Ergodic Theory, Analysis, and Efficient Simulation of Dynamical Systems

AF085812

Springer-Verlag Berlin Heidelberg GmbH

Bernold Fiedler (Editor)

Ergodic Theory, Analysis, and Efficient Simulation of Dynamical Systems

Springer

Editor:
Bernold Fiedler
Freie Universität Berlin
Institut für Mathematik I
Arnimallee 2-6
14195 Berlin, Germany
e-mail: fiedler@math.fu-berlin.de

Library of Congress Cataloging-in-Publication Data

Ergodic theory, analysis, and efficient simulation of dynamical systems / Bernold Fiedler (editor).
p. cm.
Includes bibliographical references and index.
ISBN 978-3-642-62524-4 ISBN 978-3-642-56589-2 (eBook)
DOI 10.1007/978-3-642-56589-2
1. Differentiable dynamical systems. 2. Ergodic theory. 3. Mathematical analysis. I. Fiedler, Bernold, 1956-

QA614.8 .E737 2001
515'.35--dc21

2001020854

Mathematics Subject Classification (2000): 28, 34, 35, 65, 81, 92

ISBN 978-3-642-62524-4

This work is subject to copyright. All rights are reserved, whether the whole or part of the material is concerned, specifically the rights of translation, reprinting, reuse of illustrations, recitation, broadcasting, reproduction on microfilm or in any other way, and storage in data banks. Duplication of this publication or parts thereof is permitted only under the provisions of the German Copyright Law of September 9, 1965, in its current version, and permission for use must always be obtained from Springer-Verlag. Violations are liable for prosecution under the German Copyright Law.

http://www.springer.de
© Springer-Verlag Berlin Heidelberg 2001
Originally published by Springer-Verlag Berlin Heidelberg in 2001
Softcover reprint of the hardcover 1st edition 2001

The use of general descriptive names, registered names, trademarks etc. in this publication does not imply, even in the absence of a specific statement, that such names are exempt from the relevant protective laws and regulations and therefore free for general use.

Cover design: *Erich Kirchner, Heidelberg*
Typsetting: Le-TeX Jelonek, Schmidt & Vöckler GbR, Leipzig
Printed on acid-free paper SPIN 10789119 46/3142ck-5 4 3 2 1 0

Preface

This book summarizes and highlights progress in our understanding of Dynamical Systems during six years of the German Priority Research Program "Ergodic Theory, Analysis, and Efficient Simulation of Dynamical Systems". The program was funded by the Deutsche Forschungsgemeinschaft (DFG) and aimed at combining, focussing, and enhancing research efforts of active groups in the field by cooperation on a federal level. The surveys in the book are addressed to experts and non-experts in the mathematical community alike. In addition they intend to convey the significance of the results for applications far into the neighboring disciplines of Science.

Three fundamental topics in Dynamical Systems are at the core of our research effort:

- behavior for large time
- dimension
- measure, and chaos

Each of these topics is, of course, a highly complex problem area in itself and does not fit naturally into the deplorably traditional confines of any of the disciplines of ergodic theory, analysis, or numerical analysis alone. The necessity of mathematical cooperation between these three disciplines is quite obvious when facing the formidable task of establishing a bidirectional transfer which bridges the gap between deep, detailed theoretical insight and relevant, specific applications. Both analysis and numerical analysis play a key role when it comes to building that bridge. Some steps of our joint bridging efforts are collected in this volume.

Neither our approach nor the presentations in this volume are monolithic. Rather, like composite materials, the contributions are gaining strength and versatility through the broad variety of interwoven concepts and mathematical methodologies which they span.

Fundamental concepts which are present in this volume include bifurcation, homoclinicity, invariant sets and attractors, both in the autonomous and nonautonomous situation. These concepts, at first sight, seem to mostly address *large time behavior*, most amenable to methodologies of analysis. Their intimate relation to concepts like (nonstrict) hyperbolicity, ergodicity, entropy, stochasticity and control should become quite apparent, however, when browsing through this volume.

The fundamental topic of *dimension* is similarly ubiquitous throughout our articles. In analysis it figures, for example, as a rigorous reduction from infinite-dimensional settings like partial differential equations, to simpler infinite-, finite- or even low-dimensional model equations, still bearing full relevance to the original equations. But in numerical analysis – including and transcending mere discretization – specific computational realization of such reductions still poses challenges which are addressed here.

Another source of inspiration comes from very refined *measure*-theoretic and dimensional concepts of ergodic theory which found their way into algorithmic realizations presented here.

By no means do these few hints exhaust the conceptual span of the articles. It would be even more demanding to discuss the rich circle of methods, by which the three fundamental topics of large time behavior, dimension, and measure are tackled. In addition to SBR-measures, Perron-Frobenius type transfer operators, Markov decompositions, Pesin theory, entropy, and Oseledets theorems, we address kneading invariants, fractal geometry and self-similarity, complex analytic structure, the links between billiards and spectral theory, Lyapunov exponents, and dimension estimates. Including Lyapunov-Schmidt and center manifold reductions together with their Shilnikov and Lin variants and their efficient numerical realizations, symmetry and orbit space reductions together with closely related averaging methods, we may continue, numerically, with invariant subspaces, Godunov type discretization schemes for conservation laws with source terms, (compressed) visualization of complicated and complex patterns of dynamics, and present an algorithm, GAIO, which enables us to approximately compute, in low dimensions, objects like SBR-measures and Perron-Frobenius type transfer operators. At which point our cursory excursion through methodologies employed here closes up the circle.

So much for the mathematical aspects. The range of applied issues, mostly from physics but including some topics from the life sciences, can also be summarized at most superficially, at this point. This range comprises such diverse areas as crystallization and dendrite growth, the dynamo effect, and efficient simulation of biomolecules. Fluid dynamics and reacting flows are addressed, including the much studied contexts of Rayleigh-Bénard and Taylor-Couette systems as well as the stability question of three-dimensional surface waves. The Ginzburg-Landau and Swift-Hohenberg equations appear, for example, as do mechanical problems involving friction, population biology, the spread of infectious diseases, and quantum chaos. It is the diversity of these applied fields which well reflects both the diversity and the power of the underlying mathematical approach. Only composite materials enable a bridge to span that far.

The broad scope of our program has manifested itself in many meetings, conferences, and workshops. Suffice it to mention the workshop on "Entropy" which was coorganized by Andreas Greven, Gerhard Keller, and Gerald Warnecke at Dresden in June 2000, jointly with the two neighboring DFG Priority Research Programs "Analysis and Numerics for Conservation Laws" and "Interacting Stochastic Systems of High Complexity". For further information concerning program and participants of the DFG Priority Research Program "Ergodic Theory, Analysis, and Efficient Simulation of Dynamical Systems", including a preprint server, see

– www.math.fu-berlin.de/~ danse/

For other DFG programs we refer to

– www.dfg.de
– www.dfg.de/aufgaben/Schwerpunktprogramme.html

At the end of this preface, I would like to thank at least some of the many friends and colleagues who have helped on so many occasions to make this program work. First of all, I would like to mention the members of the scientific committee who have helped initiate the entire program and who have accompanied and shaped the scientific program throughout its funding period: Ludwig Arnold, Hans-Günther Bothe, Peter Deuflhard, Klaus Kirchgässner, and Stefan Müller. The precarious conflict between great expectations and finite funding was expertly balanced by our all-understanding referees Hans Wilhelm Alt, Jürgen Gärtner, François Ledrappier, Wilhelm Niethammer, Albrecht Pietsch, Gerhard Wanner, Harry Yserentant, Eberhard Zeidler, and Eduard Zehnder. The hardships of finite funding as well as any remaining administrative constraints were further alleviated as much as possible, and beyond, by Robert Paul Königs and Bernhard Nunner, representing DFG at its best. The www-services were designed, constantly expanded and improved with unrivalled expertise and independence by Stefan Liebscher. And Regina Löhr, as an aside to her numerous other secretarial activities and with ever-lasting patience and friendliness, efficiently reduced the administrative burden of the coordinator to occasional emails which consisted of no more than "OK. BF". Martin Peters and his team at Springer-Verlag ensured a very smooth cooperation, including efficient assistance with all TEXnicalities. But last, and above all, my thanks as a coordinator of this program go to the authors of this volume and to all participants – principal investigators, PostDocs and students alike – who have realized this program with their contributions, their knowledge, their dedication, and their imagination.

Berlin,
September 2000

Bernold Fiedler

Table of Contents

Random Attractors: Robustness, Numerics and Chaotic Dynamics

Gunter Ochs*

Institut für Dynamische Systeme, Universität Bremen, Postfach 330 440, 28334 Bremen, Germany

Abstract. In this article the numerical approximation of attractors and invariant measures for random dynamical systems by using a box covering algorithm is discussed. We give a condition under which the algorithm, which defines a set valued random dynamical system, possesses an attractor close to the attractor of the original system. Furthermore, a general existence theorem for attractors for set valued random dynamical systems is proved and criteria for the robustness of random attractors under perturbations of the system are given. Our numerical algorithm is applied to the stochastically forced Duffing oscillator, which supports for certain parameter values a non-trivial random SRB measure.

1 Introduction

The qualitative behavior of an autonomous dynamical system given by the iterations of a map or the solution of an ordinary differential equation (ODE) is often characterized by invariant objects such as (global) attractors or invariant (probability) measures. Invariant measures describe the statistical behavior of a dynamical system. An attractor is a compact invariant set, which carries the asymptotic dynamics in the sense that it is approached by all trajectories under time evolution. If a global attractor exists, then it supports all invariant measures.

These "global objects" are particularly valuable in the investigation of systems with complicated ("chaotic") dynamical behavior, where reliable predictions for single trajectories are only possible for bounded time intervals.

Here we are interested in dynamics influenced by probabilistic noise. A "traditional" approach is to model this by a Markov process. In this case the evolution law of the system is given by transition probabilities instead of a deterministic map. The statistical behavior of a Markov process is often described by probability measures, which are invariant under the transition probabilities.

However, we will work in the framework of random dynamical systems (for a systematic presentation of this theory see [1]). A discrete time random dynamical system is just given by the iteration of random mappings. Stochastic differential equations (SDE's) and random differential equations (which

* Project: The Multiplicative Ergodic Theorem under Discretization and Perturbation (Ludwig Arnold and Wolfgang Krieger)

are ODE's with a randomly varying parameter) are generators of continuous time random dynamical systems.

Iterations of iid mappings and SDE's also define Markov processes. There is a notion of invariant measures for random dynamical systems, which are closely related to the invariant measures of the corresponding Markov process. However, typically a random dynamical system has more invariant measures than just these "Markov measures". The theory of random dynamical systems provides tools which allow a more detailed analysis of the dynamics than it is possible in the framework of Markov processes. In some sense it is possible to separate "noise driven dynamics" from "deterministic" dynamics.

There is a notion of a "pathwise" defined random attractor. That is, the attractor is a (compact) set valued random variable defined on the probability space which models the noise, i.e. for (almost) every realization of the stochastic process which models the noise there is a compact set. The distance between a trajectory with random initial condition and the attractor converges to zero in probability if time tends to infinity.

Such an attractor moves (in a stationary manner) in the phase space of the system under time evolution. This movement of the attractor can be interpreted as the "noise induced part" of the dynamics. In addition, there is some dynamics inside the attractor (which may be trivial). The qualitative structure of this dynamics inside the attractor is often the same for (almost) every noise realization. In this sense it can be viewed as the "deterministic part" of the dynamics.

To be more precise, let us look at two examples. Take some self mappings of a complete metric space, which are all uniform contractions. In every time step choose randomly one of these mappings. Then the asymptotic behavior of a trajectory is independent of the initial condition. This means, that the "deterministic part" of the dynamics is trivial. The attractor consists of one random point. However, if the mappings do not have a common fixed point, then an observer sees non-trivial dynamical behavior, which is in this case exclusively due to the noise.

The situation is different for the randomly forced Duffing oscillator, which will be discussed in more detail later on in this paper. There are parameter values, where the system has a positive Lyapunov exponent, which leads to sensitive dependence on initial conditions. Our numerical images of the attractor and of the "natural" measure supported by it suggest that the dynamics of the system is a combination of "deterministic dynamics" which is chaotic and "noise induced" movements in the phase space $\mathbb{R}^2$.

For many random dynamical systems the existence of a random attractor, which also implies the existence of invariant measures, is proved. However, in many interesting cases it seems to be quite hard to obtain analytical results on the structure of the attractor and the dynamics on it (the "deterministic part"). This means that one often has to rely on numerical simulations in order to obtain information about random attractors and invariant measures.

In this article we will discuss a numerical method for the approximation of random attractors and ("natural") invariant measures supported by them. We consider a "global" approach in order to approximate these global objects. We use a random version of the subdivision algorithm which was developed by Dellnitz and Hohmann [14,15] for the approximation of deterministic attractors. This algorithm produces a sequence of box coverings of the attractor.

The "random" subdivision algorithm is described in an article by Keller and Ochs [25]. Its application to the stochastically perturbed Duffing–van der Pol oscillator gave surprising new insight in the structure of the attractor.

A simplifying statement about the present paper would be that its content is essentially a new convergence proof for this algorithm. However, on the way to this proof we develop some tools which may be of interest for their own. After giving fundamental definitions (Sect. 2) we introduce in Sect. 3 the notion of a set valued random dynamical system, where the image of a point is a compact set instead of a single point. We define attractors for set valued random dynamical systems and give a general criterion for their existence based on the existence of an attracting set. This generalizes a often used existence theorem for an attractor of a standard point valued random dynamical system.

Using the notion of set valued random dynamical systems we give criteria for robustness of random attractors under perturbations. In Sect. 4 we apply these results to the "random" subdivision algorithm, which defines for a given random dynamical system φ a sequence of set valued random dynamical systems $\hat{\varphi}_k$ each of which possessing an attractor A_k (under an assumption on φ, see Corollary 26). We show that the intersection of all A_k is the global attractor for φ.

In addition (Sect. 4.2) we discuss a method for the numerical approximation of "natural" invariant measures of random dynamical systems. In the case of random dynamical systems generated by stochastic differential equations these "natural" measures are related to invariant measures of the corresponding Markov process.

In Sect. 5 we consider as an example the stochastically forced damped Duffing oscillator. We prove the existence of a global attractor. There are parameters where the corresponding random dynamical system has a positive Lyapunov exponent. This implies that the "natural" measure is a non–trivial random Sinai–Ruelle–Bowen measure. We have calculated numerical approximations of the attractor and of this measure. In addition we consider the case when the top Lyapunov exponent is negative. Then the support of the "natural" measure is a random one point set, but the global attractor seems to be a larger set carrying some sort of "transient" chaotic dynamics.

2 Preliminaries

2.1 Random Dynamical Systems

Definition 1. A (continuous) *random dynamical system* (RDS) consists of two ingredients:

- A measure preserving flow $\vartheta = (\vartheta_t)_{t\in\mathbb{T}}$ on a probability space $(\Omega, \mathcal{F}, \mathbb{P})$, which serves as a model for the noise. We always assume that ϑ is invertible, i.e. $\mathbb{T} = \mathbb{R}$ or $\mathbb{Z}$.
- A measurable mapping

$$\varphi : \mathbb{T}^{(+)} \times \Omega \times X \to X, \quad (t, \omega, x) \mapsto \varphi(t, \omega)x$$

  (with $\mathbb{T}^+ = \{t \in \mathbb{T} : t \geq 0\}$), where the *state space* X is a separable metric space (with metric d), such that
  - $(t, x) \mapsto \varphi(t, \omega)x$ is continuous for fixed ω,
  - φ satisfies the *cocycle property* $\varphi(0, \omega) = \mathrm{id}_X$ and $\varphi(t + s, \omega) = \varphi(s, \vartheta_t\omega) \circ \varphi(t, \omega)$ for $t, s \in \mathbb{T}^{(+)}$ and $\omega \in \Omega$.

We say φ is a random dynamical system on X over the *metric dynamical system* (Ω, ϑ).

Remark 2. (i) The assumption that X is separable is for technical reasons.

(ii) In some definitions continuity is only required in space (i.e. $x \mapsto \varphi(t, \omega)x$ is continuous for fixed t, ω). However, continuity in time is automatically true if $\mathbb{T} = \mathbb{Z}$ and is also satisfied for continuous time random dynamical systems generated by stochastic or random differential equations (see Arnold [1, Chapter 2]).

(iii) Throughout this paper all assertions about ω are assumed to hold on a ϑ invariant set of full measure (unless otherwise stated).

(iv) The cocycle property (which reduces to the flow property if φ is independent of the noise ω) implies, that the *skew product*

$$\Theta = \Theta_\varphi : \mathbb{T}^{(+)} \times \Omega \times X \to \Omega \times X, \quad (t, \omega, x) \mapsto \Theta(t)(\omega, x) := (\vartheta_t\omega, \varphi(t, \omega)x)$$

defines a measurable (semi-)flow on the product space $\Omega \times X$.

2.2 Generators

Iteration of Random Mappings. Let X be a metric space and $\mathbb{P}_0$ a probability measure on the space $C(X, X)$ of continuous mappings from X to itself. Set $\Omega := C(X, X)^{\mathbb{Z}}$ and let ϑ be the left shift on Ω. Define $\mathbb{P} := \mathbb{P}_0^{\mathbb{Z}}$. With $\vartheta_n = \vartheta^n$, $\varphi(\omega) = \omega_0$ and

$$\varphi(n, \omega) = \varphi(\vartheta_{n-1}\omega) \circ ... \circ \varphi(\omega)$$

for $n \geq 1$ a random dynamical system on X over (Ω, ϑ) is defined, which models the iteration of iid mappings with distribution $\mathbb{P}_0$. More generally, a random dynamical system is defined with every ϑ invariant probability measure on Ω.

If $\mathbb{P}$ is concentrated on homeomorphisms, then $\varphi(n, \omega)$ is defined for $n < 0$ via

$$\varphi(n, \omega) = \varphi(\vartheta_{-n}\omega)^{-1} \circ ... \circ \varphi(\vartheta_{-1}\omega)^{-1}.$$

In the same way a discrete time random dynamical system is generated if we have an arbitrary metric dynamical system (Ω, ϑ) and a measurable mapping $\varphi : \Omega \to C(X, X)$.

Stochastic Differential Equations. Consider the Stratonovich SDE

$$dx = f(x)\, dt + \sum_{i=1}^{m} g_i(x) \circ dW_i$$

on $\mathbb{R}^d$ with initial condition $x_0 \in \mathbb{R}^d$ and functions $f, g_i : \mathbb{R}^d \to \mathbb{R}^d$ for $i = 1, \dots, m$ and $W = (W_1, \dots, W_m)$ an m-dimensional Wiener process. If the functions f, g_i are sufficiently regular (see Theorem 2.3.36 in Arnold [1]) this equation generates a local RDS φ over (Ω, ϑ), where $\Omega = C_0(\mathbb{R}, \mathbb{R}^m)$ is the space of continuous functions ω from $\mathbb{R}$ to $\mathbb{R}^m$ with $\omega(0) = 0$ (the path space of the Wiener process) equipped with the canonical Wiener measure, and the shift ϑ is defined by $(\vartheta_t\omega)(s) = \omega(s+t) - \omega(t)$ for all $s, t \in \mathbb{R}$ and $\omega \in \Omega$. Local means that $\varphi(t, \omega)x$ is only defined for $\tau^-(\omega, x) < t < \tau^+(\omega, x)$, where $-\tau^-, \tau^+ \in (0, \infty]$ are the lifetimes of solutions before possible explosion. A sufficient condition for φ being a global RDS without explosion is global Lipschitz continuity of the functions f, g_i (see Arnold [1, Theorem 2.3.32]).

Random Differential Equations. Here the generator of a continuous time random dynamical system φ is a family of ODE's with parameter ω, which can be solved "pathwise" for each ω (in contrast to the SDE case) as a deterministic non–autonomous ODE. That is, φ satisfies the integral equation

$$\varphi(t, \omega)x = x + \int_0^t f(\vartheta_s\omega, \varphi(s, \omega)x)ds,$$

where (Ω, ϑ) is any continuous time metric dynamical system and f is a function from $\Omega \times \mathbb{R}^d$ to $\mathbb{R}^d$.

We say that φ solves the *random differential equation* $\dot{x}(t) = f(\vartheta_t\omega, x(t))$. This pathwise differential equation is fulfilled if $(t, x) \mapsto f(\vartheta_t\omega, x)$ is continuous and $x \mapsto f(\omega, x)$ is Lipschitz. In this case we speak of a *classical solution.* Otherwise the differential equation is a symbolic notation for the corresponding integral equation. For general conditions on f which are needed to generate a random dynamical system see [1, Theorems 2.2.1 and 2.2.2].

2.3 Markov Processes

A (homogeneous) Markov process (see e.g. [1, Appendix A.4]) on X is given by an initial probability distribution at time 0 and a family $(P(t,x,\cdot))_{t\in\mathbb{T}^+,x\in X}$ of *transition probabilities*, i.e. $P(t,x,\cdot)$ is a probability measure on X for every $t \geq 0$, $x \in X$. For a Borel set $A \subset X$ the value $P(x,t,A)$ is the probability to end up in A at time $s+t$ if one starts in the point x at time s. The "flow property" of a Markov process is the *Chapman–Kolmogorov equation*

$$P(s+t,x,A) = \int_X P(s,y,A)P(t,x,dy).$$

If $P(t,x,\cdot)$ is a point mass in $\varphi_t x$, then a Markov process reduces to a semi-flow.

An *invariant (or stationary) measure* is a Borel probability measure ρ on X with

$$\rho(A) = \int_X P(t,x,A)d\rho(x).$$

A random dynamical system φ has *independent increments*, if $\varphi(t,\cdot)$ and $\varphi(u,\vartheta_s\cdot)$ are stochastically independent if $0 \leq t \leq s$ and $u \geq 0$. This is the case if φ describes in discrete time the iteration of iid mappings or if φ is generated by an SDE.

Every random dynamical system φ with independent increments defines a Markov process with transition probabilities

$$P(t,x,A) = \mathbb{P}(\varphi(t,\omega)x \in A).$$

2.4 Invariant Measures for Random Dynamical Systems

An *invariant measure* for the random dynamical system φ over (Ω,ϑ) is a probability measure μ on $\Omega \times \mathbb{R}^d$ with marginal $\pi_\Omega\mu = \mathbb{P}$ on Ω, which is invariant under the skew product $\Theta(t)$.

If X is a standard measurable space (which is the case if X is a Borel subset of its completion), than μ has a disintegration $d\mu(\omega,x) = d\mu_\omega(x)d\mathbb{P}(\omega)$, where $(\mu_\omega)_{\omega\in\Omega}$ is a family of probability measures on X ($\mathbb{P}$–a.s. uniquely determined by μ). The invariance condition then means $\varphi(t,\omega)\mu_\omega = \mu_{\vartheta_t\omega}$ $\mathbb{P}$–a.s.

An ergodic invariant measure for a differentiable random dynamical system (i.e. X is a smooth finite dimensional manifold and $x \mapsto \varphi(t,\omega)x$ is differentiable) allows a "local" analysis of φ based on the multiplicative ergodic theorem of Oseledets ([33], see also Arnold [1, Theorem 4.2.6]), which provides a substitute of linear algebra. Exponential expansion rates for the linearized system (Lyapunov exponents, which generalize the real parts of eigenvalues; they are independent of ω and the initial value if $\mathbb{P}$ and μ are ergodic) and corresponding random linear subspaces, where these growth rates are realized (Oseledets spaces; they generalize eigenspaces) are defined for μ almost every (ω, x).

The following result is due to Crauel [7, Proposition 2.1.(ii)]:

Proposition 3. *Let μ be an ergodic invariant measure for a differentiable random dynamical system φ. Assume that all Lyapunov exponents are strictly negative (or φ is invertible and all Lyapunov exponents are strictly positive). Furthermore suppose that the support of μ_ω is compact $\mathbb{P}$–a.s.*

Then μ_ω is $\mathbb{P}$–a.s. a convex combination of finitely many point masses (whose number is independent of ω) with equal weights.

Another result of Crauel [8] relates invariant measures for Markov processes with invariant measures for random dynamical systems.

Proposition 4. *Let φ be a random dynamical system with independent increments and ρ an invariant measure for the corresponding Markov process.*

Then μ given by $\mu_\omega = \lim_{t\to\infty} \varphi(t, \vartheta_{-t}\omega)\rho$ (where lim *is meant in the sense of weak convergence) is an invariant measure for φ. The marginal of μ on X equals ρ.*

Measures μ which come from Proposition 4 are called *Markov measures.* In general a random dynamical system (with independent increments) has also invariant measures which are not Markov measures.

2.5 Random Attractors

Definition 5. (i) A *random set* is a family $(B(\omega))_{\omega\in\Omega}$ of subsets of X, such that its *graph* $B := \{(\omega, x) : x \in B(\omega)\} \subset \Omega \times X$ is product measurable. We identify random sets with their graphs.

(ii) A random set B is *forward invariant,* if $\Theta(t)B \subset B$ (i.e. $\varphi(t,\omega)B(\omega) \subset B(\vartheta_t\omega)$) for $t > 0$. It is *invariant,* if equality holds ($\Theta(t)B = B$).

(iii) A *random compact set* (resp. *random closed set*) is a random set A such that all $A(\omega)$ are compact (resp. closed) and

$$\omega \mapsto \operatorname{dist}(x, A(\omega)) := \inf_{y\in A(\omega)} d(x,y)$$

is measurable for every $x \in X$.

Remark 6. (i) If the probability space $(\Omega, \mathcal{F}, \mathbb{P})$ is complete, then the measurability condition in the definition of a random compact (closed) set is automatically fulfilled, i.e. every random set A with $A(\omega)$ compact (closed) is a random compact (closed) set.

(ii) We allow $A(\omega)$ to be empty, i.e. $\operatorname{dist}(x, A(\omega))$ can be infinity.

(iii) Random compact sets are random variables $\Omega \to \mathcal{K}(X)$, where $\mathcal{K}(X)$ denotes the space of all compact subsets of X (including $\emptyset$) endowed with the Hausdorff metric d_H and corresponding Borel σ–algebra.

The following facts on random closed sets are well known (see e.g. Arnold [1, Chapter 1.6] or Castaing and Valadier [6]).

The union of two and the intersection of countably many random compact (closed) sets is again a random compact (closed) set. The same is true for the image $\Theta(t)A$ of a random compact set with $t \in \mathbb{T}^{(+)}$. In the case of A a random closed set one has to take the ω–wise closure of $\Theta(t)A$. For every random closed (compact) set A there exists a *selector*, i.e. a random variable $x : \Omega \to X$ with $x(\omega) \in A(\omega)$ for every ω with $A(\omega) \neq \emptyset$. Furthermore every invariant random compact set A with $\mathbb{P}(A(\omega) \neq \emptyset) = 1$ supports an invariant measure.

Random attractors are invariant random compact sets which attract trajectories of points or of compact sets. They were defined and investigated by several authors (e.g. [4,9–12,17,23,26,32,34]).

Definition 7. Let φ be a random dynamical system over (Ω, ϑ) on a metric space X and $B \subset \Omega \times X$ be a forward invariant random set.

An invariant random compact set $A \subset B$ is called a B *point attractor*, if A attracts random points in probability, i.e.

$$\lim_{t\to\infty} \mathbb{P}\Big\{\omega : \operatorname{dist}(\varphi(t,\omega)x(\omega), A(\vartheta_t\omega)) > \varepsilon\Big\} = 0$$

for every $\varepsilon > 0$ and every random variable $x : \Omega \to X$ with $x(\omega) \subset B(\omega)$.

A is called a B *set attractor*, if it attracts random compact sets in probability, i.e.

$$\lim_{t\to\infty} \mathbb{P}\big\{\omega : \operatorname{dist}(\varphi(t,\omega)C(\omega), A(\vartheta_t\omega)) > \varepsilon\big\} = 0$$

for every random compact set $C \subset B$ and every $\varepsilon > 0$.

Here $\operatorname{dist}(K, L) := \sup_{x\in K} \operatorname{dist}(x, L)$ is the Hausdorff semi–distance.

In the sequel we will restrict our attention to set attractors and call them just attractors.

Remark 8. (i) The random variables $\operatorname{dist}(\varphi(t,\cdot)C(\cdot), A(\vartheta_t\,\cdot))$ and $\operatorname{dist}(\Theta(t)C, A) = \operatorname{dist}(\varphi(t,\vartheta_{-t}\,\cdot)C(\vartheta_{-t}\,\cdot), A(\cdot))$ have the same distribution due to the ϑ invariance of $\mathbb{P}$.

Here $\operatorname{dist}(K, L)(\omega) := \operatorname{dist}(K(\omega), L(\omega))$, which is measurable for random compact sets K, L.

(ii) Definition 7 was first given in [32]. Most authors require $\lim_{t\to\infty} \operatorname{dist}(\Theta(t)C, A) = 0$ $\mathbb{P}$–a.s. in the definition of a random attractor for C from a certain collection of random compact sets.

(iii) A *global attractor* corresponds to the case $B(\omega) \equiv X$. We speak of a *local* attractor, if the *basin* $B(\omega)$ contains an open neighborhood of $A(\omega)$. In general an attractor according to our definition need not be a global or local attractor. In this case we speak of a *relative* attractor.

(iv) If A is a B attractor and μ an invariant measure with $\mu(B) = 1$, then $\mu(A) = 1$ [32, Theorem 2].

Our next result shows, that in connection with attractors it suffices to look at discrete time systems.

Theorem 9. *Let φ be a random dynamical system on X over (Ω, ϑ). Fix $T > 0$ and define a new discrete time random dynamical system ψ over (Ω, ϑ_T) by $\psi(n, \omega) := \varphi(nT, \omega)$ for $n \in \mathbb{Z}^{(+)}$.*

Let B be a random set which is forward invariant under φ (and hence under ψ).

If A is a B set attractor for ψ, then there exists a B set attractor A^* for φ with $\mathbb{P}(A^*(\omega) = A(\omega)) = 1$.

Proof. (1) We first construct a random compact set A^*, which is invariant under φ (on a ϑ invariant set of full measure) and which coincides almost surely with A.

(1.1) Define $A' := \Theta_\varphi(t)A$ with $0 < t < T$. With $\psi(\omega) := \psi(1, \omega)$ we have $\mathbb{P}$–a.s.

$$\begin{aligned}\psi(\omega)A'(\omega) &= \varphi(T,\omega)\varphi(t,\vartheta_{-t}\omega)A(\vartheta_{-t}\omega)\\ &= \varphi(t,\vartheta_{T-t}\omega)\varphi(T,\vartheta_{-t}\omega)A(\vartheta_{-t}\omega)\\ &= \varphi(t,\vartheta_{T-t}\omega)A(\vartheta_{T-t}\omega) = A'(\vartheta_T\omega),\end{aligned}$$

i.e. A' is invariant under φ. From [32, Theorem 1] (see also the proof of Theorem 17 below) it follows that $A'(\omega) \subset A(\omega)$ $\mathbb{P}$–a.s. Analogously $A''(\omega) \subset A(\omega)$ $\mathbb{P}$–a.s. with $A'' := \Theta_\varphi(T-t)A$. Thus we have for ω from a set of full measure

$$A = \Theta_\psi A = \Theta_\varphi(T)A = \Theta_\varphi(t)A'' \subset \Theta_\varphi(t)A = A' \subset A,$$

i.e. $A = A'$, which means $\mathbb{P}\{\omega : \varphi(t,\omega)A(\omega) = A(\vartheta_t\omega)\} = 1$ for every fixed $t > 0$.

(1.2) However, we want this equality to hold simultaneously for all t and ω from a ϑ invariant set of full measure. To achieve this we need a "perfection argument" (in the nontrivial case $\mathbb{T} = \mathbb{R}$).

Let μ resp. μ^+ be a probability measure on $\mathbb{R}$ resp. $\mathbb{R}^+$ which is equivalent to Lebesgue measure. Define

$$M := \{(t,\omega) \in \mathbb{R}^+ \times \Omega : \varphi(t,\omega)A(\omega) = A(\vartheta_t\omega)\}.$$

We have

$$\begin{aligned}1 = (\mu^+ \times \mathbb{P})(M) &= \int_\Omega \int_{\mathbb{R}^+} 1_M(t,\omega)d\mu^+(t)d\mathbb{P}(\omega)\\ &= \int_{\mathbb{R}^+} \int_\Omega 1_M(t,\vartheta_{-t}\omega)d\mathbb{P}(\omega)d\mu^+(t),\end{aligned}$$

which implies $\mathbb{P}(\Omega_0) = 1$ with

$$\Omega_0 := \{\omega : (t, \vartheta_{-t}\omega) \in M \text{ for } \mu^+\text{–a.e. } t\}.$$

Since

$$(\mu \times \mathbb{P})\{(t,\omega) : \vartheta_t \omega \in \Omega_0\} = 1$$

the ϑ invariant set

$$\Omega_1 := \{\omega : \vartheta_t \omega \in \Omega_0 \text{ for } \mu\text{–a.e. } t\}$$

has also full measure.

Now assume $\omega \in \Omega_1$, $t \geq 0$ and $\vartheta_{-t}\omega \in \Omega_0$. For Lebesgue a.e. $s \geq t$ we have $(s - t, \vartheta_{-s}\omega) \in M$ and hence

$$\varphi(s, \vartheta_{-s}\omega)A(\vartheta_{-s}\omega) = \varphi(t, \vartheta_{-t}\omega)A(\vartheta_{-t}\omega) =: A^*(\omega).$$

The set $A^*(\omega)$ is uniquely defined by being equal to $\varphi(t,\omega)A(\vartheta_{-t}\omega)$ for Lebesgue almost all $t > 0$. In particular we have $A^*(\omega) = \varphi(t, \vartheta_{-t}\omega)A(\vartheta_{-t}\omega)$ for every $t \geq 0$ with $\vartheta_{-t}\omega \in \Omega_0 \cap \Omega_1$. The random compact set A^* is defined for $\omega \in \Omega_1$ and coincides with A on $\Omega_0 \cap \Omega_1$.

For $\omega \in \Omega_0 \cap \Omega_1$ we have already seen that $\varphi(t,\omega)A^*(\omega) = A^*(\vartheta_t\omega)$ for $t \geq 0$. For $\omega \in \Omega_1 \setminus \Omega_0$ choose $s > 0$ with $\vartheta_{-s}\omega \in \Omega_0$. For $t > 0$ we have then

$$\varphi(t,\omega)A^*(\omega) = \varphi(t + s, \vartheta_{-s}\omega)A(\vartheta_{-s}\omega) = A^*(\vartheta_t\omega),$$

i.e. A^* is invariant.

(2) Finally we show that A^* is an attractor for φ. Without loss of generality we assume $\Omega = \Omega_1$. Since $\mathbb{P}(A^*(\omega) = A(\omega)) = 1$ and A^* is invariant, A^* is an attractor for ψ.

Let $\varepsilon > 0$ and a random compact set $C \subset B$ be given. The continuity of $(t, x) \mapsto \varphi(t,\omega)x$ and $t \mapsto A^*(\vartheta_t\omega)$ (which follows from the invariance of A^*) implies that for every ω there exists $\delta(\omega) > 0$ with

$$\operatorname{dist}(x, A^*(\omega)) < \delta(\omega) \Rightarrow \operatorname{dist}(\varphi(t,\omega)x, A(\vartheta_t\omega)) < \varepsilon$$

for $0 \leq t < T$.

Choose $\delta_0 > 0$ with $\mathbb{P}(\delta(\omega) < \delta_0) < \frac{\varepsilon}{2}$ and $N \in \mathbb{N}$ with $\mathbb{P}(\operatorname{dist}(\Theta_\psi(n)C, A^*) \geq \delta_0) < \frac{\varepsilon}{2}$ for $n \geq N$. For $t = nT + s \geq NT$ with $n \in \mathbb{N}$ and $0 \leq s < T$ we have

$$\begin{aligned}\mathbb{P}(\operatorname{dist}(\Theta_\varphi(t)C, A^*) \geq \varepsilon) &\leq \mathbb{P}(\operatorname{dist}(\Theta_\psi(n)C, A^*) \geq \delta(\omega)) \\ &\leq \mathbb{P}(\operatorname{dist}(\Theta_\psi(n)C, A^*) \geq \delta_0) + \mathbb{P}(\delta(\omega) < \delta_0) < \varepsilon,\end{aligned}$$

which shows that A^* is a B attractor for φ. □

Remark 10. In the definition of a random attractor $\varphi(t,\omega)$ appears only for $t \geq 0$. Hence an attractor does not change if we restrict φ to $\mathbb{Z}^+ \times \Omega \times X$. However, invertibility of ϑ is needed.

3 Set Valued Random Dynamical Systems

In this section we deal (justified by Theorem 9) exclusively with discrete time $\mathbb{T} = \mathbb{Z}$. We will frequently omit the time argument in the time–one mappings, i.e. we write $\vartheta := \vartheta_1$, $\varphi(\omega) := \varphi(1, \omega)$ and $\Theta := \Theta(1)$.

The goal of this section is to investigate the robustness of random attractors under perturbations of the system. For this purpose we introduce the notion of a set valued random dynamical system and look for attractors for this type of systems. Our results can be applied to random dynamical systems defined by numerical schemes.

Definition 11. A (compact) *set valued random dynamical system* on a separable metric space X over (Ω, ϑ) is a measurable mapping

$$\varphi : \mathbb{N} \times \Omega \times X \to \mathcal{K}(X)^* := \mathcal{K}(X) \setminus \{\emptyset\}$$

with the following properties:

- φ is a cocyle, i.e.

$$\varphi(n + m, \omega)x = \varphi(m, \vartheta_n\omega)\varphi(n, \omega)x := \bigcup_{y \in \varphi(n,\omega)x} \varphi(m, \vartheta_n\omega)y$$

  for $x \in X$, $n, m \in \mathbb{N}$ and $\omega \in \Omega$,
- $\varphi(n, \omega)$ is upper semi–continuous, i.e. $\lim_{i\to\infty} x_i = x$ implies $\lim_{i\to\infty} \operatorname{dist}(\varphi(n, \omega)x_i, \varphi(n, \omega)x) = 0$.

First we derive some basic properties of set valued mappings.

Lemma 12. *Let* $K_n, L \subset X$ *be compact sets with* $\lim_{n\to\infty} \operatorname{dist}(K_n, L) = 0$. *Then the sets* $K := \overline{\bigcup_{n=1}^{\infty} K_n}$ *and* $K' := L \cup (\bigcup_{n=1}^{\infty} K_n)$ *are compact.*

Proof. Let $(x_i)_{i\in\mathbb{N}} \subset K$ be any sequence. Then there exist sequences $(n(i)) \subset \mathbb{N}$ and $(y_i) \subset X$ with $y_i \in K_{n(i)}$ and $d(x_i, y_i) < \frac{1}{i}$.

First assume that there exists an n with $n(i) = n$ for infinitely many i. Then (y_i) has an accumulation point in $K_n \subset K$, which is also an accumulation point of (x_i).

Otherwise we have $\lim_{i\to\infty} n(i) = \infty$. In this case for every i there exists $z_i \in L$ with $\lim_{i\to\infty} d(y_i, z_i) = \lim_{i\to\infty} d(x_i, z_i) = 0$. Since L is compact, there exists an accumulation point z^* of (z_i), which is also an accumulation point of (x_i). The closedness of K implies $z^* \in K$. This shows that K is compact.

The compactness of K' follows because in both cases the accumulation point lies in K'. □

Lemma 13. *Every upper semi–continuous map* $\varphi : X \to \mathcal{K}(X)^*$ *defines a mapping* $\mathcal{K}(X)^* \to \mathcal{K}(X)^*$ *also denoted by* φ *via* $\varphi(K) = \bigcup_{x\in K} \varphi(x)$, *which is also upper semi–continuous, i.e.* $\lim_{i\to\infty} \operatorname{dist}(K_i, K) = 0$ *implies* $\lim_{i\to\infty} \operatorname{dist}(\varphi K_i, \varphi K) = 0$.

Proof. In order to show that $\varphi(K)$ is compact let $(y_i) \subset \varphi(K)$ be a sequence. For every i there exists $x_i \in K$ with $y_i \in \varphi(x_i)$. Let x be an accumulation point of (x_i), w.l.o.g. $x = \lim x_i$. By Lemma 12 $\varphi(x) \cup (\bigcup_i \varphi(x_i)) \subset \varphi(K)$ is compact, which implies that (y_i) has an accumulation point in $\varphi(K)$.

The upper semi–continuity of φ on $\mathcal{K}(X)^*$ is proved by contradiction. Assume that there exist $\varepsilon > 0$ and sequences $(i(j))_j \subset \mathbb{N}$ and $(x_j) \subset X$ with $i(j) \uparrow \infty$, $x_j \in K_{i(j)}$ and $\text{dist}(\varphi(x_{i(j)}, \varphi(K)) \geq \varepsilon$ for every j.

Since $\lim_{j\to\infty} \text{dist}(K_{i(j)}, K) = 0$ the sequence (x_j) has an accumulation point $x \in K$, w.l.o.g. $x = \lim x_j$. We have

$$\text{dist}(\varphi(x_{i(j)}), \varphi(K)) \leq \text{dist}(\varphi(x_{i(j)}), \varphi(x)) \to 0,$$

which is the desired contradiction. □

Remark 14. (i) Lemma 13 implies that the composition of upper semi–continuous maps $X \to \mathcal{K}(X)^*$ defines again an upper semi–continuous map $X \to \mathcal{K}(X)^*$.

(ii) The upper semi–continuity of φ is equivalent to $\varphi\left(\bigcap_{i\in\mathbb{N}} K_i\right) = \bigcap_{i\in\mathbb{N}} \varphi(K_i)$ for every decreasing sequence $(K_i) \subset \mathcal{K}(X)^*$ (for the "$\Leftarrow$" direction consider $K_i := \{x\} \cup \{x_j : j \geq i\}$ for a sequence (x_j) with $x = \lim x_j$).

Corollary 15. *If (Ω, ϑ) is a metric dynamical system, X a separable metric space, and $\varphi : \Omega \times X \to \mathcal{K}(X)^*$ a measurable mapping which is upper semi–continuous in $x \in X$, then*

$$\varphi(n, \omega)x = \varphi(\vartheta_{n-1}\omega)...\varphi(\omega)x$$

defines a set valued random dynamical system on X over (Ω, ϑ).

Every set valued random dynamical system is generated in this way by its time–one mapping.

To proceed further we need some notations.

Definition 16. Let φ be a set valued random dynamical system on X over (Ω, ϑ).

(i) A random set $B \subset \Omega \times X$ is called *forward invariant* under φ, if $\varphi(\omega)x \subset B(\vartheta\omega)$ for every $x \in B(\omega)$.

(ii) A random compact set A is called *invariant* under φ, if $\varphi(\omega)A(\omega) = A(\vartheta\omega)$ $\mathbb{P}$–a.s.

(iii) Let $\mathcal{D}$ be a collection of random compact sets. A random compact set C is $\mathcal{D}$ *attracting*, if

$$\lim_{n\to\infty} \text{dist}(\varphi(n, \omega)D(\omega), C(\vartheta_n\omega)) = 0$$

in probability for every $D \in \mathcal{D}$.

(iv) An invariant random compact set $A \subset B$ is a B *attractor*, if

$$\lim_{n\to\infty} \text{dist}(\varphi(n, \omega)C(\omega), A(\vartheta_n\omega)) = 0$$

in probability for every random compact set $C \subset B$.

(v) A family $\mathcal{D}$ of random compact set *exhausts* a random set B, if for every random compact set $C \subset B$ and every $\varepsilon > 0$ there exists $D \in \mathcal{D}$ with $\mathbb{P}(C(\omega) \subset D(\omega)) > 1 - \varepsilon$.

Now we are prepared to prove a general criterion for the existence of an attractor for a set valued random dynamical system.

Theorem 17. *Let φ be a set valued random dynamical system, B a forward invariant random set and $\mathcal{D}$ a collection of random compact sets which exhausts B. Assume that there exists a random compact set $C \subset B$ which is $\mathcal{D}$ attracting with*

$$\lim_{n\to\infty} \operatorname{dist}(\Theta(n)C, C) = 0 \quad \mathbb{P}\text{-}a.s.$$

Then A defined by

$$A(\omega) := \bigcap_{n=1}^{\infty} \overline{\bigcup_{m\geq n} \varphi(m, \vartheta_{-m}\omega)C(\vartheta_{-m}\omega)}$$

is a B attractor.

Proof. (1) By Lemma 12 $A(\omega)$ is the intersection of a decreasing sequence of non–empty compact sets. Hence A is a non–empty random compact set. By assumption

$$\lim_{n\to\infty} \operatorname{dist}\left(\overline{\bigcup_{m\geq n} \varphi(m, \vartheta_{-m}\omega)C(\vartheta_{-m}\omega)},\ C(\omega)\right) = 0 \quad \mathbb{P}\text{–a.s.},$$

which implies $A \subset C \subset B$.

(2) Let $D \in \mathcal{D}$ and $\varepsilon > 0$ be given. The definition of A implies $\lim_{n\to\infty} \operatorname{dist}(\Theta(n)C, A) = 0$ $\mathbb{P}$–a.s., i.e. there exists $n_1 \in \mathbb{N}$ with $\mathbb{P}\left((\operatorname{dist}(\Theta(n_1)C, A) \geq \frac{\varepsilon}{2}\right) < \frac{\varepsilon}{3}$.

By the upper semi–continuity of $\varphi(n_1, \omega)$ and the compactness of $C(\omega)$ there exists $\delta(\omega) > 0$ with $\operatorname{dist}(\varphi(n_1,\omega)x, \varphi(n_1,\omega)C(\omega)) < \frac{\varepsilon}{2}$ if $\operatorname{dist}(x, C(\omega)) < \delta(\omega)$.

Choose $\delta_0 > 0$ with $\mathbb{P}(\delta(\omega) < \delta_0) < \frac{\varepsilon}{3}$. By assumption there exists $n_2 \in \mathbb{N}$ with $\mathbb{P}\left(\operatorname{dist}(\Theta(n)D, C) \geq \delta_0\right) < \frac{\varepsilon}{3}$ for $n \geq n_2$.

If $n \geq n_1 + n_2$, $\operatorname{dist}(\varphi(n - n_1, \omega)D(\omega), C(\vartheta_{n-n_1}\omega)) < \delta_0$, $\delta(\vartheta_{n-n_1}\omega) \geq \delta_0$ and $\operatorname{dist}(\varphi(n_1,\omega)C(\vartheta_{n-n_1}\omega), A(\vartheta_n\omega)) < \frac{\varepsilon}{2}$, then $\operatorname{dist}(\varphi(n,\omega)D(\omega), A(\vartheta_n\omega)) < \varepsilon$. Since this happens (for every $n \geq n_1 + n_2$) with probability $> 1 - \varepsilon$, we see that A is $\mathcal{D}$ attracting.

(3) Given $\varepsilon > 0$ and an arbitrary random compact set $E \subset B$ choose $D \in \mathcal{D}$ with $\mathbb{P}(E(\omega) \subset D(\omega)) > 1 - \frac{\varepsilon}{2}$ and $N \in \mathbb{N}$ with $\mathbb{P}(\operatorname{dist}(\Theta(n)D, A) > \varepsilon) < \frac{\varepsilon}{2}$ for $n \geq N$. Then for $n \geq N$

$$\mathbb{P}(\operatorname{dist}(\Theta(n)E, A) > \varepsilon) \leq \mathbb{P}(\operatorname{dist}(\Theta(n)D, A) > \varepsilon) + \mathbb{P}(E(\omega) \not\subset D(\omega)) < \varepsilon.$$

(4) It remains to show that A is invariant. The inclusion $A \subset \Theta A$ follows (using Lemma 12 and Remark 14.(ii)) from

$$\begin{aligned}\varphi(\omega)A(\omega) = \overline{\varphi(\omega)A(\omega)} &= \overline{\bigcap_{n=1}^{\infty} \varphi(\omega)\left(\overline{\bigcup_{m\geq n} \varphi(m,\vartheta_{-m}\omega)C(\vartheta_{-m}\omega)}\right)} \\ &\supset \overline{\bigcap_{n=1}^{\infty} \varphi(\omega)\left(\bigcup_{m\geq n} \varphi(m,\vartheta_{-m}\omega)C(\vartheta_{-m}\omega)\right)} \\ &= \bigcap_{n=1}^{\infty} \overline{\bigcup_{m\geq n} \varphi(\omega)\varphi(m,\vartheta_{-m}\omega)C(\vartheta_{-m}\omega)} = A(\vartheta\omega).\end{aligned}$$

Since $A \subset B$ and $\Theta(n)A \supset A$ for $n \geq 0$ we have for every $\varepsilon > 0$

$$\mathbb{P}(\mathrm{dist}(\Theta A, A) > \varepsilon) \leq \mathbb{P}(\mathrm{dist}(\Theta(n)A, A) > \varepsilon) \to 0 \text{ as } n \to \infty.$$

This implies $\varphi(\omega)A(\omega) \subset A(\vartheta\omega)$ with probability 1. □

Remark 18. (i) There are several versions of this theorem in the context of (standard) point valued random dynamical systems (e.g. Crauel, Debussche and Flandoli [11]). The most general one is due to Keller and Schmalfuß [27, Theorem 2.2].

There is also an extension to set valued non–autonomous systems (where the attractor is defined via "pullback convergence") by H. Keller (unpublished). The case where C is an absorbing set (also in the framework of set valued non–autonomous systems) was treated by Kloeden and Schmalfuß [30].

(ii) If there exists a forward invariant random compact set C, then $A = \bigcap_{n\geq 0} \Theta(n)C$ is a C attractor. This particular case of Theorem 17 will be important for our applications (see Theorem 19 and Corollary 23).

The following result applies to robustness of attractors under small perturbations.

Theorem 19. *Let φ_k, φ ($k \in \mathbb{N}$) be set valued random dynamical systems on X over (Ω, ϑ) with $\varphi(\omega)x \subset \varphi_{k+1}(\omega)x \subset \varphi_k(\omega)x$ for every $k \in \mathbb{N}$, $\omega \in \Omega$ and $x \in X$. Assume that there exists a random compact set B, which is forward invariant under φ_1 (and hence under all φ_k). With $\mathrm{dist}_B(\varphi_k, \varphi)(\omega) := \sup_{x\in B(\omega)} \mathrm{dist}(\varphi_k(\omega)x, \varphi(\omega)x)$ suppose $\lim_{k\to\infty} \mathrm{dist}_B(\varphi_k, \varphi) = 0$ $\mathbb{P}$-a.s.*

Then every φ_k has a B attractor A_k with $A_{k+1} \subset A_k$, and $A := \bigcap_{k\in\mathbb{N}} A_k$ is a B attractor for φ.

Proof. The existence of B attractors A_k and A' for φ_k resp. φ is clear from Theorem 17. Obviously we have $A' \subset A_{k+1} \subset A_k$ for every k, which implies $A' \subset A$. By Remark 14.(ii)

$$\varphi(\omega)A(\omega) = \bigcap_{k\in\mathbb{N}} \varphi(\omega)A_k(\omega) \subset \bigcap_{k\in\mathbb{N}} \varphi_k(\omega)A_k(\omega) = A(\vartheta\omega).$$

Since

$$\operatorname{dist}(A_k(\vartheta\omega), \varphi(\omega)A_k(\omega)) \leq \operatorname{dist}(\varphi_k, \varphi) \to 0$$

it follows $\bigcap_{k\in\mathbb{N}} \varphi(\omega)A_k(\omega) = A(\vartheta\omega)$, i.e. A is invariant under φ.

Since A' is an attractor we have

$$\mathbb{P}(\operatorname{dist}(A, A') > \varepsilon) = \mathbb{P}(\operatorname{dist}(\Theta_\varphi(n)A, A') > \varepsilon) \to 0$$

as $n \to \infty$ for every $\varepsilon > 0$, which implies $A \subset A'$ $\mathbb{P}$–a.s. □

Corollary 20. *Let φ, φ_k ($k \in \mathbb{N}$) be set valued random dynamical systems. Assume that there exists a random compact set B which is forward invariant under all φ_k.*

If $\lim_{k\to\infty} \operatorname{dist}_B(\varphi_k, \varphi) = 0$, then there exist B attractors A and A_k for φ resp. φ_k with $\lim_{k\to\infty} \operatorname{dist}(A_k, A) = 0$ $\mathbb{P}$–a.s.

Proof. First observe that the compactness of $B(\omega)$ and the forward invariance of B under all φ_k imply, that B is forward invariant under φ.

Define

$$\psi_k(\omega)x := \varphi(\omega)x \cup \left(\bigcup_{l\geq k} \varphi_l(\omega)x \right).$$

We have to show that ψ_k, φ satisfy the assumptions of Theorem 19. The inclusions $\varphi(\omega)x \subset \psi_{k+1}(\omega)x \subset \psi_k(\omega)x$ and $\psi_1(\omega)B(\omega) \subset B(\vartheta\omega)$ are clear. Furthermore, $\operatorname{dist}_B(\psi_k, \varphi) \to 0$.

By Lemma 12, $\psi_k(\omega)x$ is compact for every k, ω, x. To show that $\psi_k(\omega)$ is upper semi–continuous consider sequences $(x_n) \subset B(\omega)$ with $x = \lim x_n$ and (y_n) with $y_n \in \psi_k(\omega)x_n$ and $y = \lim y_n$. We have to show $y \in \psi_k(\omega)x$.

This follows from the upper semi–continuity of $\varphi(\omega)$ in the case that $y_n \in \varphi(\omega)x_n$ for infinitely many n. Otherwise there exist $l(n) \geq k$ (for all but finitely many n) with $y_n \in \varphi_{l(n)}(\omega)x_n$. If $\lim_{n\to\infty} l(n) = \infty$, then

$$\begin{aligned}&\operatorname{dist}(\varphi_{l(n)}(\omega)x_n, \varphi(\omega)x)\\ &\leq \operatorname{dist}(\varphi_{l(n)}(\omega)x_n, \varphi(\omega)x_n) + \operatorname{dist}(\varphi(\omega)x_n, \varphi(\omega)x) \to 0\end{aligned}$$

as $n \to \infty$, which implies $y \in \varphi(\omega)x \subset \psi_k(\omega)x$.

If $l(n) \not\to \infty$, then there exists $l \geq k$ with $l(n) = l$ for infinitely many n, which implies $y \in \varphi_l(\omega)x \subset \psi_k(\omega)x$. □

Remark 21. (i) There is some related work in the context of non–autonomous systems by Kloeden and Kozyakin [28,29].

(ii) Corollary 20 implies in particular to point valued random dynamical systems φ_k, φ with $\lim_{k\to\infty} \sup_{x\in B(\omega)} d(\varphi_k(\omega)x, \varphi(\omega)x) = 0$ $\mathbb{P}$–a.s., if there exists a random compact set B which is invariant under all φ_k.

The previous results only ensure the persistence of local ore relative attractors. However, the next result shows that the A_k are global attractors if B is an absorbing set for φ_k.

Proposition 22. *Let φ be a set valued random dynamical system, B a forward invariant random set with $B(\omega)$ open and A a B attractor.*

Then A is a $\tilde{B}$ set attractor for φ with

$$\tilde{B} := \{(\omega, x) : \varphi(n,\omega)x \subset B(\vartheta_n\omega) \text{ for some } n \geq 0\}.$$

Proof. Let $\mathcal{D}$ be the family of all random compact sets D for which $\Theta(n)D \subset B$ for some $n \geq 0$. Then A is $\mathcal{D}$ attracting.

By Theorem 17 it remains to show that $\mathcal{D}$ exhausts $\tilde{B}$. Let $C \subset \tilde{B}$ be a random compact set. For $x \in C(\omega)$ there exists a minimal $n = n(\omega, x) \geq 0$ with $\varphi(n,\omega)x \subset B(\vartheta_n\omega)$. The upper semi-continuity of $\varphi(n,\omega)$ implies $n(\omega, y) \leq n(\omega, x)$ for y from an open neighborhood of x. By compactness of $C(\omega)$ it follows $n(\omega) := \sup_{x \in C(\omega)} n(\omega, x) < \infty$.

Given $\varepsilon > 0$ there exists $n \geq 0$ with $\mathbb{P}(n(\omega) > n) < \varepsilon$. Define $D(\omega) := C(\omega)$ if $n(\omega) \leq n$ and $D(\omega) := A(\omega)$ if $n(\omega) > n$. Then $D \in \mathcal{D}$ and $\mathbb{P}(C(\omega) \subset D(\omega)) > 1 - \varepsilon$. □

Corollary 23. *Let φ be a set valued random dynamical system on X and C a random compact set with $\varphi(\omega)C(\omega) \subset \text{int}C(\vartheta\omega)$. Assume C is absorbing, i.e. for every $x \in X$ and (almost) every $\omega \in \Omega$ there exists $n \in \mathbb{N}$ with $\varphi(n,\omega)x \in C(\vartheta_n\omega)$.*

Then $A := \bigcap_{n=1}^{\infty} \Theta(n)C$ is the global attractor for φ.

Proof. Apply Remark 18.(ii) and Proposition 22 with $B(\omega) = \text{int}C(\omega)$. □

We now give a criterion for the robustness of a global attractor for a random dynamical system φ on $\mathbb{R}^d$, which imposes conditions only on φ and works for a general type of perturbations. It is based on the existence of a "Lyapunov type" function, which can be verified in many concrete examples.

Theorem 24. *Let φ be a discrete time (point valued) random dynamical system on $\mathbb{R}^d$ over (Ω, ϑ) and $V : \mathbb{R}^d \to \mathbb{R}^+$ a continuous "Lyapunov function" with $\lim_{|x|\to\infty} V(x) = \infty$. With*

$$\eta_c(x) := \sup_{|y-x|\leq c} \frac{|V(y) - V(x)|}{V(x)}$$

assume $\lim_{|x|\to\infty} \eta_c(x) = 0$ for some $c > 0$. Furthermore suppose that there exist measurable functions $a, b : \Omega \to \mathbb{R}^+$ with $\int \log^+ a \, d\mathbb{P}, \int \log^+ b \, d\mathbb{P} < \infty$ and $\int \log a \, d\mathbb{P} < 0$, such that

$$V(\varphi(\omega)x) \leq a(\omega)V(x) + b(\omega)$$

for $\omega \in \Omega$ and $x \in \mathbb{R}^d$ ($\log^+ \xi := \max\{0, \log \xi\}$).

Let $(\varphi_k)_{k\in\mathbb{N}}$ *be a sequence of set valued random dynamical systems on* $\mathbb{R}^d$ *over* (Ω,ϑ) *with* $\operatorname{dist}(\varphi_k(\omega)x,\varphi(\omega,x)) \le c$ *for* $x\in\mathbb{R}^d$ *and* $\lim_{k\to\infty}\operatorname{dist}(\varphi_k(\omega)x,\varphi(\omega,x)) = 0$ $\mathbb{P}$*-a.s. for every* $x\in\mathbb{R}^d$.

Then φ *has a global attractor* A *and every* φ_k *has a global attractor* A_k *with* $\lim_{k\to\infty}\operatorname{dist}(A_k,A) = 0$ $\mathbb{P}$*-a.s.*

If in addition $\varphi(\omega)x \in \varphi_k(\omega)x$ *for* $x\in\mathbb{R}^d$ *and* $k\in\mathbb{N}$*, then* $A\subset A_k$ *for every* k *and* $\lim_{k\to\infty} A_k = A$ $\mathbb{P}$*-a.s. in the Hausdorff metric.*

Proof. Define a set valued random dynamical system ψ by

$$\psi(\omega)x := \{y : |y-\varphi(\omega)x| \le c\}.$$

By Corollary 20 and Corollary 23 it suffices to construct an absorbing random compact set C for ψ with $\psi(\omega)C(\omega)\subset \operatorname{int}C(\vartheta\omega)$.

Choose $\varepsilon>0$ with $\int \log\left(\frac{a}{1-\varepsilon}+\varepsilon\right)d\mathbb{P} < 0$ and b_0 with $\eta_c(x)\le\varepsilon$ for $|x|\ge b_0$. Set

$$\tilde b(\omega) := \max\left\{\frac{b(\omega)}{1-\varepsilon}, b_0\right\},$$

$a_1(\omega) := \frac{a(\omega)}{1-\varepsilon}$ and $a_2(\omega) := a_1(\omega)+\varepsilon$. Then $\int\log^+ a_{1,2}\,d\mathbb{P}, \int\log^+\tilde b\,d\mathbb{P}<\infty$ and $\int\log a_{1,2}\,d\mathbb{P}<0$.

By $\chi_{1,2}(\omega)v := a_{1,2}(\omega)v+\tilde b(\omega)$ affine random dynamical systems $\chi_{1,2}$ on $\mathbb{R}$ over (Ω,ϑ) are defined. By [1, Theorem 5.6.5, Corollary 5.6.6] there exist random variables $v_1,v_2:\Omega\to\mathbb{R}$ with $\chi_i(\omega)v_i(\omega)=v_i(\vartheta\omega)$, such that $(\{v_i(\omega)\})_{\omega\in\Omega}$ is the global attractor for χ_i. This implies with $a_i(\omega)\ge 0$ and $\tilde b(\omega)\ge b_0$ that $b_0\le v_1(\omega)\le v_2(\omega)-\varepsilon b_0$

Assume $V(x)\le v_2(\omega)$, $y\in\psi(\omega)x$ and $V(y)\ge v_2(\vartheta\omega)$. Then by assumption

$$\begin{aligned} &V(y)\le V(\varphi(\omega)x)+\eta_c(y)V(y)\le a(\omega)V(x)+b(\omega)+\varepsilon V(y)\\ \Rightarrow\ &V(y)\le a_1(\omega)v_2(\omega)+\tilde b(\omega)<v_2(\vartheta\omega),\end{aligned}$$

which is a contradiction, i.e. with $C(\omega):=\{x: V(x)\le v_2(\omega)\}$ we have $\psi(\omega)C(\omega)\subset\operatorname{int}C(\vartheta\omega)$.

In order to show that C is absorbing assume $v := V(x)\ge v_2(\omega)$. It follows from [1, Corollary 5.6.6.(ii)] that there exists $n\ge 0$ with $|\chi_1(n,\omega)v - v_1(\vartheta_n\omega)|<\varepsilon b_0$. Set $x_0=0$ and for $k=1,\dots,n$ let $x_k\in\varphi(\vartheta_{k-1}\omega)x_{k-1}$. If $V(x_k)<v_2(\vartheta_k\omega)$ for some $k\le n$ then $x_n\in\operatorname{int}C(\vartheta_n\omega)$.

Now suppose $V(x_k)\ge v_1(\vartheta_k\omega)\ge b_0$ for $0\le k\le n$. Then

$$V(x_k)\le a_1(\omega)V(x_{k-1})+\tilde b(\omega).$$

Inductively it follows $V(x_n)\le\chi_1(n,\omega)v<v_2(\vartheta_n\omega)$, which implies $\psi(n,\omega)x\subset C(\vartheta_n\omega)$. □

Remark 25. If φ is a continuous time random dynamical system with $t \mapsto \varphi(t,\omega)x$ differentiable and V differentiable with

$$\frac{\partial}{\partial t} V(\varphi(t,\omega)x) \le \alpha(\vartheta_t\omega)V(\varphi(t,\omega)x) + \beta(\vartheta_t\omega)$$

with $\alpha, \beta \in L^1(\mathbb{P})$, $t \mapsto \alpha(\vartheta_t\omega), \beta(\vartheta_t\omega)$ continuous and $\int \alpha \, d\mathbb{P} < 0$, then the time–one mapping of φ meets the assumptions of Theorem 24. This follows by considering the continuous time affine random dynamical system χ generated by the random differential equation

$$\dot{v}(t) = \alpha(\vartheta_t\omega)v(t) + \beta(\vartheta_t\omega)$$

(cf. [1, Theorem 5.6.5, Corollary 5.6.6]). Then $V(\varphi(1,\omega)x) \le \chi(1,\omega)V(x)$.

4 Numerics of Random Attractors and Invariant Measures

4.1 Attractors

The *subdivision algorithm* of Dellnitz and Hohmann [14,15] was adapted to the case of random dynamical systems by Keller and Ochs [25]. Here we give only a brief description of the algorithm. For details of the implementation consult [14,15,25].

Given a discrete time random dynamical system φ on $\mathbb{R}^d$ over (Ω, ϑ) the algorithm defines a set valued random dynamical system $\hat{\varphi}$ in the following way. The state space $\mathbb{R}^d$ is divided into small (closed) boxes, which intersect only at their boundaries. The image of a point x under $\hat{\varphi}(\omega)$ consists of all those boxes, which contain $\varphi(\omega)x$. Then $\varphi(\omega)x \in \hat{\varphi}(\omega)x$ and $\mathrm{dist}(\varphi(\omega)x, \hat{\varphi}(\omega)x)$ is at most the diameter of the small boxes.

The boxes can be subdivided by bisection with respect to one coordinate (which changes cyclically between successive subdivisions). This gives a sequence $(\hat{\varphi}_k)_{k\ge 0}$ of "numerical" mappings each corresponding to a certain box size, i.e. $\hat{\varphi}_k$ describes the numerical mapping after k subdivisions. By construction

$$\lim_{k\to\infty} \sup_{\omega\in\Omega} \sup_{x\in\mathbb{R}^d} \mathrm{dist}(\varphi(\omega)x, \hat{\varphi}_k(\omega)x)) = 0.$$

With Theorem 24 this gives

Corollary 26. *Let the random dynamical system φ satisfy the assumptions of Theorem 24 (which implies the existence of a global attractor A).*

Then the set valued random dynamical systems $\hat{\varphi}_k$ defined by the numerical discretization have global attractors A_k with $A \subset A_k$ and $\lim_{k\to\infty} A_k = A$ $\mathbb{P}$–a.s. in the Hausdorff metric.

In practise it is of course not possible to treat the whole $\mathbb{R}^d$ (which consists of infinitely many boxes). We start with one "sufficiently large" box Q, which is then successively subdivided into smaller boxes. The choice of Q should ensure $A(\omega) \subset Q$ with "high" probability. We set $Q_0 = Q$ and for $n \geq 0$ we define Q_{n+1} to be the part of Q which belongs to the numerical image of Q_n under the map $\hat{\varphi}_k(\vartheta_n\omega)$ for some $k(n)$. Typically we start with $k(0)$ small (often $= 0$) and increase $k(n)$ by subdivisions in the course of the simulation. After a certain number of subdivisions we stay at a constant level, i.e. $k(n) = k$ for all $n \geq n_0$ with some $k, n_0 \in \mathbb{N}$.

If $A(\omega) \subset Q$ and

$$\hat{\varphi}_k(\vartheta_i\omega)Q_i \subset Q \qquad \text{for } i = 0, ..., n-1 \tag{$*$}$$

(note that $(*)$ is numerically checkable), then we know $A(\vartheta_n\omega) \subset Q_n$. As $n \to \infty$ our approximations Q_n should "approach" $A_k(\vartheta_n\omega)$, which is close to $A(\vartheta_n\omega)$ if k is sufficiently large.

If Q is sufficiently large, then the dynamics is expected to point inward, i.e. we have a good chance for $(*)$ to be fulfilled. In case $(*)$ is violated it is possible to change the size (and position in space) of Q adaptively, i.e. we still can ensure to have a covering of the attractor. However, we do not have general estimates on $\text{dist}(Q_n, A(\vartheta_n\omega))$ (these would involve additional assumptions on the "speed of attraction" of trajectories of φ towards A).

4.2 Invariant Measures

Dellnitz and Junge [16] approximate "natural" invariant measures for deterministic dynamical systems by calculating eigenvectors of a discretized Perron–Frobenius operator, i.e. they just discretize the equation $f\mu = \mu$ where $f : \mathbb{R}^d \to \mathbb{R}^d$ is a diffeomorphism and μ a probability measure supported by the attractor of f. Their space discretization is based on the box covering of the attractor produced by the subdivision algorithm. The discretized measure assigns a mass to each box, i.e it corresponds to a vector in $\mathbb{R}^N$ with non–negative entries, where N is the total number of boxes. The action of f on probabilities is approximated by a linear mapping on $\mathbb{R}^N$ given by a matrix $M(f)$ with non–negative entries. The entry corresponding to a pair (B, C) of boxes is the relative Lebesgue measure of that part of B which is mapped into C. That is, the space discretization gives a positive linear mapping $M(f) : \mathbb{R}^N \to \mathbb{R}^N$, which has (under an irreducibility and aperiodicity condition) by the Perron–Frobenius theorem a unique non–negative normed eigenvector corresponding to the eigenvalue one. Dellnitz and Junge show, that if f has a hyperbolic attractor A, then this eigenvector is an approximation of the Sinai–Ruelle–Bowen (SRB) measure on A. This SRB measure is the natural invariant measure for f in the sense that it describes the statistical behavior of the trajectories for Lebesgue almost all initial values.

What happens if we look for a generalization to the random case? Assume we have a differentiable discrete time random dynamical system φ (which may be the time–T mapping of a stochastic or random differential equation) with a global random attractor A. The numerical discretization gives a box covering of $A(\omega)$ with a random number $N(\omega)$ of boxes. Positive linear mappings $M(\varphi(\omega)) : \mathbb{R}^{N(\omega)} \to \mathbb{R}^{N(\vartheta\omega)}$ can be defined in the same way as in the deterministic case. There is a "random" Perron–Frobenius theorem by Gundlach and Steinkamp [21], which covers this situation. That is, under an irreducibility and an aperiodicity condition there exists a unique random vector $a(\omega) \in \mathbb{R}^{N(\omega)}$ with non–negative entries which sum up to one, and $M(\varphi(\omega))a(\omega) = a(\vartheta\omega)$. If we consider the linear random dynamical system Φ generated by $M(\varphi(\cdot))$, then $\lim_{n\to\infty} |\Phi(n,\omega)v - a(\vartheta_n\omega)| = 0$ for every probability vector $v \in \mathbb{R}^{N(\omega)}$.

However, the approach of Dellnitz and Junge is not directly transferable to the random case. Our "random eigenvector equation" $\Phi(\omega)a(\omega) = a(\vartheta\omega)$ involves ω and is thus an infinite dimensional problem, for which there is no canonical discretization, since we typically have little regularity in ω.

For this reason our approach is more primitive than the one of Dellnitz and Junge. We assign a mass to each box of the covering (starting with a uniform distribution) and just iterate this mass forward, i.e. we apply the map $\Phi(\omega)$. We expect our approximation to "converge" to $a(\vartheta_n\omega)$ for n sufficiently large. If the irreducibility and/or the aperiodicity condition of the random Perron–Frobenius theorem is violated, then Φ has still a random attractor in the space of non–negative normed vectors, which should be approached by our iterations.

However, there are no rigorous convergence results available up to now. First we have to ask which object $a(\omega)$ should approximate. There is a theory of SRB measures for random dynamical systems [5,31], which are characterized by satisfying Pesin's entropy formula and by absolute continuity on unstable manifolds. If the Markov process generated by an SDE has an absolutely continuous invariant measure, then (under an integrability condition) the corresponding Markov measure for the random dynamical system is a random SRB measure [31, Chapter IV, Theorem 1.1].

There is a possibility to prove convergence of our numerical approximation to the Markov measure using Proposition 4. However, this would not give a really satisfactory result. Therefore we will only sketch the idea instead of giving a complete proof. If the measure ρ in Proposition 4 is absolutely continuous (which is typically the case), then we can replace it by another absolutely continuous measure, namely the uniform distribution on our initial box (if $\Phi(\cdot)$ is not aperiodic, we have to be a little bit more careful). Then we can argue in the same way as in the proof of [25, Theorem 3.1]. The result would be the following:

Given ω and $\varepsilon > 0$ there exists $\delta > 0$ and $n \in \mathbb{N}$ such that the Prohorov distance between the probability measure on $\mathbb{R}^d$ represented by $\Phi(\vartheta_{-n}\omega)v$ and μ_ω is smaller than ε provided the box diameter is smaller than δ. Here v

is the probability vector representing the uniform distribution on the initial box Q.

Note that v is not concentrated on the covering of $A(\vartheta_{-n}\omega)$ but on the whole set Q. This means that (if δ is small) v has a huge number of entries, which is in practise not possible to handle.

A more satisfactory approach would be to relate the attractor of Φ (which always exists, since Φ leaves the random compact set of probability vectors on $\mathbb{R}^{N(\omega)}$ invariant) with the Markov measure (which is an attractor on a subset of the space of all probability measures).

5 The Stochastically Forced Duffing Oscillator

5.1 Introductory Remarks

The second order ODE

$$\ddot{x} = f(x) - \gamma\dot{x}, \qquad x \in \mathbb{R},$$

describes a damped oscillator with a single degree of freedom, where $f(x) = -U'(x)$ for some (differentiable) potential function $U : \mathbb{R} \to \mathbb{R}$. If there is some damping (i.e. γ is positive), the system looses energy (which is given by $E(x, \dot{x}) = U(x) + \frac{\dot{x}^2}{2}$).

This implies in the case $\lim_{x\to\pm\infty} U(x) = \infty$ and $\{x : f(x) = 0\}$ finite the existence of a global attractor for the corresponding dynamical system on $\mathbb{R}^2$. The attractor contains all equilibria which correspond to the zeros of f (and which are stable for local minima of U).

The *Duffing oscillator* has the potential $U(x) = \frac{\beta}{4}x^4 - \frac{\alpha}{2}x^2$ with $\alpha, \beta > 0$. Hence there exists a global attractor which contains 3 equilibria, two stable ones (with $x = \pm\sqrt{\frac{\alpha}{\beta}}$, $\dot{x} = 0$) and a saddle in the origin.

If a periodic forcing is introduced, the equation (written as a first order system with $y = \dot{x}$) reads

$$\begin{aligned}\dot{x} &= y \\ \dot{y} &= \alpha x - \beta x^3 - \gamma y + \sigma\cos(\mu t)\end{aligned}$$

with additional parameters $\sigma, \mu \in \mathbb{R}$. This equation can be interpreted as a random differential equation on $\mathbb{R}^2$ with $\Omega = S^1 = \{z \in \mathbb{C} : |z| = 1\}$ and $\vartheta_t\omega = \exp(2\pi i\mu t)\omega$. Then the "noise term" takes the form $\cos(\mu t) = \Re(\vartheta_t\omega_0)$ with $\omega_0 = 1 \in \Omega$.

It can be shown that this system possesses a global "random" attractor, which looks for small σ qualitatively like the attractor of the unforced system. However, if σ increases, the dynamics becomes considerably more complicated (for a detailed description see Guckenheimer and Holmes [19, Chapter 2.2]). This is due to a transversal intersection of the stable and the unstable manifold of the "perturbed origin" (the hyperbolic fixed point persists if the perturbation does not exceed a certain size).

We go beyond this periodic forcing and consider a forcing term given by a stochastic differential dW, i.e. we have the equation

$$\begin{aligned} dx &= y\,dt \\ dy &= (f(x) - \gamma y)\,dt + c(x) \circ dW \end{aligned} \tag{$*$}$$

with $f = -U'$ and $c : \mathbb{R} \to \mathbb{R}$. In particular we will focus on the case $f(x) = \alpha x - \beta x^3$ and $c(x) \equiv \sigma$.

The corresponding Markov process has a unique invariant measure ρ with density

$$c \exp\left(\frac{-1}{\eta} E(x,y)\right) = c \exp\left(\frac{-1}{\eta}\left(U(x) + \frac{y^2}{2}\right)\right)$$

with respect to Lebesgue measure, where $c > 0$ is a normalization constant and $\eta = \frac{\sigma^2}{2\gamma}$.

Note that in particular this measure is independent of γ if we choose $\sigma = \sqrt{2\eta\gamma}$ with fixed $\eta > 0$. This situation in the Markov setup was considered by Schimansky–Geier and Herzel [35]. They observed numerically, that the top Lyapunov exponent changes its sign under variation of γ. This indicates a change of the qualitative dynamical behavior, which cannot be observed in terms of the invariant measure for the Markov process. For us this served as a motivation to investigate the stochastically forced Duffing oscillator in the framework of random dynamical systems.

5.2 Existence of a Global Random Attractor

It is quite hard to get estimates in order to show the existence of a random attractor directly from an SDE. A way out is to apply a random coordinate transformation which turns the SDE into a random differential equation. Such a transformation is possible for a large number (possibly for all) SDE's on $\mathbb{R}^d$ [22]. For a random differential equation it is often possible to obtain estimates of the type of Remark 25. If the transformed equation has an attractor, then it can be transformed back to an attractor for the original system.

This technique has been used by Keller and Schmalfuß [26] to verify the existence of a global attractor for the stochastically forced Duffing–van der Pol equation. It has been generalized by Imkeller and Schmalfuß [23]. There are also applications to stochastic partial differential equations [24,27].

In [23, Sect. 3.4] the existence of a global attractor for the Duffing equation with additive noise (i.e. $c(x) \equiv \sigma$) is proved. We will present a slight generalization of this result to the case if the potential U is an arbitrary polynomial with $\lim_{|x|\to\infty} U(x) = \infty$. However, our main purpose in this part of the paper is just to demonstrate the essential techniques which can be used to prove the existence of a global attractor for an SDE. We restrict our attention to a situation which is technically not so complicated rather than proving our result in most possible generality.

The following considerations (up to Theorem 27) are mainly based on ideas of H. Keller. They are applicable in more general situations (e.g. with non-constant $c(x)$, see [24]). However, then the calculations become considerably more elaborate.

In the first part (where we stay more general) we construct the coordinate transformation. Afterwards we present an estimate which ensures the existence of an attractor for the transformed equation.

Consider (as an auxiliary tool) the SDE $dz = -\mu z + \circ dW$ with parameter $\mu > 0$. This equation generates the so called *Ornstein–Uhlenbeck process.*

The generated random dynamical system χ has a unique stationary solution $z(\omega)$ (i.e. $\chi(t,\omega)z(\omega) = z(\vartheta_t\omega)$) [1, Theorem 5.6.5], which is the global attractor for χ [1, Corollary 5.6.6]. Note that χ is defined over the path space $\Omega = \{\omega \in C(\mathbb{R},\mathbb{R}) : \omega(0) = 0\}$ of the Wiener process. The random variable z has the explicit form $z(\omega) = -\mu \int_{-\infty}^{0} e^{\mu\tau}\omega(\tau)d\tau$ [18,26]. $t \mapsto z(\vartheta_t\omega)$ is continuous and $z(\omega)$ is normally distributed with mean 0 and variance $\frac{1}{2\mu}$.

Consider now Equation (∗) and set $v := y - c(x)z + \varepsilon x$ with $0 \le \varepsilon < \gamma$. Suppose $c \in C^2(\mathbb{R},\mathbb{R})$ with uniformly bounded derivative c'. Then

$$dx = y\,dt = (v + c(x)z - \varepsilon x)\,dt$$

and (using Îto's rule [18] and going a little bit sloppy over the technical details of the stochastic calculus)

$$\begin{aligned} dv &= dy - c(x)dz - zc'(x)\,dx - \varepsilon dx \\ &= (f(x) - \gamma(v + c(x)z - \varepsilon x))\,dt + c(x)\circ dW \\ &\quad -c(x)(-\mu z\,dt + \circ dW) - (c'(x)z + \varepsilon)(v - c(x)z - \varepsilon x)\,dt. \end{aligned}$$

In particular we observe that the "stochastic" terms $\pm c(x)\circ dW$ cancel. Hence our transformed equation is a random differential equation (with right hand side continuous in (t,x) and differentiable in x)

$$\begin{aligned} \dot{x} &= v - \varepsilon x + c(x)z \\ \dot{v} &= f(x) - (\gamma - \varepsilon)(v - \varepsilon x) \\ &\quad +(\mu + \varepsilon - \gamma)c(x)z - c'(x)z(v - \varepsilon x + c(x)z). \end{aligned} \qquad (**)$$

We are free to vary the parameters ε and μ (the latter enters also into the distribution of the random variable z) in order to get some estimates.

For simplicity we assume from now on that $c(x)$ is equal to a constant $\sigma > 0$. Hence the $c'(x)$ term in (∗∗) vanishes. The equation takes the simplest possible form if we set $\varepsilon = 0$ and $\mu = \gamma$. Then the noise is "shifted" from "white noise" in the y coordinate to "real noise" in the x coordinate.

However, our purpose is to ensure that the random dynamical system φ generated by (∗∗) satisfies the assumptions of Theorem 24 (resp. Remark 25) with the the "energy" $E(x,v) = U(x) + \frac{v^2}{2}$ as a "Lyapunov function". This will only be possible if we choose ε with $0 < \varepsilon < \gamma$. For simplicity we set $\mu := \gamma - \varepsilon$. By eventual addition of a constant we can assume $E > 0$. We have

$$\frac{\partial}{\partial t} E(x(t), v(t)) = U'(x(t))\dot{x}(t) + v(t)\dot{v}(t), \text{ i.e.}$$

$$\begin{aligned}
\dot{E} &= -f(x)(v - \varepsilon x + \sigma z) + v(f(x) - (\gamma - \varepsilon)(v - \varepsilon x) + \sigma(\mu + \varepsilon - \gamma)z) \\
&= -\varepsilon x U'(x) - (\gamma - \varepsilon)v^2 + \varepsilon(\gamma - \varepsilon)xv - \sigma z f(x) \\
&=: g(x, v, z).
\end{aligned}$$

Now we make another restriction, namely that the potential U is a polynomial in x of even degree ≥ 2. Since $|ab| \leq \frac{1}{2}(a^2 + b^2)$ for real numbers a, b we have

$$\varepsilon(\gamma - \varepsilon)xv \leq \frac{1}{2}(\gamma - \varepsilon)v^2 + \varepsilon\frac{\varepsilon(\gamma - \varepsilon)}{2}x^2.$$

If we fix $\varepsilon > 0$ sufficiently small, this implies

$$\liminf_{|(x,v)| \to \infty} \frac{xU'(x) + v^2 - \varepsilon(\gamma - \varepsilon)xv}{E(x, v)} > 0.$$

Furthermore, by our assumption on U,

$$\lim_{|(x,v)| \to \infty} \frac{f(x)}{E(x, v)} = 0.$$

Hence we can find constants $R, \alpha_0, \delta > 0$ with

$$g(x, v, z) \leq (-\alpha_0 + \delta|z|)E(x, v)$$

if $|(x, v)| \geq R$. By increasing R we get δ arbitrarily small without changing α_0. Since z is normally distributed $|z|$ has finite expectation, i.e. we can choose R and δ with $\delta \int |z| \, d\mathbb{P} < \alpha_0$. Then

$$\begin{aligned}
\alpha(\omega) &:= \delta|z(\omega)| - \alpha_0 \quad \text{and} \\
\beta(\omega) &:= \sup_{|(x,v)| \leq R} (\alpha_0 E(x, v) + g(x, v, z(\omega)))
\end{aligned}$$

satisfy the conditions of Remark 25 after Theorem 24 with $V = E$. Hence we have

Theorem 27. *Let $U : \mathbb{R} \to \mathbb{R}^+$ be a polynomial of even degree ≥ 2 with $\lim_{x \to \infty} U(x) = \infty$ and $f(x) = -U'(x)$.*

Then the random dynamical system φ generated by the SDE (*) with $c(x) \equiv \sigma > 0$ (additive noise case) possesses a global random attractor.

Furthermore, φ is related to the random dynamical system ψ generated by (**) via a linear random coordinate transformation. The system ψ has for sufficiently small $\varepsilon > 0$ also a random attractor, which is robust under perturbations in the sense of Theorem 24.

Proof. It remains to show that the existence of an attractor for ψ implies the existence of an attractor for φ. This follows from [32, Theorem 4] (and is not very hard to be observed directly). □

Remark 28. The restriction on ε is due to our proof and can be removed with little technical effort.

5.3 Numerics and Discussion

Consider the random dynamical system φ generated by $(*)$ with $f(x) = \alpha x - \beta x^3$ and $c(x) \equiv \sigma$ with parameters $\alpha, \beta, \gamma, \sigma > 0$. We expect two regimes with different qualitative dynamics depending on the sign of the top Lyapunov exponent λ_1 of the Markov measure (the second exponent λ_2 as well as $\lambda_1 + \lambda_2$ are always negative since the system is volume contracting).

We have calculated λ_1 numerically (this is done by just calculating the expansion rate of a tangent vector along a long trajectory).

With the algorithm described in Sect. 4 we calculated then approximations of the attractor and the Markov measure (at least we believe that our approximation have something to do with the Markov measure) for the two regimes.

The case $\lambda_1 > 0$ is of particular interest. The randomly forced Duffing equation is (as far as we know) the first example of an SDE on $\mathbb{R}^d$, where a Markov measure with a positive Lyapunov exponent is observed. It is also possible to prove $\lambda_1 > 0$ for certain parameters analytically (Arnold and Imkeller [3]). Hence we can believe that this observation is not just a numerical artifact.

A positive Lyapunov implies the existence of (one–dimensional) unstable manifolds [31, Chapter 6], which have to be contained in the global attractor $A(\omega)$ [34, Theorem 7.3]. This implies that $A(\omega)$ has topological dimension at least one. Furthermore, $\lambda_1 > 0$ indicates sensitive dependence on initial conditions, i.e. some sort of "chaotic dynamics" on the attractor. Since the invariant measure of the Markov process is absolutely continuous, the Markov measure μ is a random SRB measure, i.e. the conditional distributions of μ_ω on the unstable manifolds are absolutely continuous and the fiber entropy is equal to $\lambda_1 > 0$ [31, Chapter 6, Theorem 1.1]. This fiber entropy [31, Chapter 1, §2] measures the complexity of the "deterministic part" of the dynamics.

Chaotic "deterministic dynamics" on a random attractor for an SDE has been observed already in numerical simulations of the attractor of the Duffing–van der Pol equation [25]. However, there the Markov measure has two negative Lyapunov exponents and its support is only a random (two–point) subset of the attractor, whereas here we have not only "transient chaos" but a "real" chaotic attractor supporting a random SRB measure.

Our numerically produced pictures (see Fig. 2.(ii)) show mass distributed over the whole attractor $A(\omega)$. Therefore we believe to "see" the SRB measure. This measure is really "random" in the sense that it is not just a small perturbation of a deterministic SRB measure.

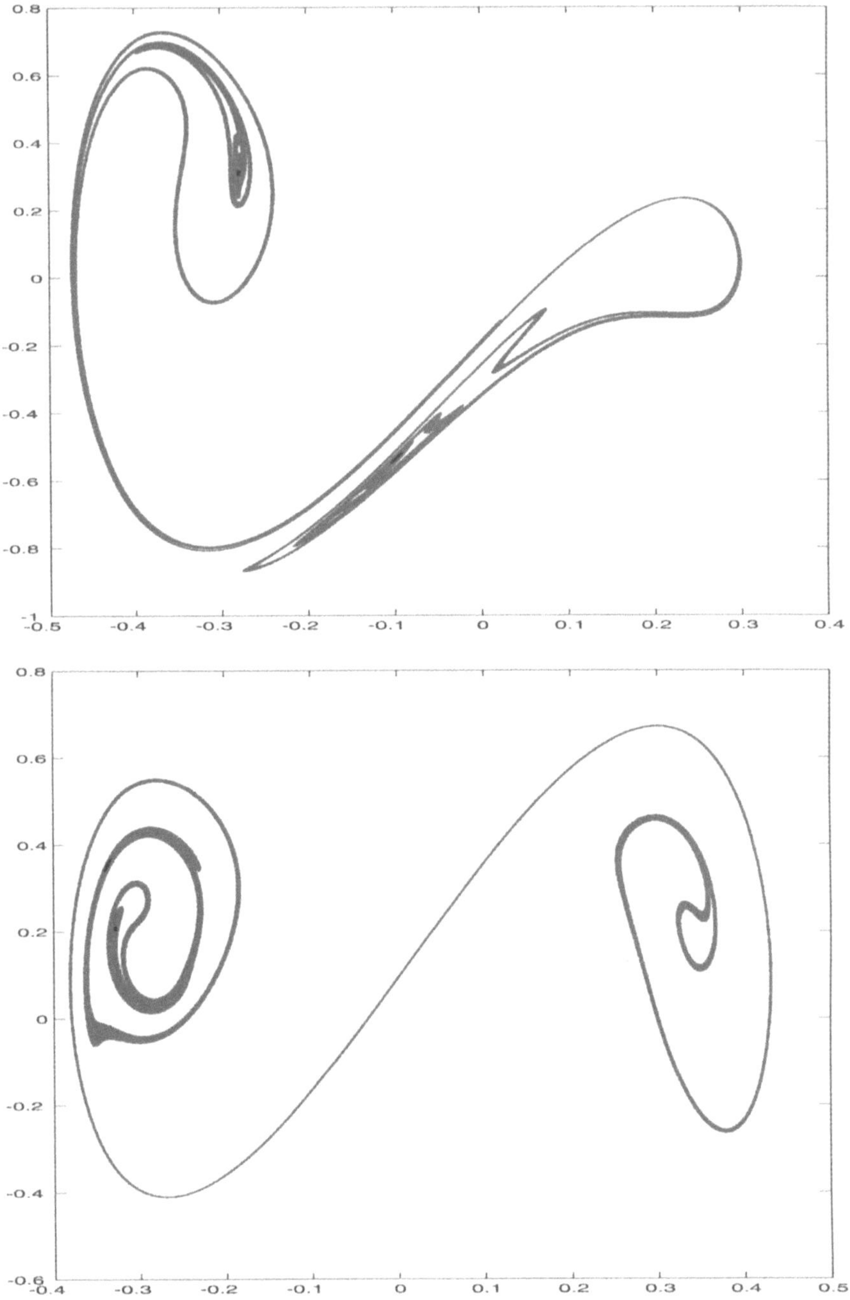

Fig. 1. Two approximations (for ω and $\theta_{2.4}\omega$) of the attractor and the Markov measure for the random Duffing oscillator in the regime $\lambda_1 < 0$. The black part has mass $1 - 10^{-6}$. The mass seems to be concentrated in two points, one of which crossing the "potential wall".

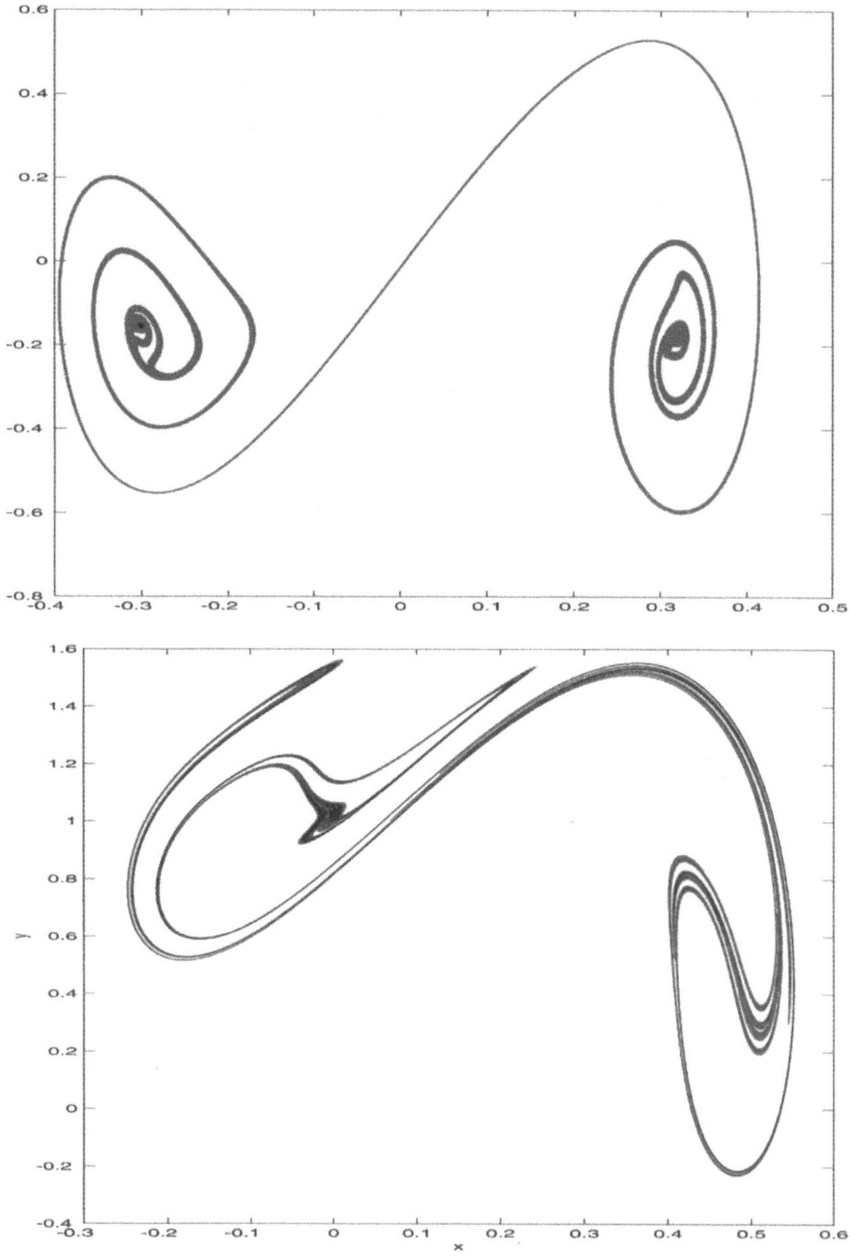

Fig. 2. (i) The final (?) picture of the random Duffing attractor in the case $\lambda_1 < 0$. The mass $(1 - 10^{-6})$ is concentrated in one (black) point.
(ii) The random Duffing attractor supporting a random SRB measure in the case $\lambda_1 > 0$. The mass distribution is indicated by different colors: 55 % green boxes with total mass 0.02, 25 % blue boxes with total mass 0.15, 16 % red boxes with total mass 0.35 and 4 % black boxes with total mass 0.5.

If we produce numerical images of $A(\vartheta_{kT}\omega)$ for $k = 0, ..., n$ and some $n \in \mathbb{N}$, $T > 0$ (i.e. we look at the time evolution of the attractor), we observe "chaotically appearing" dynamics. Frequently "noses" are growing out of the attractor and approaching another part of it.

Summarizing we come to the conclusion that stochastic noise can turn a non–chaotic deterministic system into a highly chaotic random dynamical system. This chaotic dynamics cannot be seen by only considering the corresponding Markov process (cf. the corresponding discussion by Schimansky–Geier and Herzel [35]).

Finally let us discuss the case when $\lambda_1 < 0$. By Proposition 3 the Markov measure μ_ω must have finite support. This is affirmed by our simulations.

If the algorithm runs some time, it produces a compact set which looks like a small perturbation of the deterministic attractor. The mass is concentrated in two (random) points, which are close to the stable states of the deterministic system. The shape of this set does not change significantly under further time evolution.

However, since white noise is unbounded, there should be a small but positive probability that one of these points crosses the "potential wall" near $x = 0$. This is indeed observed if we run the algorithm some more time (see Figs. 1, 2). What we see is that not the whole attractor collapses but the mass accumulates in one (random) point (which can cross the "potential wall" from time to time under further time evolution).

Hence μ_ω seems to be a Dirac measure at a point $x(\omega)$. The distribution of the random variable $x(\cdot)$ must be the (known) invariant measure for the Markov process. The invariance of μ implies $\varphi(t, \omega)x(\omega) = x(\vartheta_t\omega)$. The random set $\{x(\cdot)\}$ is a local attractor, since it has a two–dimensional stable manifold.

It is not clear (at least to the author of the present paper) whether $\{x(\cdot)\}$ is a global attractor. The numerics suggest a negative answer, but possibly there is a very small but positive probability that the whole attractor collapses to one point.

The picture suggested by the numerics is that the attractor contains besides $x(\omega)$ a compact invariant set with "horseshoe type" chaotic dynamics. "Chaos" is observed in terms of frequently appearing "noses" which move in a similar way as in the case $\lambda_1 > 0$. This phenomenon can be explained by the unboundedness of the noise and should happen for arbitrarily small $\sigma > 0$. The existence of a "horseshoe" in the case of periodic forcing is proved via transversal intersections of the stable and unstable manifold of a hyperbolic fixed point using the Smale–Birkhoff theorem, from which a version for random dynamical systems exists [20]. Anyhow, the proof of "chaos" probably cannot be carried over to the random case, because in the case of forcing by (unbounded) white noise there is no reason for the hyperbolic fixed point to persist.

The main problem here, however, is that it is not clear to which extent we can trust the numerics. Sometimes in random dynamical systems there are dynamical phenomena which happen very rarely but which are important for the asymptotic dynamics (for instance observed by Arnold et al. [2] in simulations of the stochastic Brusselator).

There is also a problem on the theoretical side. A "collapsed" random attractor (which is typical for SDE's in dimension one, see Crauel and Flandoli [13]) reflects the "very long term" behavior, but ignores important transient dynamical features. So (at least for systems with unbounded noise) a global random attractor sometimes fails to yield a satisfactory description of the qualitative dynamics.

Acknowledgment

I am very grateful to Hannes Keller, who allowed me to use his unpublished material concerning Theorem 17 and Theorem 27. He also supported me in doing the numerical work in a very helpful way.

References

1. Arnold., L,: Random Dynamical Systems. Springer, Berlin Heidelberg New York (1998)
2. Arnold, L., Bleckert, G., Schenk–Hoppé, K. R.: The stochastic Brusselator: Parametric noise destroys Hopf bifurcation. In Stochastic Dynamics, Springer, Berlin Heidelberg New York (1999) 71–92
3. Arnold, L., Imkeller, P.: Stochastic bifurcation of the noisy Duffing oscillator. In preparation
4. Arnold, L., Schmalfuß, B.: Fixed points and attractors for random dynamical systems. In IUTAM Symposium on Advances in Nonlinear Stochastic Mechanics, Kluwer, Dordrecht (1996) 19–28
5. Bahnmüller, J., Liu P.–D.: Characterization of measures satisfying the Pesin entropy formula for random dynamical systems. Journal of Dynamics and Differential Equations **10(3)** (1998) 425–448
6. Castaing, C.,Valadier, M.: Convex Analysis and Measurable Multifunctions. Volume 580 of Springer Lecture Notes in Mathematics. Springer, Berlin Heidelberg New York (1977)
7. Crauel, H.: Random Dynamical Systems: Positivity of Lyapunov Exponents, and Markov Systems. PhD thesis, Bremen (1987)
8. Crauel, H.: Extremal exponents of random dynamical systems do not vanish. Journal of Dynamics and Differential Equations **2(3)** (1990) 245–291
9. Crauel, H.: Global random attractors are uniquely determined by attracting deterministic compact sets. Ann. Mat. Pura Appl. (IV), Vol. CLXXVI (1999) 57–72
10. Crauel, H.: Random point attractors versus random set attractors. Preprint (1999)
11. Crauel, H., Debussche, A., Flandoli, F.: Random attractors. Journal of Dynamics and Differential Equations **9(2)** (1997) 307–341
12. Crauel, H., Flandoli, F.: Attractors for random dynamical systems. Probab. Theory Relat. Fields **100** (1994) 365–393
13. Crauel, H., Flandoli, F.: Additive noise destroys a pitchforf bifurcation. Journal of Dynamics and Differential Equations **10** (1998) 259–274

14. Dellnitz, M., Hohmann, A.: The computation of unstable manifolds using subdivision and continuation. In Nonlinear Dynamical Systems and Chaos, Birkhäuser (1996) 449–459
15. Dellnitz, M., Hohmann, A.: A subdivision algorithm for the computation of unstable manifolds and global attractors. Numer. Math. **75** (1997) 293–317
16. Dellnitz, M., Junge, O.: On the approximation of complicated dynamical behavior. SIAM J. Numer. Anal. **36(2)** (1998) 491–515
17. Flandoli, F., Schmalfuß, B.: Random attractors for the 3D stochastic Navier–Stokes equation with multiplicative white noise. Stochastics and Stochastics Reports **59** (1996) 21–45
18. Friedman, A.: Stochastic Differential Equations, vol. 1, Academic Press, New York (1975)
19. Guckenheimer, J., Holmes, P.: Nonlinear Oscillations, Dynamical Systems, and Bifurcations of Vector Fields, Springer, New York Berlin Heidelberg Tokyo (1983)
20. Gundlach, V. M.: Random homoclinic orbits. Random & Computational Dynamics **3** (1995) 1–33
21. Gundlach, V. M., Steinkamp, O.: Products of random rectangular matrices. Math. Nachr. **212** (2000) 51–76
22. Imkeller, P., Lederer, C.: On the cohomology of stochastic and random differential equations. Preprint (1999)
23. Imkeller, P., Schmalfuß, B.: The conjugacy of stochastic and random differential equations and the existence of global attractors. Preprint (1998)
24. Keller, H.: Attractors for hyperbolic stochastic partial differential equations. PhD thesis, in preparation
25. Keller, H., Ochs, G.: Numerical approximation of random attractors. In Stochastic Dynamics, Springer, Berlin Heidelberg New York (1999) 93–115
26. Keller, H., Schmalfuß, B.: Attractors for stochastic differential equations with nontrivial noise. Buletinul A.S. a R.M. Mathematica **1(26)** (1998) 43–54
27. Keller, H., Schmalfuß, B.: Attractors for stochastic hyperbolic equations via transformation into random equations. Report # 448, Institut für Dynamische Systeme, Bremen (1999)
28. Kloeden, P. E., Kozyakin, V.: The inflation of attractors and their discretization: the autonomous case. DANSE preprint 22/99 (1999)
29. Kloeden, P. E., Kozyakin, V.: Inflated nonautonomous pullback attractors. In preparation
30. Kloeden, P. E., Schmalfuß, B.: Asymptotic behaviour of nonautonomous difference inclusions. Preprint (1998)
31. Liu P.-D., Qian M.: Smooth Ergodic Theory of Random Dynamical Systems. Volume 1606 of Springer Lecture Notes in Mathematics, Springer, Berlin Heidelberg New York (1995)
32. Ochs, G.: Weak random attractors. Report # 449, Institut für Dynamische Systeme, Bremen (1999)
33. Oseledec, V. I.: A multiplicative ergodic theorem. Lyapunov characteristic numbers for dynamical systems. Trans. Moscow Math. Soc. **19** (1968) 197–231
34. Schenk-Hoppé, K. R.: Random attractors — general properties, existence, and applications to stochastic bifurcation theory. Discrete and Continuous Dynamical Systems **4** (1998) 99–130
35. Schimansky–Geier, L., Herzel, H.: Positive Lyapunov exponents in the Kramers oscillator. J. of Stat. Phys. **70** (1993) 141–147

Self-Similar Measures

Christoph Bandt*

Institut für Mathematik und Informatik, Arndt-Universität, 17487 Greifswald, Germany
e-mail: bandt@uni-greifswald.de

Abstract. After an introduction to self-similar deterministic and random measures, we introduce an algorithm which decides on the separation condition for given similarity mappings defining a self-similar set. We construct some new types of Sierpinski gasket and determine their exponent of interior distance.

1 Introduction

Most of the well-known examples of fractals come from the early stage of set theory and general topology at about 1900. They were designed to clarify basic concepts like "continuum", "curve" and "one-dimensional space", by demonstrating that subsets of the line or plane can have very strange, counterintuitive properties. A Cantor set is uncountable yet contains no connected pieces. A simple curve can be nowhere differentiable, it can even have positive area. The Sierpinski gasket in figure 1 has topological dimension one. Nevertheless, each point is a branching point in the sense that it can be connected by three disjoint curves to the three vertices of the triangle.

The virtue of Mandelbrot's revival of these spaces with the label "fractals" was to consider them not as pathological counterexamples, but as models for real world phenomena. This idea was eagerly accepted by many people working in the experimental sciences. In particular the Sierpinski gasket S (figure 1) was used as a universal toy model for catalysts and other porous materials. A Laplacian on S was defined and its eigenvalues determined, first by physicists ([10] gives a review of part of the literature) and later with mathematical rigour, via Dirichlet forms as well as via an appropriate Brownian motion. This theory was generalized to a class of spaces termed "post-critically finite (p.c.f.) self-similar sets with harmonic structure" [17,21,12,13].

Unfortunately, there are very few concrete examples in this class. There is a real discrepancy between the well-developed "harmonic analysis on self-similar fractals" and the small number of concrete spaces for which this theory works. Here we try to improve the situation by presenting a method which constructs plenty of new p.c.f. spaces.

---

* Project: Tangent Measures and Generalized Mandelbrot Sets (Christoph Bandt)

Fig. 1. The Sierpinski gasket S and its reversed version

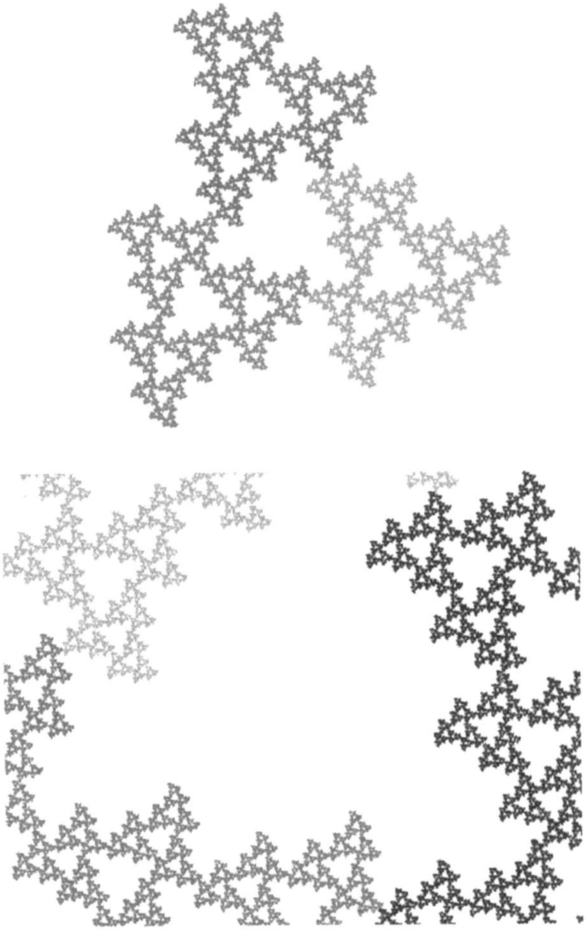

Fig. 2. A new gasket with a magnification

The set S, for instance, is still the standard example, but it has one disadvantage. Any two points of S can be connected by a curve of finite length, a kind of polygonal line. This should not be true in a proper fractal. In this paper we shall construct, by use of certain dynamical systems, the modification of S shown in figure 2. Its topological structure is very similar to that of S, but the shortest curves connecting different points are fractals themselves. They can be measured by Hausdorff measure of a certain dimension slightly larger than 1.

Before that we shall give a brief survey on self-similar deterministic and random measures and on some questions we have studied in the dynamics Schwerpunktprogramm.

2 Self-Similar Constructions

While no rigorous definition of the term "fractal" is generally accepted up to now, the intuitive idea of Mandelbrot was self-similarity: large and small pieces show the same appearance as the whole figure. A mathematical class of self-similar sets was introduced by Hutchinson in 1981, see [11,20]. A map $f : X \to X$ from a complete metric space (X, d) – usually Euclidean $\mathbb{R}^d$– into itself is called a similarity map with factor r if $d(f(x), f(y)) = r \cdot d(x, y)$ for all points x, y in X.

Given similarity maps $f_1, ..., f_m$ which are contractive, that is $r_i < 1$, a set $A \subseteq X$ is called *self-similar* with respect to the f_i if it solves the equation

$$A = f_1(A) \cup ... \cup f_m(A) .$$

The idea is that A consists of pieces which are geometrically similar to A. Substituting A on the right-hand side again by the union of pieces, we see that A consists of even smaller pieces $f_i f_j(A)$ etc. It is also rather easy to verify that among the compact subsets of X there is exactly one solution of the above equation.

Nevertheless, if one takes arbitrary similarity maps, the resulting A is not very spectacular. If the r_i are small, A will consist of tiny separate pieces – a very thin Cantor set. If the r_i are large then usually the pieces $f_i(A)$ will overlap so strongly that no pieces and no self-similarity can be recognized. Hutchinson studied a separation condition which guarantees that we can distinguish the pieces. It goes back to Moran who studied similar constructions without using mappings.

The *open set condition* says that there should exist an open set V such that the images $f_i(V)$ are pairwise disjoint and all contained in V.

One may imagine that V is like a neighbourhood of A and the $f_i(V)$ are like disjoint neighbourhoods of the pieces $f_i(A)$, which is true for Cantor sets. In general, V need not intersect A, and V is hard to find in cases like figure 3 on the left. For the Sierpinski gasket we have $f_i(x) = \frac{1}{2}(x + c_i)$ where the c_i are the three vertices of the triangle, and the interior of the triangle can be taken as V. For the other space in figure 1 we have $f_i(x) = \frac{1}{2}(-x + 3c_i)$ where the c_i denote the vertices of the big triangle in the middle, and again V can be taken as the interior of the convex hull of A, a certain hexagon. This is not true for figure 2, as can be seen in the magnification.

In figure 1 and 2, there are only finitely many points x in the intersections $A_i \cap A_j$. Moreover, when we look for preimages inside A of the form $x' = f_{i_1}^{-1}(x), x'' = f_{i_2}^{-1}(x')$ and so on, we get only finitely many points altogether. These points are considered as the boundary of the space – in figure 1 they are the vertices of V, for figure 2 they will be studied at the end of the paper. If the $A_i \cap A_j$ and the boundary are finite, the self-similar set is called post-critically finite, p.c.f. This term comes from papers on Julia sets.

The structure of self-similar sets is different from the attractors of usual dynamical systems, it rather resembles certain repellors. For example, the Julia set of the rational function $g(z) = \frac{2}{3} \cdot (z^2 - \frac{2}{z})$ is topologically identical to the Sierpinski gasket (which is not true for figure 2). On the other hand, the simplicity of the generating maps allows to study our sets in much greater detail as would be possible for hyperbolic repellors.

As for attractors and repellors, invariant measures play an important role. Given contractive similarity maps f_i and positive numbers p_i for $i = 1, ..., m$ with $p_1 + ...p_m = 1$, a Borel probability measure μ on X is called *self-similar measure* with respect to the f_i and p_i if

$$\mu = p_1 f_1(\mu) + ... + p_m f_m(\mu)$$

where $f_i(\mu)(B) = \mu(f_i^{-1}(B))$ denotes the image measure under f_i. Again there is an existence and uniqueness theorem for the solution. The choice of p_i allows to distribute the mass of μ unevenly over the pieces, which has become a standard toy model of so-called multifractals.

Since all these sets and measures look somewhat artificial, not like models of nature, there have been attempts to include randomness in the construction. The f_i and $p_i (i = 1, ..., m)$ can be chosen from a probability distribution Θ on $(Sim(X))^m \times S_m$ where $Sim(X)$ denotes the family of similarity maps and $S_m = \{(q_1, ..., q_m) | \, q_i > 0, \sum q_i = 1\}$. This should be done for each subpiece and piece of subpiece etc. in an independent and identically distributed manner. The result is a random measure – a distribution Ψ on the space $\mathcal{M}_1(X)$ of probability measures on X. Each μ is one random realization of Ψ. One example is shown in figure 3. There is again an equation describing self-similarity, not of single realizations, but of the whole random procedure:

$$\Psi = p_1 f_1(\Psi) + ... + p_m f_m(\Psi)$$

where $(f_1, ..., f_m, p_1, ..., p_m)$ is distributed with respect to Θ and $f_i(\Psi)$ means the image distribution on probability measures. This was worked out independently around 1986 by Falconer, by Graf and by Mauldin and Williams (cf. [11]). Recently, Hutchinson and Rüschendorf have simplified the existence and uniqueness theorem [16].

Fig. 3. Left: the Levy curve. Right: a random modification of S

However, there are several points which cast doubt on the usefulness of this constructive concept of random self-similarity. To get realizations with nice properties, the open set condition had to be assumed for almost all realizations, which seems a too stringent condition on the distribution Θ. So far, very few meaningful distributions Θ have been found. The randomness in the construction is unevenly distributed because the first choices decide on the overall appearance of the realization and later choices concern tiny details. Finally, one may wonder whether any fractals in nature (clouds, soil, Swiss cheese and other porous materials ...) are built by choice of random mappings.

3 Axiomatic Approach to Self-Similarity

The author has suggested an axiomatic approach to random self-similarity [4,5] which uses ideas of Mandelbrot, U. Zähle and several other authors [25,9,19,24]. Let B denote the unit ball in $\mathbb{R}^d$. A random measure on B is a function $\tilde{\mu}$ from a certain probability space Ω into the set $\mathcal{M}_1(B)$ of probability measures on B. For our purpose, we do not need this function. We shall work with the distribution of the random variable $\tilde{\mu}$ which is a probability distribution on the set $\mathcal{M}_1(B)$ and will be called Ψ. We do not want to construct Ψ by similarity maps – we just want to say under which condition Ψ is self-similar.

Let us consider an arbitrary ball C inside B. *The condition will be that an observer who sees only the restrictions to C of all realizations μ of the distribution Ψ will see the same collection of pictures as somebody who observes all complete realizations μ on B.* This will be made precise.

Let h be the homothety mapping C onto B. If C has center c and radius ϵ, then $h(y) = (y-c)/\epsilon$. For a probability measure μ on B with $\mu(C) > 0$ let $h(\mu)$ denote the image measure under h of μ restricted to C. Then

$$H(\mu) = \frac{1}{\mu(C)} \cdot h(\mu)$$

is again a probability measure. Thus we have defined a mapping H from $\mathcal{M}_1(B)$ into itself which rescales the collection of pictures of the observer of C to the size of the ball B.

Let $\mathcal{C} = \{\mu \in \mathcal{M}_1(B) | \, \mu(C) > 0\}$, and let $H(\Psi)$ denote the image measure under H of Ψ restricted to $\mathcal{C}$. In other words, we disregard all μ which give an empty view for the observer of C. Now we define

$$\Psi_C = \frac{1}{\Psi(\mathcal{C})} \cdot H(\Psi) \; .$$

Definition 3.1. (Scale-invariance of random measures)
A random measure on the unit ball B in $\mathbb{R}^d$ with distribution Ψ is called scale-invariant if $\Psi_C = \Psi$ for each ball C inside B.

This setting may seem a bit unusual but we are convinced that it is the right way to look at the intuitive concept of self-similarity. We defined a self-similarity with respect to all similarity mappings with arbitrary factors. It does not involve any exponents. Nevertheless, we obtain the existence of at least two exponents.

Proposition 3.2. (The first exponent of a scale-invariant measure)
For every scale-invariant Ψ, there is a number α between 0 and d such that for every ball $C \subseteq B$ with radius r we have

$$\Psi(\mathcal{C}) = r^{\alpha} \; .$$

The proof is simple; it was given in [5] in terms of random sets. In typical examples, $d-\alpha$ is the Hausdorff dimension of the support $supp\,\mu$ for Ψ-almost all realizations μ. However, there are a few degenerate examples consisting of discrete measures for which this does not hold.

The second exponent of a scale-invariant measure needs some more preparation since it refers to the so-called Palm measure of Ψ. Roughly speaking, if B_r denotes the ball around zero with radius r and $h(r)$ denotes the "average value" of $\mu(B_r)$ then $h(r) = r^{\beta}$. The average is taken here with respect to Ψ but we have conditioning to those μ for which *all* B_r have positive measure, and we have to take geometric averages: $h(r) = \exp(\mathbf{E}(\log \mu(B_r)))$. Thus β is the parameter called pointwise dimension for ordinary measures.

Unfortunately, not too many examples of scale-invariant measures were found so far. We can take standardized occupation time measure of Brownian motion in $\mathbb{R}^d$ with $d \geq 3$, when we start the motion with a uniformly distributed point on the surface of B. We can also consider the zero set on the time axis of one-dimensional Brownian motion, or more generally stable subordinators, with a certain standardization. There are a few other probabilistic constructions [5] but there should be many more examples connected with spatial branching processes, continuous limits of interacting particle systems, with critical percolation clusters etc. This is difficult stuff, which has motivated us to direct our work more to the field of probability.

Recently, we found that at least the deterministic self-similar Hausdorff measures with open set V also fit into the picture of scale-invariance. The idea is to take the set A not as a complete figure, but rather as a basic space from which we can sample circular "views" of part of A [24]. One can consider "centered views" around points of A which lead to the so-called tangential distribution of self-similar measures [15] but here we study views inside V which are not centered. Details will be given in a forthcoming paper.

4 An Algorithm Deciding on Separation

Now we are back to deterministic self-similar sets. The open set condition (OSC) was introduced already 1946 by Moran to prove that the Hausdorff measure $\mu^\alpha(A) > 0$ for the similarity dimension α of A. This is the unique number determined by $r_1^\alpha + ... + r_m^\alpha = 1$. Since clearly $\mu^\alpha(A) < \infty$, this means that μ^α is the natural volume function and the most simple self-similar measure on A.

Bandt and Graf [6] gave an algebraic equivalent condition. They considered the set of all maps

$$E = \{f_{i_n}^{-1} \cdot ... \cdot f_{i_1}^{-1} \cdot f_{j_1} \cdot ... \cdot f_{j_n} \mid i_1 \neq j_1, n \in N, 1 \leq i_k, j_k \leq m\}$$

Then $\mu^\alpha(A) > 0$ holds if and only if the identity map is not in the closure of E with respect to the pointwise convergence of maps on $\mathbb{R}^d$.

Schief [22] proved that this also equivalent to the OSC. Thus while originally the OSC seemed only a sufficient technical condition for positive Hausdorff measure, it has turned out that it is also necessary, and can be formulated in algebraic terms. It is the proper separation condition. While the open set is not easy to find even when you see the fractal, the algebraic condition $id \notin cl(E)$ can be used to decide on separation, given only the mappings f_i. Actually, this can be done by an algorithm indicated in [3] which can easily be implemented on a computer.

To describe the algorithm, we note that the elements of E can be generated in the form of a tree of mappings, with the identity map as root vertex. Consider the automorphisms

$$\Phi_{ij}(g) = f_i^{-1} \cdot g \cdot f_j\,, \qquad i,j = 1, ..., m$$

which act on the space of all similarity maps: $f \in Sim(\mathbb{R}^d)$. It is clear that the Φ_{ij} generate E, starting from $g = id$ and requiring $i_1 \neq j_1$ for the first step. We now want to see whether repeated action of the Φ_{ij} leads us again back to id. To simplify matters, we make the following assumption throughout the rest of the paper.

Assumption. *The factors r_i of the maps f_i are all equal.*

Let r denote the factor. The following statement is crucial.

Lemma 4.1. (The Φ_{ij} are expansive)
There is a neighbourhood U of id in $Sim(\mathbb{R}^d)$ such that the complement of U is mapped into itself by each Φ_{ij} :

$$\Phi_{ij}(g) \notin U \qquad \text{if} \quad g \notin U.$$

Proof. Let $f_i(x) = rA_i(x + a_i)$, thus $f_i^{-1}(x) = r^{-1}A_i^{-1}x - a_i$ for $i = 1, ..., m$, and $g(x) = sBx + b$, where A_i, B are orthogonal matrices and a_i, b are vectors in $\mathbb{R}^d$. Then

$$\Phi_{ij}(g) = f_i^{-1} \cdot g \cdot f_j = sCx + (sCa_j - a_i) + r^{-1}A_i^{-1}b$$

where C denotes the orthogonal matrix $A_i^{-1}BA_j$. If we let $c = \max_{i=1}^m |a_i|$ then $|sAa_j - a_i| \leq (1+s)c$ for all i, j. Moreover, $|r^{-1}A_i^{-1}b| = |b|/r$.

Thus if $|b|(\frac{1}{r} - 1) \geq (1+s)c$, then the translational part of $\Phi_{ij}(g)$ will have length at least $|b|$. The following neighbourhood U of id now fulfils the statement:

$$U = \{g \mid \ |b| \leq (1+s)c/(1-r)\}. \qquad ♣$$

Since we start with $g = id$ and $\Phi_{ij}(g)$ has the same factor as g, the number s will always remain equal to 1. In other words, we need not consider the action of the Φ_{ij} on $Sim(\mathbb{R}^d)$ but only on the subspace of isometries.

To decide whether the algebraic separation condition is true, we now can disregard all mappings which fall outside the neighbourhood U. Let us first demonstrate how our method decides whether the pieces A_i of A are disjoint, and hence A is a Cantor set.

Proposition 4.2. (Self-similar sets with disjoint pieces)
The pieces A_i of A are pairwise disjoint if and only if there exists a number n_0 such that repeated application of more than n_0 automorphisms Φ_{ij} to id, with $i_1 \neq j_1$, will always lead out of the neighbourhood U.

Proof. The mappings in E can be interpreted as "neighbour maps". Let $u = (i_1, ..., i_n)$, $v = (j_1, ..., j_n)$, and $f_v = f_{j_1} ... f_{j_n}$ and $A_v = f_v(A)$, similarly for u. Then the map $f_u^{-1}f_v$ in E is the transformation which maps A_u to A_v, renormalized to the size of A. More precisely, $f_u^{-1}f_v$ maps A to the neighbouring position $f_u^{-1}f_v(A)$ which is situated to A in a geometrically similar way as A_v is situated to A_u. See [6,3].

Now if the A_i are disjoint then there is an $\epsilon > 0$ such that any two sets A_i, A_j have distance ϵ or larger. This holds also for the subsets A_u, A_v if $i_1 \neq j_1$. So the distance between A and $f_u^{-1}f_v(A)$ is at least $r^{-n}\epsilon$. Since $f_u^{-1}f_v$ is an isometry, this means that its translational part has norm at least $r^{-n}\epsilon - 2 \cdot t$ where t denotes the maximum distance of the points of A from the origin. Thus for all n greater some n_0, the maps $f_u^{-1}f_v$ fall outside the neighbourhood U of id defined above. ♣

If we assume that the origin belongs to A – this is the case when $a_1 = 0$ – we can replace t by the diameter of A. Using the definition of U with $s = 1$, this yields the following estimate.

Corollary 4.3. (Estimate of distance between pieces)
If $a_1 = 0$ then A is a Cantor set with distance ϵ between pieces if and only if the iteration of the Φ_{ij} leads outside U after n steps, where

$$\epsilon \geq 2 \cdot r^n \cdot (\frac{c}{1-r} + \text{diam } A). \quad \clubsuit$$

In the following we shall study the graph G with (possibly infinite) vertex set $E \cap U$ where an edge is drawn from vertex g to vertex h for every (i,j) for which $h = \Phi_{ij}(g)$. Thus G is a directed graph which may have loops and multiple edges, and each edge has a corresponding mark (i,j). The vertex *id* will be considered as the root of the graph. By construction of the set E and by our lemma, every vertex can be reached from the root vertex by a finite directed path of edges. There is a very simple principle.

Principle 4.4. *For every infinite directed path of edges $e_1, e_2, ...$ starting in the root vertex id with marks $(i_1, j_1), (i_2, j_2), ...$ the two points with addresses $i_1 i_2 ...$ and $j_1 j_2 ...$ will coincide, and conversely.*

The above proposition treats the case that there is no such infinite path and in particular no directed cycle back to the root vertex *id*. Even if there were a single infinite directed path in G, then in general for $m = 3$ the set A will be a Cantor set while for $m = 2$ the set A must be connected, cf. [7].

Now, let us discuss the case that the vertex set $E \cap U$ is finite. This is true when the pieces are disjoint. When all the mappings $f_i^{-1} f_j$ are contained in a crystallographic group [14], $E \cap U$ is also finite [3]. We note that finite $E \cap U$ does not necessarily imply that the intersections $A_i \cap A_j$ are finite, see the example of the terdragon below. In the next section we shall also construct non-crystallographic examples where graph G is finite.

Proposition 4.5. *When $E \cap U$ is finite, it will be clearly generated by the Φ_{ij} in finitely many steps. If the graph contains a directed cycle back to id, then we have no separation, otherwise the OSC holds. Thus in the finite case the separation question is decidable in a finite number of steps.* $\clubsuit$

Proposition 4.6. *When $E \cap U$ is finite, existence of infinite directed paths in graph G is the same as existence of directed cycles. The intersections $A_i \cap A_j$ are finite if and only if there are only finitely many disjoint directed cycles in G. Any two directed cycles which share at least one edge will lead to an uncountable number of infinite directed paths in G. In particular this is true for a single directed cycle with a double edge.*

Proof. Let $e = (g,h)$ be a common edge of two cycles in G. Let w denote a path from *id* to g, and let w^1, w^2 be the two cycles from g over h back to g. Then for each sequence $k = k_1 k_2 k_3 ... \in \{1,2\}^\infty$, there is a corresponding infinite path $w w^{k_1} w^{k_2} w^{k_3} ...$ $\clubsuit$

5 New Sierpinski Gaskets

Clearly, the background for our algorithm comes from dynamical systems. We do not iterate a single transformation, but rather a multivalued map given by the Φ_{ij}, and the basic space is not $\mathbb{R}^d$ but the space of isometries on $\mathbb{R}^d$, with *id* as origin. Nevertheless, in the same way as for the Mandelbrot set of complex quadratic maps, it is decisive whether trajectories all go to infinite or to 0 or whether they stay in-between. Trajectories go to infinite as soon as they get out of the neighbourhood U.

In the following *we further specialize by requiring that all f_i have the same linear part,* $A_i = A$. Such systems of mappings have been considered by many authors in connection with self-affine tilings, cf. [18,3,14]. It turns out that in this case all Φ_{ij} act on the space of translations. We start with $g = id$, and if B is the identity matrix in the proof of the lemma, the matrix C of $\Phi_{ij}(g)$ will also be the identity matrix. So for $g(x) = x + b$ we can write

$$\Phi_{ij}(b) = a_j - a_i + r^{-1}A^{-1}b$$

indicating that our Φ_{ij} now act on $\mathbb{R}^d$. In dimension 2, it is more convenient to replace rA by a complex contraction factor a, and let $\lambda = 1/a$. Then $f_i(z) = a(z + a_i)$ and $f_i^{-1}(z) = \lambda z - a_i$ implies

$$\Phi_{ij}(b) = a_j - a_i + \lambda b.$$

To construct new "gaskets" like figure 2, *we now consider a self-similar set A with $m = 3$ pieces which is symmetric with respect to a 120° rotation around the origin.* More precisely, $f_3 = \overline{\omega} f_1 \omega$ and $f_2 = \omega f_1 \overline{\omega}$ where $\omega = -\frac{1}{2} + i\frac{\sqrt{3}}{2} = e^{i2\pi/3}$. If we put $a_1 = 1$, an easy calculation gives

$$f_i(z) = a(z + a_i) \quad \text{and} \quad f_i^{-1}(z) = \lambda z - a_i \quad \text{for } i = 1, 2, 3$$

with $a_1 = 1$, $a_2 = \omega$ and $a_3 = \overline{\omega}$.

Now we start with $b = 0$ and apply the Φ_{ij} repeatedly, disregarding all image points with modulus larger than $2/(1 - |a|)$ (cf. the definition of U). We wonder for which a (or λ) we get a finite set of iterates in U.

In the first step we require $i \neq j$ and obtain as possible $\Phi_{ij}(0)$ the six numbers $a_i - a_j$. These points form a regular hexagon centered at the origin, with two vertices $\pm i\sqrt{3}$ on the imaginary axis. These points are now taken as new b in the above formula for $\Phi_{ij}(b)$. We have to multiply one of the vertices of the hexagon by λ and add another of the hexagon vertices in order to obtain the result.

It seems natural to look for periodic solutions, so let us first check whether the result can again be a hexagon vertex. Since $|\lambda| > 1$, the point λb must be one of the 12 possible differences of hexagon vertices which lie outside the hexagon. This gives 12 possible values of λ, but because of the symmetry of the set A the numbers $\omega\lambda$ and $\overline{\omega}\lambda$ give exactly the same self-similar set as λ.

Thus we are left with four cases, and they are all well-known. $\lambda = 2$ gives the Sierpinski gasket, $\lambda = -2$ the "reversed version" shown in figure 1.

$\lambda = \frac{1}{2}(3 \pm i\sqrt{3})$ yield the right- and left-handed version of the terdragon [19] which tile the plane in a similar way as hexagons do. This is an example of the second case in proposition 4.6 since for each vertex b of the hexagon two Φ_{ij} can be applied successfully – one leading back to b and one to its neighbour. From this fact the symbolic description of the boundaries of the pieces A_i and of the center point $A_1 \cap A_2 \cap A_3$ could be easily derived.

Next, we look for periodic solutions where only the application of two maps Φ_{ij} will lead from one hexagon vertex to another. Thus there must be hexagon vertices w, w', w'', w''', not necessarily different, such that

$$(w\lambda - w')\lambda - w'' = w''' .$$

Dividing by w we get

$$\lambda^2 - \eta\lambda - (\psi + \xi) = 0$$

where η, ψ and ξ belong to the set of sixth roots of unity $\{\pm 1, \pm\omega, \pm\overline{\omega}\}$. Thus

$$\lambda = \eta \left(\frac{1}{2} \pm \sqrt{\frac{1}{4} + \alpha + \beta} \right)$$

where $\eta, \alpha = \psi\eta^{-2}$ and $\beta = \xi\eta^{-2}$ are again sixth roots of unity. Because of the symmetry of A, multiplication of λ by ω or $\overline{\omega}$ will lead to the same A. Hence only $\eta = \pm 1$ need to be considered. The general solution for λ now is

$$\lambda = \pm\frac{1}{2} \pm \sqrt{\frac{1}{4} + \alpha + \beta}$$

Moreover, since the complex conjugate of λ leads to a "left-handed version" of A, it suffices to study those α, β with $Im(\alpha + \beta) \geq 0$.

There are 12 possible choices left for $\alpha + \beta$ but not all of them lead to an admissible λ. It is well known that for $r < 1/2$ the set A will be a Cantor set while for $r > 1/\sqrt{3}$ the open set condition cannot hold because the similarity dimension is larger than 2. This implies

$$\sqrt{3} \leq |\lambda| \leq 2 .$$

Let us now study all possible 12 sums $\alpha + \beta$ of two sixth roots of unity in the upper half-plane. On the real line we have the values $-2, -1, 0, 1$ which all give too small $|\lambda|$, and $\alpha + \beta = 2$ which yields the ordinary Sierpinski gasket and its reversed version. Obviously, periodic orbits of period 1 of a dynamical system will also be obtained when we search for orbits of period two.

Next, on the line $Im\, z = \sqrt{3}/2$, the possible values for the real part of $\alpha + \beta$ are $\pm 1/2$ and $\pm 3/2$. While $-3/2$ and $\pm 1/2$ give too small $|\lambda|$, the value

$\alpha + \beta = 3/2 + i\sqrt{3}/2$ gives an admissible solution $\lambda \approx 1.8606 + i \cdot 0.3182$ with $|\lambda| \approx 1.8876$ and rotational angle $arg\lambda \approx 9.706^o$. See figure 4. Below we prove that in this case the number of infinite directed paths in the graph G is uncountable. Thus each intersection $A_i \cap A_j$ is a Cantor set which is also indicated by the magnification.

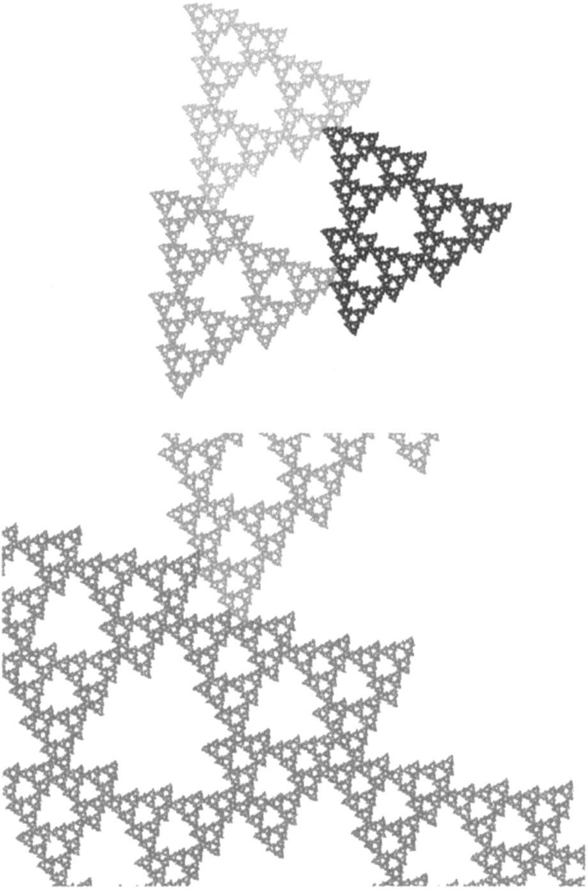

Fig. 4. A new gasket which is not p.c.f. with a magnification

Finally, on the line $Im\, z = \sqrt{3}$ we have the real part 0, which leads to the terdragon, and ± 1. For -1 we get $|\lambda| \approx 1.70$ which is a bit too small. For $\alpha + \beta = 1 + i\sqrt{3}$ we obtain figure 2. $\lambda \approx 1.8011 + i \cdot 0.6656$ has modulus about 1.92015. Thus $r \approx 0.5208$, and the angle of rotation is about 20.28^o. For $-\lambda$ we get the reversed version of figure 2, see figure 5.

For these two figures it can be easily checked that $E \cap U$ is finite, and there is exactly one cycle which alternates between the vertices of the basic hexagon and another hexagon. The point is that $\alpha = \beta$, more precisely $w = w'$,

$w'' = w''' = w(1 + i\sqrt{3})/2$. Hence there is one infinite directed path from id through each vertex of the basic hexagon, and since two paths marked $(i_1, j_1), (i_2, j_2)...$ and $(j_1, i_1), (j_2, i_2)...$ describe the same pair of addresses, these paths describe exactly three intersection points between the pieces A_i, as can be seen in the figures.

In the example of figure 4, α and β are different and can be interchanged. The same holds for w'' and w'''. From w there is an edge to $w(\lambda - 1)$, and from there we have two edges, to $w \cdot (1 + i\sqrt{3})$, and another back to w. This explains the Cantor set structure of $A_i \cap A_j$.

The methods presented here can easily be generalized in different ways. We can consider orbits of period three and higher, determined by the roots of polynomials of degree larger than 2. We can study 4 or more maps instead of 3, and we can also drop the assumption of symmetry. In each of these cases, calculations become more intricate, and a lot of new and interesting examples of fractals can be determined.

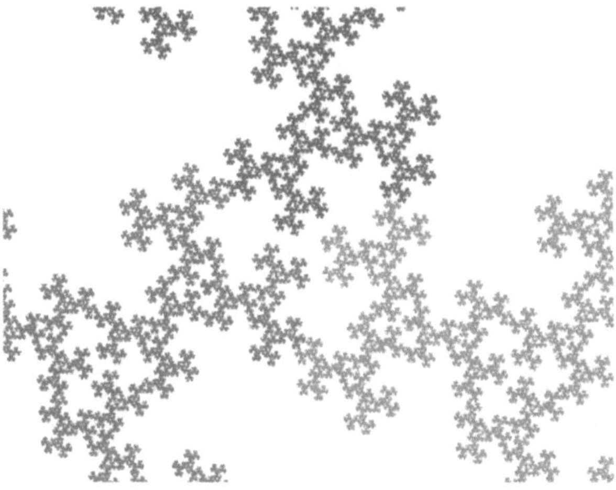

Fig. 5. Magnification of the reversed version of Fig. 2.

6 The Exponent of Interior Metric

In this section, we determine shortest paths between points in figure 2. First we study the boundary or post-critical set of A. It will turn out it consists of the 12 points labelled a_i, b_i, c_i, d_i with $i = 1, 2, 3$ in figure 6. We have $f_i(a_i) = b_i, f_i(b_i) = c_i, f(c_i) = d_i$ and $f_{i+1}(d_i) = a_{i+1}$ so that the points form a 12-cycle in counterclockwise order. Here + is mod 3, that is, 3+1=1. The two points $f_i(c_{i+1})$ and $f_{i+1}(a_i)$ coincide – this is where pieces touch each other.

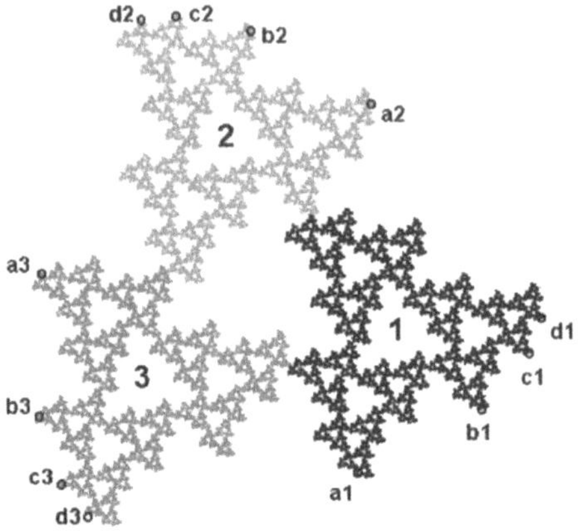

Fig. 6. The boundary of our gasket

In [8] it was shown that the system of shortest paths between boundary points is itself a system of self-similar sets, determined by a system of set equations. Because of symmetry, we need only consider 16 different shortest paths $P(x_i, y_{i+1})$, with $x, y \in \{a, b, c, d\}$. Each of these passes through one "touching point", and hence is the union of two shortest paths between boundary points in the pieces, that is, of two other shortest paths scaled down by r. Because of the open set condition, the system of these paths has a similarity dimension η. The Hausdorff measure of dimension η of shortest paths defines an interior metric on A [8]. In particular, if the length of a path P is ℓ, then the length of a scaled-down version of that path is $r^\eta \ell$.

Let $\gamma = r^\eta$, and let us introduce the notation xy for the shortest distance between x_i and y_{i+1} (more precisely, $xy = \mu^\eta(P(x_i, y_{i+1}))$ [8]) where x, y denote any two symbols in $\{a, b, c, d\}$. Then we immediately obtain the following equations which reflect the division of any path into two smaller paths.

$$aa = \gamma cd + \gamma^2 dd,\ ab = \gamma(cd + aa),\ ac = \gamma(cd + ab),\ bd = 2\gamma ac,$$

$$bc = \gamma(ac + ab),\ cc = \gamma(bc + ab),\ cd = \gamma(bc + ac),\ dd = \gamma(cc + ac).$$

To fix a multiplicative constant, let $dd = 1$. Now all variables can be expressed in terms of γ and cd.

$$aa = \gamma cd + \gamma^2,\ ab = (\gamma + \gamma^2)cd + \gamma^3,\ ac = (\gamma + \gamma^2 + \gamma^3)cd + \gamma^4,$$

$$bd = 2\gamma ac,\ ,bc = (2\gamma^2 + 2\gamma^3 + \gamma^4)cd + (\gamma^4 + \gamma^5),$$
$$cd = (\gamma^2 + 3\gamma^3 + 3\gamma^4 + \gamma^5)cd + (2\gamma^5 + \gamma^6).$$

Now cd is obtained as a rational function of γ . Substituting in

$$1 = \gamma(cc + ac) = (\gamma^2 + 2\gamma^3 + 4\gamma^4 + 2\gamma^5 + \gamma^6)cd + (2\gamma^5 + \gamma^6 + \gamma^7)$$

we get the following result.

Proposition 6.1. (Exponent of interior metric for figure 2)
γ is the real root of the polynomial $\gamma^7+\gamma^6+3\gamma^5+3\gamma^4+3\gamma^3+\gamma^2-1$. *Thus* $\gamma \approx 0.51141$. *The exponent of interior metric is* $\eta = \frac{\log \gamma}{\log r} \approx 1.028$.

This is not very far from 1. Note that the Hausdorff dimension of A is $\frac{\log 3}{-\log r} \approx 1.684$, somewhat larger than the dimension $\frac{\log 3}{\log 2} \approx 1.585$ of the Sierpinski gasket.

References

1. M. Arbeiter and N. Patzschke, Random self-similar multifractals, Math. Nachr. 181 (1996), 5-42
2. C. Bandt, Self-similarity and probability: Parameters describing the geometry of Cantor sets, Bull. London Math. Soc. 30 (1998), 1-10
3. C. Bandt, Self-similar tilings and patterns described by mappings, in: Mathematics of Aperiodic Order (ed. R. Moody) Proc. NATO Advanced Study Institute C489, Kluwer 1997, 45-83
4. C. Bandt, Random self-similarity I. Scale invariance of random sets, Preprint 2/96 of the Schwerpunktprogramm "Dynamical systems", revised version 1997
5. C. Bandt, Note on the axiomatic approach to self-similar random sets and measures, Workshop Fractals and Dynamics (eds. M. Denker et al), Mathematica Gottingensis 5 (1997), 45-49
6. C. Bandt und S. Graf, Self-similar sets 7. A characterization of self-similar fractals with positive Hausdorff measure, Proc. Amer. Math. Soc. 114 (1992), 995-1001
7. C. Bandt und K. Keller, Self-similar sets 2. A simple approach to the topological structure of fractals, Math. Nachr. 154 (1991), 27-39
8. C. Bandt and J. Stahnke, Self-similar sets 6. Interior distance on deterministic fractals, unpublished manuscript, Greifswald 1990.
9. T. Bedford and A.M. Fisher, Analogues of the Lebesgue density theorem for fractal sets of reals and integers, Proc. London Math. Soc. (3) 64 (1992), 95-124
10. A. Bunde and S. Havlin, Fractals and Disordered Systems, Springer 1991
11. K.J. Falconer, Fractal Geometry, Wiley, Chichester 1990
12. K.J. Falconer, Techniques in Fractal Geometry, Wiley, Chichester, 1997
13. K.J. Falconer, Semilinear PDEs on self-similar fractals, Preprint, St. Andrews 1999
14. G. Gelbrich, Crystallographic reptiles, Geometria Dedicata 51 (1994), 235-256
15. S. Graf, On Bandt's tangential distribution for self-similar measures, Monatshefte Math. 120 (1995), 223-246
16. J.E. Hutchinson and L. Rüschendorf, Random fractal measures via the contraction method, Indiana Univ. Math. J. 47 (1998), 471-487
17. J. Kigami, Harmonic calculus on p.c.f. self-similar sets, Trans. Amer. Math. Soc. 335 (1993), 721-755
18. J.C. Lagarias and Y. Wang, Self-affine tiles in $\mathbb{R}^n$, Advances in Math. 121 (1996), 21-49
19. B.B. Mandelbrot, The Fractal Geometry of Nature, Freeman, San Francisco, 1982.
20. P. Mattila, Geometry of Sets and Measures in Euclidean Spaces, Cambridge University Press 1995

21. U. Mosco, Lagrangian metrics on fractals, Proc. Symp. Appl. Math. 54 (1998), 301-323
22. A. Schief, Separation properties of self-similar sets, Proc. Amer. Math. Soc. 112 (1994), 111-115
23. R.S. Strichartz, A. Taylor and T. Zhang, Densities of self-similar measures on the line, Experimental Math. 4 (1995), 101-128
24. K.R. Wicks, Fractals and Hyperspaces, Lecture Notes in Math. 1492, Springer-Verlag, Berlin and New York, 1991
25. U. Zähle, Self-similar random measures I. – Notion, carrying Hausdorff dimension, and hyperbolic distribution, Probab. Th. Rel. Fields 80 (1988), 79-100

Continuation of Low-Dimensional Invariant Subspaces in Dynamical Systems of Large Dimension

Wolf-Jürgen Beyn, Winfried Kleß, and Vera Thümmler*

Fakultät für Mathematik, Universität Bielefeld
Postfach 100131, 33501 Bielefeld, Germany

Abstract. We present a continuation method for low-dimensional invariant subspaces of a parameterized family of large and sparse real matrices. Such matrices typically occur when linearizing about branches of steady states in dynamical systems that are obtained by spatial discretization of time-dependent PDE's. The main interest is in subspaces that belong to spectral sets close the imaginary axis. Our continuation procedure provides bases of the invariant subspaces that depend smoothly on the parameter as long as the continued spectral subset does not collide with another eigenvalue. Generalizing results from [32] we show that this collision generically occurs when a real eigenvalue from the continued spectral set meets another eigenvalue from outside to form a complex conjugate pair. Such a situation relates to a turning point of the subspace problem and and we develop a method to inflate the subspace at such points.

We show that the predictor and the corrector step during continuation lead to bordered matrix equations of Sylvester type. For these equations we develop a bordered version of the Bartels-Stewart algorithm which allows to reduce the linear algebra to solving a sequence of bordered linear systems.

The numerical techniques are illustrated by studies of the stability problem for traveling waves in parabolic systems, in particular for the Ginzburg-Landau and the FitzHugh-Nagumo equation.

1 Introduction

This paper is a special outcome of the DANSE project on 'Connecting orbits in dynamical systems of large dimension'. It focusses on a certain part of the project which deals with the realization of numerical techniques for dynamical systems (bifurcation, invariant manifolds, stability and spectral analysis) in higher dimensions, in particular with the efficient approximation of parameter dependent invariant subspaces. The specific results on connecting orbits will be documented elsewhere. We first summarize the main results of the paper and then relate our work to known approaches in the literature, in particular to other projects of the DANSE program.

* Project: Connecting Orbits in Dynamical Systems of Large Dimension (Wolf-Jürgen Beyn)

1.1 Outline of the Paper

Low-dimensional invariant subspaces of parametrized large matrices $A(s)$ play an important role in the numerical analysis of dynamical systems. They typically occur as Jacobians $A(s) = D_u F(u(s), \lambda(s))$ at branches $(u(s), \lambda(s))$ of steady states for a dynamical system

$$\dot{u} = F(u, \lambda).$$

If this system arises from a spatial discretization of a partial differential equation then the matrices will be large and sparse. Invariant subspaces that belong to parts of the spectrum which is close to zero or to the imaginary axis provide the information about stability, bifurcation and, more generally, about locally invariant manifolds, see e.g. [5], [17], [27], [3], [6]. In current bifurcation and continuation packages such as [11], [28] the associated eigenvalues are used for detecting and locating singular points on branches as well as for branch switching. However, these eigenvalues are usually obtained from a full resolution of the spectrum. For transferring the bifurcation techniques to large and sparse systems it will be important to incorporate the continuation of low-dimensional invariant subspaces into such software.

In this paper we develop the details of a predictor-corrector method for low-dimensional invariant subspaces of parameter-dependent large matrices. By a bordered version of the well-known Bartels-Stewart algorithm (cf. [14]) we are able to reduce the linear algebra work to solving several linear systems with bordered matrices of the type

$$\begin{pmatrix} A(s) - \tau I & \Phi \\ \hat{\Phi}^T & 0 \end{pmatrix}, \ \tau \in \mathbb{C}, A(s) \in \mathbb{R}^{m,m}, \ \Phi, \hat{\Phi} \in \mathbb{R}^{m,k}, k \ll m. \tag{1}$$

Here τ is in or close to the spectrum of $A(s)$ and k is the dimension of the invariant subspaces. Linear systems of this type occur quite frequently in bifurcation problems and several approaches have been developed for their stable and efficient solution, see [7], [15], [25], [36]. We use the method of mixed block elimination [15], [19] which requires only a black box solver for the (almost singular) principle submatrix $A(s) - \tau I$ and its transpose. This solver should be normwise backward stable in the sense of [22].

Our continuation method is independent of any multiplicities of eigenvalues that occur within the invariant subspace. However, we are forced to update the dimension when a real eigenvalue of the continued spectral subset meets an eigenvalue from outside to form a complex conjugate pair (the opposite movement presents no difficulties since real and imaginary parts of complex conjugate eigenvectors are always included in the subspace which then passes smoothly through the formation of a pair of real eigenvalues). In this case we inflate the subspace by generalizing the method from [30], [32] where the continuation of single eigenvalues is considered. In this situation the given parameter s (which is usually some arclength from the original nonlinear problem) is no longer suitable for parametrization and another arclength parameter for the subspace problem is introduced. The double real eigenvalue then appears as a turning point of the subspace problem and from this the relevant update vectors can be computed easily. Again, the linear systems can be reduced to solving with matrices of type (1) where now the size of the bordering increases by 1.

Several additional features of the current implementation of the algorithm will be discussed in section 5, such as starting procedures, reorthogonalization of the matrices in the normalizing conditions, computation of invariant subspaces for the adjoint matrices and weighted inner products for the pseudo arclength condition. A particular issue is the coupling of subspace continuation and nonlinear branch continuation, called simultaneous branch following. For systems where the matrices are not available analytically this seems to be the only approach which is conceptionally clear and at the same time avoids nested iterations and conflicts of different step size controls.

Our approach is tested on the spectra of several large matrices that occur for linearizations about traveling waves in parabolic systems. More details on the specific properties of these examples will be given below.

1.2 Related References and Projects

As mentioned above, the papers [30], [32], [20] treat the transition between real eigenvalues and complex conjugate pairs (and more general singularities of higher codimension) from the bifurcation point of view. There the focus is on the homotopy method for analyzing the spectrum of a single matrix and no precautions are taken to handle large and sparse systems.

A related approach has been developed in [37] where an extra attempt is made to ensure that the continued eigenvalue is the rightmost one which is currently not intended in our approach. To achieve such a goal it seems necessary to use repeatedly along the branch some potentially slow iterations such as inverse subspace iteration or even integration of the time-dependent system (compare the work of Lubich and others [31], [23]). In general this question is related to the so called Hopf detection problem, i.e. to detect parameter values where complex conjugate pairs (that do not belong to the current subspace) cross the imaginary axis. We refer to [13] for various approaches and an overview of this problem.

Another continuation strategy for invariant subspaces is proposed in the recent work [9] where it is used to update the boundary conditions for connecting orbits. In order to obtain a neighboring block Schur decomposition the authors take a deflation approach (apply Newton's method to the associated Riccati equation and solve a Sylvester equation in each step) in contrast to the inflation approach of the current paper (apply Newton's method to a bordered invariant subspace equation and use a bordered Bartels Stewart algorithm in each step). There is no proposal in [9] on how to handle large, sparse systems.

The current paper is an extension of an earlier version [3]. The method grew out of the thesis of the second author [26] (supported by DANSE) and was further developed and applied in the diploma thesis of the third author [42]. In [26], [2] the techniques of this paper are transferred to the problem of computing higher order Taylor terms of locally invariant manifolds. It turns out that the higher derivatives satisfy so-called multilinear Sylvester equations. Using Bartels-Stewart techniques as above these equations can be arranged in such a way that only a series of equations with bordered matrices as in (1) has to be solved. The number τ in (1) is now a sum of eigenvalues from the subspace and the well-posedness of the matrix in (1) is guaranteed by nonresonance conditions (or the stronger gap conditions) for the decomposition of the spectrum.

Our bordering approach for subspaces is strongly related to the so-called generalized Liapunov-Schmidt method pursued in Böhmer's project, see [5], [6]. There the nonlinear system itself is bordered by extra unknowns and extra equations and this is used to analyze bifurcation points numerically and to investigate approximation methods. We also mention the work in [39], [40] where Lanczos and Arnoldi procedures in Krylov spaces are discussed in the context of continuation. It is, however, not clear how these iterative methods perform for bordered systems of the type (1). Moreover, when applied directly (i.e. without a rational pretransformation) to the subspace problem, Arnoldi methods tend to produce invariant subspaces that belong to extremal eigenvalues. Therefore it may be difficult to treat some of the examples below. There the critical part of the spectrum is almost enclosed by the remaining parts and it seems inevitable to use some type of Newton's method to obtain the invariant subspace.

In our applications we consider the stability problem for traveling waves in reaction diffusion systems in 1D space. The particular difficulties in this case arise from the essential spectrum that appears for the second order linear operator on the real line. For discretizations (truncation to a finite interval with asymptotic boundary conditions plus discretization on a mesh) this creates clusters of eigenvalues that have to be separated from a few isolated eigenvalues. In [4] we investigate the influence of asymptotic boundary conditions on the point spectra of the linearized equations. More specific information on how the essential spectrum breaks up is obtained in the work of Sandstede and Scheel [38].

A model problem for traveling waves is provided by the complex Ginzburg-Landau equations the analysis of which is part of Mielke's project, see [33], [34], [35]. In fact, our project benefitted a lot from A. Mielke's suggestion to investigate the stability of the classical Hocking-Stewartson pulse [24] in certain parameter regions. Due to extra symmetries of the Ginzburg-Landau equation a double zero eigenvalue appears which even becomes triple at a certain parameter value, compare [34]. Since the corresponding eigenfunctions are all captured by the continued subspace this behavior could be resolved easily. Our numerical computations suggest that the Hocking-Stewartson pulse is always unstable, see section 3 for details.

In section 5 we apply our method to the classical equations of FitzHugh Nagumo for nerve conduction [12] . There we encounter an isolated eigenvalue in the continued invariant subspace that moves towards a cluster of eigenvalues. At such a point our continuation procedure currently terminates and an update procedure for such situations has still to be developed.

2 Smooth Branches of Invariant Subspaces

In this section we consider a family of matrices $A(s) \in \mathbb{R}^{m,m}$ that depends smoothly on a parameter $s \in \mathbb{R}$ and we ask for a smooth family of invariant subspaces of dimension $k \ll m$. Formally we try to find smooth matrices $\Phi(s) \in \mathbb{R}^{m,k}$, each of rank k, such that $\mathcal{R}(\Phi(s))$ is an invariant subspace of $A(s)$, i.e.

$$A(s)\Phi(s) = \Phi(s)\Lambda(s) \tag{1}$$

for some $\Lambda(s) \in \mathbb{R}^{k,k}$. The matrices $\Phi(s)$ will be normalized by

$$\hat{\Phi}^T\Phi(s) = \hat{\Phi}^T\Phi_0 \tag{2}$$

where $\hat{\Phi}, \Phi_0 \in \mathbb{R}^{m,k}$ are suitable rank k matrices.

First we extend the notion of a simple invariant subspace from [41] as follows.

Definition 1. For $\mathbb{K} = \mathbb{R}$ or $\mathbb{C}$ let $X \subset \mathbb{K}^m$ be an invariant subspace of $A \in \mathbb{K}^{m,m}$ and let $E \subset \mathbb{K}^m$ be the unique maximal invariant subspace of A satisfying

$$X \subset E,\ \sigma(A|_E) = \sigma(A|_X). \tag{3}$$

Then the subspace X is called *simple* if $X = E$ and it is called *multiple* in case $X \subsetneq E$ with *multiplicity*

$$\dim E - \dim X + 1. \tag{4}$$

It is clear from this definition that the multiplicity of a subspace is obtained from the Jordan normal form of A by collecting and completing all blocks that have a diagonal entry in the spectrum of $A|_X$. In the real case $A \in \mathbb{R}^{m,m}$ this spectrum is always symmetric with respect to the real axis.

The following theorem extends the well known fact that simple eigenvalues and their eigenvectors are regular solutions of a suitably normalized system of equations.

Theorem 2. *Let $A \in \mathbb{R}^{m,m}$, $\hat{\Phi}, \Phi_0 \in \mathbb{R}^{m,k}$, $\Lambda_0 \in \mathbb{R}^{k,k}$ be given such that $\hat{\Phi}^T\Phi_0$ is nonsingular.*

Then $\mathcal{R}(\Phi_0)$ is a simple invariant subspace of A with $A\Phi_0 = \Phi_0\Lambda_0$ if and only if the pair $(\Phi_0, \Lambda_0) \in \mathbb{R}^{m,k} \times \mathbb{R}^{k,k}$ is a regular solution of the equation

$$T(\Phi, \Lambda) = \begin{pmatrix} A\Phi - \Phi\Lambda \\ \hat{\Phi}^T\Phi - \hat{\Phi}^T\Phi_0 \end{pmatrix} = 0. \tag{5}$$

Proof. Regularity means that the total derivative

$$DT(\Phi_0, \Lambda_0)(H, \Delta) = \begin{pmatrix} AH - H\Lambda_0 - \Phi_0\Delta \\ \hat{\Phi}^T H \end{pmatrix} \tag{6}$$

is a nonsingular linear map in $\mathbb{R}^{m,k} \times \mathbb{R}^{k,k}$. Note that $\operatorname{rank}(\Phi_0) = k$ follows from the nonsingularity of $\hat{\Phi}^T \Phi_0$.

First, assume that $\mathcal{R}(\Phi_0)$ is a simple invariant subspace and consider $(H, \Delta) \in \mathcal{N}(DT(\Phi_0, \Lambda_0))$. Then this implies

$$A(\Phi_0, H) = (\Phi_0, H)\begin{pmatrix} \Lambda_0 & \Delta \\ 0 & \Lambda_0 \end{pmatrix} \tag{7}$$

and hence $\mathcal{R}(\Phi_0, H)$ is an invariant subspace of A with spectrum $\sigma(\Lambda_0)$. By our assumption $\mathcal{R}(\Phi_0, H) = \mathcal{R}(\Phi_0)$ and hence $H = \Phi_0 B$ for some $B \in \mathbb{R}^{k,k}$. Furthermore

$$0 = \hat{\Phi}^T H = \hat{\Phi}^T \Phi_0 B$$

implies $B = 0$ and $H = 0$, $\Delta = 0$. Thus $DT(\Phi_0, \Lambda_0)$ is nonsingular.

Conversely, if (Φ_0, Λ_0) solves (5) and $\mathcal{R}(\Phi_0)$ has multiplicity ≥ 2 then there exists an eigenvalue $\mu \in \sigma(\Lambda_0)$ and a vector $\varphi \in \mathbb{C}^m$, $\varphi \notin \Phi_0(\mathbb{C}^k)$ such that

$$(A - \mu I)\varphi = \Phi_0 c \tag{8}$$

for some $c \in \mathbb{C}^k$. Choose $Q \in \mathbb{C}^{k,k}$ such that $Q^{-1}\Lambda_0 Q$ is upper triangular with (k, k)-entry μ. Then with arbitrary $\gamma \in \mathbb{C}^k$ we define

$$H = (0, \ldots, 0, \Phi_0\gamma + \varphi)Q^{-1}$$

and obtain from (8)

$$\begin{aligned} AH - H\Lambda_0 &= [(0, \ldots, 0, A\varphi + \Phi_0\Lambda_0\gamma) - (0, \ldots, 0, \varphi + \Phi_0\gamma)Q^{-1}\Lambda_0 Q]Q^{-1} \\ &= \Phi_0(0, \ldots, 0, c + (\Lambda_0 - \mu I)\gamma)Q^{-1} = \Phi_0\Delta \end{aligned}$$

Determining γ from $\hat{\Phi}^T\Phi_0\gamma = -\hat{\Phi}^T\varphi$ we find that $(H, \Delta) \in \mathbb{C}^{m,k} \times \mathbb{C}^{k,k}$ is a nontrivial element in the (complex) nullspace of $DT(\Phi_0, \Lambda_0)$. Since $DT(\Phi_0, \Lambda_0)$ is a real operator, this proves the assertion.

We return to the parameter dependent setting and note that $(\Phi(s), \Lambda(s))$ from (1),(2) are solutions of

$$T(\Phi, \Lambda, s) = \begin{pmatrix} A(s)\Phi - \Phi\Lambda \\ \hat{\Phi}^T\Phi - \hat{\Phi}^T\Phi_0 \end{pmatrix} = 0. \tag{9}$$

If $\mathcal{R}(\Phi_0)$ is a simple invariant subspace of $A(s_0)$ then, according to Theorem 2, we can apply the implicit function theorem to (9) at (Φ_0, Λ_0, s_0) and obtain a branch of solutions $(\Phi(s), \Lambda(s))$ for small $|s - s_0|$. This branch can be continued as long as the subspaces stay simple (see the next section for the details of the continuation method).

As is well known from bifurcation theory the parametrization by the given parameter breaks down at a *turning point* of (9) and this is the only possibility in generic one-parameter systems. A turning point (Φ_0, Λ_0, s_0) of (9) satisfies

$$\dim \mathcal{N}(D_{\Phi,\Lambda}T^0) = 1, \text{ e.g. } \mathcal{N}(D_{\Phi,\Lambda}T^0) = \text{span}\{(H_0, \Delta_0)\} \tag{10}$$

$$D_s T^0 \notin \mathcal{R}(D_{\Phi,\Lambda}T^0), \tag{11}$$

where the upper index "0" indicates evaluation at (Φ_0, Λ_0, s_0). Then there is a smooth branch

$$(\Phi(t), \Lambda(t), s(t)), \ |t| < t_0, \ s(0) = s_0, \ s'(0) = 0 \tag{12}$$

passing through (Φ_0, Λ_0, s_0). Generically, the turning point will be quadratic, i.e. $s''(0) \neq 0$ which is known to be equivalent to the second order condition

$$D^2_{\Phi,\Lambda}T^0(H_0, \Delta_0)^2 \notin \mathcal{R}(D_{\Phi,\Lambda}T^0). \tag{13}$$

In the context of single eigenvalues and eigenvectors this situation (and more general ones) have been analyzed in [20], [30], [32]

The following theorem characterizes conditions (10), (13) and is a generalization of [32] to invariant subspaces. For this result we can omit the parameters.

Theorem 3. *Let the assumptions of Theorem 2 hold and let (Φ_0, Λ_0) be a solution of (5). Then the turning point conditions*

$$\mathcal{N}(DT^0) = \text{span}\{(H_0, \Delta_0)\} \text{ for some } (H_0, \Delta_0) \neq 0 \tag{14}$$

$$D^2T^0(H_0, \Delta_0)^2 \notin \mathcal{R}(DT^0) \tag{15}$$

hold if and only if $\mathcal{R}(\Phi_0)$ is an invariant subspace of multiplicity 2 and the vector needed to make $\mathcal{R}(\Phi_0)$ maximal invariant is a generalized eigenvector that belongs to a real eigenvalue $\mu \in \sigma(\Lambda_0)$ of algebraic multiplicity 2. This generalized eigenvector spans the columns of H_0 which is a rank 1 matrix.

Proof. First from (6) one calculates that

$$D^2T^0(H, \Delta)^2 = (-2H\Delta, 0), \ H \in \mathbb{R}^{m,k}, \Delta \in \mathbb{R}^{k,k}. \tag{16}$$

Therefore equations (14), (15) state that (H_0, Δ_0) is the only solution (up to constant multiples) of the system

$$AH - H\Lambda_0 - \Phi_0\Delta = 0, \qquad \hat{\Phi}^T H = 0 \tag{17}$$

and there is no solution (H, Δ) of the system

$$AH - H\Lambda_0 - \Phi_0\Delta = H_0\Delta_0, \ \ \hat{\Phi}^T H = 0. \tag{18}$$

Let us first assume that (14), (15) hold. From (17) and (7) we have that $\mathcal{R}(\Phi_0, H_0)$ is an invariant subspace of A with spectrum $\sigma(\Lambda_0)$. Since $\hat{\Phi}^T \Phi_0$ is nonsingular and $\hat{\Phi}^T H_0 = 0$ we obtain that $\mathcal{R}(\Phi_0)$ is a proper subspace of $\mathcal{R}(\Phi_0, H_0)$ and hence $\mathcal{R}(\Phi_0)$ has multiplicity $p+1 \geq 2$. Therefore we find a matrix $\Phi_1 \in \mathbb{R}^{m,p}$, $p \geq 1$ such that $\operatorname{rank}(\Phi_0, \Phi_1) = k + p$ and

$$A(\Phi_0, \Phi_1) = (\Phi_0, \Phi_1) \begin{pmatrix} \Lambda_0 & \Lambda_{01} \\ 0 & \Lambda_1 \end{pmatrix}, \quad \sigma(\Lambda_1) \subset \sigma(\Lambda_0). \tag{19}$$

By a similarity transformation with $\begin{pmatrix} Q_{00} & Q_{01} \\ 0 & Q_{11} \end{pmatrix} \in \mathbb{C}^{m+p,m+p}$ we can assume that the matrix $\begin{pmatrix} \Lambda_0 & \Lambda_{01} \\ 0 & \Lambda_1 \end{pmatrix}$ is in Jordan normal form. Then the matrices Λ_ν, Φ_ν ($\nu = 0,1$) and H_0, Δ_0, Λ_{01} become complex but the solvability conditions for (17), (18) still hold in $\mathbb{C}^{m,k} \times \mathbb{C}^{k,k}$. Moreover, by deleting columns in Φ_1 we can assume $p \leq 2$. The assertion is proved if $p = 1$ and Φ_1 extends a Jordan chain of an eigenvalue μ of algebraic multiplicity exactly 2. Therefore the following cases have to be excluded

Case 1

$p = 2$ and the vectors in $\Phi_1 = (\varphi_0, \varphi_1)$ belong to different Jordan chains in (19) (they may be eigenvectors themselves), i.e.

$$\begin{pmatrix} \Lambda_0 & \Lambda_{01} \\ 0 & \Lambda_1 \end{pmatrix} = \begin{pmatrix} \ddots & & & \\ & J_0 & & \begin{matrix} 0 & 0 \\ 1 & 0 \end{matrix} \\ & & J_1 & \begin{matrix} 0 & 0 \\ 0 & 1 \end{matrix} \\ & & & \begin{matrix} \mu_0 & 0 \\ 0 & \mu_1 \end{matrix} \end{pmatrix}, \quad J_\nu = \begin{pmatrix} \mu_\nu & 1 & & \\ & \ddots & \ddots & \\ & & \ddots & 1 \\ & & & \mu_\nu \end{pmatrix}, \quad \nu = 0,1.$$

Defining $\gamma_\nu \in \mathbb{R}^k$ by $\hat{\Phi}^T \Phi_0 \gamma_\nu = -\hat{\Phi}^T \varphi_\nu$, $\nu = 0,1$ one then verifies that

$$\begin{aligned} H &= (0, \ldots, 0, \Phi_0 \gamma_0 + \varphi_0, 0, \ldots, 0), \\ \Delta &= (0, \ldots, 0, (\Lambda_0 - \mu_0 I)\gamma_0 + e^{k_0}, 0, \ldots, 0) \\ &\text{and} \\ H &= (0, \ldots, 0, \Phi_0 \gamma_1 + \varphi_1), \\ \Delta &= (0, \ldots, 0, (\Lambda_0 - \mu_1 I)\gamma_1 + e^{k_1}) \end{aligned} \tag{20}$$

are linearly independent solutions of (17). The vector e^{k_ν}, $\nu = 0,1$ is taken to be zero if φ_ν is an eigenvector and a proper Cartesian basis vector otherwise.

Case 2

$p = 1$, $\Phi_1 = \varphi_1$, $\Lambda_1 = (\mu_1)$, φ_1 extends a Jordan block $J_1 = (\mu_1)$ in Λ_0 and μ_1 appears in another block of Λ_0 with eigenvector φ_2. This leads to the same contradiction as above because in addition to (20) we have a linearly independent solution of type

$$H = (0, \ldots, 0, \Phi_0\gamma_2 + \varphi_2, 0, \ldots, 0), \ \Delta = (0, \ldots, 0, (\Lambda_0 - \mu_1 I)\varphi_2, 0, \ldots, 0).$$

Case 3

$1 \le p \le 2$ and the columns of Φ_1 belong to a Jordan chain of length ≥ 3 for some eigenvalue μ_0. This leads to two subcases:

Case 3a $p = 1$, $\Lambda_1 = (\mu_0)$, $\Lambda_0 = \begin{pmatrix} \ddots & & \\ & \mu_0 & 1 \\ & & \mu_0 \end{pmatrix}$, $\Phi_1 = (\varphi_1)$.

Case 3b $p = 2$, $\Lambda_1 = \begin{pmatrix} \mu_0 & 1 \\ 0 & \mu_0 \end{pmatrix}$, $\Lambda_0 = \begin{pmatrix} \ddots & \\ & 0 \\ & \mu_0 \end{pmatrix}$, $\Phi_1 = (\varphi_1, \varphi_2)$.

Note that the case $p = 2$ with μ_0 being multiple in Λ_0 can be reduced to case 3a by deleting φ_2. For both subcases the unique solvability of (17) leads to

$$H_0 = (0, \ldots, 0, \Phi_0\gamma_0 + \varphi_1), \ \Delta_0 = (0, \ldots, 0, (\Lambda_0 - \mu_0 I)\gamma_0 + e^k)$$

where $\hat{\Phi}^T\Phi_0\gamma_0 = -\hat{\Phi}^T\varphi_1$. Since the last entry of $(\Lambda_0 - \mu_0 I)\gamma_0$ vanishes we obtain $H_0\Delta_0 = H_0$ in (18). One then verifies that (18) has a solution, namely

in case 3a

$$H = (0, \ldots, 0, \Phi_0\gamma_1 - \varphi_1, \Phi_0\gamma_2 + \varphi_1),$$
$$\Delta = (0, \ldots, 0, (\Lambda_0 - \mu_0 I)\gamma_1 - e^k, (\Lambda_0 - \mu_0 I)\gamma_2 + e^k - \gamma_1 - \gamma_0)$$

in case 3b

$$H = (0, \ldots, 0, \Phi_0\gamma_1 + \varphi_1, \Phi_0\gamma_2 + \varphi_2),$$
$$\Delta = (0, \ldots, 0, (\Lambda_0 - \mu_0 I)\gamma_1 + e^k, (\Lambda_0 - \mu_0 I)\gamma_2 - \gamma_0)$$

Here the vectors γ_1, γ_2 are chosen to satisfy the normalization condition.

For the converse statement assume that φ_1 is the generalized eigenvector belonging to the double eigenvalue μ and let

$$\Lambda_0 = \begin{pmatrix} \Lambda_{00} & 0 \\ 0 & \mu \end{pmatrix}, \quad (A - \mu I)\varphi_1 = \varphi_0 \tag{21}$$

Then a solution of (17) is given by

$$H_0 = (0, \ldots, 0, \Phi_0\gamma_1 + \varphi_1), \quad \Delta_0 = (0, \ldots, 0, (\Lambda_0 - \mu I)\gamma_1 + e^k) \tag{22}$$

where γ_1 is defined by $\hat{\Phi}^T\Phi_0\gamma_1 = -\hat{\Phi}\varphi_1$. Let (H_1, Δ_1) be another solution of (17), then

$$A(\Phi_0, H_0, H_1) = (\Phi_0, H_0, H_1)\begin{pmatrix} \Lambda_0 & \Delta_0 & \Delta_1 \\ 0 & \Lambda_0 & 0 \\ 0 & 0 & \Lambda_0 \end{pmatrix}.$$

Hence $\mathcal{R}(\Phi_0, H_0, H_1)$ is an invariant subspace of A with spectrum $\sigma(\Lambda_0)$. Since $\mathcal{R}(\Phi_0)$ has multiplicity 2 and $\dim \mathcal{R}(\Phi_0, H_0) = \dim \mathcal{R}(\Phi_0, \varphi_1) = k+1$ we obtain $\mathcal{R}(H_1) \subset \mathcal{R}(\Phi_0, \varphi_1)$. Therefore we can write any $\varphi \in \mathcal{R}(H_1)$ as $\varphi = \Phi_0\gamma + \alpha\varphi_1$ and from $\hat{\Phi}^T H_1 = 0$ we get $\hat{\Phi}^T \Phi_0 \gamma = -\alpha \hat{\Phi}^T \varphi_1$ and hence $\varphi = \alpha(\Phi_0\gamma_1 + \varphi_1)$ for some $\alpha \in \mathbb{R}$. Therefore we can write for some vector $\alpha^T = (\alpha_1, \dots, \alpha_{k-1})$, $\alpha_k \in \mathbb{R}$

$$H_1 = H_0 D, \quad D = \begin{pmatrix} 0 & 0 \\ \alpha^T & \alpha_k \end{pmatrix}$$

Using (17) for (H_1, Δ_1) and (H_0, Δ_0) we obtain

$$0 = AH_0D - H_0D\Lambda_0 - \Phi_0\Delta_1 = H_0(\Lambda_0 D - D\Lambda_0) + \Phi_0(\Delta_0 D - \Delta_1)$$

Now

$$(\Lambda_0 D - D\Lambda_0) = \begin{pmatrix} 0 & 0 \\ \alpha^T(\mu I - \Lambda_{00}) & 0 \end{pmatrix} \quad \text{and} \quad \varphi_1 \notin \mathcal{R}(\Phi_0)$$

imply $\alpha^T(\mu I - \Lambda_{00}) = 0$, $\Delta_0 D - \Delta_1 = 0$. Since $\mu \notin \sigma(\Lambda_{00})$ we finally have $\alpha = 0$ and $H_1 = \alpha_k H_0$, $\Delta_1 = \alpha_k \Delta_0$. Thus the nullspace is one-dimensional.

Now assume that (H_2, Δ_2) is a solution of (18). Then

$$A(\Phi_0, H_0, H_2) = (\Phi_0, H_0, H_2) \begin{pmatrix} \Lambda_0 & \Delta_0 & \Delta_2 \\ 0 & \Lambda_0 & \Delta_0 \\ 0 & 0 & \Lambda_0 \end{pmatrix},$$

and with the same argument as before $\mathcal{R}(H_2) \subset \mathcal{R}(H_0) = \operatorname{span}\{\Phi_0\gamma_1 + \varphi_1\}$. In particular $H_2 e^k = \alpha(\Phi_0\gamma_1 + \varphi_1)$ for some $\alpha \in \mathbb{R}$. Now apply the matrix $AH_2 - H_2\Lambda_0 - \Phi_0\Delta_2 = H_0$ to the vector e^k and find with (21)

$$\begin{aligned} H_0 e^k &= \Phi_0\gamma_1 + \varphi_1 = \alpha(A - \mu I)(\Phi_0\gamma_1 + \varphi_1) - \Phi_0(\Delta_2)_k \\ &= \alpha(\Phi_0\Lambda_{00}\gamma_1 + \varphi_0) - \Phi_0(\Delta_2)_k \end{aligned}$$

This contradicts $\varphi_1 \notin \mathcal{R}(\Phi_0)$.

Similar to the homotopy method for eigenvalues in [32] we suggest in this paper to apply a path-following algorithm to the invariant subspace system (9). If we encounter a turning point on the branch, Theorem 3 shows that the original parameter s will reverse its direction and a real eigenvalue from the continued spectral set collides with a real eigenvalue from outside. In fact, a pair of complex conjugate eigenvalues is created at this point if the parameter s moves beyond the turning point, see [20]. However, following the branch with the new parameter t the whole subspace passes smoothly through the multiplicity. Note that the rank of H_0 and Δ_0 is one at the turning point. In section 4 we will use this information to update the dimension of the subspace.

3 Continuation Methods for Invariant Subspaces

First we consider the case of continuing simple invariant subspaces, i.e. we compute a branch $(\Phi(s), \Lambda(s))$ for the equation (9).

3.1 The Predictor and the Corrector Step

Assume that (Φ_0, Λ_0) is a regular solution of (9) at $s = s_0$. Then we compute the tangent

$$(H_0, \Delta_0) = (\Phi'(s_0), \Lambda'(s_0))$$

to the branch $(\Phi(s), \Lambda(s))$ at $s = s_0$ from the following linear system of dimension $(m+k)k$ (cf. (6))

$$\begin{pmatrix} A(s_0)H_0 - H_0\Lambda_0 - \Phi_0\Delta_0 \\ \hat{\Phi}^T H_0 \end{pmatrix} = \begin{pmatrix} -A'(s_0)\Phi_0 \\ 0 \end{pmatrix}. \qquad (1)$$

This system contains a matrix equation for H_0 which is of Sylvester type (see [14]) and which is bordered by k^2 extra unknowns and equations. Since $\sigma(\Lambda_0) \subset \sigma(A(s_0))$ the Sylvester part $A(s_0)H_0 - H_0\Lambda_0$ is singular and it is essential to use the bordering for a stable solution (see the next subsection).

Given a stepsize δ and the solution (H_0, Δ_0) from (1) we compute the predictor from

$$(\Phi_1, \Lambda_1, s_1) = (\Phi_0, \Lambda_0, s_0) + \delta(H_0, \Delta_0, 1) \qquad (2)$$

In the corrector step we solve the system (9) with $(s, \hat{\Phi}, \Phi_0)$ replaced by (s_1, Φ_0, Φ_1), i.e. we adapt the normalization condition. Starting at (Φ_1, Λ_1), Newton's method generates the sequence (Φ_ν, Λ_ν), $\nu \geq 1$ defined by

$$\begin{pmatrix} A(s_1)\Phi_{\nu+1} - \Phi_{\nu+1}\Lambda_\nu - \Phi_\nu\Lambda_{\nu+1} \\ \Phi_0^T \Phi_{\nu+1} \end{pmatrix} = \begin{pmatrix} -\Phi_\nu\Lambda_\nu \\ \Phi_0^T\Phi_1 \end{pmatrix}. \qquad (3)$$

This system is of the same type as (1) and we use again the algorithm below. Note that the form (3) differs from the conventional realization of Newton's method (see [8], [10]) where $\Lambda_{\nu+1}$ is eliminated from the first equation in (3) with the help of the second. Our approach keeps the bordering structure.

3.2 The Bordered Bartels-Stewart Algorithm

The linear systems (1), (2) are of the form

$$\begin{pmatrix} AH - H\Lambda - \Phi\Delta \\ \hat{\Phi}^T H \end{pmatrix} = \begin{pmatrix} B \\ C \end{pmatrix} \qquad (4)$$

where $H, B, \hat{\Phi} \in \mathbb{R}^{m,k}$, $C, \Lambda, \Delta \in \mathbb{R}^{k,k}$ and $\sigma(\Lambda) \subset \sigma(A)$. We reduce the equations to systems of type (1) by the following algorithm which we call the *bordered Bartels-Stewart algorithm.* First compute the complex Schur decomposition of the matrix Λ (see [14]).

$$Q^H \Lambda Q = \tilde{\Lambda}, \quad Q^H Q = I, \quad \tilde{\Lambda} \text{ upper triangular.} \tag{5}$$

Note that this involves solving an eigenvalue problem of very small dimension $k \ll m$. Then we transform (4) into

$$\begin{pmatrix} \tilde{A}\tilde{H} - \tilde{H}\tilde{\Lambda} - \Phi\tilde{\Delta} \\ \hat{\Phi}^T \tilde{H} \end{pmatrix} = \begin{pmatrix} \tilde{B} \\ \tilde{C} \end{pmatrix} \tag{6}$$

where

$$\tilde{A} = AQ, \ \tilde{B} = BQ, \ \tilde{C} = CQ, \ \tilde{H} = HQ, \ \tilde{\Delta} = \Delta Q \tag{7}$$

Since $\tilde{\Lambda}$ is upper triangular we can compute the columns $\tilde{H}_j, \tilde{\Delta}_j$ of $\tilde{H}, \tilde{\Delta}$ similar to the Bartels-Stewart algorithm (see [14], Ch. 7.6.3) from a sequence of k bordered linear systems

$$\begin{pmatrix} \tilde{A} - \tilde{\Lambda}_{jj} I & -\Phi \\ \hat{\Phi}^T & 0 \end{pmatrix} \begin{pmatrix} \tilde{H}_j \\ \tilde{\Delta}_j \end{pmatrix} = \begin{pmatrix} \tilde{B}_j + \sum_{\nu=1}^{j-1} \tilde{\Lambda}_{\nu j} \tilde{H}_\nu \\ \tilde{C}_j \end{pmatrix}, \ j = 1, \dots, k. \tag{8}$$

Finally the solution H, Δ is obtained from $\tilde{H}, \tilde{\Delta}$ in (7). We notice that the upper left block $\tilde{A} - \tilde{\Lambda}_{jj} I$ is typically a large sparse, almost singular matrix. During the continuation of k-dimensional simple invariant subspaces we can expect that its rank drops at most by k which can be compensated for by the bordering (see [18], [16] for some estimates of condition numbers for this case).

Bordered systems of the above type occur quite frequently in bifurcation problems and various approaches have been developed for their stable and efficient solution (see [7], [19], [36]). We propose to use the mixed block elimination of Govaerts and Pryce [19] which requires only a black box solver for the principle submatrix and its transpose. If this solver is normwise backward stable in the sense of [22] and if the inverse of the principal submatrix times machine accuracy has a moderate bound then the block elimination can be shown to be forward stable [19].

3.3 The Complex Ginzburg-Landau Equation

As a first example we investigate the stability of a pulse solution due to Hocking and Stewartson [24] in the complex Ginzburg Landau equation. The latter is a well known modulation equation used in physics and chemistry [29] and we use the following form with two parameters $\alpha, \beta \in \mathbb{R}$. ([33], [34])

$$u_t = (1 + i\alpha)(u_{xx} - (1 + i\beta)^2 u + (1 + i\beta)(2 + i\beta)|u|^2 u). \tag{9}$$

The stationary solution is $\bar{u}(x) = \cosh(x)^{-(1+i\beta)} = \bar{v} + i\bar{w}$ (see [24]) where $\lim_{x\to\pm\infty} \bar{u}(x) = 0$. The linearization $L_{\alpha,\beta}$ of (9) at $\bar{u}$ is given in real and imaginary parts by the following two-dimensional system

$$L_{\alpha,\beta}\begin{pmatrix} v \\ w \end{pmatrix} = \begin{pmatrix} 1 & -\alpha \\ \alpha & 1 \end{pmatrix}\left[\begin{pmatrix} v_{xx} \\ w_{xx} \end{pmatrix} - (M_1 + M_2 M_3)\begin{pmatrix} v \\ w \end{pmatrix}\right]$$

where

$$M_1 = \begin{pmatrix} 1-\beta^2 & -2\beta \\ 2\beta & 1-\beta^2 \end{pmatrix}, \ M_2 = \begin{pmatrix} 2-\beta^2 & -3\beta \\ 3\beta & 2-\beta^2 \end{pmatrix},$$
$$M_3 = \begin{pmatrix} 3\bar{v}^2 + \bar{w}^2 & 2\bar{v}\bar{w} \\ 2\bar{v}\bar{w} & \bar{v}^2 + 3\bar{w}^2 \end{pmatrix}.$$

The operator $L_{\alpha,\beta}$ has a zero eigenvalue with geometric multiplicity at least two and corresponding eigenfunctions $\bar{u}'$ and $i\bar{u}$. The essential spectrum consists of two half-lines which cross the imaginary axis on the critical curve (cf. [4]).

$$\beta(\beta + 2\alpha) - 1 = 0. \tag{10}$$

We consider $L_{\alpha,\beta}$ in the finite interval $[x_-, x_+]$ subject to Dirichlet boundary conditions and we discretize by a centered finite difference scheme with step-size h. The changes in the spectrum caused by truncation to a finite interval have been analyzed in [4].

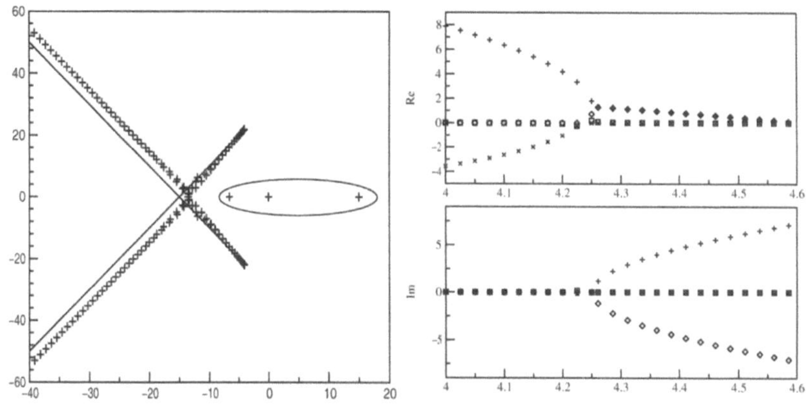

(a) Part of the essential and numerical spectrum

(b) Real and imaginary part vs. β

Fig. 1. Ginzburg-Landau equation, $\alpha = -2$

Figure 1(a) shows the two half lines of the essential spectrum and the full numerical spectrum for the case $x_\pm = \pm 10$, $h = 0.04$ and $\alpha = 2, \beta = 3$.

We continue a four-dimensional subspace that belongs to the two (almost) zero eigenvalues and the two real eigenvalues encircled in Fig. 1(a). With increasing β the stable eigenvalue passes zero (as shown by Mielke [34] this happens precisely on the curve (10) where a generalized eigenvector corresponding to $i\bar{u}$ appears), then forms a complex pair with the unstable eigenvalue which finally moves to the left half plane. For the numerical eigenvalues ($x_\pm = \pm 10, h = 0.004$) a perturbation of this appears in Figure 1(b) (see [4] for details).

A schematic drawing of the motion of the 4 eigenvalues in the infinite case and in the discretized case is shown in Fig. 2, 3. This sensitive behavior can only be revealed since the 4D-subspace stays separated from the rest of the spectrum and the remaining 4×4 eigenvalue problem can be solved very accurately. The double or even triple zero eigenvalue does not affect the continuation of the subspace which could be done with uniform step-size in the parameter β.

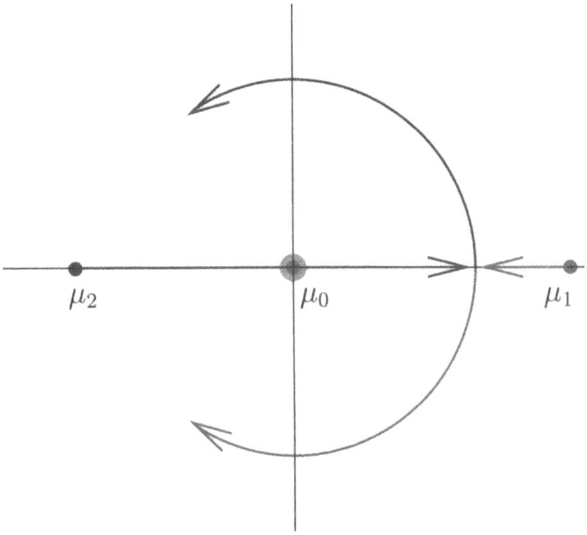

Fig. 2. Behavior of complex eigenvalues when β increases from 4 to 5

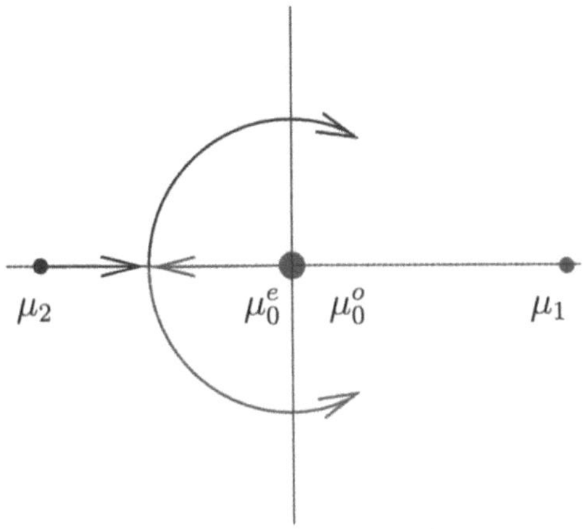

(a) $4 \leq \beta \lesssim \beta_{crit}$

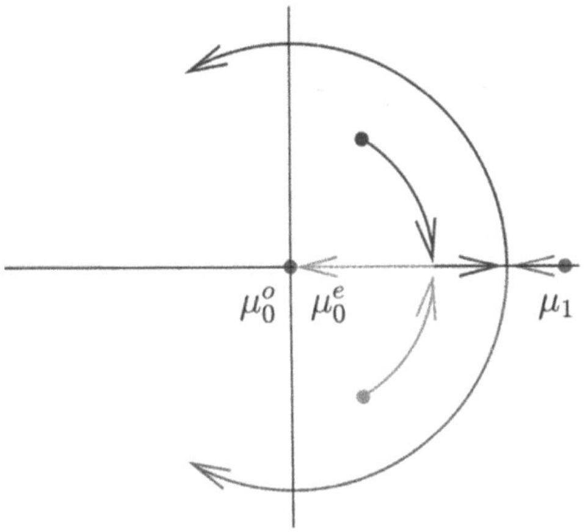

(b) $\beta_{crit} \lesssim \beta \leq 5$

Fig. 3. Behavior of numerical eigenvalues for $4 \leq \beta \leq 5$

3.4 Passing Through Nonsimple Subspaces

As explained at the end of section 2 we can apply a general continuation method to the system (9) in order to pass through double real eigenvalues. But it may also be advantageous along branches of simple invariant subspaces due to the better adaptation of step-sizes. For example in a pseudo arclength method the extra equation is of the form

$$\langle \dot{\Phi}, \Phi - \Phi_0 \rangle + \langle \dot{\Lambda}, \Lambda - \Lambda_0 \rangle + \dot{\lambda}(\lambda - \lambda_0) = \delta \tag{11}$$

where we have used the inner product $\langle A, B \rangle = \mathrm{trace}(A^T B)$ for rectangular matrices and where $\dot{\Phi}, \dot{\Lambda}, \dot{\lambda}$ are approximate tangent vectors at the previous point on the branch.

Similar to (1)-(3), both the predictor and the corrector step for (9), (11) lead to linear systems for $H \in \mathbb{R}^{m,k}$, $\Delta \in \mathbb{R}^{k,k}$ and $\mu \in \mathbb{R}$ as follows (compare (4))

$$\begin{aligned} AH - H\Lambda - \Phi\Delta + \Gamma\mu &= B \in \mathbb{R}^{m,k} \\ \hat{\Phi}^T H &= C \in \mathbb{R}^{k,k} \\ \langle \dot{\Phi}, H \rangle + \langle \dot{\Lambda}, \Delta \rangle + \dot{\lambda}\mu &= d \in \mathbb{R} \end{aligned} \tag{12}$$

where $\Gamma \in \mathbb{R}^{m,k}$. For example, $\Gamma = A'(s_0)\Phi_0$ holds in the predictor step. We can solve this system with the standard block elimination method [25]. This requires to solve two linear (H, Δ) systems with right hand sides $\binom{B}{C}$ and $\binom{\Gamma}{0}$ by the bordered Bartels-Stewart algorithm above and then form a suitable linear combination which satisfies the last equation in (12). However, very close to or at the turning point this is not reliable and the bordering by Γ is needed for stability, see Theorem 3 and (17), (18).

For this case the algorithm from 3.2 can be modified as follows. First, Λ is put into upper triangular form as in (5) and the data are transformed as in (7) where in addition $\tilde{\dot{\Phi}} = \dot{\Phi}Q$, $\tilde{\dot{\Lambda}} = \dot{\Lambda}Q$. For simplicity we drop the "~" and work with (12). For the columns H_j, Δ_j of H, Δ and for μ we use the ansatz

$$(H_j, \Delta_j, \mu) = (H_j^0, \Delta_j^0, \mu_j^0) + \sum_{i=1}^{j} \alpha_i (H_j^i, \Delta_j^i, \mu_j^i), \; j = 1, \ldots, k \tag{13}$$

in the first two equations of (12). With the matrix

$$M_j = \begin{pmatrix} A^0 - \Lambda_{jj} I_m & -\Phi & -\Gamma_j \\ \hat{\Phi}^T & 0 & 0 \\ \dot{\Phi}_j^T & \dot{\Lambda}_j^T & 0 \end{pmatrix} \tag{14}$$

we determine the unknowns in (13) for $j = 1, \ldots, k$ from

$$M_j \begin{pmatrix} H_j^0 \\ \Delta_j^0 \\ \mu_j^0 \end{pmatrix} = \begin{pmatrix} B_j + \sum_{i=1}^{j-1} \Lambda_{ij} H_i^0 \\ C_j \\ 0 \end{pmatrix}, \quad M_j \begin{pmatrix} H_j^j \\ \Delta_j^j \\ \mu_j^j \end{pmatrix} = \begin{pmatrix} 0 \\ 0 \\ 1 \end{pmatrix}$$

$$M_j \begin{pmatrix} H_j^i \\ \Delta_j^i \\ \mu_j^i \end{pmatrix} = \begin{pmatrix} \sum_{\nu=i}^{j-1} \Lambda_{\nu j} H_\nu^i \\ 0 \\ 0 \end{pmatrix} \quad \text{for } i = 1, \ldots, j-1$$

Finally the α_i and μ are calculated from the $(k+1)$-dimensional system

$$\begin{pmatrix} \mu_1^1 & & & -1 \\ \vdots & \ddots & & \vdots \\ \vdots & & \ddots & \vdots \\ \mu_k^1 & \cdots \cdots & \mu_k^k & -1 \\ 1 & \cdots \cdots & 1 & \dot{\lambda} \end{pmatrix} \begin{pmatrix} \alpha_1 \\ \vdots \\ \vdots \\ \alpha_k \\ \mu \end{pmatrix} = \begin{pmatrix} -\mu_1^0 \\ \vdots \\ \vdots \\ -\mu_k^0 \\ d \end{pmatrix} \tag{15}$$

One readily verifies that this yields the desired solution via (13). Moreover, the matrix M_j now has a bordering of width $k+1$ and can be expected to be well conditioned even at the double real eigenvalue. However, the overall method is rather expensive since one has to solve $\frac{1}{2}(k+1)(k+2)$ bordered systems.

4 Updating the Dimension

If the matrices $A(s)$ arise from linearizing about a steady state branch then one does not want the subspace continuation to reverse the direction of the parameter s (which is usually arclength for the nonlinear problem). Rather, if a real eigenvalue from the continued spectral set meets another one from outside then one should increase the dimension of the subspace by 1 and follow the complex conjugate pair in the original s-direction. This is precisely the situation that has been analyzed for single eigenvalues in [32]. We now deal with this problem in the context of general subspaces.

4.1 Inflating the Subspace

Suppose (Φ_0, Λ_0, s_0) is a quadratic turning point, i.e. (10), (11) and (13) hold. According to Theorem 3 there is a smooth branch $(\Phi(t), \Lambda(t), s(t))$ passing through (Φ_0, Λ_0, s_0) at some $t = t^*$. Therefore $\Phi_0 = \Phi(t^*), \Lambda_0 = \Lambda(t^*), s_0 = s(t^*)$ and

$$A(s(t))\Phi(t) - \Phi(t)\Lambda(t) = 0 \tag{1}$$

$$(\dot{\Phi}(t^*), \dot{\Lambda}(t^*)) = c(H_0, \Delta_0) \quad \text{for some } c \in \mathbb{R} \tag{2}$$

Since H_0 has rank 1 we obtain that after the turning point there is one new eigenvalue created by the one which becomes double, but all the other eigenvalues of $\Lambda(t)$ are identical with those that have been passed before the turning point.

Differentiating (1) at $t = t^*$ and using $s'(t^*) = 0$ leads to

$$A_0 \dot{\Phi}(t^*) = \dot{\Phi}(t^*)\Lambda_0 + \Phi_0 \dot{\Lambda}(t^*) \tag{3}$$

Since $\dot{\Phi}(t^*)$ has rank one a singular value decomposition yields

$$\dot{\Phi}(t^*) = \mu u v^T \quad \text{for some } \mu > 0,\ \|u\| = \|v\| = 1.$$

Here u is the generalized eigenvector. Insert this into (3) and multiply by v to obtain

$$\mu A_0 u = \mu u v^T \Lambda_0 v + \Phi_0 \dot{\Lambda}(t^*) v.$$

Therefore the following two steps are sufficient to update the subspace

1. SVD:
$$\dot{\Phi}(t^*) = \mu u v^T,\ \mu > 0$$
2. Update:
$$\Phi_{up} = \begin{pmatrix} \Phi(t^*) & u \end{pmatrix}, \quad \Lambda_{up} = \begin{pmatrix} \Lambda(t^*) & \frac{1}{\mu}\dot{\Lambda}(t^*)v \\ 0 & v^T \Lambda(t^*) v \end{pmatrix}$$

4.2 A Traveling Wave Example

As an example for the turning of the eigenvalues we use a variant of the Nagumo equation with an additional parameter ρ.

$$\begin{aligned} u_t &= u_{xx} + f(u,\rho), \quad u(x) : \mathbb{R} \to \mathbb{R},\ t > 0, \\ f(u,\rho) &= \rho\, u(1-u)(u-\mu),\ \mu \in (0, \frac{1}{2}) \end{aligned} \tag{4}$$

We continue traveling wave fronts of (4) i.e. solutions $\bar{u}(x,t) = \bar{v}(x - ct)$ which satisfy

$$\lim_{\xi \to \infty} \bar{v}(\xi) = 1, \qquad \lim_{\xi \to -\infty} \bar{v}(\xi) = 0$$

At $\rho = 1$ the following explicit solution is known, see [21], p. 130.

$$\bar{v}_1(x) = \frac{1}{1 + \exp(-\frac{x}{\sqrt{2}})}, \qquad c = -\sqrt{2}\,(\tfrac{1}{2} - \mu) \tag{5}$$

The function $\bar{v}$ and the parameter c satisfy the second order equation

$$F(v) = v_{xx} + c v_x + f(v,\rho) = 0, \quad -\infty < x < \infty \tag{6}$$

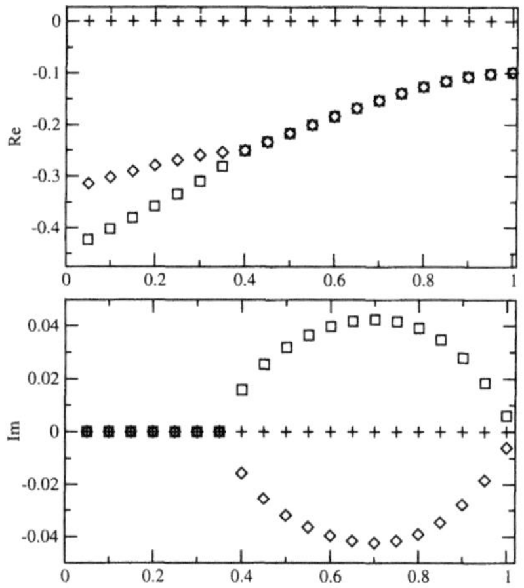

Fig. 4. Eigenvalues for $k = 3$ vs. arclength t

and the stability of the wave is governed by the spectrum of the linearized operator (see [21],[43])

$$Pu = u_{xx} + cu_x + D_v f(\bar{v}, \rho)u \tag{7}$$

We restrict to a finite interval $J = [x_-, x_+]$ and use Dirichlet boundary conditions (see [1] for the general approach).

Then the resulting boundary value problem is discretized with finite differences and the spectrum of the matrix obtained by linearizing about the solution (i.e. the discrete analog of (7)) is analyzed, for our example $J = [-10, 10]$, $h = 0.1$, $\mu = \frac{1}{4}$. We start the continuation at $\rho = 1$ with the known solution (5) and decrease ρ. We use periodic boundary conditions which tend to give good approximations of the essential spectrum but at the same time tend to produce complex eigenvalues (cf. [4]).

Figure 4 shows the result of following the three dimensional invariant subspace which belongs to the three eigenvalues with largest real part. The real and imaginary part are shown as functions of the arclength t of the subspace continuation. The eigenvalue close to zero (which always exists and is exactly zero for the continuous problem) stays real while the other two eigenvalues form a complex pair at $t = 0.4$, $\rho = 0.6$. (for the latter value see also figure 5(b))

Suppose we had started the continuation with the largest two real eigenvalues. Then the pictures in 5(a), 5(b) result where we pass a turning point at $t = 0.37$, $\rho = 0.63$.

Both eigenvalues and the parameter ρ are plotted versus arclength t and also the eigenvalue which turns is shown versus the parameter ρ. While the zero eigenvalue returns to itself the nonzero eigenvalue passes on to the third eigenvalue.

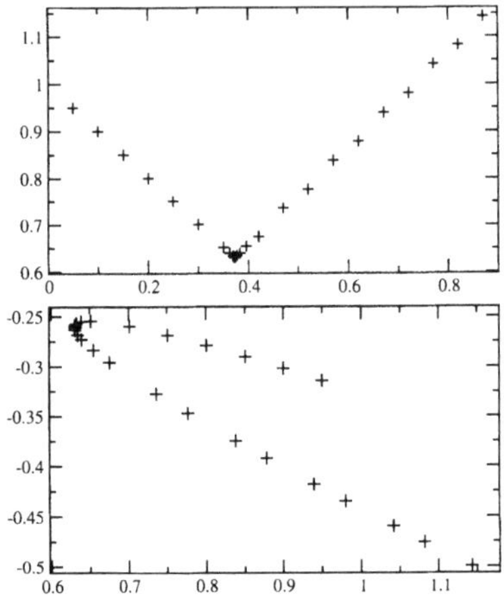

(a) Parameter ρ vs. arclength t and the second eigenvalue vs. ρ

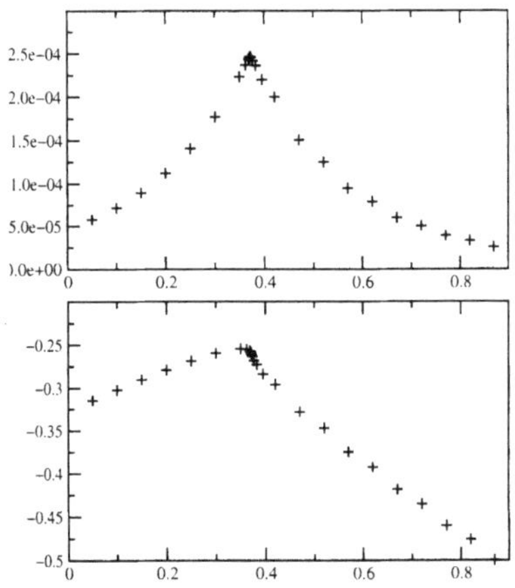

(b) Two real eigenvalues vs. arclength t for two-dimensional subspace continuation

Fig. 5. Nagumo equation, $\mu = \frac{1}{4}$, $J = [-10, 10]$

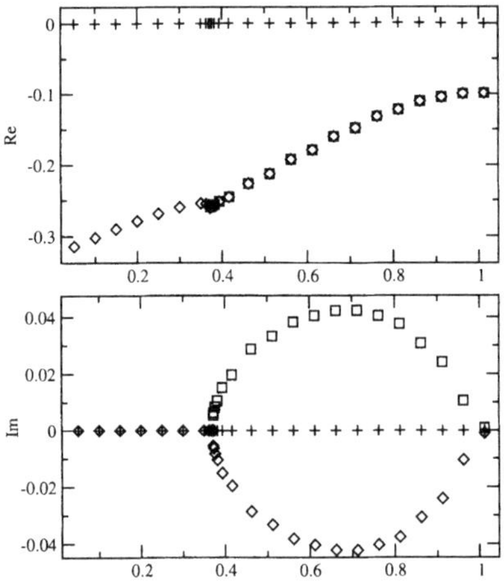

Fig. 6. Real and imaginary parts of the eigenvalues vs. arclength t

Finally, we show in Fig.6 the result of the update procedure when starting with the same two dimensional subspace as above. Near the turning point the continuation first slows down until the turn in the parameter ρ is detected. After inflating the subspace the step-size increases again and a comparison with Figures 4, 5 shows a good correspondence.

5 Application and Examples

In this section we discuss some extensions and further issues which turn out to be important for the actual implementation.

5.1 Algorithmic Details

Getting the subspace continuation started is a crucial problem. If no complete spectrum is available initially we use a combination of Cayley-transforms (see [13]) and orthogonal subspace iteration ([14]). In this way we obtain an invariant subspace that belongs to a spectral set in a prescribed vertical strip of the complex plane.

During continuation we keep the dimension of the invariant subspace minimal but above some critical number k. After each complex-real transition it is tested whether the eigenvalues with smallest real part (either a real eigenvalue or a complex conjugate pair) can be omitted without getting below dimension k. However, no device is currently implemented to guarantee that the continued spectral set contains the rightmost eigenvalues. This is a global problem and for some algorithms in this direction we refer to [13] and [37]. In order to avoid ill-conditioning of the matrices $\Phi(s)$ these matrices are constantly reorthogonalized after the successful corrector step. For example the predictor in (1) is computed with an orthogonal Φ_0 so that in the succeeding corrector step (3) the matrix $\Phi_0^T\Phi_1$ is close to the identity.

5.2 The Left Invariant Subspaces

If a solution (Φ_0, Λ_0) of (5) has been computed then the corresponding pair (Ψ_0, M_0) for the adjoint can be computed by solving just one extra linear system as the following proposition shows.

Proposition 4. *Under the assumptions of Theorem 2 let* (Φ_0, Λ_0) *be a regular solution of* (5). *Then the linear system*

$$\begin{aligned} A^T\Psi - \Psi\Lambda_0^T - \Phi_0 M &= 0 \\ \Phi_0^T\Psi &= I \end{aligned} \qquad (1)$$

has a unique solution $\Psi_0 \in \mathbb{R}^{m,k}$, $M_0 \in \mathbb{R}^{k,k}$. *This solution satisfies* $M_0 = 0$ *and* $\mathcal{R}(\Psi_0)$ *is a simple invariant subspace of* A^T *with respect to the spectral set* $\sigma(\Lambda_0)$.

Proof. Since $\mathcal{R}(\Phi_0)$ is a simple invariant subspace we have a block diagonalization of A

$$A(\Phi_0, \Phi_1) = (\Phi_0, \Phi_1) \begin{pmatrix} \Lambda_0 & 0 \\ 0 & \Lambda_1 \end{pmatrix}, \quad \sigma(\Lambda_1) \cap \sigma(\Lambda_0) = \emptyset$$

where $\Phi_1 \in \mathbb{R}^{m,m-k}$, $\Lambda_1 \in \mathbb{R}^{k,k}$ and (Φ_0, Φ_1) is nonsingular. Defining

$$(\Phi_0, \Phi_1)^{-1} = (\Psi_0, \Psi_1)^T, \quad \Psi_0 \in \mathbb{R}^{m,k}, \ \Psi_1 \in \mathbb{R}^{m,m-k}$$

we obtain that $\mathcal{R}(\Psi_0)$ is a simple invariant subspace of A^T with spectrum $\sigma(\Lambda_0^T) = \sigma(\Lambda_0)$. Since $\Phi_0^T \Psi_0 = I$ we see that $\Psi = \Psi_0$, $M = 0$ solves (1). Now, for uniqueness, suppose that (Ψ, M) solves the homogenous equation (1). Then multiply the first equation in (1) by Φ_0^T and use the second to obtain

$$0 = (A\Phi_0)^T \Psi - \Phi_0^T \Psi \Lambda_0^T - \Phi_0^T \Phi_0 M = \Lambda_0^T \Phi_0^T \Psi - \Phi_0^T \Phi_0 M = -\Phi_0^T \Phi_0 M.$$

Since $\Phi_0^T \Phi_0$ is nonsingular we find $M = 0$. Therefore, $A^T \Psi - \Psi \Lambda_0^T = 0$ and from the simplicity of $\mathcal{R}(\Psi_0)$ we obtain $\Psi = \Psi_0 B$ for some $B \in \mathbb{R}^{k,k}$. Finally, this implies

$$0 = \Phi_0^T \Psi = \Phi_0^T \Psi_0 B = B$$

as well as $\Psi = 0$.

We notice that the data Φ_0, Λ_0 of the linear system (1) are assumed to be known at this stage of the computation and that the linear system (1) can be solved by the bordered Bartels-Stewart algorithm with A replaced by A^T. Though we have not used the adjoint in our continuation method the "left invariant subspaces" provide useful information for evaluating test functions or computing singularities (see [16], [28]).

5.3 Simultaneous Branch Following

In the previous sections we always assumed that the matrices $A(s)$ are available analytically. If, however $A(s) = D_u F(u(s), \lambda(s))$ for a branch $(u(s), \lambda(s))$ of some nonlinear system $F(u, \lambda) = 0$, then any evaluation of $A(s)$ requires a new solution of the nonlinear system. This is particularly annoying when - as in section 3 - the value of s is constantly changed during the Newton iteration for the subspace problem. In order to avoid such nested iterations and to coordinate different step-size controls for the two problems we have implemented a single continuation algorithm for a very large system. It is of the form

$$S(\Phi, \Lambda, u, \lambda) = \begin{pmatrix} F(u, \lambda) \\ T(\Phi, \Lambda, u, \lambda) \end{pmatrix} = 0 \tag{2}$$

where $F : \mathbb{R}^n \times \mathbb{R} \to \mathbb{R}^n$ and

$$T(\Phi, \Lambda, u, \lambda) = \begin{pmatrix} A(u,\lambda)\Phi - \Phi\Lambda \\ \hat{\Phi}^T(\Phi - \Phi_0) \end{pmatrix}, \quad \Phi \in \mathbb{R}^{m,k}, \; \Lambda \in \mathbb{R}^{k,k}.$$

Here $A(u, \lambda) \in \mathbb{R}^{m,m}$ are matrices that depend smoothly on $(u, \lambda) \in \mathbb{R}^{n+1}$. Usually we have $n = m$ and $A(u, \lambda) = D_u F(u, \lambda)$, but other cases occur where e.g. $F(u, \lambda)$ contains boundary conditions which have to be eliminated for the spectral problem (compare the examples in this paper). We have implemented a continuation method for the large system (2) which needs only one step-size control and uses a weighted norm for $(\Phi, \Lambda, u, \lambda)$

$$\|(\Phi, \Lambda, u, \lambda)\|^2 = \frac{1}{mk}\|\Phi\|_F^2 + \frac{1}{k^2}\|\Lambda\|_F^2 + \frac{\|u\|_2^2}{n} + \lambda^2 \tag{3}$$

where $\|\Phi\|_F^2 = \mathrm{tr}(\Phi^T\Phi)$ is the Frobenius norm.

The linear systems arising during the predictor and the corrector step for (2) have a special structure (the first n equations do not depend on Φ, Λ) and they can be reduced to solving

- two linear systems with a bordering of $D_u F$
- three linear systems with a bordering of $D_{\Phi,\Lambda} T$.

The reduction is rather obvious and we do not display the details here.

During continuation turning points with respect to λ can occur for two different reasons, first because the branch of the nonlinear system $F(u, \lambda) = 0$ turns and second because a double real eigenvalue for the subspace problem occurs. The second case is indicated by

$$\dot{u}_1^T \dot{u}_0 \leq 0 \tag{4}$$

where $\dot{u}_0, \dot{u}_1$ are successive tangents to the u-part of the (u, λ)-branch. Then the update procedure for the subspace is invoked. In fact, if we pass a turning point of the (u, λ)-branch we expect $\dot{u}_1^T \dot{u}_0 > 0$ while (4) indicates that the tangent of this branch is reversed due to a turning point with respect to the parameter s as in section 4.

5.4 The FitzHugh-Nagumo System, a Final Example

The FitzHugh-Nagumo equation is a model equation for the propagation of nerve impulses [12]. We consider a two-dimensional system with a small additional diffusive term

$$v_t = v_{xx} + F(v, w), \quad w_t = \epsilon w_{xx} + G(v, w)$$
$$F(v, w) = v - \frac{1}{3}v^3 - w, \quad G(v, w) = \Phi(v + a - bw), \quad a, b, \Phi \in \mathbb{R}.$$

For the parameters $a = 0.7, b = 0.8$ there is a branch containing stable and unstable waves. We consider a specific part of the stable branch and follow again a four dimensional subspace with decreasing parameter Φ. For the actual calculation we restrict to $[x_-, x_+] = [0, 65]$, use Dirichlet boundary conditions and discretize with step-size $h = 1$ (more details can be found in [4]). Fig. 7(b) shows real and imaginary parts of the eigenvalues on the stable branch. The complex conjugate pair undergoes a transition to two real eigenvalues which is harmless for our method. However while the two largest real eigenvalues remain separated from the essential spectrum the other two move towards it. The situation for the full spectrum at the critical value $\Phi_{crit} = 0.062$ is shown in Fig. 7(a). At this point our algorithm breaks down due to stagnation of the continuation steps. The two eigenvalues can no longer be separated from the cluster that approximates the essential spectrum. Of course, concerning the stability problem for the original wave there is no need to further include these eigenvalues in the continuation. This shows that there can be reasons for deflating the subspace other than the minimality requirement discussed in 5.1.

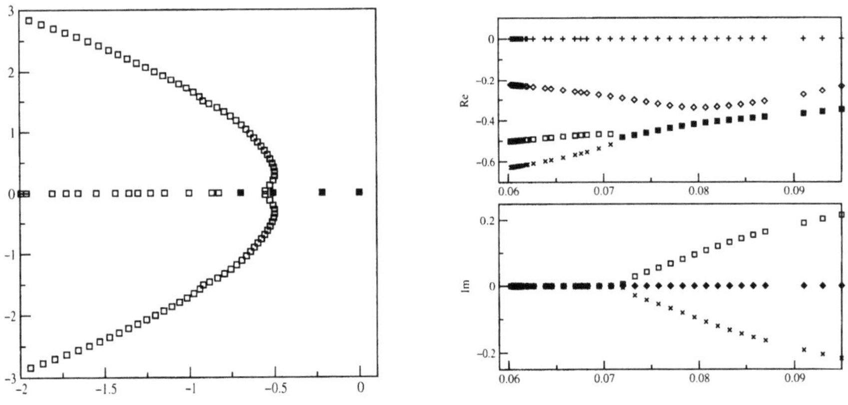

(a) Part of the numerical spectrum at Φ_{crit}

(b) Real and imaginary parts vs. Φ

Fig. 7. FitzHugh-Nagumo, stable wave, $x_- = 0, x_+ = 65, h = 1$

References

1. W.-J. Beyn. The numerical computation of connecting orbits in dynamical systems. *IMA J. Numer. Anal.*, 10:379–405, 1990.
2. W.-J. Beyn and W. Kleß. Numerical Taylor expansions of invariant manifolds in large dynamical systems. *Numerische Mathematik*, 80:1–38, 1998.
3. W.-J. Beyn, W. Kleß, and V. Thümmler. A continuation framework for invariant subspaces and its application to traveling waves. In F. Keil, W. Mackens, H. Voß, and J. Werther, editors, *Scientific Computing in Chemical Engineering II*, pages 144–151. Springer, 1999.
4. W.-J. Beyn and J. Lorenz. Stability of traveling waves: Dichotomies and eigenvalue conditions on finite intervals. *Numer. Funct. Anal. and Optimiz.*, 20:201–244, 1999.
5. K. Böhmer. On a numerical Liapunov-Schmidt method for operator equations. *Computing*, 51:237–269, 1993.
6. K. Böhmer. On hybrid methods for bifurcation studies for general operator equations. *This volume*, 2000.
7. T. F. Chan. Deflation techniques and block–elimination algorithms for solving bordered singular system. *SIAM J. Sci. Stat. Comput.*, 5:121–134, 1984.
8. F. Chatelin. *Eigenvalues of matrices.* John Wiley & Sons, 1993.
9. J. Demmel, L. Dieci, and M. Friedman. Computing connecting orbits via an improved algorithm for continuing invariant subspaces. to appear, 1999.
10. J. W. Demmel. Three methods for refining estimates of invariant subspaces. *Computing*, 38:43–57, 1987.
11. E. Doedel, T. Champneys, T. Fairgrieve, Y. Kuznetsov, B. Sandstede, and X.-J. Wang. *AUTO97 Continuation and bifurcation software for ordinary differential equations (with HomCont).* Concordia Univ. Montreal, 1997.
12. R. FitzHugh. Impulses and physiological states in theoretical models of nerve membrane. *Biophysical Journal*, 1:445–466, 1961.
13. T. J. Garratt, G. Moore, and A. Spence. A generalised Cayley transform for the numerical detection of Hopf bifurcations in large systems. In *Contributions in numerical mathematics*, pages 177–195. World Sci. Publ., 1993.
14. G. H. Golub and C. F. Van Loan. *Matrix computations.* Johns Hopkins University Press, Baltimore, MD, second edition, 1989.
15. W. Govaerts. Stable solvers and block elimination for bordered systems. *SIAM J. Matrix. Anal. Appl.*, 12:469–384, 1991.
16. W. Govaerts. Defining functions for manifolds of matrices. *Linear Algebra Appl.*, 266:49–68, 1997.
17. W. Govaerts. Mixed block elimination for linear systems with wider borders. *SIAM J. Numer. Anal.*, 34:867–880, 1997.
18. W. Govaerts and J. D. Pryce. A singular value inequality for block matrices. *Linear Algebra Appl.*, 125:141–148, 1989.
19. W. Govaerts and J. D. Pryce. Mixed block elimination for linear systems with wider borders. *IMA J. Numer. Anal.*, 13:161–180, 1993.
20. M. E. Henderson and H. B. Keller. Complex bifurcation from real paths. *SIAM J. Appl. Math.*, 50(2):460–482, 1990.
21. D. Henry. *Geometric Theory of Semilinear Parabolic Equations.* Number 804 in Lecture Notes in Mathematics. Springer, 1981.
22. N. J. Higham. *Accuracy and stability of numerical algorithms.* Society for Industrial and Applied Mathematics (SIAM), Philadelphia, PA, 1996.

23. M. Hochbruck, C. Lubich, and H. Selhofer. Exponential integrators for large systems of differential equations. *SIAM J. Sci. Comput.*, 19(5):1552–1574 (electronic), 1998.
24. L. M. Hocking and K. Stewartson. On the nonlinear response of a marginally unstable plane parallel flow to a two-dimensional disturbance. *Proc. Roy. Soc. London Ser. A*, 326:289–313, 1972.
25. H. B. Keller. The bordering algorithm and path following near singular points of higher nullity. *SIAM J. Sci. Stat. Comput.*, 4:573–582, 1983.
26. W. Kleß. *Numerische Approximation lokal invarianter Mannigfaltigkeiten in großen dynamischen Systemen.* PhD thesis, University of Bielefeld, 1997.
27. Y. Kuznetsov. *Elements of Applied Bifurcation Theory.* Springer, 1995.
28. Y. Kuznetsov and V. Levitin. *CONTENT: A multiplatform environment for analyzing dynamical systems.* GWI Amsterdam, 1997.
29. C. D. Levermore and M. Oliver. The complex Ginzburg-Landau equation as a model problem. In *Dynamical systems and probabilistic methods in partial differential equations*, pages 141–190. Amer. Math. Soc., 1996.
30. T. Li, Z. Zeng, and L. Cong. Solving eigenvalue problems of real nonsymmetric matrices with real homotopies. *SIAM J. Numer. Anal.*, 29:229–248, 1992.
31. C. Lubich. On dynamics and bifurcations of nonlinear evolution equations under numerical discretization. *This volume*, 2000.
32. S. Lui, H. Keller, and T. Kwok. Homotopy method for the large, sparse, real nonsymmetric eigenvalue problem. *SIAM J. Matrix Anal. Appl.*, 18:312–333, 1997.
33. A. Mielke. The complex Ginzburg-Landau equation on large and unbounded domains: sharper bounds and attractors. *Nonlinearity*, 10:199–222, 1997.
34. A. Mielke. The Ginzburg-Landau equation in its role as a modulation equation. Report 20/99 of DFG Schwerpunktprogramm DANSE, 1999.
35. A. Mielke, G. Schneider, and H. Uecker. Stability and diffusive dynamics on extended domains. *This volume*, 2000.
36. G. Moore. Some remarks on the deflated block elimination method. In T. Küpper, R. Seydel, and H. Troger, editors, *Bifurcation: Analysis, Algorithms, Applications*, volume 79 of *ISNM*, pages 222–234. Birkhäuser, 1987.
37. R. Neubert. Predictor-corrector techniques for detecting Hopf bifurcation points. *Internat. J. Bifur. Chaos Appl. Sci. Engrg.*, 3(5):1311–1318, 1993.
38. B. Sandstede and A. Scheel. From unbounded to bounded domains: The fate of point and essential spectrum. Report 40/99 of DFG Schwerpunktprogramm DANSE, 1999.
39. A. Schwarzer. A new class of polynomial preconditioners for large sparse eigenvalue problems. Report 5/98 of DFG Schwerpunktprogramm DANSE, to appear in SIAM J. Numer. Anal., 1998.
40. R. Sebastian. *Anwendung von Krylov-Verfahren auf Verzweigungs- und Fortsetzungsprobleme.* PhD thesis, University of Marburg, 1995.
41. G. W. Stewart and J. G. Sun. *Matrix perturbation theory.* Academic Press Inc., Boston, MA, 1990.
42. V. Thümmler. Numerische Stabilitätskriterien für wandernde Wellen. Diplomarbeit, Fakultät für Mathematik, Universität Bielefeld, 1998.
43. A. I. Volpert, V. A. Volpert, and V. A. Volpert. Traveling wave solutions of parabolic systems. In *Translations of Mathematical Monographs*, volume 140. American Mathematical Society, 1994.

On Hybrid Methods for Bifurcation and Center Manifolds for General Operators

Klaus Böhmer*

Institut für Mathematik, Philipps Universität Marburg, Marburg, Germany

Abstract. This presentation uses few basic concepts of numerical functional analysis and approximation theory as the main tools to prove convergence and stability for stationary problems. It applies to a general class of operator equations and general discretization methods. This allows an extension to numerical bifurcation studies, including Hopf bifurcation and center manifold results, for finite difference-, finite element- and spectral methods for general operators. In particular, partial differential equations (PDEs) as reaction-diffusion-systems and Navier-Stokes equations are included. The basic idea is to present an approach as simple as possible but as complex as necessary to cover all these types of problems and their discretizations with reasonably basic concepts. For the first time, the full cycle of qualitative and quantitative results, starting from PDEs via convergent discretization and post-processing back to the bifurcation scenarios in the original equation, is presented. A Γ-equi-variant example in biological pattern formation is included. Finally a C^{++}–program system with similarly general goals is indicated.

Keywords: Numerical bifurcation, numerical center manifolds, post-processing bifurcation, C^{++}–program, bordered stability, consistent differentiability.

AMS subject classification: 64F15, 15A18

1 Introduction

A wide range of projects was supported by the DFG-Schwerpunkt "DANSE". I present the main ideas of the essential projects.

The complex problems require an intense feedback between all the contributing areas, such as *Numerics, Analysis, Group Theory, Computer Algebra and Computer Science.* I have used the concept of *Hybrid Methods* to indicate this complex approach. For some problems [1], e.g., [9,4,10,12,25,64,27–29], the computations are so complex and technical, that only the old Keppler or the modern Computer Algebra is able to do it.

---

* Project: Hybrid Methods for the Dynamics of Partial Differential Equations (Klaus Böhmer)

[1] Many papers of our group were published in [9,4,10] and are referred to this way

We strive for convergence results in a wide class of operator equations and discretization methods. Only familiar concepts are used, such as projectors, approximation results,approximations to the identity, inverse estimates, monotone operators, their compact perturbations and bordered systems. On this basis, a general method is developed which allows in a first step to prove stability and convergence by a sequence of relatively easily verifiable conditions for general linear and the corresponding nonlinear problems. This technique can be extended to the full bifurcation scenarios for equally general classes of operators and discretization methods, essentially via the same argumentation, see [85,31,23,24] via Numerical Liapunov-Schmidt-and center manifold methods.

Certainly, spectacular effects in dynamical systems are of interest in the dynamical community. These effects usually will be observable in discretizations of operator equations as well as for the original operators. However, there is another and totally different community which is interested in qualitative *and* quantitative results, in particular in convergence results for discretization methods near singularities. In fact, industry needs reliable results to predict what will happen for which parameter values in the original problem. A famous example is the continuous flow stirred tank reactor. In many papers, e.g., Golubitsky/Schaeffer [52], has been studied as a model problem. In contrast to these models, in industry [69] usually many different chemicals are reacting with each other. The characteristics of the singularity have to be re-transformed to the full phase space of the reactants and parameters in the original problem in a reliable way. This paper reports how these different problems are solved.

As indicated above, we want to cover a wide class of operator equations and discretization methods with our new simplified and unified approach. Examples are elliptic, reaction-diffusion, Navier-Stokes, porous media, integral equations, and finite element , spectral and partially difference methods. Variational crimes for Petrov-Galerkin methods are only presented for the case of quadrature approximations, the more general cases are at work, [24]. The whole approach in this paper is strongly influenced by similar approaches [90–93,96,83,100,101]. [90] is mainly interested in discretization methods for initial value problems which usually are not consistently differentiable. In [91–93,83,96] a more complicated theory has to be employed to get the corresponding results for a slightly larger class of problems. Zeidler [100,101] considers, for stable methods (a main goal of our approach), a wider range of problems, however does not discuss all the different discretizations from a unified standpoint as we do it here. These approaches are presented in a more formalistic way which we want to avoid here.

1.1 Main Aim and Organization of the Paper

This paper is mainly concerned with the numerical Liapunov-Schmidt and center manifold methods. They yield bifurcation and dynamical scenarios converging to the original scenarios. Hence, the corresponding normal forms, characterizing the bifurcation scenarios of the discrete problem, converge to those for the original exact problem. We do not give proofs for the results, presented already in the many papers cited above. Rather we want to review the concepts and results.

Our approach is, w.r.t. bifurcation, more general than earlier approaches, see Section 2 and covers more complex discretization methods and operator equations, e.g. spectral methods and Navier-Stokes equations, excluded in the earlier papers. For numerical center manifolds the corresponding problems in our sense have not been discussed yet, see [71]. In Section 3 we give a short survey of the standard generalized Liapunov-Schmidt method.

The next Section 4 describes the general form and modifications of the classical discretization methods. For "regular" operator equations, with $G(u_0, \lambda_0) = 0$, the linearized operator $G_u(u_0, \lambda_0)$ is invertible. Consistency, stability and some technical conditions yield existence of a discrete solution (u_0^h, λ_0) of $G^h(u_0^h, \lambda_0) = 0$, and its convergence to the exact solution (u_0, λ_0) of the original problem. In the context of bifurcations and center manifolds, the inverse of the linear operator $G_u(u_0, \lambda_0)$ usually does not exist, and stability breaks down. Furthermore, since the nonlinearity of an operator determines the bifurcation scenarios, the usual consistency certainly will not be able to reflect this situation. Therefore we have to modify stability and consistency: We introduce the so-called ***consistent differentiability***, as the consistency of the original operator and its higher derivatives. Stability has to be replaced by ***bordered stability*** of the discretization of the bordered systems, used in Liapunov-Schmidt and center manifold methods and their numerics. The above general methods are consistently differentiable.

For stability and later for its bordered form, as presented in Section 5, we only need to consider linearized problems, see [90]. Stability and bounded invertibility of the original operator are closely related but not equivalent. For monotone or coercive operators the above concepts automatically yield stability. For compact perturbations of these operators, stability is equivalent to the existence of the inverse operator, see Theorems 8. Sometimes these conditions are not easily verified. So it is important to be able to replace them by well-known properties, monotone operators and their compact perturbations, see Theorems 7-5 or the well available discrete Brezzi-Babuska "inf-sup"condition.

Having proved consistent differentiability and bordered stability, Section 6 reports that numerical Liapunov-Schmidt methods produce numerical bifurcation scenarios converging to those of the original equations. Hence the characterizing determining equations and non- degeneracy conditions are approximately reproduced. For physically motivated, structurally stable problems (with respect to a symmetry group) we are even able to guarantee contact equivalent scenarios under an additional assumption.

In Section 7 extends these results for numerical center manifold methods. We are mainly interested in classifications of center manifolds and the influence of space discretization in contrast to Lubichs approach in [71] and [70], who discuss the approximation with Runge-Kutta-type methods.

Next we include two important examples, in Section 8 the Navier-Stokes problem on a more theoretical basis, in Section 9 we apply numerical Liapunov Schmidt methods to a structurally stable $O(3)-$ equi-variant problem.

It has turned out, that the back-transformation of the bifurcation scenarios for the low dimensional bifurcation equations back to the original operator equations with respect to its state variables and parameters is indeed a very hard problem, see Section 10. In a series of papers [64,63,27–29] this problem is solved for singularities up to co-dimension ≤ 3.

Finally we indicate in Section 11 a C^{++} – system under development, only partially documented at the moment [44,80], which allows to relatively easily do bifurcation analysis and classification for new operators and discretization methods.

1.2 Other Supported and Related Areas

Here I only want to list the other research areas which are related to the above results and have been supported by the DFG Schwerpunkt DANSE or are directly related to it.

Symmetries, group theory, symmetry respecting bases and symmetry breaking: Our and the CSU- group were working for a long time on the numerical employment of symmetries. Using characters and generalized FFT we introduced new bases, [6,4,8,73,3,10,50,11], for finite and continuous symmetry groups [13]; spectral methods mainly in [25].

Krylov-type methods for bifurcation studies: They are based on early papers of Sebastian, [4], and collected in Mei [74] and two Dissertations [87,86]. Our main interests are solution of discretized PDEs and determination of the bifurcation numerics of large systems of equations. Hence we have to determine the eigenvalues of the linearized operators close to the imaginary axis, responsible for stationary and Hopf bifurcations and center manifolds. They have to be computable with very efficient methods. The state of the art in our group essentially is described in [30]. To apply these methods to bifurcation problems we indeed need *hierarchies of pre-conditioners* to guarantee convergence of these methods. Since only few low-dimensional eigen-spaces concentrate the information for bifurcation and dynamics, these Krylow type methods are well suited. They use invariant subspace techniques in a complementary way to Beyns approach, see [19].

Extended systems play an important role for all types of singular problems. We generalize Newton's type of iterations [5] to singular operator equations and their discretization by appropriate bordering techniques, important for *symmetry breaking bifurcation points with nonlinear degeneracies,* as well, a problem studied in [26].

The computation of all bifurcating solutions has been studied by so-called scaling or blow-up techniques, for arbitrary co-dimension,see [75,86].

$k-\epsilon$ models are closely related to center manifolds. In particular, modeling turbulent floods has been studied in [84].

Homo-clinic orbits in PDEs have been studied for many years. In [13] forced symmetry breaking of homo-clinic cycles for PDEs with $O(2)-$ symmetry has been the subject. A numerical bifurcation function, based on the Lin approach of Melnikows method for homo-clinic orbits and approximations of bifurcation functions for homo-clinic orbits in large systems have been studied in [16]. In the meantime we are trying to determine homo-clinic orbits for PDEs in a more general setting [45].

Hopf bifurcations in different contexts have been intensively studied with respect to normal forms in [15] and in particular in combination with computer algebra in [4,12,15].

Crossover collisions of scroll wave filaments are studied together with Fiedler in [72].

2 Numerical Bifurcation and History

Essential phenomena in the dynamics of partial differential equations

$$\frac{\partial u}{\partial t} = G(u(t), \lambda(t)) \tag{1}$$

initially start in stationary and Hopf bifurcation points of this problem. Stationary or Hopf (time periodic) solutions bifurcation are only to be expected for purely imaginary eigenvalues $i\mu$ of $G(u_0, \lambda_0) = 0, \dim \mathcal{N}(G_u^0 - i\mu I) > 0$, and $\mu = 0$ for stationary, $\mu \neq 0$ for Hopf bifurcation. The Hopf case can be reduced by a standard transformation, see [52] and [4,12] to stationary bifurcation. So we assume

$$G(u_0, \lambda_0) = 0, \dim \mathcal{N}(G_u^0) > 0. \tag{2}$$

Collecting all the Hopf and stationary bifurcation phenomena in a specific solution point (u_0, λ_0), the local dynamics can be studied by the center manifold approach [98,99,97]. Nice examples are given in [42]. Expect [71], who do not aim for convergence of the normal forms of the discrete to the original versions, there are no convergence studies for center manifolds. We will come back to this point in Section 7.

We will mainly study stationary bifurcations and their numerical reproduction. In particular, we concentrate on the time independent form of (1). In regular points, for any specific parameter λ_0, a locally unique stationary solutions u_0 exists. Appropriate boundary conditions are indicated by $\mathcal{U}_b$.

$$G : \mathcal{D}(G) \subset \mathcal{X} = \mathcal{U} \times \mathbf{R}^q \to \mathcal{Z}', \; G(x_0) = G(u_0, \lambda_0) = 0 \text{ with } x_0 \in \mathcal{D}(G),$$
$$u_0 \in \mathcal{U}_b \subset \mathcal{U}, \mathcal{Z} \text{ Banach spaces, usually } \mathcal{U}, \mathcal{Z} \subset \{u : \Omega \subset \mathbf{R}^d \to \mathbf{R}^{\hat{d}}\}. \tag{3}$$

We have chosen $G : \mathcal{X} \to \mathcal{Z}'$, since we want to test $\mathcal{Z}'$ by the space of test functions $\mathcal{Z}^h \subset \mathcal{Z}''$ with good approximation properties. For smooth enough operators and arguments we have $G(u,\lambda) \in \mathcal{Z}$. If $\mathcal{U} = \mathbf{R}^n$, equation (1) represents an ODE and (3) a finite dimensional problem which can be solved to any given accuracy. For infinite dimensional spaces $\mathcal{U}$, usually (3) represents PDE w.r.t. spatial variables. For many important problems and specific parameter situations and boundary conditions an analytical study is able to reveal the bifurcation scenarios and the essential structural properties, see e.g. Golubitsky/Schaeffer [52], Ioos/Mielke [62], Mielke/Schneider/Uelker [77,78], Allgower/Ashwin/Böhmer/Mei [7,4], Haragus-Courcelle/Kirchgässner [59] .

However, for general problems in PDEs qualitative and quantitative analytical results are usually not available. This concerns, in particular, "real-life" problems and general parameter and boundary condition situations. Discretization methods have to be used in these cases to solve (3).

$$G^h : \mathcal{D}(G^h) \subset \mathcal{U}^h \times \mathbf{R}^q \to (\mathcal{Z}^h)', \quad \mathcal{Z}^h \subset \mathcal{Z}, \quad G^h(x_0^h) = G^h(u_0^h, \lambda) = 0, \quad (4)$$

the usual $\mathcal{U}^h \subset \mathcal{U}_b$, see Section 4, may be violated for variational crimes.

The classical convergence approach requires two essential conditions, namely stability, hence, $\|(G^h_{u^h}(u_0^h, \lambda_0^h))^{-1}\| \leq C$, and consistency, a relation of the form $G^h(u^h, \lambda^h) = G(u^h, \lambda) + \mathcal{O}(h^p), p \geq 0$. They are not applicable for bifurcation numerics. Stability breaks down since , see Section 5, the existence of $(G_u(u_0, \lambda_0))^{-1}$ is violated. Bifurcation and center manifold scenarios essentially are determined by the higher derivatives of an operator G, the above consistency certainly will hide all this essential information. The main tool to determine bifurcation and their numerics is the Liapunov-Schmidt method. The linear projectors

$$P : \mathcal{U} \to \mathcal{N}(G_u^0), I - Q : \mathcal{Z}' \to \mathcal{R}(G_u^0) \quad (5)$$

mapping the spaces $\mathcal{U}$ and $\mathcal{Z}'$ to the kernel and range of G_u^0, respectively allow to split the problem into tractable subproblems.

Bifurcation numerics have been studied in many papers: In Keller/Langford [67] perturbations of the above operators G for two-determined problems, in Jepson/Spence [65] of the projectors $P, I-Q$ in (5) are studied to allow generalizations of Liapunov-Schmidt methods. Both perturbation concepts do not cover the problems connected with discretization. Here, sequences of spaces $\mathcal{U}^h$, $\mathcal{Z}^h$ approximate the $\mathcal{U}$, $\mathcal{Z}$ respectively. In particular equi-boundedness of the discrete inverses, hence stability, and appropriate reproduction of higher order terms in discretizations are not covered. This goal is achieved for specific cases in Brezzi/Rappaz/Raviart [33,34], Crouzeix/Rappaz [41], Raugel [82] and Griewank/Reddien [56]. Using Green or, more general inverse operators, they assume G to have the form

$$G : \mathcal{U} \times \mathbf{R}^q \to \mathcal{U}, \; G(u,\lambda) = u + T(u,\lambda), \; T \text{ compact}, \; u_0 \in \mathcal{U}_b \subset \mathcal{U}. \quad (6)$$

With convergent approximation operators $P^h : \mathcal{U} \to \mathcal{U}^h \subset \mathcal{U}$, extended as $P^h(u,\lambda) := (P^h u, \lambda)$, they study discretizations of the form

$$\begin{aligned} &G^h((u^h,\lambda)) = u^h + P^h T(u^h,\lambda) \\ &\quad \text{with } P^h T = T P^h, \; G^h(u_0^h, \lambda_0) = 0, \; u_0^h \in \mathcal{U}_b^h \subset \mathcal{U}^h. \end{aligned} \quad (7)$$

These assumptions (6), (7) allow very elegantly to prove convergence of the discrete bifurcation scenarios [56]. However, the $\mathcal{U}^h \subset \mathcal{U}$ and $P^hT = TP^h$ exclude variational crimes and many important discretization methods and operators, depending on lower order derivatives, e.g., many finite element and spectral methods and Navier-Stokes equations. Furthermore, only in very few cases the numerical methods for ODEs and PDEs employ Green or discrete Green operators, see e.g., [56].

Recently, this fact has been acknowledged by [81] as well, who consider exact Galerkin methods for a very special case of (3), avoiding (6). This transition from (3) to (6) is not negligible: It is well known, that analytically equivalent numerical methods may behave very differently. So, even in [35] for nonlinear equations of the form (3) discretization method are considered and simple bifurcation is included. However, for convergence, they have to assume but do not prove, that, for their special cases, the discrete approximate the original bifurcation functions. Furthermore, the recently very popular spectral methods and finite element methods with variational crimes need different spaces of trial- and test-functions and different approximation operators P_1^h and P_2^h, see Section 4. The de-aliasing techniques, needed to avoid unwanted explicit symmetry breaking, make the situation even more complicated. This general form (3) was considered much earlier in the general theories, e.g., in [90–93,96,100,101].

To cover all these problems and to allow a direct discretization of (2), we have introduced the *Numerical Liapunov-Schmidt method* in [21,12,9,4]. Since the bifurcation effects are determined by an interplay of kernels and ranges of linear operators and nonlinear terms, we have to make sure that the discrete version of the original nonlinear problem indeed reproduces these effects. To achieve this goal we need the concepts of consistent differentiability and the stability of the bordered systems, see Sections 4, 5, 6.

3 The Generalized Liapunov-Schmidt Methods

A short version of generalized Liapunov-Schmidt methods will motivate the necessary, and essentially sufficient, conditions for convergence of the numerical bifurcation scenarios. This $\mathcal{U}^h \subset \mathcal{U}_b$, $\mathcal{Z}^h \subset \mathcal{Z}_b$ will be violated for variational crimes [24]. For operator equations only the $\lambda-$bifurcation approach is appropriate, based on G_u^0, for other approaches, see e.g. [18,55,87].

Numerically, the exact values of the solution, x_0, and the kernels $\mathcal{N}(G_u(x_0)) \subset \mathcal{U}_b$ and $\mathcal{N}((G_u(x_0))^d)$ are not available. Hence one has to use approximations and to generalize the Liapunov-Schmidt methods, see Jepson/Spence [65] and [87]. We only describe the case $\dim \mathcal{N}(G_u(x_0)) = \dim \mathcal{N}((G_u(x_0))^d) = 1$, for the general case see, e.g., [21,85,24]. Let $\mathcal{N} = \text{span}\,[\varphi]$ and $\mathcal{N}' = \text{span}\,[\psi']$ approximate [2] $\mathcal{N}(G_u(x_0))$, $\mathcal{N}((G_u(x_0))^d)$. With

[2] One can interpret the span $[\varphi]$ and span $[\psi']$ as system of $m \geq 1$ vectors.

their (closed orthogonal) complements $\mathcal{M}$ and $\mathcal{M}'$, resp., we have

$$\mathcal{U}_b = \mathcal{N} \oplus \mathcal{M}, \ \mathcal{X} = \mathcal{U}_b \times \mathbf{R}^q = (\mathcal{N} \oplus \mathcal{M}) \times \mathbf{R}^q \text{ and } \mathcal{Z}' = \mathcal{N}' \oplus \mathcal{M}'. \quad (8)$$

Accordingly, we define continuous linear projectors as

$$P : \mathcal{U}_b \to \mathcal{U}_b, \ \mathcal{R}(P) = \mathcal{N}, \ Q : \mathcal{Z}' \to \mathcal{Z}', \ \mathcal{R}(Q) = \mathcal{N}' \text{ and} \quad (9)$$

$$u = v + w := Pu + (I - P)u \in \mathcal{N} \oplus \mathcal{M} \quad (10)$$

Equivalent to the original problem $G(x_0 + x) = 0$ is the system

$$\begin{aligned} &(i)\ (I - Q)G(x_0 + x) = (I - Q)\big(G_u^0(v + w) + G_\lambda^0\lambda + R_0(v + w, \lambda)\big) = 0, \\ &(ii)\ QG(x_0 + x) = Q(G_u^0(v + w) + G_\lambda^0\lambda + R_0(v + w, \lambda)) = 0. \end{aligned} \quad (11)$$

For (8) and (9), (11)(i) is uniquely solvable, see [65], for $w = w(v, \lambda)$, if

$$\mathcal{N}(P) \cap \mathcal{N}((I - Q)G_u^0) = \{0\} \ \Leftrightarrow \ (I - Q)G_u^0(I - P) : \mathcal{M} \to \mathcal{M}' \text{ regular}. \quad (12)$$

Inserting this $w = w(v, \lambda)$ into (11)(ii), we define the bifurcation function

$$g = g(v, \lambda) := Q(G_u^0(v + w(v, \lambda)) + G_\lambda^0\lambda + R_0(v + w(v, \lambda), \lambda)) = 0. \quad (13)$$

For all $\mathcal{N}, \mathcal{N}'$, satisfying (12), these g are contact equivalent, see [65] and Section 10. Therefore, the bifurcation scenario is determined via its characteristic equations and non-degeneracy conditions, obtained by the partial derivatives of $g = g(v, \lambda)$ in the bifurcation point. Hence we need the partials with respect to v, λ for G, see e.g. [52,65,31]. The discrete bifurcation scenarios converge to the original scenarios, the partials of $g^h = g^h(v^h, \lambda^h)$ converge to those of the original g. By (11)(ii), we have to impose beyond the usual consistency the consistency of the derivatives of a discrete operator to those of the exact operators as well.

The modification of stability follows from (11)(i): We need the stability of the bordered systems. With

$$\mathcal{N} = span[\varphi] \subset \mathcal{U}_b, \ \mathcal{N}^* = span[\varphi^*], \ \mathcal{N}' = span[\psi'] \subset \mathcal{Z}', \ \mathcal{N}'^* = span[\psi], \quad (14)$$

and $< \varphi^*, \varphi > = < \psi', \psi > = 1$ we [3] consider the following bordered system

$$L := \begin{pmatrix} G_u^0 & \psi' \\ < \varphi^*, \cdot > & 0 \end{pmatrix}, \qquad L \begin{pmatrix} w \\ g \end{pmatrix} = \begin{pmatrix} f \\ 0 \end{pmatrix} \quad (15)$$

with $g, 0 \in \mathbf{R}, f = -(G_u^0 v + G_\lambda^0\lambda + R_0(v + w, \lambda))$. From (14) we obtain that system (15) is equivalent to (11)(i) and the single equation in (11)(ii), namely

$$g : \mathbf{R}^{1+q} \to \mathbf{R}, \ g(v, \lambda) := - < G_u^0(v + w) + G_\lambda^0\lambda + R_0(v + w, \lambda), \psi > = 0.$$

[3] For the general case $m > 1$ this $= 1$ is to be interpreted as identity matrix

The following *truncated Liapunov-Schmidt method*, due to Ashwin [9,4,25], generalized in [31], where not really necessary, but to simplify technicalities,

$$\text{we assume, for } \Gamma - \text{ equi-variant problems, } \quad QG_u^0 = 0. \tag{16}$$

We use the $k-$jet $j_kG(x_0+x) := \sum_{j=0}^{k} G_0^{(j)}x^j/j!$ w.r.t. x to formulate

Algorithm 1: *Truncated Liapunov-Schmidt Method:* Let $w_0(v,\lambda) = 0$.
Iteration: For $k = 2, 3, \ldots$ until determinacy do:
Define g_k, a polynomial of order k, the truncated bifurcation equation as

$$g_k : \mathcal{N}(G_u^0) \times \mathbf{R}^q \to \mathcal{N}(G_u^{0*}),\ g_k(v,\lambda) :=$$
$$Qj_k(G_u^0(v + w_{k-1}(v,\lambda)) + G_\lambda^0\lambda + R_0(v + w_{k-1}(v,\lambda),\lambda)) \tag{17}$$

Generate the next w_k, a polynomial of degree k, by

$$G_u^0 w_k(v,\lambda) := (I-Q)j_k(G_u^0(v + w_{k-1}(v,\lambda))$$
$$+G_\lambda^0\lambda + R_0(v + w_{k-1}(v,\lambda),\lambda)), w_k \in \mathcal{R}(G_u^{0*}). \tag{18}$$

For these $w_k(v,\lambda), g_k(v,\lambda)$ we have

$$w_k(v,\lambda) = j_k w(v,\lambda), \quad \text{and } g_k(v,\lambda) = j_k g(v,\lambda).$$

For $\Gamma-$equi-variant problems with continuous groups Γ, the evaluation of the nonlinear terms in G produces aliasing effects thus provoking unwanted symmetry breaking bifurcation effects. This can be avoided, by Algorithm 1 [4,12,14,25].

So, for a numerical Liapunov-Schmidt method the discretization has to be *consistently differentiable* and *bordered stable.* Both types of necessary conditions will be obtained by applying approximation results. The following two Sections Section 4 and Section 5 are concerned with these properties.

4 Consistently Differentiable Discretization Methods

4.1 Weak Formulation

Our (stationary) problems can be formulated in a weak form. For fixed λ and, with $\perp$ indicating orthogonality w.r.t.$\langle\cdot,\cdot\rangle_{\mathcal{Z}'\times\mathcal{Z}}$, e.g., $(\cdot,\cdot)_{L^2(\Omega)}$,

$$G(u_0) = 0 \Leftrightarrow G(u_0) \perp \mathcal{Z}_b, u_0 \in \mathcal{U}_b\ . \tag{19}$$

We indicate the boundary conditions, necessary for uniqueness, by $u \in \mathcal{U}_b \subset \mathcal{U}, v \in \mathcal{Z}_b \subset \mathcal{Z}$, with closed subspaces $\mathcal{U}_b, \mathcal{Z}_b$, and usually $(\mathcal{Z}_b)' = (\mathcal{Z})'$. To reduce the technical difficulties we confine the presentation for the beginning to the case of a bounded linear operator $A := G_u^0$. The evaluation and consistent differentiability of the nonlinear operators will be discussed separately, see, e.g., §4.3. For (3) and $G_u(u_0) = A \in \mathcal{L}(\mathcal{U},\mathcal{Z}')$, we solve

$$Au = f, \quad u \in \mathcal{U}, \quad f \in \mathcal{Z}'. \tag{20}$$

(20) can be re-interpreted with the continuous bilinear form $a(\cdot,\cdot)$ generated by A in the form $a(\cdot,v) :=< A\cdot, v >_{\mathcal{Z}'\times\mathcal{Z}} \quad \forall \; v \in \mathcal{Z}$ as

$$(20) \Leftrightarrow a(u,v) =< f, v >_{\mathcal{Z}'\times\mathcal{Z}} \; \forall \; v \in \mathcal{Z}. \tag{21}$$

In Galerkin methods, equation (20) is discretized by replacing the usually infinite-dimensional spaces , $\mathcal{U}, \mathcal{Z}$, by sequences of finite- dimensional subspaces, $\mathcal{U}_b^h \subset \mathcal{U}$, $\mathcal{Z}_b^h \subset \mathcal{Z}$. For variational crimes approximate boundary conditions are allowed, e.g., $u(y_j) = 0$ only in appropriate collocation points along the boundary. Here we study conforming approximations

$$\mathcal{U}_b^h = \mathcal{U}^h \subset \mathcal{U}_b, \quad \mathcal{Z}_b^h = \mathcal{Z}^h \subset \mathcal{Z}_b = \mathcal{Z} \; . \tag{22}$$

As indicated above, we restrict variational crimes to quadrature approximations, so (22) is satisfied, see [24] for the general case. Extending [100,101] we introduce

Definition 1. *Petrov-Galerkin approximating spaces:* Let $\mathcal{U}_b \subset \mathcal{U}$ be Banach spaces and $\mathcal{U}_b^h \subset \mathcal{U}_b$, finite-dimensional subspaces, assigned to every $h \in H$ with accumulation [4] point 0, see (22). We call $\mathcal{U}_b^h$ *Petrov-Galerkin approximating spaces for* $\mathcal{U}_b$ (and sometimes for $\mathcal{U}$ as well), if we have:

$$dist(u, \mathcal{U}_b^h) := \inf_{v^h \in \mathcal{U}_b^h} ||u - v^h||_{\mathcal{U}} \to 0 \quad \text{for } h \to 0 \, \forall \quad u \in \mathcal{U}_b \text{ (and } u \in \mathcal{U}), \tag{23}$$

resp. In the sequel we assume *Petrov-Galerkin approximating spaces*

$$\mathcal{U}_b^h, \;\; \mathcal{Z}_b^h \text{ for } \mathcal{U}_b, \; \mathcal{Z}_b \text{ with } \; \dim \mathcal{U}_b^h = \dim \mathcal{Z}_b^h. \tag{24}$$

If, in addition to (23) for $\mathcal{U}_b$, the $\mathcal{Z}_b^h$ satisfy(23) $\forall \;\; v \in \mathcal{Z}_b$ **and** $\forall \;\; v \in \mathcal{Z}''$, the bi-dual space of $\mathcal{Z}$, then $\mathcal{U}_b^h$, $\mathcal{Z}_b^h$ is called *a pair of bi-dual Petrov-Galerkin approximating spaces.* The bi-duality condition is usually satisfied.

We define the projector P_z^h and P_u^h, $\mathcal{Z}' \times \mathcal{Z}$ in two equivalent ways as

$$P_z^h \in \mathcal{L}(\mathcal{Z}', \mathcal{Z}_b^{h'}) \text{ by } P_z^h f := f|_{\mathcal{Z}_b^h} \Leftrightarrow < P_z^h f - f, v >_{\mathcal{Z}'\times\mathcal{Z}_b} = 0 \quad \forall \, v \in \mathcal{Z}_b^h. \tag{25}$$

The so-called Petrov-Galerkin methods for (20) determines $u^h \in \mathcal{U}_b^h$ s.t.

$$P_z^h(Au^h) = P_z^h f \;\; \Leftrightarrow \;\; < P_z^h(Au^h - f), v >_{\mathcal{Z}'\times\mathcal{Z}_b} = 0 \; \forall \;\; v \in \mathcal{Z}_b^h. \tag{26}$$

Obviously we have the equivalence

$$(26) \Leftrightarrow a(u^h, v^h) =< f, v^h >_{\mathcal{Z}'\times\mathcal{Z}} \;\; \forall \;\; v^h \in \mathcal{Z}_b^h. \tag{27}$$

Similarly, we define the projectors $\tilde{P}_z^h$, and $\tilde{P}_u^h$:

$$\tilde{P}_z^h \in \mathcal{L}(\mathcal{Z}', \mathcal{Z}_b^{h'}) \text{ by } \; < \tilde{P}_z^h f - f, v >^h_{\mathcal{Z}'\times\mathcal{Z}_b^h} = 0 \; \forall \;\; v \in \mathcal{Z}_b^h, \tag{28}$$

---

[4] We do not want to over-formalize the notation and have chosen $h \in H$

where $< \cdot, \cdot >^h_{\mathcal{Z}' \times \mathcal{Z}^h_b}$ indicates a quadrature approximation for $< \cdot, \cdot >_{\mathcal{Z}' \times \mathcal{Z}^h_b}$, only defined for smooth enough situations, e.g., $\tilde{P}^h_z f - f \in \mathcal{Z}' \cap C(\overline{\Omega})$. We omit the notational technicalities, for a detailed analysis, see [24]. The so-called *generalized Petrov-Galerkin methods* for (20) determine $u^h \in \mathcal{U}^h_b$ s.t.

$$\tilde{P}^h_z(Au^h) = \tilde{P}^h_z f \Leftrightarrow \; < \tilde{P}^h_z(Au^h - f), v >_{\mathcal{Z}' \times \mathcal{Z}_b} = 0 \;\forall \; v \in \mathcal{Z}^h_b. \tag{29}$$

Obviously we have the corresponding equivalence

$$(29) \Leftrightarrow a^h(u^h, v^h) = < Au^h, v^h >^h_{\mathcal{Z}' \times \mathcal{Z}} = < f, v^h >^h_{\mathcal{Z}' \times \mathcal{Z}_b} \;\; \forall \;\; v^h \in \mathcal{Z}^h_b. \tag{30}$$

By choosing special elements $v_i \in \mathcal{Z}^h_b, v_i(y_j) = \delta_{i,j}$ for the different quadrature points y_j, this method can often be re-interpreted as collocation method.

The main examples for this theory are finite element methods, including specific difference methods and spectral methods.

4.2 Spectral Methods

To give some concrete ideas, we collect the basic facts concerning spectral methods, see the classical and recent book and article [36,17] and, for special results, [25]. These methods are particularly appropriate for Γ- equi-variant problems with continuous groups Γ, since this equi-variance has to be reproduced in the discretization as well. We restrict the discussion to Hilbert spaces $\mathcal{U} = H^n_w(\Omega), \mathcal{Z} = L^2_w(\Omega) = \mathcal{Z}', \Omega \subset \mathbf{R}^d, w$ a weight function, $(\cdot,\cdot)_w$ a scalar product, $\|\cdot\|_n$ the corresponding norms, used in the rest of this Section. We assume, for $\mathbf{Z}^d_0 \subset \mathbf{Z}^d$, a

$$\text{complete orthogonal basis } \{\varphi_k : k \in \mathbf{Z}^d_0\} \text{ for } \mathcal{U} \text{ and } \mathcal{Z} \text{ w.r.t. } \|\cdot\|_n, \|\cdot\|_0 \tag{31}$$

with real- or complex valued $\varphi_k(x)$. We have to consider the possibly different spaces $\mathcal{U}_b \subset \mathcal{U}, \mathcal{Z}_b \subset \mathcal{Z}$, and corresponding projection operators of truncation, interpolation and orthogonal projection onto the $\mathcal{U}_b, \mathcal{Z}_b$, denoted by $T^N, I^N, P^N_u, \tilde{P}^N_u, P^N_z, \tilde{P}^N_z$. Finite dimensional approximating spaces are

$$\mathcal{U}^N = \operatorname{span}\{\varphi_k : \; k = (k_1, \ldots, k_d) \in \mathbf{K}^N\} \subset \mathcal{U}, \quad \text{finite } \mathbf{K}^N \subset \mathbf{Z}^d_0, \tag{32}$$

$u^N \in \mathcal{U}^N, \;\; N = (N_1, \ldots, N_d) \in \mathbf{N}^d_0$ the multi-index $k \in \mathbf{K}^N$ with $|k_i| \le N_i$ and $\hat{N} := |\mathbf{K}^N|, \tilde{N} := \min\{N_1, \ldots, N_d\}$. Instead of the above upper index h we therefor chose an N. Every $u \in \mathcal{U}$ (or $\mathcal{Z}$) is, alternatively, *approximated by truncation* $T^N u$ as

$$T^N u = T^N(\sum_{k \in \mathbf{Z}^d_0} \hat{a}_k \varphi_k) := \sum_{k \in \mathbf{K}^N} \hat{a}_k \varphi_k \in \mathcal{U}^N,$$

or by (unique) *interpolation* in distinct points $y_j \in \Omega, j$ multi-indices, as

$$I^N : \mathcal{U} \to \mathcal{U}^N \text{ is unique by } (I^N u - u)|_{y_j} = 0, j \in \mathbf{J}^N \subset \mathbf{Z}^d, |\mathbf{J}^N| = \hat{N} = |\mathbf{K}^N|.$$

Now, $(\cdot,\cdot)_w$ is approximated by the Gaussian quadrature rule defined as

$$(u,v)_w \approx (u,v)_w^N := \sum_{j\in\mathbf{J}^N} u(y_j)\bar{v}(y_j)w_j, \quad (\|u\|_0^N)^2 := (u,u)_w^N. \tag{33}$$

With $\rho = 0,1,2$ for Gauss, Gauss-Radau, and Gauss-Lobatto quadrature points y_j, we have $(u,v)_w = (u,v)_w^N$ in (33) for $\ell = 0,1$ and for $\ell = 2$ only for u (or v) $\in \mathcal{U}^{N-1}$, important for collocation methods. Often the $\mathcal{U}^N$, $\mathcal{Z}^N$, φ_k, y_j are chosen according to a group Γ, e.g., $\Gamma = SO(1)$ and $\varphi_k(x) = \exp(ikx)$. Then $(u,v)_w, (u,v)_w^N$ are Γ-invariant, and $T^N, I^N : \mathcal{U} \to \mathcal{U}^N$ are Γ-equi-variant orthogonal projectors w.r.t. $(\cdot,\cdot)_w$ and $(\cdot,\cdot)_w^N$, respectively, hence, the corresponding truncation and interpolation errors

$$T^N u - u \perp_w \mathcal{U}^N \text{ and } I^N u - u \perp_w^N \mathcal{U}^N. \tag{34}$$

With the Γ-invariant $\mathcal{Z}^N$ the orthogonal projectors

$$\begin{aligned} P_z^N, \tilde{P}_z^N : \mathcal{Z}_b \to \mathcal{Z}_b^N, (P_z^N z - z, v^N)_w = 0 \text{ and}\\ (\tilde{P}_z^N z - z, v^N)_w^N = 0 \;\; \forall \;\; v^N \in \mathcal{Z}_b^N, \end{aligned} \tag{35}$$

and P_u^N, $\tilde{P}_u^N$ as well, are Γ-equi-variant. Here we assume, see [25,24], $\mathcal{U}_b^N = \mathcal{U}^N$ and $\mathcal{Z}_b^N = \mathcal{Z}^N$. Then (25) shows that $T^N = P_u^N, I^N = \tilde{P}_u^N$.

Theorem 2. *Let* $0 \le n \le m$. *For a periodic function* $u \in H_w^m(\Omega)$ *choose* $\Omega = (0,2\pi)$, $w(x) \equiv 1$ *and Fourier polynomials* ($K = F$), *for non-periodic functions* $\Omega = (-1,1)$, $w(x) = 1/\sqrt{1-x^2}$ *and* $w(x) \equiv 1$ *for Chebyshev and Legendre polynomials* ($K = C$ *and* L), *respectively. Furthermore, let* $m > \tau_K(n), \iota_K(n)$ *and*

$$\begin{aligned} &\tau_F(n) := n, \tau_C(n) := \tau_L(n) &&:= 2n - (1-\delta_{n0})/2,\\ &\iota_F(n) := n, \iota_C(n) := 2n, \iota_L(n) := 2n + d/2, && \tau_K(n) \le \iota_K(n), \end{aligned} \tag{36}$$

with $\delta_{n0} = 1$ *for* $n = 0$ *and* $= 0$ *otherwise. Then the errors converge*

$$\begin{aligned} \|T_K^N u - u\|_n, &\le CN^{-m+\tau_K(n)}\|u\|_m \quad \text{for } N \to \infty\\ \| I_K^N u - u \|_n, &\le CN^{-m+\iota_K(n)} \| u \|_m \quad \text{for } N \to \infty. \end{aligned} \tag{37}$$

The quadrature errors are $\| I_N^K u \|_{\tilde{.}}$. *For functions defined on* $\Omega \subset \mathbf{R}^d, d > 1$, *the above* N *has to be replaced by* $\tilde{N}$. *If different approximations are used for different variables, the minimal* $\tau^K(n), \iota^K(n)$ *has to be chosen.*

For the parameter-dependent case we define componentwise, e.g., $(x,y)_w = ((u,\lambda),(v,\mu))_w := (u,v)_w + (\lambda,\mu)_q$ with a scalar product $(\lambda,\mu)_q$ in $\mathbf{R}^q$.

4.3 Consistent Differentiability

To avoid too many technicalities, we assume (3) in the form

$$\begin{aligned} G(x) = G(u,\lambda) &= L_0 u + \lambda R(u) = L_0 u + \lambda R_e(u, \nabla u, \int_{\Omega_0} u) \\ &= L_0 u + \lambda(u^2 + \nabla u(u + \int_{\Omega_0} u) + g) \end{aligned} \tag{38}$$

with $G(x_0) = 0$, $L_0 = G_u(x_0)$ a bounded linear operator, and R a nonlinear operator. For the general case, see [25]. With the above projectors P_z^h, $\tilde{P}_z^h$, see (25), (28), the different types of discretization methods can be formulated as

$$\begin{aligned}
&\text{determine } u_0^h \in \mathcal{U}_b^h \text{ such that } \; P_z^h G(u_0^h) = 0 \text{ or } G(u_0^h) \perp_w \mathcal{Z}_b^h && (39)\\
&\text{determine } u_0^h \in \mathcal{U}_b^h \text{ such that } \; \tilde{P}_z^h G(u_0^h) = 0 \text{ or } G(u^h) \perp_w^h \mathcal{Z}_b^h. && (40)
\end{aligned}$$

Sometimes $G(u^h)$ is replaced by an approximate operator $\tilde{G}^h(u^h)$, see [23]. For the above example (38) we have for collocation spectral methods, and using the norms of the last Subsection, to evaluate

$$\begin{aligned}
G^h(u^h)(y_j) &:= (L_0^h u^h)(y_j) + \lambda R^h(u^h)(y_j), \; L_0^h u^h := T^h\,(L_0)u^h = L_0(u^h),\\
R^h(u^h)(y_j) &:= \lambda((u^h)^2(y_j) + (\nabla u^h)(y_j)(u^h(y_j))\\
&\quad + \textstyle\sum_{i\in \mathbf{J}^N} u^h(y_i)w_i) + g(y_j)), \; j \in \mathbf{J}^N. \qquad (41)
\end{aligned}$$

In spectral methods usually the $(L_0(u^h))(y_j), (\nabla u^h)(y_j)$ are not evaluated directly, but via some linear approximation operators, e.g., the Fourier collocation derivative, see [36]. We denote these kinds of *linear* operators as

$$\begin{array}{ccccc}
L_0^h u^h & \approx & (L_0)u^h, & L_1^h u^h & \approx \nabla u^h \;\; \text{s.t.}\\
(L_0^h u^h)(y_j) & \approx & (L_0 u^h)(y_j), & (L_1^h u^h)(y_j) & \approx (\nabla u^h)(y_j)\\
l u^h & = & \int_{\Omega_0} u^h \approx & \sum_{i\in\mathbf{J}^N} u^h(y_i)w_i =: & l^h u^h.
\end{array} \qquad (42)$$

We introduce the restriction operator

$$\rho^h : C(\Omega) \to \mathbf{R}^{\hat{N}}, (\rho^h(u))(y_j) := u(y_j), \quad j \in \mathbf{J}^N. \qquad (43)$$

The $G^h(u^h)$ in (41) can now be re-interpreted as

$$\rho^h G^h(u^h) = \rho^h(L_0^h u^h) + \lambda \rho^h R^h(u^h). \qquad (44)$$

For spectral methods in the Hilbert space $L_w^2(\Omega)$ the ρ^h can, for smooth situations, equivalently be defined by the $\tilde{P}_z^h$ in (35). So we can re-write (44) as

$$\tilde{P}_z^h G^h(u^h) = \tilde{P}_z^h(L_0^h u^h) + \lambda \tilde{P}_z^h R^h(u^h) \qquad (45)$$

and are back at (40). To reveal the structure of the $R^h(u^h)$, we observe that

$$\begin{aligned}
\rho^h R^h(u^h) &:= R_e(\rho^h u^h, \rho^h(L_1 u^h), \ell^h u^h) \qquad (46)\\
&= \rho^h R_e(u^h, \nabla u^h, \ell^h u^h) + \mathcal{O}(\|I^h R(u^h) - R(u^h)\|_0)\\
&= \rho^h R(u^h) + \mathcal{O}(\|I^h R(u^h) - R(u^h)\|_0),\\
&\quad \text{and corresponding relations for the partials of } R, R^h,
\end{aligned}$$

All these operators $\rho^h, L_0^h, L_1^h, \ell^h$ are, w.r.t. appropriate norms, bounded linear operators. So the corresponding conditions for the derivatives are automatically satisfied. Studying finite element and difference methods, employing, e.g., quadrature rules and divided differences, we observe the same

structure, hence the following Theorem is valid as well. This is in contrast to methods of the Runge-Kutta type, where implicit and repeated function evaluations destroy the linearity. For the general situation we have to refer to $\mathcal{O}(\|I^h R(u^h) - R(u^h)\|_0), \mathcal{O}(\|I^h x - x\|_l)$, for spectral methods we have specified that in
(37). Then the proof for the following Theorem for spectral and the other cases is essentially identical to that in [25]:

Theorem 3. *Let the nonlinear operator G satisfy $G : H^n_w(\Omega) \to L^2_w(\Omega), G \in C^r(\mathcal{D}(G))$, $G(x_0) = 0$, $\|x_0 - x\|_n$ be small and let G^h be evaluated corresponding to (41), and satisfy (46). Then the operator G^h is* consistent *and* r-times consistently differentiable *with G, that is, for $j = 1, \ldots, r, 0 \le n \le m$,*

$$\begin{aligned}
&\|G^h(P^h_u x) - P^h_z G x\|_0 = \mathcal{O}(\|I^h x - x\|_n), \\
&\|(G^h)^{(j)}(P^h_u x) P^h_u x_1 \cdot \ldots \cdot P^h_u x_j - P^h_z G^{(j)}(x) x_1 \cdot \ldots \cdot x_j\|_0 = \\
&\quad = \mathcal{O}(\|I^h x_1 - x_1\|_n \cdot \ldots \|I^h x_j - x_j\|_n)(1 + \|I^h x - x\|_n)
\end{aligned} \tag{47}$$

for $x = (u, \lambda), x_j = (u_j, \mu) \in \mathcal{X} = \mathcal{U} \times \mathbf{R}^q$ and $u, u_1, \ldots, u_j \in H^n_w(\Omega) \cap \mathcal{U}_b$, with $\|x - x_0\|_m$ sufficiently small. Analogous results for the $\tilde{P}^h_x, \tilde{P}^h_z$ -combination are valid as well. All these operators, derivatives and $\mathcal{O}-$terms are $\Sigma-$equi-variant for $u_j \in$ $Fix(\mathcal{U}^\Sigma)$ for $\Sigma = \Gamma$ or any subgroup $\Sigma \subseteq \Gamma$.

For complicated operators an analogue form of(46) may not be satisfied. Then consistency or even consistent differentiability has to be verified directly.

5 Bordered Stability

Due to Stetters [90] results on stability of linearizations, see pp. 15 ff it is enough to prove stability for the linear discrete operator $(G^h)'(u^h)$, if G^h is continuous and some other technical properties are satisfied. In this section we again fix the parameter λ. We allow variational crimes in the form of quadrature approximations for inner products and pairings in our discussion, see [24] for the general case. We assume throughout a small enough $\|\tilde{u} - u_0\|$:

$$\text{For } A = G'(\tilde{u}),\ \|\tilde{u} - u_0\| \text{ small, choose } A^h = P^h_z A \text{ or } A^h = \tilde{P}^h_z A. \tag{48}$$

We give a sequence of results, which relatively easily allow to prove stability for the discretization methods for stationary problems including their bordered forms if certain conditions are satisfied [85,31].

Definition 4 (Stability). For the Petrov-Galerkin approximating spaces $\mathcal{U}^h_b$ and $\mathcal{Z}^h_b$ (dim $\mathcal{U}^h_b =$ dim $\mathcal{Z}^h_b\ \forall\ h \in H$) the sequence $(B^h)_{h \in H}$, of bounded operators $B^h \in \mathcal{L}(\mathcal{U}^h_b, \mathcal{Z}^{h\prime}_b)$, or often for short only B^h, is called *stable,* if there are positive constants h_0, C, independent of h, such that $(B^h)^{-1}$ exists for $h \le h_0$ and that $\|(B^h)^{-1}\|_{\mathcal{U}^h_b \leftarrow \mathcal{Z}^{h\prime}_b} \le C$ for $h \le h_0$.

We want to study the difference of the two $A^h = P_z^h\, A$ and $\tilde{P}_z^h\, A$ in (48).

Theorem 5. *Let $\mathcal{U}_b^h$, $\mathcal{Z}_b^h$ be bi-dual Petrov-Galerkin approximating spaces and $f \in \mathcal{Z}'$. Assume that the quadrature operators $\tilde{P}_z^h$ in (28) yield*

$$||P_z^h\, A - \tilde{P}_z^h\, A|_{\mathcal{U}_b^h}||_{\mathcal{Z}_b^{h\prime} \leftarrow \mathcal{U}_b^h} \to 0. \tag{49}$$

Then $P_z^h\, A$ is stable if and only if $\tilde{P}_z^h\, A$ is stable.

Stability of A^h implies the existence of $A^{-1} \in \mathcal{L}(\mathcal{Z}_b, \mathcal{U}_b)$, see [85,31]. A stable discretization possesses, in some sense, all desirable properties.

Theorem 6. *Let, for the bi-dual Petrov-Galerkin approximating spaces $\mathcal{U}_b^h$, $\mathcal{Z}_b^h$ the $A^h = P_z^h\, A$ or $A^h = \tilde{P}_z^h\, A$, see (48), be stable. Then the equations (20), (26) and (29)are uniquely solvable, more precisely:*

1. *The original equation (20) has an unique solution u.*
2. *There exists a positive constant h_0 such that, $\forall\ h \leq h_0$, the (generalized) Petrov-Galerkin equations (26) and (29) have unique solutions u^h.*
3. *The Petrov-Galerkin method converges, i.e. $u^h \to u$ for $h \to 0$, as*

$$||u - u^h||_{\mathcal{U}} \leq C\, \left(dist\ (u, \mathcal{U}_b^h) + \|Au_0 - (\tilde{P}_z^h A u_0)\|_{\mathcal{Z}_b^{h\prime}} + \|f - \tilde{f}^h\|_{\mathcal{Z}_b^{h\prime}} \right),$$

where for the generalized Petrov-Galerkin methods the exact solution u_0 has to be smooth enough, to allow the application of the quadrature formulas in $\tilde{P}_z^h A u_0$.

For an important class of operators, the so-called strongly monotone operators, [100,101,79], the stability can easily be verified with the following criterion:

Theorem 7. *Let $\mathcal{U} = \mathcal{Z}$ and $A \in \mathcal{L}(\mathcal{U}, \mathcal{U}')$ be a strongly monotone linear operator on $\mathcal{U}_b$, i.e. there exists a positive constant M such that*

$$a(u,u) \quad = \quad < Au, u >_{\mathcal{U}' \times \mathcal{U}} \quad \geq \quad M||u||_{\mathcal{U}}^2,\ \forall\ u \in \mathcal{U}_b,\ hence\ A^{-1}\ exists.$$

For $\mathcal{U}_b^h = \mathcal{Z}_b^h$ let A^h be chosen as in (48). Then A^h is stable and

$$||(A^h)^{-1}||_{\mathcal{U}_b^h \leftarrow \mathcal{U}_b^{h\prime}} \leq \varepsilon + 1/M \quad for \quad \varepsilon > 0 \quad and \quad h < h_0\ .$$

In general the existence of a bounded inverse of A is not sufficient to ensure that A^h is stable. However, the next results allows all types of elliptic equation, bordered systems and Navier-Stokes equations, see the following Sections, excluding only singular A, e.g., in bifurcation.

Theorem 8. *For the bi-dual Petrov-Galerkin approximating spaces $\mathcal{U}_b^h$, $\mathcal{Z}_b^h$ let $B \in \mathcal{L}(\mathcal{U}_b, \mathcal{Z}')$ and $A := B + C$, with $C \in C(\mathcal{U}, \mathcal{Z}')$ the set of compact operators from $\mathcal{U} \to \mathcal{Z}'$. Let A^h, B^h be chosen as in (48) and B^h be stable Then A^h is stable if and only if $A^{-1} \in \mathcal{L}(\mathcal{Z}', \mathcal{U}_b)$.*

As indicated in Section 3 we have to discretize bordered systems in order to obtain a correct discretization for bifurcation problems. We prove the stability of the linear bordered system (15) where G_u^0 is replaced by A in the linear form

$$(I - Q)Au = (I - Q)f, \quad u \in \mathcal{M}. \tag{50}$$

Theorem 9. *Let $\mathcal{U}_b^h$ and $\mathcal{Z}_b^h$ be bi-dual Petrov-Galerkin approximating spaces with and let $\hat{P}_z^h := (P_z^h, I_{\mathbf{R}^n}) = (P^h_{\mathcal{Z}\times\mathbf{R}^n})$ or $= (\tilde{P}^h_{\mathcal{Z}}, I_{\mathbf{R}^n}) = (\tilde{P}^h_{\mathcal{Z}\times\mathbf{R}^n})$: $\mathcal{Z} \times \mathbf{R}^n \to \mathcal{Z}^h \times \mathbf{R}^n$. Furthermore, let $A = B + C$ with $A, B, C \in \mathcal{L}(\mathcal{U}, \mathcal{Z}')$ and A be a compact perturbation of B with stable B^h, see Theorem8. Then the following four conditions are mutually equivalent, see (8),*

1. $((I - Q)A|_{\mathcal{M}})^{-1} \in \mathcal{L}(\hat{\mathcal{M}}, \mathcal{M})$, 2.(50) *is uniquely solvable* $\forall \; f \in \mathcal{Z}'$,
3. $L^{-1} \in \mathcal{L}(\mathcal{Z}' \times \mathbf{R}^n, \mathcal{U} \times \mathbf{R}^n)$, 4. $(\hat{P}_z^h L|_{\mathcal{U}_b^h \times \mathbf{R}^n})_{h \in H}$ *is stable.*

6 Convergent Numerical Liapunov-Schmidt Method

We want to formulate, on the basis of consistent differentiability and bordered stability, see Sections 4,5, the convergence results for the numerical Liapunov-Schmidt methods. Systematically the *operators and bilinear forms, projectors, spaces and functions* have to be replaced by their respective discrete forms:

operators G^h, L_0^h, R^h, $G_0^{h'}$, g^h, g_k^h, pairings $\langle u^h, v^h \rangle^h$,
projectors Q^h, Q^h, $I - Q^h$, T^h, I^h, P_u^h, $\tilde{P}_u^h$, P_z^h, $\tilde{P}_z^h$, $\hat{P}_z^h$,
spaces $\mathcal{X}^h$, $\mathcal{U}_b^h$, $\mathcal{Z}_b^h$, $\mathcal{N}(G_0^{h'})$, $\mathrm{im}(G_0^{h'*})$, $\mathcal{N}(G_0^{h'*})$, $\mathrm{im}(G_0^{h'})$, $\mathcal{N}(Q^h)$
and functions u^h, v^h, w^h, $w_k^h(v^h)$, $w_1^h(v^h) := 0$, g_k^h, finally , (12), (11) and Algorithm 1 have to be replaced by their discrete analogues.

We have indicated above, that the imposed conditions are either satisfied or appropriate. The next Theorem formulates the convergence results for the generalized Liapuov-Schmidt method. Then we recapitulate some basics for Γ-equi-variant problems and indicate the special results needed in Section 9. In particular, the discrete projectors Q^h have to satisfy the modified (16). Then, by [21,4,31], the following results are correct.

A discrete bifurcation function is, as for the original operators, computed from (13), where all original operators and derivatives are replaced by their discrete counterparts.

Theorem 10. *Let G^h be a $k-$ times consistently differentiable bordered stable method for G, $\;k$ the determinacy of the problem and proceed according to (51). Then, similarly to (37), the exact, discrete and the truncated k-jets $w_k^h(v) = j_k w^h(v)$, $g_k^h(v) = j_k g^h(v)$ and their derivatives $g_{v^i\lambda^j} = \partial^{i+j} g / \partial v^i \lambda^j$,* *evaluated in the origin, are related as* [5]

$$\begin{aligned} &\|w_k^h(T^h v) - T^h w_k(v)\|_n = \mathcal{O}(\|T^h v - v\|_n + \|I^h x_0 - x_0\|_n), \\ &\|g_k^h(T^h v) - T^h g_k(v)\| = \mathcal{O}(\|T^h v - v\|_n + \|I^h x_0 - x_0\|_n) \text{ and} \\ &\|g^h_{v^i\lambda^j} - g_{v^i\lambda^j}\| = \mathcal{O}(\|T^h u - u\|_n + \|T^h v - v\|_n + \|I^h x_0 - x_0\|_n). \end{aligned} \tag{51}$$

For the equi-variant problems, studied in Sections 9, we want to collect a few basics for the $\Gamma-$ equi-variant situation:

Let $E^{\nu}_{m,1}$ denote the module of germs at $(0,0)$ of C^{∞} vector or matrix-valued functions $\mathbf{R}^m \times \mathbf{R} \to \mathbf{R}^{\nu}$, over the ring $E^1_{m,1}$. We study Γ-equi-variant bifurcation problems, of the form (13),

$$\mathcal{F}_\Gamma := \{f \in E^m_{m,1} \;:\; f(0,0) = 0,\; Df(0,0) = 0,\; \gamma f(u,\lambda) = f(\gamma u, \lambda) \forall \gamma \in \Gamma\}.$$

We identify Γ- equivalent elements in $f, g \in \mathcal{F}_\Gamma$ that are merely deformations of each other we write $f \overset{\Gamma}{\sim} g$, see [53]. If $j_k g \sim g$, for minimal $k \in \mathbf{N}$, then g is said to be k-determined. With the usual pseudo-norm for $g \in \mathcal{F}_\Gamma$

$$\|g\|^0_k = \sum_{|i|+j\le k} \left| \frac{\partial^{|i|+j} g}{\partial u^i \partial \lambda^j}(0,0) \right|, \tag{52}$$

we only consider structurally stable g, the physically relevant cases, i.e., $\exists\ \varepsilon > 0$ such that $\forall\ g, f \in \mathcal{F}_\Gamma$ with $\|f - g\|^0_k < \varepsilon$ satisfy $f \overset{\Gamma}{\sim} g$. As a consequence Γ must be absolutely irreducibly represented on $\mathbf{R}^m$ (see [53]). For the problems, studied in Section 9, we consider the 1-parameter Γ-equi-variant operator equation with the singularity in $x_0 = (0,0)$ and

$$G(0,0) = 0,\; G_0' := G'(0,0) = (\partial_u G(0,0), \partial_\lambda G(0,0)) =: L =: (\partial_u G_0, \partial_\lambda G_0), \tag{53}$$

$\partial_u G_0$ a $\Gamma-$equi-variant Fredholm operator of index 1 with $dim\ \mathcal{N}(\partial_u G_0) = m+1$, $m \ge 1$. $\mathcal{F}_\Gamma$ excludes turning point bifurcations. We assume re-parameterizing if necessary, that

$$\partial_\lambda G_0 = \partial_\lambda G(0,0) = 0, \quad \partial_\lambda G_0^h = 0, \text{ hence } 0 \times \mathbf{R} \subset \mathcal{N}(G_0'),\ \mathcal{N}((G_0^h)'). \tag{54}$$

This corresponds to assuming existence of a trivial solution from which the branches bifurcate for G, and after re-parametrization, for G^h as well. Under these conditions the following convergence result is valid:

Theorem 11. *Let the conditions of the Theorem 10 be satisfied. Let the original problem be a $\Gamma-$equi-variant bifurcation problem and G^h be a $\Gamma-$equi-variant discretization based on the $\Gamma-$invariant subspaces and pairings and $\Gamma-$equi-variant operators, projectors and $\mathcal{O}-$terms with* $\dim\ \mathcal{N}(G_0^{h\prime}) = \dim\ \mathcal{N}(G_0')$. *Then we obtain Γ-equi-variant $g_k^h(T^h v) \overset{\Gamma}{\sim} g_k(v)$, determining Γ-equivalent bifurcation scenarios with Γ-equivalent perturbations $\mathcal{O}((T_v^h - v) + (I^h x_0 - x_0))$ in (51).*

[5] $\|\cdot\|$ indicates the norm in $\mathbf{R}^m$, with $m = 1$ unless by symmetry $dim\mathcal{N} = m > 1$

7 Numerical Center Manifolds

7.1 Basics for Center Manifolds

We present the basic results on center manifolds in a form which allows to later on use our numerical methods and to show stability and convergence of these methods. We start with parameter independent differential equations with appropriate boundary conditions in Banach-spaces of the form as in (3),(1)

$$\dot{u} = G(u) = L_0 u + R(u), \; G(0) = 0, \quad L_0 = G'(0). \tag{55}$$

We interpret a solution of (55) in the weak sense, see, e.g. [95]. (55) shows that w.l.o.g. we have assumed a stationary solution in $u_0 = 0$. We split the spectrum of L_0 into three parts $\sigma_s, \sigma_c, \sigma_u$, with negative, vanishing and positive $Re\,\lambda$ for the eigenvalues λ. The stable, center and unstable generalized eigenspaces are $\mathcal{N}^s, \mathcal{N} = \mathcal{N}^c, \mathcal{N}^u$. The local dynamics are well described by the dynamics on the center manifold. For the reduction, we use, as in the Liapunov-Schmidt method, subspaces and linear projectors.

Let $\mathcal{N}$ and $\mathcal{N}'$ be (closed) subspaces with $n = dim\mathcal{N} = dim\mathcal{N}' \geq 1$, spanned by the generalized eigenvectors with $Re\lambda = 0$ (or ≈ 0) of $L_0 \in \mathcal{L}(\mathcal{U}_b, \mathcal{Z}')$ and its dual $L_0^d \in \mathcal{L}(\mathcal{Z}, \mathcal{U}_b')$. As in Section (3) we define their (closed) complements $\mathcal{M}$ and $\mathcal{M}'$, the continuous linear projectors P, Q and use the splitting

$$u = v + w := Pu + (I - P)u \in \mathcal{N} \oplus \mathcal{M}. \tag{56}$$

The center stable and unstable manifolds, $\mathcal{W}^c, \mathcal{W}^s$ and $\mathcal{W}^u$ are invariant under (55) and tangential in 0 to $\mathcal{N}^c, \mathcal{N}^s, \mathcal{N}^u$, resp. Now we split (55) into

$$\dot{v} = QL_0 v + QR(v + w), \; v \in \mathcal{N}, \; w \in \mathcal{M}, \; \dot{v} \in \mathcal{N}' \text{ and} \tag{57}$$

$$\dot{w} = L_r w + (I - Q)R(v + w) \text{ with } L_r := (I - Q)L_0|_{\mathcal{M}}. \tag{58}$$

The parameterization

$$\mathcal{W}^c := \{(v, w); v \in \mathcal{N}, w = W(v) \in \mathcal{M}\} \text{ with } W(0), = 0, W'(0) = 0 \tag{59}$$

yields by inserting $\dot{w} = W'(v)\dot{v}$ into (58) the new equation, for $W(v)$ see (57),

$$\begin{aligned} C(W(v)) :&= W'(v)(QL_0 v + QR(v + W(v))) \\ &\quad - L_r W(v) - (I - Q)R(v + W(v)) = 0 \text{ and} \end{aligned} \tag{60}$$

$$\dot{v} = QL_0 v + QR(\; v + W(v)), \; v \in \mathcal{N}, \quad \text{for the center manifold} \tag{61}$$

For the above flow, (55) there exist, see [60,37,57], invariant unique stable and unstable manifolds $\mathcal{W}^u, \mathcal{W}^s \in C^r$ tangential to $\mathcal{N}^u$ and $\mathcal{N}^s$ and a (non-unique) center manifold $\mathcal{W}^c \in C^{r-1}$ tangential to $\mathcal{N}$ at 0. For most applications of (55) we have $\mathcal{N}^u = \{0\}$. Then the origin $v = 0$ of (61) and of (55) are simultaneously locally asymptotically stable and unstable, resp. We iteratively determine the Taylor coefficients of $W(v)$ via:

Theorem 12. *[60,37,99,71]: Let $\Phi(v)$ be such that $\Phi(0) = 0, \Phi'(0) = 0$ and $C(\Phi(v)) = \mathcal{O}(\|v\|^p)$, see (60), for some $p > 1$ and for $\|v\| \to 0$. Then we have the relation $W(v) = \Phi(v) + \mathcal{O}(\|v\|^p$ as $\|v\| \to 0$.*

7.2 Center Manifolds for Simple Hopf-Bifurcation

We study as concrete example a simple Hopf-bifurcation. Hence, we assume that L_0 has exactly one pair of purely imaginary simple eigenvalues, $+i\omega$, $-i\omega$ with $\omega > 0$. This implies $\mathcal{N} = [\varphi_1, \varphi_2]$, $\mathcal{N}' = [\varphi_1^*, \varphi_2^*]$ and we choose the basis such that $QL_0 = \begin{pmatrix} 0 & \omega \\ -\omega & 0 \end{pmatrix}$. We want to compute the Taylor coefficients of $W(\cdot)$ to second order accuracy. With $R(u) = F_2u^2 + F_3u^3 + \ldots$ we insert $W(v) = w_{20}v_1^2 + w_{11}v_1v_2 + w_{02}v_2^2 + \ldots = \sum_{|k|=2} w_k v^k + \ldots$ into (60) and compare the coefficients of v^k with $|k| = 2$. We end up with

$$
\begin{aligned}
(i)\; L_r w_{11} - 2\omega(w_{20} + w_{02}) &= -2QF_2\varphi_1\varphi_2 \\
(ii)\qquad L_r w_{20} - \omega_{11}\omega &= -QF_2\varphi_1\varphi_1 \\
(iii)\qquad L_r w_{02} - \omega_{11}\omega &= -QF_2\varphi_2\varphi_2.
\end{aligned}
\tag{62}
$$

We obtain with $X := w_{20} - w_{02}$ uniquely from the equation

$$
L_r X = -Q(F_2\varphi_1^2 - F_2\varphi_2^2),\; X \in \mathcal{M}, \tag{63}
$$

see [57]. Similarly, $Y := w_{20} + w_{02}$ and w_{11} are unique solutions of

$$
\begin{pmatrix} L_r & -2\omega I \\ -2\omega I & L_r \end{pmatrix} \begin{pmatrix} Y \\ w_{11} \end{pmatrix} = - \begin{pmatrix} Q(F_2\varphi_1^2 + F_2\varphi_2^2) \\ Q(F_2\varphi_1\varphi_2) \end{pmatrix}, \quad Y, w_{11} \in \mathcal{M} \tag{64}
$$

These equations (63), (64) with L_r in (58) and the X, Y, $w_{11} \in \mathcal{M}$ are bordered systems of the form (15) with two-dimensional $\varphi^* = (\varphi_1', \varphi_2')^T$, $\psi' = (\psi_1', \psi_2')$. With this known $W(v) = \Phi(v) + \mathcal{O}(\|v\|^3)$ we study the (local) dynamics of (55) via (61). Normal forms for the small dimensional center manifold system, can be obtained with the known Lie bracket operator [1] calculus. This allows, for a few cases, a classification of the dynamics, see e.g. [57].

7.3 Numerical Center Manifold Methods

There are essentially two points where discretization methods have to be used in context of center manifolds. If $w = W(v) = \Phi(v) + \mathcal{O}(\|v\|^p)$ is known, the low dimensional center manifold differential equation (61) usually has to be solved numerically. Convergence of the numerically computed center manifolds to the original center manifold solutions has been discussed for a wide class of operators and time discretization methods in Lubich/Ostermann [71]. The other context for discretization methods is the determination of the Taylor expansion for the center manifold, see Theorem 12 and (63), (64). In case of ordinary differential equations (55), this can be solved exactly. If (55) is a partial differential equation, then the w_k can be computed only approximately using some type of discretization method, see Section 4. In contrast to [71], we want to concentrate on this second aspect of space discretization.

We have seen above, that the determination of the w_k requires the solution of well understood bordered systems and their discretization. So, we obtain under the conditions of Theorems 10 (and modified 11) discretization methods with convergent $\|w_k^h - w_k\|_u \to 0$. The known constructive transformation to normal form is translated into a convergent transformation. This implies that the known construction indeed yields normal forms for the discrete problem which converge to the normal forms of the exact problem.

Finally, if instead of the original problem in (55) we have the parameter dependent problem (1) with a stationary solution $(u_0, \lambda_0) = (0,0)$, then we study the center manifold expansions in the neighborhood of this stationary point. To this end we extend, the standard approach [57], the original system into $\dot{u} = G(u, \mu), \quad \dot{\mu} = 0$.

Theorem 13. *All the above consistently differentiable and stable discretization methods define appropriate numerical center manifold methods. They yield convergent center manifold scenarios.*

8 Application to the Navier-Stokes Operator

Let $u = (u_1, \ldots, u_n)^T, p$ and f denote velocity, pressure and forcing term of an incompressible medium and v, w, q, r test functions, respectively. With the notation in Hackbusch [58], Temam [94] we assume

$$\Omega \subset \mathbf{R}^n \text{ , bounded and } \Gamma = \partial\Omega \text{ Lipschitz continuous,} \tag{65}$$

$$u, v, w \in \mathcal{U} := H_0^1(\Omega) \times \ldots \times H_0^1(\Omega), n \leq 4 \text{ factors,} \tag{66}$$

$$p, q, r \in \mathcal{W} := L_*^2(\Omega) := \{p \in L^2(\Omega) : \int_\Omega p(x)dx = 0\}.$$

Then the stationary Navier-Stokes equation has the form

$$G(u,p) := \begin{pmatrix} -\nu \Delta u + \sum_{i=1}^n u_i D_i u + \text{grad } p \\ div\ u \end{pmatrix} = \begin{pmatrix} f \\ 0 \end{pmatrix} \text{ in } \Omega$$

$$u = 0 \text{ on } \Gamma = \partial\Omega, \int_\Omega p dx = 0 \text{ guarantees a unique } p. \tag{67}$$

$G'(0,p)$, the *Stokes operator*, or the general $G'(u,p)$, applied to $(w.r)$, yields

$$G'(u,p)(w,r) = \begin{pmatrix} -\nu \Delta w + \sum_{i=1}^n (w_i D_i u + u_i D_i w) + grad\ r \\ div\ w \end{pmatrix}. \tag{68}$$

Based on the results in Sections 6, 5 there are two possible approaches: Either we use, for $\nu \approx 1$, the compact perturbation arguments relating the Stokes operator and the general derivative $G'(u,p)$. For the Stokes operator stability results are more easily available for appropriate consistently differentiable methods, see [58]. Then we get stability for an invertible $G'(u,p)$ with its discretization, see [85]. Or we employ the elaborate stability results for the general case, [94,51]. The goal is in both cases a convergenet bifurcation numeric.

With the continuous bilinear and trilinear forms $a(\cdot,\cdot), b(\cdot,\cdot)$ and $d(\cdot,\cdot,\cdot)$ and induced linear operators A, B and D (for fixed u), resp., defined as
$a(u,v) := \int_\Omega < \nabla u(x), \nabla v(x) > dx =< Au, v >_{H^{-1}(\Omega)\times H_0^1(\Omega)}$,
$b(p,v) := -\int_\Omega p(x) div\ v(x) dx =< Bp, v >_{H^{-1}(\Omega)\times H_0^1(\Omega)} \quad \forall\ p \in L_*^2(\Omega)$,
$d(u,w,v) := \sum_{i,j=1}^n \int_\Omega u_i(D_i w_j) v_j dx =< D_1(v,w), u >_{H^{-1}(\Omega)\times H_0^1(\Omega)} \quad \forall\ u$, $v, w \in \mathcal{U}$ the weak form of (67) is obtained by taking the $L^2(\Omega)$- scalar product of $G(u,p)$ with (v,q) as

$$\begin{aligned} &\nu a(u,v) + d(u,u,v) + b(p,v) = f(v) \quad \forall \quad v \in \mathcal{U} \text{ and} \\ &b(q,u) = 0 \quad \forall \quad q \in \mathcal{W}. \end{aligned} \tag{69}$$

Its linearization at (u,p) applied to (w,r) is

$$\begin{aligned} &\nu a(w,v) + d(u,w,v) + d(w,u,v) + b(r,v) = f(v) \quad \forall \quad v \in \mathcal{U} \text{ and} \\ &b(q,w) = 0 \quad \forall \quad q \in \mathcal{W}. \end{aligned} \tag{70}$$

Mind that for $u = 0$, in the Stokes case the two terms $d(u,w,v) + d(w,u,v)$ drop out. We combine $D_1, D_2 \in L(\mathcal{U},\mathcal{U}')$ as $D_1 v := d(u,\cdot,v), D_2 v := d(\cdot,u,v)$ for fixed u and the continuous and compact embedding $I : H_0^1(\Omega) \to L^2(\Omega)$. This shows, that D_1 and D_2 with, e.g., $D_1 v = D_1 I v\ \forall\ \ v \in H_0^1(\Omega)$ are, as product of a compact and a continuous operator, compact operators. This shows, that for nor too small values of $\nu > 0$, the physically not too interesting case, the $G'(u,p)$ is a compact perturbation of the Stokes operator. With these stability results we can use our machinery, to get stability via compact perturbations or directly, and then to obtain bordered stability and thus convergence for bifurcation numerics::

Theorem 14. *For all consistently differentiable and stable discretization methods for the Navier-Stokes equation, see [94,51,58,36], the numerical Liapunov-Schmidt and center manifold methods yield convergent bifurcation and center manifold scenarios.*

9 Application to Pattern Formation in Biology

9.1 The Model Equation and its Linearization

We consider a coupled system of reaction diffusion equations for on a spherical domain Ω in $\mathbf{R}^3$ of radius R

$$\frac{\partial c}{\partial t} = \tilde{L}(\beta)c + \tilde{N}(c,\beta), \quad \frac{\partial c}{\partial r} = 0 \text{ at } r = R \text{ and bounded in } 0, \tag{71}$$

with a two component mixture $c = \{c_1, c_2\}$. It arises in Turing's theory of pattern formation [61] and yields an $O(3)$-equi-variant problem. The specific model we choose was originally suggested in [88]. Here we assume a fixed $\gamma = 3$, as an effective Hill constant, the diffusion constants, following [88] as $D_1 = 0.15$ and $D_2 = 0.015$. The bifurcation parameter β represents a rate constant measuring the level of enzyme activity. The concentration vector $c = (1,1)$ is a homogeneous solution of (71) for arbitrary β. With $\partial_\beta G_0 = 0$ and the linear operator $\tilde{L} = \tilde{L}(\beta)$ we obtain

$$\tilde{L} = \begin{pmatrix} D_1\Delta - 1 & -\gamma \\ \beta & D_2\Delta + \beta(\gamma - 1) \end{pmatrix}, G_0' = (\tilde{L}, 0), \mathcal{N}(G_0') = \mathcal{N}(\tilde{L}) \times \mathbf{R}, \quad (72)$$

hence $\mathcal{N}(G_0')$ satisfies (54) and the nonlinear part is $\tilde{N}(c, \beta) = (-c_1c_2^\gamma, \beta c_1 c_2^\gamma)^T$. Now, a simple linear stability calculation shows, [25], that $C \equiv 1$ is stable for small values of β. In critical values of β the $C \equiv 1$ becomes unstable and one of the real eigenvalues passes the imaginary axis. To solve the eigenvalue problem $\tilde{L}v = \sigma v$ with boundary conditions (71), we insert the spherical harmonic ansatz

$$v = \sum_{l=0}^{\infty} \sum_{m=-l}^{l} f_l(r) Y_{lm}(\theta, \phi) = \sum_{l=0}^{\infty} \sum_{m=-l}^{l} (f_{l1}(r), f_{l2}(r))^T Y_{lm}(\theta, \phi) \in \mathcal{U}. \quad (73)$$

With $\Delta Y_{lm} = -l(l+1)Y_{lm}$, $(Y_{km}, Y_{ln})_2 = \delta_{kl} \cdot \delta_{mn}$, the linear eigenvalue problem $\tilde{L}v = \sigma v$ splits into and is reduced to the radial eigenvalue problem

$$\tilde{L} f_l Y_{lm} = Y_{lm} \tilde{L}_l f_l = Y_{lm} \sigma_l f_l, \text{ and } \tilde{L}_l f_l = \sigma_l f_l \text{ for fixed } l, \quad (74)$$

where σ_l is an eigenvalue of multiplicity $2l + 1$ and 1 for $\tilde{L}$ and $\tilde{L}_l$, resp.,

$$\text{with } d_l^2 = \frac{d^2}{dr^2} + \frac{2d}{rdr} - \frac{l(l+1)}{r^2} \text{ replacing } \Delta \text{in (72) to define} \tilde{L}_l \quad (75)$$

and boundary conditions for f_l as in (71). By[38,20] the Chebyshev ansatz

$$f_l^N(r) = \sum_{n=0}^{N-1} \hat{f}_n \, T_n(r), \hat{f}_n \in \mathbf{R}^2 \text{ for the } \hat{f}_n \in \mathbf{R}^2. \quad (76)$$

uniquely determines the numerical solution of (75)-(71) with a Chebyshev-tau collocation method in Gauss-Radau points and the exact solution in terms of spherical Bessel functions [38]. The critical or stability curves $\beta = \beta_l(R)$ in the (β, R) plane for each value of the wave number l satisfy $\Re\{\sigma_l\} = 0$. The minimal β_l value among these curves selects the critical wave number l_0 and determines the critical rate constant $\beta_c(R) = \min_l \beta_l(R)$. Generically, a unique spherical harmonic and radial eigenfunction are determined for each R for this l_0. For the parameter values in Figure 1 an intersection of two stability curves does not appear at criticality.For $R = 1.26$ we observe the minimum critical value for $l = 2$. This value is discussed here.

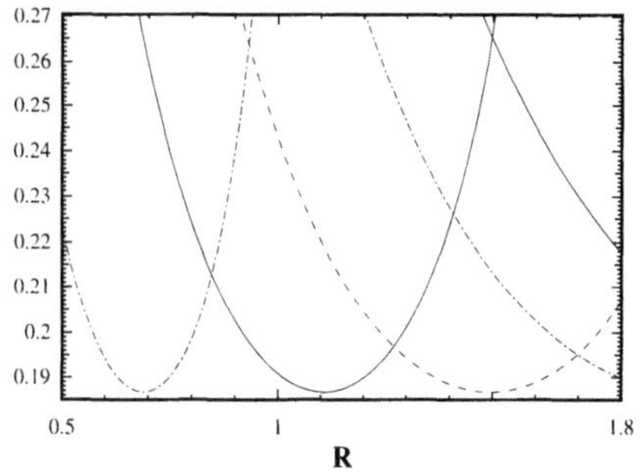

Fig. 1. Linear stability diagram, β versus R, for $l = 1$ (dot-dashed line), $l = 2$ (solid line), and $l = 3$ (dashed line).

N	$(b_1^N - b_1)/b_1$	$(b_{02-2}^N - b_{02-2})/b_{02-2}$
4	$5.7930825793 \times 10^{-02}$	$4.6691607817 \times 10^{-03}$
6	$-1.0318341612 \times 10^{-03}$	$-1.3734911834 \times 10^{-04}$
8	$1.0169306109 \times 10^{-05}$	$-2.7529791710 \times 10^{-06}$
10	$-6.2303099338 \times 10^{-08}$	$9.6679175279 \times 10^{-08}$
12	$2.3013266855 \times 10^{-10}$	$-1.9791415271 \times 10^{-09}$

Fig. 2. Relative approximation error of the coefficients b_1 and b_{02-2}, see § 9.3

9.2 Application of the Liapunov-Schmidt Reduction for $l = 2$

For the shell radius $R = 1.26$ the minimal critical value β_c occurs for $l = l_0 = 2$. With $\lambda := \beta - \beta_c$ and $u = (u_1, u_2)^T \equiv c - 1$ the critical value β_c is transformed to $\lambda_c = \lambda = 0, u = 0$. This allows the system (71) to be put into the standard form (38), for $\tilde{L}$, see (72) with $\beta = \beta_c$ and $l_0 = 2$

$$
\begin{aligned}
G : \mathcal{X} = \mathcal{U} \times \mathbf{R} \to \mathcal{Z},\ G(u, \lambda) = \tilde{L}u + R(u, \lambda) = 0, \text{ with} \quad (77)\\
\mathcal{U} = (H^1(\Omega))^2, \text{ and with identified } \mathcal{Z} = \mathcal{Z}' = (L^2(\Omega))^2 \text{ and}\\
\mathcal{N}\big(\partial_u G(0,0)\big) = \mathcal{N}(\tilde{L}) \approx \mathcal{N} = f_2^N \text{span}\{Y_{2,-2}, \dots, Y_{2,2}\}, \quad (78)\\
\mathcal{N}\big(G_0'\big) \approx \mathcal{N} \times \mathbf{R} \text{ with } \dim \mathcal{N}(\tilde{L}) = \dim \mathcal{N} = \mu = 5,
\end{aligned}
$$

and a corresponding $\mathcal{N}(\tilde{L}^*) \approx \mathcal{N}' = f_2^{N*}\text{span}\{Y_{2,-2}, \dots, Y_{2,2}\}$, $\dim \mathcal{N} = \dim \mathcal{N}' = 5$. Mind that the only difference between the exact $\mathcal{N}(\tilde{L})$ and its approximation $\mathcal{N}$ is the exact f_2 versus the discrete approximation f_2^N in (74),(76). Analogously, $\mathcal{N}(\tilde{L}^*)$ and $\mathcal{N}'$ are only distinguished by f_2^* and $f_2^{N^*}$ with $\tilde{L}^* f_l^* = 0$.
The nonlinear operator has with $\tilde{N}(c, \beta) = (-c_1 c_2^\gamma, \beta c_1 c_2^\gamma)^T$ the form

$$
R(u, \lambda) = \begin{pmatrix} -\gamma u_1 u_2 - (u_1 + 1)\sum_{k=2}^{\gamma} \gamma! u_2^k/(k!(\gamma - k)!) \\ (\lambda + \beta_c)(\gamma u_1 u_2 + (u_1 + 1)\mathcal{R}(u_2) + \lambda(u_1 + (\gamma - 1)u_2) \end{pmatrix}. \quad (79)
$$

We consider steady state solutions of (77). Bifurcating branches of solutions may then be determined by Liapunov-Schmidt reduction. With

$$
R(u, \lambda) = \sum_{i=1} \sum_{j=0} R_{ij}(u)\lambda^j, i + j \geq 2,\ R_{ij}(u) \text{ } i\text{-linear in } u,\ R_{10} \equiv 0 \quad (80)
$$

the Taylor expansion $\lambda R_{11}(u)+R_{20}(u)$ is enough, since $\lambda = \mathcal{O}(u^2)$ [39,40,46]. Following Algorithm 1, we split, see (8)

$$\mathcal{U} \ni u = v + w, \quad v \in \mathcal{N}, \quad w \in \mathcal{M}, \quad (v, \lambda) \in \mathcal{N} \times \mathbf{R} \approx \mathcal{N}(G_0') \; . \tag{81}$$

As a consequence of (78) we identify $(v,0) \in \mathcal{N}(G_0')$ and $v \in \mathcal{N}(\tilde{L})$ or $\in \mathcal{N}$

$$v = z\psi = (v_j)_{j=1}^2 = f_2^N(r)\Big(\sum_{m=-2}^{2} z_m Y_{2,m}(\theta,\phi)\Big) \in \mathcal{N} \; f_2^N = (f_{21}^N, f_{22}^N) \tag{82}$$

$z_m \in \mathbf{C}$ are complex coordinates for $\mathcal{N}$ and $\mathcal{N}'$, $z = (z_{-2}, \ldots, z_2)^T \in \mathbf{C}^5$ satisfying the reality condition $z_{-m} = (-1)^m \bar{z}_m$. (If f_2^N is replaced by f_2^{N*}).

The $O(3)$ equi-variance of (77) for G implies $T_\gamma G(u,\lambda) = G(T_\gamma u, \lambda)$, for an infinite dimensional representation $T_\gamma \;\; \forall \, \gamma \in O(3)$. T_γ induces the finite dimensional irreducible representation $T_{2,\gamma}$ on $\mathcal{N}$, see (82) via

$$T_\gamma v = T_\gamma z\psi = f_2^N \sum_{m=-2}^{2} z_m T_\gamma Y_{2,m} = f_2^N \sum_{m=-2}^{2} (T_{2,\gamma} z)_m Y_{2,m}$$

or $T_\gamma z\psi = (T_{2,\gamma} z, \psi)$ for $z\psi = v \in \mathcal{N}$ and identical $T_{z,\gamma}$ for $\mathcal{N}'$. With the $O(3)$ invariant $\mathcal{N} \approx \mathcal{N}(\tilde{L})$ and $\mathcal{N} \approx \mathcal{N}(\tilde{L}^*)$ the projectors P, Q in (9) are $O(3)$-equi-variant. Therefore, $T_\gamma P u = P T_\gamma u$ and $T_\gamma Q u = Q T_\gamma u$. The following bifurcation equations are obtained by projecting (77) with Q onto $\hat{\mathcal{N}} \approx \mathcal{N}(\tilde{L})$.

$$B(z,\lambda) = QR(z\psi + w(z,\lambda)). \tag{83}$$

The successive approximations to the bifurcation equations are calculated as a Taylor series in z and in λ until determinacy. By (17) and (82) we get

$$\begin{aligned} g_k(z,\lambda) &= Qj_k R(z\psi + w_{k-1}(z\psi,\lambda),\lambda) \\ &= Qj_k R_{20}(z\psi + w_{k-1}(z,\lambda),\lambda) + \lambda\, Qj_{k-1} R_{11}(z\psi + w_{k-1}(z,\lambda),\lambda). \end{aligned} \tag{84}$$

We only need $k=2$ and obtain for the components $-2 \le m \le 2$ the

$$\begin{aligned} g_{2,m}(z,\lambda) &= \lambda\langle R_{11}(z\psi), \psi_m' \rangle + \langle R_{20}(z\psi), \psi_m' \rangle = \lambda b_1 z_m \\ &\quad + b_2 \sum_{m'm''=-2}^{2} z_{m'} z_{m''} \int_0^\pi \int_0^{2\pi} Y_{2,m'} Y_{2,m''} \bar{Y}_{2,m} \sin\theta d\theta \, d\phi, \text{ with} \end{aligned} \tag{85}$$

$$\begin{aligned} b_1^N &= \int_0^R \big(f_{21}^N + (\gamma-1) f_{22}^N\big) f_{22}^{N*} r^2 dr \quad \text{and} \\ b_2^N &= \int_0^R \gamma \big(f_{21}^N f_{22}^N + (f_{22}^N)^2\big)\big(\beta_c f_{22}^{N*} - f_{21}^{N*}\big)\, r^2 dr. \end{aligned} \tag{86}$$

For the exact b_1, b_2 $f_2^N, f_2^{N^*}$ are replaced by f_2, f_2^*. In [25] we have indicated, how the theory of invariants can be used to decompose the series expansion at each order i in z into a finite set of m_l^i equi-variant homogeneous polynomials $Z_{lm}^{ij}(z)$ of degree i and which are independent over $\mathbf{R}$ $(1 \le j \le m_l^i)$ [53,39,40]. This results in a dramatic simplification of the computations.

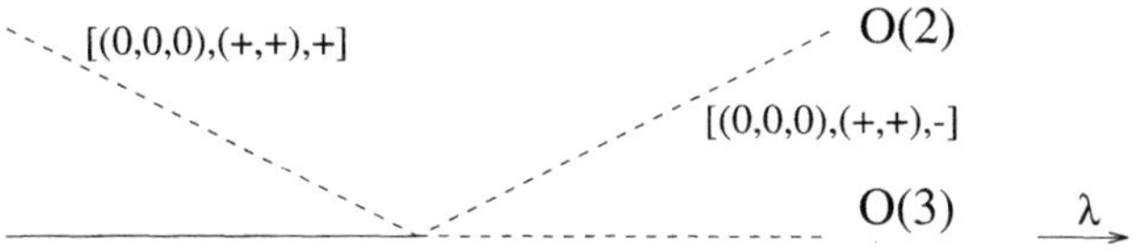

Fig. 3. Bifurcation diagram for generic bifurcations in the $l = 2$ case. The sign of the eigenvalues of the Jacobian along each solution branch is given. Multiple eigenvalues are collected into parentheses.

9.3 Approximation Errors and Solution Branches

Approximation errors in the periodic and homogeneous directions: Since we have assumed an ansatz consisting of spherical harmonics and (77) preserves spherical symmetry, the critical eigenfunction in the angular variables consists of a finite sum of harmonics.

The aliasing error which we only mentioned above may be eliminated by including sufficient collocation points in the angular variables. To second order, with $l = l_0 = 2$ critical, the aliasing error may be removed by requiring, see [25] for the many details,

$$N_{lat} \geq (3 \times 2 + 1)/2 = 7/2 \quad i.e. \quad N_{lat} \geq 4 \text{ and } N_{lon} \geq (3 \times 2 + 1) = 7. \quad (87)$$

As remarked previously, aliasing errors generate symmetry breaking perturbations. As an illustration we consider the case $N_{lat} = 4$ and $N_{lon} = 6$. The equi-variance property requires for the coefficients $b^N_{2m'm''}$ of $b^N_2 z_{m'} z_{m''}$ in (85) that $b_{222} = 0$. Here we find the value $b^N_{222} = -2.2404452355209 \times 10^{-2}$ which implies a symmetry breaking perturbation of the same order as the coefficient b_{02-2}.

Errors in the Chebyshev expansion arise from truncation. Since aliased Fourier transforms in the radial direction $r \geq 0$ do not break symmetry and no symmetry group acts in the radial direction, aliasing errors do not occur. The convergence of the Chebyshev collocation scheme is shown in Table Figure 2

Solution branches With computed coefficients in (85) the bifurcation equations are

$$g_{2,m}(z, \lambda) = 0, \quad m = -2, \ldots, 2 \quad (88)$$

Solution branches of (88) may then be computed by restricting the calculation to fixed point subspaces corresponding to a given isotropy subgroup. In order to obtain all solutions of (88) one first investigates solutions admitting the maximal isotropy subgroup $O(2)$ of $O(3)$. Solutions with the sub-maximal dihedral subgroup $D(2)$ are then investigated. Clearly, the only solution which admits the full symmetry $O(3)$ is the trivial solution $z_0 = 0$. $O(2)$ symmetric solutions can be calculated in the one dimensional fixed point subspace of $O(2)$ yielding a nontrivial solution

$$\lambda = \frac{\sqrt{5}}{7\sqrt{\pi}}\frac{b_2}{b_1}z_0 \ . \tag{89}$$

Furthermore, from [52] we know that there are no nontrivial solutions which are not conjugate to (89), i.e., there are no $D(2)$ symmetric solutions. With the numerical values b_1^N and b_{02-2}^N, computed for $N = 16$ and with $N_{lat} = 4$, $N_{lon} = 8$ quadrature points, we get $b_2/b_1 \approx 0.789 > 0$. The solution together with the eigenvalues of the Jacobian along the solution branch is shown in Figure 3. The branch is always unstable in agreement with the general results for generic bifurcations with even l representations [36]. Clearly, due to the symmetry of the problem there exists a three parameter family of conjugate solutions to (89).

10 On Imperfect Bifurcation Scenarios of co-dim ≤ 3

We discuss the classification of bifurcation scenarios with one distinguished parameter. We are interested in a bifurcation point of the highest local co-dimension, the so-called organizing center. The above Liapunov-Schmidt versions have yielded a bifurcation function g. The well-known classification process via determining equations define the corresponding normal form h^* and its universal unfolding g^* with well understood bifurcation scenarios. For applications we do need these bifurcation scenarios with respect to the original state variables and parameters. This type of *post-processing* is discussed in [55] for co-dim $= -1$ and for the cases of co-dim ≤ 3 in [64,27–29] to determine its normal form, h^*. This problem turned out to be astonishingly difficult. The main idea of our approach is as follows: We approximate a diffeomorphism that links the unfolded normal form, g^*, to that of the real bifurcation problem, g. In the case studies [64] and [27], we had verified that the idea really worked numerically.

Generalized contact equivalence A sequence of computations with different step-sizes allows to monitor the errors between the exact and the discrete problem. For sufficiently good accuracy one will discuss the bifurcation scenario for a discrete operator of the form $F : \mathbf{R}^N \times \mathbf{R}^n \to \mathbf{R}^N$. Let $F = F(u, \beta)$, $\beta = (\lambda, \alpha) \in \mathbf{R}^1 \times \mathbf{R}^k$, $n = 1 + k$. In the bifurcation context, see e.g. [52], u is the state variable, λ and α are the control and unfolding parameters. Let $(u^*, \lambda^*, \alpha^*) \in \mathbf{R}^N \times \mathbf{R}^{1+k}$ be a bifurcation point of F with co-rank $= 1$, i.e.,

$$F(u^*, \lambda^*, \alpha^*) = 0, \qquad \dim \mathcal{N}(F_u(u^*, \lambda^*, \alpha^*)) = 1\,, \quad F_u = \partial F/\partial u\ . \tag{90}$$

The point $(u^*, \lambda^*, \alpha^*)$ plays the role of an organizing center. Then the *Liapunov-Schmidt reduction* yields the bifurcation function

$$g : \mathbf{R}^1 \times \mathbf{R}^{1+k} \to \mathbf{R}^1\,, \quad g = g(x, y)\,, \quad y = (t, z)$$

of F at the point $(u^*, \lambda^*, \alpha^*)$. The solution sets $F(u, \lambda, \alpha) = 0$ and $g(x, t, z) = 0$ are locally one–to–one in neighbourhoods of $(u^*, \lambda^*, \alpha^*)$ and $0 \in \mathbf{R}^{2+k}$. The same isomorphism also links the singular roots of F and g, see e.g.[55].

Let $h : \mathbf{R}^2 \to \mathbf{R}^1$ define the perfect bifurcation scenario of the organizing center and let the unfolding be denoted by the above g such that $h(x,t) \equiv g(x, t, 0)$. Necessarily we do have, for every singularity, $h = h_x = 0$ at the origin. The classification of $h(x,t) = 0$ yields a suitable *normal form* $h^* : \mathbf{R}^2 \to \mathbf{R}^1$ with the universal unfolding g^*. The bifurcation scenario of h^* and g^* is usually well understood, e.g. [52].

Contact equivalence is defined by a smooth $M : \mathbf{R}^2 \to \mathbf{R}^1$ and a local diffeomorphism $\Psi : \mathbf{R}^2 \to \mathbf{R}^2$, $\Psi(x,t) = (\chi(x,t), \tau(t))$, such that

$$\tau, \chi = 0,\ M, \tau_t, \chi_x > 0 \ \text{ at } 0 \in \mathbf{R}^2 \text{ and } h = Mh^* \circ \Psi, \text{ denoted as } h \sim h^*. \quad (91)$$

All normal forms h^* and their *universal unfoldings* g^* for all co-dim ≤ 3 are listed and treated in [64,27–29] Here only consider the

$$\text{pitchfork:}\ \ h^*(x,t) = px^3 + qxt,\ \ g^*(x,t,z) = px^3 + qt^2 + z_1 + z_2x^2, \quad (92)$$

of $\text{co} - \text{dim} = 2$, where p, q are normalized such that $|p| = |q| = 1$. The following lemma extends (91) to the universal unfoldings g and g^*, [63].

Lemma 15. *Let M and Ψ satisfy (91) and let $g^* : \mathbf{R}^2 \times \mathbf{R}^k \to \mathbf{R}^1$ be a universal unfolding of h^*. Then there exist smooth mappings*

$$S : \mathbf{R}^{2+k} \to \mathbf{R}^1,\ \Phi : \mathbf{R}^{2+k} \to \mathbf{R}^{2+k}\,, \Phi(x,t,z) = (X(x,t,z), T(t,z), Z(z))\,,$$

such that $\Phi(\cdot,\cdot,0) = \Psi(\cdot,\cdot)$, $\quad S(\cdot,\cdot,0) = M(\cdot,\cdot)$, hence, (91) implies

$$X = T = 0,\ Z = 0,\ X_x = 1,\ T_t > 0,\ S > 0 \text{ evaluated in the origin,}$$

where the usual $X_x > 0$ has been w.o.l.o.g. normalized to $X_x = 1$, and

$$g = Sg^* \circ \Phi \quad (93)$$

in a neighborhood of $0 \in \mathbf{R}^{2+k}$, with $k = 2$ for the pitchfork.

Hence, a re-transformation of the bifurcation scenarios of the normal form of the universal unfolding g^* to the original situation g requires the determination of Φ, S. This is by far a too ambitious goal. Bifurcation scenarios present only local information. Hence it would be sufficient to determine the Frechet-derivatives, $\Phi'(0)$, of the operator Φ, linked with S, to allow a local transformation of the bifurcation scenario of the normal form universal unfolding to the unfolding of the original situation g. Via the splitting of the full problem into the bifurcating function and the complementary part see (11), this allows to determine the state variable as well.

The existence proof for the Φ and S in (93) and Lemma 15 do not allow to directly compute $\Phi'(0)$. The canonical form of Φ implies the block structure

$$\Phi'(0) = \begin{pmatrix} X_x & X_t & X_z \\ 0 & T_t & T_z \\ 0 & 0 & Z_z \end{pmatrix} \in \mathcal{L}(\mathbf{R}^4, \mathbf{R}^4), \quad Z_z \in \mathcal{L}(\mathbf{R}^2, \mathbf{R}^2); \tag{94}$$

evaluated at the origin. If $\Phi'(0)$ is known, we can solve $g = 0$ with the following linearized procedure: Choose $z \in \mathbf{R}^k$ and $t \in \mathbf{R}^1$; map them as $z \mapsto Z_z z$ and $(t, z) \mapsto (T_t t + T_z z, Z_z z)$ solve the algebraic equation $g^*(X, T_t t + T_z z, Z_z z) = 0$ for X; invert $\Phi'(0)$ and define the action of $(\Phi'(0))^{-1}$ at $(X, T_t t + T_z z, Z_x z)(x, t, z) \in \mathbf{R}^{2+k}$. Then (x, t, z) are first order approximations to the roots of $g = 0$. Similarly, all *singular* solutions of $g = 0$ in a neighborhood of the origin (e.g. limit points) can be approximated by the singular solutions of $g^* = 0$ (e.g. limit points). Implementation results for two cases were studied in [64] and [27].

Finally, the analysis of singular roots of g should be lifted from the small dimensional $\mathbf{R}^4$ to the actual state and parameter space $\mathbf{R}^{N+3}$ where the roots of F live. We achieve this by combining the roots of $g = 0$ and (11)(i), see [29].

Computing of the Frechet derivative $\Phi'(0)$ We introduce the vector of the determining equations for the pitchfork as

$$\mathbf{g} := (g, g_x, g_t, g_{xx})^T : \mathbf{R}^2 \to \mathbf{R}^4, \text{ with sgn } g_{xxx} = p, \text{ sgn } g_{xt} = q, \ p, q \neq 0.$$

evaluated in the origin. Now [52], [55] shows that g is a universal unfolding of h if and only if $\mathbf{g}'(0)$ is regular. Hence if $\mathbf{g}'(0)$, $(\mathbf{g}^*)'(0)$ are regular, we use (93) and the chain rule for differentiation to determine all the terms defining $\mathbf{g}', (\mathbf{g}^*)'$. The evaluation of the $(\mathbf{g}^*)'(0)$ on the basis of the given normal forms for the universal unfolding is trivial, see (92). The evaluation of the derivatives, defining $\mathbf{g}'(0)$ has been extensively studied in the Literature see e.g. [66]. This differentiation yields the equation

$$\mathbf{g}'(0) = \mathbf{A} \cdot (\mathbf{g}^*)'(0)\Phi'(0), \tag{95}$$

where we collect into the matrix $\mathbf{A}$ all the known and unknown higher derivatives of $\mathbf{g}, \mathbf{g}^*$, and S, Φ,respectively, which do not fit to $(\mathbf{g}^*)'(0)$ and $\Phi'(0)$. The $\mathbf{g}'(0), (\mathbf{g}^*)'(0), \Phi'(0)$ determine the size of $\mathbf{A}$.

For the case of the pitchfork we have the matrices $\mathbf{g}'(0)$, $(\mathbf{g}^*)'(0), \Phi'(0) \in \mathbf{R}^{4\times 4}$, see (94), are known, $z \in \mathbf{R}^2$. They have the form:

$$\mathbf{g}'(0) \equiv \begin{pmatrix} g_x & g_t & g_z \\ g_{xx} & g_{xt} & g_{xz} \\ g_{xt} & g_{tt} & g_{tz} \\ g_{xxx} & g_{xxt} & g_{xxz} \end{pmatrix}, \ (\mathbf{g}^*)'(0) = \begin{pmatrix} 0 & 0 & 1 & 0 \\ 0 & 0 & 0 & 1 \\ 0 & 2q & 0 & 0 \\ 6p & 0 & 0 & 0 \end{pmatrix}.$$

So we obtain for the matrix **A**

$$\mathbf{A} = \begin{pmatrix} S & 0 & 0 & 0 \\ S_x & S & 0 & 0 \\ S_t & SX_t & ST_t & 0 \\ S_{xx} & SX_{xx} + 2S_x & 0 & S \end{pmatrix}.$$

In particular we observe that the conditions $S > 0, T_t > 0$ imply the regularity of the matrix **A**.

Since $\mathbf{g}'(0), (\mathbf{g}^*)'(0)$ are known, the system (95) represents $4^2 - 1 = 13$ conditions (note that g_{xt} appear twice in $\mathbf{g}'(0)$) for 12 unknowns in $\Phi'(0)$ (taking into account that $X_x = 1$ has already been fixed) and 5 additional unknowns S, S_x, S_t, S_{xx}, X_{xx} inherited from the chain rule differentiation. Hence, four conditions are missing to determine $\Phi'(0)$ from (95). This seems to indicate that the transformation of the bifurcation scenarios back to the original situation allows a huge freedom. So we are looking for closing conditions as natural conditions added to (95) to make the system uniquely solveable and obtain

Theorem 16. *Let g be contact equivalent to the pitchfork and let the operators S and Φ satisfy the necessary contact equivalence conditions in Lemma 15. Furthermore, let the* **closing conditions** *be satisfied:*

$$T_{tt} = 0\,,\ X_{xt} = X_{xx}X_t \text{ evaluated at } 0. \tag{96}$$

Then there exist unique operators S and Φ satisfying these conditions. To determine S, Φ we need $\mathbf{g}'(0) \in \mathcal{L}(\mathbf{R}^{4\times 4})$, g_{xxxx}, g_{xxxxx}, g_{xtt}, g_{xxxt}.

11 A C^{++} Implementation for Steady-States in PDEs

There are several well established programs on the market, e.g., AUTO [43], CONTENT [68], BIFPACK [89]. In contrast to these programs the goal of our C^{++}–program is the following: for a rather general class of operators and discretizations it should provide possibilities for determining turning points, bifurcation points, Hopf points and all the bifurcating solution branches and their characteristics in these points. As we have seen above already we can restrict the discussion to a discrete operator of the form $F : \mathbf{R}^n \times \mathbf{R} \to \mathbf{R}^n$, n large. We want to solve $F(u, \lambda) = 0$ for $\lambda \in [\lambda_{min}, \lambda_{max}]$. Path-following techniques, as described e.g. in [30] with pseudo-arc length continuation based on Krylov-type methods are employed to solve the linear systems $Ax = r$ and determine the eigenvalues of A. This allows to evaluate the corresponding test functions and to get the interesting information. In this context the matrix A is either a derivative or a bordered matrix

$$A = D_uF \text{ or } A = \begin{pmatrix} D_uF & b \\ \tilde{b} & c \end{pmatrix}. \tag{97}$$

In many cases this D_uF is a sparse matrix defined via finite difference, finite element schemes or a not too large but dense matrix, as defined in spectral methods. Some efficient Krylov-type methods with successful preconditioning have been developed in [87,86] and are further developed.

Implementation. We have to solve linear systems and have to determine eigenvalues and eigenvectors. So the user has to choose the appropriate solvers and matrix-storage schemes.

We provide interfaces to the standard software packages, therefore the user can choose between direct (Lapack) and iterative solvers. To determine eigenvalues and sometimes eigenvectors of the corresponding matrices the user can choose between QR-like methods for smaller problems, ARPACK or Methods from [87,86] for larger problems.

Implemented matrix storage classes are general (full storage), sparse (row/ column/ index compressed storage), banded (Lapack-style) and algorithmically defined matrices (for iterative solvers). Algorithmically defined means that the matrix is not really stored, only the product Ax is implemented.

This C^{++}–program is intended to allow a relatively straight forward implementation of new problems and new discretization methods.

The goal of this program certainly is not to make available an end-user software, since many of these programs are available already. However to provide the chance to use for new classes of operators and discretizations a tool which provides the interesting bifurcation information on a good level.

References

1. R.H. Abraham and J.E. Marsden. *Foundation of Mechanics.* Reading, 1978.
2. E. Allgower and K. Böhmer. Resolving singular nonlinear equations. *Rocky Mountain Journal of Mathematics*, 18:225–268, 1988.
3. E.L. Allgower, K. Böhmer, K. Georg, and R. Miranda:. Exploiting symmetry in boundary element methods. *SIAM J. Numer. Anal.*, 29:534–552, 1992.
4. E.L. Allgower and K. Böhmer and M. Golubitsky. Bifurcations and Symmetry: Cross Influences between Mathematics and Applications. Volume 104 of *ISNM*, Basel, 1992. Birkhäuser Verlag.
5. E.L. Allgower, K. Böhmer, A. Hoy, and V. Janovsky. On a class of direct methods for solving nonlinear equations. *ZAMM*, 79:219–231, 1999.
6. E. L. Allgower, K. Böhmer, and Z. Mei. On a problem decomposition for semilinear nearly symmetric elliptic problems. In W. Hackbusch, editor, *Parallel Algorithms for PDE's*, volume 31 of *Notes on Numerical Fluid Mechanics*, pages 1–17, Braunschweig, Fed. Rep. Germany, 1991. Vieweg Verlag.
7. E.L. Allgower, K. Böhmer, and Z. Mei. A complete bifurcation scenario for the 2d-nonlinear Laplacian with Neumann boundary conditions on the unit square. In R. Seydel, F.W. Schneider, T. Küpper, and H. Troger, editors, *Bifurcation and Chaos: Analysis, Algorithms, Applications*, volume 97 of *ISNM*, pages 1–18, Basel, 1991. Birkhäuser Verlag.
8. E.L. Allgower, K. Böhmer, and Z. Mei. Branch switching at a corank-4 bifurcation point of semi-linear elliptic problems with symmetry. *IMA J. Numer. Anal.*, 14:161–183, 1993.

9. E. L. Allgower and K. Georg:. Nonlinear Systems of Equations. *Lectures in Applied Mathematics*, American Mathematical Society, 1990, volume 26.
10. E. L. Allgower, K. Georg and R. Miranda:. Exploiting Symmetry in Applied and Numerical Analysis. *Lectures in Applied Mathematics*, American Mathematical Society, 1993, volume 29.
11. E. L. Allgower, K. Georg, R. Miranda, and J. Tausch. Numerical exploitation of equivariance. *ZAMM*, to appear, 1998.
12. P. Ashwin, K. Böhmer, and Z. Mei. A numerical Liapunov-Schmidt method with applications to Hopf bifurcation on a square. *Math. Comp.*, 64:649–670, 1995.
13. P. Ashwin, K. Böhmer, and Z. Mei. Forced symmetry breaking of homoclinic cycles in a PDE with O(2) symmetry. *Journal of Computational and Applied Mathematics*, 70:297–310, 1996.
14. P. Ashwin and Z. Mei. A Hopf bifurcation with Robin boundary conditions. *J. Dyn. & Diff. Eq.*, 6:487–503, 1994.
15. P. Ashwin and Z. Mei. Normal form for Hopf bifurcation of partial differential equations on the square. *Nonlinearity*, 8:715–734, 1995.
16. Peter Ashwin and Z. Mei. A numerical bifurcation function for homoclinic orbits. *SIAM J. Numer. Anal.*, 35:2055-2069 1998.
17. C. Bernadi and Y. Maday. *Spectral Methods.* In P.G. Ciarlet and J.L. Lions, editors, *Handbook of Numerical Analysis*, volume V of *Techniques of Scientific Computing (Part 2)*. Elsevier Science B.V., 1998.
18. W.J. Beyn. Defining equations for singular solutions and numerical applications. In T. Küpper, H. D. Mittelmann, and H. Weber, editors, *Numerical Methods for Bifurcation Problems*, volume 70 of *ISNM*, pages 42–56, Basel, 1984. Birkhäuser Verlag.
19. W.J. Beyn, W. Kleß and V. Thümmler. Continuation of low-dimensional invariant subspaces in dynamical systems of large dimension *DFG-Schwerpunktprogramm Danse*, Preprint, University of Bielefeld, Germany 2000, this volume.
20. K. Böhmer. Über lineare Differentialgleichungen mit Lösungen von endlicher Wachstumsordnung. *Archiv der Mathematik 22,* pages394–400, 1971.
21. K. Böhmer. On a numerical Liapunov-Schmidt method for operator equations. Report University of Marburg, 1989, *Computing 51*, pages 237–269, 1993.
22. K. Böhmer. On numerical bifurcation studies for general operator equations. Proceedings Equadiff: Invited lecture, 1999.
23. K. Böhmer. Numerical center manifolds. Report, in preparation, 2000.
24. K. Böhmer. Variational crimes in Petrov-Galerkin methods for bifurcation equations. Technical Report, in preparation
25. K. Böhmer, C. Geiger, and J. Rodriguez. On a numerical Liapunov-Schmidt spectral method and applications to biological pattern formation. To appear in SIAM J. Num. Anal.
26. K. Böhmer, W. Govaerts, and V. Janovsky. Numerical detection of symmetry breaking bifurcation points with nonlinear degeneracies. *Math. Comp. (1999)*, pages 1097–1108, 1996.
27. K. Böhmer, D. Janovská, and V. Janovsky. Computer aided analysis of imperfect bifurcation diagrams, ii pitchfork bifurcation. *East-West Journal Numer. Math.*, 6(3):207–222, 1998.
28. K. Böhmer, D. Janovská, and V. Janovsky. Numerical analysis of the imperfect bifurcation diagrams. *ZAMM*, pages 445–448, 1997.

29. K. Böhmer, D. Janovská, and V. Janovsky. On the numerical analysis of the imperfect bifurcation. *submitted to SIAM*, 1999.
30. K. Böhmer, Z. Mei, A. Schwarzer, and R. Sebastian. Path-following of large bifurcation problems with iterative methods. *Numerical Methods for Bifurcation Problems and Large-Scale Dynamical Systems*, Editors E. Doedel and L.S. Tuckermann, IMA 1999, Volume 119, pages 37-67.
31. K. Böhmer and N. Sassmannshausen. Numerical Liapunov-Schmidt spectral method for k-determined problems. In *"Computational Methods and Bifurcation Theory with Applications"*, volume 170 of *Computer methods in applied mechanics and engineering*, pages 277–312, 1999.
32. S.C. Brenner and L.R. Scott. *The Mathematical Theory of Finite Element Methods*. Springer Verlag, New York, 1996.
33. F. Brezzi, J. Rappaz, and P.A. Raviart. Finite-dimensional approximation of nonlinear problems, II. Limit points. *Numer. Math.*, 37:1–28, 1981.
34. F. Brezzi, J. Rappaz, and P.A. Raviart. Finite-dimensional approximation of nonlinear problems, III. Simple bifurcation points. *Numer. Math.*, 38:1–30, 1981.
35. G. Caloz and J. Rappaz. Numerical analysis for nonlinear and bifurcation problems. In P.G. Ciarlet and J.L. Lions, editors, *Handbook of Numerical Analysis*, volume V of *Techniques of Scientific Computing (Part 2)*. Elsevier Science B.V., 1998.
36. C. Canuto, M. Y. Hussaini, A. Quarteroni, and T.A. Zang. *Spectral methods in fluid dynamics.* 3. Auflage. Springer Verlag, Berlin, 1997.
37. J. Carr. *Applications of centre manifold theory*, volume 35 of *App. Math. Sci.* Springer, New York, 1981.
38. S. Chandrasekhar. *Hygrodynamic and Hygromagnetic Stability.* Oxford University Press, Oxford, 1961.
39. P. Chossat. Branching of rotating waves in a one-parameter problem of steady-state bifurcation with spherical symmetry. *Nonlinearity*, 4:1123–1129, 1991.
40. P. Chossat. Forced symmetry breaking of an O(2)-symmetric homoclinic cycle. *Nonlinearity*, 6:723–731, 1993.
41. M. Crouzeix and J. Rappaz. *On numerical approximation in bifurcation theory.* Masson, Paris and Springer, Berlin, 1990.
42. G. Dangelmayr and L. Kramer. Mathematical tools for pattern formation. In *Evolution of Spontaneous Structures in Dissipative Continuous Systems*, volume 1, 1998.
43. E.J. Doedel, X.J. Wang, and T.F. Fairgrieve. AUTO94: Software for continuation and bifurcation in ordinary differential equations, 1994.
44. U. Garbotz. Ein C++ -Programm zum Einsatz für lokale Dynamikuntersuchungen, in Vorbeitung. Technical report, Universität Marburg, Marburg, 1999.
45. U. Garbotz. Verbindende Orbits bei partiellen Differetialgleichungen, Dissertation (laufend), Universität Marburg.
46. C. Geiger. *Strukturbildung in nichtlinearen dissipativen Systemen mit sphärischer Symmetrie.* PhD thesis, Universität Tübingen, 1994.
47. Ch. Geiger, G. Dangelmayr, D. Rodriguez, and W. Güttinger. Symmetry breaking bifurcations in the spherical Bénard convection. I: Results from singularity theory. *Fields Institute Communications*, 5:225–237, 1996.

48. C. Geiger, G. Dangelmayr, J. D. Rodriguez, and W. Güttinger. Symmetry breaking bifurcations in the spherical Bénard convection. II: Numerical results. *Fields Institute Communications*, 5:239–253, 1996.
49. K. Georg and J. Tausch. Some error estimates for the numerical approximation of surface integrals. *Math. Comp.*, 62:755–763, 1993.
50. K. Georg and J. Tausch. *User's Guide for a Package to Solve Equivariant Linear Systems.* Colorado State University, 1995. Available via anonymous ftp at `ftp.math.colostate.edu pub/georg`.
51. V. Girault and P. A. Raviart. *Finite Element Approximation of the Navier-Stokes Equations.* Lecture Notes in Math., 749. Springer-Verlag, Berlin New York, 1979.
52. M. Golubitsky and D. Schaeffer. *Singularities and Groups in Bifurcation Theory Volume 1*, volume 51 of *App. Math. Sci.* Springer, New York, 1985.
53. M. Golubitsky, I.N. Stewart, and D. Schaeffer. *Singularities and Groups in Bifurcation Theory Volume 2*, volume 69 of *App. Math. Sci.* Springer, New York, 1988.
54. W. Govaerts. Stable solvers and block elimination for bordered systems. *SIAM J. Matrix. Anal. Appl.*, 12:469–483, 1991.
55. W. Govaerts. Numerical methods for bifurcations of dynamical equlibria. In *Regional Conferences*, Philadelphia, PA, 2000. SIAM.
56. A. Griewank and G.W. Reddien. Computation of cusp singularities for operator equations and their discretization. *Journal of Computational and Applied Mathematics, North Holland*, 26:133–153, 1989.
57. John Guckenheimer and Philip Holmes. *Nonlinear Oscillations, Dynamical Systems and Bifurcations of Vector Fields.* Applied Mathematical Sciences 42. Springer Verlag, 1990. 3. Printing.
58. W. Hackbusch. *Theorie und Numerik elliptischer Differentialgleichungen.* Teubner Verlag, Stuttgart, 1986.
59. M. Haragus-Courcelle and K. Kirchgässner. *Three-dimensional steady capillary-gravity waves.* Danse report, University of Stuttgart, 1999, This volume.
60. D. Henry. *Geometric Theory of Semilinear Parabolic Equations*, volume 840 of *Lecture Notes in Mathematics.* Springer, Berlin, 1981.
61. A. Hunding. Cell to cell signaling: From experiments to theoretical models. *A. Goldbeter, Academic Press, London*, 1989.
62. G. Iooss and A. Mielke. Hydrodynamical stability problems in infinitely long cylindrical domains: existence of defects in travelling waves. In S. Rionero, editor, *Waves and Stability in Continuous Media*, Ser. Adv. Math. Appl. Sci. 4, pages 243–257. World Sci. Publishing, River Edge, NJ, 1991.
63. V. Janovský and P. Plecháč. Numerical applications of equivariant reduction techniques. In E. L. Allgower, K. Böhmer, and M. Golubitsky, editors, *Bifurcation and Symmetry*, volume 104 of *ISNM*, pages 203–214, Basel, Switzerland, 1992. Birkhäuser Verlag.
64. V. Janovský and P. Plechac. Computer aided analysis of imperfect bifurcation diagrams i. simple bifurcation point anf isola formation centre. *SIAM J. Numer. Anal.*, 21:498–512, 1992.
65. A.D. Jepson and A. Spence. On a reduction process for nonlinear equations. *SIAM J. Math. Anal.*, 22:39–56, 1989.

66. A.D. Jepson and A. Spence. The numerical solution of nonlinear equations having several parameters i: scalar equations. *SIAM J. Math. Anal.*, 20:736–759, 1989.
67. H.B. Keller and W. F. Langford. Iterations, pertubations and multiplicities for nonlinear bifurcation problems. *Arch. Rational Mech. Anal.*, 48:83–108, 1972.
68. Y.A. Kuznetsov, V.V. Levitin, and A.R. Skovoroda. Continuation of stationary solutions to evolution problems in CONTENT. Technical Report AM-R9611, Centrum voor Wiskunde en Informatica, Amsterdam, 1996.
69. HG. Kwatny and BC. Chang. Constructing linear families from parameter-dependent nonlinear dynamics. *IEEE TRANSACTIONS ON AUTOMATIC CONTROL*, 43 (8):1143–1147, 1998.
70. C. Lubich. On dynamics and bifurcations of nonlinear evolution equations under numerical discretization. *Danse preprint*, University of Tübigen, 1999, this volume.
71. C. Lubich and A. Ostermann. Hopf-bifurcation of reaction-diffusion and Navier-Stokes equations under discretization. *Numerische Mathematik*, Volume 81, Pages 53–84, 1998.
72. R. M. Mantel and B. Fiedler. Crossover collisions of scroll wave filaments. preprint, 1999.
73. Z. Mei. Exploiting symmetries in numerical analysis of bifurcations of nonlinear elliptic pdes. In *A Collection of Papers of Conference on Scientific and Engineering Computing for Young Chinese Scientists, Beijing, China 1993*, pages 312–319, Computer Center, Academia Sinica, 1993.
74. Z. Mei. Bifurcation problems in reaction diffusion equations. Philipps-Universität Marburg, 1997. Habilitationsschrift.
75. Z. Mei and A. Schwarzer. Scaling solution branches of bifurcation problems. *Journal of Mathematical Analysis and Applications*, 204:404–433, 1996.
76. A. Mielke. Saint Venant's problem and semi inverse solutions in nonlinear elasticity. *Arch. Rational Mech. Anal.*, 102:205–229, 1988.
77. A. Mielke. *Hamiltonian and Lagrangian Flows on Center Manifolds.* Lecture Notes in Mathematics, 1489. Springer-Verlag, Berlin, 1991.
78. A. Mielke. H Schneider and H. Uecker. Stability and Diffusive Dynamics on Extended Domains. *DFG-Schwerpunktprogramm Danse*, Preprint, University of Stuttgart, 1999, this volume.
79. J.T. Oden and J.N. Reddy. *An introduction to the mathematical theory of finite elements.* Wiley, New York, 1976.
80. M. Plath. C++ programm for general operators and discretizations. *Report in preparation*, University of Marburg.
81. J. Pousin and J. Rappaz. Consistency, stability, a priori and a posteriori errors for petrov-galerkin methods applied to nonlinear problems. *Numerische Mathematik*, pages 213–231, 1994.
82. G. Raugel. Approximation numérique de problemes nonlinéaires, Thèse, Université de Rennes, 1985.
83. H.J. Reinhardt. *Analysis of approximation methods for differential and integral equations.* Springer Verlag, Berlin, Heidelberg, New York, 1985.
84. A. J. Roberts Z. Mei. Equations for turbulent flood waves. In K. Kirchgässner A. Mielke, editor, *Proc. Of IUTAM/ISIMM Symposium on Structure and Dynamics of Nonlinear Waves*, World Scientific, pages 242–352, 1994.

85. N. Sassmannshausen and K. Böhmer. Petrov-Galerkin methods for projected linear operator equation, stability and convergence. *submitted to ZAMM*, 1999.
86. Andreas Schwarzer. *Skalierungstechniken für k-bestimmte Verzweigungsprobleme und Iterationsmethoden f"ur gro"se, d"unn besetzte Eigenwertprobleme.* PhD thesis, Philipps-Universität Marburg, 1997. Dissertation.
87. R. Sebastian. Anwendung von Krylov-Verfahren auf Verzweigungs- und Fortsetzungsprobleme. Dissertation am Fachbereich Mathematik, Universität Marburg, 1995.
88. E.E. Selkov. 4:79–86, 1968.
89. R. Seydel. Bifpack: A program package for calculating bifurcations. Technical report, State University of New York at Buffalo, 1983.
90. H. Stetter. *Analysis of discretization methods for ordinary differential equations.* Springer Verlag, Berlin-Heidelberg-New York, 1973.
91. F. Stummel. Diskrete Konvergenz linearer Operatoren I. *Math. Ann.*, 190:45–92, 1970.
92. F. Stummel. Diskrete Konvergenz linearer Operatoren II. *Math. Z.*, 120:231–264, 1971.
93. F. Stummel. Diskrete Konvergenz linearer Operatoren III. *Math. Z.*, 190:196–216, 1972.
94. R. Temam. Navier-stokes equations and nonlinear functional analysis. CBMS-NSF-Regional Conference Series in Applied Mathematics. SIAM, Philadelphia, 1983.
95. R. Temam. *Navier-Stokes equations, theory and numerical analysis*, volume 2 of *Studies in mathematics and its applications.* North-Holland, Amsterdam, 1984.
96. G. Vainikko. *Funktionsanalysis der Diskretisierungsmethoden.* Teubner Texte zur Mathematik. Teubner, Leipzig, 1976.
97. A. Vanderbauwhede. Center manifolds and their basic properties. an introduction. *Delft Progr. Rep.*, 12:57–78, 1988.
98. A. Vanderbauwhede and G. Iooss. Center manifolds in infinite dimensions. Technical report, University of Nice, 1989.
99. A. Vanderbauwhede and G. Iooss. Center manifold theory in infinite dimensions. In C. K. R. T. Johnes, U. Kirchgraber, and H. O. Walther, editors, *Dynamics Reported: Expositions in Dynamical Systems*, Dynam. Report. (N.S.), 1, pages 125–163. Springer-Verlag, Berlin, 1992.
100. E. Zeidler. *Nonlinear functional analysis and its applications I, fixed-point theorems.* Springer Verlag, New York, Berlin, Heidelberg, London, Paris, Tokyo, 1986.
101. E. Zeidler. *Nonlinear functional analysis and its applications II, monotone operators.* Springer Verlag, New York, Berlin, Heidelberg, London, Paris, Tokyo, 1990.

Dimension Theory of Smooth Dynamical Systems

Jörg Schmeling*

Freie Universität Berlin, Fachbereich Mathematik und Informatik, Arnimallee 2-6, 14195 Berlin, Germany

1 Introduction

One of the basic properties of dynamical systems is that local instability of trajectories gives rise to a global "chaotic" behavior. This local instability can be described as some kind of hyperbolicity. Smooth Ergodic Theory investigates the metric and stochastic properties of measures invariant under differentiable mappings or flows on manifolds. The consideration of invariant measures allows to "tame" the "chaotic" behavior from a probabilistic point of view. This transition from differentiable structures to measurable structures and vice versa makes this field fascinating and paves the way to applications far beyond this field. Due to its generality the methods and results of Smooth Ergodic Theory entered areas as Riemannian Geometry, Number Theory, Statistical Physics, Partial Differential Equations or Numerical Simulations.

In order to study the properties of an invariant measure one considers different kind of characteristics of it. This characteristical quantities entered from both the theoretical and experimental investigations. Among them the most important ones are *Lyapunov exponents* – a metric quantity measuring the local instability, *entropy* – a stochastic quantity measuring the complexity of the system, and fractal dimesions – quantities having both metric and stochastic aspects. While the first two characteristics have a commonly agreed definition the fractal dimensions collect a variety of similar but different notions according to the usefulness in applications.

In the late 70's – beginning 80's the attention of many physicists and applied mathematicians has turned to the study of dimension of strange attractors. The dimension was used to characterize the (finite) number of independent modes needed to describe the infinite-dimensional system. Several conjectures were put forward in attempts to tie together the dimension of an attractor and its other characteristics. This study has become an important breakthrough in understanding the structure of systems of evolution type.

The seminal paper [15] by Eckmann and Ruelle was one of the most serious attempts to comprehend this activity and outline a rigorous mathematical foundation for it. They discussed various concepts of dimension and

---

* Project: Ergodic and Topological Properties of Hyperbolic Dynamical Systems and Symbolic Dynamics (Hans-Günther Bothe)

ruled out the importance of the so-called pointwise dimension. The discussion in [15] can be summarized in the following conjecture (*Eckmann–Ruelle–Conjecture*):

Let M be a compact smooth Riemannian manifold without boundary, $f\colon M \to M$ a C^2–diffeomorphism, and μ an f–invariant Borel probability measure on M. Then if μ is hyperbolic, *i.e. all the Lyapunov exponents of μ are non–zero, the following limit exists*

$$d(x) := \lim_{r\to} \frac{\log \mu(B(x,r))}{\log r}$$

for a.e. $x \in M$.

The existence of this limit implies that virtually **all** the characteristics of dimension type of the measure coincide and it makes sense to talk about the fractal dimension of the measure. Moreover, the interplay of the metric and stochastic properties of the measure becomes a more clear and rigid structure.

In [5] Barreira, Pesin and Schmeling could proof the above conjecture in full generality, even the assumption that f has to be C^2 could be weakened to f belonging to the class $C^{1+\alpha}$, $\alpha > 0$. The proof of this conjecture consists of two major parts. First the work of Ledrappier and Young [25] was generalized to the class $C^{1+\alpha}$ in the case of hyperbolic measures and second it was established that arbitrary hyperbolic measures admit a very special local property: they exhibit "almost" the so-called local product structure.

Relying on this result Schmeling and Troubetzkoy [49] established a more general scaling law of hyperbolic measures for 3–dimensional systems. Instead of the ball $B(x,r)$ one can consider rectangles aligned along the invariant directions with exponentially increasing distortion. The existence of a scaling along these rectangles recovers the existence of the pointwise dimension as well as the Brin–Katok formula for the local entropies and gives a new insight in the connection in between entropy and dimension. F. Ledrappier found a similar scaling law for Takens balls (i.e. rectangles with special relations between their side–lengths) in any dimension. These general scaling laws allowed to proof the existence of the pointwise dimension for non–invertible systems with singularities.

2 Smooth Dynamical Systems

In order to develop a general theory we try to keep the assumptions on the systems considered as weak as possible. We are interested in the connection of topological (geometric), measure theoretic and other properties of these systems. In particular we want to describe the fascinating interaction of determinism and randomness. Let M be a smooth compact connected Riemannian manifold with distance d and $f\colon M \to M$ a $C^{1+\alpha}$ diffeomorphism.

A point $x \in M$ is called *Lyapunov regular* if there exists numbers

$$\lambda_1(x) < \cdots < \lambda_{k(x)}(x)$$

called *Lyapunov exponents* and a decomposition $T_xM = \bigoplus_{i=1}^{k(x)} E^i(x)$ of the tangent space, where $E^i(x)$ is the vector space

$$E^i(x) = \left\{ v \in T_xM \setminus \{0\} : \lim_{n \to \pm\infty} \frac{1}{n} \log \|d_x f^n v\| = \lambda_i(x) \right\} \cup \{0\}$$

such that

$$\lim_{n \to \pm\infty} \frac{1}{n} \log |\det d_x f^n| = \sum_{i=1}^{k(x)} \lambda_i(x) \dim E^i(x).$$

Note that $\lambda_i(f(x)) = \lambda_i(x)$ and $d_x f E^i(x) = E^i(f(x))$ for each i.

We denote by $\Lambda \subset M$ the set of all Lyapunov regular points. Below we describe some notions and results on non-uniformly hyperbolic dynamical systems. For references see [27,28,25,21].

Set $s(x) = \max\{i : \lambda_i(x) < 0\}$ and $u(x) = \min\{i : \lambda_i(x) > 0\}$. Fix $x \in \Lambda$ and $r(x) > 0$. For every $i = 1, \ldots, s(x)$ one defines the *i-th stable leaf* at x by

$$W^i(x) = \left\{ y \in B(x, r(x)) : \limsup_{n \to +\infty} \frac{1}{n} \log d(f^n(x), f^n(y)) < \lambda_i(x) \right\},$$

and for every $i = u(x), \ldots, k(x)$ the *i-th unstable leaf* at x by

$$W^i(x) = \left\{ y \in B(x, r(x)) : \limsup_{n \to -\infty} \frac{1}{n} \log d(f^n(x), f^n(y)) > \lambda_i(x) \right\}.$$

Clearly, $W^i(x) \subset W^{i+1}(x)$ if $i < s(x)$ and $W^i(x) \supset W^{i+1}(x)$ if $i \geq u(x)$. We write $W^s(x) = W^{s(x)}(x)$ and $W^u(x) = W^{u(x)}(x)$ and call them, respectively, the *stable* and *unstable manifolds* at x. One can prove that if $r(x)$ is sufficiently small than $W^i(x)$ is a $C^{1+\alpha}$ immersed submanifold.

One of the major questions in the theory is the degree of regularity of these foliations. The importance of the regularity started with Hopf's method of proving the ergodicity of the geodesic flow on a surface of negative curvature. He used the foliations to define new coordinates. The smoothness of the coordinates obviously depends on the regularity of the foliations. Another important application found its way in a series of rigidity results. As one example Hurder and Katok (see [19]) proved that an Anosov diffeomorphism of the 2–dimensional torus is C^∞– conjugated to a linear one under quite mild regularity assumption on the invariant foliations. By a number of authors (see f.i. [46,18]) it was realized that in general the stable or unstable foliations are only Hölder continuous. Optimal Hölder exponents were found and it was proved that these exponents are optimal on a set of second category in the space of all diffeomorphisms of a manifold ([40]). Hasselblatt and Wilkinson even found an open set of symplectic Anosov diffeomorphisms were the least regularity is obtained on a set of full measure on the torus. Nevertheless the situation is different if we consider strong stable foliations within a weak stable leaf.

Let us fix $c > 0$, $x \in \Lambda_c$, where Λ_c is a subset of Λ consisting of points having the same collection of Lyapunov exponents and all i–th stable leaves of inner diameter at least c, and $y' \in \Lambda_c \cap B^{i+1}(x, c)$. For each $i < s$, consider two local smooth manifolds Σ_x and $\Sigma_{y'}$ in $W^{i+1}(x)$, containing x and y', respectively and transversal to $W^i(z)$ for all $z \in \Lambda_c \cap B^{i+1}(x, c)$. The holonomy map

$$\Pi_i = \Pi_i(\Sigma_x, \Sigma_{y'})\colon\ \Sigma_x \cap \Lambda_c \cap B^{i+1}(x, c) \to \Sigma_{y'}$$

is defined by

$$\Pi_i(x') = W^i(x') \cap \Sigma_{y'}$$

with $x' \in \Sigma_x$. This map is well-defined if c is sufficiently small.

In [5] we proved the following theorem.

Theorem 1. *Let f be a $C^{1+\alpha}$ diffeomorphism. For $x \in \Lambda_c$, and $y' \in \Lambda_c \cap B^i(x, c)$ the holonomy map $\Pi_i(\Sigma_x, \Sigma_{y'})$ is Lipschitz continuous.*

This theorem was obtained by a different method by Ledrappier and Young [25] for the case of a C^2–diffeomorphism.

The natural question that arises is how large is the set we were talking about – i.e. how many regular points exist? Here the connection of geometry and measure theory comes into play. Namely, the famous theorem of Oseledec states that any invariant measure is concentrated on the set of regular points. Moreover, if we assume that the measure is ergodic it has to be concentrated on a set where the Lyapunov exponents are constant. We want to emphazise that Oseledec theorem is no longer true if we allow the system to have singularities or to be non–invertible.

For the following theory we need an additional assumption. We call a point $x \in M$ *non-uniformly hyperbolic* iff $E^0 = \{0\}$. An ergodic measure is called hyperbolic iff it is concentrated on the non–uniformly hyperbolic points.

Now we can ask what are the **interesting characteristics** of a smooth dynamical system? We encountered already the Lyapunov exponents which describe the system from the point of view of the underlying Riemannian metric. Another classical characteristic is the *entropy* which measures the inner complexity of the system. In the recent past another characteristic came into the center of interest. It is the *Hausdorff dimension* which was found to connect the previous quantities and to contain the combined information. The main role was played by the local version of the dimension, the pointwise dimension of a measure:

$$d_\mu(x) = \lim_{r \to 0} \frac{\log \mu B(x, r)}{\log r}.$$

One of the main problems is the existence of the above limit. As in the theory of entropies (the theorem of Shannon–McMillan–Breiman) the existence ensures a coincidence of different concepts of the dimension of a measure and also the connection to the other characteristics.

In an extremely important and also extremely difficult paper ([25]) Ledrappier and Young defined entropies and pointwise dimensions $d^i_\mu(x)$ w.r.t. conditional measures sitting on the i–th (un–)stable leaves. They were able to prove the a.e. existence of these quantities and stated the following formula

$$h_\mu = -\sum_{j\leq s} \chi_j (d^j_\mu - d^{j-1}_\mu)$$

This put us in the situation where we know that Lyapunov exponents, entropies and stable and unstable dimensions exist a.e. and that there is a connection between the conditional quantities. There is a connection in between the unconditioned quantities if the pointwise dimension exist. So the question is when does the pointwise dimension of a measure exist? For an hyperbolic measure (i.e. no indifferent directions) a natural guess is that the pointwise dimension exists a.e. and equals the sum of the stable and unstable dimension. Unfortunately, a simple Fubini type argument does not work since the foliations are not regular enough in general.

In [5] we prove that every hyperbolic measure invariant under a $C^{1+\alpha}$ diffeomorphism of a smooth Riemannian manifold possesses asymptotically "almost" local product structure, i.e., its density can be approximated by the product of the densities on stable and unstable manifolds up to small exponentials. This has not been known even for measures supported on locally maximal hyperbolic sets.

Using this property of hyperbolic measures we were able to prove the long-standing Eckmann–Ruelle conjecture in dimension theory of smooth dynamical systems: the pointwise dimension of every hyperbolic measure invariant under a $C^{1+\alpha}$ diffeomorphism exists almost everywhere. This implies the crucial fact that virtually all the characteristics of dimension type of the measure (including the Hausdorff dimension, box dimension, and information dimension) coincide. This provides the rigorous mathematical justification of the concept of fractal dimension for hyperbolic measures.

The problem of the existence of the pointwise dimension has a long history. In [55], Young obtained a positive answer for a hyperbolic measure μ invariant under a $C^{1+\alpha}$ surface diffeomorphism f. Moreover, she showed that in this case for almost every point x

$$\underline{d}_\mu(x) = \overline{d}_\mu(x) = h_\mu(f)\left(\frac{1}{\lambda_1} - \frac{1}{\lambda_2}\right),$$

where $h_\mu(f)$ is the metric entropy of f and $\lambda_1 > 0 > \lambda_2$ are the Lyapunov exponents of μ.

In [23], Ledrappier established the existence of the pointwise dimension for arbitrary *SRB-measures* (called so after Sinai, Ruelle, and Bowen). In [35], Pesin and Yue extended his approach and proved the existence for hyperbolic measures satisfying the so-called *semi-local product structure* (this class includes, for example, *Gibbs measures* on locally maximal hyperbolic sets).

Our main result in [5] is the following theorem.

Theorem 2. *Let f be a $C^{1+\alpha}$ diffeomorphism on a smooth Riemannian manifold M without boundary, and μ an f-invariant compactly supported ergodic Borel probability measure. If μ is hyperbolic then the following properties hold:*

- *for every $\delta > 0$ there exist a set $\Lambda \subset M$ with $\mu(\Lambda) > 1 - \delta$ and a constant $\kappa \geq 1$ such that for every $x \in \Lambda$ and every sufficiently small r (depending on x), we have*

$$r^{\delta}\mu_x^s(B^s(x, \frac{r}{\kappa}))\mu_x^u(B^u(x, \frac{r}{\kappa})) \leq \mu(B(x,r)) \leq r^{-\delta}\mu_x^s(B^s(x, \kappa r))\mu_x^u(B^u(x, \kappa r));$$

- *μ is exact dimensional and its pointwise dimension is equal to the sum of the stable and unstable pointwise dimensions, i.e.,*

$$\underline{d} = \overline{d} = d^s + d^u.$$

Let us also point out that neither of the assumptions of the Main Theorem can be omitted. Ledrappier and Misiurewicz [24] constructed an example of a smooth map of a circle preserving an ergodic measure with zero Lyapunov exponent which is not exact dimensional. In [33], Pesin and Weiss presented an example of a Hölder homeomorphism with Hölder constant arbitrarily close to 1 whose measure of maximal entropy is not exact dimensional.

Statement 1 of the Main Theorem establishes a new and non-trivial property of an arbitrary hyperbolic measure. Loosely speaking it means that every hyperbolic invariant measure possesses asymptotically "almost" local product structure. This statement has not been known even for measures supported on (uniformly) hyperbolic locally maximal invariant sets.

3 Endomorphisms with Singularities

The previous sections indicated that the dimension theory of diffeomorphisms exhibiting an hyperbolic measure is well understood. But how realistic are these systems? Often PDE's generate only semi–flows – i.e. the system is a priori non–invertible. Moreover, after reducing the system to a system of ODE's the investigation of Poincare sections induces singularities. So we are far away from the settings we studied so far. The very first question arises how to find physical meaningful measures (those measures are called *SBR–measures*)?

In this section we consider maps of a manifold which may have singularities and are non–invertible. Let $M = \bigcup K_i \cup N$ where N is the finite union of smooth lower-dimensional submanifolds. The restriction $f|_{K_i}$ is a C^2–diffeomorphism onto its image and has a smooth extension to the boundary of K_i.

We lift the map f to a map $\hat{f}$ which is invertible. Namely, we set for $I = [0,1]$ and $\tau > 0$ sufficiently small

$$\hat{M} := M \times I, \quad \hat{K}_i := K_i \times I, \quad \hat{x} := (x,\omega) \in \hat{M};$$
$$\hat{f}(x,\omega) := \left(fx, \tau\omega + \frac{i-1}{r} \right), \text{ for } x \in K_i, \ \omega \in I, \ i = 1, ..., r.$$

Let $\pi : \hat{M} \to M$ be the projection of $\hat{M}$ onto the first coordinate. Then $f\pi = \pi\hat{f}$ and $\hat{f}$ is invertible. We use this solenoidal construction rather than the inverse limit space of the map f (see [36]) since the inverse limit space is not a manifold and we cannot apply the theory of diffeomorphisms with singularities to it.

Definition 3. Let $\hat{\Lambda}$ be the set of regular points of $\hat{f}$. We say that an invariant Borel probability measure m is *regular hyperbolic* if its lift is such that $\hat{m}(\hat{\Lambda}) = 1$ and all of its Lyapunov exponents are non-zero– i.e. $s(\hat{z})+1 = u(\hat{z})$ a.e. We also call $\hat{m}$ regular hyperbolic.

One of the main examples of a regular hyperbolic measure is the **SBR-measure**. It has been shown to exist for a large class of hyperbolic endomorphisms [49].

In [50] we proved the following generalization of the main result in the previous section.

Theorem 4. *Suppose f is an endomorphism with singularities described above and that m is a regular hyperbolic measure for f. Then the limit*

$$d_m(x) := \lim_{\varepsilon \to 0} \frac{\log mB(x,\varepsilon)}{\log \varepsilon} = d^s_{\hat{M}}(\hat{x}) + d^u_{\hat{M}}(\hat{x}) - d^1_{\hat{M}}(\hat{x})$$

exists m–almost everywhere. If m is additionally ergodic then $d_m(x)$ is constant a.e.

4 Applications in 2 Dimensions

In this section we will indicate some improvement of the previous results in dimension 2.

If m is an SBR–measure of an endomorphism with singularities on a surface then its *Lyapunov dimension* is defined as

$$\dim_L m = 1 - \frac{\lambda^u}{\lambda^s}.$$

We note that an SBR–measure on a surface always has exactly one negative and one positive exponent.

We say that a system is *a.s. invertible w.r.t. m* if there is a set of full measure on which the restriction of f is invertible. This means that up to measure 0 a system has unique forward and backward trajectories on the attractor and that the semi–flow is actually a flow restricted to this set of full measure.

In [49] a class of piecewise smooth endomorphisms with hyperbolic attractors on surfaces was introduced. We developed the ergodic theory of such systems. In particular we proved the existence of the SBR-measure and gave a criterion which connects the almost sure invertibility of the map on the attractor and the dimension of the set. This result can be stated in the following way.

Theorem 5. *The system is almost surely invertible (w.r.t. the SBR-measure) on the attractor if and only if the dimension of the SBR-measure equals its Lyapunov dimension.*

In [43] we applied the general theory from [49] to a specific example. The reason is that we want to illustrate the helpfulness of the above mentioned criterion. Our example consists of a three-parameter family of piecewise smooth maps. Our main attention is paid to the question for which parameters the a priori non-invertible system is invertible on a set of full SBR–measure. From the first point of view this example seems rather special. But it contains almost all difficulties of the general case and it is the most natural one for further investigations.

We consider the three-parameter family of the Belykh map.Let us consider the square $Q = [-1, 1] \times [-1, 1] \in \mathbb{R}^2$ and the map $f : Q \to Q$ defined by

$$f(x, y) = \begin{cases} (\lambda x_1 + 1 - \lambda, \gamma x_2 + 1 - \gamma) & x_2 > k x_1 \\ (\lambda x_1 + (\lambda - 1), \gamma x_2 + (\gamma - 1)) & x_2 < k x_1 \end{cases} \tag{1}$$

with $-1 < k < 1, 1 < \gamma \leq \frac{2}{|k|+1}, 0 < \lambda \leq 1$.

The Belykh map is natural for non-invertibility and features a lot of properties of the maps in the class introduced in [49].For the parameter range where the map is invertible it was first introduced by V.P. Belykh in [9] as a simple model for phase synchronization arising as a model for the Poincare map of a system of differential equations. The ergodic properties of this system in the parameter range of invertibility were investigated by Ya.B. Pesin in [29] and this system served as an example of the investigations of E.A. Sataev in [39] and later in [1]. J. Alexander and J. Yorke [2] considered a special case of this system for parameters where it is completely non-invertible. They called it the fat baker's transformation. We generalized the above examples (including the original baker's transformation) in a natural way and obtain a family of maps which increase the amount of non-invertibility as the parameters change. We believe that this family reflects most of the properties of generic parameter families of hyperbolic piecewise smooth maps.

Using the theory of [49] we can conclude the existence of an SBR–measure and investigate its properties. In particular, we are interested in the question how the Hausdorff dimension of the measure depends on the parameters. By the above criterion this is connected to the almost sure invertibility of the map.

Theorem 6.

- *If* $\lambda \cdot \gamma^2 < 1$ *then* $f_{\lambda,\gamma,k}$ *is fully invertible on the attractor for Lebesgue a.e.* (λ, γ, k). *Hence,*

$$\dim_H m = 1 - \frac{\log \gamma}{\log \lambda}$$

- *If* $\lambda \cdot \gamma < 1$ *then* $f_{\lambda,\gamma,k}$ *is a.s. invertible on the attractor for Lebesgue a.e.* (λ, γ, k). *Hence,*

$$\dim_H m = 1 - \frac{\log \gamma}{\log \lambda}$$

- *If* $\lambda \cdot \gamma > 1$ *then* $f_{\lambda,\gamma,k}$ *for Lebesgue a.e.* (λ, γ, k)

$$\dim_H m = 2 < \dim_L m$$

Remark 7. The above theorem shows that the Lyapunov dimension equals the Hausdorff dimension of the SBR-measure for almost all Belykh maps with $\lambda \cdot \gamma < 1$. This is the well-known Kaplan-Yorke conjecture for the special case of the Belykh attractor.

5 Irregular Points

In the previous sections we investigated the behavior of the system on the sets of regular points. Our main motivation was that in many situations these points carry full invariant measure. This study gives important information about the observable properties of the dynamical system, and "typical" points with respect to different measures give complementary information.

The set of "non-typical" points, i.e., the set of points that is "typical" with respect to no measure, has rarely been considered in the literature. In [3] we show that, surprisingly, in several situations central in the theory of dynamical systems this set contains complete information about some observable properties. Namely, the set of "non-typical" points carries **full** topological entropy and **full** Hausdorff dimension. In this sense the irregular points occupy a substantial part of the phase space.

5.1 Repellers and Horseshoes

Let $f\colon M \to M$ be a C^1 map of a smooth manifold, and J an f-invariant compact subset of M. We say that f is **expanding** on J and that J is a **repeller** of f if there are constants $C > 0$ and $\beta > 1$ such that $\|d_x f^n u\| \geq C\beta^n \|u\|$ for all $x \in J$, $u \in T_x M$, and $n \geq 1$.

It is well known that repellers admit Markov partitions of arbitrarily small diameter. Each Markov partition has associated a one-sided subshift of finite type $\sigma|\Sigma$, and a **coding map** $\chi\colon \Sigma \to J$ for the repeller, which is Hölder continuous, onto, and satisfies $f \circ \chi = \chi \circ \sigma$ and $\sup\{\mathrm{card}(\chi^{-1}x)\colon x \in J\} < \infty$ (see, for example, [30] for details).

A differentiable map $f\colon M \to M$ is called **conformal** on a set J if $d_x f$ is a multiple of an isometry at every point $x \in J$. Well-known examples of conformal expanding maps include one-dimensional Markov maps and holomorphic maps. We write $a(x) = \|d_x f\|$ for each $x \in M$. For a repeller J of a conformal $C^{1+\varepsilon}$ expanding map f, the equilibrium measure m_D of $-\dim_H J \cdot \log a$ on J is called the **measure of maximal dimension** (m_D is the unique f-invariant measure μ such that $\dim_H \mu = \dim_H J$; see, for example, [30] for details). We denote by m_E the **measure of maximal entropy**, i.e., the equilibrium measure of 0.

A uniformly hyperbolic basic set (*horseshoe*) of a surface diffeomorphism can be coded by a two-sided subshift of finite type in a similar manner.

5.2 Irregular Sets for Birkhoff Averages

Let $C(\Sigma)$ be the space of continuous functions on Σ. For each function $g \in C(\Sigma)$, we define the **irregular set for the Birkhoff averages of** g by

$$\mathcal{B}(g) = \left\{ x \in \Sigma\colon \lim_{n\to\infty} \frac{1}{n} S_n g(x) \text{ does not exist} \right\},$$

where

$$S_n g(x) = \sum_{k=0}^{n} g(\sigma^k x) \tag{2}$$

for each $x \in \Sigma$ and $n \in \mathbb{N}$. By the Birkhoff Ergodic Theorem, $\mu(\mathcal{B}(g)) = 0$ for every σ-invariant measure μ.

The following statement shows that the zero measure set $\mathcal{B}(g)$ is "observable"; namely, $\mathcal{B}(g)$ carries full topological entropy for a large class of functions g.

Theorem 8. *Let $\sigma|\Sigma$ be a topologically mixing subshift of finite type, and $g_1, \ldots, g_k$ Hölder continuous functions on Σ. Then the following properties are equivalent:*

1. *the functions $g_1, \ldots, g_k$ are non-cohomologous to 0;*

2. *$\mathcal{B}(g_1) \cap \cdots \cap \mathcal{B}(g_k)$ is non-empty;*
3. *$\mathcal{B}(g_1) \cap \cdots \cap \mathcal{B}(g_k)$ is a proper dense subset;*
4. *$h(\sigma|\mathcal{B}(g_1) \cap \cdots \cap \mathcal{B}(g_k)) > 0$;*
5. *$h(\sigma|\mathcal{B}(g_1) \cap \cdots \cap \mathcal{B}(g_k)) = h(\sigma)$.*

Remark 9. If $\sigma|\Sigma$ is uniquely ergodic, then the set $\mathcal{B}$ is empty.

5.3 Irregular Sets for Local Entropies

For each probability measure μ on Σ, we define the **irregular set for the local entropies of** μ by

$$\mathcal{H}(\mu) = \left\{ x \in \Sigma \colon \lim_{n\to\infty} -\frac{\log \mu(C_n(x))}{n} \text{ does not exist} \right\},$$

where $C_n(x)$ denotes the cylinder of length n which contains the point $x \in \Sigma$. The set $\mathcal{H}(\mu)$ is σ-invariant but may not be compact.

Remark 10. 1. If μ is σ-invariant, then $\mu(\mathcal{H}(\mu)) = 0$ (using the Shannon–McMillan–Breiman Theorem).

2. If μ is a Gibbs measure, then there is a cohomology class of functions in $C(\Sigma)$ such that $\mathcal{H}(\mu) = \mathcal{B}(g)$ if and only if g belongs to this cohomology class.
3. Let μ be a σ-invariant measure of maximal entropy. If μ is a Gibbs measure, then $\mathcal{H}(\mu)$ is empty. For example, if $\sigma|\Sigma$ is a topologically mixing subshift with the specification property (in particular, if $\sigma|\Sigma$ is a subshift of finite type or a sofic subshift), then any σ-invariant measure of maximal entropy is a Gibbs measure.

We define the **irregular set for the Lyapunov exponents of** f by

$$\mathcal{L}_f = \left\{ x \in J \colon \lim_{n\to\infty} \frac{1}{n} \log\|d_x f^n\| \text{ does not exist} \right\},$$

and for each probability measure μ on J, the **irregular set for the pointwise dimensions of** μ by

$$\mathcal{D}(\mu) = \left\{ x \in J \colon \lim_{r\to 0} \frac{\log \mu(B(x,r))}{\log r} \text{ does not exist} \right\},$$

where $B(x,r) \subset J$ is the ball of radius r centered at x.

For a repeller J of a topologically mixing expanding map, if any of the invariant sets $\mathcal{B}_f$, $\mathcal{H}_f(\mu)$, $\mathcal{L}_f$, and $\mathcal{D}(\mu)$ is non-empty, then it is dense in J.

By Kingman's Subadditive Ergodic Theorem, we have $\mu(\mathcal{L}_f) = 0$ for any f-invariant probability measure μ on J. In [48] we proved that if μ is a measure invariant under an expanding map (not necessarily conformal), then $\mu(\mathcal{D}(\mu)) = 0$.

Theorem 11. *If J is a repeller of a topologically mixing $C^{1+\varepsilon}$ expanding map f, for some $\varepsilon > 0$, and f is conformal on J, then the following properties hold:*

1. $h(f|\mathcal{B}_f) = h(f|J)$ *and* $\dim_H \mathcal{B}_f = \dim_H J$;
2. $m_D \neq m_E$ *if and only if* $h(f|\mathcal{L}_f) = h(f|J)$ *and* $\dim_H \mathcal{L}_f = \dim_H J$.

If in addition $\mu \in G(f|J)$ *then*

3. $\mu \neq m_D$ *if and only if* $h(f|\mathcal{D}(\mu)) = h(f|J)$ *and* $\dim_H \mathcal{D}(\mu) = \dim_H J$;
4. $\mu \neq m_E$ *if and only if* $h(f|\mathcal{H}_f(\mu)) = h(f|J)$ *and* $\dim_H \mathcal{H}_f(\mu) = \dim_H J$;
5. *the three measures* μ, m_D, *and* m_E *are distinct if and only if*
$$h(f|\mathcal{D}(\mu) \cap \mathcal{H}_f(\mu) \cap \mathcal{L}_f) = h(f|J)$$
and
$$\dim_H(\mathcal{D}(\mu) \cap \mathcal{H}_f(\mu) \cap \mathcal{L}_f) = \dim_H J.$$

It was established by Shereshevsky in [51] that $\dim_H \mathcal{D}(\mu) > 0$, and $\overline{\mathcal{D}(\mu)} \supset \Lambda$ for a generic C^2 surface diffeomorphism possessing a locally maximal hyperbolic set Λ, and a generic Hölder continuous potential, with respect to the C^0 topology, with Gibbs measure μ. In [3] we were able to improve this result.

Theorem 12. *For a surface diffeomorphism f in a C^2 open dense set, possessing a compact locally maximal hyperbolic set Λ, then*

$$\dim_H(\mathcal{D}(\mu) \cap \mathcal{H}_f(\mu) \cap (\mathcal{L}_f \cup \mathcal{L}_{f^{-1}})) > \sup_{\nu} \dim_H \mathcal{G}(\nu).$$

for any Gibbs measure μ and where $G(\nu)$ denotes the set of generic points for the ergodic measure ν.

6 Multifractal Analysis

The multifractal analysis, i.e., the analysis of invariant sets and measures with multifractal structure, has been recently developed as a powerful tool for numerical study of dynamical systems. These spectra capture information about various dimensions associated with the dynamics. Among them are the well-known Hausdorff dimension, correlation dimension, and information dimension of invariant measures.

Another example of multifractal spectra is entropy spectra introduced in [6]. They provide an integrated information on the distribution of topological entropy associated with local entropies.

In [6,7] it is demonstrated that multifractal spectra can be used in a sense to "restore" the dynamics – the phenomenon that we call *multifractal rigidity*.

The multifractal analysis is essentially measuring the "size" (in the sense of Hausdorff dimension or topological entropy) of the level sets of special discontinuous measurable functions (such as the local entropy, the pointwise dimension, etc.) on geometrically complicated objects (supports of invariant measures). This is expressed in a function f which is called the *spectrum*. In this sense the multifractal analysis can be considered as a non-smooth Morse theory. Our main concern is the further investigation of the dimension spectrum (i.e. the Hausdorff dimension of the level sets of the pointwise dimension) and the entropy spectrum (i.e. the topological entropy of the level sets of local entropies).

This analysis was investigated in several different situations and there is a huge amount of literature on this subject (see [30] for references and more details). One of the main results in this theory is that there is an interval (α_1, α_2) on which the spectrum $f(\alpha)$ (for definition see below) is analytic, convex, and can be continuously extended to the boundary provided it is not a point spectrum.

6.1 A General Concept of Multifractality

In this section we follow an outline of [6]. The invariant sets of a dynamical system may have a complicated geometric structure. In general, these sets are not self-similar, but can often be decomposed into subsets each possessing some scaling symmetry. This decomposition is called a *multifractal decomposition* and is an essential part of the multifractal analysis of dynamical systems.

Dimension spectra are the historically first examples of more general multifractal spectra which we introduce in [6].

One of the main points is to demonstrate that multifractal spectra can be used in a sense to "restore" the dynamics – the phenomenon that we call the *multifractal rigidity*. There are two main problems related to multifractal rigidity. First, given a dynamical system of hyperbolic type there exist *finitely* many *independent* multifractal spectra which uniquely identify the main *macro-characteristics* of the system (such as its invariant measure, geometric structure of its invariant sets, their dimensions, etc.). These spectra can be viewed as a special type of *degrees of freedom*, called *multifractal degrees of freedom*, and can be effectively used in the numerical study of dynamical systems.

In particular, for subshifts of finite type and some conformal expanding maps, the dimension spectrum alone is sufficient to determine all other multifractal spectra, and thus, these systems have one multifractal degree of freedom.

Another problem is inspired by an attempt to produce a "physically meaningful" classification of dynamical systems which takes care of various aspects of the dynamics (chaoticity, instability, geometry, etc.) simultaneously.

In the theory of dynamical systems there are various types of classifications. The most prominent ones seem to be *topological classification* (up to homeomorphisms) and *measure-theoretic classification* (up to measure preserving automorphisms). From a physical point of view, these classifications trace separate "independent" characteristics of the dynamics. We suggest a new type of classification which is based upon multifractal spectra and combines features of each of the above classifications, in what we call the *multifractal classification.* The new classification has a strong physical content and identifies two systems up to a change of variables. From a mathematical point of view, we establish the smooth equivalence of two dynamical systems which are *a priori* only topologically equivalent and have the same multifractal degrees of freedom. The multifractal classification is much more rigid than the topological and measure-theoretic classifications. Besides the smooth equivalence of the two dynamical systems, it establishes the coincidence of their dimension characteristics as well as the correspondence between their invariant measures.

We begin with the general concept of multifractal spectrum.

Let X be a set and let $g\colon X \to [-\infty, +\infty]$ be a function. The *level sets* of g

$$K_\alpha^g = \{x \in X : g(x) = \alpha\}, \quad -\infty \le \alpha \le +\infty$$

are disjoint and produce a *multifractal decomposition* of X, that is,

$$X = \bigcup_{-\infty \le \alpha \le +\infty} K_\alpha^g. \tag{3}$$

Let now G be a set function, i.e., a real function that is defined on subsets of X. Assume that $G(Z_1) \le G(Z_2)$ if $Z_1 \subset Z_2$. We define the function $\mathcal{F}\colon [-\infty, +\infty] \to \mathbb{R}$ by

$$\mathcal{F}(\alpha) = G(K_\alpha^g).$$

We call $\mathcal{F}$ the *multifractal spectrum* specified by the pair of functions (g, G), or the (g, G)-multifractal spectrum. The function $\mathcal{F}$ captures an important information about the structure of the set X generated by the function g.

It often happens that the function g is defined only on a subset $Y \subset X$. In this case the decomposition (3) should be replaced by

$$X = \bigcup_{-\infty \le \alpha \le +\infty} K_\alpha^g \cup (X \setminus Y).$$

We still call this decomposition of X a *multifractal decomposition.*

6.2 Examples of Multifractal Spectra

We illustrate the general concept of multifractal spectra by studying several explicit spectra.

6.3 Dimension and Entropy Spectra

Let X be a complete separable metric space and let $f\colon X \to X$ be a continuous map. We begin with the choice of the set function G. There are two "natural" set functions on X. The first one is generated by the metric structure on X. Namely, given a subset $Z \subset X$, we set

$$G_D(Z) = \dim_H Z, \tag{4}$$

where $\dim_H Z$ is the Hausdorff dimension of Z.

The second function is generated by the dynamical system f acting on X and the metric on X. Namely,

$$G_E(Z) = h(f|_Z), \tag{5}$$

where $h(f|_Z)$ is the topological entropy of f on Z (notice that Z need not be compact nor f-invariant and one can use the definitions in [10,45]). We call the multifractal spectra generated by the function G_D *dimension spectra*, and the multifractal spectra generated by the function G_E *entropy spectra*.

We now describe some "natural" choices for the function g.

6.4 Multifractal Spectra for Pointwise Dimensions

Let μ be a Borel finite measure on X. Consider the subset $Y \subset X$ consisting of all points $x \in X$ for which the pointwise dimension $d_\mu(x)$ of the measure μ exists. We define the function g_D on Y by

$$g_D(x) = d_\mu(x).$$

We note that the corresponding multifractal decomposition consists of the sets

$$K_\alpha^{g_D} = \{x : d_\mu(x) = \alpha\}.$$

We obtain two multifractal spectra $\mathcal{D}_D = \mathcal{D}_D^{(\mu)}$ and $\mathcal{D}_E = \mathcal{D}_E^{(\mu)}$ specified by the pairs of functions (g_D, G_D) and (g_D, G_E), respectively, where the set functions G_D and G_E are given by (4) and (5). We call them *multifractal spectra for (pointwise) dimensions.*

Let us remark that the spectrum $\mathcal{D}_D$ is known in the literature as the *dimension spectrum* or *$f_\mu(\alpha)$-spectrum for dimensions.* The concept of a multifractal analysis was suggested by a group of physicists in [17] (see [30] for more references and details).

6.5 Multifractal Spectra for Local Entropies

Let X be a complete separable metric space and let $f\colon X \to X$ be a continuous map preserving a Borel probability measure μ. Consider the set $Y \subset X$

consisting of all points $x \in X$ for which the local entropy $h_\mu(x)$ exists. We define the function g_E on Y by

$$g_E(x) = h_\mu(x).$$

We note that the corresponding multifractal decomposition consists of the sets

$$K_\alpha^{g_E} = \{x : h_\mu(x) = \alpha\}.$$

We obtain two multifractal spectra $\mathcal{E}_D = \mathcal{E}_D^{(\mu)}$ and $\mathcal{E}_E = \mathcal{E}_E^{(\mu)}$ specified by the pairs of functions (g_E, G_D) and (g_E, G_E), respectively, where the set functions are given by (4) and (5). We call them *multifractal spectra for (local) entropies.*

We want to emphazise that the above definitions are dependent on the choice of a partition. But in the situations we consider later there are canonical partitions on which the spectra do not depend.

6.6 Multifractal Spectra for Lyapunov Exponents

Let X be a differentiable manifold and let $f\colon X \to X$ be a C^1 map. Consider the subset $Y \subset X$ of all points $x \in X$ for which the Lyapunov exponent $\lambda_u(x)$ exists. We define the function g_L on Y by

$$g_L(x) = \lambda(x).$$

We note that the corresponding multifractal decomposition consists of the sets

$$K_\alpha^{g_L} = \{x : \lambda(x) = \alpha\}.$$

We obtain two multifractal spectra $\mathcal{L}_D$ and $\mathcal{L}_E$ specified, respectively, by the pairs of functions (g_L, G_D) and (g_L, G_E), where the set functions G_D and G_E are given by (4) and (5). We call them *multifractal spectra for Lyapunov exponents.* The spectrum $\mathcal{L}_D$ was studied in [54] (see also the references in that paper). The spectrum $\mathcal{L}_E$ was introduced in [14].

We want to state the fundamental theorem for some spectra for expanding conformal repellers. The way we state it it is a conglomeration of results which accumulated the last years by an enormous amount of literature. For a positive Hölder continuous function $u\colon \Sigma \to \mathbb{R}^+$ we define in a unique way a metric ρ_u on Σ. This metric is specified by the condition

$$\operatorname{diam}_u(C_n(x)) = \exp\left\{-\sum_{k=0}^{n-1} u(\sigma^k x)\right\}.$$

We denote by $\dim_u(Z)$, $d_{\mu,u}(x)$ the dimensions w.r.t. the metric ρ_u. If we choose $u \equiv 1$ then the u–dimensions are simply the entropies and for $u = \log a$ we recover the dimensions on the expanding conformal repeller (see [3]).

For every real number α, set

$$K_\alpha = \{x \in \Sigma \colon d_{\mu,u}(x) = \alpha\}.$$

We set

$$\mathcal{D}_u(\alpha) = \dim_u K_\alpha.$$

The function $\alpha \mapsto \mathcal{D}_u(\alpha)$ is called the **u-dimension spectrum for u-pointwise dimensions** (with respect to the measure μ). Let φ be a continuous function on Σ. For every real number q, we define the function

$$\varphi_q = -T_u(q)u + q\varphi,$$

where the number $T_u(q)$ is chosen such that $P(\varphi_q) = 0$. We denote by ν_q and m_u, respectively, the equilibrium measures of φ_q and $-\dim_u \Sigma \cdot u$ with respect to σ.

The following is a complete multifractal analysis of the spectrum $\mathcal{D}_u$ for subshifts of finite type.

Theorem 13. *Let $\sigma|\Sigma$ be a one-sided or two-sided topologically mixing subshift of finite type, u and φ Hölder continuous functions on Σ, such that u is positive and $P(\varphi) = 0$, and μ the equilibrium measure of φ with respect to σ. Then, the following properties hold:*

1. *For μ-almost every $x \in \Sigma$, the u-pointwise dimension of μ at x exists and*

$$d_{\mu,u}(x) = -\frac{\int_\Sigma \varphi \, d\mu}{\int_\Sigma u \, d\mu} = \frac{h_\mu(\sigma)}{\int_\Sigma u \, d\mu}.$$

2. *The function $q \mapsto T_u(q)$ is real analytic on $\mathbb{R}$, and satisfies $T_u'(q) \le 0$ and $T_u''(q) \ge 0$ for every $q \in \mathbb{R}$. Moreover, $T_u(0) = \dim_u \Sigma$ and $T_u(1) = 0$.*
3. *The domain of the function $\alpha \mapsto \mathcal{D}_u(\alpha)$ is a closed interval in $[0, +\infty)$ and coincides with the range of the function $\alpha_u(q) = -T_u'(q)$. For every $q \in \mathbb{R}$, we have*

$$\mathcal{D}_u(\alpha_u(q)) = T_u(q) + q\alpha_u(q),$$

and

$$\alpha_u(q) = -\frac{\int_\Sigma \varphi \, d\nu_q}{\int_\Sigma u \, d\nu_q}.$$

4. *For every $q \in \mathbb{R}$, $\nu_q(K_{\alpha_u(q)}) = 1$, and*

$$d_{\nu_q,u}(x) = T_u(q) + q\alpha_u(q)$$

for ν_q-almost all $x \in K_{\alpha_u(q)}$. Moreover, $\overline{d}_{\nu_q,u}(x) \le T_u(q) + q\alpha_u(q)$ for every $x \in K_{\alpha_u(q)}$, and $\mathcal{D}_u(\alpha_u(q)) = \dim_u \nu_q$ for every $q \in \mathbb{R}$.

5. *If* $\mu \neq m_u$, *then* $\mathcal{D}_u$ *and* T_u *are real analytic strictly convex functions, and* $(\mathcal{D}_u, T_u)$ *is a Legendre pair with respect to the variables* α, q.
6. *If* $\mu = m_u$, *then* $d_{\mu,u}(x) = \dim_u \Sigma$ *for every* $x \in \Sigma$.

We showed in [6,7] in the case of expanding conformal repellers or hyperbolic horseshoes on surfaces that if, for instance, the spectra $\mathcal{D}_D$ and $\mathcal{E}_E$ are equivalent then all other spectra are equivalent. We call this phenomenon *multifractal rigidity*. It indicates that the spectra $\mathcal{D}_D$ and $\mathcal{E}_E$ are essentially independent. Moreover, both spectra for the Lyapunovexponent can be recovered from the entropy or dimension spectrum of an appropriately chosen measure.

7 Multifractal Rigidity

In [6] we considered another interesting phenomenon in dimension theory of dynamical systems which we regard as a multifractal rigidity phenomenon. Roughly speaking, it states that if two dynamical systems are topologically equivalent (via a homeomorphism) and some of their multifractal spectra coincide, then they are smoothly equivalent (via a diffeomorphism). This leads to a classification of maps and Gibbs measures using multifractal spectra and/or multifractal decompositions. We believe that this type of classification fits well with the "physical" interpretation of the equivalence of dynamical systems. We think that this is a non-trivial and challenging problem.

If $\mathcal{F}$ and $\widehat{\mathcal{F}}$ are two multifractal spectra for the maps f and $\widehat{f}$, respectively, the condition $\mathcal{F} = \widehat{\mathcal{F}}$ indicates the existence of a "symmetry" between the two dynamical systems. Thus, it is a requirement of "physical" nature, and this multifractal rigidity has a strong physical content: if the spectra of two dynamical systems are equal, then the systems are the same up to a change of variables, and thus should be considered the same from the physical point of view.

7.1 Preservation of the Spectra Under Markov Coding

In the theory of hyperbolic dynamical systems one often uses Markov partitions to encode the system. The coding space is a symbolic dynamical system and several useful tools from symbolic dynamics can be exploited. Among a long list of results in this direction thermodynamic formalism is one of the most fruitful tools. It is used to proof the existence of Gibbs states (see f.e. [11]) and to derive their multifractal properties (see f.e. [30,32,54,6,7]).

One of the problems of applying the coding method is that the coding map is not one–to–one on the boundaries of the Markov partitions. In particular, it is not obvious that sets of positive measure are mapped to sets of positive measure and that dimensionlike notions are preserved. The same problem occurs if one deals with entropy. In the above mentioned applications it was sufficient to establish the fact that the boundary of the Markov partition (see [11]) has zero measure for any Gibbs state. But there are problems when one has to deal with multifractal decompositions. There we have to cope with the fact that the boundaries of the Markov partitions might have positive entropy or dimensions. Therefore the coding could a priori change the spectrum. In [44] we proved:

Theorem 14. *If Λ is a horseshoe on a surface (J an expanding conformal repeller) then the map $\chi\colon \Sigma_A \to \Lambda$ ($\chi\colon \Sigma_A^+ \to J$) preserves for any subset $Z \in \Sigma_A$ ($Z \in \Sigma_A^+$) its Hausdorff dimension and its topological entropy.*

Remark 15. In [44] it is shown that the statement is no longer true for piecewise smooth systems (even with nice Markov partitions). The coding map can strictly decrease entropy and dimension.

References

1. Afraimovich, V.S., Chernov, N.I., and Sataev E.A., *Statistical properties of 2.d generalized hyperbolic attractors* Chaos, 1995, 5, 238–252.
2. J. C. Alexander, and J. A. Yorke, *Fat Baker's transformations* Erg. Th. Dyn. Syst., 1984, 4, 1–23.
3. L. Barreira and J. Schmeling, *Sets of "non-typical points have full topological entropy and full Hausdorff dimension*, Isr. J. Math. **116** (2000)
4. L. Barreira, Ya. Pesin and J. Schmeling, *On the pointwise dimension of hyperbolic measures: a proof of the Eckmann–Ruelle conjecture*, Electronic Research Announc. Amer. Math. Soc. **2** (1996), no. 1, 69–72.
5. L. Barreira, Ya. Pesin and J. Schmeling, *Dimension and product structure of hyperbolic measures*, Annals of Mathematics, Vol 149 (1999), No 3, 755–783
6. L. Barreira, Ya. Pesin and J. Schmeling, *On a general concept of multifractality: multifractal spectra for dimensions, entropies, and Lyapunov exponents. Multifractal rigidity*, Chaos **7** (1997), no. 1, 27–38.
7. L. Barreira, Ya. Pesin and J. Schmeling, *Multifractal spectra and multifractal rigidity for horseshoes*, Journal of Dynamics and Control Systems **3** (1997), no. 1, 33–49.
8. L. Barreira and J. Schmeling, *Invariant sets with zero measure and full Hausdorff dimension*, Electronic Research Announc. Amer. Math. Soc. **3** (1997), 114–118.
9. V. P. Belykh, *Models of discrete systems of phase synchronization*, in Systems of phase synchronization, V.V. Shakhildyan and L.N. Belynshina, eds. Radio i Svyaz, Moscow, 1982, 61–176.
10. R. Bowen, *Topological entropy for noncompact sets*, Trans. Amer. Math. Soc. **184** (1973), 125–136.
11. R. Bowen, *Equilibrium states and the ergodic theory of Anosov diffeomorphisms*, Springer Lecture Notes in Mathematics 470, Springer Verlag, 1975.
12. M. Brin, and A. Katok, *On local entropy* Geometric dynamics (Rio de Janeiro, 1981), 30–38, Lecture Notes in Math., 1007, Springer, Berlin-New York, 1983.
13. R. Cawley and R. D. Mauldin, *Multifractal Decompositions of Moran Fractals*, Adv.Math. **92** (1992), no. 2, 196–236.

14. J.-P. Eckmann and I. Procaccia, *Fluctuations of Dynamical Scaling Indices in Nonlinear Systems*, Phys. Rev. A (3) **34** (1986), no. 1, 659–661.
15. J.-P. Eckmann and D. Ruelle, *Ergodic theory of chaos and strange attractors*, Rev. Modern Phys. **57** (1985), no. 3, 617–656.
16. K.J. Falconer, *The Geometry of Fractal Sets*, Cambridge Tracts in Mathematics, 1985.
17. T. Halsey, M. Jensen, L. Kadanoff, I. Procaccia, and B. Shraiman, *Fractal measures and their singularities: the characterization of strange sets*, Phys. Rev. A (3) **34** (1986), no. 2, 1141–1151, errata in **34** (1986), no. 2, 1601.
18. B. Hasselblatt, *Regularity of the Anosov splitting and of horospheric foliations*, Erg. Th. Dyn. Syst., 14 no 4 (1994), 645–666
19. S. Hurder, and A. Katok, *Differentiability, rigidity and Godbillon–Vey classes for Anosov flows*, Publications Mathematiques IHES, 72 (1990), 5–61
20. A. Katok, *Lyapunov exponents, entropy and periodic orbits for diffeomorphisms*, Inst. Hautes Études Sci. Publ. Math. **51** (1980), 137–173.
21. A. Katok, B. Hasselblatt, *Introduction to the Modern Theory of Dynamical Systems*, Encyclopedia of Mathematics and its Applications, vol. 54, Cambridge University Press, 1995
22. A. Katok, and J.-M. Strelcyn, *Invariant manifolds, entropy and billiards; smooth maps with singularities*, Lecture Notes in Mathematics, 1222, Springer, Berlin, 1986.
23. F. Ledrappier, *Dimension of invariant measures*, Proceedings of the conference on ergodic theory and related topics II (Georgenthal, 1986), Teubner–texte Math, vol 94, 137–173
24. F. Ledrappier, and M. Misiurewicz, *Dimension of invariant measures for maps with exponent zero*, Erg. Th. Dyn. Syst., vol 5 (1985), 595–610
25. F. Ledrappier, and L.-S. Young, *The metric entropy of diffeomorphisms, Part I and II*, Ann. Math., 1985, 122, 509–574.
26. H. McCluskey and A. Manning, *Hausdorff dimension for horseshoes*, Ergodic Theory and Dynam. Systems **3** (1983), no. 2, 251–260, errata in **3** (1983), no. 2, 319.
27. Ya. Pesin,*Families of invariant manifolds corresponding to nonzero characteristic exponents*, Mathematics of the USSR, Isvestia, 10 no 6(1976), 1261–1305
28. Ya. Pesin,*Characteristic exponents and smooth ergodic theory*, Russian Math Surveys, 32 no 4 (1977), 55–114
29. Ya. Pesin, *Dynamical systems with generalized hyperbolic attractors: hyperbolic, ergodic and topological properties* Erg. Th. Dyn. Syst., 1992, 12, 123–151.
30. Ya. Pesin, *Dimension theory in dynamical systems: contemporary views and applications*, Chicago Lectures in Mathematics, Chicago University Press, 1997.
31. Ya. Pesin and B. Pitskel', *Topological pressure and the variational principle for noncompact sets*, Functional Anal. Appl. **18** (1984), no. 4, 307–318.
32. Ya. Pesin and H. Weiss, *A multifractal analysis of Gibbs measures for conformal expanding maps and Markov Moran geometric constructions*, J. Statist. Phys. **86** (1997), no. 1/2, 233–275.
33. Ya. Pesin, and H. Weiss, *On the dimension of deterministic and random Cantor-like sets, symbolic dynamics, and the Eckmann–Ruelle conjecture*, Comm. Math. Phys., vol 182 (1996), 105–153
34. Ya. Pesin and H. Weiss, *The Multifractal Analysis of Gibbs Measures: Motivation, Mathematical Foundation, and Examples*, Chaos: an Interdisciplinary Journal of Nonlinear Science **7** (1997), no. 3, to appear.

35. Ya. Pesin, and Ch. Yue, *The Hausdorff dimension of measures with non-zero Lyapunov exponents and local product structure*, PSU preprint
36. F. Przytycki, *Anosov endomorphisms*, Studia Math., 1976, 58, 249–185.
37. C. Rogers, *Hausdorff measures*, Cambridge University Press, London–New York, 1970
38. D. Ruelle, *Repellers for real analytic maps*, Ergodic Theory Dynam. Systems **2** (1982), no. 1, 99–107.
39. E. A. Sataev, *Invariant measures for hyperbolic maps with singularities*, Russian Math. Surveys, 1992, 47, 191–251.
40. J. Schmeling, *Hölder continuity of the holonomy maps for hyperbolic sets II*, Math. Nachr. 170 (1994), 211–225
41. J. Schmeling, *On the completeness of multifractal spectra*, Erg. Th. and Dyn. Syst. 19 (1999), 1–22
42. J. Schmeling, *Ergodic and dimension theory of hyperbolic endomorphisms with singularities* Habilitationschrift 1995.
43. J. Schmeling, *A dimension formula for endomorphisms - The Belykh family*, Erg. Th. Dyn. Sys. 18 (1998), 1283–1310
44. J. Schmeling, *Entropy preservation under Markov coding*, Preprint of the DFG priority program DANSE, no 6/99
45. J. Schmeling,*Time weighted entropies*, Coll. Math, Vol. 84/85 (2000) part 1
46. J. Schmeling, and Ra. Siegmund–Schultze, *Hölder continuity of the holonomy maps for hyperbolic systems I*, Lect. Notes Math. 1514 (1992)
47. J. Schmeling, Ra. Siegmund–Schultze, *The Singularity Spectrum of Self–affine Fractals with a Bernoulli Measure*, WIAS preprint.
48. J. Schmeling and S. Troubetzkoy, *Pointwise dimension for regular hyperbolic measures for endomorphisms*, Preprint of the DFG priority program DANSE, no 13/97
49. J. Schmeling, and S. Troubetzkoy, *Dimension and invertibility of hyperbolic endomorphisms with singularities*, Erg. Th. Dyn. Sys. 18 (1998), 1257–1282
50. J. Schmeling, and S. Troubetzkoy, *Scaling properties of hyperbolic measures*, Preprint of the DFG priority program DANSE,, no 50/98
51. M. Shereshevsky, *A complement to Young's theorem on measure dimension: the difference between lower and upper pointwise dimension*, Nonlinearity **4** (1991), no. 1, 15–25.
52. D. Simpelaere, *Dimension Spectrum of Axiom A Diffeomorphisms. I. The Bowen–Margulis Measure II. Gibbs Measures*, Journal of Statistical Physics **76** (1982), no. 5/6, 1329–1375.
53. F. Takens, *Limit capacity and Hausdorff dimension of dynamically defined Cantor sets*, Dynamical Systems (Valparaiso 1986), Eds. R. Bamón, R. Labarca and J. Palis Jr., Lecture Notes in Mathematics 1331, Springer Verlag, 1988, 196–212.
54. H. Weiss, *The Lyapunov and dimension spectra of equilibrium measures for conformal expanding maps*, preprint, 1996.
55. L.-S. Young, *Dimension, entropy and Lyapunov exponents*, Ergodic Theory Dynam. Systems **2** (1982), no. 1, 109–124.

Collision of Control Sets

Fritz Colonius[1] and Wolfgang Kliemann[2,*]

[1] Institut für Mathematik, Universität Augsburg, 86135 Augsburg, Germany
email: colonius@math.uni-augsburg.de
[2] Department of Mathematics, Iowa State University, Ames, Iowa 50011, U.S.A.
email: kliemann@iastate.edu

Abstract. Control sets describe the limit behavior of control systems. When they collide, the system behavior undergoes a qualitative change. It is shown that at a collision of an invariant control set C with another control set D there exists a ∂C-control set in the intersection of C and the closure of D.

1 Introduction

In this paper, qualitative properties of nonlinear control systems are discussed. The methods and results are a combination of geometric control theory and topological theory of dynamical systems. This approach (combined with numerical methods) has yielded a number of new insights in the last years. The basic theory is presented in the book [2]. In the present paper we analyze the collision of control sets; in particular, we describe the intersection points of an invariant control set with the closure of another control set.

We consider control-affine systems of the form

$$\dot{x} = X_0(x) + \sum_{i=1}^{m} u_i(t)X_i(x) =: X(x, u(t)), \tag{1}$$
$$u \in \mathcal{U} = \{u \in L_\infty(\mathbb{R}, \mathbb{R}^m),\ u(t) = (u_i(t)) \in U \text{ for almost all } t \in \mathbb{R}\},$$

where $X_0, ..., X_m$ are C^∞-vector fields on a Riemannian manifold M and U is a nonvoid, convex, and compact subset U of $\mathbb{R}^m$. We assume throughout, that for every initial value $x \in X$ and every $u \in \mathcal{U}$ there exists a unique solution $\varphi(t, x, u)$, $t \in \mathbb{R}$, with $\varphi(0, x, u) = x$.

The functions $u \in \mathcal{U}$ can be interpreted as controls, which are chosen to achieve a desired system behavior or as time-dependent perturbations acting on the system. The dynamical properties of system (1) are embodied in the associated control flow given by

$$\Phi : \mathbb{R} \times \mathcal{U} \times M \to \mathcal{U} \times M,\ \Phi_t(u, x) = (u(t + \cdot), \varphi(t, x, u)).$$

Then Φ becomes a continuous dynamical system if we endow $\mathcal{U}$ with the weak* topology of $L_\infty(\mathbb{R}, \mathbb{R}^m)$, and $\mathcal{U} \times M$ is a metrizable space. The limit

* Project: Analysis of Time-Dependent Perturbations of Ordinary Differential Equations (Fritz Colonius)

behavior of Φ is determined by the control sets and the chain control sets, which correspond to the topologically transitive components and the chain transitive components, respectively, of Φ. Generically, they coincide (see Section 2 for a precise statement). Qualitative changes occur when an invariant, hence closed, control set intersects the closure of another control set. Then both control sets are contained in a single chain control set and we will discuss this situation assuming local accessibility. We remark that Grünvogel [7] has described when a singular point lies in the closure of a control set with nonvoid interior. His analysis of this case where local accessibility is violated is based on Lyapunov exponents.

In Section 2 we introduce some notation and recall some facts on control sets and chain control sets. Section 3 defines control sets relative to a compact subset L of the state space and discusses their basic properties. Section 4 presents the main result, establishing the existence of a control set relative to the boundary ∂C of an invariant control set C, when C collides with another control set. Section 5 concentrates on the two dimensional case, based on results of Poincaré-Bendixson type. Section 6 presents some numerically computed examples.

2 Basic Properties of Control Sets

In this section, we cite the pertinent results from [2] on the global controllability structure of control systems.

We consider control system (1) with controls $u \in \mathcal{U}$. Sometimes it will be convenient to introduce a parameter $\rho \geq 0$, which indicates the size of the perturbations and to replace the pointwise constraint $u(t) \in U$ by

$$u(t) \in U^\rho := \rho \cdot U.$$

Then the control functions are in

$$\mathcal{U}^\rho = \{u \in L_\infty(\mathbb{R}, \mathbb{R}^m),\ u(t) \in U^\rho \text{ for almost all } t \in \mathbb{R}\}.$$

In particular, for $\rho = 0$, one recovers the nominal, unperturbed system. If the dependence on ρ does not play a role, we simply write $\mathcal{U}$. Throughout we assume that system (1) is locally accessible, i.e., for all $x \in \mathbb{R}^d$ and all $T > 0$ one has

$$\mathrm{int}\mathcal{O}^+_{\leq T}(x) \neq \emptyset \text{ and } \mathrm{int}\mathcal{O}^-_{\leq T}(x) \neq \emptyset,$$

where $\mathcal{O}^+_{\leq T}(x) = \{y \in M,\ y = \varphi(t, x, u) \text{ with } 0 \leq t \leq T \text{ and } u \in \mathcal{U}\}$ and $\mathcal{O}^-_{\leq T}(x) = \{y \in M,\ x = \varphi(t, y, u) \text{ with } 0 \leq t \leq T \text{ and } u \in \mathcal{U}\}$ are the positive and negative reachable sets (or orbits), respectively, from x.

To describe the global behavior of system (1), we recall the following concepts and results.

Definition 1. A subset D of the state space M is a control set if it is a maximal subset with the properties that (i) for all $x \in D$ the inclusion $D \subset \operatorname{cl} \mathcal{O}^+(x)$ holds and (ii) for all $x \in D$ there is $u \in \mathcal{U}$ with $\varphi(t, x, u) \in D$ for all $t \geq 0$. A control set C is an invariant control set if $\operatorname{cl} C = \operatorname{cl} \mathcal{O}^+(x)$.

If the dependence of D on ρ plays a role, we indicate this by the argument ρ in $D(\rho)$. The same remark applies to all other notions introduced. The local accessibility assumption guarantees that the invariant control sets are the closed control sets and that they have nonvoid interior.

We will also need the following weaker concept, chain control sets, which allows for arbitrarily small jumps in the trajectories. For ε, $T > 0$, a controlled (ε, T)-chain from x to y is given by $n \in \mathbb{N}$, times $T_0, ..., T_{n-1} \geq T$, points $x_0 = x, ..., x_{n-1}, x_n = y$ in M, and controls $u_0, ..., u_{n-1} \in \mathcal{U}$ such that

$$d(\varphi(T_i, x_i, u_i), x_{i+1}) < \varepsilon \text{ for all } i = 0, ..., n-1.$$

Definition 2. A *chain control set* $E \subset M$ is a maximal subset such that (i) for all $x, y \in E$ and all ε, $T > 0$ there is a controlled (ε, T)-chain from x to y; and (ii) for all $x \in E$ there is $u \in \mathcal{U}$ with $\varphi(t, x, u) \in E$ for all $t \in \mathbb{R}$.

Chain control sets are closed and every control set is contained in a chain control set. Furthermore, if two control sets D_1 and D_2 satisfy $\operatorname{cl} D_1 \cap \operatorname{cl} D_2 \neq \emptyset$, then there is a chain control set E with $D_1 \cup D_2 \subset E$.

To get a better understanding of the limit behavior we lift the control sets $D \subset M$ with nonvoid interior to sets in $\mathcal{U} \times M$ via

$$\mathcal{D} = \operatorname{cl} \{(u, x) \in \mathcal{U} \times M, \ \varphi(t, x, u) \in \operatorname{int} D \text{ for all } t \in \mathbb{R}\}, \tag{2}$$

where the closure is taken with respect to the weak* topology on $\mathcal{U}$ and the given topology on M. Observe that $\mathcal{D}$ and

$$\{(u, x) \in \mathcal{U} \times M, \ \varphi(t, x, u) \in \operatorname{int} D \text{ for all } t \in \mathbb{R}\}$$

are invariant under Φ and, if D is bounded in the metric given by the Riemannian structure on M, then $\mathcal{D}$ is compact.

Theorem 3. *Let $\mathcal{D} \subset \mathcal{U} \times M$ be a set such that the projection in M*

$$\pi_M \mathcal{D} = \{x \in M, \text{ there exists } u \in \mathcal{U} \text{ with } (u, x) \in \mathcal{D}\}$$

is compact. Then $\mathcal{D}$ is a maximal topologically mixing set if and only if there exists a control set D such that $\mathcal{D}$ is of the form (2). In this case D is unique; $\operatorname{int} D = \operatorname{int} \pi_M \mathcal{D}$, *and* $\operatorname{cl} D = \pi_M \mathcal{D}$. *Furthermore, the periodic points of Φ are dense in $\mathcal{D}$; the restricted flow $\Phi \mid \mathcal{D}$ is topologically mixing, topologically transitive, and chain transitive, and it has sensitive dependence on initial conditions.*

This is [2, Theorem 4.1.3]. An analogous interpretation in terms of the control flow can also be given for the chain control sets E. Again we consider a lift to $\mathcal{U} \times M$ by defining

$$\mathcal{E} = \{(u,x) \in \mathcal{U} \times M,\ \varphi(t,x,u) \in E \text{ for all } t \in \mathbb{R}\}. \tag{3}$$

Theorem 4. *Let $E \subset M$ be a chain control set. Then $\mathcal{E} \subset \mathcal{U} \times M$ as defined by (3) is a maximal invariant chain transitive set for the control flow $(\mathcal{U} \times M, \Phi)$. Conversely, let $\mathcal{E} \subset \mathcal{U} \times M$ be a maximal invariant chain transitive set for $(\mathcal{U} \times M, \Phi)$. Then $\pi_M \mathcal{E}$ is a chain control set.*

This is [2, Theorem 4.1.4]. In general, the control sets are proper subsets of the chain control sets. Considering systems with varying control ranges U^ρ, one finds that the relation between the control sets and chain control sets above is much closer. The chain control sets coincide generically with the closures of the control sets. Indicating the ρ-dependence by the argument ρ, we have $\operatorname{cl} D(\rho) \subset E(\rho)$ for increasing families $D(\rho)$ and $E(\rho)$, $\rho > 0$. Let

$$\mathcal{D}(\rho) := \operatorname{cl}\{(u,x) \in \mathcal{U}^\rho \times \mathbb{R}^d,\ \varphi(t,x,u) \in \operatorname{int} D(\rho) \text{ for all } t \in \mathbb{R}\}.$$

Theorem 5. *Assume that the following inner pair condition holds: for all $\rho' > \rho > 0$ and all $(u,x) \in \mathcal{D}(\rho)$ there is $T > 0$ with $\varphi(T,x,u) \in \operatorname{int} \mathcal{O}^{+,\rho'}(x)$. Then the map $\rho \mapsto \operatorname{cl} D(\rho)$ is lower semicontinuous and the map $\rho \mapsto E(\rho)$ is upper semicontinuous (with respect to Hausdorff metric). The sets of continuity points for both maps coincide, and $\rho^* \in (0,\infty)$ is a continuity point if and only if $\operatorname{cl} D(\rho^*) = E(\rho^*)$. There are at most countably many point of discontinuity.*

This theorem (see [2, Theorem 4.7.5]) shows that under the inner pair condition the chain control sets 'almost always' coincide with the closures of control sets. The exceptional cases will be of particular interest. We remark that the inner pair condition can be verified for a number of second order systems with additive controls; compare [4].

3 *L*-Control Sets

In this section, we describe control sets which are defined relative to a subset L of the state space.

Let $L \subset M$ be a subset such that every point of L is a cluster point of L, i.e., for every $x \in L$ there are $x_n \in L$ with $x \neq x_n \to x$. Furthermore we assume that L is *positively viable* (or *controlled invariant*), i.e., for every $x \in L$ there is $u \in \mathcal{U}$ such that $\varphi(t,x,u) \in L$ for all $t \geq 0$.

Definition 6. For $x \in L$, the positive L-orbit of x is given by

$$\mathcal{O}_L^+(x) = \left\{ y \in L, \begin{array}{r} \text{there are } t > 0 \text{ and } u \in \mathcal{U} \text{ such that } \varphi(t,x,u) = y \\ \text{and } \varphi(s,x,u) \in L \text{ for all } 0 < s < t \end{array} \right\}.$$

Analogously, one defines $\mathcal{O}_{L,\leq t}^+(x)$, $\mathcal{O}_L^-(x), \mathcal{O}_{L,\leq t}^-(x)$, etc.

Definition 7. An L-control set F is a subset of L with the following properties:
(i) for all $x \in F$ the inclusion $F \subset \operatorname{cl} \mathcal{O}_L(x)$ holds, (ii) for all $x \in F$ there is $u \in \mathcal{U}$ with $\varphi(t,x,u) \in F$ for all $t \geq 0$, and (iii) the set F is maximal with these properties. An L-control set F is an invariant L-control set if

$$\operatorname{cl}_L F = \operatorname{cl}_L \mathcal{O}_L^+(x),$$

where the closures are taken relative to the set L.

Thus an invariant L-control set cannot be left without leaving L.

Definition 8. The system (1) is called L-accessible if for all $x \in L$ and all $t > 0$ the sets $\mathcal{O}_{L,\leq t}^+(x)$ and $\mathcal{O}_{L,\leq t}^-(x)$ have nonvoid interiors relative to L.

For $L = M$, these definitions reduce to the standard definitions recalled earlier. We start with the following simple observation.

Proposition 9. *Let D be a control set with nonvoid interior. Then* $\operatorname{int} D = \operatorname{int} \operatorname{cl} D$ *and this set is path connected.*

Proof. Let $x \in \operatorname{int} \operatorname{cl} D$. By local accessibility, $\emptyset \neq \mathcal{O}_{\leq t}^{\pm}(x) \subset \operatorname{int} \operatorname{cl} D$ for $t > 0$, small enough. Hence, using continuous dependence on the initial value, one finds that $x \in \operatorname{int} D$. The converse inclusion is clear. Path connectedness follows from exact controllability in $\operatorname{int} D$.

Proposition 10. *Let D be a control set with nonvoid interior. Then every point of the boundary ∂D is a clusterpoint of ∂D and ∂D is viable, i.e., for every $x \in \partial D$ there exists $u \in \mathcal{U}$ such that $\varphi(t,x,u) \in \partial D$ for all $t \in \mathbb{R}$.*

Proof. The first assertion follows, because $\operatorname{int} D = \operatorname{int} \operatorname{cl} D$. For the second assertion, fix $x \in \partial D$: Then there are $y \in \operatorname{int} D$ and $v \in \mathcal{U}$ with $\varphi(t_n, y, v) \to x$ for some $t_n \to \infty$. Take a clusterpoint $u_{-1} \in \mathcal{U}$ of the sequence $(v(t_n + \cdot))$. Then $\varphi(-1, x, u_{-1}) \in \partial D$, since it is in $\operatorname{cl} D$ (because it is a clusterpoint of $\varphi(t_n - 1, y, v) \in \operatorname{int} D$); and it is not in $\operatorname{int} D$, since the boundary point x cannot be reached in finite time from the interior of D. Proceeding in this way, one constructs a control $u_- \in \mathcal{U}$ with $\varphi(t,x,u_-) \in \partial D$ for all $t \leq 0$. Applying the same procedure to the time reversed system, one finds a control $u_+ \in \mathcal{U}$ with $\varphi(t,x,u_+) \in \partial D$ for all $t \geq 0$. Thus $u(t) = u_\pm(t)$ for $t \leq 0$ and $t \geq 0$, respectively, has the desired property.

Proposition 11. *Suppose that D is a control set with nonvoid interior and compact closure. Then for every $x \in \partial D$ there exists an invariant ∂D-control set $F \subset \operatorname{cl} \mathcal{O}_{\partial D}^+(x)$.*

Proof. By Proposition 10 it follows that for every $x \in \partial D$ the positive ∂D-orbit

$$\mathcal{O}^+_{\partial D}(x) = \left\{ y \in \partial D, \begin{array}{c} \text{there are } t > 0 \text{ and } u \in \mathcal{U} \text{ such that } \varphi(t,x,u) = y \\ \text{and } \varphi(s,x,u) \in \partial D \text{ for all } 0 < s < t \end{array} \right\}$$

is nonvoid. Now we mimic the proof for the existence of invariant control sets in compact forward invariant sets; cp. [2, Theorem 3.2.8]. By compactness of $\mathrm{cl}\, D$ the boundary ∂D is compact. The family

$$\mathcal{F} = \{\mathrm{cl}\, \mathcal{O}^+_{\partial D}(x),\ x \in \partial D\}$$

is nonvoid and partially ordered by set inclusion. By compactness of ∂D, every linearly ordered subset has a lower bound. Hence $\mathcal{F}$ satisfies the assumptions of Zorn's lemma, yielding existence of a minimal element, which is an invariant ∂D-control set. Furthermore, for every $x \in \partial D$ there is an invariant ∂D-control set in $\mathrm{cl}\, \mathcal{O}^+_{\partial D}(x)$.

Proposition 12. *Suppose that the following local ∂D-accessibility condition holds:*

$$\mathrm{int}_{\partial D} \mathcal{O}^{\pm}_{\partial D, \leq t}(x) \neq \emptyset \text{ for all } x \in \partial D \text{ and all } t > 0.$$

Then every invariant ∂D-control set has nonvoid interior relative to ∂D. Furthermore, for every ∂D-control set F with nonvoid interior $\mathrm{int}_{\partial D} F$ *one has* $\mathrm{cl}\ \mathrm{int}_{\partial D} F = \mathrm{cl}\, F$, *and for all $x, y \in \mathrm{int}_{\partial D} F$ there are $t > 0$ and $u \in \mathcal{U}$ with $\varphi(t,x,u) = y$.*

Proof. This follows as the corresponding assertions for control sets; compare [2, Section 3.2].

4 Collision of Control Sets

In this section, we discuss collisions of an invariant control set C with another control set D, i.e., $C \cap \mathrm{cl}\, D \neq \emptyset$. Note that then $C \cap \mathrm{cl}\, D = \partial C \cap \partial D$. We will show that in this case the set of contact points, i.e., the points in the intersection of C with $\mathrm{cl}\, D$, must contain a control set relative to ∂C. The first result shows a weaker property.

Theorem 13. *Let C be a compact invariant control set for system (1). Suppose that $C \cap \mathrm{cl}\, D \neq \emptyset$ for some control set $D \neq C$ with nonvoid interior. Then there exist an invariant ∂D-control set F' and a ∂C-control set F with*

$$F' \subset C \cap \mathrm{cl}\, D \text{ and } F \cap C \cap \mathrm{cl}\, D \neq \emptyset.$$

Proof. Fix $x \in C \cap \mathrm{cl}\, D$. By Proposition 11, there exists an invariant ∂D-control set $F' \subset \mathrm{cl}\, \mathcal{O}^+_{\partial D}(x)$. By invariance of C, it follows that $F' \subset C \cap \mathrm{cl}\, D$. Then the definitions immediately imply that there is a ∂C-control set F containing F', and the second assertion follows.

Requiring local accessibility properties on the boundaries of the considered control sets, we obtain the following stronger result. It shows that collisions of an invariant control set with another control set occur at ∂C-control sets.

Theorem 14. *Let C be a compact invariant control set for system (1). Suppose that $C \cap \operatorname{cl} D \neq \emptyset$ for some control set $D \neq C$ with nonvoid interior.*
(i) If local ∂D-accessibility holds, then there exist an invariant ∂D-control set F' with $F' \subset \partial C \cap \partial D$ and $\operatorname{int}_{\partial D} F' \neq \emptyset$.
(ii) If, additionally, local ∂C-accessibility holds and $\operatorname{int}_{\partial C}(\partial C \cap \partial D) = \operatorname{int}_{\partial D}(\partial C \cap \partial D)$, then there exists a ∂C-control set F with $\operatorname{int}_{\partial C} F \neq \emptyset$ and

$$F \subset \partial C \cap \partial D.$$

Proof. Assertion (i) is immediate from Theorem 13 and $\emptyset \neq \operatorname{int}_{\partial D} \mathcal{O}^+_{\partial D, \leq t} \subset F'$. To see assertion (ii) observe that by Theorem 13 there exists a ∂C-control set $F \supset F'$. The assumption $\operatorname{int}_{\partial C}(\partial C \cap \partial D) = \operatorname{int}_{\partial D}(\partial C \cap \partial D)$ implies that $\emptyset \neq \operatorname{int}_{\partial D} F' = \operatorname{int}_{\partial C} F' \subset \operatorname{int}_{\partial C}(F \cap \partial C \cap \partial D)$. Hence $\emptyset \neq \operatorname{int}_{\partial C} F' \subset \operatorname{int}_{\partial C} F$ and thus F is a ∂C-control set with nonvoid interior in ∂C. By Proposition 12 this implies that $\operatorname{cl} \operatorname{int}_{\partial C} F = \operatorname{cl} F$. It remains to show that

$$\operatorname{int}_{\partial C} F \subset \partial C \cap \partial D.$$

By Proposition 12 exact controllability in $\operatorname{int}_{\partial C} F$ holds and we know that $\emptyset \neq \operatorname{int}_{\partial C} F' \cap \partial D \cap \partial C \subset \operatorname{int}_{\partial C} F \cap \partial D \cap \partial C$. Now take $x \in \operatorname{int}_{\partial C} F' \cap \partial D \cap \partial C$ and $y \in \operatorname{int}_{\partial C} F \cap \partial D \cap \partial C$. Then, by continuous dependence on the initial value and controllability from x to y, the point y can approximately be reached from $\operatorname{int} D$. Conversely, one can steer the system from some point in every neighborhood of y into the interior of D, because one can steer the system from y to x and from some point of every neighborhood of x into the interior of D.

Remark 15. In [3] the invariance radius r_{inv} of control sets is discussed which indicates when a control set around an stable equilibrium of the unperturbed system loses its invariance property for $\rho > r_{inv}$. This happens when for $\rho > r_{inv}$ the invariant control set merges with another control set D. Although this does not imply that at the invariance radius r_{inv} the invariant control set has nonvoid intersection with the closure of another control set D, this is what we see in a number of examples. In this case, the theorems above give results on the intersection of C and $\operatorname{cl} D$.

5 Poincaré-Bendixson Theory

In this section we will sketch some results of Poincaré-Bendixson type for control systems by discussing ω-limit sets for two-dimensional systems. These constructions are a modification of those in Colonius and Sieveking [5], where the limit behavior of optimal solutions in discounted optimal control problems is discussed. Then we use these results to describe ∂D-control sets in dimension $d = 2$.

First we need generalizations of transversal sections and flow boxes.

Definition 16. Let $x^0 \in \mathbb{R}^d$, $l : \mathbb{R}^d \to \mathbb{R}$ linear, and $\alpha > 0$. If $lX(x,u) > 0$ for all x in a neighborhood W of x^0 and all $u \in U$, then $S := W \cap l^{-1}(x^0)$ is a local transversal section through x^0.

We note the following observations.

Lemma 17. *(i) Suppose that $0 \notin X(x^0, U)$. Then x^0 has a local transversal section.*

(ii) A subset $L \subset \mathbb{R}^d$ either contains an equilibrium or every point in L has a transversal section.

Proof. Assertion (i) follows from the Hahn-Banach theorem, because $X(x^0, U)$ is compact and convex, and hence can be separated from $0 \notin X(x^0, U)$. Assertion (ii) is immediate from (i), because L either contains a point x^0 with $0 \in X(x^0, U)$ (i.e., an equilibrium) or $0 \notin X(x^0, U)$ for all $x^0 \in L$.

Definition 18. Let S be a local transversal section through x^0, and let $V_1 \subset V_0$ be neighborhoods of x^0. Then the triple (V_0, V_1, S) is called a flow box around x^0, if it has the following property: If $\varphi(\cdot, x^0, u)$ satisfies

$$\varphi(t_0, x^0, u) \notin V_0,\ \varphi(t_1, x^0, u) \in V_1,\ \varphi(t_2, x^0, u) \notin V_0 \text{ for some } 0 \le t_0 < t_1 < t_2,$$

then there exists $t \in (t_0, t_2)$ such that $\varphi(t, x^0, u) \in S$ and $\varphi(s, x^0, u) \in V_0$ for all s between t and t_1.

Theorem 19. *Let S be a local transversal section through x^0. Then there are neighborhoods V_0 and V_1 of x^0 such that (V_0, V_1, S) is a flow box around x^0.*

Proof. This is [5, Theorem 2.16].

We cite Jordan's curve theorem, see, e.g., Beck [1, Corollary C.23].

Theorem 20. *Let J be a Jordan curve in $\mathbb{R}^2$. Then $\mathbb{R}^2 \setminus J$ has two components, one of which is bounded (called* ins J*) and the other one (called* outs J*) is unbounded. Both have boundary* im J *and are simply connected.*

Since the orientation does not concern us, we identify J with its image. Next we discuss limit sets as time tends to infinity. For $(u, x) \in \mathcal{U} \times M$ let

$$\omega(u,x) = \{(v,y) \in \mathcal{U} \times M,\ (u(t_k + \cdot), \varphi(t_k, x, u)) \to (v,y) \text{ for some } t_k \to \infty\},$$
$$\pi_2\omega(u,x) = \{y \in M,\ \text{there is } v \in \mathcal{U} \text{ with } (v,y) \in \omega(u,x)\}.$$

If the trajectory $\varphi(\cdot, x, u)$ is bounded for positive time, then these sets are nonvoid, compact, and connected.

Lemma 21. *Let $(u,x) \in \mathcal{U} \times \mathbb{R}^2$ and suppose that the corresponding trajectory $\varphi(\cdot,x,u)$ is bounded and nonself-intersecting, i.e., there do not exist $T_2 > T_1 \geq 0$ with $\varphi(T_2,x,u) = \varphi(T_1,x,u)$. Then a local transversal section S has at most one point in common with $\pi_2\omega(u,x)$.*

Proof. This follows as [5, Lemma 4.2].

The same arguments apply to α-limit sets, which are obtained for time t tending to $-\infty$.

Proposition 22. *Let $(u,x) \in \mathcal{U} \times \mathbb{R}^2$ and suppose that the corresponding trajectory $\varphi(\cdot,x,u)$ is bounded and nonself-intersecting. Let $(v,y) \in \omega(u,x)$. Then $\pi_2\omega(v,y)$ and $\pi_2\alpha(v,y)$ consist of equilibria only or $\varphi(\cdot,y,v)$ intersects itself in a point z which has a transversal section.*

Proof. This follows as [5, Proposition 4.2].

Proposition 23. *Let $(u,x) \in \mathcal{U} \times \mathbb{R}^2$ and assume that the corresponding trajectory $\varphi(\cdot,x,u)$ is nonself-intersecting. Suppose that there are $(v,y) \in \omega(u,x)$ and $T_2 > T_1 \geq 0$ with $\varphi(T_2,x,u) = \varphi(T_1,x,u)$ such that this point has a local transversal section. Then*

$$\pi_2\omega(u,x) = \varphi([T_1,T_2],y,v).$$

Proof. This follows as [5, Proposition 4.5].

The announced result of Poincaré-Bendixson type is presented in the following theorem.

Theorem 24. *Let $(u,x) \in \mathcal{U}\times\mathbb{R}^2$ with nonself-intersecting trajectory $\varphi(\cdot,x,u)$ and suppose that $\pi_1\omega(u,x)$ contains at most finitely many equilibria. Then one of the following cases occurs:*

(i) There are $T > 0$ and $(v,y) \in \omega(u,x)$ such that

$$y = \varphi(T,y,v) \text{ and } \pi_1\omega(u,x) = \varphi([0,T],y,v).$$

(ii) There are $(v_i,y_i) \in \pi_1\omega(u,x)$ and equilibria $e_i^{\pm}$ such that for all i

$$e_i^- = \lim_{t\to-\infty} \varphi(t,y_i,v_i),\ e_i^+ = \lim_{t\to\infty} \varphi(t,y_i,v_i),$$
$$\pi_1\omega(u,x) = \bigcup_i \varphi(\mathbb{R},y_i,v_i) \cup \bigcup \{e_i^-,\, e_i^+\}.$$

Proof. This follows as [5, Theorem 4.6].

Next we will apply this result to describe boundaries of control sets. The following control sets play an exceptional role.

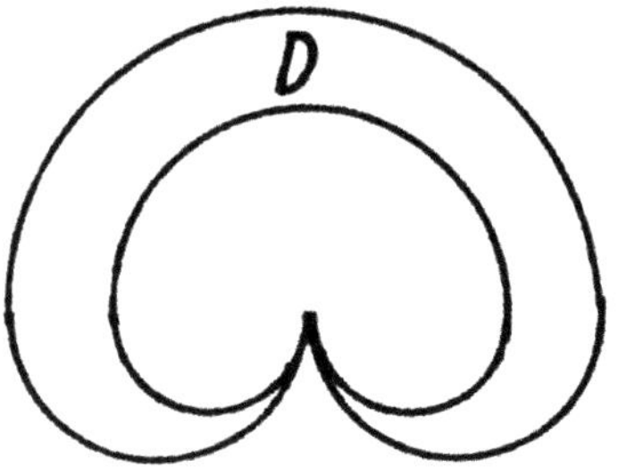

Fig. 1. Sketch of a pinched control set.

Definition 25. A control set $D \subset \mathbb{R}^2$ is pinched at a point $x \in \partial D$ if for every open neighborhood W of x the set $\partial D \cap W$ is not homeomorphic to the interval $[0, 1]$.

Although it seems difficult to give an explicit example of a pinched control set, Figure 1 gives a sketch of such a control set. The next theorem describes the ∂D-control sets for control sets that are not pinched.

Theorem 26. *Let F be a ∂D-control set of a bounded control set $D \subset \mathbb{R}^2$, which is not pinched. Suppose that local ∂D-accessibility holds and that F contains at most finitely many equilibria. Then F is a single periodic trajectory, i.e., there are $T > 0$ and $(v, y) \in \mathcal{U} \times \mathbb{R}^2$ such that $y = \varphi(T, y, v)$ and $F = \varphi([0, T], y, v)$.*

Proof. Local ∂D-accessibility implies that $F = \pi_2\omega(u, x)$ for some $(u, x) \in \mathcal{U} \times \partial D$. Suppose first that $\varphi(\cdot, x, u)$ is self-intersecting. Either there is $T > 0$ such that $\varphi(t, x, u) = \varphi(T, x, u)$ for all $t \geq T$ and the assertion follows; or there are $T_2 > T_1 \geq 0$ with $\varphi(T_1, x, u) = \varphi(T_2, x, u)$ and there is $t_0 \in (T_1, T_2)$ with $\varphi(t_0, x, u) \neq \varphi(T_1, x, u)$. If $\varphi(t, x, u) = \varphi(T_1, x, u)$ for some $t \in [T_1, T_2]$, then we can modify u such that the new control u' yields the same set of points $\{\varphi(t, x, u),\ t \geq 0\}$ and $\varphi(t, x, u') \neq \varphi(T_1, x, u)$ for all $t \in [T_1, T_2]$. Then one sees that for no neighborhood W of $\varphi(T_1, x, u)$ the set $W \cap \partial D$ is homeomorphic to $(0, 1)$. It follows that the control set D is pinched at the point $\varphi(T_1, x, u)$, contrary to our assumption. Hence $\varphi(\cdot, x, u)$ is nonself-intersecting. Then, according to Theorem 24, one has (i) $\pi_2\omega(u, x)$ is a single periodic trajectory; or (ii) there are $(v_i, y_i) \in \omega(u, x)$ and equilibria $e_i^{\pm}$ such that for all i

$$e_i^- = \lim_{t \to -\infty} \varphi(t, y_i, v_i),\ e_i^+ = \lim_{t \to \infty} \varphi(t, y_i, v_i),$$
$$\pi_2\omega(u, x) = \bigcup_i \varphi(\mathbb{R}, y_i, v_i) \cup \bigcup_i \{e_i^-, e_i^+\}.$$

In case (i) the assertion follows. In case (ii) we fix one of the equilibria $e_i^{\pm}$ say, e_1^-. Let W be a neighborhood of e_1^- such that $\partial D \cap W$ is homeomorphic to the interval $(0,1)$. Then there are $u \in \mathcal{U}$ and $t_+, \tau_+ > 0$ such that $\varphi(t_+, e_1^-, u) = \varphi(-\tau_+, y_1, v_1)$ and $\{\varphi(t, e_1^+, v_1),\ t \in [0, t_+]\} = \{\varphi(t, y_1, v_1),\ t \leq \tau_+\}$. Arguing analogously for all i one finds that $\pi_2\omega(u,x)$ consists of a single periodic trajectory. This proves the assertion.

6 Examples

In this section we discuss control sets in a simple two-dimensional model of a continuous flow stirred tank reactor (compare Poore [8] or Golubitskii/Schaeffer [6] and also the analysis in [2, Chapter 9]). The control sets in Figures 2–5 were computed with G. Häckl's program CS.

Consider the model of a continuous flow stirred tank reactor (for short: CSTR) given by the equations

$$\begin{pmatrix} \dot{x}_1 \\ \dot{x}_2 \end{pmatrix} = \begin{pmatrix} -x_1 - a(x_1 - x_c) + B\alpha(1 - x_2)e^{x_1} \\ -x_2 + \alpha(1 - x_2)e^{x_1} \end{pmatrix} + u(t) \begin{pmatrix} x_c - x_1 \\ 0 \end{pmatrix} \tag{4}$$

Here x_1 is the (dimensionless) temperature, x_2 is the product concentration, and a, α, B, x_c are positive constants. The parameter x_c is the coolant temperature, and hence the perturbation affects the heat transfer coefficient. In [8] Poore analyzes the bifurcation behavior of the nominal, i.e., $u(t) \equiv 0$, system. Here we choose parameter values such that for all constant controls $u(t) \equiv u \in [-\rho, \rho]$ the system (4) has exactly three fixed points as limit sets. Specifically, we take for our numerical analysis

$$a = 0.15,\ \ \alpha = 0.05,\ \ B = 7.0,\ \ x_c = 1.0. \tag{5}$$

First let $U = [-0.15, 0.15]$. Because of the physical constraints we have to consider the system in the set $M = [0,\infty) \times [0,1] \subset \mathbb{R}^2$. For each fixed $u \in U$ the equation (4) has three fixed points in M. Let $y^i = \alpha e^{z^i}/(1 + \alpha z^i)$, $i = 0, 1, 2$, and let $z^0 < z^1 < z^2$ be the zeros of the transcendental equation

$$-z - (a+u)(z - x_c) + B\alpha[1 - \frac{\alpha e^z}{1 + \alpha e^z}]e^z = 0.$$

Then two of these fixed points are asymptotically stable, $x^0 = (z^0, y^0)$ and $x^2 = (z^2, y^2)$, and one is hyperbolic, $x^1 = (z^1, y^1)$, i.e., the linearization about x^1 has one negative and one positive eigenvalue. The phase portrait of the nominal equation is indicated in Figure 2. There are exactly three control sets C_1, C_2, and D, containing the fixed points $x^i(u)$, $i = 0, 1, 2$, for $u \in \operatorname{int} U = (-0.15, 0.15)$ in their interior. The control sets C_1 and C_2 are invariant, the control set D is variant. The closures of these control sets are the three chain control sets of the system. Figure 2 shows the three control sets.

For different parameter values, Figure 3 shows the invariant control set C_2 as well as the positive and negative orbits from the hyperbolic equilibrium.

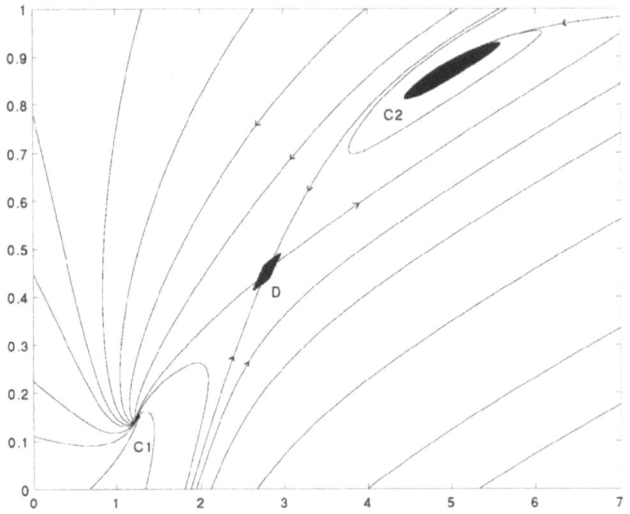

Fig. 2. Phase portrait of the unperturbed ($u(t) \equiv 0$) continuous flow stirred tank reactor and the control sets.

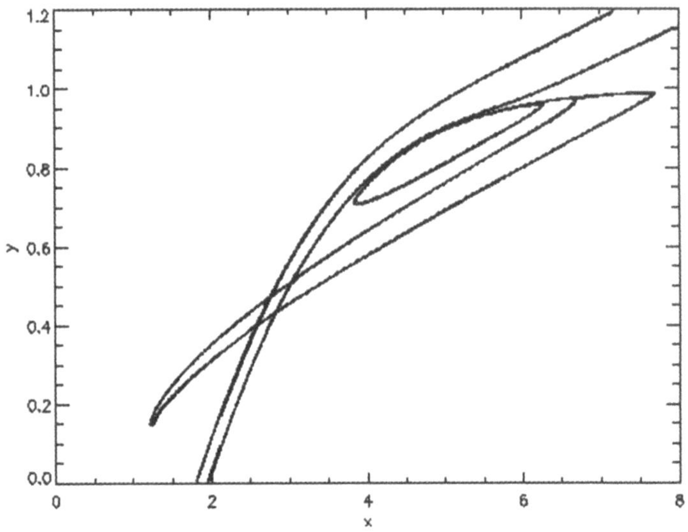

Fig. 3. Variant and invariant control sets and their domains of attraction for the CSTR

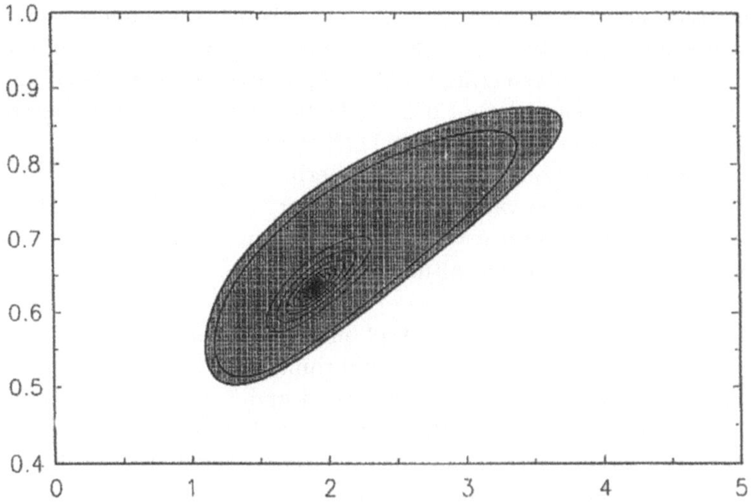

Fig. 4. Invariant control set for the CSTR with $\alpha = 0.2824$, after the Hopf bifurcation.

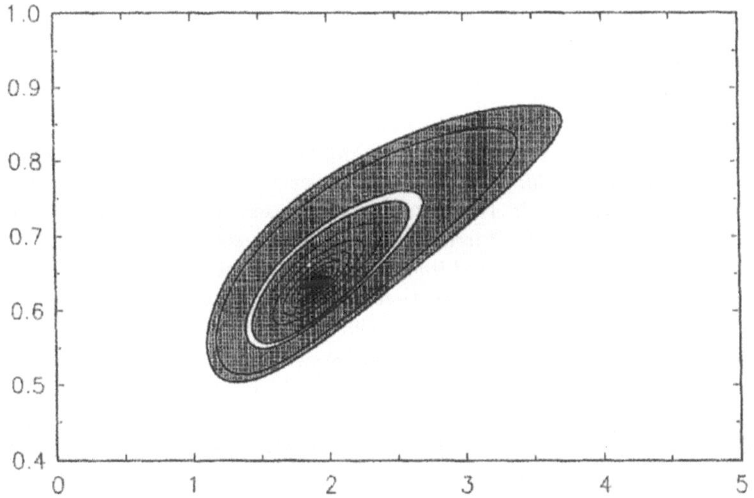

Fig. 5. Variant and invariant control sets for the CSTR with $\alpha = 0.2628$.

Their intersection is the variant control set D. Though the numerics seem to indicate that for a slightly larger control range the invariant control set C_2 loses its invariance by intersecting the domain of attraction $\mathbf{A}(D) \setminus \mathrm{cl}\, D$, the situation is different: The relevant part of the boundary of $\mathbf{A}(D)$ is the stable manifold of a hyperbolic equilibrium in $\mathrm{cl}\, D$. Hence, if $\mathbf{A}(D) \cap C_2 \neq \emptyset$, it follows from invariance of C_2 that also this hyperbolic equilibrium is in C_2. The control set C_2 touches the variant control set at the hyperbolic equilibrium corresponding to the control $u = \rho$. This one-point set is an invariant ∂C-control set. Note that this holds, although the ∂C-local accessibility condition is not satisfied here.

For different parameter values, a Hopf bifurcation (taking α as bifurcation parameter) occurs around the upper equilibrium x^2. It loses its stability and a stable periodic solution bifurcates. Figures 4 and 5 illustrate the behavior for a fixed perturbation range with varying α. Before and slightly after the Hopf bifurcation point there is only one invariant control set; see Figure 4. Then it bifurcates into an invariant control set around the stable periodic solution and a variant control set containing the equilibrium; see Figure 5. At the collision the invariant control set around the stable periodic solution touches the variant control set around the unstable equilibrium at the periodic solution which forms the inner boundary of the invariant control set (and coincides with the boundary of the variant control set).

References

1. A. Beck, *Continuous Flows in the Plane*, Springer-Verlag, 1974.
2. F. Colonius and W. Kliemann, *The Dynamics of Control*, Birkhäuser, 2000.
3. ——, *An invariance radius for nonlinear systems*, in Advances in Mathematical Systems. A Volume in Honor of D. Hinrichsen., F. Colonius, U. Helmke, D. Prätzel-Wolters, and F. Wirth, eds., Birkhäuser, 2000.
4. ——, *Invariance under bounded time-varying perturbations*, in 11th IFAC International Workshop Control Applications of Optimization (St. Petersburg, Russia, July 3-6 2000), V. Zakharov, ed., 2000, pp. 82–85.
5. F. Colonius and M. Sieveking, *Asymptotic properties of optimal solutions in planar discounted control problems*, SIAM J. Control Optim., 27 (1989), pp. 608–630.
6. M. Golubitsky and D. Schaeffer, *Singularities and Groups in Bifurcation Theory*, Springer-Verlag, 1985.
7. S. Grünvogel, *Control Sets and Lyapunov Spectrum.* Dissertation, Universität Augsburg, 2000.
8. A. B. Poore, *A model equation arising from chemical reactor theory*, Arch. Rational Mech. Anal., 52 (1974), pp. 358–388.

The Algorithms Behind GAIO - Set Oriented Numerical Methods for Dynamical Systems

Michael Dellnitz, Gary Froyland, and Oliver Junge*

Fachbereich 17 Mathematik/Informatik, Universität Paderborn, 33095 Paderborn, Germany

Abstract. In a given dynamical system there are essentially two different types of information which could be of practical interest: on the one hand there is the need to describe the behavior of single trajectories in detail. This information is helpful for the analysis of transient behavior and also in the investigation of geometric properties of dynamical systems. On the other hand, if the underlying invariant set is generated by complicated dynamics then the computation of single trajectories may give misleading results. In this case there still exists important set related information covering both topological and statistical aspects of the underlying dynamical behavior. Within the DFG-Schwerpunkt we have focussed on the development of set oriented methods for the numerical approximation of

- invariant sets (e.g. invariant manifolds, global attractors, chain recurrent sets);
- (natural) invariant measures;
- almost invariant sets.

The basic concept is a subdivision algorithm which is similar in spirit to the well known cell mapping techniques but with the crucial difference that the numerical effort mainly depends on the complexity of the dynamics rather than on the dimension of the underlying state space. First, the invariant set is covered by boxes and then the dynamical behavior on the set is approximated by a Markov chain based on transition probabilities between elements of this covering. The algorithms have been implemented in the software package GAIO (**G**lobal **A**nalysis of **I**nvariant **O**bjects), and in this article we describe both the related numerical techniques together with their theoretical foundations and how to use them within GAIO. We will also discuss details concerning the implementation such as adaptive versions of the methods.

1 Introduction and Motivation

Our aim is to capture the global structure of a given dynamical system. This will be done by using global set-oriented methods rather than by an approach based on long term computations of single trajectories. At the topological level, objects of interest are foremostly invariant *sets*, as these sets support the dynamics of the system for all time. When the dynamics has some degree of smoothness and hyperbolicity, invariant *manifolds* provide insight into the

* Project: Numerical Methods for the Investigation of Chaotic Dynamics (Michael Dellnitz)

geometric structure of the system's dynamics. At the level of statistics, invariant *measures* quantitatively describe frequencies of visitation of trajectories to different regions of phase space. These invariant measures are generalised fixed eigenfunctions of a global *transfer operator*. Other eigenfunctions of this operator provide information such as identifying *almost invariant sets* which help to fine-tune the dynamical analysis. Based on the identification of these eigenfunctions the groups of Deuflhard and Schütte and co-workers have derived a new approach to the identification of conformations of biomolecules [8,30] within this DFG research program.

The above list gives a quick run-down of *what* our set-oriented methods set out to find. *How* we do this is arguably of even greater importance. We will begin each section with a concise introduction to the mathematical object under consideration. We then give a description of the algorithm used to approximate the object, followed by rigorous results concerning convergence. The theoretical exposition will be paralleled by examples to demonstrate the efficacy of the methods.

All of the algorithms described here have been coded in the software package GAIO (**G**lobal **A**nalysis of **I**nvariant **O**bjects). It is the existence of this software package that makes the use of our techniques feasible, allowing rapid computations and informative visualisations. Many figures in this paper have been made using visualisation techniques which have been developed and implemented within the software platform GRAPE by the group of Rumpf and co-workers [29,5,7] within the DANSE program. To highlight the ease of use of the software, we present actual GAIO commands applied to each of the instructive examples. These commands will illustrate the use of the package for real problems, and collectively will provide sufficient detail so as to act as a concise tutorial for it.

2 The Computation of Invariant Sets

Our setting is that of a continuous mapping $T : M \to M$ on a compact manifold M. We will call a set $A \subset M$ *forward invariant* if $T(A) \subset A$, *backward invariant* if $T^{-1}(A) \subset A$ and *invariant* if $T(A) = T^{-1}(A) = A$. There are various collections of invariant sets that one can talk about. The largest invariant set may be defined as follows.

Definition 1. Let $\mathcal{O}^{\pm}(x) = \{\ldots, T^{-1}x, x, Tx, \ldots\}$ denote the *full orbit* of a point $x \in M$. If T is non-invertible, then $T^{-k}x = \{y \in M : T^k y = x\}$. The *maximal invariant set* contained in a set $Q \subset M$ is defined as

$$\mathrm{Inv}(Q) = \{x \in Q : \mathcal{O}^{\pm}(x) \subset Q\} \tag{1}$$

It is straightforward (see e.g. [10]) to show the following properties of maximal invariant sets.

Proposition 2. *1.* Inv(Q) *is an invariant set.*
2. If $Y \subset Q$ *is an invariant set, then* $Y \subset \text{Inv}(Q)$.
3. If Q *is forward invariant, then* $\text{Inv}(Q) = \bigcap_{k=0}^{\infty} T^k(Q)$.
4. If T *is a homeomorphism, then* $\text{Inv}(Q) = \bigcap_{k=-\infty}^{\infty} T^k(Q)$.

The maximal invariant set is defined setwise and so it contains many points which are not *recurrent* in the sense that under iteration they do not return close to themselves infinitely often. This leads to:

Definition 3. A point $q \in Q$ belongs to the *chain recurrent set* of T in Q if for every $\epsilon > 0$ there is an ϵ-pseudoperiodic orbit containing q, that is, there exists $\{q = q_0, q_1, \ldots, q_{\ell-1}\} \subset Q$ such that

$$\|T(q_i) - q_{i+1 \bmod \ell}\| \leq \epsilon \text{ for } i = 0, \ldots, \ell - 1.$$

Also the following result follows immediately from the definitions.

Proposition 4. *The chain recurrent set* $R_Q(T)$ *of* T *in* Q *is closed and invariant. Furthermore we have the following inclusion:*

$$R_Q(T) \subset \bigcap_{k\geq 0} T^k(Q). \tag{2}$$

Example 5. To illustrate the case where the inclusion of (2) is sharp, consider the map $T : [0,1] \to [0,1]$ defined by $Tx = x^2$. Here, $R_{[0,1]}(T) = \{0, 1\}$, while $\bigcap_{k\geq 0} T^k([0,1]) = [0,1]$.

This simple example illustrates the two qualitatively different types of invariant sets that we intend to approximate.

2.1 Algorithm: Relative Global Attractor

We describe numerical methods to approximate the invariant sets $\text{Inv}(Q)$ and $R_Q(T)$. These methods are based on multilevel subdivision techniques. Let us begin with an abstract algorithm of this type for the computation of the *relative global attractor* $A_Q(T) = \bigcap_{k\geq 0} T^k(Q)$, where $Q \subset M$ is an arbitrary (not necessarily forward invariant) box, i.e. a generalized rectangle $B(c, r)$ with center c and radius r.

The idea of the algorithm is to cover $A_Q(T)$ by a finite number of boxes and to recursively tighten the covering by refining appropriately selected boxes.

Algorithm 1 (Subdivision Algorithm to Compute $A_Q(T)$). We start with the collection $\mathcal{B}_0 = \{Q\}$. For $k = 1, 2, \ldots$ compute $\mathcal{B}_k$ from the collection $\mathcal{B}_{k-1}$ in two steps:

1. *Subdivision:* Bisect each box in the current collection $\mathcal{B}_{k-1}$ into two smaller boxes of equal size (for d-dimensional boxes, the cutting plane is cycled around the d coordinate directions).

2. *Selection:* Discard those refined boxes whose preimage does not intersect any box of the current (refined) collection. The remaining boxes constitute the collection $\mathcal{B}_k$.

The following result is proved in [4].

Theorem 6. *Set* $Q_k = \bigcup_{B\in\mathcal{B}_k} B$ *and* $Q_\infty = \bigcap_{k\geq 0} Q_k$.

1. $A_Q(T) \subset Q_k$ *for all* $k \geq 0$
2. $T^{-1}Q_\infty \subset Q_\infty$
3. $A_Q(T) = Q_\infty$

Since $A_Q(T)$ is the set of all points which stay inside Q under backward iteration, it follows immediately that

$$\mathrm{Inv}(Q) = A_Q(T) \cap A_Q(T^{-1}),$$

whenever T is invertible.

Remark 7. 1. Results on the speed of convergence can be obtained if $A_Q(T)$ possesses a hyperbolic structure, see [4]. Roughly speaking the stronger the contraction along the stable direction the better is the convergence behaviour.
2. The algorithm has been adapted and successfully applied to the context of *random dynamical systems* [21] by Keller and Ochs within this DFG research program (project of L. Arnold).

2.2 Practicalities: Relative Global Attractor

In GAIO, the selection criterion in step (ii) of Algorithm 1 is tested using a set of test points in each box. These test points are mapped forward one step and if at least one of these image points lies inside the box B, then B is not discarded. This procedure can be made rigorous (in the sense that $A_Q(T)$ is always covered by the box collection) through knowledge of local Lipschitz constants of T [19]; see Appendix A.

To perform one step of the algorithm with GAIO, one simply types:

```
rga(tree)
```

(the mnemonic `rga` is a short hand for "**r**elative **g**lobal **a**ttractor"), where `tree` is the data structure containing the current collection of boxes. In fact the boxes are stored in a binary tree, where the children of a box at depth k are the two boxes of half-size at depth $k+1$ formed by dividing the box at depth k into two equal pieces. This binary tree structure allows for rapid searching of which box contains the images of test points, and reduces the time for the subdivision procedure from $O(n^2)$ to $O(n \log n)$, where n denotes the number of boxes. Most of the algorithms we describe in this paper benefit significantly from the tree structure.

2.3 Example: Relative Global Attractor of a Knotted Flow

Consider a flow of a three-dimensional ordinary differential equation through an open-ended cylinder from top to bottom. On the mantle of the cylinder, the flow proceeds directly downwards. Near the center of the cylinder, some trajectories of the flow loop around to form a knot; see the red portion of Plate 4 on page 806. Using techniques from algebraic topology, it is possible to prove that there is a non-trivial invariant set contained inside the cylinder (see [7] for details). Obviously this invariant set is unstable, and therefore extremely difficult, if not impossible, to be observed numerically via simulation of single trajectories. Our set oriented approach is particularly suitable, as we cover the entire cylinder with coarse boxes and then repeatedly refine and discard boxes which are known not to contain a part of the invariant set.

In GAIO, one issues the commands

```
knot = Model('Knot')
rk4 = Integrator('RungeKutta4')
rk4.model = knot
rk4.h = 0.1
rk4.tFinal = 1.5

tree = Tree(knot.center, knot.radius)
tree.integrator = rk4
tree.domain_points = Points('Grid', knot.dim, 125)
tree.image_points = Points('Vertices', knot.dim)

rga(tree, 20)
```

The first block of commands loads the model file `Knot` into GAIO (see Appendix 5.3 on how to define your own model), and sets the parameters for the integration of the vector field defined in `Knot`. For instance here we use a fourth-order Runge-Kutta scheme in the numerical integration. The variable `rk4.h` determines the integration step-size, and `rk4.tFinal` indicates that we will treat the ODE as a 1.5-time discrete map (one iteration of the discrete map is defined by integration for 1.5 time units).

The second block of commands initialises the tree structure, and the set of test points that will be used in the selection step (ii) of Algorithm 1. Here we use a uniform grid of 125 test points (`domain_points`) in each box. Additionally we choose a set of `image_points` (given here by the vertices of a box); see Appendix 5.3 for an explanation on this.

We perform 20 subdivision steps on the initial collection given by the box with center `knot.center` and radius `knot.radius`. The covering of the corresponding backward invariant set consists of 267458 boxes; see Plates 1 and 2 on page 805. This has to be viewed as an approximation of the unstable invariant set together with its unstable manifold.

2.4 Algorithm: Chain Recurrent Sets

Often the maximal invariant set contains many "transient" points which we sometimes would like to eliminate. For instance, in Example 5, we would additionally like to know that the two points $x = 0$ and $x = 1$ are all of the recurrent points, rather than simply state the obvious fact that the entire interval $[0, 1]$ is invariant. We now show how to modify Algorithm 1 in such a way that we can approximate the chain recurrent set. As in Algorithm 1 we construct a sequence $\mathcal{B}_0, \mathcal{B}_1, \ldots$ of collections of boxes creating successively tighter coverings of the desired object.

Algorithm 2 (Subdivision Algorithm to Compute $R_Q(T)$). Set $\mathcal{B}_0 = \{Q\}$. For $k = 1, 2, \ldots$ the collection $\mathcal{B}_k$ is obtained from $\mathcal{B}_{k-1}$ in two steps:

1. *Subdivision:* Bisect each box in the current collection $\mathcal{B}_{k-1}$ into two smaller boxes of equal size (for d-dimensional boxes, the cutting plane is cycled around the d coordinate directions).
2. *Selection:* Construct a directed graph whose vertices are the boxes in the refined collection and by defining an edge from vertex B to vertex B', if

$$T(B) \cap B' \neq \emptyset. \tag{3}$$

   Compute the strongly connected components of this graph and discard all boxes which are not contained in one of these components.

Remark 8. Recall that a subset W of the nodes of a directed graph is called a *strongly connected component* of the graph, if for all $w, \tilde{w} \in W$ there is a path from w to $\tilde{w}$. The set of all strongly connected components of a given directed graph can be computed in linear time [26].

Intuitively it is plausible that the sequence of box coverings $\mathcal{B}_k$ converges to the chain recurrent set of T. Indeed, under mild assumptions on the box coverings one can prove convergence, see [11,27].

2.5 Example: Chain Recurrent Set of the Knotted Flow

We return to the previous example (the flow through an open-ended cylinder). The computations are prepared in the very same way as before. One step of Algorithm 2 is now performed by executing

```
crs(tree)
```

(where `crs` is meant to be an abbreviation for "**c**hain **r**ecurrent **s**et"). Here the `domain_points` and `image_points` of the tree are used to compute the directed graph in a way similar to the selection step of algorithm 1. The result after 18 subdivision steps is shown in Plate 3 on page 805, where the approximate chain recurrent set is shown in blue (11567 boxes), overlaying the relative global attractor. Plate 4 on page 806, which has been produced by Robert Strzodka, shows a covering of the chain recurrent set in dark blue after 30 subdivision steps.

3 Invariant Manifolds

The set oriented techniques may be applied to provide rigorous coverings of invariant manifolds within some prescribed box Q. For simplicity, we describe only the situation of unstable manifolds of hyperbolic fixed points of diffeomorphisms. However we emphasise that in principle the algorithm can be applied to general stable or unstable manifolds of arbitrary invariant sets.

Definition 9. Let x_0 be a hyperbolic fixed point for the diffeomorphism $T : M \to M$. Let U be a neighbourhood of x_0 and define the *local stable manifold* of x_0 by

$$W^s(x_0, U) = \{y : T^j y \in U \text{ for } j \in \mathbb{Z}^+ \text{ and } d(T^j y, x_0) \overset{j\to\infty}{\longrightarrow} 0\} \tag{4}$$

where $d(\cdot, \cdot)$ is a metric on M. The *local unstable manifold* is defined as

$$W^u(x_0, U) = \{y : T^{-j} y \in U \text{ for } j \in \mathbb{Z}^+ \text{ and } d(T^{-j} y, x_0) \overset{j\to\infty}{\longrightarrow} 0\}. \tag{5}$$

We have the following simplified version of the stable manifold theorem, see e.g. [28].

Theorem 10. *Let x_0 be a hyperbolic fixed point for the C^k diffeomorphism $T : M \to M$. Then there is a neighbourhood $U' \subset U$ such that the sets $W^s(U', x_0)$ and $W^u(U', x_0)$ are C^k embedded disks.*

The global stable *and* unstable manifolds *may be obtained by*

$$W^s(x_0) = \bigcup_{j\geq 0} T^{-j} W^s(x_0, U') \quad \textit{and} \quad W^u(x_0) = \bigcup_{j\geq 0} T^{j} W^u(x_0, U'), \tag{6}$$

respectively.

3.1 Algorithm: Invariant Manifolds

The rough idea behind covering the unstable manifold is as follows. Firstly, use GAIO to identify small regions containing fixed points (assuming that the fixed points are not known *a priori*). This can be done via a cycle of subdividing and throwing away all boxes whose image does not intersect itself. The remaining boxes cover all fixed points.

Once a fixed point has been located with sufficient precision, we apply Algorithm 1 to a small box containing the fixed point. Beginning with the obtained collection of boxes, these are mapped forward one iteration, and the boxes that they "hit" are added to the collection. These new included boxes are then mapped forward, and the procedure is repeated. In this way we obtain a covering (which can be made rigorous using Lipschitz estimates on the map as before; see Appendix A) of part of the global unstable manifold. Formally, the algorithm consists of two main steps:

Algorithm 3 (Continuation Algorithm to Compute $W^u(x_0)$).

1. *Initialisation:* Apply Algorithm 1 to a small box containing the hyperbolic fixed point x_0. Let the resulting collection be part of a partition of Q. Repeat the following step until no more boxes are added to the current collection:
2. *Continuation:* Map the obtained collection of boxes forward and note which other boxes of the partition are hit by these images. Add these boxes to the collection.

Remark 11.

1. It can be shown that Algorithm 3 indeed converges to part of the unstable manifold. For a detailed description of this convergence result see [3].
2. Recently there have also been results obtained on the speed of convergence in case where the unstable manifold is contained in a hyperbolic attractor, see [18]. As expected like in the case of the Subdivision Algorithm 1 the speed of convergence crucially depends on the contraction rate along the stable direction.
3. Observe that in the realisation it is not necessary to partition Q a priori into small boxes. Rather we use the same hierarchical data structure as for Algorithm 1 and just add leaves to the tree when the corresponding boxes are hit.
4. In order to cover the stable manifolds, the continuation algorithm may be applied to the inverse map T^{-1}.

3.2 Practicalities

To initialise the computations in GAIO we construct a single small box B around the fixed point x_0 within the tree data structure:

```
tree.insert(x_0, depth)
```

Here `depth` specifies at which depth of the tree the box will be generated. A higher depth corresponds to a smaller box. We may then apply Algorithm 1 in order to obtain a covering of the local unstable manifold of x_0 in B:

```
steps = 6
rga(tree, steps)
```

Finally we prepare the current collection

```
inserted = 2
tree.set_flags('all', inserted)
```

and apply several steps of Algorithm 3:

```
steps = 4
gum(tree, tree.depth, steps)
```

(here the mnemonic gum abbreviates "global unstable manifold"). The above `tree.set_flags` command (which just "marks" all boxes in the current collection with the flag "2") is necessary as a preparatory step since only newly inserted boxes are mapped forward in each step of the procedure gum.

3.3 Example: Computation of a Stable Manifold in the Lorenz System

We consider the problem of covering the two-dimensional stable manifold of the origin for the Lorenz system governed by the system of ODE's:

$$\begin{aligned}\dot{x} &= \sigma(y - x)\\ \dot{y} &= \rho x - y - xz\\ \dot{z} &= xy - \beta z\end{aligned}$$

with $\sigma = 10, \rho = 28$, and $\beta = 8/3$.

Plate 5 on page 806 was produced with the following commands:

```
lorenz = Model('lorenz')

rk4 = Integrator('RungeKutta4')
rk4.model = lorenz
rk4.tFinal = -0.1
rk4.h = -0.01

tree = Tree([0, 0, 0], [120, 120, 160])
tree.integrator = rk4
tree.domain_points = Points('Edges', 3, 100)
tree.image_points = Points('Center', 3)

x = [0; 0; 0]
depth = 21
tree.insert(x, depth)

steps = 10
gum(tree, depth, steps)
```

As usual we load the model, define the integration scheme and set up the tree object. Note that `rk4.tFinal = -0.1` and `rk4.h = -0.01`; so we integrate the ODE backwards in time. The unstable manifold of the time-reversed system is equal to the stable manifold of the forward time system. In the second block of commands we insert the box containing the origin into the tree at depth 21. Then we apply 10 steps of the continuation algorithm. Note that here we do not need to issue the `tree.set_flags` command since we inserted a single box into the tree and did not perform the subdivision algorithm on this box.

3.4 Example: Rigorous Covering of an Unstable Manifold of the Hénon System

The algorithm for generating a rigorous covering of invariant manifolds will be illustrated with the Hénon mapping $T : \mathbb{R}^2 \to \mathbb{R}^2$, given by $T(x, y) = (1 - ax^2 + y, bx)$, with $a = 1.0$ and $b = 0.54$. Using the outer box $Q = [-2.2, 3.8] \times [-2.6, 3.4]$ the lightly shaded boxes in Figure 1 were produced by the following commands:

```
henon = Model('henon')
henon.a = 1.0
henon.b = 0.54

map = Integrator('Map')
map.model = henon

tree = Tree(henon.center, henon.radius)
tree.integrator = map
tree.domain_points = Points('Lipschitz', 2)
tree.image_points = Points('Vertices', 2)

depth = 16
x = henon.fixed_point
tree.insert(x, depth)
```

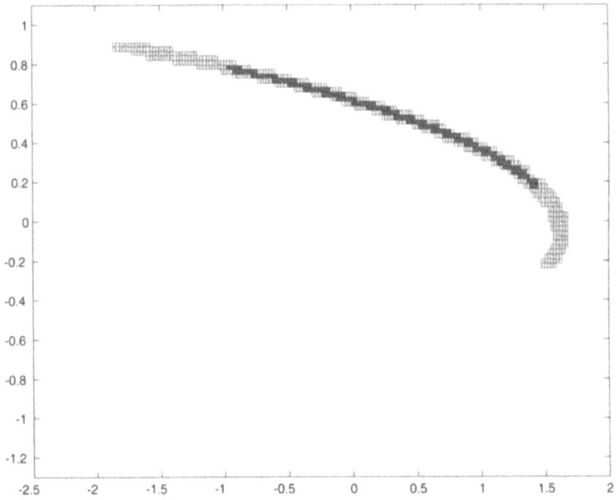

Fig. 1. Rigorous (light) and non-rigorous (dark) continuation applied to cover the unstable manifold of one of the fixed points of the Hénon map (6 continuation steps).

```
steps = 6
gum(tree, depth, steps)
```

Note the use of the point styles `Lipschitz` and `Vertices`; this option tells GAIO that the rigorous box intersection procedure (see Appendix A) is to be used. We insert the fixed point of the Hénon map into the tree at depth 16 and perform six continuation steps to extend the manifold.

The dark boxes in Figure 1 were produced by repeating the commands given above, but replacing the commands

```
tree.domain_points = Points('Lipschitz', 2)
tree.image_points = Points('Vertices', 2)
```

by

```
tree.domain_points = Points('Edges', 2, 100)
tree.image_points = Points('Center', 2)
```

yielding a non-rigorous computation of box intersections. Figure 2 shows both manifold coverings (rigorous and non-rigorous) extended by a further three steps, making a total of nine continuation steps. Finally, we perform the rigorous continuation method at depth 24 for a total of 19 steps; the result is shown in Figure 3.

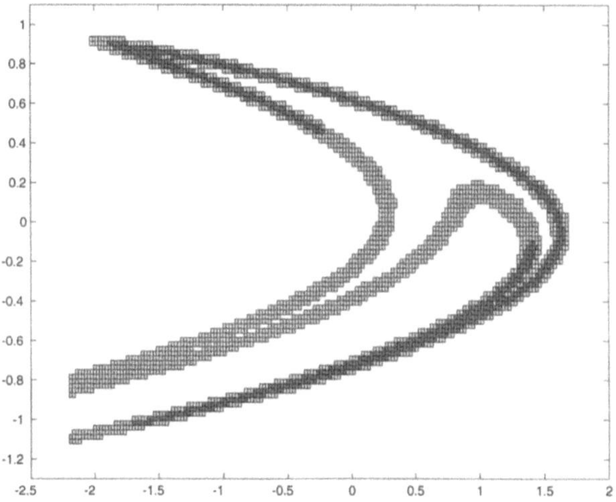

Fig. 2. Rigorous (light) and non-rigorous (dark) continuation applied to cover the unstable manifold of one of the fixed points of the Hénon map (9 continuation steps).

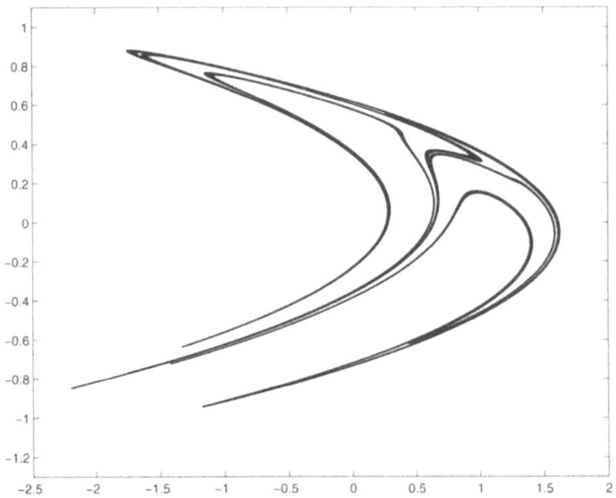

Fig. 3. The rigorous continuation algorithm applied to cover the unstable manifold of a fixed point of the Hénon map: 19 continuation steps at depth 24.

4 Invariant Measures

An invariant measure describes the distribution of points on long trajectories, with regions that are visited more often being given higher "weight" or measure. Deterministic dynamical systems typically support many invariant measures. Under mild conditions on the dynamical system, it may be shown that by adding smooth localised dynamical noise, the resulting system has a unique invariant measure. Numerically, this "noisy" measure appears to be similar to the distribution of long trajectories for the original system for a large set of initial points in a neighbourhood of the chain recurrent set. In fact there are results [23] that prove that this is true for certain types of noise added to uniformly hyperbolic diffeomorphisms.

Definition 12. A probability measure μ on M is called *T-invariant* if $\mu \circ T^{-1} = \mu$. We are particularly interested in the situation where there is an invariant measure μ with the property that

$$\#\{0 \leq k \leq N-1 : T^k x \in A\}/N \to \mu(A) \qquad \text{as } N \to \infty \tag{7}$$

for every measurable $A \subset M$ and for Lebesgue almost all x in a neighbourhood of the chain recurrent set. Such an (obviously unique) invariant measure will be called a *physical measure* or *natural invariant measure* for T.

One iteration of our noisy system will be an application of T followed by a small perturbation; that is, $x \mapsto Tx + e$, where $e \in \mathbb{R}^d$ is small and of order ϵ.

We formalise this by considering the noisy process to be a Markov chain. This Markov chain is completely defined by a transition function $\mathbf{Q}(\cdot,\cdot) : M \times \mathcal{B}(M) \to [0,1]$, where $\mathcal{B}(M)$ denotes the collection of Borel sets on M.

Example 13. 1. Let $B(x,\epsilon)$ denote the density of the uniform distribution restricted to $B_\epsilon(x)$, an ϵ ball about x. If $x \mapsto Tx + e$ with e selected from the density $B(0,\epsilon)$, then the corresponding transition function is $\mathbf{Q}_\epsilon(x,A) = \int_A B(Tx,\epsilon)\, dm(x)$, where m is Lebesgue measure. Similarly if $B(x,\epsilon)$ is replaced by the multidimensional Gaussian distribution $G(x,\epsilon)$ with mean x and variance ϵ then the corresponding transition function is $\mathbf{Q}_\epsilon(x,A) = \int_A G(Tx,\epsilon)\, dm(x)$.

2. The transition function $\mathbf{Q}(x,A) = 1$ if $Tx \in A$ and $\mathbf{Q}(x,A) = 0$ otherwise, describes the (deterministic) Markov chain corresponding to the unperturbed map T.

Definition 14. Let $\mathcal{M}(M)$ denote the space of Borel probability measures on M. Given a transition function $\mathbf{Q}_\epsilon$, we may define a linear operator on $\mathcal{M}(M)$ that describes how probability measures are "pushed forward" under one step of the Markov chain. Define

$$(\mathcal{P}_\epsilon \nu)(A) = \int_M \mathbf{Q}_\epsilon(x,A)\, d\nu(x) \tag{8}$$

where $\nu \in \mathcal{M}(M)$. In cases where we use the deterministic transition function $\mathbf{Q}$ of Example 13 (ii), we denote the deterministic operator by $\mathcal{P}$ and call it the *Perron-Frobenius operator.* The operators $\mathcal{P}_\epsilon$ will be called *noisy Perron-Frobenius operators.*

Theorem 15 ([24]). *Suppose that for all $x \in M$, $\mathbf{Q}_\epsilon(x,\cdot)$ is an absolutely continuous probability measure. Let $\mathbf{Q}_\epsilon^{(k)}$ denote the transition function for k steps of the Markov chain. If, additionally, there exists a k_0 such that the probability measure $\int_M \mathbf{Q}_\epsilon^{(k_0)}(x,\cdot)\, dm(x)$ has a strictly positive density, then $\mathcal{P}_\epsilon$ has a unique fixed point, denoted μ_ϵ. Furthermore μ_ϵ is an everywhere positive absolutely continuous probability measure, and is the unique invariant measure for the noisy system.*

Remark 16. As an example, one can choose

$$\mathbf{Q}_\epsilon(x,A) = \mathcal{N}(x) \int_A \exp(-\|Tx - y\|^2/2\epsilon)\, dm(y),$$

where $\mathcal{N}(x)$ is a normalising factor. Such a class of perturbations are considered in [31], and this provides a very readable introduction to this "noising-up" approach. In this case, $k_0 = 1$ in the above theorem.

Theorem 15 holds for very general transition functions $\mathbf{Q}_\epsilon$; they do not need to be connected in any way to a deterministic mapping T. However, we are implicitly considering only those $\mathbf{Q}_\epsilon$ which define Markov chains whose dynamics is "close" to that of T. We now make this precise.

Definition 17 ([22,23]). We will say that a family of transition functions $\{\mathbf{Q}_\epsilon\}_{\epsilon>0}$ represents a *small random perturbation* of a continuous map T if

$$\mathbf{Q}_\epsilon(x,\cdot) \to \delta_{T(x)} \quad \text{weakly as } \epsilon \to 0 \tag{9}$$

uniformly in x. Here δ_y denotes the Dirac measure at y.

Theorem 18 ([22,23]). *Suppose that $\mathbf{Q}_\epsilon$ represents a small random perturbation of T and satisfies the conditions of Theorem 15. Let $\tilde{\mu}$ be a weak limit of $\{\mu_\epsilon\}$ as $\epsilon \to 0$, where a subsequence is selected if necessary. Then $\tilde{\mu}$ is T-invariant.*

In the case where $\mathbf{Q}_\epsilon(x,\cdot)$ is absolutely continuous for all x our noisy Perron-Frobenius operators may be considered as operators on L^1. They map the space of *densities* $D = \{f \in L^1 : f \geq 0, \int_M f\, dm = 1\}$ into itself. One may think of D as representing all absolutely continuous probability measures. For technical reasons, it is often advantageous for $\mathcal{P}_\epsilon$ to be a compact operator on $L^1(M,m)$, the space of integrable functions on M. Under some further mild conditions on $\mathbf{Q}_\epsilon$, this is also true.

Theorem 19 ([18]). *Suppose that $\mathbf{Q}_\epsilon(x,\cdot)$ has a Lipschitz density for all $x \in M$. Then $\mathcal{P}_\epsilon$ is compact as an operator on L^1.*

4.1 Algorithm: Natural Invariant Measures

The box coverings introduced earlier will form the backbone of our finite-dimensional approximation of the infinite-dimensional operator $\mathcal{P}$. More precisely, for a given box collection $\{B_1,\ldots,B_n\}$ we form the (column stochastic) transition matrix

$$P_{ij} = \frac{m(B_j \cap T^{-1}B_i)}{m(B_j)}, \tag{10}$$

$i,j = 1,\ldots,n$. One computes the (assumed, unique) fixed right eigenvector p of $P = (P_{ij})$, representing the invariant distribution for the finite-state Markov chain defined by P. An approximate invariant measure μ_n is defined by assigning $\mu_n(B_i) = p_i$.

Algorithm 4 (Computation of Natural Invariant Measures).
1. Compute the matrix P above.
2. Find the (right) Perron eigenvector p of P.
3. Set $\mu_n(B_i) = p_i$, $i = 1,\ldots,n$.

In general for the deterministic case it is not clear whether or not this measure μ_n is a good approximation of the physical measure. However under certain assumptions it can be shown that the measures μ_n indeed converge to a natural invariant measure, see e.g. [25,13,14,9]. Moreover the following convergence result for the stochastically perturbed context has been shown in [6]. Here we denote by μ_n^ϵ the approximate invariant measure obtained by computing the Perron eigenvector of the transition matrix for the stochastically perturbed system.

Theorem 20. *Suppose that the diffeomorphism T has a hyperbolic attractor Λ, and that there exists an open set $U_\Lambda \supset \Lambda$ such that for the densities q_ϵ of the transition functions $\mathbf{Q}_\epsilon(x, A) = \int_A q_\epsilon(Tx, y)\, dm(y)$ we have*

$$q_\epsilon(x, y) = 0 \quad \textit{if } x \in \overline{T(U_\Lambda)} \textit{ and } y \notin U_\Lambda.$$

Then the transition function $\mathbf{Q}_\epsilon$ has a unique invariant measure μ_ϵ with support on Λ and the approximating measures μ_n^ϵ converge to the natural measure μ of T as $\epsilon \to 0$ and $n \to \infty$,

$$\lim_{\epsilon \to 0} \lim_{n \to \infty} \mu_n^\epsilon = \mu.$$

4.2 Practicalities

The crucial algorithmic step is to compute the transition probabilities between boxes, i.e. the entries of the matrix P. In GAIO, this is carried out in two ways:

Computation of P Using Test Points The first method is to select a collection of m test points within each box. The points $\{x_1, \ldots, x_m\} \in B_j$ are mapped forward by T and we set

$$P_{ij} = \#\{x \in \{x_1, \ldots, x_m\} : Tx \in B_i\}/m.$$

For example, to use a set of m points distributed randomly according to a uniform distribution in each box one would use:

```
mc = Points('MonteCarlo', dim, m)
P = tree.matrix(mc, depth)
```

where `dim` denotes the dimension of phase space and `depth` specifies on which depth of the tree the transition matrix is to be computed. The variable `depth` may also be set to -1 in which case the transition matrix is computed on the leaves of the tree. This is useful in situations where we use adapted partitions (i.e. not all of the boxes are of the same size).

Computation of P Using an Exhaustion Technique The second method is to use an approach called "exhaustion", similar to the exhaustion techniques pioneered by Eudoxus [12]. To estimate the d-dimensional volume of $B_j \cap T^{-1} B_i$, the box B_j is repeatedly subdivided into smaller boxes until the forward image of a smaller box is known to fit completely inside B_i. This criterion is tested on the basis of Lipschitz estimates on the right hand side of the underlying model. At this point, subdivision of the sub-box stops. We also stop subdividing the sub-box when its volume has decreased beyond a certain threshold. A complete description of this method can be found in [16].

In GAIO, the transition matrix is created via the command:

```
P = tree.matrix('exhaustion', depth [, err])
```

In this command, the optional integer `err` is related to the volume threshold mentioned before. The subdivision of a sub-box stops, when its volume is smaller than $2^{-\mathtt{err}}$ times the volume of B_j. The default value of `err` is 16.

Computing Eigenvalues and Eigenvectors of *P* To compute the fixed right eigenvector of the transition matrix P in GAIO, one types:

```
[v, l] = eigs(P, 1)
```

and the variable `l` will contain the largest real eigenvalue of P, with `v` containing the corresponding eigenvector.

Adaptive Partitioning Schemes So far, we have described how to compute transition matrices on a certain depth of the tree. Usually the corresponding collection will have been obtained by one of the algorithms described so far. They always lead to coverings with boxes of equal size. However, using information from the approximate invariant measure, it is possible to produce more efficient partitioning schemes. There are different strategies of how to use the information from the invariant measure, however, the basic algorithm has the following structure:

Algorithm 5 (Adaptive Subdivision Algorithm). From a box collection $\mathcal{B}_{k-1}$ and a corresponding approximate invariant measure μ_{k-1} compute $\mathcal{B}_k$ and μ_k in two steps:

1. *Subdivision*: Based on information from the approximate invariant measure, identify boxes which should be subdivided. Bisect each of those boxes into two smaller boxes of equal size (for d-dimensional boxes, the cutting plane is cycled around the d coordinate directions).
2. *Selection*: Compute the transition matrix and approximate the invariant measure μ_k for the refined box collection. Discard boxes for which the approximate measure is zero. The resulting collection constitutes $\mathcal{B}_k$.

We are now going to explain different identification procedures for step 1 of Algorithm 5 which are available in GAIO. For example one may

(a) subdivide boxes B for which $\mu_{k-1}(B) > 1/n$, where n is the number of boxes in $\mathcal{B}_{k-1}$;

(b) estimate a local approximation error from the (piecewise constant) density of the approximate invariant measure. Refine those boxes for which this estimated local error exceeds its average over all boxes (see [17] for a description of this approach).

The convergence of these adaptive schemes is analysed in detail in [18].

In GAIO, adaptive algorithm (a) is accessible via the commands

```
aim_hm(tree, method)
```

which performs one step of Algorithm 5 using subdivision procedure (a). The variable `method` determines which of the two above described methods is used in order to compute the transition matrices: it can be either a points object or the string `'exhaustion'`.

```
aim_lip(tree, method)
```

implements subdivision procedure (b). Note that for this approach to make sense we need a rather accurate result for the approximate invariant density, so in most cases one would exclusively choose the `method` to be `'exhaustion'` here.

4.3 Example: Bouncing Ball

We consider a discrete dynamical system that models a ball bouncing on a sinusoidally forced table. The approximate equations of motion are given by:

$$\begin{aligned} \phi_{t+1} &= \phi_t + v_t, \\ v_{t+1} &= \alpha v_t - \gamma \cos(\phi_t + v_t), \end{aligned}$$

with $\alpha = 0.9$ and $\gamma = 16$. Here $\phi_t \in [0, 2\pi)$ denotes the phase of the table at impact $\#t$, and $v_t \in \mathbb{R}$ denotes the exit velocity of the ball at impact $\#t$. Figure 4 shows a plot of a numerical orbit of length 10^5 with initial conditions $\phi_0 = 0$, $v_0 = 0$; that is, the points $\{(\phi_t, v_t)\}_{t=0}^{10^5}$ have been plotted on the phase space $M = S^1 \times [-100, 100]$.

Darker regions of Figure 4 contain more points, and therefore are visited more frequently by the numerical trajectory than lighter regions. Our estimate of the physical invariant measure will approximate the long term distribution of points in Figure 4 that arises in the $t \to \infty$ limit. Figure 5 is a gray scale plot of the density of an approximate invariant measure computed using the techniques described above.

Plate 6 on page 807 shows the graph of this density in three dimensions. It can clearly be seen that this invariant density has a certain spatial structure which is reminiscent of the *symmetry on average* of attractors as described in [2].

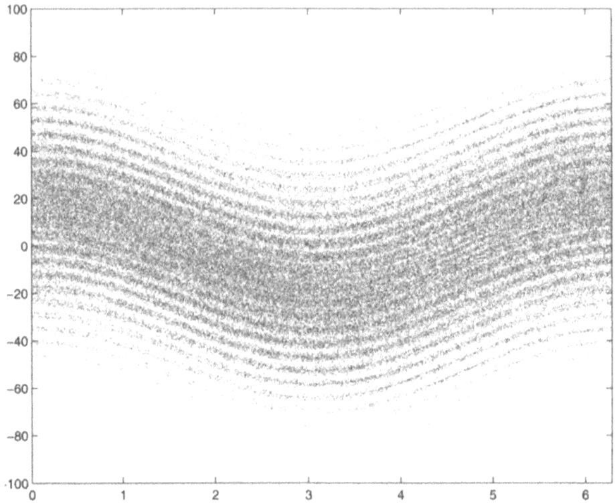

Fig. 4. A numerical trajectory of length 10^5 for the bouncing ball system.

To compute this approximation, we issued the following commands:

```
bb = Model('bouncingball')

map = Integrator('Map')
map.model = bb

tree = Tree(bb.center, bb.radius)
tree.integrator = map

to_be_subdivided = 8
for i=1:15
  tree.set_flags('all', to_be_subdivided)
  tree.subdivide(to_be_subdivided)
end

ig = Points('InnerGrid', 2, 1000)
P = tree.matrix(ig)
[v, l] = eigs(P, 1)
```

Note that because we subdivide *all* boxes at each step, we do not set the domain and image points in the first block (we don't need to worry about computing intersections of boxes). The decision to subdivide all boxes is only a little inefficient (from the data storage point of view) in this instance, as the positive density region occupies most of the phase space. We chose

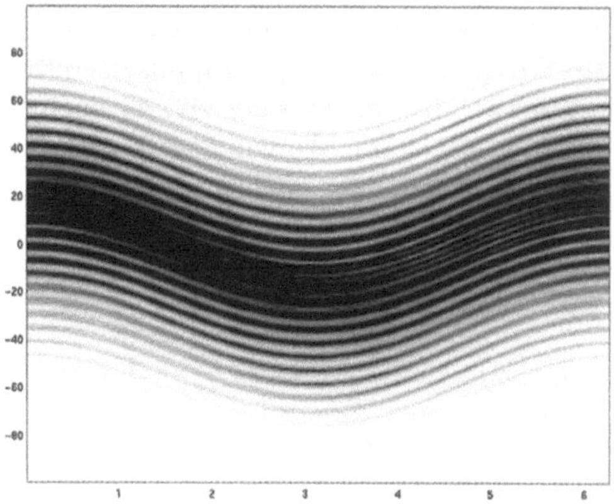

Fig. 5. Density of an approximate invariant measure using 65536 boxes (16 subdivisions). Darker areas correspond to higher density.

to subdivide all boxes purely because it is then easier to use the MATLAB visualisation function used to produce Plate 6 on page 807. Normally instead of subdividing all boxes we would use one of the Subdivision Algorithms 1, 2 or 5.

In the for-loop we repeatedly mark all boxes with the flag "8" and then subdivide all boxes which have this flag set (i.e. all boxes are subdivded). Once the box collection at depth 16 has been produced by the for-loop, we set the variable `ig` to determine how to select test points for the computation of the transition matrix P. In this example, we use 1000 points per box, arranged in a uniform grid (set slightly away from the boundary of the boxes, as indicated by `InnerGrid`). We then generate the $2^{16} \times 2^{16}$ sparse matrix P and compute its Perron eigenvector as the approximate invariant measure.

5 Almost Invariant Sets and the Isolated Spectrum

Often one observes that in transitive systems, there are regions in which orbits stay for very long times before moving to other regions, only to return some longer time later. A well-known example are the two "wings" of the Lorenz attractor. Trajectories tend to stay on each wing for quite a long time before switching to the other wing. It is not just purely of dynamical interest to identify such "almost invariant sets"; the concept of "almost invariance" has recently also successfully been used for the identification of conformations for molecules (see [8,30]).

We return to the operator $\mathcal{P}$ and consider its action on $\mathcal{M}_{\mathbb{C}}(M)$, the space of (possibly complex-valued) Borel measures (recall that $\mathcal{P}$ is defined by equation (8) using $\mathbf{Q}$ defined in Example 13 (ii)). Suppose that there is a real eigenvalue $0 < \lambda < 1$, with corresponding real eigenmeasure $\nu \in \mathcal{M}_{\mathbb{C}}(M)$ such that $\mathcal{P}\nu = \lambda\nu$. We assume that ν is normalised so that $|\nu|(M) = 1$ (this means that there are two disjoint subsets A_1, A_2 such that $\nu(A_1) = 1/2$, $\nu(A_2) = -1/2$, and $A_1 \cup A_2 = M$). We will say that the two disjoint subsets A_1, A_2 partition M into two *almost-invariant sets*. The following result [6] lends weight to this assertion.

Theorem 21. *Define*

$$\delta = \nu(A_1 \cap T^{-1}A_1)/\nu(A_1) \quad \textit{and} \quad \sigma = \nu(A_2 \cap T^{-1}A_2)/\nu(A_2).$$

Then $\delta + \sigma = \lambda + 1$.

The numbers δ and σ represent the amount of ν-mass that stays inside A_1 and A_2, respectively, under one iteration of T. Thus, if A_1 and A_2 are close to being invariant, δ and σ should be both close to one; this implies that λ is close to one also. If almost all of the ν-mass leaves A_1 in one iteration (and likewise for A_2), then both δ and σ should be close to zero; and λ should be close to -1. Thus λ close to -1 suggests that the two sets A_1 and A_2 form part of an *almost two-cycle*. Further generalisations are possible, and identical results hold for the noisy operator $\mathcal{P}_\epsilon$, see [6,15] for a detailed description.

5.1 Algorithm: Almost Invariant Sets

The transition matrix P constructed in the previous section provides a discrete approximation of the smooth dynamics. For motivational purposes let us suppose that the transition matrix is reducible, in the sense that there are two invariant subspaces V_1 and V_2 of dimension $n - r$ and r respectively, and that P restricted two each of V_1 and V_2 is irreducible. The eigenvalue one for the matrix P has geometric multiplicity two, and the spaces V_1 and V_2 can be extracted from the two fixed vectors of P (by using information on the positive and negative parts of the eigenvectors); see [6].

Suppose now that the system is perturbed, resulting in one of the eigenvalues moving away from one (one of the eigenvalues must stay at unity because the matrix remains stochastic). Then the matrix P becomes irreducible, but still close to being reducible. By continuity, the positive and negative parts of the eigenvectors corresponding to the second eigenvalue (the eigenvalue that moved away from unity) will approximate invariant sets. We call these *almost-invariant sets*.

To search for almost invariant sets, we look for eigenvalues of $\mathcal{P}$ (or $\mathcal{P}_\epsilon$) close to unity (or close to -1 when looking for almost two-cycles). We assume that the eigenmeasures and eigenvalues of $\mathcal{P}$ are well-approximated by eigenvectors and eigenvalues of the matrix P; at least for eigenvalues close to the unit circle. Such good approximation has been made precise for the noiseless operator $\mathcal{P}$ in one-dimension [20] and also for the compact operator $\mathcal{P}_\epsilon$

in higher dimensions [6]. Moreover, the existence of isolated eigenvalues and their dynamical relevance has recently been analytically studied for a certain class of one-dimensional maps, [1].

The algorithm for the computation of almost invariant sets is summarised below in the case where the second largest eigenvalue of P (in magnitude) is positive and real. The case of a large negative eigenvalue – leading to an almost invariant two-cycle – can be treated analogously.

Algorithm 6 (Computation of Almost Invariant Sets).

1. Compute the eigenvector v corresponding to the second largest real eigenvalue of P.
2. Create two index sets $\mathcal{I}_1 = \{i \in \{1,\dots,n\} : v_i \geq 0\}$ and $\mathcal{I}_2 = \{i \in \{1,\dots,n\} : v_i < 0\}$.
3. Denote the ith box by B_i. The box collections $\widetilde{A}_1 := \bigcup_{i\in\mathcal{I}_1} B_i$ and $\widetilde{A}_2 := \bigcup_{i\in\mathcal{I}_2} B_i$ approximate A_1 and A_2 in Theorem 21.

Note that the construction of $\mathcal{I}_1$ and $\mathcal{I}_2$ in step 2 is somewhat arbitrary, see [15] for a more detailed exposition on this.

5.2 Practicalities

In GAIO, we calculate the matrix P as described in §4.2. To find the s largest (in absolute value) eigenvalues, we type

```
[v, lambda] = eigs(P, s)
```

5.3 Example: Chua's Circuit

We consider the set of differential equations

$$\begin{aligned}\dot{x} &= \alpha(y - m_0 x - (1/3)m_1 x^3)\\ \dot{y} &= x - y + z\\ \dot{z} &= -\beta y\end{aligned}$$

with the parameter values $\alpha = 16, \beta = 33, m_0 = -0.2$, and $m_1 = 0.01$.

A covering of the unstable manifold of one of the fixed points is first computed using the continuation algorithm. There is numerical evidence that this leads to a covering of an attracting set of the underlying ordinary differential equation.

```
chua = Model('chua')

rk4 = Integrator('RungeKutta4')
rk4.model = chua

tree = Tree(chua.center, chua.radius)
```

```
tree.integrator = rk4
tree.domain_points = Points('Edges', 3, 100)
tree.image_points = Points('Center', 3)

depth = 21
x = chua.fixed_point
tree.insert(x, depth)

steps = 100
gum(tree, depth, steps)
```

We load the model, set the integrator to be used, and define the points to be used in the intersection tests. We then insert a single box containing one of the fixed points into the tree data structure at depth 21, and apply the continuation algorithm for 100 steps (at most; the routine gum will stop if no more boxes are added to the collection).

We compute the transition matrix

```
ig = Points('InnerGrid', 3, 1000)
P = tree.matrix(ig)
```

and using the eigs command described above, we compute the leading eigenvalues of P, and find there is a positive, real eigenvalue close to one. The corresponding eigenvector v of this second largest eigenvalue has positive and negative parts and $\sum_i v_i = 0$; we consider v as a discrete approximation of the signed measure ν introduced earlier. We choose zero as a "separator", and consider the two sets of indices $\mathcal{I}_1 = \{i \in \{1, \dots, n\} : v_i \geq 0\}$ and $\mathcal{I}_2 = \{i \in \{1, \dots, n\} : v_i < 0\}$ to partition our box-covering into two almost invariant pieces, approximating the two sets A_1 and A_2. It can be numerically verified that the two sets of indices I_1 and I_2 satisfy a probabilistic version of Theorem 21 (replacing the action of T with the matrix P, and the eigenmeasure ν with the eigenvector v). These two sets are shown in red and blue, respectively, in Plate 7 on page 807.

Acknowledgement

The authors would like to thank Michaela Schlör, Stefan Sertl, and Bianca Thiere for assistance in creating the example figures.

They gratefully acknowledge Kathrin Padberg's assistance in implementing the MATLAB interface to GAIO.

Plates 1-5 on page 805 ff. have been produced using the software platform GRAPE.

Appendix A: Rigorous Calculation of Box Intersections

Because a finite number of test points is used to compute intersections of sets in Algorithms 1, 2 and 3, it is possible to miss some intersections, and therefore possibly not have a complete covering of an invariant set or manifold. By using information on the Lipschitz constants of the mapping (for simplicity we describe only the discrete-time case), it is possible to produce a rigorous covering of an invariant set or manifold, in the sense that the invariant object is completely contained inside the resulting collection of boxes.

The problem is as follows: Given a box B in a collection $\mathcal{B}$, find all boxes in $\mathcal{B}$ that intersect $T(B)$. Roughly speaking, the solution is this: Choose a finite grid of test points $\{x_1, \dots, x_q\}$ in B in such a way that the distance between the images $T(x_i)$ and $T(x_j)$ of two neighbouring points x_i and x_j is less than the diameter of the boxes in $\mathcal{B}$. The set of all boxes in $\mathcal{B}$ intersected by the union of box-sized neighbourhoods centered at each $T(x_i)$ $(i = 1, \dots, q)$ is then guaranteed to contain the collection of all boxes intersecting $T(B)$.

In order to impose an upper bound on the distance between $T(x_i)$ and $T(x_j)$, one needs information on the Lipschitz constants of T restricted to B. Let $r \in \mathbb{R}^d$ denote the radius of the boxes in $\mathcal{B}$ (recall that all boxes are of the same size). We need to know a $d \times d$ matrix L such that $|T(x)-T(y)| \leq L|x-y|$ for $x, y \in B$. If T is differentiable, then $L_{ij} = \max_{\xi \in B} |\partial_j T_i(\xi)|$ does the job, where T_i is the ith component map of T. We now arrange test points in B lying on a d-dimensional grid $G = \{x \in B : x_i - c_i \in h_i \mathbb{Z}, i = 1, \dots, d\}$, where $c = (c_1, \dots, c_d)$ denotes the center of the box B and the vector $h \in \mathbb{R}^d$, $h > 0$, satisfies $Lh \leq 2r$.

Proposition 22. *Let* $\mathbf{B}(x,r)$ *denote a box of radius* r *centered at* $x \in M$. *Define the box collection* $\widehat{T(B)} = \{B' \in \mathcal{B} : B' \cap \bigcup_{x \in G} \mathbf{B}(Tx, r) \neq \emptyset\}$. *Then* $\{B' \in \mathcal{B} : B' \cap T(B) \neq \emptyset\} \subset \widehat{T(B)}$.

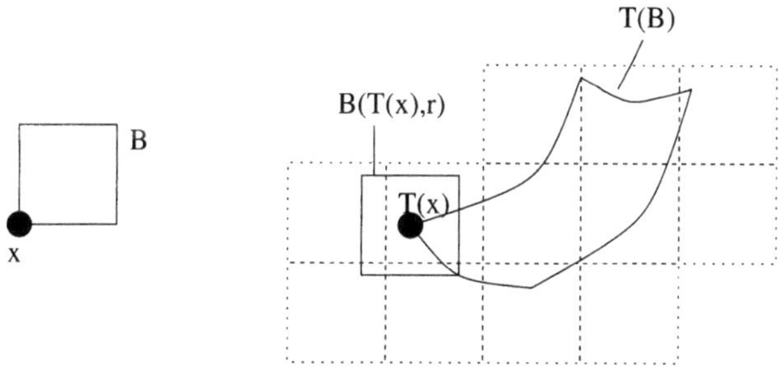

Fig. 6. Illustration of Proposition 22.

The collection $\widehat{T(B)}$ is constructed by placing a box of radius r, centered at every Tx, $x \in G$, and checking which boxes in $\mathcal{B}$ are intersected by these boxes, see also Figure 6. This check is very simple, as one only needs to include all boxes into $\widehat{T(B)}$ which contain a vertex of the boxes $B(T(x_i), r)$, $i = 1, \dots, q$.

In GAIO this construction is realized by a special choice of the test points which have to be assigned to a `Tree` object. Recall that for the non-rigorous approach one may chose

```
tree.domain_points = Points('Edges', dim, m)
tree.image_points = Points('Center', dim)
```

where `dim` denotes the dimension of phase space and m the suggested number of points. Now for the rigorous approach we instead use the commands

```
tree.domain_points = Points('Lipschitz', dim)
tree.image_points = Points('Vertices', dim)
```

An alternative, even more efficient way of constructing a grid of test points is to align the points within B along directions given by the right singular vectors of the matrix L, with the components of h directly related to the singular values. This approach reduces the required number of test points for the determination of a covering of the image of B under T. For a detailed description of the methods mentioned in this section see [18].

Appendix B: Example Model Files

Each model is defined in a single C–file which is compiled into object code by a C–compiler and then transformed into a shared object by the linker of the machine. This shared object can then be loaded into GAIO.

We give two example model files to show how one produces a model in practice. There are two main situations; the first is where our dynamical system is governed by a discrete map, and the second is where our system is a flow generated by a vector field. In the former case, the map will form the "right hand side" in our model files, while in the latter case, it is the vector field.

B.1 Map

The following is the code of `henon.c`, a C–file that produces a model file for the Hénon map.

```
char *name = "My Henon map";
char *typ = "map";
int   dim  = 2;
int   paramDim = 2;
char *paramNames[] = { "a", "b" };
double a = 1.3, b = 0.2;
double c[2] = { 0, 0 };
double r[2] = { 3, 3 };
double tFinal = 1;
```

```
void rhs(double *x, double *u, double *y) {
  y[0] = 1 - a*x[0]*x[0] + x[1];
  y[1] = b*x[0];
}
```

Except for a and b, which are obviously specific to the Hénon map, these are the variables and function(s) that have to present in each model file. The file now has to be compiled into object code by the command

```
cc -c henon.c
```

and the resulting object file henon.o has to be converted into a shared object by issuing

```
ld -shared -o henon.so henon.o
```

Note that the linker flag -shared is platform dependent. Some specific choices are given in the following table:

Linux	-shared
Solaris	-G
OSF or Irix	-shared -expect_unresolved

Additionally one has to provide a Lipschitz estimate on the right hand side if one wants to use the Lipschitz test points as mentioned in section 2.2 and explained in Appendix 5.3. The corresponding function must be called lip as in the following example for the Hénon map.

```
#define max(x,y) (x>y ? x : y)
void lip(double *c, double *r, double *L) {
  L[0] = 2.0*fabs(a)*max(fabs(c[0] + r[0]), fabs(c[0] - r[0]));
  L[1] = 1.0;
  L[2] = fabs(b);
  L[3] = 0.0;
}
```

Finally it is possible to additionally supply some special point in state space via the function fixed_point. In most cases this will be a fixed point of the underlying dynamical system. For the Hénon map we define:

```
#include <math.h>    /* defines sqrt() */

void fixed_point(double *x) {
  double t = (b-1)/(2*a);
  x[0] = t + sqrt(t*t + 1/a);
  x[1] = b*x[0];
}
```

B.2 Flow

The following is a C–file that produces a model file for the Lorenz flow, with three free parameters.

```
#include <math.h>

char *name = "The Lorenz system";
char *typ = "ode";
int   dim  = 3;
int   paramDim = 3;
char *paramNames[] = { "sigma", "rho", "beta" };
double  sigma = 10, rho = 28, beta = 2.666666666;
double c[3] = { 0, 0, 27 };
double r[3] = { 30, 30, 40 };
double tFinal = 0.2;

void rhs(double *x, double *u, double *y) {
  y[0] = sigma*(x[1]-x[0]);
  y[1] = rho*x[0] - x[1] - x[0]*x[2];
  y[2] = x[0]*x[1] - beta*x[2];
}

void fixed_point(double *x) {
  x[0] = sqrt(beta*(rho-1));
  x[1] = x[0];
  x[2] = rho-1;
}
```

Appendix C: Frequently Used GAIO Commands

Here we give a short summary of more frequently used commands. A complete description is distributed together with the software.

C.1 General Commands

`model = Model('name')`

Loads the model object from the file `name.so`.

`integ = Integrator('Map')`

Loads the `Map` integrator (always used with discrete systems). For flows there are different Runge-Kutta schemes available, including `'Euler'`, `'RungeKutta4'`, `'DormandPrince853'`, `'MidpointRule'` or `'Gauss6'`.

`tree = Tree(center, radius)`

Constructs a tree object; `center` and `radius` determine the outer box Q.

`tree.depth`

Returns the current maximal depth of `tree`.

`tree.count(depth)`

Returns the number of boxes in the `tree` at the given depth.

`points = Points('InnerGrid', dim, m)`

Loads a set of `m` test points of dimension `dim`. The type `'InnerGrid'` sets the points in a uniform grid covering the box, but with all points in the interior of the box. Other possible point types are `'Grid'`, where points are also placed around the boundary of the box, `'MonteCarlo'`, which randomly distributes points (chosen from a uniform distribution) over the box and `'Edges'` which refers to test points placed on the boundary of the boxes. To use rigorous covering algorithms, one has to use the point type `'Lipschitz'` for the domain points (this is currently restricted to maps; also note that the model has to supply the function `lip` in this case).

`tree.domain_points = points`

Defines the domain test points for each box.

`tree.image_points = points`

Defines the image test points as explained in Appendix 5.3. If one wishes to use rigorous covering algorithms, the point style `'Vertices'` has to be used for the image points.

`tree.boxes(depth)`

Returns a $(2d+2) \times n$ matrix representing the box collection on the given `depth` of the `tree`, where n is the number of boxes on this depth and d is the dimension of state space (as defined by the model). Each column of this matrix corresponds to one box, where the first d rows specify the center, the second d the radius of the box, the (2d+1)st row its flags and the last its color.

C.2 Commands for Invariant Sets and Global Attractors

`rga(tree [, steps])`

Performs `steps` (default $=$ 1) steps of the subdivision algorithm for the computation of the relative global attractor (Algorithm 1).

`crs(tree [, steps])`

Performs several steps of the subdivision algorithm for the computation of the chain recurrent set (Algorithm 2).

C.3 Commands for Invariant Manifolds

`tree.insert(x, depth)`

Inserts the box containing the point `x` at the given depth into the tree.

`gum(tree, depth [, steps])`

Performs several steps of the continuation algorithm for the computation of global unstable manifolds (Algorithm 3) at the given depth.

C.4 Commands for Transfer Operators, Invariant measures and Almost Invariant Sets

`P = tree.matrix(method, depth)`

Computes the transition matrix as a finite-dimensional approximation of the Perron-Frobenius operator on the given depth of the tree using

1. test points, if `method` is a points object;
2. the exhaustion method, if `method=='exhaustion'`.

`[v, lambda] = eigs(P, n)`

Finds the `n` eigenvectors `v` and eigenvalues `lambda` of `P` with the largest modulus. See the documentation to `eigs` for further options and details.

References

1. M. Dellnitz, G. Froyland, and S. Sertl. On the isolated spectrum of the Perron-Frobenius operator. *Nonlinearity*, 13(4):1171–1188, 2000.
2. M. Dellnitz, M. Golubitsky, and M. Nicol. *Symmetry of attractors and the Karhunen-Loéve decomposition*, pages 73–108. Number 100 in Applied Mathematical Sciences. Springer-Verlag, 1994.
3. M. Dellnitz and A. Hohmann. The computation of unstable manifolds using subdivision and continuation. In H.W. Broer, S.A. van Gils, I. Hoveijn, and F. Takens, editors, *Nonlinear Dynamical Systems and Chaos*, pages 449–459. Birkhäuser, *PNLDE* 19, 1996.
4. M. Dellnitz and A. Hohmann. A subdivision algorithm for the computation of unstable manifolds and global attractors. *Numerische Mathematik*, 75:293–317, 1997.
5. M. Dellnitz, A. Hohmann, O. Junge, and M. Rumpf. Exploring invariant sets and invariant measures. CHAOS: An Interdisciplinary Journal of Nonlinear Science, 7(2):221, 1997.
6. M. Dellnitz and O. Junge. On the approximation of complicated dynamical behavior. *SIAM J. Numer. Anal.*, 36(2):491–515, 1999.

7. M. Dellnitz, O. Junge, M. Rumpf, and R. Strzodka. The computation of an unstable invariant set inside a cylinder containing a knotted flow. In *Proceedings of Equadiff '99, Berlin*, 2000.
8. P. Deuflhard, M. Dellnitz, O. Junge, and Ch. Schütte. *Computation of essential molecular dynamics by subdivision techniques*, pages 98–115. Number 4 in Lecture Notes in Computational Science and Engineering. Springer-Verlag, 1998.
9. J. Ding and A. Zhou. Finite approximations of Frobenius-Perron operators. A solution of Ulam's conjecture to multi-dimensional transformations. *Physica D*, 92(1–2):61–68, 1996.
10. R.W. Easton. *Geometric Methods for Discrete Dynamical Systems*. Number 50 in Oxford engineering science. Oxford University Press, New York, 1998.
11. M. Eidenschink. *Exploring Global Dynamics: A Numerical Algorithm Based on the Conley Index Theory*. PhD thesis, Georgia Institute of Technology, 1995.
12. Euclid. *Elements*. Book X, (first Proposition).
13. G. Froyland. Finite approximation of Sinai-Bowen-Ruelle measures of Anosov systems in two dimensions. *Random & Computational Dynamics*, 3(4):251–264, 1995.
14. G. Froyland. Approximating physical invariant measures of mixing dynamical systems in higher dimensions. *Nonlinear Analysis, Theory, Methods, & Applications*, 32(7):831–860, 1998.
15. G. Froyland and M. Dellnitz. Detecting and locating near-optimal almost-invariant sets and cycles. In preparation.
16. R. Guder, M. Dellnitz, and E. Kreuzer. An adaptive method for the approximation of the generalized cell mapping. *Chaos, Solitons and Fractals*, 8(4):525–534, 1997.
17. R. Guder and E. Kreuzer. Control of an adaptive refinement technique of generalized cell mapping by system dynamics. *J. Nonl. Dyn.*, 20(1):21–32, 1999.
18. O. Junge. *Mengenorientierte Methoden zur numerischen Analyse dynamischer Systeme*. PhD thesis, University of Paderborn, 1999.
19. O. Junge. Rigorous discretization of subdivision techniques. In *Proceedings of Equadiff '99, Berlin*, 2000.
20. G. Keller and C. Liverani. Stability of the spectrum for transfer operators. Preprint, 1998.
21. H. Keller and G. Ochs. Numerical approximation of random attractors. In *Stochastic dynamics*, pages 93–115. Springer, 1999.
22. R.Z. Khas'minskii. Principle of averaging for parabolic and elliptic differential equations and for Markov processes with small diffusion. *Theory of Probability and its Applications*, 8(1):1–21, 1963.
23. Y. Kifer. *Random Perturbations of Dynamical Systems*, volume 16 of *Progress in Probability and Statistics*. Birkhäuser, Boston, 1988.
24. A. Lasota and M.C. Mackey. *Chaos, Fractals, and Noise. Stochastic Aspects of Dynamics*, volume 97 of *Applied Mathematical Sciences*. Springer-Verlag, New York, second edition, 1994.
25. T.-Y. Li. Finite approximation for the Frobenius-Perron operator. A solution to Ulam's conjecture. *Journal of Approximation Theory*, 17:177–186, 1976.
26. K. Mehlhorn. *Data Structures and Algorithms*. Springer, 1984.

27. G. Osipenko. Construction of attractors and filtrations. In K. Mischaikow, M. Mrozek, and P. Zgliczynski, editors, *Conley Index Theory*, pages 173–191. Banach Center Publications 47, 1999.
28. C. Robinson. *Dynamical Systems: Stability, Symbolic Dynamics, and Chaos*. CRC, Boca Raton, 1995.
29. M. Rumpf and A. Wierse. GRAPE, eine objektorientierte Visualisierungs– und Numerikplattform. *Informatik, Forschung und Entwicklung*, 7:145–151, 1992.
30. Ch. Schütte. *Conformational Dynamics: Modelling, Theory, Algorithm, and Application to Biomolecules*. Habilitation thesis, Freie Universität Berlin, 1999.
31. E.C. Zeeman. Stability of dynamical systems. *Nonlinearity*, 1:115–155, 1988.

Polynomial Skew Products

Manfred Denker[1] and Stefan-M. Heinemann[2,*]

[1] Institut für Mathematische Stochastik, Universität Göttingen, Lotzestraße 13, 37083 Göttingen, Germany

[2] Institut für Theoretische Physik, TU Clausthal, Arnold-Sommerfeld-Straße 6, 38678 Clausthal-Zellerfeld, Germany

Abstract. This is a brief overview on a new theory for the dynamical behaviour of polynomial skew products in $\mathbb{C}^n$, including some non-published facts.

1 A New Direction in Fractal Analysis and Iteration Theory of Holomorphic Maps

With the development of modern high-performance computers the 'chaos theory' in the form of fractal geometry has become a popular field of research. Simple iterative algorithms, i.e. simple repetitive step-by-step calculations, permit the composition of complicated shapes, which sometimes even appear to possess artistic value. One feature of the design underlying all these methods is self-similarity: manifestations, which when enlarged are visible to the naked eye, are repeated again and again in miniature and, given a corresponding enlargement, each step, no matter how small, will be similar to the next object as a whole. This explains how simple equations or geometrical forms can lead to complex and multilayered structures.

In recent years, the conceptual approach inherent in this theory has been extended to every realm of knowledge. Chaos theory has been applied in the attempt to explain complicated processes which cannot always be predicted, or at least not with precision. The best-known of these are the so-called Lorenz equations in physics, which describe the vertical flow of a gas. In medicine, scientists have conducted experiments aimed at modelling the generation of creativity in the brain through chaos. Electrical discharges or the creation of polymer compounds display clearly fractal structures. A further example is the modulation of population dynamics in biology, where the size of the population is subject to irregular fluctuations. The list is endless. This whole field of research has developed from problems met with when solving equations using the so-called Newton method. When employing this algorithm, initial numbers cannot just be selected at random if the aim is to achieve a useful solution to an equation. Experiments are now in train to use this system for solving equations with multiple variables. Their mathematical investigation leads towards a new theory of higher-dimensional Julia sets, which is partly reviewed and extended here. Geometric, analytic

* Project: Iterations of Maps with Several Complex Variables (Manfred Denker)

and probabilistic structures which occur under simplest dynamics of four (and higher)-dimensional space, and which comprise just these points upon which the greatest chaos reigns, are discussed (for an extended exposition see [DHP95]).

The aim is best explained by the following example. For analytic endomorphisms of the Riemann sphere S^2 it is well known that the Julia sets $J(f_c)$ of mappings of the form

$$f_c : z \mapsto z^2 + c$$

with $|c|$ small are Jordan curves (in general non-rectifiable) and $f|_{J(f_c)}$ shows similar dynamical behaviour as $f_0 : z \mapsto z^2$ restricted to S^1. This holds at least if $|c| \leq 1/4 - \varepsilon$ for some $\varepsilon > 0$. Also, as a consequence of the choice of the parameter c, the mapping has all inverse branches everywhere defined on its Julia set, hence has bounded distortion and is hyperbolic. The term is used in the sense of expanding, i.e.

$$\liminf_{\substack{x \to y \\ x \neq y \in J(f_c)}} \frac{|f_c(x) - f_c(y)|}{|x - y|} > 1.$$

This property in turn enforces probability structures by which stationary time series based on Hölder functions are approximated by Brownian motion (like in the case of independent, identically distributed discrete time processes).

One of the main points of our discussion here is concerned with similar questions in higher dimensions, in particular the analytic and geometric structures and the long time behaviour of stationary processes generated by them.

2 Another Look at the One-Dimensional Case

An analytic endomorphism (or a rational function) $f : S^2 \to S^2$ on the Riemann sphere S^2 has always a nonempty, fully invariant Julia set, defined as the set of non-normal points for the family of functions $f^n : S^2 \to S^2$ ($n \geq 0$). It is quite clear that this set constitutes the regime of greatest chaos, whence the dynamics creates large diversity in the long time behaviour of time series $X_n = \phi \circ f^n$, $n \geq 0$, as well as some disordering structure which is visible in the local structure around periodic points. In fact, such different aspects of the same phenomenon have been the centre of investigation for a long time, starting with Julia and Fatou in the beginning of this century. In case of polynomials $P : \mathbb{C} \to \mathbb{C}$ one can summarise the essential features as follows (see [Hei93] for the first time when this was observed).

Theorem 1. *For a polynomial map $P : \mathbb{C} \to \mathbb{C}$ of degree at least 2, the Julia set $J(P)$ equals each of the following sets:*

1. $\{z \in \mathbb{C} : \{P^k : k \geq 0\}$ *is not normal at* $z\}$.
2. *The boundary $\partial(K(P))$ where $K(P) = \{z \in \mathbb{C} : \sup_k |P^k(z)| < \infty\}$.*
3. *The closure $\mathcal{R}$ of the set of repelling periodic points of P.*
4. *The limit of the pull backs by P^{-k} of the boundary $\partial(\{z \in \mathbb{C} : |z| \leq r\})$ for sufficiently large $r > 0$.*
5. *The support $supp(\mu_{max})$ of the measure of maximal entropy for P.*

Before continuing it is necessary to clarify the meaning of the 5 statements in the theorem briefly. Clearly, (1.) is the description arising from geometric function theory and allows to introduce methods from complex function theory to study $J(P)$. In particular, bounded distortion properties play a fundamental role. (2.) tells us that the complement of $J(P)$ splits into connected components and hence the analysis on $J(P)$ can be studied using the theory of holomorphic functions on domains, in particular using harmonic analysis (Green's function). Certainly, this complements the description in (1.). However, as is known, this boundary is equal to the Shilov boundary $\partial_{SH}(K(P))$, which is defined to be the minimal compact set Q with the property that functions holomorphic in some neighbourhood of $K(P)$ attain their maximal modulus (over $K(P)$) in Q. Certainly, $\partial_{SH}(K(P))$ is contained in $\partial K(P)$ – it is not difficult to construct holomorphic functions which attain their maximal modulus in a given open set intersecting $K(P)$ which shows that even $\partial K(P) = \partial_{SH}(K(P))$. This is one possibility to introduce some abstract boundary theory in order to study Julia sets. Other types of boundaries are certainly well suited to approach specific problems (e.g. rigidity questions). (3.) is a dynamic description. It implies that the dynamics is essentially determined by some hyperbolic behaviour. The tracing property of repelling period points and the value of the topological entropy are an almost immediate consequence. (4.) means that we may use the well known boundary theory for large centred disks and obtain the Julia set as a pull back. Finally, we obtain the existence of a probability measure maximising entropy. Intuitively this means that in addition to (3.) we obtain a probabilistic structure which contains a maximum of randomness. Since the map P is not a homeomorphism, there is also a natural filtration given by the pull backs of the Borel σ-algebra. It is clear that this filtration can be used to introduce martingale and mixing structures which are geometrically meaningful.

Proof. (of theorem 1) First, let us remark that the set $K(P)$ is always well defined and compact. Namely, we find a radius r_P such that $|z| > r_P$ implies that $|P(z)| > |z|$ and the forward orbit of z converges to infinity. Clearly,

$$K(P) = \bigcap_{k=0}^{\infty} P^{-k}\left(\overline{B_{r_P}}\right) = \lim_{k \to \infty} P^{-k}\left(\overline{B_{r_P}}\right). \tag{1}$$

Since P is a polynomial, we get the dynamical trichotomy (with respect to complete invariance under P) of the invariant sets $\complement K(P)$, $\partial K(P)$, and $\mathring{K}(P)$ (the latter set may be empty). This also yields that $\partial K(P)$ is a perfect set. We should note here that the proof uses Montel's theorem, i.e. one makes use of the fact that $\{P^k\}$ is a normal family on some open set U once one knows that the union of the images $P^k(U)$ leaves out at least 3 points in $\mathbb{C}$. A similar result in $\mathbb{C}^n$ (i.e. providing statements about $\mathbb{C}^n$ minus a set of finitely many points) does not hold. Hence one has to develop new tools for the higher-dimensional setting.

1. $\Leftrightarrow$ 2. : On $\complement K(P)$, by definition, we obtain convergence to infinity. The open bounded set $\mathring{K}(P)$ is mapped to itself, hence $\{P^k\}$ restricted to this set is also normal. A point z in $\partial K(P)$ has a bounded forward orbit, but any neighbourhood contains an open set from $\complement K(P)$, hence we cannot get normal convergence there.

1. $\Leftrightarrow$ 3. : It is obvious that one does not get normal convergence on $\mathcal{R}$. Furthermore, let us note that the number of non-repelling periodic points is finite. Hence, it remains to show that we find a periodic point in any neighbourhood U of a point in the Julia set. We fix a point w in $U \cap J$ such that w is not a critical value of P^2. From $P^{-2}(w)$ (which has cardinality at least 4) we choose three points w_1, w_2, w_3 distinct from w. We find disjoint compact neighbourhoods Z, W_1, W_2, W_3 of w, w_1, w_2, w_3, resp., such that for each j, P^2 maps W_j homeomorphically to Z. With S_j we denote the inverses $S_j : Z \to W_j$. If, for all $n \geq 1$, all j, and all $z \in Z$, we have that $P^n(z) \neq S_j(z)$, then $\{P^n\}$ is normal in Z which contradicts $w \in Z \cap J$. But $P^n(z) = S_j(z)$ implies $P^{n+2}(z) = z$, thus we have found a periodic point in U.

2. $\Leftrightarrow$ 4. : From equation (1) we deduce that, for $r \geq r_P$, we also get that $\lim_{k\to\infty} \partial(P^{-k}\overline{B_r}) = \partial K(P)$. We should note that, for disks B_r, Shilov boundary and topological boundary coincide, whence also for $K(P)$.

1. $\Leftrightarrow$ 5. : This equivalence is non-elementary. It requires the application of Gromov's result ([Gro]) on topological entropy of holomorphic maps and of thermodynamic formalism. This makes a proof rather lengthy and technical. We refer to [Lyu83]. □

3 The Characterisation Programme in $\mathbb{C}^n$

In order to extend the one-dimensional theory to higher-dimensional complex manifolds, consider polynomial maps $P : \mathbb{C}^n \to \mathbb{C}^n$. We are interested in determining the mappings for which there exists a non-empty set $J(P)$ which can be described in the following equivalent ways:

1. $J(P) := \{z \in \mathbb{C}^n : \{P^k : k \geq 0\}$ is not weakly normal at $z\}$.
2. $J(P)$ is equal to the Shilov boundary $\partial_{SH}(K(P))$ where $K(P) := \{z \in \mathbb{C}^n : \sup_k \|P^k(z)\| < \infty\}$.
3. $J(P)$ is equal to the closure $\mathcal{R}$ of the set of repelling periodic points of P.

4. $J(P)$ is equal to the limit of the pull backs by P^{-k} of the Shilov boundary $\partial_{SH}(\{z \in \mathbb{C}^n : \|z\| \leq r\})$, for $r > 0$ sufficiently large.
5. $J(P)$ is given by the support $\text{supp}(\mu_{\text{max}})$ of the measure of maximal entropy for P.

By consideration of the example $(z_1, \ldots, z_n) \mapsto (z_1^{d_1}, \ldots, z_n^{d_n})$ where $d_j \in \mathbb{N}$, one easily sees that normality alone is too weak if one wants to detect the set of maximal randomness. Therefore, we are using the term *weak normality.* The notion is defined as follows: A family of functions $\{f_k\}$ is called *weakly normal* in a point $z \in U$ if there is

- an open neighbourhood V of z;
- a family $\mathcal{C}_x$ of at least one-dimensional (complex) analytic sets indexed by the points $x \in V$,

such that

- each x lies in the corresponding analytic set C_x;
- for each $x \in V$ the family $\{f_k\}$ restricted to $\mathcal{C}_x \cap V$ is normal (including convergence to infinity).

The condition (5.) has been used in [Bri96] to define the Julia set. It is shown there that (3.) then follows if the map is hyperbolic. The same definition has been used in [Jon97]. In the projective setting (1.) has been used in [FS94] to define a Julia set, but no results were proven. Equivalence of these five conditions has been obtained for certain skew products (see definition in section 4) in [Hei94], [Hei96], [Hei98c], [Hei98b], and [Hei98a]. In addition, these papers contain results on hyperbolicity and parameter spaces (Mandelbrot set).

It is clear that the definition selects a set of maximal randomness. It is known from [DH98] that torus like maps of the form

$$P(x, y) = (x^2 + k(y), y^2 + l(x))$$

satisfy this characterisation as long as the norms of the polynomials $k(y)$ and $l(x)$ are sufficiently small in some neighbourhood of 0. In case $k = l = 0$ one obtains the product map

$$P(x, y) = (x^2, y^2).$$

The analysis of this map is easy: The Julia set equals the torus

$$\mathbb{T}^2 = \{(x, y) : |x| = |y| = 1\}$$

and P is expanding on $\mathbb{T}^2$. For small perturbations one expects stability: The Julia set is still homeomorphic to a torus and the map is still expanding. [Hei95] shows even more: For the class of mappings for which the characterisation theorem holds (which is considerably larger than just a perturbed family) we obtain a complete analogue of the one-dimensional result mentioned in section 2. The Julia set is homotopic to a torus and P acts as an expanding transformation on $J(P)$. Hence $(J(P), P)$ forms an expanding repeller and thus defines the complete scenery of thermodynamic formalism including its probabilistic structure for stationary sequences.

A polynomial map P from $\mathbb{C}^n$ to $\mathbb{C}^n$ of the form

$$P(z_1, \dots, z_n) = \left(\prod_{l=1}^{n} z_l^{a_{1l}}, \dots, \prod_{l=1}^{n} z_l^{a_{nl}} \right)$$

leaves the n-dimensional torus $\mathbb{T}^n$ invariant. The investigation of such maps can be carried out using the theory of torus automorphisms defined by the matrix (a_{ij}) (see [Lin78] and others). However, not all of these maps have a non-empty Julia set as defined in the charactarisation programme. $\mathbb{T}^n$ carries a measure of maximal entropy for P but it may be the case that no repelling periodic point exists. Therefore, an important preliminary question is to determine whether $J(P)$ is non-empty. There is a simple and easily verifiable criterion for this fact: A polynomial map $P : \mathbb{C}^2 \mapsto \mathbb{C}^2$ satisfies the regularity condition and is called a *(p, q)-regular map* (cf. [Hei94]) if there is

$$R > 0, p \in \mathbb{Q}_+^*, q \in \mathbb{N}, k_1, k_2 > 0$$

such that, for all z with $\|z\| > R$, we have that

$$k_1 \|z\|^p \leq \|P(z)\| \leq k_2 \|z\|^q. \tag{2}$$

In particular, the left hand side of (2) shows that the dynamics of P is compatible with the simple Alexandroff-compactification, i.e. it makes sense to work on $\mathbb{C}^N := \mathbb{C}^n \cup \{\infty\}$ (cf. [Hei96]). It is important to note that the approach here is different and more successful than that of using a compactification by complex manifolds, e.g. projective space or complex solenoids ([Hei98b]).

In general, a non-empty $J(P)$ is a highly irregular fractal set as computer simulations show. The most simple ones are those where the Julia set looks like a torus. More complicated objects are discussed in the next section.

4 J^*-continuity

Polynomial skew products of $\mathbb{C}^2$ have the form

$$P(x, y) = (p_y(x), q(y)). \tag{3}$$

As just explained, the only interesting Julia sets occur for the subclasses of polynomial maps satisfying the regularity condition. We had been mentioning that for the higher-dimensional case one has to develop new tools which can replace Montel's theorem.

For skew products, we have been developing the notion of J^*-continuity (cf. [Hei98c]) which on the one hand allows us to reduce the dimension of our dynamical systems and on the other hand has turned out to be essential in establishing the equivalence in the characterisation programme. First let us note that skew products of the form (3) contain an abundance of one-dimensional subsystems. For each $y \in K(q)$ we can fix the complex line $\mathbb{C} \times \{y\}$ and investigate the dynamical behaviour of the family

$$\mathcal{P}_y := \{p_y, p_{q(y)} \circ p_y, \dots\}.$$

This yields the pre-Julia sets $J_y^* := J(\mathcal{P}_y)$ which are obtained using one-dimensional techniques.

Definition 2. ***J^**-continuity** [Hei98c] We call a skew product $P : \mathbb{C}^2 \to \mathbb{C}^2$ *J^*-continuous* if the map

$$\mathbb{J} : J(q) \to \mathbb{C}$$

which maps $y \in J(q)$ to its pre-Julia set J_y^* is continuous with respect to the Hausdorff distance d_H in $\mathbb{C}$.

If one knows that $\pi_2(J(P)) \subseteq J(q)$ then J^*-continuity yields as an immediate consequence that

$$J(P) = \bigcup_{y \in J(q)} J_y^* \times \{y\}.$$

Moreover, J^*-continuity also gives the equivalence of the characterisations (1.) to (4.) (for equivalence to (5.) cf. [Hei98a], [Hei99]).

Theorem 3. *For J^*-continuous P with $\pi_2(J(P)) \subseteq J(q)$ we have that*

$$J = \mathcal{R} = \partial_{SH} K(P) = \lim_{k \to \infty} P^{-k} \partial_{SH}(\overline{B_r}).$$

Proof. 1. $\Leftrightarrow$ 2. : For $(x^*, y^*) \in J$ and $\eta > 0$, we construct a peak function for $B_\eta(x^*, y^*)$

$$\Phi : K(P) \to \mathbb{C}.$$

There exists $\zeta > 0$, $\zeta \leq \eta/2$, such that $|y - y^*| < \zeta$ implies

$$d_H(J_y, J_{y^*}) \ < \ \eta/2.$$

We choose $y' \in \partial_{CH} J(q)$ (where ∂_{CH} denotes the Choquet boundary (which is a dense subset of the Shilov boundary) and $K_y := K(P) \cap (\mathbb{C} \times \{y\})$) with

$$|y' - y^*| \ < \ \zeta,$$

and find $x' \in \partial_{CH} K_{y'} \subseteq J_{y'}$ such that

$$|x' - x^*| \ < \ \eta/2.$$

For $0 < \delta < 1/2$, there is a peak function φ for $B_{\eta/2}(x')$ in $K_{y'}$ such that

$$\begin{aligned} \varphi(x') &= 1 = \|\varphi\|_{\pi_1(K_{y'})}, \\ \|\varphi\|_{K_{y'}\setminus B_{\eta/2}(x')} &< \delta/4. \end{aligned}$$

φ can be approximated by a polynomial $\tilde{P}$ such that

$$\|\tilde{P} - \varphi\|_{K_{y'}} < \delta/4.$$

For every $0 < \tau < \infty$, the derivative $\tilde{P}'$ of $\tilde{P}$ is bounded on the disk B_τ, i.e. for some $\omega < \infty$, we have that

$$\|\tilde{P}'\|_{B_\tau} < \omega.$$

Let us fix τ such that $\pi_1(K(P)) \subseteq B_\tau$.
It is clear that $\tilde{P}$ still is a peak function for $B_{\eta/2}(x')$ in $K_{y'}$. This follows since

$$\begin{aligned} \|\tilde{P}\|_{B_{\eta/2}(x')\cap K_{y'}} &> 1 - \delta/4, \\ \|\tilde{P}\|_{K_{y'}\setminus B_{\eta/2}(x')} &< \delta/2. \end{aligned}$$

Now we choose $0 < \xi < \eta/2$, such that $|y - y'| < \xi$ implies

$$d_H(J_y, J_{y'}) < \min\{\delta/(4\omega), \eta/2\}.$$

Since $y' \in \partial_{CH} J(q)$ we find $Q \in \mathbb{A}(J(q)$ such that

$$\begin{aligned} Q(y') &= 1 = \|Q\|_{J(q)}, \\ \|Q\|_{J(q)\setminus B_\xi(y')} &\le \delta/(1 + \delta/4 + 2\tau\omega). \end{aligned}$$

We obtain the desired peak function Φ for $B_\eta(x, y)$ by setting

$$\Phi(x, y) := \tilde{P}(x) \cdot Q(y),$$

namely

$$\|\Phi\|_{B_\eta(x^*,y^*)} \ge \|\tilde{P} \cdot Q\|_{B_{\eta/2}(x')\times B_\xi(y')} > (1 - \delta/4)\cdot 1 = 1 - \delta/4,$$

whereas

$$\|\Phi\|_{K(P)\setminus B_\eta(x^*,y^*)} \le \max\begin{cases} \|\Phi\|_{K(P)\setminus(\mathbb{C}\times(J(q)\setminus B_\xi(y')))} < \delta, \\ \|\Phi\|_{(\mathbb{C}\setminus B_\eta(x,y))\times B_\xi(y')} < 3\delta/4. \end{cases}$$

1. $\Leftrightarrow$ 3. : It is clear that we cannot have weakly normal convergence in a repelling periodic point, hence $\mathcal{R}$ is contained in J. In order to show the inclusion of J in $\overline{\mathcal{R}}$ we make use of the fact that for one-dimensional Julia sets the assertion of the theorem holds. We find $\delta > 0$ such that, for

arbitrary $(x_0, y_0) \in J$ and $\eta > 0$, $y' \in J(q)$ together with $|y' - y_0| < \delta$ implies $d_H(J_{y_0}, J_{y'}) < \eta/2$.
In $J(q)$ repelling periodic points of q are dense, we find y_1 in $J(q)$ which is k-periodic for some $k \in \mathbb{N}$ and repelling with

$$|y_1 - y_0| \ < \ \min\{\delta, \eta\}.$$

In J_{y_1} we find (x', y_1) such that

$$|x' - x_0| \ < \ \eta/2.$$

Finally, as in

$$J_{y_1} \ = \ J(\underbrace{p_{q^{k-1}(y_1)} \circ \ldots \circ p_{q(y_1)} \circ p_{y_1}}_{=: \, P_{y_1}}) \times \{y_1\}$$

repelling periodic points of P_{y_1} are dense, we find x_1 which is repelling periodic under P_{y_1} such that $|x_1 - x'| < \eta/2$. Evidently, (x_1, y_1) is a repelling periodic point of f and (using the maximum norm)

$$\begin{aligned} \|(x_1, y_1) - (x_0, y_0)\| &\leq \max\{|x_1 - x_0|, |y_1 - y_0|\} \\ &\leq \max\{|x_1 - x'| + |x' - x_0|, |y_1 - y_0|\} \\ &\leq \eta. \end{aligned}$$

1. $\Leftrightarrow$ 4. : This is an easy consequence of the one-dimensional statement as $\lim_{k\to\infty} \pi_2(P^{-k}(\overline{B_r})) = J(q)$. □

Examples of skew products where the complete characterisation programme can be verified include Cantor skews ([Hei96]), certain noodle type maps ([Hei98c]), homogeneous maps ([Hei98b]) and doughnut type maps ([Hei98a]).

5 Thermodynamical Formalism for Skew Products

Consider a polynomial skew product P of $\mathbb{C} \times \mathbb{C}^n$ where $n \geq 1$, which factors over the base $\mathbb{C}$ and the polynomial $p : \mathbb{C} \to \mathbb{C}$. The projection $\pi : \mathbb{C}^{n+1} \to \mathbb{C}$ commutes with P and p. Assume that the Julia set defined by weak normality is nonempty and completely invariant.

In this situation one can establish a relativised thermodynamic formalism. Recall that $J_z = \pi^{-1}(\{z\}) \cap J(P)$ denotes the fibre over $z \in J(p)$, and the skew product P induces canonical maps between the fibres

$$P_z : J_z \to J_{p(z)}$$

whence a relativised thermodynamic formalism for a potential $\Psi : J(P) \to \mathbb{R}_+$ is well defined by the transfer operators

$$\begin{aligned} V_z \ &: \ C(J_z) \to C(J_{p(z)}) \\ V_z\varphi(y) &= \sum_{\substack{P(y')=y; \\ \pi(y')=z}} \varphi(y')\Psi(y'). \end{aligned}$$

(In case y is a critical point, take preimages according to their multiplicity, so that the operators act on continuous functions.) A real valued function $\lambda : \mathbb{C} \to \mathbb{R}$ is called an *eigenvalue-function* if there exists a family μ_z ($z \in J(P)$) of probability measures μ_z on J_z so that

$$V_z^* \mu_{p(z)} = \lambda(z)\mu_z \qquad z \in J(p).$$

These families are called *Gibbs* and are a basic notion for the relativised thermodynamic formalism.

A weakly continuous Gibbs family (i.e. for every bounded continuous function $g : J(P) \to \mathbb{R}$ the map $z \mapsto \int g(y)\mu_z(dy)$ is continuous) can be characterised by the property that the Jacobian satisfies the equation $\frac{d\mu_{p(z)} \circ P_z}{d\mu_z} = \frac{\lambda(z)}{\Psi(y)}$. Let ν be a Gibbs measure for p and the potential $\phi : J(p) \to \mathbb{R}_+$, that is

$$V^* \nu = \nu,$$

where $Vf(z) = \sum_{p(y)=z} f(y)\phi(y)$ denotes the *Frobenius-Perron operator* for p and ϕ. Then the probability measure $\mu = \int_{\mathbb{C}} \mu_z \nu(dz)$ has two Radon-Nikodym derivatives

$$\frac{d\mu \circ P_z}{d\mu} = \frac{\lambda(z)}{\Psi(\cdot)} \qquad \mu_z \text{ a.e.}$$

$$\frac{d\mu \circ \pi^{-1} \circ p}{d\mu \circ \pi^{-1}} = \phi^{-1} \qquad \mu \circ \pi^{-1} \text{ a.e.}$$

which may be considered to be partial derivations of the measure μ with respect to the two actions on $J(P)$.

The expanding property of P depends on the metric, but equivalent metrics do not change this property. Here we choose the metric $\|x - y\| = \sum_{i=1}^{n+1} |x_i - y_i|$, and P is *expanding (hyperbolic)* if there exists $\Lambda > 1$ such that $\|P(z) - P(y)\| \geq \Lambda \|z - y\|$.

Theorem 4. *Let $J(P) \neq \emptyset$ be a bounded Julia set of the hyperbolic regular skew product $P : \mathbb{C}^{n+1} \to \mathbb{C}^{n+1}$ such that $\pi(J(P)) = J(p)$. For any Hölder continuous potential $\Psi : J(P) \to \mathbb{R}_+$ there exists a weakly Hölder continuous Gibbs family*

$$z \mapsto \mu_z,$$

and the eigenvalue-function $\lambda : J(p) \to \mathbb{R}$ is Hölder continuous.

Proof. Since $P|_{J(P)}$ is hyperbolic, there are no critical points in $J(P)$, hence the number of preimages of P_z is constant (the degree of P_z) by regularity. Also, as is easily seen after possibly passing to an equivalent metric, the fibre maps P_z are uniformly expanding and open (as maps $J_z \to J_{p(z)}$). The factor map $\pi : \mathbb{C}^{n+1} \to \mathbb{C}$ is a contraction, i.e. $|\pi(z_1) - \pi(z_2)| \leq \|z_1 - z_2\|$ holds for all $z_1, z_2 \in \mathbb{C}^{n+1}$. Since P is hyperbolic (expanding) on its Julia set, the

system $(J(P), J(p), \pi, P, p)$ is relatively Hölder continuous in the sense that for some $\alpha, b, c > 0$ the following holds:
If $y, y_0' \in \mathbb{C}^{n+1}$, $x' \in \mathbb{C}$ are such that $p(x') = \pi(y_0')$, $|\pi(y) - x'| < 2b$, $\|P(y) - y_0'\| < 2c$, then there exists a unique $y' \in \mathbb{C}^{n+1}$ such that $\pi(y') = x'$, $P(y') = y_0'$, $\|y - y'\| < 2c$, and

$$\|y - y'\| \cdot |\pi(P(y)) - \pi(P(y'))|^\alpha \leq \|P(y) - P(y')\| \cdot |\pi(y) - \pi(y')|^\alpha. \quad (4)$$

As a hyperbolic map P is open and locally invertible, there is some $c' > 0$ (small enough) such that $P : B(z, c') \to \mathbb{C}^{n+1}$ is injective, whence there exists $c > 0$ such that each inverse branch P_z^{-1} of P sending $P(z)$ to z is well defined on balls of the form $B(P(z), 2c)$. Let $P(y)$, $x \in \mathbb{C}$ and $y_0' \in B(P(y), 2c)$ with $\pi(y_0') = p(x)$ be given. Then $y' = P_y^{-1}(y_0')$ is uniquely determined by the condition $\|y' - y\| < 2c$, which also implies $\pi(y') = x$ and $|\pi(y) - x| \leq 2c$. Since p is continuous there exists a constant K such that $|p(t)-p(s)| \leq K|t-s|$ for $t, s \in J(p) = \pi J(P)$, and since P is expanding, there exists $\Lambda > 1$ such that $\|P(z) - P(z')\| \geq \Lambda\|z - z'\|$ whenever $z' \in B(z, 2c)$. Choose α so that $\Lambda^{-1}K^\alpha \leq 1$. It follows that equation (4) holds.

The theorem follows from this by application of Theorem 2.10 in [DG99]. □

This is a typical structure theorem for probability measures which are nonsingular with respect to polynomial skew products and have given conditional Jacobians. In the non-hyperbolic case, say if the polynomial is only assumed to be expansive, a change of the metric makes the map expanding (by a result essentially due to Coven and Reddy). Therefore the above theorem applies proving the existence of continuous Gibbs families. Specific examples are given in [DGH]. The reader may also notice that the above theorem strengthens the result of Jonsson in [Jon97] where a fibred structure is put on the measure of maximal entropy using a decomposition of the current defined by dd^c. Naturally, one also gets such a decomposition in more general situations.

In case of a constant potential Ψ the Gibbs family is invariant. Integrating this family with respect to the measure of maximal entropy for the base map p defines an invariant measure for P, which is called a measure of maximal entropy. In the hyperbolic case, it is unique (as a measure on $J(P)$), and possibly also as a measure on $\mathbb{C}^{n+1}$. The natural generalisation of this is the question whether invariant measures exist which are fibrewise absolutely continuous with respect to the Gibbs family for a Hölder continuous potential Ψ. This problem leads to new aspects for the theory of equilibrium states in higher dimensions. If p would be invertible, classical techniques can be used to prove the existence of such measures. However, p is not invertible and one needs to introduce new concepts. Following [DGH] a potential Ψ is called *basic* if for each $x \in J(p)$ and $z \in \pi^{-1}(\{p(x)\} \cap J(P))$

$$V_x 1(z) = \lambda(x),$$

where λ denotes the eigenvalue-function as before (see also section 3 in [DG98]). For potentials of this type one can reduce the variational problem to the relativised problem (see [LW76]) and the variational problem for p and the potential λ.

For a measure preserving transformation T of a probability space $(\Omega, \mathcal{B}, m)$ denote the metric entropy by $h_m(T)$ and the relative entropy of T given the factor S by $h_m(T|S)$. For a compact metric space Ω with Borel σ-field $\mathcal{B}$ we let

$$\mathbf{P}(T, g) = \sup\{h_m(T) + \int g dm : \; m \circ T^{-1} = m\}$$

denote the pressure of the continuous function $g \in C(\Omega)$.

In the situation of the theorem one can show that for a basic potential Ψ with eigenvalue-function λ a fibrewise absolutely continuous invariant measure m exists satisfying the relative variational principle:

$$h_m(P|p) + \int \Psi(y) m(dy) = \int \lambda dm \circ \pi^{-1},$$

and the variational problem for P reduces to that of p:

$$\sup\{h_\nu(P) + \int \Psi d\nu : \nu \circ P^{-1} = \nu\} = \sup\{h_\rho(p) + \int \lambda d\rho : \rho \circ p^{-1} = \rho\}.$$

It follows that the pressure for P and Ψ can be derived from p and λ:

$$\mathbf{P}(P, \Psi) = \mathbf{P}(p, \lambda).$$

A large class of example falls into this category: Maps of doughnut type (see [Hei99]), or Cannelloni or Spaghetti type (cf. [Hei98c]) as explained above. Details can be found in [DGH].

6 Questions and Open Problems

In dimension one the ergodic theory is completely described if the rational map $f : S^2 \to S^2$ is hyperbolic (expanding) or expansive on its Julia set $J(f)$. It is not hard to show that f is hyperbolic if and only if $J(f)$ does not contain any critical point nor any parabolic point. (Recall that a periodic point z with period p is parabolic if $(f^p)'(z)$ is a root of unity.) Also, it is known that f is expansive if and only if $J(f)$ does not contain any critical point.

In dimension ≥ 2 hyperbolicity used in a more general meaning is not the same as the expanding property. Hénon maps are invertible, and they are never expanding (their Julia sets in our sense here is always empty) but can be hyperbolic. Thus we have the first problem:

Problem 5. Which skew products $P : \mathbb{C}^n \to \mathbb{C}^n$ admit a completely invariant set on which P is expanding?

Examples when this is true were described in previous sections. Consider a polynomial skew product

$$P\begin{pmatrix} x \\ y \end{pmatrix} = \begin{pmatrix} p_y(x) \\ q(y) \end{pmatrix}.$$

A point $\binom{x}{y}$ is called parabolic if y is parabolic for q or if y is periodic (with period m say) and if x is parabolic for P_y (cf. section 4). Denote by $\mathcal{C}$ the set of critical points for P. It is not difficult to see that P is expanding if and only if $\overline{O^+(\mathcal{C})} \cap J(P) = \emptyset$, where $O^+(\mathcal{C})$ denotes the forward orbit of $\mathcal{C}$. Thus an expanding P cannot have a critical or a parabolic point.

Problem 6. Is a regular skew product $P(\binom{x}{y})) = \binom{p_y(x)}{q(y)}$ (together with its Julia set) expanding if and only if the omega-limit set of the critical locus is disjoint from $J(P)$?

As a consequence of this classification on obtains the existence of unique equilibrium states: Let $\varphi : \mathbb{C}^n \to \mathbb{C}^n$ be a Hölder continuous function. Then there exists a probability measure μ with

$$\mathbf{P}(P, \varphi) := h_\mu(P) + \int \varphi d\mu.$$

Again, in one dimension such equilibrium measures (Gibbs measures) always exist for rational functions f and are unique for Hölder continuous φ satisfying $\mathbf{P}(f, \varphi) > \sup_{z \in J(f)} \varphi(z)$.

Problem 7. If P is a regular skew product in $\mathbb{C}^n$ and φ Hölder continuous with $\mathbf{P}(P, \varphi) > \sup_{z \in J(P)} \varphi(z)$, does there exists a unique invariant Gibbs measure?

Recall that P is expansive on $J(P)$ (more precisely positively expansive) if there exists $\varepsilon > 0$ such that

$$\sup_{n \geq 0} |P^n(x) - P^n(y)| > \varepsilon$$

for all $x \neq y \in J(P)$. Again, in one dimension, a rational map is expansive if and only if its Julia set does not contain any critical point. For these maps the ergodic theory is well developed, hence one needs to characterise this class in higher dimensions.

Problem 8. Which regular polynomial $P : \mathbb{C}^n \to \mathbb{C}^n$ admits a completely invariant set in which P is open and expansive?

If one considers a regular skew product

$$P\begin{pmatrix} x \\ y \end{pmatrix} = \begin{pmatrix} p_y(x) \\ q(y) \end{pmatrix},$$

then P is not expansive on $J(P)$ if $J(P)$ contains a critical point. Conversely, there are reasons to believe that P is expansive if $J(P)$ does not contain a critical point and

$$\overline{O^+(\mathcal{C}_x)} \cap J(P)$$

is finite. Here

$$\mathcal{C}_x = \{(x,y) : P_y'(x) = 0\}$$

denotes the critical locus over the base.

Problem 9. Is a regular skew product $P\binom{x}{y} = \binom{p_y(x)}{q(y)}$ expansive (on $J(P)$) if and only if $\overline{O^+(\mathcal{C}_x)} \cap J(P)$ is finite and $J(P)$ does not contain any critical point?

If this is true, new dynamics occur for higher dimensional maps. Then there are non–expansive maps with no critical point in its Julia set. On the other hand, an open and expansive map is – after changing the metric – expanding in Ruelle's sense. Hence there exist Markov partitions for P. Since we also have an $\exp[\mathbf{P}(P,\varphi) - \varphi]$ conformal measure μ, the dynamical system $(J(P), P, \mu)$ forms a Markov fibred system (see [ADU93]).

It is not known whether skew products of $\mathbb{C}^n$ admit new probabilistic structures different from those well known arising in the one-dimensional case or in the case of group extensions. Certainly, the thermodynamic formalism as part of the dynamic behaviour is different. Thus one can expect some refined probabilistic theorems respecting canonical filtrations given by the dynamics. On the other hand, the existence of an infinite number of parabolic points may cause different spectral behaviour of the transfer operators and hence different ergodic theorems.

The fractal geometry of skew products on $\mathbb{C}^n$ for $n \geq 2$ does not fall into the category of self–similarity. It is known that even simple affine maps in higher dimension need not to have equal Hausdorff– and box–dimensions. Also for skew products

$$P\begin{pmatrix} x \\ y \end{pmatrix} = \begin{pmatrix} p_y(x) \\ q(y) \end{pmatrix}$$

one certainly obtains new phenomena when the Hausdorff dimension on fibres $J(P) \cap (\mathbb{C} \times \{y\})$ is not continuous. In fact, for the skew products as discussed previously, this can only happen for product maps. On the other hand, it is not clear whether the usual dynamic representations for the Hausdorff–dimension are valid (as the smallest zero of the pressure function). Thus we obtain the following question.

Problem 10. Characterise those regular skew products $P\binom{x}{y} = \binom{p_y(x)}{q(y)}$ for which the fibres $J_x = J(P) \cap \{x\} \times \mathbb{C}$ are continuous in the Hausdorff metric (J^*-continuity). Compute the Hausdorff dimension in this case.

References

[ADU93] J. Aaronson, M. Denker, and M. Urbański. Ergodic theory of Markov fibred systems and parabolic rational maps. *Transactions American Mathematical Society*, 337:495–548, 1993.

[Bri96] J.-Y. Briend. Exposants de Liapounoff et de points périodiques d'endomorphismes holomorphes de $\mathbb{C}P^k$. *Comptes Rendus d'Academie des Seances Paris*, 323:805–808, 1996.

[DG98] M. Denker and M. Gordin. Problems on complex dynamical systems. In S. Morosawa, editor, *RIMS Workshop*, volume 1042, pages 1–10, 1998.

[DG99] M. Denker and M. Gordin. Gibbs measures for fibred systems. *to appear: Advances in Mathematics*, 148:161–192, 1999.

[DGH] M. Denker, M. Gordin, and St. M. Heinemann. On the relative variational principle for fibre expanding maps. preprint.

[DH98] M. Denker and St. M. Heinemann. Jordan Tori as Julia sets in $\mathbb{C}^n$. *Fundamenta Mathematicae*, 157:139–159, 1998.

[DHP95] M. Denker, St. M. Heinemann, and S. J. Patterson. Erinnert stark an abstrakte Malerei. *forschung*, 4:18–21, 95.

[FS94] J. E. Fornæss and N. Sibony. Complex dynamics in higher dimension, I. *Astérisque*, 222:201–231, 1994.

[Gro] M. Gromov. On the entropy of holomorphic mappings. *Preprint d'Institute des Hautes Etudes scientifiques.*

[Hei93] St. M. Heinemann. Iteration holomorpher Abbildungen in $\mathbb{C}^n$. *Diplomarbeit Universität Göttingen*, 1993.

[Hei94] St. M. Heinemann. Dynamische Aspekte holomorpher Abbildungen in $\mathbb{C}^n$. *Dissertation Universität Göttingen*, 1994.

[Hei95] St. M. Heinemann. Jordan tori as Julia sets in $\mathbb{C}^n$. *Mathematica Gottingensis*, 44, 1995.

[Hei96] St. M. Heinemann. Julia sets for endomorphisms of $\mathbb{C}^n$. *Ergodic Theory and Dynamical Systems*, 16:1275–1296, 1996.

[Hei98a] St. M. Heinemann. Iteration of Holomorphic Endomorphisms in $\mathbb{C}^n$ – A Case Study, 1998. Habilitationsschrift, Universität Göttingen.

[Hei98b] St. M. Heinemann. Julia sets in $\mathbb{C}^n$. *Progress in holomorphic dynamics*, 387:159–185, 1998.

[Hei98c] St. M. Heinemann. Julia-sets of skew products in $\mathbb{C}^2$. *Kyushu Journal of Mathematics*, 52(2):299–329, 1998.

[Hei99] St. M. Heinemann. Shilov boundary, dynamics and entropy in $\mathbb{C}^2$. *Mathematica Gottingensis*, 6, March 1999.

[Jon97] M. Jonsson. *Dynamical Studies in Several Complex Variables.* PhD thesis, Royal Institute of Technology, Stockholm, 1997.

[Lin78] D. Lind. Ergodic group automorphisms and specification. In M. Denker and K. Jacobs, editors, *Ergodic Theory*, volume 729 of *Lecture Notes in Mathematics*, pages 93–104, Oberwolfach, 1978.

[LW76] F. Ledrappier and P. Walters. A relativised variational principle for continuous transformations. *Journal London Mathematical Society*, 10:568–576, 1976.

[Lyu83] M. Yu. Lyubich. Entropy properties of rational endomorphisms of the Riemann sphere. *Ergodic Theory and Dynamical Systems*, 3:351–385, 1983.

Transfer Operator Approach to Conformational Dynamics in Biomolecular Systems

Ch. Schütte[2], W. Huisinga[2], and P. Deuflhard[1,*]

[1] Konrad-Zuse-Zentrum Berlin (ZIB), Takustr. 7, 14195 Berlin, Germany
[2] Mathematisches Institut I, Freie Universität Berlin, Arnimallee 2-6, 14195 Berlin, Germany

Abstract. The article surveys the development of novel mathematical concepts and algorithmic approaches based thereon in view of their possible applicability to biomolecular design. Both a first deterministic approach, based on the Frobenius-Perron operator corresponding to the flow of the Hamiltonian dynamics, and later stochastic approaches, based on a spatial Markov operator or on Langevin dynamics, can be subsumed under the unified mathematical roof of the transfer operator approach to effective dynamics of molecular systems. The key idea of constructing specific transfer operators especially taylored for the purpose of conformational dynamics appears as the red line throughout the paper. Different steps of the algorithm are exemplified by a trinucleotide molecular system as a small representative of possible RNA drug molecules.

Keywords. Transfer operator, Markov process, Markov chain, molecular dynamics, biomolecular conformations, canonical ensemble, transition probability, Hamiltonian dynamics, Langevin dynamics, nearly degenerate eigenvalues, Perron cluster

Mathematics Subject Classification. 65U05, 60J25, 60J60, 47B15

1 Introduction

In recent years, biomolecular design has attracted considerable attention both in the scientific and in the economic world. A few years ago, a research group at ZIB, partly supported by the DFG research program described in this volume, has started to work in this field. The problem of biomolecular design exhibits a huge discrepancy of time scales: those relevant from the pharmaceutical point of view are in the seconds, whereas present computations reach into the nanosecond regime at most. The reason for this is twofold: First, all available numerical integrators allow stepsizes of at most some femtoseconds

---

* Project: Theoretical Analysis and Efficient Simulation of the Essential Dynamics of Molecular Systems (Peter Deuflhard)

only [39,34]. Second, trajectory-oriented simulations are ill-conditioned after, say, a few thousand integration steps [1]. As a consequence, whenever *dynamical* informations (and not only *averages* of physical observables) are wanted—which is actually the case in biomolecular design—then only short term trajectories should be exploited. This message seems to be in direct contradiction to the desired aim of long term prediction in biomolecular design.

Aware of this seemingly contradiction, the ZIB group got inspired by work of DELLNITZ and co-workers [10,9] on *almost invariant sets* of dynamical systems—within the same DFG research program. As documented in [11], the key idea was to interpret almost invariant sets in phase space as *chemical conformations*. Within chemistry, the latter term describes *metastable global states* of a molecule wherein the *large scale geometric structure* is conserved over long time spans. As it turned out, the chemists' dominant interest was anyway just in these conformations, their life spans, and their patterns of *conformational changes*. Therefore, our first approach [11] followed the line of the original paper by DELLNITZ AND JUNGE [10]: chemical conformations were identified via eigenmodes corresponding to an eigenvalue cluster of the *Frobenius-Perron operator* associated with the deterministic flow of the Hamiltonian system. However, upon keeping a clear orientation towards the design of biomolecular systems, the computational techniques based on this first approach appeared to be unsatisfactory for reasons of both lack of theoretical clarity and sheer computational complexity: The theoretical justification of the approach requires the introduction of artificial stochastic perturbations of the dynamics [10] regardless of any (physical) interpretation. Moreover, the computational techniques from [10,11] are suitable only, if the objects of interest are rather low-dimensional, whereas the search for conformations will have to include the entire high-dimensional phase space of the molecular dynamics. Therefore, an almost complete remodelling with special emphasis on both physical interpretation and dimensionality of the problem turned out to be necessary in view of biomolecular applications.

In order to define conformations as *experimentally* determinable objects, concepts of *Statistical Physics* needed to be included. In addition, the remodelling had to include the aspect that chemical conformations are purely spatial objects determined via molecular geometry. These insights gave rise to the study of "spatial" *Markov operators* beyond the Frobenius-Perron operator as well as the associated Markov chains replacing the Hamiltonian dynamics [36,35]. The thus arising special Markov operator was shown to exhibit all the desirable theoretical properties needed as a basis for efficient algorithms. Moreover, a Galerkin approximation of this Markov operator in a weighted L^2-space naturally led to the replacement of the original (expensive) subdivision techniques [9] by newly developed (cheap) Hybrid Monte Carlo (HMC) methods called reweighted adaptive temperature HMC, or short ATHMC [16]. On the basis of suggestions by AMADEI ET AL. [2], an algorithm for identifying the essential molecular degrees of freedom has been worked out that drastically reduces the eigenvalue cluster problem even in larger molecular systems [22]. With these algorithmic improvements the applicability of our approach to realistic biomolecules came into reach. By applying the above ideas to the stochastic Langevin model of molecular dynamics [37], we succeeded to show that the fruitful coupling between the concepts of Statistical Physics and the transfer operator approach to effective dynamics can be exploited in a much more general framework.

The purpose of the present article is to survey what has been achieved, and to gain further insight from that. As will be shown subsequently, we are now able to subsume both our first deterministic approach [11] and the different stochastic approaches [36,37] under the unified roof of transfer operators preserving the key idea of conformation analysis. In order to return to the original problem of biomolecular design, we illustrate the different steps of our present algorithm when applied to a small RNA molecular system.

2 Molecular Dynamics

In order to introduce our mathematical frame, we need to fix some notation. Consider a probability space $(\mathbf{X}, \mathcal{A}, \mu)$, where $X \subset \mathbf{R}^m$ for some $m \in \mathbf{N}$ denotes the state space, $\mathcal{A}$ the Borel σ–algebra on $\mathbf{X}$ and μ a probability measure on $\mathcal{A}$.

We will see below that in classical molecular dynamics the evolution of a single molecular system with initial data $x_0 \in \mathbf{X}$ is in general described by a homogeneous Markov process $\{X_t\}_{t \in \mathbf{M}}$ with $\mathbf{M} = \mathbf{R}_0^+$ or $\mathbf{M} = \mathbf{N}_0$ in continuous or discrete time, respectively. In the deterministic case, we assume that X_t is measurable and non–singular with respect to μ, i.e., $\mu({\mathbf{X}_t}^{-1}(A)) = 0$ for all $A \in \mathcal{A}$ with $\mu(A) = 0$. The evolution of a single system starting in $x(0) = x_0$ is given by $x(t; x_0) = X_t(x_0)$ for all $t \in \mathbf{M}$. We choose this more general framework to describe molecular dynamics, since it is suitable both for the deterministic case and for the stochastic situation.

Markov processes may be defined in terms of stochastic transition kernels. A function $p : \mathbf{M} \times \mathbf{X} \times \mathcal{A} \to [0, 1]$ is called a *stochastic transition kernel* [6,30], if

1. $p(t, x, \cdot)$ is a probability measure on $\mathcal{A}$ for every $t \in \mathbf{M}$, $x \in \mathbf{X}$ and furthermore, $p(0, x, \mathbf{X} \setminus \{x\}) = 0$ for every $x \in \mathbf{X}$.
2. $p(t, \cdot, A)$ is measurable for every $t \in \mathbf{M}$, $A \in \mathcal{A}$.
3. $p(\cdot, x, A)$ satisfies the Chapman–Kolmogorov equation [19,30]
$$p(t + s, x, A) = \int_{\mathbf{X}} p(t, x, \mathrm{d}y)\, p(s, y, A) \tag{1}$$
for all $t, s \in \mathbf{M}$, $x \in \mathbf{X}$ and $A \in \mathcal{A}$.

The family $\{X_t\}_{t \in \mathbf{M}}$ is called a homogeneous Markov process, if [6,30]
$$\mathbf{P}[X_t \in A \,|X_0 = x] = p(t, x, A) \tag{2}$$
for all $t \in \mathbf{M}$ and $A \in \mathcal{A}$. Thus $p(t, x, A)$ is the probability that the Markov process started in x stays in A after the time span t.

2.1 Modelling Molecular Motion

Classical models for molecular motion describe the molecular system under consideration via coupled equations of motion for the N atoms in the system (cf. textbook [1]). The most popular class of equations of motion can be written in the following general form:

$$\begin{aligned}\dot{q} &= M^{-1}p, \\ \dot{p} &= -\nabla_q V(q) \; - \; \gamma(q,p)\,p \; + \; F_{\text{ext}},\end{aligned} \tag{3}$$

where q and p are the atomic positions and momenta, respectively, M the diagonal mass matrix and $V = V(q)$ a differentiable potential energy function describing all the interactions between the atoms. The function $\gamma = \gamma(q,p)$ denotes the friction constant and F_{ext} the external forces acting on the molecular system. The state space of the system is $\Gamma \subset \mathbf{R}^{6N}$ and the solution (q_t, p_t) of (3) describes the dynamics of a *single molecular system*. In the notation introduced above, we hence have $\mathbf{X} = \Gamma$ and $X_t(q_0, p_0) = (q_t, p_t)$.

The Hamiltonian function

$$H(q,p) = \frac{1}{2} p^T M^{-1} p + V(q). \tag{4}$$

denotes the *internal energy* of the system in state $x = (q,p)$. In the following we assume $M = \mathrm{Id}_{\mathbf{R}^{3N}}$ for simplicity. In most cases, the phase space is simply given by $\Gamma = \Omega \times \mathbf{R}^{3N}$ for some $\Omega \subset \mathbf{R}^{3N}$. We will call Ω the *position space* of the system and distinguish between two fundamentally different cases:

(B) *Bounded system*: The position space Ω is unbounded, typically $\Omega = \mathbf{R}^{3N}$, and the potential energy function is smooth, bounded from below, and satisfies $V \to \infty$ for $|q| \to \infty$. Such systems are called bounded, since the energy surfaces $\{x : H(x) = E\}$ are bounded subsets of Γ.

(P) *Periodic systems*: The position space Ω is some $3N$ dimensional torus and V is continuous on Ω and thus bounded. There is an intensive discussion concerning the question of whether V can also be assumed to be smooth on Ω as we will do herein, see Sec. 2 of [35] for details.

Both cases are typical for molecular dynamics applications. Case (P) includes the assumption of periodic boundaries which is the by far the most popular modelling assumption for biomolecular systems. Subsequently, we will refer to these assumptions by referring to systems of type (B) or type (P).

Deterministic Hamiltonian Dynamics. Whenever $\gamma \equiv 0$ and $F_{\text{ext}} = 0$, equation (3) reduces to the classical Newtonian equations of motion:

$$\dot{q} = p, \qquad \dot{p} = -\nabla_q V(q). \tag{5}$$

The flow Φ^t associated with the Hamiltonian H from (4) allows to denote the solution process of (5), i.e., $x(t;x_0) = X_t(x_0) = \Phi^t x_0$ and the transition kernel is given by

$$p(t,x,C) = \chi_C\left(\Phi^t x\right), \tag{6}$$

where χ_C denotes the characteristic function of the set $C \subset \Gamma$. In this deterministic case, the equations of motion (5) model an *energetically closed system*, i.e., the Hamiltonian denotes the *total energy* of the system, which is preserved by the dynamics.

Deterministic Thermostatted Dynamics. In general, the term $\gamma(q,p)\,p$ represents the effect of some "thermostat" on the system. In "thermostatted molecular dynamics", one designs deterministic descriptions of open but conservative molecular systems contained in a heat bath by choosing $\gamma \neq 0$ and (deterministic) forces $F_{\text{ext}} \neq 0$ such that the solution of (3) conserves either kinetic or total energy [14].

Stochastic Langevin Dynamics. The most popular model for an *open system* with stochastic interaction with its environment is the so-called Langevin model [33]:

$$\dot{q} = p, \qquad \dot{p} = -\nabla_q V(q) - \gamma p + \sigma \dot{W}. \tag{7}$$

It is a special case of (3) with some constant friction $\gamma(q,p) \equiv \gamma > 0$ and an external force $F_{\text{ext}} = \sigma \dot{W}_t$ given by a $3N$-dimensional Brownian motion W_t. The external stochastic force models the influence of the Brownian motion of the heat bath surrounding the molecular system. In this case, the internal energy H is not preserved, but the interplay between stochastic excitation and damping equilibrates the internal energy as we will see in Section 3.2.

2.2 Long-Term Behavior and Conformations

In principle, a discretization of (3) permits a simulation of single system trajectories once the initial state is given. However, numerical analysis of present discretizations restricts the validity of such single system trajectory simulations to only short time spans and to comparatively small discretization steps. The reason for this is two-fold: First, numerical long-term simulation is an ill-posed problem for the Hamiltonian systems under consideration [1], and second, no numerical integrator is available that allows stepsizes larger than a few femtoseconds—neither for Hamiltonian nor for Langevin dynamics [39,34].

On the smallest time scales of about one femtosecond molecular dynamics consists of fast oscillations or fluctuations around equilibrium positions. In contrast to these fast fluctuations the term conformations describes metastable global states of the molecule, in which the *large scale geometric structure* is understood to be conserved. *Conformational changes* are therefore rare events, which will show up only in long term simulations of the dynamics, e.g., on a nano- or millisecond time scale. Thus, the effective conformational dynamics occurs on time scales not accessible via long–term simulation. We thus have to abandon the trajectory-based approach of identifying conformations via long-term simulations. Instead, we use the dynamical properties of conformations to introduce a *set-oriented concept*:

Conformations are related to geometric structure given by the atomic positions. This means that conformations are subsets of the position space. Under additional consideration of the dynamical properties, we characterize *conformations as special "almost invariant" subsets in position space* in the following sense: From an *invariant set* the dynamical process can never escape. If conformations were *invariant sets* of the molecular dynamics, then transitions between different conformations would be *impossible*. Since transitions between conformations exist but are *rare*, we have to understand conformations as *almost invariant* sets of the molecular dynamics.

In [10], DELLNITZ AND JUNGE proposed to identify almost invariant subsets of discrete dynamical systems via specific eigenvectors of corresponding transfer operators. In order to make this intriguing idea applicable to the identification of conformations, we will introduce some notation, define transfer operators for molecular motion and link them to concepts of statistical mechanics.

3 Molecular Ensembles and Transfer Operators

We *in principle* always have to accept experimental measurement uncertainties when determining the initial state—all the positions and momenta—of some molecule. As a consequence, when modelling the physical reality, we have to propagate a *statistical ensemble* of molecular systems which represents the *distribution of possible initial states determined via the initial measurement.* The distribution may be described by some *time dependent probability density* $u = u(x,t)$ in phase space. In the following, the density u is always meant with respect to the measure μ; consequently, the probability within the ensemble to encounter a system $x \in \mathbf{X}$ in a subset $C \subset \mathcal{A}$ at time t is given by

$$\mathbf{P}[X_t \in C | X_0 \sim u] = \int_C u(x,t)\, \mu(\mathrm{d}x). \tag{8}$$

Physical experiments allow for measuring relative frequencies in the ensemble, e.g., to determine the relative frequency of systems within the ensemble whose state lies in $C \subset \mathbf{X}$ at time t. The probability $\mathbf{P}_t[x \in C]$ corresponds to the relative frequency introduced above and is thus physically measurable—in contrast to the probability density $u(x,t)$. Whenever physicists use the phrase "probability density" they refer to the density from (8) *with respect to the Lebesgue measure* $\mathrm{d}x$. This means, whenever $u(x,t)$ is the density with respect to μ and, additionally, μ is absolutely continuous with respect to $\mathrm{d}x$ with density $d(x,t)$, then the *physical density* is

$f(x,t) = u(x,t)d(x,t)$. Nevertheless, it is sometimes mathematically advantageous to consider densities with respect to specific measures particularly adapted to the Markov process under investigation.

3.1 Forward and Backward Transfer Operators

The evolution of a probability density $u = u(x,t)$ in state space $\mathbf{X}$ is governed by the (micro-) dynamics $\{X_t\}_{t\in\mathbf{M}}$ of each of the identically prepared molecular systems within the ensemble. We may describe the evolution by the *propagator* or *forward transfer operator*

$$P_t u(x) = u(x,t),$$

which maps the initial probability density $u(x) = u(x,0)$ to the density $u(x,t)$ at time t. Assume for the moment that the transition kernel of the process $\{X_t\}$ is absolutely continuous with respect to the probability measure μ, i.e., $p(t,x,C) = \int_C p(t,x,y)\,\mu(\mathrm{d}y)$. Since $p(t,x,y)$ denotes the "probability" of the process to move from x to y within the time t, the propagator should have the form

$$P_t u(y) = \int_{\mathbf{X}} p(t,x,y)\,u(x)\,\mu(\mathrm{d}x). \tag{9}$$

However, since the transition kernel can not always assumed to be absolutely continuous, the propagator is in general defined according to

$$P_t u(y)\,\mu(\mathrm{d}y) = \int_{\mathbf{X}} p(t,x,\,\mathrm{d}y)\,u(x)\,\mu(\mathrm{d}x).$$

We may also proceed in a different way and define P_t via the well-known *backward transfer operator* [19]

$$T_t u(x) = \mathbf{E}_x[u(X_t)] = \int_{\mathbf{X}} u(y)\,p(t,x,\mathrm{d}y), \tag{10}$$

where $\mathbf{E}_x[u(X_t)]$ denotes the expectation of an observable $u : \mathbf{X} \to \mathbf{C}$ under the condition that the process $\{X_t\}$ has been started at $t = 0$ in x.

Consider T_t as an operator on $L^\infty_\mu(\mathbf{X})$ and P_t on $L^1_\mu(\mathbf{X})$, and let $\langle\cdot,\cdot\rangle_\mu$ denote the duality bracket between $L^\infty_\mu(\mathbf{X})$ and $L^1_\mu(\mathbf{X})$. Then, as a generalization of (9), the forward transfer operator P_t is *defined as the adjoint operator* $P_t = T_t^*$ of the backward transfer operator T_t [19], i.e.,

$$\langle T_t u, v\rangle_\mu = \langle u, P_t v\rangle_\mu, \quad \text{for all } u \in L^\infty_\mu(\mathbf{X}), v \in L^1_\mu(\mathbf{X}). \tag{11}$$

Since $p(t,x,\cdot)$ is a transition kernel, the thereby defined operator P_t is a *Markov operator* on $L^1_\mu(\mathbf{X})$. Furthermore, the semigroup property of the Markov process implies that $\{P_t\}_{t\in\mathbf{M}}$ is a semigroup of Markov operators.

In view of equations (9) and (10), the notion of "forward" and "backward" transfer operator becomes clearer. For the forward case, the state average with respect to u is taken over all initial states x, which are propagated forward in time, while for the backward case, the state average is taken over all final states y.

Invariant Measures and Stationary Densities A measure μ on $\mathbf{X}$ is called *invariant* with respect to the process $\{X_t\}$, if

$$\mu(C) = \int_X p(t,x,C)\,\mu(dx), \quad \text{for all } C \in \mathcal{A} \text{ and } t \in \mathbf{M}.$$

Due to the properties of the transition kernel and the definition of the backward transfer operator, we have—independent of the measure μ—for every $t \in \mathbf{M}$,

$$T_t \chi_{\mathbf{X}} \;=\; \chi_{\mathbf{X}}.$$

The above equality does in general not hold for the forward transfer operator, because P_t depends via (11) on the probability measure μ. However, if we assume μ to be invariant, we also get

$$P_t \chi_{\mathbf{X}} \;=\; \chi_{\mathbf{X}} \tag{12}$$

for all $t \in \mathbf{M}$. In other words, $\chi_{\mathbf{X}}$ is an invariant density of P_t, whenever μ is invariant.

Remark. Suppose additionally that μ admits a density d with respect to the Lebesgue measure. Let moreover the ensemble be distributed according to μ so that d is the stationary physical probability density of the ensemble. Then, $f(\cdot,0) = \chi_C \cdot d$ denotes the physical density of the *subensemble* of all systems being in $C \subset \mathbf{X}$ at some time $t=0$. Since P_t denotes the evolution of the ensemble in time t, the physical density of the subensemble at time t is given by $f(\cdot,t) = P_t\chi_C \cdot d$. In contrast to this, $T_t\chi_C = p(t,\cdot,C)$ denotes the probability density to *access* C at time t. This again emphasizes the difference in interpretation between P_t and T_t: P_t denotes the physically interpretable propagator of the ensemble and is defined with respect to some measure μ, while T_t denotes the transfer operator related to the Markov process (independent of the measure μ) as usually considered in stochastic theory.

3.2 Canonical Ensemble

Most experiments on molecular systems are performed under the conditions of constant temperature $\mathcal{T}$ and volume. The corresponding ensemble density (with respect to the Lebesgue measure on $\mathbf{X}$) is the *canonical density* f_{can} associated with the Hamiltonian H:

$$f_{\mathrm{can}}(x) = \frac{1}{Z}\exp\left(-\beta\, H(x)\right), \qquad Z = \int_\Gamma \exp\left(-\beta\, H(x)\right) dx, \tag{13}$$

where $\beta = 1/k_B\mathcal{T}$ denotes the inverse temperature and k_B Boltzmann's constant. Since H was assumed to be separable, f_{can} factorizes in a product of two densities $\mathcal{P}$ and $\mathcal{Q}$:

$$f_{\text{can}}(x) = \underbrace{\frac{1}{Z_p} \exp\left(-\frac{\beta}{2} p^T M^{-1} p\right)}_{=\mathcal{P}(p)} \underbrace{\frac{1}{Z_q} \exp\left(-\beta V(q)\right).}_{=\mathcal{Q}(q)} \tag{14}$$

Since we are interested in the canonical ensemble, we define the *canonical probability measure* induced by the canonical density:

$$\mu_{\text{can}}(\mathrm{d}x) = f_{\text{can}}(x)\,\mathrm{d}x.$$

It will turn out advantageous to consider transfer operators acting on weighted function spaces with respect to μ_{can}. In the sequel, we restrict our attention to potentials V, such that μ_{can} is a probability measure.

3.3 Transfer Operators and the Canonical Ensemble

In general, an equation of motion for the process $\{X_t\}$ implies an equation of motion for a probability density u. We will see below that the processes induced by both, the Hamiltonian dynamics and the Langevin dynamics, leave the canonical measure μ_{can} invariant. Since we are interested in describing fluctuations within the canonical ensemble, we thus define the forward transfer operator with respect to the canonical probability measure μ_{can}, i.e., acting on $L^1_{\mu_{\text{can}}}(\mathbf{X})$.

Langevin Dynamics. The process induced by the Langevin equation (7) leaves the canonical measure μ_{can} corresponding the the inverse temperature β invariant, if the noise and damping constants satisfy [33]:

$$\beta = \frac{2\gamma}{\sigma^2}. \tag{15}$$

The evolution of $u = u(x,t)$ with respect to μ_{can} (compare introduction to Section 3) is governed by the well-known *Fokker–Planck equation* [33]:

$$\partial_t u = \left(\underbrace{\frac{\sigma^2}{2}\Delta_p + p\cdot\nabla_q - \nabla_q V\cdot\nabla_p - \gamma p\cdot\nabla_p}_{=A} \right) u. \tag{16}$$

As a consequence, the Fokker–Planck operator A is the infinitesimal generator of the semigroup of forward transition operators $\{P_t\}_{t\in\mathbf{R}_0^+}$ acting on $L^1_{\mu_{\text{can}}}(\mathbf{X})$ with

$$P_t u = \exp(tA)u \tag{17}$$

and, since μ_{can} is invariant, we have $P_t\chi_\Gamma = \chi_\Gamma$.

Moreover, under certain conditions on the potential V (systems of type (B) with potential $V \in C^\infty(\mathbf{X})$), this is the *unique stationary density* and the semigroup $\{P_t\}_{t\in\mathbf{R}_0^+}$ is asymptotically stable [24], i.e., $P_t u \to \chi_\Gamma$ for $t \to \infty$ and every density $u \in L^1_{\mu_{\text{can}}}(\Gamma)$. Due to this property, the Langevin equation is the most prominent stochastic model for a heat bath driven relaxation of molecular ensembles to the canonical ensemble.

Hamiltonian Dynamics. The Hamiltonian equations of motion are the deterministic analogue of the Langevin equations with $\gamma = 0$ and $\sigma = 0$. As for the Langevin dynamics, the canonical probability measure μ_{can} is invariant under the dynamics. Using $\gamma = \sigma = 0$, the equation of motion (16) for the probability density u reduces to the *Liouville equation* corresponding to the Hamiltonian H:

$$\partial_t u = \Bigg(\underbrace{+p\cdot\nabla_q \; - \; \nabla_q V \cdot \nabla_p}_{=i\mathcal{L}} \Bigg) u \tag{18}$$

where $\mathcal{L}$ denotes the well-known Liouville operator [26]. The solution of (18) satisfies $u(x, t+s) = u(\Phi^{-t}x, s)$ for all $t, s \in \mathbf{R}_0^+$. Using (17), the forward transfer operator acting on $L^1_{\mu_{\text{can}}}(\Gamma)$ is given by

$$P_t u(x) \;=\; \exp(it\mathcal{L})u(x) \;=\; u(x,t) \;=\; u\left(\Phi^{-t}x\right), \tag{19}$$

which is just the definition of the *Frobenius–Perron operator* corresponding to the Hamiltonian flow Φ^t [27]. Additionally, inserting the transition kernel (6) in the definition (10) of the backward transfer operator yields

$$T_t u(x) = u\left(\Phi^t x\right), \tag{20}$$

which is simply the *Koopman operator* corresponding to Φ^t [27]. Equations (19) and (20) illustrate that P_t is the adjoint operator of T_t as discussed above.

As we have seen, the canonical density f_{can} induces the *invariant measure* μ_{can} of the deterministic Hamiltonian dynamics. However, there are infinitely many other invariant measures induced by densities of the form $f(x) = \mathcal{F}(H(x))$ for some smooth function $\mathcal{F} : \mathbf{R} \to [0,1]$ of the Hamiltonian. Furthermore, the entire discrete spectrum of the Frobenius–Perron operator P lies on the unit circle and P cannot be asymptotically stable [23]. Due to this ambiguity, pure Hamiltonian dynamics is not appropriate for modelling the relaxation of molecular ensembles to one specific ensemble, in our case the canonical ensemble. This observation corresponds to the fact that, for solving the Liouville equation, we have to specify an initial density $u(\cdot, t=0)$. Physically, the specification of an initial density corresponds to an *initial experimental preparation* of the ensemble due to (8). Thus, selecting one of the possible invariant densities means the specific initial preparation of a stationary ensemble.

4 Almost Invariant Sets of Molecular Ensembles

Assume in this section that the molecular motion is described by a Markov process $\{X_t\}_{t\in\mathbf{M}}$ that leaves the probability measure μ invariant. Moreover, assume that the Markov process is initially distributed according to μ, i.e., the probability to find the process at time $t=0$ in a subset $C \in \mathcal{A}$ is given by

$$\mathbf{P}_\mu[X_0 \in C] = \mathbf{P}[X_0 \in C | X_0 \sim \mu] \;=\; \mu(C)$$

(see introduction to Section 3).

4.1 Ensemble Transition Probabilities

The *transition probability* $p(s, C, D)$ within the ensemble from $C \in \mathcal{A}$ to $D \in \mathcal{A}$ within the time span s is defined as the conditional probability

$$p(s,C,D) \;=\; \mathbf{P}_\mu[X_s \in D \,|X_0 \in C] \;=\; \frac{\mathbf{P}_\mu[X_s \in D \text{ and } X_0 \in C]}{\mathbf{P}_\mu[X_0 \in C]}. \tag{21}$$

The similar symbols for both the transition probability $p(s, C, D)$ and for the transition kernel $p(s, x, C)$ corresponding to the process emphasizes the strong relation to (2), which, in addition to the above assumption, allows to rewrite the transition probability as

$$p(s,C,D) = \frac{1}{\mu(C)} \int_C p(s,x,D)\,\mu(\mathrm{d}x). \tag{22}$$

The transition probabilities quantify the *dynamical fluctuations within the stationary ensemble.* Using the duality bracket $\langle\cdot,\cdot\rangle_\mu$ between $L^\infty_\mu(\mathbf{X})$ and $L^1_\mu(\mathbf{X})$, the definitions of the transfer operators T_t and P_t yield

$$p(s,C,D) = \frac{\langle T_s\chi_D, \chi_C\rangle_\mu}{\langle\chi_C, \chi_C\rangle_\mu} \;=\; \frac{\langle\chi_D, P_s\chi_C\rangle_\mu}{\langle\chi_C, \chi_C\rangle_\mu}. \tag{23}$$

The above defined transition probabilities can be *measured* via the following two–step experiment on the ensemble:

1. *Pre-Selection:* Select from the ensemble all such systems with states $x \in C \in \mathcal{A}$. This selection prepares a new ensemble, which is described by the probability measure
$$\mu_C(D) = \frac{1}{\mu(C)}\,\mu(C \cap D), \quad D \in \mathcal{A}.$$
2. *Transition-Counting:* After the time span s, determine the *relative frequency* of systems in the ensemble μ_C with states in C. Since all systems evolve due to the process $\{X_t\}_{t\in\mathbf{M}}$, this relative frequency is equal to
$$\int_{\mathbf{X}} p(s,x,C)\,\mu_C(dx) = p(s,C,C).$$

4.2 Conformations as Almost Invariant Subsets

We now aim at a dynamical characterization of conformations within the ensemble; this characterization will be based on the notion of almost invariance. As already mentioned, we have to define almost invariance in terms of ensemble dynamics rather than in terms of the duration of stay of a single system.

Following [10], we call some subset $C \in \mathcal{A}$ *almost invariant*, whenever the fraction of systems within the ensemble that stay in C after some characteristic time span $s \in \mathbf{M}$ is close to 1:

$$C \text{ almost invariant} \Longleftrightarrow p(s, C, C) \approx 1.$$

This definition of almost invariance guarantees that its "degree" $p(s, C, C)$ can be *measured* via the two-step experiment introduced above.

Almost invariance may equivalently be characterized by $p(s, C, \mathbf{X} \setminus C) \approx 0$, which allows to relate it to the semigroup of forward transfer operators $\{P_t\}_{t \in \mathbf{M}}$ by the following general identity [38]:

$$\left\| P_s \frac{1}{\mu(C)} \chi_C - \frac{1}{\mu(C)} \chi_C \right\|_1 = 2\, p(s, C, \mathbf{X} \setminus C). \tag{24}$$

4.3 Identification Strategy

By definition, P_s is a Markov operator and consequently, its $L^1_\mu(\mathbf{X})$–spectrum is contained in the unit ball $\{\lambda \in \mathbf{C} : |\lambda| \leq 1\}$. Every invariant density $u \in L^1_\mu(\mathbf{X})$ of P_s satisfies $P_s u = u$ and therefore is an eigenvector of P_s corresponding to the eigenvalue $\lambda = 1$, the so–called *Perron root*. Since μ is assumed to be invariant, in particular $u = \chi_X$ is an invariant density.

Whenever a proper subset C of $\mathbf{X}$ is invariant under the Markov process, i.e., $p(t, x, \mathbf{X} \setminus C) = 0$ for all $x \in C$, the density $u = \chi_C / \mu(C)$ is an eigenvector corresponding to $\lambda = 1$.

Due to our above characterisation, the set $C \in \mathcal{A}$ is almost invariant if $p(\tau, C, \mathbf{X} \setminus C) \approx 0$, which via formula (24) implies that $\chi_C / \mu(C)$ is an approximate invariant density, i.e., an approximate normalized eigenvector associated with an eigenvalue close to the Perron root $\lambda = 1$. This motivates the following *algorithmic strategy*:

> *Invariant* sets can be identified via eigenvectors corresponding to the Perron root $\lambda = 1$, while *almost invariant* sets may be identified via eigenvectors corresponding to eigenvalues $|\lambda| < 1$ close to the Perron root $\lambda = 1$.

This strategy has first been proposed by Dellnitz and Junge [10] for discrete dynamical systems with weak random perturbations and has been successfully applied to molecular dynamics in different contexts [36,37,35]. It will be justified in more detail in Section 5.4 below, where more information about the properties of the transfer operators of interest will be available.

It is important to notice that almost invariance is defined herein with respect to some physically selected invariant probability measure μ that describes the stationary ensemble under consideration. Assume that the process $\{X_t\}$ admits another invariant measure ν, which, for the sake of simplicity, is absolutely continuous with respect to μ with density $d \in L^1_\mu(\mathbf{X})$. Then, the density $u = \chi_{\mathbf{X}} d$ is an eigenvector of P_s corresponding to $\lambda = 1$. As a consequence, one will not be able to decide in general whether some eigenvector corresponding to an eigenvalue $|\lambda| < 1$ close to the Perron root is related to an almost invariant subset of the ensemble represented by μ or rather by ν. Thus, the above algorithmic strategy requires uniqueness of the invariant measure. For its numerical realization via an eigenvalue problem we moreover need that the remaining spectrum of P_s is strictly bounded away from the Perron root, i.e., $\lambda = 1$ must be an isolated, simple eigenvalue of P_s. Additionally, the physical interpretation of the ensemble excludes other eigenvalues than $\lambda = 1$ *on* the unit circle or, equivalently, we exclude asymptotic periodicity of P_s.

We introduce the following two *fundamental conditions* on the forward transfer operator P_s that are sufficient to guarantee the desired properties:

(C1) P_s is asymptotically stable, i.e., $(P_s)^n u \to \chi_{\mathbf{X}}$ in $L^1_\mu(\mathbf{X})$ for every density $u \in L^1_\mu(\mathbf{X})$ as $n \to \infty$.

(C2) The essential spectrum of P_s in $L^1_Q(\Omega)$—and therefore in $L^2_Q(\Omega)$—is strictly bounded away from $|\lambda| = 1$.

These conditions exclude some very prominent models for molecular motion. For example, in the pure Hamiltonian case the invariant density is *not* unique in $L^1(\mathbf{X})$, and, worse, the spectrum of the Frobenius–Perron operator P_s in $L^1(\mathbf{X})$ lies *on* the unit circle[1]. Despite these fundamental problems, DEUFLHARD et al. computed almost invariant subsets of Hamiltonian systems in the above sense with quite intriguing results [11]. However, they did not use the exact Hamiltonian flow Φ^t but added small, but significant perturbations originating from time discretization errors and the related energy fluctuations. It is a widely accepted approach to model such discretization effects by small random perturbations. Under appropriate conditions, the thereby resulting transfer operator is compact and may have a unique invariant measure (see [10]). In [11] another interpretation of this approach via a sequence of nested function spaces based on subsequent coverings of the energy cell is indicated.

There are other models that satisfy our conditions without additional artificial perturbations. An example is the Langevin model introduced above. For

[1] Here, $L^1(\mathbf{X})$ may be replaced by $L^1_\mu(\mathbf{X})$ where μ may stand for μ_{can} or for any other invariant measure of the Hamiltonian flow Φ^t that is absolutely continuous with respect to the Lebesgue measure on the phase space $\mathbf{X}$.

appropriate systems (see above), its unique invariant measure is the canonical measure. Hence, application of our algorithmic strategy to the Langevin model seems to allow to attack chemically interesting systems. However, there is another condition which has to be considered and prevents the Langevin model from being a good starting point: *Chemical conformations are usually understood to be objects in position space Ω.* Therefore, a proper model needs to yield a family of forward transition operators, which are defined on $\mathbf{X} = \Omega$ rather than on the entire phase space Γ of the molecular systems.

5 Conformational Dynamics in Position Space

Since conformations are objects in position space, this section is devoted to an adequate theory of ensemble dynamics in position space, including two examples. We introduce two (reduced) Markov processes in position space and define the corresponding transfer operators. Due to physical reasons and as a consequence of (23), we restrict ourselves to the semigroup of forward transfer operators or propagators $\{P_t\}_{t\in\mathbf{M}}$ for the canonical ensemble $\mathcal{Q}$.

5.1 Positional Dynamics and Transfer Operators

Let $(\Omega, \mathcal{A}, \mu_{\mathcal{Q}})$ denote the positional probability space with $\mu_{\mathcal{Q}}(\mathrm{d}q) = \mathcal{Q}(q)\mathrm{d}q$ and refer by $L^r{}_{\mathcal{Q}}(\Omega)$ for $r = 1, 2, \ldots, \infty$ to the corresponding function spaces with respect to the canonical measure $\mu_{\mathcal{Q}}$. Note that $L^2{}_{\mathcal{Q}}(\Omega)$ is a Hilbert space with scalar product

$$\langle u, v\rangle_{\mathcal{Q}} = \int_{\Omega} u^*(q)\, v(q)\, \mathcal{Q}(q)\, \mathrm{d}q$$

and induced norm $\|u\|^2_{\mathcal{Q}} = \langle u, u\rangle_{\mathcal{Q}}$.

As a consequence of Subsection 4.3 we have to transform the *state space* dynamics into a pure *position space* dynamics. Assume that the transformed dynamics of a single system in Ω is described by a (homogeneous) Markov process $\{Q_t\}_{t\in\mathbf{M}}$ with stochastic transition kernel $p(t, q, C)$, invariant measure μ_Q and initial distribution $\mathbf{P}_\mu[Q_0 \in C] = \mu_Q(C)$. Then, the semigroup of forward transfer operators $\{P_t\}_{t\in\mathbf{M}}$ for the canonical ensemble is given by $P_t : L^r{}_{\mathcal{Q}}(\Omega) \to L^r{}_{\mathcal{Q}}(\Omega)$ such that for all $C \in \mathcal{A}$

$$\int_C P_t u(q)\mathcal{Q}(q)\, \mathrm{d}q = \int_{\Omega} u(q) p(t, q, C)\mathcal{Q}(q)\, \mathrm{d}q \tag{25}$$

under suitalbe conditions of the integrability of the transition kernel. In the following, we will consider P_t also as an operator acting on the Hilbert space $L^2{}_{\mathcal{Q}}(\Omega)$, since—as we will see below—the corresponding scalar product may reveal possible additional properties of P_t and allows to define Galerkin projections for the discretization procedure.

We conclude by stating all assumptions on the transfer operators, which result from the requirements of Subsection 4.3:

(C1) P_s is asymptotically stable, i.e., $(P_s)^n u \to \chi_\Omega$ in $L^1_\mathcal{Q}(\Omega)$ for $n \to \infty$ and every density $u \in L^1_\mathcal{Q}(\Omega)$. This implies that $\lambda = 1$ is an isolated, simple eigenvalue in $L^2_\mathcal{Q}(\Omega)$.

(C2) The essential spectrum of P_s in $L^1_\mathcal{Q}(\Omega)$—and therefore in $L^2_\mathcal{Q}(\Omega)$—is strictly bounded away from $|\lambda| = 1$.

5.2 Discrete Time Markov Chain

The first example of a reduced positional dynamics is based on the Hamiltonian equation of motion within the canonical ensemble f_{can} (14) and a characterization of conformations as special almost invariant subsets. A subset $C \subset \Omega$ of the position space is called almost invariant, if the enlarged "cylindrical" subset $C \times \mathbf{R}^d \subset \Gamma$ of the state space is almost invariant with respect to the Hamiltonian dynamics.

Let $p_\mathbf{X}(t, x, A)$ denote the stochastic transition kernel of the Markov process in state space (see (6)). Fix an observation time span $\tau > 0$. Then, $C \subset \Omega$ is almost invariant (with respect to τ), if $p_\mathbf{X}(\tau, C \times \mathbf{R}^d, C \times \mathbf{R}^d) \approx 1$. For *fixed* τ, this definition can be used to derive a reduced dynamics in position space. For two subsets $C, D \subset \Omega$, we have due to (6) and (22):

$$
\begin{aligned}
&p_\mathbf{X}(\tau, C \times \mathbf{R}^d, D \times \mathbf{R}^d) \\
&\quad = \frac{1}{\int_{C\times\mathbf{R}^d} f_{\text{can}}(x)\,\mathrm{d}x} \int_{C\times\mathbf{R}^d} \chi_{D\times\mathbf{R}^d}(\Phi^\tau(x)) f_{\text{can}}(x)\,\mathrm{d}x \\
&\quad = \frac{1}{\int_C \mathcal{Q}(q)\,\mathrm{d}q} \int_C \underbrace{\int_{\mathbf{R}^d} \chi_D(\Pi_q \Phi^\tau(q,p)) \mathcal{P}(p)\,\mathrm{d}p}_{=:p^\tau_\Omega(1,q,D)}\ \mathcal{Q}(q)\,\mathrm{d}q. \qquad (26) \\
&\quad = p^\tau_\Omega(1, C, D),
\end{aligned}
$$

where Π_q denote the projection onto the position space. A comment on the dependence of the one-step transition probability $p^\tau_\Omega(1, C, D)$ on the observation time span τ can be found in the remark below. It is easy to show, that $p^\tau_\Omega(1, q, D)$ is a transition kernel and thus defines a *discrete time* Markov process $\{Q_n\}_{n\in\mathbf{N}_0}$ on the position space Ω. Furthermore, $\{Q_n\}$ satisfies inductively for all $n \in \mathbf{N}_0$ the stochastic dynamical systems [35]

$$Q_{n+1} = \Pi_q \Phi^\tau(Q_n, P_n) \qquad (27)$$

with P_n chosen randomly according to the momenta distribution $\mathcal{P}$. The SDE (27) is the reduced positional dynamics that we were looking for. In mathematical terms, it corresponds to a Hamiltonian motion with randomly chosen momenta at discrete (physical) times $\tau, 2\tau, \ldots$. As shown in [35], $\{Q_n\}$ leaves the canonical ensemble $\mathcal{Q}$ invariant.

Via Equation (25), the transition kernel also defines a discrete time semigroup of transition operators $\{P_n\}_{n\in\mathbf{N}_0}$ on $L^r_\mathcal{Q}(\Omega)$. Exploiting that Φ^τ is a reversible, symplectic and μ_Q invariant mapping (see (19) and below, and [35]), we get

$$P_1 u(q) = \int_{\mathbf{R}^d} u(\Pi_q \Phi^{-\tau}(q,p))\mathcal{P}(p)\,\mathrm{d}p \tag{28}$$

for $u \in L^r_Q(\Omega)$. For all systems of type (P), P_t satisfies the requirements stated in Subsection 5.1 [35,23]; furthermore, it is self-adjoint on $L^2_Q(\Omega)$ due to reversibility and symplecticness of the Hamiltonian flow [35]. As a consequence, the $L^2_Q(\Omega)$-spectrum of P_t is real-valued, bounded and contained in the interval $(-1,1]$.

Remark. In (26), we have defined the *one step* transition kernel $p^\tau_\Omega(1,q,D)$ for fixed τ. Changing the observation time to σ results in a new one step transition kernel $p^\sigma_\Omega(1,q,D)$. In contrast to that, the n-term transition kernel $p^\tau_\Omega(n,q,D)$ is defined recursively by the Chapman–Kolmogorov equation (1). In general, $p^{2\tau}_\Omega(1,q,D) \neq p^\tau_\Omega(2,q,D)$ and, consequently, $P^{2\tau}_1 \neq P^\tau_2$, where the superscript indicates the corresponding observation time span (for an example, see [35, Sec. 3.7.1]). In terms of the SDE (27), this is not surprising, since $P^{2\tau}_1$ includes only one choice of momenta according to $\mathcal{P}$, while P^τ_2 does include two.

5.3 High–Friction Langevin Dynamics

The second example of a reduced positional dynamics is based on the Langevin equation. We will see that in a specific high friction limit $\gamma \to \infty$ the Langevin equation acting on the state space reduces to the so–called high–friction Langevin equation acting only on the position space.

Consider the Langevin equation (7) written in second order form

$$\ddot{q} = -\nabla_q V(q) - \gamma\dot{q} + \sigma\dot{W}. \tag{29}$$

For the high friction case, let ϵ be a small positive parameter and consider the transformed friction γ/ϵ. In order to preserve the temperature $\mathcal{T}$ of the surrounding heat bath, we simultaneously have to scale the white noise constant $\sigma \mapsto \sigma/\sqrt{\epsilon}$ due to (15). This yields

$$\ddot{q} = -\nabla_q V(q) - \frac{\gamma}{\epsilon}\dot{q} + \frac{\sigma}{\sqrt{\epsilon}}\dot{W}.$$

After rescaling the time according to $t \mapsto \epsilon\, t$ one gets

$$\epsilon^2\ddot{q} = -\nabla_q V(q) - \gamma\dot{q} + \sigma\dot{W}.$$

For systems of type (B), for which the gradient of V satisfies a global Lipschitz condition, and $0 < \epsilon \ll 1$ one may neglect the ϵ^2–term [31, Thm. 10.1] and finally get the high–friction Langevin equation

$$\dot{q} = -\frac{1}{\gamma}\nabla_q V(q) + \frac{\sigma}{\gamma}\dot{W} \tag{30}$$

modelling the high friction positional dynamics within the canonical ensemble. The stochastic differential system (30) defines a *continuous time* Markov process $\{Q_t\}_{t\in\mathbf{R}_0^+}$ on the position space Ω with corresponding transition kernel $p(t, q, C)$. The process leaves the canonical measure μ_Q invariant [33].

Remark. In contrast to the usual quasistatic approximation in mechanics, we cannot simply assume that the accelaration $\ddot{q}$ is bounded since the white noise process is unbounded. However, the investigation in [31] shows that the Langevin solution $q^{\epsilon}_{\text{Lan}}(t; q_0, p_0)$ and the solution $q_{\text{fric}}(t; q_0)$ of (30) satisfy for all p_0, with probability one: $\lim_{\epsilon\to 0} |q_{\text{fric}}(t) - q^{\epsilon}_{\text{Lan}}(t)| = 0$ uniformly for t in compact subintervals of $[0, \infty)$. However, this does not necessarily imply that the corresponding conformation coincide.

As for the general Langevin dynamics (16) in state space, the continuous time semigroup of forward transition operators $\{P_t\}_{t\in\mathbf{R}_0^+}$ may be defined in terms of its infinitesimal generator [19]:

$$A = \frac{\sigma^2}{2\gamma^2}\Delta_q - \frac{1}{\gamma}\nabla_q V(q)\cdot\nabla_q \tag{31}$$

acting on a suitable subspace of $L^r{}_{\mathcal{Q}}(\Omega)$. As a consequence, one gets

$$\begin{aligned} P_t : L^r{}_{\mathcal{Q}}(\Omega) &\to L^r{}_{\mathcal{Q}}(\Omega) \\ u \mapsto P_t u &= \exp(tA)u. \end{aligned} \tag{32}$$

Thus, every probability density $u = u(q, t)$ with respect to μ_Q evolves according to the Fokker–Planck equation $\partial_t u = Au$ and its solution is formally given by (32).

It is shown in [37,4] that for systems of type (B) the semigroup of forward transition operators satisfies the requirements of Subsection 5.1. Furthermore, $\{P_t\}_{t\in\mathbf{R}_0^+}$ is a self–adjoint semigroup in $L^2{}_{\mathcal{Q}}(\Omega)$, since the infinitesimal generator A is self–adjoint with respect to $\langle\cdot,\cdot\rangle_{\mathcal{Q}}$ [37].

Remark. The physical density $f(q, t) = u(q, t)\mathcal{Q}(q)$ (see introduction to Section 3) evolves according to the Fokker–Planck equation $\partial_t f = A^* f$, where A^* denotes the formal adjoint of A, i.e., $A^* = \sigma^2/(2\gamma^2)\Delta_q + 1/\gamma\nabla_q V(q)\cdot\nabla_q + 1/\gamma\Delta_q V(q)$ (see [19]).

Almost Invariance and First Exit Times. Recall that $\mathbf{P}_\mu[Q_0 \in C] = \mu_Q(X)$ by assumption (see Section 4). Due to experimental requirements, almost invariance of conformations is defined at *discrete points in time* (see Eq. (21)):

$$p_{\text{discr}}(t, C, C) = \frac{\mathbf{P}_\mu[Q_s \in C : s = 0 \text{ and } s = t]}{\mathbf{P}_\mu[Q_0 \in C]}. \tag{33}$$

This definition also holds for the continuous time Markov processes. However, one could alternatively want to characterize almost invariance of conformations based on *continuous time observations*:

$$p_{\mathrm{cont}}(t, C, C) = \frac{\mathbf{P}_\mu[Q_s \in C : \text{ for all } s \in [0, t]]}{\mathbf{P}_\mu[Q_0 \in C]}. \tag{34}$$

Obviously, the two definitions will in general produce different result, since the former definition does not take into account fluctuations in between the two instances. However, in contrast to the latter definition, the former one can be realized by the two-step experiment from Section 4.1.

Mathematically, both characterizations are closely related by Fokker–Planck equations on appropriate function spaces. Let τ_C^q denote the *first exit time* of the Markov process $\{Q_t\}_{t\in\mathbf{R}_0^+}$, started at time zero in $q \in C$, from an open subset $C \subset \Omega$,

$$\tau_C^q = \inf\{t \geq 0 : Q_t(q) \notin C\}. \tag{35}$$

For open, bounded subsets C with sufficiently smooth boundary ∂C the distribution of exit times $v_C(q, t) = \mathbf{P}[\tau_C^q > t] = \mathbf{P}[Q_s(q) \in C : \text{ for all } s \in [0, t]]$ for $q \in C$ is given by the Fokker–Planck equation on $C \cup \partial C$ with *Dirichlet boundary conditions*:

$$\partial_t v = Av, \qquad \begin{array}{l} v(\cdot, 0) = \chi_C \text{ on } C \text{ and} \\ v(\cdot, t) = 0 \text{ for all } t \geq 0 \text{ on } \partial C. \end{array}$$

In contrast, $u_C(q, t) = \mathbf{P}_\mu[Q_s(q) \in C : s = 0 \text{ and } s = t]$ satisfies the Fokker–Planck *Cauchy problem* on Ω:

$$\partial_t u = Au, \qquad u(\cdot, 0) = \chi_C.$$

(with implicit "transparent boundary conditions"). With respect to the above two characterization of almost invariance, we finally get

$$p_{\mathrm{discr}}(t, C, C) = \frac{1}{\mu_{\mathcal{Q}}(C)} \int_C u_C(q, t)\mathcal{Q}(q)\,\mathrm{d}q$$

and

$$p_{\mathrm{cont}}(t, C, C) = \frac{1}{\mu_{\mathcal{Q}}(C)} \int_C v_C(q, t)\mathcal{Q}(q)\,\mathrm{d}q.$$

5.4 Justification of the Algorithmic Strategy

Here, we want to pick up the algorithmic strategy presented in Section 4.3 and state more precisely how one can use eigenvectors corresponding to eigenvalues near the Perron root 1 in order to identify almost invariant subsets. In the following, we fix a time $s \in \mathbf{M}$ and—in accordance with the properties of the above two examples—we assume that the transition operator P_s is self-adjoint in $L^2_{\mathcal{Q}}(\Omega)$. Moreover, for the sake of simplicity, we restrict our considerations to the case that the Perron root is "nearly two-fold degenerate", i.e., we assume that the spectrum of P_s has the form

$$\sigma(P_s) \subset [-r, r] \cup \{\lambda_2\} \cup \{1\},$$

with $0 < r < \lambda_2 < \lambda_1 = 1$; furthermore, we assume that λ_1 and λ_2 are simple eigenvalues. The eigenvector corresponding to $\lambda_1 = 1$ is χ_Ω, while we denote the eigenvector corresponding to λ_2 by $\phi \in L^2_{\mathcal{Q}}(\Omega)$ with normalization $\|\phi\|_{\mathcal{Q}} = 1$. Note that $\langle \phi, \chi_\Omega \rangle_{\mathcal{Q}} = 0$.

Nonrigorous Approach. One intuitive idea is to interpret almost invariance as "perturbed invariance". Therefore, we assume that the above transition operator results from a *continuous* perturbation of some self-adjoint Markov operator $\tilde{P}$ with degenerate, two-fold Perron root and invariant measure μ. If the degeneracy of the Perron root is caused by the existence of two disjoint *invariant* sets, say C and $C^c = \Omega \setminus C$, the eigenspace E_1 of the Perron root is spanned by the eigenvectors χ_C and χ_{C^c}. Neither C nor C^c are invariant sets of P_s, however, $\chi_C/\mu(C)$ and $\chi_{C^c}/\mu(C^c)$ remain to be "approximative" invariant densities of P_s, in the sense that (compare Section 4.3)

$$\left\| P_s \frac{1}{\mu(C)} \chi_C - \frac{1}{\mu(C)} \chi_C \right\|_1 \approx 0.$$

By means of the *general formula* (24), this implies that C as well as C^c are *almost invariant* sets of P_s. Since χ_Ω is a common eigenvector of $\tilde{P}$ and P_s, we choose another orthonormal basis of $E_1 = \text{span}\{\chi_\Omega, u_C\}$ with

$$u_C = \sqrt{\frac{\mu(C^c)}{\mu(C)}} \chi_C - \sqrt{\frac{\mu(C)}{\mu(C^c)}} \chi_{C^c}. \tag{36}$$

Since P_s is assumed to be a continuous perturbation of $\tilde{P}$, we have to expect that the so-defined u_C is an approximation of the eigenvector ϕ of P_s corresponding to λ_2. This motivates the algorithmic strategy to identify the almost invariant sets via the second eigenvector ϕ (or $-\phi$) according to

$$C \approx \{q : \ \phi(q) > 0\} \text{ and } C^c \approx \{q : \ \phi(q) \leq 0\}. \tag{37}$$

For more details concerning the *identification algorithm* for the more general case see [13].

Rigorous Approach. Although the perturbation analysis yields an intuitive understanding of the form of the second eigenvector of P_s, we subsequently will *not* assume any kind of perturbation embedding of P_s but rather proceed in another way towards a rigorous justification of the following "equivalence":

Decomposition into almost invariant subsets $\Omega = C \cup C^c$: $p(s, C, C^c) \approx 0$, $C \approx \{q : \ \phi(q) > 0\}$	$\Longleftrightarrow$	Eigenvalue cluster $\{1, \lambda_2\}$ separated from remaining spectrum: $\epsilon = \frac{1-\lambda_2}{1-r} \ll 1$	(38)

The following rigorous statements are closely similar to the results of E.B. DAVIES [7]. To simplify reference to his results, let us denote by η_2 and ρ the positive values with

$$\lambda_2 = \exp(-s\eta_2) \quad \text{and} \quad r = \exp(-s\rho),$$

where s denotes the initially fixed time span. For the "$\Leftarrow$"-direction in (38), we assume that $\epsilon = (1-\lambda_2)/(1-r)$ is small enough, and introduce $c = \|\phi\|_\infty$ satisfying $c \geq 1$. Due to [7],[2] there exists $C \in \mathcal{A}$ given by $C = \{q : \phi(q) > 0\}$ such that $\frac{1}{2c^2} \leq \mu(C) \leq 1 - \frac{1}{2c^2}$ and

$$\|\phi - u_C\|_2 \leq 4c\sqrt{\epsilon}.$$

Furthermore, the subset C is almost invariant with

$$p(ns, C, C^c) \leq K\epsilon\,(1 + \rho n s), \quad \text{for all } n \in \mathbf{N},$$

where K depends on c and $\mu(C)$, and is independent of ϵ.

For the "$\Rightarrow$"-direction in (38), we assume that C is almost invariant with

$$p(ns, C, C^c) \leq K\,\delta\,(1 + \rho n s), \quad \text{for all } n \in \mathbf{N},$$

with $K = \frac{1-\mu(C)}{12}$ and sufficiently small $\delta > 0$. Then, we again get that u_C approximates the second eigenvector in the sense that $\|\phi - u_C\|_2 < \sqrt{2\delta}$, and that, due to Thm. 5 in [7], $0 < \eta_2/\rho < \delta$ implying

$$\epsilon < \frac{1 - r^\delta}{1 - r}.$$

Thus, the formal equivalence (38) can be taken seriously. The above statement can be generalized to the situation of more than one eigenvalue close to the Perron root, but bounded away from the remaining part of the spectrum (see [8]).

Remark. We are aware of the fact that the above assumption $\epsilon \ll 1$ on the distribution of the eigenvalues is quite restrictive. However, we observed intriguing results of the identification strategy even for situations corresponding to ϵ-values close to 1 [13].

[2] The proofs of Thms. 3 and 5 of [7] have to be adapted to our situation. In the proof of Thm. 3, the arguments using the generator H have to be replaced by analogous arguments for $1 - P_s$.

6 Spectral Approximation of Transfer Operators

We are interested in fluctuation within the canonical ensemble for some fixed observation time span τ. As a result, we restrict our consideration to the time–s transition operator P_s with $s = \tau$ or $s = 1$ for the high–friction Langevin equation or the discrete time Markov chain with the same observation time in (26), respectively. Since both associated semigroups of transfer operators are self–adjoint, we assume in this section, that P_s is a self–adjoint operator acting on $L^2_{\mathcal{Q}}(\Omega)$.

6.1 Galerkin Discretization

In order to compute the conformational subsets exploiting certain eigenvectors of P_s, we will introduce a special Galerkin procedure to discretize the eigenvalue problem $P_s u = \lambda u$.

Let $B_1, \dots, B_n \subset \Omega$ be a partition of Ω such that $B_k \cap B_l = \emptyset$ for $k \neq l$ and $\cup_{k=1}^n B_k = \Omega$. Our finite dimensional ansatz space $\mathcal{V}_n = \text{span}\{\chi_1, \dots, \chi_n\}$ is spanned by the associated characteristic functions $\chi_k = \chi_{B_k}$. Then, the Galerkin projection $\Pi_n : L^2_{\mathcal{Q}}(\Omega) \to \mathcal{V}_n$ of $u \in L^2_{\mathcal{Q}}(\Omega)$ is defined by

$$\Pi_n u = \sum_{k=1}^{n} \frac{1}{\langle \chi_k, \chi_k \rangle_{\mathcal{Q}}} \langle \chi_k, u \rangle_{\mathcal{Q}} \, \chi_k .$$

Note that $\langle \chi_k, \chi_k \rangle_{\mathcal{Q}} = \int_{B_k} \mathcal{Q}(q)\, \mathrm{d}q$ is simply the weight of the subset B_k. The resulting discretized transition operator $\Pi_n P_s \Pi_n$ induces the approximate eigenvalue problem $\Pi_n P_s \Pi_n u = \lambda \Pi_n u$ in $\mathcal{V}_n$. Using $u = \sum_{k=1}^n \alpha_k \chi_k$, the discretized eigenvalue problems in coordinate representation reads

$$\sum_{l=1}^{n} \langle P_s \chi_k, \chi_l \rangle_{\mathcal{Q}} \, \alpha_l = \lambda \, \langle \chi_k, \chi_k \rangle_{\mathcal{Q}} \, \alpha_k, \qquad \forall k = 1, \dots, n.$$

After dividing by $\langle \chi_k, \chi_k \rangle_{\mathcal{Q}} > 0$, we end up with the convenient form

$$S\alpha = \lambda \alpha \quad \text{with} \quad \alpha = (\alpha_1, \dots, \alpha_n).$$

The entries of the $n \times n$ matrix S are given by the one step transition probabilities from B_k to B_l:

$$S_{kl} = \frac{\langle P_s \chi_k, \chi_l \rangle_{\mathcal{Q}}}{\langle \chi_k, \chi_k \rangle_{\mathcal{Q}}} = p(s, B_k, B_l). \tag{39}$$

Since P_s is a Markov operator, its Galerkin discretization S is a (row) stochastic matrix, i.e., $S_{kl} \geq 0$ and $\sum_{l=1}^n S_{kl} = 1$ for all $k = 1, \dots, n$. Hence, all its eigenvalues λ satisfy $|\lambda| \leq 1$. Moreover, we have the following four important properties [36,35]:

1. The row vector $\pi = (\pi_1, \ldots, \pi_n)$ with $\pi_k = \int_{B_k} \mathcal{Q}(q)\,dq$, which represents the discretized invariant density $\mathcal{Q}$, is a left eigenvector corresponding to the eigenvalue $\lambda = 1$, i.e., $\pi S = \pi$.
2. S is *irreducible and aperiodic*, if P_s is asymptotically stable. As a consequence, the eigenvalue $\lambda = 1$ is *simple*. Hence, the discretized invariant density π is the *unique* stationary distribution of S.
3. The transition matrix S is *self–adjoint* with respect to the discrete scalar product $\langle u, v\rangle_\pi = \sum u_i v_i \pi_i$, since P_s is self–adjoint. Equivalently, S satisfies the condition of *detailed balance*:

$$\pi_k\, S_{kl} = \pi_l\, S_{lk}, \qquad \text{for all } k, l \in \{1, \ldots, n\}.$$

   Therefore, all eigenvalues of S are real–valued and, due to 2., contained in the interval $[-1, 1]$.

In other words, for an arbitrary covering $B_1, \ldots, B_n \subset \Omega$, the discretization matrix S inherits the most important properties of the transition operator P_s.

6.2 Convergence of Discrete Eigenvalues

Denote by $\sigma(P_s)$ the $L^2_{\mathcal{Q}}(\Omega)$–spectrum of P_s and by $\sigma_{\text{discr}}(P_s) \subset \sigma(P_s)$ the subset of all isolated eigenvalues of finite (algebraic) multiplicity. Then, $\sigma_{\text{discr}}(P_s)$ is called the discrete spectrum, while $\sigma_{\text{ess}}(P_s) = \sigma(P_s) \setminus \sigma_{\text{discr}}(P_s)$ is called the essential spectrum of P_s [25]. Assume that the essential spectrum is bounded away from 1 (condition C2 on page 203), i.e., there exists a constant $0 < \kappa < 1$ such that $\sigma_{\text{ess}}(P_s)$ is contained is the ball with radius κ centered a the origin.

We are interested in approximating a cluster of (real–valued) discrete eigenvalues $\lambda_c, \ldots, \lambda_1 \in \sigma_{\text{discr}}(P_s)$ near 1 "outside" the essential spectrum:

$$\kappa < \lambda_c \leq \cdots \leq \lambda_2 < \lambda_1 = 1,$$

repeated according to multiplicity. The corresponding eigenvectors $u_c, \cdots, u_1$ are assumed to be orthogonal; this is always possible, since P_s is assumed to be self–adjoint.

Assume that the sequence of Galerkin ansatz spaces $\mathcal{V}_1 \subset \mathcal{V}_2 \subset \ldots$ is dense in $L^2_{\mathcal{Q}}(\Omega)$ and the corresponding partitions are getting finer and finer, $\max_{B \in \mathcal{V}_n} \text{diam}(B) \to 0$ as $n \to \infty$. Denote by $S_{\mathcal{V}_n}$ the transition matrix (39) associated with the ansatz space $\mathcal{V}_n$ and by $\lambda_{i,\mathcal{V}_n}, u_{i,\mathcal{V}_n}$ its eigenvalues and eigenvectors, respectively, ordered in decreasing magnitude and taken into account multiplicity.

Under the above stated assumptions, the dominant eigenvalues of $S_{\mathcal{V}_n}$ are good approximations of the dominant eigenvalues of P_s, whenever the discretization is fine enough; in this case, $P_{\mathcal{V}_n}$ also has a cluster of eigenvalues

$\lambda_{c,\mathcal{V}_n} \leq \ldots \leq \lambda_{2,\mathcal{V}_n} < \lambda_{1,\mathcal{V}_n} = 1$ near 1. More precisely, for every $i = 1, \ldots, c$, we get [35]

$$\lambda_{i,\mathcal{V}_n} \longrightarrow \lambda_i \text{ and } u_{i,\mathcal{V}_n} \longrightarrow u_i \quad \text{as } n \to \infty$$

in modulus and $L^2_{\mathcal{Q}}(\Omega)$–norm, respectively.

7 Algorithmic Realization

In this section, we want to outline the basic steps for an algorithmic realization of identifying molecular conformations, their meta–stability and the transition rates between them. In doing so, we will exclusively focus on the discrete time Markov chain and the related transition operator defined in Section 5.2 due to the following two reasons. First, it is the common belief that the discrete time Markov chain approach based on Hamiltonian motion is more realistic for modelling conformational dynamics of biomolecules than the high friction Langevin approach. Second, we managed to prevent the numerical effort for solving the eigenvalue problem for the transition operator from exploding combinatorially with the number of atoms in the molecule: This was done by discretizing it by means of a special hybrid Monte Carlo method [16], such that the computational effort does not depend explicitly on the dimension of the system. The basic scheme of the resulting algorithm is illustrated in Fig. 1. We will explain the single algorithmic steps subsequently.

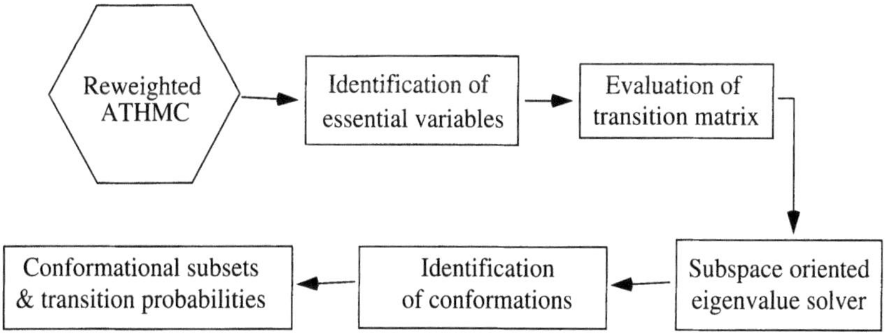

Fig. 1. Basic scheme of the algorithm.

Each step of the algorithm is illustrated by application to the triribonucleotide adenylyl(3'-5')cytidylyl(3'-5')cytidin (r(ACC)) model system in vacuum (see Fig. 2). Its physical representation ($N = 70$ atoms) is based on the GROMOS96 extended atom force field [42].

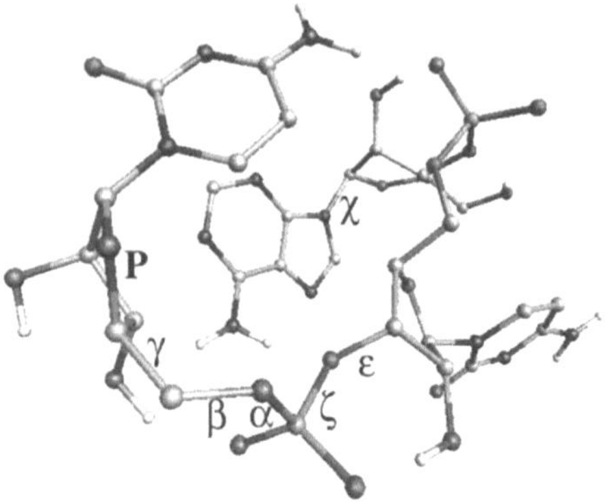

Fig. 2. Configuration of the trinucleotide $r(ACC)$ in a ball–and–stick representation. The Greek symbols indicate some of the important dihedral angles of the molecule.

7.1 Evaluation of the Transition Matrix

In order to compute the discretization matrix S of the transition operator —called the *transition matrix* in the following— we have to be able to determine transition probabilities between subsets. This task includes three subproblems: Generation of an adequate box partition of the position space Ω, sampling of the canonical ensemble $\mathcal{Q}$ and approximation of the internal dynamics within the ensemble.

Sampling of the Canonical Density. The typical approach to sample the canonical density is via Monte Carlo (MC) techniques. There is an extremely rich and varied literature on this topic (see, e.g., [5,40]) and *every* converging MC method would allow to realize this subproblem. In addition, one may also apply MD-based techniques, e.g., constant temperature sampling of the canonical density [32,3].

It is widely known, that MC simulations for ensemble averages in biomolecular systems may suffer from possible "trapping problems" [29]. As illustrated in [35], this phenomenon is related to the existence of almost invariant sets for the Monte-Carlo Markov chain.[3] We use a specific MC method, the *hybrid Monte Carlo method with adaptive choice of temperature* (ATHMC) [16], which was especially constructed to overcome this trapping problem. Moreover, ATHMC is particularly useful for linking the sampling technique with the ensemble dynamics. Future approaches will be based on a hierarchical sampling technique [15], which might be understood as a specific multilevel

[3] The trapping phenomenon occurs when the Monte-Carlo Markov chain gets trapped near a local potential energy minimum due to high energy barriers so that a proper sampling of the entire phase space within reasonable computing times is prevented.

approach to ATHMC that merges its superior sampling properties with the identification of almost invariant sets.

The result of every converging MC method is a finite sampling $\Sigma \subset \Omega$ of positions that are distributed according to the canonical ensemble.

Application to r(ACC). The simulation data were generated by means of an ATHMC sampling of the canonical density at $\mathcal{T} = 300$K. The subtrajectories of length 80 femtoseconds were computed by means of the Verlet discretization with a stepsize of 2fs. For these parameters, standard MC simulations typically require thousands of iterations only to leave the neighborhood of the initial configuration. Application of ATHMC (with adaptive temperatures between 300K and 400K) circumvents this trapping problem: one observes frequent transitions in the crucial dihedral angles of the molecule (for details see [15]). The simulation was divided into 4 Markov chains, each starting with a different state chosen from a high temperature run at 500K, which allowed the molecule to move into different conformations. The sampling took about 12h on a workstation with MIPS R10.000 processor. It was terminated by a convergence indicator [21] associated with the potential energy and all 37 dihedral angles after 320.000 steps, resulting in the sampling sequence $\Sigma = \{q_1, \dots, q_{32.000}\}$, considering only every 10th step. Since the temperature can change during the ATHMC run, each configuration is connected with a reweighting factor with respect to the canonical ensemble at 300K.

Box Partition via Essential Degrees of Freedom. Typical biomolecular systems contain hundreds or thousands of atoms. As a consequence, the number of discretization boxes, and thus the dimension of the transition matrix S, would grow exponentially with the size of the molecular system, if we would generate a box decomposition for Ω by simply partitioning every degree of freedom.

Chemical insight allows to circumvent this "curse of dimension". In the chemical literature, conformations of biomolecules are mostly described in terms of a few *essential degrees of freedom.* In the subspace of essential degrees of freedom anharmonic motion occurs that comprises most of the positional fluctuations, while in the remaining degrees of freedom the motion has a narrow Gaussian distribution and can be considered as "physically constrained".

Based on the sampling of the canonical ensemble, we may determine essential degrees of freedom either in the position space according to AMADEI ET AL. [2] or in the space of internal degrees of freedom, e.g., dihedral angles, by statistical analysis of circular data [22]. These techniques are based on the following statistical analysis of the sampling data: The correlations between atomic motions within the simulation data are expressed by the covariance matrix C.[4] Since C is symmetric, it can always be diagonalized, i.e., there is an orthonormal matrix U such that $C = U^T \Lambda U$ with Λ being the diagonal matrix whose entries are the eigenvalues of C. The matrix U defines the transformation of the original coordinates (positions or internal degrees of freedom) into the uncorrelated coordinates. The matrix Λ is connected to the systems constraints in the following way [2]: Transformed coordinates corresponding to zero or nearly zero eigenvalues behave effectively as constraints; they have narrow Gaussian distributions and do not contribute significantly to the fluctuations. In contrast to that, transformed coordinates corresponding to large eigenvalues have large deviations from their mean position, i.e., they belong to important fluctuations. Mostly, only a few coordinates exhibit such important fluctuations; these are called essential degrees of freedom. Thus, this procedure results in a tremendous reduction of the number of degrees of freedom and, consequently, in a moderate number of partition boxes when discretizing the essential variables only. [22].

Remark. As discussed in detail in [35], the transition operator can be restricted to the coordinate space spanned by the essential variables without loosing its desired spectral properties.

Application to r(ACC). Since essential degrees of freedom should solely reflect internal fluctuations of the molecule, we only consider the 37 dihedral angles of the r(ACC) molecule (see Fig. 2). The above explained transformation process based on the simulation data for r(ACC) is exemplified in Fig. 3 and Fig. 4. Figure 3 shows the circular deviations of the original and

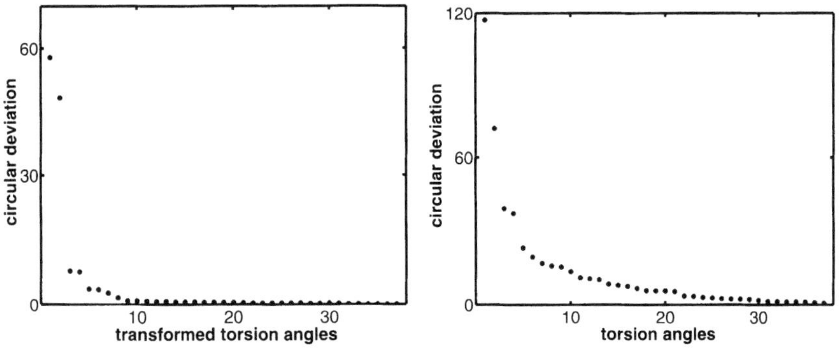

Fig. 3. Top: circular deviation of the *transformed* dihedral angles ordered by magnitude (left) and circular deviation of the original dihedral angles (right).

[4] To analyze the simulation data in terms of the dihedral angles we have to apply statistical methods for circular data [17,18]; see [22] for resulting definition of the covariance matrix.

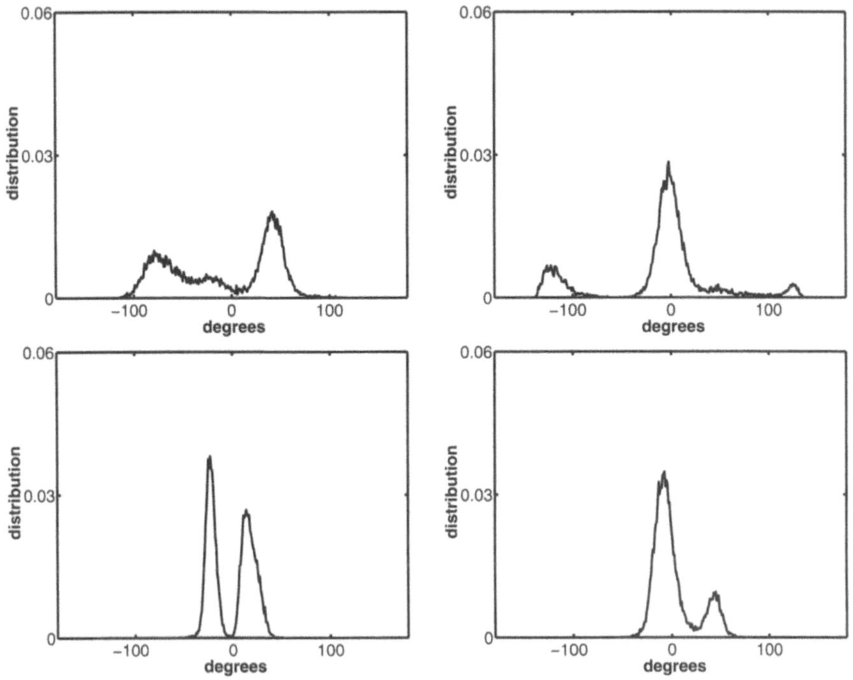

Fig. 4. Distribution of the four essential dihedral angles. The distributions at the top allow to identify three maxima each, while there are two maxima for each distribution at the bottom.

transformed dihedral angles in decreasing order of magnitude. Only the first four transformed dihedral angles have relevant circular deviation *and* are far from being Gaussian shaped (see Fig. 4), while the remaining transformed dihedral angles are Gaussian like.

The configurational space was discretized into "boxes" $B_1, \dots, B_d$, by means of all four essential degrees of freedom (see Fig. 4) resulting in $d = 36$ discretization boxes.

Approximation of Internal Dynamics. Due to equations (22) and (39) the entries of the transition matrix S with respect to the boxes B_k are given by

$$\begin{aligned} S_{kl} &= p(1, B_k, B_l) \\ &= \frac{1}{\int_{B_k} \mathcal{Q}(q)\,\mathrm{d}q} \int_{B_k} p(1, q, B_l)\,\mathcal{Q}(q)\,\mathrm{d}q. \end{aligned} \tag{40}$$

Let now $q_1, q_2, \dots$ denote an arbitrary sequence of positions generated by some ergodic Markov chain Monte Carlo method that is asymptotically distributed according to the canonical density $\mathcal{Q}$. Then, due to the law of large number for Markov chains [41], we may rewrite S_{kl} as

$$S_{kl} = \lim_{n\to\infty} \frac{\sum_{j=1}^{n} p(1, q_j, B_l)\chi_{B_k}(q_j)}{\sum_{j=1}^{n} \chi_{B_k}(q_j)}.$$

By using our particular MC sequence Σ, we thus get

$$S_{kl} \approx \frac{\sum_{q\in\Sigma} p(1, q, B_l)\chi_{B_k}(q)}{\sum_{q\in\Sigma} \chi_{B_k}(q)}.$$

Finally, we have to approximate the transition kernel $p(1, q, B_k)$ for all $q \in \Sigma$ and all B_i. For the discrete Markov chain, this can be done by applying some integration scheme to (26). A convergence analysis is presented in [35].

Application to r(ACC). The dynamical fluctuations within the canonical ensemble were approximated by integrating four short trajectories of length τ = 80fs starting from each sampling point $q \in \Sigma$. To facilitate transitions, analogous to the ATHMC sampling, the momenta were chosen according to the momenta distribution $\mathcal{P}$ for 4 different temperatures between $300K$ – 400K and reweighted afterwards. This resulted in a total of $4 \times 32.000 = 128.000$ transitions. This calculation took less than 25% of the total computing time. Then the 36×36 transition matrix S was computed based on the 128.000 transitions taking the different weighting factors into account. Since every box had been hit by sufficiently many transitions, the statistical sampling was accepted to be reliable.

7.2 Solving the Eigenvalue Problem

Once the entries of the transition matrix have been computed, we have to determine the eigenvectors corresponding to a cluster of eigenvalues near the dominant eigenvalue 1. That is, only a small part of the spectrum of S is required, *not* its full diagonalization. Actual evaluation is efficiently possible using subspace oriented iterative techniques, even if the number of discretization boxes may be about 100.000 or larger [28,20].

Application to r(ACC). The computation of the eigenvalues of S near 1 yielded a cluster of eight eigenvalues with a significant gap to the remaining part of the spectrum:

k	1	2	3	4	5	6	7	8	9	...
λ_k	1.000	0.999	0.989	0.974	0.963	0.946	0.933	0.904	0.805	...

7.3 Identification of Conformations

According to the definition of almost invariance, we are interested in unions $C = \cup_{k\in I} B_k$ of partition sets, for which $p(1, C, C) \approx 1$. In other words, we are looking for a nontrivial index set $I \subset \{1, \ldots, n\}$ such that the process Q_t almost certainly stays within $B = \cup_{k \in I} B_k$ after one step. Using the transition probabilities $p(1, B_k, B_l)$ between the partition sets, cluster algorithm can be used to identify almost invariant subsets [22]. We apply the identification strategy of Sec. 5.4 in its algorithmic realization due to [13], which exploits a certain almost constant level structure of eigenvectors corresponding to a cluster of eigenvalues near 1.

Application to r(ACC). Finally, the conformational subsets were computed based on the eigenvectors of S via the identification algorithm. This yielded eight conformations.

The conformational subsets identified turned out to be rather insensitive to further refinements of the discretization. The weighting factors within the canonical ensemble and the meta-stability of the eight identified conformations are given in the following table:

conformations	D1c	D1t	D2c	D2t	D3c	D3t	D4c	D4t
weighting factor	0.107	0.011	0.116	0.028	0.320	0.038	0.285	0.095
meta-stability	0.986	0.938	0.961	0.888	0.991	0.949	0.981	0.962

The transition probabilities between the different conformations are visualized schematically in Fig. 5. The matrix allows to define a hierarchy between

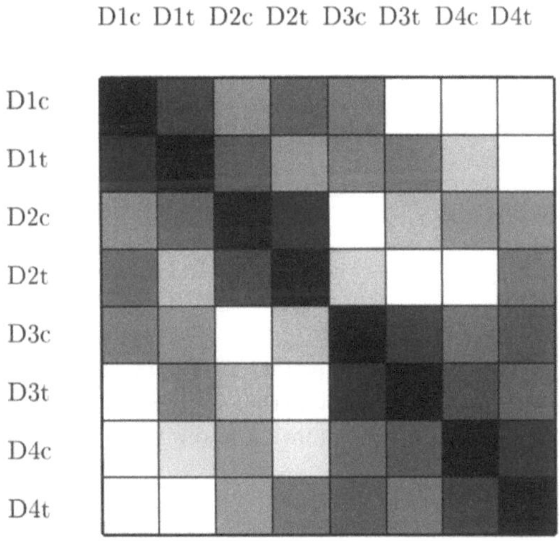

Fig. 5. Visualization of the one-step transition probabilities $p(1, D_{\text{from}}, D_{\text{to}})$ between the conformation D_{from} (row) and D_{to} (column). The colors are chosen according to the logarithm of the corresponding entries; black: $p(1, \cdot, \cdot) \approx 1$, white: $p(1, \cdot, \cdot) \approx 0$.

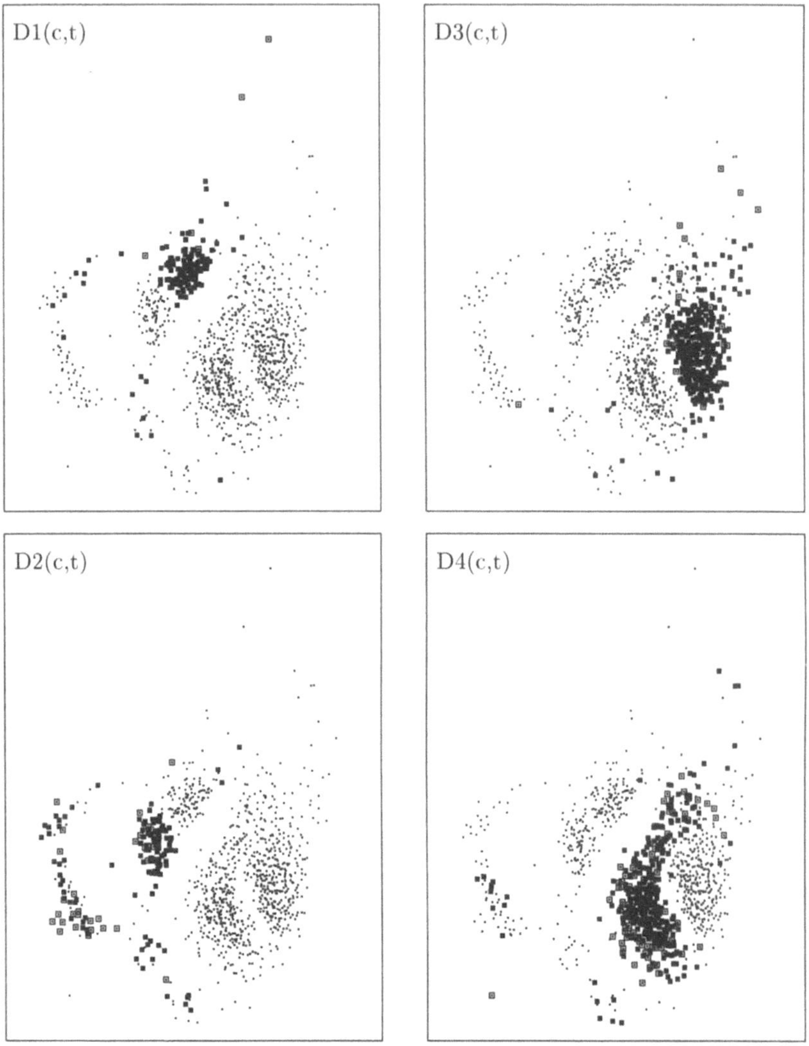

Fig. 6. 2d plot of the four conformations D1,... ,D4 (squares). The distinction between open and filled squares indicates a further splitting into eight conformations resulting from a partition into a c–conformation and a t–conformation.

the conformations, which is inherent to the algorithm. On the top level, there are two conformations, D1&D2 and D3&D4 corresponding to the two 4×4 blocks on the diagonal of S. On the next level, each of these conformations split up into two subconformations yielding D1,... ,D4. On the bottom level, each conformation is further divided into a c–part and a t–part (for interpretation see [22]). The evaluation of the transition matrix together with the execution of the identification algorithm took less than 2% of the computing time required for evaluation of the simulation data via ATHMC.

Acknowledgement

One of us (W.H.) was supported within the DFG–Schwerpunkt "Ergodentheorie, Analysis und effiziente Simulation dynamischer Systeme" under Grant De 293/2-1.

References

1. M. Allen and D. Tildesley. *Computer Simulations of Liquids*. Clarendon Press, Oxford, 1990.
2. A. Amadei, A. B. M. Linssen, and H. J. C. Berendsen. Essential dynamics of proteins. *Proteins*, 17:412–425, 1993.
3. H. Andersen. Molecular dynamics simulations at constant pressure and/or temperature. *J. Chem. Phys.*, 72:2384–, 1980.
4. A. Arnold, P. Markowich, G. Toscani, and A. Unterreiter. On logarithmic Sobolev inequalities, Csiszar–Kullback inequalities, and the rate of convergence to equilibrium for Fokker-Planck type equations. *To appear in Comm. PDE*, 2000.
5. K. Binder[Ed.]. *The Monte Carlo method in condensed matter physics*, volume Bd. 71 of *Topics in applied physics*. Springer Verlag, Berlin, Heidelberg, New York, 1992.
6. K. L. Chung. *Lectures from Markov Processes to Brownian Motion*, volume 249 of *A Series of Comprehensive Studies in Mathematics*. Springer, New York, Heidelberg, Berlin, 1982.
7. E. B. Davies. Metastable states of symmetric Markov semigroups I. *Proc. London Math. Soc.*, 45(3):133–150, 1982.
8. E. B. Davies. Metastable states of symmetric Markov semigroups II. *J. London Math. Soc.*, 26(2):541–556, 1982.
9. M. Dellnitz and A. Hohmann. A subdivision algorithm for the computation of unstable manifolds and global attractors. *Numerische Mathematik*, 75:293–317, 1997.
10. M. Dellnitz and O. Junge. On the approximation of complicated dynamical behavior. *SIAM J. Num. Anal.*, 36(2):491–515, 1999.
11. P. Deuflhard, M. Dellnitz, O. Junge, and C. Schütte. Computation of essential molecular dynamics by subdivision techniques. In *[12]*, pages 98–115, 1999.
12. P. Deuflhard, J. Hermans, B. Leimkuhler, A. E. Mark, S. Reich, and R. D. Skeel, editors. *Computational Molecular Dynamics: Challenges, Methods, Ideas*, volume 4 of *Lecture Notes in Computational Science and Engineering*. Springer, 1999.
13. P. Deuflhard, W. Huisinga, A. Fischer, and C. Schütte. Identification of almost invariant aggregates in nearly uncoupled Markov chains. Lin. Alg. Appl., 315:39–59, 2000.

14. D. Evans and G. Morriss. *Statistical Mechanics of Nonequilibrium Liquids.* Academic Press, London, 1990.
15. A. Fischer. An Uncoupling–Coupling Technique for Markov Chain Monte Carlo Methods. Report 00–04, Konrad–Zuse–Zentrum, Berlin, 2000.
16. A. Fischer, F. Cordes, and C. Schütte. Hybrid Monte Carlo with adaptive temperature in a mixed–canonical ensemble: Efficient conformational analysis of RNA. *J. Comput. Chem.*, 19:1689–1697, 1998.
17. N. I. Fisher. *Statistical Analysis of Circular Data.* University Press, Cambridge, 1993.
18. N. I. Fisher and A. J. Lee. A correlation coefficient for circular data. *Biometrica*, 70(2):327–332, 1983.
19. M. Freidlin and A. Wentzell. *Random perturbations of dynamical systems.* Springer, New York, London, 1984. Series in Comprehensive Studies in Mathematics.
20. T. Friese, P. Deuflhard, and F. Schmidt. A multigrid method for the complex Helmholtz eigenvalue problem. In C.-H. Lai, P. E. Bjørstad, M. Cross, and O. B. Widlund, editors, *Eleventh International Conference on Domain Decomposition Methods*, 1998, DDM–org Press, Bergen, pp. 18–26 (1999).
21. A. Gelman and D. Rubin. Inference from iterative simulation using multiple sequences. *Statistical Science*, 7:457–511, 1992.
22. W. Huisinga, C. Best, R. Roitzsch, C. Schütte, and F. Cordes. From simulation data to conformational ensembles: Structure and dynamic based methods. J. Comp. Chem. 20(16):1760–1774, 1999.
23. W. Huisinga. The Essential Spectral Radius and Asymptotic Properties of Transfer Operators in L^1. Submitted to Dynam. Systems Appl. Available via http://www.math.fu-berlin.de/~biocomp, 2000.
24. K. Ichihara and H. Kunita. A classification of the second order degenerate elliptic operators and its probabilistic characterization. *Z. Wahrscheinlichkeitstheorie verw. Gebiete*, 30:235–254, 1974.
25. T. Kato. *Perturbation Theory for Linear Operators.* Springer, 1995.
26. R. Kurt. *Axiomatics of Classical Statistical Mechanics.* Pergamon Press, Oxford, New York, 1980.
27. A. Lasota and M. C. Mackey. *Chaos, Fractals and Noise*, volume 97 of *Applied Mathematical Sciences.* Springer, New York, 2nd edition, 1994.
28. R. B. Lehoucq, D. C. Sorensen, and C.Yang. *ARPACK User's Guide: Solution of Large Eigenvalue Problems by Implicit Restartet Arnoldi Methods.* Rice University Houston, 1998.
29. E. Leontidis, B. Forrest, A. Widmann, and U. Suter. Monte Carlo algorithms for the atomistic simulation of condensed polymer phases. *J. Chem. Soc. Faraday Trans.*, 91(16):2355–2368, 1995.
30. S. Meyn and R. Tweedie. *Markov Chains and Stochastic Stability.* Springer, Berlin, Heidelberg, New York, Tokyo, 1993.
31. E. Nelson. *Dynamical Theories of Brownian Motion.* Mathematical Notes. Princeton University Press, 1967.
32. S. Nosé. A molecular dynamics methods for simulations in the canonical ensemble. *Mol. Phys.*, 52, 1984.
33. H. Risken. *The Fokker-Planck equation.* Springer, New York, London, 2nd edition, 1996.
34. T. Schlick. Some failures and successes of long–timestep approaches to biomolecular simulations. In *[12]*, pages 227–262, 1999.

35. C. Schütte. *Conformational Dynamics: Modelling, Theory, Algorithm, and Application to Biomolecules.* Habilitation Thesis, Fachbereich Mathematik und Informatik, Freie Universität Berlin, 1998. Available via http://www.math.fu-berlin.de/~biocomp.
36. C. Schütte, A. Fischer, W. Huisinga, and P. Deuflhard. A direct approach to conformational dynamics based on hybrid Monte Carlo. *J. Comput. Phys., Special Issue on Computational Biophysics*, 151:146–168, 1999.
37. C. Schütte and W. Huisinga. On conformational dynamics induced by Langevin processes. In EQUADIFF 99 – International Conference on Differential Equations, B. Fiedler, K. Kröger, and J. Sprekels, editors. Vol II, pp. 1247–1262 (2000).
38. G. Singleton. Asymptotically exact estimates for metatstable Markov semigroups. *Quart. J. Math. Oxford*, 35(2):321–329, 1984.
39. R. Skeel and J. Izaguirre. The five femtosecond time step barrier. In *[12]*, pages 312–325, 1999.
40. A. Sokal. Monte Carlo methods in statistical mechanics. Lecture note, Department of Physics, New York University, 1989.
41. L. Tierney. Introduction to general state-space Markov chain theory. In W. Gilks, S. Richardson, and D. Spiegelhalter, editors, *Markov chain Monte-Carlo in practice*, pages 59–74. Chapman and Hall, London, Glasgow, New York, Tokyo, 1997.
42. W. van Gunsteren, S. Billeter, A. Eising, P. Hünenberger, P. Krüger, A. Mark, W. Scott, and I. Tironi. *Biomolecular Simulation: The GROMOS96 Manual and User Guide.* vdf Hochschulverlag AG an der ETH Zürich, 1996.

Simulation and Numerical Analysis of Dendritic Growth

Michael Fried and Andreas Veeser*

Institut für Angewandte Mathematik, Universität Freiburg,
Hermann-Herder-Str. 10, 79104 Freiburg, Germany

Abstract. Dendritic growth is a nonlinear process, which falls into the category of self-organizing pattern formation phenomena. It is of great practical importance, since it appears frequently and, in the case of alloys, affects the engineering properties of the resulting solid.

In the first part of this article we report on an analysis of spatially semi-discrete approximations to the Stefan problem for two-dimensional, pure dendrites. A priori error estimates for the temperature field, the parametrization of the free boundary, relevant geometric and measuring quantities are presented and discussed.

The second part describes a new algorithm for the two-dimensional Stefan problem. Here the free boundary is represented as a level set. This allows to handle topological changes of the free boundary. The accuracy of the method is verified and various numerical simulations, including topological changes of the free boundary, are presented.

1 Introduction

Dendritic solidification is a common mode of crystallization occurring in nature and technology. For example, dendrites grow in a pure, undercooled, and thus metastable liquid from a small seed crystal. This growth is relatively fast and produces a branched morphology with preferred directions ("dendrites") of the phase boundary between liquid and solid. Dendrites, which arise in casting or welding processes are of great interest, since their size, shape, and orientation are directly related to the strength and durability of the product. Apart from its products, the dynamics of dendritic growth have attracted much interest. Surprisingly, one observes steady-state propagation of tip regions of so-called primary stems, while secondary and tertiary side branches crystallize time-dependently, e.g. [24].

Accurate numerical simulations may further our current understanding of dendritic growth. The complicated and potentially unstable (see below) dynamics of the phase boundary require efficient, reliable, and especially stable

---

* Project: Efficient Simulation and Numerical Analysis of the Dynamics of Dendrites (G. Dziuk and A. Schmidt)

numerical methods. In this context, mathematical proofs of their convergence are essential.

In this article we base on a model, which is a well-known generalization of the classical Stefan problem, cf. [4,24,26,28,46]. In order to fix notations, we recall it in the two-dimensional case. Let $\Omega \subset \mathbb{R}^2$ denote a region occupied by a pure material that may be in solid or liquid phase. The region Ω divides into the solid region ω^s, the liquid region ω^l and the one-dimensional interface γ that separates ω^s and ω^l. The partition $p = (\omega^s, \gamma, \omega^l)$ of the region Ω depends on time t. We associate the normal n, the normal velocity v and the curvature k of the interface γ with the partition p. The normal n points into the liquid region ω^l and the curvature k is positive, if the solid region ω^s is convex. Finally, $[\![\cdot]\!]_s^l$ denotes the jump across the interface γ with the sign convention "liquid" – "solid". Using this notation, the model consists of the following evolution laws for the temperature field $\theta = \theta(x,t)$ and phase distribution $p = p(t)$:

$$\begin{aligned} \theta_t - \Delta\theta &= 0 && \text{in } \omega^s(t) \cup \omega^l(t), \\ v &= -\left[\!\!\left[\frac{\partial\theta}{\partial n}\right]\!\!\right]_s^l && \text{on } \gamma(t), \\ \beta(n)\,v + a(n)\,k + \theta &= 0 && \text{on } \gamma(t), \end{aligned} \tag{1}$$

where β and a denote material quantities. The function β, defined on the sphere $\mathbb{S} := \{x \in \mathbb{R}^2 \mid |x| = 1\}$, is called kinetic coefficient and is involved in the term $\beta(n)\,v$, which models molecular attachment. The function a is the interface potential. It can be calculated from a smooth surface tension $\phi = \phi(n)$ by the formula $a(n) = D^2\phi(n)\,n^\perp \cdot n^\perp$ where $D^2\phi(n)$ is the Hessian of the positively homogeneous extension of ϕ on $\mathbb{R}^2$ and $^\perp$ denotes the counterclockwise rotation by 90°. The dependence on the normal n of the functions β and a allows to model directional anisotropies due to the crystal lattice. The last equation in model (1), a generalized Gibbs-Thomson law, makes the difference to the classical Stefan problem; it allows for a shift of the interfacial temperature due to kinetic and curvature effects. Here we have assumed that the temperature is scaled such that 0 is the melting temperature, that is the temperature of a planar interface in equilibrium.

We recall some features of (1) about its usefulness as a model for dendritic growth. Model (1) incorporates heat transport, latent heat, anisotropic surface tension and molecular attachment effects. It is widely believed that a coupling of these four effects should explain dendritic growth (see [24]). Moreover, the morphological stability analyses [5,30,40] for (1) or similar models provide a qualitative explanation of the shape of dendrites as a competition between "destabilizing" undercooling and "partially stabilizing" surface tension. Finally, model (1) is an "approximation" of a model that is consistent with thermodynamics [25].

In the sequel we restrict ourselves to the case that the material quantities satisfy

$$\beta_* := \inf_{n \in \mathbb{S}} \beta(n) > 0 \qquad \text{and} \qquad a_* := \inf_{n \in \mathbb{S}} a(n) > 0. \tag{2}$$

If we assume additionally that the material is isotropic, i.e. β and a are constant functions, then two results concerning the well-posedness of model (1) are known. Model (1), together with appropriate initial and boundary conditions, admits a unique classical solution on a short time interval [6]. There are global weak solutions [39].

In Section 2 we analyze a semi-discrete method for model (1) that has the same structure as the approach used in [2,34]. This method cannot handle "geometric singularities" such as merging or separation of phase regions. Nevertheless it can be applied to simulate important experiments in dendritic growth. The part of the method that concerns the discretization of the generalized Gibbs-Thomson law, a driven motion by anisotropic curvature, uses the same approach as [10] and [13].

In Subsection 2.1 the "derivation" of the method is presented in two steps. The first step is a reformulation of (1) introducing a parameterization $g(\cdot, t)$ of the free boundary $\gamma(t)$. A system for the parameterizations $g = g(u,t)$, which is parabolic thanks to (2), replaces the generalized Gibbs-Thomson law. The second step is an application of the standard procedure to get a finite element method from a classical formulation. The unknowns of the resulting semi-discrete problem are the approximations Θ and G to the temperature field θ and the parameterization g respectively.

In Subsection 2.2 error estimates for the described parametric method are presented and discussed. They cover the case that the free boundary γ is parameterized over the sphere $\mathbb{S}$ and does not touch the boundary $\partial\Omega$. The most important ones are

$$\left.\begin{array}{ll} \sup\limits_{(0,T)} \|\Theta - \theta\|^2_{L_2(\Omega)}, & h \displaystyle\int_0^T \|\nabla(\Theta - \theta)\|^2_{L_2(\Omega)}, \\ \sup\limits_{(0,T)} \|G - g\|^2_{H^1(\mathbb{S},\mathbb{R}^2)}, & \displaystyle\int_0^T \|G_t - g_t\|^2_{L_2(\mathbb{S},\mathbb{R}^2)} \end{array}\right\} \leq \; C h^2, \tag{3}$$

where $h \in (0, h^*]$ is the mesh size. Here the constants $h^* > 0$ and $C \geq 0$ depend on reasonable regularity (see Remark 2) and geometric properties of the continuous solution (θ, g). The H^1-estimate for the parameterization of the free boundary implies error estimates for the phase regions ω^l and ω^s. Further estimates concern approximations to the normal n, the normal velocity v and the curvature k. From these follow more estimates for approximations to the velocity and radius of a primary dendrite tip under some additional assumptions. The velocity and the radius are measured in order to investigate the steady-state propagation of a primary tip. We summarize these results by concluding that the described method is reliable and, moreover, that it can be applied in principle to compare "dendrites of model (1)" with real ones.

In Section 3 we present a new finite element algorithm for the two dimensional dendritic growth, which uses a level set method for the representation of the phase boundary $\gamma(t)$. As opposed to the parametric ansatz described in Section 2, the level set approach enables us to handle topological changes of the free boundary $\gamma(t)$. The modified Stefan problem (1) consists of determing the two quantities temperature θ and phase boundary $\gamma(t)$, which fulfill the equations (1). In order to develop a new algorithm for the whole Stefan problem, we proceed by describing separate methods for the numerical calculation of this two quantities, where we ask for one of them, assuming the other to be given. In Subsection 3.1, we outline the level set methods for the generalized mean curvature flow, which is given by the last equation in (1), the generalized Gibbs-Thomson law, while we suppose the temperature θ to be known and the interface $\gamma(t)$ to be calculated. The basic idea behind a level set method is always to interpret the interface $\gamma(t)$ as the zero level of some continuous real valued function u. In case of mean curvature flow like problems, this leads to a degenerate nonlinear pde for the function u, which has to be understood in a viscosity sense (c.f. [7], [18], [19], [20], [21], [23]). Using a regularized formulation of the inhomogeneous and anisotropic level set problem, we apply the standard procedure to get a finite element method for u, and such for the generalized evolution of $\gamma(t)$, which is defined by the zero level set of u. Time discretization by a semi-implicit Euler method finally leads to the finite element algorithm for the calculation of the free boundary $\gamma(t)$.

Now, we assume for the moment the phase boundary $\gamma(t)$, its normal velocity v and its curvature k to be known. Then a numerical method for approximating the second unknown, the temperature, is obtained by a similar finite element method as we used in Section 2, together with an implicit Euler scheme in time. The short description of this method is the subject of Subsection 3.2. This algorithm is nearly the same as introduced by Schmidt in [33], [34], except that in our case the phase boundary is understood as the zero level of some function u.

To obtain an algorithm for the whole Stefan problem, we combine the finite element methods for the free boundary and for the temperature in the following manner: for each timestep t^m, we first calculate the new finite element approximation U^m of the (regularized) level set function u, using the discrete temperature Θ^{m-1} from the previous timestep. The zero level of U^m gives us the new discrete phase boundary γ^m. With this, we apply the above mentioned finite element algorithm for the temperature, to compute the update Θ^m of the discrete temperature. Since this algorithm makes use of the curvature k of γ^m, we have to calculate k. This is done by employing an finite element approximation K of the curvature of level sets of U^m, which as well as an adaptive version of the combined algorithm is depicted in Subsection 3.3. Finally, the computation of the temperature Θ^m finishes timestep t^m.

We end by presenting various numerical results obtained by our method. They include a verification of its accuracy with the help of an example with explicitly known exact solution and an example with topological changes of the phase boundary.

2 Error Estimates for a Semi-Discrete Parametric Method

2.1 Reformulation and Semi-Discrete Problem

We employ parametrizations (here: continuous mappings into $\mathbb{R}^2$) of the free boundary to rewrite model (1). For simplicity, the domain of definition of these parametrizations is the sphere $\mathbb{S}$. We identify derivatives of mappings with their representations with respect to "canonical" bases in the involved tangential spaces. In the context of $\mathbb{R}^n$ "canonical" has the usual meaning, whereas, for example, in the case of the tangential space $T\mathbb{S}_u$ at $u \in \mathbb{S}$ it indicates the choice $u^\perp$. In order to recover the phase regions from the parametrizations we use the following "oriented distance function". Let $f : \mathbb{S} \to \mathbb{R}^2$ be an injective parametrization and define $\operatorname{sdist}(f, \cdot) : \mathbb{R}^2 \to \mathbb{R}$ by

$$\operatorname{sdist}(f, x) := \operatorname{Or}(f) \operatorname{sdist}\big(\partial \operatorname{Int}(f), x\big),$$

where $\operatorname{Or}(f) \in \{\pm 1\}$ denotes the orientation of f, $\operatorname{Int}(f)$ is the interior of the Jordan curve $f(\mathbb{S})$ (see for example Section IV.5 in [3]), and $\operatorname{sdist}(\partial A, x) := \operatorname{dist}(A, x) - \operatorname{dist}(\mathbb{R}^2 \setminus A, x)$ with $\operatorname{dist}(A, x) := \inf_{y \in A} |y - x|$ for $A \subset \mathbb{R}^2$.

Let $\Omega \subset \mathbb{R}^2$ be a bounded domain, let $T > 0$ be fixed and set $Q := \Omega \times (0, T)$ as well as $\mathbb{S}_T := \mathbb{S} \times (0, T)$. Let θ_0 be an extension on $\overline{Q}$ of the initial and boundary temperature, and let g_0 be an injective parametrization of a Jordan curve (the initial interface) in Ω.

Problem 1. Find $\theta = \theta(x, t)$ and $g = g(u, t)$ such that

(i) $g \in C^{2,1}(\mathbb{S}_T, \mathbb{R}^2)$, and $g(t) := g(\cdot, t)$ is a regular, injective parametrization with $g(\mathbb{S}, t) \subset \Omega$ for every $t \in (0, T)$,
(ii) θ is of class $C^{2,1}$ in $Q^s := \{(x, t) \in Q \mid \operatorname{sdist}\big(g(t), x\big) < 0\}$ and in $Q^l := \{(x, t) \in Q \mid \operatorname{sdist}\big(g(t), x\big) > 0\}$, and $\nabla\theta$ exists on $g(\mathbb{S}_T)$ "from both sides",
(iii) the equations

$$\begin{aligned} \theta_t &= \Delta\theta && \text{in } Q^s \cup Q^l, \\ g_t \cdot n &= -\left[\!\left[\nabla\theta\right]\!\right]_s^l (g) \cdot n && \text{on } \mathbb{S}_T, \\ \beta(n)\, g_t &= \frac{a(n)}{|g_u|^2}\, g_{uu} - \theta(g)\, n && \text{on } \mathbb{S}_T \end{aligned}$$

with $n = -g_u^\perp / |g_u|$ hold,

(iv) the initial and boundary conditions

$$\begin{aligned} g(u,0) &= g_0(u) && \text{for all } u \in \mathbb{S}, \\ \theta(x,t) &= \theta_0(x,t) && \text{for all } (x,t) \in \big[\Omega \times \{0\}\big] \cup \big[\partial\Omega \times (0,T)\big] \end{aligned}$$

are fulfilled.

Hereafter, $\theta(g)$ stands for the function $(u,t) \mapsto \theta(g(u,t),t)$. The expression $[\![\nabla\theta]\!]_s^l(g)$ is interpreted in a similar way. Condition (i) implies that Q^s and Q^l are open (see Section 20 in [44]). Moreover, it excludes that the interface touches the boundary $\partial\Omega$. The third equation in condition (iii) implies the generalized Gibbs-Thomson law (multiply by the normal n) and is a "singular", parabolic system in the sense of Petrovsky ([16]) thanks to (2).

Remark 2. Let us assume that the material is isotropic and that the data is sufficiently smooth and compatible. In [6] Chen and Reitich prove that there exists a $T > 0$ such that model (1) admits a unique solution (θ, p) on $[0,T]$, thereby representing the interface γ as "a graph over the initial interface". We find a solution (θ, g) of Problem 1 by reparametrizing the interface γ of (θ, p) in an appropriate way. This solution (θ, g) has the regularity

$$g \in C^{3+\alpha,(3+\alpha)/2}(\overline{\mathbb{S}}_T, \mathbb{R}^2), \quad g_t \in L_2\big(0,T; H^2(\mathbb{S},\mathbb{R}^2)\big),$$
$$\theta \in C^0(\overline{Q}), \quad \theta|_{Q^s} \in C^{2+\alpha,(2+\alpha)/2}(\overline{Q^s}), \quad \theta|_{Q^l} \in C^{2+\alpha,(2+\alpha)/2}(\overline{Q^l})$$

with $\alpha \in (0,1)$. Moreover, condition (i) is also satisfied for $t = 0$ (by assumption) and $t = T$. (See Section 19 in [44].)

In order to state the semi-discrete problem (and for later purposes) we define

$$\underline{\mathrm{DQ}}(f) := \inf \left\{ \frac{|f(u_1) - f(u_2)|}{\mathrm{d}_{\mathbb{S}}(u_1,u_2)} \mid u_1, u_2 \in \mathbb{S}, u_1 \neq u_2 \right\},$$

where $f : \mathbb{S} \to \mathbb{R}^2$ is a parametrization and $\mathrm{d}_{\mathbb{S}}$ denotes the standard distance on the sphere $\mathbb{S}$. The condition $\underline{\mathrm{DQ}}(f) > 0$ implies that f is injective and that the length element $|f_u|$ is bounded away from zero where it exists.

Let X and Y be finite-dimensional subspaces, e.g. (linear) finite element spaces, of $W^1_\infty(\Omega)$ and $W^1_\infty(\mathbb{S},\mathbb{R}^2)$ respectively, and set $\mathring{X} := X \cap \mathring{H}^1(\Omega)$. Let $\Theta_0 \in X$ and $G_0 \in Y$ be approximations to the initial and boundary values θ_0 and g_0 respectively. Furthermore, let G_0 satisfy $\underline{\mathrm{DQ}}(G_0) > 0$ and $G_0(\mathbb{S}) \subset \Omega$.

Problem 3. Find $\Theta : [0,T] \to X$ and $G : [0,T] \to Y$ such that

(i) $G \in H^1(0,T;Y)$, and $G(\mathbb{S},t) \subset \Omega$ as well as $\underline{\mathrm{DQ}}\big(G(t)\big) > 0$ for every $t \in (0,T)$,
(ii) $\Theta \in H^1(0,T;X)$,

(iii) the equations

$$\int_{\Omega} \Theta_t \Phi + \int_{\Omega} \nabla\Theta \cdot \nabla\Phi = -\int_{\mathbb{S}} G_t \cdot G_u^{\perp} \Phi(G) \qquad \forall \Phi \in \mathring{X},$$

$$\int_{\mathbb{S}} \frac{\beta(N)}{a(N)} G_t \cdot \Psi |G_u|^2 + \int_{\mathbb{S}} G_u \cdot \Psi_u = \int_{\mathbb{S}} \frac{\Theta(G)}{a(N)} G_u^{\perp} \cdot \Psi |G_u| \qquad \forall \Psi \in Y$$

with $N = -G_u^{\perp}/|G_u|$ hold almost everywhere in $(0,T)$,

(iv) the initial and boundary conditions

$$G(0) = G_0, \quad \Theta(0) = \Theta_0, \quad \Theta(t) - \Theta_0(t) \in \mathring{X} \text{ for all } t \in (0,T)$$

are fulfilled.

The equation for G can be obtained in the following way: let $\psi = (\psi_1, \psi_2)$ be a test function, take the scalar product of the system for g and $\psi\, |g_u|^2/a(n)$, integrate over the sphere $\mathbb{S}$, integrate by parts the principal term and then replace the solution space as well as the test space by Y. The equation for Θ can be obtained in a similar way; after integration by parts use the second equation in condition (iii) of the parametric formulation.

We can rewrite the semi-discrete Problem 3 as an initial value problem for a system of ordinary differential equations by choosing bases in X and Y. The right-hand side of that system satisfies a locally uniform Lipschitz condition. Consequently, the Theorem of Picard-Lindelöf implies the following local existence and uniqueness result.

Lemma 4. *Assume that* $\underline{\mathrm{DQ}}(G_0) > 0$, $G_0(\mathbb{S}) \subset \Omega$, *and* $\Theta_0 \in H^1(0,T;X)$. *Then there exists a* $t^* \in (0,T]$ *such that Problem 3 admits a unique solution on* $[0,t^*]$. *Moreover,* $G \in C^1([0,t^*],Y)$.

2.2 Error Estimates

For the sake of simplicity we present the results of the numerical analysis of the semi-discrete Problem 3 under simplified assumptions on data and discretization. For more general assumptions see [44] or [45].

Let Ω be a bounded, polygonal, convex domain of $\mathbb{R}^2$ and assume that the material quantities β and a satisfy $\beta, a \in W^1_\infty(\mathbb{S})$ in addition to (2). Let the initial and boundary temperature $\theta_0 = \theta_0^1 + \theta_0^2$ be split into a time-dependent part θ_0^1 and a time-independent part θ_0^2 such that

$$\begin{aligned} &\theta_0^1 \in W^1_\infty(Q) \cap L_2(0,T;H^2(\Omega)), \\ &\theta_0^2 \in W^1_\infty(\Omega) \cap \mathring{H}^1(\Omega), \quad \theta_0^2|_{\omega_0^s} \in H^2(\omega_0^s), \quad \theta_0^2|_{\omega_0^l} \in H^2(\omega_0^l), \end{aligned} \tag{4}$$

where $\omega_0^s = \{x \in \Omega \mid \mathrm{sdist}(g_0, x) < 0\}$ and $\omega_0^l = \{x \in \Omega \mid \mathrm{sdist}(g_0, x) > 0\}$. This splitting takes into account that the compatibility condition on the free boundary used in Remark 2 may imply a jump of $\nabla\theta_0$ across the initial interface. The discrete spaces X and Y we choose in the following way. Let $\mathcal{T}$ be a member of a quasi-uniform and shape-regular family of triangulations of Ω and set

$$X = \{\Phi \in C^0(\bar{\Omega}) \mid \forall S \in \mathcal{T} \;\; \Phi|_S \in P_1(S)\},$$

where $P_1(S)$ denotes the affine functions on the triangle S. Moreover, let $\mathcal{P}$ be a member of a quasi-uniform family of partitions of $\mathbb{S}$ and set

$$\begin{aligned} Z &= \{\xi \in C^0(\mathbb{S}) \mid \forall \Sigma \in \mathcal{P} \;\; \xi|_\Sigma \in P_1(\Sigma)\}, \\ Y &= Z^2, \end{aligned}$$

where $P_1(\Sigma)$ denotes the (in the arclength) affine functions on the arc Σ. We will estimate the error in terms of

$$h = \max\{ \max_{S \in \mathcal{T}} \operatorname{diam}(S), \; \max_{\Sigma \in \mathcal{P}} \operatorname{diam}(\Sigma) \}.$$

As approximations of the initial and boundary values g_0 and θ_0 we take

$$G_0 = J g_0 \quad \text{and} \quad \Theta_0 = P\theta_0^1 + R\theta_0^2,$$

where J is the Lagrange operator associated with the discrete space Y, P denotes the orthogonal projection $L_2(\Omega) \to X$, and R is the Riesz projection onto $\mathring{X}$. Finally, assume a continuous solution (θ, g) of Problem 1 as in Remark 2.

Theorem 5. *Suppose that the above assumptions are satisfied and that the solution (θ, g) of Problem 1 fulfills*

$$\inf_{t \in (0,T)} \underline{\mathrm{DQ}} \left(g(t)\right) > 0 \quad \text{and} \quad \inf_{t \in (0,T)} \operatorname{dist}\left(g(\mathbb{S}, t), \partial\Omega\right) > 0. \tag{5}$$

Then there exist $h^* > 0$ and $C \geq 0$, depending on given data Ω, T, β, a, θ_0^1, θ_0^2, g_0, on the quasi-uniformity and shape-regularity of the partitions of Ω and $\mathbb{S}$, and the exact solution (θ, g), such that for all $h \in (0, h^*]$

(i) there exists a unique solution (Θ, G) on $[0, T]$ of Problem 3,
(ii) the following error estimates for the temperature field and the parametrization of the free boundary hold:

$$\left.\begin{array}{l} \sup\limits_{(0,T)} \|\Theta - \theta\|_{L_2(\Omega)}^2, \quad h \displaystyle\int_0^T \|\nabla(\Theta - \theta)\|_{L_2(\Omega)}^2, \\ \qquad \displaystyle\int_0^T \|\Theta(G) - \theta(g)\|_{L_2(\mathbb{S})}^2, \\ \sup\limits_{(0,T)} \|G - g\|_{H^1(\mathbb{S},\mathbb{R}^2)}^2, \quad \displaystyle\int_0^T \|G_t - g_t\|_{L_2(\mathbb{S},\mathbb{R}^2)}^2 \end{array}\right\} \leq \; C h^2.$$

Note that statement (i) extends the local existence result of Lemma 4 to a global one for $[0,T]$ with $T>0$, if the corresponding continuous problem has a classical solution and the discretization is fine enough.

Let us make some remarks about the optimality of the error estimates in statement (ii). In the sequel "of optimal order" means "of the same order as the corresponding interpolation estimate". We first show that the error estimate for the heat flux $\nabla\theta$ is of optimal order. The Stefan condition implies a jump of $\big(\nabla\theta\big)(t)$, $t\in[0,T]$, across the interface $\gamma(t)$, if the interface is non-stationary. As a consequence, $\theta(t)$ has, measured in $L_2(\Omega)$, at most 3/2 derivatives (compare [37]). The interpolation error for $\nabla\theta(t)$ thus is at most of order 1/2 and so the optimal order of the error estimate for the heat flux is established. Next, we discuss the optimality of the error estimate for the interfacial temperature $\theta(g)$. We rely on the following assumption: in order to estimate the interpolation error for the interfacial temperature $\theta(g)$ one has to apply a trace theorem in L_2. We thus need at least 1/2-derivative in $L_2(\Omega)$ (see [36]) and so the interpolation error for the interfacial temperature is at most of order 1. Hence, the error estimate of the interfacial temperature is of optimal order. Since $\theta(g)$ is a coupling quantity, all the other error estimates also are of optimal order, provided the above assumption that one has to use a trace theorem in L_2.

The proof of Theorem 5 would be outside the scope of this article. Interested readers may consult [44] or [45]. Here we restrict ourselves to the following two remarks, before we present further error estimates, which can be derived from Theorem 5.

The first remark concerns an issue in estimating the error of the temperature field. It may be understood as a modification for the usual proof for the error estimate of the heat equation, which is due to the free boundary nature of (1). In the sequel "appropriately" means "in such way that the Gronwall lemma can be applied to norms that arise on the left-hand side during estimation". A key term for the "coupling" of the error estimates for $G-g$ and $\Theta-\theta$ is $\varphi(g)-\Phi(G)$, where $\Phi\in\mathring{X}$ is the test function in the equation for Θ in the semi-discrete Problem 3 and $\varphi\in\mathring{H}^1(\Omega)$ is the test function in the corresponding equation for θ. This term is connected to the second equation in (1), the so-called Stefan condition. In order to control it appropriately, one may proceed as follows: construct a mapping $H:\overline{\Omega}\times[0,t^*]\to\overline{\Omega}$ such that

$$H(\cdot,t)\text{ is bijective}\quad\text{and}\quad H\big(g(u,t),t\big)=G(u,t)$$

for all $u\in\mathbb{S}$ and $t\in[0,t^*]$. Let us denote the inverse of $H(\cdot,t)$ by $H^{-1}(\cdot,t)$. Choosing $\varphi=\Phi\circ H(\cdot,t)$, in contrary to the usual choice $\varphi=\Phi$, as the continuous test function, we calculate formally that $\hat{\theta}:=\theta\circ H^{-1}$, more precisely $\hat{\theta}(x,t)=\theta\big(H^{-1}(x,t),t\big)$, fulfills for almost every $t\in[0,t^*]$

$$\int_\Omega\hat{\theta}_t\,\Phi+\int_\Omega\nabla\hat{\theta}\cdot\nabla\Phi=-\int_{\mathbb{S}}g_t\cdot g_u^\perp\,\Phi(G)+\mathcal{E}(\Phi),\tag{6}$$

where

$$\mathcal{E}(\Phi) := \int_\Omega (\nabla\theta)\circ H^{-1}\,(H^{-1})_t\,\Phi + \int_\Omega \theta_t\,\Phi\circ H\,\big(|\det DH| - 1\big)$$
$$+ \int_\Omega (\nabla\theta)\circ H^{-1}\big(D(H^{-1}) - E\big)\cdot\nabla\Phi + \int_\Omega \nabla\theta\cdot(\nabla\Phi)\circ H\,\big(|\det DH| - 1\big)$$
$$+ \int_\Omega \nabla\theta\cdot(\nabla\Phi)\circ H\,(E - DH).$$

and E denotes the matrix of the identity in $\mathbb{R}^2$. In doing so one needs not to control $\varphi(g) - \Phi(G)$, but one has to estimate $\mathcal{E}$ and then $\Theta - \hat{\theta}$ and $\hat{\theta} - \theta$ appropriately. The needed estimates for $\mathcal{E}$ and $\hat{\theta} - \theta$ follow from a suitable construction of H, which allows to control the "distance" of $H(\cdot,t)$ and the identity in appropriate norms of $G - g$. In this way, $\Theta - \hat{\theta}$ can be estimated by the help of a trace theorem for the discrete interface and the parabolic projection (see e.g. [31]) in such manner, that one can "couple" with the error estimate for $G - g$ in terms of $\Theta(G) - \theta(g) = (\Theta - \hat{\theta})(G)$.

The second remark follows from a tracing of the dependencies of the constant C in Theorem 5. If there is a geometric singularity (i.e. a topological change of the interface) at time T_{sing}, we have $\underline{\text{DQ}}\big(g(T)\big) \downarrow 0$ when $T \uparrow T_{\text{sing}}$. This implies

$$h^* \downarrow 0 \quad \text{and} \quad C \uparrow \infty \qquad \text{as} \quad T \uparrow T_{\text{sing}};$$

in other words: one "looses" the error estimates when one approaches a geometric singularity.

The error estimates for the free boundary in Theorem 5 are rather strong. An illustration of this fact are the following consequences. The first one concerns the phase distribution $p = (\omega^s, \gamma, \omega^l)$. Let δ denote the Hausdorff distance between (arbitrary) sets in $\big(\mathbb{R}^2, |\cdot|\big)$, and define the discrete phase distribution by $\Gamma(t) := G(\mathbb{S}, t)$, $\Omega^s(t) := \{x \in \Omega \mid \operatorname{sdist}\big(G(t), x\big) < 0\}$, and $\Omega^l(t) := \{x \in \Omega \mid \operatorname{sdist}\big(G(t), x\big) > 0\}$ for $t \in [0, T]$.

Corollary 6. *Suppose that the assumptions of Theorem 5 hold and choose C and $h^* > 0$ like therein. Then the error of the phase distribution is controlled in the following way: for all $h \in (0, h^*]$*

$$\sup_{(0,T)} \delta(\omega^s, \Omega^s),\ \sup_{(0,T)} \delta(\gamma, \Gamma),\ \sup_{(0,T)} \delta(\omega^l, \Omega^l) \ \le\ C\,h.$$

A second consequence of the error estimates in Theorem 5 concerns the approximation of the geometric quantities n, v, k that appear in model (1). In the sequel we assume that v and k (like n) are defined on the sphere $\mathbb{S}$. Recall $N = -G_u^\perp/|G_u|$ and choose the approximations of the normal velocity v and curvature k in the following way:

$$V := G_t\cdot N, \quad K := -\frac{1}{a(N)}\Big(\beta(N)\,V + \Theta(G)\Big).$$

Corollary 7. *Suppose that the assumptions of Theorem 5 hold. Then the quantities N, V, K approximate n, v, k respectively, more precisely: there exists a constant $C \geq 0$ such that for all $h \in (0, h^*]$, we have*

$$\sup_{(0,T)} \|N - n\|^2_{L_2(\$,\mathbb{R}^2)}, \int_0^T \|V - v\|^2_{L_2(\$)}, \int_0^T \|K - k\|^2_{L_2(\$)} \leq C\,h^2.$$

Corollary 7 allows to approximate measuring quantities in dendritic growth experiments, namely the velocity and curvature (= inverse of the radius) of a primary tip. Let $g_{\text{tip}} : (0,T) \to \mathbb{R}^2$ with $g_{\text{tip}}(t) \in \gamma(t)$ be the position vector of a primary tip, and let $G_{\text{tip}} : (0,T) \to \mathbb{R}^2$ with $G_{\text{tip}}(t) \in \Gamma(t)$ some approximation of g_{tip}. Let Q indicate the orthogonal projection $L_2(\$) \to Z$. Define

$$v_{\text{tip}}(t) := v\Big(g^{-1}\big(g_{\text{tip}}(t), t\big), t\Big) \quad \text{and} \quad V_{\text{tip}}(t) := QV\Big(G^{-1}\big(G_{\text{tip}}(t), t\big), t\Big)$$

for $t \in [0,T]$, where $g^{-1}(\cdot, t)$ and $G^{-1}(\cdot, t)$ denote the inverse of $g(\cdot, t)$ and $G(\cdot, t)$ respectively. Define k_{tip} and K_{tip} in an analogous way. The approximation of v_{tip} and k_{tip} by V_{tip} and K_{tip} respectively is established in two steps. The first step is a reduction to the task that G_{tip} approximates g_{tip} in an appropriate way. The second step establishes such an approximation in a special situation.

Corollary 8. *Suppose that the assumptions of Theorem 5 hold. Then there exists a constant $C \geq 0$ such that for all $h \in (0, h^*]$*

$$\left.\begin{array}{l} \|V_{\text{tip}} - v_{\text{tip}}\|_{L_2(0,T)}, \\ \|K_{\text{tip}} - k_{\text{tip}}\|_{L_2(0,T)} \end{array}\right\} \leq C\left(h^{1/2} + \|G_{\text{tip}} - g_{\text{tip}}\|_{L_2((0,T),\mathbb{R}^2)}\right). \tag{7}$$

A variant of the practical construction of G_{tip} in the sequel was used in [34]. The construction relies on the following assumption on the continuous tip: suppose there is a known unit vector e (given by the anisotropy) such that

$$g_{\text{tip}}(t) \cdot e^{\perp} = 0 \quad \text{and} \quad g_{\text{tip}}(t) \cdot e = \sup\{x \cdot e \mid x \in \gamma(t)\} \tag{8}$$

for all $t \in (0,T)$.

Lemma 9. *Suppose the assumptions of Theorem 5 and (8). Choose $h^* > 0$ like in Theorem 5 and let $h \in (0, h^*]$. Then there exists a unique measurable mapping $G_{\text{tip}} : [0,T] \to \mathbb{R}^2$ with $G_{\text{tip}}(t) \in \Gamma(t)$,*

$$G_{\text{tip}}(t) \cdot e^{\perp} = 0, \quad \textit{and} \quad G_{\text{tip}}(t) \cdot e = \sup\{x \cdot e \mid x \in \Gamma(t) \textit{ and } x \cdot e^{\perp} = 0\}$$

for all $t \in [0,T]$. Moreover, we have the error estimate

$$\sup_{(0,T)} |G_{\text{tip}} - g_{\text{tip}}| \leq C\,h,$$

where $C \geq 0$ does not depend on h.

Using Lemma 9 in the Inequalities (7) we obtain the following error estimates for the velocity and curvature of the tip:

$$\|V_{\mathrm{tip}} - v_{\mathrm{tip}}\|_{L_2(0,T)},\ \|K_{\mathrm{tip}} - k_{\mathrm{tip}}\|_{L_2(0,T)} \leq C\,h^{1/2},$$

where $h \in (0, h^*]$ and C does not depend on h.

3 A Level Set Algorithm

3.1 A Level Set Formulation for the Phase Boundary

Suppose for the moment θ to be given, then the last equation in (1), the generalized Gibbs-Thomson law, leads to the following generalized mean curvature flow for the evolution of the free boundary $\gamma(t)$.

Problem 10. Find a family of interfaces $\{\gamma(t)\}_{t\in[0,T)}$ such that

$$\beta(n)v + a(n)k = -\theta \qquad \text{on} \qquad \gamma(t),\ t \in (0,T), \tag{9}$$

where $\gamma(0) = \gamma_0$.

The first part of this article dealt with a reformulation of (1) using parameterizations of the phase boundary. In this part, we turn to a level set method for the free boundary problem, as first introduced by Osher and Sethian in [32]. Their approach is this: given the initial interface γ_0, select some continuous real valued function u_0 such that

$$\gamma(0) = \{x \in \mathrm{IR}^2 \,|\, u_0(x) = 0\},\ \ u_0 > 0 \text{ in } \omega^l(0),\ \ u_0 < 0 \text{ in } \omega^s(0). \tag{10}$$

Consider the phase boundary $\gamma(t)$ to be always represented by the zero level set of a continuous function u with $u(\cdot,t) > 0$ in $\omega^l(t)$ and $u(\cdot,t) < 0$ in $\omega^s(t)$. Suppose for the moment that u is a smooth function whose spatial gradient ∇u does not vanish in some open region $\mathcal{O} \subset \mathrm{IR}^2 \times (0,T)$ of $\gamma(t)$. The normal n on $\gamma(t)$ is given by

$$n(x,t) = \frac{\nabla u(x,t)}{|\nabla u(x,t)|}, \qquad \forall x \in \gamma(t). \tag{11}$$

Then the curvature $k(x,t)$ and the normal velocity $v(x,t)$ at any point x of $\gamma(t)$ are given by

$$k(x,t) = \nabla \cdot n(x,t), \qquad v(x,t) = -\frac{u_t(x,t)}{|\nabla u(x,t)|}. \tag{12}$$

Supposing each level set of u in $\mathcal{O}$ to move according (9), we get the nonlinear degenerate and singular parabolic pde

$$\beta\left(\frac{\nabla u}{|\nabla u|}\right)\frac{u_t}{|\nabla u|} - a\left(\frac{\nabla u}{|\nabla u|}\right)\nabla \cdot \frac{\nabla u}{|\nabla u|} = \theta \text{ in } \mathcal{O} \tag{13}$$

with initial condition

$$u(\cdot,0) = u_0 \qquad \text{in} \qquad \mathbb{R}^2 \tag{14}$$

for the function u, which has to be interpreted in a viscosity sense. Conversely, if u is a viscosity solution of equation (13) in $\mathbb{R}^2$, then each level set of u evolves in a weak sense according the evolution equation (9).

In a series of publications ([18], [19], [20], [21]), Evans and Spruck investigated problem (13),(14) with $a \equiv \beta \equiv 1, \theta \equiv 0$ in spatial dimensions $m \geq 2$, and proved existence and uniqueness of the viscosity solution u of this problem. Similar results were obtained by Chen/Giga/Goto and Giga/Goto/Ishii, ([7], [23]), even for anisotropic and inhomogeneous situations.

In view of the Evolution Problem 10, the above authors showed that not only the viscosity solution u is uniquely determined by (13),(14), but also the generalized evolution of γ, defined by $\{x \in \mathbb{R}^m | \, u(x,t) = 0\}$, does only depend on γ_0, and therefore is independent of the special choice of the initial function u_0. These results allow to reformulate Problem 10 in a level set context: given some function u_0 according to (10), find the continuous function $u : \mathbb{R}^2 \times [0,T) \longrightarrow \mathbb{R}$ such that (13),(14) holds.

To overcome the numerical difficulties arising from vanishing gradients ∇u, we replace equation (13) by the regularized equation

$$\beta_\varepsilon(\nabla u_\varepsilon)\, \frac{u_{\varepsilon t}}{Q(u_\varepsilon)} - a_\varepsilon(\nabla u_\varepsilon)\, \nabla \cdot \frac{\nabla u_\varepsilon}{Q(u_\varepsilon)} = \theta \ \text{in} \ \mathbb{R}^2 \times (0,T), \tag{15}$$

where

$$Q(u_\varepsilon) = \sqrt{\varepsilon^2 + |\nabla u_\varepsilon|^2}, \qquad \varepsilon > 0, \tag{16}$$

and $a_\varepsilon : \mathbb{R}^2 \to \mathbb{R}$, $\beta_\varepsilon : \mathbb{R}^2 \to \mathbb{R}$ are suitable continuations of the functions a and β, respectively.

In case of the homogeneous and isotropic situation, Evans and Spruck proved existence and uniqueness of a solution u_ε of (15) with initial condition $u_\varepsilon(\cdot,0) = u_0$ in $\mathbb{R}^2$ (c. f. [18]). Their proof was extended to anisotropic situations, see [22]. For $\varepsilon \to 0$ the solutions u_ε converge locally uniformly to the viscosity solution of (13), (14), compare Evans and Spruck in [18]. In [9], Deckelnick proved the estimate

$$\sup_{x \in \mathbb{R}^2, 0 \leq t \leq T} |(u - u_\varepsilon)(x,t)| \leq C\varepsilon^\alpha,$$

for all $\varepsilon > 0$ and all $\alpha \in (0, \frac{1}{2})$. Hereafter, we intend to approximate the viscosity solution u by numerically solving the regularized equation (15) for u_ε. Restricting ourselves to the polygonal bounded domain $\Omega \subset \mathbb{R}^2$ of Section 2.2 with appropriate boundary conditions we state the regularized problem

Problem 11. For fixed $\varepsilon > 0$, find $u_\varepsilon \in C^0(\overline{\Omega} \times [0,T))$ such that

$$\beta_\varepsilon(\nabla u_\varepsilon)\,\frac{u_{\varepsilon t}}{Q(u_\varepsilon)} - a_\varepsilon(\nabla u_\varepsilon)\,\nabla \cdot \frac{\nabla u_\varepsilon}{Q(u_\varepsilon)} = \theta \text{ in } \Omega \times (0,T), \tag{17}$$

with initial and boundary conditions

$$u_\varepsilon = u_{\varepsilon,0} \qquad \text{on} \qquad [\overline{\Omega} \times \{0\}] \cup [\partial\Omega \times (0,T)], \tag{18}$$

where $u_{\varepsilon,0}$ is some continuous function fulfilling (10).

Remark 12. For $\varepsilon = 1$ Problem 11 describes the situation of a moving graph. By a scaling argument, we see that a solution u_ε of Problem 11 leads to a solution of the corresponding problem of a moving graph, and vice versa. Hence, results on moving graphs may be transfered to the solution of the regularized level set problem. Various results on existence and uniqueness of moving graphs on a bounded domain were obtained for instance by Huisken [27], Lieberman [29] and Veeser [43]. With particular regard to our algorithm the results of Deckelnick and Dziuk on convergence for a semi–discrete finite element method [11], [12], and an error estimate and stability results for several time discretizations obtained by Dziuk [15] are of great interest.

Our finite element method is based on the following weak formulation of the regularized Problem 11: find $u_\varepsilon(\cdot,t) \in H^1(\Omega)$ such that

$$\int_\Omega \frac{\beta_\varepsilon(\nabla u_\varepsilon)}{a_\varepsilon(\nabla u_\varepsilon)}\,\frac{u_{\varepsilon,t}\phi}{\sqrt{\varepsilon^2 + |\nabla u_\varepsilon|^2}} + \int_\Omega \frac{\nabla u_\varepsilon \nabla \phi}{\sqrt{\varepsilon^2 + |\nabla u_\varepsilon|^2}} = \int_\Omega \frac{\theta\,\phi}{a_\varepsilon(\nabla u_\varepsilon)} \qquad \forall \phi \in H_0^1(\Omega), \tag{19}$$

with initial and boundary conditions (18). In order to state the discrete problem, we define the finite element spaces V by

$$V := \{\Phi \in C^0(\overline{\Omega}) \mid \Phi \text{ is a polynomial of degree } \le 2 \text{ on each S } \in \mathcal{T}\},$$

and $\mathring{V} := V \cap H_0^1(\Omega)$, where $\mathcal{T}$ is a conforming non-degenerate simplicial triangulation of the polygonal domain Ω. Denoting by $\tau > 0$ the timestep size, and using the notations

$$\Phi^m(x) = \Phi(x,t^m), \quad t^m = m\tau \qquad (m = 0,1,\ldots,M = \left[\frac{T}{\tau}\right])$$

for a given function $\Phi : \overline{\Omega} \times [0,T] \longrightarrow \mathbb{R}$, a discretization of Problem 11 is given by:

Problem 13. For given $U_0(\cdot,t) \in V$, $t \in [0,T)$, find $U^m \in V$ $(m = 1,\ldots,M)$ such that

$$\frac{1}{\tau}\int_\Omega \frac{\beta_\varepsilon(\nabla U^{m-1})}{a_\varepsilon(\nabla U^{m-1})}\,\frac{U^m - U^{m-1}}{Q(U^{m-1})}\Phi + \int_\Omega \frac{\nabla U^m \nabla \Phi}{Q(U^{m-1})} = \int_\Omega \frac{\theta}{a_\varepsilon(\nabla U^{m-1})}\Phi \tag{20}$$

$$\forall \Phi \in \mathring{V},$$

with initial and boundary conditions

$$U^0 = U_0, \ U^m(\cdot) - U_0(\cdot, t^m) \in \mathring{V},$$

where $U_0(\cdot, t) \in V$ denotes a finite element representation of $u_0(\cdot, t)$.

Remark 14. In case of the isotropic and homogeneous situation, the Time Discretization Scheme 13 is equivalent to the semi–implicit scheme for the mean curvature flow of graphs proposed by Dziuk in [15], compare Remark 12.

3.2 Finite Element Method for the Heat Equation

Now, suppose that the free boundary $\gamma(t)$, its normal velocity v and curvature k are given. Then, together with boundary and initial conditions, the equations (1) determine the temperature θ.

Problem 15. Find the temperature θ such that

$$\theta_t - \Delta\theta = 0 \qquad \text{in } [\Omega \setminus \gamma(t)] \times (0, T), \tag{21}$$

$$v = - \left[\!\left[\frac{\partial \theta}{\partial n} \right]\!\right]_s^l \qquad \text{on } \gamma(t), \tag{22}$$

$$\beta(n)\, v + a(n)\, k + \theta = 0 \qquad \text{on } \gamma(t), \tag{23}$$

$$\theta = \theta_0 \qquad \text{on } [\overline{\Omega} \times \{0\}] \cup [\partial\Omega \times (0, T)] \tag{24}$$

In order to derive a weak formulation of Problem 15, we utilize equations (21), (22) and (23) as described in Section 2.1. Therewith we get for all $\phi \in H_0^1(\Omega)$ and all $t \in (0, T)$

$$\int_\Omega \theta_t\, \phi + \int_\Omega \nabla\theta\, \nabla\phi + \int_{\gamma(t)} \frac{1}{\beta}\, \theta\, \phi = - \int_{\gamma(t)} \frac{a}{\beta}\, k\, \phi \tag{25}$$

We discretize equation (25) with the help of an implicit Euler method in time and the finite element spaces V and $\mathring{V}$ as defined in Section 3.1. This leads to

Problem 16. Given smooth interfaces $\gamma(t^m)$ with curvature k and boundary values $\Theta_0(\cdot, t) \in V, t \in [0, T)$, find $\Theta^m \in V\, (m = 1, \ldots, M)$ such that

$$\begin{aligned} \frac{1}{\tau} \int_\Omega (\Theta^m - \Theta^{m-1})\, \Phi + \int_\Omega \nabla\Theta^m\, \nabla\Phi + \int_{\gamma(t^m)} \frac{1}{\beta}\, \Theta^m\, \Phi & \\ = - \int_{\gamma(t^m)} \frac{a}{\beta}\, k\, \Phi \qquad \forall \Phi \in \mathring{V}, & \end{aligned} \tag{26}$$

with initial and boundary conditions

$$\Theta^0 = \Theta_0, \qquad \Theta^m(\cdot) - \Theta_0(\cdot, t^m) \in \mathring{V},$$

where $\Theta_0(\cdot, t) \in V$ denotes a finite element interpolation of $\theta_0(\cdot, t)$.

In the following subsection, Discretizations 13 and 16 for the mean curvature problem and the heat equation, respectively, are combined to get a method to solve the whole Stefan problem.

3.3 Combining the Algorithms

We now turn back again to the full Stefan problem (1). In contrast to the above assumptions, neither the temperature nor the phase boundary and its curvature are known. As mentioned in Section 1, we combine the both algorithms in the following way:

In each time step t^m, given the old values Θ^{m-1} and U^{m-1}, we first compute the new free boundary by solving equation (20) for U^m, where we replaced the temperature θ on the right–hand side by the old discrete temperature Θ^{m-1}. The zero level of U^m then gives us the new discrete interface γ^m.

To compute the new temperature Θ^m, we will solve equation (26) with γ^m in place of the interface $\gamma(t^m)$. Consequently, we need the curvature k of γ^m, as it appears in equation (26). Here, γ^m is a level set of a piecewise quadratic function, and unfortunately there is no straightforward definition of the curvature of such a level set. We find a remedy by defining a discrete curvature $K \in V$ of the level sets of U^m: In case of a smooth function u with non vanishing gradient ∇u, the curvature of the level sets of u is given by

$$k(x,t) = \nabla \cdot \frac{\nabla u}{|\nabla u|}.$$

Recalling the way of regularizing the mean curvature flow equation (13), we again replace the term $|\nabla u|$ by the square root term $Q(u) = \sqrt{\varepsilon^2 + |\nabla u|^2}$, getting for all $\phi \in H_0^1(\Omega)$

$$\int_\Omega k_\varepsilon \phi = \int_\Omega \frac{\nabla u \, \nabla \phi}{Q(u)}, \tag{27}$$

with suitable boundary values $k_\varepsilon = k_0$ on $\partial\Omega$, see below. From this, we define a discrete curvature K of the level sets of the function $U^m \in V$ by

$$\int_\Omega K\Phi = \int_\Omega \frac{\nabla U^m \, \nabla \Phi}{Q(U^m)} \qquad \forall \Phi \in \mathring{V}, \tag{28}$$

with boundary values $K = K_0$ on $\partial\Omega$, where $K_0 \in V$ is an finite element approximation of k_0. A typical choice of k_0 would be $k_0(x) = |x|^{-1}$, which represents the curvature of the level sets of a rotationally symmetric function. In principle, U^m is not a rotationally symmetric function, but a discrete distance function from γ^m (compare Section 3.4), and so the level sets of U^m near the boundary $\partial\Omega$ are approximately circular, at least as long as the phase boundary γ^m is far away from $\partial\Omega$.

Now, the new temperature Θ^m is computed according to

$$\frac{1}{\tau}\int_{\Omega}(\Theta^m-\Theta^{m-1})\,\Phi+\int_{\Omega}\nabla\Theta^m\,\nabla\Phi+\int_{\gamma^m}\frac{1}{\beta}\,\Theta^m\,\Phi \tag{29}$$

$$=-\int_{\gamma^m}\frac{a}{\beta}\,K\,\Phi \qquad \forall\Phi\in\mathring{V},$$

which ends the time step t^m. Summarizing, this gives the following level set based finite element discretization of the Stefan problem (1):

Problem 17. For given $U_0(\cdot,t)$, $\Theta_0(\cdot,t)$, $K_0\in V, t\in[0,T)$, find U^m and $\Theta^m\in V\;(m=1,\dots,M)$ such that

$$\frac{1}{\tau}\int_{\Omega}\frac{\beta_\varepsilon(\nabla U^{m-1})}{a_\varepsilon(\nabla U^{m-1})}\frac{U^m-U^{m-1}}{Q(U^{m-1})}\Phi+\int_{\Omega}\frac{\nabla U^m\nabla\Phi}{Q(U^{m-1})}=\int_{\Omega}\frac{\Theta^{m-1}}{a_\varepsilon(\nabla U^{m-1})}\,\Phi \tag{30}$$

$$\forall\Phi\in\mathring{V},$$

and

$$\frac{1}{\tau}\int_{\Omega}(\Theta^m-\Theta^{m-1})\,\Phi+\int_{\Omega}\nabla\Theta^m\,\nabla\Phi+\int_{\gamma^m}\frac{1}{\beta_\varepsilon(\nabla U^m)}\,\Theta^m\,\Phi \tag{31}$$

$$=-\int_{\gamma^m}\frac{a_\varepsilon(\nabla U^m)}{\beta_\varepsilon(\nabla U^m)}\,K\,\Phi \qquad \forall\Phi\in\mathring{V},$$

with initial and boundary conditions

$$U^0=U_0,\qquad U^m(\cdot)-U_0(\cdot,t^m)\in\mathring{V},$$
$$\Theta^0=\Theta_0,\qquad \Theta^m(\cdot)-\Theta_0(\cdot,t^m)\in\mathring{V},$$

where

$$\gamma^m:=\{x\in\Omega|U^m(x)=0\},$$

and $K\in V$ is given by

$$\int_{\Omega}K\Phi=\int_{\Omega}\frac{\nabla U^m\,\nabla\Phi}{Q(U^m)} \qquad \forall\,\Phi\in\mathring{V},$$

with boundary condition

$$K-K_0\in\mathring{V}.$$

Remark 18. Though the analytical determination of the level set γ^m of U^m is straightforward, the numerical determination is costly. In practice, we therefore use a polygonal approximation Γ^m of γ^m. The basic idea for an algorithm to approximate γ^m is to find the intersections of γ^m with the edges of the triangulation $\mathcal{T}$, and then to connect them conveniently while preserving the topological type of γ^m. A detailed description of this algorithm goes beyond the scope of this article, it could be found in [22].

3.4 Numerical Algorithms

In this section we describe some of the numerical methods which were used to implement the Finite Element Method 17.

From numerical and physical experiments, it is known that the temperature θ varies rapidly in the liquid phase near the interface γ, but is flat far away from γ and inside the solid phase. In order to obtain a good finite element approximation of the temperature, a fine grid is required, at least in the region where θ varies heavily. A globally fine mesh, whose grid size is small enough to ensure an approximation as accurate as needed for the numerical simulation of dendritic growth, would lead to high computational efforts. Due to the observed behavior of θ, an adapted mesh, very fine around the phase boundary and coarser far away from it, can be a let-out. Of course, since the phase boundary moves in time, the spatially adapted mesh has to change in time, too. This requirements can be met by a time–dependent adaptive strategy, which at each time step t^m automatically adjusts the triangulation such, that an estimation of the error $\|\Theta^m - \theta(\cdot, t^m)\|_{L^2(\Omega)}$ is below a given tolerance *tol* but not much smaller than a certain fraction of *tol*. Our method consists of three parts: First, some a–posteriori error indicators, which provide information on the local (i.e. element wise) and the global error, and a prescribed tolerance for the (indicated) global error. The second part is a marking strategy. Instead of building up a completely new triangulation each time the global error indicator says that the mesh has to change, this algorithm selects elements of the actual mesh for refinement or coarsening. The last part of our method consists of refinement and coarsening algorithms for the modification of the grid, which have to make sure that the modified triangulation is also a conforming one.

As error indicators, we apply a simplified version of an a–posteriori error estimator for the Poisson equation, which was introduced by Eriksson and Johnson in [17], namely the local indicators

$$\eta_T(\Theta^m) := \left(\frac{1}{2} \sum_{e \in \partial T \cap \mathring{\Omega}} \int_e h_e^3 \left| \left[\!\left[\frac{\partial \Theta^m}{\partial n_e} \right]\!\right] \right|^2 \right)^{\frac{1}{2}},$$

where the sum is over all edges e of the triangle T, which belong to the interior of Ω, and the global error indicator

$$\eta(\Theta^m) := \left(\sum_{T \in \mathcal{T}} \eta_T^2(\Theta^m) \right)^{\frac{1}{2}} .$$

For marking, we use the *guaranteed error reduction* strategy proposed by Dörfler in [14]. The idea is to refine a subset $\mathcal{A}$ of the triangulation that produces a considerable amount of the indicated global error, and to coarsen another subset $\mathcal{B}$ such that the indicated additional error after coarsening is not larger than some fixed amount of the prescribed tolerance. A subset $\mathcal{A}$ could be found by the algorithm:

Algorithm 19 (Guaranteed error reduction strategy[14]).

```
Start with given parameters p1,p2 ∈ (0,1)

ηmax := max(ηT, T ∈ 𝒯)
sum := 0
κ := 1
while sum < (1 − p1)²η² do
   κ := κ − p2
   forall T in 𝒯 do
      if T is not marked
         if ηT > κ ηmax
            mark T for refinement
            sum := sum + ηT²
         end if
      end if
   end for
end while
```

A similar method leads to the subset $\mathcal{B}$ of triangles, which could be coarsened, compare [14]. Finally, the modification of the mesh is done by a bisectioning algorithm, which is taken from [1].

The initial triangulation is generated by an application of this adaptive procedure to the stationary problem

$$\int_{\Omega} \nabla\Theta^0 \, \nabla\Phi + \int_{\Gamma(0)} \frac{1}{\beta} \Theta^0 \, \Phi = - \int_{\Omega} \Theta_t^0 \, \Phi - \int_{\Gamma(0)} \gamma \, K \, \Phi, \;\; \forall \Phi \in \mathring{V}, \tag{32}$$

with given initial data $\Gamma(0)$ and Θ_t^0 and boundary value Θ_0. This method was proposed by Schmidt, [33]. Beside the generation of an initial adapted grid, defining Θ^0 as the solution of Equation (32) ensures that the initial temperature is compatible with the discrete interface $\Gamma(0)$ and its curvature K, compare [33].

The finite element algorithms for the approximation of the regularized level set function u_ε are analogous to the methods described for the heat equation. Namely, in both problems we apply the same finite element spaces, and consequently the same triangulation. Hence the adaptive process is completely done using the error indicator for the heat equation presented above. Differences appear as the gradients of the level set function tend to become steep or flat, which causes numerical difficulties approximating the curvature and normal at the free boundary. One way to avoid this difficulties is as follows: Remembering the equations (11) and (27) for the normal n and the weak curvature K, we see that those quantities are all depending on the gradient ∇U^m of the level set function U^m. In practice, very steep or flat gradients ∇U^m would lead to inaccurate approximations of the curvature K and the normal n, as well as this could cause stability problems like tip splitting. Therefore it is desirable to avoid having steep or flat gradients develop in U^m, whenever this would be possible. For instance, if U^m is a distance function to $\gamma(t^m)$, we will avoid such ugly gradients. However even if we choose u_0 to be the signed distance function

$$d_{\gamma_0}(x) := \begin{cases} dist(x,\gamma_0) & \text{if} \quad x \in \omega^s(0) \\ -dist(x,\gamma_0) & \text{if} \quad x \in \omega^l(0) \end{cases}$$

from γ_0, the level set function will cease to be a distance function after one timestep. This may be fixed by reinitializing the function U^m to be a distance function from the discrete phase boundary $\gamma(t^m)$ at each timestep. There are several possibilities of redistancing the level set function, see for example Chopp [8], Strain [41] or Sussman, Sekerka and Osher [42]. In view of the fact that we already have a polygonal approximation Γ_m of the interface $\gamma(t^m)$, we use a direct method, calculating a Lagrange interpolant of the distance function from the polygonal curve Γ_m.

3.5 Numerical Results

We implemented a version of the above described algorithm using the finite element tool-box ALBERT, which was developed by Schmidt and Siebert, [35]. Here, we present some numerical results obtained by this implementation.

Convergence of the Numerical Method. As a first test we compare the solution of our finite element method with the known solution in case of an isotropic test problem. This enables us to check the accuracy of our numerical method. Using the notations

$$R(t) := \sqrt{R_0^2 + 2t},\ R_0 \in \mathbb{R}^+, \qquad f(t) := \dot{W}(t) = \tfrac{2\beta}{R(t)^3}$$

$$W(t) := -\tfrac{2\beta}{R(t)}, \quad \beta \in \mathbb{R}^+, \qquad T(s) := -\sqrt{e}\int_1^s \frac{e^{-\frac{z^2}{2}}}{z}dz.$$

the problem reads as follows (c. f. Schmidt in [33]):

$$\begin{cases} \theta_t - \Delta\theta & = f \quad \text{for} \quad t \geq 0, x \in \mathbb{R}^2 \setminus \Gamma(t), \\ \left[\!\left[\dfrac{\partial\theta}{\partial n}\right]\!\right] + V_\Gamma & = 0 \quad \text{on} \quad \Gamma(t), \\ \theta + \beta C_\Gamma + \beta V_\Gamma & = 0 \quad \text{on} \quad \Gamma(t). \end{cases}$$

The difference between the modified Stefan problem (1) and the above problem is the right hand side f in the heat equation, which is chosen in such a way, that an explicit solution is known. This solution is given by

$$\theta(x,t) := \begin{cases} W(t) & \text{for } |x| \leq R(t) \\ W(t) + T\left(\frac{|x|}{R(t)}\right) & \text{else} \end{cases}$$

and the free boundary $\Gamma(t) = \{x \in \mathbb{R}^2 |\ \|x\| = R(t)\ \}$, where $v(t) = R_t(t) = \frac{1}{R(t)}$ and $k(t) = \frac{1}{R(t)}$ are the normal velocity and curvature of the free boundary.

Using the parameter $\beta = 0.1$ and the regularization parameter $\varepsilon = 10^{-6}$, we investigate the behavior of error between the computed temperature Θ and the exact temperature θ, solving the same problem with different globally refined spatial meshes and time step sizes. The error was measured by

$$E_{\infty,2}(\Theta_h) := \max_{i=0,\ldots,n} \|\, \Theta(\cdot, t_i) - \Theta_h^i \|_{L^2(\Omega)}.$$

Table 1 shows the results of these calculations.

h	1.0	0.5	0.25	0.125	0.0625
Δt	0.25	0.0625	0.0156	0.0039	0.00098
$E_{\infty,2}(\Theta_h)$	0.8319	0.3546	0.0657	0.0256	0.0133

Table 1. Exact Solution: Growing Circle. Evolution of the Error while Uniformly Refining the Grid

Numerical Experiments. In this section, we show the results of different numerical experiments in the case of the Stefan problem (1), where an explicit solution is not known. For all of the following experiments, we choose the boundary values $\Theta_0(\cdot, t) = T_0$ to be constant on $\partial\Omega$ for all $t \in [0, T]$, and the functions a and β of the form

$$\beta(n) = a(n) = \delta(1 + A\cos(k\alpha + \alpha_0)),$$

where $\alpha = \alpha_n$ denotes the angle between n and e_1– axis. The initial temperature Θ^0 is computed from Equation (32) with $\Theta_t^0 = 0$.

The first numerical experiment is a further isotropic calculation choosing the parameter $A = 0$. As opposed to the above test problem, we investigate the evolution of an initially non–convex phase boundary, namely

$$\Gamma(0) = \left\{ [0.15 + 0.4\cos(4\alpha + \frac{\pi}{4})] \begin{pmatrix} \cos(\alpha) \\ \sin(\alpha) \end{pmatrix} \mid \alpha \in [0, 2\pi] \right\}. \tag{33}$$

The investigation of such a curve was proposed by Sethian and Strain in [38]. Here, we are interested in the influence of the parameter δ. In some sense this parameter plays the role of the anisotropic functions $\beta(n)$ and $a(n)$, and in a realistic setting, it could be a very small parameter. We compare the numerical results for different values of the parameter δ. Figure 1 shows the evolution for the parameters $\delta = 0.006$ and $\delta = 0.0009$ respectively.

For larger values of δ as in the first case (figure 1 left), we found a smooth evolution of the phase boundary Γ, while the evolution for smaller $\delta \leq 0.0009$ tends to be unstable. Figure 1 (right) shows a typical pattern with so called tip splitting: the evolution becomes instable and develops small fingers and side branches. For both situations we took the spatial domain to be $\Omega = B_1(0)$ the unit circle with boundary value $T_0 = -1$, and calculated the numerical solutions using the above adaptive methods. If we choose a smaller tolerance for the adaptive method as well as finer time step sizes, the evolution is again stable, and shows qualitatively the same pattern as in the case $\delta = 0.006$. Therefore, we believe that the observed tip splitting is due to grid effects.

We now come to results obtained by simulating the anisotropic growth of a seed crystal into an undercooled melt. For the initial interface $\Gamma(0)$ given by equation (33), we calculated the evolution using the parameters

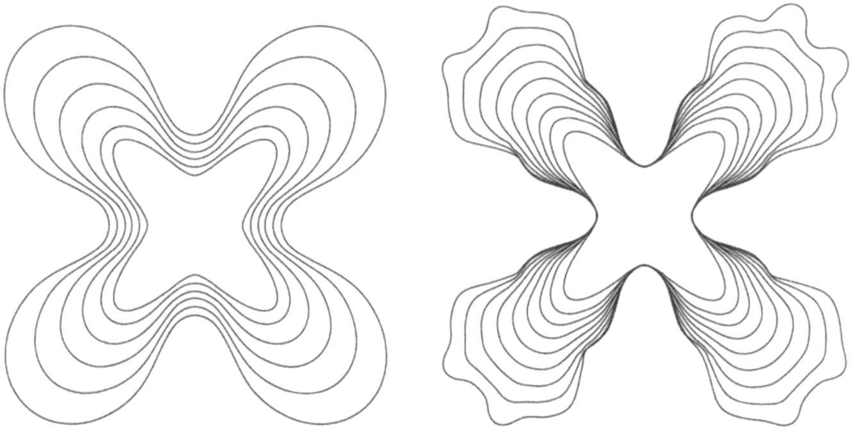

Fig. 1. Isotropic Growth for $\delta = 0.006$ (left) and $\delta = 0.0009$

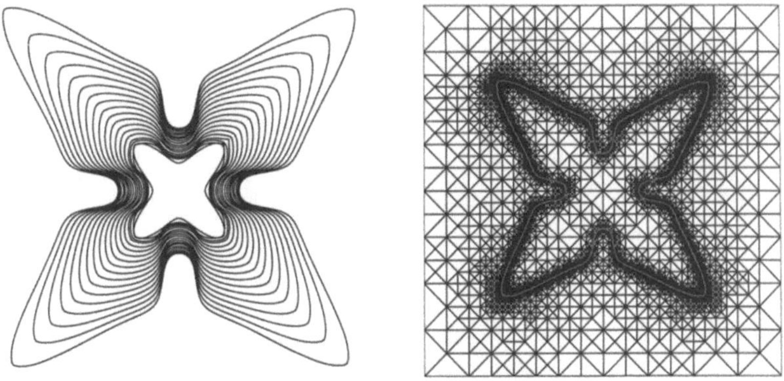

Fig. 2. Fourfold Anisotropy: Evolution of the Phase Boundary Γ and typical Adaptive Grid

$\delta = 0.006, A = 0.4$ and $k = 4$. Choosing the domain $\Omega = [-4,4]^2$ and $T_0 = -1$ we found results as depicted in figure 2 and Plate 12 on page 812.

Despite the anisotropic situation there are only small secondary dendrites. Plate 12 on page 812 shows the graph of the temperature Θ. Here we observe the expected behavior, nearly constant temperature inside the solid, but a fast decay in the liquid.

Starting with a rotational symmetric seed crystal will give us a better feeling of the anisotropy's influence on the growth. Therefore we took in a further calculation the initial curve to be circular $\Gamma(0) := \partial B_{0.1}(0)$. Figure 3 shows the result obtained with the parameters $\delta = 0.004$, $A = 0.4$, $k = 4$, $T_0 = -1$ and the domain $\Omega = [-4,4]^2$.

Using this parameters we did indeed find a transition from the initial circle to a fourfold dendrite. As in the last example, there was nearly no indication for secondary dendrites.

Choosing higher undercooling, we expect smaller temperatures inside of Ω, and such a faster growth of the solid phase. In the following experiment, we observe the evolution in case of different undercooling. We also switched to a sixfold anisotropy, fixing their parameters to be $\delta = 0.01$, $A = 0.4$ and $k = 6$. Figure 4 and 5 and Plates 13 and 14 on page 812f. illustrate the results obtained with the boundary temperatures $T_0 = -1$ and $T_0 = -2$, respectively. For both examples we choose the domain $\Omega = B_4(0)$ and the initial phase boundary $\Gamma(0) := \partial B_{0.1}(0)$.

Figure 4 and Plate 13 on page 812 show the numerical evolution of the phase boundary $\Gamma(t)$ and the graph of Θ in case of $T_0 = -1$. As in the above examples, we observe approximately constant temperatures inside the solid and steep gradients $\nabla\Theta$ in the liquid phase near the phase boundary, while the graph of Θ flattens again far away from the interface $\Gamma(t)$.

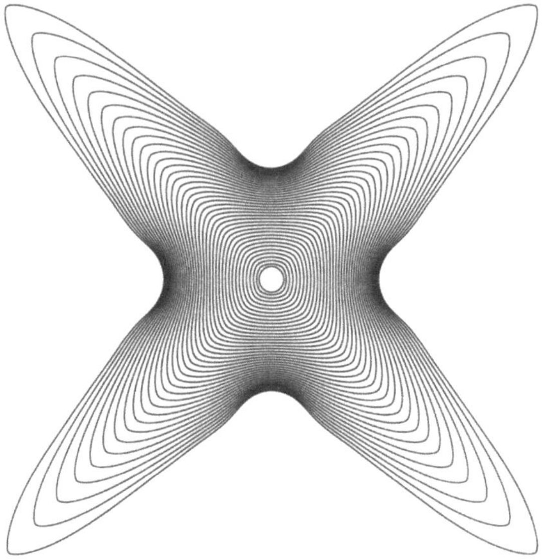

Fig. 3. Anisotropic Evolution of an Initially Circular Solid Phase

Fig. 4. Sixfold Anisotropy: Evolution of the Phase Boundary with $T_0 = -1$

Fig. 5. Sixfold Anisotropy: Evolution of the Phase Boundary with $T_0 = -2$

Reducing the boundary temperature the growth rate increases. In order to avoid an unstable evolution, we have to decrease the timestep size as well as the grid size in regions where $|\nabla\Theta|$ is big. This is done by decreasing the prescribed tolerance in the above described self adaptive methods.

The evolution of the solid phase obtained with the boundary temperature $T_0 = -2$ shows some topological changes. Different fingers touch each other and merge. In between this fingers notches are developing, while the temperature gradient near the phase boundary becomes steeper.

The last example illustrates the situation where multiple seed crystals separated by an undercooled liquid evolve and eventually merge during their growth. We found the topological change to be dependent on the the grid size of the underlying triangulation. This is demonstrated by the evolutions, which are shown in Figure 6. The only difference between the two calculations was the prescribed tolerance for the error. While in both cases we found the

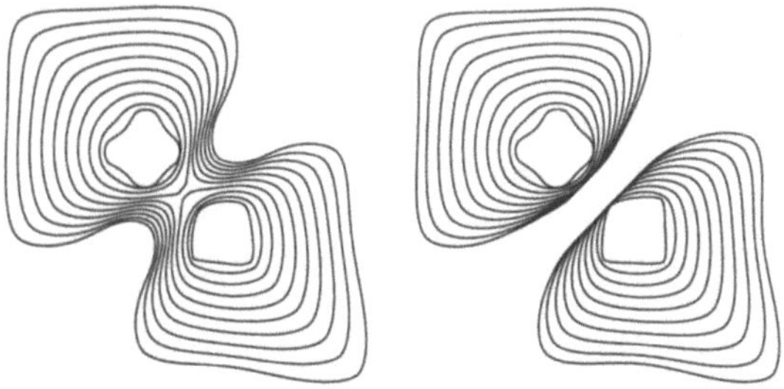

Fig. 6. Topological Changes: Dependency on the Grid size

evolution to be stable, prescribing a smaller tolerance leads to a situation without topological changes as we see in the right picture of Figure 6.

Both initially given crystals where of the form of equation (33), but the left one was rotated by 45°. The parameters of the anisotropy were $\delta = 0.006$, $A = 0.4$ and $k = 4$.

Acknowledgement

The results presented here are part of the authors' PhD theses. This research, which was founded through the Schwerpunkt DANSE of the Deutsche Froschungsgemeinschaft, was part of the project "Effiziente Simulation und Numerische Analysis der Dynamik von Dendriten", which was directed by Gerhard Dziuk and Alfred Schmidt.

References

1. E. Bänsch, *Local mesh refinement in 2 and 3 dimensions*, IMPACT Comput. Sci. Eng., (1991), pp. 181–191.
2. E. Bänsch and A. Schmidt, *A finite element method for dendritic growth*, in Computational crystal growers workshop, J. E. Taylor, ed., AMS Selected Lectures in Mathematics, 1992, pp. 16–20.
3. R. B. Burckel, *An introduction to classical complex analysis I*, vol. 64 of Lehrbücher und Monographien aus dem Gebiete der exakten Wissenschaften. Mathematische Reihe, Birkhäuser Verlag, Basel, 1979.
4. G. Caginalp, *Length scales in phase transition models: Phase field, Cahn-Hillard and blow-up problems*, in Pattern formation: symmetry methods and applications, J. Chadam and other, eds., Fields Institute for Research in Mathematical Sciences, Waterloo, Canada, 1996, American Mathematical Society, Providence, Rhode Island, pp. 67–83.
5. J. Chadam and G. Caginalp, *Stability of interfaces with velocity correction term*, Rocky Mt. J. Math., 21 (1991), pp. 617–629.
6. X. Chen and F. Reitich, *Local existence and uniqueness of solutions of the Stefan problem with surface tension and kinetic undercooling*, Journal of Mathematical Analysis and Applications, 164 (1992), pp. 350–362.
7. Y.-G. Chen, Y. Giga, and S. Goto, *Uniqueness and existence of viscosity solutions of generalized mean curvature flow equations*, J. Differ. Geom., 33 (1991), pp. 749–786.
8. D. L. Chopp, *Computing minimal surfaces via level set curvature flow*, Journal of Computational Physics, 106 (1993), pp. 77–91.
9. K. Deckelnick, *Error analysis for a difference scheme approximating mean curvature flow*, preprint, Mathematische Fakultät Freiburg, 1998.
10. K. Deckelnick and G. Dziuk, *On the approximation of the curve shortening flow*, in Calculus of Variations, Applications and Computations, C. Bandle and other, eds., Pont-à-Mousson, 1994, Longman Scientific & Technical, pp. 100–108.
11. ——, *Convergence of a finite element method for non-parametric mean curvature flow*, Numerische Mathematik, 72 (1995), pp. 197–222.

12. K. DECKELNICK AND G. DZIUK, *Discrete anisotropic curvature flow of graphs*, M2AN, Math. Model. Numer. Anal., 33 (1999), pp. 923–938.
13. K. DECKELNICK AND C. M. ELLIOTT, *Finite element error bounds for curve shrinking with prescribed normal contact to a fixed boundary*, Tech. Rep. 96/20, Centre for Mathematical Analysis and Its Applications, University of Sussex, Falmer, Brighton BN1 9QH, UK, 1996.
14. W. DÖRFLER, *A convergent adaptive algorithm for Poisson's equation*, SIAM J. Numer. Anal., 33 (1996), pp. 1106–1124.
15. G. DZIUK, *Convergence of a semi-discrete scheme for the curve shortening flow*, Mathematical Models and Methods in Applied Sciences, 4 (1994), pp. 589–606.
16. S. EIDEL'MAN, *Parabolic Systems*, North–Holland, Amsterdam, 1969.
17. K. ERIKSSON AND C. JOHNSON, *Adaptive finite element methods for parabolic problems i: A linear model problem*, SIAM J. Numer. Anal., 28 (1991), pp. 43–77.
18. L. C. EVANS AND J. SPRUCK, *Motion of level sets by mean curvature I*, J. Differ. Geom., 33 (1991), pp. 635–681.
19. ———, *Motion of level sets by mean curvature II*, Trans. Amer. Math. Soc., 330 (1992), pp. 321–333.
20. ———, *Motion of level sets by mean curvature III*, J. Geom. Ana., 2 (1992), pp. 121–150.
21. ———, *Motion of level sets by mean curvature IV*, J. Geom. Ana., 5 (1995), pp. 79–116.
22. M. FRIED, *Niveauflächen zur Berechnung zweidimensionaler Dendrite*, PhD thesis, Institut für Angewandte Mathematik, Universität Freiburg, April 1999.
23. Y. GIGA, S. GOTO, AND H. ISHII, *Global existence of weak solutions for interface equations coupled with diffusion eqautions*, Siam J. Math. Anal., 23 (1992), pp. 821–835.
24. M. E. GLICKSMAN AND S. P. MARSH, *The dendrite*, in Handbook of Crystal Growth, D. T. J. Hurle, ed., vol. 1, North Holland, Amsterdam – London – New York – Tokyo, 1993.
25. M. E. GURTIN, *Thermomechanics of evolving phase boundaries in the plane*, Clarendon Press, Oxford, 1993.
26. M. E. GURTIN AND H. M. SONER, *Some remarks on the Stefan problem with surface structure*, Quarterly of Applied Mathematics, 52 (1992), pp. 291–303.
27. G. HUISKEN, *Non-parametric mean curvature evolution with boundary conditions*, Journal of Differential Equations, 77 (1989), pp. 369–378.
28. J. S. LANGER, *Instabilities and pattern formation in crystal growth*, Reviews of Modern Physics, 52 (1980), pp. 1–28.
29. G. LIEBERMAN, *The first initial-boundary value problem for quasilinear second order parabolic equations*, Ann. Sc. Norm. Super. Pisa, Cl. Sci., IV. Ser. 13 (1986), pp. 347–387.
30. W. W. MULLINS AND R. F. SEKERKA, *Stability of a planar interface during solidification of a dilute binary alloy*, J. of Applied Physics, 35 (1964), pp. 444–451.
31. R. H. NOCHETTO AND C. VERDI, *Convergence past singularities for a fully discrete approximation of curvature-driven interfaces*, SIAM J. Numer. Anal., 34 (1997), pp. 490–512.

32. S. Osher and J. A. Sethian, *Fronts propagating with curvature-dependent speed: Algorithms based on Hamilton–Jacobi formulations*, J. Comput. Phys., 79 (1988), pp. 12–49.
33. A. Schmidt, *Die Berechnung dreidimensionaler Dendriten mit Finiten Elementen*, PhD thesis, Universität Freiburg, 1993.
34. ——, *Computation of three dimensional dendrites with finite elements*, Journal of Computational Physics, 125 (1996), pp. 293–312.
35. A. Schmidt and K. G. Siebert, *ALBERT: An adaptive hierachical finite element toolbox.* Preprint No. 06, Mathematische Fakultät Freiburg, 2000.
36. L. R. Scott, *A sharp form of the Sobolev trace theorem*, Journal of Functional Analysis, 25 (1977), pp. 70–80.
37. ——, *Applications of Banach space interpolation to finite element theory*, in Functional Analysis Methods in Numerical Analysis, M. Z. Nashed, ed., vol. 701 of Lecture Notes in Mathematics, Springer, New York, 1979.
38. J. A. Sethian and J. Strain, *Crystal growth and dendrite solidification*, J. Comp. Phys., 98 (1992), pp. 231–253.
39. H. M. Soner, *Convergence of the Phase-field equations to the Mullins–Sekerka problem with kinetic undercooling*, Archive for Rational Mechanics and Analysis, 131 (1995), pp. 139–197.
40. J. Strain, *Velocity effects in unstable solidification*, SIAM J. Appl. Math., 50 (1990), pp. 1–15.
41. J. Strain, *Fast tree-based redistancing for level set computations*, J. Comput. Phys., 152 (1999), pp. 664–686.
42. M. Sussman, P. Smereka, and S. J. Osher, *A level set method for computing solutions to incompressible two–phase flow*, Journal of Computational Physics, 114 (1994), pp. 146–159.
43. A. Veeser, *Globale Existenz von Graphen unter inhomogenen Krümmungsfluß bei Dirichlet–Randbedingungen*, Master's thesis, Institut für Angewandte Mathematik, Universität Freiburg, December 1993.
44. A. Veeser, *Fehlerabschätzungen für ein Verfahren zur Berechnung von zweidimensionalen Dendriten*, PhD thesis, Institut für Angewandte Mathematik, Universität Freiburg, April 1998.
45. ——, *Error estimates for semi-discrete dendritic growth*, Interfaces and Free Boundaries, 2 (1999), pp. 227–255.
46. A. Visintin, *Models of phase transitions*, vol. 28 of Progress in Nonlinear Differential Equations and Their Applications, Birkhäuser, Boston, 1996.

Bifurcation Phenomena and Dynamo Effect in Electrically Conducting Fluids

F. Feudel, S. Rüdiger, and N. Seehafer*

Institut für Physik, Universität Potsdam,
PF 601553, 14415 Potsdam, Germany

Abstract. Electrically conducting fluids in motion can act as self-excited dynamos. The magnetic fields of celestial bodies like the Earth and the Sun are generated by such dynamos. Their theory aims at modeling and understanding both the kinematic and dynamic aspects of the underlying processes. Kinematic dynamo models, in which for a prescribed flow the linear induction equation is solved and growth rates of the magnetic field are calculated, have been studied for many decades. But in order to get consistent models and to take into account the back-reaction of the magnetic field on the fluid motion, the full nonlinear system of the magnetohydrodynamic (MHD) equations has to be studied. It is generally accepted that these equations, i.e. the Navier-Stokes equation (NSE) and the induction equation, provide a theoretical basis for the explanation of the dynamo effect. The general idea is that mechanical energy pumped into the fluid by heating or other mechanisms is transferred to the magnetic field by nonlinear interactions. For two special helical flows which are known to be effective kinematic dynamos and which can be produced by appropriate external mechanical forcing, we review the nonlinear dynamo properties found in the framework of the full MHD equations. Specifically, we deal with the ABC flow (named after Arnold, Beltrami and Childress) and the Roberts flow (after G. O. Roberts). The appearance of generic dynamo effects is demonstrated. Applying special numerical bifurcation-analysis techniques to high-dimensional approximations in Fourier space and varying the Reynolds number (or the strength of the forcing) as the relevant control parameter, qualitative changes in the dynamics are investigated. We follow the bifurcation sequences until chaotic states are reached. The transitions from the primary flows with vanishing magnetic field to dynamo-active states are described in particular detail. In these processes the stagnation points of the flows and their heteroclinic connections play a promoting role for the magnetic field generation. By the example of the Roberts flow we demonstrate how the break up of the heteroclinic lines after the primary bifurcation leads to a complicated intersection of stable and unstable manifolds forming a chaotic web which is in turn correlated with the spatial appearance of the dynamo.

1 Introduction to the Dynamo Problem

The generation and maintenance of magnetic fields by the motion of electrically conducting fluids is the subject of dynamo theory. One of its main objectives is to explain the existence of long lasting cosmical magnetic fields,

* Project: Numerical Investigations of Symmetry Breaking Bifurcations in Spatially Continuous Systems (Fred Feudel, Jürgen Kurths, Norbert Seehafer)

as for instance, those of the Earth and the Sun. Realistic models, describing the dynamo processes, are given in the form of a complex system of nonlinear partial differential equations including the Navier-Stokes equation (NSE), the induction equation, the heat equation, and the thermodynamic equation of state. Heating causes fluid motions which in turn, notably in the presence of rotation, induce magnetic fields. For a comprehensive account of dynamo theory we refer to Refs. [1–3].

Because of the complexity of realistic models and due to restricted computer capacities, which allow only short time simulations, the dynamo processes are not completely understood yet. In order to get a tractable model, it is generally accepted that the nonlinear system of the incompressible magnetohydrodynamic (MHD) equations contains the basic elements of a dynamo [1]. It consists of the coupled system of the NSE for the flow and induction equation for the magnetic field in the form

$$\frac{\partial \mathbf{v}}{\partial t} + (\mathbf{v} \cdot \nabla)\mathbf{v} = R^{-1}\nabla^2\mathbf{v} - \nabla p - \frac{1}{2}\nabla\mathbf{B}^2 + (\mathbf{B} \cdot \nabla)\mathbf{B} + \mathbf{f}, \tag{1}$$

$$\frac{\partial \mathbf{B}}{\partial t} + (\mathbf{v} \cdot \nabla)\mathbf{B} = Rm^{-1}\nabla^2\mathbf{B} + (\mathbf{B} \cdot \nabla)\mathbf{v}, \tag{2}$$

$$\nabla \cdot \mathbf{v} = \mathbf{0}, \quad \nabla \cdot \mathbf{B} = \mathbf{0}, \tag{3}$$

where $\mathbf{v}$, p, and $\mathbf{B}$ denote fluid velocity, thermal pressure, and magnetic field, R and Rm the kinetic and magnetic Reynolds number, respectively, and $\mathbf{f}$ is a yet unspecified body force. The third and fourth terms on the right-hand side of Eq. (1) constitute the Lorentz force. Equations (3) impose the incompressibility condition on the fluid and ensure the source-free property of the magnetic field. The body force $\mathbf{f}$ on the right-hand side of the NSE has to be specified in the concrete physical context and is the sum of all forces that drive the fluid, as e.g. the buoyancy force in thermal convection, or modify the motion, like the Coriolis force in a rotating star. For simplicity we restrict our investigation to the above system of MHD equations and do not include processes generating the forces. We consider $\mathbf{f}$ as externally applied and given. It pumps energy into the fluid and we look for long lasting magnetic fields, not decaying as a result of the nonlinear coupling of NSE and induction equation. This phenomenon will be called nonlinear dynamo effect.

Traditional dynamo theory has been mainly kinematic, i.e., the induction equation, Eq. (2), is solved for a prescribed velocity field, disregarding the equation of motion, Eq. (1). In the kinematic frame the question is whether a fluid motion can amplify, or at least prevent from decaying, some weak seed magnetic field. Solving the linear induction equation positive growth rates demonstrate an instability of the zero magnetic field solution and we speak of a kinematic dynamo effect. The ABC flow $\mathbf{v}_{\mathrm{ABC}}$ [4–6] (named after Arnold, Beltrami and Childress) and the Roberts flow $\mathbf{v}_{\mathrm{R}}$ [7,8] are intensively studied examples for dynamo-effective velocity fields.

To take into account the back reaction of the magnetic field on the velocity field, the kinematic analysis has to be extended to studying the full nonlinear MHD equations, Eqs. (1)–(3). The velocity fields $\mathbf{v}_{\mathrm{ABC}}$ and $\mathbf{v}_{\mathrm{R}}$ can be produced as steady solutions of the incompressible NSE, Eq. (1), if external body forces $\mathbf{f} = -\nabla^2 \mathbf{v}_{\mathrm{ABC}}$ or $\mathbf{f} = -\nabla^2 \mathbf{v}_{\mathrm{R}}$ are applied. These forces compensate viscous losses and generate flows proportional to $\mathbf{v}_{\mathrm{ABC}}$ or $\mathbf{v}_{\mathrm{R}}$. Furthermore, together with a vanishing magnetic field, both flows are solutions of the full MHD equations, which are stable for small Reynolds numbers.

In this paper we report bifurcation studies of the MHD equations with the forcing types $\mathbf{f}_{\mathrm{ABC}}$ and $\mathbf{f}_{\mathrm{R}}$, respectively. Detailed results are published in Refs. [9–15] and are not all repeated here. We attempt to explain general properties of the nonlinear dynamo, but one has to keep in mind that there is still a large gap between the theoretical models and dynamos existing in nature. One of the most critical points is the magnitude of the Prandtl number $Pm = \nu/\eta$, the ratio between kinematic viscosity and magnetic diffusivity or, equivalently, between magnetic and kinetic Reynolds numbers. The Pm values occuring in cosmic dynamos cannot be reached by far in numerical simulations. For instance, the magnetic Prandl number in the convection zone of the Sun is of the order $10^{-4} \ldots 10^{-6}$; in contrast, due to the numerical limitations, we set $Pm = 1$.

2 Dynamo Bifurcations in the ABC Flows

The most important branch of traditional kinematic dynamo theory is the theory of the turbulent dynamo. This is a mean-field theory where one averages over turbulently fluctuating parts of velocity and magnetic field. One of its central results is that the presence of kinetic helicity,

$$H_K = \mathbf{v} \cdot (\nabla \times \mathbf{v}), \tag{4}$$

in the fluctuating part of the velocity field is favourable for a dynamo effect. Here we consider non-turbulent or laminar helical dynamos (that is, no averaging is applied). In Sec. 2.1 the stronly helical pure ABC forcing is used. In Sec. 2.2 then the role of helicity for a dynamo effect is tested by imposing a generalized ABC forcing with a varying (i.e., parameter-dependent) degree of helicity.

2.1 Pure ABC Forcing

One of the successful examples for producing a dynamo are the ABC flows, first investigated by Arnold [4],

$$\mathbf{v}_{\mathrm{ABC}} = (A\sin k_0 z + C\cos k_0 y, B\sin k_0 x + A\cos k_0 z, C\sin k_0 y + B\cos k_0 x), \tag{5}$$

where A, B, C denote constant coefficients and k_0 is a wave number. They are strongly helical Beltrami flows. Beltrami flows are flows with parallel velocity and vorticity ($\nabla \times \mathbf{v}$) vectors. The ABC flows satisfy the Beltrami condition, $\nabla \times \mathbf{v} = \gamma \mathbf{v}$, with a constant γ, namely $\gamma = k_0$, which is a necessary condition for the existence of chaotic domains in Beltrami flows [4]. For this reason, they have received much interest [5,16], notably in the kinematic context as candidates for fast dynamos [6,17] (for which the growth rate remains bounded from below by a positive constant as the magnetic diffusivity tends to zero).

Imposing an external body force

$$\mathbf{f} = -\nabla^2 \mathbf{v}_{\mathrm{ABC}} = k_0^2 \mathbf{v}_{\mathrm{ABC}} \tag{6}$$

in the NSE and applying periodic boundary conditions with period 2π in all three spatial directions, Galanti et al. [18] investigated the system of MHD equations [Eqs. (1)–(3)]. Numerically simulating the system for different Reynolds numbers and selected initial conditions, they observed that at some critical value of the Reynolds number the ABC flow with no magnetic field loses stability to a time-periodic state in which a magnetic field is excited and which, thus, represents a dynamo.

At this stage we started our study of the ABC dynamo, for the special case $A = B = C = 1$, $k_0 = 1$, and describe here further transitions which finally lead to a chaotic dynamics. We applied bifurcation-analysis techniques and paid special attention to symmetry breakings. The magnetic Prandl number was set equal to unity, $Pm = Rm/R = 1$, so that the kinetic Reynolds number R was the only remaining control parameter. R was raised from zero in small steps in order that bifurcations were detected. For small R, there exists only one stable stationary solution, namely the ABC flow [given by Eq. (5)] with a vanishing magnetic field. Its symmetry group is the full equivariance group of the ABC forced MHD equations, the octahedral group O [4,19,16].

For varying R, this steady-solution branch was traced and the Jacobian matrix and its eigenvalues were computed in each step. The primary ABC flow loses stability in a Hopf bifurcation in which a stable periodic branch with a nonvanishing magnetic field is born. Since only a single pair of complex-conjugate eigenvalues crosses the imaginary axis, the new branch, denoted as per-1, retains the full symmetry O. More precisly, the solution is no longer point symmetric with respect to all symmetry transformations since some of these produce time shifts. However, the periodic orbit as a whole is invariant. In the table in Fig. 1 the solution branches detected, their region of stability and their symmetry are compiled. The numbers in the branch designations indicate the multiplicity of the branches, i.e., the number of coexisting conjugate branches. Per-3, for instance, stands for three coexisting branches of periodic solutions, each being invariant to one of three conjugate dihedral subgroups D_4. In the right half of the figure a schematic bifurcation diagram is depicted.

We do not want to explain all details in the bifurcation diagram, but note that there are two main (stable) branches. One of them is generated by successive bifurcations from the primary ABC flow. The other one, appearing

Branch	Interval of stability for R	Symmetry
ABC flow	$0 < R < 5.7$	O
Per–1	$5.7 < R < 11.5$	O
Per–3	$7.7 < R < 16.0$	D_4
Per–4	$11.5 < R < 17.3$	D_3
Torus–3	$16.0 < R < 20.0$	
Torus–4	$17.3 < R < 17.9$	
Chaos–3	$R \geq 20.0$	
Chaos–4	$R \geq 17.9$	

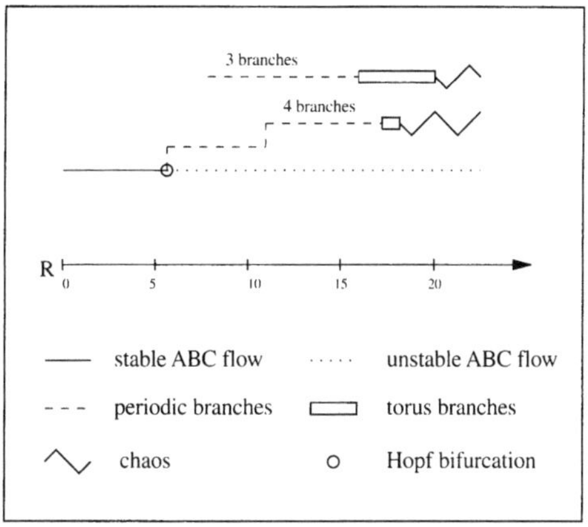

Fig. 1. Overview of the different solution branches (left). Schematic bifurcation diagram (right).

with Per-3, seems to be a result of a saddle-node bifurcation; its unstable counterpart has not been found. Starting from random initial conditions the trajectory is nearly always attracted by these solutions. Obviously, the basin of attraction of the first main branch shrinks for increasing values of R. Quasiperiodic solutions, briefly denoted as tori, and final transitions to chaos were observed for both main branches. The chaoticity of the solutions was verified by calculating the largest Lyapunov exponents. For this purpose an algorithm by Shimada and Nagashima [20] was implemented. Fig. 2 shows the cumulative values of the five largest Lyapunov exponents versus the integration time. It demonstrates good convergence of the algorithm also in applications to high-dimensional systems of ordinary differential equations as resulting from our Galerkin approximation to the original partial differential equations. It is seen that at least one of the exponents is positive.

Next we want to give an impression of the magnetic field structure in real (configuration) space and its changes under the influence of symmetry breaking bifurcations. In a kinematic dynamo study using the ABC flow with $A = B = C$, Galloway and Frisch [21] were the first to observe cigar-like concentrations of the magnetic field about velocity stagnation points. The ABC flow for the case $A = B = C$ has eight stagnation points, which are unstable fixed points of the flow $\mathbf{v}$. The corresponding eigenvalues are real and have signs $(+,-,-)$ or $(-,+,+)$. The topological structure of the ABC flow, and in particular the intersections of the stable and unstable manifolds of the stagnation points, which form a complicated web of heteroclinic lines, were comprehensively discussed by Dombre et al. [16]. Stagnation points with a two-dimensional stable manifold have been denoted as of α type and those with a two-dimensional unstable manifold as of β type. There are four stagnation

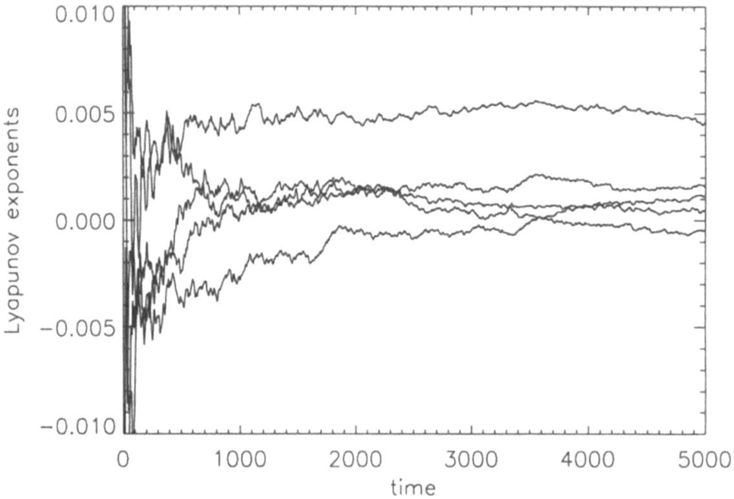

Fig. 2. The five largest Lyapunov exponents versus integration time for the branch Chaos-3 at $R = 20$.

points of each type and any two points of different type are connected by a straight line, forming a one-dimensional heteroclinic orbit. The cube diagonal through the points $(0, 0, 0)$ and $(2\pi, 2\pi, 2\pi)$, for example, is a one-dimensional invariant manifold belonging to the stagnation points $(3\pi/4, 3\pi/4, 3\pi/4)$ (β type) and $(7\pi/4, 7\pi/4, 7\pi/4)$ (α type). The rest of the ensemble of stagnation points and associated one–dimensional invariant manifolds can be obtained by applying symmetry transformations of the original equivariance group. The cigar-like structures of the magnetic field, observed by Galloway and Frisch [21] as well as by Galanti et al. [18] for the kinematic problem, are localized about the stagnation points of the α type.

Immediately after the primary Hopf bifurcation, where the ABC flow loses stability and a magnetic field appears, one expects the flow to differ only weakly from the unperturbed ABC flow and the magnetic field to resemble the unstable eigenmode of the kinematic problem. Above we already mentioned that the bifurcation retains the original symmetry for the newly created periodic branch Per–1. This symmetry has also a decisive influence on the spatial structure of the corresponding magnetic and velocity fields, here briefly discussed. A surprising feature of the new branch is that the eight stagnation points of the ABC flow survive, i.e., they remain time-independent zero-velocity points at the same location as before for the original ABC flow. In Fig. 3 isosurfaces of the magnetic field strength for a level of 65% of the maximum value are drawn. The cigar-like structures are clearly recognizable also in the nonlinear regime. Due to the temporal periodicity, the magnetic field oscillates and the shape of the isosurfaces depends also on the time at which a snapshot is made; sometimes they look much more like blobs. To give an impression of the dynamics, we consider the fields on the diagonal

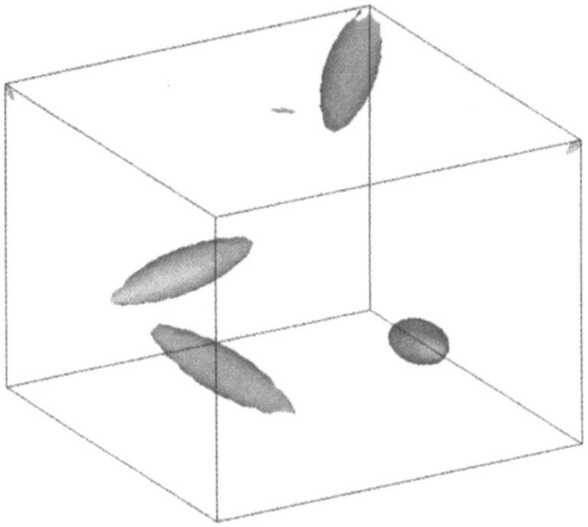

Fig. 3. Isosurfaces of the magnetic field with 65% of the maximal modulus for the symmetric branch (R=10) in the periodic box (x,y,z between 0 and 2π).

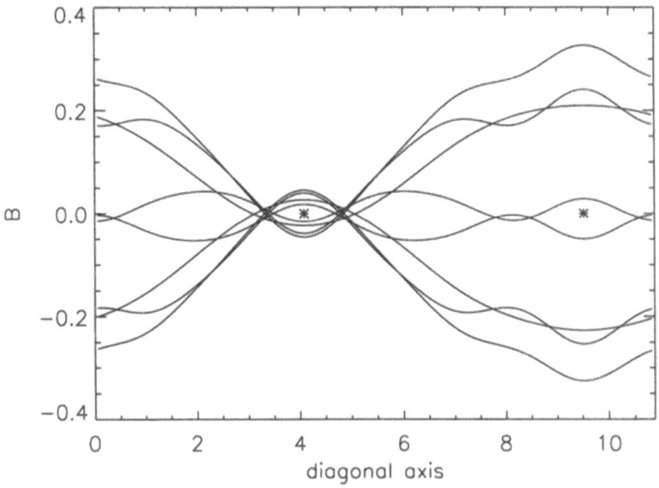

Fig. 4. Magnetic field component along the diagonal axis at different instants of time (R=10). Stagnation points are marked by asterisks.

line. As already mentioned, the stagnation points remain unchanged, but furthermore also their invariant one-dimensional manifolds are formed by the same straight lines as before. For the diagonal line, containing two stagnation points, this follows directly from the symmetry of the branch. Namely, as a consequence of the symmetry with respect to the cyclic group Z_3 as a subgroup of the full symmetry group, the velocity field has to be aligned with the diagonal direction. The same holds for the magnetic field.

In Fig. 4 the magnetic field component along the diagonal line is shown for different instants of a time period (stagnation points are marked by asterisks).

Strong oscillations with a large amplitude occur around the stagnation point of α type, and the magnetic energy is mainly located in its neighborhood. The velocity field along the diagonal line oscillates only relatively weakly about the original ABC flow.

It is not our aim here to describe the structural properties of the magnetic field for all branches presented in Fig. 1. We only mention an interesting property of the chaotic branch Chaos–3: Also for this branch the magnetic energy is on average mainly concentrated in four tube-like structures. These oscillate and move irregularly through the cube. The appearance of just four strong-field regions is still reminiscent of the four α type stagnation points of the original ABC flow.

2.2 Generalized ABC Forcing with Varying Helicity

In order to test the role of helicity for a dynamo effect we have, besides the pure ABC forcing given by Eq. (5), also applied a generalized ABC forcing of the form

$$\mathbf{f} = (1 - \lambda)\mathbf{v}_{ABC} + \lambda \mathbf{v}_{ABC}^{-}, \tag{7}$$

where we introduce

$$\mathbf{v}_{ABC}^{-} = (A \cos k_0 z + C \sin k_0 y,\ B \cos k_0 x + A \sin k_0 z,\ C \cos k_0 y + B \sin k_0 x) \tag{8}$$

and λ is a parameter varying between 0 and 0.5 (but $A = B = C = k_0 = 1$ in all calculations described here). $\mathbf{v}_{ABC}^{-}$ satisfies $\nabla \times \mathbf{v}_{ABC}^{-} = -k_0 \mathbf{v}_{ABC}^{-}$ and for a positive k_0 its helicity is thus negative whereas that of the original ABC flow is positive. The degree of helicity in the forcing varies with λ. For $\lambda = 0.5$ the addition of $\mathbf{v}_{ABC}^{-}$ in the forcing term "kills" the helicity on average in the volume while $\lambda = 0$ corresponds to the original ABC forcing.

We now present the bifurcation diagram obtained when applying the generalized ABC forcing, given by Eq. (7). For weak forcing (small R), there exists always a stable stationary, globally attracting solution (which coincides with the original ABC flow only in the special case of $\lambda = 0$). Keeping λ fixed and raising R, we have traced the steady-solution branch. For $\lambda < 0.4$ the steady state loses stability in a Hopf bifurcation, but at $\lambda = 0.4$ the type of the first bifurcation, as well as the character of the time-asymptotic states after this bifurcation, changes. While for $\lambda < 0.4$ a magnetic periodic state is the (only) new attractor, for λ between 0.4 and 0.5 new non-magnetic states appear.

The original steady state has been traced also in the region beyond the first bifurcation, where it is unstable and where it undergoes secondary bifurcations. The locations of primary and secondary bifurcations in the parameter plane are shown in Fig. 5. Thick solid and dashed lines, respectively, indicate the primary bifurcation of the original steady state. For $0.4 < \lambda < 0.49$ the Hopf bifurcation is preceded by a bifurcation in which two real eigenvalues of the Jacobian matrix become positive; these two eigenvalues are equal already

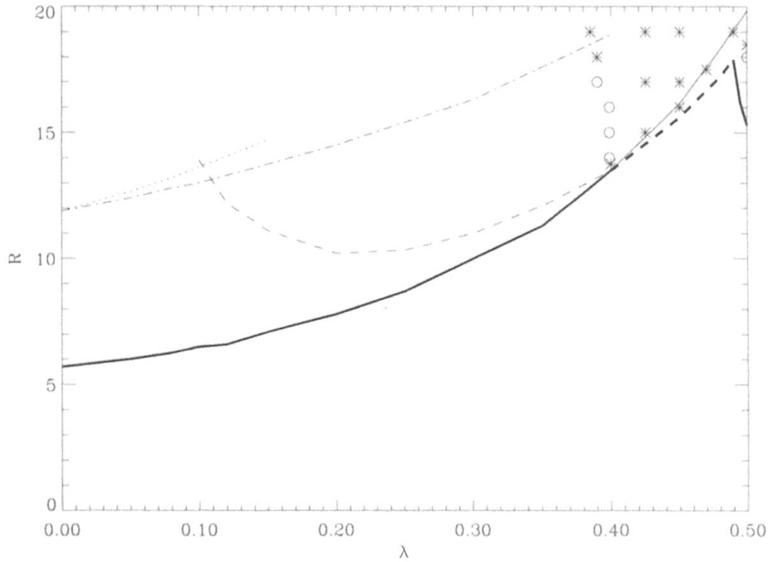

Fig. 5. Locations of primary and secondary bifurcations of the original stationary solution in the λ-R plane. Solid line and dashed-dotted line: a single pair of complex conjugate eigenvalues crosses the imaginary axis; dashed line: two real eigenvalues pass through zero; dotted line: two pairs of complex conjugate eigenvalues cross the imaginary axis. Asterisks indicate points at which, by means of simulations, non-magnetic chaotic (Shilnikov-like) time-asymptotic states have been found, while circles correspond to magnetic periodic attractors.

before the bifurcation, due to one of the symmetries in the system. The bifurcation results in three coexisting new stable stationary solutions, which can be transformed into each other by certain elements of the symmetry group of the problem (which is a subgroup of the symmetry group O in the case of the pure ABC forcing). For $\lambda = 0.4$ this bifurcation occurs simultaneously with the Hopf bifurcation.

At $\lambda = 0.49$, showing up as a kind of cusp in Fig. 5, the primary bifurcation changes its character again, namely from the real bifurcation observed for $\lambda < 0.49$ to another Hopf bifurcation for $\lambda > 0.49$. This Hopf bifurcation leads to a non-magnetic periodic state. At the cusp-like transition point the imaginary part of the pair of complex-conjugate eigenvalues responsible for the Hopf bifurcation to the right is zero, and the complex-conjugate pair coincides with the real pair responsible for the bifurcation to the left.

The new stationary and periodic solutions bifurcating from the original stationary one for $0.4 < \lambda < 0.5$ are stable only over very small intervals of the bifurcation parameter R and lose their stability directly to non-magnetic chaotic states. The presence of chaos has again been verified by calculating, for selected values of the bifurcation parameters, the largest Lyapunov exponents.

The chaotic attractors found here are strongly suggestive of Shilnikov-type homoclinic chaos. In a three-dimensional flow, this type of chaos is related to the existence of a homoclinic trajectory to a saddle point with a real eigenvalue and a pair of complex-conjugate eigenvalues, the real eigenvalue having a larger magnitude than the real part of the complex eigenvalues (cf. Ref. [22], Sec. 6.5.). The trajectory shown in Fig. 6 seems to be very close to such a homoclinic orbit. It approaches the unstable fixed point along a path apparently close to a one-dimensional stable manifold, spirals outward (thereby seemingly close to a two-dimensional unstable manifold), approaches again the fixed point, and so on. In our case the phenomenon seems to be connected with the degenerate bifurcation at $\lambda = 0.4$ (crossing of solid and dashed line in Fig. 5). In problems with two parameters, the occurence of homoclinic orbits in the vicinity of such points of degeneracy is a generic phenomenon (see e.g. Ref. [23]).

In Fig. 5, asterisks indicate points at which non-magnetic Shilnikov-like chaotic attractors have been found. The non-magnetic chaotic domain extends also to λ values less than 0.4, there causing a disappearance of the dynamo effect already present at smaller Reynolds numbers.

To repeat the main result of the studies using the generalized ABC forcing: In order that the primary bifurcation leads to a dynamo, the degree of helicity in the forcing has to exceed a threshold value.

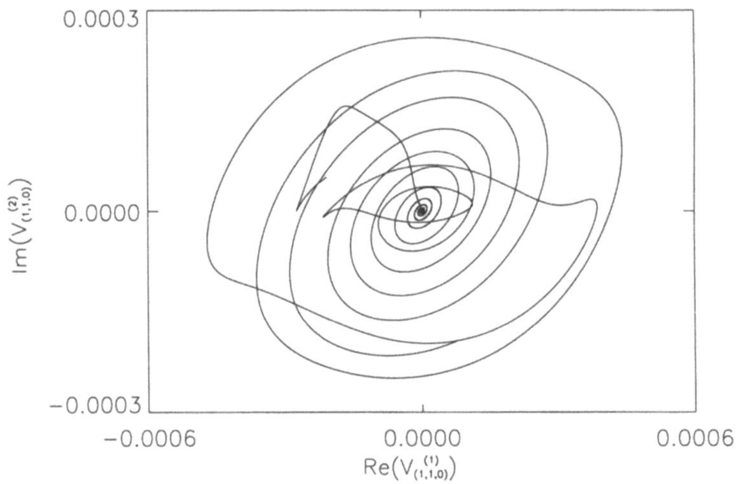

Fig. 6. Projection of a trajectory onto the plane spanned by two of the velocity components for $\lambda = 0.45$ and $R = 15.7$.

3 Dynamo Effect in Convection-Like Rolls

In this section we treat another dynamo model and compare it with the ABC dynamo in order to get further insight into the magnetic field generation mechanisms. It is based on a flow introduced by G. O. Roberts [7,8]. The Roberts flow has recently received renewed interest. On one hand it resembles the roll solutions of thermal (or solutal) convection. In the convective zones of rotating celestial bodies, for instance, convection rolls parallel to the axis of rotation tend to be formed [24]. On the other hand the Roberts flow is approximately realized in an ongoing laboratory experiment aimed at demonstrating the dynamo effect under laboratory conditions [25,26]. The experimental setup has been motivated by the kinematic dynamo effectiveness of the Roberts flow and its supposed resemblance to planetary convection. A conducting fluid (sodium) is pumped through an array of straight parallel ducts which are connected at their ends, where the flow out of a duct reverses its direction before entering a neighboring duct. The ducts contain internal guiding structures such that the flow becomes helical. All guiding structures, including the (thin) walls separating neighboring ducts, are electrically conducting.

Kinematic studies related to this experiment are due to Apel et al. [27] and Tilgner [28]. Apel et al. applied mean-field dynamo theory [2], whose central mechanism is the alpha effect (the generation of a mean electromotive force parallel to the mean magnetic field by turbulent fluctuations of velocity and magnetic field), while Tilgner used direct numerical simulation of the induction equation. In both studies the prescribed flow was the Roberts flow and system parameters most suitable for dynamo excitation were determined. But due to the kinematic nature of the models used the feedback of the magnetic field to the velocity field remained an open problem. This gave us the motivation to study bifurcations of the MHD equations for the situation where an external forcing of the Roberts type is applied.

The Roberts flow is given as a family of three-dimensional velocity fields which are independent of one of the spatial coordinates, namely,

$$\mathbf{v}_{\mathrm{R}} = (g \sin x \cos y,\ -g \cos x \sin y,\ 2f \sin x \sin y). \tag{9}$$

f and g are parameters, but we have only used $f = g = 1$ in our calculations. The flow consists of an array of rolls where the fluid spirals up and down in neighboring rolls. To give an impression of the flow structure, a projection of the velocity vectors on the x-y plane is plotted in Fig. 7. Since the flow is periodic with period 2π in the x and y directions, we consider only the four rolls shown. Applying analogously to Eq. (6) now the external forcing $\mathbf{f} = -\nabla^2 \mathbf{v}_{\mathrm{R}} = 2\mathbf{v}_{\mathrm{R}}$, the Roberts flow with vanishing magnetic field is a solution of the full MHD equations [Eqs. (1)–(3)]. It is also the only time-asymptotic steady-state for small R ($Pm = Rm/R$ is again fixed to the value 1). As in the case of the ABC forcing, we applied periodic boundary conditions with period 2π in all three spatial directions.

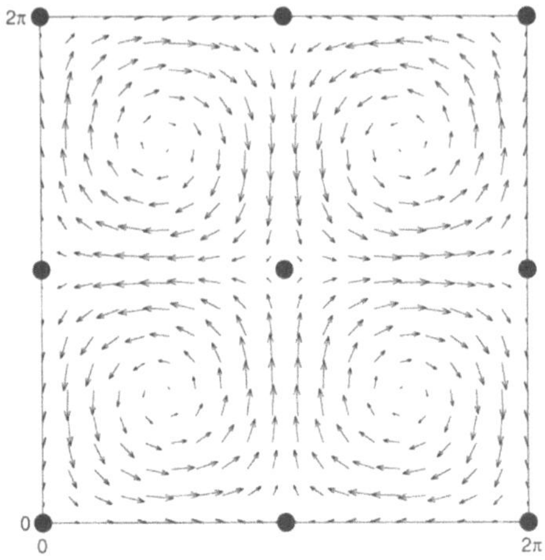

Fig. 7. Projection of the Roberts flow on the x-y plane. Black points indicate stagnation points of the flow.

For increasing Reynolds number a sequence of bifurcations occurs. Fig. 8 shows the stable solution branches that we obtained by applying continuation techniques and additional simulations. The sudden drop of the magnetic energy in the transition from periodic to quasiperiodic dynamics is conspicuous. For an explanation of this phenomenon we refer to [15] where also a discussion of the route to chaos may be found. Here we concentrate on the

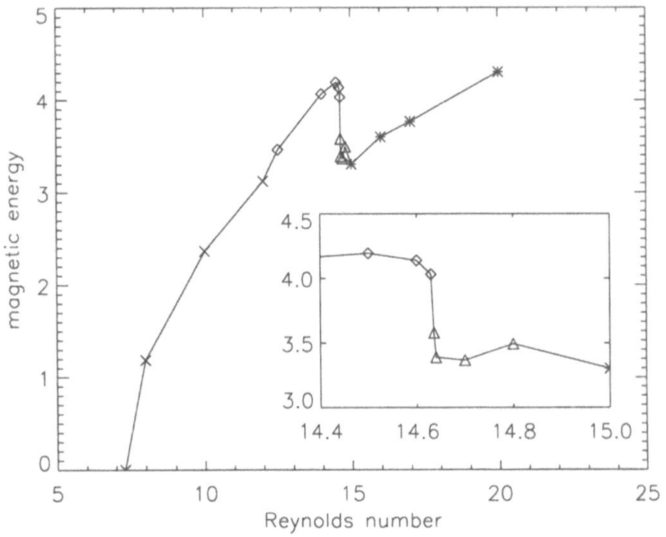

Fig. 8. Magnetic energy versus Reynolds number. Steady-state (×), periodic (◇), torus (△), and chaotic solutions (∗) are marked. The inner small box shows the zoomed region for $R = 14.4 \ldots 15.0$.

role of symmetry breaking for the onset of the dynamo and for the spatial structure of the generated magnetic field.

The external forcing, or equivalently its defining flow $\mathbf{v}_\mathrm{R}$, determines essentially, but not completely, the equivariance group of the MHD equations. The Roberts flow, given by Eq. (9), is invariant with respect to symmetry transformations forming a group which is the direct product of a discrete group G with the circle group S^1. But additionally a reflection symmetry Z_2 corresponding to the transformation $\mathbf{B} \to -\mathbf{B}$ and $\mathbf{v} \to \mathbf{v}$ leaves the MHD equations invariant. It plays an essential role in the symmetry breaking bifurcations and has to be taken into account. Thus, the whole equivariance group can be written as

$$G \times S^1 \times Z_2. \tag{10}$$

The forcing term, defined by $\mathbf{v}_\mathrm{R}$, is independent of the z coordinate. This provides together with imposed periodic boundary conditions the circle symmetry S^1. The discrete group G is also determed by the flow $\mathbf{v}_\mathrm{R}$. It consists of sixteen elements and can be characterized as a semidirect product $D_2 \times_S Z_4$. The dihedral group D_2 is the group of rotations about the x, y and z axes by π and forms a normal subgroup. The group Z_4 is generated by a rotation about the z axis by $\pi/2$ with an additional shift by π in the x direction. Applying rotations about the z axis (see Fig. 7) it has to be taken into account that the flow has a nonvanishing z component and neighboring vortices spiral in opposite directions. Thus, these rotations by $\pi/2$ have to be combined with a shift by π. Nevertheless the group is isomorphic to Z_4.

Now we analyze the first symmetry breaking bifurcation (see Fig. 8) in which the primary branch, the Roberts flow, becomes unstable and a nonvanishing magnetic field appears. At this bifurcation point two eigenvalues which belong to magnetic modes are equal to zero. The symmetry breaking pitchfork bifurcation yields new steady states representing stationary dynamos. The original symmetry is broken, namely, the solutions are no longer S^1 invariant. Now both the magnetic and the velocity fields depend on the z coordinate. The discrete symmetry $(D_2 \times_S Z_4) \times Z_2$ survives but is now formed by other transformations than before. The Z_4 symmetry generated by the rotation about the z axis and the shift in the x direction has now to be combined with an additional translation in the z direction by $\pi/2$, a remnant of the S^1 symmetry. The reflection symmetry Z_2 corresponding to the transformation $\mathbf{B} \to -\mathbf{B}$ and $\mathbf{v} \to \mathbf{v}$ has also to be combined with a translation in the z direction, namely by π.

Furthermore, the action of D_2 has become more subtle. Before giving this action, we characterize the stagnation points of the Roberts flow and the perturbed velocity field after the bifurcation. The stagnation points of $\mathbf{v}_\mathrm{R}$ are indicated in Fig. 7. Due to its independence of the z coordinate, the flow possesses continuous lines of stagnation points which are connected by a family of heteroclinic orbits. The symmetry breaking bifurcation splits up these lines into a discrete set of sixteen stagnation points. A skeleton of

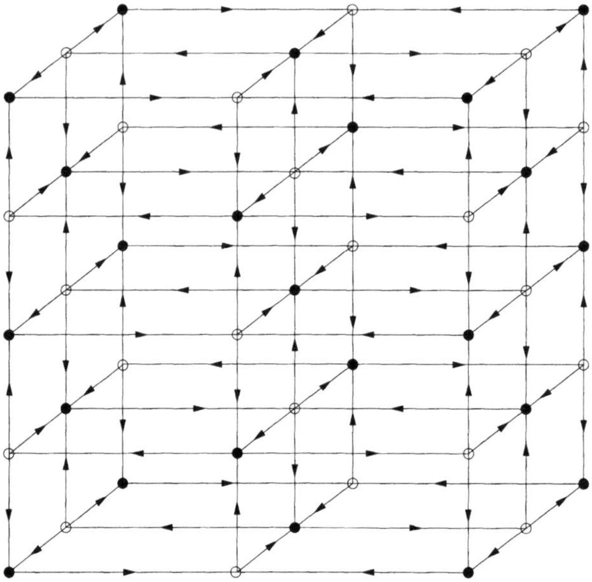

Fig. 9. Stagnation points and their connecting heteroclinic orbits after the first bifurcation. Black dots: α-type stagnation points. White dots: β-type stagnation points.

the stagnation points together with the connecting heteroclinic orbits after the bifurcation is sketched in Fig. 9 (for counting the stagnation points the periodicity has to be taken into account).

There are eight stagnation points with two negative eigenvalues and one positive eigenvalue (α type) and eight points with opposite signs of the eigenvalues (β type). The symmetry breaking bifurcation produces a family of equivalent solutions which differ only by translations in the z direction. We select one of them by choosing the coordinates $x = y = z = 0$ for one of the α-type stagnation points. Fig. 9 also helps to get a better understanding of the Z_4 symmetry described above, in particular of the necessary incorporation of translations by $\pi/2$ in the z direction as a result of the broken S^1 symmetry. But now we shall explain the actions of the surviving D_2 symmetry, one of the conjugate subgroups with respect to which the special solution branch shown in Fig. 9 is invariant. These actions are rotations by π about the x, y or z axis, combined with additional reflections $\mathbf{B} \to -\mathbf{B}$ of the magnetic field for rotations about the x or z axis (but without such a reflection for the rotation about the y axis). Obviously, these transformations form really a D_2 group.

Finally, we describe the magnetic field generated by this bifurcation. We observed that the magnetic field is strongly amplified in the neighborhood of stagnation points of the β type, which have a one-dimensional stable and

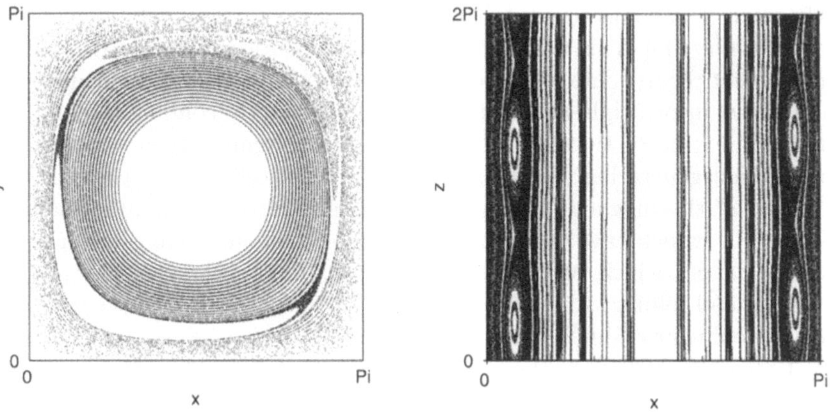

Fig. 10. Poincaré sections of the velocity field after the symmetry breaking bifurcation. Cutting planes are the x-y plane at $z=0$ (left) and the x-z plane at $y=\pi/2$ (right).

a two-dimensional unstable manifold. Plotting level surfaces of the magnetic field strength similar to Fig. 3, potato-like structures around the β-type stagnation points became visible. It is a prevalent guess that the stagnation points and, especially, the chaotic structure of the underlying flow in their vicinity, promote the dynamo effect. Therefore, we investigated the structure of the velocity field by a bunch of passive tracers which were initially randomly distributed in the periodic cube. Figure 10 shows Poincaré sections of the flow for one helical roll, with the x-y plane at $z=0$ and the x-z plane at $y=\pi/2$, respectively, as cutting planes. One recognizes the regular structure of the driven helical rolls and a chaotic region with some surviving KAM tori. Inside the helical rolls the strength of the produced magnetic field is really substantially weaker than in the chaotic region, a phenomenon which supports the above conjecture.

4 Concluding Remarks and Outlook

We have demonstrated the dynamo effect by means of two examples in which the fluid was driven by two different types of external forcing. The forcing terms in turn were motivated by the kinematic dynamo properties of two flows, the ABC flow and the Roberts flow, respectively. We want to conclude here with a discussion of some similarities and some differences between the two models.

The ABC flow is a chaotic flow from the beginning and causes a stretching, folding and twisting of the magnetic field lines, a process which is an effective drive of a dynamo [29]. Because of the chaoticity of the primary

ABC flow it is conjectured, but not proved yet, that the ABC dynamo is a fast dynamo [30]. This notion, originally defined for the kinematic problem, means that the dynamo mechanism is independent of the magnetic diffusivity and works in principle also in the limit of a perfectly conducting fluid. The first bifurcation of the ABC forced system is a symmetry preserving Hopf bifurcation generating solutions that depend periodically on time. Concentrations of the magnetic energy are found around the stagnation points of the α type, which have two-dimensional stable and one-dimensional unstable manifolds and which survive the bifurcation. The mechanism of stretching, twisting and folding, which works in their vicinity, thus continues to operate in the nonlinear regime.

Unlike the ABC flow, the Roberts flow is regular. Its stagnation points fill lines and the flow around them is non-chaotic. Therefore, the mechanism of stretching, twisting and folding is missing here and the kinematic Roberts dynamo is probably not a fast dynamo. Instead, the shearing motion between counter-rotating neighboring rolls is responsible for the kinematic dynamo effect. However, isolated stagnation points and chaotic stream lines are generated in the first bifurcation. As for the ABC flow, we got evidence that the chaoticity is caused by a complex web of heteroclinic lines formed by intersections of two-dimensional stable and unstable manifolds of the stagnation points. Furthermore, the magnetic field is concentrated in chaotic flow regions. Is then perhaps the nonlinear Roberts dynamo basically different from the kinematic one? The maximal amplification of the magnetic field occurs in the vicinity of the β-type rest points where the stretch-twist-fold mechanism is believed not to work. Thus it remains unclear whether the chaoticity of the flow is essential for the dynamo or merely a by-product. This will have to be clarified in future investigations. A next step could be a kinematic dynamo study with the stationary velocity field generated in the first bifurcation as prescribed flow, in order to see whether this flow remains dynamo effective for high magnetic Reynolds numbers. The latter would be indicative of a decisive role of chaotic stretching.

References

1. H. K. Moffatt. *Magnetic Field Generation in Electrically Conducting Fluids.* Cambridge Univ. Press, Cambridge, 1978.
2. F. Krause and K.-H. Rädler. *Mean-Field Magnetohydrodynamics and Dynamo Theory.* Akademie-Verlag, Berlin, 1980.
3. P. H. Roberts and A. M. Soward. Dynamo theory. *Ann. Rev. Fluid Mech.*, 24:459–512, 1992.
4. V. I. Arnold. Sur la topologie des êcoulements stationnaires des fluides parfaits. *C. R. Acad. Sci. Paris*, 261:17–20, 1965.
5. V. I. Arnold and E. I. Korkina. The growth of a magnetic field in a three-dimensional steady incompressible flow (in Russian). *Vest. Mosk. Univ. Mat. Mekh.*, 3:43–46, 1983.

6. D. Galloway and U. Frisch. Dynamo action in a family of flows with chaotic streamlines. *Geophys. Astrophys. Fluid Dyn.*, 36:53–83, 1986.
7. G. O. Roberts. Spatially periodic dynamos. *Phil. Trans. Roy. Soc.*, 266:535–558, 1970.
8. G. O. Roberts. Dynamo action of fluid motions with two-dimensional periodicity. *Phil. Trans. Roy. Soc.*, 271:411–454, 1972.
9. F. Feudel, N. Seehafer, and O. Schmidtmann. Fluid helicity and dynamo bifurcations. *Phys. Lett. A*, 202:73–78, 1995.
10. F. Feudel, N. Seehafer, and O. Schmidtmann. Bifurcation phenomena of the magnetofluid equations. *Math. Comput. Simulat.*, 40:235–245, 1996.
11. N. Seehafer, F. Feudel, and O. Schmidtmann. Nonlinear dynamo with ABC forcing. *Astron. and Astrophys.*, 314:693–699, 1996.
12. F. Feudel, N. Seehafer, B. Galanti, and S. Rüdiger. Symmetry-breaking bifurcations for the magnetohydrodynamic equations with helical forcing. *Phys. Rev. E*, 54:2589–2596, 1996.
13. O. Schmidtmann, F. Feudel, and N. Seehafer. Nonlinear Galerkin methods for the 3D magnetohydrodynamic equations. *Int. J. Bifurcation and Chaos*, 7:1497–1507, 1997.
14. O. Schmidtmann, F. Feudel, and N. Seehafer. Nonlinear Galerkin methods based on the concept of determining modes for the magnetohydrodynamic equations. *J. Phys. A: Math. Gen.*, 31:7141–7155, 1998.
15. S. Rüdiger, F. Feudel, and N. Seehafer. Dynamo bifurcations in an array of driven convection-like rolls. *Phys. Rev. E*, 57:5533–5538, 1998.
16. T. Dombre, U. Frisch, J. M. Greene, M. Hénon, A. Mehr, and A. M. Soward. Chaotic streamlines in the ABC flows. *J. Fluid Mech.*, 167:353–391, 1986.
17. B. Galanti, A. Pouquet, and P. L. Sulem. Influence of the period of an ABC flow on its dynamo action. In M. R. E. Proctor, P. C. Matthews, and A. M. Rucklidge, editors, *Theory of Solar and Planetary Dynamos*, pages 99–103. Cambridge University Press, Cambridge, 1993.
18. B. Galanti, P. L. Sulem, and A. Pouquet. Linear and non-linear dynamos associated with ABC flows. *Geophys. Astrophys. Fluid Dyn.*, 66:183–208, 1992.
19. V. I. Arnold. On the evolution of a magnetic field under the influence of advection and diffusion (in Russian). In V. M. Tikhomirov, editor, *Some Problems of Modern Analysis*, pages 8–21. Moscow State University, Moscow, 1984.
20. I. Shimada and T. Nagashima. A numerical approach to ergodic problem of dissipative dynamical systems. *Progr. Theor. Phys.*, 61:1605–1616, 1979.
21. D. Galloway and U. Frisch. A note on the stability of a family of space-periodic Beltrami flows. *J. Fluid Mech.*, 180:557–564, 1987.
22. J. Guckenheimer and P. Holmes. *Nonlinear Oscillations, Dynamical Systems, and Bifurcations of Vector Fields.* Springer, New York, 1983.
23. G. Nicolis and P. Gaspard. Bifurcations, chaos and self-organisation in reaction-diffusion systems. In D. Roose, B. de Dier, and A. Spence, editors, *Continuations and Bifurcations: Numerical Techniques and Applications*, pages 43–70. Kluwer, Dordrecht, 1990.
24. F. H. Busse. Convection driven zonal flows and vortices in the major planets. *Chaos*, 4:123–124, 1994.
25. R. Stieglitz and U. Müller. GEODYNAMO — Eine Versuchsanlage zum Nachweis des homogenen Dynamoeffektes. Research Report FZKA-5716, Forschungszentrum Karlsruhe, 1996.

26. F.H. Busse, U. Müller, R. Stieglitz, and A. Tilgner. Spontaneous generation of magnetic fields in the laboratory. In F.H. Busse and S.C. Müller, editors, *Evolution of Spontaneous Structures in Dissipative Continuous Systems*, Lecture Notes in Physics, pages 546–558. Springer-Verlag, Berlin, 1998.
27. A. Apel, E. Apstein, K.-H. Rädler, and M. Rheinhardt. Contributions to the theory of the Karlsruhe dynamo experiment. Research Report, Astrophysikalisches Institut Potsdam, 1996.
28. A. Tilgner. A kinematic dynamo with a small scale velocity field. *Phys. Lett. A*, 226:75–79, 1997.
29. S. Childress and A.D. Gilbert. *Stretch, Twist, Fold: The Fast Dynamo.* Lecture Notes in Physics. Springer, Berlin, 1995.
30. A.M. Soward. Fast dynamos. In M. R. E. Proctor and A.D. Gilbert, editors, *Theory of Solar and Planetary Dynamos*, pages 181–217. Cambridge University Press, Cambridge, 1994.

Cascades of Homoclinic Doubling Bifurcations

Ale Jan Homburg*

KdV Institute, University of Amsterdam, Plantage Muidergracht 24, 1018 TV Amsterdam, The Netherlands

Abstract. We present an overview of the theory of homoclinic doubling cascades, describing bifurcation theory and discussing universal scaling properties obtained from a renormalization theory.

1 Introduction

Cascades of homoclinic doubling cascades can occur in two parameter families of three dimensional vector fields and lead to the creation of homoclinic orbits with arbitrarily long arclength. Particular classes of vector fields where we think that homoclinic doubling cascades possibly occur are given by 'hooked Lorenz models', studied recently in [LuzVia99,LuzTuc99], and Shimizu-Morioka models [Shi93]. The models discussed in this review, where homoclinic doubling cascades have been shown to occur, are similar to those models.

Recall that the Lorenz model provides an example of a strange attractor for a set of differential equations that contains an equilibrium. Due to the partially hyperbolic structure of this attractor (there exists a continuous splitting of the tangent bundle along the attractor in strong stable and center unstable directions), a good description of the dynamics is possible. In contrast, the strange attractors appearing in the Hénon map where no such partial hyperbolic structure exists, are much harder to grasp. In the Lorenz model, a first return map on a cross section maps this cross section to two wedge shaped regions. Dynamics of a more complicated nature than in the Lorenz model, which still involves the equilibrium, occurs when these wedges are bended when they return to the cross section. A partially hyperbolic structure no longer exists for such hooked Lorenz maps. In fact, the resulting model is something of a combination of the Hénon model and the Lorenz model.

Where the Lorenz model possesses a (approximate) symmetry, so that all points of the cross section return, we will consider models as depicted in figure 1 where only one half of the cross section returns; we assume it returns as a hook as indicated in the figure. A fascinating phenomenon can occur in such situations when allowing to vary the differential equations with parameters: the occurrence of cascades of homoclinic doubling equations. This phenomenon is outlined in this paper. Of course, when studying symmetric vector fields the same phenomenon will also be present.

* Project: Unfoldings of Homoclinic and Heteroclinic Orbits (Bernold Fiedler)

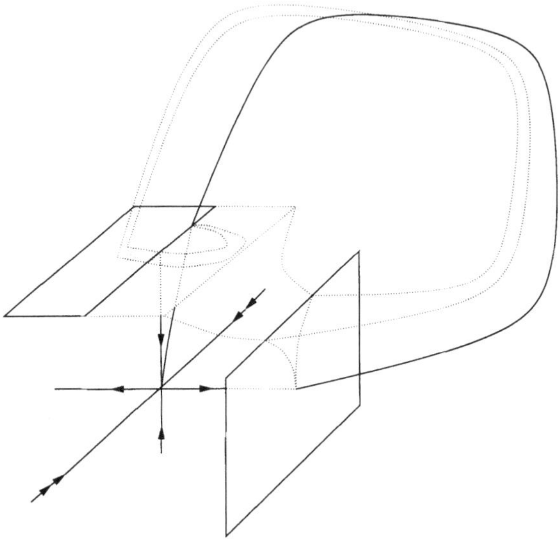

Fig. 1. We consider unfoldings of vector fields with a homoclinic orbit for which the first return map on a cross section maps a strip to a hook.

The vector fields we will consider are obtained from a bifurcation study, in which a vector field with a degenerate homoclinic orbit of codimension three is unfolded. This approach has a number of advantages. Because all the dynamics occurs near the degenerate homoclinic orbit, the depicted hook is small and near the stable manifold of the equilibrium. Since moreover in our bifurcation study the divergence of the vector fields at the equilibrium is negative, the first return map on the cross section is strongly area contracting. As a consequence the dynamics is close to one dimensional, and much can be understood from the study of a derived interval map. All of this will be explained in section 3. In the following section we introduce the homoclinic doubling bifurcation. Section 4 discusses a pathfollowing theory for homoclinic orbits; this theory is an important ingredient in existence proofs for homoclinic doubling cascades. Finally, in section 5 we study bifurcation diagrams and in particular scaling structures in the bifurcation set. This is done using renormalization theory.

A numerical study of vector fields as considered in this paper has recently been presented in [OldKraCha00]. This numerical study confirms the predictions of the theory, including the universal scaling structures.

2 Homoclinic Doubling Cascades

We will first state a precise bifurcation theorem for the homoclinic doubling bifurcation. After that we start discussing cascades of homoclinic doubling bifurcations.

2.1 Homoclinic Doubling Bifurcation

Let X be a smooth vector field on $\mathbb{R}^3$ with a hyperbolic equilibrium q. We assume that $DX(q)$ has two distinct real stable eigenvalues and one unstable eigenvalue. By a time reparametrization, we may assume the unstable eigenvalue to be equal to 1. Write $-\alpha, -\beta$ with $\alpha > \beta > 0$ for the two stable eigenvalues. Because the two stable eigenvalues are distinct, the vector field X has, contained in the stable manifold $W^{ss,s}(q)$ of q, a one dimensional strong stable manifold $W^{ss}(q)$. Its tangent space at q is the eigenspace associated to $-\alpha$. There further exists a two dimensional center unstable manifold $W^{s,u}(q)$ with tangent space at q spanned by the eigenspaces associated to the eigenvalues $-\beta$ and 1. This last invariant manifold is not unique and in general it is only C^1 [HirPugShu77]. The tangent bundle of (any) $W^{s,u}(q)$ along the unstable manifold, however, is a uniquely determined smooth bundle, see e.g. [Hom96] for a proof of this fact. All these invariant manifolds persist under perturbation of the vector field. We write $W_X^{ss,s}(q), W_X^{s,u}(q)$, etc., if we wish to stress the dependence of these manifolds on the vector field X.

Definition 1. Suppose Γ is a homoclinic orbit of X, that is, a nontrivial intersection of $W^{ss,s}(q)$ with $W^u(q)$. Γ is called an *inclination-flip* homoclinic orbit, if $W^{ss,s}(q)$ is tangent to one (and hence any) center unstable manifold $W^{s,u}(q)$ along Γ.

Inclination-flips are examples of homoclinic bifurcations of (at least) codimension two. We will not precise the notion of codimension. What is heuristically meant with a bifurcation being of codimension n, is that it is given by a collection of conditions, naturally occurring in its study, that make up a manifold of codimension n in the space of vector fields.

Consider a smooth two parameter family of vector fields $\{X_\gamma\}, \gamma \in \mathbb{R}^2$, on $\mathbb{R}^3$, satisfying the following conditions:

(BH: Basic hypothesis) The vector field X_γ has a hyperbolic equilibrium q_γ at which the linearization $DX_\gamma(q_\gamma)$ possesses two negative eigenvalues $-\alpha(\gamma) < -\beta(\gamma)$ and one positive eigenvalue 1.

(IF: Inclination-flip) The vector field X_0 possesses, at the parameter $\gamma = 0$, an inclination-flip homoclinic orbit Γ. The homoclinic orbit Γ is not contained in the strong stable manifold $W_{X_0}^{ss}(q_0)$.

(EC: Eigenvalue conditions) The eigenvalues of the linearization $DX_0(q_0)$ satisfy

$$\alpha(0) > 1 \text{ and } \frac{1}{2} < \beta(0) < 1.$$

(GU: Generic unfolding) Denote by $F_\gamma^{s,u}$ the bundle $\{T_x W_{X_\gamma}^{s,u}(q_\gamma);\ x \in W_{X_\gamma}^u(q_\gamma)\}$. The condition is then that $\bigcup_\gamma (TW_{X_\gamma}^{ss,s}(q_\gamma), \gamma)$ and $\bigcup_\gamma (F_\gamma^{s,u}, \gamma)$ intersect each other transversally along $T_\Gamma W_{X_0}^{ss,s}(q_0) \times \{0\}$ in $T\mathbb{R}^3 \times \mathbb{R}^2$.

Theorem 2. [KisKokOka93,HomKokNau98] *Let $\{X_\gamma\}$ be a two parameter family of vector fields on $\mathbb{R}^3$ as above. After a reparametrization of the parameter plane, the bifurcation diagram of $\{X_\gamma\}$ for small values of γ, is as depicted below. From the curve* $\mathbf{H}_1$ *of primary homoclinic orbits, a curve*

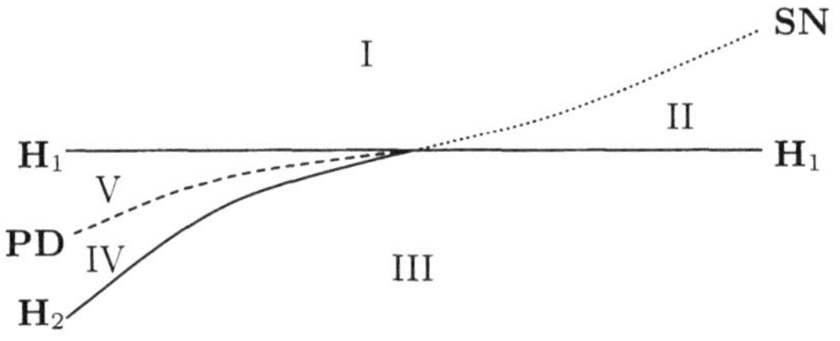

$\mathbf{H}_2$ *of doubled homoclinic orbits branches. Furthermore, a curve* **SN** *of periodic saddle-node bifurcations and a curve* **PD** *of period-doubling bifurcations branch.*

The following list describes all periodic orbits of $\{X_\gamma\}$ in a tubular neighborhood of $\overline{\Gamma}$, for parameters from the different regions I,...,V. In region I, $\{X_\gamma\}$ has no periodic orbits. In region II, $\{X_\gamma\}$ has an attracting 1-periodic orbit and a saddle 1-periodic orbit. In region III, $\{X_\gamma\}$ has an attracting 1-periodic orbit. In region IV, $\{X_\gamma\}$ has an attracting 1-periodic orbit and a saddle 2-periodic orbit. In region V, $\{X_\gamma\}$ has a saddle 1-periodic orbit and an attracting 2-periodic orbit.

2.2 Cascades of Homoclinic Doubling Bifurcations

When each doubled homoclinic orbit that is created in a homoclinic doubling bifurcation undergoes itself, at some parameter value, a homoclinic doubling bifurcation, then a cascade of homoclinic doubling bifurcations is created in which homoclinic orbits of arbitrarily long arclength occur. Though precise statements will be made in the next section, we state here informally that homoclinic doubling cascades occur persistently in two parameter families of vector fields.

Theorem 3. [HomKokNau98] *In the space of two parameter families of smooth vector fields on $\mathbb{R}^3$ there is an open set consisting of families that possess a cascade of homoclinic-doubling bifurcations.*

The following sections are devoted to a description of families of vector fields in which homoclinic doubling cascades occur persistently, plus a discussion of the resulting bifurcation set.

3 A Bifurcation Theorem; the Resonant Inclination Flip

Consider a smooth three parameter family of vector fields $\{X_\gamma\}, \gamma \in \mathbb{R}^3$, on $\mathbb{R}^3$ satisfying the following conditions:

(BH: Basic hypothesis) The vector field X_γ has a hyperbolic equilibrium q_γ at which the linearization $DX_\gamma(q_\gamma)$ possesses two negative eigenvalues $-\alpha(\gamma) < -\beta(\gamma)$ and one positive eigenvalue 1.

(IF: Inclination-flip) The vector field X_0 possesses, at the parameter $\gamma = 0$, an inclination-flip homoclinic orbit Γ. Γ is not contained in $W^{ss}_{X_0}(q)$.

(SR: Strong resonance) The eigenvalues of the linearization $DX_0(q_0)$ satisfy

$$\beta(0) = \frac{1}{2} \text{ and } \alpha(0) > 1.$$

(GU: Generic unfolding) $\bigcup_\gamma (F^{s,u}_\gamma, \beta(\gamma), \gamma)$ and $\bigcup_\gamma (TW^{ss,s}_{X_\gamma}(q_\gamma), \beta(\gamma), \gamma)$ intersect transversally along $T_\Gamma W^{ss,s}_{X_0}(q_0) \times \{1/2\} \times \{0\}$ in $T\mathbb{R}^3 \times \mathbb{R} \times \mathbb{R}^3$.

Take coordinates (x_{ss}, x_s, x_u) near q_γ in which

$$DX_\gamma(q_\gamma) = -\alpha x_{ss}\frac{\partial}{\partial x_{ss}} - \beta x_s \frac{\partial}{\partial x_s} + x_u \frac{\partial}{\partial x_u}.$$

Let $\Sigma^{in} \subset \{x_s = \delta\}$ with δ small and positive, be a cross-section transverse to Γ. By a linear rescaling, we may assume $\delta = 1$.

Proposition 5 below gives expansions for the Poincaré return map. These expansions hold in suitable smooth coordinates provided by the following normal form result. Let X_γ be given by a set of ordinary differential equations

$$\begin{aligned} \dot{x}_{ss} &= -\alpha x_{ss} + F_{ss}(x_{ss}, x_s, x_u; \gamma), \\ \dot{x}_s &= -\beta x_s + F_s(x_{ss}, x_s, x_u; \gamma), \\ \dot{x}_u &= x_u + F_u(x_{ss}, x_s, x_u; \gamma), \end{aligned} \tag{3.1}$$

where F^{ss}, F^s, F^u are quadratic and higher order terms.

Lemma 4. **[HomKra98]** *The vector field X_γ is smoothly equivalent to a vector field of the same form with*

$$\begin{aligned} F_{ss}(x_{ss}, x_s, x_u; \gamma) &= \mathcal{O}(\|(x_{ss}, x_s)\|^2), \\ F_s(x_{ss}, x_s, x_u; \gamma) &= x_u \mathcal{O}(|x_{ss}| + \|(x_{ss}, x_s)\|^2), \\ F_u(x_{ss}, x_s, x_u; \gamma) &= 0. \end{aligned}$$

Proposition 5. **[HomKra98]** *In coordinates given by lemma 4, $\Phi : \Sigma^{in} \to \Sigma^{in}$ has the following expression:*

$$\Phi(x_{ss}, x_u; \gamma) = \begin{pmatrix} Q(\gamma) + A(\gamma)x_u^\beta + B(x_{ss};\gamma)x_u^{2\beta} + R^{ss}(x_{ss}, x_u; \gamma) \\ \mu_2(\gamma) + \mu_1(\gamma)x_u^\beta + D(x_{ss};\gamma)x_u^{2\beta} + R^u(x_{ss}, x_u; \gamma) \end{pmatrix}. \tag{3.2}$$

Here μ_1, μ_2, Q, A are smooth functions of γ, with μ_1, μ_2 vanishing at $\gamma = \gamma_0$. The functions B, D are smooth in x_{ss} and γ. Furthermore, R^{ss} and R^u are smooth for $x_u > 0$; for some $\sigma > 0$, there exist constants $C_{k+l} > 0$ so that with $i = ss, s$

$$\left| \frac{\partial^{k+l}}{\partial x_u^k \partial(x_{ss}, \gamma)^l} R^i(x_{ss}, x_u) \right| \leq C_{k+l} x_u^{2\beta+\sigma-k} .$$

Define

$$\mu_3(\gamma) = 2 - \frac{1}{\beta(\gamma)}.$$

The generic unfolding condition (GU) is equivalent to

$$\frac{\partial}{\partial \gamma}(\mu_1, \mu_2, \mu_3)(\gamma)|_{\gamma=\mathbf{0}} \neq 0.$$

One may therefore, by reparametrizing the parameter space, assume that $\gamma = (\mu_1, \mu_2, \mu_3)$. Note that a 1-homoclinic orbit exists for $\mu_2 = 0$, which is an inclination-flip homoclinic orbit if also $\mu_1 = 0$.

Theorem 6. [HomKra98] *Let $\{X_\gamma\}$, $\gamma = (\mu_1, \mu_2, \mu_3) \in \mathbb{R}^3$, be a smooth three parameter family of vector fields as above. With D as in (3.2), suppose that $D > 1$ at $\gamma = 0$. For each μ_1 sufficiently small and negative, the two parameter family $\{Y_{\mu_2,\mu_3}\}$ given by $Y_{\mu_2,\mu_3} = X_{\mu_1,\mu_2,\mu_3}$, possesses a connected set of homoclinic bifurcation values in the (μ_2, μ_3)-parameter plane, containing a cascade (μ_2^n, μ_3^n) of homoclinic-doubling bifurcations (inclination-flips) in which a 2^n-homoclinic orbit is created.*

Note that the value of D in (3.2), for $\gamma = (0,0,0)$, is independent of the choice of cross section Σ. Indeed, varying Σ corresponds to conjugating Φ with a diffeomorphism. This is easily seen to leave D invariant, if β equals $\frac{1}{2}$.

3.1 Singular Rescalings to Interval Maps

A rescaling brings the Poincaré return map to a map which is a small perturbation of an interval map. Let Φ be as in (3.2) and define rescaled coordinates $(\hat{x}_{ss}, \hat{x}_u)$ by

$$x_{ss} - Q = |\mu_1| \hat{x}_{ss},$$
$$x_u = \left| \frac{\mu_1}{2D} \right|^{1/\beta} \hat{x}_u.$$

The following proposition, which gives expansions for the first return map in rescaled coordinates, is proved by a direct computation.

Proposition 7. [HomKra98] *Let $\hat{\Phi}$ be the Poincaré return map in the rescaled coordinates $(\hat{x}_{ss}, \hat{x}_u)$. Write $r = |\mu_1|^{\mu_3}/D$ and $p = r(4D\mu_2|\mu_1|^{-2}-1)$. Then*

$$\hat{\Phi}(\hat{x}_{ss}, \hat{x}_u) \to \begin{pmatrix} \frac{A}{2D}\hat{x}_u^\beta + R_1(\hat{x}_{ss}, \hat{x}_u) \\ p + r(1-\hat{x}_u^\beta)^2 + R_2(\hat{x}_{ss}, \hat{x}_u) \end{pmatrix},$$

where R_1 and R_2 converge to 0 as $\gamma \to 0$.

As $\mu_1 \to 0$, restricting $\hat{x}_u$ to a compact interval and parameters (μ_1, μ_2, μ_3) from a chart near the origin on which $|p|, |r|$ are bounded, $\hat{\Phi}$ converges to

$$\hat{x}_u \mapsto \begin{pmatrix} \frac{A}{2D}\hat{x}_u^\beta \\ p + r(1-\hat{x}_u^\beta)^2 \end{pmatrix}. \tag{3.3}$$

Ignoring the first coordinate leads to a unimodal map on an interval. Note that homoclinic orbits for the vector field correspond to periodic orbits through 0 for the interval map. Homoclinic doubling bifurcations correspond to periodic orbits that contain both 0 and the critical point.

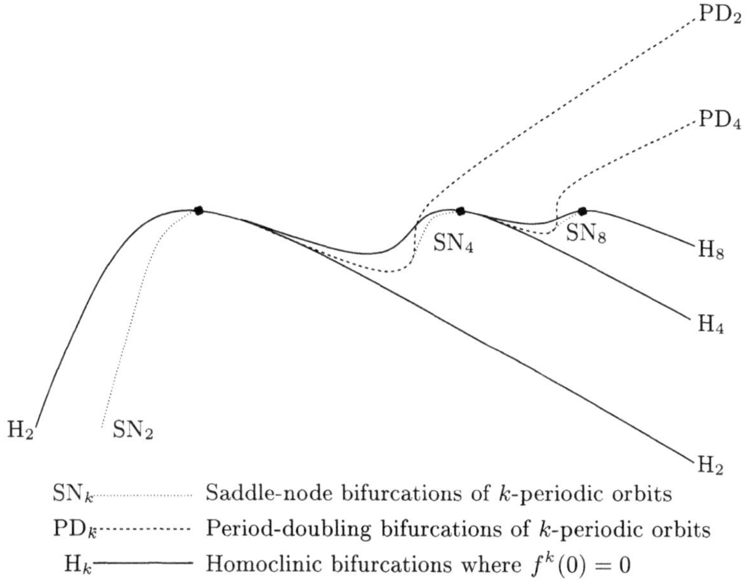

Fig. 2. Schematic impression of the expected bifurcation set of $x \mapsto p + r(1-x^\beta)^2$ in the parameter plane $\{(r,p) \in \mathbb{R}^2\}$. The thick dots indicate three subsequent 'homoclinic doubling' bifurcations.

A study of the family of interval maps (3.3) suggests what to expect for the original vector fields. It is not hard, using kneading theory, to see that the family of interval maps possesses a cascade of homoclinic doubling bifurcations.

Indeed, consider the one parameter family of interval maps $\hat{x}_u \mapsto r(1-\hat{x}_u^\beta)^2$ obtained by letting $p=0$ in (3.3). Expressed in the coordinate $u=\hat{x}_u^\beta$ this leads to the C^1 family $u \mapsto r(1-u)^{2\beta}$. Such families are known to possess a cascade of parameter values with periodic orbits of period 2^n and containing the critical point. These parameter values correspond to homoclinic doubling bifurcations.

To show that the first return maps, i.e. the vector fields, possess a cascade of homoclinic doubling bifurcations one relies on a continuation theory for homoclinic orbits. This is discussed in the next section. There exists much similarity between proving the existence of period doubling cascades in smooth families of unimodal maps (using kneading theory) and in families of planar diffeomorphisms such as dissipative Hénon families [YorAll83] (relying on a continuation theory for periodic orbits).

4 Pathfollowing Homoclinic Orbits

In this section we discuss a general pathfollowing theory for homoclinic orbits. Below, in section 4.1, we indicate how this pathfollowing theory is used to prove the existence of homoclinic doubling cascades in theorem 6. An earlier approach to pathfollowing homoclinic orbits is found in [Fie96].

Let $\mathfrak{X}$ be the set of smooth vector fields on $\mathbb{R}^3$, equipped with the weak Whitney topology. Consider the set $\mathfrak{X}_2$ of smooth two parameter families of smooth vector fields on $\mathbb{R}^3$. Let $\{X_\lambda\}$ be a two parameter family of vector fields from $\mathfrak{X}_2$, depending on a parameter $\lambda \in \mathbb{R}^2$.

Let $\mathcal{P}$ be the set of bounded closed subsets of $\mathbb{R}^3$, equipped with the Hausdorff metric. Let

$$G = \left\{ (\mu, h) \in \mathbb{R}^2 \times \mathcal{P}; \begin{array}{l} h \text{ is the union of an equilibrium and a} \\ \text{homoclinic orbit of } X_\mu \end{array} \right\}. \quad (4.4)$$

For $(\mu, h) \in G$, let $l(\mu, h)$ denote the length of h. We say that k is a virtual length of (μ, h), if there exists a sequence of perturbations $\{Y_\lambda^i\}$ of $\{X_\lambda\}$ with $\{Y_\lambda^i\} \to \{X_\lambda\}$ as $i \to \infty$ so that $\{Y_\lambda^i\}$ possesses a homoclinic orbit h_i at parameter values μ_i with $\mu_i \to \mu$, $h_i \to h$ in the Hausdorff topology and $l(\mu_i, h_i) \to k$ as $i \to \infty$. We write $\tau(\mu, h)$ for the set of virtual lengths of (μ, h).

Let $(\mu, h) \in G$ so that h is the union of a homoclinic orbit and a hyperbolic equilibrium. Write Γ for the connected component of G containing (μ, h). We call (μ, h) globally continuable if either

- $\Gamma \backslash \{(\mu, h)\}$ is connected

or each component C of $\Gamma \backslash \{(\mu, h)\}$ satisfies at least one of the following conditions:

- C is unbounded

- there exists a sequence $(\nu_i, g_i) \in C$ so that $\sup_i \tau(\nu_i, g_i) = \infty$.
- there exists a sequence $(\nu_i, g_i) \in C$ so that, as $i \to \infty$, $\nu_i \to \nu$ and g_i converges in the Hausdorff topology to a closed invariant set containing either a nonhyperbolic equilibrium, or more than two orbits.

This definition is inspired by the definition of continuable periodic orbits in one parameter families of vector fields in [AllYor84]. Note that the closure of a homoclinic orbit consists of two orbits. By a closed invariant set containing more than two orbits, one can think of a set containing two homoclinic orbits or a heteroclinic cycle.

Suppose that $(\mu, h) \in G$ is a generically unfolding codimension one homoclinic orbit with $\tau(\mu, h) = \{l(\mu, h)\}$. In such a situation, there is a sequence μ_i of parameter values converging to μ and a periodic orbit h_i of X_{μ_i} converging to h in the Hausdorff topology as $i \to \infty$. For all sufficiently large i, h_i is unique and its unstable manifold $W^u(h_i)$ is either orientable or nonorientable. Note that an unstable manifold which is either one or three dimensional is always orientable. Define

$$\phi(\mu, h) = \begin{cases} 0, & \text{if } W^u(h_i) \text{ is nonorientable for large } i, \\ 1, & \text{if } W^u(h_i) \text{ is orientable for large } i. \end{cases} \tag{4.5}$$

If $W^u(h_i)$ is two dimensional, there exists a two dimensional center manifold $W^{s,u}(h)$ of h [Hom96,San99]. Then $\phi(\mu, h) = 0, 1$ if $W^{s,u}(h)$ is nonorientable, orientable respectively. It is thus possible to define ϕ using X_μ alone.

For any $(\mu, h) \in G$, we let $\phi(\mu, h) = 1$ if the virtual lengths of (μ, h) are bounded and there exists a sequence of families $\{Y_\lambda^i\} \in \mathfrak{X}_2$ with $\{Y_\lambda^i\} \to \{X_\lambda\}$ as $i \to \infty$ and $\{Y_\lambda^i\}$ possesses a generically unfolding homoclinic orbit h_i of codimension one at parameter values μ_i with $\mu_i \to \mu$, $h_i \to h$ in the Hausdorff topology as $i \to \infty$, $\phi(\mu_i, h_i) = 1$. For all other $(\mu, h) \in G$, we let $\phi(\mu, h) = 0$. Denote

$$G_1 = \{(\mu, h) \in G;\ \ \phi(\mu, h) = 1\}. \tag{4.6}$$

The notion of continuability of homoclinic orbits (μ, h) in G_1 is as follows. Let $(\mu, h) \in G_1$ so that h is the union of a homoclinic orbit and a hyperbolic equilibrium. Write Γ_1 for the connected component of G_1 containing (μ, h). We call (μ, h) globally I-continuable if either

- $\Gamma_1 \backslash \{(\mu, h)\}$ is connected

or each component C_1 of $\Gamma_1 \backslash \{(\mu, h)\}$ satisfies at least one of the following conditions:

- C_1 is unbounded
- there exists a sequence $(\nu_i, g_i) \in C_1$ so that $\sup_i \tau(\nu_i, g_i) = \infty$ or so that $(\nu_i, g_i) \to (\nu, g) \in G$ as $i \to \infty$ with (ν, g) possessing unbounded virtual lengths.

– there exists a sequence $(\nu_i, g_i) \in C_1$ so that, as $i \to \infty$, $\nu_i \to \nu$ and g_i converges in the Hausdorff topology to a closed invariant set containing either a nonhyperbolic equilibrium, or more than two orbits.

Theorem 8. [HomKokNau98] *Let $(\mu, h) \in G_1$ be a generically unfolding codimension one homoclinic orbit of $\{X_\lambda\}$. Then (μ, h) is globally I-continuable.*

4.1 Pathfollowing Homoclinic Orbits Through Infinitely many Homoclinic Doublings

Here we briefly indicate how the above pathfollowing theory is used to prove the existence of homoclinic doubling cascades in theorem 6.

Let $(x, y) \mapsto \hat{\Phi}_{\mu_1,p,r}(x, y)$ be the rescaled Poincaré return map, see proposition 7. Recall that $\hat{\Phi}_{0,p,r}(x, y) = (\frac{A}{2D}x^\beta, f(x; p, r))$, where

$$f(x; p, r) = p + r(1 - x^\beta)^2. \tag{4.7}$$

Here we have written β for $\beta(0)$. The function f is defined on some interval $(0, A_x]$, for parameters (p, r) from a compact box $I \times J$. A computation shows that $f^2(0; p, r) = 0$ for parameter values on the curve

$$\mathbf{H}_2 = \left\{(p, r);\ \ p + r(1 - (p + r)^\beta)^2) = 0\right\}. \tag{4.8}$$

Observe that $\mathbf{H}_2$ is tangent to the diagonal at $(p, r) = (0, 1)$. Furthermore, along $\{(p, r);\ \ p + r = 2^{1/\beta}\}$, the map f satisfies $f(p + r; p, r) = p + r$.

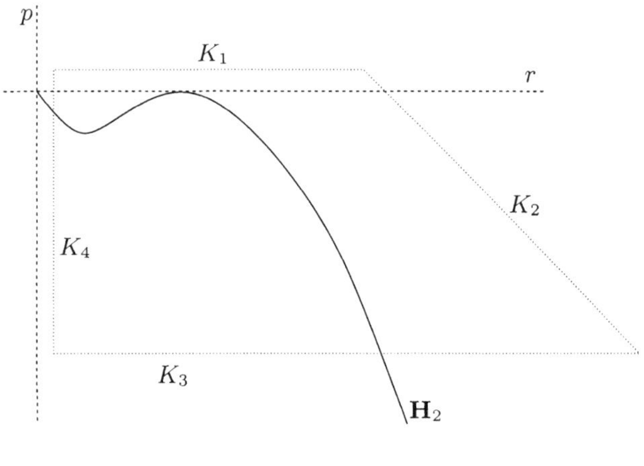

Fig. 3. The choice of the curve K.

We will restrict the parameters (p, r) to a suitably chosen domain in the parameter plane, bounded by a curve K. The curve K consists of four parts

K_1, K_2, K_3, K_4, where for some constants d_1, d_2, P_1, R_1,

$$\begin{aligned} K_1 &\subset \{(p,r);\ \ p = d_1\}, \\ K_2 &\subset \{(p,r);\ \ p + r = 2^{1/\beta} + d_2\}, \\ K_3 &\subset \{(p,r);\ \ p = P_1\}, \\ K_4 &\subset \{(p,r);\ \ r = R_1\}. \end{aligned}$$

We take $d_1 > 0$, $d_2 > 0$, $P_1 < 0$ and $R_1 < 1$.

Write X_λ for the vector field depending on $\lambda = (p,r)$. Following notation as above, let $\mu = H_2 \cap K_4$ and let h be the homoclinic orbit of X_μ. Let C_1 be the connected component of $G_1 \backslash \{(\mu, h)\}$ with parameter values that enter the domain bounded by K at μ. Let $\{Y_\lambda^i\}$ be a sequence of generic families converging to $\{X_\lambda\}$ as $i \to \infty$. Generic here means that homoclinic bifurcations unfold generically. For large i, $\{Y_\lambda^i\}$ possesses a homoclinic orbit h_i close to h at parameter values μ_i close to μ. By a combinatorial reasoning one shows the following lemma. Let C_1^i be defined as C_1 but then for Y_λ^i.

Lemma 9. [HomKokNau98] *The closure of* $\{\mu;\ \ (\mu, h) \in C_1^i\}$ *intersects* K *only in* μ_i.

The lemma says that when continuing $\{\mu;\ \ (\mu, h) \in C_1^i\}$ starting at μ_i, into the region bounded by K, one stays in this region. >From bifurcation theory and the pathfollowing theorem 8, it follows that $\{Y_\lambda^i\}$ possesses homoclinic orbits of unbounded length along C_1^i. One can now show that also $\{X_\lambda\}$ possesses homoclinic orbits of unbounded length along C_1. >From bifurcation theory it is not hard to conclude that this can only happen through a cascade of homoclinic doubling bifurcations; e.g. local and heteroclinic bifurcations can not occur since the dynamics is confined to a tubular neighborhood of the degenerate homoclinic orbit.

5 Renormalization Theory

A striking aspect of period doubling cascades is its universality; independent of details of the system, the bifurcation values scale with a fixed geometric rate. This universality was explained by renormalization theory, see [CouTre78,Fei78]. We will indicate the universal scaling structures found in the bifurcation diagram of homoclinic doubling cascades and discuss its explanation through renormalization theory. We treat renormalization for the model interval maps, it is expected that the vector fields show the same universal scalings in their bifurcation diagrams.

Let $\mathfrak{C}^\omega$ denote the set of functions $x \mapsto f(x)$, defined on a compact interval $[0, L]$, that are of the form $f(x) = g(x^\beta)$ for a real analytic function $u \mapsto g(u)$, and that are unimodal with a unique quadratic minimum at $x = 1$. We

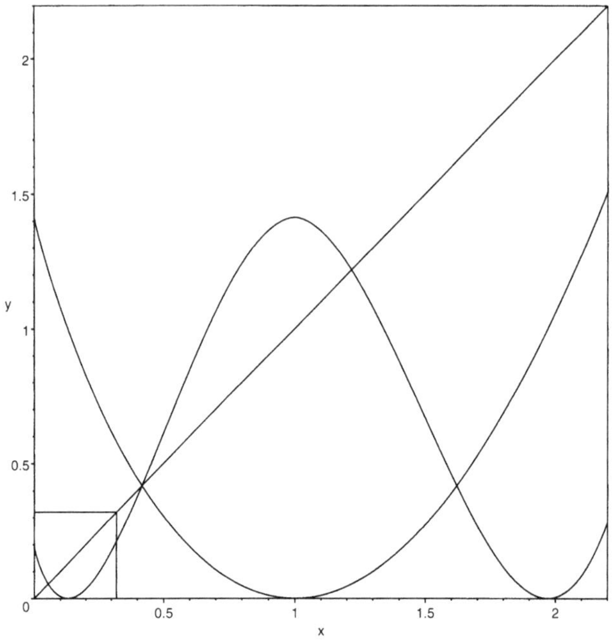

Fig. 4. The graphs of f and f^2 for $f(x) = r(1-x^{\beta})^2$ with $\beta = 0.9$ and $r = 1.413178$.

require that g extends to a bounded analytic function on a fixed complex neighborhood Ω of $[0, L^{\beta}]$. For $f \in \mathfrak{C}^{\omega}$ with $f(x) = g(x^{\beta})$, let

$$\|f\|_0 = \sup_{u \in \Omega} |g(u)|.$$

Equipped with this norm, $\mathfrak{C}^{\omega}$ is a cone in a Banach space.

We can write a function $f \in \mathfrak{C}^{\omega}$ as

$$f(x) = p + r(x^{\beta})(1 - x^{\beta})^2,$$

where p is a real number and $u \mapsto r(u)$ is a positive C^{ω} function. We are interested in such functions for which β is close to $\frac{1}{2}$, and for which p is small and r is close to 1 on a compact interval. Write

$$r(u) = 1 + \varepsilon(u).$$

The renormalization $\mathcal{R}(f)$ of f is a rescaling of the second iterate f^2. It will be defined for functions f close to $x \mapsto (1 - x^{\beta})^2$ for which $f(0) > 1$. There exists $a \in (0, 1)$ for which $f(a) = 1$. Let

$$\mathcal{R}(f)(x) = \frac{1}{a} f^2(ax).$$

The condition $f(0) > 1$ is equivalent to $p > -\varepsilon(0)$. The rescaling factor a is such that $\mathcal{R}(f)$ has its critical point at $x = 1$, so that $\mathcal{R}(f) \in \mathfrak{C}^{\omega}$.

The following theorem provides an isolated fixed point of the renormalization operator $\mathcal{R}$.

Theorem 10. [HomYou99] *For $\beta > \frac{1}{2}$ and $\beta - \frac{1}{2}$ small, the renormalization operator $\mathcal{R}$ on $\mathfrak{C}^\omega$ possesses a hyperbolic fixed point ϕ. The function ϕ depends continuously on β and converges to $x \mapsto (1-\sqrt{x})^2$ as $\beta \to \frac{1}{2}$. The linearization $D\mathcal{R}$ at ϕ has two unstable eigenvalues δ_1, δ_2. They depend continuously on β and satisfy $\delta_1 \to 2$, $\delta_2 \to \infty$ as $\beta \to \frac{1}{2}$. The remainder of the spectrum of $D\mathcal{R}(\phi)$ is strictly inside the unit disc. The unstable manifold of $\mathcal{R}$ is 2 dimensional.*

5.1 Universal Scalings

Theorem 10 provides an explanation of universal scalings in the bifurcation diagram of generic two parameter families of functions.

It follows from invariant manifold theory that $\mathcal{R}$ possesses local stable and unstable manifolds near ϕ, which are denoted by $W^s(\phi)$ and $W^{u,uu}(\phi)$, respectively. The local unstable manifold contains a strong unstable manifold $W^{uu}(\phi)$ and the weak unstable manifold $W^u(\phi) = W^{u,uu}(\phi) \cap \mathfrak{C}_0^\omega$. Note that $W^u(\phi)$ is a smooth manifold.

The next theorem is basic in finding scaling structures in bifurcation diagrams.

Theorem 11. [HomYou99] *Consider a two parameter family $\{f_{p,\varepsilon};\ (p,\varepsilon) \in \mathbb{R}^2\}$ of functions in $\mathfrak{C}^\omega$ with $f_{0,\varepsilon} \in \mathfrak{C}_0^\omega$, that intersects the local stable manifold $W^s(\phi)$ transversally at some function $f_{0,\bar\varepsilon}$. Let (p_n, ε_n) be a sequence of parameter values tending to $(0, \bar\varepsilon)$, such that $\mathcal{R}^{2^n}(f_{p_n,\varepsilon_n})$ converges to a function $\bar f$ in the local unstable manifold $W^{u,uu}(\phi)$. Then*

$$\frac{p_{n+1} - p_n}{p_n - p_{n-1}} \to \frac{1}{\delta_2} \quad \text{if } \bar f \notin W^u(\phi),$$
$$\frac{\varepsilon_{n+1} - \varepsilon_n}{\varepsilon_n - \varepsilon_{n-1}} \to \frac{1}{\delta_1} \quad \text{if } \bar f \notin W^{uu}(\phi).$$

The theorem can be applied to sequences of parameter values (p_n, ε_n) so that $\mathcal{R}^{2^n}(f_{p_n,\varepsilon_n})$ is contained in a bifurcation surface (e.g. H_2, HD_2, or S_2) and converges to a function $\bar f$ in the intersection of this bifurcation surface with $W^{u,uu}(\phi)$. The theorem shows that scalings exist in bifurcation diagrams of generic two parameter families of functions in $\mathfrak{C}^\omega$, independent of details of the family, other than the value of β. It relates the unstable eigenvalues at the fixed point ϕ of the renormalization operator $\mathcal{R}$ to scalings in such bifurcation diagrams.

In figures 5 and 6, bifurcation curves for the family $x \mapsto p + r(1 - x^\beta)^2$ with $\beta = 0.6$ are shown. These figures illustrate the scalings predicted by Theorem 11. In particular, the HD points converge exponentially at a rate δ_1 which is close to 2 for β close to $\frac{1}{2}$. As is illustrated in figure 6, the curve

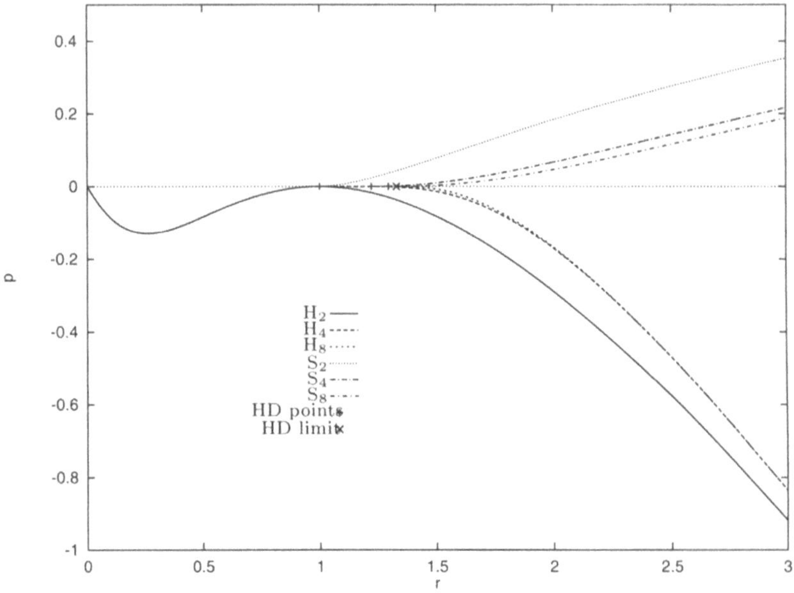

Fig. 5. Bifurcation curves for $x \mapsto p + r(1 - x^{\beta})^2$ in the (p, r) parameter plane for $\beta = 0.6$.

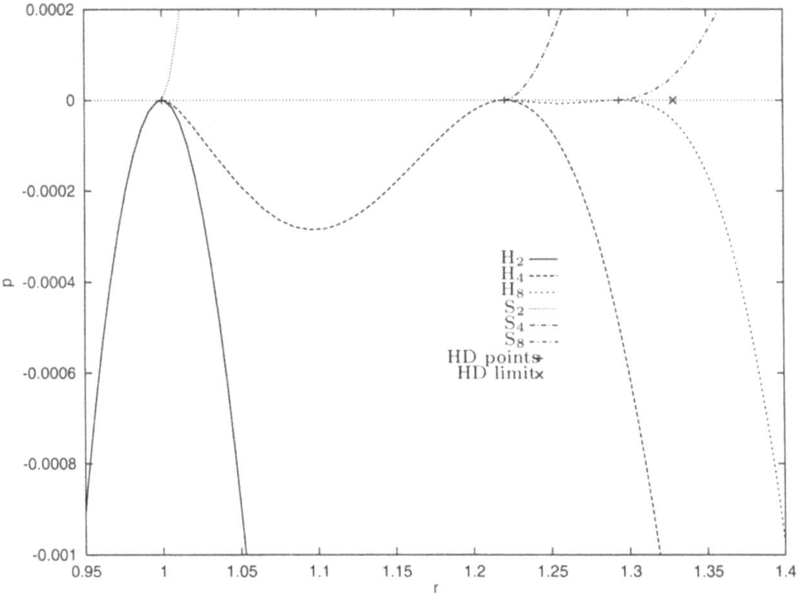

Fig. 6. Blow up of figure 5.

pieces in $\mathrm{H}_{2^{n+1}}$ between HD_{2^n} and $\mathrm{HD}_{2^{n+1}}$, accumulate on the line $\{p = 0\}$ exponentially fast at the strong rate δ_2.

It should be noted that Theorem 11 does not explain all scalings that exist in the bifurcation diagram. To see this, define $\mathrm{H}_\infty = \lim_{k\to\infty} \mathrm{H}_{2^k}$, note that H_∞ is invariant under $\mathcal{R}$. Similarly, $\mathrm{P}_\infty = \lim_{k\to\infty} \mathrm{S}_{2^k}$ is invariant under $\mathcal{R}$. H_∞ and P_∞ constitute a 'boundary of chaos'. Theorem 11 does not address scalings of sequences of bifurcation values in S_{2^n} converging to a point in P_∞ away from $W^s(\phi)$ (which are expected to scale exponentially with the usual Feigenbaum rate). Nor does it address scalings of sequences of bifurcation values in H_{2^n} converging to a point in H_∞ away from $W^s(\phi)$. We have no information on these scalings (however, [LyuPikZak89] and [Hom96] contain results on scalings of homoclinic bifurcations in similar situations).

References

[AllYor84] K.T. Alligood, J.A. Yorke, Families of periodic orbits: virtual periods and global continuability, *Journ. Diff. Eq.* **55** (1984), 59-71.

[CouTre78] P. Coullet, C. Tresser, Itérations d'endomorphismes et groupe de renormalisation, *J. de Physique* **C5** (1978), 25.

[Fei78] M.J. Feigenbaum, Quantitative universality for a class of nonlinear transformations, *Journ. Stat. Phys.* **19** (1978), 669-706.

[Fie96] B. Fiedler, Global pathfollowing of homoclinic orbits in two-parameter flows, in Pitman Res. Notes Math., Vol. 352, Longman, 1996, pp. 79-146.

[HirPugShu77] M. Hirsch, C. Pugh, M. Shub, *Invariant Manifolds*, Lect. Notes Math., Vol. 583, 1977, Springer-Verlag.

[HomKokKru94] A.J. Homburg, H. Kokubu, M. Krupa, The cusp horseshoe and its bifurcations in the unfolding of an inclination-flip homoclinic orbit, *Ergod. Th. & Dynam. Sys.* **14** (1994), 667-693.

[Hom96] A.J. Homburg, *Global Aspects of Homoclinic Bifurcations of Vector Fields*, Memoirs A.M.S., Vol. 578, A.M.S., 1996.

[HomKokNau98] A.J. Homburg, H. Kokubu, V. Naudot, Homoclinic doubling cascades, preprint (1998).

[HomKra98] A.J. Homburg, B. Krauskopf, Resonant homoclinic flip bifurcations, to appear in *J. Dynamics Differential Equations*.

[HomYou99] A.J. Homburg, T. Young, Universal scalings in homoclinic doubling cascades, preprint (1999).

[KisKokOka93] M. Kisaka, H. Kokubu, H. Oka, Bifurcations to N-homoclinic orbits and N-periodic orbits in vector fields, *J. Dynamics Differential Equations* **5** (1993), 305-357.

[KokKomOka96] H. Kokubu, M. Komuro, H. Oka, Multiple homoclinic bifurcations from orbit-flip. I. Successive homoclinic doublings. *Int. Journ. Bif. Chaos* **6** (1996), 833-850.

[LuzVia99] S. Luzatto, M. Viana, Positive Lyapunov Exponents for Lorenz-like families with criticalities, to appear in *Asterisque*.

[LuzTuc99] S. Luzatto, W. Tucker, Non-uniformly expanding dynamics for maps with critical points and singularities, to appear in *Publ. Math. I.H.E.S.*

[LyuPikZak89] D.V. Lyubimov, A.S. Pikovsky, M.A. Zaks, Universal scenarios of transitions to chaos via homoclinic bifurcations, *Sov. Sci. Rev., Sect. C, Math. Phys. Rev.* **8** (1989), 221-292.

[Nau96] V. Naudot, Strange attractor in the unfolding of an inclination-flip homoclinic orbit, *Ergod. Th. & Dynam. Syst.* **16** (1996), 1071-1086.

[OldKraCha00] B.E. Oldeman, B. Krauskopf, A.R. Champneys, Death of period-doublings: locating the homoclinic-doubling cascade, to appear in *Physica D.*

[San97] B. Sandstede, Constructing dynamical systems having homoclinic bifurcation points of codimension two, *J. Dynamics Differential Equations* **9** (1997), 269-288.

[San99] B. Sandstede, Center manifolds for homoclinic solutions, *J. Dynamics Differential Equations* **12**, (2000), 449-510.

[Shi93] A.L. Shil'nikov, On bifurcations of the Lorenz attractor in the Shimizu-Morioka model, *Physica D* **62** (1993), 338-346.

[Yan87] E. Yanagida, Branching of double pulse solutions from single pulse solutions in nerve axon equations, *Journ. Diff. Eq.* **66** (1987), 243-262.

[YorAll83] J.A. Yorke, K.T. Alligood, Cascades of period doubling bifurcations: a prerequisite for horseshoes, *Bull. A.M.S.* **9** (1983), 319-322.

Existence, Bifurcation, and Stability of Profiles for Classical and Non-Classical Shock Waves

Heinrich Freistühler[1], Christian Fries[1], and Christian Rohde[2,*]

[1] Max-Planck-Institut für Mathematik in den Naturwissenschaften, Leipzig
[2] Albert-Ludwigs-Universität Freiburg

Abstract. This paper surveys the authors' recent results on viscous shock waves in PDE systems of conservation laws with non-convexity and non-strict hyperbolicity. Particular attention is paid to the physical model of magnetohydrodynamics. The plan of the paper is as follows. Sections 1 and 2 introduce the classes of systems and the classes of shock waves we consider and recall how profiles for small-amplitude shocks are constructed via center manifold analyses of a corresponding system of ODEs. Section 3 describes the global picture, i. e., large-amplitude shock waves, for the case of magnetohydrodynamics, first the solution set of the Rankine-Hugoniot jump conditions, then a heteroclinic bifurcation occurring in the ODE system for the profiles. Section 4 presents a method for the numerical identification of heteroclinic manifolds, which is applied in Sections 5 and 6 to the case of magnetohydrodynamics. The numerical treatment confirms and details the analytical findings and, more notably, extends them considerably; in particular, it allows to study the existence / non-existence of profiles and the aforementioned heteroclinic bifurcation globally. Section 7 discusses the stability of viscous shock waves; the important nonuniformity of the vanishing viscosity limit for, in particular, non-classical MHD shock waves is *not* addressed in this paper.

1 Classification of Shock Waves

Let U be an open subset of $\mathbb{R}^n$ and $g, f : U \to \mathbb{R}^n$ smooth functions such that g maps U diffeomorphic onto its image, while $(Dg(u))^{-1}Df(u)$ is $\mathbb{R}$-diagonalizable at every $u \in U$. Consider the hyperbolic system of conservation laws

$$g(u)_t + f(u)_x = 0, \tag{1}$$

and a non-characteristic *inviscid shock wave*

$$u(x,t) = \begin{cases} u^- : x - st < 0, \\ u^+ : x - st > 0. \end{cases} \tag{2}$$

associated with (1), i. e., the triple $(u^-, u^+, s) \in U \times U \times \mathbb{R}$ with $u^- \neq u^+$ satisfies the Rankine-Hugoniot conditions

$$-s(g(u^+) - g(u^-)) + (f(u^+) - f(u^-)) = 0$$

and s is not an eigenvalue of $Df(u^-)$ nor of $Df(u^+)$.

* Project: Viscous Profiles in Magnetohydrodynamics (Dietmar Kröner, Heinrich Freistühler)

To classify such objects, introduce, for arbitrary $(u,s) \in U \times \mathbb{R}$, the spaces

$$R^-(u,s) = \sum_{\lambda<s} \ker(Df(u) - \lambda Dg(u)), \quad R^+(u,s) = \sum_{\lambda>s} \ker(Df(u) - \lambda Dg(u)).$$

The shock wave is called *Laxian*, or *classical*, if the linearized Rankine–Hugoniot conditions

$$\begin{aligned} &g(u^+) - g(u^-))\bar{\sigma}' \\ &\quad + (Df(u^+) - sDg(u^+))\bar{u}^+_+ - (Df(u^-) - sDg(u^-))\bar{u}^-_- \\ &= -(Df(u^+) - sDg(u^+))\bar{u}^+_- + (Df(u^-) - sDg(u^-))\bar{u}^-_+ \end{aligned} \tag{3}$$

have a unique solution $(\bar{u}^-_-, \bar{u}^+_+, \bar{\sigma}') \in R^-(u^-,s) \times R^+(u^+,s) \times \mathbb{R}$ for any $(\bar{u}^-_+, \bar{u}^+_-) \in R^+(u^-,s) \times R^-(u^+,s)$. Generally, let

$$l = \dim\, R^-(u^-,s) + \dim\, R^+(u^+,s) + 1$$

and

$$r = \dim\, (R^-(u^-,s) + R^+(u^+,s) + \mathbb{R}(u^+ - u^-))$$

be the number of unknowns and the rank, respectively, of (3). Let

$$\underline{\kappa} = l - r \geq 0, \qquad \overline{\kappa} = n - r \geq 0$$

denote the degrees of *under-* resp. *overdeterminacy* of this linear algebraic system. A shock wave with $\underline{\kappa} > 0$ (and $\overline{\kappa} = 0$) is called *(purely) undercompressive*; a shock wave with $\overline{\kappa} > 0$ (and $\underline{\kappa} = 0$) is called *(purely) overcompressive.* For any shock wave, call the ordered pair

$(\underline{\kappa}, \overline{\kappa})$ the *algebraic type* of the shock wave

and the integer

$$\kappa = \overline{\kappa} - \underline{\kappa} + 1$$

its *multiplicity.* Letting n^-, n^+ denote the dimensions of the spaces of "incoming" modes to the left and right of the shock wave, respectively, i. e.,

$$n^- = \dim\, R^+(u^-,s), n^+ = \dim\, R^-(u^+,s),$$

we have

$$\kappa = n - l + 1 = n - (2n - (n^- + n^+)) = n^- + n^+ - n. \tag{4}$$

Together with the 'inviscid' system (1), we consider the 'viscous' system

$$g(u)_t + f(u)_x = (B(u)u_x)_x \tag{5}$$

with some appropriate *viscosity* $B : U \to I\!R^{n\times n}$. A traveling wave solution $u(t,x) = \phi(x - st)$ of (5) corresponding to a given inviscid shock wave (2) is called its *viscous profile.* Writing $q \equiv -sg(u^{\pm}) + f(u^{\pm})$, such profile technically is a heteroclinic orbit of

$$B(\phi)\phi' = f(\phi) - sg(\phi) - q, \tag{6}$$

with end states

$$\phi(\pm\infty) = u^{\pm}.$$

Assuming for a moment that B has full rank n and the rest points $u^{\pm}$ of (6) are hyperbolic, we let

$$k^- = \dim W^u(u^-),\ k^+ = \dim W^s(u^+)$$

denote the dimensions of the unstable manifold of (6) at u^- and the dimension of the stable manifold of (6) at u^+, respectively, and define the *index* of the viscous profile ϕ as

$$k = k^- + k^+ - n.$$

Under certain conditions on B, the dimensions of $W^u(u^-)$ and $W^s(u^+)$ are equal to those of $R^+(u^-, s)$ and $R^-(u^+, s)$, i. e., $k^- = n^-, k^+ = n^+$ so that, by (4),

$$\text{multiplicity } \kappa \text{ of the shock} = \text{index } k \text{ of its profile.} \tag{7}$$

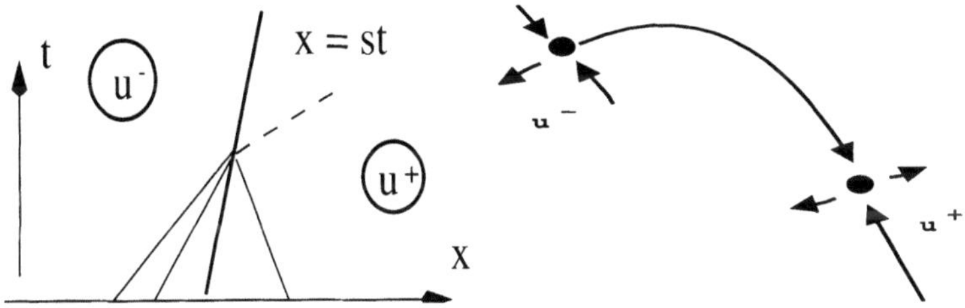

Fig. 1. Laxian shock, $(\underline{\kappa}, \overline{\kappa}) = (0, 0)$. Example with $n = 2$, $\kappa = 1 = k$.

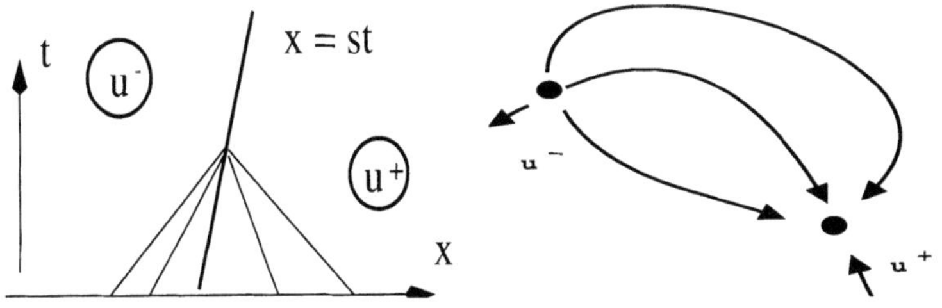

Fig. 2. Overcompressive shock, $\overline{\kappa} > 0$. Example with $n = 2$, $(\underline{\kappa}, \overline{\kappa}) = (0, 1), \kappa = 2 = k$.

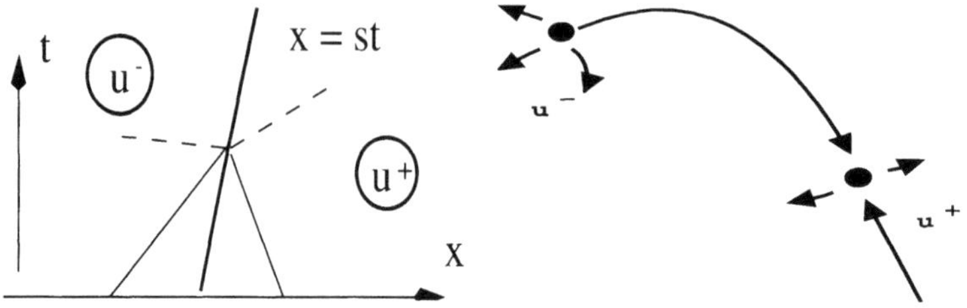

Fig. 3. Undercompressive shock, $\underline{\kappa} > 0$. Example with $n = 2$, $(\underline{\kappa}, \overline{\kappa}) = (1, 0), \kappa = 0 = k$.

This holds, e. g., for $B = I$, the identity matrix, in which case $R^+(u^-, s)$, $R^-(u^+, s)$ are the tangent spaces, at u^-, u^+, of $W^u(u^-), W^s(u^+)$.) In each of Figures 1,2,3, the left picture shows an inviscid shock wave together with characteristics ($\dot{x} = \lambda(u^\pm)$, $\lambda(u^\pm)$ eigenvalues of $Df(u^\pm)$, while the right picture sketches a corresponding phase portrait for the profile ODE (6).

2 Profiles for Small-Amplitude Shock Waves

We henceforth restrict attention to *symmetric*, canonically splitting systems, i. e., we assume that $G \equiv Dg$, $F \equiv Df$, and B are symmetric matrices with G positive definite and B positive semidefinite, and (5) decomposes as

$$\begin{aligned} &\begin{pmatrix} G_{11}(v,w) & G_{12}(v,w) \\ G_{21}(v,w) & G_{22}(v,w) \end{pmatrix} \begin{pmatrix} v \\ w \end{pmatrix}_t + \begin{pmatrix} F_{11}(v,w) & F_{12}(v,w) \\ F_{21}(v,w) & F_{22}(v,w) \end{pmatrix} \begin{pmatrix} v \\ w \end{pmatrix}_x \\ &= \left(\begin{pmatrix} 0 & 0 \\ 0 & \tilde{B}(v,w) \end{pmatrix} \begin{pmatrix} v \\ w \end{pmatrix}_x \right)_x \end{aligned} \tag{8}$$

with $\tilde{B}$ positive definite.

Theorem 1. *Consider a simple mode (λ, r), i. e.,*

$$(F(u) - \lambda(u)G(u))r(u) = 0, \tag{9}$$

where λ is real-valued and the vector field $r \neq 0$ is unique up to a scalar factor. Writing

$$F - \lambda G \equiv A \equiv \begin{pmatrix} A_{11} & A_{12} \\ A_{21} & A_{22} \end{pmatrix},$$

assume that at some state u_*,

$$A_{11}(u_*) \text{ is invertible.} \tag{10}$$

Then any small λ-shock near u_* (i. e., any shock with end states $u^-, u^+ \approx u_*$ satisfying $\lambda(u^-) > s > \lambda(u^+)$) has a viscous profile w. r. t. the viscosity B if *and only if it satisfies the strict version $(E)_s$ of Liu's entropy condition [L1]. In particular, if the mode is convex (cf. (33) below), every sufficiently small λ-shock has a profile.*

Proof of Theorem 2.1. Analogously to (8), we decompose (6) in the form

$$0 = f_1(v,w) - sg_1(v,w) - q_1 \tag{11}$$
$$\tilde{B}(v,w)w' = f_2(v,w) - sg_2(v,w) - q_2. \tag{12}$$

Consider (6) resp. (11),(12) for $u = (v,w)$ near $u_* = (v_*, w_*)$ and (q,s) near (q_*, s_*) with

$$s_* = \lambda(u_*) \text{ and } q_* = f(u_*) - s_* g(u_*).$$

Assumption (10) implies that (11) can be solved locally for v as

$$v = V(w,q,s), \tag{13}$$

i. e.,

$$0 = f_1(V(w,q,s),w) - sg_1(V(w,q,s),w) - q_1. \tag{14}$$

Plugging V into (12), we obtain the reduced system

$$\hat{B}(w,q,s)w' = h(w,q,s), \tag{15}$$

with

$$\hat{B}(w,q,s) \equiv \tilde{B}(V(w,q,s),w), \tag{16}$$
$$h(w,q,s) \equiv f_2(V(w,q,s),w) - sg_2(V(w,q,s),w) - q_2.$$

More precisely, (13) and (15) together are equivalent to (6). We claim now that any two states sufficiently close to u_* that form a λ-shock are located on a one-dimensional invariant manifold C of (6). To see this, note first that

$$D_w h|_{(q,s)=(q_*,s_*)} r_2 = 0 \tag{17}$$
$$\Leftrightarrow Ar = 0 \quad \text{with} \quad r = (r_1, r_2), r_1 = -(A_{11})^{-1} A_{12} r_2.$$

Equivalence (17) follows from

$$0 = \begin{pmatrix} A_{11} & A_{12} \\ A_{21} & A_{22} \end{pmatrix} \begin{pmatrix} r_1 \\ r_2 \end{pmatrix}$$
$$\Leftrightarrow r_1 = -(A_{11})^{-1} A_{12} r_2 \quad \text{and} \quad (-A_{21}(A_{11})^{-1} A_{12} + A_{22}) r_2 = 0$$

and

$$D_w h = A_{21} D_w V + A_{22} = -A_{21}(A_{11})^{-1} A_{12} + A_{22},$$

the latter identity coming from (14) as $0 = A_{11} D_w V + A_{12}$.

Note now that Assumption (10) implies that the w-component r_2 of the eigenvector $r = (r_1, r_2)$ does not vanish. This means that $D_w h(w_*, q_*, s_*)$, and thus $\hat{B}^{-1}(w_*, q_*, s_*) D_w h(w_*, q_*, s_*)$, have a one-dimensional kernel spanned by r_2. As $\hat{B}$ and h_w are symmetric and $\hat{B}$ is positive, $\hat{B}^{-1} h_w$ cannot have any purely imaginary eigenvalue other than 0. Applying the Center Manifold Theorem to system (15) as augmented by the further equations

$$q' = 0, s' = 0, \tag{18}$$

we see that the augmented system has, near (w_*, q_*, s_*), a center manifold with 1-dimensional w-fibres ((q, s)-sections). The left and right states u^-, u^+ of any small shock under consideration are rest points of (6). As any center manifold contains locally all rest points of the flow to which it belongs, there is precisely one fibre C (—lifting via (14), we immediately view C as lying in u-space—) that contains u^- and u^+. It is now easy to see that the open segment of C between u^- and u^+ is the desired profile if and only if there exists no other fixed point between these two. This is however equivalent to Liu's condition in its strict version $(E)_s$: For any u located between u^- and u^+ on the Hugoniot locus

$$\mathcal{H}(u^-) = \{u \in U \, : \, \exists s = s(u, u^-) : \, s(u^-, u)(g(u) - g(u^-)) = f(u) - f(u^-)\},$$

the strict inequality $s(u, u^-) < s(u^+, u^-)$ holds. Cf. [Fre2] for details. Theorem 2.1 is considered proved.

A number of important systems from continuum mechanics are of the form (8). Instances are the equations of compressible viscous, heat-conducting fluids as well as those of compressible magnetohydrodynamics in various variants corresponding to the simultaneous presence or non-presence of dissipative mechanisms associated with viscosity, heat conductivity, and electrical resistivity, when written in entropy variables. Cf. [Kw] for the identification of this class of systems and that of the mentioned physical systems as examples. Notice that in most—though not all—of these examples, the existence of viscous profiles, even for shocks of large amplitude, has been shown through *ad hoc* considerations [Gi,CS]. The purpose of the above part of the present section is to demonstrate the use of the Center Manifold Theorem in the context of degenerate viscosity which is in fact quite similar to the nondegenerate case[MP].

We now turn to non-classical shock waves. Non-classical shock waves of small amplitude arise near *umbilic points*, i. e., points near which modes, see (9), change multiplicity. For the construction of viscous profiles for small non-classical shock waves, one considers center manifolds as above, but with fibers C of dimension higher than 1. To illustrate what one can obtain in this way, we now focus on the concrete system that constitutes the primary object of our more detailed investigations. Plane waves in viscous, resistive, heat-conductive magnetohydrodynamics (MHD) satisfy the equations

$$\begin{aligned} \rho_t + (\rho v)_x &= 0 \\ (\rho v)_t + (\rho v^2 + p + \frac{1}{2}|\mathbf{b}|^2)_x &= \zeta v_{xx} \\ (\rho \mathbf{w})_t + (\rho v \mathbf{w} - a\mathbf{b})_x &= \mu \mathbf{w}_{xx} \\ \mathbf{b}_t + (v\mathbf{b} - a\mathbf{w})_x &= \nu \mathbf{b}_{xx} \\ \mathcal{E}_t + (v(\mathcal{E} + p + \frac{1}{2}(|\mathbf{b}|^2 - a^2)) - a\mathbf{w}\cdot\mathbf{b})_x &= \kappa\theta_{xx} + \zeta(v v_x)_x \\ &\quad + \mu(\mathbf{w}\cdot\mathbf{w}_x)_x + \nu(\mathbf{b}\cdot\mathbf{b}_x)_x, \end{aligned} \tag{19}$$

where $v, \mathbf{w}$ and $a, \mathbf{b}$ are the longitudinal and transverse components of the fluid's velocity $\mathbf{V} = (v, \mathbf{w}) = (v, w_1, w_2)$ and the magnetic field $\mathbf{B} = (a, \mathbf{b}) = (a, b_1, b_2)$, respectively, ($a \equiv$ *const* as *div* $\mathbf{B} = 0$) and $\mathcal{E} = \rho(\frac{1}{2}|\mathbf{V}|^2 + \epsilon) + \frac{1}{2}|\mathbf{B}|^2$, is the density of total energy. The variables $\rho, p, \theta, \epsilon$, describing density, pressure, temperature, and internal energy of the fluid, are intrinsically related with each other through the equation of state $\epsilon = \epsilon(\tau, \eta)$ and the identities $\rho = \tau^{-1}$, $p = -\epsilon_\tau(\tau, \eta)$, $\theta = \epsilon_\eta(\tau, \eta)$, where τ denotes the specific volume and η the entropy of the fluid. The internal energy ϵ is required to satisfy the conditions $-\epsilon_\tau > 0$, $\epsilon_\eta > 0$, $D^2\epsilon > 0$, $-\epsilon_{\tau\eta} > 0$, $-\epsilon_{\tau\tau\tau} > 0$; the first two of these requirements amount to the positivity of pressure and temperature, the third to the concavity of entropy η as a function of τ and e, and the fourth and fifth are known as "Weyl's conditions." The two dissipation coefficients $\mu \geq 0$ and $\zeta \geq 0$ correspond to the intrinsic viscosity of the fluid; more precisely, $\mu = \zeta_1, \zeta = \zeta_2 + \frac{4}{3}\zeta_1$ with $\zeta_1, \zeta_2 \geq 0$ the first and second viscosity coefficients of the fluid. The two remaining coefficients $\nu \geq 0$ and $\kappa \geq 0$ denote the electrical resistivity and the thermal conductivity of the fluid. We recall (e. g. from [KuLi]) some basic properties of *ideal MHD*, i. e., Eqs. (19) with $\zeta = \mu = \nu = \kappa = 0$.

The seven characteristic speeds $\lambda_{-3} \leq \lambda_{-2} \leq \lambda_{-1} \leq \lambda_0 \leq \lambda_1 \leq \lambda_2 \leq \lambda_3$ of this 7×7 hyperbolic system of conservation laws are of the form

$$\lambda_0 = v, \quad \lambda_{\pm 1} = v \pm c_-, \quad \lambda_{\pm 2} = v \pm c_A, \quad \lambda_{\pm 3} = v \pm c_+ \tag{20}$$

with the fast and slow magnetoacoustic speeds $c_+ \geq c_- \geq 0$ given by $c_\pm^2 = \frac{1}{2}[(c_s^2 + \rho^{-1}(a^2 + b^2)) \pm \sqrt{(c_s^2 + \rho^{-1}(a^2 + b^2))^2 - 4c_s^2\rho^{-1}a^2}]$ (where c_s is the sound speed, $c_s^2 = \pi_\rho(\rho, \eta)$ with $\pi(\rho, \eta) = p = -\epsilon_\tau(\tau, \eta)$) and the Alfvén speed $c_A \geq 0$ by $c_A^2 = \rho^{-1}a^2$. We assume henceforth that $a \neq 0$. Obviously, $0 < c_- < c_A < c_+$ if $\mathbf{b} \neq 0$. For $\mathbf{b} = 0$ however,

$$\begin{aligned} &0 < c_- = c_A < c_+ && \text{if } \Delta_a > 0, \\ &0 < c_- = c_A = c_+ && \text{if } \Delta_a = 0, \\ &0 < c_- < c_A = c_+ && \text{if } \Delta_a < 0, \end{aligned} \tag{21}$$

with $\Delta_a = \rho c_s^2 - a^2$. Typically, all three cases in (21) occur, with Δ_a vanishing along a smooth manifold which separates its own complement into two open sets where $\Delta_a > 0$ and $\Delta_a < 0$, respectively. E. g., $\Delta_a = \gamma p - a^2$ for a perfect gas $\epsilon(\tau, \eta) = c_v \exp(\eta / c_v) \tau^{1-\gamma}$. For shock waves, with, say w. l. o. g., $s = 0$, the Rankine-Hugoniot conditions require u^- and u^+ to satisfy

$$f(u) = q \tag{22}$$

with the same value of the relative flux q. For $q \in Q$, the set of regular values of the mapping f, Eqs. (22) have up to four solutions u_0, u_1, u_2, u_3 satisfying

$$\begin{aligned} 0 &< \pm\lambda_{\mp 3}(u_0) \\ \pm\lambda_{\mp 3}(u_1) < 0 &< \pm\lambda_{\mp 2}(u_1) \\ \pm\lambda_{\mp 2}(u_2) < 0 &< \pm\lambda_{\mp 1}(u_2) \\ \pm\lambda_{\mp 1}(u_3) < 0 &< \pm\lambda_{0}\ \ (u_3) \end{aligned} \tag{23}$$

With the two cases in (23) differing only by a direction reversal $x \mapsto -x$, we restrict attention to first one (upper signs) without loss of generality. The four states u_0, u_1, u_2, u_3 combinatorially allow for various inviscid shock waves (2), namely the twelve species $u^- = u_i, \quad u^+ = u_j, \quad i, j \in \{0, 1, 2, 3\}, i \neq j$, which are briefly referred to as being of species $i \to j$. As entropy increases with the index, i. e., $\eta(u_0) < \eta(u_1) < \eta(u_2) < \eta(u_3)$, only shocks of species $i \to j$ with $i < j$ are thermodynamically possible. One distinguishes between the classical shocks of species $0 \to 1, 2 \to 3$ which are associated with the fast and slow magnetoacoustic modes c_+, c_-, respectively, and the non-classical or "intermediate" shocks of species $0 \to 2$, $1 \to 3$, $0 \to 3$, and $1 \to 2$.

Theorem 2. *Consider an arbitrary state u_* with transverse magnetic field $\mathbf{b}_* = 0$, and an arbitrary array $\delta = (\zeta, \mu, \nu, \kappa)$ of positive dissipation coefficients. Then for any $\varepsilon > 0$, there exist shock waves, with $|u^\pm - u_*| < \varepsilon$, of types $0 \to 1$, $0 \to 2$, and $1 \to 2$ (if $c_- < c_A = c_+$ at u_*), of types $2 \to 3$, $1 \to 3$, and $1 \to 2$ (if $c_- = c_A < c_+$ at u_*), or of types $0 \to 1$, $2 \to 3$, $0 \to 2$, $1 \to 3$, $0 \to 3$, and $1 \to 2$ (if $c_- = c_A = c_+$ at u_*), which possess a viscous profile with respect to the prescribed δ. More precisely, in each of these cases, shocks of type $i \to j$ have a $(j - i)$-parameter family of profiles if $j - i > 1$, and 2 profiles if $(i, j) = (1, 2)$.*

The proof via considerations about the flow on 2- respectively 3-dimensional center manifolds can be found in [Fre1].

We conclude the section by connecting the MHD specific distinction of species "$i \to j$" with the general classification introduced in Section 1. It suffices to note that shocks of species $0 \to 1$, $2 \to 3$ have algebraic type $(0, 0)$, shocks of species $0 \to 2$, $1 \to 3$ have type $(1, 0)$, shocks of species $0 \to 3$ type $(2, 0)$, and shocks of species $1 \to 2$ are of type $(1, 1)$. Thus all intermediate MHD shock waves are overcompressive.

3 Bifurcation Analysis for MHD Shock Waves

In this section we collect first results of a bifurcation analysis for the Rankine–Hugoniot relations (22) in magnetohydrodynamics, and then recall a conjecture on a related global bifurcation occurring for viscous profiles of MHD shock waves. Attention is now restricted to a perfect gas, $p = R\rho\theta$, $\epsilon = c_v\theta$.

Equivariance and rescaling considerations entitle restriction, w. l. o. g., to the three-parameter family of cases

$$\begin{aligned} \rho v &= 1, \\ v + R\theta/v + |\mathbf{b}|^2/2 &= j, \\ \mathbf{w} - \mathbf{b} &= \mathbf{0}, \\ vb_1 - w_1 &= c, \\ vb_2 - w_2 &= 0, \\ \tfrac{v^2+|\mathbf{w}|^2}{2} + (c_v + R)\theta + v|\mathbf{b}|^2 - \mathbf{w}\cdot\mathbf{b} &= e. \end{aligned} \tag{24}$$

At first consider the case $c > 0$. It is well known that there can be up to four distinct states that solve (24). The two fast states u_0, u_1 satisfy $v_0 > v_1 > 1$ while the slow states u_2, u_3 satisfy $v_3 < v_2 < 1$. The typical configuration in the b_1v–plane is displayed in Figure 4. The subsequent lemma gives a more precise statement on the existence of *physical* solutions $u_0, \dots, u_3$, i.e. states with positive pressure.

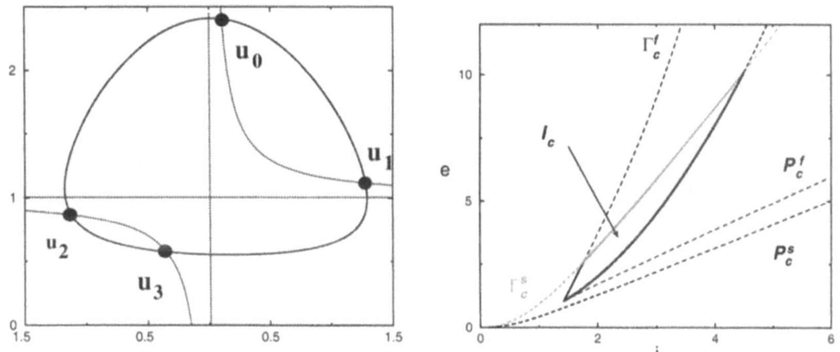

Fig. 4. Null clines of (24) in b_1v–plane and the set $\mathcal{I}_c$.

Lemma 3. *For the adiabatic coefficient $\gamma = 1 + R/c_v$, let $\bar{c}$ be the smallest (positive) solution of*

$$c^{2/3} - \sqrt{\gamma+1}\frac{2-\gamma}{3\gamma}c = \frac{2}{3\gamma}.$$

For each $c \in (0, \bar{c})$ there is a non-empty bounded open set $\mathcal{I}_c \subset (0,\infty)^2$ such that for each $(j, e) \in \mathcal{I}_c$ there exist four distinct physical states $u_0, \dots, u_3$ satisfying (24).

Lemma 3 is illustrated in the right picture of Figure 4. The curve Γ_c^s denotes the set of all points $(j,e) \in \mathbb{R}^2$ such that (24) has exactly one fast solution denoted by $u_{2=3}$. For points $(j,e) \in \mathbb{R}^2$ to the left of Γ_c^s there are no slow solutions, for $(j,e) \in \mathbb{R}^2$ to the right of Γ_c^s there are two slow solutions u_2, u_3. Γ_c^f marks the analogous partition of the je–plane for the fast solutions. The curve P_c^f consists of an upper and a lower part, ending in a cusp-type singularity for $(j,e) = (1 + \frac{3}{2}c^{2/3}, \frac{1}{2} + \frac{3}{2}c^{2/3} + \frac{3}{2}c^{4/3})$. It identifies, for the upper (lower) part, the loci where the pressure $p_0 = \rho_0 R\theta_0$ $(p_1 = \rho_1 R\theta_1)$ vanishes and changes sign. Parts of these three curves –marked with solid lines in Figure 4–form the boundary of $\mathcal{I}_c$.
The singular case $c = 0$ is of particular interest. The states u_1, u_2 degenerate to a onedimensional curve of states solving (24).

Lemma 4. *Let $c = 0$ and $\gamma = 1 + R/c_v$. There is a non-empty bounded open set $\mathcal{I}_0 \subset (0,\infty)^2$ such that $(j,e) \in \mathcal{I}_0$ if and only if*

(i) there are two physical states $u_0, u_3 \in \mathbb{R}^7$ with $v_0 > 1 > v_3$ solving (24), and
(ii) there is a set A of physical states solving (24) given by

$$A = \left\{ (\rho, v, \mathbf{w}, \mathbf{b}, \theta) \,|\, \rho = v = 1, |\mathbf{w}| = |\mathbf{b}| = r, \theta = \tfrac{1}{R}(j - \tfrac{r^2}{2} - 1)) \right\},$$
$$r = \sqrt{2\gamma j - 2(\gamma - 1)e - \gamma - 1}.$$

A is called the Alfvén circle.

Although Lemmas 3.1, 3.2—to our knowledge— cannot be found in the literature, we stress that they just refine and complement findings that trace back to the early work of Germain or Kulikovskii and Liubimov [Ge,KuLi]. For a proof of (a more detailed statement of) Lemma 3 and similar results we refer to [FreR3]. The profile ODE (6) in the MHD case, here rather (15), becomes

$$\begin{aligned}
\zeta \dot{v} &= v + p + \tfrac{1}{2} \mid \mathbf{b} \mid^2 - j,\\
\mu \dot{\mathbf{w}} &= \mathbf{w} - \mathbf{b},\\
\nu \dot{\mathbf{b}} &= v\mathbf{b} - \mathbf{w} + (c, 0)^T,\\
\kappa \dot{\theta} &= c_v \theta - \tfrac{1}{2}(|\mathbf{w}|^2 - 2\mathbf{b} \cdot \mathbf{w} + v|\mathbf{b}|^2) - \tfrac{v^2}{2} + jv + \mathbf{b} \cdot (c, 0)^T - e.
\end{aligned} \tag{Σ^6}$$

Obviously solutions of (24) are rest points of Σ^6. Conley and Smoller showed that the (Laxian) shock waves $u_0 \to u_1$ and $u_2 \to u_3$ admit a viscous profile [CS] for all $\delta \in (0,\infty)^4$ and all q such that the associated rest points exist. The situation for the intermediate waves is more complicated. The known (analytical and numerical) results from literature support the following conjecture:

There exists a threshold $\omega^* = \omega^*(q, \mu/\zeta, \kappa/\zeta) > 0$ such that the following holds for all $c \in (0, \bar{c})$, $(j,e) \in \mathcal{I}_c$, and $\delta = (\nu, \zeta, \mu, \kappa) \in (0,\infty)^4$: If $\nu/\zeta > \omega^*$, *then all intermediate shocks (for the given q) have viscous profiles (for the given δ). Conversely, if $\nu/\zeta < \omega^*$, then no intermediate shock wave has a profile.*

A proof of this conjecture for small μ and κ, following [KuLi], can be found in [FreSzm].

4 Numerical Identification of Heteroclinic Manifolds

Motivated by the dynamics of the ODE–system $\boldsymbol{\Sigma}^{\mathbf{6}}$ we are (mainly) interested in viscous profiles of shock waves that appear as several–parameter families of heteroclinic orbits. In this chapter we review a direct method to approximate general heteroclinic manifolds that has been presented in [FreR1,FreR2]. Although one key ingredient is strongly connected to the analysis of conservation laws, the method can technically be viewed as a straightforward generalization of Beyn's work for single connecting orbits [Be].

To describe the method in its general context, consider any vector field $H \in C^2(\mathbb{R}^n, \mathbb{R}^n)$, $n \in \mathbf{N}$, with two hyperbolic zeros u^- and u^+. For the ODE

$$\dot{\phi} = H(\phi), \tag{25}$$

consider a non-empty family Φ of orbits connecting the rest points u^- and u^+:

$$\Phi = \{\phi \,|\, \dot{\phi} = H(\phi) \text{ and } \phi(\pm\infty) = u^\pm\}.$$

Furthermore, we assume that the intersection of the unstable manifold of u^- and the stable manifold of u^+, given by $\{\phi(x) \,|\, \phi \in \Phi, x \in (-\infty, \infty)\}$, is a smooth manifold of dimension d for some $d \in \{1, \ldots, m\}$. In order to parametrize Φ define a mapping

$$\Omega : \Phi \to \mathbb{R}^n$$

by

$$\Omega(\phi) \equiv \int_{\mathbb{R}} A(x, \phi(x))(\phi(x) - \phi_*(x))dx,$$

with some appropriate function $A : \mathbb{R} \times \mathbb{R}^n \to \mathbb{R}^{n\times n}$ and ϕ_* either an element of Φ or given by

$$\phi_* = \begin{cases} u^- : x < 0, \\ u^+ : x > 0. \end{cases}$$

Note that, in the case $A = \mathrm{Id}$, $\Omega(\phi)$ is the relative mass of ϕ with respect to the reference object ϕ_*, a quantity with a particular natural meaning in the case of viscous profiles. The subsequent assumption means that Ω is a chart of Φ.

Assumption 5. *The mapping Ω is injective and the range $S = \Omega(\Phi)$ is a d–dimensional manifold in $\mathbb{R}^n$ allowing for a global chart $\mathbf{P} : S \to T \equiv \mathbf{P}(S) \subset \mathbb{R}^d$.*
The corresponding parameterization of Φ as $\{\phi^\tau\}_{\tau \in T}$ with ϕ^τ defined by

$$\mathbf{P}\Omega(\phi^\tau) = \tau, \tau \in T,$$

is differentiable.

For a detailed discussion of the parametrization by relative masses and the validity of Assumption 5, in particular for conservation laws, we refer to [FreR1]. Let us note that in this field the validity of Assumption 5 is a necessary condition for time–asymptotic stability (in a certain well–defined sense) of Φ as a solution of the associated PDE. Cf. partly also Section 7 of this paper.
By Assumption 5 the problem

$$\dot{\phi}^\tau = H(\phi^\tau), \quad \phi^\tau(\mp\infty) = u^\mp, \quad \mathbf{P}\int_{\mathbb{R}} \left(\phi^\tau_{\cdot} - \phi_*\right) = \tau, \tag{26}$$

has a unique solution $\phi^\tau \in C^1(\mathbb{R})$, for $\tau \in T$.
Following the work of Beyn [Be] we restrict the problem (26) to a bounded interval $I = [X_-^\tau, X_+^\tau], X_-^\tau < 0 < X_+^\tau$. The approximate solution $\phi_I^\tau \in C^1(I)$ then is supposed to fulfil

$$\dot{\phi}_I^\tau = H(\phi_I^\tau) \text{ in } I, \quad b_\mp(\phi_I^\tau(X_\mp^\tau)) = 0, \quad \mathbf{P}\int_I (\phi_I^\tau - \phi_*) = \tau. \tag{27}$$

Here the functions $b_\mp$ denote asymptotic boundary conditions, for example the spectral projections associated with the unstable/stable part of the spectrum of $DH(u^\mp)$.
Following the analysis of Beyn, as presented in [Be], it is possible to derive a rigorous convergence estimate for the error $\|\phi^\tau - \phi_I^\tau\|_{C^1(I)}$ if $|X_-^\tau|, X_+^\tau$ tend to ∞. For this sake let us assume that the d–parameter family Φ is nondegenerate in the following sense: The number $d+n$ (d dimension of the heteroclinic manifold) is given by the sum of the dimensions of the unstable subspace of $DH(u^-)$ and the stable subspace of $DH(u^+)$. Furthermore for each $\tau \in T$ we have

$$\dot{y} = DH(\phi^\tau)y, \ y(\mp\infty) = 0 \Leftrightarrow y \in \text{span}\left\{\frac{\partial\phi^\tau}{\partial\tau_1}, \dots, \frac{\partial\phi^\tau}{\partial\tau_d}\right\}.$$

Under these assumptions (and some technical requirements on $b_\mp$) we can prove

Theorem 6. *For each $\tau \in T$ there is a $\tilde{X}^\tau > 0$ such that for any $I = [X_-^\tau, X_+^\tau]$ with $|X_-^\tau|, X_+^\tau > \tilde{X}^\tau$ we have:*

(i) There is a $\delta > 0$ such that there exists a unique solution $\phi_I^\tau \in C^1(I)$ of the truncated problem (27) with $\|\phi_I^\tau - \phi^\tau\|_{C^1(I)} \le \delta$.
(ii) There is a constant $C = C(\tau) > 0$ such that:

$$\|\phi_I^\tau - \phi^\tau\|_{C^1(I)} \le C|I| \exp\left(-\min\{\lambda^- X_-^\tau, -\lambda^+ X_+^\tau\}\right), \qquad (28)$$

where λ^- (λ^+) are given by the minimal absolute value of the real parts of the unstable (stable) eigenvalues of DH at u^- (u^+).

For a detailed proof we refer to Section 4 in [FreR2]. Note that it cannot be expected that Theorem 6, in particular (28), holds uniformly for all $\tau \in T$. This issue will be further discussed in Section 6 below.

5 Numerical Study of the Heteroclinic Bifurcation in MHD

In this section we report on systematic investigations into the MHD profiles ODE system $\boldsymbol{\Sigma}^6$ using the method described in Section 4. The results illustrate dynamically interesting scenarios, in particular in regimes that could so far not be, and seem hard to be, covered analytically.
We consider the global bifurcation scenario of $\boldsymbol{\Sigma}^6$ that has been described in Section 3. While the validity of this conjecture is only proven for small values of μ and κ, numerical results that we will present in this section support the conjecture that the scenario remains *globally* true, i. e., for all $\kappa, \mu > 0$ and $(j, e) \in I_c, c \in (0, \bar{c})$. In [FreR1] we presented two methods to decide whether the global bifurcation takes place or not. We will not go into detail but mention that the methods rely on the refined conjecture that the bifurcation can be completely analyzed in an four–dimensional linear subspace E that is invariant with respect to the flow of $\boldsymbol{\Sigma}^6$. Figures 5 show some results: the bifurcation ratio ω^* for fixed $\mu = 0.01$ and different values of κ in the left picture, the bifurcation ratio ω^* for fixed $\kappa = 1$ and different values of μ in

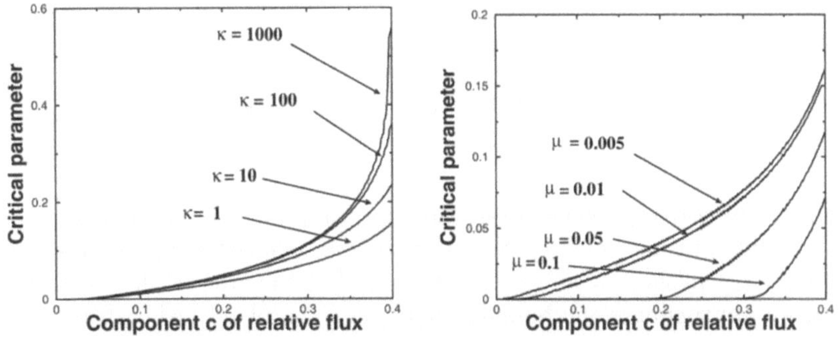

Fig. 5. Critical parameter ω^* versus c

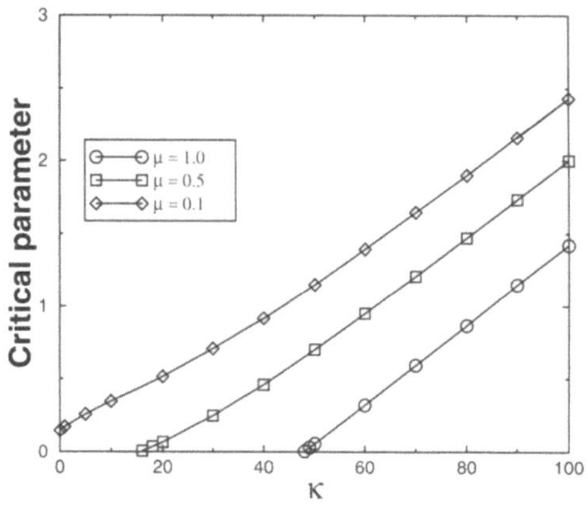

Fig. 6. Critical parameter ω^* versus κ.

the right picture. ω^* was calculated for a series of values for c, j, e such that $c \in (0, \bar{c})$ and $(j_c, e_c) \in I_c$.

We observe that ω^* vanishes for $c \to 0$ which coincides with the fact that the counterparts of the intermediate shock waves in the degenerate case $c = 0$ — the switch–on/off shock waves — have profiles for all values of δ and $(j, e) \in I_0$ (cf. Section 6 for the orbit structure in the case $c = 0$). However, it is not true, that the bifurcation parameter is uniformly bounded from above for all $\mu, \kappa > 0$ and all $c \in (0, \bar{c}), (j, e) \in I_c$ as certain partial earlier results of Wu in [W] may suggest. Figure 6 shows that the bifurcation ratio ω^* tends to ∞ as the heat conductivity κ tends to ∞, for $\mu > 0$ and q fixed.

Now we illustrate the bifurcation by a series of computations with the method described in Section 4.

Before starting let us mention some details of the implementation. The truncated problem (27) can be solved with any kind of BVP–solver, in principle. We actually use the code COLNEW [BaA] which relies on a variable step–size collocation method.

Concerning the approximation of the higher dimensional heteroclinic manifolds in $\boldsymbol{\Sigma}^6$ we will focus on the manifolds of type $u_0 \to u_2$ and $u_1 \to u_3$ and proceed as follows. Define the set T in (27) by

$$T = \{(0,0,0,\tau_1,\tau_2,0) \mid \tau_1 \in I\!R, \tau_2 \in (-\bar{\tau},\bar{\tau})\}, \quad \bar{\tau} \equiv \bar{\tau}(q,\delta) \equiv \Big|\int_{I\!R} b_2(x)\,dx\Big|. \tag{29}$$

Here the function b_2 refers to the (already computed) b_2–component of one of the orbits of type $u_1 \to u_2$. Note that τ_2 is associated to the component b_2 and that b_2 vanishes for all rest points such that the integral in (29) is finite. Now, we approximate the bounded manifolds completely when freezing the first τ–component τ_1, lets say $\tau_1 = 0$, and continuing in the parameter τ_2 starting with $\tau_2 = 0$.

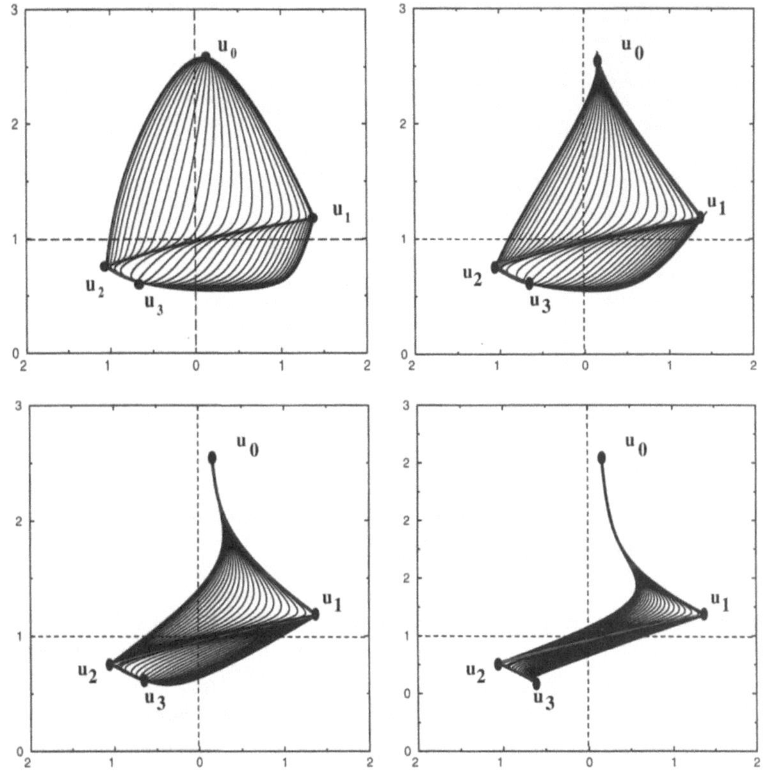

Fig. 7. Projection to b_1v–plane for $\omega = 7.5, 1.0, 0.25, 0.06$.

For the above–mentioned illustration, we fix the transverse fluid viscosity μ, heat conductivity κ and some $c \in (0, \bar{c})$, $(j, e) \in I_c$, to be specific:

$$\mu = 0.01, \quad \kappa = 1, \quad c = 0.25, \quad (j, e) = (2.68, 4.23). \tag{30}$$

By variation of the remaining free parameter, the ratio $\omega = \nu/\zeta$, we observe the global bifurcation. The numerically calculated orbits of all types except $u_0 \to u_3$ are displayed in Figure 7 as projections to the b_1v–plane. We picked out the configurations for $\omega = 7.5, 1.0, 0.25, 0.06$. For the 3D–version in b_1b_2v–space (cf. Plate 15), the colour of the surfaces which have been constructed from the orbit families of type $u_0 \to u_2$ and $u_1 \to u_3$ represent the values of the temperature θ on the corresponding heteroclinic manifold. The figures have been processed using the visualization platform GRAPE [g]. An MPEG–video showing the dynamics in more detail can be found at [w].
For the chosen set of parameters the critical value ω^* is approximately 0.0492. The graphs in Figure 8 display the situation for $\omega = \omega^*$ where only the single orbits $u_0 \to u_1$, $u_2 \to u_3$, and $u_1 \to u_2$ exist and $\omega = 0.02 < \omega_*$ where also $u_1 \to u_2$ is broken.

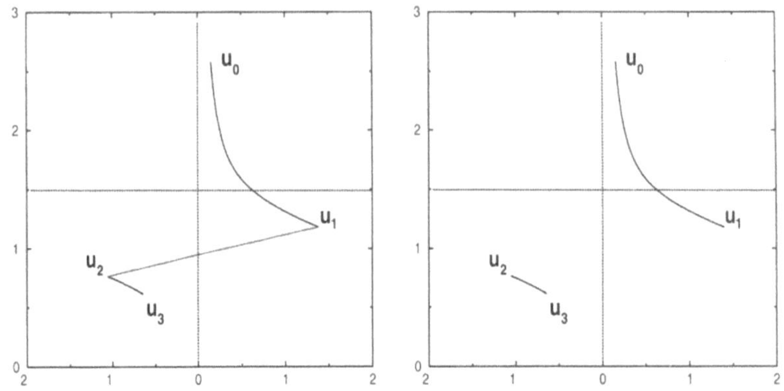

Fig. 8. Projection to b_1v–plane for $\omega = 0.0492, 0.02$.

6 Boundary Cases: (Almost-) Symmetry and Fast-Slow Dynamics

We now discuss an important special case, along with situations where our method, though reliable and robust, reaches its limitations. If the component c of the relative flux q is strictly bigger than zero the system $\boldsymbol{\Sigma}^{\mathbf{6}}$ has up to four isolated rest points $u_0, \ldots, u_3$ located in the invariant subspace E. For $c = 0$, the rest points u_0, u_3 persist in E while u_1, u_2 degenerate to a circle of rest points (Alfvén circle, cf. Section 3). In particular, for $c = 0$ the solution set of $\boldsymbol{\Sigma}^{\mathbf{6}}$ in the b_1b_2v–space is rotationally symmetric with respect to the v–axis.

In a series of pictures (Figures 9, 10) we show the orbit types $u_0 \to u_1, u_0 \to u_2, u_1 \to u_2, u_1 \to u_3, u_2 \to u_3$ for different values of c. We fixed $\nu = 0.5$, $\zeta = 1$ and μ, κ, j, e as in (30). For $c \to 0$, the phase portrait comes closer and closer to being symmetric with respect to the $b = 0$ axis, obviously.

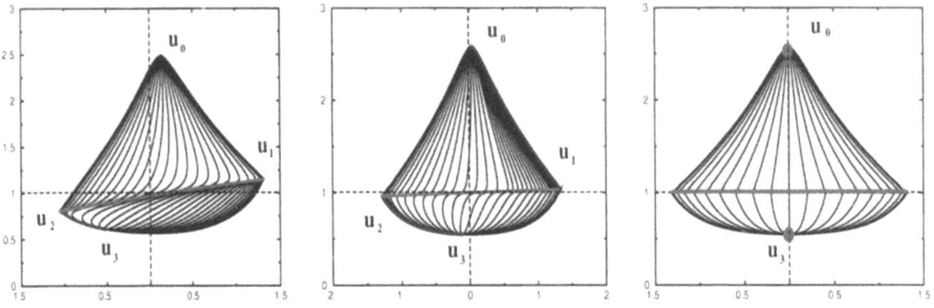

Fig. 9. Viscous profiles of type $u_0 \to u_1$, $u_0 \to u_2$, $u_1 \to u_2$, $u_2 \to u_3$ and $u_1 \to u_3$ for $c = 0.25, 0.05$ and viscous profiles connecting u_0/u_3 with the Alfvén circle for $c = 0$, projected to the b_1v–plane.

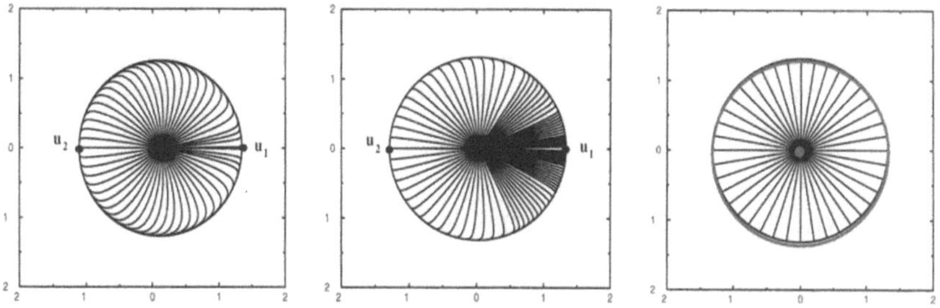

Fig. 10. The heteroclinic manifold $u_0 \to u_2$ for $c = 0.25, 0.05$ and viscous profiles connecting u_0 with the Alfvén circle for $c = 0$, projected to the $b_1 b_2$–plane.

For $c = 0$, the rest points on the Alfvén circle are not hyperbolic, but the connecting orbits reach it along the exponentially stable part of the center–stable manifold. We now comment on a numerical problem, arising at two levels. Firstly, for fixed q, the approximation of the heteroclinic manifolds of higher dimension, say $u_1 \to u_3$, is done by a continuation precedure starting with the orbit lying in the linear subspace E (for parameter $\tau = (0, \tau_2) = (0, 0)$) and ending with some orbit close to the boundary of the heterclinic manifold ($\tau = (0, \bar{\tau})$). During this continuation process, the truncation interval $[X_-^\tau, X_+^\tau]$ is enlarged, indeed exponentially, to capture the orbits in a satisfying manner. This is due to the fact that the orbits near the boundary of the manifold are closer and closer to the further rest point u_2.
Secondly, if we consider the intervals $[X_-^\tau, X_+^\tau]$, $\tau_2 \in [0, \bar{\tau})$ for decreasing c, we observe that the interval length grows even "more". This happens since in the limit $c \to 0$, the orbits of type $u_1 \to u_2$ become slower and slower, before in the limit case $c = 0$ (the closure of) their union degenerates to the Alfvén circle.
We exemplify these effects by the results of a numerical experiment, illustrated in Figure 11. We have plotted, against τ_2, the minimal intervals

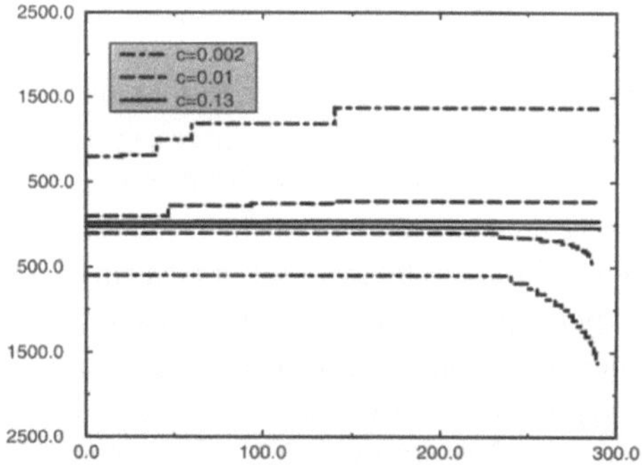

Fig. 11. Interval end points versus the continuation parameter τ_2.

$[X_-^\tau, X_+^\tau]$, such that the criterion $|\phi_\tau(X_\mp) - u_\mp| < \varepsilon$ is satisfied, for a given small tolerance $\varepsilon > 0$. As τ_2 increases and c decreases, these intervals grow to an extent which indicates that the method needs a refinement if one wishes to resolve these regimes more efficiently.

7 Stability of Viscous Shock Waves

This section discusses an analytical result on the stability of small-amplitude viscous Laxian shock waves associated with possibly nonconvex modes.

Theorem 7. *Consider a system of viscous hyperbolic conservation laws*

$$u_t + f(u)_x = \varepsilon u_{xx}, \tag{31}$$

$f \in C^3(I\!R^n, I\!R^n)$, $\varepsilon > 0$. *Let* $u_* \in I\!R^n$ *be a fixed reference state and let* $\phi : I\!R \to I\!R^n$ *denote the profile of a Laxian shock wave, near* u_*, *associated with a simple eigenvalue* λ *of* f', *i.e., the states* $u^\pm$ *are close to* u_* *and satisfy* $\lambda(u^-) > s > \lambda(u^+)$.

Let X *be the completion of* $\{\bar{u} \in C_0^\infty(I\!R, I\!R^n), \int_{-\infty}^{\infty} \bar{u}(x)dx = 0\}$ *under the norm* $\|\bar{u}\| = \|\bar{u}\|_{L_1} + \|\bar{u}\|_{H^1}$.

There exist positive constants ϵ_0, β_0 *such that if* ϕ *satisfies* $|u^+ - u^-| < \epsilon_0$ *and* $u_0 - \phi \in X$ *satisfies* $\|u_0 - \phi\| < \beta_0$, *then the solution* $u(x,t)$ *to* (31) *with data* $u(\cdot, 0) = u_0$ *exists for all times* $t > 0$ *and has*

$$\lim_{t\to\infty} \sup_{x\in I\!R} |u(x,t) - \phi(x - st)| = 0. \tag{32}$$

Briefly speaking, profiles of small-amplitude shock waves are *time-asymptotically stable.*

The result described in Theorem 7 was obtained by Goodman [Go] under the additional assumption that the eigenvalue λ be *convex*, i.e.

$$r \cdot \nabla\lambda \neq 0 \text{ where } r = \ker(f' - \lambda I). \tag{33}$$

For the general case of possibly non-convex modes, it is due to Fries [Fri1].

The stability of shock profiles had been investigated by Il'in and Oleinik [IO] for the case of a scalar equation with strictly convex flux function. Sattinger [Sa] used spectral methods to prove a stability result for travelling wave solutions of general parabolic systems which implies stability of Laxian shock profiles under perturbations of exponential decay. Goodman's proof for systems used the energy method, an approach that was introduced into the context of viscous hyperbolic conservation laws by Goodman and independently by Matsumura and Nishihara [MN1]. The result of Goodman was extended by Liu [L2,L3] and by Szepessy and Xin [SzeX] to the case of non-zero mass perturbations—still using the assumption of strict convexity. In [L3] the energy method was no longer involved, which enabled the derivation of pointwise decay-estimates. The stability of shock profiles for the non-convex scalar equation was shown by Matsumura and Nishihara [MN2] via a weigthed energy method; Gardner, Jones and Kapitula [JGKp] used a spectral approach, while Freistühler and Serre [FreSe] employed contractivity of the semigroup to establish global L_1 stability.

Recently, spectral methods have been developed to investigate the instability and stability of travelling waves for large classes of systems [GaZ,ZH]. In contrast with their enormous importance both for qualitative insight as well as for computational access, these new methods have so far not been able to provide alternative proofs for facts like the one established in Theorem 7.

We return to Theorem 7 by remarking that it has recently been extended to the non-zero mass case [Fri2]; in this note, we stay however away from the subtleties of this much more complicated case. We outline the proof of the Theorem 7 in the version it has been stated above.

We consider the solution u of (31) with initial data u_0 "close" to the profile ϕ, about which latter we assume without loss of generality that $s = 0$. Subtracting, we have

$$(u-\phi)_t + (f(u) - f(\phi))_x - \varepsilon(u-\phi)_{xx} = 0$$

i.e.

$$(u-\phi)_t + (f'(\phi)(u-\phi))_x - \varepsilon(u-\phi)_{xx} + (Q(\phi, u-\phi))_x = 0.$$

where Q is given by the Taylor expansion $f(\phi + u - \phi) - f(\phi) = f'(\phi)(u - \phi) + Q(\phi, u-\phi)$. It is thus natural to diagonalize f'. For simplicity in presentation we assume that *all* eigenvalues of f' are simple. Thus we find smooth matrix valued functions L, R such that $Lf'R = \Lambda = \mathrm{diag}(\lambda_1, \dots, \lambda_n)$ where the eigenvalues λ_k are such that $\mathrm{sign}\lambda_k = \mathrm{sign}(k-p)$ for $k \neq p$ and $\lambda_p = \lambda - s$, i.e. $\lambda_p(u_-) > 0 > \lambda_p(u_+)$. Introducing the integrated variable $U(x,t) := \int_{-\infty}^{x} u(\xi, t) - \phi(\xi) d\xi$ we have $U(\cdot, 0) \in \mathrm{H}^2(\mathbb{R})$ and obtain the *integrated equation* for U

$$U_t + f'(\phi)U_x - \varepsilon U_{xx} + Q(\phi, U_x) = 0 \tag{34}$$

Changing to characteristic coordinates $V := LU$ we obtain a diagonalized version of (34):

$$\begin{aligned} V_t + \Lambda(\phi)V_x - \varepsilon V_{xx} + LQ(\phi, RV_x) \\ + \Lambda(\phi)LR_xV - 2\varepsilon LR_xV_x - \varepsilon LR_{xx}V = 0. \end{aligned}$$

Multiplying this equation by $V^\mathsf{T}(x,t)W(x)$, where $W = \mathrm{diag}(w_1, \dots, w_n)$ is a weight matrix yet to be defined, and integrating $\int_{-\infty}^{\infty} dx$, we arrive at

$$\int_{-\infty}^{\infty} \frac{1}{2}(V^\mathsf{T}WV)_t - \frac{1}{2}V^\mathsf{T}[W\Lambda + \varepsilon W_x]_x V + \varepsilon V_x W V_x + \textit{error terms} = 0, \tag{35}$$

where integration by parts was used.

Under the assumption of a weak shock and by an appropriate choice of the weights w_i (see below) it is possible to obtain $-\frac{1}{2}[W\Lambda+\varepsilon W_x]_x \geq c|\phi_x|\mathrm{I}$ and $W \geq \mathrm{I}$ and thus

$$\frac{1}{2}\frac{\partial}{\partial t}\|V\|_{\mathrm{L}_2}^2 + c|\phi_x|\|V\|_{\mathrm{L}_2}^2 + c\varepsilon\|\sqrt{W}V_x\|_{\mathrm{L}_2}^2 + \textit{error terms} = 0. \tag{36}$$

With the assumption of a weak shock (small $|u_+ - u_-|$) and small perturbations (small β_0) we can estimate the error terms against the others. Integrating $\int_0^T dt$ and returning to the original variable U gives

$$\|U(\cdot,T)\|_{\mathrm{L}_2}^2 + \int_0^T \|U_x(\cdot,t)\|_{\mathrm{L}_2}^2 dt \leq C\|U(\cdot,0)\|_{\mathrm{L}_2}^2, \tag{37}$$

where $T>0$ denotes any time until which U exists with $U(\cdot,t)$, $0\leq t\leq T$, lying in the appropriate spaces, and $C>0$ is a constant independent of T. With this estimate it is easy to obtain the same with H^2 in place of L_2: To do so we differentiate (34) once or twice with respect to x and multiply by U_x or U_{xx} respectively. Crude estimation of terms involving h and Q then gives

$$\|U_x(\cdot,T)\|_{\mathrm{L}_2}^2 + \int_0^T \|U_{xx}(\cdot,t)\|_{\mathrm{L}_2}^2 dt \leq C\Big(\|U_x(\cdot,0)\|_{\mathrm{L}_2}^2 + \int_0^T \|U_x(\cdot,t)\|_{\mathrm{L}_2}^2 dt\Big)$$

and

$$\|U_{xx}(\cdot,T)\|_{\mathrm{L}_2}^2 \leq C(\|U_{xx}(\cdot,0)\|_{\mathrm{L}_2}^2 + \int_0^T \|U_x(\cdot,t)\|_{\mathrm{L}_2}^2 + \|U_{xx}(\cdot,t)\|_{\mathrm{L}_2}^2 dt).$$

Combining this with the above we obtain

$$\|U(\cdot,T)\|_{\mathrm{H}^2}^2 + \int_0^T \|U_x(\cdot,t)\|_{\mathrm{L}_2}^2 dt \;\leq\; C\|U(\cdot,0)\|_{\mathrm{H}^2}^2.$$

Since (34) is a uniformly parabolic system, one has a short time existence result for U. Thus the above a-priori estimate gives global existence of U, thus of u, and finally $\lim_{t\to\infty}\sup_{x\in\mathbb{R}}|u(x,t)-\phi(x-st)| = \lim_{t\to\infty}\sup_{x\in\mathbb{R}}|U_x(x,t)| = 0$.

To summarize, the two main difficulties in the whole argumentation are the choice of the weights w_k to obtain $-\frac{1}{2}[W\Lambda+\varepsilon W_x]_x \geq c|\phi_x|\mathrm{I}$ and the estimate of the error terms (especially the coupling terms involving WLR_x) to pass from (35) via (36) to (37). As regards the weights, for each k family with $k\neq p$, the choice $w_k(x) = \exp(-\int_{-\infty}^x C|\phi_x|\mathrm{sign}\lambda_k)$ gives $-\frac{1}{2}[w_k\lambda_k + \varepsilon w_{k,x}]_x \geq c|\phi_x|$ if C is sufficiently large and the shock is sufficiently weak (i.e. $|u_+ - u_-|$ sufficiently small). In the strictly convex case now, the choice $w_p \equiv 1$ yields the same property for $k=p$, allowing at the same time for straightforward treatment of the error terms [Go].

For a non-convex mode, i.e. when $\lambda_p(\phi(x))$ is *not* decreasing in x, it is still possible to obtain $-\frac{1}{2}[w_p\lambda_p + \varepsilon w_{p,x}]_x \geq c|\phi_x|$ through appropriate choice of w_p. But in this case it is non-trivial to estimate the error terms. The difficulty is easily seen from the following heuristic investigation of some properties of the weight w_p: When restricting to the scalar case, a typical non-convex flux would be $f(u) = u^3$. In this case we see that $\varepsilon\phi_x = h(\phi)$ is of the order ϵ^3 ($\epsilon := |u_+ - u_-|$). Since λ_p is then of the order ϵ^2 we see from $-\frac{1}{2}[w_p\lambda_p + \varepsilon w_{p,x}] \sim c\int|\phi_x|dx \sim \epsilon$ that w_p is of the order ϵ^{-1}. The point (of the difficulty) now is that the weight retains this order when considering a coupled system. Since the weight w_p appears in the error terms, it is not immediately clear that they can be estimated for ϵ sufficiently small. So far for the strategy and the difficulties arising in the proof of Theorem 7. For any further details we must refer the reader to [Fri1,Fri2].

We conclude by remarking that non-convex modes occur naturally in physical systems. An example is again provided by magnetohydrodynamics. Restricting the twodimensonal variables **b**, **w**, i. e., the transverse components of the magnetic field and the velocity, in (19) to a fixed line—this corresponds to restricting the full magnetic field and velocity vectors to a fixed plane: the so-called coplanar case—, one arrives at the 5 by 5 system

$$
\begin{aligned}
\rho_t + (\rho v)_x &= 0 \\
(\rho v)_t + (\rho v^2 + p + \frac{1}{2}|b|^2)_x &= \zeta v_{xx} \\
(\rho w)_t + (\rho v w - ab)_x &= \mu w_{xx} \\
b_t + (vb - aw)_x &= \nu b_{xx} \\
\mathcal{E}_t + (v(\mathcal{E} + p + \frac{1}{2}(|b|^2 - a^2)) - awb)_x &= \kappa\theta_{xx} \\
&\quad + \zeta(vv_x)_x + \mu(ww_x)_x + \nu(bb_x)_x
\end{aligned}
\tag{38}
$$

The *coplanar system* (38) retains five of the seven modes (20) of (19). The rotational modes $\lambda_{-2}, \lambda_{+2}$ being absent in (38), points (with $b = 0$) that, considered as states of the full system, have eigenvalues of multiplicity 2 by $\lambda_{\pm 2}$ coinciding with either $\lambda_{\pm 1}$ or $\lambda_{\pm 3}$, are now, for (38), points of strict hyperbolicity, and small-amplitude intermediate shocks near these points are now Laxian shocks associated with the non-convex (!) simple modes $\lambda_{\pm 1}$ or $\lambda_{\pm 3}$.

Still, Theorem 7 does of course not readily apply, as it requires the "artificial" viscosity $B = \varepsilon I$. Also, despite the conclusive study of a model problem [FreL], it is not clear how to re-proceed from the nonconvex coplanar problem to the non-classical non-coplanar problem. Work on both issues is in progress.

References

[BaA] G. BADER, U. ASCHER: A new basis implementation for a mixed order boundary value ODE solver, *SIAM J. Sci. Stat. Comput.* **8**, 483-500 (1987).

[Be] W.-J. BEYN: The numerical computation of connecting orbits in dynamical systems, *IMA J. Numer. Anal.*, **9**, 379-405 (1990).

[CS] C. Conley, J. Smoller: On the structure of magnetohydrodynamic shock waves, *Commun. Pure Appl. Math.* **28** (1974), 367-375.

[Fre1] H. Freistühler: Small amplitude intermediate magnetohydrodynamic shock waves, *Physica Scripta* **T74** (1998), 26-29.

[Fre2] H. Freistühler: Viscous profiles for Laxian shock waves in Kawashima type systems, Preprint.

[FreL] H. Freistühler, T. P. Liu: Nonlinear stability of overcompressive shock waves in a rotationally invariant system of viscous conservation laws *Commun. Math. Phys.* **153** (1993), 147-158

[FreR1] H. Freistühler, C. Rohde: Numerical Methods for Viscous Profiles of Non-Classical Shock Waves, In: Hyperbolic problems: Theory, Numerics, Applications: Seventh International Conference in Zürich 1998/ ed. by Michael Fey; Rolf Jeltsch (1999).

[FreR2] H. Freistühler, C. Rohde:, Numerical Computation of Viscous Profiles for Hyperbolic Conservation Laws, Preprint 1999, submitted to *Math. Comput.*.

[FreR3] H. Freistühler, C. Rohde:, A bifurcation analysis of the MHD Rankine-Hugoniot relations for a perfect gas, in preparation.

[FreSe] H. Freistühler, D. Serre: L^1 stability of shock waves in scalar viscous conservation laws, *Commun. Pure Appl. Math.* **51** (1998), 291-301.

[FreSzm] H. Freistühler, P. Szmolyan: Existence and bifurcation of viscous profiles for all intermediate magnetohydrodynamic shock waves, *SIAM J. Math. Anal.* **26** (1995), 112-128.

[Fri1] C. Fries: Nonlinear asymptotic stability of general small-amplitude viscous Laxian shock waves, *J. Differ. Equations* **146** (1998), 185-202.

[Fri2] C. Fries: Stability of viscous shock waves associated with non-convex modes, to appear in *Arch. Rational Mech. Anal.*

[GaZ] R.A. Gardner, K. Zumbrun: The gap lemma and geometric criteria for instability of viscous shock profiles, to appear in *Commun. Pure Appl. Math.* **51** (1998), 797-855.

[Ge] P. Germain: Contribution à la théorie des ondes de choc en magnétodynamique des fluides *Off. Nat. Etud. Aéronaut.*, Publ. **97** (1959).

[Gi] D. Gilbarg: The existence and limit behavior of the one-dimensional shock layer, *Amer. J. Math.* **73** (1951), 256-274.

[Go] J. Goodman: Nonlinear asymptotic stability of viscous shock profiles for conservation laws, *Arch. Rational Mech. Anal.* **95** (1986), 325-344.

[g] SFB 256, University of Bonn and Institut für Angewandte Mathematik, University of Freiburg. GRAPE: GRAphics Programming Environment for Mathematical Problems: http://www.mathematik.uni-freiburg.de-/IAM/Research/grape/GENERAL/index.html.

[IO] A.M. Il'In, O.A. Oleinik: Behaviour of the solutions of the Cauchy problem for certain quasilinear equations for unbounded increase of time, *Amer. Math. Soc. Translations* **42** (1964), 19-23.

[JGKp] C.K.R.T. Jones, R. Gardner, T. Kapitula: Stability of travelling waves for non-convex scalar viscous conservation laws, *Commun. Pure Appl. Math.* **46** (1993), 505-526.

[Kw] S. Kawashima: Systems of a hyperbolic-parabolic composite type, with applications to the equations of magnetohydrodynamics, Thesis Kyoto University 1983.

[KwM] S. KAWASHIMA, A. MATSUMURA: Stability of shock profiles in viscoelasticity with non-convex constitutive relations, *Commun. Pure Appl. Math.* **47** (1994), 1547-1569.

[KuLi] A. KULIKOVSKII, G. LIUBIMOV: Magnetohydrodynamics, Addison-Wesley: Reading 1965.

[L1] T.-P. Liu: The Riemann problem for general systems of conservation laws, *J. Differential Equations* **18** (1975), 218-234.

[L2] T.-P. LIU: Nonlinear stability of shock waves for viscous conservation laws, *Mem. Am. Math. Soc.* **328**, Providence: AMS 1985.

[L3] T.-P. LIU: Pointwise convergence to shock waves for viscous conservations laws, *Commun. Pure Appl. Math.* **50** (1997), 1113-1182.

[MN1] A. MATSUMURA, K. NISHIHARA: On the stability of travelling wave solutions of a one-dimensional model system for compressible viscous gas, *Japan J. Appl. Math.* **2** (1985), 17-25.

[MN2] A. MATSUMURA, K. NISHIHARA: Asymptotic stability of traveling waves for scalar viscous conservation laws with non-convex nonlinearity, *Commun. Math. Phys.* **165** (1994), 83-96.

[MP] A. MAJDA, R. PEGO: Stable viscosity matrices for systems of conservation laws, *J. Diff. Equations* **56** (1985), 229-262.

[Sa] D.H. SATTINGER: On the stability of waves of nonlinear parabolic systems, *Advances in Math.* **22** (1976), 312-355.

[SzeX] A. SZEPESSY, Z. XIN: Nonlinear stability of viscous shock waves, *Arch. Rational Mech. Anal.* **122** (1993), 53-103.

[w] http://www.mathematik.uni-freiburg.de/IAM/Research/projectskr/mhd/mhd.html

[W] C.C. WU: Formation, Structure, and Stability of MHD Intermediate Shocks, *J. Geophys. Research* **95** (1990), A6 8149-8175.

[ZH] K. ZUMBRUN, P. HOWARD: Pointwise semigroup methods and stability of viscous shock waves, *Indiana U. Math. J.* **47** (1998), 63-85.

Dynamical Systems of Population Dynamics

K.P. Hadeler and Johannes Müller*

Lehrstuhl für Biomathematik, Universität Tübingen, Auf der Morgenstelle 10, 72076 Tübingen, Germany

Abstract. After a short review of the the role of population dynamics in the formation of the theory of dynamical systems some recent developments are described, centered around work within the DANSE project: Reaction transport equations which are refinements of reaction diffusion equations, showing travelling front solutions and pattern formation, with applications to chemotaxis and the formation of polarized groups of animals or cells. Populations structured by age are modeled by conservation laws and by delay equations, the thorough investigation of the close connection between these two classes of systems leads to a solid justification of the latter as population models. Particular attention is given to models for the spread of infectious diseases: Models describing outbreaks in closed populations, optimal vaccination and control policies, the conditions for backward bifurcations and (unwanted) hysteresis phenomena in public health. Epidemic spread in space (travelling front problems), simplification of complex models to modulation equations. Modeling the behavior of single individuals connects dynamical systems to stochastic processes: Synchronous and asynchronous cellular automata, interacting particle systems, concrete applications are ring vaccination (in animal husbandry) and contact tracing (in human populations).

1 Introduction

Population dynamics, as a part of biology, is concerned with the numbers of plants and animals, with changes in population density resulting from interactions with the environment and between individuals of the same species and between species. These changes are expressed in terms of reproduction, mortality, and migration. Population dynamics is closely connected to ecology, population genetics, human demography, the theory of infectious diseases (epidemiology), parasitology, and other fields like bioeconomics. The goal of population dynamics is to describe these changes, based on field observations and experimental data, and to understand or even predict the evolution of biological systems. In population dynamics the use of mathematical models is essential.

Among all fields of biology, especially population dynamics has provided mathematical research with interesting problems in the form of dynamical systems. These are quite different from those supplied by physics and engineering where concepts like "energy" and "symmetry" play a dominant role whereas

---

* Project: Dynamical Systems of Population Dynamics, Reduction and Discretization (Karl Peter Hadeler)

"mass action kinetics" is the underlying mechanism of most biological interactions. We recall some dynamical systems approaches to population dynamics. The dynamical systems of population genetics were introduced around 1924/30 (the Fisher-Wright-Haldane model and the "fundamental law of population genetics" [9]), evolutionary game dynamics by Maynard Smith and others 1972 ff. [31], followed by adaptive dynamics in evolutionary landscapes, reaction diffusion equations for the advance of traits and population fronts were invented around 1937 (Fisher and Kolmogorov-Petrovskij-Piscunov), oscillations in interacting populations (Volterra 1926, Kolmogorov 1936), nonlinear delay equations (Hutchinson's model 1948), hyperbolic systems and renewal equations for structured populations (Sharpe-Lotka 1911, Feller 1941, McKendrick 1936); cooperative and competitive systems emerged in mathematics (M.Müller 1926, Kamke 1932), gained recognition in population biology and were systematically explored by M.Hirsch and others 1982 ff.). Modelling infectious diseases started before 1900, the first dynamical system appeared around 1927 (Kermack and McKendrick [33]). Population dynamics modeling by deterministic dynamical systems has been paralleled by stochastic approaches, e.g. in the form of birth and death processes, branching processes, stochastic epidemic models. An intermediate class of models are (deterministic and stochastic) cellular automata models and interacting particle systems [7] [5].

In the following we shall describe recent developments in the field with some bias towards problems that have been studied within the DANSE project. We hope that we nevertheless put the results in right proportions.

All topics that we have just mentioned play a major role in present day research. Here we shall report on research on spatial spread, in particular on detailed description of individual motion by reaction transport equations, on the mathematical description of aggregations of microorganisms (chemotaxis) and of schools, flocks, and herds of animals. We show how structured population models in the form of hyperbolic partial differential equations can be reduced to differential delay equations. A major topic is the study of dynamical systems modelling infectious diseases, the spread of diseases in populations and in geographical regions, vaccination and public health policies, and containment strategies based on studying individual contact patterns. Finally we report on totally discrete dynamical systems such as cellular automata, edge processes and interacting particle systems. We mention in passing results that the Tübingen DANSE group has obtained results on dynamical systems not related to population dynamics, e.g. [13] [23].

2 Spread in Space and Reactions

2.1 Reaction Transport Equations

The classical model for reaction and spatial spread is the reaction diffusion equation

$$u_t = D\Delta u + f(u). \tag{1}$$

Here $u(t,x)$ is the density of the population at the point x at time t. The ordinary differential equation $\dot{u} = f(u)$ describes the underlying reaction, and spatial spread is modeled by the diffusion equation $u_t = D\Delta u$. Reaction diffusion equations are one of the most widely used classes of dynamical systems in biology. Vector-valued reaction diffusion equations govern interactions of ecological or chemical species, propagation of nerve excitation, epidemic spread, etc. [51] [58] [49]. However, in the diffusion equation all characteristic features of moving particles are lumped into a single parameter, the diffusion coefficient D. Furthermore, the diffusion equation (and Brownian motion) show the unrealistic effect of infinitely fast transport, and the reaction diffusion equation shows the same effect. A more detailed description, where individual particles are characterized by their position in space and their individual velocity, leads to reaction transport equations. Let $u(t,x,s)$ be the density of particles at position $x \in \mathbb{R}^n$ and time t, with velocity $s \in V \subset \mathbb{R}^n$. Here V is some bounded domain, typically a ball or a sphere (in the case of constant speed). Then the linear transport equation reads

$$u_t + s \cdot \nabla_x u + \mu u = \mu \int_V K(s,\tilde{s}) u(t,x,\tilde{s}) d\tilde{s}. \tag{2}$$

Here μ is the rate at which turning events occur (changes of velocity) and $K(s,\tilde{s}) \geq 0$, with $\int_V K(s,\tilde{s}) ds = 1$, is the probability distribution for the new velocity. For $\mu = 0$ equation (2) is mere book-keeping: Each particle moves on a straight line with its own velocity. For $\mu > 0$ particles move on straight lines, stop, and choose a new direction. There are various ways in which birth and death processes can be incorporated into this system. We split the nonlinearity in the equation $\dot{u} = f(u)$ into a birth term $m(u)u$ and a death term $g(u)u$ [15] [16] [17]. Then $f(u) = m(u)u - g(u)u$. With this splitting the corresponding reaction transport equation reads

$$u_t + s \cdot \nabla_x u + \mu u = \mu \int_V K(s,\tilde{s}) u(t,x,\tilde{s}) d\tilde{s} + \frac{1}{|V|} m(\bar{u})\bar{u} - g(\bar{u})u \tag{3}$$

where

$$\bar{u}(t,x) = \int_V u(t,x,s) ds \tag{4}$$

is the population density at position x at time t and $|V|$ is the volume of V.

We consider equation (3) as the standard system for reaction and transport. The simplified system where birth and death are incorporated in one function $f(\bar{u})$ we call the "isotropic" system,

$$u_t + s \cdot \nabla_x u + \mu u = \mu \int_V K(s,\tilde{s}) u(t,x,\tilde{s}) d\tilde{s} + \frac{1}{|V|} f(\bar{u}). \tag{5}$$

A linear hyperbolic system has been introduced by Cattaneo 1948 as another system which lacks the deficiencies of the heat equation. Nonlinear Cattaneo systems can be seen as intermediates between reaction diffusion systems and "full" reaction transport equations. A nonlinear Cattaneo system has the form of a conservation law with source terms together with an equation coupling the flow to the gradient

$$\begin{aligned} u_t + \operatorname{div} v &= f(u) \\ \tau v_t + D \operatorname{grad} v + h(u)v &= 0. \end{aligned} \tag{6}$$

In the limiting case $\tau = 0$ we can solve for v in the second equation and replace v in the first equation to obtain a diffusion equation.

If the space dimension is $n = 1$ and there are only two velocities $s = \pm\gamma$ then we can write $u^{\pm}(t, x)$ instead of $u(t, x, \pm\gamma)$,

$$\begin{aligned} u_t^+ + \gamma u_x^+ &= \tfrac{1}{2}\mu(u^- - u^+) + m(u)u - g(u)u^+ \\ u_t^- - \gamma u_x^- &= \tfrac{1}{2}\mu(u^+ - u^-) + m(u)u - g(u)u^-. \end{aligned} \tag{7}$$

Equation (7), for $m = g = 0$, describes a correlated random walk. We introduce the total particle density $u = u^+ + u^-$ and the flow $v = \gamma(u^+ - u^-)$ to obtain a system of the form (6) with $\tau = 1/\mu$ and $D = \gamma^2/\mu$,

$$\begin{aligned} u_t + v_x &= f(u) \\ \tau v_t + D u_x &= -(\mu + g(u))v. \end{aligned} \tag{8}$$

Hence, in the special case of dimension 1 and constant speed, the transport equation, the correlated random walk, and the Cattaneo system are equivalent. There is no such result in dimension $n \geq 2$.

If $h(u) \equiv 1$ then we can eliminate the flow v in system (6) (the "Kac trick") and arrive at a damped wave equation or telegraph equation for the function u

$$\tau u_{tt} + (1 - \tau f'(u))u_t = D\Delta u + f(u). \tag{9}$$

For $\tau = 0$ one gets the diffusion approximation (1). Similar systems can be derived in the case of several dependent variables.

The main results obtained so far concern convergence to equilibrium (in bounded domains) [28] [55] [57], existence and stability of travelling fronts [17] [56], and Turing patterns (see below). Furthermore, approximation and scaling properties have been investigated. If time and space or, equivalently, turning rate and velocity, are appropriately scaled, then the random walk equation approximates a parabolic equation. Approximation properties have been shown for single trajectories as well as for the attractor [46] [55] [30].

2.2 Travelling Front and Waves, Emerging Patterns

Reaction diffusion equations, reaction transport equations and related integral equation models show typical bifurcation phenomena and asymptotic behavior. One of the basic phenomena is the travelling front which has been discovered as a model for a genetic trait advancing in a population by Fisher

1936. Some authors suggest that most phenomena in reaction diffusion systems can be explained in terms of interacting travelling fronts [10]. Consider a scalar reaction diffusion equation (1) in the plane with $f(u) = u(1-u)$ (often called the "Fisher equation") in the plane, with an initial datum that is positive on some compact domain and zero outside. Near the boundary of the support matter will spread outward and then increase due to the reaction term. Hence the solution will converge to 1 uniformly on compact sets. But globally the solution behaves like a circular wave. There is a critical number c_0 such that, for any unit vector e and $t \to \infty$, $u(t, x + cet) \to 0$ for $c > c_0$ and $u(t, x + cet) \to 1$ for $c < c_0$. Asymptotically the solution behaves, in a given direction, like a planar front, i.e., like a solution $u(t, x) = u(x - c_0 t)$ to $u_t = Du_{xx} + u(1-u)$ with $u(t, -\infty) = 1$, $u(t, +\infty) = 0$. Travelling front solutions have been found in many population dynamic problems. Scalar fronts describe the propagation of a single species, vector-valued fronts may describe, e.g. the joint propagation of a prey and a predator [6] [39]. In the present context the existence of traveling fronts has been shown for the correlated random walk models (7) in [17] and for the general reaction transport equation in [56].

The typical behavior of scalar reaction diffusion equations or transport equations (coefficients and source term not depending on space or time, no drift terms) in bounded domains is convergence to equilibrium. Vector-valued problems show much richer behavior. The classical example is the Turing instability. In Turing's original work the phenomenon is exhibited as a possible mechanism for morphogenesis, and it has been used to explain all kinds of coat patterns in mammals and reptiles, wing patterns in butterflies and patterns on sea shells. While some of these explanations are still lacking sufficient connection to the underlying biochemistry and genetics, it is without doubt that Turing patterns occur in chemical experiments like the Belousov-Zhabotinsky reaction and the oxygen-carbon monoxid reaction on platinum surfaces. It can be assumed that many regular patterns into which plants and animals (e.g. corals) arrange themselves are based on the Turing mechanism.

In the simplest form of the Turing model there is a system of the form (1) with two reactants, the activator u_1 and the inhibitor u_2 which coexist at a stable equilibrium under well-stirred conditions (or, which amounts to the same, in very small domains). In large domains where differential diffusion plays a role, the homogeneous state may lose its stability. Indeed, let $\bar{u}$ be the stationary state, $A = f'(\bar{u}) = (a_{jk})$ the Jacobian (a matrix of order 2), and let $a_{11} > 0$ (activator enhancement), $a_{22} < 0$ (inhibitor decay), $a_{12} < 0$ (inhibitor inhibits activator), $a_{21} > 0$ (activator enhances inhibitor), $\operatorname{tr} A = a_{11} + a_{22} < 0$, $\det A = a_{11}a_{22} - a_{12}a_{21} > 0$ (stability). With these assumptions it is true that for sufficiently distinct diffusion rates $D_2/D_1 >> 1$ and suitably large domains some modes become unstable. For the reaction diffusion case the exact conditions for this instability are well known [49] [15]. In [27] the conditions for the corresponding random walk and Cattaneo problems have been determined.

2.3 Hyperbolic Models for Chemotaxis

One common feature of all living systems is, that they sense their environment and that they respond to it. Especially moving species (most animals, many microorganisms) have strategies to find food, to avoid harmful substances or enemies, and to find mates. In many cases (bacteria, amoebae, algae, insects, mammals) chemical signals are sent out to attract or repel other individuals, a well-known example is the slime mold *Dictyostelium discoideum.* Insects release pheromones to attract potential sexual partners. The mathematical description of these processes leads to interesting problems, like the interplay of diffusion and drift terms and questions of global existence or finite time blow-up. The classical parabolic model for chemotaxis is the Patlak-Keller-Segel model. We give a brief derivation here.

We assume that in absence of any external signal the spread of the population is governed by the diffusion equation that says that the net flux j is proportional to the negative gradient, $j = -D\nabla u$. It is assumed that an external signal produces a *chemotactic velocity* β resulting in a total flux $j = -D\nabla u + u\beta$ (D has the dimension sec/cm^2). As observed in many experiments, the chemotactic velocity is proportional to the gradient ∇S of some external signal $S(t,x)$, hence $\beta = \chi(S)\nabla S$ with some *chemotactic sensitivity* $\chi(S)$. With this modified flux we arrive at the parabolic chemotaxis equation

$$u_t = \nabla(D\nabla u - \chi(S)\nabla S u). \tag{10}$$

The case $\chi(S) > 0$ is called *positive taxis*, whereas $\chi < 0$ indicates *negative taxis*. Depending on the concrete species the external signal is produced and also perceived by the individuals themselves. We assume that the spatial spread of the external signal is driven by diffusion with constant $B > 0$. Then the full system for u and S reads

$$\begin{aligned} u_t &= \nabla(D\nabla u - \chi(S)\nabla S u) \\ \tau S_t &= B\Delta S + f(S,u), \end{aligned} \tag{11}$$

where $f(S,u)$ describes production and decay of the signal and $\tau \geq 0$ indicates that the spatial spread of the organisms u and of the signal S occur on different time scales. The model (11) has first been derived by Patlak [50] and Keller, Segel [34] hence Patlak-Keller-Segel (PKS) model.

This and related models have been discussed in detail in the mathematical literature (see [30]). Lyapunov functions are known for some examples but also scenarios with blow up in finite time. However, the criticism on parabolic systems mentioned earlier applies to (PKS) as well. In [30] [29] an alternative hyperbolic model has been derived and discussed. It is based on the correlated random walk (7) in one dimension and has the form

$$\begin{aligned} u_t^+ + (\gamma(S)u^+)_x &= -\mu^+(S,S_x)u^+ + \mu^-(S,S_x)u^- \\ u_t^- - (\gamma(S)u^-)_x &= \mu^+(S,S_x)u^+ - \mu^-(S,S_x)u^- \\ \tau S_t &= BS_{xx} + f(S,u^+ + u^-). \end{aligned} \tag{12}$$

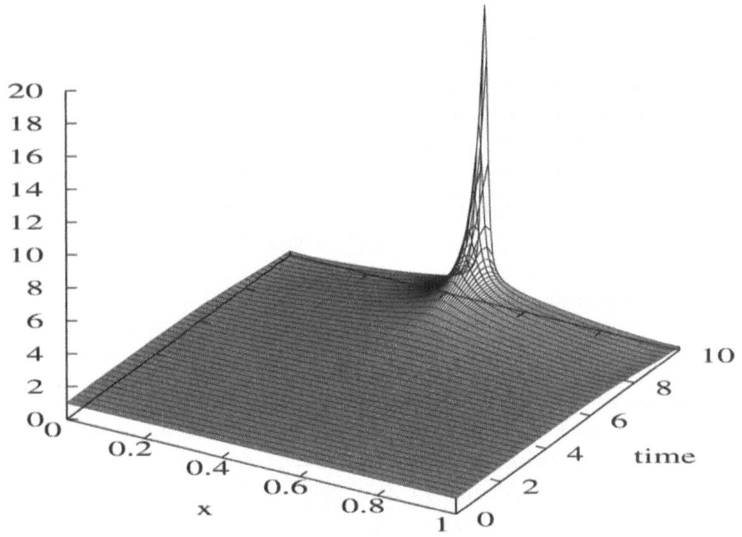

Fig. 1. Finite time blow-up for a hyperbolic chemotaxis model

The velocity $\gamma(S)$ and the turning rates $\mu^{\pm}(S, S_x)$ depend on S and its gradient. The qualitative behavior of (12) depends crucially on these functions. We have assumed linearity for S, i.e., $f(S, u) = -\beta S + \alpha u$ for some positive constants α, β.

In [30] we considered the case of constant speed γ. Under quite general assumptions we were able to prove local and global existence. It turned out that the control of the $W^{1,\infty}$-norm of the signal S is crucial for global existence. In numerical simulations we also found scenarios which show blow-up (at least numerically), see Fig.1.

In [29] the effect of non-constant velocity, $\gamma = \gamma(S)$, was investigated. For the limiting case $\tau = 0$ of (12) solutions in L^1_{loc} could be found using a vanishing viscosity method. In numerical simulations aggregation behavior and also formation of steep gradients (shocks) could be observed.

Moreover in [30] the formal relation of the general hyperbolic model (12) to the classical PKS-model has been discussed. This comparison gives an explicit relation of the microscopic parameters like turning rate μ and velocity γ to macroscopic parameters like motility D and chemotactic sensitivity χ. The microscopic parameters can be determined from experiments.

2.4 Alignment

Aggregation is a common phenomenon in many animal species. Group size and structure differ between the species and according to the situation. A striking example of individuals moving in groups are schools of fish or flocks of birds where all individuals move in the same direction. The process by which an individual adapts its direction of motion (body axis) to the one of its neighbor(s) is called alignment. It can also be observed on a microscopic level, e.g. in actin filaments.

Most detailed models for alignment in the form of discrete particle systems have been studied by simulation methods [8] [5]. In analytical models it is usually assumed that particles align but do not move, i.e., that spatial movement is slow compared to the alignment process (an exception is [40]). These models assume the form of partial integro-differential equations [8] [40] [12].

The basic ideas are as follows [12] (see also [32] for a more detailed exposition). Let $v \in V = [0, 2\pi]$ denote the angular direction and $u(t, v)$ the density of particles in direction v at time t. Particles interact with a certain rate $\eta = \eta(v_1, v_2) = \eta(v_1 - v_2)$ (interaction frequency) which depends on the (relative) orientation of the particles. Upon an interaction a particle chooses a new direction according to a function $\omega(v_0 - v_1, v_2 - v_1)$ which describes the probability that a particle in direction v_1 turns into direction v_0 after interacting with another particle with direction v_2. The master equation is a gain-loss equation for the different directions

$$\begin{aligned} \frac{\partial}{\partial t} u(t, v) &= -u(t, v) \int_V \eta(v - v_1) u(t, v_1) dv_1 \\ &\quad + \int_V \int_V \omega(v - v_1, v_2 - v_1) \eta(v_2 - v_1) u(t, v_2) u(t, v_1) dv_2 dv_1 \end{aligned} \tag{13}$$

where η, ω satisfy certain symmetry conditions. The situation where all directions are equally distributed is stationary and its stability depends on the choice of the interaction frequency and the turning probability. A detailed bifurcation analysis of a more general model is carried out in [11]. For special choices of η and ω a linear stability analysis is given in [12]. If turns to the new direction are precise enough and if the angular difference is thereby decreased then the homogeneous situation becomes unstable. Several computer simulations show a rich variety of emerging patterns. The situation of total alignment (i.e., u is a delta distribution in V) is shown to be stable.

Here a simplistic model for alignment and spatial movement has been chosen in the form of a reaction transport equation [36]. The reaction (orientation process) is modeled by comparing the velocity of a particle to the average velocity (at the point considered). We assume constant speed γ. Let $u(t, x, v)$ be the density of particles with $v \in V = \gamma S^1$ (orientation). The full transport equation reads

$$u_t(t, x, v) + v \cdot \nabla_x u(t, x, v) = \rho(\bar{u})[\bar{v} \cdot v - |\bar{v}|^2] u(t, x, v), \tag{14}$$

where $\bar{v}$ is the normalized first moment of u with respect to V, i.e. $\bar{v} = \int_V v u dv / (2\pi\gamma)$, ρ is a positive function describing the "force of alignment" depending on the total mass $\bar{u} = \int_V u dv$ and $\cdot$ is the scalar product (the turning parameter μ is chosen as zero to ease the presentation). One can generalize the model (13) slightly by allowing the turning probability ω to depend on the density [8]. Then for a special choice of η and ω model (14) is obtained. This derivation allows the following interpretation of the "macroscopic" equation (14) on the individual level: Particles interact most strongly if they meet in opposite directions and hardly ever if they are already aligned. If they turn then they choose the direction in which there are most other particles. This interpretation makes sense biologically and provides a mechanism for avoiding collisions.

In one space dimension with constant speed the model reads in its simplest form

$$\begin{aligned} u_t^+ + \gamma u_x^+ &= \frac{r}{u^3} u^+ u^- (u^+ - u^-) \\ u_t^- - \gamma u_x^- &= \frac{r}{u^3} u^+ u^- (u^- - u^+). \end{aligned} \tag{15}$$

Comparison with (7) (m=g=0) shows that the turning rates μ are now nonlinear and they change sign. For the general model in several space dimensions global existence of solutions has been shown and the limit sets have been characterized in one space dimension. The process of alignment can be studied in connection with simple birth and death processes. There are indeed situations where an aligning population survives and its non-aligning counterpart goes extinct [36].

3 Age Structure and Delays

The classical model for a population structured by age has been formulated by A.G. McKendrick in the form of a hyperbolic partial differential equation describing aging and mortality and a distributed boundary condition describing birth

$$u_t + u_a + \mu(a)u = 0, \quad u(t,0) = \int_0^\infty b(a)u(t,a)da \tag{16}$$

based on earlier work by Sharpe and Lotka. The operator $\partial/\partial a + \mu(a)$, with the given boundary condition, can be seen as the generator of a positive semigroup in $L_1(0,\infty)$. The important feature is the existence of an exponential or "persistent" solution $u(t,a) = \bar{u}(a)e^{\hat{\lambda}t}$ (under mild conditions on the coefficients μ and b). The number $\hat{\lambda}$ is the exponent of growth which is the unique solution of the characteristic equation $\int_0^\infty p(a)b(a)e^{-\lambda a}da = 1$ where $p(a) = \exp\{-\int_0^a \mu(s)ds\}$ is the survival function. Feller's renewal theorem says that most solutions approach the exponential solution for large times.

If the coefficients b and μ depend not only on age but also on population density, typically on a weighted mean $W(t) = \int_0^\infty w(a)u(t,a)da$, then we arrive at models of the Gurtin-MacCamy type

$$u_t + u_a + \mu(a,W)u = 0, \quad u(t,0) = \int_0^\infty b(a,W)uda. \tag{17}$$

With appropriate assumptions on the functions $\mu(a,W)$ and $b(a,W)$ (μ strictly increasing to ∞ as a function of W, and b strictly decreasing to 0; and a condition which ensures that the population grows at low densities such as $\int_0^\infty b(a,0)\exp\{-\int_0^a \mu(s,0)ds\}da > 1$) it follows that there is a unique nontrivial stationary state. In many situations this state describes a stable population equilibrium. However, in general, this state is not stable, and it is rather difficult to establish criteria for stability (Hopf bifurcations can occur).

On the other hand, population dynamics has been modeled by equations with retarded arguments (differential delay equations, delay equations), e.g. by Hutchinson's equation (delayed logistic equation)

$$\dot{u}(t) = au(t)(1 - u(t-\tau)/K). \tag{18}$$

This equation is equivalent with Wright's equation $\dot{v}(t) = -\alpha v(t-1)(1+v(t))$ and hence equivalent (via elementary transformations) to a special case of the equation

$$\dot{v}(t) = -\nu v(t) - \alpha f(v(t-\tau)) \tag{19}$$

which has been studied in great detail by many authors [24] [37] [38]. Hutchinson's model (18) is usually motivated by saying that somehow the population grows towards its carrying capacity K but reacts to exploited resources only with some delay due to hatching or maturation periods. In [4] a maturation period has been introduced in the McKendrick model (16). It is assumed that the mortality is a step function with a step at the age $a = \tau$ separating a juvenile from an adult age group, and that the birth rate has a delta peak at $a = \tau$ and is constant thereafter. This choice of the coefficients defines a system which is a reasonably realistic caricature of many natural populations where there is a clear division between juveniles and adults and where young adults have a relative high reproduction rate. With these assumptions the densities of the total juvenile and adult populations $U(t) = \int_0^\tau u(t,a)da$, $V(t) = \int_\tau^\infty u(t,a)da$ satisfy a set of differential delay equations. This system has the following structure. For $0 \le t \le \tau$ the variables U, V satisfy a non-autonomous system of ordinary differential equations (taking account of the history), but for $t > \tau$ they satisfy a system of neutral delay differential equations. This system separates (if juveniles do not reproduce, as we have assumed here), and the equation for adults reads

$$\dot{V}(t) = ((b_1 + b_2\mu_1)V(t-\tau) + b_2\dot{V}(t-\tau))e^{-\mu_0\tau} - \mu_1 V(t). \tag{20}$$

Here μ_i, b_i are the coefficients in $\mu(a) = \mu_0 + (\mu_1 - \mu_0)H_\tau(a)$, $b(a) = b_1 H_\tau(a) + b_2\delta_\tau(a)$ (H, δ are the Heaviside and delta function, respectively). This seems to be the first rigorous justification of a neutral differential delay equation as a population model [4].

If one applies similar considerations to the Gurtin-MacCamy model, one arrives at a nonlinear delay equation for the total adult population

$$\dot{V}(t) = g(V(t-\tau))V(t-\tau) + h(V(t-\tau))\dot{V}(t-\tau) - \mu(V(t))V(t). \tag{21}$$

In particular, for $h = 0$ the function $g(V)$ is the density-dependent birth rate and $\mu(V)$ is the death rate. The resulting equation is better motivated than equation (18).

4 Modeling Infectious Diseases

4.1 Vaccination Policies

The nucleus of most, if not all, deterministic epidemic models is the Kermack-McKendrick system

$$\dot{u} = -\beta uv/P, \quad \dot{v} = \beta uv/P - \alpha v, \quad \dot{w} = \alpha v \tag{22}$$

which describes the transition from susceptible u to infected v and to recovered (immune) w. The parameter β is the transmission rate and α is the recovery rate. The basic reproduction number R_0 is, in general, the average number of newly infecteds which are produced by one infected individual in a totally susceptible population. In the present case it is simply $R_0 = \beta/\alpha$. The basic reproduction number can be seen as a normalized bifurcation parameter. The system (22) has the distinguished stationary point $(u, v, w) = (P, 0, 0)$ describing a totally susceptible population. If $R_0 < 1$ then this point is globally stable. If $R_0 > 1$ then a trajectory is emanating from this point which describes an infected population. Along this trajectory the number of infecteds increases first and then decreases to zero, leaving a positive number of remaining susceptibles. In this case it is obvious or easy to show that all relevant quantities such as the basic reproduction number, maximal number of infecteds (maximal prevalence), total size (number of all individuals ever infected) depend in a monotone fashion on the parameters, here β and α. In more complex models for the transmission of disease, in particular in models for sexually transmitted diseases in two-sex populations with infectivity depending on time since infection, it is not obvious (and in general not true) that the basic reproduction number depends on the transmission rate. Such problems have been studied in some detail in [44] using linear complementary problems and, with simpler proofs based on coupling methods in [45].

There are various strategies to fight the spread of infectious diseases such as vaccination, quarantine, education by public health authorities. If such policies are incorporated into disease transmission models, then new groups such as vaccinated or educated individuals must be distinguished. For these more complex models the simple scenario of a transcritical bifurcation at $R_0 = 1$ need not hold any more. In fact, it seems typical for multi-group models that there are regions in the parameter space where the bifurcation is backward (i.e., subcritical). The backward bifurcation may lead to hysteresis phenomena. This mathematical observation has important consequences in public health. It can be shown that inefficient vaccination or education policies (leading to partial immunity or slightly changed behavior) may have detrimental effects [18] [25].

In a sense, vaccination is nothing else then moving individuals from the susceptible to the vaccinated class. In the formulation of a model for a population structured by age and infection, vaccination enters just as a transition term. Let u, w, v be the susceptibles, vaccinated, and infecteds,

$$\begin{aligned} u_t + u_a + \mu(a)u &= -\psi(a)u - \beta(a)uV/N \\ w_t + w_a + \mu(a)u &= \psi(a)u - \tilde{\beta}(a)uV/N \\ v_t + v_a + \tilde{\mu}(a)v &= -\alpha v + (\beta(a)u + \tilde{\beta}(a)w)V/N \end{aligned} \tag{23}$$

with boundary condition

$$u(t,0) = \int_0^\infty [b(a)(u(t,a) + w(t,a)) + \tilde{b}(a)v(t,a)]da, \; w(t,0) = 0, \; v(t,0) = 0$$

where $V = \int_0^\infty k(a)v(t,a)da$ is the total infectivity and $N = \int_0^\infty (u+w+w)da$ is the total population size. Here $\psi(a)$ is the rate at which susceptibles of age a are moved into the vaccinated class. The reproduction number $R(\psi)$ of the disease in the presence of the vaccination policy ψ can be represented in the form $R(\psi) = R_0 - F(\psi)$ where $F(\psi) \geq 0$ is some functional of the policy ψ. Then one can formulate several optimization problems, e.g. to achieve a reduction of the reproduction number $R(\psi) \leq R^* < 1$ at minimal expenses [20]. Several of these optimization problems have been discussed and solved in [20] [41] [42]. The results are somewhat astonishing (at first glance): In most of these problems, optimal policies vaccinate the population at one or two ages. There is a rather close relation to optimal harvesting or culling problems in bioeconomics. If a population structured by age or size is harvested at equilibrium and the net gain is to be maximized then optimal harvesting strategies use two age or size groups only [21].

4.2 Epidemic Spread in Space

There are basically two different ways in which the spread of an infectious disease in a spatially distributed population can be imagined. We can think of the individuals fixed at certain positions and being in contact according to some contact distribution. With this view the position of the individual is given by its residence and the contact distribution describes all contacts with individuals at different locations. In the second view the individuals move according to some stochastic process and may contact individuals which they meet at the actual position. Of course we can join both views and consider individuals which move and have contacts elsewhere. In the first view the SIR model assumes the form

$$\begin{aligned} u_t &= -\beta u \int k(x-y)v(t,y)dy \\ v_t &= \beta u \int k(x-y)v(t,y)dy - \alpha v. \end{aligned} \tag{24}$$

Here the kernel k describes the infectivity which an infected individual at position y exerts upon a susceptible at position x. In the second view we have models of the form

$$u_t = -\beta uv + D(\int K(x-y)u(t,y)dy - u(t,x))$$
$$v_t = \beta uv - \alpha v + \tilde{D}(\int \tilde{K}(x-y)v(t,y)dy - v(t,x)). \tag{25}$$

In models of the form (25) we can replace the diffusion operator by a transport operator to get a possibly more realistic model for the migration process. In [19] the following system has been studied

$$\begin{aligned} u_t(t,x,s) + su_x(t,x,s) &= \int_V K(s,\tilde{s})u(t,x,\tilde{s})d\tilde{s} - \mu u(t,x,s) \\ &\quad -\beta u(t,x,s)\int_V L(s,\tilde{s})v(t,x,\tilde{s})d\tilde{s} \\ v_t(t,x,s) + sv_x(t,x,s) &= \int_V \tilde{K}(s,\tilde{s})v(t,x,\tilde{s})d\tilde{s} - \tilde{\mu}v(t,x,s) \\ &\quad +\beta u(t,x,s)\int_V L(s,\tilde{s})v(t,x,\tilde{s})d\tilde{s} - \alpha v(t,x,s). \end{aligned} \tag{26}$$

Existence of mild solutions has been shown using a Kaniel-Shinbrot iteration and the asymptotic behavior has been explored [19].

Epidemic models of the form (24) or (25) and the corresponding diffusion approximations have typically travelling front solutions that describe the propagation of disease into a susceptible population. Existence of travelling fronts and the minimal speed have been studied in [14].

4.3 Modulation Equations

Models in mathematical biology are often very complex and not suited for analytical treatment. Numerical simulations lead to some insight, but may be not completely satisfying. However, the most interesting phenomena are connected to bifurcations. Near a bifurcation a complex system may be reduced using, for example, center manifold theory. Center manifold theory requires discrete spectra and hence its application is restricted to bounded space domains. Its analogue for unbounded domains is the theory of modulation equations or amplitude equations. This theory started with the Ginzburg-Landau equation developed in solid state physics. Only quite recently [26] [52] the underlying mathematical structure has been clarified: Modulation equations describe the evolution of the amplitude of the solution over the unstable modes, rescaled in an appropriate way, up to higher orders of a small parameter (in general the bifurcation parameter). The equation takes into account the mode-interaction (again, up to higher order terms of the small parameter). The Ginzburg–Landau equation

$$A_t = \Delta A + c_1 A + c_2 |A|^2 A \tag{27}$$

describes the generic situation, where the first unstable mode is a nonzero Fourier mode.

In epidemiology, the primary bifurcations are transcritical: A susceptible population loses its stability and the infective agent invades. A typical model for spatial spread of an epidemic has the form of equation (1) where $u(t, x)$ denotes the vector of densities for susceptibles, infecteds a.s.o. If the scalar equation $\dot{u} = f(u)$ undergoes a transcritical bifurcation, then the first unstable Fourier mode of the full system is zero. Hence the corresponding modulation equation is the classical KPP or Fisher equation

$$A_t = \Delta A + A - A^2.$$

However, most epidemic models have an additional symmetry: The vector field is homogeneous of degree one, $f(\alpha u) = \alpha f(u)$ for $\alpha > 0$. This symmetry gives additional structure, similar to the phase equation (see [35]), and the modulation equation becomes a system

$$A_t = \Delta A + 2(\nabla B)^T \nabla A + A - A^2, \quad B_t = \Delta B + |\nabla B|^2 + cA \tag{28}$$

for variables A, B governing the prevalence of infection and the total population size at the point (x, t), respectively. In many cases the solutions of these equations approach traveling fronts for large times [43]. It is an interesting point that the velocity of the spread depends only on the reproduction number and the exponent of growth of the susceptible population; whether or not the population breaks down in endemic regions has no influence on this velocity. Only if in endemic regions the population starts to *grow* much faster than in infection–free zones (a situation only of theoretical interest), then the infection may be pushed forward and spreads even faster if the growth rate of the infected population increases.

5 Models Describing Single Individuals

In the previous sections, always the density of a structured population has been considered. This density describes the behavior of an average individual; it is not possible to gain insight in the fate of a single individual. For many cases, this simplified approach gives enough information about interesting phenomena. Moreover, it leads to dynamical systems which can be analysed using a wide range of elaborated analytical tools. However, in some situations it is necessary to focus on single individuals (examples are given below). The formulation of these individual based models is often simpler than that of deterministic models: The only thing one has to do is to formulate the different behavioral patterns of single individuals and let these individuals interact. The great success of these models in recent years can partly be explained by the fact that these ideas can easily be communicated to non-mathematicians. Unfortunately, only very few tools are available to analyse individual based models. Some models and methods are presented in the following sections.

5.1 Cellular Automata and Particle Systems

A cellular automaton consists of a grid G of cells, in most cases a subset (grid domain) of the square lattice $\mathbb{Z}^2$ (or $\mathbb{Z}^1$), a neighborhood U, typically the Moore neighborhood of nine cells or the von Neumann neighborhood of five cells, a set E of elementary states, typically $E = \{0,1\}$ (empty/occupied or white/black) and finally a local function f_0. The cellular automaton (G, U, E, f_0) defines a discrete dynamical system $z^{(t+1)} = f(z^{(t)})$ on states $z : G \to E$ where the global function f is obtained from the local function as

$$f(z)(x) = f_0(z|_{U(x)}). \tag{29}$$

Cellular automata can be evaluated synchronously or asynchronously. In the asynchronous case each cell is called according to a Poisson process (with the same rate for every cell). For the same local function, trajectories of the synchronous or the asynchronous automaton look quite different. Whether synchronous or asynchronous, the "flip", i.e., the exchange of the states of two neighboring cells, cannot be realized in a standard cellular automaton. Therefore the new concept of a dimer automaton has been introduced [54]. A dimer automaton is a system (G, U, E, f_0) with a grid, a neighborhood and elementary states as before. However, the local function is a mapping $f_0 : E \times E \to E$. Hence a dimer automaton works as follows. A first cell is called according to a Poisson process, then a second cell y is called from the neighborhood $U(x) \setminus \{x\}$, then the local function is applied to the pair $(z(x), z(y))$. For a given grid domain and neighborhood, there are 256 different dimer automata. This set can be partitioned into 51 equivalence classes, and for each of these the dynamical behavior can be described qualitatively by stochastic properties such as ergodicity (see [2]) and quantitatively by mean field approximations [54]. Extensions of dimer automata can be constructed as follows. A grid domain together with a neighborhood defines a graph. The cells correspond to the vertices, an edge is running from cell x to cell y if $y \in U(x)$. Since the Moore neighborhood is symmetric, each two connected cells are connected by a pair of edges. Hence one can extend the concept to arbitrary directed graphs. Then one can consider convex combinations of dimer automata on directed graphs as follows. Define a priori a distribution on the set of all dimer automata. The graph and this distribution define an "edge process": Call an edge, inspect the states of the vertices, select a dimer automaton according to the given distribution, apply the local rule. In [3] it has been shown that, up to some rescaling of the underlying stochastic processes which amounts to monotone transformations of the time variable along realizations, edge processes can be essentially interpreted as convex combinations of seven basic processes. The travelling fronts of certain cellular automata have been studied in [53], and recently asynchronous automata have been studied which can be seen as ultimate discretizations of the Cahn-Hilliard equation [22].

5.2 Individual Based Models and Epidemiology

Infectious diseases can be seen as a problem of public health and as events in individual life. These different views are most obvious with respect to vaccination, where public health wants to prevent outbreaks and endemics whereas the individual wants protection. If the disease is not fatal and vaccination bears some risk to the vaccinated individual, then the goals of the individual and public health may diverge [42]. Deterministic models are essentially public health models; detailed information on individual fates would even be misleading. The situation is somewhat different with respect to stochastic models. There is a well developed theory of stochastic epidemics which, at least for simple transmission laws, yields explicit expressions for critical parameters, prevalence, total size of an epidemic a.s.o. If one compares the outcomes of deterministic models to the expectations in stochastic models then one finds characteristic differences that can be attributed to the variance in stochastic models.

However, there are various situations where a control policy addresses the contact structure of a single individual. In such cases models and methods based on the description of individuals are necessary. A typical example is ring vaccination, where all individuals in a neighborhood of a discovered infectious individual (the index case) are treated. In human populations the family, friends and colleagues of an infected person become protected by vaccination. In an agricultural setting, the spatial structure of contacts becomes important: A standard example is foot- and mouth disease (FMD) in cattle. Once a farm with infected animals is discovered, all farms within a certain distance (typically three to twenty kilometers) are treated. This procedure creates a ring of vaccinated farms around the index case.

Contact tracing, as opposed to mass screening, is another example of an individual based control policy. Once an infected individual is discovered (the index case), this person is interviewed about recent possibly infectious contacts. Especially for sexually transmitted diseases, these contacts are simple to characterize and at least theoretically easy to be followed up and notified. The notified persons are then examined. In this way more infected persons are discovered and again interviewed. In case these persons are asymptomatic, contact tracing is the only way to discover them.

While ring vaccination for FMD is based on spatial structure, contact tracing may be modeled in a homogeneously mixing population. Because of this difference, also the models and the methods used are slightly different.

5.3 Particle Systems and Branching Processes Modeling Ring Vaccination

The spread of FMD can be modeled as a particle system [48]. One characteristic feature of FMD is the existence of two different contact rates: short range contacts occur at a high rate, leading to many infected cases close to a primary case. Long range contacts occur at a lower rate and cause infections at farms far away from primary infected farms. In a first modeling step, one considers only local contacts by means of a cellular automaton (on a finite square grid with the Moore neighborhood U) where the state $z(x)$ at the site x can assume three values S, I and R representing a farm with animals all susceptible, some infected or all immune, respectively. The transitions $I \to R$ and $R \to S$ occur according to a Poisson process at prescribed rates. The transition $S \to I$ describes the infection process and is slightly more complex. We assume that the infection rate is a nondecreasing function of the number of infected neighbors, a cell x changes its state from S to I in a time interval of length Δt according to

$$S \to I \text{ with probability } \beta_s \#\{y \in U(x) \,|\, z(y) = I\} \Delta t + o(\Delta t).$$

The second step is modeling global contacts: At a low rate β_l, each infected particle (farm) may infect any other particle in the grid. With these global interactions, the system is not a cellular automaton any more but an interacting particle system. The total resulting probability for infection of a cell x in the state S during a small time interval Δt is

$$S \to I \quad \text{with probability}$$
$$\beta_s \#\{y \in U(x) \,|\, z(y) = I\} \Delta t + \beta_l \#\{y \in \Gamma \,|\, z(y) = I\} \Delta t + o(\Delta t).$$

In the last modeling step one introduces ring vaccination: First of all, there is a discovery rate at which index cases are observed. Once an infected site is found, all neighbors of this site are screened. If a site is susceptible, then it becomes vaccinated and hence immune. If the site is infected, it is treated (all animals at this farm are slaughtered) and recursively also around this site ring vaccination is performed.

Simulations show that local outbreaks occur in the form of several infected sites close together that are created by local contacts. These local outbreaks cause, by global contacts, other local outbreaks. Since the infected sites within local outbreaks are close together, mostly the local outbreak is completely eradicated by ring vaccination after the first index case has been discovered. This structure fits well with observed patterns of FMD.

For certain parameter ranges it is possible to approximate the dynamics of this complex nonlinear interacting particle system by a linear stochastic birth and death process on two levels. The first level, describing the dynamics of local outbreaks, is a branching process with catastrophes (at the discovery of the first member of a local outbreak the whole infectious population is locally eradicated). These local outbreaks behave like individuals which give birth to other individuals (local outbreaks) and die (due to ring vaccination or due to spontaneous recovery of all members of a local outbreak). Using the theory of branching processes on two levels [1], it is possible to derive expressions for characteristic quantities for the dynamics, e.g. for the reproduction number or the time to extinction.

Furthermore, it is possible to introduce mass vaccination, and in this way to compare mass vaccination to ring vaccination. This comparison is especially interesting, since the European Community (EU) changed its FMD policy recently: Before mass vaccination has been the rule, while now the EU relies on ring vaccination only. Since with ring vaccination herd immunity is lost, it is interesting to know how powerful certain aspects of ring- and mass vaccination are.

5.4 Nonlinear Branching Processes Modeling Contact Tracing

In contrast to ring vaccination for FMD, contact tracing can be modeled in a homogeneously mixing population [47]. Assume N individuals, S susceptibles and I infecteds. In the absence of contact tracing the infection process is given by

$$(S, I) \mapsto (S-1, I+1) \quad \text{with} \quad \text{probability} \quad \beta S I \Delta t + o(\Delta t)$$

and recovery by

$$(S, I) \mapsto (S+1, I-1) \quad \text{with} \quad \text{probability} \quad \alpha I \Delta t + o(\Delta t)$$

Now one introduces contact tracing. Like in ring vaccination, an index case has to be discovered first. Index cases are detected according to a Poisson process. Once an index case has been discovered, one starts to investigate the contacts of this person: The infected persons form a directed graph Γ. A node represents an infected person, a directed edge runs from an infector to the person infected. The graph of infected persons is not necessarily connected because recovered individuals leave the graph. Each connected component of Γ is a tree, i.e., Γ itself is a forest. If one discovers one index case in a connected component of this forest, then all neighbors of this index case can be also discovered. In general, not every neighbor will be found for sure (since people are not likely to notify e.g. all sexual partners), but for each neighbor there is a certain probability of detection, the tracing probability. In a given realization of this stochastic process, the neighbors which are actually discovered are converted into new index cases. In this way, the tracing process is running over one connected component of Γ, starting at the first index case. The size and the structure of these connected components change over time; at the onset of an outbreak, there are only very few infected individuals and the connected components are small. The components grow with time until they reach an equilibrium distribution with respect to size and structure.

The key point of the analysis is to derive the mean infectious period, since the effective reproduction number is the product of the infection rate β and the mean infectious period. Since the mean infectious period depends on the state of the infectious population as a whole, the dynamics of the forest Γ must be described.

In order to investigate the dynamics at the onset of the disease, one assumes that there are no nonlinear effects due to the infectious process, i.e., that every contact of an infected person reaches a susceptible person. The embedded Galton-Watson process, where one considers generations of infected individuals instead of time, is not stationary but asymptotically stationary. It is possible to derive recursive integro-differential equations for the probability to be infectious after time of infection a for a particle of generation i of infected persons. These functions converge with $i \to \infty$, and in this way the mean infectious period can be computed.

Heuristic time scale arguments allow for a deterministic approximation of the density of susceptible and infected individuals of the full, nonlinear process with contact tracing and infection in form of ordinary differential equations. Interesting questions that can be approached via the stochastic process are, for example: Should one look for "infectors" (i.e., the one that has infected the index case) rather than "infectees" (the persons, who are infected by the index case)? Is it sufficient to inspect the the direct neighbors of the index case or does it pay to follow the full recursivity of the dynamical process? These questions have practical implications for the design of contact tracing.

References

1. Ball, F., Mollison, D., Scalia-Tomba, G., Epidemic with two levels of mixing. Ann. Appl. Prob. 7 (1997) 46-89
2. Bandt, C., Edge processes - a simple and general version of interacting particle systems. Universität Greifswald, DANSE Preprint 1/96 (1996)
3. Bandt, C., Hadeler, K.P., Kriese, F., Particle systems acting on undirected graphs. J. Statist. Physics 91 (1998) 571-586
4. Bocharov, G., Hadeler, K.P., Structured population models, conservation laws, and delay equations. J. Diff. Equ. in print
5. Deutsch, A., Orientation-induced pattern formation: swarm dynamics in a lattice-gas automaton model. Int. J. of Bifurcation and Chaos 6 (1995) 1735-1752
6. Dunbar, S.R., Traveling waves in diffusive predator-prey equations: Periodic orbits and point-to-periodic heteroclinic orbits. SIAM J. Appl. Math. 46 (1986) 1057-1078
7. Durret, R., Ten lectures on particle systems, p. 97-201. In: P. Biane and R. Durrett (eds), Lectures on Probability Theory. Lecture Notes in Mathematics 1608, Springer Verlag, 1995
8. Edelstein-Keshet, L., Ermentrout, G.B., Models for contact mediated pattern formation. J. Math. Biol. 29 (1990) 32-58

9. Edwards, A.W.F., Foundations of mathematical genetics. Cambridge University Press 1977
10. Fife, P.C., Mathematical aspects of reacting and diffusing Systems. Lecture Notes in Biomath. 28, Springer Verlag 1979
11. Geigant, E., Nichtlineare Integro-Differential-Gleichungen zur Modellierung interaktiver Musterbildungsprozesse auf S^1. Dissertation, University of Bonn, 1999
12. Geigant, E., Ladizhansky, K., Mogilner, A., An integro-differential model for orientational distributions of F-actin in cells. SIAM J. Appl. Math. 59 (1998) 787-809
13. Hadeler, K.P., Shadowing orbits and Kantorovich's theorem. Numer. Math. 73 (1996) 65-73
14. Hadeler, K.P., Travelling epidemic waves and correlated random walks. Proc. Conf. Diff. Equ. Claremont. M. Martelli et al. (eds) Differential Equ. and Appl. to Biology and to Industry. p. 145-156. World Scientific Publ. 1996
15. Hadeler, K.P., Reaction telegraph equations and random walk systems. In: S.van Strien, S.Verduyn Lunel (eds), Stochastic and spatial structures of dynamical systems. Roy. Acad. of the Netherlands. North Holland, Amsterdam, p. 133–161 (1996)
16. Hadeler, K.P., Reaction transport systems in biological modelling. In: V.Capasso, O.Diekmann (eds), Mathematics inspired by biology, CIME Lectures 1997. p. 95-150. Florence, Lecture Notes in Mathematics 1714, Springer Verlag 1999
17. Hadeler, K.P., Nonlinear propagation in reaction transport systems. Fields Institute Communications 21 (1999) 251-257
18. Hadeler, K.P., Castillo-Chavez, C., A core group model for disease transmission. Math. Biosc. 128 (1995) 41-55
19. Hadeler, K.P., Illner, R., van den Driessche, P., A disease transport model. University of Tübingen, DANSE Preprint 49/98 (1998) To appear in: L. Weis, G. Lumer (eds), Proc. Conf. Bad Herrenalb "Evolution equations and their applications in physical and life sciences"
20. Hadeler, K.P., Müller, J., Vaccination in age structured populations. I: The basic reproduction number, II: Optimal strategies. p. 90-114. In: V.Isham, G.Medley (eds), Models for Infectious Diseases. Their structure and relation to data. Cambridge Univ. Press 1996
21. Hadeler, K.P., Müller, J., Optimal vaccination and optimal harvesting. University of Tübingen, Manuscript
22. Hadeler, K.P., Schönfisch, B., Dimer automata and edge processes. Proc. Equadiff 1999
23. Hadeler, K.P., Selivanova, E.N., On the case of Kovalevskaya and new examples of integrable conservative systems on S^2. Regular and Chaotic Dynamics 4 (1999) 45-52
24. Hadeler, K.P., Tomiuk, J., Periodic solutions of difference-differential equations. Arch. Rat. Mech. Anal. 65 (1977) 87-95
25. Hadeler, K.P., van den Driessche, P., Backward bifurcation in epidemic control Math. Biosc. 146 (1997) 15-35
26. van Harten, A., On the validity of the Ginzburg-Landau Equation, J. Nonlin. Sci. 1 (1991) 397-422
27. Hillen, T., A Turing model with correlated random walk. J. Math. Biol. 35 (1996) 49-72

28. Hillen, T., Qualitative analysis of semilinear Cattaneo systems. Math. Models and Meth. in Appl. Sci. 8 (1998) 507-519
29. Hillen, T., Rohde, C., Lutscher, F., Existence of weak solutions for a hyperbolic model of chemosensitive movement. Submitted to J. Math. Anal. Appl.
30. Hillen, T., Stevens, A., Hyperbolic models for chemotaxis in 1-D, University of Minnesota, IMA-Preprint 1592, (1998), Nonlinear Analysis, to appear
31. Hofbauer, J., Sigmund, K., The Theory of Evolution and Dynamical Systems. Cambridge U. Press 1988
32. Jäger, E., Segel, L.A. (1992) On the distribution of dominance in populations of social organisms. SIAM J. Appl. Math. 52 (1992) 1444-1468
33. Kermack, W.O., McKendrick, A.G. A Contribution to the Mathematical Theory of Epidemics. Proc. Roy. Soc. A 115 (1927) 700-721
34. Keller, L.F., Segel, L.A., Model for chemotaxis. J. Theor. Biol. 30 (1971) 225-234
35. Kuramoto, Y., Chemical oscillations, waves and turbulence. Springer Verlag 1984
36. Lutscher, F., Modelling moving polarized groups of animals and cells. Dissertation, Department of Mathematics, University of Tübingen 2000
37. Mackey, M.C., Glass, L., Oscillations and chaos in physiological control systems. Science 197 (1977) 87-89
38. Mallet-Paret, J., Nussbaum, R., Global continuation and asymptotic behaviour for periodic solutions of a differential-delay equation. Ann. Mat. Pura Appl. IV, Ser. 145 (1986) 33-128
39. Mischaikow, K., Reineck, J.F., Travelling waves in predator-prey systems. SIAM J. Math. Anal. 24 (1993) 1179-1214
40. Mogilner, A., Edelstein-Keshet, L., Spatio–angular order in populations of self-aligning objects. Physica D 89 (1996) 346-367
41. Müller, J., Optimal vaccination strategies for age structured populations. SIAM J. Appl. Math. 59 (1999) 222-241
42. Müller, J., Optimal vaccination strategies – for whom?, Math. Biosc. 139 (1997) 133-154
43. Müller, J., Scaling methods and approximative equations for homogeneous reaction–diffusion systems and applications to epidemics, J. Nonlin. Sci. 9 (1999) 149-168
44. Müller, J., Hadeler, K.P., Variable infectivity in sexually transmitted diseases. University of Tübingen, Preprint DANSE 3/97 (1997)
45. Müller, J., Hadeler, K.P., Monotonicity of the number of passages in linear chains and of the basic reproduction number in epidemic models. Zeitschr. Analysis Anwend. 19 (2000), 61-75
46. Müller, J., Hillen, T., Modulation equations and parabolic limits of reaction random-walk systems. Math. Methods Appl. Sci. 21 (1998) 1207-1226
47. Müller, J., Kretzschmar, M., Dietz, K., Contact tracing in stochastic and deterministic epidemic models. Math. Biosc. 164 (2000) 39–64
48. Müller, J., Schönfisch, B., Kirkilionis, M., Ring vaccination. J. Math. Biol., in press
49. Murray, J., Mathematical Biology. Springer Verlag 1989
50. Patlak, C.S., Random walk with persistence and external bias. Bull. Math. Biophys. 15 (1953) 311-338
51. Rothe, F., Global solutions of reaction-diffusion systems. Lect. Notes in Math. 1072, Springer Verlag 1984

52. Schneider, Guido, Error estimates for the Ginzburg-Landau approximation, Z. Angew. Math. Phys. (ZAMP) 45 (1994) 433-457
53. Schönfisch, B., Propagation of fronts in cellular automata. Physica D 80 (1995) 433-450
54. Schönfisch, B., Hadeler, K.P., Dimer automata and cellular automata. Physica D 94 (1996) 188-204
55. Schwetlick, H., Reaktions-Transportgleichungen. Dissertation, Department of Mathematics, University of Tübingen 1998
56. Schwetlick, H., Travelling fronts for multidimensional nonlinear transport equations. Ann. Inst. H. Poincaré, Analyse non lineaire 17 (2000) 523-550
57. Schwetlick, H., Limit sets for multidimensional nonlinear transport equations. J. Diff. Equ. accepted
58. Smoller, J., Shock Waves and Rection-Diffusion Equations. Springer Verlag 1982

Topological and Measurable Dynamics of Lorenz Maps

Gerhard Keller and Matthias St. Pierre*

Mathematisches Institut, Universität Erlangen-Nürnberg, Bismarckstr. $1\frac{1}{2}$,
91054 Erlangen, Germany
email: keller@mi.uni-erlangen.de

Abstract We investigate the dynamics of Lorenz maps, in particular the asymptotical behaviour of the trajectory of typical points. For Lorenz maps f with negative Schwarzian derivative we give a classification of the possible metric attractors and show that either f has an ergodic absolutely continuous invariant probability measure of positive entropy or the iterates of typical points spend most of their time shadowing the trajectory of one of the two critical values. Our main tool therefore is the construction of Markov extensions for Lorenz maps which provide a unified framework to approach both the topological and the measurable aspects of the dynamics.

We study the bifurcation diagram of a smooth two parameter family of Lorenz maps which describes the parameter dependence of the kneading invariant and show that essentially every admissible kneading invariant actually occurs if the family is sufficiently rich. Finally, we adress the problem whether the kneading invariant depends monotonously on the parameters.

1 Introduction

Lorenz maps play an important role in the study of the global dynamics of families of vector fields near homoclinic bifurcations. A typical situation where Lorenz maps are encountered is given by a family $X_\mu : \mathbb{R}^3 \to \mathbb{R}^3$ of vector fields, where X_0 has a hyperbolic saddle with eigenvalues $\lambda_{ss} < \lambda_s < 0 < \lambda_u$ and with two homoclinic orbits in the configuration of a butterfly connecting the stable and unstable direction, like in the classical Lorenz system [25]. Breaking up the homoclinic loops by changing the parameters it is possible to find vector fields with very complicated chaotic dynamics, where the trajectories of points seem to randomly follow one of two loops near the former homoclinic orbits.

The Geometric Lorenz Attractor

To explain the dynamics on the "strange" attractor of the Lorenz system, Guckenheimer [8] proposed a two-dimensional model for the flow on the

* Project: Coupled Chaotic Maps. Sinai–Bowen–Ruelle Measures, Synchronization, and Essential Dynamics (Gerhard Keller)

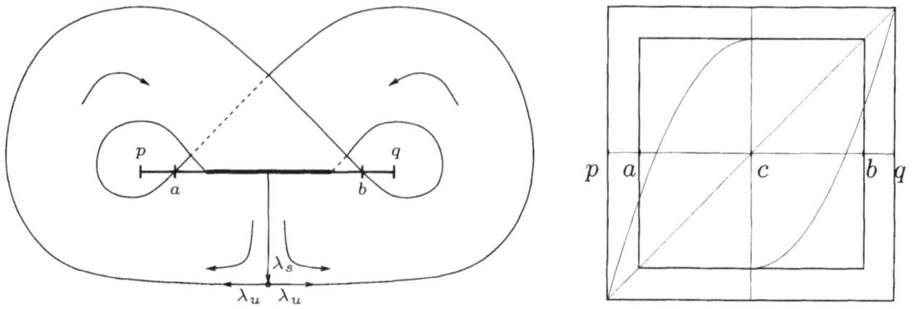

Fig. 1. The left hand side shows a phase portrait of the flow on the branched manifold and the right hand side the first return map to the cross section $\Sigma = [p, q]$.

attractor, the so-called *geometric Lorenz attractor*. It is obtained, roughly speaking, by forgetting about the strong stable direction and consists of a two-dimensional branched manifold with a hyperbolic saddle as it is shown in Figure 1. Considering the Poincaré map to the cross section $\Sigma = [p, q]$ one obtains an interval map with two monotonous branches: If c denotes the intersection of Σ with the stable manifold of the saddle then all points to the left of c follow the left loop and all points to the right of c follow the right loop until they hit the cross section again.

Hence from a topological viewpoint a Lorenz map is nothing else than an interval map with two monotonous and continuous branches and a discontinuity in between. From a metrical viewpoint it is smooth to some degree on both branches and of order $\alpha > 0$ at the discontinuity, i.e., $f(c \pm \varepsilon) \sim f(c^{\pm}) \pm \varepsilon^{\alpha}$, where the order originates from the local analysis of the linearized flow at the saddle and equals the ratio $\frac{|\lambda_s|}{|\lambda_u|}$ of the stable and the unstable eigenvalue. In practice, the smoothness of the Lorenz map is limited by the fact that in order to rigorously justify the geometric Lorenz model, the dynamics of the vector field has to be reduced by means of geometrical methods like the construction of invariant foliations or the existence of a two-dimensional invariant centre manifolds. Even if this can be done, the smoothness of the foliation or the centre manifold depends on certain gap conditions on the eigenvalues of the vector field. In the worst case it can happen that it is only of class $C^{1+\varepsilon}$, i.e., once differentiable with Hölder continuous derivative (cf. Homburg [17] for an example). Nevertheless, for some metrical results we assume that it is at least of class C^2.

If $\alpha < 1$ then the derivative of f is infinite at the discontinuity. Such maps are typically overall expanding and chaotic. Their topological type is completely determined by the *kneading invariant* of the map, a pair of binary sequences coding the orbits of the two *critical points* c^+ and c^-, (i.e., the corresponding one-sided limits of the orbits at the discontinuity c). Moreover, the set of all possible kneading invariants can be characterized by a simple combinatorial condition. Since $\alpha < 1$ holds in the situation of the classical Lorenz system, this kind of Lorenz maps has been studied by many people and their dynamics is well understood, see for example Guckenheimer [8], Rand [31], Guckenheimer & Williams [9], Williams [36], Parry [30], Hubbard & Sparrow [18], and Glendinning & Sparrow [7].

Here we are mainly concerned with Lorenz maps of exponent $\alpha > 1$. One reason for this is that in this case the derivative of f vanishes at the discontinuity which means that such maps are typically contracting in some regions and expanding in others, and due to the interplay between contraction and expansion such Lorenz maps can exhibit a much wider spectrum of behaviour. Another reason is that the condition $\alpha > 1$ is compatible with the condition of negative Schwarzian derivative,(1) whereas the condition $\alpha < 1$ is not. Under the assumption of negative Schwarzian derivative one has strong tools to control the distortion on branches of high iterates of the map, just as for smooth interval maps.

Although the theory of vector fields serves as the motivation for studying Lorenz maps, we focus here on what can be said about Lorenz maps from the viewpoint of one-dimensional dynamics. That is, we study the time discrete dynamical system (X, f), where $f : X \to X$ is a Lorenz map defined on some compact interval $X \subset \mathbb{R}$. Given an initial point $x \in X$, its time evolution under the action of f is described by the *orbit* or *trajectory* of x,

$$x_0 \to x_1 \to x_2 \to x_3 \to \dots \qquad , \text{ where } x_n := f^n(x) \ .$$

The Asymptotic Behaviour of Typical Points

A near at hand way to describe the asymptotic behaviour of the trajectory of a point x is to look at the set of its accumulation points, called the *ω-limit set* of x. In general, different points $x \in X$ can have very different asymptotic behaviour. From a physical viewpoint, a certain type of asymptotic behaviour is relevant only if it is observable with positive probability, which means that it is exhibited by a set of initial conditions which has positive Lebesgue measure. This opinion is reflected by the following definition of an attractor which goes back to Milnor [27].

Definition 1 (Attractor). A set $A \subseteq X$ is called *f-invariant* if $f(A) \subseteq A$. For a compact f-invariant set $A \subseteq X$ let $B(A) := \{x \in X \mid \omega(x) \subseteq A\}$ denote the *basin of attraction* of A. Then A is called an *attractor* if the following holds.

1. A attracts a set of positive Lebesgue measure: $m(B(A)) > 0$.
2. Every proper subset $\tilde{A}$ of A which is compact and f-invariant attracts significantly less points: $m(B(A) \setminus B(\tilde{A})) > 0$.

---

(1) For $f \in C^3$ the Schwarzian derivative is defined as $Sf := \frac{f'''}{f'} - \frac{3}{2}\left(\frac{f''}{f'}\right)^2$ where $f' \neq 0$.

An attractor A is called the *global attractor* if $B(A) = X \bmod m$. It is called *indecomposable* if it is not the disjoint union of two smaller attractors and *minimal* if it does not contain any smaller attractors.

The question about the asymptotic behaviour of typical points can now be posed as follows: What is the global attractor of f and can it be decomposed into minimal attractors? The answer as given in Theorem 16 can be summarized as follows:

- ▷ For Lorenz maps with negative Schwarzian derivative there is always a global attractor. It is minimal unless the map has two attracting periodic orbits, in which case it splits up into two minimal attractors.

A different kind of information about the asymptotic behaviour of a point can be obtained if one looks at the empirical distributions of trajectories instead of their ω-limit sets. More precisely, one considers the empirical distributions $\frac{1}{n}\sum_{k=0}^{n-1}\delta_{x_k}$, where δ_y denotes the Dirac measure in y, and asks whether they converge weakly to some f-invariant measure μ, or more generally, what the set of weak limit points is. Again, the physically most interesting measures are the ones that can be observed for a positive measured set of initial conditions.

Definition 2 (SRB-measure). A probability measure μ is called a *Sinai-Ruelle-Bowen measure*, or just an *SRB-measure*, if the set $\{x \in X \mid \mu_n(x) := \frac{1}{n}\sum_{k=0}^{n-1}\delta_{x_k} \to \mu\}$ has positive Lebesgue measure.

A trivial example of an SRB-measure is the equidistribution on an attracting periodic orbit. Much more interesting is the case where f has an absolutely continuous invariant probability measure which is ergodic on its support. In this case $\mu_n(x) \to \mu$ holds μ-almost surely by the Ergodic Theorem, whence on a set of positive Lebesgue measure.

For Lorenz maps with negative Schwarzian derivative it turns out that there are essentially two different possibilities which can be discriminated by means of upper Lyapunov-exponents $\lambda^+(x) := \limsup_{n\to\infty}\frac{1}{n}\log|(f^n)'(x)|$.

- ▷ Either $\lambda^+(x) > 0$ on a set of full Lebesgue measure. Then f has a unique (and hence ergodic) absolutely continuous invariant probability measure of positive entropy (Theorem 18).
- ▷ Or $\lambda^+(x) \le 0$ on a set of full Lebesgue measure. In this case all weak accumulation points of the sequences $(\mu_n(x))_n$ are contained in the convex closure of the union of the weak accumulation points of $(\mu_n(c^+))_n$ and $(\mu_n(c^-))_n$ for Lebesgue-almost every x (Theorem 19).

The topological conjugacy class of a Lorenz map is essentially determined by the kneading sequences of c^+ and c^-. These are 0-1-sequences which encode the topological dynamics of the map. In Theorem 22 those 0-1-sequences that can occur as kneading sequences of Lorenz maps are characterized in various purely combinatorial ways. This characterization is one of the ingredients of a bifurcation analysis for C^1-Lorenz maps that can be summarized as follows:

- ▷ Essentially all kneading sequences can be realized within any given $\mathcal{C}^1$-Lorenz family with critical order $\alpha \geq 2$, naturally parametrized by two parameters (Theorem 25).
- ▷ The finite initial pieces of the kneading sequences provide, in a natural way, a bifurcation diagram for $\mathcal{C}^1$-Lorenz families in their two-dimensional parameter space (Theorem 28 and Figure 4).

In the following we give an outline of the results obtained in [33]. Unless stated otherwise, all the theorems below are taken from there.

2 Markov Extensions

Our main tool to study the dynamics of Lorenz maps is the construction of *Markov extensions*. The great advantage of Markov extensions is that they provide a unified framework for the topological and measurable aspects of the dynamics.

The Markov Diagram

Markov extensions for piecewise monotonous interval maps were originally introduced by Hofbauer [10,11,13,12,14,15] as countable state topological Markov chains, which he called *Markov diagrams*. The state space of the Markov chain is a collection $\mathcal{D}$ of subintervals $D \subset X$ which are constructed in such a way that every interval $D \in \mathcal{D}$ is mapped by f onto the union of finitely many intervals from $\mathcal{D}$, called the *successors* of D. The successor relation defines the possible transitions for $\mathcal{D}$. Every point from X can be represented by a path in the Markov diagram which symbolically codes the sequence of intervals that are visited by the iterates of x. In this way the original system (X, f) becomes a continuous factor of the Markov chain. This fact can be used to derive statements about topological properties of f like the nature of the nonwandering set or the topological entropy.

Hofbauer Towers and Markov Extensions

Later Keller [19–23] turned the Markov diagram into an even more powerful tool by taking into account that its "states" are more than that, namely intervals which carry a smooth structure and a canonical measure, the Lebesgue measure. He introduced the so called *(canonical) Markov extension* of f, a piecewise smooth dynamical system $(\hat{X}, \hat{f})$ which has a countable Markov partition and a natural projection $\pi : \hat{X} \to X$ yielding the original system (X, f) as a countable-to-one factor. Let us describe his construction of the Markov extension in more detail.

Definition 3. i) Let $\mathcal{Z} := \{Z^-, Z^+\} := \{(p,c),(c,q)\}$ denote the partition of X into the intervals of monotonicity of f and let $\mathcal{D} \supseteq \mathcal{Z}$ be the smallest family of intervals satisfying

$$f(D \cap Z^{\pm}) \in \mathcal{D} \quad \text{whenever} \quad D \in \mathcal{D} \text{ and } D \cap Z^{\pm} \neq \emptyset \, .$$

Now take a collection $\hat{\mathcal{D}} := \{\hat{D} \mid D \in \mathcal{D}\}$ of disjoint copies $\hat{D} := D \times \{D\}$ of the intervals $D \in \mathcal{D}$. Their union $\hat{X} := \bigcup_{\hat{D} \in \hat{\mathcal{D}}} \hat{D}$ is called the *Hofbauer tower* associated to the map f and the sets $\hat{D}$ are called the *levels* of the tower.

ii) The *Markov extension* for f is the dynamical system $(\hat{X}, \hat{f})$ where $\hat{f} : \hat{X} \to \hat{X}$ is defined in the following way:

$$\hat{f}(\hat{x}) := (f(x), f(D \cap Z[x])) \quad \text{for } \hat{x} = (x, D) \in \hat{D}.$$

Here $Z[x]$ denotes the element of the partition $\mathcal{Z}$ containing x. The natural projection $\pi : \hat{X} \to X$ is given by $\pi(\hat{x}) := x$ for $\hat{x} = (x, D) \in \hat{D}$ and satisfies $\pi \circ \hat{f} = f \circ \pi$. The value $\hat{f}(\hat{c})$ is left undefined for all *critical points* $\hat{c} \in \pi^{-1}(c)$.

Note that the tower depends on the map f, although this is not indicated by the notation. The terms *tower* and *level* originate from the fact that one can imagine all these intervals piled up one upon the other, in this way forming the floors of an infinite tower built on the base $X = (p,q)$, and the map π as the vertical projection from the tower onto its base.

The possible transitions of points in the tower with respect to this Markov partition are described by Hofbauers Markov diagram.

Definition 4. The *Markov diagram* for $(\hat{X}, \hat{f})$ is the directed graph $(\hat{\mathcal{D}}, \to)$ with edges given by the following relation:

$$\hat{C} \to \hat{D} :\Longleftrightarrow \hat{D} \subseteq \hat{f}(\hat{C}) \qquad\qquad (\hat{C}, \hat{D} \in \hat{\mathcal{D}})$$

If $\hat{C} \to \hat{D}$ then $\hat{D}$ is called a *successor* of $\hat{C} \in \hat{\mathcal{D}}$.

Cutting Times and the Kneading Map

In the situation of Lorenz maps it turns out that the Markov diagram allows only a very limited set of transitions: The levels can be grouped into two different sets, denoted by $\hat{\mathcal{D}}^+ := \{\hat{D}_n^+ \mid n \in \mathbb{N}\}$ and $\hat{\mathcal{D}}^- := \{\hat{D}_n^- \mid n \in \mathbb{N}\}$, such that for every level $\hat{D}_n^{\pm}$ there is a transition $\hat{D}_n^{\pm} \to \hat{D}_{n+1}^{\pm}$ and for some special levels—called the *critical levels*—there is a second arrow $\hat{D}_n^{\pm} \to \hat{D}_{\tilde{n}+1}^{\mp}$ where $\hat{D}_{\tilde{n}}^{\mp}$ is a particular critical level $\hat{D}_{\tilde{n}}^{\mp}$ on the other side with $\tilde{n} < n$.[(2)] Graphically this means that one can either *climb up* one level in the tower

(2) Here and in the following all statements and expressions containing the symbols "$\pm$" and "$\mp$" are intended to be read twice, once using the upper signs and once using the lower signs simultaneously.

or *jump down* from a critical level to a lower or equal level on the other side of the tower.

The indices of the critical levels are called the *cutting times* and are numbered $(S_k^{\pm})_{k\geqslant 0}$ in increasing order. The Markov diagram is completely determined by the *kneading map Q*.

Proposition 5. *There exists a map* $Q = (Q^+, Q^-) : \mathbb{N} \to \mathbb{N}_\infty \times \mathbb{N}_\infty$ *such that*

$$S_k^{\pm} - S_{k-1}^{\pm} = S_{Q^{\pm}(k)}^{\mp} \qquad \textit{for all } k \geqslant 1.$$

The map Q is called the kneading map *of f.*

A characteristic feature of the kneading map is the so called *Hofbauer condition*, a combinatorial condition which is stated in Theorem 22 below.

Decomposition of the Markov Diagram

The information about the combinatorial peculiarities of the Markov diagram can be used to prove the following decomposition into irreducible components. Let us introduce some notation for the following theorem: Write $\hat{C} \twoheadrightarrow \hat{D}$ if there is a chain of transitions connecting $\hat{C}$ with $\hat{D}$. An *irreducible component* $\hat{\mathcal{X}}$ is a maximal collection of levels such that $\hat{C} \twoheadrightarrow \hat{D}$ holds for all $\hat{C}, \hat{D} \in \hat{\mathcal{X}}$ and a level $\hat{D}$ is called *transient* if $\hat{D} \not\twoheadrightarrow \hat{D}$.

Theorem 6. *The Markov diagram* $(\hat{\mathcal{D}}, \to)$ *of a Lorenz map can be decomposed as follows: The ground levels* $\hat{D}_1^+$ *and* $\hat{D}_1^-$ *are always transient. All other levels can be grouped into a finite chain* $\hat{\mathcal{T}}_1 < \hat{\mathcal{X}}_1 < \hat{\mathcal{T}}_2 < \hat{\mathcal{X}}_2 < \cdots < \hat{\mathcal{X}}_m < \hat{\mathcal{T}}_{m+1}$ *or an infinite chain* $\hat{\mathcal{T}}_1 < \hat{\mathcal{X}}_1 < \hat{\mathcal{T}}_2 < \hat{\mathcal{X}}_2 < \ldots$ *of disjoint subsets of* $\hat{\mathcal{D}} \setminus \{\hat{D}_1^+, \hat{D}_1^-\}$ *which are of the following type:*

$\hat{\mathcal{X}}_i$ *is an irreducible component*
$\hat{\mathcal{T}}_i$ *is either void or contains only transient levels*

The ordering signifies that there are paths traversing the components from left to right but no paths in the other direction.

Renormalization

We introduce two types of renormalizations for Lorenz maps, *proper* and *non-proper* renormalizations. Roughly speaking, a Lorenz map is (n, m)*-renormalizable* if one can find branches of some iterates f^n and f^m to the left and to the right of the critical point, respectively, such that the restriction of those two branches to the interval $[f^m(c^+), f^n(c^-)]$ looks like a Lorenz map which has been restricted to its dynamical interval. If the two branches extend to a larger interval $[\tilde{p}, \tilde{q}]$ such that the restriction to $[\tilde{p}, \tilde{q}]$ looks like a complete

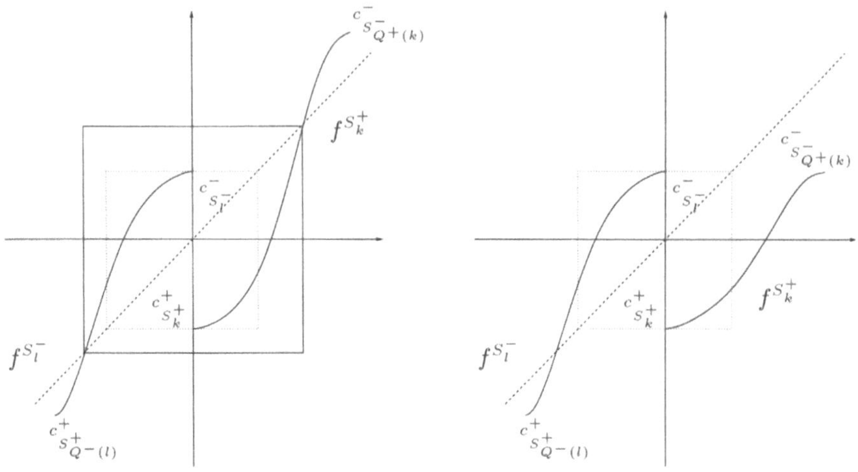

Fig. 2. A properly and a nonproperly renormalizable Lorenz map

Lorenz map again then the renormalization is *proper* (compare Figures 2 and 1, rhs.).

There is a close relation between the above decomposition of the Markov diagram and the possible renormalizations for the Lorenz map. Before stating the following theorem we remark that in the properly renormalizable case both branches are obviously critical (i.e., $m = S_k^+$ and $n = S_l^-$ for some indices k and l) and that the same can be assumed w.l.o.g. in the nonproperly renormalizable case if f has no periodic attractor.

Theorem 7. *Assume that the Lorenz map f has no periodic attractor and let $(\hat{\mathcal{D}}, \rightarrow)$ be its Markov diagram. Then the following holds.*

1. *f is (S_k^+, S_l^-)-renormalizable iff the set $\hat{\hat{\mathcal{D}}}_{k\,l} := \{\hat{D}_i^+, \hat{D}_j^- \mid i > S_k^+, j > S_l^-\}$ of levels above $\hat{D}^+_{S_k^+}$ respectively $\hat{D}^-_{S_l^-}$ is invariant in the Markov diagram.*
2. *f is properly (S_k^+, S_l^-)-renormalizable iff the set $\hat{\hat{\mathcal{D}}}_{k\,l}$ is invariant and the levels $\hat{D}^+_{S_k^+}$ and $\hat{D}^-_{S_l^-}$ belong to an irreducible component $\hat{\mathcal{X}}_i$ immediately below $\hat{\hat{\mathcal{D}}}_{k\,l}$.*

The kneading map $\mathcal{R}Q$ of the renormalized map is given by "shifting" the kneading map Q, i.e., $\mathcal{R}Q^+(j-k) = Q^+(j) - l$ for $j > k$ and $\mathcal{R}Q^-(j-l) = Q^-(j) - k$ for $j > l$.

Stated briefly, this means that proper renormalizations of f correspond to the irreducible components of the Markov diagram and nonproper renormalizations of f correspond to the transient parts in between.

3 Hopf Decompositions and Attractors

Transfer Operators

The smooth structure on the Markov extension makes it possible to study the time evolution of densities of absolutely continuous measures under the influence of $\hat{f}$ by means of the *transfer operator* $\mathrm{P}_{\hat{f}}$ (also known as the *Perron-Frobenius operator*) on the tower. It is defined with respect to a reference measure $\hat{m}$ which is the natural lift of the Lebesgue measure m onto the levels of the tower via the projection π.(3)

Definition 8. The *transfer operator* for $\hat{f}$ is the map $\mathrm{P}_{\hat{f}} : \mathrm{L}^1_{\hat{m}}(\hat{X}) \to \mathrm{L}^1_{\hat{m}}(\hat{X})$ defined by $\hat{f}(\hat{\psi} \cdot \hat{m}) = \mathrm{P}_{\hat{f}}(\hat{\psi}) \cdot \hat{m}$. Its explicit form is given by

$$\mathrm{P}_{\hat{f}}(\hat{\psi}) = \sum_{\hat{Z} \in \hat{\mathcal{Z}}} \mathrm{P}_{\hat{f}}(\hat{\psi}_{\hat{Z}}) = \sum_{\hat{Z} \in \hat{\mathcal{Z}}} (\hat{\psi} \cdot \hat{\phi}) \circ \hat{f}_{\hat{Z}}^{-1} ,$$

where $\hat{\psi}_{\hat{Z}} := \hat{\psi} \cdot 1_{\hat{Z}}$, $\hat{\phi} := \frac{1}{|\hat{f}'|}$ and $(\hat{\psi} \cdot \hat{\phi}) \circ \hat{f}_{\hat{Z}}^{-1} := (\hat{\psi} \cdot \hat{\phi})|_{\hat{Z}} \circ (\hat{f}|_{\hat{Z}})^{-1} \cdot 1_{\hat{f}(\hat{Z})}$. Using its explicit form the transfer operator can also be defined for arbitrary nonnegative measurable functions. When talking of *densities* or *mass distributions* we mean such nonnegative measurable functions, not necessarily integrable.

Similarly, from the original map $f : X \to X$ and the projection $\pi : \hat{X} \to X$ one obtains transfer operators $\mathrm{P}_f : \mathrm{L}^1_m(X) \to \mathrm{L}^1_m(X)$ and $\mathrm{P}_\pi : \mathrm{L}^1_{\hat{m}}(\hat{X}) \to \mathrm{L}^1_m(X)$, respectively. The operators P_f and $\mathrm{P}_{\hat{f}}$ are related by $\mathrm{P}_f \circ \mathrm{P}_\pi = \mathrm{P}_\pi \circ \mathrm{P}_{\hat{f}}$. By construction, the densities of absolutely continuous invariant measures are precisely the fixed points of the transfer operator: $\hat{f}(\hat{h} \cdot \hat{m}) = \hat{h} \cdot \hat{m}$ if and only if $\hat{\mathrm{P}}(\hat{h}) = \hat{h}$, and $f(h \cdot m) = h \cdot m$ if and only if $P(h) = h$. Moreover, if $\hat{h}$ is an invariant probability density for $\hat{f}$ then $h := \mathrm{P}_\pi(\hat{h})$ is an invariant probability density for f.

Regular Densities

As in the theory of countable state Markov chains the topological Markov structure already yields a lot of information about the existence and location of absolutely continuous invariant measures, since the mass can only be transported along the allowed transitions in the tower. Additionally, due to the non-linearity of the map $\hat{f}$, the mass will be compressed in some parts of the tower and thinned out in other parts. How big this effect is depends on the distortion of the map. In order to control the distortion we assume that

(3) Its explicit form is given by $\hat{m}(\hat{A}) := \sum_{\hat{D} \in \hat{\mathcal{D}}} m(\pi(\hat{A} \cap \hat{D}))$ for every measurable set $\hat{A} \subseteq \hat{X}$.

the map has negative Schwarzian derivative, i.e., that f is a $\mathcal{C}^3$-Lorenz map with $f' > 0$ and $Sf := \frac{f'''}{f'} - \frac{3}{2}\left(\frac{f''}{f'}\right)^2 < 0$ on both sides of the discontinuity.

Let us briefly state some properties of maps with negative Schwarzian derivative. From the chain rule for the Schwarzian derivative[(4)] it follows that iterates of maps with negative (resp. positive) Schwarzian derivative again have negative (resp. positive) Schwarzian derivative. A straightforward calculation yields the equality $(|f'|^{-\frac{1}{2}})'' = -\frac{1}{2}|f'|^{-\frac{1}{2}} Sf$ which shows that $Sf < 0$ (resp. $Sf > 0$) if and only if $|f'|^{-\frac{1}{2}}$ is strictly convex (resp. strictly convave). From this one immediately obtains the *Koebe Principle* which is an important tool to estimate the distortion on the branches of iterates of f.

Lemma 9 (Koebe Principle). *Let $I \subsetneq Z$ be two intervals and let $g : Z \to g(Z)$ be a diffeomorphism with negative Schwarzian derivative. If $g(Z)$ contains a δ-scaled neighbourhood of $g(I)$, i.e., if the length of both components of $g(Z) \setminus g(I)$ is at least $\delta \cdot |g(I)|$, then the distortion $\sup_{x,y \in I} |f'(x)/f'(y)|$ on I is bounded by $\left(\frac{1+\delta}{\delta}\right)^2$.*

The proof of the Koebe Principle is elementary[(5)] and based on the fact that $h := g^{-1}$ has positive Schwarzian derivative by the chain rule, whence $|h'|^{-\frac{1}{2}}$ is strictly concave on $g(Z)$. Intuitively, the Koebe Principle states that maps with negative Schwarzian derivative behave very much like linear maps on every branch if one stays away far enough from the ends of the branch, i.e., if there is some "Koebe space". That makes it sound reasonable that densities are not distorted too much when transported along such branches—at least this holds for the part of the mass that is located away from the boundary of the branch. Indeed, Keller [21] showed using results of Misiurewicz [29] that for maps with negative Schwarzian derivative there is a large positive cone of nice *regular densities* $\hat{\psi}$ which is invariant by the transfer operator and for which the quotients $\hat{\psi}(\hat{x})/\hat{\psi}(\hat{y})$ are bounded uniformly in $\hat{\psi}$ on every compact subset of the tower.

Proposition 10. *i) $\mathcal{H}$ is a positive cone which is closed in the compact open topology and $\mathcal{H} - \mathcal{H} = \mathcal{C}(\hat{X})$ is dense in $\mathrm{L}^1_{\hat{m}}(\hat{X})$. The cones $\mathcal{H} \cap \mathrm{L}^1_{\hat{m}}(\hat{X})$ and $\mathcal{H}_2 \cap \mathrm{L}^1_{\hat{m}}(\hat{X})$ are invariant by $\hat{\mathrm{P}}$.*

ii) For every $\delta > 0$ there exists a constant $C(\delta) > 0$ with the following property: Whenever $\hat{D} \in \hat{\mathcal{D}}$ contains a δ-scaled neighbourhood of $\hat{J} \subset \hat{D}$ and $\hat{\psi} \in \mathcal{H}$ is positive on $\hat{D}$ then

$$\left| \frac{\hat{\psi}(\hat{x})}{\hat{\psi}(\hat{y})} - 1 \right| \leqslant \frac{C(\delta)}{|\hat{J}|} \cdot |\hat{x} - \hat{y}| \quad \textit{for all } \hat{x}, \hat{y} \in \hat{J}.$$

(4) The chain rule for the Schwarzian derivative is $S(g \circ f) = Sg \circ f \cdot |f'|^2 + Sf$.

(5) A much more general version of the Koebe Principle is proved in [5, Section IV.1].

In particular, for every compact set $\hat{K} \subset \hat{X}$ there is a uniform bound for the Lipschitz norm of the family $\{\log \hat{\psi}|_{\hat{K}} \mid \hat{\psi} \in \mathcal{H}, \hat{\psi}|_{\hat{K}} > 0\}$ which depends only on the relative Koebe space around $\hat{K}$ in $\hat{X}$.

The Hopf Decomposition

As a consequence of the distortion bounds the transfer operator on the Markov extension behaves very much like a countable state stochastic Markov chain and it is possible to derive strong statements about the conservative and dissipative part and the existence of absolutely continuous invariant measures for $\hat{f}$. Before stating the results let us briefly recall some facts about the Hopf decomposition and introduce some notation (cf. Krengel [24]).

A measurable set $\hat{A} \subset \hat{X}$ is called a *wandering set* if $\hat{f}^n(\hat{A}) \cap \hat{A} = \emptyset$ for all $n > 0$. The map $\hat{f}$ is called *conservative* if it has no wandering sets of positive measure. If it does have wandering sets of positive measure then it is called *dissipative.*

Theorem 11 (Hopf decomposition). *There is a partition $\hat{X} = \hat{X}_c \cup \hat{X}_d$ into two disjoint measurable subsets which are characterized uniquely modulo null sets by the following properties:*

1. *$\hat{X}_c$ is invariant by $\hat{f}$ and the restriction of $\hat{f}$ to $\hat{X}_c$ is conservative.*
2. *$\hat{X}_d$ is a countable union of wandering sets.*

The sets $\hat{X}_c$ and $\hat{X}_d$ are called the conservative *and* dissipative *part of $\hat{f}$, respectively.*

The conservative part $\hat{X}_c$ can be split further into two parts $\hat{X}_+$ and $\hat{X}_0 := \hat{X}_c \setminus \hat{X}_+$, where $\hat{X}_+$ is the largest set supporting a finite absolutely continuous invariant measure. $\hat{X}_+$ is called the called the *positively recurrent* part and $\hat{X}_0$ is called the *null recurrent* part of $\hat{P}$.

Definition 12. For a nonnegative measurable function $\hat{\psi}$ let

$$\hat{\mathrm{S}}_n \hat{\psi} := \sum_{k=0}^{n-1} \hat{\mathrm{P}}^k \hat{\psi} \quad \text{and} \quad \hat{\mathrm{A}}_n \hat{\psi} := \frac{1}{n} \sum_{k=0}^{n-1} \hat{\mathrm{P}}^k \hat{\psi}$$

and denote by $\hat{\mathrm{S}}_\infty \hat{\psi}$ and $\hat{\mathrm{A}}_\infty \hat{\psi}$ the corresponding limits for $n \to \infty$ if they exist.

Roughly speaking, for a given initial distribution $\hat{\psi}$ the sum $\hat{\mathrm{S}}_n \hat{\psi}(\hat{x})$ measures the total amount of mass that passes through the point $\hat{x}$ in n time steps when the distribution evolves under the influence of $\hat{f}$, and $\hat{\mathrm{A}}_n \hat{\psi}(\hat{x})$ measures the average amount. Consequently, the behaviour of $\hat{\mathrm{S}}_n \hat{\psi}$ for $n \to \infty$ can be used to distinguish the conservative and dissipative part, and the behaviour of $\hat{\mathrm{A}}_n \hat{\psi}$ for $n \to \infty$ can be used to distinguish the null recurrent from the positively recurrent part.

Proposition 13. *The Hopf decomposition is determined uniquely modulo null sets by the following property: If* $\hat{\psi}$ *is a positive integrable function then* $\hat{X}_c = \{\hat{\mathrm{S}}_\infty \hat{\psi} = \infty\}$ *a.s. and* $\hat{X}_d = \{\hat{\mathrm{S}}_\infty \hat{\psi} < \infty\}$ *a.s.*

If one starts with an integrable regular density $\hat{\psi}$ then all partial sums $\hat{\mathrm{S}}_n \hat{\psi}$ are again integrable regular densities, so that the sequence $(\log \hat{\mathrm{S}}_n \hat{\psi})_{n\in\mathbb{N}}$ is equicontinuous on compact subsets of $\hat{X}$ by Proposition 10. Therefore, the asymptotic behaviour of the sequence $(\hat{\mathrm{S}}_n \hat{\psi}(\hat{x}_0))_{n\in\mathbb{N}}$ at a particular point $\hat{x}_0$ should determine the behaviour of the sequences $(\hat{\mathrm{S}}_n \hat{\psi}(\hat{x}))_{n\in\mathbb{N}}$ for all other points $\hat{x}$ that belong to the same irreducible component as $\hat{x}_0$. This is the core of the following theorem, which we adopt from Keller [21] to our setting.

Theorem 14. *Fix a density* $\hat{\psi} \in \mathcal{H} \cap \mathrm{L}^1_{\hat{m}}$ *and a point* $\hat{x}_0$ *from a level* $\hat{D}$ *in the Hofbauer tower, and let* $s_n := \hat{\mathrm{S}}_n \hat{\psi}(\hat{x}_0)$ *for* $n \in \mathbb{N}_\infty$*. Then* $s_n = O(n)$ *and if* $s_\infty > 0$ *then the following holds.*

1. *If* $(s_n)_{n\in\mathbb{N}}$ *is bounded then* $\hat{f}$ *is dissipative on* $\hat{D}$ *and on every level* $\hat{C} \twoheadrightarrow \hat{D}$*. The sequence* $(\hat{\mathrm{S}}_n \hat{\psi})_{\hat{C}}$ *converges uniformly on compact sets to* $(\hat{\mathrm{S}}_\infty \hat{\psi})_{\hat{C}} \in \mathcal{H}$*. If* $(s_n)_{n\in\mathbb{N}}$ *is unbounded then* $\hat{D}$ *belongs to the irreducible component* $\hat{\mathcal{X}}_{\mathrm{m}}$ *which is maximal with respect to the ordering introduced in Theorem 6*[6] *and* $\hat{f}$ *is conservative on* $\hat{X}_{\mathrm{m}} := \bigcup\{\hat{C} \mid \hat{C} \in \hat{\mathcal{X}}_{\mathrm{m}}\}$*.*
2. *If* $\hat{f}$ *is conservative on* $\hat{X}_{\mathrm{m}}$ *then it is* $\hat{m}$*-ergodic and the sequence* $\frac{1}{s_n} \hat{\mathrm{S}}_n \hat{\psi}$ *converges uniformly on compact sets to a function* $\hat{h} \in \mathcal{H}$ *which is strictly positive on* $\hat{X}_{\mathrm{m}}$ *and vanishes everywhere else. The function* $\hat{h}$ *is the unique*[7] *positive measurable function which is invariant by* $\hat{f}$*. For every* $\hat{\phi} \in \mathrm{L}^1_{\hat{m}}$
$$\lim_{n\to\infty} \frac{1}{s_n} \hat{\mathrm{S}}_n \hat{\phi} = \frac{\int \hat{\phi}\, d\hat{m}}{\int \hat{\psi}\, d\hat{m}} \cdot \hat{h} \quad \textit{holds } \hat{m}\textit{-a.e. on } \hat{X}_{\mathrm{m}}.$$
3. *If* $\hat{f}$ *is conservative on* $\hat{X}_{\mathrm{m}}$ *then* $s_n/n \to \gamma$ *for some constant* $\gamma \geqslant 0$ *and* $\hat{h}$ *is integrable if and only if* $\gamma > 0$*. In particular, for every* $\hat{\phi} \in \mathrm{L}^1_{\hat{m}}$
$$\lim_{n\to\infty} \frac{1}{n} \hat{\mathrm{S}}_n \hat{\phi} = \frac{\int \hat{\phi}\, d\hat{m}}{\int \hat{h}\, d\hat{m}} \cdot \hat{h} \quad \textit{holds } \hat{m}\textit{-a.e. on } \hat{X}_{\mathrm{m}}.$$
where the right hand side is zero if $\int \hat{h}\, d\hat{m} = \infty$*.*
   (a) *If* $\int \hat{h}\, d\hat{m} < \infty$ *then the measure preserving system* $(\hat{f}, \hat{h} \cdot \hat{m})$ *is the product of an exact system with a finite rotation.*
   (b) *If* $\int \hat{h}\, d\hat{m} = \infty$ *then* $\hat{\mathrm{P}}^n \hat{\psi}$ *tends to zero uniformly on compact sets for every* $\hat{\psi} \in \mathcal{H} \cap \mathrm{L}^1_{\hat{m}}$ *with* $\hat{\psi} \leqslant \hat{h}$*.*

If $\hat{f}$ is conservative on $\hat{X}_{\mathrm{m}}$ then $\hat{f}$ is called *essentially conservative* and if this is not the case then it is called *purely dissipative.*

---
[6] I.e., there is only a finite chain in Theorem 6 ending with $\hat{\mathcal{T}}_{m+1} = \emptyset$.

[7] In this context "unique" always means "unique up to sets of measure zero and multiplication by a constant".

Asymptotic Behaviour of Orbits on the Tower

The Hopf decomposition already provides a lot of information about the asymptotic behaviour of typical points on the tower. For example, if $\hat{f}$ is essentially conservative then the ω-limit set equals $\hat{X}_{\mathrm{m}}$ for $\hat{m}$-a.e. $\hat{x}$ and if $\hat{f}$ is purely dissipative then for $\hat{m}$-a.e. $\hat{x}$ the distance $r_n(\hat{x})$ of the nth iterate $\hat{x}_n$ to the endpoints of the current level $\hat{D}[\hat{x}_n]$ tends to zero for $n \to \infty$.

Theorem 15. *Let f be a Lorenz map with negative Schwarzian derivative and let $(\hat{X}, \hat{f})$ be its canonical Markov extension. Then the following holds.*

1. *If $\hat{f}$ is purely dissipative then $r_n(\hat{x}) \to 0$ for $\hat{m}$-a.e. $\hat{x} \in \hat{X}$.*
2. *If $\hat{f}$ is essentially conservative then $\omega(\hat{x}) = \hat{X}_{\mathrm{m}}$ for $\hat{m}$-a.e. $\hat{x} \in \hat{X}$. Moreover,*
   (a) if $\int \hat{h}\, d\hat{m} = 1$ then $\frac{1}{n}\sum_{k=0}^{n-1} \delta_{\hat{x}_k} \to \hat{\mu}$ weakly as $n \to \infty$ for $\hat{m}$-a.e. $\hat{x}$.
   (b) if $\int \hat{h}\, d\hat{m} = \infty$ then $\lim_{n\to\infty} \frac{1}{n} \operatorname{card}\{k < n \mid r_k(\hat{x}) \geqslant \varepsilon\} = 0$ holds for $\hat{m}$-a.e. $\hat{x}$ and every $\varepsilon > 0$.

The Global Attractor

The information about the asymptotic behaviour on the tower can be projected down to the original system and one obtains the following information about the global attractor of a Lorenz map f with negative Schwarzian derivative.

Theorem 16. *If f is a Lorenz map with negative Schwarzian derivative then f has a unique global attractor A which is the union of one or two minimal attractors and for m-almost every point x the ω-limit set* coincides *with a minimal attractor. More precisely, one of the following three cases applies.*

1. *If f has an attracting periodic orbit then A is the union of one or two attracting periodic orbits, each of which attracts at least one of the critical values c_1^+ and c_1^-.*

Now assume that f has no attracting periodic orbits. Then $\omega(x) = A$ for m-a.e. x and one of the following holds.

2. *If f is infinitely often renormalizable then $A = \omega(c_1^+) = \omega(c_1^-)$. If f is infinitely often properly renormalizable then A is a Cantor set.*
3. *If f is finitely often renormalizable then either*
   (a) A is a finite union of intervals
   or A is a nowhere dense set which is of one of the following types.
   (b) $A = \omega(c_1^+) = \omega(c_1^-)$.
   (c) $A = \omega(c_1^\pm) \supset \mathrm{cl}(\mathrm{orb}(c_1^\mp))$ with $c_1^\pm \in \omega(c_1^\pm)$.
   (d) $A = \mathrm{cl}(\mathrm{orb}(c_1^\pm)) \supset \mathrm{cl}(\mathrm{orb}(c_1^\mp))$ with $c_1^\pm \notin \omega(c_1^\pm)$.

The only possibility where two attractors can coexist is the case of two attracting periodic orbits.

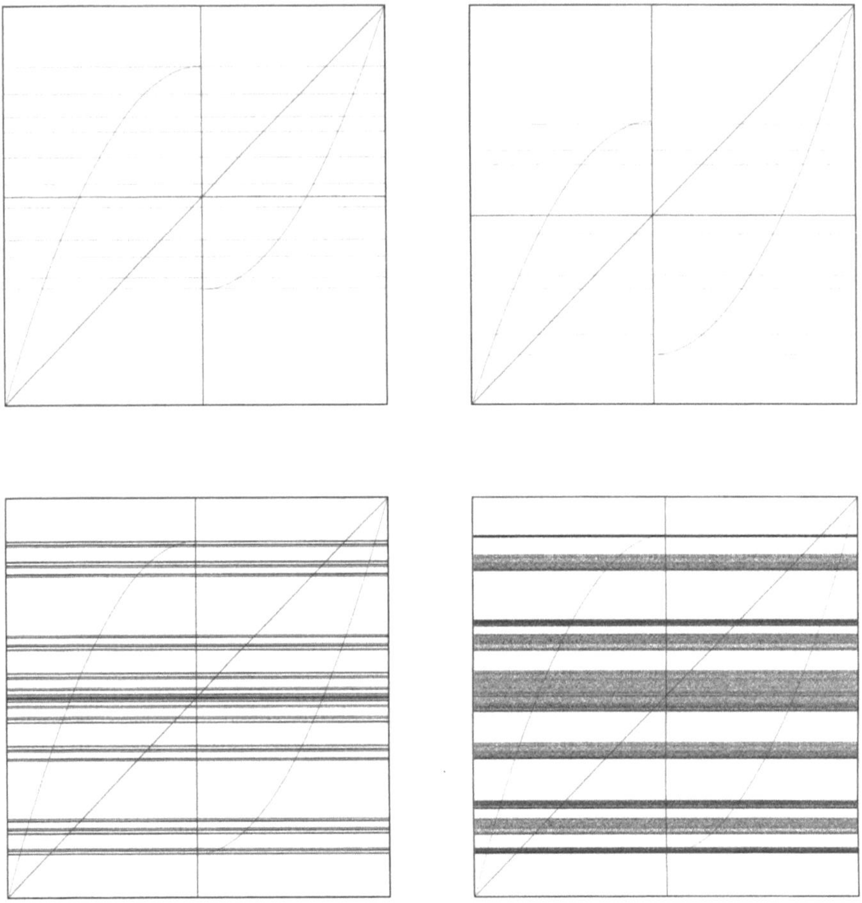

Fig. 3. Lorenz maps with one periodic attractor, two periodic attractors, an infinitely often renormalizable Cantor attractor, and an interval attractor (from left to right).

Figure 3 shows examples of Lorenz maps with attractors of types 1–3a. The attractor is visualized by plotting an approximation of $\omega(x)$ as function of x together with the map.

This theorem is very similar to the well known metric classification of attractors for unimodal maps by Blokh and Lyubich (cf. [5, Theorem V.1.3.]) except for cases 3c and 3d. In fact, there is a very close relationship between symmetric Lorenz maps and unimodal maps: A Lorenz map $f : [p,q] \to [p,q]$ is called *τ-symmetric* if there exists a continuous involution $\tau \neq \mathrm{id}$ with $\tau(c) = c$ and $f \circ \tau = \tau \circ f$. It is easily seen that a Lorenz map f is τ-symmetric iff there exists a unimodal map $g : [p,q] \to [p,q]$ with $g \circ \tau = g$ such that

$f(x) = g(x)$ for $x < c$ and $f(x) = \tau \circ g(x)$ for $x > c$. Now if g has an attractor of one of the types 1–3b from [5, Theorem V.1.3.] and if the involution τ is absolutely continuous then the symmetric Lorenz map f has an attractor of the same type.(8)

This immediately shows that all attractors 1–3a actually occur in the family $f_a : x \mapsto (-a + x^2) \cdot \mathrm{sgn}(x)$ of symmetric quadratic Lorenz maps and attractors of type 3b occur in the family $f_a : x \mapsto (-a + |x|^\alpha) \cdot \mathrm{sgn}(x)$ if $\alpha > 1$ is large enough, see Bruin et al. [3]. Also, it is not hard to construct asymmetric examples for types 1–3b which are renormalizable to one of the symmetric examples above.

Attractors of types 3c and 3d have no unimodal counterpart and it is not clear whether they actually occur. At least they are possible from a topological viewpoint, since we are able to construct examples of Lorenz maps for which the orbits of c^+ and c^- meet the required properties.

Remark 17. Attractors of type 3d appear to be a little bit exotic: If a Lorenz map has an attractor of type 3d, say, if $A = \mathrm{cl}(\mathrm{orb}(c_1^+))$, then A is the disjoint union of two sets A_α and A_ω, where A_α is a finite or countably infinite set of points that are isolated in A and $A_\omega = \omega(A_\alpha)$ is a nowhere dense set which accumulates on the left hand side but not on the right hand side of c. This implies that such maps exhibit the following kind of intermittent behaviour: The orbit of a typical point x will spend most of its time near A_ω until some iterate of x happens to land very close to c on the right hand side, where it is separated from A_ω in the next step and has to start all over again following A_α, and so on.

SRB-Measures

If the lifted map $\hat{f}$ has a finite absolutely continuous invariant measure $\hat{\mu}$ then the measure can be projected from the tower to the base and one obtains an absolutely continuous ergodic invariant measure μ of positive entropy for f.

Theorem 18. *If f is a Lorenz map with negative Schwarzian derivative then f has an absolutely continuous invariant measure μ of positive entropy if and only if $\hat{f}$ is positively recurrent on $\hat{X}_{\mathrm{m}}$. If this is the case then $\mu = h \cdot m$, where $h := \mathrm{P}_\pi \hat{h}$, μ is ergodic and $\lambda^+(x) > 0$ on a set of full Lebesgue measure.*

It was already said that such a measure is an SRB-measure. If this is not the case then still the following can be said about the weak limit sets of the empirical distributions for typical points.

(8) Observe that an orientation *reversing* n-periodic attractor for g yields one symmetric $2n$-periodic attractor for f and an orientation *preserving* n-periodic attractor for g yields two distinct n-periodic attractors for f which are mirror images of each other.

Theorem 19. *If f is a Lorenz map with negative Schwarzian derivative which has no absolutely continuous invariant probability measure of positive entropy then $\omega^*(x)$ is contained in the convex closure of $\omega^*(c_1^+) \cup \omega^*(c_1^-)$ and $\lambda^+(x) \le 0$ for a.e. $x \in X$.*

The reason for this is that the orbit of a typical point $\hat{x}$ spends more and more of its time climbing very high up in the tower, which implies that in average there are longer and longer blocks $x_{m+1}, \dots, x_{m+n}$ where the orbit of x shadows one of the critical orbits $c^\pm$ in the sense that x_{m+i} and $c_i^\pm$ lie on the same side of the critical point for $i = 1, \dots, n$. Although this does not mean that the distance $\mathrm{dist}(x_{m+i}, c_i^\pm)$ is small during the entire shadow—even more since one cannot exclude the existence of wandering intervals for Lorenz maps in general—it is possible to show that at least it is small most of the time when the length of the shadow is large enough. The dichotomy in terms of $\lambda^+(x)$ expressed by Theorems 18 and 19 is taken from [21].

4 Kneading Theory

The Kneading Invariant

The most important property that distinguishes one-dimensional dynamical systems from higher-dimensional systems is the presence of an order structure on $\mathbb{R}$. This order structure is heavily exploited by kneading theory: Since the order between two points x and y is preserved by a Lorenz map under iteration as long as the iterates x_n and y_n lie on the same side of the discontinuity, the information about the relative position of the iterates of a point x with respect to the discontinuity is of great importance. This information can be encoded in a binary sequence $\zeta(x) := (\zeta_n(x))_{n \geqslant 0}$, where $\zeta_k(x) := 0$ if $f^k(x) < c$ and $\zeta_k(x) := 1$ if $f^k(x) > c$, which is called the *itinerary* of x. Obviously, the map ζ is monotonous from the interval to the one-sided shift space and satisfies $\zeta \circ f = \sigma \circ \zeta$, where σ denotes the shift map. If one modifies the interval by doubling all preiterates of the critical point topologically and extends ζ to the resulting topological space X_f by taking one-sided limits then ζ is a semiconjugacy onto a closed invariant subshift Σ_f of the one-sided shift space. It is a conjugacy if and only if f is topologically expansive.[9]

The most important itineraries are the two limits at the critical point, together forming the *kneading invariant* $\nu := (\nu^+, \nu^-) := (\zeta(c^+), \zeta(c^-))$ of the Lorenz map f. The kneading invariant completely determines the set of all admissible itineraries of points in the dynamical interval: A sequence $\omega \in \Sigma$ occurs as the itinerary of some point $x \in X_f$ if and only if it satisfies

$$\sigma\nu^+ \leqslant \sigma^n\omega \leqslant \sigma\nu^- \qquad \forall n \geqslant 0 .$$

[9] A Lorenz map f is called *topologically expansive* if there exists some $\varepsilon > 0$ with the following property: For any two distinct points x and y there exists some n such that $d(x_n, y_n) \geqslant \varepsilon$.

Since the above equation applies in particular to the itineraries ν^+ and ν^-, any pair of binary sequences that occurs as the kneading invariant of some Lorenz map must necessarily satisfy the following *admissibility condition.*

Definition 20 (admissibility condition). A pair $\nu = (\nu^+, \nu^-)$ of binary sequences is called *admissible*, if

$$\sigma\nu^+ \leqslant \sigma^n\nu^\pm \leqslant \sigma\nu^- \quad \text{for every } n \in \mathbb{N}. \tag{AC}$$

Here and in the following we tacitly assume that $\nu = (10*, 01*)$.

Conversely, given an admissible pair we are able to construct a continuous Lorenz map with precisely this pair as kneading invariant. This yields the following theorem.

Theorem 21. *A pair $\nu = (\nu^+, \nu^-)$ of binary sequences is realizable as the kneading invariant of a Lorenz map f if and only if it satisfies condition* (AC).

This theorem completes a theorem of Hubbard & Sparrow [18] where they showed that a pair ν of binary sequences is realizable as the kneading invariant of a topologically expansive Lorenz map f if and only if it satisfies the following *expansive admissibility condition:*

$$\sigma\nu^+ \leqslant \sigma^n\nu^+ < \sigma\nu^- \quad \text{and} \quad \sigma\nu^+ < \sigma^n\nu^- \leqslant \sigma\nu^- \quad \text{for every } n \in \mathbb{N}. \tag{EAC}$$

Splitting Itineraries and Admissibility Conditions

The kneading invariant and the Markov diagram are two equivalent methods to describe the combinatorial behaviour of a Lorenz map in the sense that it is possible to translate the combinatorial information contained in the kneading invariant into the language of the Markov diagram, and vice versa. The key to this equivalence is the fact that the cutting times of the Hofbauer tower and hence the kneading map can be determined through a simple splitting technique from the kneading invariant: The first cutting$^+$ time is $S_0^+ = 1$ and if $m = S_{k-1}^+$ is a cutting$^+$ time, then $S_k^+ = n$ where $n > m$ is the minimal integer such that $c_n^+ < c < c_{n-m}^-$, i.e, the minimal integer at which the itineraries of c_{m+1}^+ and c_1^- differ:

$$\underbrace{\nu_{m+1}^+ \cdots \nu_{n-1}^+}\, \nu_n^+ = \underbrace{\nu_1^- \cdots \nu_{n-m-1}^-}\, \mathring{\nu}_{n-m}^- ,$$

where $\mathring{\nu}_k^- := 0$ if $\nu_k^- = 1$, and vice versa. Continuing in this way one obtains a splitting of ν^+ into blocks $\Delta_k^- := \nu_{S_{k-1}^+ +1}^+ \nu_{S_{k-1}^+ +2}^+ \cdots \nu_{S_k^+}^+$, i.e.,

$$\nu^+ = 1\,0\,\Delta_1^-\,\Delta_2^-\,\Delta_3^- \,\ldots ,$$

where each block Δ_k^- is a copy of the itinerary $\nu_1^- \nu_2^- \dots$ except for the last digit. (It may happen that there are only finitely many blocks and the last block has infinite length.) The cutting$^-$ times can be determined symmetrically.

The splitting technique was already used by Hofbauer and then systematized by Bruin [1,2] and Sands [32]. They introduced the so-called co-cutting times which provide additional information about the recurrence behaviour of the critical point. For Lorenz maps, the co-cutting$^+$ times can be obtained by the following modified splitting of ν^+: Starting from the smallest index greater than $S_0^+ = 1$ which is no cutting$^+$ time (i.e., the second digit that equals 1) one splits the itinerary ν^+ into blocks $\widetilde{\Delta}_k^+ := \nu^+_{\tilde{S}^+_{k-1}+1} \nu^+_{\tilde{S}^+_{k-1}+2} \dots \nu^+_{\tilde{S}^+_k}$, i.e.,

$$\nu^+ = 1\,0^{\tilde{S}_0^+ - 1} 1\, \widetilde{\Delta}_1^+ \,\widetilde{\Delta}_2^+ \,\widetilde{\Delta}_3^+ \,\dots\ ,$$

where each block $\widetilde{\Delta}_k^+$ is a copy of the itinerary $\nu_1^+ \nu_2^+ \dots$ except for the last digit. The co-cutting$^-$ times are defined symmetrically.

Note that the splitting of a kneading is obtained by comparing w.r.t. the other kneading, whereas the co-splitting is obtained by comparing w.r.t. itself. Using this technique we obtain a reformulation of the *admissibility condition* for kneading invariants in the language of the Markov diagrams, i.e., a necessary and sufficient combinatorial condition for an abstract graph to occur as the Markov diagram of some Lorenz map.

Theorem 22. *For an arbitrary pair* $(\nu^+, \nu^-) = (10*, 01*)$ *of binary sequences the following conditions are equivalent*

1. *For every* $n \in \mathbb{N}$ *the inequality* $\sigma\nu^+ \leqslant \sigma^n \nu^\pm \leqslant \sigma\nu^-$ *holds.*
2. *A cutting*$^\pm$ *is never a co-cutting*$^\pm$ *time and vice versa.*
3. *The differences between consecutive cutting*$^\pm$ *times are cutting*$^\mp$ *times and the differences between consecutive co-cutting*$^\pm$ *times are cutting*$^\pm$ *times.*
4. *The differences between consecutive cutting*$^\pm$ *times are cutting*$^\mp$ *times, i.e., there exist integers* $Q^\pm(k)$, $k \geqslant 1$, *such that*

$$S_k^\pm - S_{k-1}^\pm = S^\mp_{Q^\pm(k)}$$

and the kneading map $Q^\pm$ *satisfies the* Hofbauer condition*:*

$$\left(Q^\pm(k+j)\right)_{j\geqslant 1} \geqslant \left(Q^\pm(Q^\mp Q^\pm(k)+j)\right)_{j\geqslant 1} \quad \forall k \geqslant 1\ ,$$

where the ordering $\leqslant$ *is just the lexicographical ordering of sequences.*

Although at first sight the formulation of admissibility in terms of the kneading map (condition 4) looks much more complicated than the other three, it is in practice very powerful, since it contains the essential information about the topological properties of the Lorenz map in a condensed form.[10] This makes it very easy to construct Lorenz maps with prescribed topological properties, e.g., renormalizable maps or maps where the critical points are recurrent, respectively not recurrent, and so on.

5 Families of Lorenz Maps

Since the original motivation for studying Lorenz maps came from the analysis of homoclinic bifurcations of vector fields, it is only natural to study parametrized families of Lorenz maps and to ask how the dynamics of the maps in this family depends on its parameters.

Full Families

Given a family of Lorenz maps, we ask for sufficient conditions which guarantee that the maps of this family exhibit all different topological types of dynamical behaviour in the sense that every admissible kneading invariant occurs for a map in this family.

Definition 23. A family $\mathcal{F}$ of Lorenz maps is called a *full family* if every admissible kneading invariant ν occurs as the kneading invariant of some map $f \in \mathcal{F}$.

An obvious requirement for such a family is that should at least contain maps with "all different kinds of branches", i.e., by adjusting the parameters it should be possible to tune the length of both branches independently over the whole range—from short (i.e., not critical) to long (i.e., surjective) branches. This is the case for the following two parameter family of Lorenz maps.

Definition 24 (C^1-Lorenz family). Let Λ be an open simply connected subset of $\mathbb{R}^2$. A family $\mathcal{F} := (f_{a,b})_{(a,b)\in\Lambda}$ of maps $f_{a,b} : \mathbb{R} \to \mathbb{R}$ is called a *C^1-Lorenz family* if the following properties hold.

1. The map $(a,b,x) \mapsto f_{a,b}(x)$ is continuously differentiable for $x \neq c := 0$ and the one-sided limits $\mathrm{D}_a f_{a,b}(0^\pm) = \mathrm{D}_a c_1^\pm(a,b)$ and $\mathrm{D}_b f_{a,b}(0^\pm) = \mathrm{D}_b c_1^\pm(a,b)$ exist for all parameters $(a,b) \in \Lambda$.
2. $\mathrm{D}_x f_{a,b}(x) > 0$ and $\mathrm{D}_x f_{a,b}(x) \to 0$ for $x \to 0$.
3. The map $f_{a,b}$ has precisely two fixed points $p^-(a,b) < 0$ and $p^+(a,b) > 0$ which are hyperbolic repellers.
4. There are no neutral or repelling fixed points in $(p^-(a,b), p^+(a,b))$.

Additionally, we assume that $\mathbb{R}^2_+ \subseteq \Lambda$ and that the following properties hold.

(10) The dynamical significance of the Hofbauer condition is not so easily unveiled. We only remark here that it is trivially satisfied if both sequences Q^+ and Q^- are monotonously increasing.

5. $c_1^+(0,b) = 0$ and $c_1^-(a,0) = 0$ for all $a, b \geqslant 0$.
6. $\mathrm{D}_a\, c_1^+(a,b) < 0$ and $\mathrm{D}_b\, c_1^-(a,b) > 0$
7. $\mathrm{D}_a\, f_{a,b}(x) \leqslant 0$ and $\mathrm{D}_b\, f_{a,b}(x) \geqslant 0$ for $x \in (p^-(a,b), p^+(a,b))$.
8. There exists a parameter such that $f_{a,b}$ has full branches, i.e., $c_1^+(a,b) = p^-(a,b)$ and $c_1^-(a,b) = p^+(a,b)$.

The first four conditions ensure that the family depends C^1-smoothly on its parameters and is reasonably well behaved. The last four conditions guarantee that the branches can be adjusted arbitrarily and that the branches move monotonously as long as only one of the two parameters is modified.

Prototypes of such C^1-Lorenz families are the families

$$f_{a,b} : x \mapsto \begin{cases} -a + |x|^\alpha & , \text{if } x > 0 \\ b - |x|^\alpha & , \text{if } x < 0 \end{cases} \tag{1}$$

for fixed constant $\alpha > 1$, in particular the *quadratic Lorenz family*

$$f_{a,b} : x \mapsto \begin{cases} -a + x^2 & , \text{if } x > 0 \\ b - x^2 & , \text{if } x < 0 \end{cases} . \tag{2}$$

The parameter space Λ is chosen unnecessarily large and not every map of such a C^1-Lorenz family is really a Lorenz map in the sense that it has an invariant interval restricted to which it looks like the map in Figure 1. However, under the assumptions of the definition it can be shown that there is a bounded simply connected open region $J \subset \Lambda$ such that $p^- < c^+ < 0 < c^- < p^+$ holds for every map $f_{a,b}$, $(a,b) \in J$ and that the boundary of J is the union of four curves where the maps satisfy $c^+ = 0$, $c^+ = p^-$, $c^- = 0$, and $c^- = p^+$, respectively.

Now the answer to the above question is: Every C^1-Lorenz family is a full family—well, almost. There is an exceptional set of kneading invariants which do not necessarily occur, namely the ones where one of the critical itineraries, is a shift of the other one, say, $\nu^+ = \sigma^k \nu^-$ for some $k > 0$,[11] but ν^+ is not periodic. Whenever a Lorenz map has such a kneading invariant, a one-sided neighbourhood of the critical point is necessarily a wandering interval or contained in the basin of an inessential periodic attractor. Such kneading invariants may or may not occur in a C^1-Lorenz family. For families of maps with negative Schwarzian derivative like the quadratic Lorenz family they are certainly missing (cf. [33]). The following theorem shows that this is the only exception. We call it the *Full Family Theorem*, regardless of its little imperfections.

[11] This implies that ν is not expansive, i.e., condition (EAC) on page 349 is violated.

Theorem 25. *Let* $(f_{a,b})_{(a,b)\in J}$ *be a* C^1*-Lorenz family and let* $\nu = (\nu^+, \nu^-)$ *be an admissible kneading invariant such that*

i) ν *is expansive* *or* *ii) at least one of* ν^+ *and* ν^- *is periodic.*

Then there exists a parameter $(a, b) \in \operatorname{cl} J$ *where* $f_{a,b}$ *has the kneading invariant* ν.

We sketch a proof of this theorem which is an adaptation of the proof of the Full Family Theorem for smooth multimodal maps by de Melo & van Strien [5] (see also Martens & de Melo [26]). It is based on the so-called Thurston map which provides an elegant method for solving the following finite version of the problem posed above: Given a kneading invariant as in the theorem with the additional property that both itineraries ν^+ and ν^- are preperiodic, find a parameter where the Lorenz map is *post-critically finite*(12) and has the required kneading invariant. This problem can be reformulated as a fixed point problem for the Thurston map which is a continuous map from a finite dimensional simplex into itself. Since the fixed point is required to lie in the interior of the simplex, the Brouwer Theorem is not applicable immediately, but a closer look at its proof shows that in the given situation it can be modified to prove the existence of such a fixed point.

Although the analysis of the parameter dependence of the kneading invariant yields an independent proof of Theorem 25 as a byproduct, we included the discussion of the Thurston map approach, not only because of its great theoretical importance, but also for a very practical reason: The Thurston algorithm provides a convenient method to find Lorenz maps with specific combinatorial properties in the quadratic Lorenz family which can easily be implemented on a computer. To be honest, one has to cheat a little bit. The algorithm is based on the assumption that the Thurston map for the quadratic family (2) is a contraction with respect to some suitably chosen metric and the fixed point is found by iterating an arbitrary initial value. Unfortunately, we are not able to show that this is indeed the case, but the fact that the Thurston algorithm works very nicely and reliable in practice gives strong evidence for the following conjecture.

Conjecture 26. The Thurston map for the quadratic Lorenz family is a contraction with respect to a suitable chosen metric.

This conjecture implies that for every preperiodic admissible kneading invariant as in Theorem 25 there is precisely one parameter (a, b) such that the Lorenz map $f_{a,b}$ is post-critically finite and its kneading invariant coincides with the given one. This uniqueness problem is closely related to the question of monotonous dependence of the kneading invariant for the quadratic family which we are going to adress below.

(12) This means that the set $\operatorname{orb}(c^+) \cup \operatorname{orb}(c^-)$ is finite, i.e., both critical points are preperiodic.

The Bifurcation Diagram

Of course it is even more interesting not only to know whether all admissible sequence actually occur but also where in parameter space they can be found and how the kneading invariant changes with the parameters. In order to do this we describe the bifurcation diagram of a $\mathcal{C}^1$-Lorenz family, which consists of a refining sequence of partitions of its parameter space that is obtained by distinguishing longer and longer initial parts of the kneading invariants.

Definition 27. Let $(f_{a,b})_{(a,b)\in J}$ be a $\mathcal{C}^1$-Lorenz family. For every $n \in \mathbb{N}$ and every admissible kneading invariant $\nu = (\nu^+, \nu^-)$ let $J_n(\nu)$ be the set of parameters (a, b) such that

1. the critical points $c^+(a,b)$ and $c^-(a,b)$ of the map $f_{a,b}$ are not periodic of any period less than or equal to n, and
2. the kneading invariant $\nu(a,b)$ of $f_{a,b}$ coincides up to the nth digit with ν.

We say that the *admissible n-prefix* $(\nu_0^+ \dots \nu_n^+, \nu_0^- \dots \nu_n^-)$ is the *combinatorial type* of $J_n(\nu)$. The collection of partitions $\mathcal{J}_n := \{J_n(\nu) \mid \nu \text{ admissible }\}$ is called the *bifurcation diagram* of the Lorenz family $\mathcal{F}$.

The bifurcation diagram of the quadratic Lorenz family is shown in Figure 4. The region shaped like an almond is the parameter space J, which is contained in the square $[0,2] \times [0,2]$. Whenever the index of the partition is increased by one, some pieces of the partition break into fragments—which can be of two or four different combinatorial types—while other pieces remain unchanged. The number of possible combinatorial subtypes for a given piece is limited by the number of extensions of the given admissible $(n-1)$-prefix to an admissible n-prefix and can easily be determined using condition (2) of Theorem 22. Conversely, the following theorem shows that in a $\mathcal{C}^1$-Lorenz family every admissible subtype actually occurs.[13]

Theorem 28. *Let $(f_{a,b})_{(a,b)\in J}$ be a $\mathcal{C}^1$-Lorenz family and let $\nu = (\nu^+, \nu^-)$ be an admissible kneading invariant.*

Then there is a decreasing sequence $(G_n(\nu))_{n\in\mathbb{N}}$ of nonvoid simply connected open regions $G_n(\nu) \subseteq J_n(\nu)$. The boundary of each set $G_n(\nu)$ is contained in the union of four differentiable arcs where one of the critical points is either periodic of some period less than or equal to n (which is determined by the cutting and co-cutting times) or is mapped onto one of the repelling fixed points in the boundary.

For every parameter (a,b) in $G_\infty(\nu) := \bigcap_{n\in\mathbb{N}} G_n(\nu)$ the critical points are not periodic and the Lorenz map $f_{a,b}$ has the kneading invariant ν. If ν is expansive or periodic then $G_\infty(\nu)$ is nonvoid. In particular, this proves Theorem 25.

(13) A similar argument was used by Hofbauer & Keller [16] to prove the Full Family Theorem for unimodal maps.

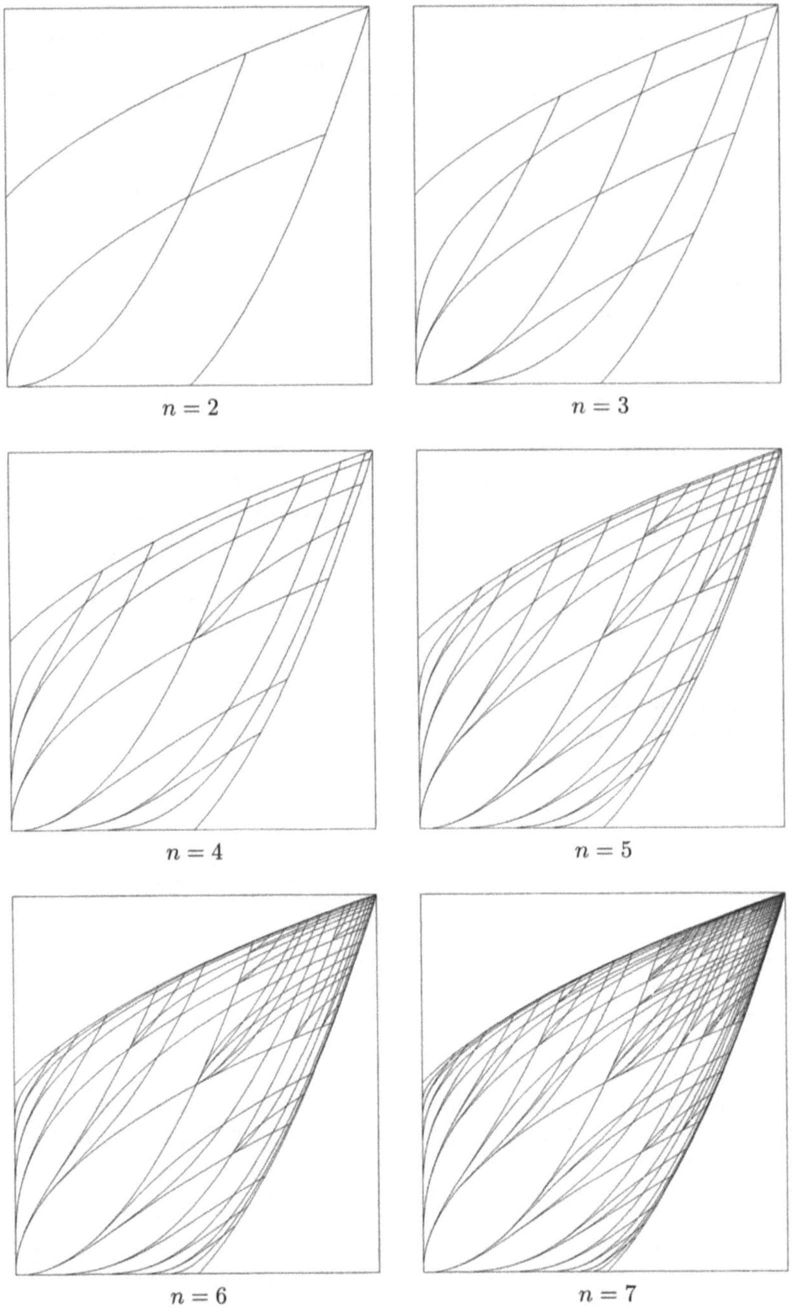

Fig. 4. The bifurcation diagram for the quadratic Lorenz family

The main tool for the proof of this theorem is a local analysis of the bifurcations at the border lines between two adjacent pieces of the partition $\mathcal{J}_n$, in particular at the intersections of these lines which form the *corners* of the pieces $J_n(\nu)$. Since the kneading invariant changes there, one or both of the critical points has to be periodic of some period less than or equal to n.

Definition 29. A periodic orbit containing one of the critical points is called a *homoclinic orbit.* Given a finite binary word $w_\pm$ of length $n^\pm$, $c^\pm$ is called *homoclinic* of type $w_\pm$ if $c^\pm_{n^\pm} = 0$, $c^\pm_k \neq 0$ for $k < n^\pm$, and $\nu^\pm = w_\pm^\infty$. The set of parameters where $c^\pm$ is homoclinic of type $w_\pm$ is denoted by $C_{w_\pm}$. Connected components of $C_{w_\pm}$ are called *hom-lines.*

In Figure 4 the lines going from left to right respectively from bottom to top are hom-lines where c^- respectively c^+ is homoclinic.

Remark 30. The term "homoclinic orbit" is motivated by the origin of Lorenz maps as first return maps for flows on the geometric Lorenz attractor, where periodic critical orbits of the Lorenz map correspond to homoclinic loops on the geometric Lorenz attractor.

At the intersection points of two hom-lines C_{w_+} and C_{w_-}, which we call *hom-points* of type (w_+, w_-), the Lorenz map has two homoclinic orbits. For nearby parameters in a quadrant northeast or souththwest of the hom-point—depending on the orientation of the intersection—the maps are renormalizable to a Lorenz map which is injective on its dynamical interval. Such maps can be considered as injective circle maps, whence they have a well defined rotation number. It can be shown that there are infinitely many hom-lines emerging from the hom-point in direction of this quadrant which form the boundaries of the frequency locked regions with rational rotation number. The combinatorial types of these hom-lines can be obtained from the original two types w_+ and w_- using concatenation rules. This type of bifurcation is known as the *gluing bifurcation*, see Coullet e.a. [4] and Gambaudo e.a. [6].

We distinguish *positively* and *negatively oriented* hom-points according to the orientation of the intersection of C_{w_+} and C_{w_-}. The local picture at a hom-point is summarized by the following proposition (cf. Figure 5).

Proposition 31. *Consider a transverse hom-point (a_0, b_0) of type (w_+, w_-) and let U be a sufficiently small neighbourhood of (a_0, b_0) which is divided into four quadrants by the hom-lines C_{w_+} and C_{w_-}. If the hom-point is positively oriented then one has the following bifurcation diagram.*

1. *Southwest of (a_0, b_0) the map f has two periodic orbits of type w_- resp. w_+ containing c^- resp. c^+ in their immediate basin. The kneading invariant equals (w_+^∞, w_-^∞).*
2. *Northwest of (a_0, b_0) the map f has a periodic orbit of type w_+ containing c^+ and $c^-_{n^-}$ in its immediate basin. The kneading invariant equals $(w_+^\infty, w_- w_+^\infty)$.*

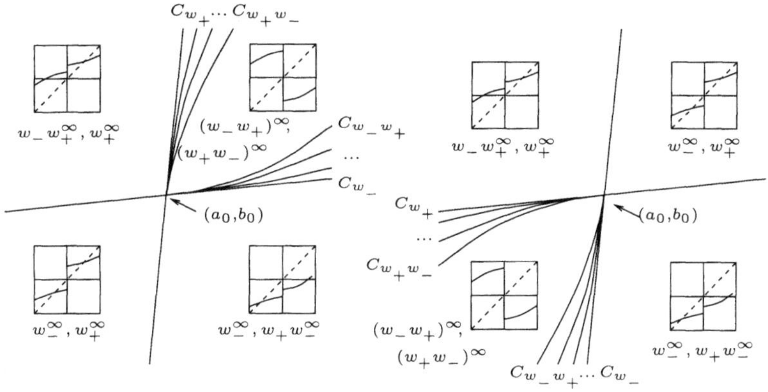

Fig. 5. If the hom-point is positively oriented (left figure) then northeast of the homoclinic value (a_0, b_0) the map f becomes renormalizable to an injective circle map and infinitely many new hom-lines are created. Similarly, infinitely many curves are annihilated in a negatively oriented hom-point (right figure).

3. *Southeast of* (a_0, b_0) *the map* f *has a periodic orbit of type* w_- *containing* c^- *and* $c^+_{n^+}$ *in its immediate basin. The kneading invariant equals* $(w_+w_-^\infty, w_-^\infty)$.
4. *Northeast of* (a_0, b_0) *the map* f *is* (w_+, w_-)*-renormalizable to an injective circle map. There are infinitely many hom-lines emerging from* (a_0, b_0) *into this direction corresponding to hom-lines of the renormalized circle map.*

If the orientation is negative then the same holds with the roles of the northeast and the southwest quadrant interchanged.

Finally, the combinatorial information obtained from the kneading invariant and the local analysis at the corners of the pieces $J_{n-1}(\nu)$ are combined in order to show that for every admissible extension of the given $(n-1)$-prefix of ν there is at least one corner of $J_{n-1}(\nu)$ where this prefix can be found (cf. Figure 6). Now Theorem 28 follows by induction.

Monotonicity of the Kneading Invariant

If $(f_a)_{a\in\mathbb{R}}$ is a C^1-family of *unimodal* maps then one says that the kneading invariant depends monotonously on the parameter, if the map $a \mapsto \nu(a)$ is a monotonous map from the real line to the shift space $\{-,+\}^{\mathbb{N}}$ endowed with the signed lexicographical order. The following definition is a natural generalization of this notion for families depending on more than one parameter.

Definition 32. Let $\mathcal{F} = (f_{a,b})_{(a,b)\in J}$ be a C^1-Lorenz family. We say that the kneading invariant of $\mathcal{F}$ *depends monotonously on the parameters* if for every admissible kneading invariant ν and every n the set $J_n(\nu)$ is simply connected.

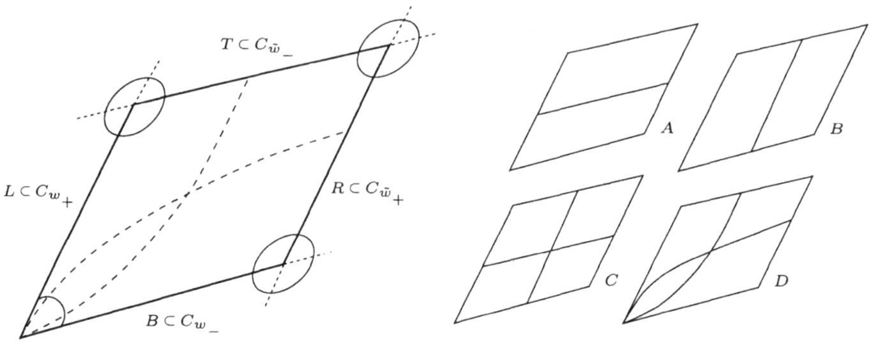

Fig. 6. The edges of a piece $J_{n-1}(\nu)$ are parts of hom-lines C_{w_+}, $C_{\tilde{w}_+}$, C_{w_-} and $C_{\tilde{w}_-}$ and the corners are hom-points. If $J_{n-1}(\nu) \notin \mathcal{J}_n$ then it splits into two or four fragments of $\mathcal{J}_n$ as in one of the figures A–D.

In the course of proving Theorem 28 it becomes evident that a sufficient condition for the sets $J_n(\nu)$ to be simply connected is the requirement that all hom-points are positively oriented.

Theorem 33. *In the situation of Theorem 28 assume that all hom-points are positively oriented. Then the kneading invariant of the Lorenz family depends monotonously on the parameters.*

The numerical bifurcation diagrams obtained for the quadratic Lorenz family (cf. Figure 4) strongly indicate that it has this property.

Conjecture 34. Every hom-point of the quadratic Lorenz family (2) is positively oriented.

For the quadratic *unimodal* family $f_a : x \mapsto a - x^2$ it was shown by Douady, Hubbard & Sullivan(14) that the kneading invariant depends monotonously on the parameter. They showed this by proving that the Thurston map is a contraction with respect to a suitable metric using deep results from Teichmüller theory. Recently, Tsujii [34,35] gave a simplified proof for the monotonicity in the quadratic family based on the following observations. First, the question whether the kneading invariant depends monotonously on the parameters can be reduced to a local problem, namely to show that

$$\frac{\mathrm{D}_a\, c_{n+1}(a)}{(f^n)'\,(c_1(a))} > 0 \tag{3}$$

whenever the critical point is periodic ("homoclinic") of period $n+1$. The interpretation of this property is that whenever the critical point is periodic and

(14) Unpublished. For a proof see the monograph of Milnor & Thurston [28].

the parameter a is increased, $c_{n+1}(a)$ moves towards the side of c corresponding to the larger kneading invariant. Second, there is a connection between the expression on the left hand side of the inequality and the Thurston map corresponding to the periodic orbit $c_0, \dots, c_n$. If $\mathrm{D}\,T(\tilde{c})$ denotes the linearization of the Thurston map at the fixed point $\tilde{c} := (c_0, \dots, c_n)$ then

$$\frac{\mathrm{D}_a\, c_{n+1}(a)}{(f^n)'\,(c_1(a))} = \det(\mathrm{I} - \mathrm{D}\,T(\tilde{c}))\ . \tag{4}$$

Since the characteristic polynomial $\det(\lambda\,\mathrm{I} - \mathrm{D}\,T(\tilde{c}))$ is a real polynomial in λ which diverges to $+\infty$ for $\lambda \to +\infty$, the above inequality follows if the spectrum of $\mathrm{D}\,T$ is contained strictly inside the unit disk, i.e., if the linearized Thurston map is a contraction.

For the hom-points of the quadratic Lorenz family we derive a two-dimensional analogue to equation (4), namely

$$\det \begin{pmatrix} \frac{\mathrm{D}_a\, c^+_{n+1}}{(f^n)'(c^+_1)} & \frac{\mathrm{D}_b\, c^+_{n+1}}{(f^n)'(c^+_1)} \\ \frac{\mathrm{D}_a\, c^-_{m+1}}{(f^m)'(c^-_1)} & \frac{\mathrm{D}_b\, c^-_{m+1}}{(f^m)(c^-_1)} \end{pmatrix} (a,b) = \det(\mathrm{I} - \mathrm{D}\,T(\tilde{c}))\ ,$$

which implies that a hom-point (a, b) is positively oriented if the linearization of the Thurston map at the corresponding fixed point $\tilde{c}$ is a contraction. In particular, Conjecture 26 implies Conjecture 34. Unfortunately, it was not possible to apply the arguments of Tsujii in order to prove the missing link, namely the local contraction of the Thurston map. Nevertheless, since we have numerical evidence for both conjectures independently, we believe very much that they are true.

References

1. H. Bruin. *Invariant measures of interval maps.* PhD thesis, Technische Universiteit Delft, 1994.
2. H. Bruin. Combinatorics of the kneading map. *Intern. J. Bif. Chaos*, 5(5):1339–1349, 1995.
3. H. Bruin, G. Keller, T. Nowicki, and S. van Strien. Wild Cantor attractors exist. *Annals of Mathematics*, 143:97–130, 1996.
4. P. Coullet, J.-M. Gambaudo, and C. Tresser. Une nouvelle bifurcation de codimension 2: le collage de cycles. *C. R. Acad. Sc. Paris Sèr. I*, 299(7):253–256, 1984.
5. W. de Melo and S. van Strien. *One-Dimensional Dynamics*, volume 25 of *Ergebnisse der Mathematik und ihrer Grenzgebiete.* Springer, Berlin Heidelberg New York, 1993.
6. J.-M. Gambaudo, P. A. Glendinning, and C. Tresser. Collage de cycles et suites de Farey. *C. R. Acad. Sc. Paris Sèr. I*, 299(14):711–714, 1984.
7. P. Glendinning and C. Sparrow. Prime and renormalisable kneading invariants and the dynamics of expanding Lorenz maps. *Physica D*, 62:22–50, 1993.

8. J. Guckenheimer. A strange, strange attractor. In J. E. Marsden and M. McCracken, editors, *The Hopf Bifurcation Theorem and its Applications*, page 368/381. Springer, 1976.
9. J. Guckenheimer and R. F. Williams. Structural stability of Lorenz Attractors. *Publ. Math. I.H.E.S.*, 50:307–320, 1979.
10. F. Hofbauer. On intrinsic ergodicity of piecewise monotonic transformations with positive entropy. *Israel Journal of Mathematics*, 34:213–237, 1979.
11. F. Hofbauer. The topological entropy of the transformation $x \mapsto ax(1-x)$. *Mh. Math.*, 90:117–141, 1980.
12. F. Hofbauer. On intrinsic ergodicity of piecewise monotonic transformations with positive entropy ii. *Israel Journal of Mathematics*, 38:107–115, 1981.
13. F. Hofbauer. The structure of piecewise monotonic transformations. *Erg. Th. Dyn. Sys.*, 1:159–178, 1981.
14. F. Hofbauer. Monotonic mod one transformations. *Studia Mathematica*, LXXX:17–40, 1984.
15. F. Hofbauer. Piecewise invertible dynamical systems. *Probab. Th. Rel. Fields*, 72:359–386, 1986.
16. F. Hofbauer and G. Keller. Quadratic maps without asymptotic measure. *Commun. Math. Phys.*, 127:319–337, 1990.
17. A. Homburg. *Some global aspects of homoclinic bifurcations of vector fields.* PhD thesis, Rijksuniversiteit Groningen, 1993.
18. J. H. Hubbard and C. T. Sparrow. The classification of topologically expansive Lorenz maps. *Commun. Pure. Appl. Math.*, XLIII:431–443, 1990.
19. G. Keller. Lifting measures to Markov extensions. *Monatshefte für Mathematik*, 108:183–200, 1989.
20. G. Keller. Markov extensions, zeta functions, and Fredholm theory for piecewise invertible dynamical systems. *Trans. Am. Math. Soc.*, 314(2):433–497, 1989.
21. G. Keller. Exponents, attractors and Hopf decompositions for interval maps. *Erg. Th. Dyn. Sys.*, 10:717–744, 1990.
22. G. Keller. Zeta functions and transfer operators for piecewise monotone transformations. *Commun. Math. Phys.*, 127:459–477, 1990.
23. G. Keller. Spectral theory, zeta functions and the distribution of periodic points for Collet-Eckmann maps. *Commun. Math. Phys.*, 149:31–69, 1992.
24. U. Krengel. *Ergodic Theorems*, volume 6 of *Studies in Mathematics.* de Gruyter, Berlin New York, 1985.
25. E. N. Lorenz. Deterministic non-periodic flow. *Journal of Atmospheric Sciences*, 20:130–141, 1963.
26. M. Martens and W. de Melo. Universal models for Lorenz maps. Preprint, Oct. 1996.
27. J. Milnor. On the concept of attractor. *Commun. Math. Phys.*, 99:177–195, 1985.
28. J. Milnor and W. Thurston. *On iterated maps of the interval*, volume 1342 of *Lecture notes in Mathematics.* Springer, Berlin New York, 1988.
29. M. Misiurewicz. Absolutely continuous measures for certain maps of an interval. *Publ. Math. I.H.E.S.*, 53, 1981.
30. W. Parry. *The Lorenz Attractor*, volume 729 of *Lecture notes in Mathematics*, pages 169–187. Springer, Berlin New York, 1979.
31. D. Rand. The topological classification of Lorenz Attractors. *Math. Proc. Camb. Phil. Soc.*, 83:451–460, 1978.

32. D. Sands. *Topological Conditions for Positive Lyapunov Exponent in Unimodal Maps.* PhD thesis, St. John's College, Cambridge, 1994.
33. M. St.Pierre. *Topological and Measurable Dynamics of Lorenz Maps.* thesis, Friedrich-Alexander Universität, Erlangen-Nürnberg, 1998. (published in *Diss. Math.* CCCLXXXII, 1999).
34. M. Tsujii. A note on Milnor and Thurston's monotonicity theorem. In *Geometry and analysis in dynamical systems*, volume 14 of *Adv. Ser. Dyn. Sys.*, Kyoto, 1993.
35. M. Tsujii. A simple proof for monotonicity of entropy in the quadratic family. *Erg. Th. Dyn. Sys.*, 20:925–934, 2000.
36. R. F. Williams. The structure of Lorenz Attractors. *Publ. Math. I.H.E.S.*, 50:321–347, 1979.

Three-Dimensional Steady Capillary-Gravity Waves

Mariana Hărăguş-Courcelle and Klaus Kirchgässner*

Mathematisches Institut A, Universität Stuttgart, Pfaffenwaldring 57, 70569 Stuttgart, Germany

Abstract. Three-dimensional steady capillary-gravity water-waves are studied in this paper. Potential flow of an ideal fluid in a layer with finite depth and upper free surface is considered. The existence of these waves is derived through bifurcation processes from the state of rest. The waves are assumed to be periodic in the direction of propagation and just bounded in the transverse direction (modulated periodic travelling waves - MPTW). Restricting the analysis to small amplitude waves, one can reduce the problem to a finite-dimensional reversible and reflectionally symmetric dynamical system. Existence and full information about the geometry of the shape of possible crests then follows via normal form analysis and persistence.

1 Introduction

Within this "Schwerpunkt DANSE" the project of the two authors of this paper has been mainly concerned with identifying scenarios of socalled dimension-breaking bifurcations for certain free-boundary value problems. These problems arise e.g. when a steady unidirectional wave on the surface of an infinitely extended ideal fluid spontaneously breaks the homogeneity in a previously homogeneous direction. Since the underlying spatial extensions are non compact, these bifurcations are by no means classical. In fact, their analysis requires a number of serious alterations of existing methods, which will become apparent during this contribution.

The technical difficulties of these problems, based on the three-dimensional Euler equations for potential flow, are well known. To overcome the obstacles, a number of model equations have been formally derived describing certain limiting situations in the parameter space. For a full justification of these equations we have still a long way to go. However a first realistic goal seems to us the proof that the results for the model equations and those for the Euler-system qualitatively do agree. This will be achieved in this paper as well.

We study analytically the existence and geometry of permanent three-dimensional (3d) water waves, in a layer of an ideal fluid with free upper boundary under gravity and surface tension. The displacement of the free

* Project: Nonlinear Waves – Global Pattern Formation, Large Time Asymptotics (Klaus Kirchgässner)

surface varies in two horizontal directions, the third direction being bounded and vertical. The velocity field is supposed to be derived from a potential.

In contrast to the case of two-dimensional (2d) water waves, where a vast literature is available for the existence problem, relatively few results are known on the existence of 3d waves. For a detailed list of references the reader can consult the review papers [Ak94], [DK98]. A large part of these results are numerical [Ro83], [MR87] or experimental [KBK99], and most of the theoretical results are formal computations of 3d waves by using perturbation expansions [Io93], [BDM98], or have been obtained for model equations such as Kadomtsev-Petviashvili (KP) or Davey-Stewartson (see e.g. [DFS97], [dBS97], [HI98], [DH99] for some recent results). For the full Euler equations, small-amplitude capillary-gravity waves have been found by Reeder & Shinbrot [RS81], who studied special geometries: periodic travelling waves whose fundamental domain is a 'symmetric diamond', and by Craig & Nicholls [CN99], who studied periodic travelling waves with general fundamental domains. Recently, Groves & Mielke [GM99] studied 3d water waves that are bounded in the direction of propagation and periodic and even in the transverse direction. We shall refer to their work frequently along this paper. The same approach has been used by Groves [Gr00] who considered waves that are periodic and even in the direction of propagation and bounded in the transverse direction.

Here we consider small-amplitude capillary-gravity waves which are bounded solutions of the free-surface Euler equations. The waves are assumed to travel at a constant speed on water of finite depth in one horizontal direction and to be periodic in this direction. We shall give a normal form classification of all possible crestforms of such waves. The method which we apply is based on ideas from the theory of dynamical systems as proposed in [Ki82]. In the generality required here we need the results in [Mi88]. Its application to the water-wave problem goes back to [Ki88] and led to existence results for 2d travelling waves such as periodic, quasi-periodic or solitary waves [IK92], [DI93], [Lo97], [BGT96], [BG99]. For 3d water waves, this method has been used in [GM99] and [Gr00]. The bounded solutions are considered as orbits of a "dynamical system" in an infinite-dimensional phase-space consisting of functions living on a bounded domain. The linearization about the trivial ground state possesses only a finite number of purely imaginary eigenvalues. Then, under suitable assumptions, a finite dimensional center manifold exists on which all solutions of small amplitude live.

For problems in domains with two or more unbounded directions, as for the 3d waves here, this method can in general not be applied, unless some additional assumptions are made. This is the reason why we assume that the waves are periodic in one horizontal direction. In general, this direction can be any horizontal direction, not only the direction of propagation or the one perpendicular to propagation. Of course, instead of assuming periodicity, one could impose boundary conditions, e.g. Dirichlet or Neumann, in one horizontal direction.

We discuss bifurcations of two types of 3d waves, which are travelling in one horizontal - namely the x - direction: Firstly, waves that are periodic in the direction transverse to the propagation and bounded. They could be called *periodically modulated travelling waves (PMTW)*. Secondly, waves periodic in the direction of propagation and just bounded otherwise: *modulated periodic travelling waves (MPTW)*. The two cases turn out to be significantly different. We treat the second case exclusively with a new method to formulate the 3d problem. Despite the wealth of solutions we shall find, we are not able, however, to construct a fully localized solution for the full 3d case.

The crosswise direction is chosen as evolutionary variable y for a pseudo-dynamical system with a reversible and an additional reflectional symmetry. For small solutions we construct a full reduction to a finite dimensional dynamical system of minimal order, inheriting all symmetries. Using normal form theory we obtain an instrument to calculate explicitly the solutions. Due to the reflectional symmetry, even waves can be found by solving a dynamical system of dimension half of the dimension of the reduced system. They all lie in an invariant subspace of the reduced system. Such waves have been considered in [GM99] and [Gr00].

This approach has previously been used for model equations to study 3d wave phenomena. The existence of PMTW e.g. has been shown in [HI98] by using a 6th order equation, being a two-dimensional generalization of the Kawahara equation. The central part of the spectrum shows qualitatively the same behavior as in [GM99] for the full Euler system. Moreover, the inherent symmetries coincide, so that, at least for the consideration of small steady waves, the model equations are qualitatively justified. Now it would be of interest to compare quantitatively the solutions in dependence of the location in the parameter space. Waves of the MPTW type have been investigated in [HK95] using the KP-equation, and in [Il99] using the 6th order equation above. They will be justified again in the above sense by the present paper.

In the analysis of the full Euler equations, one serious difficulty arises which is due to the presence of the free surface, and thus does not appear in the model equations mentioned above. The kinematic free-surface boundary condition is nonlinear and it should be linearized in order to obtain a correct formulation. For this Groves & Mielke [GM99] used the spatial Hamiltonian structure of the water wave problem to introduce suitable variables, and showed the existence of a diffeomorphism linearizing the kinematic boundary condition. However, this diffeomorphism cannot be obtained explicitly. The same idea has been used for MPTW in [Gr00]. Instead of using the Hamiltonian structure of the problem, here we overcome this difficulty by constructing a new integral (denoted by F) so that the kinematic boundary condition can be linearized with the help of a conservation law. But, in this formulation the domain for the dominating linear operator is not dense in the space where the evolution takes place, and the general reduction procedure in [Mi88] cannot be applied directly. To justify the reduction we use the ideas in [BM95]. The proof is based also on the construction of a diffeomorphism, which is similar to the one in [GM99].

In Section 2.1 we construct the conservation law that allows us to linearize the kinematic boundary condition, and introduce the new integral. In Section 2.2 the possible bifurcations of MPTW and PMTW are discussed by analyzing the central spectral part. The problem has two parameters describing the waves, λ the inverse square of the Froude number and b the Bond number, and in addition a third parameter k, namely the frequency, in the x direction in the case of MPTW, and l in the y direction in the case of PMTW. Bifurcations are found in two regions - II and III in Figure 1 - in the parameter plane (b, λ). For any (b, λ) in these regions, there is an infinite sequence of bifurcation points at critical frequencies, k resp. l. The bifurcations are of increasing complexity.

For MPTW, the first bifurcation occurs through a pair of double real eigenvalues that collide in zero and then split in two double purely imaginary eigenvalues. The center manifold is four dimensional. At the second bifurcation, another pair of real eigenvalues collide in zero and then move along the imaginary axis where we also find the two purely imaginary eigenvalues from the previous bifurcation. The center manifold is eight dimensional. In region III, the number of purely imaginary eigenvalues is increased by two after each bifurcation. All these eigenvalues are double. The situation is more complicated in region II. There we have a second (main) critical value of the frequency k beyond which all eigenvalues leave the imaginary axis successively. Resonances may occur when one pair of real eigenvalues and one pair of purely imaginary eigenvalues collide in zero simultaneously, but we do not treat this case here.

In Section 3 we discuss the first two bifurcations of MPTW. The dynamical formulation is described in Section 3.1. For the proof of the reduction we refer to [BM95] and [GM99]. The normal forms of the reduced systems for the first two bifurcations are given in Section 3.2 and justified in Appendix A. They are reversible, reflectionally symmetric, and have an additional $SO(2)$ invariance due to invariance of the full system under translations in x. The solutions of the these normal forms are discussed in Section 3.3. The first bifurcation is completely investigated. The system is similar to the one for the 1:1 resonance discussed in [IP93]. Persistence is obtained for reversible waves which, in our case, turn out to be also even waves. We find periodic and solitary waves. The persistence of asymmetric waves (not even) remains an open problem. For the second bifurcation we restrict to even solutions. They lie in an invariant subspace of dimension half of the dimension of the center manifold. On this subspace, the normal form is similar to the one studied in [IK92]. However, due to the additional $SO(2)$ invariance of the normal form, the lowest order nonlinear term is not quadratic but cubic. Persistence is again obtained for reversible solutions. In particular, we find a pair of 'generalized' solitary waves, one of depression and one of elevation with small oscillations at infinity. The lowest order approximation of the solutions is computed explicitly from the normal form, and used to plot the corresponding wave profiles.

Acknowledgements. We are grateful to Mark Groves for his competent suggestions improving this work.

2 The Euler Equations

2.1 The System

Consider nonlinear permanent waves travelling at a constant speed c on the free surface of a three-dimensional inviscid fluid layer of mean depth h and constant density ρ. Assume that both gravity and surface tension are present, and denote by g the acceleration due to gravity and by T the coefficient of surface tension. In a coordinate system (X, Y, Z) moving with constant speed c in the direction of the X-axis these waves are steady. The bottom lies at $Z = 0$ and the free surface is described by $Z = \eta(X, Y)$. The flow is supposed to be irrotational, so the velocity field has a potential $\phi = \phi(X, Y, Z)$. Introduce dimensionless variables by choosing the unit length to be h and the unit velocity to be c. The Euler equations for perturbations (Φ, η) of the state of rest $(\phi^0 = X, \eta^0 = 0)$ become

$$\Phi_{XX} + \Phi_{YY} + \Phi_{ZZ} = 0\,, \quad \text{for } 0 < Z < 1 + \eta(X,Y)\,, \tag{2.1}$$

with the boundary conditions

$$\Phi_Z = 0 \tag{2.2}$$

at the bottom $Z = 0$, and

$$\eta_X + \eta_X \Phi_X + \eta_Y \Phi_Y = \Phi_Z \tag{2.3}$$

$$\Phi_X + \frac{1}{2}(\Phi_X^2 + \Phi_Y^2 + \Phi_Z^2) + \lambda\eta - b\kappa = 0 \tag{2.4}$$

on the free surface $Z = \eta(X, Y)$, where κ is twice the mean curvature of the free surface

$$\kappa = \left(\frac{\eta_X}{\sqrt{1 + \eta_X^2 + \eta_Y^2}}\right)_X + \left(\frac{\eta_Y}{\sqrt{1 + \eta_X^2 + \eta_Y^2}}\right)_Y .$$

The dimensionless numbers

$$\lambda = gh/c^2 \quad \text{and} \quad b = T/\rho h c^2$$

are the inverse square of the Froude number and the Bond number, respectively. The analysis is done for capillary-gravity waves, so $b > 0$.

For our purposes it is convenient to replace the kinematic boundary condition at the free-surface (2.3) by the conservation law:

$$\frac{\partial}{\partial X}\left(\eta+\int_0^{1+\eta(X,Y)} \Phi_X(X,Y,Z)\,dZ\right)+\frac{\partial}{\partial Y}\left(\int_0^{1+\eta(X,Y)} \Phi_Y(X,Y,Z)\,dZ\right)=0\,. \tag{2.5}$$

A direct calculation shows that (2.5) together with (2.1)-(2.2) are equivalent with (2.1)-(2.3). Indeed,

$$\begin{aligned} 0 &= \frac{\partial}{\partial X}\left(\eta+\int_0^{1+\eta} \Phi_X\,dZ\right)+\frac{\partial}{\partial Y}\left(\int_0^{1+\eta} \Phi_Y\,dZ\right) \\ &= \eta_X+\eta_X \Phi_X\big|_{Z=1+\eta}+\eta_Y\Phi_Y\big|_{Z=1+\eta}-\Phi_Z\big|_{Z=1+\eta}\,. \end{aligned}$$

An important feature of (2.5) is that it is in divergence form. Hence, there exists a function $F=F(X,Y)$ such that

$$F_X=-\int_0^{1+\eta} \Phi_Y\,dZ \tag{2.6}$$

and

$$F_Y=\eta+\int_0^{1+\eta} \Phi_X\,dZ\,. \tag{2.7}$$

This additional variable permits us to linearize the boundary condition on the free surface (2.3). From now on, the equations (2.6)-(2.7) are used instead of (2.3).

The system (2.1)-(2.4) possesses several symmetries. They play an important role in the analysis. It is invariant under any translation in X and Y, and is Galilean invariant. It is also reversible in Y and has a reflectional symmetry with respect to X.

2.2 Bifurcations

In this section we discuss the possible bifurcations of modulated periodic travelling waves (MPTW) and periodically modulated travelling waves (PMTW).

The system for MPTW will be written in Section 3 in the form

$$\boldsymbol{V}_y+A(\lambda,b,\partial_x)\boldsymbol{V}+F(\lambda,b,\boldsymbol{V})=0,$$

where $A(\lambda,b,\partial_x)$ is a differential operator involving partial derivatives with respect to x and z, and F denotes the nonlinear terms. We seek solutions that are $2\pi/k$-periodic in x. A bifurcation from the state of rest $\boldsymbol{V}=0$ occurs at values of b and λ where the number of purely imaginary eigenvalues of the linear operator $A(\lambda, b, \partial_x)$ changes. Therefore, we have to consider, for given (b, λ), all real l, where the linear equation

$$il\boldsymbol{V} + A(\lambda, b, \partial_x)\boldsymbol{V} = 0 \tag{2.8}$$

has nontrivial solutions. These solutions can be found by using Fourier series in x, and exist for the values of b and λ for which the equation

$$il\boldsymbol{V} + A(\lambda, b, ik)\boldsymbol{V} = 0 \tag{2.9}$$

has a nontrivial solution, for some real k and l; $-il$ is the eigenvalue of $A(\lambda, b, \partial_x)$ and k the frequency in x of the solutions.

Nontrivial solutions of (2.9) correspond to solutions of the linearized equations (2.1)-(2.4) of the form

$$\Phi(x,y,z) = e^{ikx+ily}\Phi_0(z)\,, \quad \eta(x,y) = e^{ikx+ily}\eta_0\,.$$

They exist for the values of b, λ for which the dispersion relation to (2.1)-(2.4)

$$\mathcal{D}(k,l) = -k^2 + \left(\lambda + b(k^2+l^2)\right)\sqrt{k^2+l^2}\tanh\sqrt{k^2+l^2} = 0 \tag{2.10}$$

has real solutions k and l.

Similar considerations hold for PMTW, and the connection with the dispersion relation of (2.1)-(2.4) above shows that for PMTW the roles of k and l are interchanged: $-ik$ is an eigenvalue of the linear operator and l the frequency in y of the solutions. Hence, in order to detect bifurcations of MPTW, or PMTW, we have to look at the real branches of the dispersion relation (2.10). These branches are curves in the (k,l)-plane given by

$$k = \pm\sqrt{(\lambda+ba^2)a\tanh a}\,, \quad l = \pm\sqrt{a^2-(\lambda+ba^2)a\tanh a}\,, \quad a\in\mathbb{R}.$$

Their shape is shown in Figure 1 (insets) for (λ, b) in the regions of the plane indicated.

Note that due to reversibility these curves are symmetric with respect to the axis $k=0$ and $l=0$. The curves Γ and $\lambda=1$ separating the three regions I, II and III are exactly the bifurcation curves from the two-dimensional case found in [Ki88]. Along these curves the equation $\mathcal{D}(k,0)=0$ has double roots, $k=\pm k^*\neq 0$ for $(b,\lambda)\in\Gamma$ and $k=0$ for $\lambda=1$. They are determined by the parametric equations

$$\mathcal{D}(k,0)=0, \quad D_k\mathcal{D}(k,0)=0.$$

In region II the equation $\mathcal{D}(k,0)=0$ has two pairs of simple roots $\pm k_c^1$, $\pm k_c^2$, and the real branches of the dispersion relation have the limiting behavior

$$l^2 \sim \alpha_j(k-k_c^j), \qquad \alpha_j = \frac{2k_c^j\left(\sinh(2k_c^j) - 2k_c^j - 2bk_c^j\tanh k_c^j\sinh(2k_c^j)\right)}{\sinh(2k_c^j) + 2k_c^j + 2bk_c^j\tanh k_c^j\sinh(2k_c^j)}\,, \tag{2.11}$$

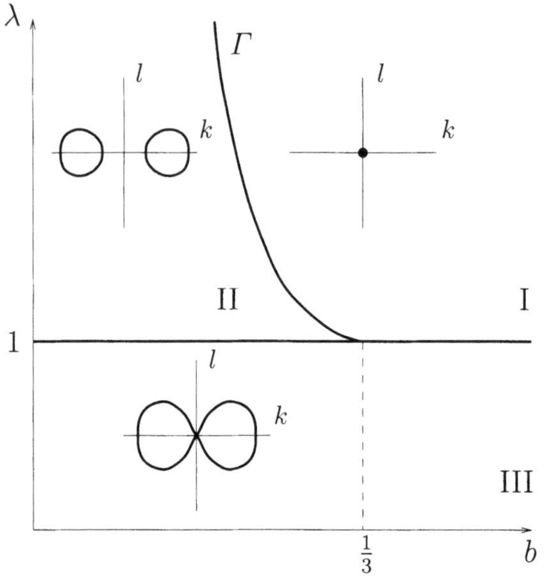

Fig. 1. Shape of the real branches of the dispersion relation.

as $k \to k_c^j$, $j = 1, 2$. In region III, the equation $\mathcal{D}(k, 0) = 0$ has only one pair of nonzero roots $\pm k_c^1$, both simple, and as $k \to k_c^1$ the real branches of the dispersion relation have the limiting behavior in (2.11). In a neighborhood of $(0, 0)$, the real branches of the dispersion curves have the limiting behavior

$$l^2 \sim (\lambda^{-1} - 1)k^2 + \left(\frac{1}{3} - \frac{b}{\lambda}\right)\lambda^{-2}k^4, \qquad \text{as } k \to 0\,. \tag{2.12}$$

The trivial solution $k = l = 0$ in region I is due to the symmetries of the Euler equations described at the end of Section 2.1. No bifurcations occur in this region. Bifurcations of MPTW and PMTW occur at any point in the regions II and III. The analysis does not apply to the boundary $b = 0$; the boundary $\lambda = 0$ is physically not relevant. The bifurcations along the curves Γ and $\lambda = 1$ are the ones from the two-dimensional problem and yield only two-dimensional solutions.

Two points of view are possible. The first one is to choose any pair (b, λ) in the region II $\cup$ III and take the frequency of the waves as bifurcation parameter. The critical frequencies at which a bifurcation occurs can be read from the picture of the real branches of the dispersion relation (Figure 1).

In the case of MPTW bifurcations occur at the values $k_c > 0$ of the frequency in x of the waves where one of the vertical lines $k = \pm nk_c$, $n = 1, 2, \ldots$, is tangent to the real dispersion curve. There, two eigenvalues of the linear operator $A(\lambda, b, \partial_x)$ collide in zero. In region III the value nk_c is equal to the maximum value k_c^1 attained by k along the dispersion curve, and in region II it equals either the maximum k_c^1 or the minimum positive value k_c^2 of k along the dispersion curve.

The first bifurcation occurs for $n = 1$, so $k_c = k_c^1$. If $k > k_c^1$, the eigenvalues of the operator $A(\lambda, b, \partial_x)$ are all bounded away from the imaginary axis and at $k = k_c$ two of the real eigenvalues, one negative and one positive, collide at zero. The multiplicity of the zero eigenvalue at bifurcation is the sum of the multiplicities of the root $l = 0$ of the dispersion relation (2.10) for $k = \pm k_c^1$. Since the lines $k = k_c$ and $k = -k_c$ are both tangent to the real dispersion curve, this root is double, so zero is an eigenvalue of multiplicity four. After bifurcation, for $k < k_c$ the zero eigenvalue splits into two double purely imaginary eigenvalues. Their imaginary parts are the values of l at which the vertical lines through $(k, 0)$ and $(-k, 0)$ intersect the real dispersion curve. There are four intersection points, two for some positive value of l and two for $-l$. These are all simple intersections, so the multiplicity of the eigenvalues il and $-il$ is two. In counting the numbers of the purely imaginary eigenvalues at bifurcation we should also take into account the intersections with the real dispersion curve of the line $k = 0$. For MPTW there is only one intersection point at $k = l = 0$. As for region I, this point is due to the symmetries of the system and, as we shall show in Section 3, its contribution can be eliminated.

By decreasing the value of k, the second bifurcation is found at $k = k_c^1/2$. Again, two real eigenvalues collide in zero which becomes an eigenvalues of multiplicity four. In addition, on the imaginary axis we find the two eigenvalues resulted from the first bifurcation. After the bifurcation there are four double purely imaginary eigenvalues. Note that since the dependence of l on k along the real dispersion curve is not monotone these imaginary eigenvalues collide on the imaginary axis at some value of k. The behavior of the critical eigenvalues during these first two bifurcations is shown in Figure 2.

In region III, the real branches of the dispersion curve exist for $|k| \in (0, k_c^1)$. Therefore, in this region we find an infinity of bifurcation points at

$$k = k_c^1, \frac{k_c^1}{2}, \ldots, \frac{k_c^1}{n}, \ldots,$$

of increasing complexity. At each bifurcation a pair of real eigenvalues collide in zero and then split into two double purely imaginary eigenvalues. After the n-th bifurcation there are $2n$ double eigenvalues on the imaginary axis.

In region II, the real branches of the dispersion curve exist for $|k| \in (k_c^2, k_c^1)$. In this case, in addition to the bifurcation points in region III, bifurcations occur also at

$$k = k_c^2, \frac{k_c^2}{2}, \ldots, \frac{k_c^2}{n}, \ldots.$$

In contrast to region III, the number of eigenvalues on the imaginary axis is not necessarily increasing with each bifurcation. This number depends on $k_c^1 - k_c^2$ and the frequency k. The eigenvalues arrive on the imaginary axis at $k = k_c^1$ and start to leave the imaginary axis at $k = k_c^2$. A new bifurcation

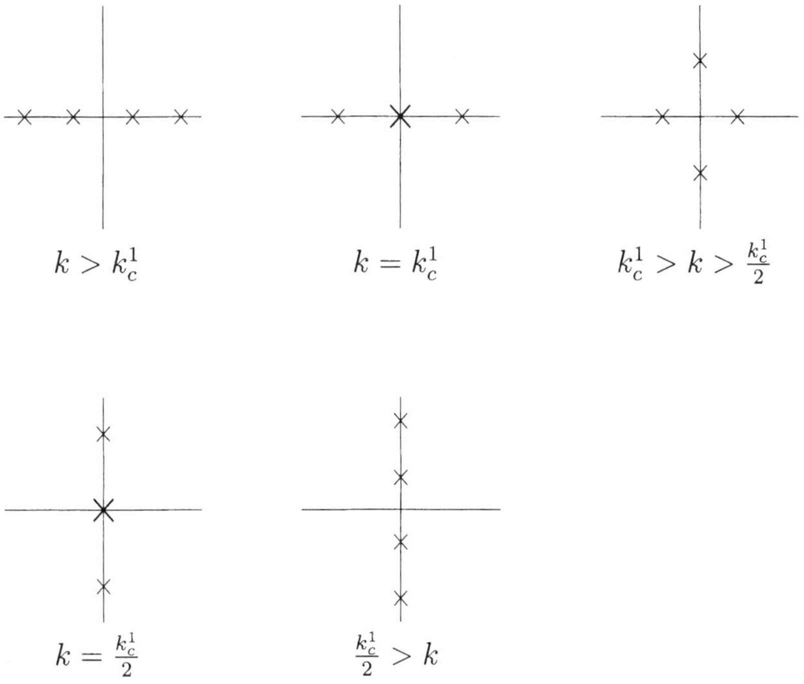

Fig. 2. MPTW: Behavior of the critical eigenvalues during the first two bifurcations. The zero eigenvalue has multiplicity four and the nonzero eigenvalues are all double.

appears when k_c^1 and k_c^2 are resonant, i.e., when $n_1 k_c^1 = n_2 k_c^2$, for some positive integers n_1, n_2. Then at $k = k_c^1/n_2 = k_c^2/n_1$ the multiplicity of the zero eigenvalue is eight, and if this is not the second bifurcation other eigenvalues may be on the imaginary axis.

In the case of PMTW, bifurcations occur at the values $l_c > 0$ of the frequency in the y-direction for which the horizontal lines $l = \pm n l_c$, $n = 1, 2, \ldots$, are tangent to the real dispersion curve. In both regions II and III, the real dispersion curve exists for $l \in (0, l_c^1)$ so, as for MPTW in region III, we find an infinity of bifurcation points at

$$l = l_c^1,\ \frac{l_c^1}{2},\ \ldots,\ \frac{l_c^1}{n},\ \ldots.$$

In contrast to the case of MPTW no collisions of eigenvalues are possible on the imaginary axis since after bifurcation the dependence of k on l is monotone. At bifurcation, two pairs of double complex eigenvalues collide in $\pm ik_1$, where $(\pm k_1, \pm l_c^1)$ are the four tangency points on the real dispersion curve. After bifurcation, they split in two pairs of purely imaginary eigenvalues that remain on the imaginary axis. Since the line $l = 0$ intersects the

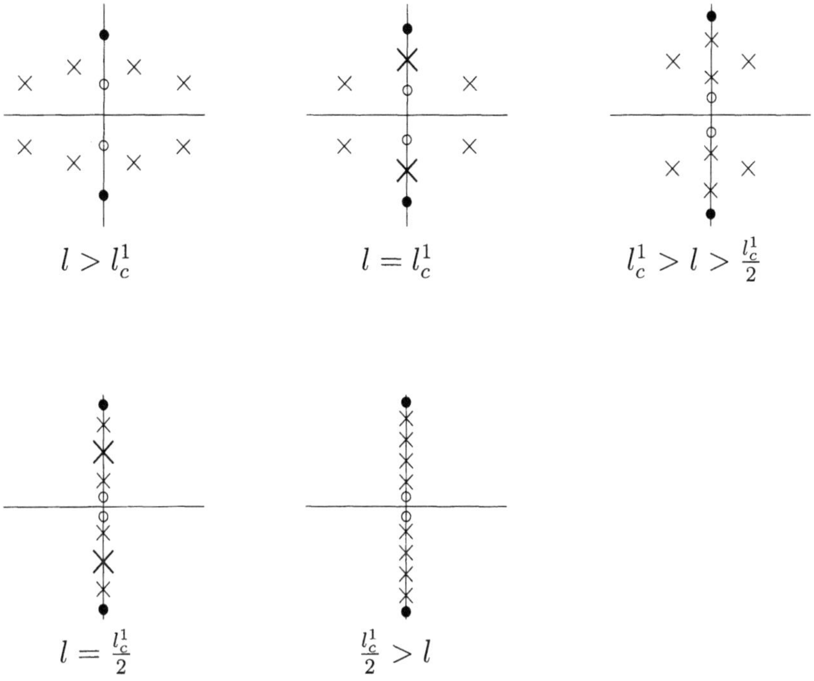

Fig. 3. PMTW: Behavior of the critical eigenvalues during the first two bifurcations. The eigenvalues denoted by '×' are double, the ones denoted by '**X**' are of multiplicity four, and the ones denoted by '○' or '•' are simple. The eigenvalues denoted by '○' are present only in region II.

real dispersion curve at $k \neq 0$, there are, in region II, four simple eigenvalues, and in region III, two simple eigenvalues that stay on the imaginary axis, for any value of l. These eigenvalues were produced by the two-dimensional bifurcation along Γ or $\lambda = 1$ and contribute to the center manifold in this case. The number of purely imaginary eigenvalues increases by four after each bifurcation so, after the n-th bifurcation there are, in region II, $4n + 4$ and in region III, $4n + 2$ eigenvalues on the imaginary axis. The behavior during the first two bifurcations of the critical eigenvalues is shown in Figure 3.

The second point of view, adopted in [HI98] and [GM99], is to choose a frequency for the wave, in x for MPTW and in y for PMTW, and to take b and λ as bifurcation parameters. Of course, the bifurcations are the same as above. This is more in the spirit of the analysis from the two-dimensional case. For each period we find, in the parameter plane (b, λ) an infinity of bifurcation curves. Their shape is shown for MPTW in Figure 4, and for PMTW in Figure 5.

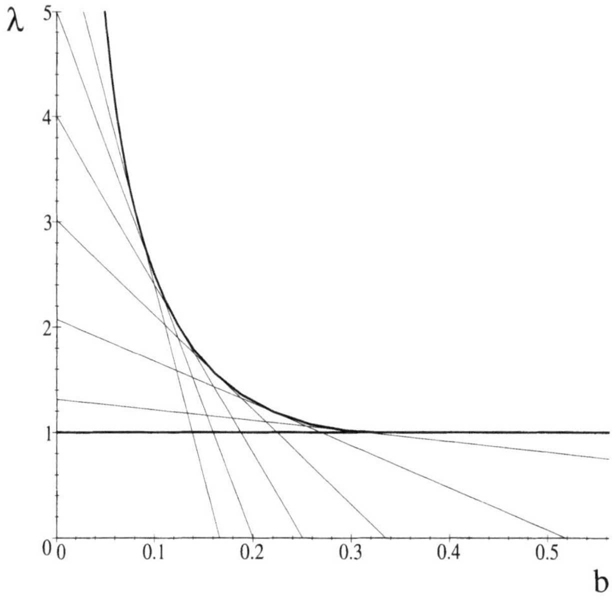

Fig. 4. MPTW: First bifurcation curves (from right to left) in the parameter plane (b, λ) for fixed k ($k = 1$).

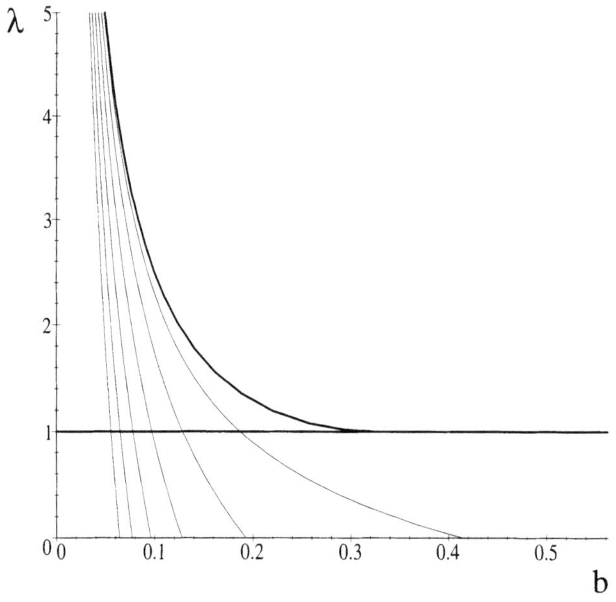

Fig. 5. PMTW: First bifurcation curves (from right to left) in the parameter plane (b, λ) for fixed l ($l = 1$).

3 Analysis of MPTW

In this section we study small-amplitude travelling waves that are periodic in the direction of propagation and just bounded in the transverse direction (MPTW). These waves are solutions of (2.1)-(2.4) that are $2\pi/k$-periodic in X.

3.1 Dynamical Formulation

It has been shown in Section 2.1 that the system (2.1)-(2.4) is equivalent with the system (2.1), (2.2), (2.4), (2.6), (2.7). We seek solutions of this system that are $2\pi/k$-periodic in X. We write the system in dynamical form by taking Y as evolutionary variable. Introduce the new variables

$$V = (1+\eta)\Phi_Y\,, \quad \beta = \frac{\eta_Y}{\sqrt{1+\eta_X^2+\eta_Y^2}}\,, \quad \varphi = kF.$$

The change of coordinates

$$x = kX\,, \quad y = Y\,, \quad z = \frac{Z}{1+\eta(X,Y)}\,, \tag{3.1}$$

maps the unknown domain $D_\eta = \{(X,Y,Z) \in \mathbb{R}^3 \,:\, 0 < Z < 1+\eta(X,Y)\}$ into $\mathbb{R}^2 \times (0,1)$. In the new coordinates (x,y,z) the system has the form

$$\boldsymbol{V}_y + A(\Lambda)\boldsymbol{V} + F(\Lambda,\boldsymbol{V}) = 0\,, \tag{3.2}$$

where $\Lambda = (\lambda, b, k)$,

$$\boldsymbol{V} = \begin{pmatrix} \Phi \\ V \\ \varphi \\ \eta \\ \beta \end{pmatrix}, \quad A(\Lambda)\boldsymbol{V} = \begin{pmatrix} -V \\ k^2\Phi_{xx} + \Phi_{zz} \\ -k\eta - k^2[\Phi_x] \\ -\beta \\ k^2\eta_{xx} - k\frac{1}{b}\Phi_x - \frac{\lambda}{b}\eta \end{pmatrix},$$

and $F = (f_1, f_2, f_3, f_4, f_5)$,

$$f_1(\Lambda,\boldsymbol{V}) = \frac{\eta V}{1+\eta} - \frac{z\beta\Phi_z}{1+\eta}\left(\frac{1+k^2\eta_x^2}{1-\beta^2}\right)^{1/2},$$

$$f_2(\Lambda,\boldsymbol{V}) = k^2\eta\Phi_{xx} - \frac{\eta\Phi_{zz}}{1+\eta} - \frac{\beta\,(V+zV_z)}{1+\eta}\left(\frac{1+k^2\eta_x^2}{1-\beta^2}\right)^{1/2} - 2zk^2\eta_x\Phi_{xz}$$
$$+ \frac{zk^2\eta_x^2\,(z\Phi_{zz}+2\Phi_z)}{1+\eta} - zk^2\eta_{xx}\Phi_z\,,$$

$$f_3(\Lambda,\boldsymbol{V}) = -k^2\eta[\Phi_x] + k^2\eta_x[z\Phi_z]\,,$$

$$f_4(\Lambda, \boldsymbol{V}) = \beta \left(1 - \left(\frac{1 + k^2 \eta_x^2}{1 - \beta^2} \right)^{1/2} \right) ,$$

$$f_5(\Lambda, \boldsymbol{V}) = -\frac{1}{2b} \left(k^2 \left(\Phi_x^{(1)} - \frac{\eta_x \Phi_z^{(1)}}{1+\eta} \right)^2 + \frac{V^{(1)^2}}{(1+\eta)^2} + \frac{\Phi_z^{(1)^2}}{(1+\eta)^2} \right) + \frac{k \eta_x \Phi_z^{(1)}}{b(1+\eta)}$$

$$+ \left(k^2 \eta_x \left(\left(\frac{1 - \beta^2}{1 + k^2 \eta_x^2} \right)^{1/2} - 1 \right) \right)_x .$$

We use the notations

$$[\Phi] = \int_0^1 \Phi \, dz \, , \quad \Phi^{(1)} = \Phi_{\big|z=1} .$$

To this system we add the boundary conditions

$$\Phi_{z\big|z=0} = 0 \, , \quad \varphi_x + [V] = 0 \, . \tag{3.3}$$

This conditions will be included in the domain of definition of the linear operator $A(\Lambda)$.

Since in (3.2) the functions Φ and F are only defined up to a constant, the linear operator $A(\Lambda)$ has at least a two-dimensional kernel, for any values of λ and b. Moreover, due to the Galilean invariance, the generalized kernel is at least three-dimensional, hence zero is an eigenvalue of $A(\Lambda)$ of multiplicity three, at least. The dimension of the reduced system is then increased by three, and since these dimensions are due to invariances they play no role in the qualitative structure of the solution set. In [HI98] and [GM99] these additional dimensions have been eliminated after the reduction to the center manifold. Here, we eliminate them before the reduction.

The periodicity in x of $\boldsymbol{V}$, allows us to decompose $\boldsymbol{V}$ into a mean-part $\boldsymbol{V}^0$ and a residual part $\widetilde{\boldsymbol{V}}$,

$$\boldsymbol{V}(x,y,z) = \boldsymbol{V}^0(y,z) + \widetilde{\boldsymbol{V}}(x,y,z) \, ,$$

where

$$\boldsymbol{V}^0(y,z) = \frac{1}{2\pi} \int_0^{2\pi} \boldsymbol{V}(x,y,z) \, dx =: [\![\boldsymbol{V}(\cdot, y, z)]\!] \, , \quad \widetilde{\boldsymbol{V}} = \boldsymbol{V} - \boldsymbol{V}^0 \, , \quad [\![\widetilde{\boldsymbol{V}}]\!] = 0 \, .$$

Accordingly, we can decompose the system (3.2),

$$\boldsymbol{V}_y^0 + A^0(\Lambda) \boldsymbol{V}^0 + F^0(\Lambda, \boldsymbol{V}^0, \widetilde{\boldsymbol{V}}) = 0 \tag{3.4}$$

$$\widetilde{\boldsymbol{V}}_y + \widetilde{A}(\Lambda) \widetilde{\boldsymbol{V}} + \widetilde{F}(\Lambda, \boldsymbol{V}^0, \widetilde{\boldsymbol{V}}) = 0 \tag{3.5}$$

where

$$F^0(\Lambda,\boldsymbol{V}^0,\widetilde{\boldsymbol{V}}) = [\![F(\Lambda,\boldsymbol{V}^0+\widetilde{\boldsymbol{V}})]\!]\,,$$
$$\widetilde{F}(\Lambda,\boldsymbol{V}^0,\widetilde{\boldsymbol{V}}) = F(\Lambda,\boldsymbol{V}^0+\widetilde{\boldsymbol{V}}) - F^0(\Lambda,\boldsymbol{V}^0,\widetilde{\boldsymbol{V}})\,.$$

Note that, if $\widetilde{\boldsymbol{V}} = 0$, the equation (3.4) is exactly the system for two-dimensional waves.

The invariances $\Phi \to \Phi + const$ and $\varphi \to \varphi + const$ from (3.2) act now only on the zero-mode components $\Phi^0 = [\![\Phi]\!]$ and $\varphi^0 = [\![\varphi]\!]$ in (3.4). In order to eliminate them, we have to eliminate φ^0 and Φ^0 from (3.4).

From the explicit formulas for (3.2) it can be seen that φ^0 appears only in the third equation of system (3.4)

$$\varphi^0_y - k\eta^0 + [\![f_3(\Lambda,\boldsymbol{V}^0+\widetilde{\boldsymbol{V}})]\!] = 0\,,$$

and thus can be eliminated.

Furthermore, Φ^0 is involved in the equations only through its derivatives Φ^0_y and Φ^0_z which determine it uniquely up to a constant. Therefore, Φ^0_y and Φ^0_z can be used in the system instead of Φ^0. Since, Φ^0_y appears only in the first equation of (3.4)

$$\Phi^0_y - V^0 + [\![f_1(\Lambda,\boldsymbol{V}^0+\widetilde{\boldsymbol{V}})]\!] = 0\,,$$

it can be eliminated as φ^0 above. Therefore, we can replace Φ^0 by the new variable

$$W^0 = \Phi^0_z\,.$$

Set $\widetilde{\boldsymbol{V}}^0 = (V^0, W^0, \eta^0, \beta^0)$. From the previous considerations follows that the system (3.4)-(3.5) is equivalent with

$$\widetilde{\boldsymbol{V}}^0_y + \widetilde{A}^0(\Lambda)\widetilde{\boldsymbol{V}}^0 + \widetilde{F}^0(\Lambda,\widetilde{\boldsymbol{V}}^0,\widetilde{\boldsymbol{V}}) = 0 \tag{3.6}$$
$$\widetilde{\boldsymbol{V}}_y + \widetilde{A}(\Lambda)\widetilde{\boldsymbol{V}} + \widetilde{F}(\Lambda,\widetilde{\boldsymbol{V}}^0,\widetilde{\boldsymbol{V}}) = 0 \tag{3.7}$$

where

$$\widetilde{A}^0(\Lambda)\widetilde{\boldsymbol{V}}^0 = \begin{pmatrix} W^0_z \\ -V^0_z \\ -\beta^0 \\ -\frac{\lambda}{b}\eta^0 \end{pmatrix}, \quad \widetilde{F}^0(\Lambda,\widetilde{\boldsymbol{V}}^0,\widetilde{\boldsymbol{V}}) = \begin{pmatrix} [\![f_2(\Lambda,\boldsymbol{V}^0+\widetilde{\boldsymbol{V}})]\!] \\ [\![f_1(\Lambda,\boldsymbol{V}^0+\widetilde{\boldsymbol{V}})]\!]_z \\ [\![f_4(\Lambda,\boldsymbol{V}^0+\widetilde{\boldsymbol{V}})]\!] \\ [\![f_5(\Lambda,\boldsymbol{V}^0+\widetilde{\boldsymbol{V}})]\!] \end{pmatrix}.$$

Here we use the fact that F depends on $\widetilde{\boldsymbol{V}}^0$, $\widetilde{\boldsymbol{V}}$, and not on $\boldsymbol{V}^0$. The boundary conditions (3.3) are included in the domains of definition of $\widetilde{A}^0(\Lambda)$ and $\widetilde{A}(\Lambda)$. We use the notations

$$\widetilde{\boldsymbol{A}}(\Lambda) = (\widetilde{A}^0(\Lambda), \widetilde{A}(\Lambda)), \quad \widetilde{\boldsymbol{F}}(\Lambda,\widetilde{\boldsymbol{V}}^0,\widetilde{\boldsymbol{V}}) = (\widetilde{F}^0(\Lambda,\widetilde{\boldsymbol{V}}^0,\widetilde{\boldsymbol{V}}), \widetilde{F}(\Lambda,\widetilde{\boldsymbol{V}}^0,\widetilde{\boldsymbol{V}}))\,.$$

Set $Q = \mathbb{R} \times (0,1)$, $I = (0,1)$, and consider the Hilbert spaces

$$\widetilde{\mathscr{X}}_s^0 = (H^s(I))^2 \times \mathbb{R}^2 , \quad \widetilde{\mathscr{X}}_s = H_{\#}^{s+1}(Q) \times H_{\#}^s(Q) \times (H_{\#}^{s+1}(\mathbb{R}))^2 \times H_{\#}^s(\mathbb{R}) ,$$

and

$$\widetilde{\mathscr{Y}}_s^0 = \widetilde{\mathscr{X}}_{s+1}^0 \cap \{(V^0, W^0, \eta^0, \beta^0) \, : \, W^0(0) = 0 \, , \ \ [V^0] = 0\} ,$$

$$\widetilde{\mathscr{Y}}_s = \widetilde{\mathscr{X}}_{s+1} \cap \{(\widetilde{\Phi}, \widetilde{V}, \widetilde{\varphi}, \widetilde{\eta}, \widetilde{\beta}) : \widetilde{\Phi}_{z}\big|_{z=0} = 0, \, \widetilde{\varphi}_x + [\widetilde{V}] = 0\},$$

for some $s \in (0, 1/2)$. The index $\#$ means that the functions are 2π-periodic and have zero mean in $x \in \mathbb{R}$. The operator $\widetilde{A}^0(\Lambda)$ (resp. $\widetilde{A}(\Lambda)$) is closed in $\widetilde{\mathscr{X}}_s^0$ (resp. $\widetilde{\mathscr{X}}_s$), with domain $\widetilde{\mathscr{Y}}_s^0$ (resp. $\widetilde{\mathscr{Y}}_s$), and $(\widetilde{F}^0, \widetilde{F})$ is a smooth map from $\widetilde{\mathscr{Y}}_s^0 \times \widetilde{\mathscr{Y}}_s$ into $\widetilde{\mathscr{X}}_s^0 \times \widetilde{\mathscr{X}}_s$.

The system (3.6)-(3.7) inherits the symmetries of the Euler equations mentioned in Section 2.1. It is invariant under translations in x, i.e. $\widetilde{\boldsymbol{A}}(\Lambda)$ and $\widetilde{\boldsymbol{F}}(\Lambda, \cdot, \cdot)$ commute with

$$\mathscr{T}_a(\widetilde{\boldsymbol{V}}^0, \widetilde{\boldsymbol{V}}(x)) = (\widetilde{\boldsymbol{V}}^0, \widetilde{\boldsymbol{V}}(x+a)) ,$$

for any $a \in \mathbb{R}$. The reversibility means that $\widetilde{\boldsymbol{A}}(\Lambda)$ and $\widetilde{\boldsymbol{F}}(\Lambda, \cdot, \cdot)$ anticommute with

$$\mathscr{R} = \mathrm{diag}\,(-1, 1, 1, -1, 1, -1, -1, 1, -1) , \tag{3.8}$$

and the reflection that they commute with

$$\mathscr{S} = \mathrm{diag}\,(-1, -1, 1, 1, -1, -1, 1, 1, 1) \circ \mathscr{S}_0 , \tag{3.9}$$

where $\mathscr{S}_0(\widetilde{\boldsymbol{V}}^0, \widetilde{\boldsymbol{V}}(x)) = (\widetilde{\boldsymbol{V}}^0, \widetilde{\boldsymbol{V}}(-x))$. These symmetries play an important role in the analysis.

3.2 Reduction and Normal Forms

The strategy of investigating the existence of periodic waves is based on a center manifold reduction performed in a neighborhood of a bifurcation point. For the system (3.6)-(3.7) the construction of the center manifold cannot be done directly by using the general reduction theorem in [Mi88] because the linear operators $\widetilde{A}^0(\Lambda)$ and $\widetilde{A}(\Lambda)$ are not densely defined. The closures of their domains of definition are proper subspaces of $\widetilde{\mathscr{X}}_s^0$ and $\widetilde{\mathscr{X}}_s$ since

$$\overline{\widetilde{\mathscr{Y}}_s^0}^{\widetilde{\mathscr{X}}_s^0} = \{(V^0, W^0, \eta^0, \beta^0) \in \widetilde{\mathscr{X}}_s^0 \, : \, [V^0] = 0\} =: \widetilde{\mathscr{Z}}_s^0$$

and

$$\overline{\widetilde{\mathscr{Y}}_s}^{\widetilde{\mathscr{X}}_s} = \{(\widetilde{\Phi}, \widetilde{V}, \widetilde{\varphi}, \widetilde{\eta}, \widetilde{\beta}) \in \widetilde{\mathscr{X}}_s \, : \, \widetilde{\varphi}_x + [\widetilde{V}] = 0\} =: \widetilde{\mathscr{Z}}_s .$$

This problem is due to the fact that the boundary conditions

$$[V^0] = 0\,, \quad \widetilde{\varphi}_x + [\widetilde{V}] = 0$$

define bounded operators on $\widetilde{\mathscr{X}}_s^0$ and $\widetilde{\mathscr{X}}_s$, respectively, so these conditions should be satisfied on the image of $\widetilde{\boldsymbol{A}}(\Lambda) + \widetilde{\boldsymbol{F}}(\Lambda,\cdot,\cdot)$. In other words, the image of $\widetilde{\boldsymbol{A}}(\Lambda)+\widetilde{\boldsymbol{F}}(\Lambda,\cdot,\cdot)$ must be in $\widetilde{\mathscr{Z}}_s^0 \times \widetilde{\mathscr{Z}}_s$. For this to hold it is required that

$$[V_y^0] = 0\,, \quad \widetilde{\varphi}_{xy} + [\widetilde{V}_y] = 0\,,$$

or, for the system (3.2) that

$$\varphi_{xy} + [V_y] = 0\,. \tag{3.10}$$

A direct calculation shows that (3.10) is equivalent with

$$\begin{aligned} \mathcal{C}(\Lambda, \widetilde{\boldsymbol{V}}^0, \widetilde{\boldsymbol{V}}) &:= k\eta_x - \frac{1}{1+\eta}\Phi_z^{(1)} + k^2\eta_x\left(\Phi_x^{(1)} - \frac{\eta_x}{1+\eta}\Phi_z^{(1)}\right) \\ &\quad + \frac{\beta}{1+\eta}\left(\frac{1+k^2\eta_x^2}{1-\beta^2}\right)^{1/2} V^{(1)} = 0\,, \end{aligned}$$

where $\eta = \eta^0 + \widetilde{\eta}$, $\Phi_x = \widetilde{\Phi}_x$, $\Phi_z = W^0 + \widetilde{\Phi}_z$, $V = V^0 + \widetilde{V}$, $\beta = \beta^0 + \widetilde{\beta}$. This is the kinematic boundary condition (2.3) written in the new coordinates (x, y, z). Hence, if $(\widetilde{\boldsymbol{V}}^0, \widetilde{\boldsymbol{V}}) \in \widetilde{\mathscr{Y}}_s^0 \times \widetilde{\mathscr{Y}}_s$ satisfy the nonlinear condition $\mathcal{C}(\Lambda, \widetilde{\boldsymbol{V}}^0, \widetilde{\boldsymbol{V}}) = 0$, then the image of $\widetilde{\boldsymbol{A}}(\Lambda)+\widetilde{\boldsymbol{F}}(\Lambda,\cdot,\cdot)$ is included in $\widetilde{\mathscr{Z}}_s^0 \times \widetilde{\mathscr{Z}}_s$. This condition can be decomposed into a mean part and a residual part

$$\mathcal{C}^0(\Lambda, \widetilde{\boldsymbol{V}}^0, \widetilde{\boldsymbol{V}}) = 0\,, \quad \widetilde{\mathcal{C}}(\Lambda, \widetilde{\boldsymbol{V}}^0, \widetilde{\boldsymbol{V}}) = 0\,, \tag{3.11}$$

where $\mathcal{C}^0 = [[\mathcal{C}]]$, $\widetilde{\mathcal{C}} = \mathcal{C} - \mathcal{C}^0$.

A center manifold in a similar situation has been constructed by Bridges & Mielke [BM95]. They construct, with the help of the compatibility condition (3.11), a new coordinate system in $\widetilde{\mathscr{Y}}_s^0$ and $\widetilde{\mathscr{Y}}_s$ such that the linear and nonlinear parts from (3.6)-(3.7) are mapped into $\widetilde{\mathscr{Z}}_s^0$ and $\widetilde{\mathscr{Z}}_s$. The general reduction method applies in these coordinate systems. It turns out that the construction of these new systems of coordinates is in fact equivalent with the construction of the diffeomorphism linearizing the boundary conditions from Groves & Mielke [GM99]. Although we are able to linearize explicitly the boundary conditions by introducing the new variable F, the problem caused by the nonlinear boundary conditions has only been postponed. In the new formulation the linear operators are not densely defined and in order to perform the reduction a map similar to the one used in [GM99] still has to be constructed.

In order to justify the reduction it is also possible to argue as in [GM99], and we adopt this approach here. To the system (3.6)–(3.7) we add the nonlinear boundary condition (3.11), and perform the reduction in $\widetilde{\mathscr{X}}_s^0 \times \widetilde{\mathscr{X}}_s$.

Set $\widetilde{\mathcal{V}} = (\widetilde{\boldsymbol{V}}_0, \widetilde{\boldsymbol{V}})$. The system (3.6)–(3.7) is of the form

$$\widetilde{\mathcal{V}}_y = \widetilde{\mathcal{N}}(\Lambda, \widetilde{\mathcal{V}}),$$

with $\widetilde{\mathcal{N}} : \mathbb{R}^3 \times \mathscr{U} \to \widetilde{\mathscr{X}}_s^0 \times \widetilde{\mathscr{X}}_s$, where $\mathscr{U}$ is a neighborhood of zero in the manifold

$$\mathscr{M}_s := \widetilde{\mathscr{X}}_{s+1}^0 \times \widetilde{\mathscr{X}}_{s+1} \cap \{W^0(0) = 0\,, \mathcal{C}^0(\Lambda, \widetilde{\mathcal{V}}) = 0\,, \widetilde{\Phi}_z\big|_{z=0} = 0\,, \widetilde{\mathcal{C}}(\Lambda, \widetilde{\mathcal{V}}) = 0\}\,.$$

We are now in a situation similar to the one in [GM99]. The linearizing diffeomorphism, as well as the resolvent estimate

$$\|(\widetilde{\boldsymbol{A}}(\Lambda) - iq)^{-1}\| \le \frac{C}{|q|}\,, \quad |q| \ge q_0\,,$$

for some $q_0 > 0$ and $C > 0$, which are essential for the reduction, are obtained following the arguments in [GM99]. The proofs are omitted, for more details about the reduction the reader can consult [GM99].

The bifurcation points in the case of MPTW have been studied in Section 2.2. In this section we analyze the first two of these bifurcations. For each of them we find the reduced system and construct its normal form.

Fix b and λ in region II $\cup$ III. The eigenvalues of $\widetilde{A}^0(\Lambda)$ are all bounded away from the imaginary axis and the bifurcations are only due to the behavior of the eigenvalues of $\widetilde{A}(\Lambda)$.

The **first bifurcation** occurs for $k = k_c^1$ satisfying

$$k = \left(\lambda + bk^2\right) \tanh k\,. \tag{3.12}$$

Set $k = k_c^1 + \mu$ and $\Lambda_c = (\lambda, b, k_c^1)$. The behavior of critical eigenvalues for small μ is shown in Figure 2. The bifurcation is due to a pair of double real eigenvalues of $\widetilde{A}(\Lambda)$ that collide in zero and then move along the imaginary axis. At bifurcation, zero is an eigenvalue of multiplicity four. To this eigenvalue there are two eigenvectors

$$\zeta_0^1 = e^{ix} \begin{pmatrix} \cosh(k_c^1 z) \\ 0 \\ 0 \\ -i \sinh k_c^1 \\ 0 \end{pmatrix}, \quad \overline{\zeta_0^1} = e^{-ix} \begin{pmatrix} \cosh(k_c^1 z) \\ 0 \\ 0 \\ i \sinh k_c^1 \\ 0 \end{pmatrix},$$

with $\widetilde{A}(\Lambda_c)\zeta_0^1 = 0$, and two generalized eigenvectors

$$\zeta_1^1 = e^{ix}\begin{pmatrix} 0 \\ -\cosh(k_c^1 z) \\ -i\frac{\sinh k_c^1}{k_c^1} \\ 0 \\ i\sinh k_c^1 \end{pmatrix}, \quad \overline{\zeta_1^1} = e^{-ix}\begin{pmatrix} 0 \\ -\cosh(k_c^1 z) \\ i\frac{\sinh k_c^1}{k_c^1} \\ 0 \\ -i\sinh k_c^1 \end{pmatrix},$$

with $\widetilde{A}(\Lambda_c)\zeta_1^1 = \zeta_0^1$.

The center manifold reduction implies that bounded solutions with small amplitude of (3.6)-(3.7) are of the form

$$\widetilde{\boldsymbol{V}}_0(y) = \mathscr{H}_0(\mu, A_j, \overline{A_j}), \tag{3.13}$$

$$\widetilde{\boldsymbol{V}}(y) = A_0(y)\zeta_0^1 + A_1(y)\zeta_1^1 + \overline{A_0}(y)\overline{\zeta_0^1} + \overline{A_1}(y)\overline{\zeta_1^1} + \mathscr{H}_1(\mu, A_j, \overline{A_j}) \tag{3.14}$$

where the reduction functions $\mathscr{H}_j$ satisfy

$$\mathscr{H}_j(\mu, A_j, \overline{A_j}) = \mathscr{O}(|A_i|(|\mu + |A_i|)), \quad j = 0, 1.$$

Substitution of (3.13)-(3.14) into (3.6)-(3.7) yields the reduced system for the amplitudes A_0 and A_1

$$\begin{aligned} A_{0y} + A_1 + \phi_0(\mu, A_0, A_1, \overline{A_0}, \overline{A_1}) &= 0 \\ A_{1y} + \phi_1(\mu, A_0, A_1, \overline{A_0}, \overline{A_1}) &= 0\,. \end{aligned} \tag{3.15}$$

This system inherits the symmetries of (3.6)-(3.7). So, the nonlinear terms ϕ_0, ϕ_1 commute with the induced operators

$$\widetilde{\mathscr{T}}_\alpha(A_0, A_1) = e^{i\alpha}(A_0, A_1), \quad \widetilde{\mathscr{S}}(A_0, A_1) = (-\overline{A_0}, -\overline{A_1}),$$

and anticommute with

$$\widetilde{\mathscr{R}}(A_0, A_1) = (A_0, -A_1).$$

In Appendix A1 we prove that the normal form of this system is

$$\begin{aligned} &B_{0y} + B_1 + iB_0 u_1 N_0(\mu, u_0, u_1^2) + \mathcal{O}(|B_j|(|\mu| + |B_j|)^k) = 0 \\ &B_{1y} + B_0 N_1(\mu, u_0, u_1^2) + iB_1 u_1 N_0(\mu, u_0, u_1^2) + \mathcal{O}(|B_j|(|\mu| + |B_j|)^k) = 0\,, \end{aligned} \tag{3.16}$$

where

$$u_0 = B_0\,\overline{B_0}\,, \quad u_1 = \frac{i}{2}\,(B_0\,\overline{B_1} - \overline{B_0}\,B_1)\,,$$

and N_0, N_1 are polynomials with real coefficients. The bounded solutions of the reduced system are studied in the next section.

The **second bifurcation** occurs for $k = k_c^1/2$. Set $k = k_c^1/2 + \mu$ and $\Lambda_c = (\lambda, b, k_c^1/2)$. For b and λ in region II assume that no resonances occur, and that $2k_c^2 < k_c^1$, so the critical eigenvalues behave as in region III. At bifurcation, zero is an eigenvalue of multiplicity four and there is a pair of complex conjugate, double eigenvalues $\pm il$ on the imaginary axis (Figure 2). To the eigenvalue zero there are two eigenvectors

$$\zeta_0^2 = e^{ix}\zeta_0^1 , \quad \overline{\zeta_0^2} = e^{-ix}\overline{\zeta_0^1} ,$$

with $\widetilde{A}(\Lambda_c)\zeta_0^2 = 0$, and two generalized eigenvectors

$$\zeta_1^2 = e^{ix}\zeta_1^1 , \quad \overline{\zeta_1^2} = e^{-ix}\overline{\zeta_1^1} ,$$

with $\widetilde{A}(\Lambda_c)\zeta_1^2 = \zeta_0^2$. To the eigenvalue il there are two eigenvectors

$$\zeta_2^2 = e^{ix}\begin{pmatrix} \cosh(\omega z) \\ -il\cosh(\omega z) \\ \frac{l\sinh\omega}{\omega} \\ -\frac{2i\omega\sinh\omega}{k_c^1} \\ -\frac{2\omega l\sinh\omega}{k_c^1} \end{pmatrix} , \quad \zeta_3^2 = e^{-ix}\begin{pmatrix} \cosh(\omega z) \\ -il\cosh(\omega z) \\ -\frac{l\sinh\omega}{\omega} \\ \frac{2i\omega\sinh\omega}{k_c^1} \\ \frac{2\omega l\sinh\omega}{k_c^1} \end{pmatrix} , \quad \omega = \sqrt{\frac{k_c^{1^2}}{4} + l^2} ,$$

and the eigenvectors associated to the eigenvalue $-il$ are

$$\overline{\zeta_2^2} , \quad \overline{\zeta_3^2} .$$

The center manifold reduction implies that bounded solutions with small amplitude of (3.6)-(3.7) are of the form

$$\widetilde{\boldsymbol{V}_0}(y) = \mathscr{H}_0(\mu, A_i, \overline{A_i}) , \tag{3.17}$$
$$\widetilde{\boldsymbol{V}}(y) = A_0(y)\zeta_0^2 + A_1(y)\zeta_1^2 + A_2(y)\zeta_2^2 + A_3(y)\zeta_3^2 + c.c. + \mathscr{H}_1(\mu, A_i, \overline{A_i}) \tag{3.18}$$

where $c.c.$ denote the complex conjugated terms and the reduction functions $\mathscr{H}_j$ satisfy

$$\mathscr{H}_j(\mu, A_i, \overline{A_i}) = \mathscr{O}(|A_i|(|\mu + |A_i|)) , \quad j = 0, 1 .$$

Substitution of (3.17)-(3.18) into (3.6)-(3.7) yields the reduced system for the amplitudes A_j, $j = 0, \ldots, 3$

$$\begin{aligned} A_{0y} + A_1 + \phi_0(\mu, A_j, \overline{A_j}) &= 0 \\ A_{1y} + \phi_1(\mu, A_j, \overline{A_j}) &= 0 \\ A_{2y} + ilA_2 + \phi_2(\mu, A_j, \overline{A_j}) &= 0 \\ A_{3y} + ilA_3 + \phi_3(\mu, A_j, \overline{A_j}) &= 0 . \end{aligned} \tag{3.19}$$

This system inherits the symmetries of (3.6)-(3.7). The nonlinear terms ϕ_j commute with the induced operators

$$\widetilde{\mathscr{T}_\alpha}(A_0, A_1, A_2, A_3) = (e^{-2i\alpha}A_0, e^{-2i\alpha}A_1, e^{-i\alpha}A_2, e^{i\alpha}A_3) ,$$

$$\widetilde{\mathscr{S}}(A_0, A_1, A_2, A_3) = (-\overline{A_0}, -\overline{A_1}, -A_3, -A_2) ,$$

and anticommute with

$$\widetilde{\mathscr{R}}(A_0, A_1, A_2, A_3) = (A_0, -A_1, \overline{A_3}, \overline{A_2}) .$$

In Appendix A2 we prove that the normal form of this system is

$$\begin{aligned}
&B_{0y} + B_1 + B_0 N_0(\mu, B_0, \overline{B_0}, u_1, u_2, u_3, u_4) + \mathcal{O}(|B_j|(|\mu| + |B_j|)^k) = 0 \\
&B_{1y} + B_1 N_0(\mu, B_0, \overline{B_0}, u_1, u_2, u_3, u_4) + N_1(\mu, B_0, \overline{B_0}, u_1, u_2, u_3, u_4) \\
&\qquad\qquad + \mathcal{O}(|B_j|(|\mu| + |B_j|)^k) = 0 \qquad\qquad (3.20) \\
&B_{2y} + ilB_2 + iB_2 N_2(\mu, B_0, \overline{B_0}, u_1, u_2, u_3, u_4) + \mathcal{O}(|B_j|(|\mu| + |B_j|)^k) = 0 \\
&B_{3y} + ilB_3 + iB_3 N_2(\mu, -\overline{B_0}, -B_0, -u_1, u_3, u_2, \frac{u_3 \overline{u_4}}{u_2}) \\
&\qquad\qquad + \mathcal{O}(|B_j|(|\mu| + |B_j|)^k) = 0 ,
\end{aligned}$$

where

$$u_1 = \frac{i}{2}\left(B_0 \overline{B_1} - \overline{B_0} B_1\right), \quad u_2 = B_2 \overline{B_2}, \quad u_3 = B_3 \overline{B_3}, \quad u_4 = \frac{B_2}{B_3} .$$

Here, N_0, N_1, N_2 are polynomials in $B_0, \overline{B_0}, u_1, u_2, u_3$ and rational functions in u_4 such that the terms $B_0 N_0$, $B_1 N_0$, N_1, $B_2 N_2$, $B_3 N_2$ in (3.20) are polynomials in B_j, $\overline{B_j}$, $j = 0, \ldots, 3$. Further properties of the (3.20) can be obtained from the fact that the nonlinear terms commute with $\widetilde{\mathscr{T}}_\alpha$, $\widetilde{\mathscr{S}}$ and anticommute with $\widetilde{\mathscr{R}}$. For our purposes it is enough to do this up to terms of order 3. We find

$$\begin{aligned}
&B_{0y} + B_1 + iB_0(d_{01}u_1 + d_{02}u_2 - d_{02}u_3) = 0 \\
&B_{1y} + iB_1(d_{01}u_1 + d_{02}u_2 - d_{02}u_3) + B_0(d_1 + d_{10}u_0 + d_{12}u_2 + d_{12}u_3) \\
&\qquad\qquad + id_{14}B_2\overline{B_3} = 0 \\
&B_{2y} + ilB_2 + iB_2(d_2 + d_{20}u_0 + d_{21}u_1 + d_{22}u_2 + d_{23}u_3) + d_{24}B_0 B_3 = 0 \\
&B_{3y} + ilB_3 + iB_3(d_2 + d_{20}u_0 - d_{21}u_1 + d_{23}u_2 + d_{22}u_3) - d_{24}\overline{B_0} B_2 = 0 \\
&\qquad\qquad\qquad\qquad\qquad\qquad\qquad\qquad (3.21)
\end{aligned}$$

where $u_0 = B_0\overline{B_0}$, and d_j, d_{ij} are real coefficients.

3.3 Solutions of the Reduced System

In this section we investigate the set of bounded solutions of (3.16) and (3.20). The fact that the system (3.6)-(3.7) commutes with the symmetry (3.9) implies that the analysis can be also done in the subspace of $\widetilde{\mathscr{Z}_s^0} \times \widetilde{\mathscr{Z}_s}$ consisting of solutions satisfying $\mathscr{S}(\widetilde{\boldsymbol{V}}^0, \widetilde{\boldsymbol{V}}) = (\widetilde{\boldsymbol{V}}^0, \widetilde{\boldsymbol{V}})$. This is an invariant subspace for (3.6)-(3.7). The surface profiles η that correspond to solutions in this subspace are even in x. We call these solutions *even solutions*. For the reduced systems (3.16), (3.20), this restriction means that we look only for solutions in the invariant subspace

$$\mathscr{B}_e = \{\widetilde{\mathscr{S}}(B_j) = (B_j)\}\,.$$

In this way, half of the modes are eliminated and the dimension of the reduced systems is divided by 2. Thus, in order to determine even solutions of (3.6)-(3.7), we have to solve, for the first bifurcation, a two-dimensional, and for the second bifurcation, a four-dimensional reduced system. For this reason in each case we start by investigating the set of even solutions.

First bifurcation. Even solutions of (3.6)-(3.7) are of the form (3.13)-(3.14) with

$$A_0 + \overline{A_0} = 0\,, \quad A_1 + \overline{A_1} = 0\,.$$

They correspond to solutions of (3.16) that satisfy

$$B_0 + \overline{B_0} = 0\,, \quad B_1 + \overline{B_1} = 0\,,$$

so $\Re B_0 = \Re B_1 = 0$.

Set $b_0 = \Im B_0$ and $b_1 = \Im B_1$. Then b_0 and b_1 verify the system

$$\begin{aligned} &b_{0,y} + b_1 = 0 \\ &b_{1,y} + b_0 N_1(\mu, b_0^2, 0) = 0\,. \end{aligned} \tag{3.22}$$

The lowest order terms in the expansion of N_1 are

$$N_1(\mu, b_0^2, 0) = c_0\mu + c_2 b_0^2 + \dots\,.$$

For our problem the coefficients c_0 and c_2 can be calculated explicitly. We find

$$c_0 = -\frac{2{k_c^1}^2 + k_c^1 \sinh(2k_c^1)}{\sinh(2k_c^1)} < 0\,,$$

and

$$c_2 = -\frac{{k_c^1}^3}{\lambda}\,\frac{2{k_c^1}^2 + \lambda k_c^1 \sinh(2k_c^1)(3-4\cosh^2 k_c^1) + 3b\lambda {k_c^1}^2 \sinh^2 k_c^1 - \alpha\lambda \sinh(2k_c^1)(2\cosh^2 k_c^1 + 1)}{k_c^1 + \sinh k_c^1 \cosh k_c^1 + 2bk_c^1 \sinh^2 k_c^1}\,,$$

with

$$\alpha = -\frac{3k_c^1}{2}\,\frac{1 + bk_c^1 \sinh(2k_c^1)}{3bk_c^1 \sinh k_c^1 \cosh k_c^1 - \sinh^2 k_c^1}\,.$$

For values of λ, b and k_c^1 satisfying (3.12) it can be verified that $c_2 < 0$.

The qualitative behavior of the set of bounded solutions of (3.22) is shown in Figure 6. The persistence of these phase portraits for the full reduced system can be proved by arguing as in [Ki88].

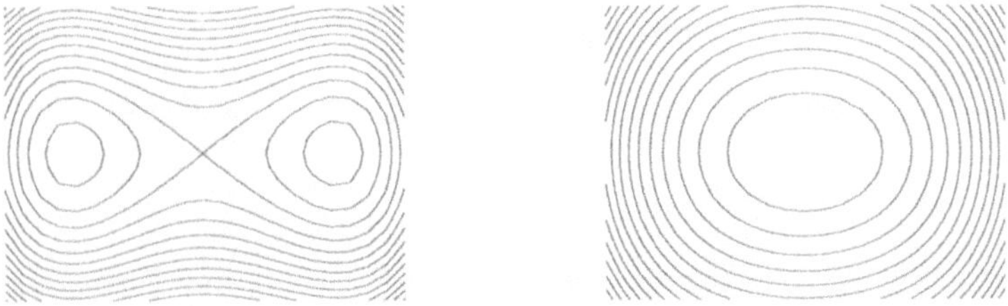

Fig. 6. Qualitative behavior of the bounded solutions of (3.22) in the (b_0, b_1)-plane.

As a solution of the Euler equations (2.1)–(2.4), we find for the free surface in lowest order approximation

$$\eta(X,Y) = 2\sinh k_c^1 \sqrt{\mu}\, B_0(\sqrt{\mu}\, Y)\cos(k_c^1 X) + \mathcal{O}(|\mu|)\,,$$

where B_0 is a bounded solution of the equation

$$B_{0yy} = c_0 \operatorname{sign}\mu B_0 + c_2 B_0^3\,.$$

The surface profiles are even and periodic in x, the direction of propagation. Periodic solutions of (3.22) in Figure 6 correspond to surface profiles that are periodic in both x and y (Figure 7 (a) and (b)). The two homoclinic solutions for $\mu < 0$ in Figure 6 correspond to solitary waves (Figure 7 (c)). We find for the homoclinic solutions

$$B_0(y) = \pm\sqrt{-\frac{2c_0}{c_2}}\operatorname{sech}(\sqrt{c_0}\, y)\,. \tag{3.23}$$

General case. The normal form (3.16) is similar to the normal form in the case of the 1:1 resonance studied by Iooss and Pérouème [IP93]. The system is integrable with integrals

$$K = \frac{i}{2}\left(B_0\overline{B}_1 - \overline{B}_0 B_1\right)\,,$$

and

$$H = |B_1|^2 - G(\mu, u_0, K)\,, \quad G(\mu, u_0, K) = \int\limits_0^{u_0} N_1(\mu, s, K^2)\, ds\,.$$

Set

$$B_0(y) = r_0(y)e^{i\psi_0(y)}\,, \quad B_1(y) = r_1(y)e^{i\psi_1(y)}\,.$$

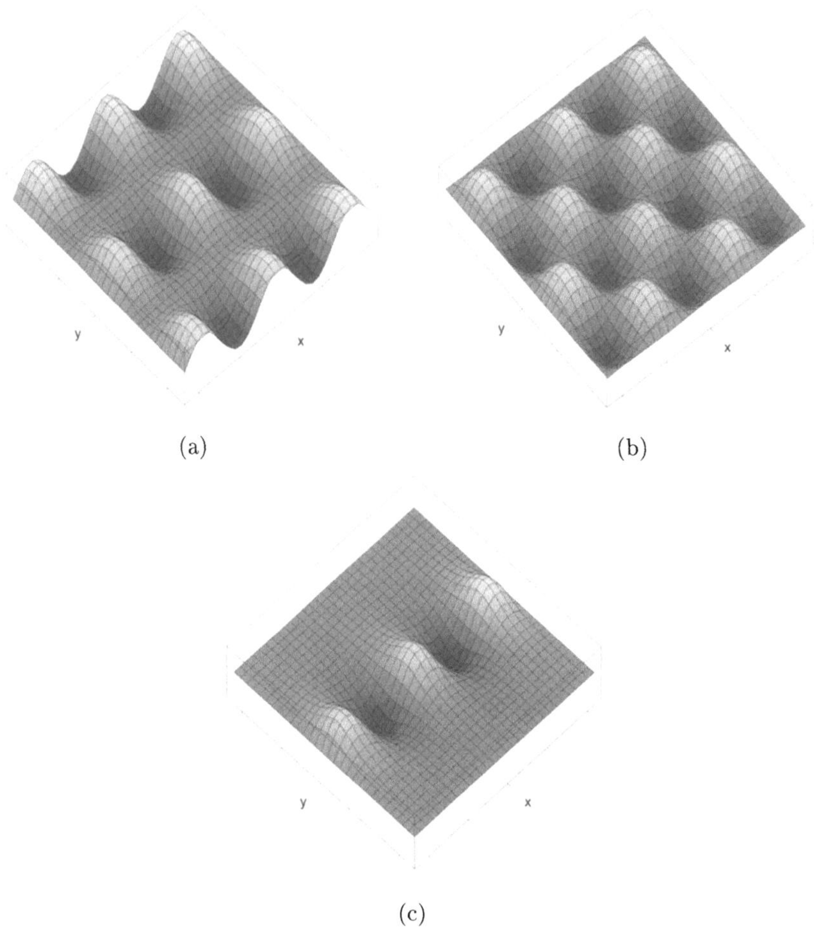

Fig. 7. Surface profiles of even MPTW: (a) and (b) periodic waves, (c) solitary wave.

Then $u_0 = r_0^2$, and the truncated system (3.16) reduces to

$$\left(\frac{du_0}{dy}\right)^2 = 4\left(u_0(G(\mu, u_0, K) + H) - K^2\right), \tag{3.24}$$

$$\frac{d(\psi_1 - \psi_0)}{dy} = K(u_0 r_1^2)^{-1} \frac{\partial}{\partial u_0}\left(u_0(G(\mu, u_0, K) + H) - K^2\right). \tag{3.25}$$

Consider the scaling

$$H = \mu^2 h, \quad K = |\mu|^{3/2} k, \quad u_0 = |\mu| U_0 .$$

Steady solutions of (3.24)-(3.25) correspond to double roots of the polynomial

$$f(u_0) = u_0\left(G(\mu, u_0, K^2) + H\right) - K^2 .$$

They lie on the curve in the (h, k) plane given parametrically by

$$h = -\frac{3}{2}c_2 a^2 - 2c_0 \operatorname{sign}\mu a + \mathcal{O}(|\mu|) ,$$
$$k = \pm\sqrt{-c_2 a^3 - c_0 \operatorname{sign}\mu a^2} + \mathcal{O}(|\mu|) .$$

For $c_2 < 0$, this curve is unbounded. Bounded solutions of (3.24)-(3.25) exist on one side of the curve and are periodic. In addition, if $\mu < 0$ at the origin $h = k = 0$ there are two homoclinic solutions with

$$r_0(y) = \pm\sqrt{-\frac{2\mu c_0}{c_2}}\operatorname{sech}(\sqrt{\mu c_0}\, y) + \mathcal{O}(|\mu|^{3/2}), \quad \psi_0(y) = \gamma_0 . \tag{3.26}$$

Going back to the Euler equations, we find for the lowest order approximation of the free surface

$$\begin{aligned}\eta(X, Y) &= 2\Re\left(-i \sinh k_c^1 e^{ik_c^1 X} B_0(Y)\right) + \mathcal{O}(|\mu|) \\ &= 2\sinh k_c^1\, r_0(Y) \sin(k_c^1 X + \psi_0(Y)) + \mathcal{O}(|\mu|) .\end{aligned}$$

Now is not hard to see that the homoclinic solutions in (3.26) with $\gamma_0 = \pi/2$ give in fact the two even solitary waves found before. The solitary waves obtained for $\gamma_0 \neq \pi/2$ are just translations in x of the even solitary waves.

In fact, the equalities

$$\frac{d\psi_0}{dy} = -K\left(u_0^{-1} + N_0(\mu, u_0, K^2)\right) ,$$

and (3.25) imply that solutions of (3.16) with $K = 0$ have

$$\psi_0 = \psi_1 = const.$$

By setting the constant to $\pi/2$ we obtain the even solutions described at the beginning of this section. A different choice for ψ_0 means just that the solutions are translated in x by a constant.

Asymmetric surface profiles are found if $K \neq 0$. In Figure 8 we show the shape of an asymmetric wave that is periodic in y. Note that there are no asymmetric solitary waves. As for the persistence of these solutions for the full system, this remains an open problem. Persistence can be proved for reversible solutions [IP93], i.e. for solutions satisfying

$$B_0(y) = B_0(-y) , \quad B_1(y) = -B_1(-y) .$$

Then $K(y) = -K(-y)$, and since K is a first integral $K = 0$. The argument above shows that solutions with $K = 0$ are even, or translations by a constant of an even solution, so reversible solutions are always even and can be obtained by solving the two-dimensional system (3.22).

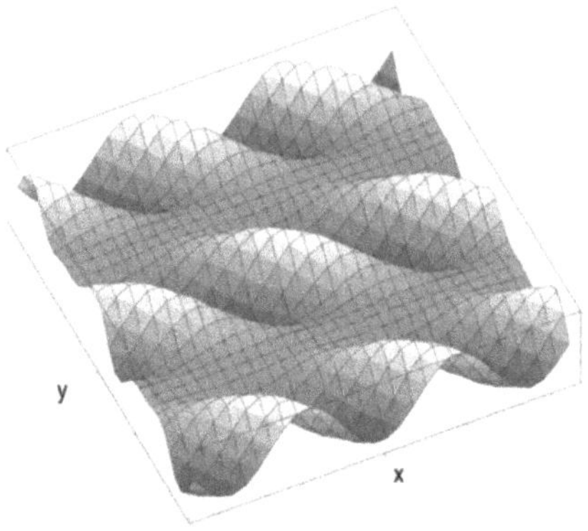

Fig. 8. Surface profiles of asymmetric MPTW that are periodic in y.

Second bifurcation. Even solutions of (3.20) satisfy

$$B_0 + \overline{B_0} = 0\,, \quad B_1 + \overline{B_1} = 0\,, \quad B_2 + B_3 = 0\,.$$

Set $b_0 = \Im B_0$ and $b_1 = \Im B_1$. The normal form (3.20) becomes

$$\begin{aligned} &b_{0y} + b_1 + b_0 n_0(\mu, b_0, u_2) = 0 \\ &b_{1y} + b_1 n_0(\mu, b_0, u_2) + n_1(\mu, b_0, u_2) = 0 \\ &B_{2y} + ilB_2 + iB_2 n_2(\mu, b_0, u_2) = 0 \end{aligned}$$

where n_j are polynomials, n_0, n_1 with real coefficients,

$$\begin{aligned} n_{0,2} &= N_{0,2}(\mu, ib_0, -ib_0, 0, u_2, u_2, -1)\,, \\ n_1 &= -iN_1(\mu, ib_0, -ib_0, 0, u_2, u_2, -1)\,. \end{aligned}$$

The reversibility $\widetilde{\mathscr{R}}$ implies $n_0 = 0$ and that n_2 has real coefficients. We conclude

$$\begin{aligned} &b_{0y} + b_1 = 0 \\ &b_{1y} + n_1(\mu, b_0, u_2) = 0 \\ &B_{2y} + ilB_2 + iB_2 n_2(\mu, b_0, u_2) = 0\,. \end{aligned} \tag{3.27}$$

With the notations from (3.21) we have the expansions of n_j up to order 3:

$$\begin{aligned} n_1(\mu, b_0, u_2) &= d_1 \mu b_0 - d_{14} u_2 + d_{10} b_0^3 + \dots\,, \\ n_2(\mu, b_0, u_2) &= d_2 \mu - d_{24} b_0 + \dots\,. \end{aligned}$$

Moreover, we find $d_1 = c_0$ and $d_{10} = c_2$. This is due to the invariance of the system (3.6)-(3.7) with respect to

$$x \to \alpha x\,, \qquad k \to \frac{k}{\alpha}\,. \tag{3.28}$$

This invariance was introduced by the change of coordinates (3.1).

Set $B_2(y) = r_2(y)e^{i\psi_2(y)}$. Before we start with the analysis of the normal form (3.27) we note the formula for the free surface corresponding to a bounded solution of (3.27)

$$\eta = 2\sinh k_c^1\, b_0(Y)\cos(k_c^1 X) + \frac{4\omega\sinh\omega}{k_c^1} r_2(Y)\sin\psi_2(Y)\cos\left(\frac{k_c^1 X}{2}\right) + \mathcal{O}(|\mu|)\,. \tag{3.29}$$

The system (3.27) is similar with the one discussed in [IK92]. The difference is that the coefficient of the quadratic term b_0^2 in the expansion of n_1 is here zero and it has been non zero in [IK92]. However, the question of the persistence of the solutions of (3.27) for the full reduced system can be treated in the same way. Persistence can be proved for reversible solutions, i.e. for solutions with

$$b_0(y) = b_0(-y)\,, \quad b_1(y) = -b_1(-y)\,, \quad B_2(y) = -\overline{B_2}(-y)\,.$$

The last equality implies

$$r_2(y) = r_2(-y)\,, \quad \Psi_2(y) = \frac{\pi}{2} + \widetilde{\psi}_2(y)\,, \quad \widetilde{\psi}_2(y) = -\widetilde{\psi}_2(-y)\,.$$

The system is integrable, with integrals

$$K = u_2\,, \quad H = b_1^2 - 2G(\mu, b_0, K)\,, \quad G(\mu, b_0, K) = \int_0^{b_0} n_1(\mu, s, K)\,ds\,.$$

Then

$$r_2^2 = K\,, \quad \psi_{2y} = -l - n_2(\mu, b_0, K)\,,$$

and

$$\left(\frac{db_0}{dy}\right)^2 = H + 2G(\mu, b_0, K) = H - 2d_{14}Kb_0 + d_1\mu b_0^2 + \frac{1}{2}d_{10}b_0^4 + \dots \tag{3.30}$$

Consider the scaling

$$b_0(y) = |\mu|^{1/2}\beta(|\mu|^{1/2}y)\,, \quad H = \mu^2 h\,, \quad K = |\mu|^{3/2}k\,.$$

Then (3.30) reads

$$(\beta')^2 = p(\beta) = h - 2d_{14}k\beta + d_1\,\mathrm{sign}\,\mu\beta^2 + \frac{1}{2}d_{10}\beta^4 + \mathcal{O}(|\mu|^{1/2})\,.$$

The qualitative behavior of the polynomial $p(\beta)$ determines the different types of bounded solutions of (3.27). Depending on the sign of μ, we have two possible cases.

Steady solutions of (3.30) are double roots of $p(\beta)$. They lie on the curve in the half plane $(h, k \geq 0)$ given by

$$h = d_1 \operatorname{sign} \mu \beta^2 + \frac{3}{2} d_{10} \beta^4 + \mathcal{O}(|\mu|^{1/2})$$

$$k = \frac{d_1 \operatorname{sign} \mu}{d_{14}} \beta + \frac{d_{10}}{d_{14}} \beta^3 + \mathcal{O}(|\mu|^{1/2}) .$$

Note that this curve does not depend on the sign of d_{14} (for $-d_{14}$ change $\beta \to -\beta$). The shape of this curve is indicated in Figure 9. For (h, k) along this curve (3.30) has steady solutions. To these solutions correspond periodic solutions of (3.27) of the form

$$b_0 = b_0^*, \quad b_1 = 0, \quad B_2 = \sqrt{K} e^{-i(l+n_2(\mu, b_0^*, H))y + i\theta_2} .$$

Reversible solutions have $\theta_2 = \pi/2$. Substitution into (3.29) gives the corresponding free surface: a periodic wave, both in x and y.

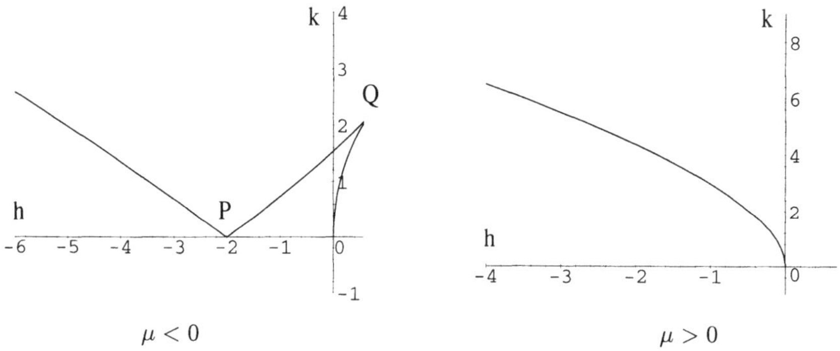

Fig. 9. Location of steady solutions of (3.30) in the $(h, k \geq 0)$ half plane.

For $\mu < 0$ the equation (3.30) has bounded solutions for any $(h, k \geq 0)$ outside the small angle opened to the left in P. For each such (h, k) it has periodic solutions. Those with $k = 0$ yield periodic solutions of (3.27), hence surface profiles that are periodic both in x and y. The ones with $k \neq 0$, correspond to quasiperiodic solutions of (3.27) and the corresponding surface profiles given by (3.29) are quasiperiodic in y and periodic in x. In addition, for (h, k) along the curve between the singular point Q and the origin, (3.30) has a pair of homoclinic solutions. They are homoclinic to a nontrivial steady solution, and, since $k \neq 0$, they correspond to a pair of 'generalized' solitary waves, one of depression and one of elevation. They have oscillations of amplitude $\sqrt{K}$ at infinity. The fact that the reversible ones persist for the full reduced system can be proved as in [IK92]. In Figure 10 we show the surface profile of a generalized solitary wave. At the origin $(0,0)$, we find a pair of homoclinic to zero solutions. However, these solutions do not persist for the full system, they develop exponentially small oscillations at infinity (see e.g. [Lo97], [Lo99a], [Su99], [Lo99b]).

If $\mu > 0$, the equation (3.30) has only periodic solutions for (h,k) in the interior of the half parabola in Figure 9 (b). The ones with $k=0$ correspond again to periodic waves, and those with $k \neq 0$ to quasiperiodic waves.

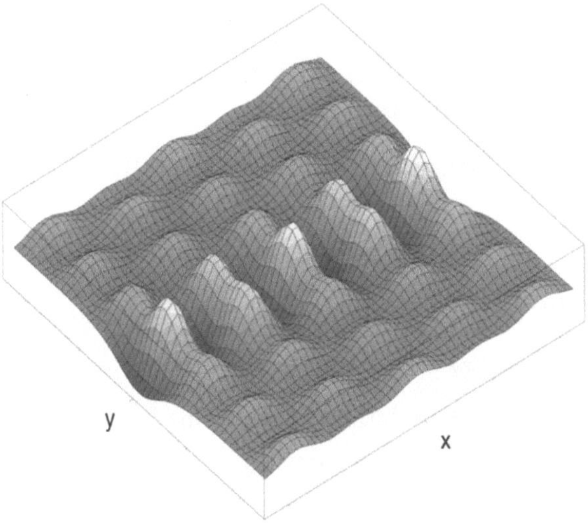

Fig. 10. MPTW: even generalized solitary wave.

General case. For the normal form (3.20) the situation is more difficult since it is not clear whether the system is integrable or not. Besides the even solutions described above, other particular solutions of (3.21) can be found by restricting the analysis to some invariant subspaces for the flow of the truncated normal form.

The normal form (3.20) has at least three invariant subspaces

$$\mathscr{B}_1 = \{B_2 = B_3 = 0\}\,, \quad \mathscr{B}_2 = \{B_0 = B_1 = B_3 = 0\}\,,$$

$$\mathscr{B}_3 = \{B_0 = B_1 = B_2 = 0\}\,.$$

These subspaces are invariant for (3.20) but not for the full reduced sytem. Of course it is easier to find solutions in these spaces since they are of lower dimension. However, we shall see that the solutions in $\mathscr{B}_1$ are the ones already obtained at the first bifurcation, and that the solutions in $\mathscr{B}_2$ or $\mathscr{B}_3$ are genuinely one-dimensional. Since the analysis of the normal form is identical on $\mathscr{B}_2$ and $\mathscr{B}_3$, we shall only consider solutions in $\mathscr{B}_1$ and in $\mathscr{B}_2$.

Solutions in $\mathscr{B}_1$. Set $B_2 = B_3 = 0$. Due to the equivariance under $\widetilde{\mathscr{T}_\alpha}$, $\widetilde{\mathscr{S}}$ and to reversibility $\widetilde{\mathscr{R}}$, the normal form (3.20) restricted to $\mathscr{B}_1$ is

$$B_{0y} + B_1 + iB_0 u_1 n_0(\mu, u_0, u_1) + \mathcal{O}(|B_j|(|\mu| + |B_j|)^k) = 0$$
$$B_{1y} + iB_1 u_1 n_0(\mu, u_0, u_1) + B_0 n_1(\mu, u_0, u_1) + \mathcal{O}(|B_j|(|\mu| + |B_j|)^k) = 0\,,$$

where $u_0 = B_0\overline{B_0}$. This is exactly the normal form found for the first bifurcation. Moreover, due to the invariance (3.28), the solutions of this normal form and those obtained at the first bifurcation correspond to the same solutions of the Euler equations.

Solutions in $\mathscr{B}_2$. Set $B_0 = B_1 = B_3 = 0$. Then (B_0, B_1, B_2, B_3) is a solution of (3.20) if B_2 satisfies

$$B_{2y} + ilB_2 + iB_2 n_2(\mu, u_2) = 0\,, \tag{3.31}$$

where

$$n_2(\mu, u_2) = d_2\mu + d_{22}u_2 + \ldots\,.$$

Set $B_2(y) = r_2(y)e^{i\psi_2(y)}$. Then

$$r_{2y} = 0\,, \quad \psi_{2y} = -l - n_2(\mu, r_2^2)\,,$$

so (3.31) has only periodic solutions of the form

$$B_2(y) = He^{-i(Ly+K)}\,, \quad L = l + n_2(\mu, H)\,,$$

for any real numbers H and K. The corresponding free surface is

$$\eta(X,Y) = \frac{4\omega\sinh\omega}{k_c^1} H \sin\left(\frac{k_c^1 X}{2} - Ly - K\right) + \mathcal{O}(|\mu|)\,.$$

This is an 'oblique' periodic wave and is genuinely one-dimensional.

Appendix A: Normal Forms

A1: First Bifurcation of MPTW

The normal form of (3.15) can be found by following the method in [IA92]. The normal form theorem in [IA92] shows that there exists a nonlinear change of coordinates $(A_0, A_1) \to (B_0, B_1)$ close to identity such that in the new coordinates (B_0, B_1) the system becomes

$$B_{0,y} + B_1 + P_0(\mu, B_0, B_1, \overline{B_0}, \overline{B_1}) + \mathcal{O}(|B_j|(|\mu| + |B_j|)^k) = 0$$
$$B_{1,y} + P_1(\mu, B_0, B_1, \overline{B_0}, \overline{B_1}) + \mathcal{O}(|B_j|(|\mu| + |B_j|)^k) = 0$$

with P_0, P_1 polynomials of degree k satisfying

$$\mathscr{D}^* P_0 = 0\,, \quad \mathscr{D}^* P_1 = P_0\,,$$

where $\mathscr{D}^*$ is the differential operator

$$\mathscr{D}^* = B_0 \frac{\partial}{\partial B_1} + \overline{B_0} \frac{\partial}{\partial \overline{B_1}} .$$

Moreover, the normal form inherits the symmetries of (3.6)-(3.7). Since

$$\mathscr{T}_a \zeta_0^1 = e^{ik_c^1 a} \zeta_0^1 , \quad \mathscr{T}_a \zeta_1^1 = e^{ik_c^1 a} \zeta_1^1 ,$$
$$\mathscr{R} \zeta_0^1 = \zeta_0^1 , \quad \mathscr{R} \zeta_1^1 = -\zeta_1^1 ,$$
$$\mathscr{S} \zeta_0^1 = -\overline{\zeta_0^1} , \quad \mathscr{S} \zeta_1^1 = -\overline{\zeta_1^1} ,$$

the polynomials P_0, P_1 commute with the induced operators

$$\widetilde{\mathscr{T}_\alpha}(B_0, B_1) = e^{i\alpha}(B_0, B_1) , \quad \widetilde{\mathscr{S}}(B_0, B_1) = (-\overline{B_0}, -\overline{B_1}) ,$$

and anticommute with

$$\widetilde{\mathscr{R}}(B_0, B_1) = (B_0, -B_1) .$$

In the next calculations we can set $\mu = 0$. The equation $\mathscr{D}^* u = 0$ has three independent first integrals

$$B_0 , \quad \overline{B_0} , \quad u_1 = \frac{i}{2} (B_0 \overline{B_1} - \overline{B_0} B_1) .$$

Since $\mathscr{D}^* P_0 = 0$ it follows that

$$P_0(B_0, B_1, \overline{B_0}, \overline{B_1}) = Q_0(B_0, \overline{B_0}, u_1) ,$$

and Q_0 is a polynomial in its arguments. Since P_0 commutes with $\widetilde{\mathscr{T}_\alpha}$, the expression

$$e^{-i\alpha} Q_0(e^{i\alpha} B_0, e^{-i\alpha} \overline{B_0}, u_1)$$

is independent of α. Hence

$$Q_0(B_0, \overline{B_0}, u_1) = B_0 R_0(u_0, u_1) , \quad u_0 = B_0 \overline{B_0} .$$

Moreover, P_0 anticommutes with $\widetilde{\mathscr{R}}$, so R_0 is odd in u_1. Then

$$P_0(B_0, B_1, \overline{B_0}, \overline{B_1}) = i B_0 u_1 N_0(u_0, u_1^2) .$$

and since P_0 commutes with $\widetilde{\mathscr{S}}$ the polynomial N_0 has real coefficients.

A direct calculation shows that $i B_1 u_1 N_0(u_0, u_1^2)$ is a solution of $\mathscr{D}^* u = P_0$. Hence

$$P_1(B_0, B_1, \overline{B_0}, \overline{B_1}) = i B_1 u_1 N_0(u_0, u_1^2) + \widetilde{P_1}(B_0, B_1, \overline{B_0}, \overline{B_1}) ,$$

where $\mathscr{D}^* \widetilde{P_1} = 0$. By arguing as above we find

$$\widetilde{P_1}(B_0, B_1, \overline{B_0}, \overline{B_1}) = Q_1(B_0, \overline{B_0}, u_1) ,$$

and by using the symmetries $\widetilde{\mathscr{T}_\alpha}$, $\widetilde{\mathscr{R}}$, $\widetilde{\mathscr{S}}$ we conclude

$$\widetilde{P_1}(B_0, B_1, \overline{B_0}, \overline{B_1}) = B_0 N_1(u_0, u_1^2) ,$$

where N_1 is a polynomial with real coefficients.

A2: Second Bifurcation of MPTW

We compute here the normal form of (3.19). Set $\mu = 0$ in the following calculations. By the normal form theorem in [IA92] there exists, for any positive integer k, a nonlinear change of coordinates $(A_0, A_1) \to (B_0, B_1)$ close to identity such that the system (3.19) is transformed to

$$\begin{aligned}
&B_{0,y} + B_1 + P_0(B_j, \overline{B_j}) + \mathcal{O}(|B_j|(|\mu| + |B_j|)^k) = 0\\
&B_{1,y} + P_1(B_j, \overline{B_j}) + \mathcal{O}(|B_j|(|\mu| + |B_j|)^k) = 0\\
&B_{2,y} + ilB_2 + P_2(B_j, \overline{B_j}) + \mathcal{O}(|B_j|(|\mu| + |B_j|)^k) = 0\\
&B_{3,y} + ilB_3 + P_3(B_j, \overline{B_j}) + \mathcal{O}(|B_j|(|\mu| + |B_j|)^k) = 0
\end{aligned}$$

with P_j, $j = 0, \ldots, 3$, polynomials of degree k satisfying

$$\mathscr{D}^* P_0 = 0\,, \quad \mathscr{D}^* P_1 = P_0\,, \quad \mathscr{D}^* P_2 = -ilP_2\,, \quad \mathscr{D}^* P_3 = -ilP_3\,,$$

where $\mathscr{D}^*$ is the differential operator

$$\mathscr{D}^* = B_0 \frac{\partial}{\partial B_1} - ilB_2 \frac{\partial}{\partial B_2} - ilB_3 \frac{\partial}{\partial B_3} + \overline{B_0} \frac{\partial}{\partial \overline{B_1}} + il\overline{B_2} \frac{\partial}{\partial \overline{B_2}} + il\overline{B_3} \frac{\partial}{\partial \overline{B_3}}\,.$$

Moreover, the normal form inherits the symmetries of (3.6)-(3.7). The polynomials P_j commute with the induced operators

$$\widetilde{\mathscr{T}}_\alpha(B_0, B_1, B_2, B_3) = (e^{-2i\alpha} B_0, e^{-2i\alpha} B_1, e^{-i\alpha} B_2, e^{i\alpha} B_3)\,,$$

$$\widetilde{\mathscr{S}}(B_0, B_1, B_2, B_3) = (-\overline{B_0}, -\overline{B_1}, -B_3, -B_2)\,,$$

and anticommute with

$$\widetilde{\mathscr{R}}(B_0, B_1, B_2, B_3) = (B_0, -B_1, \overline{B_3}, \overline{B_2})\,.$$

We compute now P_0. The equation $\mathscr{D}^* u = 0$ has seven independent first integrals

$$B_0\,, \quad \overline{B_0}\,, \quad u_1 = \frac{i}{2}\,(B_0\overline{B_1} - \overline{B_0}B_1)\,, \quad u_2 = B_2\overline{B_2}\,, \quad u_3 = B_3\overline{B_3}\,,$$

$$u_4 = \frac{B_2}{B_3}\,, \quad u_5 = il\frac{B_1}{B_0} + \ln B_3\,.$$

Since $\mathscr{D}^* P_0 = 0$ it follows that

$$P_0 = Q_0(B_0, \overline{B_0}, u_1, u_2, u_3, u_4, u_5)\,.$$

The equalities

$$\frac{\partial P_0}{\partial \overline{B_1}} = B_0 \frac{\partial Q_0}{\partial u_1}\,, \quad \frac{\partial P_0}{\partial \overline{B_2}} = B_2 \frac{\partial Q_0}{\partial u_2}\,, \quad \frac{\partial P_0}{\partial \overline{B_3}} = B_3 \frac{\partial Q_0}{\partial u_3}\,,$$

imply that Q_0 is a polynomial in u_1, u_2 and u_3. Then, from

$$\frac{\partial P_0}{\partial \overline{B_0}} = \frac{\partial Q_0}{\partial \overline{B_0}} - B_1 \frac{\partial Q_0}{\partial u_1}, \quad \frac{\partial P_0}{\partial B_1} = -\overline{B_0} \frac{\partial Q_0}{\partial u_1} + il \frac{1}{B_0} \frac{\partial Q_0}{\partial u_5},$$

we deduce that Q_0 is also a polynomial in $\overline{B_0}$ and u_5. Since P_0 is a polynomial, it follows now that Q_0 is a rational function in the other two arguments B_0 and u_4. A comparison of the asymptotic behavior of P_0 and Q_0 for $B_3 \to \infty$ shows that Q_0 does not depend on u_5. Then, since Q_0 is a polynomial in u_1 and is independent of u_5, we conclude that Q_0 is a polynomial in B_0, too.

Hence,

$$P_0 = Q_0(B_0, \overline{B_0}, u_1, u_2, u_3, u_4),$$

with Q_0 a polynomial in the first five arguments and a rational function in u_4, such that it yields a polynomial in B_j, $\overline{B_j}$.

In order to compute P_1 we need a particular solution of the equation $\mathscr{D}^* P_1 = P_0$. Set $P_0 = B_0 R_0$. Then

$$\mathscr{D}^*(B_1 R_0) = B_0 R_0 + B_1 \mathscr{D}^*(R_0) = P_0,$$

so

$$P_1 = B_1 R_0(B_0, \overline{B_0}, u_1, u_2, u_3, u_4) + Q_1(B_0, \overline{B_0}, u_1, u_2, u_3, u_4),$$

where $\mathscr{D}^* Q_1 = 0$. The arguments above show that Q_1 is a polynomial in B_0, $\overline{B_0}$, u_1, u_2, u_3 and a rational function in u_4. Moreover, since $B_1 R_0$ should be a polynomial in B_j, $\overline{B_j}$ and, since $R_0 = Q_0/B_0$, R_0 is also a polynomial in B_0, $\overline{B_0}$, u_1, u_2, u_3 and a rational function in u_4.

We look for P_2 in the form $P_2 = B_2 R_2$. Then $\mathscr{D}^* R_2 = 0$ and, by arguing as for P_0, we obtain

$$P_2 = B_2 R_2(B_0, \overline{B_0}, u_1, u_2, u_3, u_4),$$

with R_2 a polynomial in the first five arguments and a rational function in u_4, such that it yields a polynomial in B_j, $\overline{B_j}$. Finally,

$$P_3 = B_3 R_2(-\overline{B_0}, -B_0, -u_1, u_3, u_2, \frac{u_3 \overline{u_4}}{u_2}),$$

due to the commutativity with $\widetilde{\mathscr{S}}$.

References

[Ak94] T. Akylas, Three-dimensional long water-wave phenomena, *Ann. Rev. Fluid Mech.* **26** (1994), 191-210.

[dBS97] A. de Bouard & J.-C. Saut, Solitary waves of generalized Kadomtsev-Petviashvili equations, *Ann. Inst. H. Poincaré Anal. Non Linéaire* **14** (1997), 211-236.

[BDM98] T. Bridges, F. Dias & D. Menasce, Steady three-dimensional finite-depth patterns on the ocean surface: a new characterization of short-crested Stokes waves interacting with a mean flow, (1998), preprint.

[BM95] T.J. Bridges & A. Mielke, A proof of the Benjamin-Feir instability, *Arch. Rat. Mech. Anal.* **133** (1995), 145-198.

[BG99] B. Buffoni & M.D. Groves, A multiplicity result for solitary gravity-capillary waves in deep water via critical-point theory, *Arch. Rat. Mech. Anal.* **146** (1999), 183-220.

[BGT96] B. Buffoni, M.D. Groves & J.F. Toland, A plethora of solitary gravity-capillary water waves with nearly critical Bond and Froude numbers, *Phil. Trans. Roy. Soc. Lond. A* **354** (1996), 575-607.

[CN99] W. Craig & D.P. Nicholls, Traveling two and three dimensional capillary gravity water waves, (1999), preprint.

[DH99] F. Dias & M. Hărăguş-Courcelle, On the transition from two-dimensional to three-dimensional water waves, *Stud. Appl. Math.*, to appear.

[DI93] F. Dias & G. Iooss, Capillary-gravity solitary waves with damped oscillations, *Physica D* **65** (1993), p. 399-423.

[DK98] F. Dias & C. Kharif, Nonlinear gravity and capillary-gravity waves, *Ann. Rev. Fluid Mech.* **31** (1999), 301-346.

[DFS97] B. A. Dubrovin, R. Flickinger & H. Segur, Three-phase solutions of the Kadomtsev-Petviashvili equation, *Stud. Appl. Math.* **99** (1997), 137-203.

[Gr00] M.D. Groves, An existence theory for three-dimensional periodic travelling gravity-capillary water waves with bounded transverse profiles, *Physica D*, to appear.

[GM99] M.D. Groves & A. Mielke, A spatial dynamics approach to three-dimensional gravity-capillary steady water waves, *Proc. Roy. Soc. Edinburgh A*, to appear.

[HK95] M. Hărăguş & K. Kirchgässner, Breaking the dimension of a steady wave: some examples, in "Nonlinear dynamics and pattern formation in the natural environment", *Res. Notes in Math.* **335**, Pitman (1995), 119-129.

[HI98] M. Hărăguş & A. Il'ichev, Three dimensional solitary waves in the presence of additional surface effects, *Eur. J. Mech. B/Fluids* **17** (1998), 739-768.

[Il99] A. Il'ichev, Self-channelling of surface water waves in the presence of an additional surface pressure, in "Three-dimensional aspects of air-sea interaction", *Eur. J. Mech. B/Fluids* **18** (1999), 501-510.

[IA92] G. Iooss & M. Adelmeyer, *Topics in bifurcation theory and applications.* World Scientific (1992).

[IK92] G. Iooss & K. Kirchgässner, Water waves for small surface tension: an approach via normal form, *Proc. Roy. Soc. Edinb.* **122** A (1992), 267-299.

[IP93] G. Iooss & M. C. Pérouème, Perturbed homoclinic solutions in reversible 1:1 resonance vector fields, *J. Diff. Eq.* **102** (1993), 62-88.

[Io93] M. Ioualalen, Fourth order approximation of short-crested waves, *C. R. Acad. Sci. Paris, Série II* **316** (1993), 1193-1200.

[KBK99] O. Kimmoun, H. Branger & C. Kharif, On short-crested waves: experimental and analytical investigations, *Europ. J. Mech. B/Fluids*, (1999), to appear.

[Ki82] K. Kirchgässner, Wave solutions of reversible systems and applications, *J. Diff. Eq.* **45** (1982), 113-127.

[Ki88] K. Kirchgässner, Nonlinearly resonant surface waves and homoclinic bifurcation, *Adv. Appl. Mech.* **26** (1988), 135-181.

[Lo97] E. LOMBARDI, Orbits homoclinic to exponentially small periodic orbits for a class of reversible systems. Application to water waves, *Arch. Rat. Mech. Anal.* **137** (1997), 227-304.

[Lo99a] E. LOMBARDI, Non-persistence of homoclinic connections for perturbed integrable reversible systems, *J. Dyn. Diff. Eq.* **11** (1999), 129-208.

[Lo99b] E. LOMBARDI, Oscillatory integrals and phenomena beyond any algebraic order with applications to homoclinic orbits in reversible systems. Book in preparation.

[MR87] T. R. MARCHANT & A. J. ROBERTS, Properties of short-crested waves in water of finite depth. *J. Austral. Math. Soc. Ser.* **B 29** (1987), 103-125.

[Mi88] A. MIELKE, Reduction of quasilinear elliptic equations in cylindrical domains with applications, *Math. Meth. Appl. Sci.* **10** (1988), 51-66.

[RS81] J. REEDER & M. SHINBROT, Three-dimensional, nonlinear wave interaction in water of constant depth, *Nonlinear Analysis TMA* **5** (1981), 303-323.

[Ro83] A. J. ROBERTS, Highly nonlinear short-crested water waves, *J. Fluid Mech.* **135** (1983), 301-321.

[Su99] S. M. SUN, Non-existence of truly solitary waves in water with small surface tension, *Proc. Roy. London* **455** (1999), 2191-2228.

MIX
Papier aus verantwortungsvollen Quellen
Paper from responsible sources
FSC® C105338

If you have any concerns about our products,
you can contact us on
ProductSafety@springernature.com

In case Publisher is established outside the EU,
the EU authorized representative is:
Springer Nature Customer Service Center GmbH
Europaplatz 3, 69115 Heidelberg, Germany

Printed by Libri Plureos GmbH
in Hamburg, Germany

Ergodic Theory, Analysis,
and Efficient Simulation of Dynamical Systems

Springer-Verlag Berlin Heidelberg GmbH

Bernold Fiedler (Editor)

Ergodic Theory, Analysis, and Efficient Simulation of Dynamical Systems

Springer

Editor:
Bernold Fiedler
Freie Universität Berlin
Institut für Mathematik I
Arnimallee 2-6
14195 Berlin, Germany
e-mail: fiedler@math.fu-berlin.de

Library of Congress Cataloging-in-Publication Data

Ergodic theory, analysis, and efficient simulation of dynamical systems / Bernold Fiedler (editor).
p. cm.
Includes bibliographical references and index.
ISBN 978-3-642-62524-4 ISBN 978-3-642-56589-2 (eBook)
DOI 10.1007/978-3-642-56589-2
1. Differentiable dynamical systems. 2. Ergodic theory. 3. Mathematical analysis. I. Fiedler, Bernold, 1956-

QA614.8 .E737 2001
515'.35--dc21

2001020854

Mathematics Subject Classification (2000): 28, 34, 35, 65, 81, 92

ISBN 978-3-642-62524-4

This work is subject to copyright. All rights are reserved, whether the whole or part of the material is concerned, specifically the rights of translation, reprinting, reuse of illustrations, recitation, broadcasting, reproduction on microfilm or in any other way, and storage in data banks. Duplication of this publication or parts thereof is permitted only under the provisions of the German Copyright Law of September 9, 1965, in its current version, and permission for use must always be obtained from Springer-Verlag. Violations are liable for prosecution under the German Copyright Law.

http://www.springer.de
© Springer-Verlag Berlin Heidelberg 2001
Originally published by Springer-Verlag Berlin Heidelberg in 2001
Softcover reprint of the hardcover 1st edition 2001

The use of general descriptive names, registered names, trademarks etc. in this publication does not imply, even in the absence of a specific statement, that such names are exempt from the relevant protective laws and regulations and therefore free for general use.

Cover design: *Erich Kirchner, Heidelberg*
Typsetting: Le-TeX Jelonek, Schmidt & Vöckler GbR, Leipzig
Printed on acid-free paper SPIN 10789119 46/3142ck-5 4 3 2 1 0

Preface

This book summarizes and highlights progress in our understanding of Dynamical Systems during six years of the German Priority Research Program "Ergodic Theory, Analysis, and Efficient Simulation of Dynamical Systems". The program was funded by the Deutsche Forschungsgemeinschaft (DFG) and aimed at combining, focussing, and enhancing research efforts of active groups in the field by cooperation on a federal level. The surveys in the book are addressed to experts and non-experts in the mathematical community alike. In addition they intend to convey the significance of the results for applications far into the neighboring disciplines of Science.

Three fundamental topics in Dynamical Systems are at the core of our research effort:

- behavior for large time
- dimension
- measure, and chaos

Each of these topics is, of course, a highly complex problem area in itself and does not fit naturally into the deplorably traditional confines of any of the disciplines of ergodic theory, analysis, or numerical analysis alone. The necessity of mathematical cooperation between these three disciplines is quite obvious when facing the formidable task of establishing a bidirectional transfer which bridges the gap between deep, detailed theoretical insight and relevant, specific applications. Both analysis and numerical analysis play a key role when it comes to building that bridge. Some steps of our joint bridging efforts are collected in this volume.

Neither our approach nor the presentations in this volume are monolithic. Rather, like composite materials, the contributions are gaining strength and versatility through the broad variety of interwoven concepts and mathematical methodologies which they span.

Fundamental concepts which are present in this volume include bifurcation, homoclinicity, invariant sets and attractors, both in the autonomous and nonautonomous situation. These concepts, at first sight, seem to mostly address *large time behavior*, most amenable to methodologies of analysis. Their intimate relation to concepts like (nonstrict) hyperbolicity, ergodicity, entropy, stochasticity and control should become quite apparent, however, when browsing through this volume.

The fundamental topic of *dimension* is similarly ubiquitous throughout our articles. In analysis it figures, for example, as a rigorous reduction from infinite-dimensional settings like partial differential equations, to simpler infinite-, finite- or even low-dimensional model equations, still bearing full relevance to the original equations. But in numerical analysis – including and transcending mere discretization – specific computational realization of such reductions still poses challenges which are addressed here.

Another source of inspiration comes from very refined *measure*-theoretic and dimensional concepts of ergodic theory which found their way into algorithmic realizations presented here.

By no means do these few hints exhaust the conceptual span of the articles. It would be even more demanding to discuss the rich circle of methods, by which the three fundamental topics of large time behavior, dimension, and measure are tackled. In addition to SBR-measures, Perron-Frobenius type transfer operators, Markov decompositions, Pesin theory, entropy, and Oseledets theorems, we address kneading invariants, fractal geometry and self-similarity, complex analytic structure, the links between billiards and spectral theory, Lyapunov exponents, and dimension estimates. Including Lyapunov-Schmidt and center manifold reductions together with their Shilnikov and Lin variants and their efficient numerical realizations, symmetry and orbit space reductions together with closely related averaging methods, we may continue, numerically, with invariant subspaces, Godunov type discretization schemes for conservation laws with source terms, (compressed) visualization of complicated and complex patterns of dynamics, and present an algorithm, GAIO, which enables us to approximately compute, in low dimensions, objects like SBR-measures and Perron-Frobenius type transfer operators. At which point our cursory excursion through methodologies employed here closes up the circle.

So much for the mathematical aspects. The range of applied issues, mostly from physics but including some topics from the life sciences, can also be summarized at most superficially, at this point. This range comprises such diverse areas as crystallization and dendrite growth, the dynamo effect, and efficient simulation of biomolecules. Fluid dynamics and reacting flows are addressed, including the much studied contexts of Rayleigh-Bénard and Taylor-Couette systems as well as the stability question of three-dimensional surface waves. The Ginzburg-Landau and Swift-Hohenberg equations appear, for example, as do mechanical problems involving friction, population biology, the spread of infectious diseases, and quantum chaos. It is the diversity of these applied fields which well reflects both the diversity and the power of the underlying mathematical approach. Only composite materials enable a bridge to span that far.

The broad scope of our program has manifested itself in many meetings, conferences, and workshops. Suffice it to mention the workshop on "Entropy" which was coorganized by Andreas Greven, Gerhard Keller, and Gerald Warnecke at Dresden in June 2000, jointly with the two neighboring DFG Priority Research Programs "Analysis and Numerics for Conservation Laws" and "Interacting Stochastic Systems of High Complexity". For further information concerning program and participants of the DFG Priority Research Program "Ergodic Theory, Analysis, and Efficient Simulation of Dynamical Systems", including a preprint server, see

- www.math.fu-berlin.de/~ danse/

For other DFG programs we refer to

- www.dfg.de
- www.dfg.de/aufgaben/Schwerpunktprogramme.html

At the end of this preface, I would like to thank at least some of the many friends and colleagues who have helped on so many occasions to make this program work. First of all, I would like to mention the members of the scientific committee who have helped initiate the entire program and who have accompanied and shaped the scientific program throughout its funding period: Ludwig Arnold, Hans-Günther Bothe, Peter Deuflhard, Klaus Kirchgässner, and Stefan Müller. The precarious conflict between great expectations and finite funding was expertly balanced by our all-understanding referees Hans Wilhelm Alt, Jürgen Gärtner, François Ledrappier, Wilhelm Niethammer, Albrecht Pietsch, Gerhard Wanner, Harry Yserentant, Eberhard Zeidler, and Eduard Zehnder. The hardships of finite funding as well as any remaining administrative constraints were further alleviated as much as possible, and beyond, by Robert Paul Königs and Bernhard Nunner, representing DFG at its best. The www-services were designed, constantly expanded and improved with unrivalled expertise and independence by Stefan Liebscher. And Regina Löhr, as an aside to her numerous other secretarial activities and with ever-lasting patience and friendliness, efficiently reduced the administrative burden of the coordinator to occasional emails which consisted of no more than "OK. BF". Martin Peters and his team at Springer-Verlag ensured a very smooth cooperation, including efficient assistance with all TEXnicalities. But last, and above all, my thanks as a coordinator of this program go to the authors of this volume and to all participants – principal investigators, PostDocs and students alike – who have realized this program with their contributions, their knowledge, their dedication, and their imagination.

Berlin,
September 2000

Bernold Fiedler

Table of Contents

Discretization, Inflation and Perturbation of Attractors

L. Grüne and P. E. Kloeden*

Fachbereich Mathematik, Johann Wolfgang Goethe Universität
60054 Frankfurt am Main, Germany
e-mail: gruene@math.uni-frankfurt.de,
kloeden@math.uni-frankfurt.de

Abstract. The discretization of attractors for autonomous and nonautonomous systems is considered. Unlike the autonomous case, where most basic issues are now well understood, the nonautonomous case still has many open questions, which will be discussed here.

1 Introduction

The basic issues concerning the effect of discretization or perturbation on autonomous attractors are now quite well understood. For nonautonomous systems matters are, however, considerably more complicated as solutions now depend explicitly on both the initial and the current time, so limiting objects need not exist in current time or be invariant, the semigroup evolutionary property no longer holds, and the concept of an attractor for autonomous systems is generally too restrictive.

Nonautonomous systems are ubiquitous. They are easily obtained by including time variation in the vector field of an autonomous differential equation and also arise naturally without an underlying autonomous model. Moreover, they cannot be entirely avoided when one is interested primarily in a particular autonomous system, since perturbations and noise terms are more realistically time dependent, while numerical schemes with variable step size are essentially nonautonomous difference equations even when the underlying differential equation is autonomous.

This Chapter begins with a brief review of results for the autonomous case and more recent ideas on inflated autonomous attractors. The cocycle formalism for a nonautonomous system and the concepts of pullback convergence and pullback attractors in such systems are then outlined. Results on the existence of pullback attractors and of Lyapunov functions characterizing pullback attractors are presented, the formulation of a numerical scheme with variable time steps as a discrete time cocycle system is discussed and the comparison of numerical and original pullback attractors considered, at least in special cases, along with the inflation of pullback attractors. Finally, some open questions and desirable future developments are mentioned.

* Project: Discretization of Nonautonomous Attractors (Peter E. Kloeden)

2 Autonomous Dynamical Systems

The solution $x(t) = x(t; x_0)$ with initial value $x(0; x_0) = x_0$ of an autonomous differential equation

$$\dot{x} = \frac{dx}{dt} = F(x), \qquad x \in I\!R^d, \tag{1}$$

generates a continuous time semigroup $\phi = \{\phi_t\}_{t \in I\!R^+}$ on $I\!R^d$ defined by $\phi_t(x_0) := x(t; x_0)$ for each $t \geq 0$ and $x_0 \in I\!R^d$ under assumptions on the vector field F that ensure the existence, uniqueness and global extendability of all such solutions. In particular, to simplify the exposition, it will be assumed here that F satisfies a uniform Lipschitz condition on $I\!R^d$ with Lipschitz constant K.

Recall that the Hausdorff separation of nonempty compact subsets A and B of $I\!R^d$ is defined by

$$H^*(A, B) := \max_{a \in A} \operatorname{dist}(a, B) = \max_{a \in A} \min_{b \in B} \|a - b\|$$

and that the Hausdorff metric is defined by

$$H(A, B) := \max\{H^*(A, B), H^*(B, A)\}.$$

The long term dynamical behaviour of a semidynamical system ϕ often occurs in or near its *maximal attractor*, that is, a nonempty compact subset A_0 of $I\!R^d$ which is *ϕ–invariant*, i.e. with $\phi_t(A_0) = A_0$ for all $t \geq 0$, and *attracting*, i.e. with

$$\lim_{t \to \infty} H^*(\phi_t(D), A_0) = 0 \quad \text{for any bounded subset } D \subset \mathbf{R}^d. \tag{2}$$

The existence of a maximal attractor follows from that of geometrically simpler and more easily found absorbing sets. A positively invariant compact subset B of $\mathbf{R}^d$, i.e. with $\phi_t(B) \subseteq B$ for all $t \geq 0$, is called an *absorbing set* for the semidynamical system ϕ on $I\!R^d$ if for every bounded subset D of $I\!R^d$ there exists a $t_D \in I\!R^+$ such that $\phi_t(D) \subset B$ for all $t \geq t_D$. The maximal attractor is then given uniquely by

$$A_0 = \bigcap_{t \geq 0} \phi_t(B). \tag{3}$$

A maximal attractor is uniformly asymptotically stable [29] and, as shown by Yoshizawa [31], there then exists a *Lyapunov function* $V : I\!R^d \to [0, \infty)$ satisfying

1. *V is uniformly Lipschitz on* $\mathbb{R}^d$, *i.e. there exists a constant* $L > 0$ *such that*
$$|V(x) - V(y)| \leq L\,\|x - y\| \qquad \textit{for all} \quad x, y \in \mathbb{R}^d;$$
2. *there exist continuous strictly increasing functions* α, β : $\mathbb{R}^d \mapsto [0, \infty)$ *with* $\alpha(0) = \beta(0) = 0$ *and* $0 < \alpha(r) < \beta(r)$ *for all* $r > 0$ *such that*
$$\alpha(\mathrm{dist}(x, A_0)) \leq V(x) \leq \beta(\mathrm{dist}(x, A_0)) \qquad \textit{for all} \quad x \in \mathbb{R}^d;$$
3. *there exists a constant* $c > 0$ *such that*
$$V(\phi_t(x_0)) \leq e^{-ct}\, V(x_0) \qquad \textit{for all} \quad t \geq 0,\ x_0 \in \mathbb{R}^d. \tag{4}$$

Such Lyapunov functions are a very convenient tool for discretization and perturbation investigations as they do not require explicit knowledge of the solutions of the differential equation. For example, the inequality
$$V(x_1) \leq e^{-ch}\, V(x_0) + L\, C_p\, h^{p+1}. \tag{5}$$
is satisfied [19] by a pth–order one–step numerical scheme (possibly implicit)
$$x_{n+1} = x_n + hF(h, x_n, x_{n+1}) \tag{6}$$
with constant step size $h > 0$ applied to the differential equation (1), where
$$\|x_{n+1} - \phi_h(x_n)\| \leq C_p\, h^{p+1}$$
is the local discretization error with constant C_p. Similarly, the inequality
$$V(y(t; y_0)) \leq e^{-ct}\, V(y_0) + L\, K^{-1} t e^{Kt}\, h \tag{7}$$
is satisfied by a solution $y(t; y_0)$ of the perturbed differential equation
$$\frac{dy}{dt} = f(y) + h\, g(y) \tag{8}$$
with uniformly bounded continuously differentiable perturbations g satisfying $\|g(y)\| \leq 1$ on $\mathbb{R}^d$. These Lyapunov inequalities can be used to show the existence of absorbing sets for the the discrete time semidynamical system generated by the numerical scheme (6) and for the continuous time semidynamical system generated by the perturbed differential equation (7). From this follows the existence of a maximal numerical attractor A^h_{num} and maximal perturbed attractor A^h_{pert}, which converge upper semicontinuously to A_0, i.e.
$$H^*(A^h_{num}, A_0) \to 0 \qquad \text{as} \quad h \to 0$$
and similarly for the perturbed attractor A^h_{pert} [12,19,29,24].

In general, the Hausdorff separation H^* above cannot be replaced by Hausdorff metric H, so the numerical attractor A^h_{num} or the perturbed attractor A^h_{pert} may approximate in the near limit only a proper subset of the original attractor A_0, which represents a collapse of the original attractor A_0 under discretization or perturbation. For example, the closed unit disc $A_0 = \{(x,y) \in I\!R^2 : x^2+y^2 \leq 1\}$ is the maximal attractor of the two–dimensional system

$$\frac{dx}{dt} = y - x(1-x^2-y^2)^2, \quad \frac{dy}{dt} = -x - y(1-x^2-y^2)^2 \tag{9}$$

and a disc of radius slightly larger than 1 is the maximal attractor of the explicit Euler scheme applied to (9), while the singleton set $A^h_{num} = \{(0,0)\}$ is the maximal attractor of the corresponding implicit Euler scheme when the step size h is sufficiently small.

2.1 Inflation of the Attractor

The totality of possible elements of such discretized or perturbed attractors can be determined [17] by inflating the vector field of the differential equation (1) to form a *differential inclusion* or setvalued differential equation

$$\frac{dx}{dt} \in F_\epsilon(x) := \{y \in I\!R^d : \|y - F(x)\| \leq \epsilon\} \tag{10}$$

The set $F_\epsilon(x)$ here is nonempty, compact and convex, and depends continuously on ϵ, while the mapping $x \mapsto F_\epsilon(x)$ satisfies a uniform Lipschitz condition on $I\!R^d$ with the same Lipschitz constant K as the function F. These properties ensure the existence [2] of an absolutely continuous solution with initial value $x(0) = x_0$ satisfying

$$x(t) \in x_0 + \int_0^t F_\epsilon(x(s))\, ds \qquad \text{for all} \quad t \geq 0.$$

Moreoever, the setvalued mapping $(t, x_0) \mapsto \Phi^\epsilon_t(x_0)$, where $\Phi^\epsilon_t(x_0)$ is the attainability set formed by all such solutions, is continuous with respect to the Hausdorff metric, while $\Phi^\epsilon_t(x_0)$ is a nonempty compact connected subset of $I\!R^d$ with

$$H\left(\{\phi_t(x_0)\}, \Phi^\epsilon_t(x_0)\right) \leq K^{-1} t e^{Kt}\, \epsilon,$$

where $\phi_t(x_0)$ is the solution of the singlevalued differential equation (1). The solutions can also be shown to satisfy a Lyapunov inequality like (7) with the parameter h replaced by ϵ, which can then be used to construct an absorbing set and hence to establish the existence of a maximal attractor A^ϵ_{infl} for the setvalued semidynamical system [30] generated by the Φ^ϵ_t on $I\!R^d$. The attractor A^ϵ_{infl} was called the ϵ*–inflated attractor* [17] of the original singlevalued semidynamical system ϕ. By the construction, A^ϵ_{infl} contains A_0 and converges continuously rather than just upper semicontinuously to A_0, i.e.

$$\lim_{\epsilon \to 0} H\left(A^\epsilon_{infl}, A_0\right) = 0.$$

If the step size or perturbation parameter h in the numerical scheme (6) or perturbed differential equation (8) is chosen small enough compared with ϵ, then the numerical and the perturbed dynamics will be contained within and carried along by the inflated setvalued dynamics $\Phi_t^\epsilon(x_0)$ and hence the numerical attractor A_{num}^h and the perturbed attractor A_{pert}^h will be contained in the ϵ–inflated attractor A_{infl}^ϵ. The effects of roundoff error, which usually vary from step to step so the actual numerical dynamical system generated within the computer will be nonautonomous even if a constant time step h is used, will similarly be contained in the inflated attractor A_{infl}^ϵ provided ϵ is larger than the machine precision. The inflated attractor A_{infl}^ϵ is thus the smallest set containing all possible limiting behaviour or approximate autonomous attractors or nonautonomous attractor components (to be defined later) resulting from all possible perturbations and approximations of appropriate magnitude of the original semidynamical system ϕ. In particular, there is no loss of information in the inflated attractor about the original asymptotic dynamics as may occur with certain approximate systems for which the approximate attractors converge only upper semicontinuously to the original maximal attractor A_0.

2.2 Convergence Rates

The theorems used above giving the upper semi continuous convergence of the numerical and perturbed attractors to the original one have a rate of the form $\alpha^{-1}(h^p)$, where α the strictly increasing function that bounds the Lyapunov function V from below and is usually not known explicitly in practice. To be able to say something more specific about the convergence rate, one needs to know or assume something more about the attractor A_0.

For example, the ϵ–inflated attractor $A_{infl}^\epsilon = [-\epsilon^{1/\rho}, \epsilon^{1/\rho}]$ of the scalar differential equation $\dot{x} = -x|x|^{\rho-1}$, where $\rho \geq 1$, converges to the maximal attractor $A_0 = \{0\}$ with order $1/\rho$. Essentially, the rate of convergence here depends on how fast the unperturbed attractor attracts its neighbourhoods. This is, in fact, typical of the general situation, as was shown in [9] (see also [10]) using a different kind of perturbation that is, however, equivalent to the inflated dynamics of [17].

Let A_0 be the maximal attractor of the semidynamical system generated by (1). A family of forward invariant compact sets $\{B_\mu, \mu \geq 0\}$ that depend continuously on μ with respect to the Hausdorff metric H and satisfy $A_0 \subset \text{int } \mathrm{B}_0$ is called a *contracting family of neighbourhoods* if there exist a $T > 0$ with

(i) $A_0 = \bigcap_{\mu \in I\!R_0^+} B_\mu$,
(ii) $B_{\mu'} \subseteq B_\mu$ for all $\mu, \mu' \in I\!R_0^+$, $\mu' \geq \mu$,
(iii) $\phi_T(B_\mu) \subseteq B_{\mu+T}$ for all $\mu \in I\!R_0^+$.

Furthermore, let $d_{\min}(A, B) := \inf_{x \notin B} \min_{y \in A} d(x, y)$ and let $\gamma : I\!R_0^+ \to I\!R_0^+$ be a strictly increasing and continuous function with $\gamma(0) = 0$ such that for all postive s_0 and r there exist positive $\tilde{r} \leq \bar{r}$ for which $\tilde{r}\gamma(s) \leq \gamma(rs) \leq \bar{r}\gamma(s)$ for all $s \in [0, s_0]$. Then, a contracting family of neighbourhoods B_μ is said to be *contracting with rate* γ if there exists $C > 0$ such that

$$H(B_\mu, A_0) \leq C\gamma(d_{\min}(\phi_T(B_\mu), B_\mu)) \quad \text{for all } \mu \in I\!R_0^+.$$

The existence of a contracting family of neighbourhoods with rate of contraction γ is both necessary and sufficient for the rate of convergence γ of the inflated attractor [9].

Theorem 1. *Let B be an absorbing set for a maximal attractor A_0 for which $A_0 \subset \operatorname{int} B$. Then A_0 admits a contracting family of neighbourhoods B_μ with $B_0 = B$ and contraction rate γ if and only if there is an $\epsilon^* > 0$ such that there exists an inflated attractor $A^\epsilon_{infl} \subset \operatorname{int} B$ for each $\epsilon \in (0, \epsilon^*]$ with*

$$H\left(A^\epsilon_{infl}, A_0\right) \leq K\gamma(\epsilon).$$

for some constant K.

In the simple example above, $\gamma(s) = s^{1/\rho}$.

3 Nonautonomous Dynamical Systems

Suppose that a unique solution $x(t) = x(t; t_0, x_0)$ of a nonautonomous differential equation

$$\dot{x} = \frac{dx}{dt} = F(t, x), \qquad x \in I\!R^d,\ t \in I\!R \tag{11}$$

with initial value $x(t_0; t_0, x_0) = x_0$ at time t_0 exists for all $x_0 \in I\!R^d$ and $t \geq t_0 \in I\!R$. The semigroup property of solutions of an autonomous differential equation now becomes

$$x(t_2; t_0, x_0) = x(t_2; t_1, x(t_1; t_0, x_0)) \tag{12}$$

for all $x_0 \in I\!R^d$ and all $t_0 \leq t_1 \leq t_2$ in $I\!R$, which is called a cocycle property.

An abstract nonautonomous dynamical system that is sometimes called a *process* [12] can be defined in terms of the solution mapping $(t, t_0, x_0) \to x(t; t_0, x_0)$ with this generalized semigroup property together with initial condition and continuity properties. An alternative formulation due to Sell [28] that retains the semigroup representation is somewhat more abstract but includes more information about how they evolve in time. It is based on the fact that whenever $x(t)$ is a solution of the differential equation (11), then the $x_\tau(t) := x(\tau + t)$ with fixed τ satisfies the nonautonomous differential equation

$$\frac{d}{dt}x_\tau(t) = F_\tau(t, x_\tau(t)) := F(\tau + t, x(\tau + t)).$$

Denote by $\mathcal{F}$ a set of functions $F : I\!R \times I\!R^d \to I\!R^d$ such that $F_\tau(\cdot,\cdot) := F(\tau + \cdot,\cdot) \in \mathcal{F}$ for all $\tau \in I\!R$; for example, $\mathcal{F}$ is a compact metric space for almost periodic differential equations. Then introduce a group of shift operators $\theta_\tau : \mathcal{F} \mapsto \mathcal{F}$ by $\theta_\tau F := F_\tau$ for each $\tau \in I\!R$, define $\mathcal{X} = I\!R^d \times \mathcal{F}$ and write $x(t; x_0, F)$ for the solution of (11) with initial value x_0 at initial time $t_0 = 0$. Finally define $\Psi_t : \mathcal{X} \mapsto \mathcal{X}$ by $\Psi_t(x_0, F) := (x(t; x_0, F), \theta_t F)$. Then the family of mappings $\{\Psi_t, t \in I\!R\}$ is a continuous–time semigroup on the state space $\mathcal{X}$ and with an appropriate topology on $\mathcal{F}$ so that $(t, x_0, F) \to x(t; x_0, F)$ is continuous it forms an autonomous semidynamical system on the extended state space $I\!R^d \times \mathcal{F}$ which is called the *skew–product flow.* To see this observe that the first component of the semigroup identity $\Psi_{t+s}(x_0, F) = \Psi_t \circ \Psi_s(x_0, F)$ expands out as

$$x(t + s; x_0, F) = x(t; x(s; x_0, F), \theta_s F), \tag{13}$$

which is also a cocycle property.

3.1 Cocycle Formalism

The shift operators in the cocycle property (13) can be considered as a driving mechanism that indicates how the dynamics of the nonautonomous system changes with time. This motivates the following definition of an abstract nonautonomous dynamical system. Let $\theta = \{\theta_t, t \in I\!R\}$ be a group of mappings on a nonempty parameter set P, that is, $\theta_t : P \mapsto P$ with $\theta_0 = id.$ and $\theta_t \circ \theta_s = \theta_{t+s}$ for all t, $s \in I\!R$, and write $\theta_t p$ for $\theta_t(p)$.

Definition 2. A family of mappings $\phi_{(t,p)} : I\!R^d \to I\!R^d$ for $t \in I\!R^+$ and $p \in P$ is called a cocycle on $I\!R^d$ with respect to a group θ of mappings on P if

$$(i) \quad \phi_{(0,p)} = id, \qquad \text{and} \qquad (ii) \quad \phi_{(t+s,p)} = \phi_{(t,\theta_s p)} \circ \phi_{(s,p)}$$

for all t, $s \in I\!R^+$ and $p \in P$.

The use of a general parameter set P here may seem an unnecessary abstraction, but in fact allows for broader applicability and richer dynamical behaviour, particularly when P is a compact metric space. In the skew–product formalism above P is the function space $\mathcal{F}$, while for a periodic differential equation (i.e. with $F(t + T, x) = F(t, x)$ in (11)) a circle $S^1 \cong I\!R \pmod T$ representing the fundamental periodic interval can be used as P. A control system $\dot{x}(t) = f(x(t), u(t))$ can be formulated as a cocyle with a compact metric space of all measurable control functions taking values in a given compact and convex set as the parameter set P. A general nonautonomous differential equation (11) can also included in this new formalism with P being the set $I\!R$ of initial times and the shift operators θ_t by $\theta_t t_0 := t_0 + t$, but the parameter space P is now no longer compact. The parameter space P may not even be a topological space, as happens with random dynamical systems for which a canonical probabilistic sample space is used as the parameter space [1,16].

4 Nonautonomous Attraction and Attractors

A nonautonomous differential equation (11) can sometimes have an attractor as defined for autonomous systems. For example, $\bar{\phi}(t) \equiv 0$ is a solution of (11) if the vector field satisfies $F(t, 0) = 0$ for all $t \in I\!R$ and could be asymptotically stable in the sense of Lyapunov [31]. However, even in this simple case, the rate of attraction and absorbing sets need not be uniform in time, as can be seen from the example $\dot{x} = -2tx$, which has the asymptotically stable solution $\bar{u}(t) \equiv 0$ and general solutions $x(t; t_0, x_0) = x_0 e^{-t^2+t_0^2}$. The situation is more complicated for a nonzero asymptotically stable solution $\bar{\phi}$. Of course, the time varying change of coordinates $z(t) = x(t) - \bar{\phi}(t)$ will convert this to the preceding situation provided $\bar{\phi}$ is known explicitly. If not, how can a specific point $\bar{\phi}(t) \in \{\bar{\phi}(s), s \in I\!R\}$ be determined analytically or numerically for a given finite $t \in I\!R$? The example $\dot{x} = -x + g(t)$ for a continuous function $g : I\!R \to I\!R$, which has the general solution

$$x(t; t_0, x_0) = x_0 e^{-t+t_0} + e^{-t} \int_{t_0}^{t} e^s g(s)\, ds,$$

gives some insight here. Holding t fixed and letting the initial time $t_0 \to -\infty$ gives the limit

$$x(t; t_0, x_0) \longrightarrow \bar{\phi}(t) := e^{-t} \int_{-\infty}^{t} e^s g(s)\, ds \qquad \text{as} \quad t_0 \to -\infty, \quad t, x_0 \ \text{fixed}, \tag{14}$$

provided that the improper integrals here exist and are finite for each $t \in I\!R$; see Figure 1.

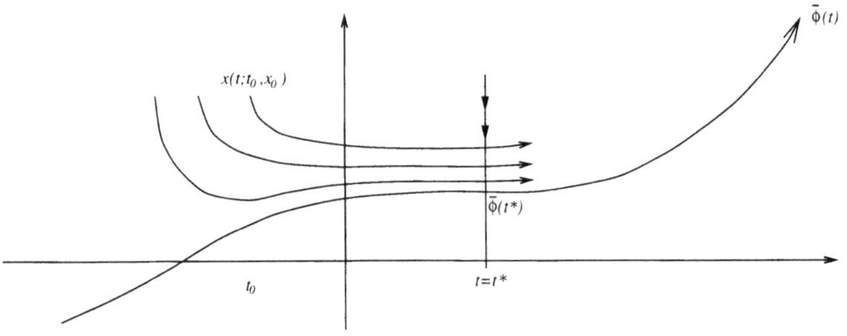

Fig. 1. Convergence to $\bar{\phi}(t)$ at $t = t^*$ as $t_0 \to -\infty$

Note that $\bar{\phi}$ is a solution of the differential equation here that exists for all $t \in I\!R$ and although the geometric trajectory set $\{\bar{\phi}(s), s \in I\!R\}$ is not invariant under the nonautonomous dynamics, the solution satisfies a dynamical invariance of the form

$$x(t; t_0, \bar{\phi}(t_0)) = \bar{\phi}(t) \qquad \text{for all} \quad t \geq t_0 \quad \text{in} \quad I\!R. \tag{15}$$

The convergence in (14) can be rewritten as

$$\lim_{t_0 \to -\infty} \|x(t; t_0, x_0) - \bar{\phi}(t)\| = 0 \quad \text{with} \quad t, x_0 \text{ fixed}$$

and is called *pullback convergence* to distinguish it from the usual *forwards convergence* given by

$$\lim_{t \to \infty} \|x(t; t_0, x_0) - \bar{\phi}(t)\| = 0 \quad \text{with} \quad t_0, x_0 \text{ fixed.}$$

The idea of pullback convergence has been used in other contexts for many years, for example by Mark Krasnosel'skii [25] in the 1960s to establish the existence of solutions of (11) that remain bounded for all $t \in I\!R$. To help understand what it means, recall that in an autonomous system convergence with time $t \to \infty$ gives the same result as convergence with the elapsed time $t - t_0 \to \infty$ with t fixed and $t_0 \to -\infty$, since autonomous dynamics depend only on the elapsed time and the attractor or limit set exists for all time and is invariant. In a nonautonomous system pullback convergence involves essentially $t - t_0 \to \infty$ with t fixed and $t_0 \to -\infty$, and thus differs from the usual forward convergence with $t - t_0 \to \infty$ for fixed t_0. In general, pullback convergence and forwards convergence are independent concepts in nonautonomous systems, as the examples $\dot{x} = -2tx$ and $\dot{x} = 2tx$ show, since, as the Figures 2 and 3 indicate, the first is forwards but not pullback convergent, whereas the latter is pullback but not forwards convergent.

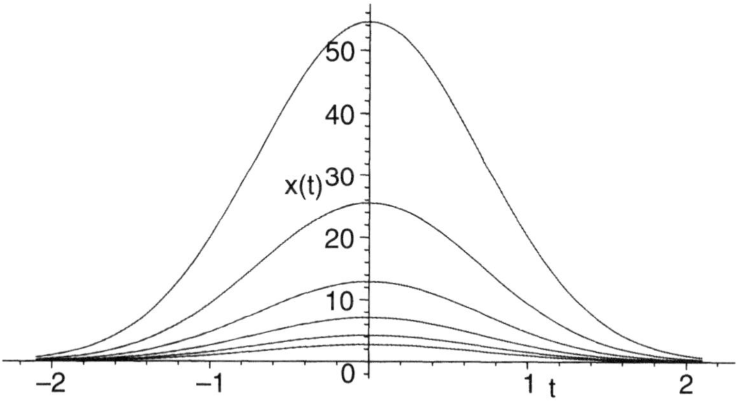

Fig. 2. Trajectories for $\dot{x} = -2tx$

4.1 Pullback Attractors

The above observations suggest that a nonautonomous attractor could be defined in terms of pullback convergence, with such a *pullback attractor* consisting of a family of compact subsets that are mapped into each other under the forward action of the cocycle mappings.

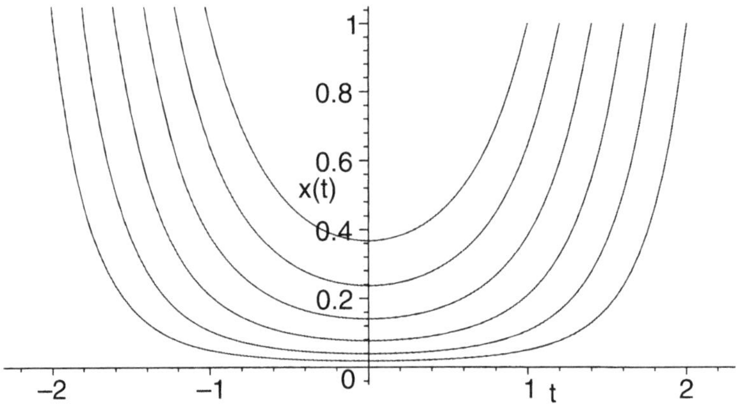

Fig. 3. Trajectories for $\dot{x} = 2tx$

Definition 3. A family $\widehat{A} = \{A_p, p \in P\}$ of compact subsets of $I\!R^d$ is called a *pullback attractor* of a cocycle $\{\phi_{(t,p)}, t \in I\!R^+, p \in P\}$ on $I\!R^d$ if it is *invariant* in the sense that

$$\phi_{(t,p)}\left(A_p\right) = A_{\theta_t p},\ t \in I\!R^+,\ p \in P, \tag{16}$$

and *pullback attracting* in the sense that

$$\lim_{t\to\infty} H^*\left(\phi_{(t,\theta_{-t}p)}(D), A_p\right) = 0 \quad \text{for any bounded subset } D \text{ of } I\!R^d. \tag{17}$$

For example, in (14) above with $P = I\!R$, $p = t_0$ and $\theta_t t_0 = t_0 + t$, the component sets $A_{t_0} = \{\bar{\phi}(t_0)\}$ for each $t_0 \in I\!R$ form a pullback attractor. The definition also includes the usual autonomous attractor by representing the autonomous semigroup as a cocycle with respect to a singleton parameter set $P = \{p\}$.

The existence of a pullback attractor also follows from that of more easily found absorbing sets, but these are now defined in terms of the pullback action and families of parametrized sets are used to allow for nonuniformities that are ubiquitous in nonautonomous systems. A family $\widehat{B} = \{B_p, p \in P\}$ of compact subsets of $I\!R^d$ is called a *pullback absorbing family* for a cocycle $\{\phi_{(t,p)}, t \in I\!R^+, p \in P\}$ on $I\!R^d$ if for each $p \in P$ and every bounded subset D of $I\!R^d$ there exists a $t_D(p) \in I\!R^+$ such that

$$\phi_{(t,\theta_{-t}p)}(D) \subseteq B_p \qquad \text{for all } t \geq t_D(p). \tag{18}$$

$\widehat{B}$ is said to be uniformly absorbing if the $t_D(p)$ here do not depend on p.

Theorem 4. *Let $\{\phi_{(t,p)}, t \in I\!R^+, p \in P\}$ be a cocycle of continuous mappings on $I\!R^d$ with a pullback absorbing family $\widehat{B} = \{B(p), p \in P\}$. Then* *there exists a pullback attractor* $\widehat{A} = \{A_p, p \in P\}$ *with components uniquely determined by*

$$A_p = \bigcap_{\tau \geq 0} \overline{\bigcup_{t \geq \tau} \phi_{(t,\theta_{-t}p)}(B_{\theta_{-t}p})}. \tag{19}$$

Proofs of various versions of this theorem in a number of different contexts can be found in [6,8,20,21,24,27].

Although pullback convergence does not in general imply forwards convergence, additonal continuity and compactness assumptions allow one to conclude that all forward limiting behaviour is contained in the union of all of the pullback attractor component sets.

Corollary 5. *Suppose in addition to the assumptions of Theorem 4 that the mappings* $\phi_{(t,\cdot)}(\cdot) : P \times I\!R^d \mapsto I\!R^d$ *are continuous,* P *is a compact metric space, the* θ_t *are continuous and* $\widehat{B}$ *is uniformly absorbing. Then*

$$\lim_{t\to\infty} \sup_{p\in P} H^*\left(\phi_{(t,p)}(D), A(P)\right) = 0 \tag{20}$$

for any bounded subset D *of* $I\!R^d$, *where* $A(P) := \bigcup_{p\in P} A_p$.

Furthermore [3], if $\widehat{B}$ consists of just a single absorbing set for all $p \in P$, then the autonomous skew–product flow $\Psi_t(x_0,p) := \phi_{(t,p)}(x_0), \theta_t p)$ on the extended state space $I\!R^d \times P$ has a maximal autonomous attractor $\mathcal{A}$ in $I\!R^d \times P$ with the sectional structure $\mathcal{A} = \bigcup_{p\in P} A_p \times \{p\}$.

4.2 Lyapunov Functions for Pullback Attractors

A pullback attractor can also be characterized by a Lyapunov function [14]. Suppose that the cocycle dynamical system (ϕ,θ) is generated by a nonautonomous differential equation

$$\dot{x} = \frac{dx}{dt} = F(p,x), \qquad p \in P,\ x \in I\!R^d, \tag{21}$$

i.e. with

$$\frac{d}{dt}\phi_{(t,p)}(x_0) = F\left(\theta_t p, \phi_{(t,p)}(x_0)\right),$$

where for simplicity $(p,x) \mapsto f(p,x)$ is assumed to be continuous in $(p,x) \in P \times I\!R^d$, $x \mapsto f(p,x)$ to be uniformly Lipschitz continuous on $I\!R^d$ with Lipschitz constant $L(p)$ for each $p \in P$, and $(t,p) \mapsto \theta_t p$ to be continuous.

Let $\widehat{A}$ be a pullback attractor for (ϕ,θ). Then there exists a pullback neighbourhood system $\widehat{B}$ with $A_p \subset \mathrm{int}B_p$ for each $p \in P$ such that the function $V : P \times I\!R^d \to I\!R^d$ defined by

$$V(p,x_0) := \sup_{t\geq 0} e^{-T_{p,t}} \mathrm{dist}\left(x_0, \Phi(t,\theta_{-t}p, B_{\theta_{-t}p})\right),$$

where $T_{p,t} := t + \int_0^t L(\theta_{-s}p)\,ds$ with $T_{p,0} := 0$, satifies the following properties:

(1) *For each $p \in P$ there exists a function $a(p,\cdot) : I\!R^+ \to I\!R^+$ with $a(p,0) = 0$ and $a(p,r) > 0$ for all $r > 0$ which is monotonic increasing in r such that*

$$a(p, \mathrm{dist}(x_0, A_p)) \le V(p, x_0) \le \mathrm{dist}(x_0, A_p); \tag{22}$$

for all $x_0 \in I\!R^d$;

(2) *V is uniformly Lipschitz on $I\!R^d$ with Lipschitz constant 1 for all $p \in P$;*

(3) *For all $p \in P$ and any bounded set D in $I\!R^d$*

$$\mathrm{limsup}_{t\to\infty} \sup_{x_0 \in D} V(p, \phi_{t,\theta_{-t}p}(x_0) = 0;$$

In addition, it can be shown that there exists a family $\widehat{N} = \{N_p, p \in P\}$ of nonempty compact sets of $I\!R^d$ which are positively invariant w.r.t. ϕ in the sense that $\phi_{t,p}(N_p) \subseteq N_{\theta_t p}$ for all $t \ge 0$, $p \in P$, and satisfying $A_p \subset \mathrm{int} N_p$ for each $p \in P$ such that

$$V(\theta_t p, \phi_{t,p}(x_0)) \le e^{-t} V(p, x_0) \tag{23}$$

for all $x_0 \in N_p$ and $t \ge 0$, which in turn implies that

$$a(\theta_t p, \mathrm{dist}(\phi_{t,p}(x_0), A_{\theta_t p})) \le e^{-t} V(p, x_0),$$

However, this does not imply Lyapunov stability or asymptotic stability, since there is no guarantee (without additional assumptions) that $\inf_{j \ge 0} a(\theta_t p, r) > 0$ for $r > 0$, so $\mathrm{dist}(\phi_{t,p}(x_0), A_{\theta_t p})$ need not become small as $t \to \infty$.

This is in fact what happens with the differential equation $\dot{x} = 2tx$ with solutions

$$x(t; t_0, x_0) = x_0 e^{t^2 - t_0^2}.$$

The pullback attractor here has components $A_{t_0} = \{0\}$ for each $t_0 \in P = I\!R$ and a Lyapunov function meeting the above requirements is given by

$$V(t_0, x_0) = |x_0| e^{-t_0 - t_0^2 - \frac{1}{4}}.$$

Property **(1)** with $a(t_0, x_0) = |x_0| e^{-|t_0| - t_0^2 - \frac{1}{4}}$ and Property **(2)** are immediate, while Property **(3)** follows from

$$\begin{aligned} V(t_0, x(t_0; t_0 - t, x_0)) &= \left| x_0 e^{t_0^2 - (t_0 - t)^2} \right| e^{-t_0 - t_0^2 - \frac{1}{4}} \\ &= e^{-(t_0 - t)^2 - t_0 - \frac{1}{4}} |x_0| \to 0 \quad \text{as} \quad t \to \infty. \end{aligned}$$

In addition, V satisfies inequality (23), since

$$\begin{aligned} V(t_0 + t, x(t_0 + t; t_0, x_0)) &= \left| x_0 e^{(t_0 + t)^2 - t_0^2} \right| e^{-(t_0 + t) - (t_0 + t)^2 - \frac{1}{4}} \\ &= e^{-t} V(t_0, x_0) \to 0 \quad \text{as} \quad t \to \infty, \end{aligned}$$

although from Figure 3 the zero solution is clearly not Lyapunov stable.

5 Approximation of Pullback Attractors

A nonautonomous dynamical system arises if variable stepsizes h_n are used in the numerical scheme (6) or a timedependent perturbation $g(t,y)$ in the perturbed differential equation (8), even though the original dynamical system generated by the differential equation (1) is autonomous. However, in both cases the Lyapunov inequalites (5) and (7) remain valid and can be used to construct a single uniform absorbing set about the autonomous maximal attractor A_0 for each of the resulting nonautonomous dynamical systems. The nonautonomously perturbed ordinary differential equation obviously generates a cocycle with respect to the parameter set $P = \mathbb{R}$ of initial times t_0, for which there thus exists a pullback attractor $\widehat{A}^h_{pert} = \{A^h_{pert,t_0}, t_0 \in \mathbb{R}\}$ and the individual component sets converge upper semi continuously to A_0 [24], i.e.

$$H^*\left(A^h_{pert,t_0}, A_0\right) \longrightarrow 0 \quad \text{as} \quad h \to 0+, \quad \text{for all} \quad t_0 \in \mathbb{R}.$$

An analogous result holds for the numerical scheme with variable stepsizes [21,23], but the formulation of such a numerical scheme as a discrete time cocycle is not as obvious.

5.1 Numerical Schemes as Discrete Time Cocycles

Consider an explicit one–step numerical scheme (6) with variable time–steps,

$$x_{n+1} = F_{h_n}(x_n) := x_n + h_n F(h_n, x_n), \tag{24}$$

applied to the autonomous differential equation (1). Define $\mathcal{H}^\delta$ to be the set of all two sided sequences $\mathbf{h} = \{h_n\}_{n\in\mathbb{Z}}$ satisfying

$$\frac{1}{2}\delta \le h_n \le \delta, \qquad \text{for all} \quad n \in \mathbb{Z} \tag{25}$$

for $\delta > 0$ (the particular factor 1/2 here is chosen just for convenience) and define the shift operator $\tilde{\theta} : \mathcal{H}^\delta \to \mathcal{H}^\delta$ by $\tilde{\theta}\mathbf{h} = \tilde{\theta}\{h_n\}_{n\in\mathbb{Z}} := \{h_{n+1}\}_{n\in\mathbb{Z}}$. The set $\mathcal{H}^\delta$ is a compact metric space with the metric

$$\rho_{\mathcal{H}^\delta}\left(\mathbf{h}^{(1)}, \mathbf{h}^{(2)}\right) = \sum_{n=-\infty}^{\infty} 2^{-|n|}\left|h_n^{(1)} - h_n^{(2)}\right|$$

and the shift operator $\tilde{\theta}$ is a homeomorphism on this metric space, so its iterations form a discrete time group. It then follows that the numerical scheme (24) with variable time steps generates a discrete time cocycle ψ on $\mathbb{R}^d$ with the parameter space $\mathcal{H}^\delta$ and shift operator group defined by

$$\psi_{0,\mathbf{h}}(x_0) = x_0, \qquad \psi_{n,\mathbf{h}}(x_0) = x_n = F_{h_{n-1}} \circ \cdots \circ F_{h_{n_0}}(x_{n_0})$$

for any $n \in I\!N$, $x_0 \in I\!R^d$ and $\mathbf{h} = \{h_n\}_{n\in\mathbb{Z}} \in \mathcal{H}^\delta$. As mentioned above, it then has a pullback attractor $\widehat{A}^h_{num} = \{A^\delta_{num,\mathbf{h}}, \mathbf{h} \in \mathcal{H}^\delta\}$ for which the components converge upper semicontinuously to the autonomous maximal attractor A_0 uniformly in the sense that

$$\sup_{\mathbf{h}\in\mathcal{H}^\delta} H^*\left(A^\delta_{num,\mathbf{h}}, A_0\right) \longrightarrow 0 \quad \text{as} \quad \delta \to 0+.$$

The situation is somewhat more complicated for the discretization of a nonautonomous differential equation of the form (21) that generates a cocycle on $I\!R^d$ with respect to the given parameter space P and group θ. An explicit one–step numerical scheme with variable step size applied to (20) now takes the form

$$x_{n+1} := x_n + h_n\, F\left(h_n, \theta_{t_n} p, x_n\right), \tag{26}$$

where the times t_n are related to a sequence of stepsizes $\mathbf{h} \in \mathcal{H}^\delta$ by $t_0 = 0$ and define $t_n = t_n(\mathbf{h}) := \sum_{j=0}^{n-1} h_j$ and $t_{-n} = t_{-n}(\mathbf{h}) := -\sum_{j=1}^{n} h_{-j}$ for $n \geq 1$. Define a mapping $\psi : \mathbb{Z}^+ \times \mathcal{Q}^\delta \times I\!R^d \to I\!R^d$ by

$$\psi(0, q, x_0) := x_0, \qquad \psi(n, q, x_0) = \psi(n, (\mathbf{h}, p), x_0) := x_n \qquad n \geq 1,$$

where $\mathcal{Q}^\delta := \mathcal{H}^\delta \times P$ and x_n is the nth iterate of the numerical scheme (26) with initial value $x_0 \in I\!R^d$, initial parameter $p \in P$ and stepsize sequence $\mathbf{h} \in \mathcal{H}^\delta$. Then ψ is a discrete time cocycle on $I\!R^d$ with the extended parameter space $\mathcal{Q}^\delta$ and the group $\Theta = \{\Theta_n\}_{n\in\mathbb{Z}}$ on $\mathcal{H}^\delta \times P$ with $\Theta_n : \mathcal{Q}^\delta \to \mathcal{Q}^\delta$ for $n \in \mathbb{Z}$ defined by iteration of the component shift operators,

$$\Theta_0 := \mathrm{id}_{\mathcal{Q}^\delta}, \quad \Theta_1(\mathbf{h}, p) := \left(\tilde{\theta}_1 \mathbf{h}, \theta_{h_0} p\right), \quad \Theta_{-1}(\mathbf{h}, p) := \left(\tilde{\theta}_{-1} \mathbf{h}, \theta_{-h_{-1}} p\right).$$

A numerical pullback attractor now has the form

$$\widehat{A}^\delta_{num} = \{A^\delta_{num,(\mathbf{h},p)}, (\mathbf{h}, p) \in \mathcal{Q}^\delta\},$$

if it exists. The existence of both continuous time and discrete time numerical pullback attractors were established in [3] under very strongly uniform structural assumptions on the vector field of the nonautonomous differential equation (21). Here the upper semicontinuous convergence reads

$$\sup_{\mathbf{h}\in\mathcal{H}^\delta} H^*\left(A^\delta_{num,\mathbf{h},p}, A_p\right) \longrightarrow 0 \quad \text{as} \quad \delta \to 0+, \qquad \text{for each} \quad p \in P.$$

A practical complication here is that a numerical scheme (26) applied to a differential equation of the form (21) may have have a lower order than the scheme on which it is based (e.g. a Runge-Kutta scheme) since the mapping $t \mapsto F(\theta_t p, x)$ may not be sufficiently smooth to justify the usual error estimations. The original higher order may still be retained if one first averages the vector field over each discretization subinterval with an appropriately chosen sampling step [11].

5.2 Inflated Pullback Attractors

Inflating the vector field of the differential equation (21) leads to a nonautonomous differential inclusion or setvalued differential equation of the form

$$\frac{dx}{dt} \in F_{\epsilon_p}(p,x) := \{y \in I\!R^d \; : \; \|y - F(p,x)\| \leq \epsilon_p\}, \tag{27}$$

where the use of a family $\widehat{\epsilon} := \{\epsilon_p, p \in P\}$ of inflation parameters is to handle nonuniformities in the nonautonomous vector field. Solutions of this equation are interpreted as absolutely continuous functions $x(t)$ satisfying

$$x(t) \in x_0 + \int_0^t F_{\epsilon_{\theta_s p}}(\theta_s p, x(s))\, ds,$$

which requires that the mappings $t \mapsto \epsilon_{\theta_t p}$ must satisfy some kind of continuity property to ensure that the resulting attainability sets $\Phi^{\widehat{\epsilon}}_{t,p}(x_0)$ generate a setvalued cocycle mapping [23].

Consider the uniform inflation of the differential equation $\dot{x} = 2tx$, that is, the differential inclusion

$$\frac{dx}{dt} \in [2tx - \epsilon, 2tx + \epsilon]$$

which generates the setvalued cocycle

$$\begin{aligned} &\Phi^{\epsilon}_{t_0+t,t_0}(x_0) \\ &= x_0 e^{(t+t_0)^2 - t_0^2} + \left[-\epsilon e^{(t+t_0)^2} \int_{t_0}^{t+t_0} e^{-s^2}\, ds,\; \epsilon e^{(t+t_0)^2} \int_{t_0}^{t+t_0} e^{-s^2}\, ds \right] \end{aligned}$$

over the parameter set $P = I\!R$ with shift $\theta_t t_0 = t_0 + t$. The setvalued or ϵ–inflated pullback attractor $\widehat{A}^{\epsilon}_{infl} = \left\{ A^{\epsilon}_{infl,t_0}, t_0 \in I\!R \right\}$ here has components

$$A^{\epsilon}_{infl,t_0} = \left[-\epsilon e^{t_0^2} \int_{-\infty}^{t_0} e^{-s^2}\, ds,\; \epsilon e^{t_0^2} \int_{-\infty}^{t_0} e^{-s^2}\, ds \right]$$

for $t_0 \in I\!R$, but requires a restriction on the regions of pullback attraction to subsets $D^{\epsilon}_{t_0} = \{x_0 \in I\!R^1 \; : \; |x_0| \leq e^{t_0^2} \epsilon \sqrt{\pi}\}$ for each $t_0 \in I\!R$, see Figure 4, where the positive part of the pullback attractor is given by the shaded region and the upper curve indicates the upper bound on $D^{\epsilon}_{t_0}$.

Such a restriction on the regions of pullback attraction is typical in many examples and the theory of pullback attractors has been extended to handle it [16]. The component sets of any perturbed or numerical pullback attractor for sufficiently close perturbation or numerical approximations will lie within the corresponding component of the inflated pullback attractor. Their regions of pullback attraction will, in general, also need to be parameter dependent or the magnitude of the error need to be made increasingly smaller with increasing t_0.

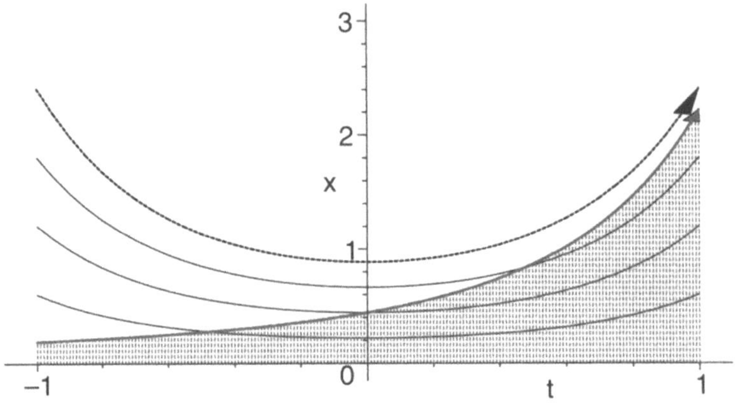

Fig. 4. Positive part of the inflated pullback attractor for $\dot{x} = 2tx$, $\epsilon = 1/2$

6 Scope of the Project and Future Work

This article surveys the progress that has been made as well as sketching the background and basic issues involved with a project that has been funded as part of the DANSE research program during the final two years of its six year existence. The original broad aim of the project was to investigate the effects of discretization and perturbation on attractors of nonautonomous dynamical systems and, more specifically, to generalize the 1986 result of Kloeden and Lorenz [19] on the discretization of autonomous attractors, for which the Lyapunov inequality (5) was used to construct an absorbing set for the discretized system. The characterization of pullback attractors by a Lyapunov function was thus seen as the crucial initial step in the project. The longer term motivation for the project was to understand the effects of discretization and perturbation on random dynamical systems, which are intrinsically nonautonomous and, moreover, highly nonuniform.

Results on the discretization of attractors are essentially perturbation results, if for rather atypical types of perturbations, for which the methods that are traditionally applied usually require for some kind of uniformity in the assumed behaviour under consideration. For cocycle systems with compact parameter sets and other nice topological and structural properties, as is assumed for the skew-product flow formalism of nonautonomous deterministic differential eqautions, reasonable progress can be expected and has been made [3]. The example of an inflated pullback attractor in the previous section is only uniform over the past up to any present time rather than for all times so just how much of uniformity with respect to the parameter space is required to give the sought result remains to be seen. Convergence rates for approximations of pullback and inflated pullback attractors also still need to be carefully investigated as does the apparently strong connection between the pullback attractors and controllability properties of the control systems that are representable as cocycle systems.

Measurability rather than continuity is the dominate characteristic of random dynamical systems, at least with respect to the parameter in the cocycle mapping, so some very deep and challenging theoretical analysis seems to be required. At present only simple special cases have been investigated, e.g. [16]. Results of numerical simulations of random dynamical systems, again only for special cases, using the subdivision algorithm of Dellnitz and Hohmann [7] reveal very interesting dynamical behaviour and suggest that such an investigation will be well worth the effort [26].

References

1. L. Arnold, *Random Dynamical Systems.* Springer–Verlag, Heidelberg, 1998.
2. J.P. Aubin and A. Cellina, *Differential Inclusions.* Springer–Verlag, Berlin, 1984.
3. D.N. Cheban, P.E. Kloeden and B. Schmalfuß, Pullback attractors in dissipative nonautonomous differential equations under discretization, *J. Dyn. Syst. Diff. Eqns.* (to appear)
4. D.N. Cheban, P.E. Kloeden and B. Schmalfuß, The Relationship between pullback, forwards and global attractors of nonautonomous dynamical systems, *DANSE*-Preprint 27/00, FU Berlin, 2000.
5. F. Colonius and W. Kliemann, *The Dynamics of Control.* Birkhäuser, Boston, 1999.
6. H. Crauel and F. Flandoli, Attractors for random dynamical systems, *Probab. Theory Relat. Fields,* **100** (1994), 365–393.
7. M. Dellnitz and A. Hohmann, A subdivision algorithm for the computation of unstable manifolds and global attractors, *Numer. Math.* **75** (1997), 293–317.
8. F. Flandoli and B. Schmalfuß, Random attractors for the 3d stochastic Navier Stokes equation with multiplicative white noise, *Stochatics and Stochastic Reports,* **59** (1996), 21–45.
9. L. Grüne, Convergence rates of perturbed attracting sets with vanishing perturbations, *J. Math. Anal. Appl.* **244** (2000), 369–392.
10. L. Grüne, Persistence of attractors for one-step discretizations of ordinary differential equations, *IMA J. Numer. Anal.* (to appear)
11. L. Grüne and P.E. Kloeden, Pathwise approximation of random ordinary differential equations. *BIT* (submitted)
12. J. Hale, *Asymptotic Behavior of Dissipative Dynamical Systems*, Amer. Math. Soc., Providence, RI, 1988.
13. P.E. Kloeden, Lyapunov functions for cocycle attractors in nonautonomous difference equation, *Izvestiya Akad Nauk RM. Mathematika* **26** (1998), 32–42.
14. P.E. Kloeden, A Lyapunov function for pullback attractors of nonautonomous differential equations, *Electronic J. Differential Eq.* Conference 05 (2000), 91–102.
15. P.E. Kloeden, Pullback attractors in nonautonomous difference equations, *J. Difference Eqns. Applns.* **6** (2000), 33–52.

16. P. E. Kloeden, H. Keller and B. Schmalfuß, Towards a theory of random numerical dynamics, in *Stochastic Dynamics*. Editors: H. Crauel and V. M. Gundlach, Springer–Verlag, 1999, pp. 259–282.
17. P.E. Kloeden und V.S. Kozyakin, Inflation of pullback attractors and discretization: The autonomous case, *Nonlinear Analysis TMA* **40** (2000), 333–343.
18. P.E. Kloeden and V.S. Kozyakin, The inflation of nonautonomous systems and their pullback attractors, *Transactions of the Russian Academy of Natural Sciences, Series MMMIU*. **4**, No. 1-2, (2000), 144-169.
19. P.E. Kloeden and J. Lorenz, Stable attracting sets in dynamical systems and their one–step discretizations, *SIAM J. Numer. Anal.* **23** (1986), 986–995.
20. P.E. Kloeden and B. Schmalfuß, Lyapunov functions and attractors under variable time–step discretization, *Discrete & Conts. Dynamical Systems* **2** (1996), 163–172.
21. P.E. Kloeden and B. Schmalfuß, Cocycle attractors sets of variable time–step discretizations of Lorenzian systems, *J. Difference Eqns. Applns.* **3** (1997), 125–145.
22. P.E. Kloeden and B. Schmalfuß, Nonautonomous systems, cocycle attractors and variable time–step discretization, *Numer. Algorithms* **14** (1997), 141–152.
23. P.E. Kloeden and B. Schmalfuß, Asymptotic behaviour of nonautonomous difference inclusions, *Systems & Control Letters* **33** (1998), 275–280.
24. P.E. Kloeden and D.J. Stonier, Cocycle attractors in nonautonomously perturbed differential equations, *Dynamics of Discrete, Continuous and Impulsive Systems* **4** (1998), 211–226.
25. M.A. Krasnosel'skii, *The Operator of Translation along Trajectories of Differential Equations*, Translations of Mathematical Monographs, Volume 19. American Math. Soc., Providence, R.I., 1968.
26. G. Ochs and H. Keller, Numerical approximation of random attractors, in *Stochastic Dynamics*. Editors: H. Crauel and V. M. Gundlach, Springer–Verlag, 1999, pp. 93–115.
27. B. Schmalfuß, The stochastic attractor of the stochastic Lorenz system, in *Nonlinear Dynamics: Attractor Approximation and Global Behaviour*, Proc. ISAM 92 (Editors: N. Koksch, V. Reitmann and T. Riedrich), TU Dresden, 1992, 185–192.
28. Sell, G.R., *Lectures on Topological Dynamics and Differential Equations*. Van Nostrand–Reinbold, London, 1971.
29. A.M. Stuart and A.R. Humphries, *Numerical Analysis and Dynamical Systems*. Cambridge University Press, Cambridge 1996.
30. G.P. Szegö and G. Treccani, *Semigruppi di Trasformazioni Multivoche*, Springer Lecture Notes in Mathematics, Vol. 101, 1969.
31. T. Yoshizawa, *Stability Theory by Lyapunov's Second Method*. The Mathematical Society of Japan, Tokyo, 1966

Aspects on Data Analysis and Visualization for Complicated Dynamical Systems

J. Becker[1], D. Bürkle[2], R.-T. Happe[1], T. Preußer[3], M. Rumpf[3], M. Spielberg[3], and R. Strzodka[3,*]

[1] Mathematische Fakultät, Universität Freiburg
[2] Institut für Angewandte Mathematik, Universität Freiburg
[3] Institut für Angewandte Mathematik, Universität Bonn

Abstract. Flow visualization is an indispensible tool for the understandig of complex flow phenomena in computational fluid dynamics and the analysis of dynamical systems. In this note we will present several ways for an effective post processing of fluid flows and flows on invariant manifolds of dynamical systems. Feature extraction techniques will be present which reduce the informational content of large timedependent data sets to its mainly intersesting essence. Futhermore, we present visualization approaches which are based on partial differential equations. Similar to the modelling of physical phenomena by partial differential equations, in the postprocessing of data such equations naturally arise as well. Finally, the method for the dense covering of an invariant manifold with streamlines is outlined, which enables us to represent the geometry of the objects, statistical information on it, and the local flow properties at the same time.

1 Introduction

The understanding of complex structures in dynamical systems is a challenging subject not only from the analytical or numerical point of view. Visualization serves as a tool to get insight in solution structures and their dynamical behaviour. Frequently, standard methods for a graphical representation break down almost at the beginning. For instance the visualization of timedependent vector fields by arrow icons leads to visual clutter, or drawing single orbits on invariant manifolds often hides important features of this object. Furthermore, for timedependent problems especially in 3D, a drawing of complicated geometric pattern often hides essential information, e. g. in terms of critical points, heteroclinic and homoclinic orbits. We will described recent approaches from different fields of dynamical systems to avoid these shortcomings. Therefore, we gather brief descriptions of several methods and algorithms in this note. Detailed discussions of these techniques can be found in publications, where they have been presented first [3,19,7,4,10,14,2]. Here, our main intention is to describe them as bricks of diverse origin but with the same aim to enable a better understanding of complex flow phenomena

* Project: Visualization of Topological and Geometric Characteristics of Time Dependent Partial Differential Equations (Dietmar Kröner, Martin Rumpf)

in computational fluid dynamics and the theory of dynamical systems. Our goal is to graphically represent flow data in an intuitively understandable and precise way.
At first, we will discuss feature extraction techniques for fluid flow. If we are interested in the topology of flow fields, we may focus on the evolution of critical points, connecting orbits, or vortex cores. Iconic visualization techniques will be presented, which help to extract such features and to visualize them in an intuitively receptable way.
Furthermore, two methods based on a modelling with partial differential equations are described which allow an easy perception of flow data. The texture transport method especially applies to timedependent velocity fields. Lagrangian coordinates are computed solving the corresponding linear transport equations numerically. Choosing an appropriate texture on the reference frame the coordinate mapping can be used as a suitable texture mapping. Alternatively, the nonlinear diffusion methods serves as an appropriate scale space method for the visualization of complicated flow patterns. It is closely related to nonlinear diffusion methods in image analysis where images are smoothed while still retaining and enhancing edges. Here an initial noisy image is smoothed along streamlines, whereas the image is sharpened in the orthogonal direction. The two methods have in common that they are based on a continuous model and discretized only in the final implementational step. Therefore, many important properties are naturally established already in the continuous model.
Concerning invariant manifolds of dynamical systems, a novel visualization approach is presented. It is based on research concerning efficient and robust set oriented computational methods, which were introduced by M. Dellnitz, T. Hohmann, and O. Junge [8,9]. Thereby the manifolds are covered with leaf boxes of a binary tree of boxes. The visualization technique to be presented here allow an interactive manipulation and inspection of these sets and an accompanying invariant measure density. Furthermore to struggle out the local dynamics, a covering of the leaf boxes with a dense set of short integral lines is considered. These line segments can then be shaded and animated.
To introduce the general topic, let us briefly recall the principle setting of flow visualization. The visualization of field data, especially of velocity fields from CFD computations is one of the fundamental tasks in scientific visualization. The simplest method to draw vector plots at nodes of some overlayed regular grid in general produces visual clutter, because of the typically different local scaling of the field in the spatial domain, which leads to disturbing multiple overlaps in certain regions, whereas in other areas small structures such as eddies can not be resolved adequately. The central goal is to obtain a denser, intuitively better receptible method. Furthermore it should be closely related to the mathematical meaning of field data, which is mainly expressed in its one to one relation to the corresponding flow. Single particle lines only very partially enlighten features of a complex flow field. Thus, we ask for an *automatic selection procedure of interesting particle lines and features* or

alternatively a *suitable dense pattern which represents the flow* ϕ globally on the computational domain.

2 Iconic Visualization of Flow Phenomena

Complex physical phenomena can be simulated and resolved with large scale computations based on recent numerical methods, in particular adaptive, time–dependent, two and three dimensional finite element or finite volume algorithms based on unstructured grids. Characteristics of the solution, which are topologically invariant and globally describe the physical phenomena, are in general hidden in enormous masses of information. Instead of an "overall" visualization, concepts to display selected important aspects are required. We are forced to carefully depict these features of interest, which characterize the global solution. A couple of selection techniques has recently been studied. Globus et al. [13] propose to extract critical points from flow data sets. At these locations they graphically represent the eigenspaces. On boundary shapes, Helman and Hesselink [15] construct topological skeletons for vector fields. In [11] Demarcelle and Hesselink give a complete analysis of second order tensor field topology on two dimensional domains. Post et al. [18] apply methods based on mathematical morphology to locate interesting regions in large data sets. To represent the local solution in regions of interest graphically, icons have been investigated. An icon is a geometric object which acts as a symbolic representation for specific data quantities and features of the solution. DeLeeuw and van Wijk [16] have developed an iconic flow probe. Post et. al. [18] give several glyphs for various simulation features.
In this section we contribute new icons and criteria for point selection and apply them in different stationary and especially time–dependent applications. At first let us consider the linearization of a flow close to a particle path. We pick up the above one–to–one relation between a velocity field and the induced flow φ defined by the ordinary differential equation

$$\dot{\varphi}(X,t) = v(\varphi(X,t),t)$$

where $\varphi(X,0) = X$ describes the motion of particles initially located at positions X driven by the velocity v in Eulerian coordinates. Therefore the above equation can be rewritten as $\dot{x} = v(x,t)$. Now we ask for the acceleration $\ddot{x}$ of a particle. By applying the chain rule we obtain the material derivative Dv/dt ($Dv/dt = \partial_t v + v \cdot \nabla v$ by definition) of the velocity v:

$$\ddot{x} = \frac{\partial}{\partial t} v + v \cdot \nabla v = \frac{D}{dt} v$$

If this derivative vanishes on a particle path, the corresponding particle is up to first order in a constant motion. Now let us assume we have selected some particle of interest denoted x_0 at time t_0. Its path can be expanded in terms of v and $\frac{D}{dt}v$

$$x_0(t) = x_0(t_0) + v(x_0, t_0)(t - t_0) + \frac{1}{2}\frac{D}{dt}v(x_0, t_0)(t - t_0)^2 + O\left(\frac{\partial}{\partial t}^3\right)$$

We will study the motion of nearby particles moving along the latter path more closely and expand the offset

$$\begin{aligned}(x - x_0)(t) &= (x - x_0)(t_0) + (v(x, t_0) - v(x_0, t_0))(t - t_0) + \\ &\quad \frac{1}{2}\left(\frac{D}{dt}v(x, t_0) - \frac{D}{dt}v(x_0, t_0)\right)(t - t_0)^2 + O((t - t_0)^3) \\ &= (x - x_0)(t_0) + \nabla v(x_0, t_0)(t - t_0)(x - x_0) + \\ &\quad \frac{1}{2}\nabla\frac{D}{dt}v(x_0, t_0)(t - t_0)^2(x - x_0) + O((t - t_0)^3 + (t - t_0)(x - x_0)^2)\end{aligned}$$

Linearizing this equation we obtain

$$\dot{\delta} = \nabla v(x_0, t_0)\delta \qquad \delta(0) = \delta_0$$

To summarize, the first order motion in a neighbourhood of a specific particle x_0 at time t_0 is described by the velocity $v(x_0, t_0)$ and the velocity gradient $\nabla v(x_0, t_0)$. Now we ask for a graphical representation of this offset motion. Therefore let us look closer onto the induced linear field. We will restrict ourselves to the three dimensional case. The considerations in two dimensions then are a straightforward consequence. ∇v has at least one real eigenvalue, which we will suppose to be the third. The others might be real as well or conjugate complex. If the real parts of all three eigenvalues are positive, respectively negative, x is a moving source respectively sink. The flow of an incompressible medium in a closed system is source and sink free. To facilitate the exposition, let us restrict to this case. For differentiable velocities incompressibility is equivalent to vanishing divergence. By that assumption in the nondegenerate case there is a two dimensional subspace of $\mathbb{R}^3$ spanned by the eigenvectors corresponding to the eigenvalues of equal sign (resp. to the complex eigenvalue and its conjugate) and one remaining direction corresponding to the third eigenvalue. The induced flow is hyperbolic, particles stream in along the plane and they stream out of x in the direction of the third eigenvector, or vice versa. Graphically the direction of the third eigenvalue is represented by two opposite vectors positioned at x and pointing in or out, depending on the sign of the corresponding eigenvalue. If the other two eigenvalues of opposite sign are real, we display the restricted sink or source type flow by a disk centered at x, scale the two eigenvectors by the eigenvalues and place them on the disk. In the complex conjugate case the restricted flow is swirling in or out on the plane. To support an intuitive understanding we partition the above disk into 4 segments with alternating colour. The real part of the eigenvalue $\lambda = \alpha + i\beta$ drives the particles into the center or away from it proportionally to $e^{\alpha t}$. That determines the period

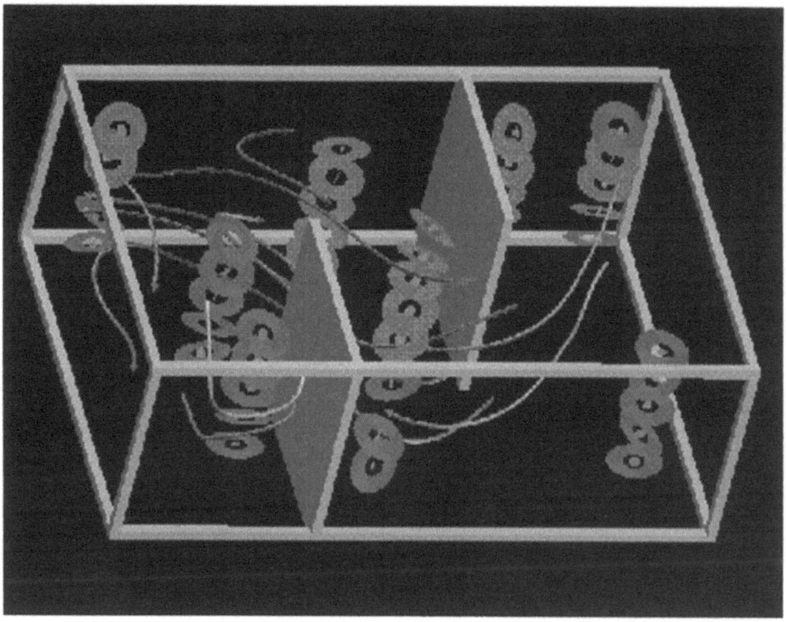

Fig. 1. Rotation icons in an incompressible flow.

of time τ the particles need to traverse the disk. Afterwards, they'll have been swirled around the angle $\beta\tau$. The rim of the disk is twisted according to that angle, and the disk is deformed linearly as indicated by the real and imaginary parts of the complex eigenvector. This leads to spiral shaped segments.

The separation lines between the segments can be interpreted as first order relative particle paths.

Icons can be released at positions related to the domain geometry and yield first insights in solution aspects. Fig. 1 depicts an example, where the incompressible Navier–Stokes equations are solved in a rectangular box with walls inside, one outlet and one inlet model the flow in a water reservoir. Some particle traces in the stationary flow field indicate the principal motion. The three–dimensional structure of the recirculation zones is visualized by placing columns of icons in the volume. Each icon shows the rotation of the velocity evaluated at its center. Especially for flow problems icons can be aligned to particle lines [16]. But one has to be very careful in finding appropriate paths which give significant insight in interesting qualitative aspects of the underlying phenomena. Critical points, characterized by $v = 0$, are of specific interest in velocity fields, in particular in the stationary case. They are topological invariants of the underlying flow [1] and can be taken as seed points to reconstruct a topological skeleton. Plate 16 on page 815 shows icons visualizing the local flow at critical points extracted automat-

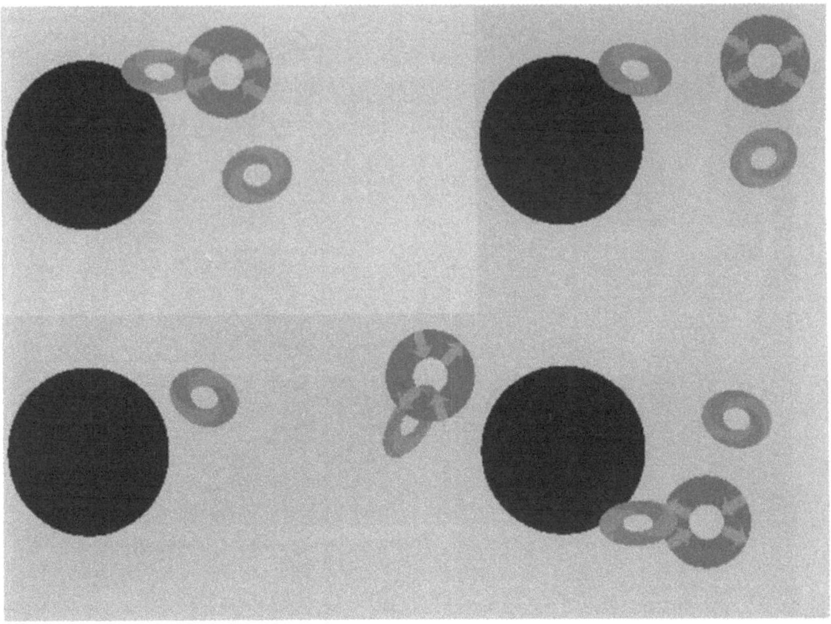

Fig. 2. Icons placed at critical points in a non stationary, incompressible flow

ically from a three dimensional volume. The underlying interpretation has already been discussed above.In the non stationary case critical points do not have the same meaning as for stationary velocity fields. But nevertheless they are still topologically invariant and give insight in qualitative aspects of the flow. Fig. 2 shows several snapshots of the incompressible, nonstationary flow behind an obstacle in two dimensions. It enlightens part of the process responsible for the formation of a Karman vortex street. Finally, Plate 17 on page 815 shows icons and streamlines on the homoclinic, respectively heteroclinic orbits in a convective flow.

3 Vector Field Aligned Nonlinear Diffusion

Let us now discuss a first PDE based method. Here, nonlinear anisotropic diffusion applied to some initial random noisy image will enable an intuitive and scalable visualization of complicated flow fields. Therefore, we pick up the idea of line integral convolution, where a strong correlation in the image intensity along streamlines is achieved by convolution of an initial white noise along the streamlines. As proposed already by Cabral and Leedom [5,21] a suitable choice for the convolution kernel is a Gaussian kernel. On the other hand an appropriately scaled Gaussian kernel is known to be the fundamental solution of the heat equation. Thus, line integral convolution is nothing else than solving the heat equation in 1D on a streamline parametrized with respect to arclength. If we ask for a wellposed *continuous* diffusion problem with similar properties, we are lead to some anisotropic diffusion, now controlled by a suitable diffusion matrix. In the direction perpendicular to the flow field we incorporate a sharpening process known from scale space methods in image processing [17]. In detail we consider the following parabolic differential equation problem:

$$\begin{aligned} \tfrac{\partial}{\partial t}\rho - \operatorname{div}\left(A(\nabla\rho_\epsilon)\nabla\rho\right) = f(\rho)\,, &\quad \text{in } I\!R^+ \times \Omega\,, \\ \rho(0,\cdot) = \rho_0 \quad , &\quad \text{on } \Omega\,, \\ \tfrac{\partial}{\partial \nu}\,\rho = 0 \quad , &\quad \text{on } I\!R^+ \times \partial\Omega \end{aligned}$$

for given initial density $\rho_0 : \Omega \to [0,1]$. Here $\rho_\epsilon = \chi_\epsilon * \rho$ is a mollification of the current density. This ensures the wellposedness of the above parabolic, boundary and initial value problem. In our setting we interprete the density as an image intensity, a scalar greyscale or – with a slight extension to the vector valued case – as a vector valued color. Thus, the solution $\rho(\cdot)$ can be regarded as a family of images $\{\rho(t)\}_{t\in I\!R^+}$, where the time t serves as a scaling parameter.

Let us now focus on the anisotropic diffusion matrix A. For a given vector field $v : \Omega \to I\!R^n$ we consider linear diffusion in the direction of the vector field and a Perona Malik type diffusion orthogonal to the field. If we suppose that v is continuous and $v \neq 0$ on Ω, then there exists an family of continuous orthogonal mappings $B(v) : \Omega \to SO(n)$ such that $B(v)v = e_0$, where $\{e_i\}_{i=0,\cdots,n-1}$ is the standard base in $I\!R^n$. Thus we define

$$A(v,d) = B(v)^T \begin{pmatrix} \alpha(\|v\|) & \\ & G(d)\,\mathrm{Id}_{n-1} \end{pmatrix} B(v)$$

where $\alpha : I\!R^+ \to I\!R^+$ is a supposed to be monoton, controlling the linear diffusion in vector field direction, i. e. along streamlines, and $G(\cdot)$ acts as an edge enhancing diffusion coefficient in the orthogonal directions (cf. [23,17]), e. g. $G(d) = \beta\,(1 + \|d\|^2)^{-1}$. As initial data ρ_0 we choose some random noise of an appropriate frequency range.

Hence pattern will grow upstream and downstream, whereas the edges tangential to these patterns are successively enhanced. Still there is some diffusion perpendicular to the field which supplies us for evolving time with a scale of progressively coarser representation of the flow field. If we run the evolution for vanishing right hand side f the image contrast will unfortunately decrease due to the diffusion along streamlines. Therefore, we strengthen the image contrast during the evolution, selecting an appropriate function $f : [0,1] \to I\!R^+$, with $f(0) = f(1) = 0$, $f > 0$ on $(0.5, 1)$, and $f < 0$ on $(0, 0.5)$.

If we ask for pointwise asymptotic limits of the evolution, we expect an almost everywhere convergence to $\rho(\infty,\cdot) \in \{0,1\}$ due to the choice of the contrast enhancing function $f(\cdot)$ (cf. Fig. 3). The space of asymptotic lim-

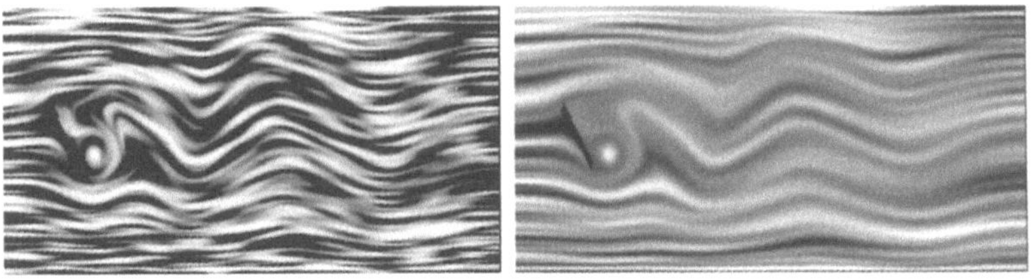

Fig. 3. A single timestep is depicted from the nonlinear diffusion method applied to the vector field describing the flow around an obstacle at a fixed time. A discrete white noise is considered as initial data. We run the evolution on the left for a small and on the right for a large constant diffusion coefficient α.

its significantly influences the richness of the developing vector field aligned structures. To enrich the set of asymptotic states settled by anisotropic diffusion we can consider a vector valued $\rho : \Omega \to [0,1]^m$ for some $m \geq 1$ and a corresponding system of parabolic equations. Finally we end up with the method of nonlinear anisotropic diffusion to visualize complex vector fields [20] (cf. Plate 18 on page 816).

4 Texturing Lagrangian Coordinates

The second method based on a modelling with partial differential equations consists in the numerical calculation of the flux ϕ itself [3]. We adopt the idea of the implicit streamsurfaces presented by J. van Wijk [24] and discuss the corresponding transport problem for timedependent data. Inflow time and inflow coordinates may be regarded as a Lagrangian frame. The method we propose here displays Lagrangian coordinates by texture mapping, which map a certain pattern from a Lagrangian coordinates system, i. e. from texture space, to the Eulerian frame. Let us assume $\Omega \subset \mathbb{R}^n$ to be a domain describing a fluid container with an inlet boundary $\Gamma^+ \subset \partial\Omega$ and an outlet boundary $\Gamma^- \subset \partial\Omega$. Furthermore we suppose the fluid velocity $v : \Omega \times [0, \hat{T}] \to \mathbb{R}^n$ to be given for a fixed time $\hat{T}$. In the application this velocity will be delivered by a numerical simulation, which runs simultaneously or has stored its results in files on disk. Let us now interpret the coordinates X on the inlet boundary Γ^+, respectively the inflow time T as depending variables, which are transported with the fluid. Then they are described by the following transport equation for a density ρ

$$\begin{aligned} \partial_t \rho + v \cdot \nabla \rho &= 0 \quad \text{in } \Omega\,, \\ \rho &= \rho_\Gamma \quad \text{on } \Gamma^+\,. \end{aligned}$$

Thus we obtain $\rho = X$ for $\rho_\Gamma = X$ on Γ^+, respectively $\rho = T$ for $\rho_\Gamma = T$ on Γ^+. On the outlet Γ^- no boundary condition has to be described if $v \cdot \nu \geq 0$ for all times, where ν is the outer normal of the domain Ω. This transport can be interpreted as a simultaneous and global particle tracing. On a particle path $x(t)$ the solution ρ of the above transport equation is constant, because $\dot{x}(t) = v(x(t), t)$ and

$$\frac{d}{dt}\rho(x(t), t) = \partial_t \rho(x(t), t) + \dot{x}(t) \cdot \nabla \rho(x(t), t) = 0\,.$$

Therefore points of constant X value are located on the particle line starting at position X on Γ^+. Analogously a constant T value indicates points on a surface which is the image of a corresponding surface on the inlet under the flow $\phi(\cdot, T)$. In this sense X, T as functions on $\Omega \times [0, \hat{T}]$ can be regarded as Lagrangian coordinates describing the motion of particles which pass through Γ^+. Particles which have earlier entered the fluid container are not considered so far.
The transport equation becomes a wellposed problem by prescribing suitable initial conditions. If every particle path starting at a position in Ω has left the domain, the solution ρ no longer depends on these initial conditions. For moderate values of $\hat{T}$ this might not be the case and for certain applications especially the initial phase of the physical simulation is of great importance. Therefore we suppose that $\tilde{X}$ and $\tilde{T}$ are extensions of $X|_{\Gamma^+}$ respectively 0 on Ω and choose them as initial conditions for the two transport problems. E. g. if $\Omega \subset \mathbb{R}^+ \times \mathbb{R}$ and $\Gamma_+ \subset 0 \times \mathbb{R}$ we choose $\tilde{X}(x_1, x_2) = (0, x_1)$, $\tilde{T}(x_1, x_2) = 0$.

Finally, we have to define an appropriate pattern in the texture space $\Gamma^+ \times [0, \hat{T}]$. There are several desirable features which should be realized by the textural representation of the Lagrangian coordinates. It should simultaneously code time and inlet coordinates. Furthermore, to enable long time animation of moving fluids the pattern in the texture space should be periodic in T and the zooming into detailed areas has to be supported by a scalability property. Thus, we use a periodic color coding of T and a periodic scalable 1D texture for X (cf. Plate 20 on page 817, Fig. 4).

5 Streamlines on Invariant Manifolds of Dynamical Systems

In this final section we will deal with a dense coverage of approximations of invariant manifolds with streamlines. They display the local flow on the manifold in an intuitively understandable way. This method is an analog of the techniques presented before, but now on geometrically complex objects. To begin with, let us consider a dynamic system. If time is assumed to evolve continuously then this system is frequently given by an ordinary differential equation of the form

$$\frac{dx}{dt}(t) = g(x), \tag{1}$$

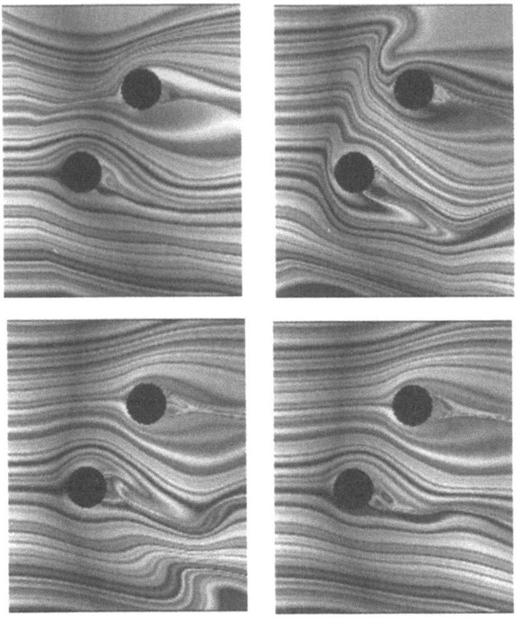

Fig. 4. Texture transport applied to a compressible Euler flow arround two obstacles.

where $g : I\!R^n \rightarrow I\!R^n$. Alternatively, we may study a discrete dynamical system of the form

$$x_{j+1} = f(x_j), \quad j = 0, 1, \ldots, \tag{2}$$

where $f : I\!R^n \rightarrow I\!R^n$. Observe that this type of dynamical system naturally arises when an ordinary differential equation is integrated by an explicit numerical scheme. Topological information on the long term behavior of the dynamical system is given by invariant sets: a set $A \subset I\!R^n$ is *invariant* if

$$f(A) = A.$$

In this section we present a visualization techniques based on recent numerical approximation methods by D. Dellnitz, H. Hohmann, O. Junge. The central object which is approximated by the subdivision algorithm developed in [9] is the so-called *relative global attractor*,

$$A_Q = \bigcap_{j \geq 0} f^j(Q), \tag{3}$$

where $Q \subset I\!R^n$ is a compact subset. Roughly speaking, the set A_Q should be viewed as the union of invariant sets inside Q together with their unstable manifolds. In particular, A_Q may contain subsets of Q which cannot be approximated by direct simulation. A subdivision algorithm for the approximation A_Q^k of A_Q generates a sequence $\mathcal{B}_0, \mathcal{B}_1, \cdots, \mathcal{B}_k, \ldots$ of finite collections of boxes which contain A_Q and approximate this relative invariant set for increasing values of k. In the concrete implementation the boxes are generalized rectangles which build up a binary tree, generated by successive bisection [8]. As an example we study here invariant set in the Lorenz system [7]. Once a box covering $\mathcal{B}$ of the attractor A_Q has been computed, one can approximate the statistics of the dynamics on A_Q by the computation of a corresponding natural invariant measure. I. e. the the transition probabilities

$$p_{ij} = \frac{m(f^{-1}(B_i) \cap B_j)}{m(B_j)}, \quad i,j = 1, \ldots, N$$

can be approximated, where m denotes Lebesgue measure. Then an eigenvector for the eigenvalue 1 is computed numerically and serves as an approximation of the invariant measure. We will use this measure for coloring of the flow lines.

The set oriented algorithm is not restricted to approximations of attracting sets which are smooth submanifolds of $\mathbb{R}^n$. The attractor under consideration may have a Hausdorff dimension which is not an integer but is of dimension between two and three. Nevertheless, frequently attractors are contained in the closure of unstable manifolds which are locally m dimensional surfaces within $\mathbb{R}^n$. Unfortunately, this surface structure is hidden in our discrete box approach. In terms of a surface interpretation the fundamental question is how to define a tangent space, or, equivalently, how to give a suitable definition of normals. Here we apply a method related to Nielson's approach in the interpolation of scattered data [6]. For given ϵ we consider the neighbourhood $U_\epsilon(c_B)$ of the center point c_B of every box $B \in \mathcal{B}_k$. (The distance is measured in the maximum norm.) Then we define the local center of gravity $\bar{x}_B$ and the first momentum matrix S_B by

$$\bar{x}_B = \int_{U_\epsilon(c_B)\cap A_Q^k} \mathrm{d}x\,, \quad S_B = \frac{3}{\epsilon^2} \int_{U_\epsilon(c_B)\cap A_Q^k} (x - \bar{x}_B)(x - \bar{x}_B)^T \,\mathrm{d}x.$$

The scaling of S_B obviously guarantees that $S_B = Id$, if the invariant set completely covers $U_\epsilon(c_B)$. Concerning the implementation we can avoid the exact evaluation of the integrals and approximate them by a counting measure over box centers. We thereby take into account values of ϵ which are multiples of the length of box edges, e. g. in the applications we consider a factor of 10. The momentum matrix S_B is symmetric. Thus, there exists an orthonormal system of eigenvectors $v_1, \cdots, v_n$ and corresponding real eigenvalues $\lambda_1 \leq \lambda_2 \leq \cdots \leq \lambda_n$. By construction the approximate set A_Q^k is locally more extended in directions of eigenvectors with relatively large eigenvalues λ_j and vice versa. If the actual invariant set A_Q is locally an m dimensional surface, then it is reasonable to assume that A_Q^k reflects this property in the sense that there are $(n - m)$ small eigenvalues, i. e.

$$\lambda_1 \leq \cdots \leq \lambda_{n-m} \ll \lambda_{n-m+1} \leq \cdots \leq \lambda_n .$$

We make use of this fact and require in the algorithm that $\frac{\lambda_{n-m}}{\lambda_{n-m+1}} < \delta$ for a small constant δ, in our case 0.1. Then the eigenvectors $v_1, \cdots, v_{n-m}$ are interpreted as approximate normals and $v_{n-m+1}, \cdots, v_n$ as approximate tangent vectors. Concerning the visualization, in particular for $n = 3$ and $m = 2$ the definition of a normal allows an appropriate shading and thereby supports the visual reception of the streamline coverage of a complicated invariant manifolds.
Our streamline visualization approach is related to the method of illuminated streamlines introduced by Stalling et al. [22]. Here, a coverage of the frequently lower dimensional invariant manifolds is attained similar to the line art illustration method by Elber [12].
We use streamlines to emphasize the local dynamics on the invariant set A_Q, i. e. the direction and velocity of the continuous flow according to the underlying ODE. Streamlines are suitable tools to visualize such information. We will now describe an algorithm which generates a coverage of A_Q^k with streamlines at a prescribed density in a preprocessing step. Then, for the lateron interactive rendering we use transparent illuminated streamlines and color them according to the invariant measure. For our case we make use of the approximate surface normals and therefore shade the individual streamlines with respect to these normals. We thus ensure the graphical representation of the *global geometry* and the *local dynamics* of the dynamical system at the same time while still retaining the *surface type appearance.*
Our coverage will be of equal density all over A_Q^k in the sense that the ratio $\gamma(B){:=}L(B)/m(B)$ of the sum $L(B)$ of the length of streamline segments in the boxes B of the binary tree and the local volume $m(B)$ is balanced. We achieve this by an interative insertion process of streamlines of fixed length 2δ. We successively select starting positions x_0, compute streamline segments $x : [-\delta, \delta] \rightarrow I\!R^n$ as numerical solutions of the ODE problems

$$\dot{x} = g(x)\,; \qquad x(0) = x_0$$

and distribute the local segments onto the corresponding boxes. Simultaneously we update the current densities $\gamma(B)$ on the involved boxes.
Plate 21 on page 817 shows results for the Lorenz system..

Acknowledgement

The authors would like to thank M. Dellnitz and O. Junge for the ongoing cooperation and exchange of ideas and to acknowledge Karol Mikula, Jarke van Wijk for inspiring discussions and many useful comments on flow visualization and image processing. Furthermore they thank Eberhard Bänsch from Bremen University for providing the incompressible flow data sets.

References

1. D. Asimov: Notes on the topology of vector fields and flows, Tutorial Notes, IEEE Visualization '95, 1995
2. J. Becker, T. Preußer, M. Rumpf: PDE Methods in Flow Simulation Post Processing, Computing and Visualization in Science, to appear
3. J. Becker, M. Rumpf: Visualization of time-dependent velocity fields by Texture Transport, Proceedings of the Eurographics Workshop on Scientific Visualization '98, Blaubeuren, 1998
4. D. Bürkle, M. Dellnitz, O. Junge, M. Rumpf, M. Spielberg: Visualizing Complicated Dynamics, Late Braking Hot Topics, Visualization '99
5. B. Cabral and L. Leedom. Imaging vector fields using line integral convolution. In J. T. Kajiya, editor, Computer Graphics (SIGGRAPH '93 Proceedings), volume 27, pages 263–272, Aug. 1993.
6. W. L. F. Degen and V. Milbrandt. The geometric meaning of nielson's affine invariant norm. CAGD, 15:19–25, 1997.
7. M. Dellnitz, A. Hohmann, O. Junge, and M. Rumpf. Exploring invariant sets and invariant measures. CHAOS: An Interdisciplinary Journal of Nonlinear Science, **7**(2):221, 1997.
8. M. Dellnitz and A. Hohmann. The computation of unstable manifolds using subdivision and continuation, in Nonlinear Dynamical Systems and Chaos (H.W. Broer, S.A. van Gils, I. Hoveijn und F. Takens eds.), PNLDE 19 (Birkhäuser, 1996), 449-459.
9. M. Dellnitz and A. Hohmann, A subdivision algorithm for the computation of unstable manifolds and global attractors. Numerische Mathematik 75, 293-317, 1997.
10. M. Dellnitz, O. Junge, M. Rumpf, R. Strzodka: The computation of an unstable invariante set inside a cylinder containing a knotted flow, Proceedings of the EquaDiff '99
11. T. Delmarcelle, L. Hesselink: The Topology of Symmetric, Second–Order Tensor Fields, IEEE Visualization '94, 140–147, 1994
12. G. Elber. Line Art Illustrations of Parametric and Implicit Forms. IEEE Transactions on Visualization and Computer Graphics, 4 (1):71–81, 1998.
13. A. Globus, C. Levit, T. Lasinski: A Tool for Visualizing the Topology of Three–Dimensional Vector Fields, IEEE Visualization '91, 33–40, 1991
14. R.-T. Happe, M. Rumpf: Characterizing gobal features of simulation data by selected local icons, M. Göbel, J. David, P. Slavik, J.J. van Wijk (eds.): Virtual Environments and Scientific Visualization'96, Springer, Vienna 1996
15. J.L. Helman, L. Hesselink: Visualizing Vector Field Topology in Fluid Flows, IEEE CG&A 11, No. 3, 36–46, May 1991
16. W.C. Leeuw, J.J. van Wijk: A Probe for Local Flow Field Visualization, IEEE Visualization '93, 39–45, 1993
17. J. Malik, P. Perona: Scale space and edge detection using anisotropic diffusion, IEEE Computer Society Workshop on Computer Vision, 1987
18. F.J. Post, T. van Walsum, F.H. Post, D. Silver: Iconic Techniques for Feature Visualization, IEEE Visualization '95,288–295, 1995
19. U. Diewald, T. Preußer, M. Rumpf: Anisotropic Nonlinear Diffusion in Vector Field Visualization on Euclidean Domains and Surfaces, Tans. Vis. and Comp. Graphics 2000, to appear

20. T. Preußer, M. Rumpf: An Adaptive Finite Element Method for Large Scale Image Processing, Journal of Visual Comm. and Image Repres., 11:183–195, 2000
21. D. Stalling, C. Hege: Fast and Resolution Independent Line Integral Convolution, Proceedings SIGGRAPH '95, 1995
22. D. Stalling, M. Zöckler, and H.-C. Hege. Fast display of illuminated field lines. IEEE Transactions on Visualization and Computer Graphics, 3(2), Apr.–June 1997. ISSN 1077-2626.
23. J. Weickert: Anisotropic diffusion in image processing, Teubner, Stuttgart 1998
24. J. J. van Wijk. Flow visualization with surface particles. IEEE Computer Graphics and Applications, 13(4):18–24, July 1993.

Non-Smooth Dynamical Systems: An Overview

Markus Kunze and Tassilo Küpper*

Mathematisches Institut der Universität zu Köln, Weyertal 86, 50931 Köln, Germany
e-mail: mkunze@mi.uni-koeln.de
kuepper@mi.uni-koeln.de

Abstract. We review mathematical results on non-smooth dynamical systems, i.e., systems that incorporate effects of friction and/or impacts.

1 Introduction

Differential equations with non-smooth components occur in various situations. For example they arise in mechanical systems if the effects of dry friction are included into the model, or they occur in the case of impacts. They are present in electrical circuits or in biological systems, if non-smooth characteristics are used to represent switches. In control theory they frequently appear when discontinuous controls are involved.

In the beginning of the theory of non-smooth systems they used to be replaced by smooth approximations whose properties are much better understood. While this approach is indeed appropriate in many situations, in the course of a growing demand concerning a refined understanding of the technical models it is no longer justified to neglect effects giving rise to non-smooth phenomena and to overcome this difficulty by smoothing. Therefore to gain a better understanding of the true effects of the non-smooth parts a direct treatment is required, and correspondingly a more general concept of differential equations is needed. Such an extension has been developed within the frame of differential inclusions that permit typical behaviour such as a lack of uniqueness, non-discrete sets of equilibria, or only temporal sticking to such "rest points". While in this, by now traditional, research area the main focus has been directed towards a theory of differential inclusions that deals with standard questions like existence of solutions, properties of solution sets, etc., see [4,6,11], the aim of our project "Non-Smooth Dynamical Systems" within the DFG priority research program "Ergodentheorie, Analysis und effiziente Simulation dynamischer Systeme" was mainly to analyze the qualitative behaviour of such non-smooth models. As standard methods of dynamical systems theory rely heavily on linearization (and this requires smoothness) explicitly extensions of classical methods had to be found. Our

* Project: Non-Smooth Dynamical Systems (Markus Kunze, Tassilo Küpper)

studies were motivated by experimental and numerical investigations [35,36] on the friction oscillator. While those results clearly indicated the presence of attractors, a more reliable picture of the dynamics was not at hand. For such systems Lyapunov exponents provide an useful tool to determine the long-term behaviour. Our initial results based on Lyapunov exponents computed in a strictly formal way suggested that indeed the dynamical behaviour could be characterized by such formally computed quantities. It was our first goal to find out how the notion of Lyapunov exponents could rigorously be extended to non-smooth systems, and in case they do exist, if conclusions can be drawn from the Lyapunov exponents to the dynamics of the system. While still preliminary, mostly formal work was carried out in [41,31,13,14], the new concept has been developed in [18], see also [23] for some preliminary results. In [18] it was also shown that it is possible to verify the hypotheses at least for important prototype models. Further results on numerically computed Lyapunov exponents, in particular for multiple mass friction oscillators, are in [28]. Moreover, implications concerning the asymptotic stability of periodic solution were drawn. In Section 2 we summarize the results on Lyapunov exponents. The importance of the theoretical investigations is underlined by the fact that there are situations where "Lyapunov exponents" can be calculated in a formal way numerically, although they can be shown to exist not at all. The misleading information due to numerical shortcomings indicate the danger of any naive way of arguing only on the basis of formal manipulations without mathematical background. Concerning Lyapunov exponents in general, we took much profit from L. Arnold and his group being as well a member of the priority research program; see especially [2].

Another important tool to prove existence of non-trivial solutions or to study bifurcations are topological methods like the Conley index. In [21] it has been shown how this concept can used for non-smooth systems by extending the classical Conley index to differential inclusions. This approach relies on proper selections of single-valued maps approximating the multivalued right-hand side of the differential inclusion. In this way bifurcation results are obtained which are illustrated by an application to a model that describes the motion of bird wings, see Section 3.

Section 4 explains what are the problems encountered when one tries to use Kolmogorov-Arnold-Moser (KAM) theory for conservative non-smooth oscillators. The related classical question is if all solutions of forced oscillators like $\ddot{x} + f(x) = p(t)$ are bounded in the phase plane. Those classical results always rely on the application of a suitable twist theorem. We then tried to understand the influence of additional discontinuous terms of type $a\text{sgn}(x)$ that model e.g. state-dependent kicks. It turns out that the discontinuity helps to keep the solutions bounded, since for a sufficiently large discontinuity all solutions remain bounded. On the other hand, if a is below a certain threshold (which is sharp), then the solutions are unbounded. From a mathematical point of view, the main problem was how to apply (a variant of) the twist theorem that requires high smoothness assumptions.

For non-smooth dynamical systems, planar models are easy to handle but at the same time do offer many interesting problems. The simple geometric constellation permits special arguments which cannot be applied in higher dimensions. In Section 5 the situation of two linear systems, each given in a half plane, is described. It results in a highly non-linear problem which is investigated in [12] with respect to the onset of periodic orbits. Using geometrical arguments about the form of the Poincaré return map to the discontinuity line, a complete description of the possible bifurcation scenarios could be obtained. Further studies are concerned with a generalized concept of Hopf bifurcation [30] for piecewise linear systems.

Finally in Section 6 we describe recent work on the generalization of Melnikov's method to non-smooth systems. It can be used to analyze the perturbations of planar systems with a homoclinic orbit. In [43] it is shown that for systems where sliding motion is excluded, the "non-smooth" Melnikov function contains two additional terms compared to the usual smooth one which result from impulses when the trajectory crosses the discontinuity line.

The paper is concluded by a short outlook in Section 7. Finally, the items of the bibliography are understood only as a selection. A more exhaustive list of references concerning non-smooth dynamical systems can be found at

`http://www.mi.uni-koeln.de/mi/Forschung/Kuepper/Forschung1.htm`

These references also include a lot of papers from engineering that often focus upon numerical results, a topic which mostly has been omitted in the present paper; see also [20,41,24] for some work of our group in this direction.

2 Lyapunov Exponents for Non-Smooth Systems

The idea of Lyapunov exponents is to define characteristic numbers for a dynamical system that allow to classify the behaviour of the system is a concise manner. These numbers should account for exponential convergence or divergence of trajectories that start close to each other. To describe this in more detail, for a general smooth system $\dot{x} = f(t, x)$ consider a reference solution trajectory $x(t)$ emerging from some initial value $x_0 \in \mathbb{R}^n$. Let $\bar{x}(t)$ denote the solution with initial value $\bar{x}_0 \in \mathbb{R}^n$ close to x_0. We are interested in the time evolution of the initial difference $z_0 = \bar{x}_0 - x_0$ and write for this purpose

$$\bar{x}(t) - x(t) \approx \frac{\partial \varphi}{\partial x}(t, x_0)(\bar{x}_0 - x_0) = \frac{\partial \varphi}{\partial x}(t, x_0) z_0, \tag{1}$$

where φ denotes the flow of the system i.e., $x(t) = \varphi(t, x_0)$ and $\bar{x}(t) = \varphi(t, \bar{x}_0)$. Hence the quantity

$$\lambda^+(x_0, z_0) := \limsup_{t \to \infty} \left(\frac{1}{t} \log \left| \frac{\partial \varphi}{\partial x}(t, x_0) z_0 \right| \right) \tag{2}$$

somehow describes the evolution of small perturbations of the initial value, and it is called an upper Lyapunov exponent; "upper" referring to the lim sup. These numbers can be used to make reliable predictions only in case that $\limsup_{t\to\infty}$ does exist as a $\lim_{t\to\infty}$. A criterion for this to hold is provided by the famous ergodic theorem of Oseledets [32] that relies on some *a priori* probability measure μ, related in a natural way to the system and invariant under the flow, i.e., under the flow the μ-measure of sets in phase space remains constant. Additionally, μ is assumed to be ergodic, meaning that every flow-invariant set has μ-measure either zero or one. The theorem reads as follows.

Theorem. *Let μ be an ergodic probability measure which is invariant w.r. to the flow. Then there exists a set $\Gamma \subset \mathbb{R}^n$ of full μ-measure and $k \leq n$ distinct numbers $\lambda^{(1)}, \dots, \lambda^{(k)}$, with $\lambda^{(1)} = -\infty$ possible, such that for all initial points $x_0 \in \Gamma$ and all $z_0 \in \mathbb{R}^n$ the quantity $\lambda^+(x_0, z_0)$ from (2) equals one of the $\lambda^{(i)}$, and additionally the* $\limsup_{t\to\infty}$ *is in fact the* $\lim_{t\to\infty}$.

We will only use this part of the theorem, which in its complete version contains much more additional information.

To see the difficulty with the definition of Lyapunov exponents for non-smooth systems, we consider the example

$$\ddot{x} + x + \operatorname{sgn}(\dot{x}) = \gamma \sin(\eta t). \tag{3}$$

This friction oscillator is one of the simplest dry friction models, cf. Fig. 1.

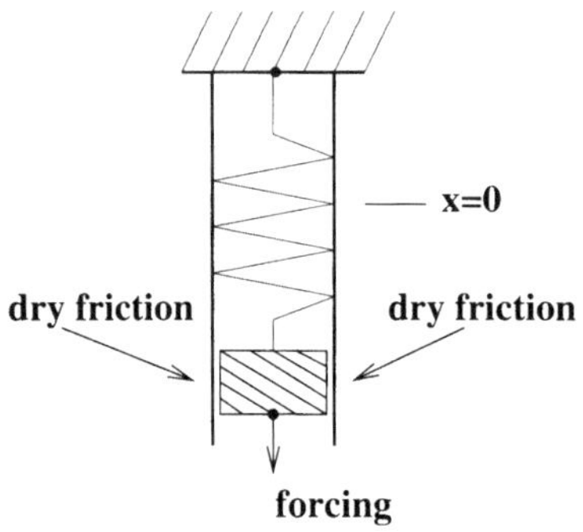

Fig. 1. A pendulum with dry friction

Equation (3) describes a pendulum that consists of a mass being attached to a spring. The mass is sinusoidally forced, it moves in a straight tube and has contact with the wall of the tube. Depending on the size of the dry friction between mass and wall, and depending on the strength of the forcing, the mass moves up or down ("slip phase") or it may stick to the wall. The term $\operatorname{sgn}(\dot{x})$ accounts for the dry friction, by Coulomb's friction law, since the friction force is proportional to the negative of the normal force (constant, since the tube is straight) multiplied by the sign of the velocity. In realistic applications this simple characteristic would have to be replaced by a more refined friction characteristics. Note, too, that in (3) we have set all physical constants equal to one, besides amplitude γ and frequency $(2\pi)/\eta$ of the forcing.

Solution trajectories of (3) may stay either in $\{\dot{x} > 0\}$ or in $\{\dot{x} < 0\}$. In that case we just have a simple oscillator. There is a third possibility that turns (3) non-linear, namely a solution trajectory may stick to $\{\dot{x} = 0\}$ for some time, corresponding to sticking of the mass to the wall. In that case we have $x(t) \equiv x_0$ in some time interval $t \in [t_1, t_2]$. Hence from (3) we find that for sticking to happen we need to have $x_0 + \text{sgn}(0) = \gamma \sin(\eta t)$ for $t \in [t_1, t_2]$, and this is impossible if we simply defined $\text{sgn}(0) := 0$. The proper idea is to take $\text{sgn}(0)$ as the whole interval $[-1, 1]$ that "connects" $+1$ to -1. Therefore we arrive at the condition $x_0 \in [\gamma \sin(\eta t) - 1, \gamma \sin(\eta t) + 1]$ for $t \in [t_1, t_2]$. From this criterion we can decide whether a solution that hits the surface $\{\dot{x} = 0\}$ in phase space at time t_1 will just pass from one to the other side, or whether it has to rest there up to some time $t_2 > t_1$. Note that for amplitudes $\gamma \leq 1$ there are some solutions sticking forever $(t_2 = \infty)$, e.g. the one with $x_0 = 0$. In particular, solutions in general do *not* cross the discontinuity surface transversely.

The mathematical theory of differential equations where the values of the right-hand side are intervals (like here $\text{sgn}(0) := [-1, 1]$), or more general sets, is well-established within the frame of differential inclusion, see [4,6]. From these results we know that (3) defines a semi-flow, i.e., a flow for positive times. (Note that we cannot expect solutions to be unique in backward time, since the discontinuity surface can be left in a non-unique way in backward time.) We denote this semi-flow by $\varphi(t, y_0)$ $(t \geq 0)$ where, as (3) is non-autonomous, the initial value $y_0 = (x_0, \dot{x}_0, t_0)$ has three components: a starting time t_0, the position x_0, and the velocity $\dot{x}_0$ at time t_0, respectively. The semi-flow can be shown to be locally Lipschitz continuous w.r. to both t and y_0.

Our general goal is to define and estimate Lyapunov exponents for systems like (3), and it is already here that we encounter the first main problem. As (3) is non-smooth, the semi-flow will not be smooth w.r. to the initial value y_0, and the corresponding $\lambda^+(y_0, z_0)$ from (2) is not well-defined, since it requires $\frac{\partial \varphi}{\partial y}(t, y_0)$ to exist. To see this, imagine a trajectory that touches the discontinuity surface D tangentially ("grazing"), with a nearby initial value $\bar{y}_0$ leading to a trajectory just crossing from, say, $\{\dot{x} > 0\}$ to $\{\dot{x} < 0\}$; see Fig. 2 below.

Thus $\frac{\partial \varphi}{\partial y}(t, y_0)$ is not defined at all y_0, as would be necessary to apply Oseledets' theorem in the above version in order to have a well-defined Lyapunov spectrum. Here it helps that the theorem in fact is much more general as stated. Instead of the linearization it is only necessary to have a cocycle, i.e., a map $T : [0, \infty[\times \mathbb{R}^n \to M(n \times n)$ (the $(n \times n)$-matrices) satisfying

$$T(t+s, y_0) = T(t, \varphi(s, y_0)) \circ T(s, y_0) \quad \text{for} \quad t, s \in [0, \infty[\quad \text{and} \quad y_0 \in \mathbb{R}^n. \quad (4)$$

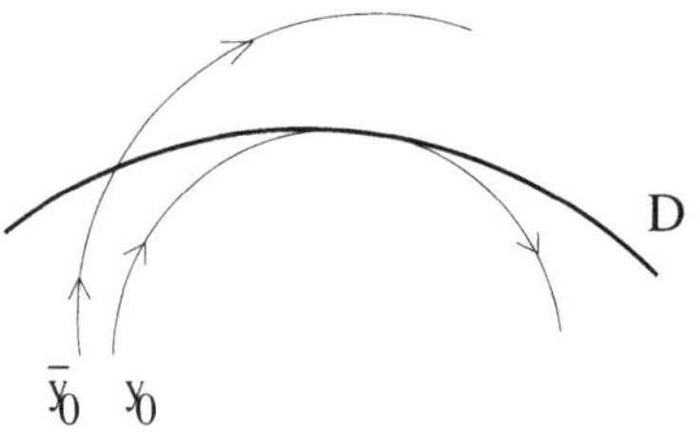

Fig. 2. A trajectory hits some discontinuity surface D tangentially

The corresponding generalization of (2) is

$$\lambda^+(y_0, z_0) := \limsup_{t\to\infty} \left(\frac{1}{t}\log|T(t, y_0)z_0|\right), \tag{5}$$

and the ergodic theorem of Oseledets holds accordingly. Observe that the identity in (4) follows from the flow equation $\varphi(t+s, y_0) = \varphi(t, \varphi(s, y_0))$ in the smooth case by differentiation w.r. to the initial value y_0.

Hence our problem has been shifted to defining a suitable cocycle for a non-smooth system as in (3). Of course we will set $T(t, y_0) = \frac{\partial\varphi}{\partial y}(t, y_0)$ whenever this is possible, i.e., as long as the trajectory stays in either $\{\dot{x} > 0\}$ or $\{\dot{x} < 0\}$. In the definition of $T(t, y_0)$ one certainly will have to exclude initial values y_0 like above that lead to grazing. Further, trajectories hitting the discontinuity surface infinitely often in a finite time do pose problems; this occurs e.g. with a gyroscopic pendulum that can have contact to a wall while rotating. However, it can be rigorously proved that it does not happen for examples like (3), and it can also be shown that the set of initial values which will *not* lead to grazing at some later time do form an invariant set G ("good initial values") of full Lebesgue measure in phase space. Therefore it is reasonable to define $T(t, y_0)$ only for $y_0 \in G$, and it is sufficient to clarify what happens when the flow changes from one of the regions $\{\dot{x} > 0\}$, $\{\dot{x} < 0\}$, or the "sticking region", to another, with the trajectory arriving non-tangentially. Therefore we have to derive a formula that describes how the linearization of the semi-flow is transformed while crossing a certain discontinuity surface. So we consider an initial point $y_0 \in G$ that leads to consecutive switching times $0 = t_0(y_0) < t_1(y_0) < t_2(y_0) < \ldots$ from one of the regions to another. Assume that at time $t_{i+1} = t_{i+1}(y_0)$ and at the point $\xi_{i+1} = \xi_{i+1}(y_0)$ the trajectory hits some discontinuity surface $\mathcal{D}$ which can be described as the level set $\{y : h_l(y) = 0\}$ for a smooth function h_l. Moreover, we suppose that just before the hitting the governing equation is $\dot{y} = f_l(y)$ with flow φ_l, while shortly afterwards it becomes $\dot{y} = f_p(y)$ with flow φ_p. We will derive the form of the transition matrix $A_{i+1} = A_{i+1}(y_0)$ of the linearization, i.e., A_{i+1} will satisfy

$$\lim_{t\to t_{i+1}^+} \frac{\partial\varphi}{\partial y}(t, y_0) = A_{i+1} \circ \left(\lim_{t\to t_{i+1}^-} \frac{\partial\varphi}{\partial y}(t, y_0)\right). \tag{6}$$

To get (6), note $\varphi(t, y_0) = \varphi_p(t - t_{i+1}, \xi_{i+1})$ for $t \in [t_{i+1}, t_{i+2}[$, hence

$$\begin{aligned}\frac{\partial \varphi}{\partial y}&(t, y_0)\\ &= -\frac{\partial \varphi_p}{\partial t}(t - t_{i+1}, \xi_{i+1})\frac{dt_{i+1}}{dy}(y_0) + \frac{\partial \varphi_p}{\partial y}(t - t_{i+1}, \xi_{i+1})\frac{d\xi_{i+1}}{dy}(y_0)\\ &= -f_p(\varphi_p(t - t_{i+1}, \xi_{i+1}))\frac{dt_{i+1}}{dy}(y_0) + \frac{\partial \varphi_p}{\partial y}(t - t_{i+1}, \xi_{i+1})\frac{d\xi_{i+1}}{dy}(y_0).\end{aligned}$$

Therefore

$$\lim_{t \to t_{i+1}^+} \frac{\partial \varphi}{\partial y}(t, y_0) = -\, f_p(\xi_{i+1})\, \frac{dt_{i+1}}{dy}(y_0) + \frac{d\xi_{i+1}}{dy}(y_0). \tag{7}$$

Denote $\Gamma(t, y) = \varphi_l(t - t_i(y), \xi_i(y))$. Then $\Gamma(t_{i+1}(y), y) = \xi_{i+1}(y)$, thus

$$\frac{d\xi_{i+1}}{dy}(y_0) = f_l(\xi_{i+1})\, \frac{dt_{i+1}}{dy}(y_0) + \frac{\partial \Gamma}{\partial y}(t_{i+1}, y_0). \tag{8}$$

By differentiation of $0 = h_l(\xi_{i+1}(y)) = h_l(\Gamma(t_{i+1}(y), y))$ we additionally obtain

$$\frac{dt_{i+1}}{dy}(y_0) = -\, \frac{1}{\langle \nabla h_l,\, f_l\rangle(\xi_{i+1})}\, \langle \nabla h_l(\xi_{i+1}),\, \frac{\partial \Gamma}{\partial y}(t_{i+1}, y_0)\,\cdot\rangle. \tag{9}$$

As $\frac{\partial \varphi}{\partial y}(t, y_0) \to \frac{\partial \Gamma}{\partial y}(t_{i+1}, y_0)$ for $t \to t_{i+1}^-$, (6) follows from (7), (8), and (9), with the matrix

$$\begin{aligned}&A_{i+1}\\ &:= \frac{1}{\langle \nabla h_l,\, f_l\rangle(\xi_{i+1})}\Big(\big[f_p(\xi_{i+1}) - f_l(\xi_{i+1})\big]_\alpha \cdot \big[\nabla h_l(\xi_{i+1})\big]_\beta\Big)_{1 \le \alpha, \beta \le n} + \mathrm{id}_{\mathbb{R}^n}\end{aligned} \tag{10}$$

for a general problem in $\mathbb{R}^n$, $[\ldots]_\alpha$ denoting the α's component of a n-vector.

As an example, we consider (3) and a switching from $\{\dot{x} > 0\}$ to $\{\dot{x} < 0\}$ at time $t_{i+1} = t_{i+1}(y_0)$, for a trajectory starting at some $y_0 = (x_0, \dot{x}_0, t_0)$. With $\xi_{i+1} := (x_{i+1}, \dot{x}_{i+1}, t_{i+1})$, $h_l(x, v, t) = v$, $f_p(x, v, t) = (v, -x + \gamma \sin(\eta t) + 1, 1)$ [the "1" in the third component due to $\dot{t} = 1$ for (3) written as an autonomous system], $f_l(x, v, t) = (v, -x + \gamma \sin(\eta t) - 1, 1)$, it results that here

$$A_{i+1} = \begin{pmatrix} 1 & 0 & 0 \\ 0 & \theta_{i+1}^+ & 0 \\ 0 & 0 & 1 \end{pmatrix}, \quad \theta_{i+1}^+ = 1 - \frac{2}{x_{i+1} - \gamma \sin(\eta t_{i+1}) + 1}. \tag{11}$$

Returning to the general set-up, we can define

$$T(t, y_0) = \begin{cases} \frac{\partial \varphi}{\partial y}(t, y_0) & : \ y_0 \in G,\ t \text{ is no switching time} \\ A_{i+1}(y_0)\Big(\lim\limits_{s \to t_{i+1}(y_0)^-} T(s, y_0)\Big) & : \ y_0 \in G,\ t = t_{i+1}(y_0) \\ 0 & : \ y_0 \notin G \end{cases}$$

and it can be verified that $T(t, y_0)$ is a cocycle, in the sense that (4) holds. Hence we have shown

Lemma 1. *For systems like (3), a canonical cocycle can be defined. Hence Oseledets' ergodic theorem applies (if there is a suitable invariant ergodic measure), and it ensures that the corresponding Lyapunov spectrum is well-defined.*

See [18] for full details. With Lemma 1 at hand, it will be the next task to estimate the Lyapunov exponents in analytically accessible cases as (3). Here we have the following theorem.

Theorem 2. *For the pendulum with dry friction (3) we have*

$$\lambda^+(y_0, z_0) \leq 0$$

for all $y_0, z_0 \in \mathbb{R}^3$. *In particular, there are no positive Lyapunov exponent. This holds for every choice of amplitude* $\gamma \geq 0$ *and frequency* $(2\pi)/\eta \geq 0$.

The proof of this result is as follows. By (5) and due to the sublinearity of the exponents, it will be enough to estimate

$$|T(t, y_0)e_j|, \quad y_0 \in \mathbb{R}^3, \quad j = 1, 2, 3,$$

with the unit vectors e_j in $\mathbb{R}^3$. Due to the form of T we only need to consider "good" initial values $y_0 \in G$. Such y_0 defines a sequence of switching times $0 = t_0 < t_1 < t_2 < \ldots$ as above, and we have $t_i \to \infty$, since there is no accumulation point of switching times. At each t_{i+1} the cocycle is multiplied by the corresponding transition matrix A_{i+1}, cf. (10), and it can be shown by analyzing all possible cases of switchings that each A_{i+1} will be of the form (11), where possibly θ^+_{i+1} is replaced by

$$\theta^-_{i+1} = 1 - \frac{2}{\gamma \sin(\eta t_{i+1}) + 1 - x_{i+1}}$$

or by $\theta^0_{i+1} = 0$, if the trajectory switches to the sticking region. Then $\theta^+_{i+1} \in [0, 1]$ and $\theta^-_{i+1} \in [0, 1]$, as is obtained by an analysis of the vector field. Hence all transition matrices like (11) are of norm less or equal to unity. Note that this nicely reflects the dissipative nature of (3), and the largest dissipation occurs when the pendulum sticks to the wall for a certain time.

To estimate $|T(t, y_0)e_j|$, according to the definition of T we have

$$T(t, y_0) = B_{i+1}(t) \circ \Big(\lim_{s \to t^-_{i+1}} T(s, y_0) \Big)$$

for $t \in [t_{i+1}, t_{i+2}[$ with a suitable matrix $B_{i+1}(t)$ arising from the linearization of the semi-flow. This may be iterated to yield for integers i and $t \in [t_{i+1}, t_{i+2}[$

$$T(t, y_0) = B_{i+1}(t) \circ B_i(t_{i+1}) \circ \ldots \circ B_1(t_2) \circ T^-_1 \tag{12}$$

with $T_1^- = \lim_{s \to t_1^-} T(s, y_0)$. The proper estimation of (12) is rather technical, but it can be shown, mainly because of $\theta_{i+1}^+ \in [0,1]$ and $\theta_{i+1}^- \in [0,1]$, that with each additional multiplication of a matrix $B_{i+1}(t)$ the norm in (12) can increase at most as $\approx (t - t_{i+1})$; see [17]. Hence we obtain the bound

$$|T(t, y_0)e_j| \approx \left((t - t_{i+1}) + \sum_{k=0}^{i}(t_{k+1} - t_k)\right) = t,$$

and therefore by (5),

$$\lambda^+(y_0, e_j) \le \limsup_{t \to \infty} \left(\frac{\log t}{t}\right) = 0.$$

This concludes a short sketch of the proof of Theorem 2.

For the specific example of the pendulum in (3) it turns out that the Lyapunov exponents do reflect the dynamical behaviour of the system, as can be checked in the resonant case $\eta = 1$ where the situation is completely clarified. See [17], also for further references, and [28] for corresponding numerical results. In a general situation there can be derived some conclusions from the Lyapunov spectrum about the stability of the underlying system.

Theorem 3. *Under some reasonable assumption for the non-smooth system: If for a periodic orbit all Lyapunov exponents are negative (besides one being zero), the periodic solution is orbitally asymptotically stable.*

See [28, Kor. 4.2] for the details.

3 Conley Index Theory for Non-Smooth Systems

In this section we outline extensions to differential inclusions of the Conley index which is a powerful tool to determine the occurrence of bifurcations. The change of some kind of index usually is a sign for bifurcations in a nonlinear dynamical system $\dot{x} = f_\lambda(x)$ depending on some parameter λ. Since we are interested in properties of the flow, the index should be connected to changes of the flow. In Ch. Conley's approach, some index is assigned to special sets (as the equilibria), and this index will account for changes in the phase portrait when λ is varied. The key notions are "isolating neighborhoods" and "isolated invariant sets". An isolating neighborhood is a set such that no boundary point stays in the set forever (in forward and backward time). In a 1D-example, every interval having no equilibrium as an endpoint will serve as an isolating neighborhood. Equivalently, an isolating neighborhood is a set U which contains its maximal invariant subset $\text{inv}(U)$ in the interior. An isolated invariant set I is determined through the requirement that it is the maximal invariant subset of some isolating neighborhood.

To define the Conley index of an isolated invariant set, suppose that for a given isolated invariant set I there exists a special isolating neighborhood B (called an isolating block) such that each boundary point leaves B immediately either in forward or in backward time, and denote b^+ the set of all boundary points that leave B immediately in forward time. Then we can determine the homotopy type of the quotient B/b^+ and define an index as $h(I) = [B/b^+]$, this homotopy type; this only amounts to identifying all points in b^+, they are collapsed to a single distinguished point. It can be shown that in general an isolating block exists, and, moreover, that the index does not depend on the special choice of the isolating block, i.e., if B_1, B_2 are isolated blocks with exit sets b_1^+, b_2^+ and $\operatorname{inv}(B_1) = I = \operatorname{inv}(B_2)$, then $[B_1/b_1^+] = [B_2/b_2^+]$.

The usefulness of this index rests on its properties.

(a) (Existence Principle.) If $I = \operatorname{inv}(U)$ and $h(I) \neq \overline{0}$ (the pointed one-point space), then $\operatorname{inv}(U) \neq \emptyset$, i.e., there exists an initial value in the isolating neighborhood U such that the full forward and backward trajectory through this initial value is contained in the isolating neighborhood.

(b) (Continuation Theorem.) In the flow of a parameter-dependent problem $\dot{x} = f_\lambda(x)$ is continuous in $\lambda \in [a, b]$, and if U is an isolating neighborhood on all λ-levels between a and b, the Conley index of the generated isolated invariant set is independent of λ, i.e., $h(\operatorname{inv}(U); f_\lambda)$ is constant.

As an application of (b), we derive a criterion for bifurcation. Suppose that zero is a fixed-point of all the systems $\dot{x} = f_\lambda(x)$ for $\lambda \in [a, b]$ (i.e., $f_\lambda(0) = 0$) such that $\{0\}$ is an isolated invariant set. Then we have

(c) (Criterion for Bifurcation.) In the above set-up, if $h(\{0\}; f_a) \neq h(\{0\}; f_b)$, then some $\lambda_0 \in [a, b]$ is a bifurcation point for small non-trivial solutions of $\dot{x} = f_\lambda(x)$. This means that for $\varepsilon > 0$ there exists $\lambda \in [a, b]$ and a global non-trivial solution x of $\dot{x} = f_\lambda(x)$ such that $|\lambda - \lambda_0| + |x(t)| \leq \varepsilon$ for all times $t \in \mathbb{R}$.

If λ varies from a to b, a jump of the Conley index causes bifurcation. To prove this, it follows directly from (b) and the definition of a bifurcation point that in case there would be no bifurcation point the function $\lambda \mapsto h(\{0\}; f_\lambda)$ were (locally) constant on $[a, b]$, a contradiction. The Conley index also enjoys an addition property, i.e., the index of a disjoint union of two isolated invariant sets is the "sum" of the respective homotopy types. See [29,37] and others for much more information about classical Conley index theory.

Now we turn to a generalization of this useful apparatus to non-smooth systems. For illustration we consider the example

$$\ddot{x} + \lambda \dot{x} + x^3 + \varphi(x, \dot{x}) \operatorname{sgn}(\dot{x}) = 0, \tag{13}$$

with some function $\varphi(x, \dot{x})$ and the bifurcation parameter $\lambda \in \mathbb{R}$. Equations of this type are used to model wings of insects; see [38]. Here, as in the last section, sgn(0) should be considered to be the whole interval $[-1,1]$. Thus if we rewrite (13) with $x' = v$ as the first-order system

$$(x,v)' = F_\lambda(x,v) = \Big(v,\ -\lambda v - x^3 - \varphi(x,v)\,\mathrm{sgn}(v)\Big), \tag{14}$$

then the right-hand side $F_\lambda(x,v)$ for $v=0$ is given by

$$F_\lambda(x,0) = \{0\} \times [-\varphi(x,0) - x^3, \varphi(x,0) - x^3]. \tag{15}$$

Since this is a set, we will more exactly write $\dot{y} \in F_\lambda(y)$, but note this reduces to (14), with $y = (x,v)$, for $v \neq 0$. As such differential inclusions $\dot{y} \in F_\lambda(y)$ can have several solutions, we first need to adapt the definition of an isolated invariant set. We introduce

$$\begin{aligned}\mathrm{inv}(U;F_\lambda) := \big\{y_0 : \text{there exists } \textit{some} \text{ solution of } \dot{y} \in F_\lambda(y), y(0) = y_0,\\ \text{that remains in } U \text{ for all times}\big\}.\end{aligned}$$

Therefore it is only required that at least one of all possible solutions stayed in U. Observe that in general the system $\dot{y} \in F_\lambda(y)$ will not generate a semi-flow, not even what is called a "multi-valued (semi-) flow".

We call I an isolated invariant set, if $I = \mathrm{inv}(U) \subset U$ for some isolating neighborhood U, and we are going to define a (generalized) Conley index $H(I;F)$ that reduces to the Conley index $h(I;f)$ introduced before, if restricted to usual dynamical systems $\dot{y} = f(y)$ instead of the multi-valued $\dot{y} \in F(y)$.

The key to such extension of Conley index theory is provided by the continuation theorem, cf. (b) above, that may also be seen as follows: if a smooth right-hand side can be deformed continuously to another smooth right-hand side, the Conley index does not change along the deformation. Now it is known that set-valued maps like $F_\lambda(x,v)$ in (14) can be approximated through smooth ones; just think of e.g. approximating

$$\mathrm{sgn}(v) = \begin{cases} -1 & : \quad v < 0 \\ [-1,1] & : \quad v = 0 \\ 1 & : \quad v > 0 \end{cases} \tag{16}$$

by e.g. $(2/\pi)\arctan(v/\varepsilon)$, $\varepsilon \to 0^+$. Hence given an error bound ε we will find, roughly speaking, a smooth function $f : \mathbb{R}^n \to \mathbb{R}^n$ such that

$$f(y) \in F(y) + \text{ "small error of order } \varepsilon\text{"}, \tag{17}$$

and for this f we already do have a Conley index $h(I;f)$, by the classical theory. Therefore it is tempting to define

$$H(I;F) := h(I;f). \tag{18}$$

Now (18) is indeed justified, since the continuation theorem can be used to prove that this assignment is independent of the choice of the approximation f, if the error bound ε is sufficiently small. In fact, for two approximations f_0 and f_1 it turns out that the deformation $\lambda \mapsto f_\lambda = (1-\lambda)f_0 + \lambda f_1$ of f_0 into f_1 is appropriate for an application of the continuation theorem, and thus $h(I; f_0) = h(I; f_1)$. Hence $H(I; F)$ is well-defined. It also can be shown that an isolated invariant set I for F is an isolated invariant set for the approximation f; see [21] or [17, Sec. 6] for the details.

To summarize, we now have defined a Conley index $H(I; F)$ for problems of the form $\dot{y} \in F(y)$, and it can be seen that all the essential properties of the classical Conley index h are transferred to H via (18). Hence we have

Lemma 4. *The existence principle (a), the continuation theorem (b), and the criterion for bifurcation (c) also hold for the generalized Conley index H.*

As an illustration, we show how the criterion for bifurcation (c) can be used in (13) with some continuous $\varphi(x, v)$ being superlinear in v, i.e., we require $|\varphi(x,v)| \le C|v|^{1+\alpha}$ for some constants $\alpha > 0$ and $C > 0$. Written in the form (14), we first observe that $(x, v) = (0,0)$ is an equilibrium for all $\lambda \in \mathbb{R}$, since $\varphi(0,0) = 0$, and therefore $F_\lambda(0,0) = (0,0)$, cf. (15). Hence we have to verify that $\{(0,0)\}$ is an isolated invariant set for F_λ and that $H(\{(0,0)\}, F_\lambda)$ jumps when λ crosses zero.

We first consider the case $\lambda > 0$. Define the Lyapunov function $V(x,v) = \frac{1}{2}v^2 + \frac{1}{4}x^4$, and $U_r = \{(x,v) \in \mathbb{R}^2 : V(x,v) < r\}$. Then we have the bounds

$$2C|v|^\alpha \le \lambda \quad \text{and} \quad |\varphi(x,v)| \le C|v|^{1+\alpha}, \quad (x,v) \in U_r, \tag{19}$$

if r (depending on λ) is chosen small enough. We claim that $\mathrm{inv}(U_r; F_\lambda) = \{(0,0)\}$, i.e., $\{(0,0)\}$ is an isolated invariant set for F_λ with isolating neighborhood U_r. To see this, assume that there is a solution $X = (x, \dot{x}) : \mathbb{R} \to U_r$ of (14) with $X(0) = (x_0, v_0) \ne (0,0)$. We write (13) in the form

$$\ddot{x} + \lambda\dot{x} + x^3 + \varphi(x,\dot{x})\, w(t) = 0,$$

with some function $w : \mathbb{R} \to [-1,1]$ that realizes $\mathrm{sgn}(\dot{x}(t))$, i.e., $w(t) \in \mathrm{sgn}(\dot{x}(t))$, the multi-valued signum function from (16). By definition of $w(t)$ and (19) we obtain

$$\begin{aligned}\frac{d}{dt} V(X(t)) &= \dot{x}\,[\ddot{x} + x^3] = \dot{x}\,[-\lambda\dot{x} - \varphi(x,\dot{x})\, w(t)] \\ &= -\lambda\,|\dot{x}|^2 - \varphi(x,\dot{x})\,|\dot{x}| \le -\lambda\,|\dot{x}|^2 + C\,|\dot{x}|^{2+\alpha} \\ &\le -(\lambda/2)\,|\dot{x}|^2.\end{aligned}$$

It may be shown that this enforces the solution to be identically zero; in particular $(x_0, v_0) = (0,0)$, a contradiction. Hence $\{(0,0)\}$ is an isolated invariant set for F_λ when $\lambda > 0$. On the other hand, in case that $\lambda < 0$ a similar argument may be used, since then with analogous notation

$$\frac{d}{dt} V(X(t)) = -\lambda\, |\dot{x}|^2 - \varphi(x,\dot{x})\, |\dot{x}| \geq -\lambda\, |\dot{x}|^2 - C\, |\dot{x}|^{2+\alpha} \geq -(\lambda/2)\, |\dot{x}|^2.$$

Next we will calculate $H(\{(0,0)\}; F_\lambda) = \Sigma^0$ in the case $\lambda > 0$, with Σ^0 being the pointed zero-sphere. For that, we have to use the definition of H in (18) and approximate the multi-valued map. Let $w_\varepsilon(v) = (2/\pi)\arctan(v/\varepsilon)$ be the standard approximations of $\mathrm{sgn}(v)$, see (16) above; whence

$$f_\varepsilon(x,v) = \Big(v,\ -\lambda v - x^3 - \varphi(x,v)\, w_\varepsilon(v)\Big)$$

serves as an approximation of F_λ in the (somewhat rough) sense of (17). Thus

$$H(\{(0,0)\}; F_\lambda) = h(\mathrm{inv}(U_r); f_\varepsilon)$$

for $\varepsilon > 0$ small by definition (18). Since $|w_\varepsilon(v)| \leq 1$, $V(x,v)$ can also be used as Lyapunov function for the system $(x,v)' = f_\varepsilon(x,v)$. As above we obtain $\mathrm{inv}(U_r; f_\varepsilon) = \{(0,0)\}$. Therefore $B = U_r$ is a corresponding isolating block with immediate exit set $b^+ = \emptyset$. Thus

$$H(\{(0,0)\}; F_\lambda) = h(\mathrm{inv}(U_r); f_\varepsilon) = [B/b^+] = \Sigma^0,$$

because $B = U_r$ can be contracted to a single point.

In case $\lambda < 0$, we may proceed similarly. Then $b^+ = \partial B$ is the whole boundary of B, and identifying this to a point we obtain the pointed two-sphere in $\mathbb{R}^3$ denoted by Σ^2. Consequently we have shown

$$H(\{(0,0)\}; F_\lambda) = \Sigma^0 \ (\lambda > 0) \quad \text{and} \quad H(\{(0,0)\}; F_\lambda) = \Sigma^2 \ (\lambda < 0).$$

Using the bifurcation criterion (c) this jump of the index implies

Theorem 5. *If $|\varphi(x,v)| \leq C|v|^{1+\alpha}$, then $\lambda_0 = 0$ is a bifurcation point for (13).*

Of course this theorem also holds for other systems, e.g., x^3 can be replaced by x^{2k+1} in (13) for some integer k. In principal what goes on in the example is a Hopf bifurcation of small periodic orbits; see [30].

More sophisticated topological arguments, mainly resting on [40], allow to prove a global bifurcation theorem for problems like (13), see [21] and [17], in the following way: the bifurcation point λ_0 with corresponding solution through the initial value $x_0 = 0 \in \mathbb{R}^n$ lies in some connected component $\mathcal{C}$ of the space $[a,b] \times \mathbb{R}^n$, where in "horizontal" direction the parameter space $\lambda \in [a,b]$ is drawn, and in "vertical" direction the initial values to solutions. Then the component $\mathcal{C}$ has two possibilities, namely either it is unbounded in $[a,b] \times \mathbb{R}^n$, or it has to meet the boundary $\{a,b\} \times \mathbb{R}^n$.

4 Application of KAM Theory to Non-Smooth Systems

Contrary to the systems we dealt with in Sects. 2 and 3 which are of a "dissipative" nature, there are also some non-smooth problems that lead to "conservative" equations. An example for this is provided by

$$\ddot{x} + x + a\,\mathrm{sgn}(x) = p(t) \tag{20}$$

with periodic forcing p and a parameter $a > 0$. Equations of this type are models for oscillators with state-dependent kicks, and it may be seen for (20) that the discontinuity line $\{x = 0\}$ is crossed transversely; hence here is no need to generalize the signum function to $\mathrm{sgn}(0) := [-1, 1]$, as was necessary in Sect. 2. It is a classical question about such forced planar oscillators, initiated in [26] some time ago, whether all solutions are bounded in the $(x, \dot{x})$-phase plane, or whether there do exist unbounded solutions. Although the equations under investigation were as simple as $\ddot{x}+2x^3 = p(t)$, the answers even nowadays require the application of tools as advanced as KAM theory. See [17] for a more detailed historical account. The main idea is to find, for a suitably transformed system, invariant curves with arbitrary high energy thus preventing solutions to escape to infinity. This additionally yields for free the existence of infinitely many periodic and quasi-periodic solutions, the former by the Poincaré-Birkhoff theorem.

Usually the application of KAM theory requires high regularity assumptions, and even without going into detail it may be clear that (20) will not give rise to such a smooth system; in action-angle variables (for $p = 0$) the system will only be continuous in the angle for $x = 0$. It turns out, however, that by means of a suitable canonical transformation at a certain stage of the proof the role of time t and the angle variable φ can be interchanged. Under the hypothesis that p is sufficiently smooth, it was proved in [22] that indeed KAM theory applies to give the desired boundedness of all solutions, at least for $a > 0$ sufficiently large. The main technical difficulty is to keep track of the errors through all the transformations. As a special case of a more general oscillator problem with linear growth at infinity (as is satisfied for (20), the result was generalized in [27] to hold even for $|\int_0^{2\pi} p(t)e^{it}\,dt| < 4a$. See [19] for a somewhat streamlined version of the proof in [27]. Since it is known from [15] that the bound for a is sharp, we can summarize the results as

Theorem 6. *(a) For $|\int_0^{2\pi} p(t)e^{it}\,dt| < 4a$ all solutions of (20) are bounded in the phase plane. In addition, there exist infinitely many periodic and quasiperiodic solutions.*

(b) For $|\int_0^{2\pi} p(t)e^{it}\,dt| > 4a$ all solutions of (20) are unbounded.

The case $|\int_0^{2\pi} p(t)e^{it}\,dt| = 4a$ is degenerate.

5 Non-Smooth Planar Systems

A particular accessible case of non-smooth systems are planar linear systems of the form

$$\dot{q} = Aq + \operatorname{sgn}(w \cdot q)v, \tag{21}$$

with $q = (x, y) \in \mathbb{R}^2$, A a real (2×2)-matrix, and given vectors $v, w \in \mathbb{R}^2$ with $w \neq 0$. Such systems play a role in electrical circuits with a twin triode [1, p. 344] or in control systems with a two-point relay characteristic [25, p. 82]. Equation (21) has a discontinuity line $\{q \in \mathbb{R}^2 : w \cdot q = 0\}$, and the problem is to classify the dynamical behaviour of the system in dependence of the 8 parameters (4 for A, and 2 for w and v, respectively). Following [12,33], we make the following assumptions.

$$\operatorname{trace}(A) \neq 0, \quad \operatorname{trace}(\mathrm{A})^2 < 4\det(A). \tag{22}$$

It is then possible to detect the number of periodic solutions of (21) and their stability. Some of those solutions stick to the discontinuity for some time (sliding motion).

As a first reduction step, the number of parameters can be decreased to three, since by a transformation (21) is seen to be equivalent to

$$\dot{q} = A_\sigma q + \operatorname{sgn}(x)b, \quad A_\sigma = \begin{pmatrix} -1 & 1 \\ -\sigma & 0 \end{pmatrix}, \quad \sigma = \det(A)/\operatorname{trace}(\mathrm{A})^2 > 1/4, \tag{23}$$

where

$$b_1 = -(w \cdot v)/\operatorname{trace}(\mathrm{A}), \quad b_2 = -[(Aw^\perp) \cdot v^\perp]/\operatorname{trace}(\mathrm{A})^2,$$

with $w^\perp = (-w_2, w_1)$. This also rotates the discontinuity to $\{(x, y) \in \mathbb{R}^2 : x = 0\}$. In (23) there are only three parameters left, σ, b_1, and b_2.

Equation (23) has to be considered as the corresponding differential inclusion $\dot{q} \in A_\sigma q + \operatorname{sgn}(x)b$, with $q = (x, y)$ and $\operatorname{sgn}(0) = [-1, 1]$, the latter being necessary since here once more solutions can stick to the discontinuity line. Then an analysis can be carried out for the resulting semiflow to determine the critical points, i.e., such $q_0 = (x_0, y_0)$ with $(0, 0) \in A_\sigma q_0 + \operatorname{sgn}(x_0)b$, in all the different possible cases; see [12]. As the existence of periodic solutions is concerned, $b_1 \geq 0$ is a necessary condition as may be seen from the phase portraits. For $b_1 > 0$ the closed orbits that cross the discontinuity line transversely (no sticking) can be found as fixed points of the return map $\Pi = \Pi^- \circ \Pi^+$, where e.g. Π^+ maps a point $(0, y_0)$ along a trajectory that is contained in $\{(x, y) : x > 0\}$ to the first return point $(0, y_1)$ to the y-axis. Under certain circumstances, it may be verified that Π, considered as a map on the y-axis, is strictly increasing and concave, and hence has at most two fixed points. Thus there are at most two closed orbits that do not stick to the discontinuity line. With some more effort, the precise number of periodic solutions can be detected, and also their stability. Moreover, the existence of periodic solutions with sliding motion and of homoclinic orbits can be studied in detail; see [12] for the complete results. In particular, it turns out that there are at most three periodic solutions for (23).

6 Melnikov's Method for Non-Smooth Systems

A further important tool to obtain global analytical results for smooth dynamical systems is Melnikov's method, cf. [3,42]. This method allows to detect subharmonic solutions of arbitrary period and homoclinic orbits in certain systems $\dot{q} = f(q) + \epsilon g(q,t)$ that are small perturbations of systems possessing a homoclinic orbit $q_0(t)$. It turns out that this method carries over to particular non-smooth systems which are of the type $\dot{q} \in f(q) + \epsilon G(q,t)$, with a multi-valued function $G(q,t)$ arising as a model of dry friction as in (3). Thus, the choice of $\epsilon G(q,t)$ means that the friction effect has to be small; see [9,10,5] for analytical and numerical results. A second situation where Melnikov's method can be used in non-smooth systems are planar systems where it is not possible for trajectories to stick to the discontinuity surface for some time. To be definite, we deal with the following example taken from [43]. With $q = (x,y) \in \mathbb{R}^2$ let $f^+(x,y) = (1,-2x)$, $f^-(x,y) = (-y-1,-x)$, and

$$f(x,y) = f^+(x,y), \quad y > 0, \quad \text{and} \quad f(x,y) = f^-(x,y), \quad y < 0.$$

So we consider

$$\begin{matrix} \dot{x} = & 1 \\ \dot{y} = & -2x \end{matrix}, \quad y > 0 \qquad \text{and} \qquad \begin{matrix} \dot{x} = & -y-1 \\ \dot{y} = & -x \end{matrix}, \quad y < 0. \tag{24}$$

The system has a homoclinic orbit

$$q_0(t) = \begin{cases} (-e^{t+1}, e^{t+1} - 1) & : \ t \in]-\infty, -1], \\ (t, -t^2 + 1) & : \ t \in [-1, 1], \\ (e^{-t+1}, e^{-t+1} - 1) & : \ t \in [1, \infty[, \end{cases}$$

connecting the hyperbolic fixed point $p_0 = (0,-1)$ to itself. The solution has been adjusted in phase such that $q_0(0) = (0,1)$, and it transversely intersects the discontinuity line $\{(x,y) \in \mathbb{R}^2 : y = 0\}$ twice, at time $t_d^- = -1$ in $P_d^- = q_0(t_d^-) = (-1,0)$ and at $t_d^+ = 1$ in $P_d^+ = q_0(t_d^+) = (1,0)$; see Fig. 3 below.

We also note that here

$$\text{trace}(Df^+) = \text{trace}(Df^-) = 0 \tag{25}$$

in the respective domains of definition. It is for such situations and small perturbations

$$\dot{q} = f(q) + \epsilon g(q,t) \tag{26}$$

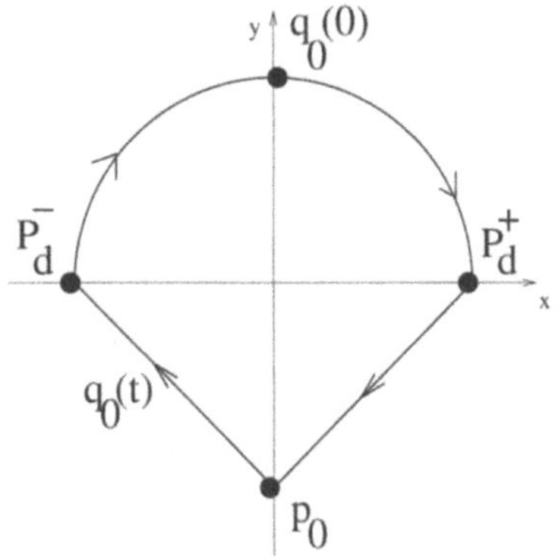

Fig. 3. The homoclinic orbit for (24)

that Melnikov's method can be used equally well for non-smooth system to prove e.g. the existence of a homoclinic orbit for ϵ small. Compared to the usual Melnikov function

$$M_0(\theta_0) = \int_{\mathbb{R}} f(q_0(t-\theta_0)) \wedge g(q_0(t-\theta_0), t)\, dt \tag{27}$$

for smooth systems, two additional terms will be picked up through crossing the discontinuity. To see this, we first have to re-develop some parts of the classical theory, cf. [3, Sect. 3.8], always arguing on a quite heuristic rather than a rigorous level. We firstly expect that the fixed point p_0 perturbs to a fixed point p_ϵ of (26) that is of distance $\mathcal{O}(\epsilon)$ of p_0. For the augmented system

$$\dot{q} = f(q) + \epsilon g(q, \theta), \quad \dot{\theta} = 1, \tag{28}$$

this means that there exists a hyperbolic periodic orbit γ_ϵ close to $\gamma_0 = \{(p_0, \theta) : \theta \in S^1)\}$. Fix $\theta_0 \in S^1 \cong [0, 2\pi]$ and denote L a line segment in the $\theta = \theta_0$ plane $\Sigma_{\theta_0} = \mathbb{R}^2 \times \{\theta_0\}$ that is perpendicular to the homoclinic orbit $\{q_0(t) : t \in \mathbb{R}\}$ at $q_0(0)$, and hence has the direction of $f^\perp(q_0(0))$. Denote p_{ϵ,θ_0} the intersection of γ_ϵ with Σ_{θ_0}, and let $q^{u,s}(t;\theta_0,\epsilon)$ be the unique trajectories of (28) that lie in the unstable ($=u$) resp. stable ($=s$) manifold of p_{ϵ,θ_0} and cross L at shortest distance to $q_0(0)$; see Fig. 4 below.

Then we consider the function

$$\Delta_\epsilon(t, \theta_0) = f(q_0(t-\theta_0)) \wedge [q^u(t;\theta_0,\epsilon) - q^s(t;\theta_0,\epsilon)] \tag{29}$$

which for $t = \theta_0$ measures the distance between the unstable and stable manifolds of p_{ϵ,θ_0}. Hence to find a homoclinic orbit for (26) one needs to study the zeroes of $\theta_0 \mapsto \Delta_\epsilon(\theta_0, \theta_0)$; see [3, p. 173], [42, Sect. 4.5]. We have

$$q^{u,s}(t;\theta_0,\epsilon) = q_0(t-\theta_0) + \epsilon q_1^{u,s}(t,\theta_0) + \mathcal{O}(\epsilon^2), \tag{30}$$

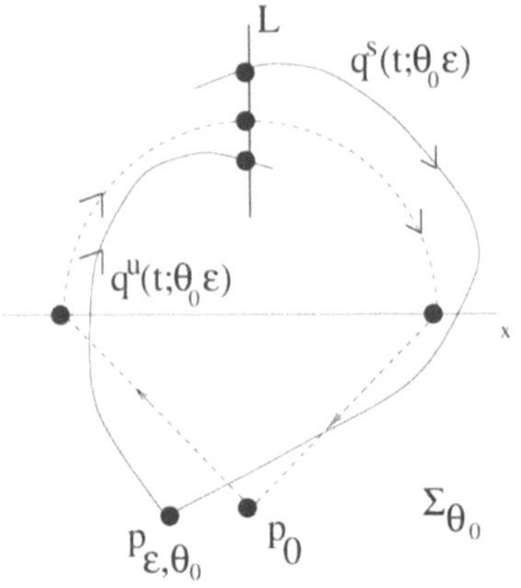

Fig. 4. Unstable and stable manifolds of p_{ϵ,θ_0}

with $q_1^{u,s}$ solutions to

$$\dot{q}_1^{u,s}(t,\theta_0) = Df(q_0(t-\theta_0))q_1^{u,s}(t,\theta_0) + g(q_0(t-\theta_0),t). \tag{31}$$

Let $\Delta_\epsilon^{u,s}(t,\theta_0) = \epsilon f(q_0(t-\theta_0)) \wedge q_1^{u,s}(t,\theta_0)$. Since $f(q_0(t))$ is orthogonal to $q_0(t)$, it follows from (29) and (30) that

$$\Delta_\epsilon(t,\theta_0) = \Delta_\epsilon^u(t,\theta_0) - \Delta_\epsilon^s(t,\theta_0) + \mathcal{O}(\epsilon^2). \tag{32}$$

According to the trace condition (25) it then can be shown as in the smooth case that

$$\dot{\Delta}_\epsilon^u(t,\theta_0) = \epsilon f(q_0(t-\theta_0)) \wedge g(q_0(t-\theta_0),t). \tag{33}$$

To integrate this w.r. to t we have to know at which times $\theta_0 + T^{u,s}(\theta_0,\epsilon)$ (on the level Σ_{θ_0}) the trajectories $q^{u,s}(t;\theta_0,\epsilon)$ will cross the discontinuity surface. According to the above construction we expect $\theta_0 + T^{u,s}(\theta_0,\epsilon) \cong \theta_0 + t_d^{\pm} + \mathcal{O}(\epsilon)$, and hence we will omit the error $\mathcal{O}(\epsilon)$, since we expect this to give a further term $\mathcal{O}(\epsilon^2)$ in (32). The intersection points of $q^{u,s}(t;\theta_0,\epsilon)$ with the discontinuity are ϵ-close to P_d^- and P_d^+, respectively. Integration of (33) over $t \in]-\infty, \theta_0 + t_d^-]$ yields, taking into account that $\Delta_\epsilon^u(-\infty,\theta_0) = 0$ due to $f(p_0) = 0$,

$$-\Delta_\epsilon^{u,-}(\theta_0 + t_d^-, \theta_0) = \epsilon \int_{-\infty}^{\theta_0 + t_d^-} f^-(q_0(t-\theta_0)) \wedge g(q_0(t-\theta_0),t)\,dt;$$

here the "$-$" in $\Delta_\epsilon^{u,-}$ should indicate that one needs to choose f^- in the definition of Δ_ϵ^u, also to determine q_1^u. Similarly, observing $t_d^- < 0$, we have

$$\Delta_\epsilon^{u,+}(\theta_0,\theta_0) - \Delta_\epsilon^{u,+}(\theta_0 + t_d^-, \theta_0) = \epsilon \int_{\theta_0 + t_d^-}^{\theta_0} f^+(q_0(t-\theta_0)) \wedge g(q_0(t-\theta_0),t)\,dt,$$

and analogously, omitting the arguments in the integral,

$$\Delta_\epsilon^{s,+}(\theta_0 + t_d^+, \theta_0) - \Delta_\epsilon^{s,+}(\theta_0, \theta_0) = \epsilon \int_{\theta_0}^{\theta_0 + t_d^+} f^+ \wedge g \, dt,$$

$$-\Delta_\epsilon^{s,-}(\theta_0 + t_d^+, \theta_0) = \epsilon \int_{\theta_0 + t_d^+}^{\infty} f^- \wedge g \, dt.$$

According to (32), and since $q_0(0)$ is in the upper half plane, this leads to

$$\begin{aligned} \Delta_\epsilon(\theta_0, \theta_0) &= \Delta_\epsilon^{u,+}(\theta_0, \theta_0) - \Delta_\epsilon^{s,+}(\theta_0, \theta_0) + \mathcal{O}(\epsilon^2) \\ &= M_0(\theta_0) + \Big(\Delta_\epsilon^{u,+}(\theta_0 + t_d^-, \theta_0) - \Delta_\epsilon^{u,-}(\theta_0 + t_d^-, \theta_0) \\ &\quad + \Delta_\epsilon^{s,-}(\theta_0 + t_d^+, \theta_0) - \Delta_\epsilon^{s,+}(\theta_0 + t_d^+, \theta_0) \Big) + \mathcal{O}(\epsilon^2), \end{aligned}$$

with the standard Melnikov integral $M_0(\theta_0)$ from (27). The Δ-terms are contributions due to the discontinuity, and they are also of order $\mathcal{O}(\epsilon)$. Thus according to the definition of $\Delta_\epsilon^{u,s}$ the relevant "non-smooth" Melnikov function is

$$\begin{aligned} M(\theta_0) = M_0(\theta_0) + \Big(& f^+(q_0(t_d^-)) \wedge q_1^{u,+}(\theta_0 + t_d^-, \theta_0) \\ & - f^-(q_0(t_d^-)) \wedge q_1^{u,-}(\theta_0 + t_d^-, \theta_0) \\ & + f^-(q_0(t_d^+)) \wedge q_1^{s,-}(\theta_0 + t_d^+, \theta_0) \\ & - f^+(q_0(t_d^+)) \wedge q_1^{s,+}(\theta_0 + t_d^+, \theta_0) \Big). \end{aligned}$$

The functions $q_1^{u/s,\pm}$ are determined by (31), e.g. $\varphi(s) = q_1^{u,+}(\theta_0 + s, \theta_0)$ solves the equation $\frac{d}{ds}\varphi = Df^+(q_0(s))\varphi + g(q_0(s), \theta_0 + s)$ for s near $s = t_d^-$. Again the zeroes of $M(\theta_0)$ determine the (transversal) intersection of stable and unstable manifolds, whence a simple zero gives rise to a homoclinic orbit for (26).

Returning to the particular example (24) from the beginning of this section, with $g(q,t) = (0, \sin t)$ the corresponding Melnikov function was calculated in [43], and numerically simple zeroes of M have been found.

7 Further Directions

Of course there is still a lot to do for non-smooth systems. Since the models that are mathematically accessible are often by far too simplified, it will be of interest to detect a "small" system within some larger system, say in the spirit of a center manifold, where all the relevant dynamics will take place. The problem is then to investigate how that dynamics of the "small" non-smooth system is related to the full higher-dimensional system. Such ideas will be followed in [34]. There are further tools from smooth dynamical system theory that are useful also for non-smooth systems, like e.g. results on Hopf bifurcation of periodic orbits. This has been studied in [30], and the problem is once more that a non-smooth system in general does not have a linearization. Finally, on the more numerical side, one should try to get a better understanding of the non-smooth dynamics, in particular of real-world systems, through simulation of the corresponding attractors and invariant measures. Here a suitable approach to be taken is through box-division algorithms developed within the priority research program by the group of M. Dellnitz; see [7,8], and also [39] for a first step regarding applications of this algorithm to handle non-smooth problems.

References

1. Andronov A.A., Vitt A.A., Khaikin S.E. (1966) Theory of Oscillators. Dover Publications, New York
2. Arnold L. (1998) Random Dynamical Systems. Springer, Berlin Heidelberg New York
3. Arrowsmith D.K., Place C.M. (1990) An Introduction to Dynamical Systems. Cambridge University Press, Cambridge New York
4. Aubin J.P., Cellina A. (1984) Differential Inclusions. Springer, Berlin Heidelberg New York
5. Awrejcewicz J., Holicke M.M. (1999) Melnikov's method and stick-slip chaotic oscillations in very weakly forced mechanical systems. Int J Bifurcation and Chaos 9: 505-518
6. Deimling K. (1992) Multivalued Differential Equations. de Gruyter, Berlin New York
7. Dellnitz M., Hohmann A. (1997) A subdivision algorithm for the computation of unstable manifolds and global attractors. Numer Mathematik 75: 293-317
8. Dellnitz M., Froyland G., Junge O. (2000) The algorithms behind GAIO-set oriented numerical methods for dynamical systems. In this DANSE book volume
9. Fečkan M. (1996) Bifurcations from homoclinic to periodic solutions in ordinary differential equations with multivalued perturbations. J Differential Equations 130: 415-450
10. Fečkan M. (1999) Chaotic solutions in differential inclusions: chaos in dry friction problems. Trans Amer Math Soc 351: 2861-2873
11. Filippov A.F. (1988) Differential Equations with Discontinuous Right-Hand Sides. Kluwer, Dordrecht Boston London
12. Giannakopoulos F., Kaul A., Pliete K. (1999) Qualitative analysis of a planar system of piecewise linear differential equations with a line of discontinuity. Preprint
13. Hubbuch F. (1995) Die Dynamik des periodisch erregten Reibschwingers. Z Ang Math Mech 75, Supplement 1: S51-S52
14. Hubbuch F., Müller A. (1995) Lyapunov Exponenten in dynamischen Systemen mit Unstetigkeiten. Z Ang Math Mech 75, Supplement 1: S91-S92
15. Kunze M. (1998) Unbounded solutions in non-smooth dynamical systems at resonance. Z Angew Math Mech 78, Supplement 3: S985-S986

16. Kunze M. (1999) Periodic solutions of conservative non-smooth dynamical systems. Z Angew Math Mech 79, Supplement 1: S97-S100
17. Kunze M. (1999) Non-Smooth Dynamical Systems. Habilitation Thesis, Universität Köln
18. Kunze M. (2000) On Lyapunov exponents for non-smooth dynamical systems with an application to a pendulum with dry friction. J Dynamics Differential Equations 12: 31-116
19. Kunze M. (2000) Remarks on boundedness of semilinear oscillators. In: Sanchez L. (Ed.) Proc. Autumn School Nonlinear Analysis and Differential Equations, Lisbon 1998. Birkhäuser, Basel Boston
20. Kunze M., Küpper T. (1997) Qualitative analysis of a non-smooth friction-oscillator model. Z Angew Math Phys 48: 1-15
21. Kunze M., Küpper T., Li J. (2000) On the application of Conley index theory to non-smooth dynamical systems. Differential Integral Equations 13: 479-502
22. Kunze M., Küpper T., You J. (1997) On the application of KAM theory to discontinuous dynamical systems. J Differential Equations 139: 1-21
23. Kunze M., Michaeli B. (1995) On the rigorous applicability of Oseledets' ergodic theorem to obtain Lyapunov exponents for non-smooth dynamical systems. To appear in: Arino O. (Ed.) Proc 2nd Marrakesh International Conference on Differential Equations
24. Kunze M., Neumann J. (1997) Linear complementary problems and the simulation of the motion of rigid body systems subject to Coulomb friction. Z Angew Math Mech 77: 833-838
25. Lefschetz S. (1965) Stability of Nonlinear Control Systems. Academic Press, New York London
26. Littlewood J. (1968) Some Problems in Real and Complex Analysis. Heath. Lexington, Massachusetts
27. Liu B. (1999) Boundedness in nonlinear oscillations at resonance. J Differential Equations 153: 142-174
28. Michaeli B. (1998) Lyapunov-Exponenten bei nichtglatten dynamischen Systemen. Ph D Thesis, Universität Köln
29. Mischaikow K. (1995) Conley Index Theory. In: Johnson R. (Ed.) Dynamical Systems, Montecatini Terme 1994. Lecture Notes in Mathematics Vol 1609. Springer, Berlin Heidelberg New York, 119-207
30. Moritz S. (2000) Hopf-Verzweigung bei unstetigen planaren Systemen. MA Thesis, Universität Köln
31. Müller A. (1994) Lyapunov Exponenten in nicht-glatten dynamischen Systemen. MA Thesis, Universität Köln
32. Oseledets V.I. (1968) A multiplicative ergodic theorem. Ljapunov characteristic numbers for dynamical systems. Trans Moscow Math Soc 19: 197-231
33. Pliete K. (1998) Über die Anzahl der geschlossenen Orbits bei unstetigen stückweise linearen dynamischen Systemen in der Ebene. MA Thesis, Universität Köln
34. Pliete K. Ph D Thesis, in preparation
35. Popp K., Stelter P. (1990) Nonlinear oscillations of structures induced by dry friction. In: Schiehlen W. (Ed.) Nonlinear Dynamics in Engineering Systems-IUTAM Symposium Stuttgart 1989. Springer, Berlin Heidelberg New York, 233-240
36. Popp K., Stelter P. (1990) Stick-slip vibrations and chaos. Phil Trans Roy Soc London A 332: 89-105

37. Rybakowski K.P. (1987) The Homotopy Index and Partial Differential Equations. Springer, Berlin Heidelberg New York
38. Scharstein H. (1998) Kräfte- und Leistungsbilanz bei der künstlichen Schlagbewegung einzelner Insektenflügel. In: Nachtigall W., Wisser A. (Eds.) Technische Biologie und Bionik 4, München 1998. Gustaf Fischer Verlag, Stuttgart
39. Voßhage Ch. (2000) Visualisierung von Attraktoren und invarianten Maßen in nichtglatten dynamischen Systemen. MA Thesis, Universität Köln
40. Ward J.R., Jr. (1998) Global bifurcation of periodic solutions to ordinary differential equations. J Differential Equations 142: 1-16
41. Wiederhöft A. (1994) Der periodisch erregte Einmassenreibschwinger. MA Thesis, Universität Köln
42. Wiggins S. (1990) Introduction to Applied Nonlinear Dynamical Systems and Chaos. Springer, Berlin Heidelberg New York
43. Zou Y.-K., Küpper T. (2000) Melnikov method and detection of chaos for nonsmooth systems. Submitted to SIAM J Math Anal

Forced Symmetry Breaking and Relative Periodic Orbits

Frédéric Guyard[1] and Reiner Lauterbach[2,*]

[1] DMA, École Polytechnique Fédéral de Lausanne, 1015 Lausanne, Switzerland
[2] Fachbereich Mathematik, Universität Hamburg, 20146 Hamburg, Germany

Abstract. We would like to discuss some aspects of forced symmetry breaking in equivariant systems. This topic is of significance for applications since in real life systems symmetries occur only as approximations. This is in particular relevant for the applications of genericity theories. There are certain dynamical features (heteroclinic cycles) which are generic within the class of symmetric systems, however they are of high codimension in the context of general dynamical systems. Therefore one might expect that forced symmetry breaking destroys such behavior. However it has been observed that weak perturbations of symmetric systems can introduce new complicated dynamical properties. Therefore it is interesting to study the influence of weak symmetry breaking from a purely mathematical point of view as well as from an applied point of view. In this paper we focus on the mathematical part and we try to describe some of our ideas in geometrical terms.

1 The Setting and Basic Results

1.1 Basic Definitions

In this paper we study perturbations of differential equation which break the symmetry of these equations. So we start with an equivariant vectorfield $f : \mathbb{R}^n \to \mathbb{R}^n$ (or more generally $f : M \to TM$) and the corresponding equation

$$\dot{x} = f(x). \tag{1}$$

Equivariance means that the trajectories are mapped onto trajectories under the linear action of a Lie group G. Precise definitions will be given shortly. If small terms of the form $\varepsilon h(x)$ perturb this equation and h transforms the same way as f but only with respect to a subgroup $H < G$, then we call this *forced symmetry breaking*. Besides some basic notions from group theory which are needed to describe the setting of equivariant systems we will need some terminology to describe the geometry induced by group actions. It turns out that this geometry provides a lot of structure which makes it easier to understand such complicated dynamical systems.

* Project: Numerical Methods for the Spherical Bénard Problem (Eberhard Bänsch, Reiner Lauterbach)

Let us begin with a vector space V and a compact Lie group G acting on V. The first and simplest notions are the *isotropy subgroup of a point* $x \in V$

$$G_x = \left\{ g \in G \;\middle|\; gx = x \right\}$$

and the *group orbit*

$$Gx = \left\{ gx \;\middle|\; g \in G \right\}.$$

Since we are interested in special time orbits, which intersect a group orbit either for all $t \in \mathbb{R}$ or for a certain subset $\mathbb{T} \subset \mathbb{R}$, we find it convenient to work on the *orbit space* V/G. This orbit space is the topological space obtained from identifying group orbits to points. Since we need to do analysis on such orbit spaces we have to introduce a topology, which is the finest one, such that the natural projection, ie the map

$$\pi : V \to V/G : x \mapsto Gx$$

is continuous. Note, that although we have a smooth structure on V and although we consider smooth actions, the orbit space need not be a smooth manifold. In fact in many cases V/G is not a smooth manifold. The simplest such case is the $\mathbb{Z}_2$ action on $\mathbb{R}$, which is given by $x \mapsto -x$. The orbit space can be represented by the set of points $\left\{ x \in \mathbb{R} \;\middle|\; x \geq 0 \right\}$. Obviously the set of nonnegative reals is not a manifold and we need some additional structure to account for this bad behavior.

In Figure 1 we indicate an orbit space and its relation to dynamics.

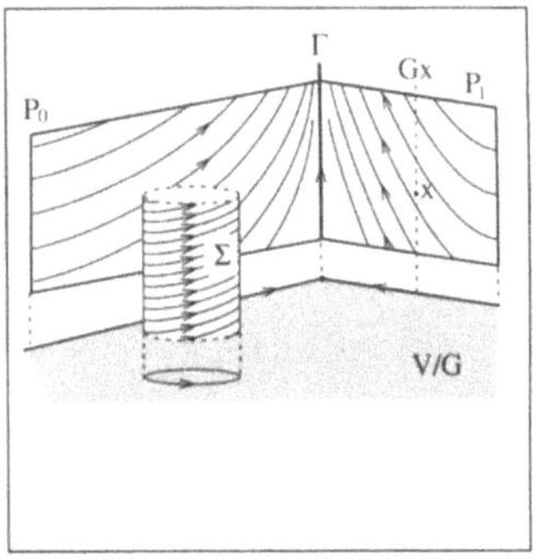

Fig. 1. A schematic view of an orbit space.

An important observation which is fundamental for all to come is, that the isotropy subgroups of points on one group orbit are conjugate, ie if $y = Gx$, then

$$G_y = gG_xg^{-1}.$$

The proof of this fact is elementary and left to the reader, however this observation allows us to speak of the orbit type: for a point x we say its orbit type $\tau(x)$ is the equivalence class of groups which are conjugate to G_x, ie $\tau(x) = [G_x] = \left\{ gG_xg^{-1} \,\middle|\, g \in G \right\}$. Since the above observation tells us, that the orbit type is constant along group orbits, we can introduce a map $\bar{\tau}$ from V/G to the set of orbit types, ie to the set of conjugacy classes of subgroups of G, by mapping $Gx \to \tau(x) = [G_x]$. So we can decompose the topological space V/G into sets of points of the same orbit type. So let $S_\tau = \left\{ Gx \in V/G \,\middle|\, \bar{\tau}(Gx) = \tau \right\}$.

T *he connected component of S_τ is a smooth manifold. The set of all connected components of all S_τ, where τ runs through the set of conjugacy classes of subgroups is a stratification of V/G into smooth strata.*

We call this construction the stratification of V and of V/G into isotropy types. Note, that sometimes in the literature the orbit type G/G_x is used instead of the isotropy type G_x. SCHWARZ [28] shows that one can define a smooth structure on V/G such that the projection of a smooth equivariant vectorfields is smooth.

Another important concept, which is in some sense dual to the isotropy subgroup is the fixed point space, ie given a subgroup $H < G$, we write

$$\text{Fix}(H) = \left\{ v \in V \,\middle|\, hv = v \text{ for all } h \in H \right\}.$$

Closely related to the fixed point space is the normalizer of H, which is denoted by $\mathrm{N}(()H)$. The normalizer acts naturally on $\text{Fix}(H)$.

Now we want to begin to study dynamics in terms of these geometric notions. We consider equivariant ODE's, ie equations of the type

$$\dot{x} = f(x), \tag{2}$$

where f is equivariant, ie $f(gx) = gf(x)$ for all $g \in G$. The orbit space allows to define *relative objects* for such equations in a natural manner. Observe, that due to the above observation there is a way to project vectorfield on the orbit space and to define a flow for such a projected field. Let us write $\pi^* f$ for the image of f under this mapping. Details of such a construction may be found in [17,18,23,26,25,27,28]. Let $\pi^*\Phi$ be the flow on V/G which is the projection of the flow Φ on V. Observe that equivariance implies that strata are flow invariant. This is the first observation in terms of geometric restrictions for equivariant flows.

The first example of a relative object is the relative equilibrium, which we define as follows.

Definition 1. A relative equilibrium is the preimage $\pi^{-1}(y_0)$ of an equilibrium $y_0 \in V/G$ with respect to the flow $\pi^*\Phi$.

In the literature one finds other descriptions of relative equilibria, for example that a relative equilibrium is a group orbit which is invariant under the flow, ie that the time orbit is part of the group orbit. If we take a one-parameter subgroup of G which corresponds to a time orbit on Gx, we observe a steady state solution in this particular moving frame. We shall see later that this construction of relative objects applies also to more complicated objects like relative periodic orbits or other relative invariant sets.

Examples for relative equilibria are equilibria, of course and more generally periodic motions of bodies moving with constant angular velocity around one of their axis of symmetry.

1.2 Forced Symmetry Breaking: a First Look

Forced symmetry breaking means the systematic study of equivariant equations which are perturbed slightly so that they loose part of their symmetry. Assume f is a G-equivariant vectorfield and $H \subset G$ is a closed subgroup. Consider an H-equivariant vectorfield g and the resulting equation on V

$$\dot{x} = f(x) + \varepsilon g(x).$$

Of course changing the symmetry group of an equation has severe consequences. If V is a sum of a certain number of irreducible representations of G it will decompose into irreducible representations of H. In general this decomposition is rather complicated. Even if V is an absolutely irreducible representation we might find many representations of H which interact in a complicated manner. As an example we recall the $\mathbf{SO}(3)$ situation. Irreducible representations are in any odd dimension and for each dimension we find up to equivalence precisely one such representation. If H equals $\mathbf{SO}(2)$ its irreducible representations are of dimension at most two and if V has dimension $2\ell + 1$ it decomposes into $\ell + 1$ representations of $\mathbf{SO}(2)$. As a consequence local bifurcation points split into several bifurcation points and we may find many connecting orbits, most of them will approach the underlying steady state when ε tends to 0. Depending on the precise nature of the perturbation g this splitting can be different and we get a complicated zoo of bifurcating solutions. Even in simple cases there is (or seems to be) no hope to get a complete picture of what is going on. However, if we refrain from describing the global situation by following all local bifurcations we might be able to get some reasonable statements. So let us assume that we are in a reasonable distance d to all local bifurcation points, where reasonable means that this distance should be in a relation to the size of the perturbation. In fact we need a bit more. To describe the necessary ingredients, let us look at the stability of a relative equilibrium. Of course it is described by the eigenvalues of linearization of the vector field on the orbit space. Alternatively we use the group action to view the relative equilibrium as an equilibrium in suitable coordinates, see FIELD [8]. The size of the eigenvalues gives the rate of contraction in directions transversal to the group orbit. If all these rates are nonzero then the group orbit is *normally hyperbolic*, see HIRSCH, PUGH

& SHUB [15]. Normal hyperbolicity guarantees that the underlying manifolds persist as flow-invariant manifolds, see [15], Theorem 4.1. Generically near a bifurcation point these eigenvalues are proportional to certain powers of the distance to the bifurcation point. In order to ensure the persistence of the invariant manifold we allow only such perturbations which are sufficiently small to not destroy normal hyperbolicity, ie we require

$$|\varepsilon| < Cd^{\alpha}$$

for an appropriate choice of α. This will be our standing hypothesis. Observe that so far, the theory is the same for relative equilibria and for *relative periodic orbits.*

Definition 2. A relative periodic orbit is the preimage of a periodic orbit in the orbit space.

Observe that relative periodic orbits can be periodic themselves or they can be quasiperiodic.

If we want to describe briefly the effects of the perturbation on the dynamics on this object, then differences come up. Basically there are three cases. The most elementary setting is the description of the behavior on a perturbed relative equilibrium. Similar ideas apply to relative periodic orbits, which are periodic. The most difficult case is the one of an aperiodic relative periodic orbit.

First we want to describe briefly the situation for relative equilibria. Since, in Subsection 3.2 we do the same for periodic relative periodic orbits we just outline the procedure. We will discuss this last case in the main part of this paper.

Now we concentrate on the description of the most simple case: let G be a compact Lie group, $L = G_x$ be the isotropy subgroup of a point $x \in V$ and $H \subset G$ be the symmetry of the perturbed equation. Then we know that Gx is diffeomorphic to G/L and we consider H-equivariant vector fields on G/L. Important for the discussion are isotropy subgroups for this action and their fixed point subspaces. The details of these computations may be found in LAUTERBACH & ROBERTS [22]. We call a subgroup $K \subset H$ subconjugate to L, if there exists a $g \in G$ with $gKg^{-1} \subset L$. Then the isotropy subgroups for the action of H on G/L are just the subgroups which are subconjugate to L and the fixed point spaces are parameterized by the set of elements $N(K, L) = \left\{ g \in G \,\middle|\, gKg^{-1} \subset L \right\}$. So we have a complete description of the geometric properties of this action, for specific examples one can compute the relevant data, see also [22]. The main observation in this context is the appearance of heteroclinic cycles for specific choices of G, L, K ([22]). A systematic study when such cycles occur was carried out in [21].

2 Periodic and Quasi-Periodic Solutions in Equivariant Systems

We are interested in trajectories of (2) that are periodic orbits *up to the group action*, ie corresponding to non-trivial periodic orbits $\widetilde{\Sigma}$ for the smooth projection of the vector field $\tilde{f}$ of f in the orbit space V/G and we denote its period by T. Since $\tilde{f}$ is smooth, $\widetilde{\Sigma}$ is diffeomorphic to S^1 and is contained in the stratum $\widetilde{(H)}$ of an isotropy subgroup H for the action of G. If $\pi : V \mapsto V/G$ is the canonical projection on the orbit space, we denote by Σ the preimage $\pi^{-1}(\widetilde{\Sigma})$ and by Σ_H the intersection $\Sigma \cap V^{\mathrm{H}}$. Σ is called a *relative periodic orbit* for (2) and T its *relative period.* The group $\mathrm{N} = \mathrm{N}(H)/H$ acts freely on Σ_H and $\pi\Sigma_H = \widetilde{\Sigma}$. Let x_0 be a point in Σ_H and we denote its time orbit by $x(t)$ with $x(0) = x_0$. There is a point $x \in \Sigma_H$ and an element n in N such that $x = x(T) = nx_0$. The element n generates a monogenic subgroup $\widetilde{K}$ of N. It can be shown ([9,19]) that

$$\tilde{K}x_0 = \overline{\{x(\mathbb{R}) \cup Gx_0\}}$$

and furthermore that generically a connected component of Σ_H is a torus of dimension rank $\mathrm{N}(H)/H + 1$. In fact, there exist an element $k_0 \in \mathrm{N}(H)$ such that $x(T) = k_0 x_0$ and it generates a monogenic subgroup group $K \subset \mathrm{N}(H)$ that projects down on $\widetilde{K}$ under the canonical projection $\mathrm{N}(H) \to \mathrm{N}(H)/H$; $K \subset \mathrm{N}(H)$ is a cyclic extension of H in $\mathrm{N}(H)$ and $K/H = \widetilde{K}$.

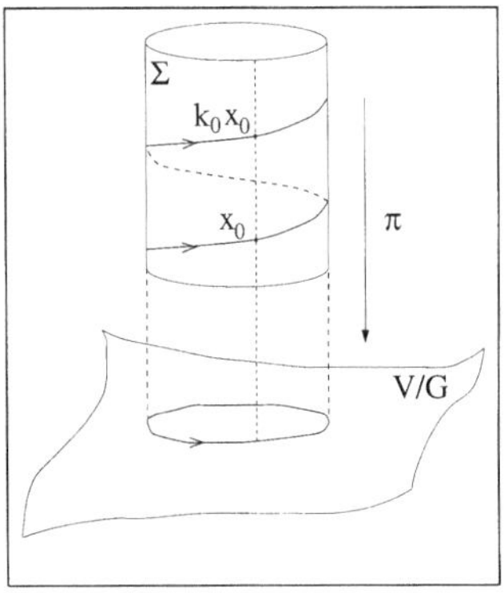

Fig. 2. A schematic representation of the structure of the (relative) periodic orbit Σ

Geometrically, H is the symmetry group of the point x_0 in Σ. Hence, all points in Σ have $[H]$ as isotropy type. The group generated by k_0 (whose closure is K) fixes the integral line of (2) passing through x_0 as a whole. Remark that if x_1 is a point in the group orbit of x_0 with $x_1 = gx_0$ then the group generated with gk_0g^{-1} fixes the integral line of (2) passing through x_1

as a whole. The conjugacy type of K corresponds to the spatio-temporal part of the symmetry. The relative periodic orbit is usually called a *standing wave* if $K = \{\mathbb{1}\}$ and a *discrete rotating wave* when $K = \mathbb{Z}_m$ for a given integer m.

3 Periodic Orbits and Their Forced Symmetry Breaking

The simplest case of forced symmetry breaking for relative periodic orbits occurs when some trajectories in V are periodic orbits. However, the methods used to study forced symmetry breaking are geometric in nature since one can only get informations about the geometric structures that are forced to persist under the perturbation. In this respect, it turns out that forced symmetry breaking for relative periodic orbits that are not periodic orbits is very similar. We recall in this section the results obtained in [13] concerning the periodic case.

3.1 The Geometry of Periodic Orbits

We assume now that Σ is a relative periodic orbit associated to a periodic orbit P through a point x_0 with relative period T. Of course, $x(\mathbb{R})$ project down onto $\widetilde{\Sigma}$ and $\Sigma = \{gx(t)|g \in G, t \in \mathbb{R}\}$. By periodicity, there is an integer n such that the period of P is $T_P = nT$. This is also the period of all trajectories in Σ. Rescaling the time in (2), we assume that $T_P = 2\pi$. This allow us to define a transitive action of $G \times \mathrm{S}^1$ on Σ as follows

$$(g, \alpha) \in G \times \mathrm{S}^1, y \in \Sigma, (g, \alpha) \circ y = g \cdot y(\alpha)$$

Since K acts on P, we can define a group homomorphism $\theta : K \mapsto \mathrm{S}^1$ such that

$$kx_0 = x(\theta(k))$$

with $\ker \theta = H$ and $\operatorname{im} \theta = \mathbb{Z}_m$. The isotropy subgroup Γ_{x_0} of the distinguished point x_0 for this action is then

$$\Gamma_{x_0} = \{(k, \theta(k^{-1}))|k \in K_0\}$$

and the relative periodic orbit Σ is diffeomorphic to $(G \times \mathrm{S}^1)/\Gamma_{x_0}$. On the other hand, one can consider the free right action of K on G making G a right principal K-bundle and the left K-action on S^1 given by

$$k \cdot \alpha = \alpha + \theta(k)$$

Now the quotient manifold of $G \times_K S^1$ under the diagonal action of K is the twisted product $\mathrm{E} = G \times_K \mathrm{S}^1$.

Any element in E can be uniquely represented as

$$[g, \alpha], \ g \in G, \alpha \in \mathrm{S}^1$$

and the projection $p : \mathrm{E} \mapsto G/K$ defines with $p([g,\alpha]) = [g]_K$ is a bundle projection for a bundle with total space E, base G/K, fiber S^1 and structure group K associated to the principal bundle $G \mapsto G/K$.

There is a G-equivariant diffeomorphism μ_{x_0} between E and Σ provided by (see [13], Propositions 3.7 and 3.8)

$$mu_{x_0}([g,\alpha]) = gx(\alpha)$$

Remark that with this parameterization, the time orbit of a point $y \in \Sigma$ project down on a unique point through the projection onto G/K. Indeed, any point y is uniquely written as $y = [g, \alpha]$ for $g \in G$ and $\alpha \in \mathrm{S}^1$. Now any real number $t \in \mathbb{R}$ can be written as $t = qT + \beta$ with $q \in \mathbb{Z}$ and $\beta \in [0, 2\pi[$. Then $y(t) = k_0^q y(\beta)$ is represented as $[gk_0^q, \beta]$ and projects down on $[g]_K$ (k_0 is a generator of K). The flow induced from the flow on Σ by the diffeomorphism μ_{x_0} is then a vertical flow in E ([13] Theorem 4.19) ie it preserves the fibres of the bundle. More precisely, each fibre of the bundle is a single trajectory of the flow. Since E is locally diffeomorphic to $G/K \times \mathrm{S}^1$ (over an open set $\mathcal{U} \in G/K$), the flow associated to (2) is locally represented as

$$\left.\begin{array}{ll} \dot{[g]}_{\mathrm{K}} & = 0 \\ \dot{\alpha} & = r([g], \alpha) \end{array}\right\} \tag{3}$$

where r is a smooth vectorfield without fixed point on S^1.

Finally, one can show that, up to a G-equivariant diffeomorphism, the construction given above does not depend on the distinguished point x_0 on Σ.

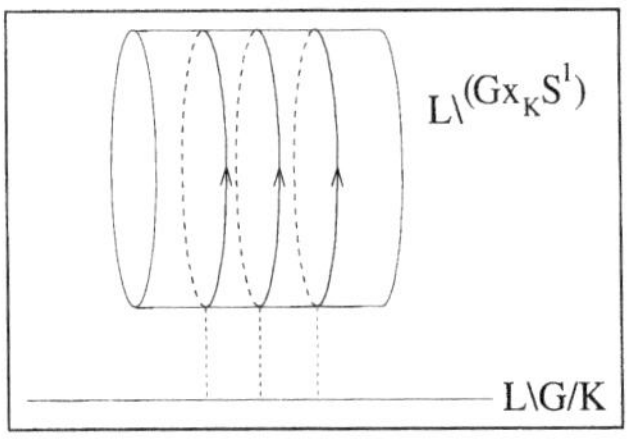

Fig. 3. A local representation of the geometry of the bundle $\mathrm{L}\backslash\mathrm{G} \times_{\mathrm{K}} \mathrm{S}^1 \mapsto \mathrm{L}\backslash\mathrm{G}/\mathrm{K}$ with the induced periodic flow for the unperturbed system.

3.2 Forced Symmetry Breaking for Periodic Orbits

The information we can obtain concerning forced symmetry breaking using the methods developed in [22] are based on the characterization of the geometric structure that persist under the perturbation. We now assume that the relative periodic orbit Σ in (2) is normally hyperbolic and one considers the new system

$$\dot{x} = f(x) + \epsilon h(x) \tag{4}$$

with ϵ small and h a C^r L-equivariant vector field with $r \geq 1$. Then Proposition 1.1 in [22] ensures that if ϵ is small enough, there exists a unique C^r-manifold Σ_ϵ near Σ which is invariant under the flow associated to (4) and which is L-equivariantly C^r-diffeomorphic to Σ. Classifying the geometric structure corresponding to fixed point subspaces for the isotropy subgroups of L on Σ_ϵ can allow to obtain the necessary existence of fixed points for the flows, or possibility of heteroclinic connections or heteroclinic cycles etc$\cdots$ (see [22,20,21,24]). Since Σ_ϵ and Σ are L-equivariantly diffeomorphic, we can undertake this characterizations on Σ itself for which we have an explicit algebraic representation.

Since G acts on a point $g_0 x_0(t_0)$ as $g \circ g_0 x_0(t_0) = g g_0 x_0(t_0)$, the G action induced on E is given by

$$g \circ [g_0, t_0] = [g g_0, t_0]$$

The subgroup $L \subset G$ of symmetry of the perturbation acts on E by restriction of the action of G. Let H_1 and H_2 two subgroups of G and let $\mathrm{N}(H_1, H_2)$ be defined by

$$\mathrm{N}(H_1, H_2) = \{g \in G \mid H_1 \subset g H_2 g^{-1}\}.$$

If $\mathrm{N}(H_1, H_2) \neq \emptyset$ then H_1 is said to be *subconjugated* to H_2. Then fixed points subspaces for the L-action on E are characterized by (see [13] Proposition 4.13):

Proposition 3. *Let L' be a subgroup of L. We have*

(a) $\mathrm{Fix}(L') \neq \emptyset \Leftrightarrow \mathrm{N}(L', H) \neq \emptyset$
(b) $\mathrm{Fix}(L') = [\mathrm{N}(L', H), \mathrm{S}^1]$

In particular, part (*b*) of the proposition shows that the fixed point subspaces are vertical spaces relatively to the bundle structure of E and all points in a periodic orbit on Σ will still be in the same strata for the action of L after the perturbation.

Since L acts on the left E, on can perform a orbit space projection on the orbit space $L\backslash\mathrm{E}$. The fact that the left action of L on E and indeed on G commutes with the right K-action on G inside the twisted product $G \times_K \mathrm{S}^1$ allows to show the following ([13] Theorem 4.17) :

T *he mapping* $\pi_0 : L\backslash(G\times_K \mathrm{S}^1) \mapsto L\backslash G/K$ *defines a fiber bundle with fiber* S^1 *and structure group* K.

Furthermore, using the form (3) of the unperturbed vector field (ie $\epsilon = 0$), one can show the following ([13] Theorem 4.19 and Proposition 4.20)

T *he projection on* $L\backslash\mathrm{E}$ *of the vector field on* E *induced by* $f|_\Sigma$ *is a vertical vector field relatively to the bundle structure on* $L\backslash\mathrm{E}$. *Furthermore, all solutions of* $\widetilde{f}$ *are* T/n *periodic with* n *the order of the cyclic group* $\theta(L\cap K)$.

Remark that since the fibers in $L\backslash\mathrm{E}$ are isomorphic to S^1, the last result implies that each fiber in $L\backslash\mathrm{E}$ is a periodic orbit and then there is a one-to-one correspondence between the points in the base $L\backslash G/K$ and the periodic orbits of the unperturbed vector field. The restriction of the initial flow to $L\backslash\mathrm{E}$ can be thought of as a global cross-section of the set of periodic orbits in Σ (remark however that such global cross-section does not necessarily exists as a bundle-section since $\mathrm{E} \mapsto L\backslash\mathrm{E}$ has no reason to be a trivial bundle). In conclusion, for the perturbed system (4) with ϵ small enough one has :

- 0-dimensional strata in $L\backslash G/K$ correspond to periodic orbits persisting to the perturbation
- 1-dimensional strata with boundaries indicate the possibility of homoclinic or heteroclinic trajectories
- 1-dimensional strata, compact and without boundary, indicate the possibility of relative periodic orbits

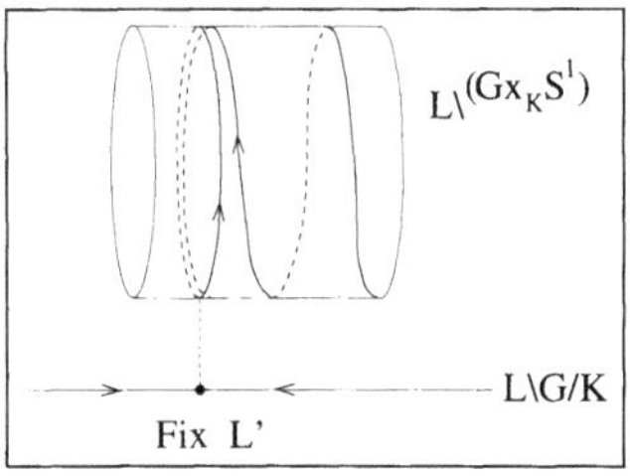

Fig. 4. A local representation of a flow for the perturbed system. Here there is a persisting periodic orbit associated with the point Fix(L') in $L\backslash G/K$. A typical trajectory outside this fixed point is also sketched.

4 Forced Symmetry Breaking for Relative Periodic Orbits

In this section, we show that the problem of forced symmetry breaking for relative periodic orbits can be brought back to the same problem for periodic orbits.

4.1 The Geometry of Relative Periodic Orbits

In order to describe the geometric structure of relative periodic orbits, we go back to their definition : a relative periodic Σ of isotropy type H is the preimage by the orbit map $\pi : V \mapsto V/G$ of a smooth non-trivial periodic orbit $\widetilde{\Sigma}$ in the strata $\widetilde{(H)}$ of H. When no confusion is possible, we denote with π the canonical orbit map as well as it restriction on Σ. Since the action of G is smooth, one has the following result ([12]) :

T *he mapping* $\pi \ : \Sigma \mapsto \widetilde{\Sigma}$ *defines a fiber bundle with base* S^1, *fiber* G/H *and group* $\mathrm{N} = \mathrm{N}(H)/H$ *associated to the principal* N*-bundle* $\pi \ : \Sigma_H \mapsto \widetilde{\Sigma}$, *ie*

$$\Sigma = (G/H) \times_{\mathrm{N}} \Sigma_H$$

The classification of possible relative periodic orbit with isotropy type H is then equivalent to the classification of principal N-bundles over S^1. According to the general scheme of classification for bundles over spheres ([29]), on can show the following ([14]) :

Proposition 4. *The set* $\mathrm{P}(\mathrm{N}, \mathrm{S}^1)$ *of isomorphism of principal* N*-bundles over* S^1 *is in one-to-one correspondence with the set of conjugacy classes in* N/N_0 *with* N_0 *the connected component of the identity in* N.

If x_0 is a distinguished point in Σ such that $x(T) = n_0 x_0$ with $n_0 \in \mathrm{N}$, then n_0 project down to an element $\widetilde{n}_0$ in N/N_0 whose conjugacy class characterizes the isomorphism type of the corresponding relative periodic orbit Σ.

In fact, more informations can be found concerning the geometry if one notice that the vector field on Σ is smooth and non-singular and thus provides a smooth 1-foliation on Σ. Therefore, the fiber bundle $\Sigma \mapsto \widetilde{\Sigma}$ is an example of foliated bundle and is *flat*, meaning that its structure groups can be reduced to a discrete one. If $\widetilde{n}_0$ is the element of N/N_0 defined above, then its generates a cyclic subgroup $\widetilde{\Gamma}(\widetilde{n}_0) \subset \mathrm{N}/\mathrm{N}_0$ of a given order q. It can be shown (Lemma 4.2 [9]), that there is a cyclic subgroup Γ of order q in N projecting down on $\widetilde{\Gamma}$ in N/N_0. If γ is a generator of Γ then one has $x(T) = n_0 x_0 = \gamma n_0' x_0$ with $n_0' \in \mathrm{N}_0$. One has the following result ([14]) :

Lemma 5. *The structure group of the bundle* $\pi : \Sigma_H \mapsto \widetilde{\Sigma}$ *can be reduced to* $\Gamma(n_0)$ *and* Σ_H *can be written as*

$$\Sigma_H = \mathrm{N} \times_\Gamma \mathrm{S}^1$$

and is the total space of a smooth fiber bundle with fiber N *associated to the smooth principal* Γ*-bundle* $p : \mathrm{S}^1 \mapsto \mathrm{S}^1$ *corresponding to the q-covering of* S^1.

Using [9] Lemma 4.2 (replacing G_0 by H), there is a subgroup $\Gamma' \subset \mathrm{N}(H)$ isomorphic to Γ and covering it for the mapping $\mathrm{N}(H) \mapsto \mathrm{N}(H)/H$. We denote with K the product $\Gamma' H$. Since Γ' acts on the right on G/H, one can define an action of Γ on G/H with

$$[g]_H\gamma = [g\gamma']_H$$

where γ' projects down on γ. Then we can identify Γ and Γ' when there are no possible confusions. As a consequence, the relative periodic orbit Σ can be seen as the total space of the Γ-principal bundle $(G/H) \times_\Gamma \mathrm{S}^1$. The elements of Σ can be uniquely represented as

$$[[g]_H, \alpha], \ g \in G, \ \alpha \in S^1$$

with the additional condition

$$[[g]_H, \alpha + 2\pi/q] = [[g\gamma]_H, \alpha]$$

We can use the flow associated to (2) to define a connection on Σ, ie an smooth and equivariant distribution of "horizontal" tangent spaces transversal to the fibers. Here this connection is given by tangent vectors to the flow. Let $\omega : [0, 2\pi] \mapsto \widetilde{\Sigma}$ be a smooth loop in $\widetilde{\Sigma}$ with base point $\pi(x_0)$ in $\widetilde{\Sigma}$. It can be lifted in a unique integral flow line s_ω of (2) on Σ with $s_\omega(0) = x_0$. Let $\Omega\widetilde{\Sigma}$ be the space of smooth loops in $\widetilde{\Sigma}$. Then, the mapping $h : \Omega\widetilde{\Sigma} \mapsto \mathrm{N}(H)/H$ define with $\omega \longrightarrow g$ such that $s_\omega(1) = gx_0$ is called the *holonomy* of the bundle. It clearly depends on the flow associated to (2). However, the conjugacy class (under inner automorphism of $\mathrm{N}(H)/H$) of the homotopy class of h denoted $h(\Sigma)$ and called the *holonomy map* is independent of this flow and is a geometric invariant of the group action (see [16] Theorem 1.1). With the previous construction, this holonomy map assigns γ to the generator of $\Omega\widetilde{\Sigma} = \Omega\mathrm{S}^1$. We call Γ the *holonomy group* of Σ. The (conjugacy class of the) holonomy group completely characterizes the type of the bundle corresponding to Σ (compare with Proposition 4). Using properties of fiber bundles, on can show the following ([12]):

1. *If $\widetilde{\Sigma} \subset V/G$ is contractible in the strata of H, then the holonomy map $h(\Sigma)$ is trivial and $\Sigma \simeq G/H \times \mathrm{S}^1$.*
2. *If Σ and Σ' are two relative periodic orbits with isotropy H for the system (2) such that there respective projection $\widetilde{\Sigma}$ and $\widetilde{\Sigma}'$ are homotopically equivalent within $(\widetilde{H})$, then Σ and Σ' have conjugated holonomy group.*

4.2 Forced Symmetry Breaking for Relative Periodic Orbits

Since the principal Γ-bundle $\mathrm{S}^1 \mapsto \mathrm{S}^1$ defined in Lemma 5 is smooth, one can find a smooth $2\pi q$-periodic flow covering the non trivial 2π periodic flow on

S^1. It induces a G-equivariant $2\pi q$-periodic flow on Σ covering the 2π-periodic flow on $\widetilde{\Sigma}$:

$$\forall t \in \mathbb{R}, \forall [[g]_H, \alpha] \in (G/H) \times_\Gamma S^1, \ t \circ [[g]_H, \alpha] = [[g]_H, \alpha + t \bmod 2\pi/q]$$

(recall that $\gamma^q = e$). Here $[[e]_H, 0]$ is identified with the distinguish point x_0. Details of this construction are given in [14]. As a conclusion, one can always find a periodic flow on a relative periodic orbit. Now the subgroup $\Gamma \subset N(H)$ is precisely the subgroup of G fixing the periodic orbit through $[[e]_H, 0]$ as a set. Indeed, this periodic orbit is given with $\{[[e]_H, \alpha], \alpha \in [0, 2\pi q]\}$ and if γ' generates Γ' and $h \in H$, one has

$$\gamma' h[[e]_H, \alpha]] = [[\gamma']_H, \alpha]] = [[e]_H \gamma, \alpha]] = [[e]_H, \alpha + 2\pi]$$

One then obtain the following :

The problem of forced symmetry breaking for relative periodic orbits with holonomy group Γ is "equivalent" to the problem of forced symmetry breaking for periodic orbits with spatio-temporal symmetry $K = \Gamma H$ (or $\widetilde{K} = \Gamma$). More precisely we have:

From the point of view of the persistence of geometric structures, the problem of forced symmetry breaking for relative periodic orbits with holonomy group Γ can be brought back to the problem of forced symmetry breaking for periodic orbits with spatio-temporal symmetry $K = \Gamma H$ (or $\widetilde{K} = \Gamma$).

Another way to see this point is that the total space $(G/H) \times_\Gamma S^1$ of the bundle associated to Σ is G-equivariantly and smoothly isomorphic to $G \times_{\Gamma H} S^1$ through the mapping

$$[[g]_H, \alpha] \longrightarrow [g, \alpha]$$

The proof that this mapping is well defined is straightforward and only uses the definitions and the fact the Γ and H commute.

5 An Application

In this section we want to discuss some related results for special dynamical systems without going into any detail. The details can be found in the literature. Observe that the dynamics close to a heteroclinic cycle is characterized by sudden bursts being followed by long periods of rest. A typical example for this type of motion is the change of polarity of the earth's magnetic field. Measurements (of magnetism in solidified lava) indicate that the orientation of the field is constant on time scales of 10^4 to 10^6 years and that a reversal takes several hundred years. We do not want to go into details of this fascinating subject (see for example CHILDRESS [2], CHILDRESS & GILBERT [3] or GHIL & CHILDRESS [11] for further background and more references). The problem for the underlying mechanism is still unsolved. Here, we want to describe some recent research on the occurrence of heteroclinic cycles in magnetohydrodynamics. The underlying equations are the Maxwell equations coupled with the Navier-Stokes equations. A first (very crude) approximation of the problem might be to look at these equations in a spherical shell and to neglect the force on the fluid motion due to the magnetic field. If we do this we might consider the Navier-Stokes equations in a shell and take solutions of these equations to compute the resulting magnetic field (Observe that there are some problems due to the fact that this is a linear equation.) from the induction equation.

Performing a center manifold reduction one reduces the problem to a low-dimensional ordinary differential equation. Restricting to the the spherical case this kind of reduction has been done at various places, see CHOSSAT [4,5]. In [7] it has been shown that near the onset of convection no complicated dynamics can develop if the representation of the group $\mathbf{O}(3)$, the symmetry group of the sphere, acts absolutely irreducibly on the critical modes. Therefore one might ask, whether there are parameter values where we see mode interactions generically. This question has been answered positively by ARMBRUSTER & CHOSSAT [1]. They also proved the existence of a heteroclinic cycle in the nonrotating case for the eight-dimensional representation of $\mathbf{O}(3)$ which is a sum of the three and five-dimensional irreducible representation. This cycle involves two types of axisymmetric solutions: one type is characterized by an inflow along the equator and an outflow at the poles and the other way around at the other type. This burst-like dynamics was already predicted by a numerical study of FRIEDRICH & HAKEN [10]. However, the solutions found numerically do not completely agree with the analysis in [1]. There are qualitative differences, see [6]. In [6] it was also shown that the burst-like behavior found by FRIEDRICH & HAKEN [10] can be explained by the existence of a large flow-invariant set involving heteroclinic cycles. This set is crucial for the stability analysis of the heteroclinic cycles found by ARMBRUSTER & CHOSSAT [1]. All this applies to the nonrotating sphere. Rotation can be modelled as a forced symmetry breaking. This implies that we are looking at very slow angular velocities. Then it is shown that we still find heteroclinic behavior, however it involves not only equilibria but also rotating waves. Of course the rotation in the case of the earth is not slow, so these results are not applicable to the question of pole reversals.

References

1. D. ARMBRUSTER & P. CHOSSAT. Heteroclinic cycles in a spherically invariant system. *Physica D*, 50, 155-176, 1991.
2. S. CHILDRESS. Convective dynamos. In Spiegel & Zahn, editors, *Problems of stellar convection.* Springer Verlag, 1976. Lecture Notes in Physics 71.
3. S. CHILDRESS & A. D. GILBERT. *Stretch, Twist, Fold: The Fast Dynamo.* Springer, 1995.
4. P. CHOSSAT. Bifurcation and stability of convective flow in a rotating or nonrotating spherical shell. *SIAM J. Appl. Math.*, 37, 624-647, 1979.

5. P. Chossat. *Le Problème de Bénard dans une Couche Sphérique*. PhD thesis, Nice, 1981.
6. P. Chossat, F. Guyard & R. Lauterbach. Heteroclinic sets in spherically invariant systems and their perturbations. *J. Nonl. Sc.*, 9, 479-524, 1999.
7. P. Chossat, R. Lauterbach & I. Melbourne. Steady-state bifurcation with **O**(3)-symmetry. *Arch. Rat. Mech. Anal.*, 113(4), 313-376, 1991.
8. M. Field. Equivariant dynamical systems. *Trans. Am. Math. Soc.*, 259(1), 185-205, 1980.
9. M. Field. Local structure of equivariant dynamics. In M. Roberts & I. Stewart, editors, *Singularity Theory and its Applications, Warwick 1989,* Part II. Springer Verlag, 1991. Lecture Notes in Mathematics 1463.
10. R. Friedrich & H. Haken. Static, wavelike, and chaotic thermal convection in spherical geometries. *Physical Rev. A*, 34, 2100-2120, 1986.
11. M. Ghil & S. Childress. *Topics in Geophysical Fluid Dynamics: Athmospheric Dynamics, Dynamo Theory, and Climate Dynamics.* Number 39 in Applied Math. Sciences. Springer Verlag, New York, 1986.
12. F. Guyard. to be published. *private communication*, 1999.
13. F. Guyard & R. Lauterbach. Forced symmetry breaking perturbations for periodic solutions. *Nonlinearity*, 10, 291-310, 1997.
14. F. Guyard & R. Lauterbach. Forced symmetry breaking: theory and applications. In *Pattern Formation in Continuous and Coupled Systems*, 121-135. IMA, 1999.
15. M. W. Hirsch, C. C. Pugh & M. Shub. *Invariant Manifolds*, volume 583 of *Lecture Notes in Mathematics*. Springer Verlag, 1977.
16. F. Kamber & P. Tondeur. Flat bundles and characteristic classes of group-representation. *Amer. J. Math*, 89, 857-886, 1967.
17. M. Kœnig. Linearization of vector fields on the orbit space of the action of a compact Lie group. *Math. Proc. Cambridge Phil. Soc.*, 121, 401-424, 1996.
18. M. Kœnig & P. Chossat. Caractérisation des bifurcations pour les champs de vecteur éequivariants sous l'action d'un groupe de Lie compact. *C. R. Acad. Sci. Paris,* Ser. I, (to appear), 1997.
19. M. Krupa. Bifurcations of relative equilibria. *SIAM J. Math. Anal.*, 21(6), 1453-1486, 1990.
20. R. Lauterbach. Forced symmetry breaking from **O**(3). In K. B. Eugene Allgower & M. Golubitsky, editors, *Bifurcation and Symmetry,* Int. Ser. Num. Math. 104, 253-262. Birkhäuser, 1992.
21. R. Lauterbach, S. Maier & E. Reissner. A systematic study of heteroclinic cycles in dynamical system with broken symmetries. *Proc. Roy. Soc. Edinburgh*, 126A, 885-909, 1996.
22. R. Lauterbach & M. Roberts. Heteroclinic cycles in dynamical systems with broken spherical symmetry. *J. Diff. Equat.*, 100, 428-448, 1992.
23. R. Lauterbach & J. Sanders. Bifurcation analysis for spherically symmetric systems using invariant theory. *J. Dyn. and Diff. Equat.*, 9(4), 535-560, 1997.
24. S. Maier-Paape & R. Lauterbach. Heteroclinic cycles for reaction diffusion systems by forced symmetry breaking. *Trans. Am. Math. Soc.*, (to appear), 1998.
25. M. Rumberger. Finitely differentiable invariants. *Math. Zeitschr.*, 229, 675-694, 1998.

26. M. RUMBERGER & J. SCHEURLE. Invariant C^j-functions and center manifold reduction. In H. W. Broer, S. van Gils, I. Hoveijn & F. Takens, editors, *Nonlinear Dynamical Systems and Chaos*. Birkhäuser, 1996.
27. G. SCHWARZ. Smooth functions invariant under the action of a compact Lie group. *Topology*, 14, 63-68, 1975.
28. G. SCHWARZ. Lifting smooth homotopies. *IHES*, 51, 37-135, 1980.
29. N. STEENROD. *Topology of Fibre Bundles*. Princeton Mathematical Series. Princeton University Press, 1952.

On Dynamics and Bifurcations of Nonlinear Evolution Equations Under Numerical Discretization

Christian Lubich*

Mathematisches Institut, Universität Tübingen, Auf der Morgenstelle 10,
72076 Tübingen, Germany
e-mail: lubich@na.uni-tuebingen.de

Abstract. This article reviews recent results on long-time behaviour, invariant sets and bifurcations of evolution equations under discretization by numerical methods. The emphasis is on time discretization. Finite-time error bounds of low order for non-smooth data, of high order for smooth data, and attractive invariant manifolds are tools that pervade large parts of the article. To illustrate the mechanisms, the following combinations of dynamics/equations have been selected for a detailed discussion:

1. Shadowing near hyperbolic equilibria of singularly perturbed ODEs
2. Hyperbolic periodic orbits of delay differential equations
3. Hopf bifurcation of semilinear parabolic equations
4. Inertial manifolds of semilinear parabolic equations
5. Attractors of damped wave equations.

Introduction

This article was written under the premise that it should

(a) survey the subject area,
(b) review work by the author in the DANSE project,
(c) give one or the other new result.

These goals are not necessarily compatible. Clearly, (b) inflicts a strong bias on (a). With (a) and/or (b) achieved, (c) can only be rudimentary within the assigned pages. This caveat notwithstanding, may the reader find this article useful!

What are the mechanisms that make a numerical discretization capture the long-time dynamics of the differential equation? How good can the approximation be? These are the questions underlying the present article. Answers are different for different classes of equations:

(i) ordinary differential equations for which the Jacobian of the right-hand side has a bound that is much smaller than the inverse of the step size needed to approximate solutions over some finite time ("nonstiff" ODEs);

* Project: Large Time Behavior of Numerical Methods for Parabolic Partial Differential Equations (Christian Lubich)

(ii) evolution equations for which the linearization involves unbounded operators or operators of arbitrarily large norm, e.g., partial differential equations, singularly perturbed ODEs, and (less obviously) delay differential equations.

The present paper is concerned with class (ii). Reference to the more developed subject of dynamics of discretized nonstiff ODEs is made only for a comparison of the arguments. A further distinction needs to be made between "dissipative" and "conservative" problems, e.g., damped wave equations versus Hamiltonian wave equations. The dynamic properties considered here will be for the "dissipative" type.

Inhomogeneous and far-spread as the class of equations is, it nevertheless turns out that there are a few tools which go a long way with many different evolution equations and dynamic phenomena. Of these tools, I would like to single out the roles played by low-order nonsmooth-data error bounds and high-order smooth-data error bounds over finite times, and the reduction of the dynamics to the nonstiff case via attractive invariant manifolds. It is a strange fact that two of the papers that have influenced me most in this area, have never been published: Larsson's report [35] which highlights the importance of error bounds for general initial data that do not admit a smooth solution; and Nipp and Stoffer's report [47] on an extremely useful version of an invariant manifold theorem. That theorem is recapitulated in Appendix A.

The sampling in the three-dimensional lattice of Dynamics $\times$ Equations $\times$ Numerical Methods, as listed in the abstract, has been guided by the objective of showing relationships as well as differences among a wide variety of problems, and admittedly by (b) above. Section 5 was chosen to comply with (c). Each section presents the analytical framework and the numerical method, and states just one theorem. The result and its background are explained, and related results are indicated. The sections can be read independently, but they share common ground.

Stuart [55] gives a good survey of the field as of ~1995. That article concerns sectorial evolution equations, which is where most of the work has been done. Stuart and Humphries [56] is a basic reference on the dynamics of nonstiff ODE discretizations, Hairer, Nørsett and Wanner [19,20] on time discretization methods.

1 Shadowing Near Hyperbolic Equilibria of Singularly Perturbed Ordinary Differential Equations

Following [41], we consider numerical solutions to singularly perturbed differential equations in the neighbourhood of a saddle point. We apply an implicit Runge-Kutta method with a step size larger than the small perturbation parameter. The main result is a shadowing property: After an initial transient, the numerical solution in a neighbourhood of the hyperbolic stationary point remains for all times close to some exact solution of the differential equation. Conversely, after the elapse of a short time also every exact solution in the neighbourhood remains close to some numerical solution. In both cases, the approximation takes place with the order of approximation of smooth solutions on finite intervals. We will see that the main difficulty, as compared to the analogous problem for nonstiff ordinary differential equations, is that finite-time errors are not uniformly small for arbitrary initial data in an open set. This difficulty, which we will again encounter with all the other classes of evolution equations considered in this article, is here overcome by using strongly attractive invariant manifolds for both the continuous and the discrete problem.

1.1 Analytical Framework

We consider the singularly perturbed problem

$$\begin{aligned} \frac{dy}{dt} &= f(y,z) \\ \varepsilon\,\frac{dz}{dt} &= g(y,z)\,, \qquad 0<\varepsilon\ll 1\,, \end{aligned} \tag{1}$$

in the neighbourhood of a stationary point at the origin. The functions f and g are arbitrarily differentiable. With subscripts denoting partial derivatives, we assume the following.

$$\text{All eigenvalues of } g_z(0,0) \text{ have negative real part.} \tag{2}$$

$$(f_y - f_z g_z^{-1} g_y)(0,0) \;\; \text{has no eigenvalues on the imaginary axis.} \tag{3}$$

The first condition is familiar in the theory of singularly perturbed problems and yields the uniform well-posedness of the system as $\varepsilon \to 0$; see, e.g., [50]. The second assumption is equivalent to stating that for small ε the fixed point of (1) is *hyperbolic*, that is, the Jacobian of the system (1) at the stationary point has no eigenvalues on the imaginary axis.

1.2 Numerical Method

We consider an implicit Runge-Kutta method applied to (1) with step size $h \geq \varepsilon$:

$$y_{n+1} = y_n + h\sum_{j=1}^{m} b_j Y'_{nj}\,, \qquad z_{n+1} = z_n + h\sum_{j=1}^{m} b_j Z'_{nj}\,,$$

with internal stages ($i = 1,\ldots,m$)

$$Y_{ni} = y_n + h\sum_{j=1}^{m} a_{ij} Y'_{nj}\,, \qquad Z_{ni} = z_n + h\sum_{j=1}^{m} a_{ij} Z'_{nj}$$

satisfying relations of the form of (1):

$$\begin{aligned} Y'_{ni} &= f(Y_{ni}, Z_{ni}) \\ \varepsilon\, Z'_{ni} &= g(Y_{ni}, Z_{ni}) \,. \end{aligned}$$

The method is determined by its coefficents a_{ij} and b_j.

We assume that the Runge-Kutta method is *strongly A-stable*, that is, the stability function

$$R(w) = 1 + wb^T(I - w\mathcal{O})^{-1}1\!\!1 \,,$$

(where $b^T = (b_1, \ldots, b_m)$, $\mathcal{O} = (a_{ij})_{i,j=1}^m$, $1\!\!1 = (1, \ldots, 1)^T$) satisfies

$$|R(w)| \le 1 \qquad \text{for } \operatorname{Re} w \le 0 \,,$$

all eigenvalues of the Runge-Kutta matrix $\mathcal{O} = (a_{ij})_{i,j=1}^m$ have positive real part, and $R(\infty) = 1 - b^T\mathcal{O}^{-1}1\!\!1$ satisfies

$$|R(\infty)| < 1 \,.$$

We require the following approximation properties: the method has *classical order p*, that is, the error of the method applied to nonstiff ordinary differential equations is $O(h^p)$ on bounded time intervals. The approximation properties for the singularly perturbed problem (1) depend in addition on the *stage order q*, which is determined by the condition

$$\sum_{j=1}^m a_{ij} c_j^{k-1} = \frac{c_i^k}{k} \qquad \text{for } k = 1, \ldots, q \quad \text{and all } i \,. \tag{4}$$

Here c_i is defined by (4) with $k = 1$.

A well-known and widely used class of Runge-Kutta methods satisfying the above assumptions are the Radau IIA methods [20], which for each stage number $m \ge 1$ have $p = 2m - 1$, $q = m$, and satisfy $a_{mj} = b_j$ $(j = 1, \ldots, m)$, which in particular implies $R(\infty) = 0$.

1.3 Statement of the Result

Theorem 1. [41] *Under the above assumptions, there are positive constants r and h_0 such that the following holds for $0 < \varepsilon \le h \le h_0$.*

(A) For every Runge-Kutta solution with $\|(y_n, z_n)\| \le r$ for $0 \le n \le N$, there exists a solution $(y(t), z(t))$ of (1) for $0 \le t \le T = Nh$, such that for $0 \le n \le N$

$$\|y_n - y(nh)\| \le C \cdot (h^p + \varepsilon h^{q+1} + \varepsilon \rho^n)$$

$$\|z_n - z(nh)\| \le \begin{cases} C \cdot (h^p + \varepsilon h^q + \rho^n) & \text{if } a_{mj} = b_j \text{ for } j = 1, \ldots, m \,, \\ C \cdot (h^{q+1} + \rho^n) & \text{else} \,. \end{cases}$$

Here $\rho < 1$ and C depend only on f, g, r and h_0, and in particular are independent of ε, h, and N. Moreover, for $\varepsilon \ll h$ we have $\rho = |R(\infty)| + O(\varepsilon/h)$.

(B) Conversely, for every solution of (1) with $\|(y(t), z(t))\| \le r$ for $0 \le t \le T = Nh$, there exists a Runge-Kutta solution (y_n, z_n), $0 \le n \le N$, such that the difference to $(y(nh), z(nh))$ satisfies the above bounds with $\rho = e^{-\kappa h/\varepsilon}$, with C and $\kappa > 0$ independent of ε, h, and N.

1.4 Discussion and Comparison with the Nonstiff ODE Case

Theorem 1 is an analogue of a shadowing result of Beyn [4] for numerical solutions near a hyperbolic stationary point of the ordinary differential equation

$$\frac{dx}{dt} = f(x) , \tag{5}$$

where sufficiently many derivatives of the nonlinearity f are assumed to be bounded. The following is shown in [4, Theorem 3.1] for pth order methods applied with step size h:

For every numerical solution (x_n), $0 \le n \le N$, which stays in a sufficiently small neighborhood U of the stationary point, there is a solution $x(t)$ of (5) satisfying $\|x_n - x(nh)\| \le Ch^p$. Conversely, for every solution $x(t)$ in U, $0 \le t \le T = Nh$, there is a Runge-Kutta solution (x_n) with $\|x_n - x(nh)\| \le Ch^p$. The constant C is independent of h and N.

As in Theorem 1, it is essential that the estimates remain uniform on time intervals that can become arbitrarily large. We review briefly the considerations that lead to Beyn's result.

Let R^h denote the Runge-Kutta map, so that $R^h(x)$ is the result of one step of the method applied with step size h, starting from the point x. A numerical solution sequence (x_n) thus satisfies

$$x_{n+1} = R^h(x_n) .$$

Further, let S^h denote the flow map of the differential equation (5) over time h, so that for a solution $x(t)$ of (5) we have

$$x(t+h) = S^h(x(t)) .$$

The *local error* $R^h(x) - S^h(x)$ then satisfies

$$R^h(x) - S^h(x) = O(h^{p+1}) , \qquad \frac{\partial R^h}{\partial x}(x) - \frac{\partial S^h}{\partial x}(x) = O(h^{p+1}) \tag{6}$$

uniformly for x in an arbitrary compact set.

The first estimate is obtained by comparing the Taylor expansions of the exact and the numerical solution, the second estimate follows by interpreting the two error expressions combined as the local error of the Runge-Kutta method applied to the system composed of (5) and its variational equation $dv/dt = \partial f/\partial x(x)v$.

By the variation-of-constants formula, a sequence of exact solution values $\widetilde{x}_n = x(nh)$ satisfies a recurrence relation

$$\widetilde{x}_{n+1} = e^{hA}\widetilde{x}_n + h\phi(\widetilde{x}_n) , \tag{7}$$

where A is the Jacobian of (5) at the stationary point $x^* = 0$, and where ϕ is an h-dependent function with $\phi(0) = 0$ and Lipschitz constant of size $O(r)$ in an r-neighbourhood $B(r)$ of the stationary point. Similarly, by (6), every numerical solution sequence satisfies a recursion of the form

$$x_{n+1} = e^{hA}x_n + h\psi(x_n) , \tag{8}$$

where ψ has a Lipschitz constant of size $O(r) + O(h^p)$ in $B(r)$, and

$$\psi(x) - \phi(x) = O(h^p) \qquad \text{uniformly in } B(r).$$

By the assumption of hyperbolicity, A has the form (up to a similarity transform)

$$A = \begin{pmatrix} A^- & 0 \\ 0 & A^+ \end{pmatrix}$$

where all eigenvalues of A^- have negative real part and those of A^+ have positive real part. We now apply the discrete variation-of-constants formula to the stable components in (7) and (8) in the forward direction, and to the unstable components in the backward direction:

$$x_n^- = \exp(nhA^-)x_0^- + h\sum_{j=0}^{n-1} \exp((n-j-1)hA^-)\psi^-(x_j) ,$$
$$x_n^+ = \exp((n-N)hA^+)x_N^+ - h\sum_{j=n}^{N-1} \exp((n-j-1)hA^+)\psi^+(x_j) .$$

As the Lipschitz constants of ϕ and ψ can be made arbitrarily small by reducing the radius r of the neighbourhood, the Banach contraction principle yields the following: *Let* (x_n) *be a solution of (8) with* $\|x_n\| \le r$ $(n = 0, \ldots, N)$*. If* r *is sufficiently small, there is a unique solution of (7) with boundary values* $\widetilde{x}_0^- = x_0^-$, $\widetilde{x}_N^+ = x_N^+$. *This solution satisfies*

$$x_n - \widetilde{x}_n = O(h^p) \qquad \textit{uniformly for } n = 0, \ldots, N .$$

Evidently, the same holds with the roles of x_n and $\widetilde{x}_n$ interchanged. This yields Beyn's shadowing result as stated above. We emphasize that this construction depends crucially on the uniform approximation estimate (6).

Let us now return to the singularly perturbed problem (1) and its Runge-Kutta discretization. Here we face the difficulty that an approximation estimate (6) does not hold when $\varepsilon \le h$. Since general solutions of (1) undergo rapid initial changes, the best possible local error estimate valid in a neighbourhood of the stationary point is only $R^h(x) - S^h(x) = O(1)$!

More favourable error bounds exist only for initial values (y_0, z_0) which are such that the corresponding solution $(y(t), z(t))$ of (1) is *"smooth"* in the sense that arbitrarily many derivatives are bounded independently of ε. For such initial data, the following sharp finite-time error bounds were shown in [18] to hold for numerical solutions obtained by strongly A-stable Runge-Kutta methods of classical order p and stage order q:

$$\|y_n - y(nh)\| \leq C\,(h^p + \varepsilon h^{q+1})$$

$$\|z_n - z(nh)\| \leq \begin{cases} C\,(h^p + \varepsilon h^q) & \text{if } a_{mj} = b_j \text{ for } j = 1, \ldots, m\,, \\ Ch^{q+1} & \text{else}\,. \end{cases}$$

The constants depend on bounds for the derivatives of the solution and on the length T of the time interval, but are independent of ε, h and n with $nh \leq T$. We note that these orders of approximation are the same as stated in Theorem 1.

The way to circumvent the missing uniformity in the error bounds, is to make use of *attractive invariant manifolds.* As is known from the geometric theory of singular perturbation problems [14], [46], there is a manifold $\mathcal{M}_\varepsilon = \{(y, z) : \; z = s_\varepsilon(y)\}$ (locally near the stationary point, which itself is on $\mathcal{M}_\varepsilon$), such that solutions of (1) starting on $\mathcal{M}_\varepsilon$ remain on $\mathcal{M}_\varepsilon$ and are smooth in the above sense. Also the function s_ε defining the manifold has arbitrarily many derivatives bounded independently of ε. An arbitrary solution $(y(t), z(t))$ of (1) near $(0,0)$ rapidly approaches a solution on $\mathcal{M}_\varepsilon$: There is a solution $(\widetilde{y}(t), \widetilde{z}(t))$ on $\mathcal{M}_\varepsilon$ such that

$$\|y(t) - \widetilde{y}(t)\| + \varepsilon \cdot \|z(t) - \widetilde{z}(t)\| \leq C\,\varepsilon e^{-\kappa t/\varepsilon}\,, \quad 0 \leq t \leq T\,,$$

with some constants C and $\kappa > 0$ which do not depend on ε and T ("property of asymptotic phase"). If $(y(t), z(t)) = (y(t), s_\varepsilon(y(t)))$ is a solution of (1) on $\mathcal{M}_\varepsilon$, then $y(t)$ is a solution of the differential equation with smooth right-hand side

$$\frac{dy}{dt} = f(y, s_\varepsilon(y))\,, \tag{9}$$

which has $y = 0$ as a stationary point. There, the Jacobian is of the form $(f_y - f_z g_z^{-1} g_y)(0,0) + O(\varepsilon)$. By condition (3), this matrix has no eigenvalues on the imaginary axis for small ε.

Not only the continuous system (1), but also its Runge-Kutta discretization admits an invariant manifold. A combination of results and techniques of [47] and [18] yields the following result; cf. Nipp and Stoffer [48].

For $0 < \varepsilon \leq h \leq h_0$, *there is a local attractive invariant manifold* $\mathcal{M}_{\varepsilon,h} = \{(y, z) |\; z = s_{\varepsilon,h}(y)\}$ *for the Runge-Kutta discretization.* $\mathcal{M}_{\varepsilon,h}$ *is close to* $\mathcal{M}_\varepsilon$*:*

$$\|s_{\varepsilon,h}(y) - s_\varepsilon(y)\| \leq \begin{cases} C\,\varepsilon\, h^q & \textit{if } b_i = a_{si} \textit{ for all } i\,, \\ C\, h^{q+1} & \textit{else}\,. \end{cases}$$

There is a property of asymptotic phase: for every (y_0, z_0) in an h- and ε-independent open set that contains $\mathcal{M}_{\varepsilon,h}$, there exists $(\widetilde{y}_0, \widetilde{z}_0) \in \mathcal{M}_{\varepsilon,h}$ such that the corresponding Runge-Kutta solutions satisfy

$$\|y_n - \widetilde{y}_n\| + \varepsilon \cdot \|z_n - \widetilde{z}_n\| \leq C\,\varepsilon\rho^n \ , \quad 0 \leq n \leq N \ ,$$

where $\rho < 1$ and C do not depend on ε, h, and N. For $\varepsilon \ll h$ we have $\rho = |R(\infty)| + O(\varepsilon/h)$.

With these results at hand, the construction of the shadowing solution in [41] proceeds in several steps. Starting from a given Runge-Kutta solution (y_n, z_n), $0 \leq n \leq N$, of (1) staying in an r-neighborhood of $(0,0)$, with r sufficiently small (but independent of ε), a solution $(y(t), z(t))$ of (1) shadowing the numerical solution is constructed as follows.

1. Take the Runge-Kutta solution $(\widetilde{y}_n, \widetilde{z}_n)$ on $\mathcal{M}_{\varepsilon,h}$ with the same asymptotic phase, so that $y_n - \widetilde{y}_n = O(\varepsilon\rho^n)$ for $0 \leq n \leq N$.
2. Construct a Runge-Kutta solution (η_n) of (9) shadowing $(\widetilde{y}_n)$: $\widetilde{y}_n - \eta_n = O(\varepsilon h^{q+1})$ for $0 \leq n \leq N$.
3. Apply Beyn's result to (9) to obtain a shadowing solution $y(t)$ with $\eta_n - y(nh) = O(h^p)$ for $0 \leq n \leq N$. Take $z(t) = s_\varepsilon(y(t))$.

This yields part (A) of Theorem 1. For part (B), one proceeds similarly from a given solution $(y(t), z(t))$, $0 \leq t \leq T = Nh$, of (1).

1'. Take the solution $(\widetilde{y}(t), \widetilde{z}(t))$ on $\mathcal{M}_\varepsilon$ with the same asymptotic phase, so that $y(t) - \widetilde{y}(t) = O(\varepsilon e^{-\kappa t/\varepsilon})$ for $0 \leq t \leq T$.
2'. Apply the converse direction of Beyn's result to (9) to obtain a Runge-Kutta solution (η_n) of (9) shadowing $(\widetilde{y}(nh))$: $\widetilde{y}(nh) - \eta_n = O(h^p)$ for $0 \leq n \leq N$.
3'. Construct a Runge-Kutta solution (y_n, z_n) of (1) on $\mathcal{M}_{\varepsilon,h}$ shadowing $(\eta_n, s_\varepsilon(\eta_n))$: $\eta_n - y_n = O(\varepsilon h^{q+1})$ for $0 \leq n \leq N$, and $z_n = s_{\varepsilon,h}(y_n)$.

Steps 2 and 3' are the technically demanding steps in these constructions; see [41] for the details.

1.5 Related Results

In the nonstiff ODE case, Beyn [4] further shows that the local stable and unstable manifolds near the hyperbolic stationary point are approximated with the order of the method. For the Runge-Kutta discretization of the singularly perturbed problem (1), the techniques of [41,47] yield also that there are local stable and unstable submanifolds of the attractive invariant manifold $\mathcal{M}_{\varepsilon,h}$ of the discretization which approximate the stable and unstable manifolds of the reduced equation (9), again with the order of the finite-time smooth-data error bounds.

The results of [41] permit to relate the dynamics of Runge-Kutta discretizations of the singularly perturbed problem (1) to those of the discretization of the nonstiff differential equation (9). This reduction makes it possible to transfer many of the known results on the long-time behaviour of discretizations of nonstiff ODEs to singularly perturbed problems — at least as long as trajectories stay away from regions where some eigenvalue of g_z has non-negative real part. This restriction excludes, for example, relaxation oscillations.

For BDF-like multistep methods applied to singularly perturbed problems, optimal-order finite-time smooth-data error bounds were derived in [40,49], and their attractive invariant manifolds were studied by Nipp and Stoffer [49]. A combination of [49] and [41] extends Theorem 1 to multistep discretizations.

There is a rich literature on "numerical shadowing", mainly for nonstiff ordinary differential equations. (Zentralblatt lists over 40 articles, e.g., by Chow, Coomes/Koçak/Palmer, Corless, Eirola, Hadeler, Kloeden, Pilyugin, Sauer/Yorke, Van Vleck.) Less has been done for stiff differential equations or partial differential equations, and apparently nothing for delay differential equations.

Alouges and Debussche [1] extend Beyn's shadowing result near hyperbolic equilibria to implicit Euler time discretizations of semilinear parabolic problems, Larsson and Sanz-Serna [37,38] to finite element space discretizations and full discretizations. Ostermann and Palencia [51] derive a shadowing result for an implicit Euler time discretization of non-autonomous parabolic problems.

The shadowing lemma of Chow, Lin and Palmer [9] combined with non-smooth-data error bounds for finite element and Runge-Kutta discretizations of semilinear parabolic equations [35,43] yields numerical shadowing near general hyperbolic invariant sets of such equations.

Larsson and Pilyugin [36] investigate numerical shadowing near the attractor for finite element/implicit Euler discretizations of reaction-diffusion equations in one space dimension, using a reduction to known finite-dimensional results on Morse-Smale systems via inertial manifolds. It is shown that every numerical trajectory shadows some exact solution of the problem after the elapse of a finite time.

Long-time error bounds of numerical discretizations of semilinear parabolic problems near asymptotically stable stationary points were obtained earlier [25,34,53].

2 Hyperbolic Periodic Orbits of Delay Differential Equations

Following [29], we study the persistence of stable hyperbolic periodic orbits of delay differential equations under numerical discretization. We show the existence of attractive closed curves for Runge-Kutta discretizations, which approximate the periodic orbit with the full order of the method. The proof requires an infinite-dimensional analytical/numerical framework and combines finite-time error bounds for both smooth and non-smooth data with an invariant manifold theorem.

2.1 Analytical Framework

We consider a delay differential equation with fixed delay $\tau > 0$,

$$\frac{dx}{dt}(t) = f(x(t), x(t-\tau)) \tag{10}$$

where f is bounded and sufficiently often differentiable with bounded derivatives. With d denoting the dimension of the system (10), we let

$$\mathcal{C} = \mathcal{C}([-\tau, 0], \mathbf{R}^d)$$

be the Banach space of $\mathbf{R}^d$-valued continuous functions on $[-\tau, 0]$ equipped with the maximum norm, which we denote by $\|\cdot\|$. For a given initial function $x^0 \in \mathcal{C}$, Eq. (10) has a unique solution $x : [-\tau, \infty) \to \mathbf{R}^d$. For $t \geq 0$, we define

$$x^t \in \mathcal{C} \qquad \text{via} \quad x^t(\theta) = x(t+\theta) \;\; \text{for} \;\; \theta \in [-\tau, 0] \; .$$

To indicate the dependence of the solution section x^t on the initial function x^0, we write

$$x^t = S^t(x^0) \; .$$

This gives a semigroup on $\mathcal{C}$. Further, $S^t : \mathcal{C} \to \mathcal{C}$ is a Fréchet differentiable map. We denote its derivative at $x^0 \in \mathcal{C}$ by $DS^t(x^0)$.

We assume that (10) has a *stable hyperbolic periodic orbit*, that is, (10) has a nonconstant periodic solution

$$\overline{x} : \mathbf{R} \to \mathbf{R}^d \qquad \text{of period} \;\; \omega > 0 \; ,$$

and the derivative of the period map, $DS^\omega(\overline{x}^0)$, has 1 as a simple eigenvalue, whereas the remaining part of the spectrum of $DS^\omega(\overline{x}^0)$ is bounded in modulus by a number strictly smaller than 1. Let

$$\Gamma = \{\overline{x}^t : t \in \mathbf{R}\}$$

denote the periodic orbit of (10) in $\mathcal{C}$.

2.2 Numerical Method

We restrict our attention to step sizes $h > 0$ for which the delay τ is an integer multiple:

$$\tau = \nu h \; , \qquad \text{with integer } \nu \; .$$

A Runge-Kutta discretization of (10) reads as follows [19]: Given an initial function $x^0 \in \mathcal{C}$, we define starting values (for $i = 1, \ldots, m$)

$$\begin{aligned} X_{ni} &= x^0(nh + c_i h) \; , \qquad n = -\nu, \ldots, -1 \\ x_0 &= x^0(0) \end{aligned}$$

and set recursively for $n = 0, 1, 2, \ldots$

$$x_{n+1} = x_n + h \sum_{i=1}^{m} b_i X'_{ni}$$

with internal stage relations ($i = 1, \ldots, m$)

$$X_{ni} = x_n + h \sum_{j=1}^{m} a_{ij} X'_{nj}$$

and

$$X'_{ni} = f(X_{ni}, X_{n-\nu,i}) \ .$$

The method is determined by the real coefficients a_{ij}, b_j, c_i ($i, j = 1, \ldots, m$), where $c_i \in [0,1]$. It is explicit if $a_{ij} = 0$ for $i \leq j$. We assume that the Runge-Kutta method has classical order p.

For a formulation of the result on the persistence of the periodic orbit under the discretization, we need to interpolate the discrete solution values to functions in $\mathcal{C}$. For $t = nh$ with integer $n \geq 0$, we construct

$$R_h^t(x^0) = x_h^t \in \mathcal{C}$$

by setting $x_h^0 = x^0$ and defining $x_h^t(\theta)$ for $-h \leq \theta \leq 0$ by polynomial interpolation through $x_n, x_{n-1}, \ldots, x_{n-p}$ (with $x_k = x^0(kh)$ for negative k), and we set recursively $x_h^t(\theta) = x_h^{t-h}(\theta + h)$ for $-\tau \leq \theta \leq -h$, for $t = h, 2h, 3h, \ldots$.

2.3 Statement of the Result

Theorem 2. [29] *In the above situation, for any given $L > 0$, there are positive constants r, c, C, C_0 and h_0 such that the following holds for $0 < h \leq h_0$. There is a closed curve $\Gamma_h \subset \mathcal{C}$ which attracts numerical solutions at an exponential rate,*

$$\mathrm{dist}(R_h^t(x^0), \Gamma_h) \leq C\, e^{-ct} \ , \qquad t = nh > 0 \ ,$$

whenever the initial function x^0 is in an r-neighbourhood of the periodic orbit, viz. $\mathrm{dist}(x^0, \Gamma) \leq r$, *and x^0 has a Lipschitz constant not exceeding L. The Hausdorff distance to the periodic orbit Γ is bounded by*

$$\mathrm{dist}_{\mathrm{H}}(\Gamma_h, \Gamma) \leq C_0\, h^p \ .$$

2.4 Discussion and Comparison with the ODE Case

Theorem 2 is related to results by Braun and Hershenov [7], Beyn [6], and Eirola [12] on invariant curves of numerical discretizations of (nonstiff) smooth ordinary differential equations with a hyperbolic periodic orbit. Theorem 2 considers only the stable case, but similar to ODEs, the result could be extended to general hyperbolic periodic orbits. In the stable case, the result for pth-order Runge-Kutta discretizations of smooth ordinary differential equations with sufficiently small step size h reads as follows.

There is a closed curve which is invariant under the numerical method and $O(h^p)$ close to the stable hyperbolic orbit of the ordinary differential equation. Locally, it attracts numerical solutions exponentially with an h-independent rate.

We now indicate how this result for ordinary differential equations follows directly from the attractive invariant manifold theorem of Kirchgraber, Nipp and Stoffer [32,47] restated here in the appendix. Consider the differential equation (5) and denote its flow by $S^t(x_0) = x(t)$. Let the equation have the periodic orbit $\overline{x}(t)$ with period ω. By definition, the orbit is stable hyperbolic if the derivative of the period map, $\partial S^\omega(\overline{x}(0))/\partial x$, has 1 as a simple eigenvalue, and all other eigenvalues are strictly smaller than 1 in modulus. We now use normal coordinates, in terms of which every point in a neighbourhood of the periodic orbit is written as $x = \overline{x}(\alpha) + \beta$, where $\alpha \in \mathbf{R}$ is unique up to integer multiples of the period ω, and β is unique in the maximal invariant subspace of $\partial S^\omega(\overline{x}(\alpha))/\partial x$ that does not contain the eigenvector to the eigenvalue 1. Written in these coordinates, the flow map S^t satisfies, for sufficiently large t, the conditions of Theorem A.1 on a strip $\mathbf{R} \times B$, with B a ball. The estimate (6) for the local error implies that, on finite time intervals, there is a uniform error estimate between the Runge-Kutta solution $R_h^t(x) = x_n$ at $t = nh$ and the exact solution $S^t(x)$,

$$R_h^t(x) - S^t(x) = O(h^p) , \qquad \frac{\partial R_h^t}{\partial x}(x) - \frac{\partial S^t}{\partial x}(x) = O(h^p) \tag{11}$$

uniformly for x in an arbitrary compact set.

Hence, the Runge-Kutta map R_h^t is a small Lipschitz perturbation of the flow map. R_h^t thus still satisfies the assumptions of Theorem A.1 for sufficiently small h, and the existence of an attractive invariant closed curve of the numerical discretization follows. Corollary A.2 provides the $O(h^p)$ bound for the distance to the periodic orbit.

We return to the delay differential equation (10). As in the previous section, a principal difficulty in extending the ODE result is that the uniform error bounds (6) or (11) are no longer valid. For the above Runge-Kutta discretization, the following finite-time *nonsmooth-data error bound* is known [29]: If $x^0, v \in \mathcal{C}$ are Lipschitz bounded by L, and $\|v\| \leq 1$, then the difference between the Runge-Kutta solution $x_h^t = R_h^t(x^0)$ and the exact solution $x^t = S^t(x^0)$ corresponding to the initial function x^0, is bounded by

$$\|R_h^t(x^0) - S^t(x^0)\| \leq CLh , \qquad \|DR_h^t(x^0)v - DS^t(x^0)v\| \leq CLh$$

with a constant C which depends on t, but is independent of x^0 and v with the stated properties. Without assuming additional regularity of $x^0, v \in \mathcal{C}$, such as Lipschitz continuity, there is no uniform convergence on bounded sets of $\mathcal{C}$ as $h \to 0$. On the other hand, for sufficiently differentiable initial data x^0, there is the full-order *smooth-data error bound* [19, Sec. II.17]

$$\|R_h^t(x^0) - S^t(x^0)\| \leq C\, h^p \ ,$$

where C depends on bounds of the first p derivatives of x^0, and on t.

A further difficulty not present in the ODE case results from the fact that the numerical method incorporates the delay via past internal stages, not via past solution values. As a consequence, a Runge-Kutta step must be viewed as a mapping

$$(x_{n+k}, X_{n+k,1}, \ldots, X_{n+k,m})_{k=-\nu}^{0} \mapsto (x_{n+1+k}, X_{n+1+k,1}, \ldots, X_{n+1+k,m})_{k=-\nu}^{0}$$

or, extending to continuous functions via polynomial interpolation, as a mapping $\mathcal{C} \times \mathcal{C}^m \to \mathcal{C} \times \mathcal{C}^m$. The attractive curve Γ_h in $\mathcal{C}$ of Theorem 2 is in general *not* an invariant curve of the numerical method, but instead it consists of the projection to the first component of an attractive invariant curve in $\mathcal{C} \times \mathcal{C}^m$ of the Runge-Kutta map. This leads to additional problems in bringing the invariant manifold theorem into play; see [29] for details. Here, for simplicity, we continue the discussion with a Runge-Kutta method for which all internal stages are linear combinations of numerical solution values, such as the explicit or implicit Euler method or the trapezoidal rule or the implicit midpoint rule. In this case, the numerical one-step map R^h can indeed be viewed as a map on $\mathcal{C}$, and for $t = nh$, the time-t numerical solution map R_h^t is the n-fold composition $R_h^t = (R^h)^n$.

For this special case, we now outline the arguments in the proof of Theorem 2. Analogous to ordinary differential equations, there exist normal coordinates near the periodic orbit $\Gamma = \{\overline{x}^\alpha : \alpha \in \mathbf{R}\}$. As is known from Hale [21, Ch. 10], every function $x \in \mathcal{C}$ in some neighbourhood of Γ can be written as $x = \overline{x}^\alpha + \beta$, where $\alpha \in \mathbf{R}$ is unique up to integer multiples of the period ω, and β is unique in the maximal invariant subspace of $DS^\omega(\overline{x}(\alpha))$ that does not contain the eigenfunction to the eigenvalue 1. It can be verified that the exact flow map S^t written in normal coordinates, $(\alpha, \beta) \mapsto (\widehat{\alpha}, \widehat{\beta})$, satisfies the conditions of the attractive invariant manifold theorem Theorem A.1 in a strip of $\mathbf{R} \times \mathcal{C}$, clearly with the periodic orbit Γ as attractive invariant curve. Consider now the closed bounded set in $\mathcal{C}$,

$$B_{r,L} = \{\beta \in \mathcal{C} : \|\beta\| \leq r,\ \beta \text{ is Lipschitz bounded by } L\} \ .$$

Using the above *nonsmooth-data error bound* and a uniform Lipschitz bound of numerical solutions $x_h^t \in \mathcal{C}$, it is seen, for appropriately chosen r and L, that together with S^t also the Runge-Kutta map R_h^t written in normal coordinates satisfies the conditions of Theorem A.1 on $\mathbf{R} \times B_{r,L}$ for sufficiently small step size h. This yields the existence of an exponentially attractive invariant curve Γ_h for R_h^t, and subsequently also for the one-step map R^h. Since the periodic orbit $\overline{x}$ is arbitrarily differentiable, the *smooth-data error bound* yields, uniformly for $\alpha \in \mathbf{R}$,

$$\|R_h^t(\overline{x}^\alpha) - S^t(\overline{x}^\alpha)\| \leq C\, h^p \ .$$

Corollary A.2 requires just this bound of the difference of R_h^t and S^t on the periodic orbit Γ, and hence it yields the optimal-order distance estimate $\mathrm{dist_H}(\Gamma_h, \Gamma) = O(h^p)$.

2.5 Related Results

The persistence of (not necessarily stable) hyperbolic periodic orbits under discretization has been studied also for semilinear parabolic differential equations. For implicit Euler time discretizations of such problems, Alouges and Debussche [2] show the existence of invariant closed curves approximating the hyperbolic periodic orbit with a sub-optimal order smaller than 1. In [43], it is shown for Runge-Kutta time discretizations that the approximation order of the invariant curve is actually that of high-order finite-time error bounds for the Runge-Kutta approximation of smooth solutions of the parabolic problem. As in the case of delay differential equations, the result relies on both low-order nonsmooth-data and high-order smooth-data finite-time error bounds.

3 Hopf Bifurcation of Semilinear Parabolic Differential Equations

Following [44], we study the long-time behaviour of numerical discretizations in a situation where hyperbolicity gets lost, in the neighbourhood of a bifurcation point. We consider Runge-Kutta time discretization of semilinear parabolic equations near a generic, supercritical Hopf bifurcation. The phase portrait is shown to persist under the discretization, and in particular, the bifurcation point and the Hopf orbits are approximated with the high order of finite-time approximations to smooth solutions of the parabolic equation. The analysis uses a reduction to two-dimensional center manifolds of both the continuous problem and its discretization, and a comparison of the dynamics on the center manifolds via normal forms. The existence, smoothness, and approximation properties of the center manifold of the discretization are obtained by studying the discretization, by the same numerical method, of a boundary value problem on the negative half-line.

3.1 Analytical Framework

We consider reaction-diffusion equations and incompressible Navier-Stokes equations in the abstract setting of sectorial evolution equations in a Banach space as given in Henry's book [24]. We let the equation be parametrized by a real bifurcation parameter λ,

$$\frac{du}{dt} + A(\lambda)u = F(u,\lambda) \ . \tag{12}$$

We assume that the system can be transformed, via suitable spectral projections, to a block-diagonal form

$$\begin{aligned} \frac{dy}{dt} + B(\lambda)y &= f(y,v,\lambda) \\ \frac{dv}{dt} + L(\lambda)v &= g(y,v,\lambda) \end{aligned} \tag{13}$$

with the following specifications. The real 2×2-matrix $B(\lambda)$ has a pair of complex conjugate eigenvalues which cross the imaginary axis at the parameter value λ^* with non-vanishing speed, i.e.,

$$\sigma(-B(\lambda)) = \{\alpha(\lambda) \pm i\omega(\lambda)\}, \quad \text{with} \ \ \alpha(\lambda^*) = 0, \ \frac{d\alpha}{d\lambda}(\lambda^*) > 0, \ \omega(\lambda^*) > 0 \ .$$

We further assume, with $\omega^* = \omega(\lambda^*)$,

$$B(\lambda^*) = \begin{pmatrix} 0 & -\omega^* \\ \omega^* & 0 \end{pmatrix} .$$

The linear operator $L(\lambda)$ in (13) is a densely defined closed operator on a Banach space X, with domain $D(L)$ independent of λ. The spectrum of $-L(\lambda)$ is in a sector lying strictly in the left half-plane, and $L(\lambda)$ satisfies the sectorial resolvent bound, with an angle $\phi < \pi/2$ and a positive abscissa $\ell > 0$,

$$\|(z + L(\lambda))^{-1}\|_{\mathcal{L}(X)} \leq \frac{K}{|z+\ell|} \qquad \text{for all complex } z \text{ with } |\arg(z+\ell)| \geq \phi$$

uniformly for λ in an interval Λ around λ^*. The functions

$$\left.\begin{aligned} \lambda &\mapsto B(\lambda) \in \mathbf{R}^{2\times 2} \\ \lambda &\mapsto L(\lambda) \in \mathcal{L}(D(L),X) \end{aligned}\right. \qquad \text{are arbitrarily differentiable.}$$

The nonlinearities f and g, taking values in $\mathbf{R}^2$ and X, respectively, are arbitrarily differentiable on $\mathbf{R}^2 \times V \times \Lambda$, where $V = D(L(\lambda)^\alpha)$ for some $\alpha < 1$. (Note, V is independent of λ because of the uniform resolvent condition.) We assume that the left-hand side of (13) represents the linearization of the equation at the stationary point $(0,0)$, i.e.,

$$\begin{gathered} f(0,0,\lambda) = 0 \ , \quad g(0,0,\lambda) = 0 \\ D_y f(0,0,\lambda) = 0 \ , \quad D_v f(0,0,\lambda) = 0 \ , \quad D_y g(0,0,\lambda) = 0 \ , \quad D_v g(0,0,\lambda) = 0 \ . \end{gathered}$$

It is known [24,59] that the system (13) has a *center manifold*

$$\mathcal{M}_\lambda = \{(y, s(y,\lambda)) : y \in \mathbf{R}^2\} ,$$

with a defining function $s : \mathbf{R}^2 \times \Lambda \to V$ which, for arbitrary integer k, is k times continuously differentiable in a k-dependent neighbourhood of $(0, \lambda^*)$. The center manifold is locally invariant and attracts solutions at an exponential rate. The dynamics of the system (13) in a neighbourhood of the stationary point $(0,0)$ is determined by the equation reduced to the center manifold,

$$\frac{dy}{dt} + B(\lambda)y = \varphi(y,\lambda) , \tag{14}$$

where $\varphi(y,\lambda) = f(y, s(y,\lambda), \lambda)$ satisfies $\varphi(0,\lambda) = 0,\ D_y\varphi(0,\lambda) = 0$. A near-identity change of coordinates transforms (14) to *normal form* [60, Sections 2.2B and 3.1B], which in polar coordinates reads for $\lambda = \lambda^* + \rho^2$

$$\begin{aligned} \frac{dr}{dt} &= (a\,\rho^2 + c\,r^2)r + O(r(\rho^4 + r^4)) \\ \frac{d\theta}{dt} &= \omega^* + b\,\rho^2 + d\,r^2 + O(\rho^4 + r^4) , \end{aligned} \tag{15}$$

where $a = d\alpha/d\lambda\,(\lambda^*) > 0$, $b = d\omega/d\lambda\,(\lambda^*)$, and where the coefficients c and d depend on second and third derivatives of φ at the bifurcation point. The normal form yields that the dynamical behaviour passes for λ growing across λ^* from an asymptotically stable equilibrium to an asymptotically stable periodic orbit if $c < 0$, and such a change in the dynamics is possible only if $c \le 0$. In the following we assume

$$c < 0 .$$

The periodic orbits of the reduced equation (14) are close to circles centered at 0 with radius $r = \rho\sqrt{-a/c}$, where again $\rho = \sqrt{\lambda - \lambda^*}$. It will be convenient to parametrize the periodic orbits by ρ via $\lambda(\rho) = \lambda^* + \rho^2$. We denote the periodic orbit of Eq. (12) with parameter $\lambda(\rho)$ by $\Gamma(\rho)$. This periodic orbit lies on the center manifold to the parameter $\lambda(\rho)$.

3.2 Numerical Method

A Runge-Kutta time discretization of (12) reads

$$u_{n+1} = u_n + h\sum_{i=1}^m b_i U'_{ni} , \quad U_{ni} = u_n + h\sum_{j=1}^m a_{ij} U'_{nj} ,$$

$$U'_{ni} + A(\lambda)U_{ni} = F(U_{ni},\lambda) \qquad (i = 1,\dots,m) .$$

The stability condition we need here is a weakened form of the strong A-stability defined in Section 1.2. For an angle $\theta \le \frac{1}{2}\pi$, the method is called *strongly $A(\theta)$-stable* if all eigenvalues of the Runge-Kutta matrix $\mathcal{A}$ lie outside the closed complex sector $|\arg(-z)| \leq \theta$, and if the stability function of the method satisfies

$$|R(z)| \leq 1 \quad \text{for } |\arg(-z)| \leq \theta$$

and

$$|R(\infty)| < 1 \, .$$

We assume that the method is strongly A(θ)-stable with $\theta > \phi$, where ϕ is the angle in the sectorial resolvent condition. As in Section 1.2, we assume that the method has classical order p and stage order q. We let

$$k = \min(p, q+1) \, .$$

3.3 Statement of the Result

Theorem 3. [44] *In the above situation, there exist $h_0 > 0$, $\lambda_0 < \lambda^*$, $\rho_0 > 0$, and constants C and C^*, such that for all positive time steps $h \leq h_0$, there is a parameter value λ_h^* with*

$$|\lambda_h^* - \lambda^*| \leq C^* h^p$$

such that the following holds:

(i) For every $\lambda \in [\lambda_0, \lambda_h^*)$, the Runge-Kutta discretization has the asymptotically stable equilibrium point 0.

(ii) For every $\rho \in (0, \rho_0]$, the Runge-Kutta discretization with parameter $\lambda_h(\rho) = \lambda_h^* + \rho^2$ has an attractive invariant closed curve $\Gamma_h(\rho)$. Its Hausdorff distance to the periodic orbit $\Gamma(\rho)$ with parameter $\lambda(\rho) = \lambda^* + \rho^2$ satisfies

$$\mathrm{dist_H}\big(\Gamma_h(\rho), \Gamma(\rho)\big) \leq C \, \rho \, h^k \, .$$

3.4 Discussion and Comparison with the ODE Case

For ordinary differential equations

$$\frac{dx}{dt} = f(x, \lambda)$$

the question of the behaviour of Hopf bifurcation under numerical discretization was first considered by Brezzi, Ushiki and Fujii [8]. They state an analogue of Theorem 3 for the explicit Euler method and give an outline of a proof for the two-dimensional case.

Hairer and Lubich [16] study the behaviour of Runge-Kutta methods near a Hopf bifurcation of ordinary differential equations with real-analytic right-hand side via a *backward analysis* of numerical integrators. It is shown in [3,16] that there exists a modified differential equation

$$\frac{d\widetilde{x}}{dt} = f_h(\widetilde{x}, \lambda)$$

such that the numerical solution map R_h^t departs from the flow S_h^t of the modified differential equation by only exponentially small terms in the step size: for a fixed $t = nh$,

$$R_h^t(x) - S_h^t(x) = O(e^{-\gamma/h})$$

uniformly for x in any complex compact subset of the domain of analyticity. The constant $\gamma > 0$ is inversely proportional to a Lipschitz constant of f. The real-analytic function f_h is $O(h^p)$ close to f, again uniformly on complex compact sets. The modified differential equation undergoes a Hopf bifurcation at a parameter $\lambda_h^* = \lambda^* + O(h^p)$. Although the hyperbolicity of the Hopf orbits deteriorates as λ approaches the bifurcation point, the exponential smallness of the error allows us to use the approximation result of hyperbolic periodic orbits (see Sect. 2.4) down to parameters that are exponentially close to the bifurcation point: $\lambda \geq \lambda_h^* + e^{-\gamma/(2h)}$. For such a λ, we thus obtain the existence of an attractive invariant closed curve of the discretization which is exponentially close to the periodic orbit of the modified equation for the same parameter value.

Backward analysis is a powerful tool for studying numerical discretizations even in non-hyperbolic situations in nonstiff ordinary differential equations. However, it is not applicable to partial differential equations due to the unboundedness of the operator. (See, however, [45] for some partial results in that direction when very strong smoothing properties such as Gevrey regularity are available.)

The proof of Theorem 3 follows a standard procedure in the analysis of bifurcations [33,58,59] outlined already in Section 3.1: reduction of the dynamics to a center manifold and analysis of the normal form of the reduced equation. The first step is to construct a center manifold of the discretization, and to study its relationship to the center manifold of the continuous problem. In [44], the center manifold is constructed from a boundary value problem on the negative half-line,

$$\frac{dy}{dt} + B(\lambda)y = f(y, v, \lambda) , \qquad y(0) = \eta$$

$$\frac{dv}{dt} + L(\lambda)v = g(y, v, \lambda) , \qquad \limsup_{t \to -\infty} \|v(t)\| < \infty ,$$

from which the function $s : \mathbf{R}^2 \times \Lambda \to V$ defining the center manifold $\mathcal{M}_\lambda$ is obtained by setting

$$s(\eta, \lambda) = v(0) .$$

Similarly, the center manifold of the Runge-Kutta discretization is obtained by formally applying the Runge-Kutta method to this boundary value problem. Existence, attractivity, smoothness and approximation properties of the

discrete center manifold then follow from studying the numerical discretization of the boundary value problem. Using the convolution quadrature interpretation of the Runge-Kutta method [42] and exploiting the temporal smoothness of the solution of the boundary value problem, which follows from the smoothness of the function s, the following result is obtained.
There exists a center manifold

$$\mathcal{M}_{\lambda,h} = \{(y, s_h(y,\lambda)) : y \in \mathbf{R}^2\}$$

which is invariant under the numerical method and uniformly in h and λ exponentially attractive. The defining function $s_h : \mathbf{R}^2 \times \Lambda \to V$ has the same regularity properties as s, with derivatives bounded uniformly in h. There is the approximation estimate

$$\|s_h(y,\lambda) - s(y,\lambda)\| \leq C\,|y|\,h^k$$

in the norm of the space V, and the same order of approximation is valid for any fixed number of derivatives of $s_h - s$ with respect to y and λ.

The next step is to compare the normal forms of the time-h flow map of the reduced differential equation on $\mathbf{R}^2$ and of the Runge-Kutta map reduced to the discrete center manifold, giving another map on $\mathbf{R}^2$. This comparison uses the above estimates for s_h and leads to a situation to which Theorem A.1 and Corollary A.2 can be applied. This yields the existence of invariant curves and the approximation properties as stated in Theorem 3.

We remark that this construction can of course be carried out also in the ODE situation. In that case, the approximation estimate in Theorem 3 improves to the full order $k = p$.

A related construction can be used also for finite element space discretizations of reaction-diffusion equations and subsequently for full discretizations [joint work by S. Larsson and the author, in preparation]. There are additional difficulties due to the fact that the method is then no longer invariant under the transition from (12) to (13), and that the equilibrium point of the differential equation need not lie in the finite element space. Multistep time discretization is studied by H. Selhofer [doctoral thesis in preparation].

4 Inertial Manifolds of Semilinear Parabolic Equations

Following [11], we show high-order approximation of the inertial manifold of a semilinear parabolic equation by inertial manifolds of full discretizations combined of a spectral Galerkin method in space and a Runge-Kutta method in time. The result follows by a combination of low-order nonsmooth-data error bounds and high-order error bounds for smooth data over finite time intervals, and once more with the invariant manifold theorem of the appendix. The smoothness of solutions on the inertial manifold, as implied by time analyticity and Gevrey regularity, renders the high order of approximation possible.

4.1 Analytical Framework

The following applies to the complex Ginzburg-Landau and Kuramoto-Sivashinsky equations and to classes of reaction-diffusion equations with analytic nonlinearities in one or two space dimensions; cf. [57] and references therein. The limitation of the space dimension is due to a spectral gap condition, which is needed in the proof of existence of an inertial manifold and which is not satisfied by the three- (or higher-) dimensional Laplacian.

We consider the evolution equation

$$\frac{du}{dt} + Au = F(u) \tag{16}$$

under the following assumptions: the operator A is of the form $A = (1+ia)L$, where $a \in \mathbf{R}$ and L is a self-adjoint, densely defined linear operator on a separable Hilbert space H, with a compact inverse and eigenvalues

$$0 < \lambda_1 \leq \lambda_2 \leq \ldots \nearrow +\infty \,.$$

For some $\alpha < 1$, the nonlinearity is defined on $V = D(L^\alpha)$. The function $F : V \to H$ is at least twice continuously Fréchet differentiable.

We denote the norm on H by $|\cdot|$, that on V by $\|\cdot\|$. Let P_m denote the orthogonal projection on the space spanned by the first m eigenfunctions of L, and let $Q_m = I - P_m$ be the projection on the orthogonal complement. $B(\rho)$ denotes the ball of radius ρ in V centered at the origin. As in previous sections, S^t denotes the time-t flow of the evolution equation.

We are interested in an *inertial manifold* $\mathcal{M}$ for (16) as introduced by Foiaş, Sell and Temam [15]. This is a positively invariant set (more precisely, for a fixed $\rho > 0$, $\mathcal{M}$ satisfies $S^t(\mathcal{M} \cap B(\rho)) \subset \mathcal{M}$ for all $t \geq 0$) defined through a Lipschitz continuous function $s : P_m V \to Q_m V$ via

$$\mathcal{M} = \{v \in V : Q_m v = s(P_m v)\} \,.$$

$\mathcal{M}$ is exponentially attracting: there exist $\nu > 0$ and a constant C (depending on ρ) such that for all $u_0 \in V$ with $\|u_0\| \leq \rho$,

$$\operatorname{dist}(S^t(u_0), \mathcal{M}) \leq C\, e^{-\nu t} \qquad \text{for all} \quad t \geq 0 \,.$$

The dynamics of the infinite-dimensional evolution equation (16) is then determined by its restriction to the inertial manifold, which is a finite-dimensional ordinary differential equation.

The existence of an inertial manifold is known under the following conditions: (16) has an *absorbing ball*, i.e., there exist $r > 0$ and, for every $\rho > 0$, a $\tau(\rho) > 0$ such that

$$\|v\| \leq \rho \quad \text{implies} \quad \|S^t(v)\| \leq r \quad \text{for all} \quad t \geq \tau(\rho) \,.$$

The second condition is a *spectral gap condition*: for a sufficiently large constant K,

$$\lambda_{m+1} - \lambda_m \geq K\, \lambda_{m+1}^\alpha \,.$$

Finally, it is needed that λ_m is sufficiently large compared to a Lipschitz constant of F. Under these conditions, the existence of an inertial manifold of dimension m actually follows directly from Theorem A.1, used with $\Phi = S^t$ for $t = c/\lambda_{m+1}$ with a sufficiently small constant c.

For our approximation results, we make use of strong regularity results in time and space, which are known to hold for the differential equations mentioned in the beginning of this subsection; cf. [52].

Analyticity in time: for every $v \in V$ of norm bounded by ρ, the function $t \mapsto S^t(v)$ is analytic on the intersection of a complex sector $|\arg t| < \phi$ with a strip $|\mathrm{Im}\, t| < c$, and is bounded there by $\|S^t(v)\| \leq C$ uniformly for $\|v\| \leq \rho$.

Gevrey regularity: for a given $\rho > 0$, there exist a constant C and a time $\bar{t} > 0$ such that $S^t(v)$ is in the domain of $\exp(+(tA)^{1/2})$ and

$$\| \exp((tA)^{1/2}) S^t(v)\| \leq C \quad \text{for} \quad \|v\| \leq \rho,\ 0 \leq t \leq \bar{t}\ .$$

4.2 Numerical Method

We consider a spectral Galerkin discretization in space combined with Runge-Kutta discretization in time. The spectral Galerkin method yields an approximation to the solution of (16) in the space V_N spanned by the first N eigenfunctions of A. With P_N denoting the orthogonal projection on V_N, the method solves

$$\frac{du_N}{dt} + Au_N = P_N F(u_N)\ , \qquad u_N(0) = P_N u_0\ .$$

This problem is discretized in time by a strongly $A(\theta)$-stable Runge-Kutta method of order p and stage order q; cf. Section 3.2. The angle θ should be larger than $\arg(1 + ia)$. We set again

$$k = \min(p, q+1)\ .$$

The time step is again denoted by h, and we let for brevity

$$\Delta = (N, h)\ .$$

The numerical approximation at time $t = nh$ is written as $R^t_\Delta(u_0)$.

4.3 Statement of the Result

Theorem 4. [11] *In the above situation, for given $\rho > 0$, there exist positive constants C_0, C_1, C_2, c, ℓ, κ (independent of the dimension m of the inertial manifold $\mathcal{M}$) such that the following holds for N and h with $\lambda_m/\lambda_{N+1} \leq \kappa$ and $h\lambda_m \leq \kappa$. There exists a manifold $\mathcal{M}_\Delta$ that is positively invariant under the numerical method (more precisely, $R^t_\Delta(\mathcal{M}_\Delta \cap B(\rho)) \subset \mathcal{M}_\Delta$ for $t = nh \geq$*

0). It is defined by a function $s_\Delta : P_m V \to Q_m V$, which is Lipschitz bounded by ℓ, via

$$\mathcal{M}_\Delta = \{v \in V : Q_m v = s_\Delta(P_m v)\} .$$

$\mathcal{M}_\Delta$ attracts all numerical solutions starting with $\|u_0\| \le \rho$ exponentially,

$$\text{dist}(R_\Delta^t(u_0), \mathcal{M}_\Delta) \le C_0 \, e^{-t\nu/2} \quad \textit{for all } \; t = nh \ge 0 .$$

The Hausdorff distance to the inertial manifold of (16) is bounded by

$$\text{dist}_{\text{H}}(\mathcal{M}_\Delta, \mathcal{M}) \le C_1 \, e^{-c\sqrt{\lambda_{N+1}/\lambda_m}} + C_2 \left(\lambda_m h\right)^k .$$

4.4 Discussion

Theorem 4 gives exponential convergence in space and high order in time. We have included the dependence on the dimension m because the spectral gap condition is usually satisfied for infinitely many m, leading to a nested sequence of inertial manifolds with growing attractivity exponents ν. Moreover, even the smallest possible m may be quite large in applications. The way the distance estimate depends on m shows that only the time and length scales of the differential equation reduced to the inertial manifold need to be resolved properly by the discretization for an accurate approximation of the inertial manifold.

We outline a proof of Theorem 4 that uses Theorem A.1 and Corollary A.2, whereas [11] employs a Hadamard graph transform adopted from [30].

The existence of an inertial manifold of the discretization can be proved using Theorem A.1 and a *nonsmooth-data* error bound. Combining the results of [43] and [55, Sect. 3.3] on time and space discretizations, respectively, the following finite-time error bound is obtained: uniformly for $\|u_0\| \le \rho$ and $0 < t \le T$,

$$\|R_\Delta^t(u_0) - S^t(u_0)\| \le C \left(t^{-\alpha} h |\log h| + (h/t)^k + (t\lambda_{N+1})^{\alpha-1}\right) .$$

The same bound holds also for the derivative, $\|DR_\Delta^t(u_0) - DS^t(u_0)\|_{\mathcal{L}(V)}$. The error bound implies the existence of an absorbing ball of the discretization for sufficiently small h and large N. Consequently, the dynamics is not changed if the scheme is modified outside a sufficiently large ball in V, say of radius $\widehat{r}$. We can then achieve that the above error bounds hold globally on V. In the light of Theorem A.1, R_Δ^t is thus a small Lipschitz perturbation of S^t. Hence, for $t = c/\lambda_{m+1}$ with sufficiently small c, together with S^t also R_Δ^t satisfies the conditions of that theorem. This yields the existence of the inertial manifold of the discretization.

The distance estimate between the inertial manifolds $\mathcal{M}$ and $\mathcal{M}_\Delta$ can be based on Corollary A.2. This requires to estimate $R_\Delta^t(u_0) - S^t(u_0)$ only for $u_0 \in \mathcal{M}$, which is achieved via a study of the regularity of solutions on $\mathcal{M}$ and using *smooth-data* finite-time error bounds. Theorem 4.1 of [42] gives the following error bound. If the solution $u(t) = S^t(u_0)$ satisfies, with $d(t) = Q_N F(u(t)) = Q_N(du/dt + Au(t))$,

$$\|u^{(k)}(0)\| + \int_0^T \|u^{(k+1)}(t)\| \, dt \le \mu \ ,$$

$$\|A^{-1}d(0)\| + \int_0^T \|A^{-1}d'(t)\| \, dt \le \delta_N \ ,$$

(where $u^{(k)}$ denotes the kth time derivative of u, and d' the time derivative of d), then

$$\|R^t_\Delta(u_0) - S^t(u_0)\| \le C(\mu h^k + \delta_N)$$

for $0 \le t \le T$. Solutions on the inertial manifold $\mathcal{M} \cap B(r)$ can be continued backward in time, and they stay in $\mathcal{M} \cap B(\widehat{r})$ at least for a time that is inverse proportional to λ_m (recall that the equation and the scheme were modified outside $\widehat{r}$). With a time $t \sim \lambda_m^{-1}$, we can thus employ the time analyticity and Gevrey regularity estimates to obtain

$$\mu = O(\lambda_m^k) \ , \quad \delta_N = O(e^{-c\sqrt{\lambda_{N+1}/\lambda_m}})$$

uniformly for $u_0 \in \mathcal{M} \cap B(r)$. Corollary A.2 then yields the desired high-order distance bound.

4.5 Related Results

The existence of inertial manifolds for spectral discretizations in space was studied by Foiaş, Sell and Temam [15] and Jones and Stuart [30]. The distance estimate of Theorem 4.1 of [15], in a framework which corresponds to the case $\alpha = 1/2$, is $O(\lambda_{N+1}^{-1/4})$. The bound of [30] is $O(\lambda_{N+1}^{\alpha-1})$, which corresponds to the nonsmooth-data finite-time error bound. (The dependence on m is considered explicitly in these papers, but is not reproduced here.)

The first results on the existence of inertial manifolds of time discretizations were given by Demengel and Ghidaglia [10]. Full discretizations were subsequently studied by Jones and Stuart [30]. Those authors consider linearly implicit Euler and fractional step time discretizations and obtain distance estimates of the low order of nonsmooth-data error bounds. Shardlow [54] studies multistep time discretizations and also obtains low-order distance estimates of an order not exceeding 1.

Lord [39] establishes discrete Gevrey regularity for finite difference methods for the complex Ginzburg-Landau equation and uses it to study approximation of the inertial manifold.

C^1-approximation of inertial manifolds has been studied by Jones, Stuart and Titi [31].

5 Attractors of Damped Wave Equations

For a specially constructed time discretization combined with a spectral Galerkin space discretization of a dissipative wave equation, we show the existence of an attractor of the discretization which lies close to that of the partial differential equation. The result relies on a nonsmooth-data finite-time error bound, which is not available for usual time discretizations.

5.1 Analytical Framework

We consider the abstract damped wave equation

$$\frac{d^2u}{dt^2} + \alpha \frac{du}{dt} + Au = g(u) \tag{17}$$

with a positive damping parameter $\alpha > 0$. The linear operator A is assumed to be a self-adjoint, densely defined operator on a Hilbert space H, with a compact inverse and eigenvalues $0 < \lambda_1 \leq \lambda_2 \leq \dots$. We let $V = D(A^{1/2})$ and set $X = V \times H$. The norm on X is denoted by $\|\cdot\|$, viz. $\|(u,v)\|^2 = |A^{1/2}u|^2 + |v|^2$, where $|\cdot|$ is the norm on H.

For the nonlinearity we assume that for some $\gamma > 0$,

$$\begin{aligned} g &: V \to D(A^{\gamma/2}) \\ g &: D(A^{(1-\gamma)/2}) \to H \end{aligned} \qquad \text{are continuously Fréchet differentiable,}$$

and that it permits unique solutions $S^t(u_0, v_0) = (u(t), du/dt(t))$ of (17) in X for all times t and initial values $(u_0, v_0) \in X$.

We further assume that (17) has a (global) *attractor*, that is, a compact set $\mathcal{A} \subset X$ which is invariant under the flow, viz. $S^t(\mathcal{A}) = \mathcal{A}$ for all t, and which attracts bounded sets in X. The latter means that for every bounded set B in X and for every $\varepsilon > 0$, there is a t_0 (which depends on ε and B), such that $S^t(B)$ is in an ε-neighbourhood of $\mathcal{A}$ for $t \geq t_0$.

We refer to Temam [57, Ch. IV] for conditions which ensure the existence of an attractor, and for concrete examples of damped wave equations having an attractor.

5.2 Numerical Method

The choice of time discretization requires more care than in the preceding sections. In contrast to the parabolic case, standard numerical integration schemes for (17) do not admit convergent nonsmooth-data error bounds, which are needed here. Instead we consider a time-stepping method in the spirit of [17,28]. To motivate the method, we start from the variation-of-constants formula

$$\begin{pmatrix} A^{1/2}u(t+h) \\ v(t+h) \end{pmatrix} = E(hA^{1/2}) \begin{pmatrix} A^{1/2}u(t) \\ v(t) \end{pmatrix}$$

$$+ h \int_0^1 E((1-\theta)hA^{1/2}) \begin{pmatrix} 0 \\ -\alpha v(t+\theta h) + g(u(t+\theta h)) \end{pmatrix} d\theta ,$$

where $v = du/dt$ and

$$E(\xi) = \begin{pmatrix} \cos\xi & \sin\xi \\ -\sin\xi & \cos\xi \end{pmatrix} .$$

Expressing the term $v(t + \theta h)$ under the integral once more by the same formula and dropping terms of formal order $O(h^2)$ leads to the following method:

$$\begin{pmatrix} A^{1/2}u_{n+1} \\ v_{n+1} \end{pmatrix} = \Big(E(hA^{1/2}) - h\alpha\Phi(hA^{1/2})\Big)\begin{pmatrix} A^{1/2}u_n \\ v_n \end{pmatrix} + h\Psi(hA^{1/2})\begin{pmatrix} 0 \\ g(u_n) \end{pmatrix}$$

with

$$\Phi(\xi) = \int_0^1 E((1-\theta)\xi) \begin{pmatrix} 0 & 0 \\ 0 & 1 \end{pmatrix} E(\theta\xi)\, d\theta$$

$$\Psi(\xi) = \int_0^1 E((1-\theta)\xi)\, d\theta .$$

The integrals can be evaluated analytically. They are such that the entries of the transformed matrices $D(\xi)^{-1}(\Phi(\xi), \Psi(\xi))D(\xi)$ with $D(\xi) = \mathrm{diag}(\xi, 1)$ are entire functions of ξ^2.

As space discretization we take a standard spectral Galerkin method. With P_N denoting the orthogonal projection on the space H_N (or V_N) spanned by the first N eigenfunctions, and with A_N the restriction of A to H_N, this space discretization is obtained by replacing A by A_N and g by $P_N g$ in the above formula.

The numerical solution starting from $(u_0, v_0) \in X_N = V_N \times H_N$ is denoted by $R^t_\Delta(u_0, v_0) = (u_n, v_n)$ at $t = nh$, where again Δ symbolizes the pair (N, h) of discretization parameters.

Remark. The above method is just a particular example of a class of methods that are exact solvers for $d^2u/dt^2 + Au = 0$. It is a first-order method when applied to initial value problems (17) that admit a smooth solution. A second-order method could be constructed along similar lines.

Remark. Even in cases where an eigendecomposition of the matrix arising from the space discretization is computationally not feasible, as in finite element methods, the method can be efficiently implemented using superlinearly convergent Krylov subspace approximations of matrix function times vector products; cf. [28].

5.3 Statement of the Result

Theorem 5. *If the discretization is sufficiently fine, it has a (local) attractor $\mathcal{A}_\Delta$. The semi-distance to the attractor $\mathcal{A}$ converges to 0 as $h \to 0$, $N \to \infty$. More precisely, for every $\varepsilon > 0$ and every bounded set B in X, there exist positive h_0, N_0, t_0 such that*

$$\operatorname{dist}(R_\Delta^t(u_0, v_0), \mathcal{A}) < \varepsilon \qquad \text{for } t > t_0$$

whenever $(u_0, v_0) \in B \cap X_N$ and $N > N_0$, $h < h_0$.

5.4 Discussion and Related Results

Theorem 5 is apparently the first result on the dynamics of a fully discretized dissipative wave equation. An analogue of Theorem 5 for spectral Galerkin semi-discretization in space was obtained already by Hale, Lin and Raugel [22]. However, the derivation of a corresponding result for time discretizations was hampered by the fact that standard time discretization methods admit no convergent error bounds where the error is measured in the same norm in which bounds for the initial data are specified. Such a *nonsmooth-data error bound* does exist for the numerical method of Section 5.2:

For every $\rho > 0$ and $T > 0$, there exists C such that

$$\|R_\Delta^t(u_0, v_0) - S^t(u_0, v_0)\| \le C\left(h^\gamma + \lambda_{N+1}^{-\gamma}\right)$$

for all $(u_0, v_0) \in X_N$ with $\|(u_0, v_0)\| \le \rho$ and for $0 \le t = nh \le T$.

This is proved below. With this error bound, the distance estimate of Theorem 5 then follows by an argument given by Larsson [35]: Let $\varepsilon > 0$ and let B be a bounded set in X, of which we may assume that it contains the ε-neighbourhood of $\mathcal{A}$. We choose T such that

$$\operatorname{dist}(S^t(B), \mathcal{A}) := \sup_{x \in B} \operatorname{dist}(S^t(x), \mathcal{A}) < \tfrac{1}{2}\varepsilon \quad \text{for } t \ge \tfrac{1}{2}T .$$

If the discretization is sufficiently fine, the above error bound yields for $B_N = B \cap X_N$

$$\operatorname{dist}(R_\Delta^t(B_N), \mathcal{A}) < \varepsilon \quad \text{for } \tfrac{1}{2}T \le t \le T ,$$

and in particular, $R_\Delta^t(B_N) \subset B_N$ for such t. Hence, the bound must hold *for all* $t \ge \frac{1}{2}T$. This proves the distance estimate of Theorem 5. The existence of an attractor $\mathcal{A}_\Delta$ of the discretization then follows from the discrete-time version of Theorem I.1.1 in [57].

Like other results in the spirit of [22], such as [13,26,35,39] on parabolic problems, Theorem 5 gives no estimate for the distance between $\mathcal{A}_\Delta$ and $\mathcal{A}$, nor for the dimension of $\mathcal{A}_\Delta$, and no information about the discrete flow on $\mathcal{A}_\Delta$. It also does not ensure that $\mathcal{A}_\Delta$ lies close to every point on the attractor $\mathcal{A}$. The latter has been shown for gradient flows by Hale and Raugel

[23], but is not true in general. Hill and Süli [27] study set convergence $\mathcal{A}_\Delta \to \mathcal{A}_0$, where $\mathcal{A}_0$ is a compact invariant subset of $\mathcal{A}$. The shadowing results cited in Section 1.5 relate the dynamics on the discrete attractor to those of the partial differential equation in situations where the attractor is a hyperbolic invariant set or where the system has Morse-Smale structure.

5.5 Proof of the Nonsmooth-Data Error Bound

By the variation-of-constants formula, the solution $(u(t), v(t)) = S^t(u_0, v_0)$ satisfies

$$\begin{pmatrix} A^{1/2}u(t+h) \\ v(t+h) \end{pmatrix} = \left(E(hA^{1/2}) - h\alpha\Phi(hA^{1/2})\right)\begin{pmatrix} A^{1/2}u(t) \\ v(t) \end{pmatrix}$$

$$+ h\Psi(hA^{1/2})\begin{pmatrix} 0 \\ g(u(t)) \end{pmatrix} + d(t)$$

with the defect

$$d(t) = h\int_0^1 E((1-\theta)hA^{1/2})\begin{pmatrix} 0 \\ g(u(t+\theta h)) - g(u(t)) \end{pmatrix} d\theta + O(h^2) .$$

By our assumption on g we have

$$|g(u(t+\theta h)) - g(u(t))| \le L|A^{(1-\gamma)/2}(u(t+\theta h) - u(t))| = O(h^\gamma) ,$$

where the last estimate follows again from the variation-of-constants formula of Section 5.2. Hence, $|d(t)| = O(h^{1+\gamma})$. By a standard stability estimate, this yields the error bound for the semi-discretization in time,

$$\|R_h^t(u_0, v_0) - S^t(u_0, v_0)\| \le C\,h^\gamma ,$$

uniformly for $\|(u_0, v_0)\| \le \rho$ and $0 \le t \le T$. The difference between the semi-discrete and the fully discrete numerical solution is estimated similarly, using

$$|(I - P_N)g(u)| = O(\lambda_{N+1}^{-\gamma}) .$$

This yields

$$\|R_\Delta^t(u_0, v_0) - R_h^t(u_0, v_0)\| \le C\,\lambda_{N+1}^{-\gamma}$$

uniformly for $\|(u_0, v_0)\| \le \rho$ and $0 \le t \le T$, and the desired error bound follows.

A Attractive Invariant Manifolds

Attractive invariant manifold theorems can be traced back to Hadamard a hundred years ago, and their usefulness has been rediscovered and re-established ever since. Here we give a version due to Kirchgraber, Lasagni,

Nipp, and Stoffer [32], which is particularly useful in applications because of its explicit handling of constants. A proof is contained in the report [47].

Consider a map $\Phi : A \times B \to A \times B$ defined on the Cartesian product of a Banach space A and a closed bounded subset B of another Banach space. We write $\Phi(\alpha, \beta) = (\widehat{\alpha}, \widehat{\beta})$ with

$$\begin{aligned} \widehat{\alpha} &= \alpha + F(\alpha, \beta) \\ \widehat{\beta} &= \qquad G(\alpha, \beta) \, . \end{aligned}$$

We assume that F and G are Lipschitz bounded, with Lipschitz constants $L_{\alpha\alpha}$, $L_{\alpha\beta}$ and $L_{\beta\alpha}$, $L_{\beta\beta}$ with respect to α, β. If these Lipschitz constants are sufficiently small, then the map Φ has an attractive invariant manifold. More precisely, the following holds.

Theorem A.1. [32,47] *If $L_{\alpha\alpha} + L_{\beta\beta} + 2\sqrt{L_{\alpha\beta}L_{\beta\alpha}} < 1$, then there is a function $s : A \to B$, Lipschitz bounded by $\ell < 2L_{\beta\alpha}/(1 - L_{\alpha\alpha} - L_{\beta\beta})$, such that*

$$\mathcal{M} = \{(\alpha, s(\alpha)) : \alpha \in A\}$$

is invariant under Φ. $\mathcal{M}$ attracts orbits of Φ with rate $r = \ell L_{\alpha\beta} + L_{\beta\beta} < 1$, i.e., the inequality $\|\widehat{\beta} - s(\widehat{\alpha})\| \le r \, \|\beta - s(\alpha)\|$ holds for all $(\alpha, \beta) \in A \times B$.

Remark. If $A = \mathbf{R}$, and if F, G are periodic in α with period ω, then s is again ω-periodic.

Corollary A.2. *Consider maps Φ, $\widetilde{\Phi} : A \times B \to A \times B$. Assume Φ, $\widetilde{\Phi}$ satisfy the conditions of Theorem A.1 with the same Lipschitz constants $L_{\alpha\alpha}, L_{\alpha\beta}, L_{\beta\alpha}, L_{\beta\beta}$. Let s and $\widetilde{s}$ be the functions defining the attractive invariant manifolds $\mathcal{M}$ and $\widetilde{\mathcal{M}}$, respectively. If the bound*

$$\|\widetilde{\Phi}(\alpha, \beta) - \Phi(\alpha, \beta)\| \le \delta \qquad \text{for all} \;\; (\alpha, \beta) \in \mathcal{M}$$

holds in the norm $\|(\alpha, \beta)\| = \ell \, \|\alpha\| + \|\beta\|$ on $A \times B$, then

$$\|\widetilde{s}(\alpha) - s(\alpha)\| \le \frac{\delta}{1 - r} \qquad \text{for all} \;\; \alpha \in A \, .$$

For our applications of this result it is essential that a bound of $\widetilde{\Phi} - \Phi$ is needed only on the invariant manifold $\mathcal{M}$, not on all of $A \times B$ (as is required in the formulation of [47]). The result as stated follows by tracing the proof in [47], or with the alternative proof in [29].

Acknowledgement

I thank K.P. Hadeler and A. Ostermann for helpful comments on an earlier draft of this article.

References

1. F. Alouges, A. Debussche, On the qualitative behavior of the orbits of a parabolic partial differential equation and its discretization in the neighborhood of a hyperbolic fixed point. Numer. Funct. Anal. Optimiz. 12 (1991), 253–269.
2. F. Alouges, A. Debussche, On the discretization of a partial differential equation in the neighborhood of a periodic orbit. Numer. Math. 65 (1993), 143–175.
3. G. Benettin, A. Giorgilli, On the Hamiltonian interpolation of near to the identity symplectic mappings with application to symplectic integration algorithms. J. Statist. Phys. 74 (1994), 1117–1143.
4. W.-J. Beyn, On the numerical approximation of phase portraits near stationary points. SIAM J. Numer. Anal. 24 (1987), 1095–1113.
5. W.-J. Beyn, On invariant closed curves for one-step methods. Numer. Math. 51 (1987), 103–122.
6. W.-J. Beyn, Numerical methods for dynamical systems, in Advances in Numerical Analysis, Vol. I, Nonlinear partial differential equations and dynamical systems, W. Light, ed., Clarendon Press, Oxford, 1991, 175–236.
7. M. Braun, J. Hershenov, Periodic solutions of finite difference equations. Quart. Appl. Math. 35 (1977), 139–147.
8. F. Brezzi, S. Ushiki, H. Fujii, Real and ghost bifurcation dynamics in difference schemes for ordinary differential equations. In: T. Küpper, H.D. Mittelmann, H. Weber, eds., Numerical Methods for Bifurcation Problems. Birkhäuser, Boston, 1984, 79–104.
9. S.-N. Chow, X.-B. Lin, K.J. Palmer, A shadowing lemma with applications to semilinear parabolic equations. SIAM J. Math. Anal. 20 (1989), 547–557.
10. F. Demengel, J.-M. Ghidaglia, Inertial manifolds for partial differential evolution equations under time discretization: existence, convergence, and applications. J. Math. Anal. Appl. 155 (1991), 177–225.
11. J.L.M. van Dorsselaer, Ch. Lubich, Inertial manifolds of parabolic differential equations under higher-order discretizations. IMA J. Numer. Anal. 19 (1999), 455–471.
12. T. Eirola, Invariant curves of one-step methods. BIT 28 (1988), 113–122.
13. C.M. Elliott, A.M. Stuart, The global dynamics of discrete semilinear parabolic equations. SIAM J. Numer. Anal. 30 (1993), 1622–1663.
14. N. Fenichel, Geometric singular perturbation theory for ordinary differential equations. J. Diff. Eq. 31 (1979), pp. 53–98.
15. C. Foiaş, G.R. Sell, R. Temam, Inertial manifolds for nonlinear evolutionary equations. J. Diff. Eq. 73 (1988), 309–353.
16. E. Hairer, Ch. Lubich, The life-span of backward error analysis for numerical integrators. Numer. Math. 76 (1997), 441–462.
17. E. Hairer, Ch. Lubich, Long-time energy conservation of numerical methods for oscillatory differential equations. Preprint, 1999.
18. E. Hairer, Ch. Lubich, M. Roche, Error of Runge-Kutta methods for stiff problems studied via differential-algebraic equations. BIT 28 (1988), 678–700.
19. E. Hairer, S.P. Nørsett, G. Wanner, Solving Ordinary Differential Equations I. Nonstiff Problems. Springer-Verlag, 2nd ed., 1993.
20. E. Hairer, G. Wanner, Solving Ordinary Differential Equations II. Stiff and Differential-Algebraic Problems. Springer-Verlag, 2nd ed., 1996.

21. J. Hale, Theory of Functional Differential Equations. Springer-Verlag, New York, 1977.
22. J.K. Hale, X.-B. Lin, G. Raugel, Upper semicontinuity of attractors for approximations of semigroups and partial differential equations. Math. Comp. 50 (1988), 89–123.
23. J.K. Hale, G. Raugel, Lower semicontinuity of attractors of gradient systems and applications. Ann. Mat. Pura Appl. 154 (1989), 281–326.
24. D. Henry, Geometric Theory of Semilinear Parabolic Equations. LNM 840, Springer-Verlag, 1981.
25. J.G. Heywood, R. Rannacher, Finite element approximation of the nonstationary Navier-Stokes problem II. Stability of solutions and error estimates uniform in time. SIAM J. Numer. Anal. 23 (1986), 750–777.
26. A.T. Hill, E. Süli, Upper semicontinuity of attractors for linear multistep methods approximating sectorial evolution equations. Math. Comp. 64 (1995), 1097–1122.
27. A.T. Hill, E. Süli, Set convergence for discretizations of the attractor. IMA J. Numer. Anal. 16 (1996), 289–296.
28. M. Hochbruck, Ch. Lubich, H. Selhofer, Exponential integrators for large systems of differential equations. SIAM J. Sci. Comp. 19 (1998), 1552–1574.
29. K. in 't Hout, Ch. Lubich, Periodic orbits of delay differential equations under discretization. BIT 38 (1998), 72–91.
30. D.A. Jones, A.M. Stuart, Attractive invariant manifolds under approximation. Inertial manifolds. J. Diff. Eq. 123 (1995), 588–637.
31. D.A. Jones, A.M. Stuart, E.S. Titi, Persistence of invariant sets for dissipative evolution equations. J. Math. Anal. Appl. 219 (1998), 479–502.
32. U. Kirchgraber, F. Lasagni, K. Nipp, D. Stoffer, On the application of invariant manifold theory, in particular to numerical analysis. Internat. Ser. Numer. Math. 97, Birkhäuser, Basel, 1991, 189–197.
33. Yu.A. Kuznetsov, Elements of Applied Bifurcation Theory. Springer-Verlag, New York, 1995.
34. S. Larsson, The long-time behaviour of finite element approximations of solutions to semilinear parabolic problems. SIAM J. Numer. Anal. 26 (1989), 348–365
35. S. Larsson, Nonsmooth data error estimates with applications to the study of the long-time behavior of finite element solutions of semilinear parabolic problems. Report 1992–36, Dept. of Mathematics, Chalmers Univ. Göteborg, 1992.
(http://www.math.chalmers.se/~stig/papers/index.html)
36. S. Larsson, S. Yu. Pilyugin, Numerical shadowing near the global attractor for a semilinear parabolic equation. Preprint 1998-21, Department of Mathematics, Chalmers University of Technology.
37. S. Larsson, J.M. Sanz-Serna, The behaviour of finite element solutions of semilinear parabolic problems near stationary points. SIAM J. Numer. Anal. 31 (1994), 1000–1018.
38. S. Larsson, J.M. Sanz-Serna, A shadowing result with applications to finite element approximation of reaction-diffusion equations. Math. Comp. 68 (1999), 55-72.
39. G.J. Lord, Attractors and inertial manifolds for finite-difference approximations of the complex Ginzburg-Landau equation. SIAM J. Numer. Anal. 34 (1997), 1483–1512.

40. Ch. Lubich, On the convergence of multistep methods for nonlinear stiff differential equations. Numer. Math. 58 (1991), 839–853.
41. Ch. Lubich, K. Nipp, D. Stoffer, Runge-Kutta solutions of stiff differential equations near stationary points. SIAM J. Numer. Anal. 32 (1995), 1296–1307.
42. Ch. Lubich, A. Ostermann, Runge-Kutta methods for parabolic equations and convolution quadrature. Math. Comp. 60 (1993), 105–131.
43. Ch. Lubich, A. Ostermann, Runge-Kutta time discretization of reaction-diffusion and Navier-Stokes equations: nonsmooth-data error estimates and applications to long-time behaviour. Appl. Numer. Math. 22 (1996), 279–292.
44. Ch. Lubich, A. Ostermann, Hopf bifurcation of reaction-diffusion and Navier-Stokes equations under discretization. Numer. Math. 81 (1998), 53–84.
45. K. Matthies, Time averaging of parabolic partial differential equations: exponential estimates. Doctoral dissertation, FU Berlin, 1999.
46. K. Nipp, Invariant manifolds of singularly perturbed ordinary differential equations. ZAMP 36 (1985), 309–320.
47. K. Nipp, D. Stoffer, Attractive invariant manifolds for maps: Existence, smoothness and continuous dependence on the map. Report 92-11, SAM, ETH Zürich, 1992.
(http://www.sam.math.ethz.ch/Reports/prepr92.html)
48. K. Nipp, D. Stoffer, Invariant manifolds of numerical integration schemes applied to stiff systems of singular perturbation type. I: RK-methods. Numer. Math. 70 (1995), 245–257.
49. K. Nipp, D. Stoffer, Invariant manifolds and global error estimates of numerical integration schemes applied to stiff systems of singular perturbation type. II: Linear multistep methods. Numer. Math. 74 (1996), 305–323
50. R.E. O'Malley, Jr., Singular Perturbation Methods for Ordinary Differential Equations. Springer-Verlag, New York, 1991.
51. A. Ostermann, C. Palencia, Shadowing for nonautonomous parabolic problems with applications to long-time error bounds. Preprint 2–1999, IMG, Univ. Innsbruck. To appear in SIAM J. Numer. Anal.
52. K. Promislow, Time analyticity and Gevrey regularity for solutions of a class of dissipative partial differential equations. Nonl. Anal., Theory, Methods and Applic. 16 (1991), 959–980.
53. J.M. Sanz-Serna, A.M. Stuart, A note on uniform in time error estimates for approximations to reaction-diffusion equations. IMA J. Numer. Anal. 12 (1992), 457–462
54. T. Shardlow, Inertial manifolds and linear multi-step methods. Numer. Algor. 14 (1997), 189–209.
55. A.M. Stuart, Perturbation theory for infinite dimensional dynamical systems. In: M. Ainsworth, J. Levesley, W.A. Light, M. Marletta (eds.), Theory and Numerics of Ordinary and Partial Differential Equations. Advances in Numerical Analysis IV, Clarendon Press, Oxford 1995, 181–290.
56. A.M. Stuart, A.R. Humphries, Dynamical Systems and Numerical Analysis. Cambridge Univ. Press, 1996.
57. R. Temam, Infinite-Dimensional Dynamical Systems in Mechanics and Physics. Springer, New York, 2nd ed., 1997.
58. A. Vanderbauwhede, Centre manifolds, normal forms and elementary bifurcations. In: U. Kirchgraber, H.O. Walther, eds., Dynamics Reported 2, Teubner, Stuttgart, and J. Wiley, Chichester, 1989.

59. A. Vanderbauwhede, G. Iooss, Center manifold theory in infinite dimensions. In: C.K.R.T. Jones, U. Kirchgraber, H.O. Walther, eds., Dynamics Reported 1, New series, Springer-Verlag, Berlin, 1992.
60. S. Wiggins, Introduction to Applied Nonlinear Dynamical Systems and Chaos. Springer-Verlag, New York, 1990.

Evolution of Microstructure: an Example

Felix Otto*

Institut für Angewandte Mathematik, Universität Bonn, Wegelerstr. 10,
53115 Bonn, Germany

Abstract. We consider a continuum model for the flow of two phases of different mobility and density in a Hele–Shaw cell or a porous medium. As a consequence of the Saffman–Taylor instability, the phase distribution is thought to develop a microstructure, so that its evolution is effectively unpredictable.
We identify the constraints on the macroscopic quantities, like the averaged volume fraction of the phases, and show that these constraints allow to derive some predictions on how the macroscopic quantities change over time. Furthermore, we investigate a class of closure hypothesis, which complement these constraints and thereby determine an evolution of the macroscopic quantities themselves, by analyzing the stability of this evolution.
Our analysis uses a combination of tools from nonlinear scalar conservation laws and ideas from the theory of effective moduli.

1 Generation of Microstructure

In this paper, we investigate a simple continuum model [12] for a saturated immiscible two-phase flow in a Hell-Shaw cell or porous medium. The two phases (indexed with h and l) have different mobilities, w.l.o.g.

$$0 < m_h < m_l,$$

and different densities. Let the open set $\Omega \subset (0,\infty) \times \mathbb{R}^N$ denote the distribution of the h-phase in space-time $(0,\infty) \times \mathbb{R}^N$ ($N = 2$ for the Hell-Shaw cell and $N = 3$ for the porous medium, where one should think of Ω as the mesoscopic phase distribution, as averaged over many pores). We introduce the following notation

- $\bar{\Omega}^c$ for the complement of the closure of Ω in $(0,\infty) \times \mathbb{R}^N$, that is, the distribution of the l-phase,
- $\partial\Omega$ for the boundary of Ω, that is, the interface,
- ν for the inner spatial normal to $\partial\Omega$,
- V for the normal velocity of $\partial\Omega$.

The velocity u, an $\mathbb{R}^N$-valued function of $(t,x) \in (0,\infty) \times \mathbb{R}^N$, satisfies the kinematic conditions

* Project: Dynamics of Phase Transitions and Microstructure (Stefan Luckhaus and Stefan Müller)

$$\nabla \cdot u \;=\; 0 \quad \text{in} \quad \Omega \cup \bar{\Omega}^c, \tag{1}$$

$$u \cdot \nu \text{ is continuous across } \partial\Omega \text{ and coincides with } V. \tag{2}$$

Darcy's law states that there exists a pressure p, an $\mathbb{R}$-valued function of $(t,x) \in (0,\infty) \times \mathbb{R}^N$, such that

$$u \;=\; \left\{ \begin{array}{ll} m_h\,(-\nabla p - e) & \text{in } \Omega \\ m_l\,(-\nabla p) & \text{in } \bar{\Omega}^c \end{array} \right\}, \tag{3}$$

$$p \text{ is continuous across } \partial\Omega. \tag{4}$$

Here the given vector $e \in \mathbb{R}^N$ models the effect of density difference.
We observe that (1), (2), (3) & (4) constitute an elliptic problem for p with discontinuous coefficients and right hand side, both depending on the phase distribution. If p or the normal component of u is prescribed on the boundary of the physical domain, we see that at any time t, (1), (2), (3) & (4) determine $p(t)$ – up to an additive constant in space – and thus $u(t)$, given the spatial phase distribution $\Omega(t)$ at that time. Hence (1), (2), (3) & (4) formally define an evolution of the spatial phase distribution. By evolution we mean that we can predict the spatial phase distribution Ω at a later time if we know Ω at an earlier time. As we shall see now, this is only formally the case – effectively, Ω turns out to be unpredictable.
The effective unpredictability of Ω becomes apparent by investigating the stability of a moving planar front. A moving planar front is a solution (Ω_*, u_*, p_*) of (1), (2), (3) & (4) of the form

$$\Omega_* \;=\; \{\nu_* \cdot x > V_*\,t\}, \quad \left\{ \begin{array}{l} c := u_* \cdot \nu_* \\ h := (-\nabla p_*) - (-\nabla p_*) \cdot \nu_*\,\nu_* \end{array} \right\} \;=\; const.,$$

for given $c \in \mathbb{R}$ and $\nu_*,\ h \in \mathbb{R}^N$ with $|\nu_*| = 1$ and $h \cdot \nu_* = 0$. The normal velocity V_* is determined by (2)

$$V_* \;=\; c, \tag{5}$$

and the pressure is determined by (3)

$$-p_*(t,x) \;=\; h \cdot x \;+\; (\nu_* \cdot x - V_*\,t) \left\{ \begin{array}{ll} \frac{c}{m_h} + \rho & \text{for } \nu_* \cdot x \geq V_*\,t \\ \frac{c}{m_l} & \text{for } \nu_* \cdot x \leq V_*\,t \end{array} \right\}$$

where $\rho := e \cdot \nu_*$.
By stability investigation we mean the answer to the question: Do initially nearby solutions (Ω, u, p) of (1), (2), (3) & (4) stay close to (Ω_*, u_*, p_*), or do deviate from (Ω_*, u_*, p_*) over time? Since (1) & (3) imply that $-\nabla^2 p = 0$ in $\Omega \cup \bar{\Omega}^c$, we make the following Ansatz for a perturbation (Ω, u, p):

$$\Omega = \{\nu_* \cdot x \;>\; V_*\,t + A\,\cos\}, \tag{6}$$

$$\begin{aligned} -p(t,x) = h \cdot x \;+\; (\nu_* \cdot x - V_*\,t) &\left\{ \begin{array}{ll} \frac{c}{m_h} + \rho & \text{for } \nu_* \cdot x \;>\; V_*\,t + A\,\cos \\ \frac{c}{m_l} & \text{for } \nu_* \cdot x \;<\; V_*\,t + A\,\cos \end{array} \right\} \\ + A &\left\{ \begin{array}{ll} a_h\,\exp_h\,\cos \;+\; b_h\,\exp_h\,\sin & \text{for } \nu_* \cdot x \;>\; V_*\,t + A\,\cos \\ a_l\,\exp_l\,\cos \;+\; b_l\,\exp_l\,\sin & \text{for } \nu_* \cdot x \;<\; V_*\,t + A\,\cos \end{array} \right\} \end{aligned} \tag{7}$$

where we use the abbreviations

$$\left.\begin{aligned} A &:= A_0 \exp\left(\frac{\mu_\nu t}{l}\right), \\ \cos &:= \cos\left(\frac{\tau\cdot x-\mu_\tau t}{l}\right), \quad \sin := \sin\left(\frac{\tau\cdot x-\mu_\tau t}{l}\right), \\ \exp_h &:= \exp\left(-\frac{\nu_*\cdot x-V_* t}{l}\right), \exp_l := \exp\left(\frac{\nu_*\cdot x-V_* t}{l}\right). \end{aligned}\right\} \qquad (8)$$

We think of the direction τ (with $\tau\cdot\nu_*=0$), the initial amplitude A_0 and the wavelength $2\pi l$ of the perturbation as given. As we shall see, the six other parameters (μ_ν, μ_τ, a_h, a_l, b_h and b_l) are determined by the requirement that (Ω,u,p) (with u defined as in (3)) satisfies (1), (2), (3) & (4) to first order in $A/l << 1$. The parameter of interest is the growth rate μ_ν/l of the amplitude. This linear stability analysis is standard [12,3] and is reproduced here only for the convenience of the reader and since we use a similar calculus in Chapter 4.

The two traces of the function (7) on the interface $\partial\Omega$, one coming from Ω and the other from $\bar\Omega^c$, are given by

$$\begin{aligned} -\tfrac{1}{l}\,p_{|\Omega} &\doteq \tfrac{1}{l}\,h\cdot x+\tfrac{A}{l}\left\{\left(\tfrac{c}{m_h}+\rho\right)\cos+a_h\cos+b_h\sin\right\}, \\ -\tfrac{1}{l}\,p_{|\bar\Omega^c} &\doteq \tfrac{1}{l}\,h\cdot x+\tfrac{A}{l}\left\{\tfrac{c}{m_l}\cos+a_l\cos+b_l\sin\right\}, \end{aligned}$$

where "$\doteq$" denotes equality up to first order in A/l. (4), that is, the continuity of p across $\partial\Omega$ therefore imposes

$$\tfrac{c}{m_h}+\rho+a_h = \tfrac{c}{m_l}+a_l \quad\text{and}\quad b_h = b_l.$$

Since the perturbed normal ν satisfies

$$\nu \doteq \nu_* + \tfrac{A}{l}\,\tau\sin,$$

we have for the two traces of the normal derivative of the function in (7)

$$\begin{aligned} -\nabla p\cdot\nu_{|\Omega} &\doteq \tfrac{c}{m_h}+\rho+\tfrac{A}{l}\left\{h\cdot\tau\sin-a_h\cos-b_h\sin\right\}, \\ -\nabla p\cdot\nu_{|\bar\Omega^c} &\doteq \tfrac{c}{m_l}+\tfrac{A}{l}\left\{h\cdot\tau\sin+a_l\cos+b_l\sin\right\}. \end{aligned}$$

Together with

$$-e\cdot\nu_{|\Omega} \doteq -\rho-\tfrac{A}{l}\,e\cdot\tau\sin,$$

we obtain from (3)

$$\left.\begin{aligned} u\cdot\nu_{|\Omega} &\doteq c+\tfrac{A}{l}\,m_h\left\{h\cdot\tau\sin-a_h\cos-b_h\sin-e\cdot\tau\sin\right\}, \\ u\cdot\nu_{|\bar\Omega^c} &\doteq c+\tfrac{A}{l}\,m_l\left\{h\cdot\tau\sin+a_l\cos+b_l\sin\right\}. \end{aligned}\right\} \qquad (9)$$

Hence (2), that is, the continuity of $u\cdot\nu$ across $\partial\Omega$ imposes

$$-m_h\,a_h = m_l\,a_l \quad\text{and}\quad m_h\,(h\cdot\tau-b_h-e\cdot\tau) = m_l\,(h\cdot\tau+b_l).$$

Thus we are forced to choose

$$\begin{aligned} -m_h\,a_h = m_l\,a_l &= \tfrac{m_l-m_h}{m_l+m_h}\,c \qquad + \tfrac{m_l\,m_h}{m_l+m_h}\,\rho, \\ -b_h \quad = \quad -b_l &= \tfrac{m_l-m_h}{m_l+m_h}\,h\cdot\tau + \tfrac{m_h}{m_l+m_h}\,e\cdot\tau. \end{aligned}$$

Since the normal velocity V of the boundary of (6) is to first order given by

$$V \doteq V_* + \tfrac{A}{l}\,\{\mu_\nu \cos + \mu_\tau \sin\},$$

the kinematic condition $V = u\cdot\nu$ in (2) enforces in view of (9) and (5) that

$$\mu_\nu = m_l\,a_l \quad \text{and} \quad \mu_\tau = m_l\,h\cdot\tau + m_l\,b_l.$$

Hence we obtain

$$\mu_\nu = \frac{1}{\frac{1}{m_h}+\frac{1}{m_l}}\left((\tfrac{1}{m_h}-\tfrac{1}{m_l})\,c+\rho\right),$$

so that it is the sign of

$$\lambda := (\tfrac{1}{m_h}-\tfrac{1}{m_l})\,c+\rho$$

which determines the stability ($\lambda > 0$ means instable).
We see that the more mobile phase displacing the less mobile phase ($c > 0$) and the heavier phase above the lighter phase ($\rho > 0$) creates instabilities. These are the celebrated Saffman-Taylor instabilities [12], which are seen as the origin of the numerically [2] and experimentally [4] observed "fingering" of the phase distribution. We also observe that the growth rate of the amplitude

$$\frac{\mu_\nu}{l} = \frac{1}{\frac{1}{m_h}+\frac{1}{m_l}}\,\frac{\lambda}{l}$$

is inversely proportional to the wave length $2\pi l$. This reflects the scale invariance of (1), (2), (3) & (4) under the hyperbolic scaling

$$t \to \alpha\,t, \qquad x \to \alpha\,x. \tag{10}$$

Hence in the unstable case, small wave length perturbations of the phase distribution grow very fast — which we interpret as effective unpredictability. Only the modelling of an additional physical effect, which breaks the scale invariance (10), can be mathematically regularizing in the sense that it provides a lower cut-off l_- for the band of unstable wave lengths (we think of surface tension or molecular diffusion). An upper cut-off l_+ is for instance provided by the dimensions of the physical flow domain. Unfortunately, $l_- << l_+$ in the applications of interest. Hence we expect the generation and evolution of a fine scale structure of the phase distribution. The goal of this paper is to convince the reader that despite this broad band of active modes, certain aspects of the evolution of the phase distribution can be predicted. Our starting point is the unregularized set of equations (1), (2), (3) & (4), even if the corresponding evolution problem is mathematically ill-posed. It is believed that additional physical effects, like molecular diffusion, dispersion or surface tension, act as a singular perturbation and constrain the evolving microstructure beyond the predictions of this paper. We plan to analyze these further constraints in the future — this will require new methods.

The analysis presented here is an extension of earlier work by the author contained in [9–11]. Analysis from a similar point of view has been performed by Alt [1] and Luckhaus & Plotnikov [6].

2 Containment of the Microstructured Zone

We continue to investigate perturbations (Ω, u, p) of the moving planar front (Ω_*, u_*, p_*) in the unstable case, that is,

$$\lambda = (\tfrac{1}{m_h} - \tfrac{1}{m_l})\, c + \rho > 0,$$

this time in the nonlinear regime. We restrict ourselves to the case where the front normal ν_*, the velocity and the gravity vector are all aligned:

$$u_* = c\,\nu_* \quad \text{and} \quad e = \rho\,\nu_*. \tag{11}$$

The linear stability analysis of Chapter 1 showed that the amplitude $A(t)$ of a perturbation (Ω, u, p) of wave length $2\pi l$ grows like

$$A(t) \approx \exp\left(\frac{1}{\frac{1}{m_h} + \frac{1}{m_l}} \frac{\lambda t}{l} \right) A(0) \quad \text{as long as} \quad A(t) << l.$$

Loosely speaking, we will show in Proposition 1 (see Remark 2 thereafter), that in an average sense, the growth of the amplitude is estimated as follows

$$A(t) \leq A(0) + (m_l + m_h)\, \lambda t \quad \text{for} \quad \lambda t >> l. \tag{12}$$

The interpretation of this estimate is the following: The zone where the phase distribution is characterized by a finger–like microstructure can at most grow linearly in time, with a rate $(m_l + m_h)\,\lambda$. Observe that it is the sum $\frac{1}{m_h} + \frac{1}{m_l}$ of the resistancies which enters in the linear result and the sum of the mobilities $m_l + m_h$ which appears in the nonlinear result.

In Proposition 1, we show that the deviation from (12) for any perturbation is small. In this nonlinear context, by a perturbation (Ω, u, p) of wave length $2\pi l$ we understand a solution of (1), (2), (3) & (4) which is periodic of period $2\pi l$ in the tangential variables and approaches (Ω_*, u_*, p_*) for the normal variable $\to \pm\infty$. To be more precise, it is convenient to describe the phase distribution Ω by the characteristic function

$$\chi := \left\{ \begin{matrix} 1 & \text{in} & \Omega \\ 0 & \text{in} & \bar{\Omega}^c \end{matrix} \right\}$$

and to formulate (1), (2), (3) & (4) in the weak form

$$\partial_t \chi + \nabla \cdot (\chi u) = 0 \quad \text{distributionally,} \tag{13}$$

$$\nabla \cdot u = 0 \quad \text{distributionally,} \tag{14}$$

$$u = (m_h \chi + m_l (1-\chi))(-\nabla p - \chi e) \quad \text{a.e..} \tag{15}$$

By periodicity of period $2\pi l$ and asymptotic behavior like (Ω_*, u_*, p_*) we understand

$$\chi, u, p \quad \text{are periodic with period } 2\pi l \text{ in} \quad y_1 := \tau_1 \cdot x, \ldots, y_{N-1} := \tau_{N-1} \cdot x,$$

$$\int_{Q \cap \{z>0\}} (1-\chi) + \int_{Q \cap \{z<0\}} \chi + \int_Q |u - c\,\nu_*|^2 < \infty \quad \text{where } z := \nu_* \cdot x$$

at any given time, where $\{\tau_1, \ldots, \tau_{N-1}, \nu_*\}$ is an orthonormal basis of $\mathbb{R}^N$ and $Q := (0, 2\pi l)^{N-1} \times \mathbb{R}$ the periodic box.
The main idea is to compare the fraction of phase h at height z and time t, that is,

$$s(t,z) := \frac{1}{(2\pi l)^{N-1}} \int_{(0,2\pi l)^{N-1}} \chi(t,y,z)\, dy \in [0,1],$$

with the (entropy) solution s_* of the following Riemann problem of the scalar conservation law

$$(\partial_t + c\,\partial_z) s_* + \lambda\, \partial_z [f(s_*)] = 0 \quad \text{and} \quad s_*(t=0,z) = \left\{ \begin{matrix} 1 & \text{for} & z > 0 \\ 0 & \text{for} & z < 0 \end{matrix} \right\}.$$

Here the flux function f

$$f(s) = -\frac{m_h s\, m_l (1-s)}{m_h s + m_l (1-s)}$$

is strictly convex. For Riemann problems for a scalar conservation law with strictly convex flux function we refer the reader for instance to [13, Chapter 16]. There is an explicit formula for s_*, which we shall introduce now. It is convenient to pass to rescaled variables

$$\hat{t} = \lambda t \quad \text{and} \quad \hat{z} = z - ct. \tag{16}$$

Since s_* solves

$$\partial_{\hat{t}} s_* + \partial_{\hat{z}} [f(s_*)] = 0 \quad \text{and} \quad s_*(\hat{t}=0,\hat{z}) = \left\{ \begin{matrix} 1 & \text{for} & \hat{z} > 0 \\ 0 & \text{for} & \hat{z} < 0 \end{matrix} \right\}, \tag{17}$$

it is of self-similar form

$$s_* = s_*(\hat\zeta) \quad \text{with} \quad \hat\zeta = \hat z/\hat t,$$

where the profile is implicitly given by

$$f'(s_*(\hat\zeta)) = \left\{ \begin{array}{lll} f'(1) & \text{for} & f'(1) \le \hat\zeta \\ \hat\zeta & \text{for} & f'(0) \le \hat\zeta \le f'(1) \\ f'(0) & \text{for} & \hat\zeta \le f'(0) \end{array} \right\}. \tag{18}$$

A straightforward but tedious calculation shows that s_* is explicitly given by

$$s_*(\hat\zeta) = \left\{ \begin{array}{lll} 1 & \text{for} & m_l \le \hat\zeta \\ \frac{m_h^{-1} - (m_h\, m_l + (m_l - m_h)\,\hat\zeta)^{-1/2}}{m_h^{-1} - m_l^{-1}} & \text{for} & -m_h \le \hat\zeta \le m_l \\ 0 & \text{for} & \hat\zeta \le -m_h \end{array} \right\}. \tag{19}$$

Proposition 1. *Assume that $z_{h,0} \le z_{l,0}$ are so that initially,*

$$\int_{-\infty}^{z_{h,0}} s(0,z)\,dz = 0 \quad \text{and} \quad \int_{z_{l,0}}^{+\infty} (1-s)(0,z)\,dz = 0.$$

Then we have for all t with $\frac{m_h \lambda t}{l} \ge 2$ and all $\hat\zeta \in \mathbb{R}$ the two estimates

$$\frac{1}{(m_l\, m_h)^{1/2}} \left\{ \begin{array}{l} \frac{1}{\lambda t} \int_{-\infty}^{z_{h,0}+ct+\hat\zeta\lambda t} s\,(t,z)\,dz - \int_{-\infty}^{\hat\zeta} s_*\,(\hat\zeta')\,d\hat\zeta' \\ \frac{1}{\lambda t} \int_{z_{l,0}+ct+\hat\zeta\lambda t}^{+\infty} (1-s)(t,z)\,dz - \int_{\hat\zeta}^{+\infty} (1-s_*)(\hat\zeta')\,d\hat\zeta' \end{array} \right\}$$

$$\le C \left(\frac{l}{m_h\, \lambda t} \right)^{1/2} \log \frac{m_h\, \lambda t}{l}, \tag{20}$$

where $C < \infty$ is a universal constant.

Remark 2. In view of (19), the choices $\hat\zeta = -m_h$ and $\hat\zeta = m_l$ in Proposition 1 yield

$$\frac{1}{(m_l\, m_h)^{1/2} \lambda t} \left\{ \int_{-\infty}^{z_{h,0}+ct-m_h\lambda t} s(t,z)\,dz + \int_{z_{l,0}+ct+m_l\lambda t}^{+\infty} (1-s)(t,z)\,dz \right\}$$

$$\le C \left(\frac{l}{m_h\, \lambda t} \right)^{1/2} \log \frac{m_h\, \lambda t}{l}$$

for all t with $\frac{m_h \lambda t}{l} \ge 2$. The left hand side measures the deviation from (12), in the sense that it measures the volume of the part of the phase distribution which violates (12). It measures the normalized volume; normalized by the reference volume of $(0, 2\pi l)^{N-1} \times (-m_h \lambda t, m_l \lambda t)$. The estimate states that this normalized volume decreases as $t^{-1/2}$ as time goes to infinity and as $l^{1/2}$ as the perturbation length goes to zero.

Evoking the underlying symmetry, we restrict ourselves to the proof of the first part of Proposition 1. By translational symmetry, we may assume $z_{h,0} = 0$. We will reformulate the first part of Proposition 1 in terms of the distribution functions of s and s_*, that is

$$V(t,z) \;=\; \int_{-\infty}^{z} s(t,\tilde z)\,d\tilde z \quad \text{resp.} \quad V_*(t,z) \;=\; \int_{-\infty}^{z} s_*(t,\tilde z)\,d\tilde z$$

and the rescaled variables (16). We observe that V_* inherits the self similarity of s_*

$$V_*(\hat t,\hat z) \;=\; \hat t\int_{-\infty}^{\hat\zeta} s_*(\hat\zeta')\,d\hat\zeta \quad \text{where} \quad \hat\zeta \;=\; \hat z/\hat t, \tag{21}$$

so that

$$\frac{1}{\hat t}\left(V(\hat t,\hat z) \;-\; V_*(\hat t,\hat z)\right) \;=\; \frac{1}{\lambda t}\int_{-\infty}^{ct+\hat\zeta\lambda t} s(t,\tilde z)\,d\tilde z \;-\; \int_{-\infty}^{\hat\zeta} s_*(\hat\zeta')\,d\hat\zeta'.$$

On the other hand, we have according to $z_{0,h} = 0$

$$V(0,\hat t) - V_*(0,\hat t) \;=\; V(0,\hat t) - \hat t_+ \;\le\; 0,$$

so that the first part of Proposition 1 is a consequence of

Proposition 3. *There exists a universal constant $C < \infty$, s. t. for all $\hat t$ with $\frac{m_h\,\hat t}{l} \ge 2$*

$$\frac{1}{(m_h\,m_l)^{1/2}\,\hat t}\left\{\sup_{\hat z\in\mathbb{R}}(V(\hat t,\hat z) - V_*(\hat t,\hat z))_+ \;-\; \sup_{\hat z\in\mathbb{R}}(V(0,\hat z) - V_*(0,\hat z))_+\right\}$$
$$\le C\left(\frac{l}{m_h\,\hat t}\right)^{1/2}\log\frac{m_h\,\hat t}{l}.$$

For the proof of Proposition 3 we observe that

$$\begin{aligned} s(t,z) &= \frac{1}{(2\pi l)^{N-1}}\int_{(0,2\pi l)^{N-1}} \chi(t,y,z)\,dy,\\ c &= \frac{1}{(2\pi l)^{N-1}}\int_{(0,2\pi l)^{N-1}} u(t,y,z)\cdot\nu_*\,dy \end{aligned} \tag{22}$$

and notice that

$$j(t,z) := \frac{1}{\lambda}\,\frac{1}{(2\pi l)^{N-1}}\int\{\chi(t,y,z)\,u(t,y,z)\cdot\nu_* - s(t,z)\,c\}\,dy \tag{23}$$

is such that (13) in the rescaled variables projects onto

$$\partial_{\hat t}s + \partial_{\hat z}j \;=\; 0 \quad \text{distributionally.}$$

In view of (17), we have to relate j to $f(s)$ in order to compare s to s_*. As can be seen from (23) j has an interesting interpretation: For the quantities χ and $u \cdot \nu_*$, j measures the deviation of the horizontal average of the product from the product of the horizontal averages. If one is willing to give up complete localization in z, one obtains a fairly sharp estimate of j in terms of $f(s)$.

Lemma 4. *We have* $\int_{-\infty}^{+\infty} |j|\, dz < \infty$ *and if* $\varphi = \varphi(z)$ *is positive and slowly varying on scale L in the sense of*

$$L\left|\frac{d\varphi}{dz}\right| \leq \varphi, \tag{24}$$

then we have for all $t \in (0, \infty)$

$$\left|\int_{-\infty}^{+\infty} j\,\varphi\, dz\right| \leq \frac{1}{1-\frac{l}{L}(\frac{m_l}{m_h})^{1/2}} \int_{-\infty}^{+\infty} (-f(s))\, \varphi\, dz. \tag{25}$$

At the heart of Lemma 4 is the following elliptic estimate

Lemma 5. *Let* $m, \tilde{p}, \tilde{u}$ *and* $\tilde{f}$ *be periodic in* y *with* $\nabla\tilde{p},\ \tilde{u},\ \tilde{f} \in L^2(Q)^N$. *Assume that*

$$\begin{aligned} \nabla \cdot \tilde{u} &= 0 \quad \textit{distributionally}, \\ \tilde{u} &= m\,(-\nabla\tilde{p} - \tilde{f}) \quad \textit{a.e.}, & (26)\\ m &\in [m_h, m_l] \quad \textit{a.e.}. & (27)\end{aligned}$$

Then for any φ *as in Lemma 4 we have*

$$\int_Q m\,|\nabla\tilde{p}|^2\,\varphi + \int_Q \frac{1}{m}\,|\tilde{u}|^2\,\varphi \leq \frac{1}{1-\frac{l}{L}(\frac{m_l}{m_h})^{1/2}} \int_Q m\,|\tilde{f}|^2\,\varphi. \tag{28}$$

We now use the conservation laws

$$\left.\begin{aligned} \partial_{\hat{t}} s \;+ \partial_{\hat{z}} j \qquad &= 0 \\ \partial_{\hat{t}} s_* + \partial_{\hat{z}}[f(s_*)] &= 0 \end{aligned}\right\} \tag{29}$$

and Lemma 4 to compare s and s_* directly.

Lemma 6. *Let* $\varphi = \varphi(\hat{\zeta}), \hat{\zeta} = \hat{z}/\hat{t}$, *be bounded, decreasing with derivative slowly varying on scale* Λ *in the sense of*

$$\Lambda\left|\frac{d^2\varphi}{d\hat{\zeta}^2}\right| \leq -\frac{d\varphi}{d\hat{\zeta}}. \tag{30}$$

Let $\hat{\tau} = \hat{z}(\hat{t})$ *be an adapted reparametrization of time given by*

$$\left\{\begin{aligned} \frac{d\hat{\tau}}{d\hat{t}} &= 1 - \frac{l}{\Lambda\hat{t}}(\frac{m_l}{m_h})^{1/2} \quad \textit{for}\ \hat{t} > \frac{l}{\Lambda}(\frac{m_l}{m_h})^{1/2} \\ \hat{\tau} &= 0 \qquad\qquad\qquad\quad\ \ \textit{for}\ \hat{t} = \frac{l}{\Lambda}(\frac{m_l}{m_h})^{1/2} \end{aligned}\right\}. \tag{31}$$

Then

$$\frac{d}{d\hat{t}} \int_{-\infty}^{+\infty} (s \circ \hat{\tau} - s_*)\,\varphi\, d\hat{z} \leq 0 \qquad \textit{for} \quad \hat{t} > \frac{l}{\Lambda}(\frac{m_l}{m_h})^{1/2}.$$

We now choose φ appropriately to localize the result of Lemma 6

Lemma 7.

$$\sup_{\hat z\in\mathbb{R}}\left(V(\hat\tau,\hat z)-V_*(\hat t,\hat z)\right)_+ \;-\; \sup_{\hat z\in\mathbb{R}}\left(V(0,\hat z)-V_*(\tfrac{l}{\Lambda}(\tfrac{m_l}{m_h})^{1/2},\hat z)\right)_+ \;\le\; \Lambda\,\hat t.$$

We now pass from $V_*(\hat t,\hat z)$ to $V_*(\hat\tau,\hat z)$ using the Lipschitz continuity of V_* in time.

Lemma 8.

$$\sup_{\hat z\in\mathbb{R}}(V-V_*)_+(\hat\tau,\hat z) \;-\; \sup_{\hat z\in\mathbb{R}}(V-V_*)_+(0,\hat z) \;\le\; \frac{1}{4}\,(m_h\,m_l)^{1/2}\,(\hat t-\hat\tau)+\Lambda\,\hat t.$$

We finally optimize the parameter Λ for a given $\hat\tau$.

Lemma 9. . *There exists a universal constant $C<\infty$ s. t. for all $\hat\tau$ with $\frac{m_h\,\hat\tau}{l}\ge 2$,*

$$\sup_{\hat z\in\mathbb{R}}(V-V_*)_+(\hat\tau,\hat z) \;-\; \sup_{\hat z\in\mathbb{R}}(V-V_*)_+(0,\hat z) \;\le\; C\,(m_l\,l\,\hat\tau)^{1/2}\,\log\frac{m_h\,\hat\tau}{l}.$$

3 Coarse Graining of the Equations

In Chapter 1, we saw that although (13), (14) & (15) formally define an evolution of the microscopic spatial phase distribution χ, χ is effectively unpredictable from the knowledge of χ at an earlier time. In Chapter 2, we worked out that certain aspects of macroscopic averages s of χ are predictable from the knowledge of s at an earlier time. In this chapter, we continue to investigate the constraints on the macroscopic average s of χ, but in a more general, albeit qualitative fashion.

We cannot expect to predict s from its knowledge at earlier time, since we encounter the following closure problem: Taking macroscopic averages, denoted by "-", does not commute with nonlinearities. More precisely, although the linear differential operators commute with taking macroscopic averages,

$$\left.\begin{aligned}\partial_t\bar\chi+\nabla\cdot(\overline{\chi u})&=0\\ \nabla\cdot\bar u&=0\end{aligned}\right\}\quad\text{distributionally,}$$

the nonlinear pointwise relations are not preserved

$$\left.\begin{aligned}\overline{\chi u}&\neq\bar\chi\,\bar u\\ \bar u&\neq(m_h\bar\chi+m_l(1-\bar\chi))\,(-\nabla\bar p-\bar\chi\,e)\end{aligned}\right\}\quad\text{a. e. .}\qquad(32)$$

The first defect in (32) has been estimated in Lemma 4 of Chapter 2. In Proposition 11 we point out that the only object required to describe both defects in (32) is the effective mobility M of the microscopic mobility $m_h\chi + m_l(1-\chi)$.

Loosely speaking, the effective mobility M is characterized by the following property: whenever

$$\nabla \cdot v \;=\; 0 \quad \text{and} \quad v \;=\; (m_h\chi + m_l(1-\chi))(-\nabla q),$$

then

$$\nabla \cdot \bar{v} \;=\; 0 \quad \text{and} \quad \bar{v} \;=\; M\,(-\nabla \bar{q}).$$

M is a symmetric $N \times N$-matrix (an element of $\mathrm{Sym}(\mathbb{R}^N)$) with

$$\left(\tfrac{1}{m_h}\bar{\chi} + \tfrac{1}{m_l}(1-\bar{\chi})\right)^{-1}\mathrm{id} \;\leq\; M \;\leq\; (m_h\bar{\chi} + m_l(1-\bar{\chi}))\,\mathrm{id} \tag{33}$$

in the sense of ordering between symmetric matrices. It is well-known that M need not to be isotropic (that is, a multiple of the identity matrix id).

After this motivation, we take a more systematic approach: We embed the space of solutions (χ, u, p) of (13), (14) & (15) into a set of solutions (s, u, p, M) of a new collection of linear differential equations and nonlinear pointwise relations which are stable under the operation of taking macroscopic averages (for s, u and p) and passing to effective moduli (for M). Of course, the smaller this set of solutions (s, u, p, M), the better. We propose to consider the set of all (s, u, p, M), where s is a $[0,1]$-valued, u an $\mathbb{R}^N$-valued, p an $\mathbb{R}$-valued and M a $\mathrm{Sym}(\mathbb{R}^N)$-valued function of (t,x), which satisfy

$$\partial_t s \,+\, \nabla \cdot (G\,u + F\,e) \;=\; 0 \quad \text{distributionally,} \tag{34}$$

$$\nabla \cdot u \;=\; 0 \quad \text{distributionally,} \tag{35}$$

$$u \;=\; M\,(-\nabla p - G\,e) \quad \text{a. e. ,} \tag{36}$$

$$\left(\tfrac{1}{m_h}s + \tfrac{1}{m_l}(1-s)\right)^{-1}\mathrm{id} \;\leq M \leq\; (m_h s + m_l(1-s))\,\mathrm{id} \quad \text{a. e. ,} \tag{37}$$

where the $\mathrm{Sym}(\mathbb{R}^N)$-valued G and F are given by

$$G \;:=\; \frac{M^{-1} - m_l^{-1}\,\mathrm{id}}{m_h^{-1} - m_l^{-1}}, \qquad F \;:=\; \frac{G - s\,\mathrm{id}}{m_h^{-1} - m_l^{-1}}. \tag{38}$$

The mathematically convenient way of expressing stability of (34), (35), (36) & (37) under the operation of taking averages (for s, u and p) and passing to the effective moduli (for M) in a qualitative sense is the following: The set of solutions (s, u, p, M) of (34), (35), (36) & (37) is closed under weak convergence (for s, u and p) and H-convergence (for M). For the convenience of the reader, we recall the definition of H-convergence here — in particular since we need a slightly modified notion due to the time dependence of the fields. For the unmodified notion, we refer the reader to Murat & Tartar [8] and [14].

Definition 10. A sequence $\{M_n\}_{n\uparrow\infty}$ of $\mathrm{Sym}(\mathbb{R}^N)$–valued functions of (t,x) is said to H-converge to a $\mathrm{Sym}(\mathbb{R}^N)$-valued M if the following holds: For any $\mathbb{R}^N$–valued v_n, v, and $\mathbb{R}$–valued q_n, q with

$$\begin{aligned} v_n &\rightharpoonup v \quad \text{weakly in } L^2, \qquad \nabla q_n \rightharpoonup \nabla q \quad \text{weakly in } L^2, \\ \nabla\cdot v_n &= \nabla\cdot v \quad \text{distributionally,} \end{aligned}$$

we have

$$v_n = M_n\,(-\nabla q_n) \quad \text{a.e.} \quad \Longrightarrow \quad v = M\,(-\nabla q) \quad \text{a.e..}$$

Proposition 11. *i) (34), (35), (36) & (37) turns into (13), (14) & (15) for* $s=\chi\in\{0,1\}$ *since*

$$\left.\begin{aligned} \frac{m_h\,s}{m_h\,s+m_l\,(1-s)}\,\mathrm{id} \le G \le s\,\mathrm{id} \\ -\frac{m_h\,s\;m_l\,(1-s)}{m_h\,s+m_l\,(1-s)}\,\mathrm{id} \le F \le 0 \end{aligned}\right\} \quad a.e.\ . \tag{39}$$

ii) (34), (35), (36) & (37) is closed under the operation of taking macroscopic averages (for s,u,p) and passing to effective moduli (for M) in the following sense: If the elements of the sequence $\{(s_n,u_n,p_n,M_n)\}_{n\uparrow\infty}$ satisfy (34), (35), (36) & (37), and the sequence converges to (s,u,p,M) in the sense of

$$\left.\begin{aligned} s_n &\rightharpoonup s && \textit{weak–}\ast\textit{ convergence in } L^\infty, \\ (u_n,\nabla p_n) &\rightharpoonup (u,\nabla p) && \textit{weak convergence in } L^2, \\ M_n &\to M && H-\textit{convergence in the sense of Definition 10} \end{aligned}\right\} \tag{40}$$

then (s,u,p,M) satisfies (34), (35), (36) & (37).

Remark 12. Besides mimicking the operations of taking the macroscopic average and passing to the effective modulus, the convergences (40) are also natural in the following sense: The set of solutions (s,u,p,M) of (34), (35), (36) & (37) is not only closed, but also "almost" compact in this topology. Indeed, according to $s_n\in[0,1]$ a.e., $\{s_n\}_{n\uparrow\infty}$ is bounded in L^∞. Controlled boundary conditions on the pressure automatically yield boundedness of $\{u_n\}_{n\uparrow\infty}$ and $\{\nabla p_n\}_{n\uparrow\infty}$ in L^2, according to the elliptic equation (35) & (36) with controlled coefficients (37). If it weren't for the time-dependence of M_n, the control (37) would ensure the compactness of $\{M_n\}_{n\uparrow\infty}$ w.r.t. H-convergence [8, Theorem 2].

PROOF OF PROPOSITION 11. The upper bounds $G\le s\,\mathrm{id}$ and $F\le 0$ are immediate from (37) in the form of

$$M^{-1} \le \left(\tfrac{1}{m_h}s+\tfrac{1}{m_l}(1-s)\right)\,\mathrm{id}$$

and the definitions (38) if G and F. The lower bounds in (39) follow from (37) in the form of

$$M^{-1} \ge \tfrac{1}{m_h s+m_l(1-s)}\,\mathrm{id}$$

and the identities

$$\frac{1}{\frac{1}{m_h}-\frac{1}{m_l}}\left(\frac{1}{m_h s+m_l(1-s)}-\frac{1}{m_l}\right) = \frac{m_h s}{m_h s+m_l(1-s)},$$
$$\frac{1}{\frac{1}{m_h}-\frac{1}{m_l}}\left(\frac{m_h s}{m_h s+m_l(1-s)}-s\right) = -\frac{m_h s\, m_l(1-s)}{m_h s+m_l(1-s)}.$$

We now address part ii). The fact that (34), (35), (36) are preserved in the limit follows from two algebraic observations: (36) is preserved since it can be rewritten as

$$u+\frac{1}{\frac{1}{m_h}-\frac{1}{m_l}}\,e \;=\; M\left(-\nabla p+\frac{1}{m_l}\,\frac{1}{\frac{1}{m_h}-\frac{1}{m_l}}\,e\right), \tag{41}$$

which is an immediate consequence of the definition of G in (38). By definition of H-convergence, (41) is stable, since $u+\frac{1}{\frac{1}{m_h}-\frac{1}{m_l}}\,e$ is divergence–free and $\nabla p-\frac{1}{m_l}\,\frac{1}{\frac{1}{m_h}-\frac{1}{m_l}}\,e$ is a gradient. (34) is preserved since the flux can be rewritten as

$$\begin{aligned}
&G\,u+F\,e\\
&\overset{(38)}{=} G\,(u+\frac{1}{\frac{1}{m_h}-\frac{1}{m_l}}\,e) \;-\; \frac{1}{\frac{1}{m_h}-\frac{1}{m_l}}\,s\,e\\
&\overset{(38)}{=} \frac{1}{\frac{1}{m_h}-\frac{1}{m_l}}\,M^{-1}\,(u+\frac{1}{\frac{1}{m_h}-\frac{1}{m_l}}\,e)-\frac{1}{m_l}\,\frac{1}{\frac{1}{m_h}-\frac{1}{m_l}}\,(u+\frac{1}{\frac{1}{m_h}-\frac{1}{m_l}}\,e)-\frac{1}{\frac{1}{m_h}-\frac{1}{m_l}}\,s\,e\\
&\overset{(41)}{=} \frac{1}{\frac{1}{m_h}-\frac{1}{m_l}}\,(-\nabla p+\frac{1}{m_l}\,\frac{1}{\frac{1}{m_h}-\frac{1}{m_l}}\,e)-\frac{1}{m_l}\,\frac{1}{\frac{1}{m_h}-\frac{1}{m_l}}\,(u+\frac{1}{\frac{1}{m_h}-\frac{1}{m_l}}\,e)-\frac{1}{\frac{1}{m_h}-\frac{1}{m_l}}\,s\,e,
\end{aligned}$$

which is a linear expression in (s,u,p).

The argument that the inequalities (37) are preserved under weak convergence in s and H-convergence in M is only a slight modification of the one given in [14, Theorem 4 & 5].

4 A Closure for the Coarse Grained Equations

We return to the closure problem: (34), (35), (36) & (37) defines an evolution for s only if one adds a relation which allows to recover M from s, locally in time. In this chapter and the next, we will investigate the simplest form of this closure, namely

$$M \;=\; m(s)\,\mathrm{id}, \tag{42}$$

where m is a function of $s\in[0,1]$ which in view of (37) necessarily satisfies

$$\left(\tfrac{1}{m_h}\,s+\tfrac{1}{m_l}\,(1-s)\right)^{-1} \;\le\; m(s) \;\le\; m_h\,s+m_l\,(1-s) \quad \text{for all } s\in[0,1].$$

Then (34), (35) & (36) turn into

$$\partial_t s \;+\; \nabla\cdot(g(s)\,u+f(s)\,e)=0 \quad \text{distributionally}, \tag{43}$$
$$\nabla\cdot u=0 \quad \text{distributionally}, \tag{44}$$
$$u=m(s)\,(-\nabla p-g(s)\,e) \quad \text{a.e.}, \tag{45}$$

where

$$g(s) \;=\; \frac{(\frac{1}{m})(s)-\frac{1}{m_l}}{\frac{1}{m_h}-\frac{1}{m_l}} \quad \text{and} \quad f(s) \;=\; \frac{g(s)-s}{\frac{1}{m_h}-\frac{1}{m_l}}. \tag{46}$$

One important remark w.r.t. assumption (42) is in order: It would seem that (42) is based on the implicit assumption that the microscopic mobility $m_h\,\chi+m_l\,(1-\chi)$ is such that the effective mobility M is isotropic. In fact, (42) only implies that the macroscopic quantity $(\frac{1}{m_h}-\frac{1}{m_l})\,u+e$ is an eigenvector of the effective mobility M with eigenvalue depending only on the macroscopic variable s, as can be seen from (41) in the proof of Proposition 11.

In this chapter, we derive an important consistency property for the class of models (42): In Proposition 15, we show that longitudinal and transversal stability of a moving planar front coincide for this class of models, provided $\frac{1}{m}$ is strictly convex.

By a moving planar front we understand a solution (s_*, u_*, p_*) of (43), (44) & (45) of the form

$$\left.\begin{array}{c} s_* \;=\; \left\{\begin{array}{ll} s_h & \text{for } x\cdot\nu_* > V_*\,t \\ s_l & \text{for } x\cdot\nu_* < V_*\,t \end{array}\right\} \\ \left\{\begin{array}{lcl} c &:=& u_*\cdot\nu_* \\ h &:=& (-\nabla p_*)-(-\nabla p_*)\cdot\nu_*\,\nu_* \end{array}\right\} \;=\; const., \end{array}\right\} \tag{47}$$

where $0\le s_l<s_h\le 1$, $c\in\mathbb{R}$ and ν_*, $h\in\mathbb{R}^N$ with $|\nu_*|=1$ and $\nu_*\cdot h=0$ are given. The normal velocity V_* ist then determined by (43):

$$V_* \;=\; \frac{(c\,g+\rho\,f)(s_h)-(c\,g+\rho\,f)(s_l)}{s_h-s_l}, \tag{48}$$

where $\rho=e\cdot\nu_*$. The pressure p_* is determined by (44) & (45):

$$-p_*(t,x) \;=\; h\cdot x \;+\; (\nu_*\cdot x-V_*\,t)\left\{\begin{array}{ll} \frac{c}{m(s_h)}+\rho\,g(s_h) & \text{for } x\cdot\nu_*>V_*\,t \\ \frac{c}{m(s_l)}+\rho\,g(s_l) & \text{for } x\cdot\nu_*<V_*\,t \end{array}\right\}.$$

We claim that longitudinal and transversal stability of s_* coincide.

By stability we mean that initial perturbations do not grow over time. By longitudinal stability, we mean that (s_*, u_*, p_*) is stable under all one-dimensional perturbations (s,u,p), that is, solutions of (43), (44) & (45) of the form

$$s \;=\; s(t,x\cdot\nu_*), \quad u\cdot\nu_* \;=\; c \quad \text{and} \quad -\nabla p-(-\nabla p)\cdot\nu_*\,\nu_* \;=\; h.$$

Then s is a weak solution of the scalar conservation law

$$\partial_t s \;+\; \partial_z(c\,g(s)+\rho\,f(s)) \;=\; 0, \tag{49}$$

It is well-known that the special weak solution s_* of (49), a "shock", is stable under the formation of "rarefaction waves", also solutions s of (49), if and only if the following condition is fulfilled

$$(c\,g+\rho\,f)(s) \;\ge\; (c\,g+\rho\,f)(s_l)+V_*\,(s-s_l) \quad \text{for all } s\in[s_l,s_h], \tag{50}$$

see for instance [7, Chapter 2, Example (3.11) and (3.13)] and [13, Chapter 15, Example 6] for this admissibility criterion for shocks. In view of (48) and (50) we say

Definition 13. (s_*, u_*, p_*) is longitudinally stable if the following "chord"–condition is fulfilled

$$(c\,g+\rho\,f)(s) \;\geq\; \tfrac{s-s_l}{s_h-s_l}\,(c\,g+\rho\,f)(s_h) + \tfrac{s_h-s}{s_h-s_l}\,(c\,g+\rho\,f)(s_l) \quad \text{for all } s \in [s_l, s_h]. \tag{51}$$

By transversal stability we mean that (s_*, u_*, p_*) is stable under all perturbations (s, u, p) with $s \in \{s_l, s_h\}$. In fact we mean linear stability as in Chapter 1. As there, we make the following Ansatz for the perturbation (s, u, p):

$$\begin{aligned}
s(t,x) &= \left\{\begin{array}{ll} s_h & \text{for } \nu_*\cdot x > V_*\,t + A\,\cos \\ s_l & \text{for } \nu_*\cdot x < V_*\,t + A\,\cos, \end{array}\right\} \\
-p(t,x) &= h\cdot x \\
&\quad + (\nu_*\cdot x - V_*\,t)\left\{\begin{array}{lll} \frac{c}{m(s_h)} + \rho\,g(s_h) & \text{for} & \nu_*\cdot x \;>\; V_*\,t + A\,\cos \\ \frac{c}{m(s_l)} + \rho\,g(s_l) & \text{for} & \nu_*\cdot x \;<\; V_*\,t + A\,\cos \end{array}\right\} \\
&\quad + A\left\{\begin{array}{lll} a_h\,\exp_h\,\cos + b_h\,\exp_h\,\sin & \text{for} & \nu_*\cdot x \;>\; V_*\,t + A\,\cos \\ a_l\,\exp_l\,\cos + b_l\,\exp_l\,\sin & \text{for} & \nu_*\cdot x \;<\; V_*\,t + A\,\cos, \end{array}\right\},
\end{aligned}$$

where we use the same abbrevations $A, \cos, \sin, \exp_h, \exp_l$ as in Chapter 1, see (8). As in Chapter 1, we consider the direction τ (with $\tau\cdot\nu_* = 0$), the initial amplitude A_0 and the wavelength $2\pi l$ of the perturbation as given and determine the remaining six parameters μ_ν, μ_τ, a_h, a_l, b_h, b_l such that (s, u, p) (with u defined to satisfy (45)) satisfies (43), (44) & (45) to first order in $A/l << 1$. As in Chapter 1, we see that continuity of p and $\nabla u\cdot\nu$ across the interface $\{\nu_*\cdot x = V_*\,t + A\,\cos\}$ imposes

$$\left.\begin{array}{rclcl}
-m(s_h)\,a_h &=& m(s_l)\,a_l &=& \frac{m(s_l)-m(s_h)}{m(s_l)+m(s_h)}\,c \;+\; \frac{m(s_h)\,m(s_l)\,(g(s_h)-g(s_l))}{m(s_l)+m(s_h)}\,\rho, \\
-b_h &=& -b_l &=& \frac{m(s_l)-m(s_h)}{m(s_l)+m(s_h)}\,h\cdot\tau \;+\; \frac{m(s_h)\,g(s_h)-m(s_l)\,g(s_l)}{m(s_l)+m(s_h)}\,e\cdot\tau.
\end{array}\right\} \tag{52}$$

The normal velocity of the interface $\{\nu_*\cdot x = V_*\,t + A\,\cos\}$ is

$$V \;\doteq\; V_* \;+\; \frac{A}{l}\,\{\mu_\nu\cos + \mu_\tau\sin\}$$

and should according to (43) be given by

$$\begin{aligned}
V \;&=\; \frac{(g(s_h)\,u\cdot\nu + f(s_h)\,e\cdot\nu) - (g(s_l)\,u\cdot\nu + f(s_l)\,e\cdot\nu)}{s_h - s_l} \\
&\overset{(48)}{=}\; V_* \;+\; \frac{g(s_h)-g(s_l)}{s_h-s_l}\,(u\cdot\nu - c) \;+\; \frac{f(s_h)-f(s_l)}{s_h-s_l}\,(e\cdot\nu - \rho).
\end{aligned}$$

This enforces

$$\mu_\nu \;=\; \frac{g(s_h)-g(s_l)}{s_h-s_l}\, m(s_l)\, a_l, \tag{53}$$

and a more complicated formula for μ_τ. In view of (52) and (53) we say

Definition 14. (s_*,u_*,p_*) is transversally stable if

$$\frac{1}{(\frac{1}{m})(s_h)+(\frac{1}{m})(s_l)}\,\frac{g(s_h)-g(s_l)}{s_h-s_l}\,\left((c\,\tfrac{1}{m}+\rho\, g)(s_h)-(c\,\tfrac{1}{m}+\rho\, g)(s_l)\right)\;\le\;0. \tag{54}$$

Observe that Definition 13 involves g and f, but not m, whereas Definition 14 involves m and g, but not f. The relations (46) are just such that we have

Proposition 15. *Provided $\frac{1}{m}$ is a strictly convex function of s, longitudinal stability is equivalent to transversal stability in the sense that (51) and (54) are equivalent to the same condition*

$$\lambda \;=\; (\tfrac{1}{m_h}-\tfrac{1}{m_l})\,c+\rho\;\le\;0. \tag{55}$$

Remark 16. . It is well–known (see for instance [7, Chapter 2, (6.18), (6.19) and Remark 5.7]), that the weak formulation together with the shock admissibility criterion (50) for the scalar conservation law (49) are equivalent to its "entropy formulation":

$$\partial_t\eta(s)\;+\;\partial_z(c\,g_\eta(s)+\rho\, f_\eta(s))\;\le\;0\quad\text{distributionally} \tag{56}$$

for all convex functions η ("entropies") and g_η, f_η ("entropy fluxes") related by

$$g_\eta' \;=\; \eta'\,g' \qquad\text{and}\qquad f_\eta' \;=\; \eta'\,f'. \tag{57}$$

Kruzkov [5] has shown that the entropy formulation renders the initial value problem of a scalar conservation law well–posed. In view of this and Proposition 15, it is natural to pass from (43), (44) & (45) to

$$\partial_t\eta(s)\;+\;\nabla\cdot(g_\eta(s)\,u+f_\eta(s)\,e)\le 0\quad\text{distributionally}, \tag{58}$$

$$\nabla\cdot u=0\quad\text{distributionally}, \tag{59}$$

$$u=m(s)\,(-\nabla p-g(s)\,e)\quad\text{a. e. }, \tag{60}$$

for all convex η and g_η, f_η with (57). We stress that this entropy condition is not motivated by "viscosity" (which would have no physical meaning in our coarse grained setting), but by stability considerations!

Proof of Proposition 15. We just use the definitions (46) to show that (51) and (54) are equivalent to (55). Indeed, we infer from the definition of g (46) that μ_ν can be rewritten as

$$\mu_\nu \;=\; \frac{1}{(\frac{1}{m})(s_h)+(\frac{1}{m})(s_l)}\left(\frac{(\frac{1}{m})(s_h)-(\frac{1}{m})(s_l)}{\frac{1}{m_h}-\frac{1}{m_l}}\right)^2\frac{1}{s_h-s_l}\left((\tfrac{1}{m_h}-\tfrac{1}{m_l})\,c+\rho\right),$$

so that, since $s_h > s_l$, we have $\mu_\nu \leq 0$ if and only if (55) holds. On the other hand, (50) can be written as

$$h(s) \geq \tfrac{s-s_l}{s_h-s_l}\, h(s_h) + \tfrac{s_h-s}{s_h-s_l}\, h(s_l) \quad \text{for all } s \in [s_l, s_h], \tag{61}$$

where according to the definition of g and f in (46)

$$\begin{aligned} h(s) &:= c\, g(s) + \rho\, f(s) \\ &= \frac{1}{(\frac{1}{m_h} - \frac{1}{m_l})^2} \left((\tfrac{1}{m_h} - \tfrac{1}{m_l})\, c + \rho \right) (\frac{1}{m})(s) \; + \; \text{linear term in } s. \end{aligned}$$

Since by assumption, $\frac{1}{m}$ is strictly convex and $s_h > s_l$, (61) holds if and only (55) is true.

5 Stability of the Coarse Grained Model

In this chapter, we would like to convince the reader that within the class of closures considered in Chapter 4, the model

$$m(s) \;=\; m_h\, s + m_l\, (1-s) \tag{62}$$

enjoys good stability properties. We observe that in this model f and g have the form already familiar from Chapter 2

$$\left.\begin{aligned} g(s) &= \frac{(\frac{1}{m})(s) - \frac{1}{m_l}}{\frac{1}{m_h} - \frac{1}{m_l}} &&= \frac{m_h\, s}{m_h\, s + m_l\, (1-s)} \\ f(s) &= \frac{g(s) - s}{\frac{1}{m_h} - \frac{1}{m_l}} &&= \frac{m_h\, s\, m_l\, (1-s)}{m_h\, s + m_l\, (1-s)} \end{aligned}\right\}. \tag{63}$$

An important remark with respect to the assumption (62) is in order: As for (42), (62) implies that the macroscopic quantity $(\frac{1}{m_h} - \frac{1}{m_l})\, u + e$ is an eigenvalue of the effective mobility M with eigenvalue given this time given by the average $m_h s + m_l(1-s)$ of the microscopic mobility. Loosely speaking, this is equivalent to the assumption that the microscopic mobility $m_h\chi + m_l(1-\chi)$ can only oscillate perpendicular to $(\frac{1}{m_h} - \frac{1}{m_l})\, u + e$. Hence the choice of the specific closure (62) amounts to the assumption that the microscopic phase distribution organizes itself such that it is mostly aligned with $(\frac{1}{m_h} - \frac{1}{m_l})\, u + e$. As shown in Chapter 1, the only layering of the microscopic phase distribution which is not (linearly) unstable is indeed a layering parallel to $(\frac{1}{m_h} - \frac{1}{m_l})\, u + e$. Hence the assumption is at least not unplausible.

Our argument in favor of (62) is that all one–dimensional solutions of (58), (59) & (60) are stable. We recall that by a one–dimensional solution we understand a solution (s_*, u_*, p_*) of the form

$$s_* \;=\; s_*(t, \nu_* \cdot x) \quad \text{and} \quad \left\{ \begin{aligned} c &:= u_* \cdot \nu_* \\ h &:= (-\nabla p_*) - (-\nabla p_*) \cdot \nu_*\, \nu_* \end{aligned} \right\} \;=\; const.$$

for some $c \in \mathbb{R}$, $\nu_*, h \in \mathbb{R}^N$ with $|\nu_*| = 1$ and $h \cdot \nu_* = 0$. Then (58) forces s_* to be an entropy solution of the scalar conservation law

$$\partial_t s_* + \partial_z[h(s_*)] \;=\; 0, \tag{64}$$

where $z := \nu_* \cdot x$ and the flux function h is given by

$$\begin{aligned} h(s) &:= (g(s)\,u_* + f(s)\,e)\cdot\nu_* \\ &= c\,g(s) + \rho\,f(s) \\ &\overset{(63)}{=} \left\{ \begin{array}{l} c\,s + \lambda\,f(s) \\ \frac{\lambda}{(\frac{1}{m_l}-\frac{1}{m_h})^2}\left((\frac{1}{m})(s) - \frac{1}{m_l}\right) - \frac{\rho}{\frac{1}{m_h}-\frac{1}{m_l}}\,s \end{array} \right\} \end{aligned} \tag{65}$$

with $\rho = \nu_* \cdot e$. Furthermore, (59) & (60) force p_* to be of the form

$$-p_* \;=\; h\cdot x + \int_0^z [c\,(\tfrac{1}{m})(s_*) + \rho\,g(s_*)]. \tag{66}$$

We restrict ourselves to initial data of the form

$$s_*(t=0,z) \;=\; \left\{ \begin{array}{ll} s_h & \text{for } z>0 \\ s_l & \text{for } z<0 \end{array} \right\} \quad \text{with} \quad 0 \;\le\; s_l \;<\; s_h \;\le\; 1.$$

Then s_* is uniquely determined and must be of selfsimilar form

$$s_* \;=\; s_*(\zeta) \quad \text{where} \quad \zeta \;=\; z/t.$$

In case of

$$\lambda \;=\; (\tfrac{1}{m_l} - \tfrac{1}{m_h})\,c + \rho \;\le\; 0,$$

the entropy solution of (64) is a shock

$$s_* \;=\; \left\{ \begin{array}{ll} s_h & \text{for } \zeta>0 \\ s_l & \text{for } \zeta<0 \end{array} \right\}$$

and we have already established in Proposition 15 that this (s_*, u_*, p_*) is nonlinearly stable under longitudinal perturbations and linearly stable under transversal perturbations (observe that $\frac{1}{m}$ is indeed strictly convex for our model (62)). In case of

$$\lambda \;=\; (\tfrac{1}{m_l} - \tfrac{1}{m_h})\,c + \rho \;>\; 0,$$

the entropy solution of (64) is a rarefaction wave implicitly given by

$$h'(s_*(\zeta)) \;=\; \left\{ \begin{array}{lll} h'(s_h) & \text{for} & h'(s_h) \;\le\; \zeta \\ \zeta & \text{for} & h'(s_l) \;\le\; \zeta \le h'(s_h) \\ h'(s_l) & \text{for} & \zeta \;\le\; h'(s_l) \end{array} \right\} \tag{67}$$

and we will show in Proposition 17 that this (s_*, u_*, p_*) is nonlinearly stable under all perturbations. We restrict ourselves to the geometry of Chapter

2, that is the alignment of velocity, gravity and front normal (11) and to $s_h = 1, s_l = 0$. As in Chapter 2, perturbations (s, u, p) of (s_*, u_*, p_*) admitted in Proposition 17 are solutions of (58), (59) & (60), which are periodic in the tangential variables and approach (s_*, u_*, p_*) for normal variable $\to \pm\infty$. More precisely, ν_* can be extended to an orthonormal basis $\{\tau_1, \ldots, \tau_{N-1}, \nu_*\}$ such that

$$s, u, p \quad \text{are periodic with period } 2\pi l \text{ in} \quad y_i \;:=\; \tau_i \cdot x, \tag{68}$$

$$\int_Q |s - s_*|^2 + \int_Q |u - u_*|^2 \;<\; \infty, \tag{69}$$

for some $l > 0$ (which turns out to be irrelevant for our result), where $Q := (0, 2\pi l)^{N-1} \times \mathbb{R}$ is the periodic box.

Proposition 17.

$$\frac{d}{dt}\int_Q \left\{\lambda\, m(s)\left|\frac{(\frac{1}{m})(s) - (\frac{1}{m})(s_*)}{\frac{1}{m_h} - \frac{1}{m_l}}\right|^2 \right.$$
$$\left. + \,(\zeta - h'(1))_+\,(1 - s) \;+\; (h'(0) - \zeta)_+\, s\right\} \;\le\; 0.$$

Remark 18. Proposition 17 identifies an adapted distance between s and s_* which decreases over time. This distance dominates the $L^2(Q)$-distance between s and s_*. The distance does not depend on the period $2\pi l$.

PROOF OF PROPOSITION 17. We fix a ζ_* and use the special Kruzkov entropy / entropy flux pairs, that is

$$\eta(s) \;=\; (s - s_*(\zeta_*))_+, \qquad \left\{\begin{array}{l} g_\eta(s) = (g(s) - g(s_*(\zeta_*)))_+, \\ f_\eta(s) = (f(s) - f(s_*(\zeta_*)))_+, \end{array}\right\}$$

with the parameter choosen to be $s_*(\zeta_*)$; see for instance [7, Chapter 2, Remark 5.7]. The entropy inequality (58) turns into

$$\partial_t(s - s_*(\zeta_*))_+ + \nabla\cdot[(g(s) - g(s_*(\zeta_*)))_+\, u + (f(s) - f(s_*(\zeta_*)))_+\, e] \;\le\; 0 \tag{70}$$

in the distributional sense. According to the periodicity (68) and the behavior at infinity (69) we obtain from (70) by integrating over $\{z < \zeta_*\, t\} \cap Q$

$$\frac{d}{dt}\int_{\{z<\zeta_* t\}\cap Q} (s - s_*(\zeta_*))_+\, d\mathcal{L}^N$$
$$= \zeta_* \int_{\{z=\zeta_* t\}\cap Q} (s - s_*(\zeta_*))_+\, d\mathcal{H}^{N-1} \;+\; \int_{\{z<\zeta_* t\}\cap Q} \partial_t(s - s_*(\zeta_*))_+\, d\mathcal{L}^N$$
$$\le \int_{\{z=\zeta_* t\}\cap Q} \Big\{\zeta_*\,(s - s_*(\zeta_*))_+ \;-\; (g(s) - g(s_*(\zeta_*)))_+\, u\cdot\nu_*$$
$$- \,(f(s) - f(s_*(\zeta_*)))_+\, \rho\Big\} d\mathcal{H}^{N-1}.$$

Likewise, we have

$$\begin{aligned}&\frac{d}{dt}\int_{\{z>\zeta_* t\}\cap Q}(s_*(\zeta_*)-s)_+\,d\mathcal{L}^N\\&\le\int_{\{z=\zeta_* t\}\cap Q}\Big\{-\zeta_*\,(s_*(\zeta_*)-s)_+\;+\;(g(s_*(\zeta_*))-g(s))_+\,u\cdot\nu_*\\&\qquad+\;(f(s_*(\zeta_*))-f(s))_+\,\rho\Big\}d\mathcal{H}^{N-1}.\end{aligned}$$

We add both

$$\begin{aligned}&\frac{d}{dt}\Big[\int_{\{z<\zeta_* t\}\cap Q}(s-s_*(\zeta_*))_+\,d\mathcal{L}^N+\int_{\{z>\zeta_* t\}\cap Q}(s_*(\zeta_*)-s)_+\,d\mathcal{L}^N\Big]\\&\le-\int_{\{z=\zeta_* t\}\cap Q}\Big\{-\zeta_*(s-s_*(\zeta_*))\;+\;(g(s)-g(s_*(\zeta_*)))\,u\cdot\nu_*\\&\qquad+\;(f(s)-f(s_*(\zeta_*)))\,\rho\Big\}d\mathcal{H}^{N-1}\end{aligned}$$

and integrate over $\zeta_*\in\mathbb{R}$

$$\begin{aligned}&\frac{d}{dt}\int_Q\left\{\int_\zeta^{+\infty}(s-s_*(\zeta_*))_+\,d\zeta_*+\int_{-\infty}^{\zeta}(s_*(\zeta_*)-s)_+\,d\zeta_*\right\}d\mathcal{L}^N\\&\le-\frac{1}{t}\int_Q\{-\zeta\,(s-s_*)\;+\;(g(s)-g(s_*))\,u\cdot\nu_*\;+\;(f(s)-f(s_*))\,\rho\}\,d\mathcal{L}^N,\end{aligned}$$

where $\zeta=z/t$. The result now follows from the following three lemmas.

Lemma 19. *For $\zeta\in\mathbb{R}$ and $s\in[0,1]$ we have*

$$\begin{aligned}&\int_\zeta^{+\infty}(s-s_*(\zeta_*))_+\,d\zeta_*+\int_{-\infty}^{\zeta}(s_*(\zeta_*)-s)_+\,d\zeta_*\\&=\lambda\,m(s)\left|\frac{(\frac{1}{m})(s)-(\frac{1}{m})(s_*(\zeta))}{\frac{1}{m_h}-\frac{1}{m_l}}\right|^2+(\zeta-h'(1))_+\,(1-s)+(h'(0)-\zeta)_+\,s.\end{aligned}$$

Lemma 20. *For $\zeta\in\mathbb{R}$, $s\in[0,1]$ and $u\in\mathbb{R}^N$ we have*

$$\begin{aligned}&-\zeta\,(s-s_*(\zeta))\;+\;(g(s)-g(s_*(\zeta)))\,u\cdot\nu_*\;+\;(f(s)-f(s_*(\zeta)))\,\rho\\&\qquad\ge\frac{1}{\lambda}\left(m(s)\,|\tilde f|^2\;+\;\tilde f\cdot\tilde u\right),\end{aligned}$$

where

$$\tilde u\;:=\;u-c\,\nu_*\quad\textit{and}\quad\tilde f\;:=\;\lambda\,\frac{(\frac{1}{m})(s)-(\frac{1}{m})(s_*(\zeta))}{\frac{1}{m_h}-\frac{1}{m_l}}\,\nu_*.\qquad(71)$$

Lemma 21. *(59) and (60) imply*

$$\int_Q \left\{ m(s)|\tilde{f}|^2 + \tilde{f} \cdot \tilde{u} \right\} \geq 0, \tag{72}$$

where $\tilde{u}$ and $\tilde{f}$ are defined like in (71).

The only place where (62) enters is through the following identity, which is needed in Lemma 19 and 20. It quantifies the strict convexity of $\frac{1}{m}$.

Lemma 22. *For $s, s_* \in [0,1]$ we have*

$$(\tfrac{1}{m})(s) - (\tfrac{1}{m})(s_*) - (\tfrac{1}{m})'(s_*)\,(s - s_*) \;=\; m(s)\,\left((\tfrac{1}{m})(s) - (\tfrac{1}{m})(s_*)\right)^2. \tag{73}$$

Acknowledgment

The author started working on this subject as a post–doc under Stephan Luckhaus, instigated by him. At that time, the research was funded through the Schwerpunktprogramm DANSE of the Deutsche Forschungsgemeinschaft. The author acknowledges partial funding through the National Science Foundation and the A. P. Sloan Research Foundation.

References

1. H. –W. Alt, unpublished notes.
2. H. Aref & G. Tryggvason, Numerical experiments on Hele–Shaw flow with sharp interface, *J. Fluid Mech.* **136** (1983), pp. 1 – 30
3. R. L. Chouke, P. van Meurs & C. van der Poel, The instability of slow, immiscible, viscous liquid–liquid displacements in permeable media, *Trans. AIME* **216** (1958), pp. 188–194
4. G. M. Homsy, Viscous fingering in porous media, *Ann. Rev. Fluid Mech.* **19** (1987), pp. 272–311.
5. S.N. Kružkov, First order quasilinear equations in several independent variables, *Math. USSR-Sb* **10** (1970), pp. 217 - 243
6. S. Luckhaus, P. I. Plotnikov, Entropy solutions to the Buckley-Leverett equations, *Siberian Math J.* **41 (2)** (2000), pp. 329–348.
7. J. Malek, J. Nečas, M. Rockyta and M. Ručička, *Weak and Measure–valued solutions to evolutionary partial differential equations*, Chapman & Hall, (1996)
8. F. Murat, L. Tartar,*H*–convergence, in *Topics in the mathematical modelling of composite materials*, R. Kohn & A. Cherkaev (Edts.), Birkhäuser, (1997), pp. 21–45
9. F. Otto, Stability investgation of planar solutions of the Buckley–Leverett equations, preprint no. 345 of the Sonderforschungbereich 256, University of Bonn (1994)
10. F. Otto, Viscous fingering: bound on the growth rate of mixing zone, *SIAM J. Appl. Math.* **57** (1997), pp. 982 – 990

11. F. Otto, Evolution of microstructure in unstable porous media flow: a relaxational approach, *Comm. Pure Appl. Math.* **52** (1999), pp. 873 – 915
12. P. G. Saffman & G. I. Taylor, The penetration of a fluid into porous medium or Hele–Shaw cell containing a more viscous liquid, *Proc. R. Soc. Lond. A* **245** (1958), pp. 312–329
13. J. Smoller, *Shock waves and reaction–diffusion equations*, Springer, (1994)
14. L. Tartar, Estimations of homogenized coefficients, in *Topics in the mathematical modelling of composite materials*, R. Kohn & A. Cherkaev (Edts.), Birkhäuser, (1997), pp. 9–21

An Extension of the Thermodynamic Formalism Approach to Selberg's Zeta Function for General Modular Groups

Cheng-Hung Chang and Dieter H. Mayer*

Institut für Theoretische Physik, TU Clausthal, Arnold-Sommerfeld-Straße 6, 38678 Clausthal-Zellerfeld, Germany
e-mail: ch.chang@tu-clausthal.de
dieter.mayer@tu-clausthal.de

Abstract. In the framework of the thermodynamic formalism for dynamical systems [Rue78] Selberg's zeta function [Sel56] for the modular group $PSL(2,\mathbb{Z})$ can be expressed through the Fredholm determinant of the generalized Ruelle transfer operator for the dynamical system defined by the geodesic flow on the modular surface [May91b]. In the present paper we generalize this result to modular subgroups Γ of $PSL(2,\mathbb{Z})$ with finite index. The corresponding surfaces of constant negative curvature with finite hyperbolic volume are in general ramified covering surfaces of the modular surface for $PSL(2,\mathbb{Z})$. Selberg's zeta function for these modular subgroups can be expressed through the generalized transfer operator for $PSL(2,\mathbb{Z})$ belonging to the representation of $PSL(2,\mathbb{Z})$ induced by the trivial representation of the subgroup Γ. The decomposition of this induced representation into its irreducible components leads to a decomposition of the transfer operator for these modular groups in analogy to a well known factorization formula of Venkov and Zograf for Selberg's zeta function for modular subgroups [VZ83]. The transfer operator can be continued into the entire complex β plane and its Fredholm determinant defines a meromorphic function in this plane.

1 Introduction

The thermodynamic formalism in ergodic theory is concerned with the investigation of dynamical systems and their properties by employing ingredients and techniques from statistical mechanics of lattice spin systems [Rue78]. Besides other quantities like partition functions, the free energy, different entropies, the transfer operator is playing a very special role in this approach. This operator was originally devised to determine the partition functions of periodic configurations of the 1-dimensional Ising lattice-spin model [Isi25] of statistical mechanics. Based on the analogy between the dynamics of the shift operator on the configuration space of such spin models and the symbolic dynamics of discrete time dynamical systems the thermodynamic formalism for hyperbolic dynamical systems was introduced by D. Ruelle, Y. Sinai and

* Project: Divisors of Fredholm Determinants of Generalized Perron-Frobenius Operators and Dynamical Zeta Functions (Dieter Mayer)

R. Bowen to study ergodic properties of such discrete time and, through the Poincaré map, also continuous time dynamical systems.

For very special systems like the geodesic flow on the modular surface $PSL(2,\mathbb{Z})\backslash\mathbb{H}$, this approach can be made quite explicit and leads to an interesting new approach to Selberg's zeta function for this group. This function, which counts in a certain way the length spectrum of the geodesic flow on this surface, is a special case of a dynamical zeta function introduced in the sixties by S. Smale and D. Ruelle. Contrary to the general case, for the geodesic flow on the modular surface the symbolic dynamics and hence the transfer operator can be written down explicitly so that many questions about this operator and hence Selberg's zeta function can be answered in great detail [May91b].

Selberg's classical approach to his zeta function through the trace formula shows that this function encodes through its zeros and poles interesting spectral properties of the Laplace-Beltrami operator on the modular surface respectively topological properties of the surface. In this sense this zeta function relates properties of a chaotic classical Hamiltonian system, namely the geodesic flow on a surface of constant negative curvature, to properties of the corresponding quantum system, namely the free motion of a quantized particle on this surface. This answers at least for a very special case Einstein's by now famous question in [Ein17] which is at the basis of all modern developments in the field of 'quantum chaos' [Gut90].

Up to now properties of the distribution of the eigenvalue spectrum of quantum systems with chaotic classical limit and their relation to random matrix models respectively the topography of their eigenfunctions have been studied mostly numerically. Rigorous results have been obtained only for very special systems with nice arithmetic properties like the geodesic flow on arithmetic surfaces of constant negative curvature [Sar95]. Results for the eigenfunctions are even harder to get since the trace formulas are not of great help in this case [see the contribution of Steiner et al. to this volume]

The transfer operator approach to Selberg's zeta function where this function gets expressed as the Fredholm determinant of this operator could open a new possibility to attack such questions. In this approach spectral properties of an operator which is completely determined by properties of the underlying classical system like its length spectrum describe the analytic properties of the Selberg function and hence also the quantum spectrum of the quantized dynamical system. This surprising relation is achieved simply by analytic continuation of the classical transfer operator into the complex β plane. For a special value of this parameter namely $\beta = 1$ the transfer operator is just the well known Perron-Frobenius operator from ergodic theory.

Rather surprising is furthermore a result known up to now only for the group $PSL(2,\mathbb{Z})$: the eigenfunctions of the transfer operator with eigenvalue 1 at the β values corresponding to eigenvalues of the Laplace Beltrami operator $-\triangle$ can be explicitly related to the corresponding eigenfunctions of $-\triangle$, the so called Maass wave forms for $PSL(2,\mathbb{Z})$ [CM99], [CM98]. The explicit form of this transformation was derived by J. Lewis in [LZ97]. This is the first time where such a direct relation between quantum states and classical objects has been established for a chaotic system where the usual construction of quasimodes [Laz93] based on invariant Lagrangian manifolds cannot be applied. In the present case the quantum mechanical eigenfunctions are just related to eigenfunctions of the classical transfer operator. An example is the ground state of our quantum system on the modular surface which is related through the Lewis transformation to the invariant density of the Gauss map which on the other hand is closely related to the Poincare map of the classical flow. One should therefore expect that also the other eigenfunctions of the transfer operator related to the Maass wave forms should have an interesting interpretation for the classical system.

In this paper however we address another question namely how far is it possible to extend the transfer operator approach to Selberg's zeta function to more general Fuchsian groups than $PSL(2,\mathbb{Z})$. Hence we study subgroups Γ of $PSL(2,\mathbb{Z})$ with finite index. Since the geodesic flow on the corresponding surface $\Gamma\backslash\mathbb{H}$ is just a lift of the one for $PSL(2,\mathbb{Z})$ to a finite, in general ramified covering surface of $PSL(2,\mathbb{Z})\backslash\mathbb{H}$, its symbolic dynamics and hence also its generalized transfer operator are closely related to the ones for $PSL(2,\mathbb{Z})$. Indeed, its transfer operator coincides with the generalized transfer operator for $PSL(2,\mathbb{Z})$ when one considers the geodesic flow with some representation of this group , corresponding in fact to a lift of the geodesic flow to some vector bundle [Fri86], [BO95] over the modular surface.

In detail the paper is organized as follows: in chapter 2 we recall the congruence subgroups of $PSL(2,\mathbb{Z})$ and introduce the special groups Γ_2, $\Gamma_0(2)$, $\Gamma^0(2)$, Γ_ϑ and $\Gamma(2)$ whose transfer operators we will discuss later in great detail. In chapter 3 we deduce the symbolic dynamics of the geodesic flow on the covering surfaces $\Gamma\backslash\mathbb{H}$ and give an explicit expression for the Poincaré map for these flows. In chapter 4 we discuss the transfer operators $\tilde{\mathcal{L}}_\beta$ for subgroups Γ, their analyticity properties in the 'temperature' β and the relation to transfer operators for $PSL(2,\mathbb{Z})$ with a representation, especially the one induced from a representation of Γ. We show that $\tilde{\mathcal{L}}_\beta$ is nuclear and that Selberg's zeta function for Γ can be simply expressed as the Fredholm determinant of $\tilde{\mathcal{L}}_\beta$. We derive a decomposition of $\tilde{\mathcal{L}}_\beta$ into a direct sum corresponding to the decomposition of the induced representation into its irreducible components and the resulting product formula of Venkov and Zograf for zeta functions for subgroups of $PSL(2,\mathbb{Z})$.

In a forthcoming second part to this paper we will discuss further spectral properties of the transfer operators for the special groups Γ_2, $\Gamma_0(2)$, $\Gamma^0(2)$, Γ_ϑ and $\Gamma(2)$ and how the theory of Lewis and Zagier [LZ97] of period functions for $PSL(2,\mathbb{Z})$ can be extended to these groups. We will also show how the transfer operator approach for these subgroups reflects the well known theory of Atkin-Lehner of old and new forms for cofinite Fuchsian groups [AL70].

2 The Hyperbolic Surfaces M_Γ

The Poincaré upper half-plane $\mathbb{H}$ is the half-plane

$$\mathbb{H} = \{x + iy \mid x,\ y \in \mathbb{R}, y > 0\} \tag{1}$$

equipped with the Poincaré metric (arc length)

$$d\,s^2 = y^{-2}\,(\,d\,x^2 + d\,y^2\,). \tag{2}$$

The surface $\mathbb{H}$ is a hyperbolic surface with constant negative Gaussian curvature -1 [Ter85].

The geodesics γ on $\mathbb{H}$ are the shortest paths through two points with respect to the Poincaré metric (2). They are either perpendicular half lines based on the real axis $\mathbb{R}$ or semi-circles based on two basepoints $\gamma_{-\infty}$ and $\gamma_{+\infty}$ in $\mathbb{R}$. A free particle on $\mathbb{H}$ moves along the geodesics with constant velocity. This motion is known to be hyperbolic; the geodesic flow on $\mathbb{H}$ is an Anosov flow but completely integrable. When introducing however the action of Fuchsian groups and their fundamental domains the induced geodesic flow on the corresponding quotient surface becomes highly chaotic and not predictable.

2.1 The Modular Groups

Consider first the group

$$PSL(2,\mathbb{R}) = SL(2,\mathbb{R})/\{\pm\, id\}\,,$$

where

$$SL(2,\mathbb{R}) := \{g = \begin{pmatrix} a\ b \\ c\ d \end{pmatrix} \mid a,b,c,d \in \mathbb{R},\ ad - bc = 1\},$$

and its action on $\mathbb{H}$ by Möbius-transformations

$$g\,z = \frac{az+b}{cz+d}, \tag{3}$$

for $z \in \mathbb{H}$. The group $PSL(2,\mathbb{R})$ is the isometry group of analytic actions on the surface $\mathbb{H}$. The discrete subgroups of $SL(2,\mathbb{R})$ are called Fuchsian groups. Examples are the full modular group

$$\Gamma(1) := PSL(2,\mathbb{Z}) = \{\begin{pmatrix} a\ b \\ c\ d \end{pmatrix} \mid a,\ b,\ c,\ d, \in \mathbb{Z},\ ad - bc = 1\}/\{\pm\, id\} \tag{4}$$

and its subgroups. The subgroups $\Gamma \subseteq \Gamma(1)$ with finite index $[\Gamma(1) : \Gamma]$ are called the modular groups.

The group $\Gamma(1)$ has two generators, e.g.

$$Q := \begin{pmatrix} 0 & 1 \\ -1 & 0 \end{pmatrix} \quad \text{and} \quad T := \begin{pmatrix} 1\ 1 \\ 0\ 1 \end{pmatrix} \tag{5}$$

acting as $Qz = -\frac{1}{z}$ and $Tz = z+1$ for $z \in \mathbb{H}$, which fulfill the following two relations

$$Q^2 = (QT)^3 = id. \tag{6}$$

The elements $\sigma \neq \pm id$ of $\Gamma(1)$ fall into three classes [Ter85]:

(i) σ is parabolic $\Leftrightarrow$ $|\mathrm{trace}\,\sigma| = 2$ and σ has Jordan normal form $\pm \begin{pmatrix} 1 & a \\ 0 & 1 \end{pmatrix}$ with $a \neq 0$.

(ii) σ is elliptic $\Leftrightarrow$ $|\mathrm{trace}\,\sigma| < 2$ and σ has Jordan normal form $\begin{pmatrix} a & 0 \\ 0 & 1/a \end{pmatrix}$ with $a \notin \mathbb{R}$ and $|a| = 1$.

(iii) σ is hyperbolic $\Leftrightarrow$ $|\mathrm{trace}\,\sigma| > 2$ and σ has Jordan normal form $\begin{pmatrix} a & 0 \\ 0 & 1/a \end{pmatrix}$ with $a \in \mathbb{R}$ and $|a| > 1$.

An element $\sigma \in \Gamma$ is primitive if σ is not a power of another element in Γ.

The principal congruence subgroups $\Gamma(N)$ of level N, defined as

$$\Gamma(N) := \{ \begin{pmatrix} a\,b \\ c\,d \end{pmatrix} \in \Gamma(1) \,|\, \begin{pmatrix} a\,b \\ c\,d \end{pmatrix} \equiv \begin{pmatrix} 1\,0 \\ 0\,1 \end{pmatrix} \; mod\, N,\; N \in \mathbb{N} \}, \tag{7}$$

are subgroups of the modular group $\Gamma(1)$. They are normal subgroups with index [Sch74]

$$\mu = [\Gamma(1) : \Gamma(N)] = \begin{cases} 6 & N = 2\,, \\ \frac{1}{2} N^3 \prod_{p\,|N} (1 - \frac{1}{p^2}) & N > 2\,, \end{cases}$$

where the devisor p of N is a prime number. The quotient groups $\Gamma(N)\backslash\Gamma(1)$ with $N = 2, 3, 4, 5$ are isomorphic to symmetric groups S_ν respectively alternating groups A_ν [KF66], [Sch74]:

$$\Gamma(2)\backslash\Gamma(1) \cong S_3,\; \Gamma(3)\backslash\Gamma(1) \cong A_4,\; \Gamma(4)\backslash\Gamma(1) \cong S_4 \text{ and } \Gamma(5)\backslash\Gamma(1) \cong A_5 \tag{8}$$

A subgroup Γ of $\Gamma(1)$ with the property $\Gamma(N) \subseteq \Gamma \subseteq \Gamma(1)$ for some $N \in \mathbb{N}$ is called a congruence subgroup. The following are examples of congruence subgroups:

$$\Gamma_0(N) := \{ \begin{pmatrix} a\,b \\ c\,d \end{pmatrix} \in \Gamma(1) |\; c = 0 \; mod\, N,\; N \in \mathbb{N} \}, \tag{9}$$

$$\Gamma^0(N) := \{ \begin{pmatrix} a\,b \\ c\,d \end{pmatrix} \in \Gamma(1) |\; b = 0 \; mod\, N,\; N \in \mathbb{N} \}. \tag{10}$$

These groups are not normal subgroups of $\Gamma(1)$ and have index μ in the full modular group $\Gamma(1)$ with [Sch74]

$$\mu = [\Gamma(1) : \Gamma_0(N)] = [\Gamma(1) : \Gamma^0(N)] = N \prod_{p|N} (1 + \frac{1}{p}). \tag{11}$$

Hence the groups $\Gamma_0(2)$ and $\Gamma^0(2)$ both have index three. Another congruence subgroup of index three is the theta group Γ_ϑ with

$$\Gamma_\vartheta := \Gamma(2) \cup \Gamma(2)Q\,. \tag{12}$$

The three groups $\Gamma_0(2)$, $\Gamma^0(2)$ and Γ_ϑ are conjugate to each other [Ran77]:

$$\Gamma^0(2) = G_1\,\Gamma_0(2)\,G_1^{-1} \text{ respectively } \Gamma_\vartheta = G_2\,\Gamma_0(2)\,G_2^{-1}\,,$$

with

$$G_1 = \begin{pmatrix} 0 & 1 \\ -1 & -1 \end{pmatrix} \text{ and } G_1^{-1} = \begin{pmatrix} -1 & -1 \\ 1 & 0 \end{pmatrix} \text{ respectively}$$
$$G_2 = \begin{pmatrix} 1 & 1 \\ -1 & 0 \end{pmatrix} \text{ and } G_2^{-1} = \begin{pmatrix} 0 & -1 \\ 1 & 1 \end{pmatrix}.$$

Besides the principal congruence subgroups $\Gamma(N)$ the group $\Gamma(1)$ has other normal subgroups, e.g. [Sch74]

$$\Gamma_2 := \Gamma(2) \cup \Gamma(2)QT \cup \Gamma(2)(QT)^2 \tag{13}$$

with index two. Indeed all the subgroups of $\Gamma(1)$ with index $\mu \leq 6$ are congruence subgroups. Non-congruence subgroups appear only for $\mu \geq 7$ [Ran77].

The following table collects some of the subgroups Γ of $\Gamma(1)$, their generators and representatives of the quotient sets $\Gamma\backslash\Gamma(1)$:

Table 1. Generators and representatives of different subgroups of $\Gamma(1)$

Group	Generators	Representatives of $\Gamma\backslash\Gamma(1)$
$\Gamma(1)$	Q, T	id
Γ_2	QT, QT^{-1}	id, Q
$\Gamma_0(2)$	$T, QT^{-2}Q, T^{-1}QT^{-2}Q$	id, Q, QT
$\Gamma^0(2)$	$T^2, QT^{-1}Q, TQT$	id, Q, T
Γ_ϑ	$T^2, Q, TQT^{-1}QT^{-1}$	id, T, TQ
$\Gamma(2)$	$T^2, QT^{-2}Q$	$id, T, Q, TQ, QT^{-1}, TQT$
$\Gamma^0(3)$	T^3, QTQ, TQT	id, T, T^{-1}, Q

2.2 The Modular Surfaces M_Γ

Let Γ be a subgroup of $\Gamma(1)$ with finite index $\mu = [\Gamma(1) : \Gamma] < \infty$. Two points $z_1, z_2 \in \mathbb{H}$ are Γ-equivalent if there exists an element $\sigma \in \Gamma$ with $z_1 = \sigma\, z_2$ as given in (3). The modular surface $M_\Gamma := \Gamma\backslash\mathbb{H}$ is the quotient surface defined by the Γ-equivalent points on $\mathbb{H}$.

A fundamental domain for Γ is a connected region $\mathcal{F}_\Gamma$ in $\mathbb{H}$ such that $\cup_{g\in\Gamma}\, g(\mathcal{F}_\Gamma) = \mathbb{H}$ with the property

$$g(\mathcal{F}_\Gamma^0) \cap g'(\mathcal{F}_\Gamma^0) = \emptyset\,,\ \ \forall\, g, g' \in \Gamma \text{ and } g \neq g'\,,$$

where $\mathcal{F}_\Gamma^0$ denotes the interior of $\mathcal{F}_\Gamma$. Conventionally one selects connected regions bounded by geodesics as fundamental domains. An example for such a domain for $\Gamma = \Gamma(1)$ is

$$\mathcal{F}_{\Gamma(1)} = \left\{ z \,\middle|\, |\Re z| \leq \frac{1}{2},\ |z| \geq 1,\ z \in \mathbb{H} \right\}. \tag{14}$$

For $\Gamma \subseteq \Gamma(1)$ one can then choose as a fundamental domain

$$\mathcal{F}_\Gamma = \cup_{\{g\} \in \Gamma\backslash\Gamma(1)}\, g(\mathcal{F}_{\Gamma(1)}), \tag{15}$$

where the union runs over representatives of the μ different equivalence classes of the quotient $\Gamma\backslash\Gamma(1)$.

The modular surface M_Γ can be constructed by identifying boundary points of the fundamental domain $\mathcal{F}_\Gamma$ related by the generators of the group Γ [Ran77]. Topologically, the surface M_Γ is a sphere with finitely many handles and finitely many cusps located at ∞ and at the inequivalent rational points in $\mathbb{R}$. The group Γ is not cocompact but cofinite, i.e., its fundamental region is not compact but has a finite hyperbolic area.

The Poincaré upper half-plane $\mathbb{H}$ is the universal covering space of the quotient space M_Γ. The quotient space M_Γ is an in general ramified μ-fold covering space of the modular surface $M_{\Gamma(1)} = \Gamma(1)\backslash\mathbb{H}$.

3 The Geodesic Flow on M_Γ

3.1 Coding of the Geodesics on M_Γ

We recall briefly the symbolic description of geodesics on the modular surfaces [Ser85]. For this, let us consider a special class of geodesics in $\mathbb{H}$, namely the set $\mathcal{A}$ of all oriented semi-circle running from $\gamma_{-\infty}$ to $\gamma_{+\infty}$ with $\gamma_{-\infty}$, $\gamma_{+\infty} \in \mathbb{R}$ such that

$$\mathcal{A} := \{\gamma \,|\, 0 < |\gamma_{-\infty}| \leq 1 \leq |\gamma_{+\infty}|,\ \gamma_{-\infty}\gamma_{+\infty} < 0\}. \tag{16}$$

The last condition $\gamma_{+\infty}\gamma_{-\infty} < 0$ makes sure that the geodesics in $\mathcal{A}$ cut the imaginary axis I. Every geodesic γ in $\mathbb{H}$ can be brought to $\mathcal{A}$ by means of an element in $g \in \Gamma(1)$ such that $g\gamma \in \mathcal{A}$. To describe the geodesics in $\mathcal{A}$ we consider the following two representations:

The first representation makes use of the continued fraction expansion. Since E. Artin [Art24] it is known that the geodesics on the modular surface $M_{\Gamma(1)}$ are closely related to the continued fraction transformation (Gauß transformation). Suppose $\gamma \in \mathcal{A}$ is a geodesic in $\mathbb{H}$ with the two basepoints $\gamma_{\pm\infty}$. Since $\gamma_{\pm\infty} \in \mathbb{R}$ these two points can be expressed by continued fractions:

$$\gamma_{-\infty} = -\varepsilon \left(\cfrac{1}{n_{-1} + \cfrac{1}{n_{-2} + \cdots}} \right) := -\varepsilon \, [n_{-1}, n_{-2}, n_{-3}, \cdots] , \tag{17}$$

$$\gamma_{+\infty} = \varepsilon \left(n_0 + \cfrac{1}{n_1 + \cfrac{1}{n_2 + \cdots}} \right) := \varepsilon \, [n_0, n_1, n_2, \cdots]^{-1} , \tag{18}$$

with $n_i \in \mathbb{N}$, where $\varepsilon = 1$ respectively $\varepsilon = -1$ denote the orientation $\gamma_{+\infty} > \gamma_{-\infty}$ respectively $\gamma_{+\infty} < \gamma_{-\infty}$. Hence every geodesic γ in $\mathcal{A}$ is determined by

$$\gamma := ([n_0, n_1, n_2, \cdots], [n_{-1}, n_{-2}, n_{-3}, \cdots], \varepsilon) \tag{19}$$

$$= (\varepsilon \, \gamma_{+\infty}^{-1}, -\varepsilon \, \gamma_{-\infty}, \varepsilon). \tag{20}$$

The second representation for the geodesics in $\mathcal{A}$ uses the so-called cutting sequences [Ser85]. To define this representation one needs the Farey tessellation $\mathbb{F}$ of the Poincaré upper half-plane $\mathbb{H}$. This tessellation $\mathbb{F}$ is defined by the hyperbolic triangle

$$\triangle_{\mathbb{F}} := \{ z \in \mathbb{H} \mid 0 \leq \Re z \leq 1, \, \frac{1}{2} \leq |z - \frac{1}{2}| \} \tag{21}$$

and its translations under the group $\Gamma(1)$, hence $\mathbb{H} = \cup_{g \in \Gamma(1)} g \triangle_{\mathbb{F}}$. The boundary of the tessellation $\mathbb{F}$ consists of all boundaries of all the hyperbolic triangles $g \triangle_{\mathbb{F}}$ with $g \in \Gamma(1)$. They are either the half-lines $\Re z \in \mathbb{Z}$ with $z \in \mathbb{H}$ or semi-circles with rational basepoints p/q and p'/q' with p, q, p', $q' \in \mathbb{Z}$, q, $q' \neq 0$ and $pq' - qp' = \pm 1$ [Ser85].

Almost every oriented geodesic $\gamma \in \mathcal{A}$ intersects infinitely many hyperbolic triangles in $\mathbb{F}$ (up to those that run along the boundaries or run into the cusps of the hyperbolic triangles) and hence are cut by the boundaries of the triangles into infinitely many pieces of arcs. The three vertices of the triangle that the geodesic γ crosses are separated into the two sides of γ. Depending on which side (left respectively right) the single vertex of the triangle is located with respect to the oriented geodesic, the piece of the arc of γ in this triangle will be denoted by L respectively R. Therefore the geodesic γ intersecting the imaginary axis I in y, i.e., $y = \gamma \cap I$, can be described by the symbol sequence:

$$\gamma := \begin{cases} \cdots R^{n_{-3}} L^{n_{-2}} R^{n_{-1}} y L^{n_0} R^{n_1} L^{n_2} \cdots & \text{for } -1 \leq \gamma_{-\infty} < 0, \ \varepsilon = 1 , \\ \cdots L^{n_{-3}} R^{n_{-2}} L^{n_{-1}} y R^{n_0} L^{n_1} R^{n_2} \cdots & \text{for } \ 0 < \gamma_{-\infty} \leq 1, \ \varepsilon = -1 , \end{cases} \tag{22}$$

where $n_i \in \mathbb{N}$ denotes the number of consecutive arcs L respectively R.

Comparing the representations (20) and (22) for γ shows that the numbers n_i in the continued fraction expansions (17) respectively (18) coincide exactly with the numbers n_i of the consecutive arcs L respectively R in (22) [Ser85]. The sequence (22) stops in case the geodesic runs into a cusp of $\mathbb{H}$, namely the rational numbers $\mathbb{Q}$, or starts out in such a cusp. Its continued fraction expansion in (18) respectively (17) then is finite. In this case $\gamma_{-\infty}$ respectively $\gamma_{+\infty}$ is rational. The representation (22) of γ is called a cutting sequence [Ser85]. Later we will use it to determine the Poincaré return map for the geodesic flow in $M_{\Gamma(1)}$.

3.2 The Geodesics on M_Γ and the Geodesic Flow

As mentioned earlier the Poincaré upper half-plane $\mathbb{H}$ is an infinite covering of the quotient space M_Γ for $\Gamma \subseteq \Gamma(1)$ and $[\Gamma(1) : \Gamma] < \infty$. Two different geodesics $\gamma^{(1)}$ and $\gamma^{(2)}$ in $\mathbb{H}$ are identified on the surface M_Γ if $\gamma^{(1)}$ and $\gamma^{(2)}$ are related by $\sigma\gamma^{(1)} = \gamma^{(2)}$ with some element $\sigma \in \Gamma$. Denote by

$$\pi : \mathbb{H} \to M_\Gamma \tag{23}$$

the projection of $\mathbb{H}$ onto M_Γ. Every geodesic γ in $\mathbb{H}$ is projected by π to a geodesic $\hat{\gamma}$ in M_Γ. Conversely a geodesic $\hat{\gamma}$ in M_Γ can be lifted to infinitely many γ in $\mathbb{H}$. Almost all geodesics in M_Γ cut the imaginary axis I, more precisely the projection of I in M_Γ, infinitely often.

A free particle on the surface $\mathbb{H}$ moves along a geodesic with constant velocity v, say $|v| = 1$. The physical phase space for this free particle is hence the unit tangent bundle $T_1\mathbb{H}$ of $\mathbb{H}$ of dimension three. The geodesic flow on $T_1\mathbb{H}$ is

$$\begin{aligned} \phi_t : T_1\mathbb{H} &\to T_1\mathbb{H} \\ (\gamma_{\mathbb{H}}(0), \dot{\gamma_{\mathbb{H}}}(0)) &\mapsto (\gamma_{\mathbb{H}}(t), \dot{\gamma_{\mathbb{H}}}(t)), \quad t \in \mathbb{R}, \end{aligned} \tag{24}$$

where $\gamma_{\mathbb{H}} : \mathbb{R} \to \mathbb{H}$ denotes the geodesic in $\mathbb{H}$ parameterized by the arc length respectively the time t through the initial position $\gamma_{\mathbb{H}}(0)$ with the initial tangent vector $\dot{\gamma_{\mathbb{H}}}(0)$. A (geodesic) orbit of the geodesic flow ϕ_t is a path in $T_1\mathbb{H}$ through the initial point $(\gamma_{\mathbb{H}}(0), \dot{\gamma_{\mathbb{H}}}(0)) \in T_1\mathbb{H}$.

Let $T_1 M_\Gamma$ be the unit tangent bundle of the configuration space of a free particle on M_Γ . Analogous to (24), the geodesic flow reads

$$\begin{aligned} \hat{\phi}_t : T_1 M_\Gamma &\to T_1 M_\Gamma \\ (\hat{\gamma}_{M_\Gamma}(0), \dot{\hat{\gamma}}_{M_\Gamma}(0)) &\mapsto (\hat{\gamma}_{M_\Gamma}(t), \dot{\hat{\gamma}}_{M_\Gamma}(t)) \end{aligned} \tag{25}$$

where $\hat{\gamma}_{M_\Gamma} : \mathbb{R} \to M_\Gamma$ denotes the geodesic in M_Γ through the point $\hat{\gamma}_{M_\Gamma}(0)$ with tangent vector $\dot{\hat{\gamma}}_{M_\Gamma}(0)$. The projection π in (23) relates the orbits of the geodesic flows on $\mathbb{H}$ and M_Γ in the obvious way:

$$\begin{aligned} \pi_* : T_1\mathbb{H} &\to T_1 M_\Gamma \\ (\gamma_{\mathbb{H}}(t), \dot{\gamma_{\mathbb{H}}}(t)) &\mapsto (\pi\gamma_{\mathbb{H}}(t), \pi\dot{\gamma_{\mathbb{H}}}(t)). \end{aligned} \tag{26}$$

Every periodic orbit of ϕ_t of period l in $T_1 M_\Gamma$ can be identified with a closed geodesic in M_Γ of length l, and vice versa [Pol91].

3.3 The Poincaré Map for the Geodesic Flow on $M_{\Gamma(1)}$

To understand the construction of a Poincaré map for the geodesic flow on an arbitrary modular surface M_Γ we recall briefly the case of the full modular group $\Gamma(1)$. We use the approach by Series [Ser85].

To construct a Poincaré section for the geodesic flow $\hat{\phi}_t : T_1 M_{\Gamma(1)} \to T_1 M_{\Gamma(1)}$ for the modular surface for $\Gamma(1)$, consider first the half line $S = [i, i\infty) \in \mathbb{H}$ on the imaginary axis I. Suppose a point $y \in S$ is given. Obviously not all the geodesics through y change their type L or R in the cutting representation at the point y. Define $C(y)$ as the set of all unit vectors $v(y)$ at the point y such that the corresponding geodesic belongs to $\mathcal{A}$ and hence changes its type at y. One can then choose the Poincaré section X for the geodesic flow on $T_1 M_{\Gamma(1)}$ as [Ser85]

$$X := \bigcup_{y \in S,\, v(y) \in C(y)} \pi_*(y, v(y)) . \tag{27}$$

To find the Poincaré return map for the Poincaré section X in (27) we consider a geodesic $\hat{\gamma}$ in $M_{\Gamma(1)}$ and lift $\hat{\gamma}$ to a geodesic γ_0 in $\mathcal{A}$. Suppose γ_0 has the cutting sequence representation

$$\gamma_0 = \cdots R^{n_{-3}} L^{n_{-2}} R^{n_{-1}} y_0 L^{n_0} R^{n_1} L^{n_2} \cdots , \quad \varepsilon = +1 \tag{28}$$

with $y_0 = \gamma_0 \cap S$ and has basepoints

$$\gamma_{-\infty} = -\left[n_{-1}, n_{-2}, n_{-3}, \cdots\right] \quad \text{and} \quad \gamma_{+\infty} = \left[n_0, n_1, n_2, \cdots\right]^{-1} .$$

The corresponding orbit in $T_1 M_{\Gamma(1)}$ of the geodesic $\hat{\gamma}$ in $M_{\Gamma(1)}$ has then the crossing point

$$\pi_*(y_0, v(y_0)) \in X$$

with the Poincaré section X where $v(y_0)$ denotes the unit tangent vector of γ_0 at y_0.

To find the next crossing point of this orbit with X, we follow the oriented geodesic γ_0. The next change of the type along γ_0 takes place on the axis $\Re z = n_0 := n$ which can be identified with the imaginary axis by means of the elements $QT^{-n} \in \Gamma(1)$ or $T^{-n} \in \Gamma(1)$, where Q and T are defined in (5). The geodesic $T^{-n}\gamma_0$ however doesn't belong to $\mathcal{A}$, because its basepoints $|\gamma_{+\infty}| \leq 1$ and $|\gamma_{-\infty}| \geq 1$ violate the conditions imposed on $\mathcal{A}$. Therefore, the point $\pi_*(y_1', v(y_1'))$ with $y_1' = T^{-n}\gamma_0 \cap S$ cannot be the next crossing point in X. In contrast, the element QT^{-n} brings the geodesic $\gamma_0 \in \mathcal{A}$ in (28) into the set $\mathcal{A}$ again. The action of Q however leads to a change of the orientation of the geodesic γ_0, namely the basepoints $\gamma_{\pm\infty}'$ of $QT^{-n}\gamma_0$ have $\gamma_{-\infty}' > \gamma_{+\infty}'$

instead of $\gamma_{-\infty} < \gamma_{+\infty}$. The geodesics $\gamma_1 = QT^{-n}\gamma_0$ and γ_0 are equivalent under $\Gamma(1)$, i.e., γ_1 and γ_0 define the same geodesic in $M_{\Gamma(1)}$. The basepoints of the equivalent geodesic $\gamma = \gamma_1 = QT^{-n}\gamma_0$ are

$$\gamma'_{-\infty} = [n_0, n_{-1}, n_{-2}, n_{-3}, \cdots] \quad \text{and} \quad \gamma'_{+\infty} = -[n_1, n_2, n_3 \cdots]^{-1}.$$

The corresponding cutting sequence representation on the other hand is

$$\gamma_1 = QT^{-n_0}\gamma_0 = \cdots R^{n_{-3}} L^{n_{-2}} R^{n_{-1}} L^{n_0} y_1 R^{n_1} L^{n_2} \cdots, \quad \varepsilon = -1,$$

where $y_1 = \gamma_1 \cap S$ denotes the point of intersection of γ_1 with S. Let $v(y_1)$ be the tangent vector of γ_1 at y_1. Then the first return point of γ_0 to X is $\pi_*(y_1, v(y_1))$. The Poincaré return map $P : X \to X$ transforms the initial crossing point $\pi_*(y_0, v(y_0)) \in X$ into the next crossing point $\pi_*(y_1, v(y_1)) \in X$. Obviously the above arguments are valid also for the geodesics in $\mathcal{A}$ with $\varepsilon = -1$.

Combining the cases $\varepsilon = 1$ and $\varepsilon = -1$, suppose $\gamma \in \mathcal{A}$ is described as

$$\gamma = (\varepsilon\gamma_{+\infty}^{-1}, -\varepsilon\gamma_{-\infty}, \varepsilon) = ([n_0, n_1, n_2, \cdots], [n_{-1}, n_{-2}, n_{-3}, \cdots], \varepsilon), \quad \varepsilon = \pm 1. \tag{29}$$

Then the Poincaré return map $P : X \to X$ transforms $\pi_*(y, v(y)) \in X$ into $\pi_*(y', v(y')) \in X$ with $y = \gamma \cap S$ and $y' = QT^{-n\varepsilon}\gamma \cap S$ where $n = n_0$ is the integer part of $\varepsilon\gamma_{+\infty}$. However it is known [Ser85] that the map

$$\rho : X \to \mathcal{A}$$

is bijective up to the two geodesics with basepoints $\gamma_{\pm\infty} = \pm 1$ respectively $\gamma_{\pm\infty} = \mp 1$ which are mapped into each other under Q and hence correspond to a single point in X [Ser85]. The map ρ allows to describe the points of X through the geodesics in $\mathcal{A}$ and hence by their basepoints $\gamma_{-\infty}$ and $\gamma_{+\infty}$. As in (29) the geodesics γ in $\mathcal{A}$ can be described as

$$\gamma = (\varepsilon\gamma_{+\infty}^{-1}, -\varepsilon\gamma_{-\infty}, \varepsilon)$$

with $\varepsilon = \pm 1$ and $\varepsilon\gamma_{+\infty}^{-1}$, $-\varepsilon\gamma_{-\infty}$ two real numbers in $[0, 1]$. In this coordinate system the Poincaré return map P has the following explicit form:

$$\begin{aligned} P : [0,1] \times [0,1] \times \mathbb{Z}_2 &\to [0,1] \times [0,1] \times \mathbb{Z}_2 \\ (x_1, x_2, \varepsilon) &\mapsto (-T^n Q x_1, -QT^n x_2, -\varepsilon) \end{aligned} \tag{30}$$

where for $x := (x_1, x_2, \varepsilon)$ the number $n = n(x) = [\frac{1}{x_1}]$ denotes the integer part of $\frac{1}{x_1}$ and the change of the orientation $\varepsilon \to -\varepsilon$ reflects the action of Q.

Since the first entry in the map (30) is nothing but the Gauß transformation

$$T_G : [0,1] \to [0,1] \quad \text{with} \quad T_G : z \mapsto \frac{1}{z} - [\frac{1}{z}], \tag{31}$$

we can rewrite (30) as

$$\begin{aligned} P(x_1, x_2, \varepsilon) &= (\frac{1}{x_1} - [\frac{1}{x_1}], \frac{1}{[\frac{1}{x_1}] + x_2}, -\varepsilon) \\ &= (T_G x_1, \frac{1}{[\frac{1}{x_1}] + x_2}, -\varepsilon). \end{aligned} \tag{32}$$

Obviously the x_1-direction is the expanding and the x_2-direction the contracting direction of the map P. The ergodic properties of the map P are determined by the behavior in the expanding directions of P [Rue78], [Bow75] which are (x_1, ε) with $\varepsilon = \pm 1$.

Apparently the geodesics in $\mathbb{H}$ are not closed. In contrast, due to the identifications through the group Γ there exist closed geodesics $\hat{\gamma}$ in M_Γ. Obviously this $\hat{\gamma}$ must define a periodic point of the Poincaré map P. An easy calculation then shows that for such a closed geodesic one has

$$x_1 = [\overline{n_0, n_1, \cdots, n_{m-2}, n_{m-1}}],\ x_2 = [\overline{n_{m-1}, n_{m-2}, \cdots, n_1, n_0}] \text{ and } \varepsilon = \pm 1\,,$$

where the bar denotes the periodic repetition of the sequence. For $\Gamma = \Gamma(1)$, suppose $\hat{\gamma}$ is a periodic geodesic in $M_{\Gamma(1)}$. Then the lifted geodesic γ in $\mathbb{H}$ belonging to $\mathcal{A}$ can be expressed in the coordinates as used in (19):

$$\gamma = ([\overline{n_0, n_1, \cdots, n_{m-2}, n_{m-1}}], [\overline{n_{m-1}, n_{m-2}, \cdots, n_1, n_0}], \varepsilon)\,.$$

Every hyperbolic element $\sigma \in \Gamma$ determines a closed geodesic in M_Γ by fixing the two basepoints $\gamma_{-\infty}$ and $\gamma_{+\infty}$ on $\mathbb{R}$, namely $\sigma\gamma_{\pm\infty} = \gamma_{\pm\infty}$. The length $l(\gamma)$ of γ is known to be related to the trace of σ [Hej76]:

$$2\cosh\frac{l(\gamma)}{2} = (e^{l(\gamma)})^{1/2} + (e^{l(\gamma)})^{-1/2} = |\mathrm{trace}(\sigma)|\,. \tag{33}$$

3.4 The Poincaré Map for the Geodesic Flow on M_Γ

Consider now any subgroup $\Gamma \subseteq \Gamma(1)$ with finite index $\mu = [\Gamma(1) : \Gamma] < \infty$. The surface $M_\Gamma = \Gamma\backslash\mathbb{H}$ is then a finite, in general ramified covering surface of $M_{\Gamma(1)}$. Denote by $\pi_\Gamma : M_\Gamma \to M_{\Gamma(1)}$ and $\pi : \mathbb{H} \to M_\Gamma$ the corresponding projection maps. To generalize the construction of the Poincaré map to arbitrary subgroups Γ of $\Gamma(1)$ consider the set $\mathcal{A}_\Gamma = \cup_{i=1}^{\mu} g_i\mathcal{A}$ where $\mathcal{A}$ was defined in (16) and the g_i are representatives of the μ different equivalence classes $\{g_i\}$ in $\Gamma\backslash\Gamma(1)$. As a natural Poincaré section for the geodesic flow on M_Γ one can choose

$$X_\Gamma = \pi_*\Big(\bigcup_{i=1}^{\mu} g_i \bigcup_{y\in S,\, v(y)\in C(y)} (y, v(y))\Big)\,. \tag{34}$$

which obviously is identical to ${\pi_{\Gamma,*}}^{-1}(X)$, where X is the Poincaré section in $M_{\Gamma(1)}$ and hence the lift of X to the covering sphere bundle $T_1 M_\Gamma$ of

$T_1 M_{\Gamma(1)}$. There is again a bijective map of $\mathcal{A}_\Gamma$ onto X_Γ, apart from possibly a finite set of geodesics in $\mathcal{A}$: for $\gamma \in \mathcal{A}_\Gamma$ there exists a $\{g_i\} \in \Gamma\backslash\Gamma(1)$ such that $g_i^{-1}\gamma \in \mathcal{A}$. To this $g_i^{-1}\gamma$ there corresponds exactly one point $(y, v(y))$ with $y \in S$ and $v(y) \in C(y)$ and hence there corresponds to γ exactly one point $g_i(y, v(y))$ with $y \in S$ and $v(y) \in C(y)$ which on the other hand defines exactly one point in X_Γ. If Q belongs to Γ then there exist again geodesics in $\mathcal{A}_\Gamma$ which correspond to the same point in the Poincaré section namely the geodesics γ with basepoints $\gamma_{\pm\infty} = \pm 1$ or ∓ 1 and perhaps some or all of their images $g_i\gamma$.

To determine the explicit form of the Poincaré map $P_\Gamma : X_\Gamma \to X_\Gamma$ we proceed as follows: consider an orbit intersecting X_Γ and the geodesic γ in $\mathcal{A}_\Gamma$ corresponding to this point in X_Γ. Then there exists a $\{g_i\} \in \Gamma\backslash\Gamma(1)$ with $g_i^{-1}\gamma \in \mathcal{A}$. >From our discussion of the group $\Gamma(1)$ we know that the geodesic $QT^{-n\varepsilon}g_i^{-1}\gamma \in \mathcal{A}$ describes the action of the Poincaré map for the group $\Gamma(1)$. Obviously, the geodesics γ and $QT^{-n\varepsilon}g_i^{-1}\gamma$ do not describe in general the same orbit for the geodesic flow on M_Γ. But there exists a unique $\{g_j\} \in \Gamma\backslash\Gamma(1)$ such that $g_jQT^{-n\varepsilon}g_i^{-1} \in \Gamma$: consider namely $g := hg_iT^{n\varepsilon}Q$ with $h \in \Gamma$ arbitrary, then $gQT^{-n\varepsilon}g_i^{-1} = h \in \Gamma$ and g defines a unique g_j with $\{g\} = \{g_j\} \in \Gamma\backslash\Gamma(1)$. Hence $g_jQT^{-n\varepsilon}g_i^{-1}\gamma \in g_j\mathcal{A} \in \mathcal{A}_\Gamma$ defines an unique point on the Poincaré surface X_Γ. That this point is the first point in X_Γ to which the orbit returns follows immediately from our discussion for $\Gamma(1)$. We can introduce now on X_Γ coordinates in analogy to the section X, namely if $\gamma' = g_i\gamma$ with $\{g_i\} \in \Gamma\backslash\Gamma(1)$ and $\gamma \in \mathcal{A}$ described as in (29) as $\gamma = (\varepsilon\gamma_{+\infty}^{-1}, -\varepsilon\gamma_{-\infty}, \varepsilon)$, then we can describe γ' as

$$\gamma' = (\varepsilon\gamma_{+\infty}^{-1}, -\varepsilon\gamma_{-\infty}, \varepsilon, \{g_i\}) .$$

In these coordinates the Poincaré map $P_\Gamma : X_\Gamma \to X_\Gamma$ then can be written as

$$\begin{aligned} P_\Gamma : [0,1] \times [0,1] \times \mathbb{Z}_2 \times \Gamma\backslash\Gamma(1) &\to [0,1] \times [0,1] \times \mathbb{Z}_2 \times \Gamma\backslash\Gamma(1) \\ (x_1, x_2, \varepsilon, \{g\}) &\mapsto (T_G x_1, \frac{1}{[\frac{1}{x_1}] + x_2}, -\varepsilon, \{gT^{n\varepsilon}Q\}) , \end{aligned} \tag{35}$$

where $n = n(x) = [\frac{1}{x_1}]$ is the integer part of $\frac{1}{x_1}$ for $x := (x_1, x_2, \varepsilon, \{g\})$. It is just an extension of the Poincaré map for $\Gamma(1)$ by the set $\Gamma\backslash\Gamma(1)$.

Obviously we get back from P_Γ the Poincaré map P in the case $\Gamma = \Gamma(1)$ since then $\Gamma\backslash\Gamma(1)$ consists just of one element. The obvious interpretation of the Poincaré map P_Γ in (35) is as the lift of P to the different sheets of $T_1 M_\Gamma$ as a μ-fold covering of the space $T_1 M_{\Gamma(1)}$ characterized by the μ classes $\{g_i\}$, $i = 1, 2, \cdots, \mu$ of $\Gamma\backslash\Gamma(1)$. For the construction of the generalized Ruelle transfer operator for the geodesic flow on M_Γ we need the action of P_Γ along the expanding directions. Obviously they are the x_1 directions in the different sheets described by $\varepsilon = \pm 1$ and $\{g\} \in \Gamma\backslash\Gamma(1)$. Hence we get

$$\begin{aligned} P_\Gamma|_{ex} &: [0,1] \times \mathbb{Z}_2 \times \Gamma\backslash\Gamma(1) \to [0,1] \times \mathbb{Z}_2 \times \Gamma\backslash\Gamma(1) \\ P_\Gamma|_{ex}(z, \varepsilon, \{g\}) &= (T_G\, z, -\varepsilon, \{gT^{n\varepsilon}Q\}) = (\frac{1}{z} - [\frac{1}{z}], -\varepsilon, \{gT^{n\varepsilon}Q\}), \end{aligned} \tag{36}$$

with $n = n(x) = [\frac{1}{z}]$ and $x = (z, \varepsilon, \{g\})$. This allows us now to construct the transfer operator for the geodesic flow on M_Γ.

4 The Transfer Operator for the Geodesic Flow on M_Γ

4.1 The Transfer Operator for Modular Subgroups $\Gamma \subseteq \Gamma(1)$

The generalized Ruelle transfer operator for an expanding map $\tau : M \to M$ with inverse temperature β and weight function $A : M \to \mathbb{R}$ is defined in [Rue78], [May91a] as

$$\mathcal{L}_{\tau,\beta} f(x) = \sum_{y \in \tau^{-1}x} \exp\left(-\beta\, A(y)\right) f(y),$$

where the sum runs over all preimages of x under the map τ and where f is some function on M, which has to be chosen appropriately. The weight function in the case of the Poincaré map $P_\Gamma|_{ex}$ which describes the ergodic properties of the geodesic flow is the function $A_{P_\Gamma|_{ex}}(x) = log|T'_G(z)|$ with $x = (z, \varepsilon, \{g\})$, quite similar to the one for $\Gamma(1)$ [May91a]. Furthermore, the preimages $P_\Gamma|_{ex}^{-1} x$ of the point $x = (z, \varepsilon, \{g\})$ are just

$$\left\{ (\frac{1}{z+n}, -\varepsilon, \{gQT^{n\varepsilon}\}) \mid n \in \mathbb{N} \right\}, \tag{37}$$

because

$$\begin{aligned} P_\Gamma|_{ex}(\frac{1}{z+n}, -\varepsilon, \{gQT^{n\varepsilon}\}) &= (z, \varepsilon, \{gQT^{n\varepsilon}\, T^{-n(x')\varepsilon}Q\}) \\ &= (z, \varepsilon, \{g\}), \end{aligned} \tag{38}$$

where we used $n(x') = n((\frac{1}{z+n}, -\varepsilon, \{gQT^{n\varepsilon}\})) = [z+n] = n$. The transfer operator $\tilde{\mathcal{L}}_\beta^{\Gamma,\chi}$ for a subgroup $\Gamma \subseteq \Gamma(1)$ with finite index $\mu = [\Gamma(1) : \Gamma] < \infty$ and representation $\chi : \Gamma \to end\, V$ can be defined as follows

$$\tilde{\mathcal{L}}_\beta^{\Gamma,\chi} \underline{f}(z, \varepsilon, \{g\}) = \sum_{n=1}^{\infty} (\frac{1}{z+n})^{2\beta} \chi(gQT^{n\varepsilon}g'^{-1}) \underline{f}(\frac{1}{z+n}, -\varepsilon, \{gQT^{n\varepsilon}\}), \tag{39}$$

where g' is the unique element among $g_1, \cdots, g_\mu$, such that $\{g'\} = \{gQT^{n\varepsilon}\}$, that means $\Gamma g' = \Gamma gQT^{n\varepsilon}$. Therefore $gQT^{n\varepsilon}g'^{-1} \in \Gamma$ and $\chi(gQT^{n\varepsilon}g'^{-1})$ is well defined as acting on the function $\underline{f}$ if $\underline{f}$ takes values in the space V. The exact properties of the functions in the variable z will be discussed later. Before doing this we will give a simple interpretation of the above transfer operator in terms of a generalized transfer operator for the group $\Gamma(1)$ with a certain representation.

4.2 Transfer Operators for $\Gamma(1)$ with Representation U^χ

If $\psi : \Gamma(1) \to end W$ is a finite dimensional representation on the vector space W then the transfer operator for the geodesic flow on the modular surface for $\Gamma(1)$ with representation ψ has the following form

$$\tilde{\mathcal{L}}_\beta^{\Gamma,\psi} \underline{f}(z,\varepsilon) = \sum_{n=1}^{\infty} (\frac{1}{z+n})^{2\beta} \psi(QT^{n\varepsilon}) \underline{f}(\frac{1}{z+n}, -\varepsilon), \tag{40}$$

where the function $\underline{f}$ takes values in the space W. This operator generalizes the transfer operator used in [CM98] to describe the Selberg zeta function for the geodesic flow on the modular surface for $\Gamma(1)$. Indeed, this operator will be shown to describe the generalized Selberg zeta function for the group $\Gamma(1)$ with representation ψ [VZ83]. To understand the form of the transfer operator in (39) for Γ we briefly recall the definition of the representation of a group G induced by a representation χ of a subgroup $G_1 \subseteq G$ with $[G : G_1] = m < \infty$.

Let $\chi : G_1 \to V$ be a representation of the subgroup G_1 of G. Then the representation U^χ of G induced from the representation χ of G_1 can be defined as follows [VZ83]: for $\underline{v} = (v_i)_{i=1}^m \in V^m$ one defines for $g \in G$

$$(U^\chi(g)\underline{v})_i = \chi(g_i\, g\, g_j^{-1})\, v_j \tag{41}$$

where the $g_i \in G$, $i = 1, \cdots, m$ are fixed by the condition that $G = \bigcup_{i=1}^m G_1 g_i$ with $g_1 = id$ the unit element in G, and g_j in (41) is uniquely determined by the condition $g_i g g_j^{-1} \in G_1$ that means $G_1 g_i g = G_1 g_j$. Coming back now to our transfer operator $\tilde{\mathcal{L}}_\beta^{\Gamma,\chi}$ in (39) and identifying the functions $\underline{f}(z,\varepsilon,\{g_i\}) \in V$, $i = 1, \cdots, \mu$ with the elements v_i in (41) and writing $(\underline{\underline{f}}(z,\varepsilon))_{\{g_i\}} = \underline{f}(z,\varepsilon,\{g_i\})$ we find

$$(\tilde{\mathcal{L}}_\beta^{\Gamma,\chi} \underline{\underline{f}}(z,\varepsilon))_{\{g\}} = \sum_{n=1}^{\infty} (\frac{1}{z+n})^{2\beta} \chi(gQT^{n\varepsilon}g'^{-1}) \underline{f}(\frac{1}{z+n}, -\varepsilon, \{g'\}) \tag{42}$$

and hence

$$\tilde{\mathcal{L}}_\beta^{\Gamma,\chi} \underline{\underline{f}}(z,\varepsilon) = \sum_{n=1}^{\infty} (\frac{1}{z+n})^{2\beta} U^\chi(QT^{n\varepsilon}) \underline{\underline{f}}(\frac{1}{z+n}, -\varepsilon) \tag{43}$$

Hence we have shown

Theorem 1. *For $\Gamma \subseteq \Gamma(1)$ a subgroup of $\Gamma(1)$ with finite index and $\chi : \Gamma \to end V$ a finite dimensional representation of Γ the transfer operator $\tilde{\mathcal{L}}_\beta^{\Gamma,\chi}$ of the geodesic flow on $\Gamma\backslash\mathbb{H}$ with representation χ of Γ is identical up to isomorphy to the transfer operator $\tilde{\mathcal{L}}_\beta^{\Gamma(1),U^\chi}$ of the geodesic flow on $\Gamma(1)\backslash\mathbb{H}$ with representation U^χ of $\Gamma(1)$ induced by the representation χ of Γ.*

For the special case when χ is the trivial 1-dimensional representation of Γ the transfer operator (43) can be reduced to

$$\tilde{\mathcal{L}}_\beta^{\Gamma,\chi}\underline{f}(z,\varepsilon)=\sum_{n=1}^{\infty}(\frac{1}{z+n})^{2\beta}U^\chi(QT^{n\varepsilon})\underline{f}(\frac{1}{z+n},-\varepsilon) \tag{44}$$

where the representation U^χ is defined as $(U^\chi(g)\underline{v})_i=v_j$ where j is determined again by the condition $g_igg_j^{-1}\in\Gamma$. Employing the function $\omega:\Gamma(1)\to\mathbb{R}$ with

$$\omega(g)=\begin{cases}1 & g\in\Gamma\,,\\ 0 & g\notin\Gamma\,,\end{cases}$$

one can also write

$$(U^\chi(g)\underline{v})_i=\sum_{j=1}^{\mu}\omega(g_igg_j^{-1})\,v_j\,.$$

Consequently the linear operator U^χ in this case has the following matrix representation $\chi^\Gamma(g)$:

$$\chi^\Gamma(g)=\begin{pmatrix}\omega(g_1gg_1^{-1}) & \omega(g_1gg_2^{-1}) & \cdots & \omega(g_1gg_\mu^{-1})\\ \omega(g_2gg_1^{-1}) & \omega(g_2gg_2^{-1}) & \cdots & \omega(g_2gg_\mu^{-1})\\ \cdots & \cdots & \cdots & \cdots\\ \omega(g_\mu gg_1^{-1}) & \omega(g_\mu gg_2^{-1}) & \cdots & \omega(g_\mu gg_\mu^{-1})\end{pmatrix}. \tag{45}$$

The matrix χ^Γ is a μ-dimensional permutation matrix. In every row respectively column there is only one entry different from zero and its value is 1. For fixed μ, there exist only $\mu!$ different matrices of this type. Thus, for every $g\in G$ we can always find integers $r_1,\,r_2\in\mathbb{N}$ with $r_1<r_2$ such that $\chi^\Gamma(g^{r_1})=\chi^\Gamma(g^{r_2})$ and consequently $\chi^\Gamma(g^{r_2-r_1})=1$. That is, there always exists $r\in\mathbb{N}$ such that $0<r<\infty$ and $\chi^\Gamma(g^r)=1$.

As an example let us consider the group $\Gamma_0(2)$. According to (11) it is a subgroup of $\Gamma(1)$ with index 3 and has furthermore the following coset decomposition

$$\Gamma(1)=\Gamma_0(2)g_1\cup\Gamma_0(2)g_2\cup\Gamma_0(2)g_3\,,$$

where $g_1=id$, $g_2=Q$ and $g_3=QT$ are representatives of the three different equivalence classes in $\Gamma_0(2)\backslash\Gamma(1)$. The representations $\chi^{\Gamma_0(2)}$ of Q and T read

$$\chi^{\Gamma_0(2)}(Q)=\begin{pmatrix}\omega(id\,Q\,(id)^{-1}) & \omega(id\,Q\,(Q)^{-1}) & \omega(id\,Q(QT)^{-1})\\ \omega(QQ(id)^{-1}) & \omega(QQ(Q)^{-1}) & \omega(QQ(QT)^{-1})\\ \omega(QT\,Q(id)^{-1}) & \omega(QT\,Q(Q)^{-1}) & \omega(QT\,Q(QT)^{-1})\end{pmatrix}=\begin{pmatrix}0&1&0\\1&0&0\\0&0&1\end{pmatrix}$$

respectively

$$\chi^{\Gamma_0(2)}(T)=\begin{pmatrix}\omega(id\,T(id)^{-1}) & \omega(id\,T(Q)^{-1}) & \omega(id\,T(QT)^{-1})\\ \omega(Q\,T(id)^{-1}) & \omega(Q\,T(Q)^{-1}) & \omega(Q\,T(QT)^{-1})\\ \omega(QT\,T(id)^{-1}) & \omega(QT\,T(Q)^{-1}) & \omega(QT\,T(QT)^{-1})\end{pmatrix}=\begin{pmatrix}1&0&0\\0&0&1\\0&1&0\end{pmatrix}.$$

The induced representations χ^{Γ} of Q and T for the groups Γ_2, $\Gamma_0(2)$, $\Gamma^0(2)$, Γ_ϑ and $\Gamma^0(3)$ are summarized in the following table:

Table 2. The induced representations χ^{Γ} of Q and T for different subgroups Γ of $\Gamma(1)$

group Γ	$\chi^{\Gamma}(Q)$	$\chi^{\Gamma}(T)$	representatives of $\Gamma\backslash\Gamma(1)$
$\Gamma(1)$	1	1	id
Γ_2	$\begin{pmatrix}0&1\\1&0\end{pmatrix}$	$\begin{pmatrix}0&1\\1&0\end{pmatrix}$	$g_1 = id$ $g_2 = Q$
$\Gamma_0(2)$	$\begin{pmatrix}0&1&0\\1&0&0\\0&0&1\end{pmatrix}$	$\begin{pmatrix}1&0&0\\0&0&1\\0&1&0\end{pmatrix}$	$g_1 = id$ $g_2 = Q$ $g_3 = QT$
$\Gamma^0(2)$	$\begin{pmatrix}0&1&0\\1&0&0\\0&0&1\end{pmatrix}$	$\begin{pmatrix}0&0&1\\0&1&0\\1&0&0\end{pmatrix}$	$g_1 = id$ $g_2 = Q$ $g_3 = T$
Γ_ϑ	$\begin{pmatrix}1&0&0\\0&0&1\\0&1&0\end{pmatrix}$	$\begin{pmatrix}0&1&0\\1&0&0\\0&0&1\end{pmatrix}$	$g_1 = id$ $g_2 = T$ $g_3 = TQ$
$\Gamma^0(3)$	$\begin{pmatrix}0&0&0&1\\0&0&1&0\\0&1&0&0\\1&0&0&0\end{pmatrix}$	$\begin{pmatrix}0&0&1&0\\1&0&0&0\\0&1&0&0\\0&0&0&1\end{pmatrix}$	$g_1 = id$ $g_2 = T^{-1}$ $g_3 = T$ $g_4 = Q$

In the following we restrict our discussion to the trivial 1-dimensional representation χ of Γ, but our results can be extended without problem to the more general case of non-trivial χ's. The corresponding transfer operator with the representation (45) reads

$$\tilde{\mathcal{L}}_\beta^{\Gamma,\chi}\underline{f}(z,\varepsilon) = \sum_{n=1}^{\infty}(\frac{1}{z+n})^{2\beta}\chi^{\Gamma}(QT^{n\varepsilon})\underline{f}(\frac{1}{z+n},-\varepsilon)\,. \tag{46}$$

4.3 Decomposition of the Transfer Operator $\tilde{\mathcal{L}}_\beta^{\Gamma}$

In general, the induced representation U^χ will be reducible, that is U^χ can be decomposed into its irreducible components

$$\chi^{\Gamma} = \oplus_i \chi_i^{\Gamma}\,. \tag{47}$$

In the special case of a normal subgroup $\Gamma \subseteq \Gamma(1)$ the quotient set $\Gamma\backslash\Gamma(1)$ is itself a group and the induced representation of the group $\Gamma\backslash\Gamma(1)$ is isomorphic to the right regular representation of the group $\Gamma\backslash\Gamma(1)$. Since $U^\chi(h) = 1$ for $h \in \Gamma$ the representation U^χ defines also a representation of $\Gamma\backslash\Gamma(1)$ by $U^\chi(\Gamma g) = U^\chi(g)$. Then one finds [Vin89]

$$\chi^{\Gamma} = \bigoplus_{\chi_i^{\Gamma}\in\chi^*(\Gamma\backslash\Gamma(1))} n_i\chi_i^{\Gamma}\,, \tag{48}$$

where $\chi^*(\Gamma\backslash\Gamma(1))$ denotes the set of all inequivalent unitary irreducible representations of $\Gamma\backslash\Gamma(1)$ and n_i the dimension of χ_i^Γ.

According to (48) the transfer operator (46) can then be decomposed as

$$\begin{aligned}
&\tilde{\mathcal{L}}_\beta^{\Gamma(1),\chi^\Gamma} \underline{f}(z,\varepsilon) \\
&= \sum_{n=1}^\infty (\frac{1}{z+n})^{2\beta} \Big(\bigoplus_{\chi_i^\Gamma \in \chi^*(\Gamma\backslash\Gamma(1))} n_i\, \chi_i^\Gamma(QT^{n\varepsilon})\, \underline{f}_i(\frac{1}{z+n}, -\varepsilon)\Big) \qquad (49) \\
&= \Big(\bigoplus_{\chi_i^\Gamma \in \chi^*(\Gamma\backslash\Gamma(1))} n_i\, \tilde{\mathcal{L}}_\beta^{\Gamma(1),\chi_i^\Gamma} \underline{f}_i(z,\varepsilon)\Big) \qquad (50)
\end{aligned}$$

with

$$\tilde{\mathcal{L}}_\beta^{\Gamma(1),\chi_i^\Gamma} \underline{f}_i(z,\varepsilon) = \sum_{n=1}^\infty (\frac{1}{z+n})^{2\beta} \chi_i^\Gamma(QT^{n\varepsilon})\, \underline{f}_i(\frac{1}{z+n}, -\varepsilon),$$

where the function $\underline{f}_i$ takes values in the representation space of χ_i^Γ and $\underline{f} = \oplus_i \underline{f}_i$.

The simplest example is the transfer operator for $\Gamma(1)$ with the trivial representation $\chi^\Gamma = 1$:

$$\tilde{\mathcal{L}}_\beta^{\Gamma(1),1} f(z,\varepsilon) = \sum_{n=1}^\infty (\frac{1}{z+n})^{2\beta} f(\frac{1}{z+n}, -\varepsilon)\,. \qquad (51)$$

A non-trivial example is the transfer operator for Γ_2. The induced representations χ^{Γ_2} of the generators Q and T in table 2 read in this case

$$\chi^{\Gamma_2}(Q) = \begin{pmatrix} 0 & 1 \\ 1 & 0 \end{pmatrix} \quad \text{and} \quad \chi^{\Gamma_2}(T) = \begin{pmatrix} 0 & 1 \\ 1 & 0 \end{pmatrix}. \qquad (52)$$

Applying the matrices $M = \begin{pmatrix} \frac{1}{2} & \frac{1}{2} \\ -\frac{1}{2} & \frac{1}{2} \end{pmatrix}$ respectively $M^{-1} = \begin{pmatrix} 1 & -1 \\ 1 & 1 \end{pmatrix}$ one gets

$$M\chi^{\Gamma_2}(Q)M^{-1} = M\chi^{\Gamma_2}(T)M^{-1} = \begin{pmatrix} 1 & 0 \\ 0 & -1 \end{pmatrix}.$$

That is, under the above basis transformation the 2-dimensional representation χ^{Γ_2} in (52) can be decomposed into two 1-dimensional unitary irreducible representations $\chi^{\Gamma_2} = \chi_1^{\Gamma_2} \oplus \chi_{-1}^{\Gamma_2}$ with

$$\chi_1^{\Gamma_2} = 1 \quad \text{and} \quad \chi_{-1}^{\Gamma_2}(Q) = \chi_{-1}^{\Gamma_2}(T) = -1\,. \qquad (53)$$

Therefore the transfer operator for Γ_2 can be decomposed as

$$\tilde{\mathcal{L}}_\beta^{\Gamma(1),\chi^{\Gamma_2}} \underline{f}(z,\varepsilon) = \sum_{n=1}^\infty (\frac{1}{z+n})^{2\beta} \begin{pmatrix} \chi_1^{\Gamma_2}(QT^{n\varepsilon}) & 0 \\ 0 & \chi_{-1}^{\Gamma_2}(QT^{n\varepsilon}) \end{pmatrix} \underline{f}(\frac{1}{z+n}, -\varepsilon)\,. \qquad (54)$$

Another non-trivial example of a normal subgroup of $\Gamma(1)$ is the principal congruence subgroup $\Gamma(2)$. The quotient group $\Gamma(2)\backslash\Gamma(1)$ has index 6 and is isomorphic to the permutation group S_3 in (8). The induced representation $\chi^{\Gamma(2)}$ of $\Gamma(2)$ can be decomposed as $\chi^{\Gamma(2)} = \chi_1^{\Gamma(2)} \oplus \chi_{-1}^{\Gamma(2)} \oplus \chi_2^{\Gamma(2)} \oplus \chi_2^{\Gamma(2)}$ with the following three unitary irreducible representations $\chi_i^{\Gamma(2)}$, $i = 1, -1, 2$ [Vin89]: $\chi_1^{\Gamma(2)}$ is the trivial 1-dimensional representation, i.e., $\chi_1^{\Gamma(2)}(Q) = \chi_1^{\Gamma(2)}(T) = 1$. $\chi_{-1}^{\Gamma(2)}$ is the non-trivial 1-dimensional representation with

$$\chi_{-1}^{\Gamma(2)}(Q) = \chi_{-1}^{\Gamma(2)}(T) = -1. \tag{55}$$

The representation $\chi_2^{\Gamma(2)}$ is 2-dimensional and is usually realized by the matrices

$$\hat{\chi}_2(Q) = \begin{pmatrix} -\cos\frac{2\pi}{3} & \sin\frac{2\pi}{3} \\ \sin\frac{2\pi}{3} & \cos\frac{2\pi}{3} \end{pmatrix} \quad \text{and} \quad \hat{\chi}_2(T) = \begin{pmatrix} -1 & 0 \\ 0 & 1 \end{pmatrix}.$$

An alternative choice more convenient for our later discussion is the following:

$$\tilde{\chi}_2(Q) = \begin{pmatrix} 1 & 0 \\ -1 & -1 \end{pmatrix} \quad \text{and} \quad \tilde{\chi}_2(T) = \begin{pmatrix} 0 & 1 \\ 1 & 0 \end{pmatrix}, \tag{56}$$

which is related to the former by $\tilde{\chi}_2 = M\,\hat{\chi}_2\,M^{-1}$ with

$$M = \begin{pmatrix} \frac{1}{2} & \frac{\sqrt{3}}{6} \\ -\frac{1}{2} & \frac{\sqrt{3}}{6} \end{pmatrix} \quad \text{and} \quad M^{-1} = \begin{pmatrix} 1 & -1 \\ \sqrt{3} & \sqrt{3} \end{pmatrix}.$$

In the following we choose $\chi_2^{\Gamma(2)} = \tilde{\chi}_2$. The transfer operator for $\Gamma(2)$ can then be decomposed as

$$\tilde{\mathcal{L}}_\beta^{\Gamma(1),\chi^{\Gamma(2)}} \underline{f}(z,\varepsilon) = \sum_{n=1}^{\infty} (\frac{1}{z+n})^{2\beta} \tag{57}$$

$$\times \begin{pmatrix} \chi_1^{\Gamma(2)}(QT^{n\varepsilon}) & & & 0 \\ & \chi_{-1}^{\Gamma(2)}(QT^{n\varepsilon}) & & \\ & & \chi_2^{\Gamma(2)}(QT^{n\varepsilon}) & \\ 0 & & & \chi_2^{\Gamma(2)}(QT^{n\varepsilon}) \end{pmatrix} \underline{f}(\frac{1}{z+1}, -\varepsilon).$$

The representation U^χ for non-normal subgroups Γ is in general also reducible. As examples we consider the groups $\Gamma_0(2)$, $\Gamma^0(2)$ and Γ_ϑ. The induced representations χ^Γ in table 2 for the groups $\Gamma \in \{\Gamma_0(2), \Gamma^0(2), \Gamma_\theta\}$ can be decomposed into the following irreducible components:

$$M_\Gamma \chi^\Gamma(Q) M_\Gamma^{-1} = \begin{pmatrix} 1 & 0 & 0 \\ 0 & 1 & 0 \\ 0 & -1 & -1 \end{pmatrix} = \begin{pmatrix} \chi_1^\Gamma(Q) & 0 \\ 0 & \chi_2^\Gamma(Q) \end{pmatrix},$$

$$M_\Gamma \chi^\Gamma(T) M_\Gamma^{-1} = \begin{pmatrix} 1 & 0 & 0 \\ 0 & 0 & 1 \\ 0 & 1 & 0 \end{pmatrix} = \begin{pmatrix} \chi_1^\Gamma(T) & 0 \\ 0 & \chi_2^\Gamma(T) \end{pmatrix}, \tag{58}$$

with

$$\chi_1^\Gamma(Q) = 1,\ \chi_1^\Gamma(T) = 1 \quad \text{and}$$
$$\chi_2^\Gamma(Q) = \begin{pmatrix} 1 & 0 \\ -1 & -1 \end{pmatrix},\ \chi_2^\Gamma(T) = \begin{pmatrix} 0 & 1 \\ 1 & 0 \end{pmatrix}. \tag{59}$$

The matrices M_Γ and M_Γ^{-1} for the different groups Γ have the form

$$M_{\Gamma_0(2)} = \begin{pmatrix} -\frac{1}{9} & -\frac{1}{9} & -\frac{1}{9} \\ -1 & -1 & 2 \\ -1 & 2 & -1 \end{pmatrix},\quad M_{\Gamma_0(2)}^{-1} = \begin{pmatrix} -3 & -\frac{1}{3} & -\frac{1}{3} \\ -3 & 0 & \frac{1}{3} \\ -3 & \frac{1}{3} & 0 \end{pmatrix},$$
$$M_{\Gamma^0(2)} = \begin{pmatrix} \frac{1}{9} & \frac{1}{9} & \frac{1}{9} \\ 1 & 1 & -2 \\ -2 & 1 & 1 \end{pmatrix},\quad M_{\Gamma^0(2)}^{-1} = \begin{pmatrix} 3 & 0 & -\frac{1}{3} \\ 3 & \frac{1}{3} & \frac{1}{3} \\ 3 & -\frac{1}{3} & 0 \end{pmatrix},$$
$$M_{\Gamma_\vartheta(2)} = \begin{pmatrix} \frac{1}{9} & \frac{1}{9} & \frac{1}{9} \\ -2 & 1 & 1 \\ 1 & -2 & 1 \end{pmatrix},\quad M_{\Gamma_\vartheta(2)}^{-1} = \begin{pmatrix} 3 & -\frac{1}{3} & 0 \\ 3 & 0 & -\frac{1}{3} \\ 3 & \frac{1}{3} & \frac{1}{3} \end{pmatrix}.$$

According to (58) the transfer operators $\tilde{\mathcal{L}}_\beta^{\Gamma(1),\chi^\Gamma}$ for $\Gamma \in \{\Gamma_0(2),\ \Gamma^0(2),\ \Gamma_\theta\}$ can then be decomposed as follows:

$$\tilde{\mathcal{L}}_\beta^{\Gamma(1),\chi^\Gamma} \underline{f}(z,\varepsilon) = \sum_{n=1}^\infty \left(\frac{1}{z+n}\right)^{2\beta} \begin{pmatrix} \chi_1^\Gamma(QT^{n\varepsilon}) & 0 \\ 0 & \chi_2^\Gamma(QT^{n\varepsilon}) \end{pmatrix} \underline{f}\left(\frac{1}{z+n}, -\varepsilon\right). \tag{60}$$

Comparing (55) with (53) and (59) with (56) one concludes

$$\chi_{-1}^{\Gamma_2} = \chi_{-1}^{\Gamma(2)} \text{ and } \chi_2^{\Gamma(2)} = \chi_2^\Gamma \text{ for } \Gamma \in \{\Gamma_0(2),\ \Gamma^0(2),\ \Gamma_\theta\}.$$

Using the notations

$$\chi_1(Q) = 1, \qquad \chi_1(T) = 1, \tag{61}$$
$$\chi_{-1}(Q) = -1, \qquad \chi_{-1}(T) = -1, \tag{62}$$
$$\chi_2(Q) = \begin{pmatrix} 1 & 0 \\ -1 & -1 \end{pmatrix}, \chi_2(T) = \begin{pmatrix} 0 & 1 \\ 1 & 0 \end{pmatrix}, \tag{63}$$

one hence finds for the decompositions of the different induced representations χ^Γ of $\Gamma(1)$:

$$\begin{aligned} \chi^{\Gamma(1)} &= \chi_1, \\ \chi^{\Gamma_2} &= \chi_1 \oplus \chi_{-1}, \\ \chi^{\Gamma} &= \chi_1 \oplus \chi_2, \qquad\qquad \Gamma \in \{\Gamma_0(2), \Gamma^0(2), \Gamma_\vartheta\}, \\ \chi^{\Gamma(2)} &= \chi_1 \oplus \chi_{-1} \oplus \chi_2 \oplus \chi_2 \end{aligned} \tag{64}$$

and for the corresponding decompositions of the transfer operators $\tilde{\mathcal{L}}_\beta^{\Gamma(1),\chi^\Gamma}$:

$$\tilde{\mathcal{L}}_\beta^{\Gamma(1),1} f(z,\varepsilon) = \sum_{n=1}^{\infty} (\frac{1}{z+n})^{2\beta} f(\frac{1}{z+n}, -\varepsilon),$$

$$\tilde{\mathcal{L}}_\beta^{\Gamma(1),\chi^{\Gamma_2}} \underline{f}(z,\varepsilon) = \sum_{n=1}^{\infty} (\frac{1}{z+n})^{2\beta} \begin{pmatrix} \chi_1(QT^{n\varepsilon}) & 0 \\ 0 & \chi_{-1}(QT^{n\varepsilon}) \end{pmatrix} \underline{f}(\frac{1}{z+n}, -\varepsilon),$$

$$\tilde{\mathcal{L}}_\beta^{\Gamma(1),\chi^{\Gamma}} \underline{f}(z,\varepsilon) = \sum_{n=1}^{\infty} (\frac{1}{z+n})^{2\beta} \begin{pmatrix} \chi_1(QT^{n\varepsilon}) & 0 \\ 0 & \chi_2(QT^{n\varepsilon}) \end{pmatrix} \underline{f}(\frac{1}{z+n}, -\varepsilon),$$

$$\tilde{\mathcal{L}}_\beta^{\Gamma(1),\chi^{\Gamma(2)}} \underline{f}(z,\varepsilon) = \sum_{n=1}^{\infty} (\frac{1}{z+n})^{2\beta}$$

$$\times \begin{pmatrix} \chi_1(QT^{n\varepsilon}) & & & 0 \\ & \chi_{-1}(QT^{n\varepsilon}) & & \\ & & \chi_2(QT^{n\varepsilon}) & \\ 0 & & & \chi_2(QT^{n\varepsilon}) \end{pmatrix} \underline{f}(\frac{1}{z+1}, -\varepsilon).$$

As we shall see later, these decompositions of the transfer operators are closely related to the Venkov-Zograf factorization of the Selberg zeta function for the different subgroups $\Gamma \subseteq \Gamma(1)$. Before doing this we will first discuss spectral properties of the transfer operators.

4.4 Spectral Properties of the Transfer Operator

In the following we write the transfer operator $\tilde{\mathcal{L}}_\beta^{\Gamma(1),\chi^\Gamma}$ in (46) for an arbitrary subgroup $\Gamma \subseteq \Gamma(1)$ with finite index simply as $\tilde{\mathcal{L}}$:

$$\tilde{\mathcal{L}}\, \underline{f}(z,\varepsilon) = \sum_{n=1}^{\infty} (\frac{1}{z+n})^{2\beta} \chi^\Gamma(QT^{n\varepsilon}) \underline{f}(\frac{1}{z+n}, -\varepsilon). \tag{65}$$

Combining the functions $\underline{f}(z,\varepsilon)$ for $\varepsilon = \pm 1$ into the vector function

$$\underline{\underline{f}}(z) := \begin{pmatrix} \underline{f}(z,+1) \\ \underline{f}(z,-1) \end{pmatrix} = \begin{pmatrix} \underline{f}_+(z) \\ \underline{f}_-(z) \end{pmatrix}$$

the operator $\tilde{\mathcal{L}}$ can be written as

$$\tilde{\mathcal{L}}\, \underline{\underline{f}}(z) = \begin{pmatrix} 0 & \mathcal{L}_- \\ \mathcal{L}_+ & 0 \end{pmatrix} \begin{pmatrix} \underline{f}_+(z) \\ \underline{f}_-(z) \end{pmatrix} \tag{66}$$

with

$$(\mathcal{L}_+ \underline{f}_+)(z) = \sum_{n=1}^{\infty} (\frac{1}{z+n})^{2\beta} \chi^\Gamma(QT^n) \underline{f}_-(\frac{1}{z+n}), \tag{67}$$

$$(\mathcal{L}_- \underline{f}_-)(z) = \sum_{n=1}^{\infty} (\frac{1}{z+n})^{2\beta} \chi^\Gamma(QT^{-n}) \underline{f}_+(\frac{1}{z+n}), \tag{68}$$

with μ-dimensional vectors $\underline{f}_{\pm}$. Inserting (67) and (68) into (66) gives

$$\tilde{\mathcal{L}}\,\underline{\underline{f}}(z) = \sum_{n=1}^{\infty}\left(\frac{1}{z+n}\right)^{2\beta}\begin{pmatrix} 0 & \chi^{\Gamma}(QT^{-n}) \\ \chi^{\Gamma}(QT^{n}) & 0 \end{pmatrix}\underline{\underline{f}}\left(\frac{1}{z+n}\right) \tag{69}$$

with a 2μ-dimensional vector $\underline{\underline{f}}$.

Due to this structure trace $\tilde{\mathcal{L}}^n$ vanishes for odd n. The dynamical reason for this is that the orbits of the geodesic flow have at consecutive points of intersections with the Poincaré section different orientations described by $\varepsilon \to -\varepsilon$. Hence the Poincaré map P_Γ cannot have any fixed point. Only at the next intersection the orientation is the same and there can exist fixed points P_Γ^2 and so on.

An operator of the form (66) has very special spectral properties: Suppose $\tilde{\lambda}$ is an eigenvalue of the transfer operator $\tilde{\mathcal{L}}$ with eigenfunction $\underline{\underline{f}}(z) = \begin{pmatrix} \underline{f}_+(z) \\ \underline{f}_-(z) \end{pmatrix}$, i.e.,

$$\begin{pmatrix} 0 & \mathcal{L}_- \\ \mathcal{L}_+ & 0 \end{pmatrix}\begin{pmatrix} \underline{f}_+(z) \\ \underline{f}_-(z) \end{pmatrix} = \tilde{\lambda}\begin{pmatrix} \underline{f}_+(z) \\ \underline{f}_-(z) \end{pmatrix}. \tag{70}$$

Then $-\tilde{\lambda}$ is also an eigenvalue of $\tilde{\mathcal{L}}$ with eigenfunction $\begin{pmatrix} \underline{f}_+(z) \\ -\underline{f}_-(z) \end{pmatrix}$ as one can verify easily. Furthermore, equation (70) is equivalent to the equations

$$\mathcal{L}_+\underline{f}_+(z) = \tilde{\lambda}\underline{f}_-(z) \quad \text{and} \quad \mathcal{L}_-\underline{f}_-(z) = \tilde{\lambda}\underline{f}_+(z)\,.$$

They imply immediately

$$\mathcal{L}_-\mathcal{L}_+\underline{f}_+(z) = \tilde{\lambda}\mathcal{L}_-\underline{f}_-(z) = \tilde{\lambda}^2\underline{f}_+(z)$$

and

$$\mathcal{L}_+\mathcal{L}_-\underline{f}_-(z) = \tilde{\lambda}\mathcal{L}_+\underline{f}_+(z) = \tilde{\lambda}^2\underline{f}_-(z)\,.$$

That is, both the operators $\mathcal{L}_-\mathcal{L}_+$ respectively $\mathcal{L}_+\mathcal{L}_-$ have eigenvalue $\tilde{\lambda}^2$ with corresponding eigenfunction $\underline{f}_+(z)$ respectively $\underline{f}_-(z)$. Indeed the two operators $\mathcal{L}_-\mathcal{L}_+$ and $\mathcal{L}_+\mathcal{L}_-$ have the same eigenvalues: if $\underline{g}(z)$ is an eigenfunction with eigenvalue λ of $\mathcal{L}_-\mathcal{L}_+$ respectively $\mathcal{L}_+\mathcal{L}_-$, then $\mathcal{L}_+\underline{g}(z)$ respectively $\mathcal{L}_-\underline{g}(z)$ is an eigenfunction of $\mathcal{L}_+\mathcal{L}_-$ respectively $\mathcal{L}_-\mathcal{L}_+$ with the same eigenvalue, as long as the two operators $\mathcal{L}_+$ and $\mathcal{L}_-$ have trivial kernel. For the Fredholm determinant of $\tilde{\mathcal{L}}$ one finds

$$\begin{aligned} \det(1-\tilde{\mathcal{L}}) &= \det\left(1-\begin{pmatrix} 0 & \mathcal{L}_- \\ \mathcal{L}_+ & 0 \end{pmatrix}\right) \\ &= \det(1-\mathcal{L}_+\mathcal{L}_-) = \det(1-\mathcal{L}_-\mathcal{L}_+)\,, \end{aligned} \tag{71}$$

under the assumption that the operator $\tilde{\mathcal{L}}$ is for instance nuclear in some Banach space in the sense of Grothendieck, which will be shown later.

The only difference in the definition of the two operators $\mathcal{L}_+$ in (67) and $\mathcal{L}_-$ in (68) is the sign of the power of T in the representation χ^Γ. Hence $\mathcal{L}_+$ and $\mathcal{L}_-$ are identical, iff $\chi^\Gamma(QT^n) = \chi^\Gamma(QT^{-n})$ for all $n \in \mathbb{N}$, i.e., iff $\chi^\Gamma(T^2) = 1$ holds. For example this is true for the groups $\Gamma(1)$, $\Gamma(2)$, $\Gamma_0(2)$ and $\Gamma^0(2)$. In this case we set $\mathcal{L}_+ = \mathcal{L}_- := \mathcal{L}$ and relation (71) then says

$$\det(1 - \tilde{\mathcal{L}}) = \det(1 - \mathcal{L}^2) = \det(1 + \mathcal{L})\, \det(1 - \mathcal{L})\,. \tag{72}$$

To understand in this case the spectrum of $\tilde{\mathcal{L}} = \begin{pmatrix} 0 & \mathcal{L} \\ \mathcal{L} & 0 \end{pmatrix}$, it's obviously enough to study the spectrum of the operator

$$\mathcal{L}\, \underline{f}(z) = \sum_{n=1}^{\infty} (\frac{1}{z+n})^{2\beta} \chi^\Gamma(QT^n)\, \underline{f}(\frac{1}{z+n}) \tag{73}$$

where, compared to (65), the dependence on the parameter ε has disappeared. Assuming again nuclearity of $\tilde{\mathcal{L}}$ its spectrum consists of all numbers $+\lambda$ respectively $-\lambda$ with λ an eigenvalue of $\mathcal{L}$. The corresponding eigenfunctions of $\tilde{\mathcal{L}}$ are $\begin{pmatrix} \underline{f} \\ \underline{f} \end{pmatrix}$ respectively $\begin{pmatrix} \underline{f} \\ -\underline{f} \end{pmatrix}$ with $\mathcal{L}^2 \underline{f} = \lambda^2 \underline{f}$. For $\Gamma = \Gamma(1)$ and χ^Γ the trivial representation (73) is nothing but the transfer operator for the Gauß transformation [May91a], [CM99]

$$\mathcal{L}_\beta f(z) = \sum_{n=1}^{\infty} (\frac{1}{z+n})^{2\beta} f(\frac{1}{z+n})\,.$$

4.5 The Analytic Continuation of the Transfer Operator

We have still to find an appropriate Banach space of functions on which the transfer operators we have discussed up to now in a formal way are well defined nuclear operators. Consider vector valued functions $\underline{\underline{f}}$ given by

$$\underline{\underline{f}}(z) = \bigoplus_{\varepsilon=\pm 1,\, i=1\cdots\mu} f_{\varepsilon\, i}(z) = \begin{pmatrix} f_{+1}(z) \\ \vdots \\ f_{+\mu}(z) \\ f_{-1}(z) \\ \vdots \\ f_{-\mu}(z) \end{pmatrix} \tag{74}$$

where $\mu = [\Gamma(1) : \Gamma]$ is the dimension of the induced representation χ^Γ. For general subgroups $\Gamma \subseteq \Gamma(1)$ we choose in analogy to the case $\Gamma(1)$

the functions $f_{\varepsilon,i}$ to belong to the Banach space $B(D)$ of all holomorphic functions on the disk [May91a]

$$D := \{z | z \in \mathbb{C}, |z-1| < \frac{3}{2}\} \tag{75}$$

which are continuous on the closure $\bar{D}$. Consider then the transfer operator $\tilde{\mathcal{L}}_\beta$ as acting on the space $\oplus_{i=1\cdots 2\mu} B(D)$:

$$\tilde{\mathcal{L}}_\beta : \bigoplus_{i=1\cdots 2\mu} B(D) \to \bigoplus_{i=1\cdots 2\mu} B(D) . \tag{76}$$

Due to the contraction property [May91a] of the transformation $\psi_n(z) = \frac{1}{z+n}$ in $\tilde{\mathcal{L}}_\beta$ in (65) for all $n \in \mathbb{N}$ the transfer operator $\tilde{\mathcal{L}}_\beta$ is well defined on this space for $\Re\beta > \frac{1}{2}$ and even holomorphic in this half plane. We will show next, that the operator $\tilde{\mathcal{L}}_\beta$ can be analytically continued to a meromorphic family of operators in the entire complex β plane as follows:

In a first step we select a number $\kappa \in \mathbb{N}_0 = \mathbb{N} \cup \{0\}$. According to our discussion in paragraph 4.3 there exists a smallest number r with $0 < r < \infty$ and $\chi^\Gamma(T^r) = 1$. Using this property and writing $n = r(n'-1)+m$, $1 \le m \le r$ and $n' \ge 1$ the transfer operator (65) can be written as

$$\begin{aligned}
\tilde{\mathcal{L}}_\beta \underline{f}(z,\varepsilon) &= \sum_{n=1}^{\infty} (\frac{1}{z+n})^{2\beta} \chi^\Gamma(QT^{n\varepsilon})\, \underline{f}(\frac{1}{z+n}, -\varepsilon) \\
&= \sum_{n=1}^{\infty}\sum_{m=1}^{r} \Big(\frac{1}{z+m+r(n-1)}\Big)^{2\beta} \chi^\Gamma(QT^{(m+r(n-1))\varepsilon})\, \underline{f}\Big(\frac{1}{z+m+r(n-1)}, -\varepsilon\Big) \\
&= \Big\{ \sum_{n=1}^{\infty}\sum_{m=1}^{r} \Big(\frac{1}{z+m+r(n-1)}\Big)^{2\beta} \chi^\Gamma(QT^{m\varepsilon}) \\
&\quad \times \Big[\underline{f}\Big(\frac{1}{z+m+r(n-1)}, -\varepsilon\Big) - \sum_{l=0}^{\kappa} \frac{\underline{f}^{(l)}(0,-\varepsilon)}{l\,!} \Big(\frac{1}{z+m+r(n-1)}\Big)^{l}\Big]\Big\} \\
&\quad + \sum_{l=0}^{\kappa}\sum_{m=1}^{r} \chi^\Gamma(QT^{m\varepsilon}) \frac{\underline{f}^{(l)}(0,-\varepsilon)}{l\,!} (\frac{1}{r})^{2\beta+l} \sum_{n=1}^{\infty} \Big(\frac{1}{n+(\frac{z+m}{r}-1)}\Big)^{2\beta+l} .
\end{aligned} \tag{77}$$

The last sum can be expressed by the Hurwitz zeta function $\zeta(s, z+1) := \sum_{n=1}^{\infty} (\frac{1}{n+z})^s$ as follows:

$$\sum_{n=1}^{\infty} \Big(\frac{1}{n+(\frac{z+m}{r}-1)}\Big)^{2\beta+l} = \zeta(2\beta+l, \frac{z+m}{r}) . \tag{78}$$

Due to well known analyticity properties of this function and the property [MOS66]

$$\lim_{s\to 1} \Big[\zeta(s,z) - \frac{1}{s-1}\Big] = -\psi(z) = -\frac{\Gamma'(z)}{\Gamma(z)} , \tag{79}$$

where $\Gamma(z)$ is the gamma function and $\psi(z)$ is the psi function, the function (78) is meromorphic in the entire complex β-plane with poles of first order at those β values where $2\beta + l = 1$ with $l \in \mathbb{N}_0$. Substituting (78) into (77) shows that the operator $\tilde{\mathcal{L}}_\beta$ can be written as

$$\tilde{\mathcal{L}}_\beta = \tilde{\mathcal{A}}_\beta^{(\kappa)} + \tilde{\mathcal{L}}_\beta^\kappa\,, \tag{80}$$

with

$$\tilde{\mathcal{A}}_\beta^{(\kappa)} \underline{f}(z,\varepsilon) := \sum_{l=0}^{\kappa} (\frac{1}{r})^{2\beta+l} \sum_{m=1}^{r} \chi^\Gamma(QT^{m\varepsilon}) \frac{\underline{f}^{(l)}(0,-\varepsilon)}{l\,!} \zeta(2\beta+l, \frac{z+m}{r}) \tag{81}$$

and

$$\tilde{\mathcal{L}}_\beta^\kappa \underline{f}(z,\varepsilon) := \sum_{n=1}^{\infty} \sum_{m=1}^{r} \chi^\Gamma(QT^{m\varepsilon}) \Big(\frac{1}{z+m+r(n-1)}\Big)^{2\beta} \tag{82}$$
$$\times \Bigg[\underline{f}\Big(\frac{1}{z+m+r(n-1)}, -\varepsilon\Big) - \sum_{l=0}^{\kappa} \frac{\underline{f}^{(l)}(0,-\varepsilon)}{l\,!} \Big(\frac{1}{z+m+r(n-1)}\Big)^l\Bigg].$$

The operator $\tilde{\mathcal{L}}_\beta^\kappa$ is obviously holomorphic in the region $\Re\beta > -\frac{\kappa}{2}$ and the operator $\tilde{\mathcal{A}}_\beta^{(\kappa)}$ is meromorphic in $\mathbb{C}$ with possible poles only at $\beta = \beta_l = \frac{1-l}{2}$, $l = 0, 1, \dots, \kappa$. Consequently, the operator $\tilde{\mathcal{L}}_\beta$ is meromorphic in the region $\Re\beta > -\frac{\kappa}{2}$ with possible poles at the points $\beta = \beta_l$. Since $\kappa \in \mathbb{N}_0$ was arbitrary, $\tilde{\mathcal{L}}_\beta$ is meromorphic in the entire β-plane.

In the limit $\beta \to \beta_\kappa = \frac{1-\kappa}{2}$ the term $l = \kappa$ in the sum of $\tilde{\mathcal{A}}_\beta^{(\kappa)}$ in (81) is the only one which can become singular: due to (79) we have

$$\zeta(2\beta + \kappa, z) = \frac{1}{2}\frac{1}{\beta - \beta_\kappa} + O(1) \qquad \text{for } \beta \to \beta_\kappa\,. \tag{83}$$

Hence the operator $\tilde{\mathcal{A}}_\beta^{(\kappa)}$ for $\beta \to \beta_\kappa$ behaves like

$$\tilde{\mathcal{A}}_\beta^{(\kappa)} \underline{f}(z,\varepsilon) = \Big(\sum_{m=1}^{r} \chi^\Gamma(QT^{m\varepsilon})\Big) \Big[\underline{a}_\kappa \frac{\underline{f}^{(\kappa)}(0,-\varepsilon)}{\beta - \beta_\kappa} + O(1)\Big], \tag{84}$$

with

$$\underline{a}_\kappa := \frac{1}{2\,r\,\kappa\,!}. \tag{85}$$

Whether the operator $\tilde{\mathcal{A}}_\beta^{(\kappa)}$ has in $\beta = \beta_\kappa$ really a singularity, depends on the representation χ^Γ. For $r = 2$ and $\chi^\Gamma(Q) = \chi^\Gamma(T) = -1$ for example the operator $\mathcal{A}_\beta^{(\kappa)}$ is regular and therefore the operator $\tilde{\mathcal{L}}_\beta$ is holomorphic in the entire β-plane. The transfer operator for the group Γ_2 in the irreducible representation $\chi_2^{\Gamma_2}$ of the induced representation of χ^{Γ_2} belongs to this case.

4.6 Nuclearity of the Transfer Operator for $\Gamma \subseteq \Gamma(1)$

It is known that the transfer operator for the group $\Gamma(1)$ in the Banach space $B(D)$ is a nuclear operator [May91a] in the sense of Grothendieck [Gro55]. Here we will show that also the transfer operator for an arbitrary subgroup $\Gamma \subseteq \Gamma(1)$ with $[\Gamma(1) : \Gamma] < \infty$ is a nuclear operator in the Banach space $\oplus_{n=1}^{2\mu} B(D)$.

For this consider the analytically continued transfer operator $\tilde{\mathcal{L}}_\beta$ as defined in (80). Obviously the operator $\tilde{\mathcal{A}}_\beta^{(\kappa)}$ for $\beta \neq \beta_l$, $l = 1, \cdots, \kappa$ is a nuclear operator of order zero, because $\tilde{\mathcal{A}}_\beta^{(\kappa)}$ has finite rank $\kappa + 1$. To show the operator $\tilde{\mathcal{L}}_\beta^\kappa$ to be nuclear, we write $\tilde{\mathcal{L}}_\beta^\kappa$ analogous to $\tilde{\mathcal{L}}_\beta$ in (69) as follows:

$$\tilde{\mathcal{L}}_\beta^\kappa \underline{\underline{f}}(z) := \sum_{n=1}^{\infty} \sum_{m=1}^{r} \begin{pmatrix} 0 & \chi^\Gamma(QT^{-m}) \\ \chi^\Gamma(QT^m) & 0 \end{pmatrix} \left(\frac{1}{z+m+r(n-1)} \right)^{2\beta} \times \left[\underline{\underline{f}}\left(\frac{1}{z+m+r(n-1)} \right) - \sum_{l=0}^{\kappa} \frac{\underline{\underline{f}}^{(l)}(0)}{l\,!} \left(\frac{1}{z+m+r(n-1)} \right)^l \right]. \tag{86}$$

Suppose $\underline{\underline{f}} \in \mathbb{C}^{2\mu} \otimes B(D)$ with

$$\underline{\underline{f}} = \sum_{i=1}^{2\mu} \underline{e}_i \otimes f_i \,, \tag{87}$$

where $\{\underline{e}_1, \underline{e}_2, \cdots, \underline{e}_{2\mu}\}$ is a basis in $\mathbb{C}^{2\mu}$ and $f_i \in B(D)$ for $i = 1, 2, \cdots, 2\mu$. Then the operator $\tilde{\mathcal{L}}_\beta^\kappa$ in (82) can be written as

$$\begin{aligned} &\tilde{\mathcal{L}}_\beta^\kappa \underline{f}(z) \\ &= \sum_{i=1}^{2\mu} \sum_{m=1}^{r} \begin{pmatrix} 0 & \chi^\Gamma(QT^{-m}) \\ \chi^\Gamma(QT^m) & 0 \end{pmatrix} \underline{e}_i \otimes \sum_{n=1}^{\infty} \left(\frac{1}{z+m+r(n-1)} \right)^{2\beta} \\ &\times \left[f_i\left(\frac{1}{z+m+r(n-1)} \right) - \sum_{l=0}^{\kappa} \frac{f_i^{(l)}(0)}{l\,!} \left(\frac{1}{z+m+r(n-1)} \right)^l \right]. \end{aligned} \tag{88}$$

Using the notations $\chi_m : \mathbb{C}^{2\mu} \to \mathbb{C}^{2\mu}$

$$\chi_m := \begin{pmatrix} 0 & \chi^\Gamma(QT^{-m}) \\ \chi^\Gamma(QT^m) & 0 \end{pmatrix},$$

and

$$\mathcal{L}_{\beta,m} f(z) := \sum_{n=1}^{\infty} \left(\frac{1}{z+m+r(n-1)} \right)^{2\beta} \times \left[f\left(\frac{1}{z+m+r(n-1)} \right) - \sum_{l=0}^{\kappa} \frac{f^{(l)}(0)}{l\,!} \left(\frac{1}{z+m+r(n-1)} \right)^l \right] \tag{89}$$

the operator (88) reads as

$$\tilde{\mathcal{L}}_\beta^\kappa \underline{f}(z) = \sum_{i=1}^{2\mu} \sum_{m=1}^{r} \chi_m \, \underline{e}_i \otimes \mathcal{L}_{\beta,\,m} f_i(z) \tag{90}$$

and hence we find

$$\tilde{\mathcal{L}}_\beta^\kappa = \sum_{m=1}^{r} \chi_m \otimes \mathcal{L}_{\beta,\,m} \,. \tag{91}$$

Since the operator $\mathcal{L}_{\beta,\,m}$ for $\Re\beta > -\frac{\kappa}{2}$ is a nuclear operator of order zero [May91a], the operator $\tilde{\mathcal{L}}_\beta^\kappa$ is also a nuclear operator of order zero for all $\Re\beta > -\frac{\kappa}{2}$, $\beta \neq \beta_\kappa$ and consequently of trace class. The same holds then for the transfer operator $\tilde{\mathcal{L}}_\beta$.

5 The Thermodynamic Formalism Approach to Selberg's Zeta Function for $\Gamma \subseteq \Gamma(1)$

Having established the trace class property for $\tilde{\mathcal{L}}_\beta$ we can now apply this operator to dynamical zeta functions and especially the Selberg zeta function for $\Gamma \subseteq \Gamma(1)$.

5.1 Generalized Selberg Zeta Functions for $PSL(2, \mathbb{Z})$

The dynamical zeta function $\zeta_{RS}(\beta; \chi)$ of Ruelle and Smale for the group $\Gamma(1)$ with some representation $\chi : \Gamma(1) \to GL(V)$ is defined as [Rue94]

$$\zeta_{RS}(\beta; \chi) = \prod_\gamma \left[\det \left(1 - \chi(\sigma_\gamma)\, e^{-\beta\, l(\gamma)} \right) \right]^{-1}, \tag{92}$$

where the product runs over all primitive periodic orbits[1] γ of the geodesic flow (25) on the modular surface, $l(\gamma)$ denotes the period of γ and σ_γ the hyperbolic element in the group $\Gamma(1)$ which fixes the geodesic γ in $\mathbb{H}$, i.e., $\sigma_\gamma \gamma = \gamma$. Obviously σ_γ is determined only up to conjugation with some element in $\Gamma(1)$.

We briefly recall the description of the periodic orbits for the geodesic flow for $\Gamma(1)$ since for general subgroups $\Gamma \subseteq \Gamma(1)$ we will heavily rely on this. Consider the Poincaré map $P : X \to X$ in (32) given as

$$Px = P(x_1, x_2, \varepsilon) = (T_G x_1, \frac{1}{[\frac{1}{x_1}] + x_2}, -\varepsilon\,), \tag{93}$$

[1] Notice that geodesics on $M_{\Gamma(1)}$ and geodesic orbits on $T_1 M_{\Gamma(1)}$ are not the same. But we use γ to denote both of them, since they are closely related to each other. $l(\gamma)$ stands for the period of the periodic geodesic orbit respectively the length of the geodesic.

where $x = \gamma \cap X$ denotes the point of intersection of γ with the Poincaré section X. The orbit γ is closed, iff there exists some $m \geq 1$ such that $P^m x = x$. This implies

$$T_G^m x_1 = x_1 \quad \text{and} \quad (-1)^m \varepsilon = \varepsilon \tag{94}$$

and therefore

$$x_1 = [\overline{n_0, n_1, \cdots, n_{m-2}, n_{m-1}}] \quad \text{with even } m\,.$$

A simple calculation shows that x_2 in (93) must then be of the form:

$$x_2 = [\overline{n_{m-1}, n_{m-2}, \cdots, n_1, n_0}].$$

That means up to conjugation a periodic orbit γ exactly corresponds to a geodesic with basepoints

$$\gamma_{-\infty} = -\varepsilon\,[\overline{n_{m-1}, n_{m-2}, \cdots, n_1, n_0}], \quad \gamma_{+\infty} = \varepsilon\,[\overline{n_0, n_1, \cdots, n_{m-2}, n_{m-1}}]^{-1}$$

in the representation (20). Since

$$QT^{n_{m-1}\varepsilon} QT^{-n_{m-2}\varepsilon} \cdots QT^{n_1\varepsilon} QT^{-n_0\varepsilon} \gamma_{\pm\infty} = \gamma_{\pm\infty},$$

this closed geodesic γ is fixed by the hyperbolic element

$$\sigma_\gamma = QT^{n_{m-1}\varepsilon} QT^{-n_{m-2}\varepsilon} \cdots QT^{n_1\varepsilon} QT^{-n_0\varepsilon} \quad \text{with even } m\,. \tag{95}$$

The period of such an orbit respectively the length of the corresponding closed geodesic can be expressed in terms of the recurrence time function (roof function) $r(x)$ [Pol86] for the geodesic flow

$$l(\gamma) = \sum_{k=0}^{m-1} r(P^k x)$$

where $x \in \gamma \cap X$ and $r(x)$ denotes the recurrence time respectively the length of the geodesic orbit γ from x to Px. For the geodesic flow for $\Gamma(1)$, this function $r(x)$ is equal to $r(x) = r((x_1, x_2, \varepsilon)) = log|T_G'(x_1)| = -log\, x_1^2$ [May91a]. Since the hyperbolic length is invariant under $PSL(2, \mathbb{R})$, this function is also the recurrence time function for the geodesic flow of any subgroup, $\Gamma \subseteq \Gamma(1)$ with respect to the Poincaré section X_Γ.

To apply this to the generalized Ruelle-Smale function in (92) we use a result of Ruelle [Rue94]:

Lemma 2. *Let $\tau : M \to M$ be a discrete time dynamical system and $\Phi : M \to GL(\mu, \mathbb{C})$ a matrix-valued function. Then the following identity holds*

$$\zeta_R(z, \Phi) = \prod_\gamma \left[\det \left(1 - z^{n(\gamma)} \prod_{k=0}^{n(\gamma)-1} \Phi(\tau^k \xi_\gamma)\right)\right]^{-1} \tag{96}$$

$$= \exp \sum_{n=1}^{\infty} \frac{z^m}{m} Z_m(\Phi) \tag{97}$$

where the first product runs over all primitive periodic orbit $\gamma \in M$ *with* $\xi_\gamma \in \gamma$ *and* $n(\gamma) \in \mathbb{N}$ *is the period of* γ, *i.e.*, $\tau^{n(\gamma)}\xi_\gamma = \xi_\gamma$, *and* $Z_m(\Phi)$ *denotes the partition function*

$$Z_m(\Phi) = \sum_{\xi \in Fix\, \tau^m} \operatorname{trace}\Big(\prod_{k=0}^{m-1} \Phi(\tau^k \xi) \Big) \tag{98}$$

where the sum runs over all fixed point of τ^m.

The dynamical zeta function (92) is a special case of $\zeta_R(z, \Phi)$ in (96) with $z = 1$, $\tau = P$ and

$$\Phi(x) = \chi\,(QT^{N(x)}) \exp(-\beta\, r(x))\,, \tag{99}$$

where $N(x) = -\varepsilon\, n(x)$ with $n(x) = [\frac{1}{x_1}]$ for $x = (x_1, x_2, \varepsilon) \in X$: inserting (99) into definition (96) namely gives

$$\begin{aligned}
&\zeta_{RS}(\beta; \chi) = \zeta_R(1, \Phi) \\
&= \prod_\gamma \Big[\det\, \big[1 - \prod_{k=0}^{m-1} \chi(QT^{N(P^k x)}) \exp\big(-\beta r(P^k x)\big)\big]\Big]^{-1}
\end{aligned}$$

with $P^m x = x$ and therefore

$$\begin{aligned}
\zeta_{RS}(\beta; \chi) &= \prod_\gamma \Big[\det \Big(1 - \chi(\sigma_\gamma) \exp\big(-\beta \sum_{k=0}^{m-1} r(P^k x)\big)\Big)\Big]^{-1} \\
&= \prod_\gamma \Big[\det \Big(1 - \chi(\sigma_\gamma)\, e^{-\beta\, l(\gamma)}\Big)\Big]^{-1}.
\end{aligned} \tag{100}$$

Due to (97) the dynamical zeta function (92) hence can be expressed as

$$\zeta_{RS}(\beta; \chi) = \sum_{m=1}^{\infty} \frac{Z_m(\beta; P; \chi)}{m}\,, \tag{101}$$

with Z_m the partition function in (98) and Φ defined in (99):

$$\begin{aligned}
Z_m(\beta; P; \chi) &= \sum_{x \in Fix\, P^m} \operatorname{trace}\Big[\prod_{k=0}^{m-1} \Phi(P^k x)\Big] \\
&= \sum_{x \in Fix\, P^m} \operatorname{trace}\Big[\prod_{k=0}^{m-1} [\chi\,(QT^{N(P^k x)}) \exp(-\beta\, r(P^k x))]\Big] \\
&= \sum_{x \in Fix\, P^m} \operatorname{trace} \chi\,(\sigma_\gamma) \exp\Big(-\beta \sum_{k=0}^{m-1} r(P^k x)\Big).
\end{aligned} \tag{102}$$

Z_m obviously vanishes for odd integers m, because according to (94) P^m doesn't have fixed points for odd m.

The generalized Selberg zeta function for Γ with representation χ is defined as [Ven90]

$$Z_S(\beta;\Gamma(1);\chi) = \prod_{\gamma}\prod_{k=0}^{\infty} \det\left(1 - \chi(\sigma_\gamma)\, e^{-(\beta+k)\, l(\gamma)}\right). \tag{103}$$

Comparing with expression (92) shows that Selberg's zeta function can be expressed as

$$Z_S(\beta;\Gamma(1);\chi) = \prod_{k=0}^{\infty} \zeta_{RS}(\beta+k;\chi)^{-1} \tag{104}$$

in terms of the Ruelle-Smale dynamical zeta function with representation χ.

5.2 Dynamical Zeta Functions and Fredholm Determinants of Transfer Operators

As in statistical mechanics the transfer operator in the thermodynamic formalism of dynamical systems is used to determine partition functions as in (98) [May80]. In the case of the geodesic flow on modular surfaces this can be done as follows:

Using the notation $Uz = \frac{1}{z}$ the transfer operator $\tilde{\mathcal{L}}_\beta$ in (69) can be expressed as

$$\begin{aligned}
\tilde{\mathcal{L}}_\beta \underline{f}(z) &= \sum_{n_1=1}^{\infty} \begin{pmatrix} 0 & \chi^\Gamma(QT^{-n_1}) \\ \chi^\Gamma(QT^{n_1}) & 0 \end{pmatrix} \left(\frac{1}{n_1+z}\right)^{2\beta} \underline{f}\left(\frac{1}{n_1+z}\right) \\
&= \sum_{n_1=1}^{\infty} \begin{pmatrix} 0 & \chi^\Gamma(QT^{-n_1}) \\ \chi^\Gamma(QT^{n_1}) & 0 \end{pmatrix} (UT^{n_1}z)^{2\beta} \underline{f}(UT^{n_1}z),
\end{aligned}$$

where for short $\underline{\underline{f}}(z)$ has been replaced by $\underline{f}(z)$. The m-th iterate of the operator $\tilde{\mathcal{L}}_\beta$ is then equal to

$$\begin{aligned}
\tilde{\mathcal{L}}_\beta^m \underline{f}(z) &= \underbrace{(\tilde{\mathcal{L}}_\beta \circ \tilde{\mathcal{L}}_\beta \circ \cdots \circ \tilde{\mathcal{L}}_\beta)}_{m-times} \underline{f}(z) \\
= & \sum_{n_1,n_2,\dots,n_m=1}^{\infty} \chi_{n_1,n_2,\dots,n_m} \\
& \times \left[(UT^{n_m}z)(UT^{n_{m-1}}UT^{n_m}z)\cdots(UT^{n_1}UT^{n_2}\cdots UT^{n_{m-1}}UT^{n_m}z)\right]^{2\beta} \\
& \times \underline{f}(UT^{n_1}UT^{n_2}\cdots UT^{n_{m-1}}UT^{n_m}z),
\end{aligned} \tag{105}$$

with

$$\chi_{n_1,n_2,\dots,n_m} = \begin{cases} \begin{pmatrix} \chi^{\Gamma}(\sigma_{n_1,n_2,\dots,n_m,-1}) & 0 \\ 0 & \chi^{\Gamma}(\sigma_{n_1,n_2,\dots,n_m,+1}) \end{pmatrix} & \text{for } m \text{ even} \\ \begin{pmatrix} 0 & \chi^{\Gamma}(\sigma_{n_1,n_2,\dots,n_m,+1}) \\ \chi^{\Gamma}(\sigma_{n_1,n_2,\dots,n_m,-1}) & 0 \end{pmatrix} & \text{for } m \text{ odd} \end{cases} \tag{106}$$

and

$$\sigma_{n_1,n_2,\dots,n_m,\varepsilon} = QT^{(-1)^m n_m \varepsilon} QT^{(-1)^{m-1} n_{m-1}\varepsilon} \cdots QT^{n_2\varepsilon} QT^{-n_1\varepsilon}, \quad \varepsilon = \pm 1 . \tag{107}$$

Since $\tilde{\mathcal{L}}_\beta$ is nuclear, also $\tilde{\mathcal{L}}_\beta^m$ is nuclear for all $m \in \mathbb{N}$. With the abbreviations

$$\psi_{n_1,n_2,\dots,n_m}(z) = UT^{n_1}UT^{n_2}\cdots UT^{n_{m-1}}UT^{n_m}z, \tag{108}$$

$$\varphi_{n_1,n_2,\dots,n_m}(z) = \Big[\prod_{k=1}^{m}(UT^{n_k}UT^{n_{k+1}}\cdots UT^{n_{m-1}}UT^{n_m}z)\Big]^{2\beta},$$

$$\Phi_{n_1,n_2,\dots,n_m}(z) = \chi_{n_1,n_2,\dots,n_m}\,\varphi_{n_1,n_2,\dots,n_m}(z), \tag{109}$$

(105) can be rewritten as

$$\tilde{\mathcal{L}}_\beta^m \underline{f}(z) = \sum_{n_1,n_2,\dots,n_m=1}^{\infty} \Phi_{n_1,n_2,\dots,n_m}(z)\,\underline{f}(\psi_{n_1,n_2,\dots,n_m}(z)). \tag{110}$$

In analogy we find

$$(-\tilde{\mathcal{L}}_{\beta+1})^m \underline{f}(z) = \sum_{n_1,n_2,\dots,n_m=1}^{\infty} \Phi_{n_1,n_2,\dots,n_m}(z)\,\psi'_{n_1,n_2,\dots,n_m}(z)\,\underline{f}(\psi_{n_1,n_2,\dots,n_m}(z)). \tag{111}$$

To calculate the trace of the operators $\tilde{\mathcal{L}}_\beta^m$ and $(-\tilde{\mathcal{L}}_{\beta+1})^m$, we follow the arguments in [May91a] :

The operator (110) has the form

$$L\underline{f}(z) = \sum_i L_i\underline{f}(z), \tag{112}$$

where i stands for the multi-index $(n_1, n_2, \dots, n_m)$ and L_i is a composition operator [May91a] of the form

$$L_i\underline{f}(z) = \Phi_i(z)\,\underline{f}(\psi_i(z)), \tag{113}$$

with Φ_i a matrix-valued function. To determine trace L_i consider the eigenfunction equation

$$L_i \underline{f}(z) = \Phi_i(z) \underline{f}(\psi_i(z)) = \lambda \underline{f}(z) . \tag{114}$$

Since ψ_i in (108) is holomorphic in the disk D in (75) and maps the disk D strictly inside itself, ψ_i has exactly one fixed point z_i^* in D [Rue76]. At the point $z = z_i^*$ equation (114) reads

$$\Phi_i(z_i^*) \underline{f}(z_i^*) = \lambda \underline{f}(z_i^*) ,$$

i.e., if $\underline{f}(z_i^*) \neq \underline{0}$, then λ must be an eigenvalue of $\Phi(z_i^*)$ with corresponding eigenvector $\underline{f}(z_i^*)$. If $\underline{f}(z_i^*) = \underline{0}$, then we differentiate equation (114) with respect to z and get

$$D\Phi_i(z) \underline{f}(\psi_i(z)) + \Phi_i(z) \underline{f}'(\psi_i(z)) \psi_i'(z) = \lambda \underline{f}'(z) . \tag{115}$$

Since $\underline{f}(z_i^*) = \underline{0}$, setting $z = z_i^*$ in (115) one gets the equation

$$\psi_i'(z_i^*) \Phi_i(z_i^*) \underline{f}'(z_i^*) = \lambda \underline{f}'(z_i^*) \tag{116}$$

and therefore $\lambda = \rho \psi_i'(z_i^*)$, with ρ an eigenvalue of the matrix $\Phi_i(z_i^*)$ with eigenvector $\underline{f}'(z_i^*) \neq \underline{0}$. If also $\underline{f}'(z_i^*) = \underline{0}$, one differentiates equation (115) again. Repeating the argument shows that the eigenvalues of L_i must belong to the set $\{\rho \psi_i'(z_i^*)^n\}$, $n \in \mathbb{N}_0$, where ρ runs over all eigenvalues of the matrix $\Phi_i(z_i^*)$.

Conversely, to show that every one of the numbers $\lambda_n := \rho \psi_i'(z_i^*)^n$ really belongs to the spectrum of L_i, one must show $(L_i - \rho \psi_i'(z_i^*)^n)$ is not invertible, that means there is no solution $\underline{f}(z)$ for the equation

$$(L_i - \rho \psi_i'(z_i^*)^n) \underline{f}(z) = \underline{g}(z) \tag{117}$$

respectively

$$\Phi_i(z) \underline{f}(\psi_i(z)) - \rho \psi_i'(z_i^*)^n \underline{f}(z) = \underline{g}(z) \tag{118}$$

for a certain function $\underline{g}(z)$ in $\oplus_{i=1}^{2\mu} B(D)$. Let us choose a function $\underline{g}(z)$ with the properties

$$\underline{g}^{(k)}(z_i^*) = 0 \quad \text{for } 0 \leq k \leq n-1 \quad \text{and} \quad \underline{g}^{(n)}(z_i^*) \neq \underline{0} . \tag{119}$$

Then for $n = 0$ it follows from (118) that

$$(\Phi_i(z_i^*) - \rho) \underline{f}(z_i^*) = \underline{g}(z_i^*) \neq \underline{0} . \tag{120}$$

One then chooses $\underline{g}(z_i^*)$ such that this matrix equation doesn't have a solution $\underline{f}(z_i^*)$. This is possible since ρ is an eigenvalue of $\Phi_i(z_i^*)$ and thus $(\Phi_i(z_i^*) - \rho)$ is not invertible. Therefore $\lambda_0 = \rho$ is in the spectrum of L_i.

For $n \geq 1$ we have to consider two cases. First, suppose $\rho\psi_i'(z_i^*)^k$ are not eigenvalues of $\Phi_i(z_i^*)$ for all $k \in \mathbb{N}$. Equation (118) at $z = z_i^*$ then reads:

$$(\Phi_i(z_i^*) - \rho\psi_i'(z_i^*)^n)\underline{f}(z_i^*) = \underline{g}(z_i^*) = \underline{0}. \tag{121}$$

This implies immediately $\underline{f}(z_i^*) \equiv \underline{0}$, because $\rho\psi_i'(z_i^*)^n$ is not an eigenvalue of $\Phi_i(z_i^*)$. Differentiating equation (118) once and setting $z = z_i^*$ gives

$$D\Phi_i(z_i^*)\underline{f}(z_i^*) + \Phi_i(z_i^*)\psi_i'(z_i^*)\underline{f}'(z_i^*) - \rho\psi_i'(z_i^*)^n\underline{f}'(z_i^*) = \underline{g}'(z_i^*) = \underline{0}\,.$$

Since $\underline{f}(z_i^*) = \underline{0}$ we get

$$\psi_i'(z_i^*)\,(\,\Phi_i(z_i^*) - \rho\psi_i'(z_i^*)^{n-1}\,)\,\underline{f}'(z_i^*) = g'(z_i^*) = \underline{0}$$

and therefore $\underline{f}'(z_i^*) \equiv \underline{0}$, because $\rho\psi_i'(z_i^*)^{n-1}$ is not an eigenvalue of $\Phi_i(z_i^*)$. Repeating this argument n-times one finds

$$\psi_i'(z_i^*)^n(\Phi_i(z_i^*) - \rho)\underline{f}^{(n)}(z_i^*) = \underline{g}^{(n)}(z_i^*) \neq \underline{0}\,. \tag{122}$$

Choose now the vector $\underline{g}^{(n)}(z_i^*)$ such that equation (122) doesn't have a solution $\underline{f}^{(n)}(z_i^*)$. This is possible, because ρ is an eigenvalue of $\Phi_i(z_i^*)$. This shows that for a function $\underline{g}(z)$ obeying the conditions (119) equation (117) does not have a solution and therefore all $\rho\psi_i'(z_i^*)^n$ with $n \in \mathbb{N}_0$ are eigenvalues of L_i. The fact that the eigenvalues $\rho\psi_i'(z_i^*)^n$ for a given ρ are indeed simple follows from arguments similar to the ones given in [May91a]. The trace of L_i is therefore the sum of the geometrical series $\rho\psi_i'(z_i^*)^n$ summed over all eigenvalues ρ of $\Phi_i(z_i^*)$, i.e.,

$$\operatorname{trace} L_i = \frac{\operatorname{trace} \Phi_i(z_i^*)}{1 - \psi_i'(z_i^*)}\,. \tag{123}$$

In the second case suppose some of the numbers $\rho\psi_i'(z_i^*)^k$ for $k \in \mathbb{N}$ are eigenvalues of $\Phi_i(z_i^*)$. Then we consider a new transformation $\psi_{i,\delta} : D \to D$ slightly deformed in a δ-neighborhood of the transformation ψ_i. Obviously $\psi_{i,\delta}$ will have new fixed points $z_{i,\delta}^*$, slightly different from the z_i^*. We choose $\psi_{i,\delta}$ such that all $\rho\psi_{i,\delta}'(z_{i,\delta}^*)^k$ for $k \in \mathbb{N}_0$ don't belong to the spectrum of $\Phi_i(z_i^*)$. This is possible, because $\Phi_i(z_i^*)$ in (109) is a finite-dimensional permutation matrix and has only finitely many non-vanishing eigenvalues and $\rho\psi_{i,\delta}'(z_{i,\delta}^*)^n$ converges to zero for large n. Repeating the arguments of the first case shows

$$\operatorname{trace} L_{i,\delta} = \frac{\operatorname{trace} \Phi_i(z_{i,\delta}^*)}{1 - \psi_{i,\delta}'(z_{i,\delta}^*)}$$

Since the trace is continuous in ψ_i, taking the limit $\delta \to 0$ one finds

$$\lim_{\delta \to 0} \operatorname{trace} L_{i,\delta} = \operatorname{trace} L_i \;=\; \frac{\operatorname{trace} \Phi_i(z_i^*)}{1 - \psi_i'(z_i^*)}$$

as in (123).

Summarizing, we have therefore shown for the operator L in (112):

$$\operatorname{trace} L = \sum_i \operatorname{trace} L_i = \sum_i \frac{\operatorname{trace} \Phi_i(z_i^*)}{1 - \psi_i'(z_i^*)} . \tag{124}$$

With the definition

$$L_i^{(s)} f(z) = \Phi_i(z) \left(\psi_i'(z)\right)^s f(\psi_i(z)), \quad s = 0, 1$$

and $L^{(s)} = \sum_i L_i^{(s)}$ one finally gets the relation

$$\begin{aligned} \operatorname{trace} L^{(0)} - \operatorname{trace} L^{(1)} &= \sum_i \left(\operatorname{trace} L_i^{(0)} - \operatorname{trace} L_i^{(1)} \right) \\ &= \sum_i \operatorname{trace} \Phi_i(z_i^*) . \end{aligned} \tag{125}$$

Applying this relation to the operators (110) and (111) yields

$$\begin{aligned} &\operatorname{trace} \tilde{\mathcal{L}}_\beta^m - \operatorname{trace}(-\tilde{\mathcal{L}}_{\beta+1})^m \\ &= \sum_{n_1, n_2, \dots, n_m = 1}^{\infty} \operatorname{trace}\left(\chi_{n_1, n_2, \dots, n_m}\right) \varphi_{n_1, n_2, \dots, n_m}(z^*_{n_1, n_2, \dots, n_m}) \end{aligned} \tag{126}$$

with $z^*_{n_1, n_2, \dots, n_m}$ the fixed point of $\psi_{n_1, n_2, \dots, n_m}(z)$ in (108), i.e.,

$$UT^{n_1} UT^{n_2} \cdots UT^{n_{m-1}} UT^{n_m} z^*_{n_1, n_2, \dots, n_m} = z^*_{n_1, n_2, \dots, n_m}$$

and consequently

$$z^*_{n_1, n_2, \dots, n_m} = [\overline{n_1, n_2, \cdots, n_{m-1}, n_m}] . \tag{127}$$

This however is also a fixed point of T_G^m in (94). For this fixed point one has

$$\begin{aligned} \varphi_{n_1, n_2, \dots, n_m}(z^*_{n_1, n_2, \dots, n_m}) &= \left(\prod_{k=1}^{m} [\overline{n_k, n_{k+1}, \cdots, n_m, n_1, \cdots, n_{k-1}}] \right)^{2\beta} \\ &= \left(\prod_{k=0}^{m-1} T_G^k \, z^*_{n_1, n_2, \dots, n_m} \right)^{2\beta} . \end{aligned} \tag{128}$$

Furthermore, it follows from (106) that

$$\operatorname{trace}(\chi_{n_1, n_2, \dots, n_m}) = \begin{cases} 0 & m \text{ odd}, \\ \sum_{\varepsilon = \pm 1} \operatorname{trace} \chi^{\Gamma}\left(\sigma_{n_1, n_2, \dots, n_m, \varepsilon}\right) & m \text{ even}. \end{cases} \tag{129}$$

Inserting (128) and (129) in (126) one gets for odd m

$$\operatorname{trace} \tilde{\mathcal{L}}_\beta^m - \operatorname{trace}(-\tilde{\mathcal{L}}_{\beta+1})^m = 0$$

and for even m

$$\operatorname{trace} \tilde{\mathcal{L}}_\beta^m - \operatorname{trace}(-\tilde{\mathcal{L}}_{\beta+1})^m = \sum_{z \in Fix\, T_G^m} \sum_{\varepsilon=\pm 1} \operatorname{trace} \chi^\Gamma(\sigma_{z,\varepsilon}) \Big(\prod_{k=0}^{m-1} T_G^k z \Big)^{2\beta}, \tag{130}$$

where the fixed points $z^*_{n_1,n_2,\dots,n_m}$ of $\psi_{n_1,n_2,\dots,n_m}(z)$ are replaced by the fixed points z of T_G^m and the element $\sigma_{n_1,n_2,\dots,n_m,\varepsilon}$, which depends on z, is abbreviated as $\sigma_{z,\varepsilon}$. The summation on the right-hand side of (130) gives nothing but the partition function $Z_m(\beta; P; \chi)$ in (102) for $\chi = \chi^\Gamma$. Notice that σ_γ in formula (102) coincides with $\sigma_{n_1,n_2,\dots,n_m,\varepsilon}$ in (107) and can be expressed in the matrix form (106). Hence we find

$$\operatorname{trace} \tilde{\mathcal{L}}_\beta^m - \operatorname{trace}(-\tilde{\mathcal{L}}_{\beta+1})^m = Z_m(\beta; P; \chi^\Gamma), \tag{131}$$

which is valid not only for even but also for m odd, because for m odd both sides of (131) vanish.

Finally we can express the dynamical zeta function in (101) in terms of the transfer operator $\tilde{\mathcal{L}}_\beta$ as follows:

$$\begin{aligned} \zeta_{RS}(\beta; \chi^\Gamma) &= \exp \sum_{m=1}^\infty \frac{Z_m(\beta; P; \chi^\Gamma)}{m} \\ &= \exp \sum_{m=1}^\infty \frac{1}{m} (\operatorname{trace} \tilde{\mathcal{L}}_\beta^m - \operatorname{trace}(-\tilde{\mathcal{L}}_{\beta+1})^m) \\ &= \frac{\exp \sum_{m=1}^\infty \frac{1}{m} \operatorname{trace} \tilde{\mathcal{L}}_\beta^m}{\exp \sum_{m=1}^\infty \frac{1}{m} \operatorname{trace}(-\tilde{\mathcal{L}}_{\beta+1})^m}. \end{aligned}$$

Since $\operatorname{trace} \tilde{\mathcal{L}}_\beta^m = 0$ for m odd, this yields immediately

$$\zeta_{RS}(\beta; \chi^\Gamma) = \frac{\exp \sum_{m=1}^\infty \frac{1}{m} \operatorname{trace} \tilde{\mathcal{L}}_\beta^m}{\exp \sum_{m=1}^\infty \frac{1}{m} \operatorname{trace} \tilde{\mathcal{L}}_{\beta+1}^m}. \tag{132}$$

Applying the identity [Gro56]

$$\begin{aligned} \det(1 - z\mathcal{L}) &= \exp(\operatorname{trace} log(1 - z\mathcal{L})) = \exp\Big(- \operatorname{trace} \sum_{m=1}^\infty \frac{(z\mathcal{L})^m}{m} \Big) \\ &= \frac{1}{\exp \sum_{m=1}^\infty \frac{1}{m} \operatorname{trace}(z\mathcal{L})^m} \end{aligned}$$

for nuclear operators $\mathcal{L}$ of order zero, we obtain from (132) the formula

$$\zeta_{RS}(\beta; \chi^\Gamma) = \frac{\det(1 - \tilde{\mathcal{L}}_{\beta+1})}{\det(1 - \tilde{\mathcal{L}}_\beta)}. \tag{133}$$

5.3 The Selberg Zeta Function for Subgroups $\Gamma \subseteq \Gamma(1)$

Equality (133) gives a connection between the transfer operator $\tilde{\mathcal{L}}_\beta$ and the dynamical zeta function $\zeta_{RS}(\beta;\chi^\Gamma)$. Since on the other hand the dynamical zeta function is related to Selberg's zeta function (104) through [Ven90], [Rue94], [Cha99]

$$Z_S(\beta;\Gamma(1);\chi^\Gamma) = \prod_{k=0}^{\infty} \zeta_{RS}(\beta+k;\chi^\Gamma)^{-1}\,, \tag{134}$$

Selberg's zeta gets expressed in terms of the transfer operator by inserting (133) into (134):

Theorem 3. *Let Γ be a subgroup of $\Gamma(1)$ of finite index and χ^Γ be the representation of $\Gamma(1)$ induced from the trivial representation of the subgroup Γ. Furthermore, let $\tilde{\mathcal{L}}_\beta^\Gamma$ respectively $\tilde{\mathcal{L}}_\beta^{\Gamma(1),\chi^\Gamma}$ be the transfer operators for the geodesic flows on $\Gamma\backslash\mathbb{H}$ respectively $\Gamma(1)\backslash\mathbb{H}$ with representation χ^Γ. Then the Selberg zeta functions $Z_S(\beta;\Gamma)$ respectively $Z_S(\beta;\Gamma(1);\chi^\Gamma)$ are related to the Fredholm determinants of these transfer operators by*

$$Z_S(\beta;\Gamma(1);\chi^\Gamma) = \det(1-\tilde{\mathcal{L}}_\beta^{\Gamma(1),\chi^\Gamma}) = det(1-\tilde{\mathcal{L}}_\beta^\Gamma) = Z_S(\beta;\Gamma)\,. \tag{135}$$

In the framework of the traditional approach to Selberg's zeta function by means of Selberg's trace formula the following relation between Selberg's zeta functions for subgroups of different Fuchsian groups with representations is well known [VZ83]:

Theorem 4. *Let G be an arbitrary Fuchsian group of the first kind and G_1 be a subgroup of G of finite index. Let χ be an arbitrary finite-dimensional unitary representation of G_1 and U^χ the representation of G induced from χ. Then one has*

$$Z_S(\beta;G_1;\chi) = Z_S(\beta;G;U^\chi)\,. \tag{136}$$

Our Theorem 2 is just a dynamical proof of this result for the special case $G=\Gamma(1)$, $G_1=\Gamma$ and χ the trivial representation.

Remark: Obviously, as soon as one can establish the transfer operator approach for a general Fuchsian group, Theorem 3 follows exactly along the line of arguments we have given in the case $G=\Gamma(1)$.

5.4 Factorization of the Selberg Zeta Functions

As mentioned in paragraph 4.3 the decomposition of the induced representation χ^Γ in its irreducible components implies also a decomposition of the transfer operator $\tilde{\mathcal{L}}_\beta^{\Gamma(1),\chi^\Gamma}$. Combined with (135) this implies a factorization of the Fredholm determinant $\det(1-\tilde{\mathcal{L}}_\beta^{\Gamma(1),\chi^\Gamma})$ and hence the factorization of the Selberg zeta function $Z_S(\beta;\Gamma(1);\chi^\Gamma)$.

Let χ^Γ be the representation of $\Gamma(1)$ induced from the trivial representation of the subgroup Γ of finite index which decomposes as $\chi^\Gamma=\oplus_i\chi_i^\Gamma$. Due to (47) the Fredholm determinant of $\tilde{\mathcal{L}}_\beta^{\Gamma(1),\chi^\Gamma}$ can then be factorized as

$$\det(1-\tilde{\mathcal{L}}_\beta^{\Gamma(1),\chi^\Gamma})=\prod_i\det(1-\tilde{\mathcal{L}}_\beta^{\Gamma(1),\chi_i^\Gamma}).$$

This implies immediately a factorization of the Selberg zeta function as

$$Z_S(\beta;\Gamma(1);\chi^\Gamma)=\prod_i Z_S(\beta;\Gamma(1);\chi_i^\Gamma).$$

If Γ is a normal subgroup of $\Gamma(1)$, the decomposition of the induced representation χ^Γ in (48) implies

$$\det(1-\tilde{\mathcal{L}}_\beta^{\Gamma(1),\chi^\Gamma})=\prod_{\chi_i^\Gamma\in\chi^*(\Gamma\backslash\Gamma(1))}\det(1-\tilde{\mathcal{L}}_\beta^{\Gamma(1),\chi_i^\Gamma})^{dim\,\chi_i^\Gamma}, \tag{137}$$

where $\chi^*(\Gamma\backslash\Gamma(1))$ denotes the set of all inequivalent irreducible unitary representations of the group $\Gamma\backslash\Gamma(1)$. This leads to the factorization of the Selberg zeta function as

$$Z_S(\beta;\Gamma(1);\chi^\Gamma)=\prod_{\chi_i^\Gamma\in\chi^*(\Gamma\backslash\Gamma(1))}Z(\beta;\Gamma(1);\chi_i^\Gamma)^{dim\,\chi_i^\Gamma}, \tag{138}$$

which is just a special case of the result of Venkov and Zograf in [VZ83]:

Proposition 5. *Let G be an arbitrary Fuchsian group of the first kind and G_1 be a normal subgroup of finite index in G. Let χ be an arbitrary finite dimensional unitary representation of G_1 and U^χ be the representation of G induced from χ. Then one has*

$$Z_S(\beta;G;U^\chi)=\prod_{\chi_i^\Gamma\in\chi^*(G_1\backslash G)}Z_S(\beta;G;\chi_i^\Gamma)^{dim\,\chi_i^\Gamma}.$$

If the subgroup Γ of $\Gamma(1)$ is not normal, the corresponding transfer operator $\tilde{\mathcal{L}}_\beta^{\Gamma(1),\chi^\Gamma}$ is in general also reducible. It follows from (64) that for example for the subgroups $\Gamma_0(2)$, $\Gamma^0(2)$ and Γ_ϑ the induced representation χ^Γ always contains the trivial representation $\chi^{\Gamma(1)}$. We will see in the second part of our paper that this decomposition of the transfer operators is closely related to the theory of new and old automorphic forms for subgroups of $\Gamma(1)$. That the transfer operator $\mathcal{L}_\beta^{\Gamma(1)}$ is contained in the transfer operator for any subgroup just reflects the fact that any automorphic form for $\Gamma(1)$ is also an automorphic form for any of its subgroups.

According to the decomposition of the induced representations in (64), Selberg's zeta function $Z_S(\beta, \Gamma, 1)$ for these groups can be factorized as followed:

$$\begin{aligned} Z_S(\beta, \Gamma(1), 1) &= \det(1 - \tilde{\mathcal{L}}_\beta^{\chi_1}), \\ Z_S(\beta, \Gamma_2, 1) &= \det(1 - \tilde{\mathcal{L}}_\beta^{\chi_1}) \det(1 - \tilde{\mathcal{L}}_\beta^{\chi_{-1}}), \\ Z_S(\beta, \Gamma, 1) &= \det(1 - \tilde{\mathcal{L}}_\beta^{\chi_1}) \det(1 - \tilde{\mathcal{L}}_\beta^{\chi_2}), \quad \Gamma \in \{\Gamma_0(2), \Gamma^0(2), \Gamma_\vartheta\}, \\ Z_S(\beta, \Gamma(2), 1) &= \det(1 - \tilde{\mathcal{L}}_\beta^{\chi_1}) \det(1 - \tilde{\mathcal{L}}_\beta^{\chi_{-1}}) \det(1 - \tilde{\mathcal{L}}_\beta^{\chi_2}) \det(1 - \tilde{\mathcal{L}}_\beta^{\chi_2}), \end{aligned} \tag{139}$$

where $\tilde{\mathcal{L}}_\beta^{\chi_i}$ denotes the transfer operator for $\Gamma(1)$ with representation χ_i, $i = 1, -1, 2$ as given in (61), (62) and (63). Relation (72) implies the further factorization

$$\det(1 - \tilde{\mathcal{L}}_\beta^{\chi_i}) = \det(1 - \mathcal{L}_\beta^{\chi_i}) \det(1 + \mathcal{L}_\beta^{\chi_i}).$$

Obviously, the relations in (139) imply also the following identities:

$$\begin{aligned} Z_S(\beta, \Gamma(1), 1) &= \det(1 - \tilde{\mathcal{L}}_\beta^{\chi_1}), \\ Z_S(\beta, \Gamma_2, 1) &= Z_S(\beta, \Gamma(1), 1) \det(1 - \tilde{\mathcal{L}}_\beta^{\chi_{-1}}), \\ Z_S(\beta, \Gamma_0(2), 1) &= Z_S(\beta, \Gamma(1), 1) \det(1 - \tilde{\mathcal{L}}_\beta^{\chi_2}). \end{aligned}$$

6 Conclusion

In this paper we have discussed the transfer operator for the dynamical system of a free particle moving on the modular surfaces $M_\Gamma = \Gamma\backslash\mathbb{H}$ belonging to the subgroups $\Gamma \subseteq \Gamma(1)$ of finite index $\mu = [\Gamma(1) : \Gamma]$, which are μ-fold covering surfaces of the modular surface $M_{\Gamma(1)} = \Gamma(1)\backslash\mathbb{H}$. Topologically these surfaces are spheres with a finite number of handles with cusps at infinity. Starting from the well understood case of the modular surface $\Gamma(1)\backslash\mathbb{H}$ we construct Poincaré surfaces X_Γ and the Poincaré return maps P_Γ from which the transfer operators $\tilde{\mathcal{L}}_\beta^\Gamma$ for Γ can be determined in a standard way. We show that the operators $\tilde{\mathcal{L}}_\beta^\Gamma$ are meromorphic in $\beta \in \mathbb{C}$, determine their spectral properties as nuclear operators and show how they can be decomposed into their basic components. By using trace formulas similar to the Atiyah-Bott formula the dynamical zeta functions of Ruelle and Smale for the geodesic flows on the surfaces of constant negative curvature given by Γ can be expressed in the thermodynamic formalism through Fredholm determinants of the transfer operator. The aforementioned decomposition of the transfer operator leads to a factorization of the Selberg zeta function for Γ

and gives a new dynamical interpretation of the Venkov-Zograf factorization proved originally within the classical approach by the Selberg trace formula. Finally we discuss some simple groups where the approach developed in this paper can be performed explicitly.

It would be interesting to compare the transfer operators we have constructed for subgroups Γ of $PSL(2,\mathbb{Z})$ with the operators Morita defined in his recent work on general cofinite Fuchsian groups in [Mor97]. In a second paper we will discuss the spectral properties of the transfer operators in more detail especially their eigenfunctions to the eigenvalue $\lambda = 1$ and their relation to automorphic forms of the groups Γ.

References

[AL70] A. O. L. Atkin and J. Lehner. Hecke operators on $\Gamma_0(m)$. *Math. Ann.*, 185:134–160, 1970.

[Art24] E. Artin. Ein mechanisches System mit quasi-ergodischen Bahnen. *Abh. Math. Sem. d. Hamburgischen Universität*, 3:170–175, 1924.

[BO95] U. Bunke and M. Olbrich. *Selberg Zeta and Theta Functions.* Akademie Verlag, 1995.

[Bow75] R. Bowen. *Equilibrium States and the Ergodic Theory of Anosov Diffeomorphisms.* Springer-Verlag, 1975. L.N.in Mathematics 470.

[Cha99] C.-H. Chang. *Die Transferoperator-Methode für Quantenchaos auf den Modulflächen* $\Gamma\backslash\mathbb{H}$. Papierflieger, Clausthal-Zellerfeld, January 1999. ISBN 3-89720-253-0.

[CM98] C.-H. Chang and D. Mayer. The period function of the nonholomorphic Eisenstein series for $PSL(2,\mathbb{Z})$. *Math. Phys. Elec. J.*, 4(6), 1998.

[CM99] C.-H. Chang and D. Mayer. The transfer operator approach to Selberg's zeta function and modular and Maass wave forms for $PSL(2,\mathbb{Z})$. In D. Hejhal and M. Gutzwiller et al, editors, *IMA Volumes 109 'Emerging applications of number theory'*, pages 72–142. Springer-Verlag, 1999.

[Ein17] A. Einstein. Zum Quantensatz von Sommerfeld und Epstein. *Verhandlungen der Deutschen Physikalischen Gesellschaft*, 19:82–92, 1917.

[Fri86] D. Fried. The zeta functions of Ruelle and Selberg I. *Ann. scient. éc. norm. sup.* 4e *Série*, 19:491–517, 1986.

[Gro55] A. Grothendieck. Produits tensoriels topologiques et espaces nucléaires. *Mem. Am. Math. Soc.*, 16, 1955.

[Gro56] A. Grothendieck. La theorie de Fredholm. *Bull. Soc. Math. France*, 84:319–384, 1956.

[Gut90] M. Gutzwiller. *Chaos in classical and quantum mechanics.* Springer-Verlag, 1990.

[Hej76] D. Hejhal. The Selberg trace formula and the Riemann zeta function. *Duke Math. J.*, 43(3):441–482, Sep. 1976.

[Isi25] E. Ising. Beitrag zur Theorie des Ferromagnetismus. *Zeitschrift für Physik*, 31:253–258, 1925.

[KF66] F. Klein and R. Fricke. *Vorlesungen über die Theorie der elliptischen Modulfunktionen. Ausgearbeitet und vervollständigt von Robert Fricke. Vol 2.* Teubner, Leipzig, 1966.

[Laz93] V. Lazutkin. *KAM Theory and Semiclassical Approximations to Eigenfunctions.* Springer-Verlag, 1993.

[LZ97] J. Lewis and D. Zagier. Period functions and the Selberg zeta function for the modular group. In *The Mathematical Beauty of Physics*, Adv. Series in Math. Physics 24, pages 83–97. World Scientific, Singapore, 1997.

[May80] D. Mayer. *The Ruelle-Araki transfer operator in classical statistical mechanics.* L. N. in Phys. 123. Springer-Verlag, Berlin, 1980.

[May91a] D. Mayer. Continued fractions and related transformations. In *Ergodic Theory, Symbolic Dynamics and Hyperbolic Spaces*, chapter 7, pages 175–222. Oxford Univ. Press, Oxford, 1991.

[May91b] D. Mayer. The thermodynamic formalism approach to Selberg's zeta function for $PSL(2,\mathbb{Z})$. *Bull. Am. Math. Soc.*, 25:55–60, 1991.

[Mor97] T. Morita. Markov systems and transfer operators associated with cofinite Fuchsian groups. *Ergodic Theory Dyn. Syst.*, 17(5):1147–1181, 1997.

[MOS66] W. Magnus, F. Oberhettinger, and R. Soni. *Formulas and Theorems for the Special Functions of Mathematical Physics.* Springer-Verlag, 1966.

[Pol86] M. Pollicott. Distribution of closed geodesics on the modular surface and quadratic irrationals. *Bull. Soc. math. France*, 114:431–446, 1986.

[Pol91] M. Pollicott. Closed geodesics and zeta functions. In *Ergodic Theory, Symbolic Dynamics and Hyperbolic Spaces*, chapter 6, pages 153–173. Oxford Univ. Press, Oxford, 1991.

[Ran77] R. Rankin. *Modular Forms and Functions.* Cambridge University Press, 1977.

[Rue76] D. Ruelle. Zeta-functions for expanding maps and Anosov flows. *Invent. Math.*, 34:231–242, 1976.

[Rue78] D. Ruelle. *Thermodynamic formalism.* Addison Wesley, Reading MA, 1978.

[Rue94] D. Ruelle. *Dynamical Zeta Functions for Piecewise Monotone Maps of the Interval.* AMS, 1994.

[Sar95] P. Sarnak. Arithmetic chaos. In *Israel Math. Conf. Proc.*, volume 8, pages 183–236, 1995.

[Sch74] B. Schoeneberg. *Elliptic Modular Functions.* Springer-Verlag, 1974.

[Sel56] A. Selberg. Harmonic analysis and discontinuous groups in weakly symmetric Riemannian spaces with applications to Dirichlet series. *J. Indian Math. Soc.*, 20:47–87, 1956.

[Ser85] C. Series. The modular surface and continued fractions. *J. London Math. Soc.*, (2), 31:69–80, 1985.

[Ter85] A. Terras. *Harmonic Analysis on Symmetric Spaces and Applications I.* Springer-Verlag, 1985.

[Ven90] A.B. Venkov. *Spectral Theory of Automorphic Functions and Its Applications.* Kluwer Academic Publishers, 1990.

[Vin89] E. Vinberg. *Linear Representation of Groups.* Birkhäuser, Basel, 1989.

[VZ83] A. B. Venkov and P. G. Zograf. On analogues of the Artin factorization formulas in the spectral theory of automorphic functions connected with induced representations of Fuchsian groups. *Math. USSR Izvestiya*, 21(3):435–443, 1983.

Stability and Diffusive Dynamics on Extended Domains

Alexander Mielke[1], Guido Schneider[2], and Hannes Uecker[2,*]

[1] Mathematisches Institut A, Universität Stuttgart, Pfaffenwaldring 57, 70569 Stuttgart, Germany
[2] Mathematisches Institut, Universität Bayreuth, 95440 Bayreuth, Germany

Abstract. We consider dissipative systems on the real axis in situations when the evolution is dominated by a dynamics similar to the one of a linear diffusion equation. It is surprising that such a diffusive behavior occurs in relatively complicated systems.
After a discussion of the linear and nonlinear diffusion equation, we give a brief introduction into the methods which are available to describe diffusive behavior in nonlinear systems. These are L^1-L^∞ estimates, Lyapunov functions and discrete and continuous renormalization groups.
In the second part of the paper we show examples, where such a diffusive dynamics can be seen. For the Ginzburg–Landau equation we consider the nonlinear stability of Eckhaus-stable equilibria and the diffusive mixing of two different Eckhaus-stable equilibria. Diffusive dynamics also occurs in pattern forming systems as the Swift–Hohenberg equation or hydrodynamical stability problems as Bénard's problem. In such cases the method of reduced instability allows us to analyze the linearized problem.
We close with an outlook on situations, where diffusive behavior is expected, but where a proof is still missing.

1 Introduction

The dynamics of dissipative partial differential equations (PDEs) on extended domains differs significantly from that on bounded domains. Many new solution types appear, e.g., traveling waves, fronts and pulses, [DFKM96]. Besides studies of particular solution classes, an existence theory for attractors for PDEs on unbounded domains was developed in [BV90,Fei96,MS95,Mie97a, ES99b,EZ99]. The inherent lack of compactness enforces a parallel use of a uniform and a localized topology; the attractors can be characterized quantitatively by Kolmogorov's ε-entropy and the dimension per unit volume, [CE99b,CE99a,Zel99].

For PDEs on bounded domains the bifurcation theory is very well developed: the center-manifold theory and the Liapunov–Schmidt reduction allow for a finite-dimensional description. On unbounded domains a reduction may still be possible by the theory of modulation equations or, for special solution

* Project: Analysis and Modelling of Pattern Formation Processes (Alexander Mielke, Guido Schneider)

classes, by the Kirchgässner reduction (spatial center-manifold reduction). However, the reduced problem remains infinite dimensional and is given by a simple partial differential equation, e.g., the (complex) Ginzburg–Landau equation. Although the multiple scaling ansatz of modulation theory has been used formally for more than 30 years, a mathematical justification was only obtained in [vH91,KSM92,Eck93,Sch94b,Sch94c,MS95,Sch98a,MSZ00] for model problems and in [Sch94a,Sch99b] for the Navier–Stokes equation; for a survey see [Mie99].

Here we survey one particular subject in the theory of PDEs on extended domains, namely the stability of spatially homogeneous or spatially periodic steady states. Moreover, we address the phenomenon of diffusive mixing of such steady states.

A basic concept in stability theory is the stability induced by the linearization alone. This means that the nonlinear terms can be controlled if the linearized problem dissipates energy with an exponential rate. Then stability can be achieved by considering the linearized problem alone. For dissipative problems on unbounded domains the linearization possesses continuous spectrum up to the imaginary axis in the complex plane. For these problems the linearized problem shows a dynamics similar to the one of a linear diffusion equation and so by an interplay of norms very often polynomial decay rates of the linearized problem can be obtained. As a consequence of the polynomial decay not all nonlinear terms can be controlled by the linearized problem. But if the low order terms are absent, again stability can be established with the help of the linearized problem. Such nonlinearities are called irrelevant.

On unbounded domains diffusive behavior occurs as a new aspect in the theory of stability. In fact, it turned out that for many interesting problems the nonlinear terms are irrelevant, such that this method is widely applicable. For instance, it has been applied successfully for proving the nonlinear stability of Taylor vortices in infinite cylinders with respect to spatially localized perturbations.

After a discussion of the linear and nonlinear diffusion equation, we give a brief introduction into the methods which are available to describe the diffusive behavior when the basic state is spatially homogeneous. These are L^1-L^∞ estimates, Lyapunov functions and discrete and continuous renormalization groups.

Next we provide typical examples, where such a diffusive dynamics can be seen.

For the Ginzburg–Landau equation we consider the nonlinear stability of Eckhaus-stable equilibria and the diffusive mixing of different steady state solutions.

We use Bloch's theory to generalize the spectral theory from spatially homogeneous steady states to spatially periodic ones. For steady states, which bifurcate from a homogeneous state, the linearized stability can be investigated by the theory of reduced instability (cf. [Mie95]). Nonlinear diffusive stability is obtained by showing the irrelevance of the nonlinear terms by proving appropriate convolution identities. For the Swift–Hohenberg equation the stability of the Eckhaus stable roll patterns is shown with respect to one- and two-dimensional perturbations. Applications in hydrodynamics include the roll solutions in Rayleigh–Bénard convection and the Taylor vortices in the Taylor–Couette experiment.

We close with an outlook on situations, where diffusive behavior is expected, but a proof is still missing.

Here, we restrict ourselves to the case of diffusive repair and diffusive mixing of equilibria. There are also results for diffusive stability for traveling front solutions [BK92,Gal94,EW94a,RK98,GR97,GR98]. The transfer of these last stability results to modulated front solutions ([HCS99]) will be found in [ES99a].

Solutions which are localized perturbations of an exponentially homogeneous state (traveling pulses or modulated pulses) [BL99b,KS98,Sch00,Uec00] may be exponentially stable. The stability question of localized perturbations of diffusively stable states was attacked in [SS99] where a spectral stability result was obtained.

Acknowledgment: This work was partially supported by the DFG-Schwerpunktprogramm DANSE under the grant Mi 459/2. The research has benefited from the interaction with the DANSE research groups in Berlin, Bielefeld, Stuttgart and Tübingen.

2 Diffusive Repair and Diffusive Mixing in the Linear Case

2.1 Diffusive Repair

In this section we consider the linear diffusion equation

$$\partial_t u = \partial_x^2 u, \qquad u|_{t=0} = u_0 \tag{1}$$

with $x \in \mathbb{R}$, $t \geq 0$ and $u(t,x) \in \mathbb{R}$. The solution can be written explicitly as

$$u(t,x) = \tfrac{1}{\sqrt{2\pi t}} \int_{\mathbb{R}} \mathrm{e}^{-(x-y)^2/(4t)} u_0(y)\,\mathrm{d}y = \int_{\mathbb{R}} G(x-y,t) u_0(y)\,\mathrm{d}y. \tag{2}$$

Spatially constant functions stay constant in time, but by Young's inequality for convolutions with $p > q$ we obtain

$$\|u(t)\|_{L^p} \leq C t^{-1/(2r)} \|u_0\|_{L^q}, \quad \text{where} \quad 1/p = 1/q - 1/r,\ 1 \leq p, r \leq \infty, \tag{3}$$

for some constant C independent of time. Thus, spatially localized initial conditions give rise to solutions with polynomial decay rates. Moreover, the solutions become flatter and flatter, since we have for instance

$$\|\partial_x^n u\|_{L^\infty} \leq \|\partial_x^n G\|_{L^\infty} \|u_0\|_{L^1} \leq C t^{-(n+1)/2} \|u_0\|_{L^1}.$$

There is some additional structure which can be seen by looking at the Fourier transform

$$\hat{u}(k) = (Fu)(k) = \tfrac{1}{2\pi} \int_{\mathbb{R}} u(x) \mathrm{e}^{-ikx}\,\mathrm{d}x$$

of (1). The equation $\partial_t \hat{u} = -k^2 \hat{u}$ possesses the solution $\hat{u}(t,k) = \mathrm{e}^{-k^2 t} \hat{u}_0(k)$. Renormalizing this solution gives

$$\hat{u}(t, k/\sqrt{t}) = \mathrm{e}^{-k^2} \hat{u}_0(k/\sqrt{t}) = \mathrm{e}^{-k^2} \left(\sum_{j=0}^{n} t^{-j/2} k^j \hat{u}_0^{(j)}(0) + o(t^{-n/2}) \right) \tag{4}$$

if $\hat{u}_0$ is n-times differentiable. Since smoothness in Fourier space corresponds to decay properties in x-space, solutions to spatially localized initial conditions decay in a universal way to 0. Loosely speaking, if the initial conditions spatially decay like $|x|^{-n}$, we obtain

$$u(t,x) = \sum_{j=0}^{n-1} t^{-(j+1)/2} \hat{u}_0^{(j)}(0) H_j(x/\sqrt{t}) + O(t^{-n/2})$$

for $t \to \infty$, where H_j is a multiple of the jth Hermite polynomial.

To be more precise we use the fact that Fourier transform is an isomorphism from $H^n(m)$ to $H^m(n)$, where

$$H^m(n) = \{u : \mathbb{R} \to \mathbb{C} \mid \|u\|_{H^m(n)} = \|u\rho^n\|_{H^m} < \infty\} \quad \text{and} \quad \rho(x) = (1+x^2)^{1/2}. \tag{5}$$

For the lowest order terms we obtain $\|\hat{u}(t, k/\sqrt{t}) - \mathrm{e}^{-k^2} \hat{u}_0(0)\|_{H^2(2)} \leq C t^{-1/2}$, i.e.,

$$\|\sqrt{t}\, u(t, \sqrt{t}\, x) - \sqrt{\pi}\, \hat{u}_0(0) \mathrm{e}^{-x^2/4}\|_{H^2(2)} \leq C t^{-1/2}. \tag{6}$$

The result is based on the fact that the linear evolution operator $\mathrm{e}^{-k^2 t}$ concentrates the Fourier modes at the wave number $k = 0$. Depending on the differentiability of the initial conditions $\hat{u}_0$ the local behavior of $\hat{u}_0$ at the wavenumber $k = 0$ is extracted by the linear evolution operator $\mathrm{e}^{-k^2 t}$ for $t \to \infty$.

Remark 1. Most of the above theory also holds if we have a linear evolution operator $\mathrm{e}^{t\lambda(k)}$ with eigenvalues $\lambda(k) \sim -k^2$ for $k \to 0$. This is the reason why diffusive behavior can be observed in a big variety of problems.

From (6), i.e., $u(t,x) = \sqrt{\pi/t}\, \hat{u}_0 \mathrm{e}^{-x^2/(4t)} + O(t^{-1})$, it is obvious that stability of $u = 0$ in $H^2(2)$ does not hold, but again we have

$$\sup_{x \in \mathbb{R}} |u(t,x)| \leq C t^{-1/2}$$

for initial conditions $u|_{t=0} = u_0 \in H^2(2)$. Thus, it makes sense to introduce the following definition.

Definition 2. Let $\mathcal{B}_1, \mathcal{B}_2$ be Banach spaces and let S_t be an evolution operator. A fixed point $u_0 = S_t u_0$ is called $(\mathcal{B}_1, \mathcal{B}_2)$-stable under S_t if the following holds: For all $\varepsilon > 0$ there exists a $\delta > 0$ such that from $\|v - u_0\|_{\mathcal{B}_1} < \delta$ it follows that $\|S_t v - u_0\|_{\mathcal{B}_2} < \varepsilon$ for all $t \geq 0$. The point u_0 is called asymptotically $(\mathcal{B}_1, \mathcal{B}_2)$-stable if additionally $\lim_{t\to\infty} S_t v = u_0$ in $\mathcal{B}_2$.

The proof of (6) for some nonlinear problem also establishes asymptotic $(H^2(2), L^\infty)$-stability of $u = 0$. The usage of two different norms is very common for problems posed on unbounded domains, cf. [BV90,MS95].

2.2 Diffusive Mixing

Above we have shown that spatially localized perturbations decay diffusively to 0. Another interesting question is the behavior of solutions to initial conditions

$$u_0(x) = \begin{cases} a \text{ for } x < -\ell, \\ b \text{ for } x \geq \ell, \end{cases}$$

for some $\ell > 0$, i.e., we prescribe two different constants on the left and on the right. Since $\partial_x u_0$ is again spatially localized and since $\partial_x u$ also satisfies (1) we obtain the results from above for $\partial_x u$. Integration with respect to x leads to

$$\sup_{x\in\mathbb{R}} |u(t,x) - [a+(b-a)\mathrm{Erf}(x/\sqrt{t})]| \leq C/\sqrt{t},$$

where $\mathrm{Erf}(x) = 1/\sqrt{4\pi}\int_{-\infty}^{x} \mathrm{e}^{-y^2/4}\,\mathrm{d}y$. For the linear diffusion equation all spatially constant functions are stable equilibria. Thus, we have some diffusive mixing of the stable states.

The question arises whether such a behavior also occurs in more complicated systems if we have different diffusively stable equilibria for $x \to \pm\infty$.

3 Irrelevant Nonlinearities and Nonlinear Diffusive Stability

In this section we explain how the polynomial decay rates for the linearized problem can be used to control the nonlinear terms. As an example we consider a nonlinear diffusion equation

$$\partial_t u = \partial_x^2 u + c\,u^p, \quad u|_{t=0} = u_0 \tag{7}$$

with $t \geq 0$, $x \in \mathbb{R}$, $p \in \mathbb{N}$ and $c \in \mathbb{R} \setminus \{0\}$.

In case $p = 2$ we have blowup of the solution for most initial conditions, in case $p = 3$ the sign of c decides about stability ($c < 0$) and instability ($c > 0$), but for $p \geq 4$ the sign of c doesn't play any role and small spatially localized perturbations vanish for $t \to \infty$ with the same polynomial decay rate as in the linear case.

There are essentially three methods to prove the last assertion, namely a) L^1-L^∞ estimates, b) the construction of Lyapunov functions, and c) the discrete and continuous renormalization approach.

3.1 L^1-L^∞ Estimates

This method relies on the L^q-L^p estimate (3), the variation-of-constants formula and suitable estimates of the nonlinearity.

Lemma 3. *Let $p > 3$. For all $C > 0$ there exists $\varepsilon > 0$ such that solutions u of (7) with $\|u_0\|_{L^1} + \|u_0\|_{L^\infty} \leq \varepsilon$ satisfy*

$$\|u(t)\|_{L^1} \leq C \quad \textit{and} \quad \|u(t)\|_{L^\infty} \leq C/(1+t)^{\frac{1}{2}}$$

for all $t \geq 0$.

Proof. We consider the variation-of-constants formula

$$u(t) = \mathrm{e}^{t\partial_x^2} u_0 + c \int_0^t \mathrm{e}^{(t-s)\partial_x^2} u^p(s)\,\mathrm{d}s$$

for (7). With $\|u^p\|_{L^\infty} \leq \|u\|_{L^\infty}^p$ and $\|u\|_{L^1}^p \leq \|u\|_{L^\infty}^{p-1} \|u\|_{L^1}$, the abbreviations

$$a(t) = \sup_{0\leq s\leq t} \|u(s)\|_{L^1} \quad \text{and} \quad b(t) = \sup_{0\leq s\leq t} \|(1+s)^{1/2} u(s)\|_{L^\infty}$$

and the estimates of Section 2 we obtain

$$\begin{aligned}(1+t)^{1/2}\left\|\int_0^t \mathrm{e}^{(t-s)\partial_x^2} u^p(s)\,\mathrm{d}s\right\|_{L^\infty} &\leq (1+t)^{1/2}\int_0^t \|\mathrm{e}^{(t-s)\partial_x^2}\|_{L^1\to L^\infty} \|u^p\|_{L^1}\,\mathrm{d}s \\ \leq (1+t)^{1/2}\int_0^t (t-s)^{-1/2}(1+s)^{-(p-1)/2}\,\mathrm{d}s \cdot b(t)^{p-1} a(t) &\leq C_1\, b(t)^{p-1} a(t)\end{aligned}$$

with a constant C_1 independent of t for $p > 3$. Furthermore, we have

$$\begin{aligned}\left\|\int_0^t \mathrm{e}^{(t-s)\partial_x^2} u^p\,\mathrm{d}s\right\|_{L^1} &\leq \int_0^t \|e^{(t-s)\partial_x^2}\|_{L^1\to L^1} \|u^p\|_{L^1}\,\mathrm{d}s \\ \leq \int_0^t (1+s)^{-(p-1)/2}\,\mathrm{d}s \cdot b(t)^{p-1} a(t) &\leq C_2 b(t)^{p-1} a(t).\end{aligned}$$

Together we obtain

$$a(t) \leq a(0) + |c| C_1 b(t)^{p-1} a(t) \quad \text{and} \quad b(t) \leq a(0) + |c| C_2 b(t)^{p-1} a(t).$$

If $a(0) + b(0) < \varepsilon$ with $\varepsilon > 0$ sufficiently small we have the existence of $C > 0$ such that $a(t), b(t) \leq C$ for all $t \geq 0$.

3.2 Lyapunov Functions

The usage of Lyapunov functions is well established in nonlinear stability problems if the nonlinearity has some sign as in $\partial_t u = \partial_x^2 u - u^3$. This has been used for diffusive stability problems in [EW94b] and [GR97] and in [GM98] for the proof of diffusive mixing. But also if the nonlinearity has the wrong sign, Lyapunov functions can be used. We do not obtain the optimal power p of the irrelevant nonlinearities, but the method is also applicable on unbounded domains $\Omega \subset \mathbb{R}^d$ with $\Omega \neq \mathbb{R}^d$, cf. [ES98]. As an example we consider again

$$\partial_t u = \partial_x^2 u + c\, u^p, \quad u|_{t=0} = u_0.$$

We introduce the functionals $I(u) = \int_{\mathbb{R}} u^2 \, dx$, $J(u) = \int_{\mathbb{R}} (\partial_x u)^2 \, dx$ and $K(u) = \int_{\mathbb{R}} (\partial_x^2 u)^2 \, dx$. This fixes the L^q-power to $q = 2$, and due to the L^2-L^∞ estimate (3) we can only handle nonlinearities for $p \geq 5$.

With $\|u\|_{L^\infty}^2 \leq I^{1/2} J^{1/2}$ and $J^2 \leq I\,K$ we obtain

$$\begin{aligned} \tfrac{1}{2}\tfrac{\mathrm{d}}{\mathrm{d}t} I &= \int u \partial_t u \, dx = \int u \partial_x^2 u + c u^{p+1} \, dx \\ &= \int -(\partial_x u)^2 + c u^{p+1} \, dx \;\leq\; -J + |c| \|u(x)\|_{L^\infty}^{p-1} \int u^2 \, dx \\ &\leq -J + |c| I^{(p+3)/4} J^{(p-1)/4} = -J \, [1 - |c| I^{(p+3)/4} J^{(p-5)/4}] \end{aligned}$$

and

$$\begin{aligned} \tfrac{1}{2}\tfrac{\mathrm{d}}{\mathrm{d}t} J &= \int (\partial_x u)(\partial_t \partial_x u) \, dx \\ &= \int (\partial_x u)(\partial_x^3 u) + c p u^{p-1} (\partial_x u)^2 \, dx \;\leq\; -\int (\partial_x^2 u)^2 \, dx + |c| p \|u\|_{L^\infty}^{p-1} J \\ &\leq -K + |c| p I^{(p-1)/4} J^{(p+3)/4} \;\leq\; -K \, [1 - |c| p I^{(p+3)/4} J^{(p-5)/4}]. \end{aligned}$$

Hence, we have $\dot{I} \leq 0$ and $\dot{J} \leq 0$ if $I(u_0)^{(p+3)/4} J(u_0)^{(p-5)/4}$ is sufficiently small. Thus, we have proved the following result.

Lemma 4. *Let $p \geq 5$. Then, there exists $\varepsilon_0 > 0$ such that for all $\varepsilon \in (0, \varepsilon_0)$ and all solutions u of (7) with $\|u_0\|_{H^1} \leq \varepsilon$ we have* $\sup_{t \geq 0} \|u(t)\|_{H^1} \leq \varepsilon$.

3.3 The Discrete and Continuous Renormalization Process

In addition to some stability result this method gives the asymptotics of the decay to 0. It relies on formula (6). By a fixed point argument we prove also for the nonlinear system (7) that the renormalized solution $\sqrt{t}\, u(t, \sqrt{t}\, x)$ converges towards a multiple of the Gaussian $\mathrm{e}^{-x^2/4}$. There are two approaches, a discrete and a continuous one. We sketch the first one very briefly and the second one in a little more detail.

The discrete approach In the discrete approach (cf. [BK92,BKL94,Gal94, Sch96]) a sequence of problems is considered which converges towards the linear diffusion equation. Define $u_n(\tau, x) = L^n u(L^{2n}\tau, L^n x)$ with $L > 1$ and $n \in \mathbb{N}$. Then u_n satisfies

$$\partial_\tau u_n = \partial_x^2 u_n + L^{n(3-p)} u_n^p \quad \text{for } \tau \in [L^{-2}, 1]; \qquad u_n(L^{-2}, x) = L u_{n-1}(1, Lx). \tag{8}$$

Obviously, for $p > 3$ the influence of the nonlinear terms tends exponentially to 0 as $n \to \infty$ and in the limit we obtain the linear diffusion equation. Solving the sequence of problems (8) is equivalent to solving (7). By a fixed point argument it then follows (see for instance [BK92,Sch96]) that for spatially localized initial conditions u_0 the functions $u_n|_{\tau=1} = L^n u(L^{2n}, L^n \cdot)$ converge towards a multiple of the Gaussian $\mathrm{e}^{-x^2/4}$.

The continuous approach Here the system satisfied by the renormalized solution is considered directly, where additionally a logarithmic time scale is taken to transfer the polynomial decay rates into exponential ones. We follow the lines of [Way97] and introduce the new variable w and the new coordinates ξ and τ by

$$u(t,x) = t^{-1/2} w(\log t, x/\sqrt{t}) = \mathrm{e}^{-\tau/2} w(\tau,\xi). \tag{9}$$

The transformed equation is given by

$$\partial_\tau w = w/2 + (\xi/2)\partial_\xi w + \partial_\xi^2 w + \mathrm{e}^{(3-p)\tau/2} w^p \tag{10}$$

with $w|_{\tau=0} \in H^2(2)$. The linearization around $w = 0$ leads to the spectral problem which reads in Fourier space

$$-k^2\hat{w} - (k/2)\partial_k \hat{w} - \lambda\hat{w} = \hat{f}, \tag{11}$$

with $\hat{f} \in H^2(2)$. The eigenfunctions $\hat{\psi}_s(k) = k^s \mathrm{e}^{-k^2}$ to the real eigenvalues $\lambda = -s/2$ are parameterized with $s \in \mathbb{R}$. See also (4). Since $\partial_k^j \hat{\psi}_s \in L^2$ is required for $j = 0, 1, 2$ (for which the possible singularity of $\partial_k^j \hat{\psi}_s$ at $k = 0$ plays the crucial role) this leads in (11) to two discrete eigenvalues 0 and $-1/2$ and to essential spectrum $\{\lambda \in \mathbb{C} \mid \mathrm{Re}\lambda < -3/4\}$ due to Sobolev's embedding theorem. Since the solutions of the linearized problem are uniformly bounded and since the nonlinear terms vanish with an exponential rate there exist $\varepsilon, C > 0$ such that $\sup_{\tau\geq 0} \|w(\tau)\|_{H^2(2)} < C$ for the solutions w of (10) if $\|w|_{\tau=0}\|_{H^2(2)} < \varepsilon$ and $p > 3$. If we denote with w_0 the part of w belonging to the eigenvalue 0 and with w_1 the rest of w we can conclude by integration of the variation of constants formula with respect to time that

$$w_0(\tau) = w_{\lim}\mathrm{e}^{-(\cdot)^2/4} + O(\mathrm{e}^{\frac{3-p}{2}\tau}) \quad \text{and} \quad w_1(\tau) = O(\mathrm{e}^{\max\{-\frac{1}{2},\frac{3-p}{2}\}\tau})$$
for $\tau \to \infty$.

Herein, $\psi_0 = \mathrm{e}^{-(\cdot)^2/4}$ is the eigenvector to 0 in physical space, and $w_{\lim} \in \mathbb{R}$ is a constant only depending on the initial conditions. This leads to the following convergence result, cf. [Way97].

Theorem 5. *Let $p > 3$. Then there exist $\varepsilon, C > 0$ such that the following holds. Let u be a solution of (7) with $\|u|_{t=0}\|_{H^2(2)} \leq \varepsilon$. Then there exists a $w_{\lim} \in \mathbb{R}$ such that*

$$\|\sqrt{t}\,u(t,\cdot\sqrt{t}) - w_{\lim}\mathrm{e}^{-(\cdot)^2/4}\|_{H^2(2)} \leq C(1+t)^{-\max\{1/2,(3-p)/2\}} \quad \textit{for all } t \geq 0.$$

3.4 Some Remarks

Clearly, the above theory can be adjusted to more general nonlinearities as well as to higher space dimensions. If we call the exponent p of the nonlinear term u^p degree of irrelevance then $u^{p_1}(\partial_x u)^{p_2}(\partial_x^2 u)^{p_3}$ has the degree of irrelevance $p = p_1 + 2p_2 + 3p_3$. Thus, nonlinear terms with derivatives give some additional irrelevance. For instance, Lemma 3 holds for all nonlinearities with $p > 3$. In higher space dimensions the solutions of the linear diffusion equation $\partial_t u = \Delta u$, $x \in \mathbb{R}^d$ satisfy the estimate

$$\|u(t)\|_{L^\infty} \leq Ct^{-d/2}\|u\|_{L^1}.$$

From the above analysis it is easy to see, that all nonlinear terms with degree of irrelevance $p>1+2/d$ are irrelevant. As a consequence, in dimensions $d \geq 3$ all sufficiently smooth nonlinear terms are irrelevant. Physically this means that there are enough directions in which the energy can diffuse away before the quadratic nonlinear terms have time to act. See also [BL99a] for a generalization to larger classes of initial data.

4 Diffusive Behavior in the Ginzburg–Landau Equation

The (real) Ginzburg–Landau equation

$$\partial_t u = \partial_x^2 u + u - |u|^2 u, \quad u = u(t,x) \in \mathbb{C}, \quad u|_{t=0} = u_0 \tag{12}$$

occurs as an amplitude equation for bifurcation problems on infinitely long cylindrical domains, cf. for instance [CE90b,Sch94b,MS96,Mie99,Sch99b]. It possesses so-called stationary roll solutions, i.e. spatially periodic steady states of the form $u_{q,\beta}(x) = \sqrt{1-q^2}\,\mathrm{e}^{\mathrm{i}(qx+\beta)}$. Letting $u(t,x) = u_{q,\beta}(x) + \mathrm{e}^{-\mathrm{i}(qx+\beta)}v$ and linearizing in v one obtains that a roll $u_{q,\beta}$ is linearly stable if and only if $q^2 \leq 1/3$, cf. [Eck65]. For $q^2 > 1/3$ a roll $u_{q,\beta}$ is sideband- or Eckhaus-unstable. This means that $u_{q,\beta}$ is unstable with respect to perturbations $\mathrm{e}^{\mathrm{i}\tilde{q}x}$ with a slightly different wavenumber $\tilde{q} \approx q$, $\tilde{q} \neq q$.

In this section we review results from [CEE92,BK92] and [GM98] concerning the nonlinear diffusive stability and the diffusive mixing of rolls, respectively.

4.1 Diffusive Stability of Equilibria

To understand the Ginzburg–Landau equation (12) in the vicinity of a roll we introduce coordinates $u(t,x) = r(t,x)\,\mathrm{e}^{\mathrm{i}\phi(t,x)}$. Then

$$\partial_t \phi = \partial_x^2 \phi - 2\frac{\partial_x r}{r}\partial_x \phi, \quad \partial_t r = \partial_x^2 r + r[1-r^2-(\partial_x \phi)^2]. \tag{13}$$

The roll solution $u_{q,\beta}$ now takes the form $(\phi_{q,\beta}, r_q) = (qx+\beta, \sqrt{1-q^2})$. On the linear level we see that $\phi-\phi_{q,\beta}$ behaves diffusively while $r-r_q$ is linearly exponentially damped with rate $-2(1-q^2)$. Moreover, ϕ itself does not appear on the right-hand side.

This means roughly that the amplitude is slaved to the local wave length $\eta = \partial_x \phi$. Heuristically, we obtain asymptotically $r = \sqrt{1-(\partial_x \phi)^2} + \text{h.o.t.}$.

Neglecting the higher order terms and inserting the relation in the equation for ϕ we arrive at the *phase diffusion equation*

$$\partial_t \phi = \frac{1-3(\partial_x \phi)^2}{1-(\partial_x \phi)^2} \partial_x^2 \phi. \tag{14}$$

Writing $\phi = \phi_{q,\beta} + \psi$ allows us to study the question of diffusive stability

$$\partial_t \psi = a(q)\partial_x^2 \psi + [a(q+\partial_x \psi) - a(q)]\partial_x^2 \psi \quad \text{where } a(q) = (1-3q^2)/(1-q^2).$$

Clearly, we need $a(q) > 0$ which characterizes the Eckhaus-stable domain. The lowest order nonlinear terms are a multiple of $(\partial_x \psi)^{p_2} \partial_x^2 \psi$ for a $p_2 \geq 1$. Following the remarks given in Section 3.4 we have $p_3 = 1$ and hence $p = 2p_2 + 3 > 3$. Thus, for the phase diffusion equation the nonlinearity is irrelevant.

Of course, the proof of diffusive stability (in the sense of $(H^2(2), L^\infty)$ stability) of Eckhaus-stable rolls $u_{q,\beta}$ for the full Ginzburg–Landau equation (12) is much more involved as the coupling between $\partial_x \phi$ and r has to be studied precisely. See [BK92] for a proof using renormalization theory as described in subsection 3.3 and see [Kap94] for a proof which is based on Lyapunov functions and L^2-L^∞ estimates.

4.2 Diffusive Mixing

The next question is the evolution of u for an initial condition $u_0(x) = r_0(x)e^{i\phi_0(x)}$ that converges to two different stable rolls $u_{q_\pm,\beta_\pm}$ for $x \to \pm\infty$, where $(\beta_+ - \beta_-)^2 + (q_+ - q_-)^2 \neq 0$. Clearly, the associated solution will, for all $t > 0$, satisfy the same boundary conditions at infinity:

$$u(t,x) - u_{q_\pm,\beta_\pm}(x) \to 0 \quad \text{for} \quad x \to \pm\infty.$$

The question is how the solutions behave in the intermediate regime. Diffusive mixing is seen, if the initial condition u_0 is chosen properly. Then, for large t the solution develops an intermediate wave length q_* which solely depends on q_- and q_+. However, in general the phase β does not converge but grows as $\sqrt{t}$.

For $q_- = q_+ = q \in (-1/\sqrt{3}, 1/\sqrt{3})$ and small $\delta = \beta_+ - \beta_- \neq 0$ the problem has been treated in [CEE92]. It is shown that, if $r_0 - \sqrt{1-q^2}$ and $\phi_0(x) - qx$ are small and spatially localized, then the solution (r, ϕ) of (13) satisfies

$$\|r(t) - \sqrt{1-q^2}\|_{L^\infty} + \|\partial_x \phi(t) - q\|_{L^\infty} \to 0 \quad \text{as } t \to \infty. \tag{15}$$

The extension to the case $q_- \neq q_+$ with small $|q_\pm|, |\beta_\pm|$ and more detailed asymptotics have been obtained in [BK92] using the discrete renormalization approach. In [GM98] these results are generalized to arbitrary $q_-, q_+ \in (-1/\sqrt{3}, 1/\sqrt{3})$. In the case $q_+ \neq q_-$ we may assume $\beta_- = \beta_+ = 0$

without loss of generality (use the translation $x \mapsto x+y$ and the phase invariance $\phi \mapsto \phi+\alpha$).

The first step in the analysis is the construction of a limiting profile $\tilde{\eta}$ for the local wave vector. It is obtained as the unique similarity solution $\phi(t,x) = \tilde{\eta}(x/\sqrt{t})$ of the phase diffusion equation (14) which satisfies

$$[a(\tilde{\eta})\tilde{\eta}']' + \frac{\xi}{2}\tilde{\eta}' = 0 \text{ for } \xi \in \mathbb{R} \quad \text{and} \quad \tilde{\eta}(\xi) \to q_\pm \text{ for } \xi \to \pm\infty.$$

Then we introduce

$$\tilde{\phi}(x) = q_- x + \textstyle\int_{-\infty}^{x} [\tilde{\eta}(\xi) - q_-]\, \mathrm{d}\xi$$

and define the limiting profile

$$\widetilde{U}(t,x) = \sqrt{1-\tilde{\eta}(x/\sqrt{t})^2}\, \mathrm{e}^{\mathrm{i}\sqrt{t}\,\tilde{\phi}(x/\sqrt{t})}.$$

The following result was proved by using a nonlinear change of variables for the amplitude r and the phase ϕ, the continuous renormalization approach from Subsection 3.3 combined with energy estimates in the sense of Section 3.2 and the theory of nonlinear monotone operators.

Theorem 6. *[GM98, Theorem 4.2] For all $q_-, q_+ \in (-1/\sqrt{3}, 1/\sqrt{3})$ there exist $t_0 > 0$ and $\varepsilon > 0$ such that for all $\nu \in (0,1)$ and for all $u_0 \in H^2_{\mathrm{ul}}(\mathbb{R})$ satisfying $\|u_0 - \widetilde{U}(t_0,\cdot)\|_{H^2} \le \varepsilon$ the unique solution of (12) in $H^2_{\mathrm{ul}}(\mathbb{R})$ with $u(0,\cdot) = u_0$ satisfies*

$$\|u(t,\cdot) - \widetilde{U}(t_0+t,\cdot)\|_{H^1_{\mathrm{ul}}} = O(t^{-\nu/4}), \quad \|\,|u(t,\cdot)| - |\widetilde{U}(t_0+t,\cdot)|\,\|_{H^1_{\mathrm{ul}}} = O(t^{-3\nu/4}) \tag{16}$$

as $t \to \infty$.

Here the spaces H^m_{ul} are the uniformly local Sobolev spaces introduced in [MS95]. Using (16) and simple properties of $\tilde{\eta}$ one obtains, for all fixed ℓ,

$$\sup_{|x|\le \ell} \left| u(t,x) - \sqrt{1-q^{*2}}\, \mathrm{e}^{\mathrm{i}(q^* x + \sqrt{t}\,\phi^*)} \right| = O(t^{-\nu/4}),$$

where $q^* = \tilde{\eta}(0)$ and $\phi^* = \tilde{\phi}(0)$ with $\mathrm{sign}(\phi^*) = \mathrm{sign}(q_+ - q_-)$. The weaker decay rate compared to [BK92] is due to less restrictive conditions on u_0.

5 General Pattern Forming Systems

So far we have studied the diffusive stability and the diffusive mixing of special periodic states of the Ginzburg–Landau equation which has the phase invariance as an SO(2) symmetry. The rotating waves $u_{q,\beta}$ are in fact relative equilibria (in the stationary problem) with respect to this symmetry. As a consequence it is possible to factor out the phase and treat the problem as a spatially homogeneous one. In the following we want to indicate, that it is also possible to apply the above techniques to general spatially periodic solutions in parabolic systems. A basic tool for this situation is Bloch analysis which generalizes Fourier analysis.

5.1 The Bloch Analysis

For simplicity we only treat the one-dimensional case. For the general case we refer to the standard reference [RS78] as well as to the recent work [Mie97b,Sca99], see also [OZ00]. We consider a differential operator

$$\mathcal{L}v = M\partial_x^2 v + N(x)\partial_x v + P(x)v$$

where $u \in \mathbb{R}^m$, $M \in \mathbb{R}^{m\times m}$ is invertible and N, P are periodic matrices with period p. The basic observation is that $\mathcal{L}$ maps functions of the form $\mathrm{e}^{\mathrm{i}lx}V(x)$ with $V(x+p) = V(x)$ into itself. Functions of this form are called Bloch waves and $l \in \mathcal{T}_p = \mathbb{R}/p\mathbb{Z}$ is called their Bloch wave number.

Defining the Hilbert spaces

$$\mathbb{H}_l^k(\mathbb{R}) = \{v \in H_{\mathrm{loc}}^k(\mathbb{R}) \mid v(x+p) = \mathrm{e}^{\mathrm{i}lp}v(x)\},$$

the operator $\mathcal{L}$ restricts to a closed unbounded operator $\mathcal{L}_l$ with compact resolvent from $\mathbb{H}_l^0(\mathbb{R})$ into itself. Its domain is $D(\mathcal{L}_l) = \mathbb{H}_l^2(\mathbb{R})$. In fact, the transformation $V = T_l v$ with $V(x) = \mathrm{e}^{-\mathrm{i}lx}v(x)$ maps $\mathbb{H}_l^k(\mathbb{R})$ into $H_{\mathrm{per}}^k((0,p)) := \mathbb{H}_0^k(\mathbb{R})$. This transformation defines the *Bloch operator*

$$B_l = T_l\mathcal{L}_l T_l^{-1} \quad \text{with} \quad B_l V = MV'' + [N(x)+2\mathrm{i}lM]V' + [P(x)+\mathrm{i}lN(x)-l^2M]V.$$

All these operators map $H_{\mathrm{per}}^2((0,p))$ into $L_{\mathrm{per}}^2((0,p))$ and they depend polynomially on l. Thus classical perturbation arguments apply.

The main tool of the Bloch analysis is the decomposition of $L^2(\mathbb{R})$ into the orthogonal, direct sum of the spaces $\mathbb{H}_l^0(\mathbb{R})$:

$$L^2(\mathbb{R}) = \bigoplus_{l\in\mathcal{T}_p} \mathbb{H}_l^0, \quad v(x) = \int_{l\in[0,p)} v_l(x)\,\mathrm{d}l,$$

where the integral is called the *direct integral* which has to be understood in the $L^2(\mathbb{R})$ sense. We have $\|v\|_{L^2(\mathbb{R})}^2 = \int_0^p \|v_l\|_{L^2(\mathbb{R})}^2\,\mathrm{d}l$ and v_l can be expressed through the Fourier transform $\hat{u}$ as $v_l(x) = (2\pi)^{-1/2}\sum_{m\in\mathbb{Z}} \mathrm{e}^{\mathrm{i}(l+mp)}\hat{v}(l+mp)$. Thus, the Bloch decomposition is a partial Fourier transform which is exactly adjusted to the periodicity of the underlying problem.

Moreover, the operator $\mathcal{L}$ takes the form $\oplus_{l\in\mathbb{S}^1}\mathcal{L}_l$ with respect to this decomposition. As a result the exponential and the resolvent take the form

$$\mathrm{e}^{t\mathcal{L}} = \bigoplus_{l\in\mathcal{T}_p} \mathrm{e}^{t\mathcal{L}_l}, \quad (\lambda-\mathcal{L})^{-1} = \bigoplus_{l\in\mathcal{T}_p} (\lambda-\mathcal{L}_l)^{-1}.$$

The orthogonality of the decomposition implies the norm identities

$$\|e^{t\mathcal{L}}\|_{H^k(\mathbb{R})\to H^k(\mathbb{R})} = \sup_{l\in\mathcal{T}_p} \|e^{tB_l}\|_{\mathbb{H}^k_l\to\mathbb{H}^k_l},$$
$$\|(\lambda-\mathcal{L})^{-1}\|_{H^k(\mathbb{R})\to H^k(\mathbb{R})} = \sup_{l\in\mathcal{T}_p} \|(\lambda-B_l)^{-1}\|_{\mathbb{H}^k_l\to\mathbb{H}^k_l}.$$

As a consequence we have a useful result for the spectrum of $\mathcal{L}$:

$$\mathrm{spec}_{L^2(\mathbb{R})}\mathcal{L} = \bigcup_{l\in\mathcal{T}_p} \mathrm{spec}_{\mathbb{H}^0_l(\mathbb{R})}\mathcal{L}_l = \bigcup_{l\in\mathcal{T}_p} \mathrm{spec}_{L^2_{\mathrm{per}}((0,p))} B_l .$$

We refer to [RS78] for the self-adjoint case and to [Mie97b] for the case of general elliptic operators. In [Sca99] the theory is developed for the Navier–Stokes equation where slight deviations arise, cf. also [Mie97c] for a concrete treatment of the operator arising in the Rayleigh–Bénard problem.

5.2 The Swift–Hohenberg Equation in one and two Dimensions

A widely studied model problem for the pattern formation over unbounded domains is the Swift–Hohenberg equation [SH77,CE90a],

$$\partial_t u = -(1+\Delta)^2 u + \varepsilon^2 u - u^3, \quad t \geq 0, \quad x \in \mathbb{R}^d, \quad u = u(t,x) \in \mathbb{R}. \tag{17}$$

First we consider the case $d = 1$ and below the case $d = 2$. For small $\varepsilon > 0$ there exist stationary roll solutions $u_{\varepsilon,\kappa}$ of (17) with $\kappa \in (-\varepsilon, \varepsilon)$, which bifurcate from $(\varepsilon, u) \equiv (0,0)$. These rolls have an amplitude $r \approx \widetilde{a} = \widetilde{a}(\varepsilon,\kappa) = \sqrt{4(\varepsilon^2-\kappa^2)/3}$, are even in x and periodic with period $2\pi/k$, where $k = \sqrt{\kappa+1}$. They may be expanded as

$$u_{\varepsilon,\kappa}(x) = a_1 \cos(kx) + a_3 \cos(3kx) + O(\widetilde{a}^5), \tag{18}$$

where $a_1 = \widetilde{a} + \widetilde{a}^3/512 + O(\widetilde{a}^5)$ and $a_3 = O(\widetilde{a}^3)$, see e.g. [CE90a,Mie95].

The nonlinear diffusive stability of rolls $u_{\varepsilon,\kappa}$ with respect to spatially localized perturbations has been shown in [Sch96]. Here we outline the ideas. Letting $u(t,x) = u_{\varepsilon,\kappa}(x) + v(t,x)$, the perturbations v have to satisfy

$$\partial_t v = \mathcal{L}v + F(v), \tag{19}$$

where $\mathcal{L}v = -(1+\partial_x^2)^2 v + \varepsilon^2 v - 3u_{\varepsilon,\kappa}^2 v$ and $F(v) = -3u_{\varepsilon,\kappa}v^2 - v^3$. The operator $\mathcal{L}$ can be treated with the Bloch analysis of Section 5.1. The associated Bloch operators are

$$B(\varepsilon,\kappa,l)V \stackrel{\mathrm{def}}{=} -(1+(\partial_x+\mathrm{i}l)^2)^2 V + (\varepsilon^2 - 3u_{\varepsilon,\kappa}^2)V, \tag{20}$$

which are unbounded operators on $L^2_{\mathrm{per}}(\mathcal{T}_{2\pi/k})$ with domain $H^4_{\mathrm{per}}(\mathcal{T}_{2\pi/k})$. For every fixed $l \in \mathcal{T}_k$ the eigenvalue problem (20) is self-adjoint with a discrete set of real eigenvalues $\{\lambda_j^{(\varepsilon,\kappa)}(l) \in \mathbb{R} : j \in \mathbb{N}\}$, $\lambda_j^{(\varepsilon,\kappa)}(l) \geq \lambda_{j+1}^{(\varepsilon,\kappa)}(l) \to -\infty$ for $j \to \infty$.

For $(\varepsilon,\kappa) = (0,0)$ the eigenvalues are $\lambda_j^{(0,0)}(l) = -(1-(kj+l)^2)^2$, and by perturbation arguments $\lambda_j^{(\varepsilon,\kappa)}(l)$ can be calculated for all small (ε,κ). The stability of $u_{\varepsilon,\kappa}$ is then determined from the behavior of the smooth function $l \mapsto \lambda_1^{(\varepsilon,\kappa)}(l)$ for l close to 0. In fact, we always have $\lambda_1^{(\varepsilon,\kappa)}(0) = 0$. This eigenvalue 0 comes from the fact that we have a family (w.r.t. β) of stationary solutions and that translations along this family lead to the associated eigenvector $\partial_x u_{\varepsilon,\kappa}$. For small l we have the expansion

$$\lambda_1^{(\varepsilon,\kappa)}(l) = -c_1(\varepsilon,\kappa)l^2 + O(l^4), \tag{21}$$

where expansions for the coefficient $c_1(\varepsilon,\kappa)$ are given in [CE90a,Mie95, Mie97b]. If $c_1 < 0$ then $u_{\varepsilon,\kappa}$ is linearly unstable (Eckhaus' sideband instability) which occurs for $\varepsilon^2 < E_{\mathrm{Eckh}}(\kappa) = 3\kappa^2 + O(|\kappa|^3)$. If ε^2 lies above the Eckhaus stability boundary $E_{\mathrm{Eckh}}(\kappa)$, then the rolls are linearly stable.

Moreover, the parabolic expansion (21) of the critical eigenvalue suggests that solutions to the linear problem $v_t = \mathcal{L}v$ decay like solutions to the linear diffusion equation (1).

Regarding the nonlinear equation (19) the nonlinearity doesn't seem to be irrelevant by naive power counting in the sense of Section 3. However, in the spatially periodic case a more elaborate way of power counting involving the Bloch analysis is necessary. This can be understood by relating the problem on the unbounded domain to the center-manifold theory for (17) with periodic boundary conditions. Then via normal form transformations and appropriate convolution identities it turns out that the nonlinearity in Bloch space vanishes up to a sufficiently high order. For details see Section 3 of [Sch98b]. Using the renormalization group approach from Subsection 3.3 one obtains the following theorem.

Theorem 7. *[Sch96] There exists an $\varepsilon_1 > 0$ such that for all $\varepsilon \in (0,\varepsilon_1)$, all Eckhaus-stable rolls $u_{\varepsilon,\kappa}$ and all $p \in (0,1/2)$ the following holds. There exist $\delta, C > 0$ such that for all initial conditions v_0 satisfying*

$$v_0(x) = \textstyle\sum_n \varepsilon A_n(\varepsilon x)\mathrm{e}^{\mathrm{i}nx} + \text{c.c.} \quad \textit{with} \quad \textstyle\sum_n n^2\|A_n\|_{H^2(2)} < \delta \tag{22}$$

we have a constant $\alpha = O(1)$ such that the solution v of (19) with $v|_{t=0} = v_0$ satisfies

$$\|v(t,x) - \tfrac{\alpha}{\sqrt{t}}\,\mathrm{e}^{-x^2/(4c_1t)}\partial_x u_{\varepsilon,\kappa}(x)\|_{L^\infty(\mathbb{R})} \le C\varepsilon^{-1+2p}t^{-1+p} \quad \textit{for } t \to \infty,$$

with $c_1 = c_1(\varepsilon,\kappa)$ from (21).

Remark 8. From (22) one obtains that the attracted neighborhood $\mathcal{U}$ of $u_{\varepsilon,\kappa}$ is of order ε in $L^\infty(\mathbb{R})$. Defining

$$\mathcal{B}_1 = \{u : u(x) = \sum_n \varepsilon A_n(\varepsilon x)\mathrm{e}^{\mathrm{i}nx} \text{ with } \sum_n n^2\|A_n\|_{H^2(2)} < \infty\}$$

we have asymptotic $(\mathcal{B}_1, L^\infty)$ stability of $u_{\varepsilon,\kappa}$. A more refined characterization of $\mathcal{U}$ is given in [Sch96].

The above results have been transfered to the two-dimensional Swift–Hohenberg equation in [Uec99]. Here the rolls considered are $u_{\varepsilon,\kappa}(x_1,x_2) = u^{1\mathrm{d}}_{\varepsilon,\kappa}(x_1)$, i.e., they are independent of x_2. Inserting $u(t,x) = u_{\varepsilon,\kappa}(x_1)+v(t,x)$ into (17) one obtains

$$\partial_t v = \mathcal{L}v + F(v) \tag{23}$$

with $F(v) = -3u_{\varepsilon,\kappa}v^2 - v^3$ as before, but now $\mathcal{L}v = -(1+\Delta)^2 v + \varepsilon^2 v - 3u^2_{\varepsilon,\kappa}v$. The eigenvalue problem $\mathcal{L}v = \lambda v$ can still be treated by the Bloch analysis with $v(x) = \mathrm{e}^{\mathrm{i}l\cdot x}V(x_1)$ where $l = (l_1,l_2) \in \mathcal{T}_k \times \mathbb{R}$ is now a two-dimensional Bloch wave vector. The Bloch operators read

$$B(\varepsilon,\kappa,l)V \stackrel{\mathrm{def}}{=} -(1+(\partial_x+\mathrm{i}l_1)^2 - l_2^2)^2 V + (\varepsilon^2 - 3u^2_{\varepsilon,\kappa})V = \lambda V. \tag{24}$$

Again, the stability properties of $u_{\varepsilon,\kappa}$ are determined by the behavior of the spectral surface $(l_1,l_2) \mapsto \lambda_1^{(\varepsilon,\kappa)}(l_1,l_2)$. Rigorous stability results and a complete characterization of the set of unstable Bloch wave vectors $\{l \in \mathcal{T}_k \times \mathbb{R} \mid \lambda_1^{(\varepsilon,\kappa)}(l) > 0\}$ in dependence of (ε,κ) have been obtained in [Mie97b]. In addition to the Eckhaus instability for $\varepsilon^2 < E_{\mathrm{Eckh}}(\kappa)$ there occurs a second instability mechanism called zigzag-instability. This means instability with respect to Bloch waves $\mathrm{e}^{\mathrm{i}l\cdot x}V(x_1)$ with wave vectors $l = (0,l_2)$ with small $l_2 \neq 0$. The linearized stability results may be summarized as follows.

Theorem 9. *[Mie97b] There exist an $\varepsilon_1 > 0$ and curves $E_{\mathrm{Eckh}}, K_{\mathrm{zigzag}}$ with expansions $E_{\mathrm{Eckh}}(\kappa) = 3\kappa^2 - \kappa^3 + O(\kappa^4)$ and $K_{\mathrm{zigzag}}(\varepsilon) = -\varepsilon^4/512 + O(\varepsilon^6)$ such that a roll $u_{\varepsilon,\kappa}$ with $\varepsilon \in (0,\varepsilon_1]$ is linearly stable if and only if*

$$\varepsilon^2 > E_{\mathrm{Eckh}}(\kappa) \quad \text{and} \quad \kappa > K_{\mathrm{zigzag}}(\varepsilon).$$

In this case the surface $\lambda_1^{(\varepsilon,\kappa)}$ of the largest eigenvalue of (24) has the expansion

$$\lambda_1^{(\varepsilon,\kappa)}(l) = -c_1(\varepsilon,\kappa)l_1^2 - c_2(\varepsilon,\kappa)l_2^2 + O(|l|^4) \quad \text{for } l \to 0 \tag{25}$$

with $c_j(\varepsilon,\kappa) > 0$ for $j = 1,2$.

In [Uec99] these results are combined with the techniques developed in [Sch96] to show the nonlinear diffusive stability of marginally stable rolls. In order to obtain estimates on the size of the domain of attraction we need to take care of the dependence of (c_1,c_2) on (ε,κ). Here it is useful to consider the special parameterization $\kappa = \beta\varepsilon$ with $\beta \in (0,1/\sqrt{3})$ and the transformed parameter set $\mathcal{P} = (0,1/\sqrt{3}) \times (0,\varepsilon_1)$. With the definition (5) adapted to $H^m(k) = \{u : \mathbb{R}^2 \to \mathbb{C} : \|u\|_{H^m(k)} = \|u\rho^k\|_{H^m(\mathbb{R}^2)} < \infty\}$, $\rho(x) = (1+|x|^2)^{1/2}$, we have the following result.

Theorem 10. *[Uec99] There exist continuous functions $\delta, C : (0,1/\sqrt{3}) \to \mathbb{R}_+$ and a continuous function $A : \mathcal{P} \times H^2(3) \to \mathbb{R}$ such that for all $(\beta,\varepsilon) \in \mathcal{P}$*

the following holds. Let $\kappa = \beta\varepsilon$ *and let* $v = v(t,x)$ *be the solution to (23) with the initial condition* v_0 *satisfying*

$$v_0(x) = \varepsilon^{3/2} \sum_{n\geq 0} A_n(\varepsilon x_1, \sqrt{\varepsilon} x_2) \mathrm{e}^{\mathrm{i}nx_1} + \mathrm{c.c.} \tag{26}$$

$$\textit{with} \quad \sum_{n\geq 0} (1+n^2) \|A_n\|_{H^2(3)} \leq \delta(\beta). \tag{27}$$

Then we have

$$\|v(t,x) - \tfrac{\alpha}{\sqrt{c_1 c_2}\ t}\ \mathrm{e}^{-[d_1 x_1^2 + d_2 x_2^2]/(4t)} \partial_{x_1} u^{\mathrm{1d}}_{\varepsilon,\kappa}(x_1)\|_{L^\infty(\mathbb{R}^2)} \leq C(\beta)\varepsilon^{-3/2} t^{-3/2},$$

with $\alpha = A(\beta, \varepsilon, v_0)$ *and* $d_j = 1/c_j(\varepsilon, \beta\varepsilon)$ *from (25),* $j = 1, 2$.

Remark 11. Similar to Remark 8 one obtains from (26) that in two dimension the attracted neighborhood $\mathcal{U}$ of $u_{\varepsilon,\kappa}$ is of order $\varepsilon^{3/2}$ in $L^\infty(\mathbb{R}^2)$, see [Uec99].

5.3 Application to Hydrodynamical Stability Problems

With the above methods it was possible to solve a class of hydrodynamical stability problems on unbounded domains which have been open for almost 30 years, namely the nonlinear stability of linearly Eckhaus-stable spatially periodic equilibria in infinite cylinders with respect to spatially localized perturbations. The linear stability analysis leads to a similar situation as for the Swift–Hohenberg equation, i.e. continuous spectrum up to the imaginary axis and no obvious sign for the nonlinear terms.

The physical problems which we have in mind are the Taylor–Couette problem and Rayleigh–Bénard's problem. The Taylor–Couette problem consists in finding the flow of a viscous incompressible fluid filling the domain between two rotating infinitely extended cylinders. Bénard's problem consists in finding the flow of a viscous incompressible fluid filling an infinitely extended strip subjected to some heating from below.

In both problems the velocity field is governed by the Navier–Stokes equations, and in both problems there exist a spatially homogeneous flow, the Couette flow and pure heat conduction, respectively, which gets unstable and bifurcates into a family of spatially periodic equilibria, the Taylor vortices and roll solutions, respectively.

The result for the Taylor–Couette problem is formulated in [Sch98b]. For completeness we will give here an explicit formulation of the nonlinear diffusive stability result for roll solutions in the Rayleigh–Bénard problem. The linear Bloch theory with one unbounded direction was first studied in [KvW97] and that with two unbounded directions in [Mie97c]. Here we only treat the two-dimensional problem in the strip $\Omega = \mathbb{R} \times (0, \pi)$. The velocity field $u = (u_1, u_2)$, the temperature T and the pressure p satisfy

$$\begin{aligned} \partial_t u &= \nu \Delta u - \nabla p - T\boldsymbol{e}_2 - (u \cdot \nabla) u \\ \partial_t T &= \kappa \Delta T + u_2 - (u \cdot \nabla) T \\ 0 &= \nabla \cdot u \end{aligned}$$

for all $(x,y) \in \Omega$, with $\Delta = \partial_x^2 + \partial_y^2$, the mean flux condition $\int_0^\pi u_1 \, dy = 0$ and the boundary conditions

$$\partial_y u_1 = u_2 = 0 \text{ for } y = 0, \pi, \quad T = T_0 \text{ for } y = 0, \text{ and } T = T_1 \text{ for } y = \pi$$

where $T_0 < T_1$.

There exists a trivial spatially homogeneous solution $u = 0$, $T = T_0 + y(T_1 - T_0)/\pi$, which becomes unstable when the parameter $\mu = T_1 - T_0$ is sufficiently large. Then, a one-dimensional family of spatially periodic equilibria $(u_{q,\mu}, T_{q,\mu})$ with

$$(u_{q,\mu}, T_{q,\mu})(x,y) = (u_{q,\mu}, T_{q,\mu})(x+2\pi/q, y),$$

bifurcates, where the horizontal wave number q lies in the interval $(q_{\text{ex}}^-(\mu), q_{\text{ex}}^+(\mu))$. The linear stability analysis leads to a similar situation as for the Swift–Hohenberg equation. We have the linear Eckhaus-stability for all $(u_{q,\mu}, T_{q,\mu})$ with

$$q \in (q_{\text{stab}}^-(\mu), q_{\text{stab}}^+(\mu)) \subset (q_{\text{ex}}^-(\mu), q_{\text{ex}}^+(\mu)).$$

In [Sch98b] the following result has been proved.

Theorem 12. *There exist $\varepsilon, C > 0$ such that the following holds. Let the initial condition $(u,T)|_{t=0} = (u_{q,\mu}, T_{q,\mu}) + (v,\theta)|_{t=0}$ satisfy $q \in (q_{\text{stab}}^-(\mu), q_{\text{stab}}^+(\mu))$ and $\|(v,\theta)\|_{H^2(2)} \leq \varepsilon$. Then the associated solution $(u,T) = (u_{q,\mu}, T_{q,\mu}) + (v,\theta)$ satisfies*

$$\left\| \begin{pmatrix} v \\ \theta \end{pmatrix} - \frac{A^*}{\sqrt{t}} e^{-dx^2/t} \partial_x \begin{pmatrix} u_{q,\mu} \\ T_{q,\mu} \end{pmatrix} \right\|_{L^\infty} \leq \frac{C}{t} \quad \textit{for all } t > 0,$$

where $d = d(q,\mu) > 0$ and $A^* \in \mathbb{R}$ is a constant depending on the initial condition.

In a different situation the diffusive stability method has been used to prove the nonlinear stability of Kolmogorov flow (cf. [Sch99a]).

5.4 An Open Problem: Diffusive Mixing in Pattern Forming Systems

In the last section we explained how diffusive behavior occurs near spatially periodic equilibria in pattern forming systems. If such steady state solutions are perturbed in a spatially localized manner, the perturbations are repaired diffusively and decay to 0 algebraically in t as $t \to \infty$.

For the Ginzburg–Landau equation we additionally know that diffusive mixing takes place if two different stable steady states are prescribed for $x \to -\infty$ and $x \to \infty$, see Section 4.2. Although a proof of diffusive mixing of steady states in general pattern forming systems is still missing, it is widely expected that such a behavior occurs. For the one dimensional Swift–Hohenberg equation this conjecture is as follows.

Conjecture: *Fix* $\varepsilon > 0$ *sufficiently small and let* $u_{\varepsilon,k_-}, u_{\varepsilon,\kappa_+}$ *be two stable rolls with* $\kappa_- \neq \kappa_+$. *Then there exist limiting profiles* $\tilde{k} \in C_b^4(\mathbb{R})$ *and* $\tilde{U} \in C^4(\mathbb{R}_+ \times \mathbb{R}, \mathbb{R})$ *with* $\tilde{k}(\xi) \to k_\pm = \sqrt{\kappa_\pm + 1}$ *for* $\xi \to \pm\infty$ *and*

$$\|\tilde{k}(\cdot) - (k_- + (k_+ - k_-)\mathrm{Erf}(\cdot))\|_{C^4} \leq C\varepsilon^2$$
$$\|\tilde{U}(t,\cdot) - a_1(\varepsilon, \tilde{k}(\cdot/\sqrt{t}))\cos(\sqrt{t}\,\tilde{\phi}(\cdot/\sqrt{t}))\|_{C^4} \leq C\varepsilon^3,$$

where $\tilde{\phi}(\xi) = k_-\xi + \int_{-\infty}^{\xi}[\tilde{k}(s) - k_-]\,\mathrm{d}s$ *and* $a_1(\varepsilon, k)$ *from (18), such that the following holds. There exist* $t_0 > 0$, $\delta > 0$ *such that for all* $u_0 \in H^2_{\mathrm{lu}}(\mathbb{R})$ *with* $\|u_0 - \tilde{U}(t_0,\cdot)\|_{H^2(2)} \leq \delta$ *we have*

$$\|u(t,\cdot) - \tilde{U}(t,\cdot)\|_{L^\infty} \leq Ct^{-1/3}.$$

In particular, this implies that for all fixed $\ell > 0$,

$$\sup_{|x|\leq\ell} |u(t,x) - u_{\varepsilon,k_*^2-1}(x + \sqrt{t}\,\phi_*)| \leq C_\ell t^{-1/3},$$

where $k_* = \tilde{k}(0) = \frac{1}{2}(k_+ + k_-) + O(\varepsilon^2)$ *and* $\phi_* = \tilde{\phi}(0) = (k_+ - k_-)/\sqrt{\pi} + O(\varepsilon^2)$.

For Bénard's problem a similar conjecture can be stated. The proof of the diffusive repair of the steady states heavily relies on the Bloch wave analysis. So far it is not known how to transfer it to the diffusive mixing case, where we have to combine the Bloch wave analysis of the spatially periodic steady states with a local analysis in x-space to prove these conjectures.

References

[BK92] J. Bricmont and A. Kupianen. Renormalization group and the Ginzburg–Landau equation. *Comm. Math. Phys.*, 150:287–318, 1992.

[BKL94] J. Bricmont, A. Kupianen, and G. Lin. Renormalization group and asymptotics of solutions of nonlinear parabolic equations. *Comm. Pure Appl. Math.*, 6:893–922, 1994.

[BL99a] S. Benachour and P. Laurençot. Global solutions to viscous Hamilton-Jacobi equations with irregular initial data. *Comm. Partial Differential Equations*, 24(11-12):1999–2021, 1999.

[BL99b] W.-J. Beyn and J. Lorenz. Stability of traveling waves: dichotomies and eigenvalue conditions on finite intervals. *Numer. Funct. Anal. Optimization*, 20(3-4):201–244, 1999.

[BV90] A.V. Babin and M.I. Vishik. Attractors of partial differential equations in an unbounded domain. *Proc. Royal Soc. Edinb.*, 116A:221–243, 1990.

[CE90a] P. Collet and J.-P. Eckmann. *Instabilities and Fronts in Extended Systems.* Princeton University Press, 1990.

[CE90b] P. Collet and J.-P. Eckmann. The time dependent amplitude equation for the Swift–Hohenberg problem. *Comm. Math. Phys.*, 132:139–152, 1990.

[CE99a] P. Collet and J.-P. Eckmann. The definition and measurement of the topological entropy per unit volume in parabolic PDEs. *Nonlinearity*, 12:451–473, 1999.

[CE99b] P. Collet and J.-P. Eckmann. Extensive properties of the complex Ginzburg–Landau equation. *Comm. Math. Physics*, 200:699–722, 1999.

[CEE92] P. Collet, J.-P. Eckmann, and H. Epstein. Diffusive repair for the Ginzburg–Landau equation. *Helv. Phys. Acta*, 65:56–92, 1992.

[DFKM96] G. Dangelmayr, B. Fiedler, K. Kirchgässner, and A. Mielke. *Dynamics of Nonlinear Waves in Dissipative Systems: Reduction, Bifurcation and Stability*, volume 352 of *Pitman Research Notes in Mathematics*. Addison Wesley Longman, 1996.

[Eck65] W. Eckhaus. *Studies in Non-Linear Stability Theory.* Springer Tracts in Nat. Phil. Vol.6, 1965.

[Eck93] W. Eckhaus. The Ginzburg–Landau manifold is an attractor. *J. Nonlinear Science*, 3:329–348, 1993.

[ES98] J. Escher and B. Scarpellini. Stability properties of parabolic equations in unbounded domains. *Arch. Math.*, 71(1):31–45, 1998.

[ES99a] J.-P. Eckmann and G. Schneider. Nonlinear stability of modulating fronts in the Swift–Hohenberg equation. In M. Otani, editor, *Nonlinear evolution equations and application*, Volume 1105, pages 81–90, Kyoto, 1999, RIMS Kokyuroku.

[ES99b] M. Efendiev and A. Scheel. Upper and lower bounds for the Hausdorff dimension of the attractor for reaction–diffusion equations in $\mathbb{R}^n$. Preprint, Universität Bayreuth, 1999.

[EW94a] J.-P. Eckmann and C.E. Wayne. The nonlinear stability of front solutions for parabolic partial differential equations. *Comm. Math. Phys.*, 161(2):323–334, 1994.

[EW94b] J.-P. Eckmann and C.E. Wayne. The nonlinear stability of front solutions for parabolic partial differential equations. *Comm. Math. Phys.*, 161(2):323–334, 1994.

[EZ99] M. Efendiev and S. Zelik. The attractor for a nonlinear reaction–diffusion system in an unbounded domain. Preprint no. 505, WIAS Berlin, 1999.

[Fei96] E. Feireisl. Bounded, locally compact global attractors for semilinear damped wave equations on $\mathbb{R}^n$. *J. Diff. Integral Eqns.*, 9:1147–1156, 1996.

[Gal94] T. Gallay. Local stability of critical fronts in parabolic partial differential equations. *Nonlinearity*, 7:741–764, 1994.

[GM98] T. Gallay and A. Mielke. Diffusive mixing of stable states in the Ginzburg–Landau equation. *Comm. Math. Phys.*, 199:71–97, 1998.

[GR97] T. Gallay and G. Raugel. Stability of travelling waves for a damped hyperbolic equation. *Z. Angew. Math. Phys.*, 48(3):451–479, 1997.

[GR98] T. Gallay and G. Raugel. Scaling variables and asymptotic expansions in damped wave equations. Orsay Preprint no. 97–66, 1997.

[HCS99] M. Haragus-Courcelle and G. Schneider. Bifurcating fronts for the Taylor–Couette problem in infinite cylinders. *ZAMP*, 50:120–151, 1999.

[Kap94] T. Kapitula. On the nonlinear stability of plane waves for the Ginzburg–Landau equation. *Comm. Pure Appl. Math.*, 47:831–841, 1994.

[KS98] T. Kapitula and B. Sandstede. Stability of bright and dark solitary-wave solutions to perturbed nonlinear Schrödinger equations. *Physica D*, 124:58–103, 1998.

[KSM92] P. Kirrmann, G. Schneider, and A. Mielke. The validity of modulation equations for extended systems with cubic nonlinearities. *Proc. Royal Soc. Edinburgh*, 122A:85–91, 1992.

[KvW97] Y. Kagei and W. von Wahl. The Eckhaus criterion for convection roll solutions of the Oberbeck–Boussinesq equations. *Int. J. Non-Linear Mechanics*, 32:563–620, 1997.

[Mie95] A. Mielke. A new approach to sideband instabilities using the principle of reduced instability. In A. Doelman and A. van Harten, editors, *Nonlinear dynamics and pattern formation in the natural environment*, pages 206–222. Longman UK, 1995.

[Mie97a] A. Mielke. The complex Ginzburg–Landau equation on large and unbounded domains: sharper bounds and attractors. *Nonlinearity*, 10:199–222, 1997.

[Mie97b] A. Mielke. Instability and stability of rolls in the Swift–Hohenberg equation. *Comm. Math. Phys.*, 189:829–853, 1997.

[Mie97c] A. Mielke. Mathematical analysis of sideband instabilities with applicaton to Rayleigh–Bénard convection. *J. Nonlinear Science*, 7:57–99, 1997.

[Mie99] A. Mielke. The Ginzburg–Landau equation in its role as a modulation equation. In B. Fiedler, G. Iooss, N. Kopell, and F. Takens, editors, *Handbook of Dynamical Systems III. Towards Applications.* Springer–Verlag, 1999. To appear.

[MS95] A. Mielke and G. Schneider. Attractors for modulation equations on unbounded domains — existence and comparison. *Nonlinearity*, 8(5):743–768, 1995.

[MS96] A. Mielke and G. Schneider. Derivation and justification of the complex Ginzburg–Landau equation as a modulation equation. In Percy Deift, C. David Levermore, and C. Eugene Wayne, editors, *Lect. Appl. Math. 31*, pages 191–216. American Mathematical Society, 1996.

[MSZ00] A. Mielke, G. Schneider, and A. Ziegra. Comparison of inertial manifolds and application to modulated systems. *Math. Nachrichten*, 214:53–69, 2000.

[OZ00] J. H. Ortega and E. Zuazua. Large time behavior in $\mathbb{R}^n$ for linear parabolic equations with periodic coefficients. *Asymptot. Anal.*, 22(1):51–85, 2000.

[RK98] G. Raugel and K. Kirchgässner. Stability of fronts for a KPP-system. II. The critical case. *J. Diff. Eqns.*, 146(2):399–456, 1998.

[RS78] M. Reed and B. Simon. *Methods of Modern Mathematical Physics IV.* Academic Press, New York, 1978.

[Sca99] B. Scarpellini. *Stability, Instability, and Direct Integrals*, volume 402 of *Pitman Research Notes in Mathematics.* Chapman & Hall/CRC, 1999.

[Sch94a] G. Schneider. Error estimates for the Ginzburg–Landau approximation. *ZAMP*, 45:433–457, 1994.

[Sch94b] G. Schneider. Global existence via Ginzburg–Landau formalism and pseudo–orbits of Ginzburg–Landau approximations. *Comm. Math. Phys.*, 164:157–179, 1994.

[Sch94c] G. Schneider. A new estimate for the Ginzburg–Landau approximation on the real axis. *J. Nonlinear Science*, 4:23–34, 1994.

[Sch96] G. Schneider. Diffusive stability of spatial periodic solutions of the Swift–Hohenberg equation. *Comm. Math. Phys.*, 178:679–702, 1996.

[Sch98a] G. Schneider. Hopf–bifurcation in spatially extended reaction–diffusion systems. *J. Nonlinear Science*, 8:17–41, 1998.

[Sch98b] G. Schneider. Nonlinear stability of Taylor–vortices in infinite cylinders. *Arch. Rational Mech. Analysis*, 144(2):121–200, 1998.

[Sch99a] G. Schneider. Cahn–Hilliard description of secondary flows of a viscous incompressible fluid in an unbounded domain. *ZAMM*, 79:615–626, 1999.

[Sch99b] G. Schneider. Global existence results for pattern forming processes in infinite cylindrical domains – applications to 3d Navier–Stokes problems –. *J. Mathématiques Pures Appliquées*, 78:265–312, 1999.

[Sch00] G. Schneider. Existence and stability of modulating pulse solutions in disspative systems. *Physica D*, 140(3–4):283–293, 2000.

[SH77] J. Swift and P.C. Hohenberg. Hydrodynamic fluctuations at the convective instability. *Physical Review A*, 15(1):319–328, 1977.

[SS99] B. Sandstede and A. Scheel. Spectral stability of modulated travelling waves bifurcating near essential instabilities. *Proc. Roy. Soc. Edinb. A*, 129:1263–1290, 1999.

[Uec99] H. Uecker. Diffusive stability of rolls in the two–dimensional real and complex Swift–Hohenberg equation. *Comm. PDE*, 24(11–12):2109–2146, 1999.

[Uec00] H. Uecker. Stable modulating multi–pulse solutions for dissipative systems with resonant spatially periodic forcing. *J. Nonlin. Sci.*, 2000 To appear.

[vH91] A. van Harten. On the validity of Ginzburg–Landau's equation. *J. Nonlinear Science*, 1:397–422, 1991.

[Way97] C.E. Wayne. Invariant manifolds for parabolic partial differential equations on unbounded domains. *Arch. Rational Mech. Analysis*, 138(3):279–306, 1997.

[Zel99] S. Zelik. The attractor for nonlinear reaction-diffusion system in the unbounded domain and Kolmogorov's ε–entropy. *Mathem. Nachr.*, 1999. Submitted (DANSE Preprint 34/99).

Dimension Estimates for Invariant Sets of Dynamical Systems

Volker Reitmann*

Technische Universität Dresden, Fachrichtung Mathematik, Mommsenstr. 13, 01062 Dresden, Germany

Abstract We investigate the relationships between various types of dimension like characteristics (e.g. Hausdorff, box, and Lyapunov dimensions, topological entropy) of dynamical systems and the geometric complexity of invariant and limit sets (e.g. isolated equilibria, global attracting or unstable cycles, products of Cantor-type sets and regular objects). We consider both flows and discrete-time dynamical systems of smoothness C^1 on n-dimensional smooth Riemannian manifolds which possess compact invariant sets. We formulate our results in terms of the singular values of the linearized evolution operator and consider additionally global informations such as the homology group and curvature properties of the manifold, natural Lyapunov functions and Losinskii norms, and the degree of non-injectivity of the generating map.

1 Introduction

The first general result for upper Hausdorff dimension estimates of flow invariant sets in $\mathbb{R}^n$ in terms of singular values of the linearization are given in [8]. This approach was extended in [47,31,4,51] to map-invariant sets on Riemannian manifolds and in [32,34] by including Lyapunov functions into the contraction conditions for outer Hausdorff measures. In [10,55] the results by Douady-Oesterlé were extended to estimates for invariant sets of evolution systems in Hilbert spaces. Hausdorff dimension estimates for invariant sets of general flows using the eigenvalues of the symmetric part of the linearization are deduced in [54] for the $\mathbb{R}^n$ and in [47,35,37] for manifolds. Douady-Oesterlé-type estimates for piecewise smooth maps on manifolds are given in [51,46] (see also [53]). The hyperbolic or quasi-hyperbolic structure of the invariant set was used in dimension estimates in [12,22,16] where also an entropy term was introduced into the dimension bound.

Several dimension upper bounds of invariant sets allow conclusions on the dynamical behaviour of the system. The key step in the papers [54,33,34,47] is to use properties of sets with Hausdorff dimension less than two. One can show ([54,5]) that such a compact invariant set contains no simple closed piecewise smooth invariant curves. On the basis of Hausdorff dimension estimates

* Project: Dimension Estimates of Attractors via Feedback-Systems and Lyapunov-Functions (Volker Reitmann)

a generalization of certain global stability results for ODE's of Hartman-Olech and Borg ([25]), but also of other types of classical results from the Bendixson-Poincaré theory were derived in [34,41,42].
Parallel to Hausdorff dimension estimates a number of upper and lower bounds for the box dimensions of invariant sets were deduced in [4,60,29,2,36,43,55,26]. The box dimension of a set is never smaller than the Hausdorff dimension and gives important informations about the possibility to use embedding strategies. Recently it was shown ([14]) that certain embedding homeomorphisms can be chosen generically with Hölder-Lipschitz continuous inverse which enable conclusions for dimension estimates.
Hausdorff and box dimension estimates for flow invariant sets are very effective if appropriate types of local, global and uniform Lyapunov exponents are introduced ([10,31,32,55,60]). On the basis of such Lyapunov exponents various types of Lyapunov dimensions of a dynamical system with respect to an invariant set were defined (Kaplan-Yorke formula [18]) and it was conjectured that in typical cases these dimensions coincide with the Hausdorff dimension of the invariant set.
Parallel to the dimension and stability investigation of invariant sets of discrete-time and continuous-time dynamical systems various types of dimensions of invariant measures have been developed ([31,48,59]). Defining for the invariant ergodic measure of a dynamical system the Lyapunov exponents one can introduce the Lyapunov dimension of this measure which is an upper bound of the Hausdorff dimension of the measure. As in the measure free case various stability and convergence properties of the underlying dynamical system may be derived from the magnitude of the Lyapunov exponents of the measure. It is shown in [9] that if the invariant measure of a flow is ergodic and all Lyapunov exponents of this measure are negative, the support of the measure is a stable equilibrium point. If exactly one exponent of such a measure is zero and the remaining ones are negative, the support of the measure is a stable equilibrium or a stable limit cycle.
An important class of invariant sets of dynamical systems have locally the structure of a product of a smooth (often one-dimensional) submanifold directed "along the attractor" and a set, "transversal" to the attractor. Thus, it is naturally to investigate the dimension of such sets and stability of orbits of such attractors considering surfaces which are transversal to the flow-lines.
The use of transversal intersections is well-known in stability investigations of flow lines: contracting or expanding behaviour in sections transversal to the flow line directions is the main reason for the stability or instability of the orbit ([25,34,37]).
The paper is organized as follows. In Section 2 we present a short review on attractors and Lyapunov functions. In Section 3 we introduce singular values of differentials, the covariantly written variational equation and basic facts on Losinskii measures and measure-free Lyapunov exponents. In Section 4 we give the definitions of various types of dimensions and of the topological entropy. A number of upper and lower dimension estimates (especially of

Hausdorff dimension) of invariant sets are contained in Section 5. In Section 6 upper Hausdorff dimension bounds for flow-negatively invariant sets are formulated in terms of the eigenvalues of the symmetric part of an associated system in normal variations. In Section 7 we give upper Hausdorff dimension estimates for systems with an equivariant tangential bundle splitting. As a special case flows on manifolds with negative curvature are considered. In the last Section 8 the cardinality of pre-image sets is used to get dimension estimates for non-injective maps.

2 Attractors, Dissipativity and Lyapunov Functions

Suppose M is an n-dimensional C^k-manifold with Riemannian metric g. Let on M be defined a dynamical system $\{\varphi^t\}_{t\in\Gamma}$ with $\Gamma \in \{\mathbb{Z}_+, \mathbb{Z}, \mathbb{R}, \mathbb{R}_+\}$. In the discrete-time case ($\Gamma \in \{\mathbb{Z}_+, \mathbb{Z}\}$) we assume that the dynamical system $\{\varphi^t\}_{t\in\Gamma}$ is generated by a sufficiently smooth map $\varphi \equiv \varphi^1 : M \to M$. In the continuous-time case ($\Gamma \in \{\mathbb{R}_+, \mathbb{R}\}$) we assume a smooth vector field $f : M \to TM$ as generator of $\{\varphi^t\}_{t\in\Gamma}$. If $\Gamma = \mathbb{R}$ we speak also about a *flow*. If $\Gamma \in \{\mathbb{R}, \mathbb{Z}\}$ we call the system *invertible*. Further we use the notion $\Gamma_+ := \Gamma \cap \mathbb{R}_+$.
A subset $A \subset M$ is called *invariant* with respect to $\{\varphi^t\}_{t\in\Gamma}$ if $\varphi^t(A) = A$ for all $t \in \Gamma$ and *positively invariant* if $\varphi^t(A) \subset A$ for all $t \in \Gamma_+$. A compact set $B_0 \subset M$ is *absorbing* for $\{\varphi^t\}_{t\in\Gamma}$ if the following properties hold:

1) $\varphi^t(B_0) \subset B_0$ for all $t \in \Gamma_+$.
2) For every bounded set $B \subset M$ there exists $t_B \in \Gamma_+$ such that $\varphi^t(B) \subset B_0$ for all $t \in \Gamma_+$ with $t \geq t_B$.

A system with a compact absorbing set is also called *dissipative*. A set $E \subset M$ is called *minimal* for $\{\varphi^t\}_{t\in\Gamma}$ if E is non-empty, closed, and invariant, and has no proper subsets with the same property.
Suppose $\{\varphi^t\}_{t\geq 0}$ is dissipative with respect to the bounded set B_0. Then B_0 contains a minimal set. To see this, note that since B_0 is bounded, the non-wandering set $\Omega(\varphi)$ of $\varphi = \{\varphi^t\}_{t\geq 0}$ is non-empty, closed, φ^t-invariant, and $\Omega(\varphi) \subset B_0$. Moreover, $\Omega(\varphi)$ is bounded. But any bounded, closed invariant set contains a minimal set.
A set $A \subset M$ is called *attractor* for $\{\varphi^t\}_{t\in\Gamma}$ if A is compact, invariant and if there exists an open set $U \supset A$ such that for any $\varepsilon > 0$ there is a time $t_\varepsilon \in \Gamma_+$ with $\varphi^t(U) \subset N_\varepsilon(A)$ for all $t \in \Gamma_+$ with $t \geq t_\varepsilon$. The set $W^s(A) = \{p \in M : \varphi^t(p) \to A \text{ for } t \to \infty\}$ is called *domain of attraction*. A set A is a *global attractor* for $\{\varphi^t\}_{t\in\Gamma}$ if it is an attractor and for any bounded set $B \subset M$ holds $\operatorname{dist}(\varphi^t(B), A) \to 0$ for $t \to \infty$. If $\{\varphi^t\}_{t\in\Gamma}$ has a compact absorbing set B_0 then $A = \bigcap_{t\in\Gamma_+} \varphi^t(B_0)$ is the global attractor for $\{\varphi^t\}_{t\in\Gamma}$.
An effective tool for the characterization of attractors or dissipativity regions are Lyapunov functions. For a given dynamical system $\varphi = \{\varphi^t\}_{t\in\Gamma}$ on M a smooth function $v : U \subset M \to \mathbb{R}$ is called *Lyapunov function* in $U_1 \subset U$

if $\mathcal{L}_\varphi v(p) \leq 0$ for all $p \in U_1$. Here $\mathcal{L}_\varphi v$ denotes the *Lie operator* of v with respect to $\{\varphi^t\}_{t\in\Gamma}$, defined as

$$\mathcal{L}_\varphi v(p) = \begin{cases} v(\varphi(p)) - v(p) & \text{if } p, \varphi(p) \in U_1 \text{ (discrete-time case)}, \\ \dot{v}(p) = \dfrac{d}{dt} v(\varphi^t p)_{|t=0} & \text{if } \varphi^t p \in U_1, |t| < \varepsilon \text{ (continuous-time case)}. \end{cases}$$

Lyapunov functions are important in the description of attractors and absorbing domains. It is well-known ([3]), that if A is an attractor for the flow $\{\varphi^t\}_{t\in\Gamma}$ on M, then there exists a C^k-Lyapunov function $v : W^s(A) \to \mathbb{R}_+$ such that $v(p) = 0$ on A, $v(p) > 0$ and $\dot{v}(p) < 0$ on $W^s(A)\backslash A$ and $v(p) \to +\infty$ for $p \to \partial W^s(A)$.
Hence, any attractor of a given flow may be described by means of a Lyapunov function v as $\{p \in M : v(p) = 0\}$, which shows ([3,21])that any attractor has the shape of a polyhedron.
We now describe a useful method for verifying the existence of Lyapunov functions for a dynamical system $\{\varphi^t\}_{t\in\Gamma}$ which shall be used in the following. In the continuous-time case we assume for this that the flow is given by a vector field in *feedback-control* form

$$\dot{x} = Ax + bu, \;\; u = \phi(c^* x) \;, \tag{1}$$

where $A : E_n \to E_n$, $b : E_m \to E_n$, and $c : E_l \to E_n$ are linear operators, $\phi : U \subset E_l \to E_m$ is a C^1-function (E_k denotes an Euclidean space of dimension k).
In the discrete-time case we consider the system

$$x \mapsto Ax + bu, \;\; u = \phi(c^* x) \;, \tag{2}$$

where A, b, c, and ϕ are as above. We say that the pair (A, b) of (1), respectively (2), is *stabilizable* if there exists an operator h such that the spectrum of $A + b\,h^*$ is in the continuous-time case contained in $\{z \in \mathbb{C} : \operatorname{Re} z < 0\}$ and in the discrete-time case contained inside the complex unit circle.
The following theorem is proved by V.A.Yakubovich ([58]) and is called *frequency-domain lemma.*

Theorem 1. *Consider a dynamical system $\{\varphi^t\}_{t\in\Gamma}$ of (1) or (2) and assume that (A, b) is stabilizable. Let $F : E_n \times E_m \to \mathbb{R}$ be a quadratic form. Then there exists a Lyapunov function $v(x) = \langle x, Hx\rangle$ generated by an operator $H = H^* : E_n \to E_n$ and satisfying $\mathcal{L}_\varphi v(x, u) < 0$ for $x \in E_n, u \in E_m$ with $F(x, u) \geq 0$ and $\|x\| + \|u\| > 0$, if and only if there exists a $\delta > 0$ such that $F^c(x, u) \leq -\delta(\|x\|^2 + \|u\|^2)$ for all $x \in E^c_u$, $u \in E^c_m$ such that $\lambda x = Ax + bu$ for all $\lambda \in S$, where $S = \{i\omega : \omega \in \mathbb{R}\}$ in the continuous-time case and $S = \{\lambda \in \mathbb{C} : |\lambda| = 1\}$ in the discrete-time case. (Here is $\mathcal{L}_\varphi v(x, u) = 2x^* H(Ax + bu)$ in the continuous-time case and $\mathcal{L}_\varphi v(x, u) = (Ax + bu)^* H(Ax + bu) - x^* Hx$ in the discrete-time case. F^c denotes the Hermitean extension.)*

Remark 2. Using the frequency-domain lemma and a procedure which is called *averaging over the group* we can construct Lyapunov function for factor manifolds. Consider a dynamical system $\{\varphi^t\}_{t\in\Gamma}$ on M which is equivariant with respect to a finite group G of homeomorphisms of M, i.e. for any $t \in \Gamma$ and $g \in G$ we have $\varphi^t \circ g = g \circ \varphi^t$. Suppose also that $v : M \to \mathbb{R}$ is a Lyapunov function for φ on M, i.e. $\mathcal{L}_\varphi v(p) \le 0$ on M. We want to consider the dynamical system on the factor-set M/G, which is assumed to have the structure of a factor manifold. We define a new scalar-valued function on M by $w(p) := \frac{1}{|G|}\sum_{g\in G} v(gp)$. It is easy to see that

$$w(gp) = \frac{1}{|G|}\sum_{g'\in G} v(g'gp) = w(p) \quad \text{for any } g \in G \text{ and}$$

$$w(\varphi^t p) - w(p) = \frac{1}{|G|}\sum_{g\in G} v(g\varphi^t p) - v(gp) = \frac{1}{|G|}\sum_{g\in G} \underbrace{v\left(\varphi^t(gp)\right) - v(g(p))}_{\le 0}$$

for $p \in M$ and $t \in \Gamma$. Thus w has all properties of a Lyapunov function on M/G.

3 Multilinear Operators Acting on Bundles

3.1 Singular Values of Operators Acting Between Different Spaces

Let E and E' be two n-dimensional Euclidean spaces with the scalar products $\langle\cdot,\cdot\rangle_{\mathrm{E}}$ and $\langle\cdot,\cdot\rangle_{\mathrm{E}'}$, respectively. If $L : E \to E'$ is a linear map, we denote by $L^{[*]} : E' \to E$ the unique linear operator satisfying $\langle L^{[*]}u, v\rangle_{\mathrm{E}} = \langle u, Lv\rangle_{\mathrm{E}'}$, for all $u \in E'$ and $v \in E$. $L^{[*]}$ is called $\langle\cdot,\cdot\rangle_{\mathrm{E}}$, $\langle\cdot,\cdot\rangle_{\mathrm{E}'}$-*adjoint*, or shortly *adjoint* ([20]). If $L = L^{[*]}$ the operator is called *self-adjoint*. Using the definition it is easy to see that $L^{[*]}L$ is self-adjoint.

Example 3. Consider the special case $L : (\mathbb{R}^n, \langle\cdot,\cdot\rangle_{\mathrm{G}_1}) \to (\mathbb{R}^n, \langle\cdot,\cdot\rangle_{\mathrm{G}_2})$. We identify L with its matrix in the canonical basis and assume scalar products given by positive-definite matrices G_1 and G_2. Then $L^{[*]} = G_1^{-1}L^*G_2$, where L^* denotes the usual transpose of the matrix L. It follows that $L^{[*]}L = G_1^{-1}L^*G_2L$ and $(L^{[*]}L)^{[*]} = L^{[*]}L$. The matrix $L^{[*]}L$ is non-negative, by which we understand that all eigenvalues of the generalized eigenvalue problem $L^*G_2L = \lambda G_1$ are non-negative.

The *singular values* of a linear operator $L : E \to E'$, denoted by $\sigma_1(L) \ge \ldots \ge \sigma_n(L)$ are defined to be the eigenvalues of the positive part of L, i.e. of the positive semi-definite operator $(L^{[*]}L)^{\frac{1}{2}} : E \to E$. (They are ordered with respect to their size and algebraic multiplicity). The definition of the numbers $\sigma_i(L)$ agrees with the min-max characterization (Courant-Hilbert theorem) of the eigenvalues of the operator $(L^{[*]}L)^{\frac{1}{2}}$: For every integer $m \in \{1, \dots, n\}$ we have

$$\sigma_m(L) = \sup_{\substack{F \subset E \\ \dim F = m}} \inf_{\substack{v \in F \\ \|v\|_{\mathrm{E}}=1}} \|Lv\|_{\mathrm{E}'} \ . \tag{3}$$

Using the singular values of $L : E \to E'$ we can define the absolute value of the determinant by $|\det L| = [\det L^{[*]}L]^{\frac{1}{2}}$. The geometric meaning of the singular values is the following (see [55]). If $L : E \to E'$ has the singular values $\sigma_1(L) \geq \ldots \geq \sigma_n(L) > 0$ and B is a ball in E of radius r then LB is an ellipsoid in E' with the length of semi-axes $\sigma_i r$, $i = 1, \dots, n$.
With $E^{\wedge k}$ or $\bigwedge^k E$ we denote the k-th exterior power of the n-dimensional Euclidean space E and we define the k-th *exterior power* of L by $L^{\wedge k} : E^{\wedge k} \to E'^{\wedge k}$ and the k-th *additive compound operator* by $L_k := \frac{d}{ds}(\mathrm{id} + sL)^{\wedge k}|_{s=0}$.
It is easy to see that $(L^{\wedge k})^{[*]} = (L^{[*]})^{\wedge k}$ for $k = 1, \dots, n$. If L is unitary, i.e. L is bijective and $L^{-1} = L^{[*]}$, then $L^{\wedge k}$ is also unitary.
By means of the singular values $\sigma_1(L) \geq \ldots \geq \sigma_n(L)$ for an arbitrary integer $k \in \{0, 1, \dots, n\}$ we define

$$\omega_k(L) := \begin{cases} \sigma_1(L) \cdot \ldots \cdot \sigma_k(L) = \|L^{\wedge k}\|_{\mathrm{op}} & \text{for } k > 0 \ , \\ 1 & \text{for } k = 0 \ . \end{cases}$$

Further, for a number $d \in (0, n]$ written as $d = d_0 + s$ ($d_0 \in \{0, 1, \dots, n-1\}$, $s \in (0, 1]$) we put

$$\omega_d(L) := \omega_{d_0}^{1-s}(L) \cdot \omega_{d_0+1}^{s}(L) = \|L^{\wedge d_0}\|_{\mathrm{op}}^{1-s} \|L^{\wedge (d_0+1)}\|_{\mathrm{op}}^{s} \ .$$

We indicate that the generalized *Horn's inequality* holds ([46,47]):
Suppose that E, E' and E'' are n-dimensional Euclidean spaces, $L : E \to E'$ and $L' : E' \to E''$ are linear operators, and $d \in [0, n]$ is an arbitrary number. Then $\omega_d(L'L) \leq \omega_d(L') \cdot \omega_d(L)$.

3.2 The Vertical Part of the Variational Equation and Liouville-Inequality

For a given smooth dynamical system $\{\varphi^t\}_{t \in \Gamma}$ on the smooth n-dimensional Riemannian manifold (M, g) let us consider the linearization, i.e. the first order Taylor expansion

$$\varphi^t(\exp_p v) = \exp_{\varphi^t p}((d_p \varphi^t) v + o(v)) \tag{4}$$

with $p \in M$, $v \in T_p M$, $t \in \Gamma$, $|t| \leq c$. ($\exp_p$ denotes the exponential map around p.) If the system is time-continuous the differential of the time t-map is described by the covariantly written variational equation with respect to the infinitesimal generator f of $\{\varphi^t\}_{t \in \mathbb{R}}$,

$$f(x) := \frac{d}{dt}\varphi^t(x)|_{t=0} \in T_x M \ .$$

The variational equation with respect to the motion $t \mapsto \varphi^t p$ is given by

$$\frac{Dy}{dt} = \nabla f(\varphi^t p) y \ , \tag{5}$$

where the linear operator $\nabla f(p) : T_pM \to T_pM$ is the *covariant derivative* of f in p, defined in local coordinates with the help of the Christoffel symbols Γ_{ij}^k by $v = v^i \partial_i(p) \in T_pM \mapsto \nabla_i f^k v^i \partial_k(p)$ with

$$\nabla_i f^k = \frac{\partial f^k}{\partial x^i} + \Gamma_{ij}^k f^j \ . \tag{6}$$

$\frac{Dy}{dt}$ denotes the absolute derivative along the integral curve $t \mapsto \varphi^t(p)$ and is in local coordinates of a chart x given by $\frac{Dy^i}{dt} = \nabla_l f^i y^l$. In order to investigate the stability of orbits and the deformation of k-forms in the tangent space we consider the operator $S\nabla f(p) := \frac{1}{2}[\nabla f(p) + \nabla f(p)^{[*]}] : T_pM \to T_pM$, which in a chart x is given by the matrix $S_m^i = \frac{1}{2}[g^{ij}\nabla_j f^k g_{km} + \nabla_m f^i]$. The eigenvalues of the operator $S\nabla f(p)$ are defined as the eigenvalues of the symmetric quadratic form *(deformation tensor)*

$$e_{si} = \tfrac{1}{2}[f_{s,i} + f_{i,s}] \ \text{ with } f_{s,i} = g_{st}\nabla_i f^t \ , \tag{7}$$

i.e. λ is a zero of the equation $\det[e_{si} - \lambda g_{si}] = 0$. In general, for an arbitrary number $k \in \{1, \dots, n\}$, we consider the *k-th compound variational equation*

$$\frac{Dw}{dt} = [\nabla f(\varphi^t p)]_k w \ , \tag{8}$$

where $[\nabla f(u)]_k : \bigwedge^k T_uM \to \bigwedge^k T_uM$ is the k-th additive compound of $\nabla f(u)$.

If $y_1(\cdot), \dots, y_k(\cdot)$ are solutions of the variational equation (5) with respect to $\varphi^t(p)$ then $y_1(\cdot) \wedge \dots \wedge y_k(\cdot)$ is a solution of the k-th compound variational equation (8) since

$$\frac{D}{dt}[y_1(t) \wedge \dots \wedge y_k(t)] = \frac{Dy_1(t)}{dt} \wedge \dots \wedge y_k(t) + \dots + \ y_1(t) \wedge \dots \wedge \frac{Dy_k(t)}{dt} \ .$$

Suppose that in the time-continuous case $\sigma_i(d_u\varphi^t)$, $i = 1, 2, \dots, n$, are the singular values of $d_u\varphi^t$ for $u \in M$ and $t \in \mathbb{R}_+$. Assume further that $\alpha_1(u) \geq \dots \geq \alpha_n(u)$ are the eigenvalues of the operator $S\nabla f(u)$. Then for any $t \geq 0$ the generalized Liouville inequalities have the form ([47])

$$\underbrace{\sigma_1(d_u\varphi^t) \cdot \ldots \cdot \sigma_k(d_u\varphi^t)}_{\omega_k(d_u\varphi^t)} \leq \exp \int_0^t [\alpha_1(\varphi^\tau u) + \dots + \alpha_k(\varphi^\tau u)] d\tau \tag{9}$$

and

$$\sigma_n(d_u\varphi^t)\cdot\ldots\cdot\sigma_{n-k+1}(d_u\varphi^t)\geq \exp\int_0^t[\alpha_n(\varphi^\tau u)+\ldots+\alpha_{n-k+1}(\varphi^\tau u)]d\tau\ . \tag{10}$$

As a special case we get the equality $|\det(d_u\varphi^t)|=\exp\int_0^t \operatorname{div} f(\varphi^\tau u)d\tau$.

3.3 Losinskii Measures on Bundles

In order to improve the contraction properties of the singular value function with respect to a given metric we also consider on M, parallel to the Riemannian structure, a Banach space structure in the tangent spaces T_uM, given by a family $\{\|\cdot\|_u\}$ of norms on TM.
Let us consider at first the general situation $T:E\to E$, where T is a linear operator on the n-dimensional space E with norm $\|\cdot\|_{\mathrm{E}}$. The functional $m:L(E,E)\to\mathbb{R}$ defined by

$$m(T):=\lim_{s\to 0+0}\frac{\|id_E+sT\|_{\mathrm{op}}-1}{s}$$

is called ([6]) *Losinskii measure* (*Losinskii norm* or *logarithmic norm*). The Losinskii measure has several basic properties which follow directly from its definition:

$$\begin{aligned}&(1)\quad m\,(S+T)\leq\ m(S)+m(T)\quad \text{for}\quad S,T\in L(E,E)\ .\\ &(2)\qquad m\,(\alpha T)=\ \alpha m(T)\quad \text{for}\quad \alpha\geq 0\ .\\ &(3)\qquad m\ (T)\leq\ \|T\|_{\mathrm{op}}\ .\end{aligned}$$

Instead of the generalized Liouville inequalities, using the supplementary Finsler structure on the Riemannian manifold (i.e. the family of norms $\{\|\cdot\|_u\}$ on the tangent bundle TM), we get the following inequalities: Suppose that $m(\cdot)$ is a Losinskii measure on the tangent bundle. Then there exists (see [5]) a continuous function $\rho:M\to\mathbb{R}^+$ such that for any $k\in\{1,\ldots,n\}$, $t\geq 0$, and $u\in M$ the inequalities

$$\sigma_1(d_u\varphi^t)\cdot\ldots\cdot\sigma_k(d_u\varphi^t)\frac{\rho(\varphi^t u)}{\rho(u)}\leq\exp\int_0^t m[\nabla f(\varphi^\tau u)]_k d\tau \tag{11}$$

and

$$\frac{\rho(\varphi^t u)}{\rho(u)}\sigma_n(d_u\varphi^t)\cdot\ldots\cdot\sigma_{n-k+1}(d_u\varphi^t)\geq\exp\int_0^t -m[-\nabla f(\varphi^\tau u)]_k d\tau \tag{12}$$

are satisfied.
Note that in the special case, where the norms are generated by the Riemannian structure, it holds $m([\nabla f(\varphi^t u)]_k)=\alpha_1(d_u\varphi^t)+\ldots+\alpha_k(d_u\varphi^t)$, and choosing $\rho(u)\equiv 1$, we get the inequalities (9) resp. (10).

3.4 Lyapunov Exponents

Suppose that $\{\varphi^t\}_{t\in\Gamma}$ is a smooth injective dynamical system on the Riemannian manifold M. Let $\sigma_1(d_u\varphi^t) \geq \ldots \geq \sigma_n(d_u\varphi^t) > 0$ denote the singular values of $d_u\varphi^t : T_uM \to T_{\varphi^t u}M$, i.e. the eigenvalues of the positive definite operator $([d_u\varphi^t]^{[*]}[d_u\varphi^t])^{\frac{1}{2}} : T_uM \to T_uM$. Let $K \subset M$ be a compact invariant set of $\{\varphi^t\}_{t\in\Gamma}$. Define for any $j \in \{1,\ldots,n\}$ and $t \in \Gamma_+$

$$\omega_{j,K}(\varphi^t) = \sup_{u\in K} \omega_j(d_u\varphi^t) \text{ and } \omega_{0,K}(\varphi^t) = 1 . \tag{13}$$

For any $j \in \{1,\ldots,n\}$ and $u \in M$ the *local Lyapunov exponents* $\nu_j(u)$ of $\{\varphi^t\}_{t\in\Gamma}$ are defined iteratively by [10,55,56]

$$\nu_1(u) + \ldots + \nu_j(u) = \limsup_{t\to\infty} \frac{1}{t} \ln \omega_j(d_u\varphi^t) . \tag{14}$$

The *global Lyapunov exponents* $\nu_1,\ldots,\nu_n$ are defined by

$$\nu_1 = \lim_{t\to\infty} \frac{1}{t} \ln \omega_{1,K}(\varphi^t) \text{ and } \nu_j = \lim_{t\to\infty} \frac{1}{t} \ln \frac{\omega_{j,K}(\varphi^t)}{\omega_{j-1,K}(\varphi^t)} \tag{15}$$

for $j = 2,3,\ldots,n$. Note that these limits exist because of the subexponential property ([55]) of the singular value function. The *upper Lyapunov exponents* are given by

$$\overline{\nu}_k = \limsup_{t\to\infty} \frac{1}{t} \ln \sup_{u\in K} \ln \sigma_k(d_u\varphi^t) \quad (k = 1,2,\ldots,n) .$$

4 Basic Definitions from Dimension Theory

4.1 Topological, Hausdorff- and Box Dimension

Let X be a topological space. We say that X is *finite dimensional* if there exists an integer $n \geq -1$ such that, for every open covering α of X, there exists another open covering β refining α such that every point of X belongs to at most $n+1$ sets of β. (Recall that a covering $\alpha = \{A\}$ of X is *open* if all sets A of α are open. A covering β of X *refines* another covering α of X if for any $B \in \beta$ there exists an $A \in \alpha$ such that $B \subset A$.) In this case the *topological dimension* of $X, \dim_{\mathrm{T}} X$, is defined as the minimal number n satisfying this property (see [30]). Then $\dim_{\mathrm{T}} \emptyset = -1, \dim_{\mathrm{T}} \mathbb{R}^n = n$, and if K is a compact finite dimensional space, it is homomorphic to a subset of $\mathbb{R}^n$ with $n = 2\dim_{\mathrm{T}} K + 1$.
Suppose that (X,ϱ) is a metric space, E is an arbitrary subset of X, $d \geq 0$ and $\varepsilon > 0$ are real numbers. Let us cover E by at most countable many balls of radii $r_j < \varepsilon$ and define

$$\mu_{\mathrm{H}}(E,d,\varepsilon) = \inf \sum_j r_j^d ,$$

where the infimum is taken over all such countable ε-covers of E under the convention that $\inf \emptyset = +\infty$. It is obvious that $\mu_{\mathrm{H}}(E, d, \varepsilon)$ increases with decreasing ε. Thus, there exists the limit (which may be infinite)

$$\mu_{\mathrm{H}}(E, d) = \lim_{\varepsilon \to 0+0} \mu_{\mathrm{H}}(E, d, \varepsilon) \ ,$$

which is called the outer *Hausdorff-d-measure* of E. For any fixed $E \subset X$ there exists a critical value $d_{\mathrm{cr}} \in [0, +\infty]$ such that

$$\mu_{\mathrm{H}}(E, d) = \begin{cases} +\infty & \text{for all } 0 \le d < d_{\mathrm{cr}}, \\ 0 & \text{for all } d > d_{\mathrm{cr}} \ . \end{cases}$$

This critical value is defined as the *Hausdorff dimension* $\dim_{\mathrm{H}} E$ of E.
Let (X, ϱ) be a metric space, $K \subset X$ a compact subset. For any $\varepsilon > 0$ denote by $N_\varepsilon(K)$ the minimal number of balls of radius ε which are necessary to cover K. Define the *box dimension* (or *upper box, fractal, entropic dimension*) of K, and the *lower box dimension* of K by

$$\overline{\dim_{\mathrm{B}}} K = \limsup_{\varepsilon \to 0+0} \frac{\ln N_\varepsilon(K)}{-\ln \varepsilon} \quad \text{and} \quad \underline{\dim_{\mathrm{B}}} K = \liminf_{\varepsilon \to 0+0} \frac{\ln N_\varepsilon(K)}{-\ln \varepsilon} \ ,$$

respectively.
It is well-known ([48]) that for any compact set K it holds

$$\dim_{\mathrm{T}} K \le \dim_{\mathrm{H}} K \le \underline{\dim_{\mathrm{B}}} K \le \overline{\dim_{\mathrm{B}}} K \ .$$

4.2 Topological Pressure and Topological Entropy

Suppose that M is an n-dimensional Riemannian manifold and $\{\varphi^t\}_{t \in \Gamma}$ is a dynamical system on M. For $g \in C(M)$, $p \in M$ and $t \ge 1$ define $S_t g(p) = \sum_{i=0}^{t-1} g(\varphi^i p)$ in the discrete-time case, and $S_t g(p) = \int_0^t g(\varphi^s p) ds$ in the continuous-time case.
In the discrete-time case a set $E \subset M$ is (ε, T)*-separated* with respect to $\{\varphi^k\}_{k \in \Gamma}$ if $p, q \in E, p \ne q$, imply that $\varrho(\varphi^k p, \varphi^k q) > \varepsilon$ for some integer $k \in \{0, \dots, T-1\}$. In the continuous-time case a set $E \subset M$ is (ε, T)-*separated* with respect to $\{\varphi^t\}_{t \in \Gamma}$ if $p, q \in E, p \ne q$, imply $\varrho(\varphi^t p, \varphi^t q) > \varepsilon$ for some $t \in [0, T]$. For a dynamical system $\varphi = \{\varphi^t\}_{t \in \Gamma}$ on M, $g \in C(M)$, $\varepsilon > 0$ and $T \in \Gamma, T > 0$, define

$$Z_{\mathrm{T}}(\varphi, g, \varepsilon) = \sup_{E : (\varepsilon, T)-\text{separated}} \left\{ \sum_{p \in E} \exp S_T g(p) \right\}$$

$$\text{and} \qquad P(\varphi, g, \varepsilon) = \limsup_{T \to \infty} \frac{1}{T} \ln Z_{\mathrm{T}}(\varphi, g, \varepsilon) \ .$$

Then there exists the limit $P(\varphi, g) = \lim_{\varepsilon \to 0} P(\varphi, g, \varepsilon)$ which is called ([48,57]) *topological pressure* of g.

In particular, taking $g \equiv 0$, we get the *topological entropy* $h_{\mathrm{top}}(\varphi) = P(\varphi, 0)$. For expanding and hyperbolic systems which are both expansive ([52,57]) we have the following property. There is some $\varepsilon_0 > 0$ such that for any $\varepsilon < \varepsilon_0$ and any continuous function g

$$P(\varphi, g) = \limsup_{T \to \infty} \frac{1}{T} \ln Z_{\mathrm{T}}(\varphi, g, \varepsilon) \ .$$

5 Dimension Estimates of Invariant Sets

5.1 Upper Hausdorff and Box Dimension Estimates

Suppose there is given a C^1-map $\varphi : U \subset M \to M$ on an open set U of an n-dimensional Riemannian C^k-manifold. Further we assume that $K \subset U$ is a φ-invariant compact set. In order to estimate various types of dimension of this set we consider the singular values $\sigma_1(d_u\varphi) \geq \ldots \geq \sigma_n(d_u\varphi) \geq 0$ of the differential $d_u\varphi : T_uM \to T_{\varphi(u)}M$ in any point $u \in U$ and for a fixed $d = d_0 + s$ ($d_0 \geq 0$ integer, $s \in (0,1]$) the singular value function

$$\omega_{d,K}(\varphi) = \sup_{u \in K} \omega_d(d_u\varphi), \text{ where } \omega_d(d_u\varphi) = \omega_{d_0}^{1-s}(d_u\varphi)\omega_{d_0+1}^{s}(d_u\varphi) \ .$$

For $p \in U$ the *local Lyapunov dimension of* φ at p denoted by $\dim_{\mathrm{L}}(\varphi, p)$ is defined to be the largest number $d \in [0, n]$ such that $\omega_d(d_p\varphi) \geq 1$. If $\sigma_1(d_p\varphi) < 1$ let $\dim_{\mathrm{L}}(\varphi, p) = 0$. For a compact set $K \subset U$ let $\dim_{\mathrm{L}}(\varphi, K) = \max \dim_{\mathrm{L}}(\varphi, p)$ be the *Lyapunov dimension* of φ on K ([18,29]).
A basic result for various upper dimension bounds is the following theorem ([47,46,31,4,51]). Note that the introduction of a Lyapunov function in the dimension estimate for the phase space $M = \mathbb{R}^n$ was first done in [33].

Theorem 4. *Suppose that $\varphi : U \subset M \to M$ is as above, $K \subset U$ is a φ-invariant compact set, $d \in (0, n]$ is a fixed number and $\rho : U \to \mathbb{R}_+$ is a continuous function such that*

$$\sup_{u \in K} \frac{\rho(\varphi(u))}{\rho(u)} \omega_d(d_u\varphi) < 1 \ . \tag{16}$$

Then $\dim_{\mathrm{H}} K \leq d$.

Remark 5. The basic constructions of the proof of Theorem 4 are the following. There exists an $\varepsilon > 0$ such that the exponential map $\exp_p$ is an embedding of the closed ε-ball $\overline{B_\varepsilon(0_p)} = \{v \in T_pM : \|v\| \leq \varepsilon\}$ into M for every point $p \in K$. This gives closed neighbourhoods $V_p = \overline{B_\varepsilon(p)}$ of $p \in K$. The local coordinates describing V_p depend continuously on p. If $\varrho(p, q) < \delta$ we can use the radial isometry of $\exp_p$, i.e. we can pass to coordinates on V_p in which $q \in V_p \subset M$ corresponds to a point $z \in T_pM$ with $\|z\|_{T_pM} = \varrho(p, q)$.

Using these properties of the exponential map we can fix a $D \in [0,n]$ and show that for any $b_0 > 2\sqrt{n}$ and $C_0 > 2^D n^{D/2}$ there is an $\varepsilon_0 > 0$ such that for all $u \in K$ and all $A \subset B_{\varepsilon_0}(u)$

$$\mu_{\mathrm{H}}(A, D, b\varepsilon) \le C\mu_{\mathrm{H}}(\varphi(A), D, \varepsilon)$$

for all $\varepsilon < \varepsilon_0$, where $C = C(C_0, n, D)$ and $b = b(C_0, n, D)$.

Remark 6. It is useful to compare condition (16) with the following situation. If $\varphi : M \to M$ is a diffeomorphism on the manifold M and $\rho : M \to \mathbb{R}_+$ is the density of a φ-invariant measure with respect to the canonical volume form then

$$\frac{\rho(\varphi(u))}{\rho(u)} \omega_n(d_u\varphi) \equiv 1 \ .$$

To illustrate how the frequency-domain lemma (Theorem 1) can facilitate the application of Theorem 4 we consider the special case of a map in the feedback-control form (2). The following result was shown by A. Noack in [45].

Corollary 7. *Consider the map (2) and assume that $K \subset D \subset E_n$ (D open) is a compact invariant set for (2). Assume further that the pairs (A,b) and (A^*,c) from (2) are stabilizable and that for certain real numbers $d \in (0,n]$, $\kappa > 0$ and $\lambda > 0$ the following conditions hold:*

(1) *All eigenvalues of $\frac{1}{\kappa}A$ lie outside the complex unit circle.*

(2) $\frac{1}{2}\eta^*([\phi'(c^*x)]^* + \phi'(c^*x))\eta \le \lambda\|\eta\|^2$ *for all* $\eta \in E_m$ *and* $x \in D$.

(3) $\mathrm{Re}\ \chi(\kappa z) + \lambda\chi^*(\kappa z)\chi(\kappa z) \le 0$ *for all* $z \in \mathbb{C}$ *with* $|z| = 1$.

(4) $\frac{1}{\kappa^{n-d}} \det[A + b\phi'(c^*x)c^*] < 1$ *for all* $x \in D$.

Then $\dim_{\mathrm{H}} K < d$.

We now formulate a direct consequence of Theorem 4 for the Hausdorff dimension estimation of flow-invariant sets (see [47,5]). Note that using the results of [19] the following bound is also true for the box dimension.

Corollary 8. *Suppose that the flow $\{\varphi^t\}_{t\in\mathbb{R}}$ on the n-dimensional manifold M is generated by the C^1-vector field f and $\alpha_1(u) \ge \ldots \ge \alpha_n(u)$ are the eigenvalues of the symmetrized covariant operator $S\nabla f(u) : T_uM \to T_uM$. Suppose that there is an integer $d_0 \in (0, n-1]$ and a real number $s \in (0,1]$ such that for a $t > 0$*

$$\sup_{u\in K} \int_0^t [\alpha_1(d_u\varphi^\tau) + \ldots + \alpha_{d_0}(d_u\varphi^\tau) + s\ \alpha_{d_0+1}(d_u\varphi^\tau)]d\tau < 0 \ . \tag{17}$$

If K is a compact set and $\varphi^t(K) \supset K$, then $\dim_{\mathrm{H}} K \le d_0 + s$.

Proof. For small t the singular values of the operator $d_u\varphi^t : T_uM \to T_{\varphi^t u}M$ are the eigenvalues of the matrix

$$\sqrt{[\mathrm{id}_{T_uM} + tB(x^k) + o\,(t,x^k)]^{[*]}\,[\mathrm{id}_{T_uM} + tB\,(x^k) + o\,(t,x^k)]},$$

where $B(x^k) = \nabla_i f^i(x^k)$ is the representation of the covariant derivative in a chart x. Since

$$\left[\mathrm{id}_{T_uM} + tB(x^k) + o\,(t,x^k)\right]^{[*]} \left[\mathrm{id}_{T_uM} + tB(x^k) + o\,(t,x^k)\right] =$$
$$= (\mathrm{id}_{T_uM} + \tfrac{1}{2}t\,[B^{[*]}(x^k) + B(x^k)])^2 + o\,(t,x^k)$$

we get for $i = 1, 2, \ldots, n$

$$\sigma_i(d_u\varphi^t) = \sqrt{(1 + t\alpha_i(u))^2 + o\,(t,u)} = 1 + t\alpha_i(u) + o\,(t,u)$$

and, consequently,

$$\omega_d(d_u\varphi^t) = (1 + t\alpha_1(u)) \cdot \ldots \cdot (1 + t\alpha_{d_0}(u))(1 + t\alpha_{d_0+1}(u))^s + o\,(t,u) =$$
$$1 + t[\alpha_1(u) + \ldots + \alpha_{d_0}(u) + s\,\alpha_{d_0+1}(u)] + o\,(t,u)\ .$$

□

Remark 9. Let us consider the special case of a smooth vector field f on the manifold $M = \mathbb{R}^n$ with the metric $(g_{ij}(x)) = \rho^2(x)I$, where $\rho : \mathbb{R}^n \to \mathbb{R}_+$ is a C^1-function. To verify condition (17) we have to compute the eigenvalues of the deformation tensor e_{ij}, defined in (7), which is in this case $\frac{1}{2}\,[J + J^*] + \frac{\rho'}{\rho}I$, where $\frac{1}{2}\,[J + J^*]$ is the symmetrized Jacobi matrix. If we take $\rho(x) = e^{\frac{v(x)}{d_0+s}}$, where v is a C^1-function, we have $\frac{\rho'}{\rho} = \frac{\dot v}{d_0+s}$. It follows that if $\alpha_i(x)$ are the eigenvalues of $\frac{1}{2}\,[J(x) + J(x)^*] + \frac{\dot v(x)}{d_0+s}$ and $\widetilde{\alpha}_i(x)$ are the eigenvalues of $\frac{1}{2}\,[J(x) + J(x)^*]$, then $\alpha_i(x) = \widetilde{\alpha}_i(x) + \frac{\dot v(x)}{d_0+s}$ for $i = 1, 2, \ldots, n$.

For embedding techniques it is very important to have not only upper bounds for the Hausdorff dimension but also for the box dimension. The next theorem gives such an upper bound ([4]) under the additional condition (1), which, as was shown later in [29,2,19], is not necessary .

Theorem 10. *Suppose that $\varphi : U \subset M \to M$ is defined as in Theorem 4, $K \subset U$ is a compact φ-invariant set, and $\rho : U \to \mathbb{R}_+$ is a continuous function. Suppose further that the following conditions are satisfied:*

(1) $\omega_n(d_u\varphi) = \mathrm{const} \neq 0$ *for all* $u \in K$.
(2) *There exists a number* $s \in (0,1]$ *such that* $\sup\limits_{u\in K} \frac{\rho(\varphi(u))}{\rho(u)}\omega_{n-1+s}(d_u\varphi) < 1$.

Then $\overline{\dim}_{\mathrm{B}}K \leq n - 1 + s$. *Further if* $\rho(u) \equiv 1$ *then* $\overline{\dim}_{\mathrm{B}}K \leq \dim_{\mathrm{L}}(\varphi, K)$.

The full proof of Theorem 10 is given in [4]. Let us only note here that for the proof of the theorem the new metric $\widetilde{g}_{ij}(x) = \rho^2(x) g_{ij}(x)$ on M is considered. Note that the transition to an equivalent metric in M does not alter the box dimension. Thus, the singular-value function which is taken with respect to a fixed Riemannian metric, can be replaced by the singular-value function with respect to an adapted metric $\widetilde{g}$.

Example 11. Consider a mapping which describes a model for repeated impacts of a bouncing ball on a sinusoidally vibrating table and which is given by the map $\varphi_{\alpha,\gamma} : \mathbb{R}^2 \to \mathbb{R}^2$ through

$$\begin{pmatrix} x \\ y \end{pmatrix} \mapsto \begin{pmatrix} x + y \\ \alpha y - \gamma \cos(y + x) \end{pmatrix} . \tag{18}$$

(See also [7].) Here γ plays the role of force amplitude for the table motion and $0 < \alpha \leq 1$ is the dissipation. An approach of the dynamics can be found in [23].

For any pair of parameters (α, γ) the map $\varphi_{\alpha,\gamma}$ is a diffeomorphism. The determinant of its Jacobian matrix is equal to α, which means that $\varphi_{\alpha,\gamma}$ generates a dissipative dynamical system for $\alpha < 1$. For that case all orbits enter and remain within the strip bounded by $y = \pm\gamma/(1-\alpha)$. Since (18) is invariant under coordinate change $x \mapsto x + 2\pi n$ for $n = \pm 1, \pm 2, \ldots$ the phase space can be interpret to be the cylinder $S^1 \times \mathbb{R}$. Furthermore, there is an compact attracting set K within the absorbing set ($\varepsilon > 0$ small) $D \subset \{(x, y) : |y| \leq \varepsilon + \gamma/(1-\alpha)\} \subset S^1 \times \mathbb{R}$. With the standard metric for factor manifolds given by $(g_{ij}(x)) = I$ in any chart, $S^1 \times \mathbb{R}$ is a 2-dimensional smooth Riemannian manifold, the flat cylinder. The largest singular value of $d_p \varphi_{\alpha,\gamma}$ for $p = (x, y) \in S^1 \times \mathbb{R}$ with respect to this metric is given by

$$\sigma_1(d_p \varphi_{\alpha,\gamma}) = \frac{1}{2} \left(\sqrt{(\alpha + 1)^2 + r(x, y)} + \sqrt{(\alpha - 1)^2 + r(x, y)} \right) ,$$

where $r(x, y) = 2\gamma^2 \sin^2(x + y) + 2\gamma \sin(x + y)\alpha + 1$, and $\sigma_1(d_p \varphi_{\alpha,\gamma}) > 1$ for all $p \in S^1 \times \mathbb{R}$ and all $\alpha \in (0, 1)$. Estimating the box dimension Theorem 10 yields

$$\overline{\dim}_{\mathrm{B}} K \leq 1 + \left(1 - \frac{\ln \alpha}{\ln \left(\sqrt{(1 + \alpha)^2 + b} + \sqrt{(1 - \alpha)^2 + b} \right) - \ln 2} \right)^{-1} ,$$

with $b = 2\gamma^2 - 2\gamma + 1$. For parameter values $\alpha = 0.5, \gamma = 10$ considered in [23] we obtain $\overline{\dim}_{\mathrm{B}} K \leq 1.790$.

Example 12. We consider the Hénon map $\varphi_{\alpha,\beta} : \mathbb{R}^2 \to \mathbb{R}^2$ given by ([27])

$$\begin{pmatrix} x \\ y \end{pmatrix} \mapsto \begin{pmatrix} \alpha + \beta y - x^2 \\ x \end{pmatrix} \tag{19}$$

with parameters $0 < \alpha, 0 < \beta < 1$. The map has the two fixed points $p_\pm = (x_\pm, y_\pm)$ with $y_\pm = x_\pm = \frac{1}{2}(\beta - 1 \pm \sqrt{(\beta-1)^2 + 4\alpha})$. Define a function $v : \mathbb{R}^2 \to \mathbb{R}$ by $v(x, y) = \gamma(x + \beta y)$ for $\gamma = \left[(-2x_- + \beta - 1)\left(\sqrt{x_-^2 + \beta}\right)\right]^{-1}$ and introduce the metric tensor g in the standard chart around $p = (x, y)$ as $e^{2(1-s)v(x,y)} \begin{pmatrix} \sqrt{\beta} & 0 \\ 0 & 1 \end{pmatrix}$ for some $s \in (0, 1)$. The singular values with respect to this metric are given for $p = (x, y)$ by

$$\sigma_{1/2}(d_p\varphi_{\alpha,\beta}) = e^{(1-s)\gamma(\alpha+(\beta-1)x-x^2)}\left(\sqrt{x^2 + \beta} \pm |x|\right) .$$

Consider the case $\beta = 0.3,\ \alpha = 1.4$. For any $\varphi_{\alpha,\beta}$-invariant compact set $K \subset \mathbb{R}^2$ Theorem 10 gives $\overline{\dim}_{\mathrm{B}} K \leq 1.496$. The result agrees with the one obtained in [36], where the estimate is obtained using a coordinate transform and the function v as a Lyapunov-type function instead of making use of a certain metric. For every compact invariant set K which is located in the square $-1.8 \leq x, y \leq 1.8$ using the flat metric on $\mathbb{R}^2$ gives the estimate $\overline{\dim}_{\mathrm{B}} K \leq 1.523$ (see [29]). We refer to [45] for another application on choosing Riemannian metrics which implies even smaller bounds. Following the observation in [36] it can be seen that the global Lyapunov dimension $\dim_{\mathrm{L}}(\varphi_{\alpha,\beta}, K)$ is equal to the local Lyapunov dimension $\dim_{\mathrm{L}}(\varphi_{\alpha,\beta}, p_-)$.

Example 13. We consider the *complex Lorenz equation* which is introduced in [15] and which may be written as

$$\begin{aligned} \dot{x} &= -c\,(x - y), \\ \dot{y} &= -xz + rx - ay, \\ \dot{z} &= -bz + \tfrac{1}{2}\,(x^* y + x y^*)\ . \end{aligned} \tag{20}$$

In this system c and b are positive parameters and $a = 1 - ie$ resp. $r = r_1 + ir_2$ are complex parameters ($e, r_1, r_2 \in \mathbb{R}$).
For $r_2 = e = 0$ we have the *real Lorenz equation.* If either Im $(r - a) \neq 0$ or Im $(r - a) = 0$ and Re $(r - a) \leq 0$ system (20) has the only equilibrium point $(0, 0, 0)$. For Im $(r - a) = 0$ and Re$(r - a) > 0$ the system has a continuum of equilibrium points ([39,38]).
Writing $x = x_1 + ix_2$ and $y = y_1 + iy_2$ we obtain from (20) the real ODE system in $\mathbb{R}^5$

$$\begin{aligned} \dot{x}_1 &= -cx_1 + cy_1, \\ \dot{x}_2 &= -cx_2 + cy_2, \\ \dot{y}_1 &= -x_1 z + r_1 x_1 - r_2 x_2 - y_1 - e y_2, \\ \dot{y}_2 &= -x_2 z + r_2 x_1 + r_1 x_2 + e y_1 - y_2, \\ \dot{z} &= -bz + x_1 y_1 + x_2 y_2\ . \end{aligned} \tag{21}$$

Note that if f denotes the right-hand side of (21) then $\mathrm{div} f \equiv -2c - 2 - b < 0$. In order to prove the dissipativity of (21) we use a Lyapunov function as in [39].

Put $\kappa_0 = \min\{1, b, c\}$ and take real auxiliary constants $\kappa_1 \in (0, \kappa_0)$, $\kappa_2 > 0$ and κ_3 satisfying $\kappa_2 (c - \kappa_1)(1 - \kappa_1) - \frac{1}{4} |c - \kappa_2 r - \kappa_3|^2 \geq 0$. Define also constants κ_4 and κ_5 by

$$\kappa_4 > \frac{\kappa_3^2 (b - 2\kappa_1)^2}{8\kappa_1 \kappa_2 (b - \kappa_1)} \quad \text{and} \quad \kappa_5 > \frac{\kappa_3^2}{\kappa_2 (1 + \kappa_2)} \left(\frac{(b - 2\kappa_1)^2}{4\kappa_1 (b - \kappa_1)} + 1 \right) .$$

Introduce now the function $v : \mathbb{R}^5 \to \mathbb{R}$ by

$$v(x_1, x_2, y_1, y_2, z) = \frac{1}{2}[x_1^2 + x_2^2 + \kappa_2(y_1^2 + y_2^2 + z^2)] - \kappa_3 z .$$

A straightforward calculation shows (see [39,38]) that for any bounded set $B \subset \mathbb{R}^5$ there exists a time $t = t_B$ such that for $t \geq t_B$

$$v(x_1(t), x_2(t), y_1(t), y_2(t), z(t)) < \kappa_4 \quad \text{and} \quad x_1^2(t) + x_2^2(t) < \kappa_5$$

holds provided $(x_1(0), x_2(0), y_1(0), y_2(0), z(0)) \in B$ and $t \geq t_B$.

It follows that the global attractor A of the Lorenz system (21) is contained in the ellipsoid

$$\mathcal{E} = \Big\{ (x_1, x_2, y_1, y_2, z) \in \mathbb{R}^5 : x_1^2 + x_2^2 + \kappa_2(y_1^2 + y_2^2) + \\ + \kappa_2 (z - \frac{\kappa_3}{\kappa_2})^2 < 2\kappa_4 + \frac{\kappa_3^2}{\kappa_2} \Big\} .$$

From this we get the estimates

$$y_l^2 \leq \frac{1}{\kappa_2} \left(2\kappa_4 + \frac{\kappa_3^2}{\kappa_2} \right), \quad l = 1, 2, \quad \text{and} \quad \left(z - \frac{\kappa_3}{\kappa_2} \right)^2 \leq \frac{1}{\kappa_2} \left(2\kappa_4 + \frac{\kappa_3^2}{\kappa_2} \right) .$$

Using the global Lyapunov exponents $\nu_1, \ldots, \nu_5$ with respect to A and the estimate of A given by $\mathcal{E}$ we obtain for the box dimension on the basis of Theorem 10 (see also [26])

$$\overline{\dim}_{\mathrm{B}} A \leq 4 + \frac{\nu_1 + \nu_2 + \nu_3 + \nu_4}{|\nu_5|} \leq 4 + \frac{C}{-\operatorname{div} f + C} ,$$

where $\nu_1 + \nu_2 + \nu_3 + \nu_4 \leq C$ and $|\nu_5| = -\nu_5 = \nu_1 + \nu_2 + \nu_3 + \nu_4 - \operatorname{div} f$.

The bound C can be estimated by means of the variational equation, using the ellipsoid $\mathcal{E}$. In [24] for parameter values $(r_1, r_2, e, c, b) = (60, 0.02, 0.06, 2, 0.8)$ the estimate $\dim_{\mathrm{H}} A \leq 4.46$ for the global attractor A of (21) was given. On the basis of Theorem 10 this number bounds also the box dimension of the attractor A.

5.2 Dimension Estimates and the Spanning Surface Principle

In this subsection we use upper dimension estimates of invariant sets of dynamical systems for the characterization of the underlying dynamics. Using such an approach it is possible to get higher-dimensional generalizations of certain theorems of Bendixson-Poincaré type and global convergence results. The key step is to exclude the existence of closed orbits of the flow by using a spanning surface argument. In the ODE case for $M = \mathbb{R}^n$ such an approach is developed in several papers ([41,33,34]). In [47] the analog problem is considered for vector fields on manifolds. In the 1-dimensional version of invariant objects the "spanning surface argument" means that for any closed invariant curve we consider spanning surfaces and argue that 2-dimensional measures of such surfaces are bounded from below. We suppose that the n-dimensional smooth manifold M has this property, i.e. if there is a smooth closed curve then there exists a two-dimensional surface in M with finite 2-dimensional Lebesque-measure such that it spans the given curve. We can use in this situation the generalization of a result from [41] which can be formulated as follows. Suppose that $D = \{(u,v) \in \mathbb{R}^2 : u^2 + v^2 < 1\}$ is the open disc. On Lip $(D \to M)$, i.e. on the space of Lipschitz continous functions from D in M, we consider the area functionals

$$A(h) = \text{mes}_2\, h\,(D) \ .$$

If x is a chart in $h\,(D)$, $x(u,v)$ is the coordinate representation in this chart of $x \circ h$ and $g_{ij}(x)$ is the local realization of the metric tensor, then this area functional is given by

$$\iint_D \sqrt{EG - F^2}\, dudv \ ,$$

where $E = g_{ij}(x)x_u^i x_u^j$, $G = g_{ij}(x)x_v^i x_v^j$ and $F = g_{ij}(x)x_u^i x_v^j$. If c is a simple closed rectifiable curve in M and $\Sigma(c, M)$ is the set of all Lipschitz continuous maps h from $\overline{D} \subset \mathbb{R}^2$ in M satisfying $h(\partial D) = c(\partial D)$ it can be shown that there exists a $\delta > 0$ such that $A(h) \geq \delta$ for all $h \in \Sigma(c, M)$. (For the $\mathbb{R}^n$ case this is done in [41], the manifold case is investigated in [5]).
The next theorem can be considered as generalized negative Bendixson theorem for manifolds ([47]). Other versions of such generalizations are given in [5,33,34].

Theorem 14. *Suppose that on the simply-connected compact n-dimensional Riemannian C^m-manifold M $(m \geq 4)$ there is defined a flow $\{\varphi^t\}_{t\in\mathbb{R}}$, generated by a smooth vector field f. Suppose also that $\alpha_1(u) \geq \alpha_2(u)$ are the two largest eigenvalues of the symmetrized covariant derivative $S\nabla f(u)$ such that $\alpha_1(u) + \alpha_2(u) < 0$ on M. Then $\{\varphi^t\}_{t\in\mathbb{R}}$ has on M no closed invariant curve.*

Proof. Suppose the opposite, i.e. suppose that there exists a closed invariant curve γ of $\{\varphi^t\}_{t\in\mathbb{R}}$. By assumption we can construct a surface given by a parametrization h, which spans γ, i.e. $h(\partial D) = \gamma$. We can now use the fact, as mentioned above, that there is a $\delta > 0$ such that $A(h) \geq \delta > 0$ for any such γ-spanning surface. Since $\varphi^t(\gamma) = \gamma$ for all $t \in \mathbb{R}$, we conclude by the smoothness of the flow that $h^t := \varphi^t \circ h$ is for any t also an integrable parametrization of a γ-spanning surface. This gives a contradiction, since by $\alpha_1(u) + \alpha_2(u) < 0$ on M it follows that $\mathrm{mes}_2(\varphi^t \overline{D}) \to 0$ as $t \to +\infty$. □

Remark 15. An analogous result is possible if we consider the existence of k-dimensional $(2 \leq k \leq n)$ invariant submanifolds for a given flow $\{\varphi^t\}_{t\in\mathbb{R}}$. If we require that for the largest eigenvalue $\alpha_1^{(k)}(u)$ of the symmetrized k-th compound operator $[S\nabla f]_k(u)$ the inequality $\alpha_1^{(k)}(u) < 0$ on M is satisfied and any that k-dimensional invariant submanifold may be written as boundary of a $(k+1)$-dimensional surface, then our dynamical system $\{\varphi^t\}_{t\in\mathbb{R}}$ can not have any such k-dimensional invariant submanifold on M. A basic idea for the proof is the use of Stokes theorem for the integration of k-forms ω^k on $(k+1)$-dimensional submanifolds U^{k+1} on M by $\int_{U^{k+1}} d\omega^k = \int_{\partial U^{k+1}} \omega^k$.

In the following we want to derive conditions for the global convergence of motions of a given flow to equilibrium points on the basis of Theorem 14. There we apply the C^1-closing lemma of Pugh ([49]), which we formulate for completeness in the following version.

Lemma 16. *Suppose that $\varphi = \{\varphi^t\}_{t\in\Gamma}$ is a C^1-dynamical system on an n-dimensional Riemannian C^m-manifold M. Let p_0 be a non-wandering point of φ such that the iterates $\{\varphi^t(p_0)_{t\in\Gamma}\}$ of p_0 are contained in a compact subset of M. Then $\{\varphi^t\}_{t\in\Gamma}$ can be C^1-approximated by a dynamical system $\{\psi^t\}_{t\in\Gamma}$ which has a periodic orbit through p_0.*

Now we can state a convergence theorem for flows on manifolds. (See also [46,47].)

Theorem 17. *Suppose that on a simply-connected compact n-dimensional Riemannian C^m-manifold M a smooth flow $\{\varphi^t\}_{t\in\mathbb{R}}$ is given, generated by the vector field f such that the conditions of Theorem 14 are satisfied. Suppose further that $\{\varphi^t\}_{t\in\mathbb{R}}$ has equilibrium points and these points are all isolated. Then any motion of $\{\varphi^t\}_{t\in\mathbb{R}}$ tends for $t \to +\infty$ to an equilibrium.*

Proof. Suppose that $t \mapsto \varphi^t(u)$ is an arbitrary motion of the flow. Since M is compact the ω-limit set $\omega(u) \neq \emptyset$. Suppose that there exists a point $p \in \omega(u)$ which is not an equilibrium point. Since such a point is non-wandering, due to Pugh's lemma there exists in any C^1-neighborhood of f a smooth vector field $\widetilde{f}$ such that the associated flow $\{\widetilde{\varphi}^t\}_{t\in\mathbb{R}}$ has a periodic motion through p. Since M is compact, and the conditions of Theorem 14 are satisfied, we can choose $\widetilde{f}$ such that the two largest eigenvalues $\widetilde{\alpha}_1(u) \geq \widetilde{\alpha}_2(u)$ of $S\nabla\widetilde{f}(u)$ satisfy the inequality $\widetilde{\alpha}_1(u) + \widetilde{\alpha}_2(u) < 0$ on M. On the basis of Theorem 14 we conclude that the vector field $\widetilde{f}$ cannot generate closed motions. It follows that p is an equilibrium point. Since $\omega(u)$ is connected and the equilibrium points are by assumption isolated, it follows that $\omega(u) = \omega(p)$, which shows that the motion through u converges. □

5.3 Lower Estimates for the Lower Box Dimension by Use of Lyapunov Functions

In a general not necessarily hyperbolic situation lower estimates of dimensions are very rare in the literature. Only for the lower box dimension sufficiently general estimates are known. A classical result goes back to G. Kushnirenko ([12]).

Theorem 18. *Suppose that M is an n-dimensional C^m-manifold ($m \geq 3$). Suppose further that $\varphi : M \to M$ is a C^1-map, $K \subset M$ is a compact φ-invariant set, and $\sup_{u \in K} \|d_u\varphi\| > 1$. Then*

$$\underline{\dim}_B K \geq \frac{h_{\text{top}}(\varphi_{|K})}{\ln \sup_{u \in K} \|d_u \varphi\|} \ .$$

In the following two corollaries we give some technical realizations of such a lower bound, which are due in parts to A. Noack ([46]).

Corollary 19. *Assume that on the manifold M a smooth flow $\varphi = \{\varphi^t\}_{t \in \mathbb{R}}$ is given, which is generated by the vector field f and has a compact invariant set K. Suppose further that $v : D \to \mathbb{R}$ ($D \subset M$ open with $K \subset D$) is a C^1-function and $\alpha_1(u)$ is the largest eigenvalue of the symmetrized covariant derivative of f in u such that*

$\sup_{u \in K}\ [\alpha_1(u) + \dot{v}(u)] > 0$. *Then* $\underline{\dim}_B K \geq \dfrac{h_{\text{top}}(\varphi_{|K})}{\sup_{u \in K}[\alpha_1(u) + \dot{v}(u)]}$ *holds.*

In the next corollary we consider again a general dynamical system, but we assume that this system is injective.

Corollary 20. *Suppose that $\varphi = \{\varphi^t\}_{t \in \Gamma}$ is an injective dynamical system on M, having a compact invariant set K such that the first upper Lyapunov exponent $\overline{\nu}_1$ on K is positive. Then*

$$\underline{\dim}_B K \geq \frac{h_{\text{top}}(\varphi^1_{|K})}{\overline{\nu}_1} \ .$$

Proof. For any positive integer k we have

$$h_{\text{top}}({\varphi^1}_{|K}) = \frac{1}{k} h_{\text{top}}({\varphi^k}_{|K}) \leq \frac{1}{k} \ln \sup_{u \in K} \sigma_1(d_u \varphi^k)\, \underline{\dim}_B K \ .$$

This implies

$$h_{\text{top}}({\varphi^1}_{|K}) \leq \limsup_{k \to \infty} \frac{1}{k} \ln \sup_{u \in K} \sigma_1(d_u \varphi^k)\, \underline{\dim}_B K \ .$$

□

Remark 21. If μ is a Borel-measure, the *L-box dimension* of μ (named after F. Ledrappier ([31])), is

$$\dim_{\mathrm{L}}(\mu) = \lim_{\varepsilon \to 0} \limsup_{\delta \to 0} \frac{\ln N(\delta;\varepsilon)}{\ln 1/\varepsilon} ,$$

where $N(\delta;\varepsilon)$ is the minimal number of δ-balls needed to cover a compact set up to a μ-measure ε. It is easy to see that there is an analogy to the box dimension of a set ([48]): Suppose φ is C^1 on the compact Riemannian manifold M and μ is an invariant ergodic probability measure with $\lambda_1(\mu) > 0$ and metric entropy h_μ. Then $\dim_{\mathrm{L}}(\mu) \geq \frac{h_\mu}{\lambda_1(\mu)}$.

5.4 Topological Dimension Estimates of Minimal Sets of Flows

In difference to the great number of Hausdorff and box dimensions estimates analog results for the topological dimension of invariant sets are rare (of course, we always have the estimates $\dim_{\mathrm{T}} K \leq \dim_{\mathrm{H}} K \leq \underline{\dim}_{\mathrm{B}} K$). One of the few results in the literature concerning only the topological dimension goes back to G. F. Hilmy ([28]; see also [5]). We formulate and prove an adapted version for Riemannian manifolds.

Theorem 22. *Suppose that $\{\varphi^t\}_{t\in\Gamma}$ is a smooth flow on an n-dimensional Riemannian C^m-manifold M and $\Sigma \subsetneqq M$ is a minimal set of $\{\varphi^t\}_{t\in\Gamma}$. Then* $\dim_{\mathrm{T}} \Sigma \leq n-1$.

In order to prove this theorem we need two auxiliary results.

Lemma 23. *If $E \subset M$ is an invariant set for $\{\varphi^t\}_{t\in\Gamma}$ then the boundary ∂E is also invariant.*

Proof. Suppose $p \in \partial E$ is an arbitrary point, $t \in \mathbb{R}$ and $\varepsilon > 0$ are arbitrary numbers. Using the continuity of the map $\varphi^t : M \to M$ we can find a $\delta > 0$ such that $\varphi^t(B_\delta(p)) \subset B_\epsilon(\varphi^t(p))$. Since $B_\delta(p)$ contains points of E as well as points of $M\backslash E$, it follows that $\varphi^t(B_\delta(p))$ and, consequently, $B_\varepsilon(\varphi^t(p))$ has the same property. But $\varepsilon > 0$ is arbitrary, therefore $\varphi^t(p) \in \partial E$. □

Lemma 24. *If Σ is a minimal set of $\{\varphi^t\}_{t\in\Gamma}$ all its points are either boundary or inner points.*

Proof. Suppose that Σ has boundary points as well as inner points. Since Σ is closed, $\partial\Sigma \subset \Sigma$. By Lemma 23 the boundary $\partial\Sigma$ is a proper closed and invariant subset of Σ. But this contradicts the minimality of Σ. □

Proof (of Theorem 22). Suppose that $\dim_{\mathrm{T}} \Sigma = n$. Then ([30]) Σ contains an inner point. Thus, according to Lemma 24, all points of Σ are inner points. But this implies that Σ is open. Thus, $\Sigma \notin \{M, \emptyset\}$ is open and closed. But this is not possible, which shows that $\dim_{\mathrm{T}} \Sigma < n$. □

Remark 25. It is interesting to compare the estimate of the topological dimension given for Σ on the base of Theorem 22 with Theorem VIII 3' from [30]. In the latter theorem it is proved that in order that $\dim_{\mathrm{T}} C \le n-1$ for a compact space X of a finite dimension it is necessary and sufficient that for the given closed subset C of X only the zero element of the homology group $H_{n-1}(C)$ bounds in X. (This means that the natural homomorphism of $H_{n-1}(C)$ in $H_{n-1}(X)$ must be an isomophism of $H_{n-1}(C)$ in $H_{n-1}(X)$.)

6 Upper Dimension Estimates for Semi-Flows in Normal Coordinates

6.1 The Variational System in Normal Coordinates

The aim of this subsection is to introduce a type of reparametrization of the neighborhood of flow lines, which goes back to [35] and which is very useful in stability and dimension theory.

Consider a semi-flow $\{\varphi^t\}_{t\ge 0}$ on an n-dimensional Riemannian C^m-manifold M generated by a vector field f of class C^2. For every regular point $u \in M$ with $f(u) \ne O_u$ (the origin of T_uM), we introduce the linear subspace of T_uM

$$T^{\perp}(u) := \{w \in T_uM : \langle w, f(u)\rangle = 0\} \ .$$

To describe how the orthogonal deviation of a perturbation in the initial conditions evolves in time we split a solution $y(\cdot)$ of (5) into orthogonal components $y(t) = z(t) + \mu(t) f(\varphi^t p)$, with $z(t) \in T^{\perp}(\varphi^t p)$ and $\mu(t)$ a time-dependent factor. Then $z(\cdot)$ is a solution of the *system in normal* or *transversal* variations

$$\frac{Dz}{dt} = A(\varphi^t p) z \tag{22}$$

where the linear operator $A(u) : T_uM \to T_uM$ is defined by

$$A(u)v = \nabla f(u)v - \frac{2\langle f(u), S\nabla f(u)v\rangle}{\|f(u)\|^2} f(u) \tag{23}$$

for all $v \in T_uM$. In local coordinates of a chart x around u the operator is given by the tensor

$$A_i^k = \nabla_i f^k - \frac{2}{g_{rs} f^r f^s} f^k g_{jl} f^l S_i^j \ ,$$

where f^k and g_{jl} are the coordinates of the vector field f and the metric tensor in the chart x, respectively, and $S_i^j = \frac{1}{2}(g^{ik}\nabla_k f^l g_{li} + \nabla_i f^j)$ is the representation of $S\nabla f(u)$ in this chart.

To formulate the following central result on the reparametrization of orbits in a general not necessarily hyperbolic situation, we need some notations. Recall

that since M is $C^m (m > 3)$ for any $p \in M$ there exists an open set $\Omega_p \subset T_pM$ around O_p such that the exponential map $\exp_p : \Omega_p \to \exp_p(\Omega_p) \subset M$ is a C^{m-2}-diffeomorphism. For $\delta > 0$ sufficiently small such that $B_\delta(O_p) \subset \Omega_p$ we consider the $(n-1)$-dimensional submanifold $B_\delta^\perp(p) := \exp_p(B_\delta(O_p)) \cap T^\perp(p)$ of M through p, where $B_\delta(O_p) \subset T_pM$ denotes a ball of radius δ centered in the origin of T_pM. Since every point $u \in B_\delta^\perp(p)$ is uniquely determined by the geodesic connecting u and p the point u can be described by the coordinates (r, v) such that $u = \exp_p(r, v)$, where $v \in T^\perp(p)$ is a vector of length 1 and $r \in [0, \delta)$ measures the arc length along the geodesic.
To investigate the stability of a fixed orbit or the dimension properties of a flow invariant set it suffices to consider stability or deformation properties of the projection of the flow lines transversally to the considered orbits or a fixed orbit of the invariant set, resp..
Keeping this in mind several methods of reparametrization of flow lines are of great importance ([25,34]). The following result goes back to [37]. (See also [46].)

Lemma 26. *Suppose $\varphi^{(\cdot)}(p)$ with $p \in M$ to be a non-constant solution of f with bounded positive semi-orbit $\gamma_+(p)$ such that for a certain $C_0 > 0$ the inequality $\|f(u)\| > C_0$ is satisfied for all $u \in \gamma_+(p)$. Then it holds: For any finite time $T_0 > 0$ there exists a $\delta = \delta(T_0) > 0$ such that for any pair $(r, v) \in [0, \delta] \times (T^\perp(p) \cap \partial B_1(O_p))$ we may find a C^{m-1}-diffeomorphism $s(\cdot, r, v) : \mathbb{R}_+ \to \mathbb{R}_+$ with $s(t, r, v) = t$ for all $t \in \mathbb{R}_+$ such that near $\gamma_+(p)$ the reparametrized flow $\phi(t, r, v) = \varphi^{s(t,r,v)}(\exp_p(rv))$ of f satisfies the condition*

$$\langle D_2\phi(t, r, v), f(\phi(t, r, v))\rangle = 0$$

for all $t \in [v, T_0]$.

6.2 Upper Hausdorff Dimension Estimates in Normal Coordinates

In this subsection we give Hausdorff dimension bounds of flow invariant sets of $\{\varphi^t\}_{t \geq 0}$, generated by f, which are formulated in terms of the eigenvalues of the symmetric part of the operator which generates the associated system in normal variations with respect to the direction of the vector field.
In the following we denote at any regular point p of f the eigenvalues of the operator $SA(p)$ restricted to the subspace $T^\perp(p)$ by $\beta_1(p) \geq \ldots \geq \beta_{n-1}(p)$. By $Z(t, p)$ we denote the operator solution of (22) satisfying $Z(0, p) = \mathrm{id}_{T^\perp(p)}$. For every $t \in \mathbb{R}_+$ the linear operator $Z(t, p) : T^\perp(p) \to T^\perp(\varphi^t p)$ maps between the subspaces $T^\perp(p)$ and $T^\perp(\varphi^t p)$ being orthogonal to the vector field in p and $\varphi^t p$, respectively.
The next lemma which can be proved analogously to the inequality (9) (see [37]) shows the distortion of the d-dimensional ellipsoid-measure under the linearized flow of f.

Lemma 27. *Suppose that $p \in M$ is a regular point of f and $Z(\cdot, p)$ is the operator solution of the equation in transversal variations (22). If $d \in (0, n-1]$ is an arbitrary number, written as $d = d_0 + s$ with integer d_0 and $s \in (0,1]$ for $t \geq 0$ it holds*

$$\omega_d(Z(t,p)) \leq \exp\left\{\int_0^t [\beta_1(\varphi^\tau p) + \ldots + \beta_{d_0}(\varphi^\tau p) + s\beta_{d_0+1}(\varphi^\tau p)]d\tau\right\} .$$

Remark 28. For a fixed regular point $p \in M$, small $r > 0$ and $t \geq 0$ the set $\mathcal{E}(t) := Z(t,p)B_r^\perp(O_p)$ is an ellipsoid in the subspace $T^\perp(\varphi^t p)$. If the numbers $\sigma_1(\mathcal{E}(t)) \geq \ldots \geq \sigma_{n-1}(\mathcal{E}(t))$ are the lengths of the semi-axes of $\mathcal{E}(t)$ and $d \in (0, n-1]$ is an arbitrary number we have $\omega_d(\mathcal{E}(t)) = \omega_d(Z(t,p))r^d$.

The next theorem which goes back to K. Gelfert (see [35]) is the realization of the Douady-Oesterlé estimate of Theorem 4 adapted to our special system of normal coordinates.

Theorem 29. *Let K and $\widetilde{K}$ be two compact sets in M satisfying the inclusions $K \subset \varphi^t(K) \subset \widetilde{K}$ for all $t \geq 0$, where K does not contain equilibrium points of $\{\varphi^t\}$. Denote by Λ the set of equilibrium points of $\{\varphi^t\}$ in M. For $p \in \widetilde{K}\backslash\Lambda$ let $\beta_1(p) \geq \ldots \geq \beta_{n-1}(p)$ be the eigenvalues of the symmetric part $SA(p)$ restricted to the subspace $T^\perp(p)$, where $A(p)$ is the operator from (23). Let $v : M\backslash\Lambda \to \mathbb{R}$ be a C^1-function and suppose that there exist real numbers $d \in (0, n-1]$ $(d = d_0 + s$, $s \in (0,1]$, d_0 integer$)$, $\Theta > 0$ and $T > 0$ such that*

$$\sup_{p\in\widetilde{K}\backslash\Lambda} \int_0^T [\beta_1(\varphi^\tau p) + \ldots + \beta_{d_0}(\varphi^\tau p) + s\beta_{d_0+1}(\varphi^\tau p) + \dot{v}(\varphi^\tau p)]d\tau \leq -\Theta .$$

Then $\dim_H K \leq d + 1$. Further if $d = 1$ we have $\dim_H K \leq 1$.

6.3 Transversal Lyapunov Exponents for Semi-Flows

We now introduce Lyapunov exponents with respect to the variational equation in normal coordinates. Suppose again that $K \subset M$ is a compact invariant set of the semi-flow $\{\varphi^t\}_{t\geq 0}$ and in K there are no equilibrium points of this flow. For points in K we introduce the tangent bundle splitting $T_K M = E^1 \oplus E^2$ with $E_p^1 = T^\perp(p)$ and $E_p^2 = T''(p) = \mathrm{span}\{f(p)\}$ and require that the equivariance property $E^i_{\varphi^t p} = d_p\varphi^t E^i_p$ $(i = 1, 2,\ p \in K,\ t \geq 0)$ is satisfied. Denote the *transversal singular* values of the differential $d_p\varphi^t{}_{|E_p^1}$ by $\sigma_1^\perp(d_p\varphi^t) \geq \ldots \geq \sigma_{n-1}^\perp(d_p\varphi^t)$ and the *transversal singular value function* with respect to $\sigma_i^\perp(d_p\varphi^t)$ by $\omega_k^\perp(d_p\varphi^t)$. Analogously to the definitions in Section 3.4 let us introduce global transversal Lyapunov exponents by the following procedure. Define for $t \geq 0$ and $j = 1, 2, \ldots, n-1$ the numbers

$$\omega_{j,K}^{\perp}(\varphi^t) = \sup_{p \in K} \omega_j^{\perp}(d_p\varphi^t) \text{ and } \omega_{0,K}^{\perp}(t)(\varphi^t) \equiv 1 .$$

Then the *global transversal Lyapunov exponents* are

$$\nu_j^{\perp} = \lim_{t \to \infty} \frac{1}{t} \ln \frac{\omega_{j,K}^{\perp}(\varphi^t)}{\omega_{j-1,K}^{\perp}(\varphi^t)}, \quad j = 1, \ldots, n-1 . \tag{24}$$

With respect to Theorem 29 we have now the following

Corollary 30. *Suppose under the above assumptions that M is an n-dimensional Riemannian manifold ($n \geq 2$), $\{\varphi^t\}_{t \geq 0}$ is a semi-flow on M with invariant compact set K, and the global transversal Lyapunov exponents are defined by (24). Suppose further that there is a number $d_0 \in \{1, \ldots, n-2\}$ such that $\nu_1^{\perp} + \ldots + \nu_{d_0}^{\perp} < 0$ and $\nu_1^{\perp} + \ldots + \nu_{d_0+1}^{\perp} > 0$.*

Then $\dim_H K \leq d_0 + 1 + \dfrac{\nu_1^{\perp} + \ldots + \nu_{d_0}^{\perp}}{|\nu_{d_0+1}^{\perp}|}$.

7 Upper Hausdorff Dimension Estimates for Dynamical Systems with an Equivariant Tangent Bundle Splitting

We introduce now a structure which is naturally given for hyperbolic systems. But in contrast to these systems we do not require any norm contraction or expansion properties with respect to a bundle of invariant subspaces. On an n-dimensional Riemannian manifold we consider an invertible dynamical system $\{\varphi^t\}_{t \in \Gamma}$ having a compact invariant set K. We say that $\{\varphi^t\}_{t \in \Gamma}$ possesses an *equivariant tangent bundle splitting* $T_K M = E^+ \oplus E^- \oplus E^0$ if for any $p \in K$ the set $E_p^+ := E^+ \cap T_p M$ is an n^+-dimensional subspace of $T_p M$, the set $E_p^- := E^- \cap T_p M$ is an n^--dimensional subspace of $T_p M$ and $E_p^0 := E^0 \cap T_p M$ is an n^0-dimensional subspace of $T_p M$ such that $n^- + n^+ + n^0 = n$ and $d_p\varphi^t(E_p^{\xi}) = E_{\varphi^t p}^{\xi}$ for $t \in \Gamma$ and $\xi \in \{+, -, 0\}$.

Example 31. Hyperbolic sets of flows allow a specific type of equivariant splitting of the tangent bundle, if we consider the stable part E^s as E^+, the unstable part E^u as E^- and the central part E^c as E^0.

We want to consider systems with a certain contraction property of the singular values of the tangent map. In order to use the splitting structure of the tangent bundle we define a singular function in a modified form. Suppose that there is an equivariant tangent bundle splitting as above, the numbers $\sigma_1^+(d_p\varphi^t), \ldots, \sigma_{n^+}^+(d_p\varphi^t)$ are the singular values of $d_p\varphi^t_{|E_p^+}$ and the numbers

$\sigma_1^-(d_p\varphi^{-t}), \dots, \sigma_{n^-}^-(d_p\varphi^{-t})$ are the singular values of $d_p\varphi^{-t}_{|E_p^-}$. For arbitrary $p, q \in K$ and $t > 0$ relabel the set of singular values

$$\{\sigma_1^+(d_p\varphi^t), \dots, \sigma_{n^+}^+(d_p\varphi^t), \sigma_1^-(d_q\varphi^{-t}), \dots, \sigma_{n^-}^-(d_q\varphi^{-t})\}$$

as $\{\sigma_1(t,p,q), \sigma_2(t,p,q), \dots, \sigma_{n^++n^-}(t,p,q)\}$ with $\sigma_1(t,p,q) \geq \dots \geq \sigma_{n^++n^-}(t,p,q)$ and define the *singular value function* with respect to the splitting (E^+, E^-, E^0) and $d = d_0 + s \in (0, n^+ + n^-]$ by

$$\omega_{d,K}^{E^+, E^-, E^0}(\varphi^t) = \begin{cases} 1 \\ \sup_{p,q\in K} \sigma_1(t,p,q) \cdot \ldots \cdot \sigma_{d_0}(t,p,q) \sigma_{d_0+1}^s(t,p,q) \end{cases}$$

for $d = 0$ resp. $d > 0$.
We say that φ^t is *d-contractive* with respect to the splitting $E^+ \oplus E^- \oplus E^0$ and $t \in \Gamma$ if it holds $\omega_{d,K}^{E^+, E^-, E^0}(\varphi^t) < 1$.
Since $\omega_{d,K}^{E^+, E^-, E^0}(\varphi^t)$ is subexponential in t, the limit

$$\overline{\nu}_d = \lim_{t\to\infty} \frac{1}{t} \ln \omega_{d,K}^{E^+, E^-, E^0}(\varphi^t)$$

exists for all $d \in [0, n^- + n^+]$. The numbers $\nu_1, \dots, \nu_{n^++n^-}$, computed by the formula $\nu_i = \overline{\nu}_i - \overline{\nu}_{i-1}$ $(i = 1, 2, \dots, n^+ + n^-)$ with $\overline{\nu}_0 = 0$, are the *global Lyapunov exponents of the splitting* (E^+, E^-, E^0). The following theorem goes back to A. Franz ([16]) and generalizes results of [12,22].

Theorem 32. *Let M be an n-dimensional Riemannian C^m-manifold, $\{\varphi^t\}_{t\in\Gamma}$ a smooth invertible dynamical system, and $K \subset M$ a compact φ^t-invariant set. Suppose that there exist an equivariant splitting $T_KM = E^+ \oplus E^- \oplus E^0$ of dimensions (n^+, n^-, n^0) and a number $d \in (0, n^+ + n^-]$ such that φ^t is d-contractive with respect to the splitting. Let us assume that there exists a time $t \in \Gamma$ such that*

$$\omega_{d,K}^{E^+, E^-, E^0}(\varphi^t) < e^{-2h_{\text{top}}(\varphi^t{}_{|K})} .$$

Then $\dim_H K \leq d + n_0$.

Corollary 33. *Let M be an n-dimensional Riemannian C^m-manifold, $U \subset M$ be an open set, $\{\varphi^t\}_{t\in\mathbb{R}}$ a C^1-flow on U and $K \subset U$ a compact φ^t-invariant set which admits an equivariant splitting $T_KM = E^+ \oplus E^- \oplus E^0$ of dimensions (n^+, n^-, n^0). Let $\nu_1, \dots, \nu_{n^++n^-}$ be the global Lyapunov exponents of $\{\varphi^t\}_{t\in\mathbb{R}}$ on K with respect to this splitting and let $d_0 < n^+ + n^-$ be the smallest natural number with*

$$2h_{\text{top}}(\varphi^1{}_{|K}) + \nu_1 + \ldots + \nu_{d_0} + \nu_{d_0+1} < 0 .$$

Then $$\dim_H K \leq d_0 + \frac{2h_{\text{top}}(\varphi^1{}_{|K}) + \nu_1 + \ldots + \nu_{d_0}}{|\nu_{d_0+1}|} + n_0.$$

Example 34. We consider the geodesic flow on a smooth n-dimensional Riemannian manifold, i.e. the flow defined on the unit tangent bundle $T^1M = \{v \in TM : \|v\| = 1\}$ such that the projection π of any flow line $t \mapsto \varphi^t v$ with $v \in TM$ onto M is the geodesic $\gamma_v(t)$ which is determined by $\gamma_v(0) = \pi(v)$ and $\dot{\gamma}_v(0) = v$. It follows $\dot{\gamma}_v(t) = \varphi^t(v)$ for any $v \in T^1M$ and $t \in \mathbb{R}$.
Let $K \subset T^1M$ be a compact φ^t-invariant set. It is well-known that there exists an equivariant splitting of the tangent bundle $T_K T^1 M$ with a one-dimensional subbundle E^0 tangential to the flow lines and a $(2n-2)$-dimensional subbundle transversal to the flow lines. Assume that all 2-planes Π which are tangent to the geodesic have strictly negative sectional curvature $S(\Pi)$. Since M is smooth and K is compact there are positive constants k_1 and k_2 with $-k_1^2 \le S(\Pi) \le -k_2^2$ for all 2-planes Π tangent to the geodesic in a point of $\pi(K)$. The $(2n-2)$-dimensional subbundle $T_K T^1 M$ which is transversal to the flow can equivariantly be splitted into a stable subbundle E^+ and an unstable subbundle E^- such that for any $u \in K$ and $t \in \mathbb{R}$ the relations

$$\frac{k_2}{k_1}\|\xi\|e^{-k_1 t} \le \|d_u\varphi^t\xi\| \le \frac{k_1}{k_2}\|\xi\|e^{-k_2 t} \quad \text{for any } \xi \in E_u^+$$

and

$$\frac{k_1}{k_2}\|\xi\|e^{k_1 t} \ge \|d_u\varphi^t\xi\| \ge \frac{k_2}{k_1}\|\xi\|e^{k_2 t} \quad \text{for any } \xi \in E_u^-$$

hold. It follows that the global Lyapunov exponents of the flow $\{\varphi^t\}$ on K with respect to the splitting $E^- \oplus E^+ \oplus E^0$ satisfy for $i = 1, \dots, 2n-2$ the inequalities $-k_1 \le \nu_i \le -k_2$.
If $(n-1)k_2 > h_{\text{top}}(\varphi^1{}_{|K})$ there is a natural number $d_0 < 2n-1$ with $2h_{\text{top}}(\varphi^1{}_{|K}) + \nu_1 + \ldots + \nu_{d_0} + \nu_{d_0+1} < 0$. Using now Corollary 33 we get the dimension estimate

$$\begin{aligned}\dim_{\mathrm{H}} K &\le d_0 + 1 + \frac{2h_{\text{top}}(\varphi^1{}_{|K}) + \nu_1 + \ldots + \nu_{d_0}}{|\nu_{d_0+1}|} \\ &\le d_0 + 1 + \frac{2h_{\text{top}}(\varphi^1{}_{|K}) - d_0 k_2}{k_1} \le 1 + \frac{2h_{\text{top}}(\varphi^1{}_{|K})}{k_1} + d_0\left(1 - \frac{k_2}{k_1}\right),\end{aligned}$$

provided that $(n-1)k_2 < h_{\text{top}}(\varphi^1_{|K})$ holds.

8 Upper Hausdorff Dimension Estimates for Non-Injective Maps

Whereas the majority of the previous sections were concerned with dimension estimates for injective dynamical systems here we want to consider time-discrete dynamical systems generated by a non-injective map. In this case the conditions of the Douady-Oesterlé dimension estimate of Theorem 4 can be weakened using the cardinality of the pre-image sets, described by the multiplicity function (see for instance [13]). For a map $\varphi : M_1 \to M_2$ between two arbitrary sets M_1 and M_2 the *multiplicity function* $N(\varphi, K, u)$ of φ with respect to a set $K \subset M_1$ in a point $u \in M_2$ is defined as the cardinality of the set $\{v \in K | \varphi(v) = u\}$. The following theorems are proved by A.Franz ([17]).

Theorem 35. *Let M be a n-dimensional Riemannian C^m-manifold, $U \subset M$ an open set, $\varphi : U \to M$ a C^1-map and $K \subset U$ a compact φ-invariant set with $|\det(d_u\varphi)| > 0$ for all $u \in K$. If there exist numbers $d \in (0, n]$ and $\nu \in (0,1)$ with*

$$\omega_d(d_u\varphi)^{\frac{1}{d}} \leq \nu N(\varphi, K, \varphi(u))^{\frac{1}{n}} \quad \textit{for all } u \in K,$$

then $\dim_{\mathrm{H}} K \leq d$.

Note that the condition of this theorem contains the singular-value function ω_d with an exponent $\frac{1}{d}$, since in contrast to ω_d the function $\omega_d^{1/d}$, as a function of d, is always monotone decreasing in the whole interval $(0, n]$.
The following theorem uses the local inverse maps, therefore a few additional assumptions are needed ($\mathrm{cl}(\cdot)$ denotes the closure of a set). The notation $\overline{\omega}_d$ refers to the inverse singular-value function defined as the product not of the largest but of the smallest d singular values.

Theorem 36. *Let M be a n-dimensional Riemannian C^m-manifold, $U \subset M$ an open set, $\varphi : U \to M$ a C^1-map and $K \subset U$ a compact φ-invariant set and $\widetilde{U} \supset K$ an open set with $\mathrm{cl}(\widetilde{U}) \subset U$ compact, $|\det(d_u\varphi)| > 0$ for all $u \in \mathrm{cl}(\widetilde{U})$ and $\varphi^{-1}(\widetilde{U}) \subset \widetilde{U}$. If there exist numbers $d \in (0, n]$ and $\nu \in (0,1)$ with*

$$\nu\, \overline{\omega}_d(d_u\varphi) \geq N(\varphi, \widetilde{U}, \varphi(u)) \quad \textit{for all } u \in K,$$

then $\dim_{\mathrm{H}} K \leq d$.

The proof of the last two theorems is based on transformation formulas for Hausdorff outer integrals (see [17]), which have, applied to the characteristic function of a compact set, the same critical value as the outer Hausdorff-d-measure and can therefore be used in a similar way to obtain dimension estimates.

Example 37. Both theorems of this section can be applied to a wide class of non-invertible maps and to iterated function systems. Some classes of injective maps like horseshoe maps can be split into partial non-invertible maps, or the state space (for instance for Belykh maps ([1])) can be factorized such that the resulting non-injectivity of the map can be used to obtain dimension bounds for an invariant set. A well known class of non-injective maps are polynomials in the complex plane. We want to illustrate the dimension estimate by Theorem 36 for Julia sets of quadratic polynomials $\varphi : \mathbb{C} \to \mathbb{C}$ in the form

$$\varphi(z) = z^2 + c,$$

where $c \in \mathbb{C}$ is a parameter. The Julia set K is the closure of the set of repelling periodic points (see for instance [11,52]). This set is compact, non-empty and invariant under φ. For different parameters c this set has different shapes. Therefore it is interesting to estimate the Hausdorff dimension of this set.

Let us consider $\mathbb{C}$ as two-dimensional Riemannian C^∞-manifold with a constant metric tensor giving the metric

$$\varrho(z_1, z_2) = |z_1 - z_2| = \sqrt{(\mathrm{Re}z_1 - \mathrm{Re}z_2)^2 + (\mathrm{Im}z_1 - \mathrm{Im}z_2)^2}\ .$$

Under this metric the singular values of $d_z\varphi$ are

$$\sigma_1(d_z\varphi) = \sigma_2(d_z\varphi) = 2\sqrt{(\mathrm{Re}z)^2 + (\mathrm{Im}z)^2} = 2|z|.$$

Thus for $d \in (0, 2]$ and any $z \in \mathbb{C}$ we have $\omega_d(d_z\varphi) = \overline{\omega}_d(d_z\varphi) = (2|z|)^d$. It can be shown (see [11]) that the absolute value of the points of K can be estimated by

$$\begin{aligned} &\frac{1}{2} + \sqrt{\frac{1}{4} - |c|} < |z| < \frac{1}{2} + \sqrt{\frac{1}{4} + |c|} \quad \text{for } |c| < \frac{1}{4}, \\ &\sqrt{|c| - \sqrt{|2c|}} \le |z| \le \sqrt{|c| + \sqrt{|2c|}} \quad \text{for } |c| > 2. \end{aligned}$$

The multiplicity function of φ with respect to K and any open neighbourhood $\widetilde{U}$ of K satisfies $N(\varphi, K, \varphi(z)) = N(\varphi, \widetilde{U}, \varphi(z)) = 2$ for all $z \in K$ under the condition, that the origin is not contained in K. Thus the condition of Theorem 36 has the form $(2|z|)^d > 2$ for all $z \in K$ which is satisfied if

$$\begin{aligned} &\left(1 + \sqrt{1 - 4|c|}\right)^d > 2 \quad \text{for } |c| < \frac{1}{4}, \\ &2\sqrt{|c| - \sqrt{|2c|}} > 2 \qquad \text{for } |c| > 2. \end{aligned}$$

Thus Theorem 36 gives, similar to [11], the dimension estimate

$$\dim_{\mathrm{H}} K \le \begin{cases} \frac{\ln 2}{\ln(1+\sqrt{1-4|c|})} & \text{for } |c| < \frac{\sqrt{2}-1}{2}, \\ \frac{\ln 2}{\ln\left(2\sqrt{|c|-\sqrt{|2c|}}\right)} & \text{for } |c| > \frac{3}{2} + \sqrt{2}, \end{cases}$$

In the case $|c| > \frac{3}{2} + \sqrt{2}$ results of [12] can be applied as well and yield the same dimension bound.

Acknowledgment

I would like to thank Dr. Astrid Franz from the Chemnitz University of Technology and Katrin Gelfert and Dr. Antje Noack from the University of Technology Dresden for fruitful discussions, support and help.

References

1. Belykh, V.N., Shelesnyak, I.L.: Dimension estimates for a strange attractor of a two-dimensional discontinuous map. Methods of Qualitative Theory and Theory of Bifurcations, Nishny Novgorod, University of Nishny Novgorod, (1992) 12–17 (Russian)
2. Blinchevskaya, M. A., Ilyashenko, Yu. S.: Estimate for the entropy dimension of the maximal attractor for k-contracting systems in an infinite-dimensional space. J. of Math. Phys. **6** 1 (1999) 20–27 (Russian)
3. Bogatyj, S.A., Gutsu, V.I.: On the structure of attracting compacta. Differential'nye Uravneniya **25** (1989) 907–909 (Russian)
4. Boichenko, V.A., Franz, A., Leonov, G.A., Reitmann, V.: Hausdorff and fractal dimension estimates for invariant sets of non-injective maps. Z. Anal. Anw. (ZAA) **17** 1 (1998) 207–223
5. Boichenko, V.A., Leonov, G.A., Reitmann, V.: Dimension Theory for Ordinary Differential Equations. Preliminary version (2000)
6. Daleckii, Yu.L., Krein, M.G.: Stability of Solutions of Differential Equations in Banach Spaces. Am. Math. Soc., Providence, 1974
7. Dellnitz M., Froyland G., Junge O.: The algorithms behind GAIO - set oriented numeric methods for dynamical systems. (This volume)
8. Douady, A., Oesterlé, J.: Dimension de Hausdorff des attracteurs. C. R. Acad. Sci. Paris Ser. A **290** (1980) 1135–1138
9. Eckmann, J.-P., Ruelle, D.: Ergodic theory of chaos and strange attractors. Rev. Mod. Phys. **57** (1985) 617–656
10. Eden, A., Foias, C., Temam, R.: Local and global Lyapunov exponents. J. Dyn. Diff. Equ. **3** (1991) 133–177
11. Falconer, K.J.: Fractal Geometry: Mathematical Foundations and Applications. Wiley, Chichester, (1990)
12. Fathi, A.: Expansiveness, hyperbolicity and Hausdorff dimension. Commun. Math. Phys. **126** (1989) 249–262
13. Federer, H.: Geometric Measure Theory. Springer, New York, (1969)
14. Foias, C., Olsen, E.: Finite fractal dimension and Hölder-Lipschitz Parametrization. Indiana University Math. J. **65** 3 (1995) 603–616
15. Fowler, A.C., Gibbon, J.D., Guiness, Mc.: The real and complex equations and their relevance to physical systems. Physica 7 D, (1983) 126–134
16. Franz, A.: Hausdorff dimension estimates for invariant sets with an equivariant tangent bundle splitting. Nonlinearity **11** (1998) 1063–1074
17. Franz, A.: Hausdorff dimension estimates for non-injective maps using the cardinality of the pre-image sets. Nonlinearity **13** (2000) 1425–1438
18. Fredrickson, P., Kaplan, J. L., Yorke, E. D., Yorke, J. A.: The Lyapunov dimension of strange attractors. J. Diff. Equ. **49** (1983) 185–207

19. Gelfert, K.: Estimates of the box-counting dimension of map-invariant sets on Riemannian manifolds. Proc. of the Workshop "Dimensions of attractors", St. Petersburg, (1999)
20. Gohberg, I.C., Krein, M.G.: Introduction to the Theory of Linear Nonselfadjoint Operators. Translations of mathematical monographs 18, Amer. Math. Soc., Providence, RI, (1969)
21. Günther, B., Segal, J.: Every attractor of a flow on a manifold has the shape of a finite polyhedron. Proc. of the Amer. Math. Soc. **119** 1 (1999) 321–329
22. Gu, X.: An upper bound for the Hausdorff dimension of a hyperbolic set. Nonlinearity **4** (1991) 927–934
23. Guckenheimer, J., Holmes, P.: Nonlinear Oscillation, Dynamical Systems, and Bifurcation Theory of Vector Fields. New York, Springer, (1983)
24. Hakamada, T., Imai, H., Ishimura, N.: Analytical approach to estimating the dimension of attractors. Appl. Math. Opt. **34** (1996) 29–36
25. Hartman, P., Olech, C.: On global asymptotic stability of solutions of ordinary differential equations. Trans. Amer. Math. Soc. **104** (1962) 154–178
26. Heineken, W., Reitmann, V.: Fractal dimension estimates for flow invariant sets on manifolds. To appear in Z. Anal. Anw.(ZAA)
27. Hénon, M.: A two-dimensional mapping with a strange attractor. Commun. Math. Phys. **50** (1976) 69–76
28. Hilmy, G.F.: On a property of minimal sets. Dokl. Akad. Nauk SSSR **14** (1937) 261–262 (Russian)
29. Hunt, B.R.: Maximal local Lyapunov dimension bounds the box dimension of chaotic attractors. Nonlinearity **9** (1996) 845–852
30. Hurewicz, W., Wallman, H.: Dimension Theory. Princeton University Press, Princeton, (1948)
31. Ledrappier, F.: Some relations between dimension and Lyapunov exponents. Commun. Math. Phys. **81** (1981) 229–283
32. Leonov, G.A.: Lyapunov Exponents and Problems of Linearization. From Stability to Chaos. St. Petersburg Univ. Press., St. Petersburg, (1997)
33. Leonov, G.A., Boichenko, V.A.: Lyapunov's direct method in the estimation of the Hausdorff dimension of attractors. Acta Appl. Math. **26** 1 (1992) 1–60
34. Leonov, G.A., Burkin, I.M., Shepelyawyi, A.I.: Frequency Methods in Oscillating Theory. Kluwer Academic Publishers, Ser. Mathematics and Its Applications, Dordrecht/Boston/London **357** (1996)
35. Leonov, G.A., Gelfert, K., Reitmann, V.: Hausdorff dimension estimates by use of a tubular Carathéodory structure and their application to stability theory. Advances of Stability Theory at the End of XXth Century, A. A. Martynyuk (Ed.), Stability and Control: Theory, Methods and Applications, Gordon and Breach Science Publishers, London, **13** (2000) 41–66
36. Leonov, G.A., Lyashko, S.A.: Lyapunov's direct method in estimates of fractal dimension of attractors. Diff. Urav. **30** 1 (1997) 68–74 (Russian)
37. Leonov, G.A., Noack, A., Reitmann, V.: Asymptotic orbital stability conditions for flows by estimates of singular values of the linearization. Nonlinear Anal., Theory Methods Appl. 9 (2000) (to appear)
38. Leonov, G.A., Reitmann, V.: Attraktoreingrenzung für Nichtlineare Systeme. Teubner-Texte zur Mathematik, Bd. 97, Teubner-Verlag, Leipzig, (1986)
39. Leonov. G.A., Reitmann, V.: Dissipativität und globale Stabilität des komplexen Lorenz-Systems. Z. Anal. Anw. (ZAA) **6** (1987) 575–582

40. Leonov, G.A., Reitmann, V., Smirnova, V.B.: Non-Local Methods for Pendulum-Like Feedback Systems. Teubner-Texte zur Mathematik, Bd. 132, B.G. Teubner Stuttgart-Leipzig, (1992)
41. Yi Li, M., Muldowney, J.S.: On Bendixson's criterion. J. Diff. Equ. **106** (1993) 27–39
42. Yi Li, M., Muldowney, J.S.: Lower bounds for the Hausdorff dimension of attractors. J. Dyn. Diff. Equ. **7** 3 (1995) 457–469
43. Mané, R.: On the dimension of the compact invariant sets of certain non-linear maps. Springer, Lectures Notes in Math. **898** (1981) 230–241
44. Morrey, Ch.B.: Multiple Integrals in the Calculus of Variations. Springer Berlin, Heidelberg, (1966)
45. Noack, A.: Hausdorff Dimension Estimates for time-discrete Feedback Control Systems. Z. Angew. Math. Mech. (ZAMM), **77** 12 (1997) 891–899
46. Noack, A.: Dimensions- und Entropieabschätzungen sowie Stabilitätsuntersuchungen für nichtlineare Systeme auf Mannigfaltigkeiten. Ph. D. thesis, TU Dresden (1998)
47. Noack, A., Reitmann, V.: Hausdorff dimension of invariant sets of time-dependent vector fields. Z. Anal. Anw. (ZAA), **15** 2 (1996) 457–473
48. Pesin, Ya.B.: Dimension Theory in Dynamical Systems. Chicago Lectures in Mathematics. The University of Chicago Press, Chicago and London, (1997)
49. Pugh. C.C.: The closing lemma. Amer. J. Math. **89** (1967) 956–1021
50. Reitmann, V.: Hausdorff dimension bounds, homology group properties and the global behaviour of flows on manifolds. Proc. Third Intern. Conf. "Differential Equations and Applications", St. Petersburg (2000) 87–88
51. Reitmann, V., Schnabel, U.: Hausdorff dimension estimates for invariant sets of piecewise smooth maps. Z. Angew. Math. Mech. (ZAMM), **80** 9 (2000) 623–632
52. Ruelle, D.: Repellers for real analytic maps. Ergodic Th. Dynam. Sys. **2** (1982) 99–108
53. Schmeling, J.: A dimension formula for endomorphisms - The Belykh family, Ergodic Th. Dyn. Sys. **18** (1998) 1283–1309
54. Smith, R.A.: Some applications of Hausdorff dimension inequalities for ordinary differential equations. Proc. Roy. Soc. Edinburgh **104A** 3–4 (1986), 235–259
55. Temam, R.: Infinite-Dimensional Dynamical Systems in Mechanics and Physics. Springer, New York-Berlin, (1988)
56. Thieullen, P.: Entropy and the Hausdorff dimension for infinite-dimensional dynamical systems. J. Dyn. Diff. Equ. **4** 1 (1992) 127–159
57. Walters, P.: An Introduction to Ergodic Theory. Springer, Berlin, (1982)
58. Yakubovich, V.A.: The frequency theorem in control theory. Sibirsk. Math. Zh. **14** 2 (1973) 384–420
59. Young, L.-S.: Capacity of attractors. Ergodic Th. Dynam. Sys. **1** (1981) 381–388
60. Zhi-Min Chen: A note on Kaplan-Yorke-type estimates on the fractal dimension of chaotic attractors. Chaos, Solitons Fractals **3** (1993) 575–582

On the Inverse Problem of Fractal Compression

Hannes Hartenstein[1], Matthias Ruhl[2], Dietmar Saupe[3], and
Edward R. Vrscay[4,*]

[1] Computer & Communication Research Lab, NEC Europe Ltd., Heidelberg, Germany, E-mail: Hannes.Hartenstein@ccrle.nec.de
[2] Laboratory of Computer Science, Massachusetts Institute of Technology, USA E-mail: ruhl@mit.edu
[3] Institut für Informatik, Universität Leipzig, Germany E-mail: saupe@informatik.uni-leipzig.de
[4] Department of Applied Mathematics, University of Waterloo, Canada E-mail: ervrscay@links.uwaterloo.ca

Abstract. The inverse problem of fractal compression amounts to determining a contractive operator such that the corresponding fixed point approximates a given target function. The standard method based on the *collage coding* strategy is known to represent a suboptimal method. Why does one not search for optimal fractal codes? We will prove that optimal fractal coding, when considered as a discrete optimization problem, constitutes an NP-hard problem, i.e., it cannot be solved in a practical amount of time. Nevertheless, when the fractal code parameters are allowed to vary continuously, we show that one is able to improve on collage coding by fine-tuning some of the fractal code parameters with the help of differentiable methods. The differentiability of the attractor as a function of its luminance parameters is established. We also comment on the approximating behavior of collage coding, state a lower bound for the optimal attractor error, and outline an annealing scheme for improved fractal coding.

1 Introduction

Fractal compression seeks to approximate a target function f with a function $\bar{f}_p$ which is the fixed point, or *attractor*, of a 'simple' contractive operator T_p that acts on a suitable metric space $(\mathcal{F}, d_{\mathcal{F}})$ of functions. The parameter vector p (also called the *fractal code*) that defines T_p (also called *fractal transform operator*) is then used as a (lossy) representation of the target function f. The fixed point $\bar{f}_p$ is generated by iterating the operator T_p on an arbitrary function of the space $\mathcal{F}$; this is the decoding step. The encoding problem of fractal compression lies in finding in a suitable class of operators the one whose corresponding fixed point gives the best approximation of the target function. Of course, the class of operators considered for fractal coding purposes has to be constrained to 'simple' operators that can be coded compactly in order to lead to data compression. The encoding problem is

* Project: Partitioned Iterated Function Systems. The Inverse Problem of Image Coding and the Complexity of Associated Dynamical Systems (Dietmar Saupe)

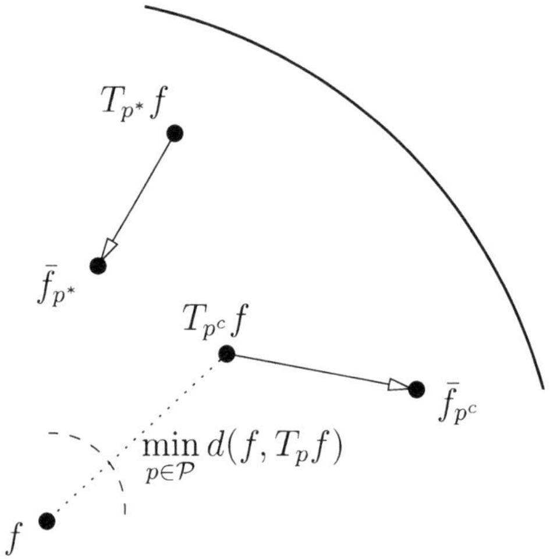

Fig. 1. Schematic presentation of the relationship between the target f, an optimal collage $T_{p^c}f$, the fixed point $\bar{f}_{p^c} = T_{p^c}\bar{f}_{p^c}$ corresponding to T_{p^c}, as well as the collage $T_{p^*}f$ of an optimal fractal code and the corresponding optimal fixed point $\bar{f}_{p^*} = T_{p^*}\bar{f}_{p^*}$. The boldfaced arc indicates the upper bound on the feasible attractor error provided by the collage theorem while the dotted arc indicates a lower bound on the optimal attractor error. This lower bound is given in Section 5. Clearly, a collage optimal fractal code does not in general coincide with an optimal fractal code.

also called the inverse problem of fractal compression since it involves the determination of 'causes', i.e., the determination of the operator parameters, based on a desired 'effect', i.e., the desired fixed point.

In practice, fractal coding algorithms rely upon the method of *collage coding.* Given a target function f and a suitable parameter space $\mathcal{P}$ one determines a fractal transform operator $T_{p^c}, p^c \in \mathcal{P}$, that minimizes the *collage error* $d_{\mathcal{F}}(f, Tf)$. This procedure is motivated by the *collage theorem* [1,3], a corollary of the contraction mapping principle. The collage theorem states that the attractor error $d_{\mathcal{F}}(f, \bar{f}_{p^c})$ is bounded from above by a multiple of the collage error $d_{\mathcal{F}}(f, T_{p^c}f)$. Thus, with the collage coding method one minimizes a bound on the actual attractor error. However, with this approach one generally does not find an *optimal* fractal code for the target function f, i.e., a fractal code $p^* \in \mathcal{P}$ such that $d_{\mathcal{F}}(f, \bar{f}_{p^*}) = \min_{p\in\mathcal{P}} d_{\mathcal{F}}(f, \bar{f}_p)$ (see Figure 1).

It is, therefore, natural to address the question: why does one not search for an optimal fractal code? In this paper we will show the following result: *Optimal fractal coding is NP-hard.* Thus, fractal coding —when considered as an optimization problem— represents an intractable problem, i.e., it is the computational complexity that prevents us from determining optimal fractal codes.

Given that optimal fractal coding is intractable, can one at least improve upon collage coding or is collage coding essentially the best one can do? We will show that one is able to improve on collage coding by fine-tuning some of the fractal code parameters with the help of differentiable methods.

The above short outline of the results indicates that we tackle the inverse problem of fractal compression from two different directions using two different mathematical methodologies. For the NP-hardness proof we consider fractal coding as a discrete optimization problem, whereas for the improvements over collage coding some of the parameters are assumed to be continuous. Since it is easier to state the discrete problem by reference to the 'continuous' problem, the first part of the paper will deal with the question of how to improve on collage coding with the use of differentiable methods, while the second part presents the NP-hardness result —in contrast to the 'logical' order of arguments. This paper summarizes the main results from our conference publications [27,28].

The use of contractive transforms and their corresponding attractors for the compression of signals and images was proposed by Barnsley and Jacquin in the late 1980s [2,19]. Before the birth of fractal compression and without technical applications in mind, Williams [29] and Hutchinson [18] had published mathematical studies of compositions of contractions and iterated function system. During the last 10 years about 400 papers were published in the field of fractal compression, as well as four books [4,10,20,11]. Several studies have attempted to find attractor functions $\bar{f}$ that are better approximations to a target f than the "collage attractors" $\bar{f}_{p^c}$. Indeed, these studies have typically employed the collage attractor $\bar{f}_{p^c}$ as a starting point. For example, Barthel [5] and then Lu [20] have devised "annealing schemes" that produce sequences of attractors $\bar{f}^{(n)}$ that are then used to "collage" the target f. The sequences $\bar{f}^{(n)}$ are observed to provide better approximations to the target. However, there is still no rigorous theoretical basis for this method. On the other hand, Dudbridge and Fisher [9], using the Nelder-Mead simplex algorithm, searched the fractal code space $\mathcal{P}$ in the vicinity of the collage attractor to locate (local) minima of the approximation error $d_{\mathcal{F}}(f, \bar{f})$. Their method was applied to a restricted class of (separable) fractal transforms, in which four 4×4 pixel range blocks shared a common domain block [22]. Withers [30] has derived differentiability properties of Iterated Functions Systems with probabilities whose attractors model graphs of 1D functions. Newton's method is used to compute parameters.

The paper is organized as follows: below notations and basic definitions are introduced. In Section 3 the differentiability of the attractor functions with respect to the luminance parameters is proven and results obtained by using gradient methods are presented. In Section 4, the problem of optimal fractal coding is stated as an combinatorial problem, and the computational complexity of this problem is analyzed. Further results are surveyed in Section 5.

2 Mathematical and Notational Preliminaries

What is the form of a fractal transform operator T? Let (X,d) denote the support or *base space*, assumed to be a metric space, e.g. $X = [0,1)$ or $X = [0,1)^2$. Let $\mathcal{F}(X) = \{f : X \to \mathbb{R}\}$ denote a suitable complete space of functions with metric $d_{\mathcal{F}}$. Now let $R_k \subset X, k = 1,2,\ldots,n_{\mathcal{R}}$ denote a set of *range blocks* that partition X, i.e., (1) $\bigcup_{k=1}^{n_{\mathcal{R}}} R_k = X$ and (2) $R_i \cap R_j = \emptyset$ for $i \neq j$. With each range block are associated the following:

1. a *domain block* $D_k \subset X$ and a one-to-one contraction map $w_k : D_k \to R_k$ with a contraction factor $c_k \in [0,1)$.
2. an affine map $\phi : \mathbb{R} \to \mathbb{R}, \phi_k(t) = s_k t + o_k$, where $s_k, o_k \in \mathbb{R}$.

In the language of [13], the above ingredients comprise an (affine) $n_{\mathcal{R}}$-map Iterated Function System with Grey Level Maps (IFSM). The fractal transform operator $T : \mathcal{F}(X) \to \mathcal{F}(X)$ associated with such a (nonoverlapping) IFSM is defined as follows. Given a function $f \in \mathcal{F}(X)$ then for all $x \in R_k$, $k = 1,2,\ldots,N$,

$$\begin{aligned}(Tf)(x) &= \phi_k(f(w_k^{-1}(x)) \\ &= s_k f(w_k^{-1}(x)) + o_k. \end{aligned} \tag{1}$$

The maps w_k incorporate some form of *self-reference.* When considering the function values $f(x)$ as *luminance values* one can view the parameters s_k and o_k as control parameters for *contrast* and *brightness* (s stands for *scaling factor*, o for *offset*). They are also called *luminance parameters.* Figure 2 illustrates a fractal transform operator for the case of image coding.

It is well known that if $|s_k| < 1, 1 \le k \le n_{\mathcal{R}}$, then the operator T is contractive in the complete metric space of functions $\mathcal{L}^\infty(X)$. In the complete metric space of functions $\mathcal{L}^2(X)$, a straightforward calculation shows that

$$\|Tf_1 - Tf_2\|_2 \le C\|f_1 - f_2\|_2, \quad \forall f_1, f_2 \in \mathcal{L}^2(X),$$

where

$$C = \sum_{k=1}^{n_{\mathcal{R}}} c_k |s_k|. \tag{2}$$

Therefore, the condition $C < 1$ is sufficient (but not necessary) for contractivity of T in $\mathcal{L}^2(X)$. An example of the iterative application of an contractive fractal transform operator is given in Figure 3.

In the first part of this paper, we assume that one is given a target function, a range partition as well as a range-domain assignment. Then we examine a systematic method to perform attractor optimization using the partial derivatives of attractor functions with respect to the luminance parameters, $\partial \bar{f}_p / \partial s_k, \partial \bar{f}_p / \partial o_k, k = 1,2,\ldots,n_{\mathcal{R}}$. To this end, we use $\mathcal{F}(X) = \mathcal{L}^2(X)$, the

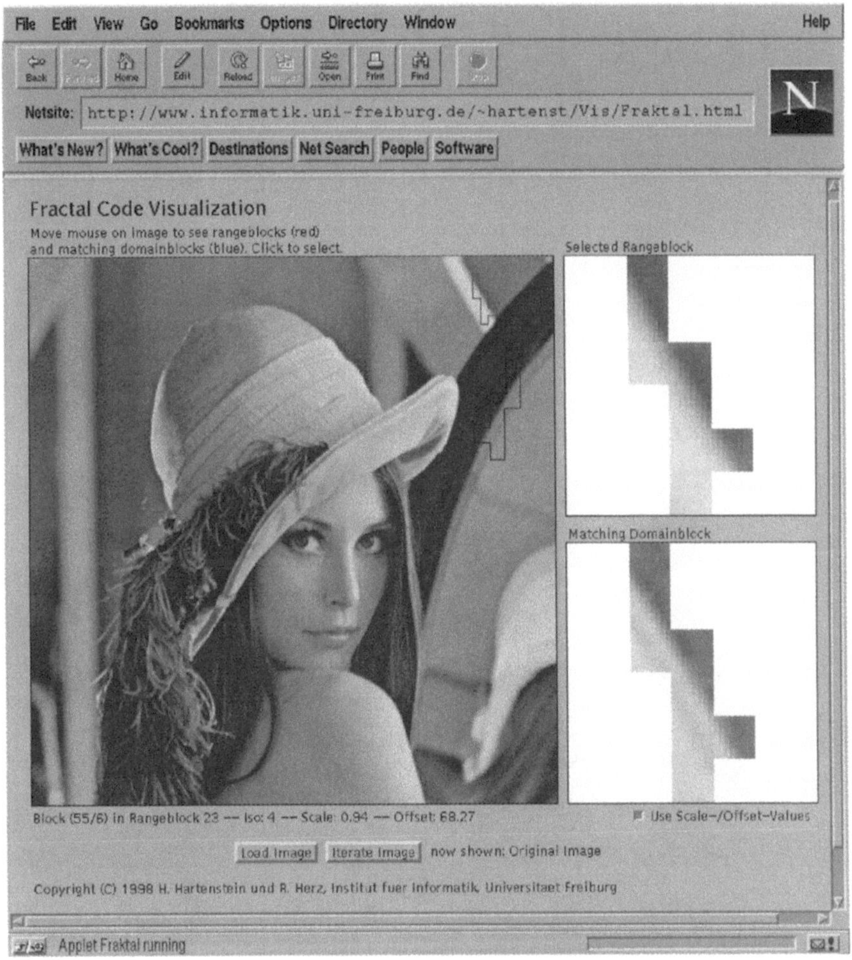

Fig. 2. A screenshot of the *Fractal Code Visualizer* (a Java applet) that is available from `http://www.informatik.uni-leipzig.de/cgip/`. As input the visualizer takes an original image and a fractal code thereof. By moving the mouse pointer over the image, the borders of the range to which one is pointing are drawn as well as the borders of the corresponding domain. To the right the selected range and domain are depicted; here, the (reflected) domain is viewed with the corresponding luminance transformation ϕ applied.

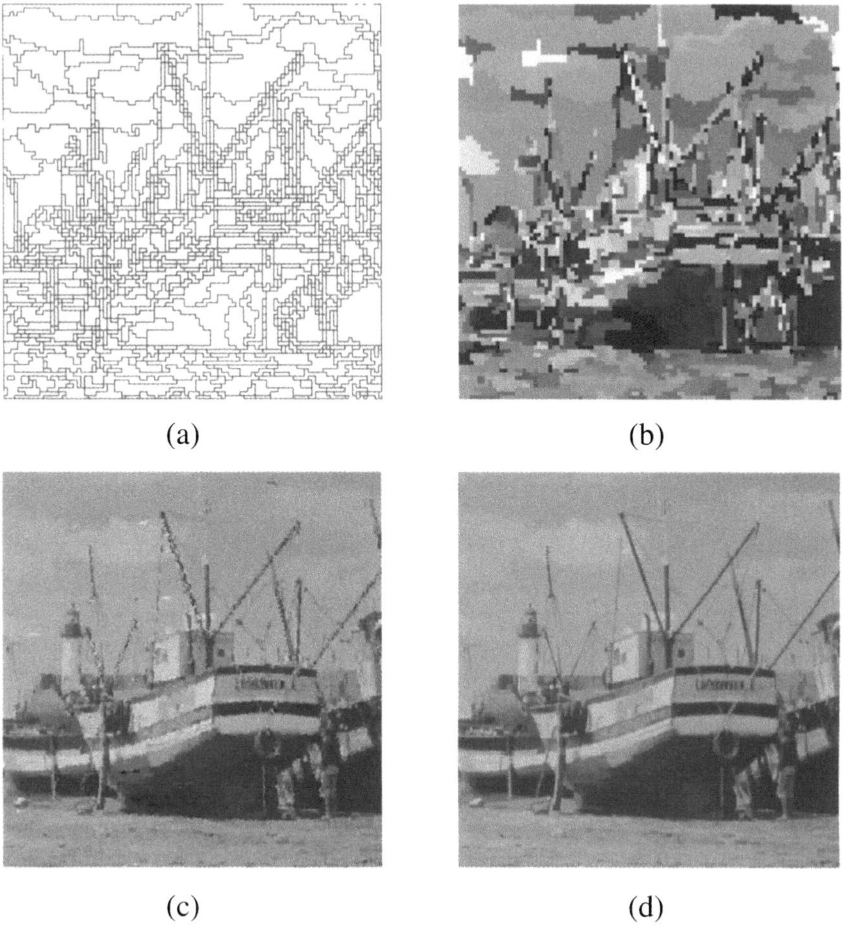

Fig. 3. Decoding of a fractal code for the standard test image 'Boat'. (a) Range partition, (b) first iteration, i.e., the operator applied to an all-black image, (c) third iteration, (d) 10th iteration.

space of square integrable functions on X with the usual metric, and set the parameter space to

$$\mathcal{P}^0 = \{p = ((s_1, o_1), \ldots, (s_{n_\mathcal{R}}, o_{n_\mathcal{R}}))|s_i, o_i \in \mathbb{R}, 1 \leq i \leq n_\mathcal{R}, \\ \text{s.t. } T_p \text{ is contractive in } \mathcal{L}^2(X)\}.$$

We first establish the existence of these derivatives and show that they are attractor functions of "vector fractal transform" operators (in the sense of hierarchical IFS [25, Chapter 5]). A knowledge of these derivatives permits the computation of the gradient vector of the error function $d_{\mathcal{F}}(f, \bar{f}_p)$ which, in turn, allows the use of gradient descent algorithms.

For the second part, we assume that one is given a target function and a range partition but no range-domain assignments. We now view the problem of optimal fractal coding as a combinatorial optimization problem, i.e., we model the space of feasible fractal codes via

$$\mathcal{P}^1 = \{p = ((z_1, s_1, o_1), \ldots, (z_{n_\mathcal{R}}, s_{n_\mathcal{R}}, o_{n_\mathcal{R}})) | 1 \leq z_i \leq n_D, \\ s_i \in Q(s), o_i \in Q(o), 1 \leq i \leq n_\mathcal{R}\}.$$

Here, each z_i represents an *address* for a domain block and $n_\mathcal{D}$ gives the (finite) number of domain choices per range. The sets $Q(s)$ and $Q(o)$ represent finite sets of feasible values for the scaling parameters and offsets, respectively. For practical applications one can assume that one is acting on a function space that is a finite-dimensional vector space. Thus, in order to guarantee convergence of the sequence of iterates $T^j f$, the constraint $|s_k| < 1, 1 \leq k \leq n_\mathcal{R}$ can be employed. We will show that the problem of determining in the parameter space a fractal code whose corresponding fractal transform operator gives the minimal attractor error is NP-hard, and, therefore, it cannot be solved in a practical amount of time.

3 Direct Attractor Optimization Based on Gradient Methods

3.1 Partial Derivatives of IFSM Attractor Functions with Respect to Luminance Parameters

Let us assume that the range partition and the range-domain assignments are given, i.e., the IFS maps $w_i, 1 \leq i \leq n_\mathcal{R}$, are fixed. Thus, the corresponding fractal transform operators T that are contractive in the space $\mathcal{F}(X) = \mathcal{L}^2(X)$ are parameterized via $\mathcal{P}^0$. We now consider the corresponding attractor functions $\bar{f}_p$ as functions not only of position but also of the luminance parameters p, i.e., we will write $\bar{f}(x, p)$ instead of $\bar{f}_p$.

Then, from Eq. (1) and with $p = (s_1, o_1, \ldots, s_{n_\mathcal{R}}, o_{n_\mathcal{R}}) \in \mathcal{P}^0$,

$$\bar{f}(x, p) = s_k \bar{f}(w_k^{-1}(x), p) + o_k, \quad x \in R_k. \tag{3}$$

Proposition 1. *The attractor $\bar{f}$ is continuous w. r. t. the fractal parameters s_l, o_l, $l = 1, 2, \ldots, n_\mathcal{R}$.*

The continuity of IFSM attractors with respect to grey level maps ϕ_l was proved in [12], using the methods described in [6]. It is straightforward to establish the continuity in terms of the luminance parameters s_l and o_l.

Proposition 2. *The set $\mathcal{P}^0$ is open.*

Proof. We prove that $\overline{\mathcal{P}^0} = \mathbb{R}^{2n_\mathcal{R}} - \mathcal{P}^0$ is closed. Let $p^{(n)} \in \overline{\mathcal{P}}$, $n = 1, 2, \ldots$, be a convergent sequence (in the topology of $\mathbb{R}^{2n_\mathcal{R}}$) with limit p. Each (infeasible) fractal code vector $p^{(n)} \in \overline{\mathcal{P}^0}$ defines a noncontractive fractal transform operator $T^{(n)} : \mathcal{L}^2(X) \to \mathcal{L}^2(X)$ with associated factor (cf. Eq. (2)) $C^{(n)} = \sum_{k=1}^{n_\mathcal{R}} c_k |s_k^{(n)}|$. Now, for each operator $T^{(n)}$, define its "optimal" Lipschitz factor as follows,

$$L^{(n)} = \sup_{y_1 \neq y_2} \frac{\| T^{(n)} y_1 - T^{(n)} y_2 \|_2}{\| y_1 - y_2 \|_2}. \tag{4}$$

From this definition and the noncontractivity of the $T^{(n)}$, it follows that $1 \leq L^{(n)} \leq C^{(n)}$ for all n. From the convergence of the code vectors $p^{(n)}$, it also follows that $\lim_{n\to\infty} C^{(n)} = C \geq 1$. So, from Proposition 1, the fractal transform T defined by the limit code vector p has associated factor C and Lipschitz factor $L \geq 1$. Therefore T is not contractive, implying that $p \notin \mathcal{P}^0$. Thus $\overline{\mathcal{P}^0}$ is closed, proving the proposition. □

Theorem 3. *The partial derivatives of the attractor $\bar{f}$ with respect to the fractal parameters s_l, o_l, $l = 1, 2, \ldots, n_\mathcal{R}$, exist at any point $p \in \mathcal{P}^0$.*

In the proof we need a special type of IFSM/fractal transform that involves "condensation" [21]. For a function $f \in \mathcal{F}(X)$, define Tf as follows: For all $x \in R_k$, $k = 1, 2, \ldots, n_\mathcal{R}$,

$$(Tf)(x) = s_k f(w_k^{-1}(x)) + \theta_k(x). \tag{5}$$

The functions $\theta_k(x)$ are known as *condensation* functions. Note that condensation functions do *not* affect the contractivity of T. The following result, which establishes the continuity of attractor functions with respect to condensation functions, is a simple consequence of Proposition 1.

Proposition 4. *Let T_1 and T_2 be contractive $n_\mathcal{R}$-map IFSM operators as in Eq. 5, with condensation functions $\theta^{(1)}(x)$ and $\theta^{(2)}(x)$, respectively, and identical scaling parameters s_k. Let $\bar{f}^{(1)}$ and $\bar{f}^{(2)}$, respectively, denote the fixed points of these operators. Then given an $\epsilon > 0$, there exists a $\delta > 0$ such that $\| \theta^{(1)} - \theta^{(2)} \|_2 < \delta$ implies that $\| \bar{f}^{(1)} - \bar{f}^{(2)} \|_2 < \epsilon$.*

Proof of Theorem 3: For any $p \in \mathcal{P}^0$, the associated fractal transform T is contractive. This implies that for any $f^{(0)} \in \mathcal{L}^2$, the sequence of functions defined by $f^{(n+1)} = Tf^{(n)}$ converges to $\bar{f}$, that is, $\| f^{(n)} - \bar{f} \|_2 \to 0$ as $n \to \infty$. Let $f^{(0)} = \theta$, where

$$\theta(x) = \sum_{k=1}^{n_\mathcal{R}} o_k I_{R_k}(x) \tag{6}$$

and $I_S(x)$ is the characteristic function of a subset $S \subset X$. Then, for $M \geq 0$, $f^{(M)} = T^{\circ M} f^{(0)}$ is given by

$$f^{(M)}(x, p) = \theta(x) + \sum_{n=1}^{M} \sum_{i_1, \ldots, i_n = 1}^{n_\mathcal{R}} s_{i_1} \cdots s_{i_n} \theta(w_{i_n}^{-1} \circ \cdots \circ w_{i_1}^{-1}(x)), \tag{7}$$

with the standard convention that $\theta(w_{i_n}^{-1} \circ \cdots \circ w_{i_1}^{-1}(x))$ equals zero if $w_{i_n}^{-1} \circ \cdots \circ w_{i_1}^{-1}(x)$ does not exist. The $f^{(M)}$ are partial sums of an infinite series that converge, in the $\mathcal{L}^2$ metric, to $\bar{f}$. Thus we can write

$$\bar{f}(x,p) = \theta(x) + \sum_{n=1}^{\infty} \sum_{i_1,\ldots,i_n=1}^{n_{\mathcal{R}}} s_{i_1} \cdots s_{i_n} \theta(w_{i_n}^{-1} \circ \cdots w_{i_1}^{-1}(x)), \tag{8}$$

where the equation is understood in the $\mathcal{L}^2$ sense.

Now consider an $x \in R_k$ for some $k \in \{1, 2, \ldots, n_{\mathcal{R}}\}$. Then the index i_1 in Eq. (7) must equal k (in order for $w_{i_1}^{-1}(x)$ to be defined). Therefore, Eq. (7) becomes

$$f^{(M)}(x,p) = \theta(x) + s_k f^{(M-1)}(w_k^{-1}(x), p), \qquad x \in R_k. \tag{9}$$

For a given $l \in \{1, 2, \ldots, n_{\mathcal{R}}\}$, we partially differentiate the terms in this equation with respect to s_l:

$$\frac{\partial f^{(M)}}{\partial s_l}(x,p) = s_k \left[\frac{\partial f^{(M-1)}}{\partial s_l}(w_k^{-1}(x), p)\right] + [f^{(M-1)}(w_k^{-1}(x), p)]\delta_{kl}. \tag{10}$$

Define the following $n_{\mathcal{R}}$-map IFSM operator T_l with condensation:

$$(T_l f)(x) = s_k f(w_k^{-1}(x)) + \xi_k(x), \qquad x \in R_k, \quad 1 \leq k \leq n_{\mathcal{R}}, \tag{11}$$

where $\xi_k(x) = [\bar{f}(w_k^{-1}(x))]\delta_{kl}$ with $\delta_{kl} = 1$ if $k = l$ and zero otherwise. Since T is contractive, it follows that T_l is contractive in $\mathcal{L}^2$. (T and T_l have identical IFS maps and fractal parameters s_k.) Let $\bar{v}_l$ denote the fixed point of T_l. From Propositions 1 and 2, $\bar{v}_l$ is continuous with respect to the parameters s_k, in particular, s_l. We now show that $\bar{v}_l = \partial\bar{f}/\partial s_l$. (In what follows, for simplicity of notation, only x and s_l will be written explicitly in the list of independent variables.)

Note that Eq. (10) does not correspond to a single IFSM operator with condensation. However, since the functions $f^{(M)}$ converge to $\bar{f}$, it follows, from Proposition 4, that the sequence of functions $\partial f^{(M)}/\partial s_l$ converges to $\bar{v}_l$. That is, for a given $p \in \mathcal{P}^0$ and $\epsilon_1 > 0$, there exists an $M_1 > 0$ such that

$$\left\|\frac{\partial f^{(M)}}{\partial s_l}(x, s_l) - \bar{v}(x, s_l)\right\|_2 < \epsilon_1, \qquad \forall M > M_1. \tag{12}$$

It is convenient to denote our reference point as

$$p^0 = (s_1^0, \ldots, s_{n_{\mathcal{R}}}^0, o_1^0, \ldots, o_{n_{\mathcal{R}}}^0) \in \mathcal{P}^0.$$

Let $N_l(\delta)$, $\delta > 0$, be a restricted neighborhood of the point p^0 in which only the element s_l is allowed to vary, i.e., $s_l \in I_\delta = [s_l^0 - \delta, s_l^0 + \delta]$, such that the corresponding vectors p lie in $\mathcal{P}^0$. (The existence of such a neighborhood is guaranteed since $\mathcal{P}^0$ is open.) Let $h \in \mathbb{R}$, with $|h| < \delta$. Then for each $x \in X$

there exists, by the Mean Value Theorem, a $\gamma^{(M)} \in I_h = [s_l^0 - h, s_l^0 + h]$, such that

$$f^{(M)}(x, s_l^0 + h) - f^{(M)}(x, s_l^0) = \frac{\partial f^{(M)}}{\partial s_l}(x, \gamma^{(M)})h. \tag{13}$$

Therefore,

$$\begin{aligned}
\| f^{(M)}(x, s_l^0 + h) &- f^{(M)}(x, s_l^0) - h\bar{v}(x, s_l^0) \|_2 \\
&= h \left\| \frac{\partial f^{(M)}}{\partial s_l}(x, \gamma^{(M)}) - \bar{v}(x, s_l^0) \right\|_2 \\
&\leq h \left\| \frac{\partial f^{(M)}}{\partial s_l}(x, \gamma^{(M)}) - \bar{v}(x, \gamma^{(M)}) \right\|_2 \\
&\qquad + h \, \| \bar{v}(x, \gamma^{(M)}) - \bar{v}(x, s_l^0) \|_2 \\
&\leq h \left\| \frac{\partial f^{(M)}}{\partial s_l}(x, \gamma^{(M)}) - \bar{v}(x, \gamma^{(M)}) \right\|_2 \\
&\qquad + \max_{s_l \in I_h} h \, \| \bar{v}(x, s_l) - \bar{v}(x, s_l^0) \|_2 \,.
\end{aligned} \tag{14}$$

Since I_δ is closed, there exists an $\overline{M} > 0$ such that the inequality in (12) is satisfied for all $M > \overline{M}$ at all $p \in N_l(\delta)$. Therefore, for a fixed $h \in (-\delta, \delta)$, we may take the limit $M \to \infty$ on both sides of (14) to yield

$$\left\| \frac{\bar{f}(x, s_l^0 + h) - \bar{f}(x, s_l^0)}{h} - \bar{v}(x, s_l^0) \right\|_2 \leq \max_{s_l \in I_h} \| \bar{v}(x, s_l) - \bar{v}(x, s_l^0) \|_2 \,. \tag{15}$$

Since $\bar{v}$ is continuous with respect to s_l, the right side term may be made arbitrarily small by choosing h sufficiently small, thus establishing the differentiability of $\bar{f}$ with respect to s_l at p^0.

The differentiability of $\bar{f}$ with respect to the o_l may be derived in a similar fashion. □

Remark: From Eq. (10) (and its analogue for differentiation with respect to o_l), the partial derivatives of $\bar{f}$ with respect to the fractal parameters s_l and o_l may be obtained by formally differentiating both sides of Eq. (3). For a fixed $x \in R_k$:

$$\frac{\partial \bar{f}}{\partial s_l}(x, p) = s_k \left[\frac{\partial \bar{f}}{\partial s_l}(w_k^{-1}(x), p) \right] + [\bar{f}(w_k^{-1}(x), p)]\delta_{kl}, \tag{16}$$

$$\frac{\partial \bar{f}}{\partial o_l}(x, p) = s_k \left[\frac{\partial \bar{f}}{o_l}(w_k^{-1}(x), p) \right] + \delta_{kl}. \tag{17}$$

Eqs. (3), (16) and (17) may be considered to define a $(2n_{\mathcal{R}}+1)$-component "vector IFSM with condensation" that may be written in the following compact form:

$$\bar{\mathbf{f}} = \mathbf{T}\bar{\mathbf{f}}, \tag{18}$$

where

$$\bar{\mathbf{f}}(\mathbf{x}, \mathbf{p}) = \left[\bar{f}(x,p), \frac{\partial \bar{f}}{\partial p_1}(x,p), \ldots, \frac{\partial \bar{f}}{\partial p_{2n_{\mathcal{R}}}}(x,p) \right]^t . \tag{19}$$

Now define the space $\mathcal{F}^{2n_{\mathcal{R}}+1}(X) = \{\mathbf{f} = (f_1, f_2, \ldots, f_{2n_{\mathcal{R}}+1}) \mid f_j \in \mathcal{F}(X)\}$ with associated metric $d_{\mathcal{F}^{2n_{\mathcal{R}}+1}}(\mathbf{f}, \mathbf{g}) = \max_{1 \leq j \leq 2n_{\mathcal{R}}+1} d_{\mathcal{F}}(f_j, g_j)$. Then $\mathbf{T} : \mathcal{F}^{2n_{\mathcal{R}}+1}(X) \to \mathcal{F}^{2n_{\mathcal{R}}+1}(X)$. For an $f \in \mathcal{F}^{2n_{\mathcal{R}}+1}(X)$,

$$(\mathbf{Tf})(x) = s_k \mathbf{f}(w_k^{-1}(x)) + \mathbf{e}_k \cdot \mathbf{f}(w_i^{-1}(x)) + \Theta_k(x), \quad x \in R_k. \tag{20}$$

The vector $[\mathbf{e}_k]^t = (0, 0, \ldots, 1, \ldots, 0)$, where the "1" occurs in the $(k+1)$st entry, represents the only "mixing" of components of $\mathbf{f}$ under the action of $\mathbf{T}$. The function $\Theta_k(x)$ represents a condensation vector composed of constant functions: $[\Theta_k(x)]^t = (o_k, 0, 0, \ldots, 1, \ldots, 0)$, where the "1" occurs in the $(n_{\mathcal{R}} + 1 + k)$th entry.

Proposition 5. *Suppose that T is contractive in $(\mathcal{F}(X), d_{\mathcal{F}})$. Then $\mathbf{T}$ is contractive in $\mathcal{F}^{2n_{\mathcal{R}}+1}(X)$. Its fixed point $\bar{\mathbf{f}}$ is given by Eq. (19), where $\bar{f}$ is the fixed point of T, see Eq. (3).*

From Banach's Fixed Point Theorem, contractivity of T allows the computation of its fixed point function $\bar{f}$ by means of iteration. The above proposition implies that all partial derivatives $\partial \bar{f} / \partial p_l$ may also be computed by iteration: Begin with a "seed" $\mathbf{f}^{(0)} \in \mathcal{F}^{2n_{\mathcal{R}}+1}(X)$ and construct the sequence of vector functions $\mathbf{f}^{(n+1)} = \mathbf{Tf}^{(n)}$, $n \geq 0$. The calculations are very complex: Except in special cases, $\bar{f}$ and its partial derivatives will have to be computed for all $x \in X$. This will be discussed in more detail below.

3.2 Experimental Image Coding Results

Let $f \in \mathcal{L}^2(X)$ again denote the target function we seek to approximate. For a given fractal code $p \in \mathcal{P}^0$ we will consider the squared $\mathcal{L}^2$ error

$$E(p) = \| f - \bar{f}_p \|_2^2 .$$

We now employ the attractor $\bar{f}_p$, in particular the attractor $\bar{f}_{p^c}$ where p^c again denotes a collage error optimal fractal code, as a starting point and vary the fractal code parameters p in an attempt to decrease the error function $E(p)$ as much as possible. This was also the strategy of Dudbridge and Fisher [9], who employed the Nelder-Mead simplex algorithm. In their scheme, the error function $E(p)$ is computed at strategic points.

A knowledge of the partial derivatives of $\bar{f}_p$ with respect to the fractal parameters p permits the computation of elements of the gradient vector of E:

$$\frac{\partial E}{\partial p_l}(p) = -2 \left\| f - \bar{f}_p - \frac{\partial \bar{f}_p}{\partial p_l} \right\|_2^2 , \quad l = 1, 2, \ldots, 2n_{\mathcal{R}}. \tag{21}$$

This allows us to employ gradient-descent and related methods to search for local minima.

Practically speaking, however, the partial derivatives $\partial\bar{f}/\partial p_l(x,p)$ must be computed at all points (pixels) $x \in X$. In addition to an $n \times n$ matrix required to store an image, an additional $2n_{\mathcal{R}}$ $n \times n$ matrices are needed, in general, to store the derivatives at all pixels. Borrowing from the terminology of quantum chemists, this "full configuration interaction" will compute the total rate of change of the attractor — hence the approximation error — with respect to changes in all fractal parameters p_l *for a fixed set of domain-range pair assignments*. When applying a gradient descent method to minimize the error function E less storage is required. It suffices to provide one additional $n \times n$ matrix to sequentially compute each component of the gradient $(\partial E/\partial p_1, \dots, \partial E/\partial p_{2n_{\mathcal{R}}})$.

We apply our method to the fractal transform scheme examined by Dudbridge and Fisher [9], designed to minimize the interdependency of range blocks. The following four 512×512 pixel images (8 bpp), used in [9], were also used in this study: *Lena*, *Boat*, *Mandrill* and *Peppers*.[1] Each image was partitioned into 4×4 pixel range blocks, with four neighboring range blocks sharing a common 8×8 pixel domain block, namely the one that consists of the four ranges. Therefore, for each image, the inverse problem separates into 64^2 independent problems, each involving an 8×8 pixel image with four range blocks R_k, hence 8 fractal parameters (four scaling and four offset values).

As in [9], for each test image we first used collage coding to determine a fractal code p^c that minimizes the collage error. We then used this code as a starting point for a gradient-descent method. The NAG [23] subroutine E04DKF, which performs a quasi-Newton conjugate gradient minimization, was used. It was also desirable to compare these results with the non-gradient calculations of [9]. However, since some of our collage error results differed from those of [9], we have independently carried out attractor optimization using the Nelder-Mead simplex algorithm. The NAG subroutine E04CCF was used.

In all cases, the simplex and gradient methods yielded almost identical improvements. A comparison with [9] reveals some nonnegligible differences, not only in the collage errors but also in the improvements obtained by the simplex method. In all cases, we improved on the results of [9]. In both the simplex as well as the gradient algorithms, the results are quite sensitive to the settings of the tolerance/accuracy parameters as well as the maximum number of iterations (*maxiter*) allowed. Generally the best performance was obtained when the tolerance parameters for the simplex and gradient subroutines were set to 10^{-5} and 10^{-6}, respectively. The parameter *maxiter* was set to 2000, which is virtually infinity.

---

[1] These 512×512 images may be retrieved by anonymous ftp from the Waterloo Fractal Compression Project site `links.uwaterloo.ca` in the appropriate subdirectories located in `ftp/pub/BragZone/GreySet2`.

In Table 1 we present the *peak-signal-to-noise-ratio* (PSNR) values associated with collage coding and subsequent simplex and gradient optimized attractor coding, along with the improvements in PSNR. The numbers in brackets represent the CPU time required for each calculation. (We emphasize that these numbers are presented for the purpose of comparison, since the computer codes themselves are not optimized.)

Table 1. Results of (a) collage coding and attractor optimization using (b) simplex and (c) gradient methods, the latter two using collage coding as a starting point. All results are expressed in PSNR (dB). The final two columns list the improvement in PSNR achieved by the simplex method obtained in this study and Ref. [9], respectively.

	Collage attractor	Attractor optimization		ΔPSNR	ΔPSNR [9]
		Simplex	Gradient		
Lena	29.25	29.87 (301)	29.87 (229)	0.62	0.35
Boat	26.66	27.42 (300)	27.42 (299)	0.56	0.41
Mandrill	21.52	22.11 (532)	22.08 (1500)	0.59	0.33
Peppers	29.34	30.02 (277)	29.94 (591)	0.68	0.33

In an attempt to understand how good the initial estimate provided by collage coding actually is, we have performed simplex and gradient optimization calculations for another set of initial conditions, namely, piecewise constant approximations to the images. In this case, all s_l are initially set to zero and the o_l are simply the mean values of the range block. (Of course, in more general problems than the one studied here, there would remain the problem of assigning a domain block to each range block.) In Table 2, we present the results of these calculations. The first column gives the error associated with the initial piecewise constant approximation. The next two columns list the PSNR values of the optimized attractors obtained from the simplex and gradient methods along with the CPU times. The final column gives the PSNR improvement yielded by the better of the two methods.

Table 2. Results of (a) piecewise constant approximation (PCA) and attractor optimization using (b) simplex and (c) gradient methods, the latter two using the PCA as a starting point. All results are expressed in PSNR (dB). The final column lists the improvement in PSNR achieved by the better of methods (b) and (c).

	PCA	Simplex	Gradient	ΔPSNR
Lena	26.93	29.73 (421)	29.74 (288)	2.81
Boat	25.08	27.30 (452)	27.32 (618)	2.24
Mandrill	20.85	22.00 (663)	21.97 (3333)	1.15
Peppers	25.97	29.76 (420)	29.56 (2888)	1.79

We observe that the simplex and gradient methods, using such suboptimal initial conditions, i.e., piecewise constant approximations, yield approximations that are almost as good as those found from collage attractors. The worst case is *Peppers*, for which a 0.26 dB difference is found. For the others, the discrepancy is on the order of 0.1 dB.

Results of the gradient descent algorithms applied to fractal image encodings based on quadtree partitions can be found in our paper [28]. In these quadtree experiments we used the conjugate gradient algorithm from [26]. The major computational burden is the computation of the gradients required in each step, which allowed us to do experiments only with images of size 256 × 256. The gain obtained by the gradient descent method varied between 0.16 and 0.25 dB PSNR. However, the necessary quantization destroyed a large part of these gains. Thus, the achievable gains for fractal coding with the quadtree method are negligible.

4 On the Computational Complexity of Optimal Fractal Coding

In this section we will analyze the inverse problem of fractal coding from the computational complexity point of view,[2] i.e., we will consider optimal fractal coding as a discrete optimization problem. Thus, the support X is now given by $\{1, \ldots, n\}$,[3] and the space of functions $\mathcal{F}(X)$ equals $\mathbb{R}^n$. Instead of directly defining fractal transform operators acting on functions $f \in \mathbb{R}^n$, we simply interpret a function $f \in \mathbb{R}^n$ as a function on $[0, 1)$ that is constant on each interval of $\mathcal{I} = \{[\frac{j}{n}, \frac{j+1}{n}) | 0 \leq j < n\}$. We will make the following assumptions that will allow us to easily translate 'back and forth' between discrete and continuous settings:

- Each range is a (connected) union of elements of $\mathcal{I}$.
- The affine mappings w_k^{-1} are of the form $w_k^{-1}(x) = \left(2x + \frac{j}{n}\right) \bmod 1, j \in \mathbb{Z}$; thus, the contraction factor of the mappings $w_k, 1 \leq k \leq n_{\mathcal{R}}$, is fixed to 0.5, and each domain is a union of elements of $\mathcal{I}$.

A fractal transform operator whose action is defined for $x \in I \subset R_k$, $I \in \mathcal{I}$, by

$$(Tf)(x) = s_k \cdot n \int_I f(w_k^{-1}(u))du + o_k \tag{22}$$

will again output a function that is constant on each interval of $\mathcal{I}$. Thus, an operator satisfying the above conditions and (22) can be regarded as a fractal operator acting on $\mathbb{R}^n$. Its basic difference to the original definition is the averaging over neighboring samples. Therefore, we will not distinguish

[2] For an introduction into the topic of computational complexity see, e.g., [14,24].

[3] For simplicity we restrict ourselves to the one-dimensional case. It is straightforward to extend all results and discussions to higher dimensions.

between the above operator and a 'truly' discrete operator, and write Tf also for functions $f \in \mathbb{R}^n$. Using the translation mechanism between piecewise constant functions and discrete functions we can now use the terms range, domain etc. also for the discrete case.

For the analysis of the computational complexity of optimal fractal coding we assume for simplicity that one is given a function $f \in \mathbb{R}^{m \cdot n_{\mathcal{R}}}$ that is uniformly partitioned into $n_{\mathcal{R}}$ ranges with m components each, i.e., $R_i = \{im, \ldots, (i+1)m - 1\}, 1 \leq i \leq n_{\mathcal{R}}$. The domains are non-overlapping and have twice the size of the ranges, i.e., the domains are given by $D_j = \{j \cdot 2m, \ldots, (j+1) \cdot 2m - 1\}, 1 \leq j \leq n_{\mathcal{D}} = \lfloor \frac{n_{\mathcal{R}}}{2} \rfloor$. We require the scaling parameters to have an absolute value smaller than 1 in order to guarantee convergence in the decoding. Thus, the set of feasible fractal codes for function f is given by

$$\mathcal{P}^1_{n_{\mathcal{R}}} = \{p = ((z_1, s_1, o_1), \ldots, (z_{n_{\mathcal{R}}}, s_{n_{\mathcal{R}}}, o_{n_{\mathcal{R}}}))| 1 \leq z_i \leq n_D, \\ s_i \in Q(s), o_i \in Q(o), 1 \leq i \leq n_{\mathcal{R}}\},$$

where $Q(s)$ and $Q(o)$ are finite sets of real values, and $|s| < 1$ for $s \in Q(s)$. The number of fractal codes in $\mathcal{P}^1_{n_{\mathcal{R}}}$ with different range-domain assignments is $(n_{\mathcal{D}})^{n_{\mathcal{R}}} = \lfloor \frac{n_{\mathcal{R}}}{2} \rfloor^{n_{\mathcal{R}}}$, since for each range one of the $n_{\mathcal{D}}$ domains is chosen. Thus, the number of feasible fractal codes grows exponentially with the number of ranges.

A fractal code $p^* \in \mathcal{P}^1_{n_{\mathcal{R}}}$ is called an optimal fractal code for function f (uniformly partitioned into $n_{\mathcal{R}}$ ranges) if

$$\|f - \bar{f}_{p^*}\|_2^2 \leq \min_{p \in \mathcal{P}^1_{n_{\mathcal{R}}}} \|f - \bar{f}_p\|_2^2,$$

where $\bar{f}_p$ denotes again the attractor corresponding to the fractal code p.

Let us now formally define FRACCODE as the decision problem associated with the problem of optimal fractal coding.

FRACCODE
INSTANCE: Function $f \in \mathbb{Z}^n$ uniformly partitioned into $n_{\mathcal{R}}$ ranges with m components each, quantization levels $Q(s), Q(o)$, positive number Δ.
QUESTION: Is there an element p in $\mathcal{P}^1_{n_{\mathcal{R}}}$ (as defined above) whose attractor $\bar{f}_p$ satisfies $\|f - \bar{f}_p\|_2^2 \leq \Delta$?

We will now prove that FRACCODE represents an NP-hard problem, thus, optimal fractal coding is NP-hard. Particularly, we will show that solving the FRACCODE problem is at least as hard as solving an instance of (unweighted) MAXCUT, i.e., we will give a polynomial transformation from MAXCUT to FRACCODE. The MAXCUT (decision) problem is defined as follows:

MAXCUT
INSTANCE: Undirected graph $\mathcal{G} = (\mathcal{V}, \mathcal{E})$ with $n_{\mathcal{V}}$ vertices and $n_{\mathcal{E}}$ edges, positive integer k.

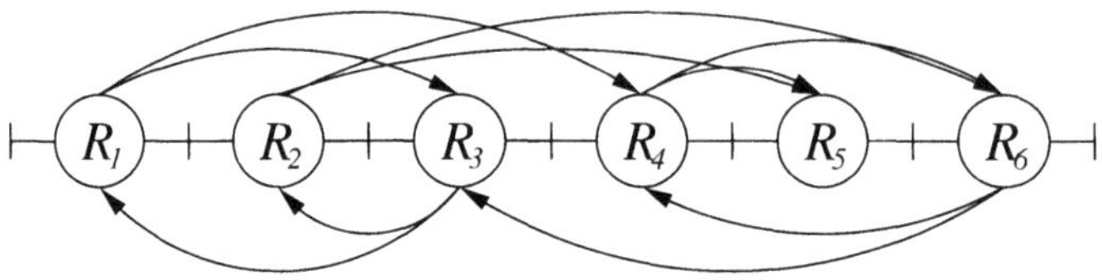

Fig. 4. Example of a dependency graph. Here, range R_5 is simply coded by an offset value.

QUESTION: Is there a partition of $\mathcal{V}$ into disjoint sets $\mathcal{V}_1$ and $\mathcal{V}_2$ such that the number of edges that have one endpoint in $\mathcal{V}_1$ and one endpoint in $\mathcal{V}_2$ is at least k?

Since MAXCUT is known to be NP-hard (to be precise, it is NP-complete, cf. [14, problem ND16] or [7, problem ND11]), it then follows that FRACCODE is also NP-hard.

Before proceeding to the formal proof, let us first explain intuitively what makes the problem of optimal fractal coding a hard problem.

The reconstruction quality for a function f on range R_i depends on how well the function is reconstructed on the domain for that range. Therefore, it depends on the reconstruction quality of the function on the ranges that are contained in the domain for range R_i, and so on. Those dependencies can be represented using a *dependency graph* as proposed in [8]. The dependency graph of a fractal code consists of the set of ranges $\{R_1, \dots, R_{n_\mathcal{R}}\}$ as the set of vertices, and the set of edges is given by

$$\{(R_i, R_j) | R_j \text{ overlaps, fully or partially, with domain assigned to range } R_i\}.$$

An example is given in Figure 4. With the collage coding approach, each range is coded separately in a greedy fashion; the dependencies of the interference of the various range-domain maps are ignored by the collage coder which is the reason why collage coding is a suboptimal strategy. These dependencies are the reason why the determination of the optimal fractal code represents a computationally hard problem.

4.1 The Main Theorem

The reduction from MAXCUT will proceed as follows: given a graph $\mathcal{G} = (\mathcal{V}, \mathcal{E})$ with $n_\mathcal{V}$ vertices and $n_\mathcal{E}$ edges, we will construct in polynomial time a signal $f(\mathcal{G}) \in \mathbb{R}^n$ with $n_\mathcal{R}$ ranges, sets $Q(s), Q(o)$ and a function $\Delta(\mathcal{G}, k)$ monotonically decreasing in k, $k \in \mathbb{N}$, such that the following holds:

Theorem 6. *$\mathcal{G}$ has a cut of size $\geq k \iff \exists p \in \mathcal{P}^1_{n_\mathcal{R}}$ such that $\|f(\mathcal{G}) - \bar{f}_p\|_2^2 \leq \Delta(\mathcal{G}, k)$.*

Thus, the question whether there exists a cut of a given cardinality k is reduced to the question whether there is an attractor $\bar{f}_p$ that approximates the signal $f(\mathcal{G})$ with an error of at most $\Delta(\mathcal{G}, k)$. To prove Theorem 6 we will proceed in three steps. First, the construction of $f(\mathcal{G})$ and $\Delta(\mathcal{G}, k)$ will be given. From the construction the $\Rightarrow$-direction will follow immediately:

Lemma 7. *$\mathcal{G}$ has a cut of size $\geq k \Rightarrow \exists p \in \mathcal{P}^1_{n\mathcal{R}}$ such that $\|f(\mathcal{G}) - \bar{f}_p\|_2^2 = \Delta(\mathcal{G}, k)$.*

As the last step we show the $\Leftarrow$-direction of Theorem 6 which is equivalent to the statement

$$\begin{gathered}\mathcal{G} \textit{ has a maximal cut of size smaller than } k\\ \Rightarrow \nexists p \in \mathcal{P}^1_{n\mathcal{R}} \textit{ such that } \|f(\mathcal{G}) - \bar{f}_p\|_2^2 \leq \Delta(\mathcal{G}, k).\end{gathered}$$

This in turn is equivalent to the following lemma:

Lemma 8. *$\mathcal{G}$ has a maximal cut of size $k \Rightarrow \nexists p \in \mathcal{P}^1_{n\mathcal{R}}$ such that $\|f(\mathcal{G}) - \bar{f}_p\|_2^2 \leq \Delta(\mathcal{G}, k+1)$.*

In Subsection 4.2 we give the construction of $f(\mathcal{G})$. The function Δ is given together with Lemma 7 in Subsection 4.3. Lemma 8 is shown in Subsection 4.4. Note that, for simplicity, in the following we will also call $f \restriction R_i$ a *range* and $f \restriction D_i$ a *domain*.

4.2 Construction of $f(\mathcal{G})$

In order to satisfy Theorem 6 we have to construct a signal $f(\mathcal{G})$ such that the approximation error resulting from the optimal attractor indicates whether or not the graph $\mathcal{G}$ has a cut of size at least k. The signal $f(\mathcal{G})$ will consist of five segments $S_0, \ldots, S_4$ that are designed as follows.

First of all, we assign to each vertex $\mathfrak{v} \in \mathcal{V}$ of the graph $\mathcal{G}$ a distinct signal, the *vertex ID*. IDs pertaining to different vertices will differ significantly from each other. The segments $S_1, \ldots, S_4$ are constructed as follows:

- Signal segment S_1 contains for each vertex $\mathfrak{v} \in \mathcal{V}$ four ranges as shown in Figure 5 a). The first and the third range contain the vertex ID for $\mathfrak{v}$, the second and the fourth range contain signals that are complementary to each other. The two complementary signals are used as binary *flags* and are denoted by B_1 and B_2.
- Segment S_2 contains two ranges for each vertex $\mathfrak{v}$ (cf. Figure 5 b)). The first half of the first range is again the vertex ID of $\mathfrak{v}$, shrunk to half its width. The rest of the two ranges equals zero.
- In the third segment, S_3, for each edge $(\mathfrak{v}_i, \mathfrak{v}_j) \in \mathcal{E}$ we have the following two ranges (cf. Figure 5 c)): The first quarter of the first range is the appropriately shrunk vertex ID of $\mathfrak{v}_i$, the first quarter of the second range contains the vertex ID of $\mathfrak{v}_j$. The rest of the two ranges is zero.

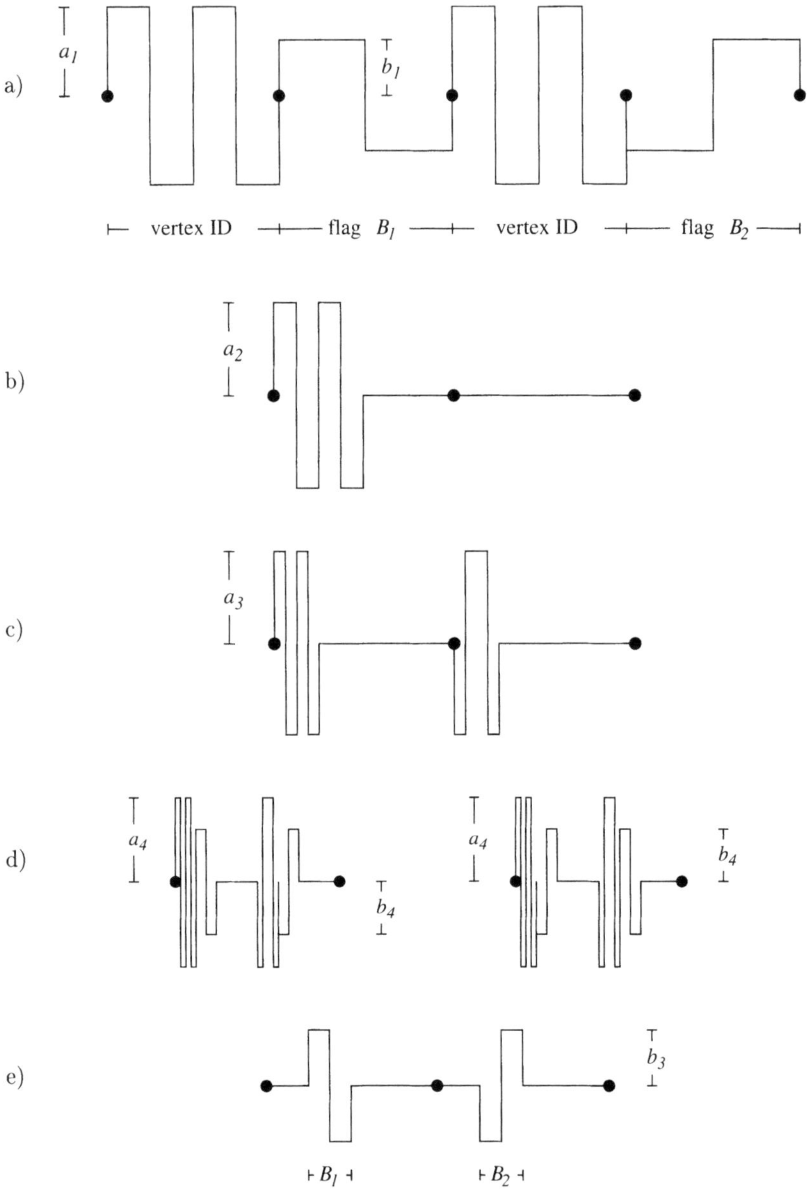

Fig. 5. Design of $f(\mathcal{G})$

- The fourth segment, S_4, contains two ranges for every edge in $(\mathfrak{v}_i, \mathfrak{v}_j) \in \mathcal{E}$ (cf. Figure 5 d)). Both ranges contain the vertex IDs of $\mathfrak{v}_i$ and $\mathfrak{v}_j$. Next to the vertex IDs copies of the flags are placed. In the first range, these are the flags B_1 and B_2, in the second range B_2 and B_1 (in this order).

The heights $a_1, \ldots, a_4$ of the signal are related by $a_2 = \sigma_1 \cdot a_1$, $a_3 = \sigma_2 \cdot a_2$, $a_4 = \sigma_3 \cdot a_3$. Furthermore, we set $b_i = \frac{a_i}{\sqrt{2}}$ for $i = 1, \ldots, 4$.[4] We set a_4 to some arbitrary, but fixed, constant. Thus, all parameters are completely determined by $a_4, \sigma_1, \sigma_2, \sigma_3$. The values of the σ_i will be determined in subsection 4.4. Note that due to this definition the signal does not necessarily consist of integer values. The assumption is that the parameters can be scaled by some sufficiently large factor and then rounded.

To motivate the construction, let us assume for the moment that the ranges of S_i have to be coded by domains from S_{i-1} for $i = 2, 3, 4$ and S_1 is given as side information. The vertex IDs will be designed in such a way that an ID mismatch will be *very* costly, i.e., when a fractal code assigns a domain to a range with a different vertex ID, this will result in a large reconstruction error for that range. Thus, for each range in S_2 the only 'possible' domains are the two domains with corresponding ID in S_1. Both contribute the same distortion in the attractor. Selecting one of them for each range corresponds to the partitioning of $\mathcal{V}$ into $\mathcal{V}_1$ and $\mathcal{V}_2$. The flag (B_1 or B_2, respectively) associated to the vertex $\mathfrak{v}$, therefore, indicates to which set of the partition $\mathfrak{v}$ belongs ($\mathcal{V}_1$ resp. $\mathcal{V}_2$). Again, each range of S_3 has to be coded by the domain of S_2 with the same vertex ID. In the attractor this third segment contains the information about which edges of the graph $\mathcal{G}$ belong to the cut. The segment S_4 will be used to *count* the number of these edges. An edge in the cut consists of a pair of vertices to which different flags (B_1 and B_2, or vice versa) have been assigned. In that case, we can find an exact match for one of the ranges in S_4 belonging to that edge. By doing so, the error of the attractor is coupled with the size of the cut.

In fact, the signal is hard to code since at segment S_2 it does not make any difference which of the two domains in S_1 with the same ID is chosen for each range, but the effect of the choice will affect the reconstruction error in segment S_4. The collage coder cannot decide which domain should be chosen because it does not take into account the implications of such a decision. Therefore, it simply uses some kind of tie breaking rule.

To make things explicit, we now give the remaining details of the construction of the signal $f(\mathcal{G})$. The IDs are built using the following lemma:

Lemma 9. *For each $\kappa \in \mathbb{N}$ there exists a binary code with κ codewords $\mathfrak{c}_1, ..., \mathfrak{c}_\kappa$, each of length $\ell = \mathcal{O}(\kappa)$, such that for $i \neq j$ the Hamming distances $d_{\mathcal{H}}(\mathfrak{c}_i, \mathfrak{c}_j)$ and $d_{\mathcal{H}}(\mathfrak{c}_i, \overline{\mathfrak{c}_j})$ equal $\ell/2$. $\overline{\mathfrak{c}_i}$ denotes the binary complement of $\mathfrak{c}_i$.*

[4] By following the proof backwards one can derive the feasible ratios between b_i and a_i; our choice facilitates calculations. More details are given in [17].

Proof. We will show by induction that the lemma holds for $\kappa = \ell = 2^q$, for all $q \in \mathbb{N}$. For all other κ simply choose κ of the codewords constructed for size $2^{\lceil \log_2 \kappa \rceil}$. To begin the induction, $\mathfrak{c}_1 = 0$ is such a code for $\kappa = 2^0$. For $\kappa = 2^{q+1}$ take the set $\{\mathfrak{c}_i\mathfrak{c}_i, \mathfrak{c}_i\overline{\mathfrak{c}_i} | 1 \leq i \leq 2^q\}$, where the $(\mathfrak{c}_i)_{1 \leq i \leq 2^q}$ form a code of the desired type of length 2^q. This gives a new binary code of size 2^{q+1} that is easily shown to have the desired property. □

Let $(\mathfrak{c}_i)_{1 \leq i \leq n_\mathcal{V}}$ be a binary code of $n_\mathcal{V}$ codewords of length ℓ constructed as in Lemma 9. From $(\mathfrak{c}_i)_{1 \leq i \leq n_\mathcal{V}}$ we build the binary code $\mathfrak{C}$ with codewords of length 2ℓ:

$$\mathfrak{C} := \{\mathfrak{c}_i\overline{\mathfrak{c}_i} | 1 \leq i \leq n_\mathcal{V}\}$$

$\mathfrak{C}$ has the property that two different codewords differ in half their bits and —as a consequence of Lemma 9— has the following features, which we will use in our calculations:

- Every codeword consists of ℓ 0s and ℓ 1s.
- For codewords $\mathfrak{c}_i\overline{\mathfrak{c}_i}, \mathfrak{c}_j\overline{\mathfrak{c}_j} \in \mathfrak{C}$ the following holds:
  - there are exactly $\frac{\ell}{2}$ positions where $\mathfrak{c}_i\overline{\mathfrak{c}_i}$ has a 0 and $\mathfrak{c}_j\overline{\mathfrak{c}_j}$ has a 1.
  - there are exactly $\frac{\ell}{2}$ positions where $\mathfrak{c}_i\overline{\mathfrak{c}_i}$ has a 1 and $\mathfrak{c}_j\overline{\mathfrak{c}_j}$ has a 0.

From the code $\mathfrak{C}$ we obtain the vertex IDs for segment S_1 as follows. Essentially, we interpret the 0s and 1s of the binary codewords as $-a_1$ and a_1. But in order to have unaliased geometrically shrunk versions for the vertex IDs in segments S_2, S_3, S_4, each value has to be repeated 8 times. Thus, the size m of a range has to be $m = 16\ell$. Therefore, the range size depends linearly on the number of vertices $n_\mathcal{V}$. We remark that the vertex IDs shown in Figures 5 and 7 are chosen for their simple shapes and are not constructed with the above approach. Also note that the vertex IDs have to be distinct from the binary flags. The above properties of the code $\mathfrak{C}$ guarantee that when for a range containing the vertex ID for $\mathfrak{v}_i$ a domain is assigned such that the vertex ID for $\mathfrak{v}_i$ is approximated by a vertex ID for $\mathfrak{v}_j, i \neq j$, this results in a large approximation error (cf. Figure 6). The proof in Subsection 4.4 depends heavily on this property.

In order to have all ingredients for coding segment S_1 without any distortion, we add a construction segment S_0 to the signal. For example, S_0 contains the signal parts that represent geometrically scaled copies of length $2m, 4m, \ldots, m^2$ of the ranges in S_1. Thus, we add construction segments $S_{0,0}, \ldots, S_{0,\log_2 m-1}$ where $S_{0,i}$ is built by repeating each component (sample) of S_1 $\frac{m}{2^i}$ times, $0 \leq i < \log_2 m$. We set $S_0 := S_{0,0} \ldots S_{0,\log_2 m-1}$. Clearly, the length of S_0 depends polynomially on the number of vertices $n_\mathcal{V}$.

For the edge counting to work we also need an extra block in S_3 of the shape sketched in Figure 5 e). Of course, this also leads to the addition of some construction blocks in segments S_0, S_1, S_2.

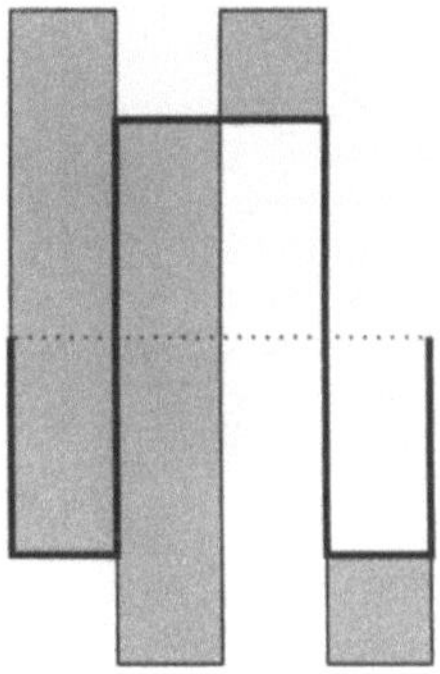

Fig. 6. Schematic representation of an ID mismatch; the grey-shaded area indicates the error.

4.3 Constructing an Attractor for $f(\mathcal{G})$

For the signal $f(\mathcal{G}) = S_0S_1S_2S_3S_4$ as described above we now give a transformation T_p that will later be shown to generate the optimal attractor. In the fractal code p a range in S_i is assigned a domain in $S_{i-1}, i = 1,\ldots,4$. We will be able to determine easily the attractor of T_p, since there will be no need for iterating the operator T_p, i.e., the dependency graph corresponding to T_p will not contain any cycles.

First of all, the segments S_0, S_1 can be coded without any distortion. By hypothesis, $\mathcal{G} = (\mathcal{V}, \mathcal{E})$ has a cut of cardinality k by partitioning $\mathcal{V}$ into $\mathcal{V}_1$ and $\mathcal{V}_2$. For a range in S_2 containing a (geometrically shrunk) vertex ID we choose the domain in S_1 with the same ID and the flag set in accordance to the graph partition. The scaling parameter and offset are set to $s = \frac{2}{3}\sigma_1$ and $o = 0$. In this way the maximal height of the attractor on S_2 is $\frac{2}{3}$ of the maximal height of the original signal on S_2 (cf. Figure 7(a)). On the first half of the range an error of $\frac{m}{2}(a_2 - \frac{2}{3}a_2)^2 = \frac{m}{18}a_2^2$ occurs, on the other half of the range the error is $\frac{m}{2}(\frac{2}{3}\frac{a_2}{\sqrt{2}})^2 = \frac{m}{9}a_2^2$. Therefore, on each range of segment S_2 that contains a vertex ID an error of $\frac{m}{6}a_2^2$ occurs. Thus, the total distortion of Ω_F in segment S_2 is

$$n_{\mathcal{V}} \cdot \frac{m}{6}a_2^2.$$

For each range in S_3 we choose the corresponding domain of S_2, i.e., the one with the same vertex ID, and scale it using $s = \sigma_2, o = 0$. The error introduced in segment S_3 then is

$$2n_{\mathcal{E}} \cdot \frac{m}{12}a_3^2.$$

The distortion in segment S_4 depends on the size of the cut k. For each edge there are two ranges in S_4 differing only in the flags. Depending on whether or not an edge belongs to the cut, we proceed as follows:

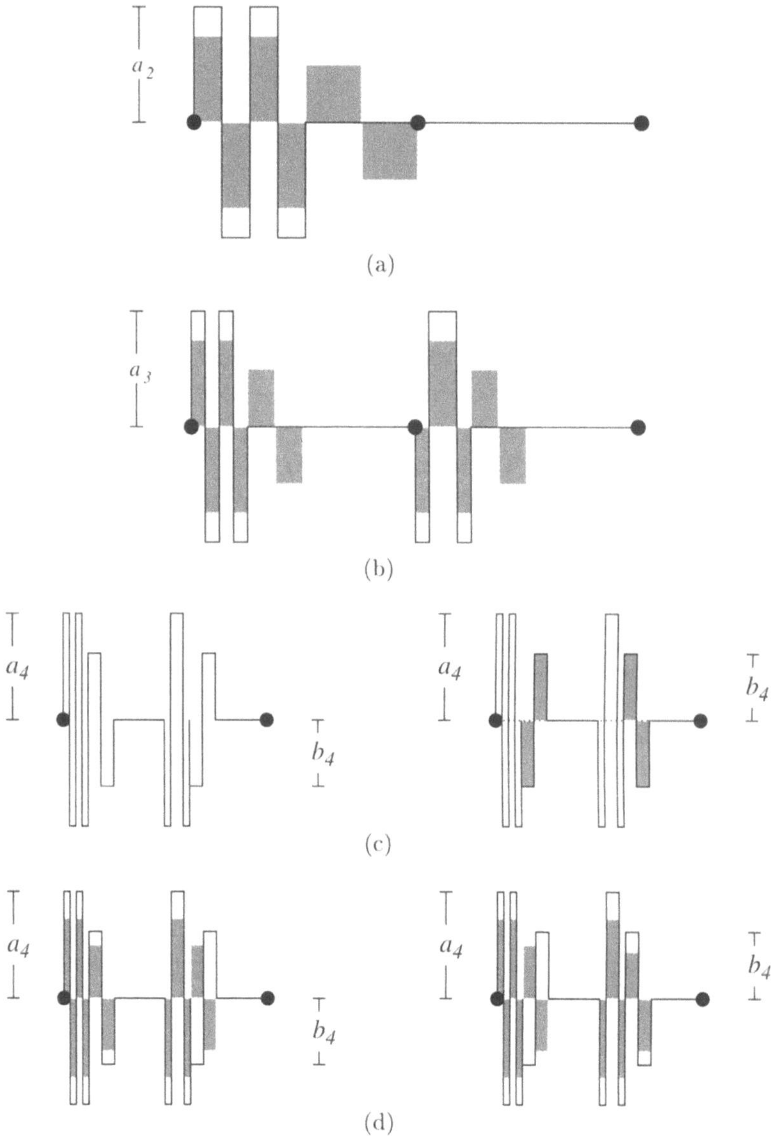

Fig. 7. Schematic representation of the attractor of F as defined in Section 4.3 (the attractor is given by the grey-shaded signal); (a) ranges of S_2, (b) ranges of S_3, (c) the two ranges in S_4 corresponding to an edge when the edge is in the cut, and (d) when the edge is not in the cut.

- *Edge belongs to cut.* In this case, one of the two ranges can be coded without any distortion by the corresponding domain in S_3. For this mapping the luminance parameters are $s = \frac{3}{2}\sigma_3$ and $o = 0$. The second range will be coded by the extra block as exemplified in Figure 5 e) $(s = \sigma_3, o = 0)$ yielding a distortion of $\frac{m}{4}a_4^2$.
- *Edge does not belong to cut.* In this case, we code both of the ranges in S_4 with the corresponding domain in S_3 $(s = \sigma_3, o = 0)$ obtaining a total error of

$$2m\left[\frac{1}{4}\left(a_4 - \frac{2}{3}a_4\right)^2 + \frac{1}{8}\left(b_4 - \frac{2}{3}b_4\right)^2 + \frac{1}{8}\left(b_4 + \frac{2}{3}b_4\right)^2\right] = m \cdot \frac{5}{12}a_4^2.$$

Therefore, the error introduced in S_4 by an edge that is not in the cut is $\frac{5m}{12}a_4^2 = \frac{m}{4}a_4^2 + \frac{m}{6}a_4^2 > \frac{m}{4}a_4^2$. Thus, the total error introduced in segment S_4 is $(\frac{1}{4}n_{\mathcal{E}} + \frac{1}{6}(n_{\mathcal{E}} - k))ma_4^2$. We define $\Delta(\mathcal{G}, k)$ as the distortion made by the attractor in all segments of the signal:

$$\Delta(\mathcal{G}, k) := m\left[\frac{1}{6}n_{\mathcal{V}}a_2^2 + \frac{1}{6}n_{\mathcal{E}}a_3^2 + (\frac{1}{4}n_{\mathcal{E}} + \frac{1}{6}(n_{\mathcal{E}} - k))a_4^2\right].$$

With this definition we (trivially) obtain Lemma 7. Now, using this definition we have to show the correctness of Lemma 8.

4.4 Proof of Lemma 8

We have to show that when $\mathcal{G}$ has a maximal cut of size k then no fractal code in $\mathcal{P}^1_{n_\mathcal{R}}$ leads to a distortion smaller or equal to $\Delta(\mathcal{G}, k+1)$. Let assume, on the contrary, that the graph $\mathcal{G}$ has a maximal cut of size k, but there exists a fractal code $p' \in \mathcal{P}^1_{n_\mathcal{R}}$ such that the attractor $\bar{f}_{p'}$ satisfies $\|f(\mathcal{G}) - \bar{f}_{p'}\|_2^2 \leq \Delta(\mathcal{G}, k+1)$. From Lemma 7 we know that there is an attractor $\bar{f}$ with $\|f(\mathcal{G}) - \bar{f}\|_2^2 = \Delta(\mathcal{G}, k)$. Obviously, $\bar{f}_{p'}$ is closer to the original signal $f(\mathcal{G})$ than $\bar{f}$. Consequently it must approximate $f(\mathcal{G})$ better on at least one of the segments of the signal, S_0, S_1, S_2, S_3, S_4. By setting $\sigma_1, \sigma_2, \sigma_3$ — depending only on the input graph $\mathcal{G}$— we will enforce that $\bar{f}_{p'}$ cannot be a better approximation than $\bar{f}$ on any part of the signal. Thus, our hypothesis is false and the lemma is proven.

Let us first assume that the ranges of S_2 have to be coded by domains from S_1, ranges from S_3 by domains from S_2 and ranges from S_4 by domains from S_3. At the end of the proof we will indicate how to remove this restriction.

Case 1: *$\bar{f}_{p'}$ is better than $\bar{f}$ on S_0 or S_1*
Since the difference of $\bar{f}$ and $f(\mathcal{G})$ is zero on S_0 and S_1, no improvement is possible, and, therefore, case 1 cannot occur.

Case 2: *$\bar{f}_{p'}$ is better than $\bar{f}$ on S_2*
For simplicity, first assume that $\bar{f}_{p'}$ is identical to $f(\mathcal{G})$ on part S_1. For a range R of segment S_2 that contains a vertex ID there are two possibilities for choosing a domain D:

1. When selecting a domain with a fitting vertex ID, the incurred error is (depending on scaling factor s and offset o)

$$E_{D,R}(s,o) = \frac{m}{4}\Big((a_2 - (a_1 \cdot s + o))^2 + (-a_2 - (-a_1 \cdot s + o))^2 + (b_1 \cdot s + o)^2 + (-b_1 \cdot s + o)^2\Big).$$

Solving this equation for the optimal values of s and o yields $\tilde{s} = \frac{2}{3}\sigma_1, \tilde{o} = 0$. This leads to an error of $E_{D,R}(\frac{2}{3}\sigma_1, 0) = \frac{m}{6}a_2^2$ for the range R.
2. When selecting a domain with an incorrect vertex ID, the error will be

$$E_{D,R}(s,o) = \frac{m}{8}\Big((a_2 - (a_1 \cdot s + o))^2 + (-a_2 - (-a_1 \cdot s + o))^2 + (a_2 - (-a_1 \cdot s + o))^2 + (-a_2 - (a_1 \cdot s + o))^2 + 2 \cdot (b_1 \cdot s + o)^2 + 2 \cdot (-b_1 \cdot s + o)^2\Big).$$

Again, solving for optimal s, o yields $\tilde{s} = \tilde{o} = 0$ with an error of $E_{D,R}(0,0) = \frac{m}{2}a_2^2$, three times the error incurred when matching correct IDs. Here, we have used the properties of the binary code $\mathfrak{C}$ of Subsection 4.2.

Thus, the error of $\bar{f}_{p'}$ on S_2 is at least $(n_{\mathcal{V}} + 2l) \cdot \frac{1}{6}a_2^2 \cdot m$, where l is the number of incorrect ID assignments. We choose σ_2 so small that the error made by one ID mismatch is larger than the error made by $\bar{f}$ in segments S_3 and S_4:

$$\begin{aligned}
2 \cdot \frac{1}{6}a_2^2 &> \frac{1}{12}n_{\mathcal{E}}a_3^2 + (\frac{1}{4}n_{\mathcal{E}} + \frac{1}{6}(n_{\mathcal{E}} - 0))a_4^2 \\
\iff \frac{1}{3}a_2^2 &> \frac{1}{6}n_{\mathcal{E}}\sigma_2^2 a_2^2 + \frac{5}{12}n_{\mathcal{E}}\sigma_2^2\sigma_3^2 a_2^2 \\
\iff \frac{1}{3} &> \sigma_2^2\left(\frac{1}{6} + \frac{5}{12}\sigma_3^2\right)n_{\mathcal{E}} \\
\iff \frac{2}{\sqrt{(2+5\sigma_3^2)n_{\mathcal{E}}}} &> \sigma_2
\end{aligned} \tag{23}$$

Therefore, l must equal zero, since otherwise the error incurred in segment S_2 alone would be larger than $\Delta(\mathcal{G}, k)$.

Let us now deal with the assumption that $\bar{f}_{p'}$ equals $f(\mathcal{G})$ on segment S_1. Note that the difference between $\bar{f}_{p'}$ and $f(\mathcal{G})$ on S_1 has to be less than $\Delta(\mathcal{G}, k)$, and this value does not depend on σ_1. By choosing σ_1 sufficiently small, we can assure that the error of $\Delta(\mathcal{G}, k)$ is very small relative to a_1. This relative error will then change our calculations slightly. But by scaling σ_1 we can make these differences arbitrarily small, in fact, significantly smaller than $\Delta(\mathcal{G}, k) - \Delta(\mathcal{G}, k+1)$.

Case 3: *$\bar{f}_{p'}$ is better than $\bar{f}$ on S_3*

First, we can assume that $\bar{f}_{p'}$ looks essentially like $\bar{f}$ on S_2. This is because by choosing σ_2 small enough, any difference that is noticeable after scaling down a domain from S_2 would mean a large additional error in the domain, larger than any potential savings in S_3 and S_4. Thus, we can assume that $\bar{f}_{p'}$ is identical to $f(\mathcal{G})$ on part S_2. For a range R of segment S_3 there are two possibilities for choosing a domain D:

1. When selecting a domain with a fitting vertex ID, the incurred error is

$$E_{D,R}(s,o) = \frac{m}{8}\Big((a_3 - (\frac{2}{3}a_2 \cdot s + o))^2 + (-a_3 - (-\frac{2}{3}a_2 \cdot s + o))^2 + (\frac{2}{3}b_2 \cdot s + o)^2 + (-\frac{2}{3}b_2 \cdot s + o)^2\Big).$$

   The factor $\frac{2}{3}$ comes into play, since we compare the range against the *reconstructed* domain, i.e., against the attractor $\bar{f}_{p'}$ on S_2. Solving this equation for the optimal values of s and o yields $\tilde{s} = \sigma_1, \tilde{o} = 0$. This leads to an error of $E_{D,R}(\sigma_1, 0) = \frac{m}{12}a_3^2$ for the range R.
2. When selecting a domain with an incorrect vertex ID, the error will be

$$E_{D,R}(s,o) = \frac{m}{16}\Big((a_3 - (\frac{2}{3}a_2 \cdot s + o))^2 + (-a_3 - (-\frac{2}{3}a_2 \cdot s + o))^2 + (a_3 - (-\frac{2}{3}a_2 \cdot s + o))^2 + (-a_3 - (\frac{2}{3}a_2 \cdot s + o))^2 + 2 \cdot (\frac{2}{3}b_2 \cdot s + o)^2 + 2 \cdot (-\frac{2}{3}b_2 \cdot s + o)^2\Big).$$

   Solving for optimal s and o yields $\tilde{s} = \tilde{o} = 0$ with an error of $E_{D,R}(0,0) = \frac{m}{4}a_3^2$.

The error of $\bar{f}_{p'}$ on S_3 is at least $(2n_\mathcal{E} + 2l) \cdot \frac{m}{12}a_3^2$, where l is the number of incorrect ID assignments. Again, by choosing σ_3 small enough, we can assure that if we use incorrect IDs, the error will be larger than the error made by $\bar{f}$ in S_4 (which does not depend on σ_3); thus, the total error would be larger than the error of $\bar{f}$:

$$\begin{aligned} & \frac{1}{6}a_3^2 > \left(\frac{1}{4}n_\mathcal{E} + \frac{1}{6}n_\mathcal{E}\right) a_4^2 \\ \Longleftrightarrow\ & \frac{1}{6}a_3^2 > \left(\frac{1}{4}n_\mathcal{E} + \frac{1}{6}n_\mathcal{E}\right) \sigma_3^2 a_3^2 \\ \Longleftrightarrow\ & \frac{1}{6} > \frac{5}{12}n_\mathcal{E}\sigma_3^2 \\ \Longleftrightarrow\ & \sqrt{\frac{2}{5n_\mathcal{E}}} > \sigma_3. \end{aligned}$$

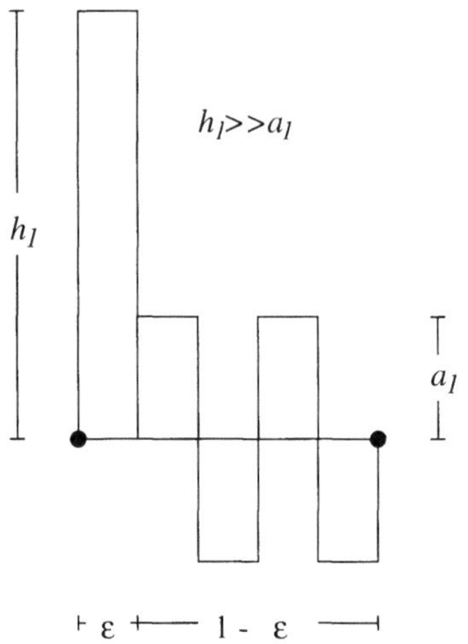

Fig. 8. Adding a peak to the ID

Case 4: *$\bar{f}_{p'}$ is better than $\bar{f}$ on S_4*
Again, we can assume that $\bar{f}_{p'}$ and $\bar{f}$ look 'the same' on segment S_3.

We will now examine what error the possible domain-range pairings will incur. To this end, we distinguish two cases: first, if an edge belongs to the cut, i.e., the flags of the two vertices are different, and second, if the flags are the same.

1. *Edge belongs to cut.* In this case one of the two edge copies in S_4 can be mapped with error zero. As for the other copy, by computing the optimal transformation parameters for all possible domains in S_3, we see that mapping the extra block on the range yields the minimum error of $\frac{1}{4}a_4^2 \cdot m$.
2. *Edge does not belong to cut.* In this case there exists no exact matching domain in S_3 for the two edge copies in S_4. Thus, these ranges can only be coded with a mismatched ID, wrong flag or the extra block. By computing the optimal transformation parameters for all possible domains we see that the error for each of the two edge copies is at least $\frac{5}{24}a_4^2 \cdot m$ for a total of $\frac{5}{12}a_4^2 \cdot m$ (we omit the details of the calculations) .

Thus, the error of $\bar{f}_{p'}$ in S_4 is at least $(n_{\mathcal{E}} \cdot \frac{1}{4}a_4^2 + (n_{\mathcal{E}} - k) \cdot \frac{1}{6}a_4^2)m$, which is exactly the error of $\bar{f}$.

It remains to be shown how one can assure that ranges from segment S_i are only coded by ranges from segment S_{i-1} for $i = 2, 3, 4$. This can be achieved by a slight modification of the signal $f(\mathcal{G})$. The basic idea is the following: We add at the left end of each ID in segment S_1 a peak of height h_1 and width ϵ as depicted in Figure 8 (the rest of the IDs are shrunk accordingly); we also add peaks of height h_i and width $\epsilon/2^{i-1}$ to the IDs of segments S_i for $i = 2,3,4$. Now, by appropriately choosing the heights $h_1,\ldots,h_4$ we can achieve that range-domain assignments other than those considered in our proofs above lead to arbitrarily large errors: when a range of $S_i, i \in \{1,2,3,4\}$, containing an ID is assigned a domain from a segment other than S_{i-1}, the corresponding peak will differ with respect to the peak support in such a way that a large error occurs no matter what the rest of the domain and range looks like. Thus, we exploit the fact that the *geometric* scaling factor is fixed. At the same time, all argumentations of this and the previous subsection can be easily translated to the case of the 'peak-added' signal.

This consideration concludes our proof of Lemma 8. □

5 Further Results

In this section we survey further results related to optimal fractal compression and collage coding. We state that collage coding does not constitute an approximating algorithm for the problem of optimal fractal coding, derive a lower bound on the optimal attractor error, and devise an 'annealing' scheme for improved fractal coding.

5.1 Approximability

The NP-hardness result of the last section poses the question of whether the problem FRACCODE admits an approximating algorithm. In [27] we have shown that the method of collage coding does not represent an approximating algorithm, i.e., we have shown that there exits no (finite) constant $\rho > 0$ such that

$$\frac{d_{\mathcal{F}}(f, \bar{f}_{p^c})}{d_{\mathcal{F}}(f, \bar{f}_{p^*})} < \rho \quad \forall f \in \mathcal{F},$$

where again the attractor for a collage error optimal code is denoted by $\bar{f}_{p^c}$, and a truly optimal attractor is given by $\bar{f}_{p^*}$. In other words, it is possible that the ratio of the collage attractor error to the optimal attractor error can be arbitrarily large (for more details, cf. [17]). While this result shows that the collage coding strategy has fundamental shortcomings, it remains open whether *near-optimal* fractal coding is possible.

5.2 Anti-Collage-Theorem

Another interesting result concerning collage coding is given by the "Anti-Collage-Theorem" [28] that provides a lower bound for the collage attractor error in terms of the collage error. The generalized formulation is as follows:

Proposition 10. *Given (Y, d_Y) a complete metric space. Let $T : Y \to Y$ be Lipschitz, i.e., there exists an $L_T \geq 0$ such that $d_Y(Ty_1, Ty_2) \leq L_T d_Y(y_1, y_2)$ for all $y_1, y_2 \in Y$. As well, assume that $\bar{y}$ is a fixed point of T. Then for any $y \in Y$,*

$$d_Y(y, \bar{y}) \geq \frac{1}{1 + L_T} d_Y(y, Ty). \tag{24}$$

Proof. From the triangle inequality:

$$\begin{aligned} d_Y(y, Ty) &\leq d_Y(y, \bar{y}) + d_Y(\bar{y}, Ty) \\ &\leq d_Y(y, \bar{y}) + L_T(\bar{y}, y), \end{aligned} \tag{25}$$

from which the desired result follows.

Thus, for nonzero collage error, there is no chance that the error can be small "by accident."

5.3 Annealing

The result of Section 4 has shown that optimal fractal coding represents an intractable problem. However, the results obtained with the gradient-based direct attractor optimization (Section 3) indicate that improvements over collage coding are feasible. But with this approach only the luminance parameters are modified. As a next step one would like to include the domain addresses in the optimization as well. Unfortunately, even when the domain addresses would be considered as continuously varying entities, they cannot be included into the gradient-based optimization method for two reasons: the complexity would be prohibitive and the chance to get trapped in a local optimum would also be very high. Thus, in order to include the domain addresses in the optimization, one has to deal with the discrete problem.

In [15] a method is proposed for *local iterative improvements* of a fractal code. The basic idea is as follows: one tries to improve a given fractal code by selecting a single range and modifying the corresponding domain address and luminance parameters. The new 'candidate code' is then decoded, and one checks whether an improvement over the original code has occurred. When this is the case, the 'candidate code' is used as the new original code and the procedure is repeated with another range. The way the fractal code for a range is modified follows the proposal of Barthel [5] and Lu [20]: the domain search is performed in the attractor of the original code (not in the image to be coded as it is the case for the collage coding strategy).

For a practical application computational efficiency is crucial since the naive straightforward implementation leads to unacceptably long compute times. Here are some of the issues that are addressed in [16]:

- In what order should the range blocks be processed?
- Can one restrict the search for a matching domain to a few candidate blocks which are determined a priori in a preprocessing step?

- The decoder, which is an integral component of the coder, must be accelerated by exploiting the fact that the code changes only locally during one iteration of the algorithm.

The results (cf. Table 3) show significant improvements of about 0.6 dB over standard collage coding.

Table 3. Results for quadtree encodings of the 512 × 512 Boat image using the annealing procedure. Shown are the number of ranges of the partition, the compression ratio, the attractor error obtained using the collage coding strategy, the attractor error obtained with the additional annealing procedure, and the observed gain in PSNR.

No. of ranges	Comp. ratio	Collage coding (dB PSNR)	Annealing (dB PSNR)	Δ PSNR
4510	17:1	31.94	32.54	0.60
3352	23:1	30.19	30.77	0.58
2560	30:1	28.89	29.51	0.62
1972	38:1	27.83	28.51	0.68

6 Summary

In this paper we have reported on approaches to the inverse problem of fractal compression from two different directions using two different mathematical methodologies. In the first part we have derived the theoretical foundations necessary for any application of differentiable methods for attractor error reduction in fractal compression, namely

- the establishment of the differentiability of the attractor as a function of its (real valued) scaling and offset parameters, and
- the feasibility of gradient computation by iteration of a properly defined vector Iterated Function System with gray level Maps.

Moreover, we have implemented gradient descent algorithms for the problem and reported computational results for a few test cases. While the computer programs have demonstrated that the methods work in practice, the outcomes, however, are not promising. Although gains for the encoding based on the method of Dudbridge and Monro are around one half of a dB in PSNR, the conceptually less complex method using a simplex hill climbing algorithm performs just as well at the same cost in terms of computation time.

In the second part of the paper we have analyzed the computational complexity of the inverse problem, i.e., we have considered fractal compression as a discrete optimization problem and have analyzed the complexity of determining for a given function the fractal code —out of a class of feasible fractal codes— that achieves the least attractor error. Here, we have been able to prove that the problem is inherently intractable, i.e., NP-hard. This explains the predominant use of the suboptimal collage coding strategy, and, unfortunately, limits the prospects of improving fractal compression by searching for optimal fractal codes.

References

1. Barnsley, M. F., *Fractals Everywhere*, Academic Press, New York, 1988.
2. Barnsley, M. F., *Fractal modeling of real world images*, in: *The Science of Fractal Images*, H.-O. Peitgen and D. Saupe (eds.), Springer-Verlag, New York, 1988.
3. Barnsley, M.F., Ervin, V., Hardin, D., and Lancaster. J., *Solution of an inverse problem for fractals and other sets*, Proc. Nat. Acad. Sci. USA **83** (1985) 1975–1977.
4. Barnsley, M. F., Hurd, L., *Fractal Image Compression*, AK Peters, Wellesley, 1993.
5. Barthel, K. U., *Festbildcodierung bei niedrigen Bitraten unter Verwendung fraktaler Methoden im Orts- und Frequenzbereich* (Dissertation, Technische Universität Berlin), Wissenschaft & Technik Verlag, Berlin, 1996.
6. Centore, P., Vrscay, E.R., *Continuity of attractors and invariant measures for Iterated Function Systems*, Canad. Math. Bull. **37**(3) (1994) 315–329.
7. Crescenzi, P., Kann, V., *A compendium of NP optimization problems*, Technical report SI/RR-95/02, Universita di Roma "La Sapienza", 1995.
8. Domaszewicz, J., Vaishampayan, V. A., *Graph-theoretical analysis of the fractal transform*, Proc. IEEE Int. Conf. on Acoustics, Speech, and Signal Processing, vol. 4, Detroit, 1995.
9. Dudbridge, F., Fisher, Y., *Attractor optimixation in fractal image encoding*, Proc. Third Fractals in Engineering Conference, Arcachon, France, May, 1997.
10. Fisher, Y., *Fractal Image Compression, Theory and Application*, Springer-Verlag, New York, 1995.
11. Fisher, Y. (ed.), *Fractal Image Coding and Analysis*, Proc. NATO ASI, Springer-Verlag, New York, 1998.
12. Forte, B., Vrscay, E.R., *Solving the inverse problem for function and image approximation using Iterated Function Systems*, Dyn. Cont. Disc. Imp. Syst. **1** (1995) 177–231.
13. Forte, B., Vrscay, E.R., *Inverse problem methods for generalized fractal transforms*, in *Fractal Image Encoding and Analysis*, ed. Y. Fisher, Springer Verlag, Heidelberg, 1998.
14. Garey, M. R., Johnson, D. S., *Computers and Intractability: A Guide to the Theory of NP-Completeness,* Freeman, New York, 1979
15. Hamzaoui, R., Hartenstein, H., Saupe, D., *Local iterative improvement of fractal image codes*, Image & Vision Computing **18** (2000) 565–568.
16. Hamzaoui, R., Saupe, D., Hiller, M., *Fast code enhancement with local search for fractal image compression,* Proc. IEEE Int. Conf. on Image Processing, Vancouver, Sept. 2000.
17. Hartenstein, H., *Topics in Fractal Image Compression and Near-Lossless Image Coding*, Doctoral Dissertation, Institut für Informatik, Universität Freiburg, 1998.

18. Hutchinson, J., *Fractals and self-similarity*, Indiana Univ. Math. J. **30** (1981) 713–714.
19. Jacquin, A. E., *A Fractal Theory of Iterated Markov Operators with Applications to Digital Image Coding*, Doctoral Dissertation, Georgia Institute of Technology, August 1989.
20. Lu, N., *Fractal Imaging*, Academic Press, New York, 1997.
21. Mendivil, F., Vrscay, E.R., *Correspondence between fractal-wavelet transforms and Iterated Function Systems with Grey-level Maps*, in *Fractals in Engineering*, ed. J. Lévy Véhel, E. Lutton and C. Tricot, Springer, New York, 1997.
22. Monro, D. M., Dudbridge, F., *Fractal block coding of images*, Electronics Letters **28**,11 (1992) 1053–1055.
23. NAG Fortran Library, The Numerical Algorithms Group Ltd, Oxford, UK.
24. Papadimitriou, C. H., *Computational Complexity*, Addison-Wesley, Reading, Mass., 1994
25. Peitgen, H.-O., Jürgens, H., Saupe, D., *Chaos and Fractals*, Springer-Verlag, New York, 1992.
26. Press, W. H., Teukolsky, S. A., Vetterling, W. T., Flannery, B. P., *Numerical Recipes in C,* Second Edition, Cambridge University Press, 1992.
27. Ruhl, M., Hartenstein, H., *Optimal fractal coding is NP-hard*, in Proc. IEEE Data Compression Conference, ed. J. Storer and M. Cohn, Snowbird, Utah, 1997, 261–270.
28. Vrscay, E. R., Saupe, D., *Can one break the "collage barrier" in fractal image coding?*, in: *FRACTALS: Theory and Applications in Engineering*, Proceedings, Delft, June 1999, M. Dekking, J. Levy Vehel, E. Lutton, C. Tricot (eds.), p. 289–305, Springer-Verlag, Heidelberg, 1999.
29. Williams, R. F., *Compositions of contractions*, Bol. Soc. Brasil. Mat. **2** (1971) 55–59.
30. Withers, D., *Newton's method for fractal approximation*, Constructive Approximation **5** (1989) 151-170.

The Orbit Space Method: Theory and Application

Matthias Rumberger and Jürgen Scheurle*

Zentrum Mathematik, Technische Universität München, 80290 München, Germany
e-mail: rumberger@mathematik.tu-muenchen.de
scheurle@mathematik.tu-muenchen.de

1 Introduction

If a system does not change its structure, while it is exposed to some transformations such as translations, rotations, reflections, or even more complicated transformations, it possesses symmetry. This is a feature of a host of phenomena. A good mathematical description of them must take the symmetry into account. Further symmetry may come about by simplifications. Often, deep statements are possible just by symmetry arguments. On the other hand, the presence of symmetry helps to analyze a given system. For example, one can find new solutions knowing a special one. Or, the symmetry may exclude some situations or force some phenomena to appear, which usually do not occur in generic nonsymmetrical systems.

We consider dynamical systems with symmetry being given by a group action on the phase space. Our aim is to use the symmetry to simplify the system in order to obtain information about its dynamical behavior, e.g. about bifurcation and stability properties. To this end, we propose to use the orbit space method which factors out the symmetry as follows. Due to symmetry, the dynamics of a system maps each (group) orbit onto an orbit. Therefore, the dynamics of such a system can be viewed as dynamics on the set of orbits, i.e., on the so-called orbit space (reduced dynamics). This projection reduces the complexity of dynamical objects. For example, a dynamically invariant set which is contained in a group orbit reduces to an equilibrium on the orbit space; therefore, its points are called relative equilibria. Similarly, relative periodic points are projected onto periodic points.

By means of this projection, the dynamics along any orbit is faded out, and in turn the directions transversal to the orbits are emphasized. In fact, all trajectories meeting one fixed orbit are projected onto the same trajectory on the orbit space, and show the same behavior. Modulo the symmetry, they cannot be distinguished. Thus, it seems reasonable to investigate them in their entirety. This is the basic idea of the orbit space method. For example, the stability of such a collection of trajectories as a union, i.e., orbital

* Project: Constructive Methods for Stability and Bifurcation Analysis of High-Dimensional Nonlinear Dynamical Systems (Jürgen Scheurle)

stability which is equivalent to stability on the orbit space, appears to be the appropriate notion of stability in this context. Throughout this paper, stability refers to this notion.

By the way, the term 'trajectory' is to be understood in the dynamical sense here and subsequently, while 'orbit' will exclusively be used in the context of group actions.

To handle the orbit space, it is convenient to embed it into some vector space. In finite dimensions, the embedding can be performed by a Hilbert basis; this is a set of generators of the algebra of polynomials on the phase space that are invariant under the action of the symmetry group. For compact or even reductive groups, the existence of a Hilbert basis is guaranteed by a famous theorem due to Hilbert and Weyl. Here embedding means to introduce coordinates which are invariant with respect to the group action. The new coordinates are nonlinear. Thus, degeneracies of the dynamics forced by the symmetry are converted into singularities of the orbit space. We are not going to exclude this kind of singularities later on. Rather, a key feature of this contribution is to show how to apply the orbit space method at such singularities (singular reduction).

There are several papers using orbit space reduction for dynamical systems with symmetry, e.g. [2], [25], [7], [8], [9], [26], [36], [35]. See also the contribution by F. Guyard and R. Lauterbach to the present volume. Also, there is an extensive literature about more sophisticated versions for particular systems such as mechanical systems (see, for example, [1] and the references therein). However, this is far beyond the theory discussed here. For the approach of restriction to fixed point spaces of isotropy subgroups we refer to [16,17].

In the present paper, we give a theoretical justification of the orbit space method for general dynamical systems with symmetry, and outline fundamental techniques for its application. Most topics in this paper concern both continuous and discrete dynamics. In Sect. 2, we start with a suitable description of the symmetry, introduce the orbit space, and develop the concept of reduction to the orbit space. An extension ([34]) of the fundamental theorem of Schwarz ([37]) allows to consider not only smooth, but also finitely differentiable systems, which arise, for instance, from a center manifold reduction. In this sense, infinite-dimensional systems are included in the theory developed here. In fact, our main example will be an application to a system of Kuramoto-Sivashinsky type of partial differential equations which is of that nature (Sect. 7).

The main purpose of the orbit space method is to draw conclusions from the reduced dynamics to the dynamics of the original system. Here we concentrate on dynamical features such as bifurcation and stability properties that can be characterized by eigenvalues (exponents) of appropriately constructed linearized systems. For example, since the orbit space will be embedded into a vector space, we extend the reduced dynamics to this vector space and consider the linearization of the extended system at points of the orbit space. In particular, this linearization is well defined at singular points. Since the process of reduction uses nonlinear coordinates, at first glance it is by no means obvious how eigenvalues (exponents) on the orbit space relate to corresponding eigenvalues (exponents) on the original phase space, and vice versa. Of course, only the modes in directions transversal to group orbits in the original phase space play a role in this context. Moreover, since the orbit space will be embedded into a vector space of higher dimension, in general, not all eigenvalues (exponents) inside this vector space affect the dynamics on the orbit space. We are going to address both these issues for equilibria of maximal symmetry (isotropy) in Sect. 3, for general relative equilibria in Sect. 4, for relative periodic points in Sect. 5, and finially, for even more general trajectories in Sect. 6. In all cases it turns out that there are certain algebraic expressions for the eigenvalues (exponents) on the orbit space in terms of the ones on the original phase space depending on the representation of the symmetry group. If the dynamics is given by a vector field, then eigenvalues of the original system appear as linear combinations over $\mathbb{N}_0$ on the orbit space (cf. [23]). The same is true for exponents. For discrete systems, products of eigenvalues of the original system appear on the orbit space. In particular, for compact symmetry groups the relevant eigenvalues (exponents) on the orbit space are sufficient to determine the spectral (orbital) stability and instability of trajectories of the original system.

Recall from above that the orbit space is not a smooth manifold, in general. Rather, it is a semialgebraic variety. To determine the relevant eigenvalues on the orbit space, we use a concept of tangent space (tangent bundle) which is adopted from algebraic geometry, and restrict the linearized system to that. The tangent space at equilibria of maximal symmetry always covers the whole vector space the orbit space is embedded in, if the Hilbert basis is minimal. Hence, at such points the computations are easier than at general points.

To deal with general relative equilibria we use the slice theorem. Together with a splitting of the dynamics into parts tangent to the orbits and transversal to them due to M. Field [14] and M. Krupa [24] (cf. also [12]), it allows to reduce an arbitrary point to a point with maximal symmetry. Furthermore, it provides a suitable stratification for the corresponding representation space. This partition into sets of 'equal' symmetry is preserved during the reduction process.

Our results for relative equilibria also apply to relative periodic points. To show this we construct an invariant hypersurface transversal to the flow and a first return map on this hypersurface (cf. [13]). This map is group invariant and possesses a relative equilibrium corresponding to the relative periodic point. It reduces to a Poincaré map on the orbit space. In [31] this result is generalized to Lyapunov exponents. In Sect. 6, we just indicate the main differences between the cases of eigenvalues and Lyapunov exponents. For details concerning the latter, we refer the interested reader to [31].

We point out that the concept of a tangent space as indicated above is not yet sophisticated enough to get rid of any spurious eigenvalue on the orbit space. For instance, at a point corresponding to a hyperbolic dynamically invariant orbit of saddle type the linearization of the reduced system restricted to the tangent space may have critical eigenvalues besides stable and unstable ones. In other words, hyperbolicity on the original phase space does not necessarily imply hyperbolicity on the orbit space in that context. To accomplish that this implication is true, the concept of a tangent cone which is a subset of the tangent space has been introduced by M. Kœnig [23]. In Sect. 8, we invoke this concept to refine our theory as to reproduce all eigenvalues transverse to the group orbits in the original phase space from the eigenvalues on the orbit space. Thus, we establish a kind of converse to the results of the previous sections.

For a Hartman-Grobman type of linearization theory on the orbit space, see [10].

Finally, Sect. 9 is devoted to a discussion of the case of reductive symmetry groups. These are groups whose finite-dimensional representations are completely reducible. They are not compact, in general. Examples show that the orbit space method may fail in that case.

2 Reduction of Symmetric Systems

First we introduce some notation. Let Γ be a compact Lie group acting orthogonally on the real vector space $V = \mathbb{R}^m$. Let $P : V \to V$ be a Γ-equivariant map. This means that P commutes with the group action: $\forall x \in V, \gamma \in \Gamma : P(\gamma x) = \gamma P(x)$. Therefore, the image of each point γx of the orbit Γx is known, if $P(x)$ is known. Let F be a Γ-equivariant vector field on V. Then, due to the uniqueness, the flow Φ_t of F is a Γ-equivariant map $V \to V$ for each t. Assume P and F, and therefore Φ_t, to be qn-times continuously differentiable, where the integers q and n are to be fixed later.

The trajectories of all points of an orbit are conjugated by the symmetry group. Moreover, an orbit Γx is mapped onto the orbit $\Gamma P(x) = P(\Gamma x)$ or $\Gamma \Phi_t(x)$ by P or Φ_t, respectively. Thus, it is reasonable to regard the dynamics as a system on the set $\{\Gamma x \mid x \in V\}$ of orbits, called the *orbit space*. It is identified with the quotient V/Γ. The canonical projection π_Γ : $\begin{cases} V \longrightarrow V/\Gamma \\ x \longmapsto \Gamma x \end{cases}$ induces a topology on V/Γ.

The map $\pi_\Gamma \circ P$ is Γ-invariant, i.e., constant on all orbits. Therefore, there exists a map $\overline{P} : V/\Gamma \to V/\Gamma$ with

$$\overline{P} \circ \pi_\Gamma = \pi_\Gamma \circ P \ .$$

In this way, the symmetry is divided out of P, and a possibly simpler map comes up. If Γx is a dynamically invariant orbit, meaning that no trajectory of P leaves the set Γx, then $\Gamma x = \pi_\Gamma(x)$ is an equilibrium of $\overline{P}$. We use the word 'equilibrium' also for a fixed point of a map to distinguish from a 'fixed point' of the group action. In the sense of orbital stability, the stability of Γx with respect to P agrees with the stability of $\pi_\Gamma(x)$ with respect to $\overline{P}$. In a similar way, other dynamical objects corresponding to P reduce to simpler objects corresponding to $\overline{P}$.

The embedded orbit space To handle the orbit space and the reduced dynamics, it is convenient to embed the orbit space into a vector space. To do so, we introduce a kind of invariant coordinates.

The action of Γ on V induces an action on the functions $f : V \to N$, where N is some set:

$$\gamma f : x \longmapsto f(\gamma^{-1}x) \ .$$

The fixed point space of this action contains exactly the Γ-invariant functions. As usual, we mark the fixed point space by a superscript Γ, e.g. $\mathbb{R}[V]^\Gamma$ is the $\mathbb{R}$-algebra of Γ-invariant polynomials. A well-known theorem by Hilbert and Weyl states that this algebra is finitely generated ([42]). A set of generators $\rho_1, \ldots, \rho_k$ is called a *Hilbert basis*. Each Γ-invariant polynomial f can be written as a polynomial g in $\rho_1, \ldots, \rho_k$:

$$f(x) = g(\rho_1(x), \ldots, \rho_k(x)) \ .$$

To simplify notations, we define the *Hilbert map* $\rho(x) = (\rho_1(x), \ldots, \rho_k(x))$. (It is quite difficult to compute a Hilbert basis for a given representation, but there exist algorithms for finite groups (see Sturmfels [40], Kemper [22]) and for compact and reductive groups (see Gatermann [15] and Derksen [11]).)

The polynomial $R : x \mapsto \langle x, x\rangle$ is Γ-invariant, due to the orthogonality of the action. Therefore, there is a polynomial g with $R = g \circ \rho$. Let $K \subset \mathbb{R}^k$ be a compact set and L a compact interval containing $g(K)$. Then $R^{-1}(L)$ is closed and bounded, and contains $\rho^{-1}(K) \subset R^{-1}(g(K))$. Thus, the preimage $\rho^{-1}(K)$ itself is compact. This means that a Hilbert map of an orthogonal representation is proper. As a consequence, ρ is closed and even open as a map from V into $Y := \rho(V)$.

Since the action of Γ on $\mathbb{R}[V]$ preserves the degree, we can choose homogeneous generators of $\mathbb{R}[V]^\Gamma$. In this case, we say the Hilbert basis and the Hilbert map are *homogeneous*. A homogeneous Hilbert basis and the corresponding Hilbert map are called *minimal*, if each proper subset of the basis does not generate $\mathbb{R}[V]^\Gamma$. In this paper ρ is assumed to be a homogeneous Hilbert map with image $Y \subset \mathbb{R}^k$.

Assuming a decomposition $V = V_1 \oplus \cdots \oplus V_l$ with coordinates $x = (x_1, \ldots, x_l)$ we can generalize the notion of degree in the following way. For a multiindex $\beta \in \mathbb{N}_0^l$, denote by $\mathbb{R}[V]_\beta$ the subspace of polynomials which are homogeneous of degree β_i in x_i, $i = 1, \ldots, l$. These polynomials are called *multihomogeneous*, and β is their *multidegree*. If the spaces V_i are Γ-invariant, then $\mathbb{R}[V]_\beta$ is Γ-invariant for all β, and a multihomogeneous Hilbert basis exists.

Remark 1. The number of elements of a minimal Hilbert basis and their (multi-)degrees are unique up to permutation.

Using the Weierstraß approximation theorem it is easy to see that a Hilbert map separates the Γ-orbits, i.e., the images $\rho(x)$ and $\rho(y)$ differ for $x,\ y \in V$ with $x \notin \Gamma y$. Hence, the map $\overline{\rho} : V/\Gamma \to \mathbb{R}^k$ with $\overline{\rho} \circ \pi_\Gamma = \rho$ is one-to-one. Moreover, it is a homeomorphism. Therefore, we identify the orbit space V/Γ with $Y = \rho(V)$, and call Y the *(embedded) orbit space.*

A relation of ρ is a polynomial $r \in \mathbb{R}[\mathbb{R}^k]$ with $r \circ \rho = 0$. The relations form an ideal I in $\mathbb{R}[\mathbb{R}^k]$. If a given Γ-invariant polynomial f is represented by $g \in \mathbb{R}[\mathbb{R}^k]$, then all polynomials $h \in g + I$ represent f by $f = h \circ \rho$. Thus, in contrast to $g|_Y$, g is not unique, if I is nontrivial.

The variety $Z := \{y \in \mathbb{R}^k \mid \forall r \in I : r(y) = 0\}$ associated with this ideal is the smallest variety containing Y. The subset Y of Z is described by some polynomial inequalities obtained as follows. Define the $k \times k$-matrix $G(x)$ by $G_{i,j}(x) := \operatorname{grad} \rho_i(x) \cdot \operatorname{grad} \rho_j(x)$. Then G is Γ-invariant, and there exists a map $H : \mathbb{R}^k \to \mathbb{R}^{k\times k}$ with $H(\rho(x)) = G(x)$. If $y \in Z$ lies in Y, then $H(y)$ is positive semi-definite. It is proved in [30] that the converse is also valid. Therefore, the inequalities which describe the definiteness of $H(y)$ determine the shape of $Y \subset Z$. Thus, Y is a semialgebraic set.

The differentiability of the reduced system Since Γ-invariant polynomials can be represented on the orbit space by polynomials, one may ask, whether a differentiable Γ-invariant function can be represented by a differentiable function. For example, the even (i.e., $\mathbb{Z}_2$-invariant) function $f : x \mapsto |x|^3$ is C^2, but all the functions g with

$$f = g \circ \rho\,, \tag{1}$$

where $\rho(x) = x^2$, are only C^1, since $g : y \to |y|^{3/2}$ on $Y = \mathbb{R}_0^+$. However, there is a C^1-function g satisfying (1). In general, a loss of differentiability is expected, but this loss is controllable:

Proposition 2 ([37,27,34]). *There exists an integer q such that for each $n = 0, \ldots, \infty$ a continuous map*

$$\mathcal{J} : C^{qn}(V)^\Gamma \longrightarrow C^n(\mathbb{R}^k)$$

exists satisfying $\mathcal{J}(f) \circ \rho = f$ for $f \in C^{qn}(V)^\Gamma$.
The Taylor polynomial of degree n of $\mathcal{J}(f)$ depends only on the Taylor polynomial of degree qn of f.

The integer q depends only on the representation, and is at least the degree of a minimal Hilbert basis.

We use this result to find differentiable representations of P and F on the orbit space. $\rho \circ P$ is a Γ-invariant C^{qn}-map $V \to \mathbb{R}^k$. We apply the proposition to each component, and get a C^n-map $Q : \mathbb{R}^k \to \mathbb{R}^k$ such that the diagram

$$\begin{array}{ccc} V & \xrightarrow{P} & V \\ \rho \big\downarrow & & \big\downarrow \rho \\ \mathbb{R}^k & \xrightarrow{Q} & \mathbb{R}^k \end{array} \tag{2}$$

commutes. Denote the space of C^n-maps from V to W by $C^n(V, W)$, and indicate the subspace of equivariant maps by a subscript Γ: $C^n_\Gamma(V, W)$. Then there is a continuous map

$$\mathcal{J}_{\mathrm{m}} : C^{qn}_\Gamma(V, V) \longrightarrow C^n(\mathbb{R}^k, \mathbb{R}^k)$$

which satisfies $\mathcal{J}_{\mathrm{m}}(P) \circ \rho = \rho \circ P$ for all $P \in C^{qn}_\Gamma(V)$.

Also vector fields can be reduced by factoring out the symmetry. The space of vector fields of class C^n on V is denoted by $\mathrm{Vec}^n(V)$. The subspace of Γ-equivariant vector fields is denoted by $\mathrm{Vec}^n_\Gamma(V)$. There is a continuous and linear map

$$\mathcal{J}_{\mathrm{v}} : \mathrm{Vec}^{qn}_\Gamma(V) \longrightarrow \mathrm{Vec}^n(\mathbb{R}^k)$$

with $\mathcal{J}_{\mathrm{v}}(F) \circ \rho(x) = D\rho(x)F(x)$ for all $F \in \mathrm{Vec}^{qn}_\Gamma(V)$. Denote the flow of $\mathcal{J}_{\mathrm{v}}(F)$ by Ψ_t, then we obtain $\rho \circ \Phi_t = \Psi_t \circ \rho$.

This approach to symmetric systems works on Γ-manifolds, too. Let M be a smooth manifold without boundary, and suppose there is a smooth action of Γ on M. The only further condition on M is that the action has at most finitely many orbit types. Then there is a Γ-equivariant embedding of M into a representation space V of Γ. Since a Hilbert map ρ is proper, an open set W exists such that $\rho(M) = \rho(V) \cap W$. Now $\rho : V \to \mathbb{R}^k$ may be replaced by $\rho : M \to W$ in the following sections.

Reducing linear maps Even if the Γ-equivariant map P is linear, the reduced map Q need not to be linear. Since relations can be added to Q, it is interesting to ask, whether one can choose a linear Q obeying (2). We can restrict our investigations to the case of minimal Hilbert bases.

Consider the effect of P as an action on $\mathbb{R}[V]^\Gamma$. The maximal ideal J of polynomials vanishing at 0 is invariant with respect to P. Each Hilbert basis generates J. If the Hilbert basis is minimal, then it is projected by the canonical surjection to a vector space basis of J/J^2. Therefore, a minimal Hilbert basis is a vector space basis of a complement C of J^2 in J, and vice versa.

Suppose C is invariant with respect to P. Then $\rho_i \circ P$ is in C, and there is a linear function Q_i with $\rho_i \circ P = Q_i \circ \rho$. The question arises whether a P-invariant complement of J^2 exists.

Example 3. Consider the natural action of $\mathbb{Z}_4$ on $\mathbb{R}^2$. A Hilbert basis is given by $u(x,y) = x^2+y^2$, $v(x,y) = x^2y^2$, $w(x,y) = x^3y-xy^3$. There is the relation $w^2 = v(u^2-4v)$. A $\mathbb{Z}_4$-equivariant linear map has the form $A := \begin{pmatrix} \lambda & -\mu \\ \mu & \lambda \end{pmatrix}$. We have:

$$\begin{aligned} u(A(x,y)) &= u(\lambda,\mu)u(x,y) \qquad u(A(x,y))^2 = u(\lambda,\mu)^2u(x,y)^2 \\ v(A(x,y)) &= v(\lambda,\mu)u(x,y)^2 + 2w(\lambda,\mu)w(x,y) + (u(\lambda,\mu)^2 - 8v(\lambda,\mu))v(x,y) \\ w(A(x,y)) &= (u(\lambda,\mu)^2 - 8v(\lambda,\mu))w(x,y) + w(\lambda,\mu)(u(x,y)^2 - 8v(x,y)) \end{aligned}$$

Take $v'(A(x,y)) = u(x,y)^2 - 8v(x,y) = x^4 - 6x^2y^2 + y^4$ instead of v ($v'^2 + 16w^2 = u^4$). Then

$$\begin{aligned} v'(A(x,y)) &= v'(\lambda,\mu)v'(x,y) - 16w(\lambda,\mu)w(x,y) \\ w(A(x,y)) &= w(\lambda,\mu)w(x,y) + v'(\lambda,\mu)w(x,y) \ . \end{aligned}$$

Thus, all linear maps can be reduced to linear maps on the orbit space, using the Hilbert basis u, v', and w. This is also true for vector fields:

$$\begin{aligned} \operatorname{grad} u \cdot A &= 2\lambda u & \operatorname{grad} v \cdot A &= 4\lambda v + 2\mu w \\ \operatorname{grad} w \cdot A &= \mu(u^2 - 8v) + 4\lambda w = \mu v' + 4\lambda w & \operatorname{grad} v' \cdot A &= 4\lambda v' - 16\mu w \end{aligned}$$

Proposition 4. *There is a Hilbert basis such that for all Γ-equivariant automorphisms P a linear map Q on $\mathbb{R}^k$ satisfying (2) exists.*

Proof. $\Delta := \mathrm{GL}(V)^\Gamma$ is the group of Γ-equivariant automorphisms of V. It is the centralizer of Γ in $\mathrm{GL}(V)$. If the i-th isotypical component of V has type $\mathbb{K}_i$ and multiplicity n_i, then $\Delta \cong \bigotimes_i \mathbb{K}_i \otimes_{\mathbb{R}} \mathrm{GL}(\mathbb{R}^{n_i})$ is a reductive group. Thus, the representation of Δ on $J \cap \mathbb{R}[V]^\Gamma_{\leq d}$, where $d := \max_i \deg \rho_i$, is completely reducible, and the invariant space $J^2 \cap \mathbb{R}[V]^\Gamma_{\leq d}$ has an invariant complement. □

The same is valid for vector fields, see [23].

3 Eigenvalues at Equilibria of Maximal Isotropy

We consider a Γ-equivariant map $P : V \to V$ which has an equilibrium invariant under the the full symmetry group Γ. Without loss of generality we assume this equilibrium to be 0. The eigenvalues of the linearization $DP(0)$ indicate the stability of the equilibrium, if they lie inside the unit circle. The reduced map Q has an equilibrium at $0 = \rho(0)$. The question arises, how the eigenvalues of the linearizations $DP(0)$ and $DQ(0)$ are related. In particular, can the stability of $0 = P(0)$ be read off from the eigenvalues of $DQ(0)$?

If a relation $r \in I$ with non-vanishing linear terms exists, then $DQ(0)$ is not unique. Moreover, then we can choose Q in such a way that an arbitrarily given number λ is an eigenvalue of $DQ(0)$. However, the following straight forward lemma provides a condition for uniqueness.

Lemma 5. *A homogeneous Hilbert basis is minimal, if and only if the linear part* $\operatorname{grad} r(0)$ *of all relations* $r \in I$ *vanishes. In this case, the linear part of* $Q \in C^n(\mathbb{R}^k, \mathbb{R}^k)$ *satisfying (2) depends uniquely on* $P \in C^{qn}_{\Gamma}(V, V)$.

Theorem 6 ([32]). *Let* P *and* Q *be as in (2). Suppose* $P(0) = 0$, *and assume the Hilbert map* ρ *to be minimal. Then the eigenvalues of* $DQ(0)$ *are products of eigenvalues of* $DP(0)$, *a factor may occur several times in each product.*

Proof. Assume P to be polynomial. The map $Q = \mathcal{J}_{\mathsf{m}}(P) : \mathbb{R}^k \to \mathbb{R}^k$ satisfies (2). Decompose P into a sum $P = P_\ell + P_{\mathrm{r}}$, where P_ℓ contains the linear terms of P, and P_{r} consists of the terms with degree greater than 1. Since the action preserves the degree, P_ℓ and P_{r} are Γ-equivariant.

Suppose that the Hilbert basis is homogeneous and ordered by increasing degree. Then $DQ(0)$ has the block form

$$\begin{pmatrix} A_1 & * & \cdots & * \\ 0 & A_2 & \ddots & \vdots \\ \vdots & \ddots & \ddots & * \\ 0 & \cdots & 0 & A_j \end{pmatrix}, \tag{3}$$

where the matrix A_1 belongs to the ρ_i's with the lowest degree, and so on. The vanishing entries in the lower triangle result from the increasing degrees of the elements of the Hilbert basis: Since ρ is minimal, the terms of $\rho_i \circ P$ have at least degree $\deg \rho_i$. Here we use the last assertion of Proposition 2. A similar argument yields that the entries above the diagonal depend only on P_{r}, and P_{r} has no effect on the A_i's. Therefore, we can ignore P_{r} without changing the eigenvalues of $DQ(0)$.

First we assume $P_\ell = P$ to be semisimple. Let $\lambda_1, \ldots, \lambda_l$ be the eigenvalues of P_ℓ. Since P_ℓ is Γ-equivariant, the corresponding eigenspaces V_i are Γ-invariant: If $v \in V_i$, then $P_\ell(\gamma v) = \gamma P_\ell(v) = \gamma \lambda_i v = \lambda_i \gamma v$, thus $\gamma v \in V_i$. We have, changing the coordinates in V,

$$P = \begin{pmatrix} \lambda_1 \operatorname{id}_{V_1} & & 0 \\ & \ddots & \\ 0 & & \lambda_l \operatorname{id}_{V_l} \end{pmatrix}.$$

Denote the coordinates in V_i by x_i, and let $\sigma_1, \ldots, \sigma_k$ be a minimal Hilbert basis consisting of multihomogeneous polynomials. If there are complex eigenvalues, it is understood that we pass to the complexification $V_{\mathbb{C}} = \mathbb{C} \otimes_{\mathbb{R}} V$ of the representation space V. However, a Hilbert basis on V is a real Hilbert basis on $V_{\mathbb{C}}$, and vice versa.

According to the classical theory, there is a polynomial map $Q' : \mathbb{R}^k \to \mathbb{R}^k$ with $Q' \circ \sigma = \sigma \circ P$. As a consequence of the following lemma, $DQ'(0)$ and $DQ(0)$ have the same spectrum.

Lemma 7. *The spectrum of the reduced map is independent of the choice of the minimal Hilbert basis. Passing to another minimal Hilbert basis means a polynomial change of coordinates, and conjugating the differentials.*

Proof. Let ρ and σ be minimal Hilbert maps. Obviously, polynomial maps $\Lambda, \mathrm{M} : \mathbb{R}^k \to \mathbb{R}^k$ with $\rho = \Lambda \circ \sigma$ and $\sigma = \mathrm{M} \circ \rho$ exist. From the minimality of the Hilbert bases, $D\Lambda(0)^{-1} = D\mathrm{M}(0)$ follows. Here, we use the fact that the differentials are uniquely determined by those relations (cf. Lemma 5).

Let Q' be the reduced map with respect to σ: $\sigma \circ P = Q' \circ \sigma$. Then we have $\Lambda \circ Q' \circ \mathrm{M} = Q$ on Y, and $DQ(0) = D\Lambda(0)DQ'(0)D\Lambda(0)^{-1}$. □

Write

$$\sigma_j(x) = \sum_\alpha c_\alpha \prod_{i=1}^{l} x_i^{\alpha_i} ,$$

where the sum runs over the l-tuple of multiindices $\alpha_i \in N_0^{\dim V_i}$. According to our assumption, there exists a multiindex $\beta \in N_0^l$ such that $c_\alpha \neq 0$, only if $|\alpha_1| = \beta_1, \ldots, |\alpha_l| = \beta_l$. Now compute

$$Q_j'(\sigma(x)) = \sigma_j(P(x)) = \sum_\alpha c_\alpha \prod_{i=1}^{l} (\lambda_i x_i)^{\alpha_i} = \sigma_j(x) \prod_{i=1}^{l} \lambda_i^{|\alpha_i|} \tag{4}$$

Thus, Q' is linear and diagonal, and its eigenvalues are some products of eigenvalues of P. Since $DQ(0)$ and $Q' = DQ'(0)$ are conjugated, we have calculated the eigenvalues of $DQ(0)$ for a polynomial map $P : V \to V$ with semisimple linear part.

Now, take an arbitrary $P \in C_\Gamma^{qn}(V, V)$. Then there is a sequence of polynomials $P_i \in C_\Gamma^{qn}(V, V)$ with semisimple linear parts, which converges to P. Since $\widetilde{\mathcal{J}} : P \mapsto D\mathcal{J}_{\mathrm{m}}(P)(0) \in \mathbb{R}^{k \times k}$ is continuous, and the eigenvalues depend in some sense continuously on a matrix (see [21, II Thm. 5.14]), the claim holds for all maps P. □

Corollary 8. *In the situation of the last theorem, let λ be an eigenvalue of $DP(0)$, then $DQ(0)$ has the eigenvalue λ or $|\lambda|^2$.*

Proof. Resume the notations used in the proof of the theorem. Denote the eigenspace of λ by V_λ. Let $W \subset V_\lambda$ be an irreducible subrepresentation. If W is the trivial representation, then the coordinate σ_1 in $W \cong \mathbb{R}$ is Γ-invariant, and there is a minimal multihomogeneous Hilbert basis containing σ_1. According to (4), λ is an eigenvalue of $DQ(0)$.

Now suppose that W is nontrivial. If λ is real, then the eigenspace of λ contains a nontrivial (real) representation W'. This representation is orthogonal, hence the square σ_1 of the Euclidean norm on W' is Γ-invariant. Moreover, there are no linear invariants. Therefore, there is a multihomogeneous Hilbert basis containing σ_1. And according to (4), $\lambda^2 = |\lambda|^2$ is an eigenvalue of $DQ(0)$.

If λ is complex, then $\bar{\lambda}$ is also an eigenvalue, and the eigenspace $V_{\bar{\lambda}}$ contains the conjugated representation $\overline{W}$. If x_W and $x_{\overline{W}}$ are the coordinates of W and $\overline{W}$, respectively, then $\sigma_1(x) := \langle x_W, \mu x_{\overline{W}} \rangle$ is a Γ-invariant polynomial of degree 2, where $\mu : \overline{W} \to W$ is a anti-linear Γ-equivariant isomorphism. Since there are no linear invariants on W and $\overline{W}$, a minimal multihomogeneous Hilbert basis containing σ_1 exists. And again according to (4), $\lambda\bar{\lambda} = |\lambda|^2$ is an eigenvalue of $DQ(0)$. □

More details concerning the relation between the eigenvalues of $DP(0)$ and $DQ(0)$ will be given in Sect. 8. By Theorem 6 and Corollary 8 we conclude that the eigenvalues of $DP(0)$ lie inside the (closed) unit disc, if and only if all eigenvalues of $DQ(0)$ have this property. In particular, it follows by the principle of linearized stability that 0 is an asymptotically stable equilibrium of P, if all eigenvalues of $DQ(0)$ are strictly inside the unit disc; it is an unstable equilibrium, if at least one eigenvalue of $DQ(0)$ lies outside the closed unit disc. Also, if $DP(0)$ has a critical eigenvalue, then the same holds for $DQ(0)$. In general, the converse is not true. However, it is true provided that the spectrum of $DQ(0)$ is contained in the closed unit disc.

Remark 9. (Cf. Sect. 8) Let R be the subspace of $\mathbb{R}^k$ associated with the linear and quadratic components of ρ. Then $DQ(0)$ keeps R invariant. To determine the stability of the equilibrium 0, it suffices to check the eigenvalues of $DQ(0)|_R$.

Example 10. Let $\Gamma = \mathbb{Z}_4$ act freely on $\mathbb{R}^2$. A minimal Hilbert basis is given by $u(x,y) = x^2 + y^2$, $v(x,y) = x^2y^2$, and $w(x,y) = x^3y - xy^3$. A $\mathbb{Z}_4$-equivariant map P has the equilibrium $P(0) = 0$ and the linear part $\begin{pmatrix} \lambda & -\mu \\ \mu & \lambda \end{pmatrix}$. The reduced map Q has the differential $DQ(0) = \begin{pmatrix} \lambda^2+\mu^2 & * & * \\ 0 & \lambda^4 - 6\lambda^2\mu^2 + \mu^4 & 2\lambda^3\mu - 2\lambda\mu^3 \\ 0 & -8\lambda^3\mu - 8\lambda\mu^3 & \lambda^4 - 6\lambda^2\mu^2 + \mu^4 \end{pmatrix}$.
The eigenvalues of $DP(0)$ are $\lambda + \mathrm{i}\mu$ and $\lambda - \mathrm{i}\mu$, while $DQ(0)$ has the eigenvalues $\lambda^2 + \mu^2$, $(\lambda + \mathrm{i}\mu)^4$, and $(\lambda - \mathrm{i}\mu)^4$. The eigenvalue $\lambda^2 + \mu^2$, which corresponds to the quadratic invariant u, suffices to determine asymptotic stability or instability of the equilibrium $0 = P(0)$.

Example 11. Let $\Gamma = \mathbb{Z}_2$ act on $\mathbb{R}^2$ as $\pm\,\mathrm{id}$. A minimal Hilbert basis is given by $\rho_1(x,y) = x^2$, $\rho_2(x,y) = y^2$, and $\rho_3(x,y) = xy$. They have the relation $r(\rho_1,\rho_2,\rho_3) = \rho_1\rho_2 - \rho_3^2$. The orbit space is contained in the variety $r = 0$, and given by $\rho_1 \geq 0$.

Let P be a $\mathbb{Z}_2$-equivariant map. Then P has the equilibrium 0 and the linear part $\begin{pmatrix} a & b \\ c & d \end{pmatrix}$. The eigenvalues of $DP(0)$ are $\lambda_{1,2} := \frac{a+d}{2} \pm \frac{1}{2}\sqrt{(a-d)^2 + 4bc}$.

The reduced map Q has the linear part $DQ(0) = \begin{pmatrix} a^2 & b^2 & 2ab \\ c^2 & d^2 & 2cd \\ ac & bd & ad+bc \end{pmatrix}$ with eigenvalues $\mu_1 = \lambda_1^2$, $\mu_2 = \lambda_2^2$ and $\mu_3 = \lambda_1\lambda_2$.

Observe, that if $\lambda_1 = \lambda_2^{-1}$ and $\lambda_1 > 1$, then the eigenvalue μ_3 of $DQ(0)$ lies on the unit circle, but no one of $DP(0)$ does. This phenomenon might occur, if P and hence Q have unstable eigenvalues.

Here we can use the eigenvectors corresponding to μ_3 in order to exclude that $DP(0)$ has a critical eigenvalue. An eigenvector of μ_3 is given by $v := (-b, c, \frac{a-d}{2})$. The line $\mathbb{R}v$ meets the orbit space nontrivially, if and only if the eigenvalues λ_1 and λ_2 coincide, i.e., $(a-d)^2 - 4bc = 0$. In this case, μ_3 is equal to μ_1 and μ_2. Thus, $DP(0)$ has a critical eigenvalue, if and only if μ_3 is critical and has an eigenvector in the orbit space. We show in Sect. 8 how to handle such situations more generally.

As a final conclusion from Theorem 6 we state the following corollary.

Corollary 12. *If P is a local diffeomorphism at 0, then the same holds true for Q.*

Proof. No eigenvalue of $DP(0)$ vanishes, hence all eigenvalues of $DQ(0)$ differ from zero. Therefore, $DQ(0)$ is a local diffeomorphism. □

Of course, analogous results are valid for vector fields (cf. [23,33]).

Theorem 13. *Let $F \in \mathrm{Vec}^q_\Gamma(V)$ and $G = \mathcal{J}_\rho(F)$. Suppose that $F(0) = 0$, and ρ is a minimal Hilbert map. Then the eigenvalue of $DG(0)$ are sums of the eigenvalues of $DF(0)$.*
If λ is an eigenvalue of $DF(0)$, then $DG(0)$ has λ or $\lambda + \bar{\lambda}$ in its spectrum.

4 Eigenvalues at a Relative Equilibrium

In this section the meaning of Γ, V, ρ, Y, P, and Q is as before. Next we study dynamically invariant orbits. Assume the orbit Γx to be invariant with respect to P or, equivalently, $P(x) \in \Gamma x$. Such a point x is called a *relative equilibrium.* Then the orbit Γx of x is an *invariant orbit.*

As mentioned above, an invariant orbit Γx reduces to an equilibrium $\rho(x)$ on the orbit space. Moreover, the stability of Γx is equivalent to the stability of $\rho(x)$ with respect to $Q|_Y$. Since Q does not know anything about the dynamical drift along the orbits, we cannot expect to find statements about the stability of the single trajectory of x using Q, but only for the whole orbit Γx as a set. This is a natural point of view, because all points of the orbit show exactly the same dynamical behavior.

We extend Theorem 6 to points x with submaximal symmetry, and exploit the eigenvalues of $DQ(\rho(x))$ to determine the stability of the invariant orbits Γx. For this we need to have informations about the local structure of the orbit space near $\rho(x)$. The slice theorem states that the orbit space locally looks like the origin of the orbit space of the isotropy group Γ_x. Furthermore, it is necessary to introduce a kind of linearization of P along an orbit. We begin by discussing the relevant concepts.

Local structure of the orbit space: The slice theorem An important concept in this context is the *isotropy subgroup* Γ_x *of a point* x. This group of all $\gamma \in \Gamma$ which fix x ($\Gamma_x := \{\gamma \in \Gamma \mid \gamma x = x\}$) describes the local symmetry.

In our simple situation tangent spaces inherit their inner products from the vector space V. If we work with manifolds, we use a Γ-invariant Riemannian metric, which exists always. The exponential map $\exp_x : T_x V \to V$ is a local Γ_x-equivariant diffeomorphism. In our case we can identify $T_x V$ with V, and $\exp_x$ is just the translation by x.

Let N_x be the orthogonal complement of $T_x \Gamma x$ in $T_x V$. N_x is the space of vectors normal to the orbit, and is called the *normal space* at x. The dimension of N_x depends on the dimension of the orbit, and is constant on orbits. The isotropy subgroup Γ_x acts naturally on the tangent space $T_x V$. This action emerges from taking the derivation of the action of Γ. Note that this action is orthogonal, too, and that the tangent space $T_x \Gamma x$ of the orbit and, therefore, N_x are Γ_x-invariant.

Since a group element γ maps $T_x \Gamma x$ onto $T_{\gamma x} \Gamma x$, we have $\gamma : N_x \to N_{\gamma x}$. For a small $\varepsilon > 0$ and $y \in \Gamma x$ we call the ε-neighborhood S_y of 0 in N_y a *slice at* y, which is Γ_y-invariant. We identify S_y with its image $\exp_y(S_y)$, and consider N_y as a subspace of V.

If $\varepsilon > 0$ is small enough, an ε-neighborhood U_x of the orbit Γx carries the structure of a twisted product or of a fiber bundle with fiber S_x over the orbit. The isotropy group Γ_x acts on the product $\Gamma \times S_x$ by

$$(\delta, (\gamma, v)) \longmapsto (\gamma\delta^{-1}, \delta v) \ .$$

Let $[\gamma, v]$ be the equivalence class $\Gamma_x(\gamma, v)$. The quotient $(\Gamma \times S_x)/\Gamma_x$ of this free action, denoted by $\Gamma \times_{\Gamma_x} S_x$, is a Γ-bundle with fiber S_x over $\Gamma x \cong \Gamma / \Gamma_x$. Γ acts on $\Gamma \times_{\Gamma_x} S_x$ by $(\delta, [\gamma, v]) \mapsto [\delta\gamma, v]$.

Proposition 14 (Differential slice theorem). *If* ε *is small enough, then*

$$\lambda_x : \begin{cases} \Gamma \times_{\Gamma_x} S_x \longrightarrow U_x \\ \quad [\gamma, v] \quad \longmapsto \gamma v \end{cases}$$

is a Γ*-equivariant diffeomorphism.*

Proof. See [20,5], and for the case of a reductive group [39]. □

We call a bundle $S \to \Gamma x$ satisfying the slice theorem a *slice bundle*. S is accompanied by a *tubular neighborhood* $U_x = \lambda_x(S)$, also called a *tube*. There are some characterizing properties of a slice bundle: If the intersection of $\lambda_x(S_y)$ and $\lambda_x(S_{\gamma y}) = \lambda_x(\gamma S_y) = \gamma \lambda_x(S_y)$ is non-empty, then $\gamma \in \Gamma_y$. In this case both sets coincide. Of course, the converse holds true, too. Furthermore, $\lambda_x(S_y)$ is closed in $U_x = \Gamma \lambda_x(S_y)$.

From the slice theorem follows that the orbit space is locally diffeomorphic to the orbit space of Γ_x, meaning that λ_x induces a homeomorphism

$$\lambda_x^* : \begin{cases} C^\infty(U_x)^\Gamma \longrightarrow C^\infty(S_x)^{\Gamma_x} \\ \quad f \quad \longmapsto \ f \circ \lambda_x \end{cases} .$$

As the following corollary implies, this is also valid for the embedded orbit spaces.

Corollary 15. *Let $\rho_1, \ldots, \rho_k$ be a Hilbert basis of V, and $\sigma_1, \ldots, \sigma_l$ a minimal Hilbert basis for the action of Γ_x on N_x. Denote the local orbit space $\sigma(N_x) \subset \mathbb{R}^l$ by Y_x. Then a C^∞-map $H_x : \mathbb{R}^l \to \mathbb{R}^k$ exists with*

1. *$H_x \circ \sigma = \rho \circ \lambda_x$ on S_x.*
2. *There are open sets $U \subset \mathbb{R}^l$ and $W \subset \mathbb{R}^k$ such that $H_x(Y_x \cap U) = H_x(\sigma(S_x)) = \rho(U_x) = Y \cap W$.*
3. *The differential $DH_x(0)$ is one-to-one.*
4. *There are open sets $U \subset \mathbb{R}^l$ and $W \subset \mathbb{R}^k$ and a C^∞-map $\widetilde{H}_x : W \to \mathbb{R}^l$ such that $\widetilde{H}_x \circ H_x|_U = \mathrm{id}_U$.*

Obviously, $l \leq r$ and $\max_i \deg \sigma_i \leq \max_i \deg \rho_i$.

We will prove this together with the next lemma.

At first, we introduce an appropriate notion of tangent spaces of the orbit space Y. Since it is possible to add relations to Q without destroying (2), P determines Q uniquely on Y, only. If the relations possess nontrivial linear parts at $y \in Y$, then $DQ(y)$ is not uniquely determined by P and (2). However, this is true for the restriction of $DQ(y)$ to the tangent space of Y defined as follows.

As above, denote the ideal of relations by I, and the variety of I by Z. The tangent space of Z at $z \in Z$ is given in terms of I in the following way: Let $r_1, \ldots, r_j$ be a set of generators of I. The differentials $Dr_1(z), \ldots, Dr_j(z)$ at z are linear functions on $T_z\mathbb{R}^k$, and define a subspace T_zZ called the *(Zariski) tangent space of Z at z*. For $y \in Y$ we define the *tangent space of Y at y* as $T_yY := T_yZ$. This definition is independent of the choice of the generators of I (cf. [19]).

If $f \in C^{qn}(V)^\Gamma$ and $g \in C^n(\mathbb{R}^k)$ with $g \circ \rho = f$, then $Dg(y)|_{T_yY}$ is uniquely determined by f. Also $DQ(y)|_{T_yY}$ is uniquely determined by P and (2), and respects the tangent spaces: $DQ(y)T_yY \subset T_{Q(y)}Y$.

If the Hilbert basis is minimal, then no relation possesses a nontrivial linear part at 0, and therefore, the tangent space of Y at $\rho(0)$ has dimension k (cf. Lemma 5).

There is a supplement to Corollary 15:

Lemma 16. *For the map H_x in Corollary 15 we have* range $DH_x(0) = T_yY$, *where $y := \rho(x)$.*

Proof. We prove Corollary 15 and Lemma 16 together. Since $\rho \circ \lambda_x : N_x \to \mathbb{R}^k$ is a Γ_x-invariant C^∞-map, there is a C^∞-map $H_x : \mathbb{R}^l \to \mathbb{R}^k$ with $\rho \circ \lambda_x = H_x \circ \sigma$. Since ρ and σ are proper, the second claim of the Corollary follows.

Furthermore, there is a map $\tilde{\sigma} \in C^\infty(U_x, \mathbb{R}^l)^\Gamma$ with $\tilde{\sigma} \circ \lambda_x = \sigma$. According to Proposition 2 we can find a C^∞-map Λ with $\tilde{\sigma} = \Lambda \circ \rho$. Altogether, it gives

$$\sigma = \Lambda \circ \rho \circ \lambda_x = \Lambda \circ H_x \circ \sigma \ .$$

Since σ is minimal, $\Lambda \circ \rho \circ \lambda_x$ determines the linear part of $\Lambda \circ H_x$ at 0 uniquely. Thus,

$$\mathrm{id}_{\mathbb{R}^l} = \mathrm{id}_{T_0 \sigma(N_x)} = D(\Lambda \circ H_x)(0) = D\Lambda(\rho(x)) DH_x(0) \ .$$

Therefore, $DH_x(0)$ is one-to-one. This proves claims 3 and 4 of the corollary.

The same arguments work for Λ, and yield $DH_x(0)D\Lambda(y)|_{T_yY} = \mathrm{id}_{T_yY}$. Therefore, $D\Lambda(y) : T_yY \to T_0Y_x \cong \mathbb{R}^l$ is one-to-one.

Let r be a relation of ρ, then $r \circ H_x$ is a relation of σ, and has a vanishing linear part at $0 \in \mathbb{R}^l$. Hence, range $DH_x(0) \subset \ker Dr(H_x(0))$. Since this holds true for all relations r of ρ, $DH_x(0)$ maps into $T_yY = \bigcap_r Dr(y)$. We have the sequence of maps

$$T_0Y_x \stackrel{DH_x(0)}{\longrightarrow} T_yY \stackrel{D\Lambda(y)}{\longrightarrow} T_0Y_x$$

which are one-to-one. From this the assertion of the lemma follows. □

By Remark 1, the dimension of T_0Y_x is independent of the choice of σ. Since the map H_x establishes a local diffeomorphism between the orbit space Y near y and the local orbit space Y_x, the dimension of T_yY is independent of the choice of σ and ρ.

Lemma 17. *Define $R_y := DH_x(0)R_0$, where $y := \rho(x)$ and $R_0 = R$ as in Remark 9. Let Q be the reduced dynamics of $P \in C^q_\Gamma(V,V)$. Then $DQ(y)$ maps R_y into $R_{Q(y)}$.*

Proof. This follows from Remark 9 and Corollary 15. □

For later use we mention another definition of the tangent space T_yY of the orbit space. Denote the maximal ideal of Γ-invariant polynomials which vanish at $x \in V$ by J_x. The tangent space of the orbit space V/Γ at Γx is the dual space of J_x/J_x^2 ([27, p. 147]). Let $\rho_1, \ldots, \rho_k$ be a minimal Hilbert basis. It generates a complement of J_0^2 in J_0. Let $p = \mathcal{J}(p) \circ \rho$ be a Γ-invariant polynomial. Due to the minimality of ρ, the linear part of $\mathcal{J}(p)$ is unique. We identify the unit vector $e_i \in T_0Y$ with the linear function which sends a polynomial p to $\partial_{\rho_i}\mathcal{J}(P)(0)$, the coefficient of the ρ_i-term in $\mathcal{J}(p)$. This function is well-defined on J_0/J_0^2, too. This identification depends on the choice of the Hilbert basis. The slice theorem allows to define such a duality also for $x \neq 0$.

Remark 18. Let K_i be the subset of J_x of homogeneous polynomials of degree i, then $(K_1 \oplus K_2)/J_x^2$ is dual to $R_{\rho(x)}$.

Remark 19. Using the alternative definition of T_yY, one can view the differential $DQ(y)|_{T_yY}$, $y := \rho(x)$, of the reduced map $Q = \mathcal{J}_{\mathrm{m}}(P)$ as the dual map of

$$P^* : \begin{cases} J_{P(x)}/J_{P(x)}^2 & \longrightarrow J_x/J_x^2 \\ p & \longmapsto p \circ P \end{cases} .$$

Suppose $P(0) = 0$ and that $V_1 \oplus V_2$ is a Γ-invariant decomposition of V, and V_1 is invariant with respect to $DP(0)$, where 0 is an equilibrium of P. Denote by $L_{2,0}$, $L_{1,1}$, and $L_{0,2}$ the sets of Γ-invariant, multihomogeneous, quadratic polynomials. Then $L_{0,2}$ and $L_{1,1} \oplus L_{0,2}$ are $DP(0)$-invariant, and $(L_{2,0}/J_0^2)^*$ and $((L_{2,0} \oplus L_{1,1})/J_0^2)^*$ are invariant with respect to $S := D\mathcal{J}_m(DP(0))$. These facts can be adopted for a relative equilibrium of P using the map H_x.

Stratification by orbit type is another useful concept to describe the structure of the orbit space. The origin of a representation has the full group as isotropy group. This is, of course, also valid for the fixed points. All other points have smaller isotropy groups. The slice theorem shows that the isotropy group Γ_x cannot increase suddenly, if the point x varies. Modulo conjugation, the isotropy groups of points near an orbit Γx are subgroups of Γ_x.

The conjugation class of the isotropy subgroup is constant on an orbit ($\Gamma_{\gamma x} = \gamma \Gamma_x \gamma^{-1}$). The map $\mathcal{C} : x \mapsto [\Gamma_x]$ is Γ-invariant, where $[\Gamma_x]$ is the conjugacy class of Γ_x in Γ, called the *orbit type of* x. The fibers of $\mathcal{C}$ are Γ-invariant, homogeneous ($\lambda \neq 0 \Rightarrow \Gamma_{\lambda x} = \Gamma_x$), and are called *strata*. Sometimes the connected components of the fibers of $\mathcal{C}$ bear this name. The stratum of x is a bundle with fiber $\Gamma/\Gamma_x \cong \Gamma x$.

The strata have some nice properties following from the slice theorem:

1. The strata are locally closed smooth manifolds.
2. The boundary of a stratum is a union of other strata.
3. Let Σ_1 and Σ_2 be two distinct strata such that $\Sigma_1 \cap \overline{\Sigma_2} \neq \emptyset$, then Σ_1 is already contained in the boundary of Σ_2,
4. and there are points $x \in \Sigma_1$ and $y \in \Sigma_2$ such that Γ_y is a proper subgroup of Γ_x.
5. Let $N_x^0 := N_x^{\Gamma_x}$ be the space of fixed points in N_x. If $S_x^0 := S_x \cap N_x^0$, then $\Gamma \times_{\Gamma_x} N_x^0$ and $\Gamma \times_{\Gamma_x} S_x^0$ are trivial bundles over the orbit. We have $\lambda_x(\Gamma \times_{\Gamma_x} S_x^0) = U_x \cap \Sigma_x$, where Σ_x is the stratum containing x.
6. There is a unique stratum which is open and dense in V, called the *principal stratum*.
7. There are at most finitely many strata in V.

$\mathcal{C}$ and the Hilbert map ρ induce a stratification of the orbit space Y. The ρ-images of the strata of V are strata of Y with 1.–3. and 6. This stratification coincides with the canonical stratification of Y as a semialgebraic set ([4]).

Denote by Σ_x the stratum containing x. The image of the map $D\rho(x) : T_xV \to T_y\mathbb{R}^k$ is the tangent space $T_y\Upsilon$ of the stratum $\Upsilon := \rho(\Sigma_x) \subset Y$ at $y := \rho(x)$. Indeed, $D\rho(x)(T_x\Sigma_x) = T_y\Upsilon$. If Σ_x is the principal stratum, then Υ is the principal stratum of Y, and the tangent spaces of Υ and Y coincide: $T_y\Upsilon = T_yY$. Otherwise $T_y\Upsilon$ is a proper subset of T_yY.

$T_y\Upsilon$ covers the directions inside the stratum Υ, while T_yY contains also the directions out of Υ but inside the orbit space.

The stratification can be useful for investigating equivariant dynamics. A Γ-equivariant vector field F is tangent to the strata, i.e., $F(x) \in T_x\Sigma_x$. The reduced vector field $G := \mathcal{J}(F)$ is also tangent to the strata of Y, because

$$G \circ \rho = D\rho(x)F(x) \tag{5}$$

and $D\rho(x)T_x\Sigma_x = T_y\Upsilon$. A result by Bierstone [4] and Schwarz [38] states that each smooth vector field G on $\mathbb{R}^k$ which is tangent to the strata of Y can be lifted to a smooth Γ-equivariant vector field F on V such that (5) holds.

Similar things hold for a Γ-equivariant map P. The isotropy group of the image $P(x)$ of a point x contains Γ_x, because of $\gamma P(x) = P(\gamma x) = P(x)$ for $\gamma \in \Gamma_x$. Therefore, fixed point spaces are invariant, and $P(x)$ lies in the closure of Σ_x. If P is one-to-one, then the strata are invariant.

The map H_x, defined in Corollary 15, sends the strata of the local orbit space to strata of Y, locally. As a byproduct we make the remark

Remark 20. Over each stratum of the orbit space there are two continuous bundles with fibers T_zY and R_z, respectively. They are invariant with respect to reduced diffeomorphisms.

Linearizations along an invariant orbit and eigenvalues at a relative equilibrium First we consider an invariant orbit Γx of the vector field $F \in \mathrm{Vec}^q_\Gamma(V)$, i.e., $F(x) \in T_x\Gamma x$. The slice theorem provides coordinates near Γx. Locally, the space is split into directions tangent to the orbit and normal directions. This yields a decomposition of the vector field $F = F_{\mathrm{T}} + F_{\mathrm{N}}$ in a tangent part F_{T} and a normal part F_{N} such that $F_{\mathrm{T}}(z) \in T_z\Gamma z$ and $F_{\mathrm{N}}(z) \in S_y$ ([24]), for $z \in S_y$ and $y \in \Gamma x$. Moreover, this decomposition can be achieved in such a way that F_{N} is a Γ_y-equivariant vector field on S_y.

Proposition 21. *Let Γx be an invariant orbit of $F \in \mathrm{Vec}^q_\Gamma(V)$. Then there are a slice bundle $S \to \Gamma x$ with tube $\lambda_x(S) = U_x$ and two vector fields F_{T}, $F_{\mathrm{N}} \in \mathrm{Vec}^q_\Gamma(U_x)$ with*

1. $F|_{U_x} = F_{\mathrm{T}} + F_{\mathrm{N}}$.
2. $\forall z \in U_x : F_{\mathrm{T}}(z) \in T_z\Gamma z$.
3. $\forall y \in \Gamma x, z \in S_y : F_{\mathrm{T}}(z) \in N_y$.

If $y \in \Gamma x$, then $F_{\mathrm{N}}|_{S_y} \in \mathrm{Vec}^q_{\Gamma_y}(S_y)$ has an equilibrium at $0 \in S_x$.

Proof. The proofs for C^∞-vector fields in [14,24] also work in the present case. □

For a given relative equilibrium x we consider the vector field $\widetilde{F} := F_{\mathrm{N}}|_{S_x} \in \mathrm{Vec}^q_{\Gamma_x}(S_x)$. Since F_{T} influences only the flow inside the orbits, it does not affect the stability of the invariant orbit Γx. But F_{N}, and therefore $\widetilde{F}$, describe whether the distance between Γx and a nearby trajectory of F increases or decreases, and thus reflect the stability of the invariant orbit (cf. the discussion for maps below). Thus, we use $\widetilde{F}$ to describe the stability of the relative equilibrium x, and regard its linearization at 0 as the linearization of F along Γx.

Note that F_{T} and F_{N} are not unique, in general. However, if G is the reduced vector field, then $G \circ \rho = D\rho F = D\rho(F_{\mathrm{N}} + F_{\mathrm{T}}) = D\rho F_{\mathrm{N}}$. Now we are ready to apply Theorem 13 to $\widetilde{F}$ (cf. [23]).

Theorem 22. *Let $F \in \mathrm{Vec}^q_\Gamma(V)$, and $G \in \mathrm{Vec}^1(\mathbb{R}^k)$ be a reduced vector field with (5). Suppose Γx is an invariant orbit of F, and, therefore, $y := \rho(x)$ is an equilibrium of G. Define $\widetilde{F}$ be as above. Then the eigenvalues of $DG(y)|_{T_yY}$ are sums of the eigenvalues of $D\widetilde{F}(0)$. Moreover, if λ is an eigenvalue of $D\widetilde{F}(0)$, then $DG(y)|_{T_yY}$ has the eigenvalue λ or $\lambda + \bar{\lambda}$.*

Remark 23. This statement remains true, if $DG(y)|_{R_y}$ replaces $DG(y)|_{T_yY}$, where R_y is as in Lemma 17.

Proof. Assume a minimal Hilbert map σ on N_x. There is a reduced vector field $G_\sigma \in \mathrm{Vec}^1(W)$ with $G_\sigma \circ \sigma = D\sigma\widetilde{F}$, where W is an open set with $W \cap \sigma(N_x) = \sigma(S_x)$. The relation between G_σ and G is given by

$$G \circ H_x \circ \sigma = G \circ \rho \circ \lambda_x = D\rho F = DH_x\, D\sigma F = (DH_x\, G_\sigma) \circ \sigma\ ,$$

where H_x is as in Corollary 15. As σ is minimal, this yields $D(G \circ H_x)(0) = D(DH_x\, G_\sigma)(0) = DH_x(0)DG_\sigma(0) + DDH[G_\sigma(0), v]$. Since the last term vanishes, we obtain $DG(y)DH_x(0) = DH_x(0)DG_\sigma(0)$.

Since $DH(0)$ is one-to-one onto T_yY, the spectra of $DG(y)|_{T_yY}$ and $DG_\sigma(0)$ coincide. Now applying Theorem 13 to G_σ proves the claims. □

Finally we carry out the analogous analysis for a Γ-equivariant map P. The splitting of maps is a little bit harder than for vector fields. We are searching for a map $\widetilde{P}$ on the slice S_x such that P is the combination of $\widetilde{P}$ and a drift along the orbits. A way to define $\widetilde{P}$ is to shift $P(y)$ along the orbit $\Gamma P(y)$. If y is close enough to the invariant orbit, then $\Gamma P(y)$ meets the slice S_x, and we may take a point of intersection for $\widetilde{P}(y)$. Unfortunately, we cannot guarantee that $\widetilde{P}$ obtained this way is Γ_x-equivariant. Therefore, we have to be more careful.

Proposition 24. *Let Γx be an invariant orbit of the map $P \in C^q_\Gamma(V, V)$. There is a smallest number a such that an element c in the centralizer $C(\Gamma_x)$ of Γ_x exists with $P^a(x) = cx$. Then a slice bundle $S \to \Gamma x$ with tube U_x exists, and there are maps $\widetilde{P} : S_x \to N_x$ and $\gamma : S_x \to \Gamma$ with*

1. *$\forall y \in S_x : P^a(y) = \gamma(y)(\widetilde{P}(y))$.*
2. *$\widetilde{P} \in C^q_{\Gamma_x}(S_x, N_x)$.*
3. *$\forall y \in S_x : \gamma(y) \in C^q(\Gamma_x)$.*

Proof. The proofs in [14] or [28] for the C^∞ case remain valid in the present case. □

The stability of the invariant orbit Γx with respect to P is equivalent to the stability of the equilibrium 0 with respect to $\widetilde{P}$: Given a neighborhood U of Γx, then there is a neighborhood U' of 0 in S_x such that $\Gamma(U') \subset U$. Suppose the equilibrium $0 = \widetilde{P}(0)$ is stable, i.e., a neighborhood V' exists such that $\widetilde{P}^n(y) \in U'$ for all $n \in \mathbb{N}_0$ and $y \in V'$. Then $P^{an}(\Gamma(V')) = \Gamma P^{an}(V') = \Gamma \widetilde{P}^n(V') \subset U$. Choose a neighborhood V of Γx such that $P^n(V) \subset \Gamma V'$ for all $n \leq a$, then $P^n(V) \subset U$ for all $n \in \mathbb{N}$, i.e., Γx is stable. Also the converse is true.

Hence, $\widetilde{P}$ provides an appropriate description of the stability of the relative equilibrium x. We take its linearization at 0 for the linearization of P along Γx. If x is even an equilibrium of P, then the eigenvalues of $DP(x)|_{T_x \Gamma_x}$ have modulus 1. And the a-th power of the other eigenvalues of $DP(x)$ are the eigenvalues of $D\widetilde{P}(0)$.

Theorem 25 ([32]). *Let P and Q be as in (2). Suppose Γx is an invariant orbit of P, and therefore $y := \rho(x)$ is an equilibrium of Q. Let $\widetilde{P}$ be as in Proposition 24.*
Let U be either the open or the closed unit disc or the complement of one of them. Then the eigenvalues of $DQ(y)|_{T_y Y}$ lie in U, if and only if the same holds true for $D\widetilde{P}(0)$. This statement remains valid, if $DQ(y)|_{R_y}$ replaces $DQ(y)|_{T_y Y}$, where R_y is taken from Lemma 17.

Proof. The line of arguments is very similar to the proof of Theorem 22. Using Theorem 6 and its Corollaries, and Corollary 15, we establish a correspondence between the eigenvalues of $D\widetilde{P}(0)$ and $DQ(y)|^a_{T_y Y}$. Note that the eigenvalues of $DQ(y)^a$ are a-th powers of the eigenvalues of $DQ(y)$. □

This theorem provides a criterion for asymptotic stability and instability of a relative equilibrium in terms of the reduced map. We point out that $\widetilde{P}$ is used only to prove, but not to apply the criterion. It can also be used to detect, when an asymptotically stable invariant orbit becomes unstable. We refer also to the Section 8 and the discussion after Corollary 8.

5 Relative Periodic Points and Poincaré Maps

In this section we use the last theorem to discuss relative periodic points. We consider a vector field $F \in \mathrm{Vec}^q_\Gamma(V)$ and its reduction to the orbit space $G := \mathcal{J}_\ast(F)$. Let $y \in Y$ be a periodic point of G, i.e., the trajectory through y is closed. The preimage Ω of its trajectory is a compact manifold in V. It is invariant with respect to both the flow Φ_t induced by F and the action of Γ. The points of Ω are called *relative periodic points.* Ω is contained in some stratum.

To investigate the stability of a periodic trajectory, one usually constructs a Poincaré map. This is the first return map Q on a hypersurface E transversal to the trajectory. The point y is an equilibrium of Q on E, whose stability reflects the stability of the periodic trajectory.

For the manifold Ω we construct a similar first return map (see [13]), and relate it to Q. The orthogonal complement $M\Omega$ of the tangent bundle $T\Omega$ in $TV|_\Omega$ has the fibers $M_z = T_z\Omega^\perp$. $M\Omega$ is Γ-invariant.

Take $x \in \Omega$. Let $S \to \Gamma x$ be a slice bundle with the tubular neighborhood U_x. The subbundle $R \to \Gamma x$ with fibers $R_z := S_z \cap M_z$ gives a Γ-invariant manifold $D := \lambda_x(R)$, where λ_x is as in the slice theorem.

Lemma 26. *If the slices are small enough, D is a Γ-invariant hypersurface transversal to F.*

Proof. Take the orthogonal projection v of $F(x)$ into $N_x = (T_x\Gamma x)^\perp$. Since M_x is a codimension one subspace of N_x, and is orthogonal to $F(x)$, v is also orthogonal to M_x, and is not the zero vector. As $F(x)$ is Γ_x-invariant, the vector v and, therefore, the function $\varphi(w) := \langle w, v\rangle$ are also Γ_x-invariant. Using λ^*, φ can be extended to a Γ-invariant function near Γx. Since, for $w \in N_x$, $\varphi(w) = 0$ is equivalent to $w \in M_x$, the Γ-invariance of φ yields $\forall z \in U_x : \varphi(z) = 0 \Leftrightarrow z \in D$. Furthermore, $\langle \operatorname{grad}\varphi(x), F(x)\rangle = \langle v, F(x)\rangle \neq 0$. □

D intersects Ω along the orbit Γx transversally. Furthermore, $T_zD = M_z \oplus T_z\Gamma z$.

Let θ be the period of y. Then $\Phi_\theta(x)$ meets the orbit at some γx. Since $\varphi(\gamma x) = 0$ and

$$\frac{d}{dt}\varphi(\Phi_\theta(x)) = \langle \operatorname{grad}\varphi(\gamma x), \dot{\Phi}_\theta(x)\rangle = \langle (T_x\gamma)v, (T_x\gamma)F(x)\rangle = \langle v, F(x)\rangle \neq 0\,,$$

applying the implicit function theorem to $(t, z) \mapsto \varphi(\Phi_t(z))$ gives a Γ-invariant function τ in a neighborhood of Γx, so that $\varphi(\Phi_{\tau(z)}(z)) = 0$.

If D' is the intersection of the domain of τ with D, then define the map

$$P : \begin{cases} D' \longrightarrow & D \\ z \longmapsto & \Phi_{\tau(z)}(z) \end{cases}.$$

P is Γ-equivariant and as smooth as F. Furthermore, Γx is an invariant orbit of P. We call (D, D', P, τ) *Poincaré system.*

The next proposition establishes a correspondence between a Poincaré map Q to the periodic point y of G and P.

Proposition 27. *Assume x to be a relative periodic point of $F \in \mathrm{Vec}^q_\Gamma(V)$ with Poincaré system (D, D', P, τ). At the periodic point $y := \rho(x)$ of the reduced system G, there is a transversal section E and a Poincaré map Q with $\rho^{-1}(E) \subset D$ and $\rho \circ P = Q \circ \rho$.*

Proof. By Proposition 2, a C^∞-function ψ exists in a neighborhood of y with $\psi \circ \rho = \varphi$. Define $E := \{z \in \mathbb{R}^k \mid \psi(z) = 0 \wedge \operatorname{grad} \psi(z) \neq 0\}$. Since

$$D\psi(y)G(y) = D\psi(y)D\rho(x)F(x) = D(\psi \circ \rho)(x)F(x) = D\varphi(x)F(x) \neq 0\,,$$

$\operatorname{grad} \psi(y)$ is not orthogonal to $G(y)$, and E is a section transversal to the trajectory of y.

Let $z \in V$ with $\rho(z) \in E$, then $\psi(\rho(z)) = \varphi(z) = 0$ and, therefore, $z \in D$. So $\rho^{-1}(E) \subset D$.

The flow of G is denoted by Ψ_t, and has the property $\Psi_t \circ \rho = \rho \circ \Phi_t$. Let Q be the Poincaré map on E. The trajectory of $\rho(z) \in E$, $z \in D$, meets E at the time $\tau(z)$, because $\rho(\Phi_{\tau(z)}(z)) \in \rho(D) \subset E$, but not before $\tau(z)$. So $Q(\rho(z)) = \rho(\Phi_{\tau(z)}(z)) = \rho(P(z))$. □

Theorem 25 applies to P. In particular, we have

Theorem 28. *In the situation of the last proposition, set $T := T_yY \cap T_yE$. For $\widetilde{P}$ as in Proposition 24, the eigenvalues of $D\widetilde{P}(0)$ are contained in $\{z \mid |z| < 1\}$, ($\{z \mid |z| \leq 1\}$, $\{z \mid |z| > 1\}$, or $\{z \mid |z| \geq 1\}$), if and only if the eigenvalues of $DQ(0)|_T$ do so.*

Using this theorem, the stability of relative periodic points is determinable in terms of eigenvalues of the linearization of any Poincaré map on the orbit space. Recall that Poincaré maps are related by conjugacy.

6 Lyapunov Exponents

Lyapunov exponents are a generalization of eigenvalues at equilibria. One averages the separation of neighboring trajectories during time evolution. The Lyapunov exponents are given by the average rate of expansion of distance vectors $v \neq 0$, e.g.

$$\lambda_v(x) := \lim_{n\to\infty} \frac{1}{n} \log \|DP^n(x)v\|$$

is a Lyapunov exponent of a map P at x. By Oseledec's multiplicative ergodic theorem, this limit exists almost everywhere with respect to an invariant measure ([3]), and at most $\dim V$ many different values occur, if the measure is ergodic.

Concerning orbit space reduction, for Lyapunov exponents the same results hold as presented above for eigenvalues of vector fields. In [31] it is proved that the Lyapunov exponents of a reduced map Q are $\mathbb{N}_0$-linear combinations of the exponents of the original map P. Moreover, if λ is an exponent of P, then λ or 2λ is an exponent of Q. Again, only those Lyapunov exponents of the reduced system are relevant which belong to the tangent space of the orbit space. By the way, here the invariant measure on the orbit is supposed to be the measure induced by the invariant measure of the original system via the Hilbert map.

Of course, eigenvalues and Lyapunov exponents have a lot of common aspects. However, the case of Lyapunov exponents is much more involved. Concerning eigenvalues, our argument in Sect. 3 is more or less elementary due to the fact that the general eigenvalue case can be reduced to the case, where the differential of the reduced map is diagonal and, therefore, self adjoint. It is not obvious how to achieve such a diagonalization in the general case of Lyapunov exponents. However, using the duality principle indicated in Remark 19 and the bundle picture as outlined throughout the paper, the argument in Sect. 3 can be carried over to the case of Lyapunov exponents. In particular, estimates are needed in order to exclude the influence of nonlinear terms in the original system, i.e., for the block matrix in (3) the asymptotic growth of the blocks above the diagonal as $t \to \infty$ must be estimated along the basic trajectory. They must not grow exponentially fast with respect to some norm.

7 Example

Let $P := \{\cos(kx), \sin(kx) \mid k \in \mathbb{N}_0\}$, and denote the span of P in the Sobolev space $H^m(]0, 2\pi[)$ by H^m_{per}. Set $Z := (H^0_{\text{per}})^4$, and let $D(C) := (H^4_{\text{per}})^4$ be the domain of definition of the closed operator C in Z given by the linearization at $u = 0$ of the right-hand side of the partial differential equation

$$u_t = -\partial_x^4 u - D\Delta u + \Lambda u + f(u, u_x) \ . \tag{6}$$

Here $u : \mathbb{R} \times \mathbb{R} \to \mathbb{R}^4$ is supposed to fulfill periodic boundary conditions $u(t, x) = u(t, x + 2\pi)$, and $f(u, u_x) = o(|u| + |u_x|)$ to be sufficiently smooth. Furthermore, we assume that (6) is invariant with respect to the flip $x \mapsto -x$ and the following action of the torus group T^2:

$$\begin{cases} T^2 \times \mathbb{R}^4 & \longrightarrow \qquad \mathbb{R}^4 \\ ((\vartheta_1, \vartheta_2), u) & \longmapsto (R(-\vartheta_1)(u_1, u_2)^t, R(-\vartheta_2)(u_3, u_4)^t) \end{cases}$$

Therefore, restrictions for D, Λ, and f arise. We consider the constant, real matrices

$$\Lambda = \begin{pmatrix} \lambda & -\omega_1 & 0 & 0 \\ \omega_1 & \lambda & 0 & 0 \\ 0 & 0 & \sigma & -\omega_2 \\ 0 & 0 & \omega_2 & \sigma \end{pmatrix}$$

and $D = \operatorname{diag}(d_1, d_1, d_2, d_2)$. The T^2-invariants and T^2-equivariants are

$$\begin{aligned} e_1 &= u_1^2 + u_2^2 & \qquad e_3 &= u_3^2 + u_4^2 \\ e_2 &= u_{x1}^2 + u_{x2}^2 & e_4 &= u_{x3}^2 + u_{x4}^2 \end{aligned}$$

$$\begin{aligned} e_5 &= u_1 u_{x1} + u_2 u_{x2} & \qquad e_7 &= u_3 u_{x3} + u_4 u_{x4} \\ e_6 &= -u_1 u_{x2} + u_2 u_{x1} & e_8 &= -u_3 u_{x4} + u_4 u_{x3} \end{aligned}$$

$$E_1 = \begin{pmatrix} u_1 \\ u_2 \\ 0 \\ 0 \end{pmatrix} \quad E_2 = \begin{pmatrix} -u_2 \\ u_1 \\ 0 \\ 0 \end{pmatrix} \quad E_3 = \begin{pmatrix} 0 \\ 0 \\ u_3 \\ u_4 \end{pmatrix} \quad E_4 = \begin{pmatrix} 0 \\ 0 \\ -u_4 \\ u_3 \end{pmatrix}$$
$$E_5 = \begin{pmatrix} u_{x1} \\ u_{x2} \\ 0 \\ 0 \end{pmatrix} \quad E_6 = \begin{pmatrix} -u_{x2} \\ u_{x1} \\ 0 \\ 0 \end{pmatrix} \quad E_7 = \begin{pmatrix} 0 \\ 0 \\ u_{x3} \\ u_{x4} \end{pmatrix} \quad E_8 = \begin{pmatrix} 0 \\ 0 \\ -u_{x4} \\ u_{x3} \end{pmatrix}$$

Changing the sign of x keeps $e_1, \ldots, e_4$ and $E_1, \ldots, E_4$ invariant, and switches the sign of $e_5, \ldots, e_8$ and $E_5, \ldots, E_8$. Therefore, the general form of the nonlinearity f is $\sum f_i(e_1, \ldots, e_8)E_i$, where $e_5, \ldots, e_8$ and $E_5, \ldots, E_8$ appear only in pairs.

In particular, $u = 0$ is always an equilibrium of (6). The eigenvalues of C are $\alpha_\pm^k = \lambda - k^4 + d_1k^2 \pm \mathrm{i}\omega_1$ and $\beta_\pm^k = \sigma - k^4 + d_2k^2 \pm \mathrm{i}\omega_2$, $k \in \mathbb{N}_0$, with eigenvectors

$$w_1^k = \begin{pmatrix} \mathrm{i}e^{ikx} \\ e^{ikx} \\ 0 \\ 0 \end{pmatrix} \quad w_2^k = \begin{pmatrix} \mathrm{i}e^{-ikx} \\ e^{-ikx} \\ 0 \\ 0 \end{pmatrix} \quad w_3^k = \begin{pmatrix} -\mathrm{i}e^{ikx} \\ e^{ikx} \\ 0 \\ 0 \end{pmatrix} \quad w_4^k = \begin{pmatrix} -\mathrm{i}e^{-ikx} \\ e^{-ikx} \\ 0 \\ 0 \end{pmatrix}$$
$$w_5^k = \begin{pmatrix} 0 \\ 0 \\ \mathrm{i}e^{ikx} \\ e^{ikx} \end{pmatrix} \quad w_6^k = \begin{pmatrix} 0 \\ 0 \\ \mathrm{i}e^{-ikx} \\ e^{-ikx} \end{pmatrix} \quad w_7^k = \begin{pmatrix} 0 \\ 0 \\ -\mathrm{i}e^{ikx} \\ e^{ikx} \end{pmatrix} \quad w_8^k = \begin{pmatrix} 0 \\ 0 \\ -\mathrm{i}e^{-ikx} \\ e^{-ikx} \end{pmatrix},$$

where w_1^k and w_2^k belongs to α_+^k, and so on. The eigenvalues are double for $k \in \mathbb{N}$, and simple for $k = 0$. We introduce $v_1^k = (w_1^k + w_4^k)/2$, $v_2^k = (w_1^k - w_4^k)/2\mathrm{i}$, $v_3^k = (w_2^k + w_3^k)/2$, $v_4^k = (w_2^k - w_3^k)/2\mathrm{i}$, $v_5^k = (w_5^k + w_8^k)/2$, $v_6^k = (w_5^k - w_7^k)/2\mathrm{i}$, $v_7^k = (w_6^k + w_7^k)/2$, and $v_8^k = (w_6^k - w_8^k)/2\mathrm{i}$. These vectors form a Hilbert space basis of both $D(C)$ and Z.

We have the following action of $\Gamma := \mathrm{O}(2) \times T^2$ on Z: $\varphi \in \mathrm{SO}(2)$ acts on u by $\varphi \cdot u : (t, x) \mapsto u(t, x - \varphi)$. Fix $\kappa \in \mathrm{O}(2) \setminus \mathrm{SO}(2)$, and let its action be $\kappa u : (t, x) \mapsto u(t, 2\pi - x)$. For $k \in \mathbb{N}$, set $v^k = (v_1^k, v_2^k, v_3^k, v_4^k, v_5^k, v_6^k, v_7^k, v_8^k)^t$, then

$$\varphi v^k = \begin{pmatrix} R(-k\varphi) & & & \\ & R(k\varphi) & & \\ & & R(-k\varphi) & \\ & & & R(k\varphi) \end{pmatrix} v^k \quad \text{and} \quad \kappa v^k = \begin{pmatrix} & I & & \\ I & & & \\ & & & I \\ & & I & \end{pmatrix} v^k,$$

with $R(\varphi) = \begin{pmatrix} \cos\varphi & -\sin\varphi \\ \sin\varphi & \cos\varphi \end{pmatrix}$ and $I = \begin{pmatrix} 1 & 0 \\ 0 & 1 \end{pmatrix}$, and

$$(\vartheta_1, \vartheta_2)v^k = \begin{pmatrix} R(-\vartheta_1) & & & \\ & R(-\vartheta_1) & & \\ & & R(-\vartheta_2) & \\ & & & R(-\vartheta_2) \end{pmatrix} v^k,$$

and similarly for $k = 0$.

Up to degree 5, f has the general form

$$\begin{aligned} f(u) &= \sum_{i,j=1}^{4} f_{ij} e_j(u) E_i(u) + \sum_{i,j=5}^{8} f_{ij} e_j(u) E_i(u) \\ &\quad + \sum_{j,k=1}^{4} \sum_{i=1}^{j} f_{kij} e_i(u) e_j(u) E_k(u) + \sum_{k=1}^{4} \sum_{j=5}^{8} \sum_{i=5}^{j} f_{kij} e_i(u) e_j(u) E_k(u) \\ &\quad + \sum_{j,k=5}^{8} \sum_{i=1}^{4} f_{kij} e_i(u) e_j(u) E_k(u) \ . \end{aligned}$$

We write

$$u = \sum_{i=1}^{4} b_i^{(0)} v_i^0 + \sum_{k=1}^{\infty} \sum_{i=1}^{8} b_i^{(k)} v_i^k \ .$$

The action of Γ on the coordinates is the transpose of the action on the basis vectors v.

Equations of type (6) appear as mathematical models in various contexts of pattern formation (see e.g. [41, Ch. III. 4]). Here the SO(2) component of the symmetry group Γ is related to the periodic boundary condition imposed above, while the remaining symmetries are either of geometric nature or normal form symmetries. In the following we are going to study symmetry breaking bifurcations from the basic equilibrium $u = 0$ of this problem. Of course, $u = 0$ has the symmetry of the full group Γ. We use λ and σ as bifurcation parameters, while the remaining parameters in the problem are supposed to be fixed. A few restrictions for the latter will be stated where needed.

We do not intend to analyze the bifurcation behavior of this problem as complete as possible here. In order to illustrate our general theory, we rather restrict ourselves to a specific bifurcation scenario including many aspects of the theory developed before. To this end we consider a critical point ($\lambda_0 = 1 - d_1$, $\sigma_0 = 1 - d_2$) in the parameter plane, where the basic equilibrium $u = 0$ becomes unstable simultaneously with respect to all modes $v_1^1, \ldots, v_8^1$ with wave number $k = 1$, while it remains stable with respect to all other modes v_i^k ($1 < d_1, d_2 < 5$). Also we assume $\omega_1 \omega_2 \neq 0$, and $\omega_1 \neq \omega_2$.

Near such a critical point in the parameter space and for u near 0 the original problem can be reduced to a Γ-equivariant first order system of ordinary differential equations on an 8-dimensional local center manifold.

The map h which describes the center manifold as a graph in Z over the center space $\operatorname{span}\{v_1^1, \ldots, v_8^1\}$ of C is Γ-equivariant. Due to the T^2-action, h has only terms of odd degree. Therefore, the reduced system up to degree 3 can be computed without knowledge of h. Since the terms of f are only of odd degree, the next non-vanishing terms of the reduced system have degree 5. To compute them the cubic terms of h are needed. The action of SO(2) implies that the image of h is contained in the space generated by $\{v_i^k \mid k \text{ odd}, k > 1\}$. Moreover, the image of the cubic terms is contained in $\operatorname{span}\{v_1^3, \ldots, v_8^3\}$.

The coordinates of the center space are $b_1^{(1)}, \ldots, b_8^{(1)}$, $\lambda' = \lambda + d_1 - 1$, $\sigma' = \sigma + d_2 - 1$. The vector field on the center manifold is Γ-equivariant. Therefore, we can perform a further reduction by the orbit space method. For the sake of simplicity, we use complex notation $z_1 = b_1^{(1)} + ib_1^{(1)}$, $z_2 = b_3^{(1)} + ib_4^{(1)}$, and so on. A minimal Hilbert basis is given by

$$\begin{aligned} \rho_1 &= z_1\bar{z}_1 + z_2\bar{z}_2 & \rho_5 &= \delta_1\delta_2 \\ \rho_2 &= z_3\bar{z}_3 + z_4\bar{z}_4 & \rho_6 &= \tfrac{1}{2}(z_1\bar{z}_2\bar{z}_3z_4 + \bar{z}_1z_2z_3\bar{z}_4) \\ \rho_3 &= \delta_1^2 & \rho_7 &= \tfrac{1}{2}\delta_1(z_1\bar{z}_2\bar{z}_3z_4 - \bar{z}_1z_2z_3\bar{z}_4) \\ \rho_4 &= \delta_2^2 & \rho_8 &= \tfrac{1}{2}\delta_2(z_1\bar{z}_2\bar{z}_3z_4 - \bar{z}_1z_2z_3\bar{z}_4) \end{aligned}$$

where $\delta_1 := z_1\bar{z}_1 - z_2\bar{z}_2$ and $\delta_2 := z_3\bar{z}_3 - z_4\bar{z}_4$. This Hilbert basis has several relations. The orbit space Y is five dimensional and has the following strata (The hierarchy of the strata reflects the fact that the components (u_1, u_2) and (u_3, u_4) are interchangeable in (6) from the qualitative point of view.):

Stratum	Isotropy*	$\dim \Upsilon_i$	$\dim T_yY$	in the closure of**
$\Upsilon_0 := \{0\}$	$O(2) \times T^2$	0	8	$\Upsilon_1, \Upsilon_2, \Upsilon_4, \Upsilon_5$
$\Upsilon_1 := \{\rho_1 > 0, \rho_2 = \cdots \rho_8 = 0\}$	$\vartheta_2 \in S^1, \kappa$	1	8	$\Upsilon_3, \Upsilon_7, \Upsilon_8$
$\Upsilon_2 := \{\rho_1 > 0, \rho_3 = \rho_1^2, \rho_i = 0\}$	$\vartheta_1 = -\varphi, \vartheta_2 \in S^1$	1	6	$\Upsilon_3, \Upsilon_9, \Upsilon_{10}$
$\Upsilon_3 := \{\rho_1 > 0, \rho_1^2 > \rho_3 > 0, \rho_i = 0\}$	$\vartheta_2 \in S^1$	2	6	Υ_{11}
$\Upsilon_4 := \{\rho_2 > 0, \rho_i = 0\}$	$\vartheta_1 \in S^1, \kappa$	1	8	$\Upsilon_6, \Upsilon_7, \Upsilon_8$
$\Upsilon_5 := \{\rho_2 > 0, \rho_4 = \rho_2^2, \rho_i = 0\}$	$\vartheta_1, \vartheta_2 = -\varphi \in S^1$	1	6	$\Upsilon_6, \Upsilon_9, \Upsilon_{10}$
$\Upsilon_6 := \{\rho_2 > 0, \rho_2^2 > \rho_4 > 0, \rho_i = 0\}$	$\vartheta_1 \in S^1$	2	6	Υ_{11}
$\Upsilon_7 := \{\rho_1, \rho_2 > 0, 4\rho_6 = \rho_1\rho_2, \rho_i = 0\}$	κ	2	8	Υ_{11}
$\Upsilon_8 := \{\rho_1, \rho_2 > 0, 4\rho_6 = -\rho_1\rho_2, \rho_i = 0\}$	$\kappa(0, 0, \pi)$	2	8	Υ_{11}
$\Upsilon_9 := \{\rho_1, \rho_2 > 0, \rho_3 = \rho_1^2, \rho_4 = \rho_2^2, \rho_5 = \rho_1\rho_2, \rho_i = 0\}$	$\vartheta_1 = \vartheta_2 = -\varphi \in S^1$	2	6	Υ_{11}
$\Upsilon_{10} := \{\rho_1, \rho_2 > 0, \rho_3 = \rho_1^2, \rho_4 = \rho_2^2, \rho_5 = -\rho_1\rho_2, \rho_i = 0\}$	$\vartheta_1 = -\vartheta_2 = -\varphi \in S^1$	2	6	Υ_{11}
$\Upsilon_{11} :=$ rest	*	5	5	

* This column contains generators of one representative of the conjugacy class of the stratum. κ is a fixed element of $O(2) \setminus SO(2)$, φ parametrizes $SO(2)$, and $(\vartheta_1, \vartheta_2)$, $\vartheta_i \in S^1$, parametrizes T^2. Since all isotropy subgroups contain the element $\varphi = \vartheta_1 = \vartheta_2 = \pi$, this is omitted in the table.

** If a stratum lies in the closure of another one, then it is contained in the closure of any stratum which contains the latter.

We computed the reduced vector field using the computer algebra system Singular [18]. The part which arises from the linear and cubic terms of the right hand side of (6) is the following:

$$
\left(\begin{array}{l}
2f_{68}y_5 + (4f_{13} + 4f_{57} - 4f_{14})y_6 + (f_{12} + 3f_{11} + f_{55})y_1^2 + (2f_{14} + 2f_{13})y_2y_1 \\
\qquad + (-f_{11} + 2f_{66} - f_{55} + f_{12})y_3 + 2y_1(\lambda - d_1) \\
2y_5f_{86} + (-4f_{32} + 4f_{31} + 4f_{75})y_6 + (2f_{31} + 2f_{32})y_2y_1 + (3f_{33} + f_{77} + f_{34})y_2^2 \\
\qquad + 2y_2(\sigma - d_2) + (-f_{33} + 2f_{88} - f_{77} + f_{34})y_4 \\
(-8f_{23} - 8f_{67} + 8f_{24})y_7 + (4f_{14} + 4f_{13})y_3y_2 + 4y_1y_5f_{68} \\
\qquad + (4f_{11} + 4f_{66} + 4f_{12})y_3y_1 + 4y_3(\lambda - d_1) \\
(8f_{41} + 8f_{85} - 8f_{42})y_8 + 4y_2y_5f_{86} + (4f_{32} + 4f_{31})y_4y_1 \\
\qquad + (4f_{33} + 4f_{88} + 4f_{34})y_4y_2 + 4y_4(\sigma - d_2) \\
(2(\sigma - d_2) + 2(\lambda - d_1))y_5 + (4f_{41} - 4f_{42} + 4f_{85})y_7 + (4f_{24} - 4f_{23} - 4f_{67})y_8 \\
\qquad + 2y_2y_3f_{86} + (2f_{11} + 2f_{66} + 2f_{31} + 2f_{12} + 2f_{32})y_5y_1 \\
\qquad + (2f_{34} + 2f_{88} + 2f_{14} + 2f_{13} + 2f_{33})y_5y_2 + 2y_1y_4f_{68} \\
(2(\sigma - d_2) + 2(\lambda - d_1))y_6 + (f_{22} - f_{65} - f_{21} + 2f_{76} - 2f_{56})y_7 \\
\qquad + (2f_{78} - 2f_{58} + f_{87} - f_{44} + f_{43})y_8 \\
\qquad + (-\frac{1}{4}f_{75} - \frac{1}{4}f_{31} + \frac{1}{4}f_{32})y_3y_2 + (-\frac{1}{4}f_{13} + \frac{1}{4}f_{14} - \frac{1}{4}f_{57})y_4y_1 \\
\qquad + (2f_{31} + 3f_{11} + f_{12} + f_{55} + 2f_{32})y_6y_1 + (\frac{1}{4}f_{57} + \frac{1}{4}f_{13} - \frac{1}{4}f_{14})y_2^2y_1 \\
\qquad + (f_{77} + f_{34} + 2f_{13} + 2f_{14} + 3f_{33})y_6y_2 + (\frac{1}{4}f_{31} - \frac{1}{4}f_{32} + \frac{1}{4}f_{75})y_2y_1^2 \\
(2(\sigma - d_2) + 4(\lambda - d_1))y_7 + (4f_{23} + 4f_{67} - 4f_{24})y_6^2 + (\frac{1}{2}f_{67} - \frac{1}{2}f_{24} + \frac{1}{2}f_{23})y_3y_2^2 \\
\qquad + (\frac{1}{4}f_{85} - \frac{1}{4}f_{42} + \frac{1}{4}f_{41})y_5y_3 + (-2f_{78} - f_{43} + f_{44} - f_{87} + 2f_{58})y_6y_5 \\
\qquad + (2f_{32} + 3f_{12} + 2f_{66} + 5f_{11} + f_{55} + 2f_{31})y_7y_1 + (-\frac{1}{4}f_{41} + \frac{1}{4}f_{42} - \frac{1}{4}f_{85})y_5y_1^2 \\
\qquad + (-\frac{1}{4}f_{23} - \frac{1}{4}f_{67} + \frac{1}{4}f_{24})y_2^2y_1^2 + (\frac{1}{4}f_{67} - \frac{1}{4}f_{24} + \frac{1}{4}f_{23})y_4y_1^2 + 2y_1y_8f_{68} \\
\qquad + (-\frac{1}{2}f_{23} - \frac{1}{2}f_{67} + \frac{1}{2}f_{24})y_5^2 + (f_{65} - f_{22} + 2f_{56} - 2f_{76} + f_{21})y_6y_3 \\
\qquad + (f_{34} + f_{77} + 4f_{13} + 4f_{14} + 3f_{33})y_7y_2 \\
(4(\sigma - d_2) + 2(\lambda - d_1))y_8 + (4f_{42} - 4f_{41} - 4f_{85})y_6^2 + (\frac{1}{4}f_{42} - \frac{1}{4}f_{41} - \frac{1}{4}f_{85})y_3y_2^2 \\
\qquad + (f_{65} - f_{22} + 2f_{56} - 2f_{76} + f_{21})y_6y_5 + (\frac{1}{4}f_{85} - \frac{1}{4}f_{42} + \frac{1}{4}f_{41})y_2^2y_1^2 \\
\qquad + (-\frac{1}{2}f_{41} + \frac{1}{2}f_{42} - \frac{1}{2}f_{85})y_4y_1^2 + (3f_{11} + 4f_{31} + f_{12} + f_{55} + 4f_{32})y_8y_1 \\
\qquad + (-\frac{1}{2}f_{42} + \frac{1}{2}f_{41} + \frac{1}{2}f_{85})y_5^2 + 2y_2y_7f_{86} \\
\qquad + (\frac{1}{4}f_{67} - \frac{1}{4}f_{24} + \frac{1}{4}f_{23})y_5y_2^2 + (5f_{33} + 2f_{13} + 2f_{14} + f_{77} + 3f_{34} + 2f_{88})y_8y_2 \\
\qquad + (-\frac{1}{4}f_{23} - \frac{1}{4}f_{67} + \frac{1}{4}f_{24})y_5y_4 + (-2f_{78} - f_{43} + f_{44} - f_{87} + 2f_{58})y_6y_4
\end{array}\right)
$$

We will focus on the 2-dimensional stratum Υ_7, which is the image of the fixed point space of κ. The boundary of Υ_7 is the union of Υ_0, Υ_1, and Υ_4. We consider the restriction to $\overline{\Upsilon_7}$. If we introduce new coordinates ($x = \rho_1 \geq 0$ and $y = \rho_2 \geq 0$) and scale all variables appropriately by powers of ε, then the general system on $\overline{\Upsilon_7}$ becomes

$$
\begin{aligned}
\dot{x} &= 2x(\lambda' + a_1y + a_2x + \varepsilon(a_3y^2 + a_4xy + a_5x^2 + \varepsilon(\ldots))) \\
\dot{y} &= 2y(\mu' + b_1x + b_2y + \varepsilon(b_3x^2 + b_4xy + b_5y^2 + \varepsilon(\ldots)))\ ,
\end{aligned} \tag{7}
$$

where a_i and b_i are constants. The first four constants are given in terms of the cubic terms of f as follows:

$$
\begin{aligned}
a_1 &= \tfrac{1}{2}(3f_{13} + f_{14} + f_{57}) \qquad & a_2 &= \tfrac{1}{2}(3f_{11} + f_{12} + f_{55}) \\
b_1 &= \tfrac{1}{2}(3f_{31} + f_{32} + f_{75}) \qquad & b_2 &= \tfrac{1}{2}(3f_{33} + f_{34} + f_{77})
\end{aligned}
$$

The next six constants involve the quintic terms of f:

$$
\begin{aligned}
a_3 = {} & \tfrac{1}{2}(f_{134} + 5f_{133} + f_{144} + f_{177} + f_{537} + f_{547}) \\
& +(\tfrac{3}{4}f_{34}f_{14} - \tfrac{3}{4}f_{33}f_{14} + \tfrac{3}{4}f_{77}f_{14} + \tfrac{1}{4}f_{77}f_{13} + \tfrac{1}{4}f_{57}f_{34} - \tfrac{1}{4}f_{57}f_{33} \\
& \quad +\tfrac{1}{4}f_{34}f_{13} - \tfrac{1}{4}f_{33}f_{13} - \tfrac{1}{2}f_{87}f_{58} - \tfrac{1}{2}f_{58}f_{44} + \tfrac{1}{2}f_{58}f_{43} + \tfrac{1}{4}f_{77}f_{57})h_4 \\
& +(\tfrac{1}{4}f_{57}f_{14} - \tfrac{1}{4}f_{57}f_{13} - \tfrac{1}{4}f_{24}f_{23} + \tfrac{1}{4}f_{14}f_{13} - \tfrac{1}{8}f_{13}^2 - \tfrac{1}{8}f_{14}^2 + \tfrac{1}{8}f_{23}^2 \\
& \quad +\tfrac{1}{8}f_{24}^2 + \tfrac{3}{8}f_{57}^2 - \tfrac{3}{8}f_{67}^2 - \tfrac{1}{4}f_{67}f_{24} + \tfrac{1}{4}f_{67}f_{23})h_2 \\
a_4 = {} & +\tfrac{1}{2}(f_{545} + f_{124} + f_{123} + f_{535} + 5f_{113} + f_{157} + f_{517} + f_{527} + f_{114}) \\
& +(-\tfrac{1}{4}f_{75}f_{13} - \tfrac{3}{4}f_{75}f_{14} + \tfrac{3}{4}f_{31}f_{14} - \tfrac{3}{4}f_{32}f_{14} + \tfrac{1}{4}f_{57}f_{31} + \tfrac{1}{2}f_{85}f_{58} \\
& \quad -\tfrac{1}{4}f_{57}f_{32} + \tfrac{1}{2}f_{58}f_{42} + \tfrac{1}{4}f_{31}f_{13} - \tfrac{1}{4}f_{75}f_{57} - \tfrac{1}{4}f_{32}f_{13} - \tfrac{1}{2}f_{58}f_{41})h_3 \\
& +(-\tfrac{1}{8}f_{24}f_{21} - \tfrac{5}{8}f_{13}f_{12} - \tfrac{3}{8}f_{13}f_{11} + \tfrac{3}{8}f_{65}f_{23} - \tfrac{3}{8}f_{67}f_{65} - \tfrac{1}{8}f_{23}f_{22} \\
& \quad -\tfrac{3}{8}f_{65}f_{24} + \tfrac{5}{8}f_{14}f_{12} - \tfrac{5}{8}f_{55}f_{13} - \tfrac{1}{8}f_{67}f_{21} - \tfrac{1}{2}f_{67}f_{56} + \tfrac{1}{8}f_{67}f_{22} \\
& \quad +\tfrac{5}{8}f_{57}f_{12} + \tfrac{3}{8}f_{57}f_{11} + \tfrac{5}{8}f_{57}f_{55} + \tfrac{1}{8}f_{24}f_{22} + \tfrac{1}{2}f_{56}f_{23} + \tfrac{5}{8}f_{55}f_{14} \\
& \quad +\tfrac{1}{8}f_{23}f_{21} - \tfrac{1}{2}f_{56}f_{24} + \tfrac{3}{8}f_{14}f_{11})h_2 \\
& +(-\tfrac{1}{8}f_{65}f_{23} + \tfrac{1}{8}f_{55}f_{13} + \tfrac{1}{8}f_{65}f_{24} + \tfrac{1}{8}f_{24}f_{22} - \tfrac{1}{8}f_{23}f_{22} + \tfrac{3}{8}f_{57}f_{12} \\
& \quad -\tfrac{1}{8}f_{24}f_{21} + \tfrac{1}{8}f_{14}f_{11} + \tfrac{3}{8}f_{57}f_{55} - \tfrac{1}{8}f_{14}f_{12} + \tfrac{1}{8}f_{13}f_{12} + \tfrac{3}{8}f_{67}f_{21} \\
& \quad -\tfrac{3}{8}f_{57}f_{11} - \tfrac{1}{8}f_{13}f_{11} - \tfrac{1}{8}f_{55}f_{14} - \tfrac{3}{8}f_{67}f_{65} + \tfrac{1}{8}f_{23}f_{21} - \tfrac{3}{8}f_{67}f_{22})h_1 \\
a_5 = {} & \tfrac{1}{2}(f_{515} + 5f_{111} + f_{112} + f_{155} + f_{525} + f_{122}) + (-\tfrac{3}{8}f_{11}^2 + \tfrac{5}{8}f_{55}^2 + \tfrac{1}{8}f_{21}^2 \\
& \quad +\tfrac{1}{8}f_{22}^2 + \tfrac{5}{8}f_{12}^2 - \tfrac{3}{8}f_{65}^2 - \tfrac{1}{4}f_{22}f_{21} + \tfrac{5}{4}f_{55}f_{12} - \tfrac{1}{4}f_{65}f_{22} \\
& \quad +\tfrac{1}{4}f_{65}f_{21} - \tfrac{1}{2}f_{65}f_{56} - \tfrac{1}{2}f_{56}f_{22} + \tfrac{1}{2}f_{56}f_{21} - \tfrac{1}{4}f_{55}f_{11} - \tfrac{1}{4}f_{12}f_{11})h_1 \\
b_3 = {} & +\tfrac{1}{2}(f_{312} + 5f_{311} + f_{322} + f_{355} + f_{715} + f_{725}) \\
& +(\tfrac{1}{4}f_{85}f_{42} - \tfrac{1}{4}f_{32}f_{31} - \tfrac{3}{8}f_{75}^2 - \tfrac{1}{4}f_{85}f_{41} + \tfrac{1}{4}f_{75}f_{31} + \tfrac{1}{4}f_{42}f_{41} \\
& \quad +\tfrac{3}{8}f_{85}^2 - \tfrac{1}{8}f_{42}^2 + \tfrac{1}{8}f_{31}^2 + \tfrac{1}{8}f_{32}^2 \\
& \quad -\tfrac{1}{8}f_{41}^2 - \tfrac{1}{4}f_{75}f_{32})h_3 \\
& +(\tfrac{3}{4}f_{55}f_{32} + \tfrac{1}{4}f_{75}f_{55} - \tfrac{3}{4}f_{32}f_{11} + \tfrac{1}{4}f_{75}f_{12} - \tfrac{1}{4}f_{75}f_{11} + \tfrac{1}{4}f_{55}f_{31} \\
& \quad +\tfrac{1}{4}f_{31}f_{12} - \tfrac{1}{4}f_{31}f_{11} + \tfrac{1}{2}f_{76}f_{21} - \tfrac{1}{2}f_{76}f_{65} - \tfrac{1}{2}f_{76}f_{22} + \tfrac{3}{4}f_{32}f_{12})h_1 \\
b_4 = {} & \tfrac{1}{2}(5f_{313} + f_{314} + f_{324} + f_{323} + f_{717} + f_{727} + f_{357} + f_{735} + f_{745}) \\
& +(\tfrac{1}{8}f_{87}f_{42} - \tfrac{3}{8}f_{87}f_{85} - \tfrac{1}{8}f_{33}f_{31} - \tfrac{3}{8}f_{85}f_{44} + \tfrac{3}{8}f_{75}f_{34} + \tfrac{1}{8}f_{43}f_{41} \\
& \quad -\tfrac{1}{8}f_{77}f_{32} - \tfrac{1}{8}f_{43}f_{42} + \tfrac{1}{8}f_{77}f_{31} - \tfrac{1}{8}f_{44}f_{41} - \tfrac{1}{8}f_{34}f_{32} - \tfrac{1}{8}f_{87}f_{41} \\
& \quad +\tfrac{1}{8}f_{34}f_{31} + \tfrac{3}{8}f_{77}f_{75} + \tfrac{1}{8}f_{33}f_{32} + \tfrac{3}{8}f_{85}f_{43} + \tfrac{1}{8}f_{44}f_{42} - \tfrac{3}{8}f_{75}f_{33})h_4 \\
& +(-\tfrac{3}{8}f_{87}f_{41} - \tfrac{1}{8}f_{43}f_{41} + \tfrac{1}{8}f_{44}f_{41} + \tfrac{1}{2}f_{85}f_{78} - \tfrac{5}{8}f_{34}f_{32} + \tfrac{3}{8}f_{87}f_{42} \\
& \quad +\tfrac{5}{8}f_{34}f_{31} + \tfrac{1}{2}f_{78}f_{42} + \tfrac{3}{8}f_{87}f_{85} - \tfrac{1}{8}f_{85}f_{44} - \tfrac{3}{8}f_{33}f_{32} - \tfrac{5}{8}f_{77}f_{32} \\
& \quad +\tfrac{1}{8}f_{43}f_{42} - \tfrac{5}{8}f_{77}f_{75} - \tfrac{5}{8}f_{75}f_{34} - \tfrac{3}{8}f_{75}f_{33} + \tfrac{3}{8}f_{33}f_{31} + \tfrac{5}{8}f_{77}f_{31} \\
& \quad -\tfrac{1}{8}f_{44}f_{42} + \tfrac{1}{8}f_{85}f_{43} - \tfrac{1}{2}f_{78}f_{41})h_3 \\
& +(\tfrac{3}{4}f_{32}f_{14} - \tfrac{1}{4}f_{75}f_{13} + \tfrac{1}{4}f_{31}f_{14} - \tfrac{1}{4}f_{31}f_{13} + \tfrac{1}{4}f_{75}f_{57} - \tfrac{1}{2}f_{76}f_{24} \\
& \quad +\tfrac{1}{2}f_{76}f_{23} + \tfrac{1}{4}f_{75}f_{14} + \tfrac{3}{4}f_{57}f_{32} - \tfrac{3}{4}f_{32}f_{13} + \tfrac{1}{4}f_{57}f_{31} - \tfrac{1}{2}f_{76}f_{67})h_2 \\
b_5 = {} & \tfrac{1}{2}(5f_{333} + f_{377} + f_{344} + f_{737} + f_{747} + f_{334}) + (\tfrac{5}{8}f_{34}^2 - \tfrac{3}{8}f_{87}^2 + \tfrac{5}{8}f_{77}^2 \\
& \quad +\tfrac{1}{8}f_{44}^2 + \tfrac{1}{8}f_{43}^2 + \tfrac{5}{4}f_{77}f_{34} - \tfrac{1}{4}f_{77}f_{33} + \tfrac{1}{4}f_{87}f_{43} - \tfrac{1}{2}f_{78}f_{44} \\
& \quad -\tfrac{3}{8}f_{33}^2 - \tfrac{1}{2}f_{87}f_{78} + \tfrac{1}{2}f_{78}f_{43} - \tfrac{1}{4}f_{44}f_{43} - \tfrac{1}{4}f_{34}f_{33} - \tfrac{1}{4}f_{87}f_{44})h_4
\end{aligned}
$$

J. Menck investigated the system (7) in [29]. In the limit as $\varepsilon \to 0$, and under the conditions $A_1 := a_2 b_2 < 0$ and $A_2 := a_2 b_2 - a_1 b_1 > 0$ a branch of equilibria emerges from the origin inside Υ_1 or Υ_4, depending on the sign of a_2. A secondary symmetry breaking bifurcation arises into the stratum Υ_7. Furthermore, the new branch finally undergoes a Hopf bifurcation. Between consecutive bifurcations, all equilibria are asymptotically stable inside $\overline{\Upsilon_7}$. This scenario persists for small $\varepsilon > 0$.

The Hopf bifurcation is degenerate for $\varepsilon = 0$. But, if ε becomes positive, the direction and stability of the branch of periodic points can be computed. The determining coefficient in the corresponding normal form has the expansion

$$\begin{aligned}
\alpha &= 0 + \varepsilon \tfrac{d}{d\varepsilon}\alpha(0) + O(\varepsilon^2) \\
&= \frac{a_2 b_1}{A_2}\Big((b_2 a_2 b_1^2 - b_1 b_2 a_2^2) a_3 + (-b_2^2 a_2^2 + A_2 b_2 a_2 + b_1 b_2^2 a_2 + 2 b_2 b_1 A_2) a_4 \\
&\qquad + (-a_1 b_2^2 a_2 + 2 A_2 b_2 a_1 + b_2^3 a_2 - b_2^2 A_2) a_5 \\
&\qquad + (a_1^2 a_2 b_2 - a_1 b_2^2 a_2) b_3 \\
&\qquad + (-b_2 a_1 a_2^2 - b_2^2 a_2^2 + A_2 b_2 a_2 + 2 b_1 b_2^2 a_2 + 2 b_2 b_1 A_2) b_4 \\
&\qquad + (a_2^3 b_2 - A_2 a_2^2 - b_1 b_2 a_2^2 + 2 A_2 b_1 a_2) b_5 \Big)\varepsilon + O(\varepsilon^2) \ .
\end{aligned}$$

If $\frac{d}{d\varepsilon}\alpha(0)$ is negative, then the bifurcation is supercritical, and locally the periodic points are stable inside the stratum Υ_7.

It remains to be shown that the solutions in question are also stable with respect to the directions transversal to $\overline{\Upsilon_7}$. To this end, we note that the gradient of all relations of ρ vanishes at points of $\overline{\Upsilon_7}$; thus, the tangent spaces of the orbit space at these points have dimension 8, and all eigenvalues of the differential of the reduced vector field in $\mathbb{R}^8$ are relevant to determine the stability. That differential has in suitable coordinates block structure (cf. (3)).

At the origin there are the following blocks on the diagonal of the differential: A 2×2-matrix corresponding to the quadratic elements of the Hilbert basis, a 4×4-matrix corresponding to the invariants $\rho_3, \ldots, \rho_6$, and another 2×2-matrix corresponding to ρ_7 and ρ_8. By Remark 9, only the first block must be considered. This is already done by analyzing the restriction of the dynamics to $\overline{\Upsilon_7}$.

At points x of the one-dimensional strata Υ_1 or Υ_4 the representation of the isotropy subgroup $\Gamma_x \cong S^1 \times \mathbb{Z}_2$ on the 6-dimensional normal space decomposes into a sum of four non-equivalent irreducible subrepresentations: The first one is trivial and corresponds to the stratum of x, the kernel of the second one-dimensional subrepresentation is S^1, the kernel of the third one is $\mathbb{Z}_2$, and the last one has a trivial kernel. Therefore, a local minimal Hilbert basis on N_x has one linear and three quadratic elements. Thus, here $\dim R = 4$, and the restriction of the differential to R has in suitable coordinates the form

$$\begin{pmatrix} a_1 & * & * & * \\ 0 & a_2 & 0 & 0 \\ 0 & 0 & a_3 & 0 \\ 0 & 0 & 0 & a_4 \end{pmatrix}$$

where $a_i \in \mathbb{R}$ corresponds to the i-th subreprestation.

Since the stratum Υ_7 is two-dimensional, on the diagonal of the differential there are a 2×2-block corresponding to the stratum and a 6×6-block, whose eigenvalues control the stability transversal to the stratum. This 6×6-matrix provides a criterion for the stability of the solutions in the stratum with respect to the transversal directions. If its eigenvalues are negative at the point of the Hopf bifurcation, then we can conclude that the Lyapunov exponent (Floquet exponents) of the periodic trajectories corresponding to those directions are negative near the bifurcation point.

For example, if we demand $f_{23} = f_{24}$ and $f_{41} = f_{42}$, and assume that all other terms are independent of u_x, i.e., $f_{\bullet 2}$, $f_{\bullet 4}$, and $f_{5\bullet}, \ldots, f_{8\bullet}, f_{5\bullet\bullet}, \ldots, f_{8\bullet\bullet}$ vanish, then we have the following sufficient conditions for the asymptotic stability of the branching periodic trajectories.

$$\begin{aligned} f_{11} &> 0 \\ f_{33} f_{11} &< 0 \\ f_{33} f_{11} - f_{31} f_{13} &> 0 \\ -f_{31} f_{33} + f_{13} f_{11} &> 0 \\ (-f_{31} f_{13} + f_{31} f_{33} + f_{11} f_{13} - f_{33} f_{11}) f_{33} f_{11} &> 0 \end{aligned}$$

and $\beta\, b_1 a_2 = -\frac{32}{81}\alpha'(0) A_2 > 0$ with

$$\begin{aligned} \beta := 5(&(-f_{31} f_{13} f_{33}^2 - f_{13} f_{33}^2 f_{11} + 2 f_{13}^2 f_{31} f_{33}) f_{111} \\ &+(-3 f_{31} f_{33}^2 f_{11} + f_{31} f_{13} f_{33} f_{11} + 2 f_{13} f_{31}^2 f_{33}) f_{113} \\ &+(f_{31} f_{33} f_{11}^2 - f_{33} f_{11} f_{31}^2) f_{133} + (-f_{13}^2 f_{11} f_{33} + f_{13} f_{33}^2 f_{11}) f_{311} \\ &+(f_{31} f_{13} f_{33} f_{11} + 2 f_{13} f_{31}^2 f_{33} + f_{33} f_{13} f_{11}^2 - 4 f_{31} f_{33}^2 f_{11}) f_{313} \\ &+(-f_{31} f_{13} f_{11}^2 + 2 f_{13} f_{11} f_{31}^2 - f_{31} f_{33} f_{11}^2) f_{333}) \\ +(&\tfrac{3}{4} f_{11}^3 f_{13} f_{33}^2 - \tfrac{1}{4} f_{11} f_{13} f_{33}^2 f_{21}^2 - \tfrac{1}{4} f_{13} f_{31} f_{33}^2 f_{21}^2 + f_{11}^2 f_{13} f_{31} f_{33}^2 + \tfrac{1}{2} f_{13}^2 f_{31} f_{33} f_{21}^2 \\ &\qquad - \tfrac{1}{2} f_{11} f_{13}^2 f_{31}^2 f_{33} - \tfrac{5}{4} f_{11}^2 f_{13}^2 f_{31} f_{33}) h_1 \\ +(&-\tfrac{3}{2} f_{11}^2 f_{13}^2 f_{31} f_{33} - f_{13}^2 f_{31}^3 f_{33} + 2 f_{11} f_{13} f_{31}^2 f_{33}^2 \\ &- \tfrac{7}{4} f_{11} f_{13}^2 f_{31}^2 f_{33} + \tfrac{9}{4} f_{11}^2 f_{13} f_{31} f_{33}^2) h_2 \\ +(&\tfrac{1}{4} f_{11} f_{13}^2 f_{31}^2 f_{33} + f_{13}^2 f_{31}^3 f_{33} + \tfrac{3}{2} f_{13} f_{31}^3 f_{33}^2 - 3 f_{11} f_{31}^2 f_{33}^3 \\ &+ \tfrac{3}{4} f_{11}^2 f_{13} f_{31} f_{33}^2 - \tfrac{1}{2} f_{11} f_{13} f_{31}^2 f_{33}^2) h_3 \\ +(&f_{11} f_{31}^2 f_{33}^3 - \tfrac{1}{2} f_{13} f_{31}^3 f_{33}^2 - \tfrac{1}{4} f_{11}^2 f_{31} f_{33} f_{43}^2 + \tfrac{1}{2} f_{11} f_{13} f_{31}^2 f_{43}^2 \\ &\qquad + \tfrac{3}{4} f_{11}^2 f_{31} f_{33}^3 - \tfrac{1}{4} f_{11}^2 f_{13} f_{31} f_{43}^2 - \tfrac{5}{4} f_{11} f_{13} f_{31}^2 f_{33}^2) h_4 \,, \end{aligned}$$

where $h_1 = \frac{1}{3d_1 - 39 - \lambda_0}$, $h_2 = \frac{1}{-39 + 4d_1 - d_2 - \sigma_0}$, $h_3 = \frac{1}{d_1 + 39 + \lambda_0 - 4d_2}$, $h_4 = \frac{1}{3d_2 - \sigma_0 - 39}$. For example, this is fulfilled, if $f_{11} = 1$, $f_{13} = -\frac{3}{4}$, $f_{31} = 2$, $f_{33} = -1$, and all other coefficients vanish. In this situation the equilibria of the branches in Υ_4 and Υ_7 are asymptotically stable in the full space $\mathbb{R}^8$ up to the bifurcations. There are similar conditions in the case $f_{11} < 0$.

By the general theory we finally conclude that under the previous conditions the following bifurcations occur in the original problem (6) near the critical parameter value considered. The trivial solution becomes unstable in Z, and a branch of stable, invariant 2-tori emerges, which bifurcate again to stable 3-dimensional invariant orbits. These relative equilibria become unstable at a "Hopf point", where finally a branch of stable 4-dimensional invariant manifolds of relative periodic points emerges.

8 Reconstruction of Eigenvalues

This section is devoted to the question, whether the eigenvalues of $DP(0)$ can be reconstructed with the aid of the spectrum of $DQ(0)$, where $P : V \to V$ is a Γ-equivariant map with equilibrium 0 and $Q = \mathcal{J}_{\mathsf{m}}(P)$ is the reduced map. By Theorem 6, the eigenvalues of $DQ(0)$ are products of the eigenvalues of $DP(0)$. The fixed point space V^Γ is invariant with respect to $DP(0)$, and since $D\rho(0)|_{V^\Gamma}$ is one-to-one, the eigenvalues of $DP(0)$ which have an eigenvector in V^Γ appear on the orbit space. Therefore, an eigenvalue of $DQ(0)$ with an eigenvector tangent to the stratum of $\rho(0)$ is also an eigenvalue of $DP(0)$.

Due to the block structure (3) of $DQ(0)$ there is the invariant subspace R of T_0Y which is dual to space of the linear and quadratic invariants. We will only use the restriction $DQ(0)|_R$ to reconstruct the eigenvalues of $DP(0)$. It may happen that the blocks A_1 and A_2 corresponding to the spaces of linear and quadratic invariants, respectively, have a common eigenvalue, then its multiplicity depends on the block above A_2, which is determined by nonlinear terms of P. As we do not care so much about multiplicity of eigenvalues here, we occasionally assume that the right upper block of the matrix $DQ(0)|_R$ is zero.

Remark 29. Let $V^\Gamma \oplus \bigoplus_{i=1}^{j} V_i$ be a decomposition of V into irreducible subrepresentations. Let $\rho_1, \ldots, \rho_k$ be a multihomogeneous minimal Hilbert basis with respect to this decomposition. Let α be the multidegree of a quadratic element ρ_i. Then either $\alpha_j = 2\delta_{j,k}$ for some k or $\alpha_j = \delta_{j,k} + \delta_{j,l}$ for some pair (k, l), where V_k and V_l are equivalent. This is because $\operatorname{grad} \rho_i$ is a Γ-equivariant linear map.
Thus, each linear and quadratic element of the Hilbert basis belongs to a unique isotypical component. Since R is dual to the space of these invariants (in the sense of Remark 18), the subspace R of T_0Y decomposes uniquely into $\bigoplus_i R_{X_i}$, where R_{X_i} corresponds to the isotypical component X_i and is $DQ(0)$-invariant.
The map $\begin{pmatrix} A_1 & 0 \\ 0 & A_2 \end{pmatrix} : R \to R$ keeps this decomposition invariant, because there are no equivariant linear maps between nonequivalent representations but the trivial one. Of course, this is not correct for $DQ(0)|_R$, in general.

Remark 30. Using the techniques of Sect. 4 the last remark can be carried over to relative equilibria.

As Example 11 shows, there are eigenvalues in the tangent space of the orbit space which have no meaning to the original system P. Especially, these eigenvalues may be critical, even if $DP(0)$ has no critical eigenvalue.

The generalized eigenspace of the stable (unstable) eigenvalues is tangent to the stable (unstable) manifold. A Hilbert map maps the stable (unstable) manifolds into stable (unstable) manifolds on the orbit space.

Proposition 31. *If $P \in C_\Gamma^{qn}(V,V)$ is a diffeomorphism, and $Q := \mathcal{J}_m(P)$ the reduced map, let $W^s(\Gamma x)$ ($W^u(\Gamma x)$) be the stable (unstable) manifold of the invariant orbit Γx of P, and $W^s(y)$ ($W^u(y)$) be the stable (unstable) manifold of the equilibrium $y := \rho(x)$ of Q.*
Then $\rho(W^s(\Gamma x)) = W^s(y) \cap Y$ and $\rho(W^u(\Gamma x)) = W^u(y) \cap Y$.

Proof. Let $z \in W^s(\Gamma x)$, then the distance $d(P^t(z), \Gamma x)$ of the iterates of x to the invariant orbit shrinks exponentially for $t \to \infty$. Also $d(Q^t(\rho(z)), y)$ shrinks exponentially for $t \to \infty$, i.e., $\rho(z) \in W^s(y)$. Also the converse holds. □

This suggests that only eigenvalues whose eigenspaces are 'tangent' to Y in a narrower sense should be used to determine the noncritical eigenvalues at a relative equilibrium. Here, a vector v is 'tangent' to $W^s(y) \cap Y$ means that there is a path in Y starting at y in direction v. This concept leads to the tangent cone.

The tangent cone of the orbit space The *tangent cone* C_yY of a set $Y \subset \mathbb{R}^k$ at $y \in Y$ is the limit of the secants of Y through y:

$$s \in C_yY :\Longleftrightarrow \exists \lambda_i > 0 : \exists y_i \in Y : y_i \to y \wedge \lambda_i(y_i - y) \to s \qquad \text{as} \quad i \to \infty$$

C_yY is a union of rays emerging from y.

First, we consider the tangent cone of the embedded orbit space $Y = \rho(V)$ at 0, where ρ is a minimal Hilbert map. The representation space V decomposes Γ-invariantly into the fixed point space $V_1 = V^\Gamma$ and the orthogonal complement V_2. Let ρ_1 be a minimal Hilbert map on V_1, i.e., coordinates on V_1, and (ρ_2, ρ_3) a minimal homogeneous Hilbert map on V_2, where ρ_2 is a quadratic map and ρ_3 contains the higher order terms. Then $\rho = (\rho_1, \rho_2, \rho_3)$ is a Hilbert map on V, and $Y = \rho_1(V_1) \times (\rho_2, \rho_2)(V_2) = V_1 \times (\rho_2, \rho_3)(V_2)$.

Proposition 32 ([23]). *The tangent cone of Y at 0 is given by $C_0Y = V_1 \times \rho_2(V_2) \times \{0\}$.*

Proof. In the last paragraph we proved that $C_0Y = V_1 \times C_0(\rho_2, \rho_3)(V_2)$. We proceed by showing $C_0(\rho_2, \rho_3)(V_2) = C_0\rho_2(V_2) \times \{0\}$. Take $s = (s^{(2)}, s^{(3)}) \in C_0(\rho_2, \rho_3)(V_2)$. There are sequences $\lambda_i > 0$ and $y_i \in (\rho_2, \rho_3)(V_2)$ with $y_i \to 0$ and $\lambda_i y_i \to s$.

Choose $x_i \in V_2$ with $y_i = (\rho_2(x_i), \rho_3(x_i)) =: (y_i^{(2)}, y_i^{(3)})$. x_i converges to 0. Since $R : x \mapsto \langle x, x\rangle$ is Γ-invariant, there is linear function g with $R = g \circ \rho_2$. If the component r of ρ_3 has degree $d > 2$, then $|r(x_i)| \leq CR(x_i)^{d/2} = Cg(y_i)^{d/2} \leq C'\|y_i\|^{d/2}$, where C and C' are suitable constants. This yields $|\lambda_i r(x_i)| \leq C'\|\lambda_i y_i\|\,\|y_i\|^{d/2-1} \to C'\|s\|0 = 0$. Therefore, $s^{(3)} = 0$.

We have $\lambda_i y_i^{(2)} = \rho_2(\sqrt{\lambda_i}x_i)$. Since ρ_2 is proper, $\sqrt{\lambda_i}x_i$ is bounded, and has a limit $v \in V_2$, eventually pass to a subsequence. Hence, $\rho_2(v) = \rho_2(\lim \sqrt{\lambda_i}x_i) = \lim \lambda_i \rho_2(x_i) = s^{(2)}$, and $s \in \rho(V_2) \times \{0\}$. □

Now we consider the tangent cone at an arbitrary point $y \in Y$. Let $x \in V$ such that $\rho(x) = y$. We regard the tangent cone as a subset of $T_y\mathbb{R}^k$. To achieve this we identify the tangent space $T_y\mathbb{R}^k$ with $\mathbb{R}^k$ in the canonical way. Obviously, the tangent cone C_yY contains the tangent space $T_y\Upsilon_y$ of the stratum Υ_y of y. And it is a subset of the tangent space T_yY of the orbit space: Let r be a relation of ρ, and write $r(z) = Dr(y)(z-y) + r_2(z-y)$. Let $s \in C_yY$ and $y_i \in Y$ be a sequence with limit y, and $\lambda_i > 0$ a sequence with $\lambda_i(y_i - y) \to s$. Setting $d_i := (y_i - y) - s/\lambda_i$, we have

$$\begin{aligned} 0 &= r(y_i)\lambda_i = r(y + s/\lambda_i + d_i)\lambda_i = Dr(y)s + Dr(y)d_i\lambda_i + r_2(s/\lambda_i + d_i)\lambda_i \\ &= Dr(y)s + Dr(y)d_i\lambda_i + o(1/\lambda_i) + o(d_i) \longrightarrow Dr(y)s\,, \end{aligned}$$

because $d_i\lambda_i \to 0$.

Proposition 33. *Let ρ and σ be two Hilbert maps for the same representation. Denote the embedded orbit spaces by $Y_\rho = \rho(V)$ and $Y_\sigma = \sigma(V)$, respectively. There is a polynomial Λ with $\sigma = \Lambda \circ \rho$. The isomorphism $D\Lambda(\rho(x)) : T_{\rho(x)}Y_\rho \to T_{\sigma(x)}Y_\sigma$ induces a bijection between $C_{\rho(x)}Y_\rho$ and $C_{\sigma(x)}Y_\sigma$, $x \in V$.*
For $x \in V$ and $y := \rho(x)$, the isomorphism $DH_x(0) : T_0Y_x \to T_yY$ from Lemma 16 induces a bijection between the tangent cones C_0Y_x and C_yY, where Y_x is the local orbit space.

Proof. We proof the second statement. Let $s \in C_0Y_x$, $y_i \in Y_x$, and $\lambda_i > 0$ with $y_i \to 0$ and $\lambda_i y_i \to s$ for $i \to \infty$, and $d_i := (y_i - y) - s/\lambda_i$, then

$$\begin{aligned} \lambda_i(H_x(y_i) - y) &= \lambda_i(H_x(s/\lambda_i + d_i) - y) \\ &= DH_x(0)s + DH_x(0)\lambda_i d_i + o(1/\lambda_i) + o(d_i) \longrightarrow DH_x(0)s\,. \end{aligned}$$

Thus, $DH_x(0)$ maps C_0Y_x into C_yY. The converse is shown in the same way. □

Remark 34. For a given point $x \in V$ write $V_2 := (V^{\Gamma_x})^\perp$. Expand ρ at x according to Taylor, and denote the linear part of the expansion by $\rho_\mathrm{l} = D\rho(x)$ ($V_2 = \ker \rho_\mathrm{l}$), the quadratic part of $\rho|_{V_2}$ by ρ_q, and the other terms by ρ_r. Then $\rho = \rho(x) + \rho_\mathrm{l} + \rho_\mathrm{q} + \rho_\mathrm{r}$, and $C_{\rho(x)}Y = \rho_\mathrm{l}(V) + \rho_\mathrm{q}(V) \subset T_{\rho(x)}Y$ (see Propositions 32 and 33).
Moreover, for $x = 0$, $C_0Y = \rho_\mathrm{l}(V) + \sum_j \rho_\mathrm{q}(X_j)$, where $X_1, \ldots$ are the isotypical components of V. Note that $\rho_\mathrm{l}(V) = \rho_\mathrm{l}(V^\Gamma)$ and $\rho_\mathrm{q}(V^\Gamma) = \{0\}$.

Remark 35. If the Hilbert basis $\rho_1, \ldots, \rho_k$ is homogeneous and ordered by degree, then the differential of the reduced map has the form (3). The diagonal part is obtained by reducing $DP(0)$:

$$S := D\mathcal{J}_\mathrm{m}(DP(0)) = \begin{pmatrix} A_1 & 0 \\ 0 & \ddots \end{pmatrix}$$

Moreover, $(\rho_\mathrm{l} + \rho_\mathrm{q}) \circ DP(0) = S \circ (\rho_\mathrm{l} + \rho_\mathrm{q})$.

Lemma 36. *Suppose that* $V_1 \subset V$ *is invariant with respect to both the group action and* $DP(0)$. *Then there is a subspace* $R' \subset R_0$ *which is* S*-invariant, and contains* $(\rho_l + \rho_q)(V_1) \subset C_0 Y$.
If V_1 *is irreducible, then we can choose* R' *to be one-dimensional. If* V_1 *is the sum of two equivalent nontrivial irreducible representations of type* $\mathbb{R}$, *then a three-dimensional* R' *can be chosen.*
If V_1 *being the sum of* m *equivalent nontrivial irreducible representations has a cyclic element* v, *i.e.,* $V_1 := \operatorname{span}\{\Gamma v, \ldots, \Gamma DP(0)^{m-1} v\}$, *then the subspace* $R'' = \operatorname{span}\{S^i \rho_q(v) \mid i \in \mathbb{N}_0\} \subset R'$ *is at least* m*-dimensional.*

Proof. If X is a isotypical component, then $X \cap V_1$ is $DP(0)$-invariant. Therefore, we can assume V_1 to be contained in an isotypical component. If X is the fixed point space V^Γ, then nothing remains to show, because $D\rho(0)$ is an isomorphism between V^Γ and the tangent space of the stratum of the origin.

Now, suppose V_1 lies in a nontrivial isotypical component. There is a Γ-invariant complement of V_1, say V_2. Remark 19 applies to this situation. The image $\rho_q(V_1)$ is a subset of $R' := (L_{2,0}/J_0^2)^*$: Choose a multihomogeneous Hilbert basis $\sigma_1, \ldots, \sigma_k$ ordered in such a way that $\sigma_i \in L_{2,0}$, if and only if $i \le l$, for some l. For $v \in V_1$ we have $\sigma_i(v) = 0$ provided that $i > l$ and $\deg \sigma_i = 2$. It follows $\sigma_q(v) = (a_1, \ldots, a_l, 0, \ldots, 0)$, where $a_i \in \mathbb{R}$ and, therefore, $\sigma_q(v) \in R'$. A change to the Hilbert map ρ does not affect the statement.

If V_1 is irreducible, then $l = 1$. Therefore, $\dim L_{2,0}/J_0^2 = 1$. In the second case, we have $l = 3$.

With respect to a certain basis of $V_1 = \operatorname{span}\{\Gamma v\} \oplus \operatorname{span}\{\Gamma DP(0) v\} \oplus \cdots$ we obtain the block matrix $\begin{pmatrix} 0 & \cdots & 0 & C \\ \mathrm{id} & \ddots & \vdots & * \\ & \ddots & 0 & \vdots \\ 0 & & \mathrm{id} & * \end{pmatrix}$ for $DP(0)|_{V_1}$, where $\det C \neq 0$. Let $\sigma_i \in L_{2,0}$ be the radius on $\operatorname{span}\{\Gamma DP^{i-1} v\}$, $i = 1, \ldots, m$. Then σ_1 is a cyclic element of a subspace of $L_{2,0}$ with dimension $\ge m$. If at least one of the $*$'s stands for a non-zero matrix, then this space has dimension $> m$. The dual $R'' \subset R'$ of this space has the same dimension and possesses $\sigma_q(v)$ as a cyclic element. □

Theorem 37. *Let* ρ *be a homogeneous Hilbert basis, and* P *and* Q *be as in (2). Suppose* $P(0) = 0$. *Let* S *be as in Remark 35. Let* λ *be an eigenvalue of* $DP(0)$.

1. *If the eigenspace of* λ *contains a trivial subrepresentation, then* λ *is an eigenvalue of* S, *and there is an eigenvector of* λ *tangent to the stratum of* $\rho(0)$.
2. *If* λ *is real, and there is a nontrivial representation in the eigenspace of* λ, *then* $\mu := \lambda^2$ *is an eigenvalue of* S *with an eigenvector in the tangent cone* $C_0 Y$.

3. If λ is not real, and there is an irreducible subrepresentation of type $\mathbb{C}$ or $\mathbb{H}$ in the eigenspace of λ, then $\mu := \lambda\bar{\lambda}$ is an eigenvalue of S, and there exists an eigenvector of μ in C_0Y.

4. If the eigenspace of $\lambda \notin \mathbb{R}$ contains a nontrivial irreducible subrepresentation of type $\mathbb{R}$, then S possesses the eigenvalues λ^2, $\lambda\bar{\lambda}$ and $\bar{\lambda}^2$ and there are eigenvectors w_1, w_2, and w_3 corresponding to these eigenvalues such that the real part of $\operatorname{span}\{w_1, w_2, w_3\}$ *intersects the tangent cone C_0Y nontrivially.*

Proof. 1. Since $D\rho(0) : V^\Gamma \to T_0\Upsilon_0$ is an isomorphism, we obtain the claim from $DQ(0)D\rho(0) = D\rho(0)DP(0)$.

2. Suppose V_1 is a nontrivial irreducible subrepresentation in the eigenspace of λ. By Lemma 36 there is a one-dimensional subspace R' of R_0, which contains $\rho_\mathrm{q}(V_1) \subset C_0Y$. Since R' is an invariant space for S, it must lie in an eigenspace. According to the proof of Theorem 6, the corresponding eigenvalue of S is $\mu := \lambda^2$.

3. Let W_1 be an irreducible (complex) subrepresentation of type $\mathbb{C}$ or $\mathbb{H}$ in the eigenspace of λ. Of course, $\bar{\lambda}$ is an eigenvalue of $DP(0)$, and its eigenspace contains $\overline{W_1}$. The real part V_1 of $W_1 \oplus \overline{W_1}$ is an irreducible subrepresentation. As above there is an S-invariant line R'. Note that there is a Γ-invariant quadratic polynomial on V_1 which is of degree 1 on W_1 and $\overline{W_1}$. Therefore, the eigenvalue of S on R' is $\lambda\bar{\lambda}$.

4. Denote the mentioned representation by W_1. Then the real part V_1 of $W_1 \oplus \overline{W_1}$ is the sum of two equivalent irreducible representations $V_1^1 \oplus V_1^2$. Let $L_{2,0}$ the space of Γ-invariant polynomials of multidegree $(2,0)$ as in Remark 19. By Lemma 36 we have a three-dimensional S-invariant subspace $R' = (L_{2,0}/J_0^2)^*$ of R.

$L_{2,0} \otimes_\mathbb{R} \mathbb{C}$ contains polynomials $\sigma_1, \ldots, \sigma_3$ with multidegree $(2,0)$, $(1,1)$, and $(0,2)$ with respect to $W_1 \oplus \overline{W_1}$. According to the proof of Theorem 6 there are eigenvalues λ^2, $\lambda\bar{\lambda}$, and $\bar{\lambda}^2$ of S with eigenvectors in $R' \otimes_\mathbb{R} \mathbb{C}$. Therefore, these eigenvectors generate $R' \otimes_\mathbb{R} \mathbb{C}$. □

Remark 38. If we replace S by $DQ(0)$, the statements of the theorem remains true, due to the structure of the tangent cone (Remark 34).
Note that if A_1 and A_2 have a common eigenvalue, then all of its eigenvectors may be tangent to the stratum of the origin. In part 4, then the intersection of the real part of $\operatorname{span}\{w_1, w_2, w_3\}$ and the tangent cone is contained in $\mathbb{R}w_2$.

Example 39. Consider the action of $\mathbb{Z}_2$ on $\mathbb{R}^3$ given by $-\mathrm{id}(x,y,z) = (x,-y,-z)$, and the equivariant map $P(x,y,z) = (2x + y^2 + z^2, y - z, y + z)$. The eigenvalues of $DP(0)$ are 2 and $1 \pm \mathrm{i}$. If we choose the Hilbert basis x, y^2, z^2, yz, then $DQ(0)$ has the shape

$$\begin{pmatrix} 2 & 1 & 1 & 0 \\ 0 & 1 & 1 & -2 \\ 0 & 1 & 1 & 2 \\ 0 & 1 & -1 & 0 \end{pmatrix}$$

and the eigenvalues 2, 2i, and $-2\mathrm{i}$ with eigenvectors $(1,0,0,0)$, $(0,\mathrm{i},-\mathrm{i},1)$, and $(0,-\mathrm{i},\mathrm{i},1)$, respectively. The first eigenvalue is double. The span of the eigenvectors intersects the tangent cone $C_0Y = Y = \{(u_1,u_2,u_3,u_4) \mid u_2u_3 = u_4^2 \wedge u_2,u_3 \geq 0\}$ at the line $\mathbb{R}(1,0,0,0)$.

The eigenvalue 2 of S has the linear independent eigenvectors $(1,0,0,0)$ and $(0,1,1,0)$. The intersection of $\mathrm{span}\{(a,1,1,0),(0,\mathrm{i},-\mathrm{i},1),(0,-\mathrm{i},\mathrm{i},1)\}$, $a \in \mathbb{R}$, and C_0Y contains no eigenvector of S.

We state a converse of the last theorem, which allows to reconstruct important facts about the eigenvalues of $DP(0)$.

Proposition 40. *Resume the situation of the last Theorem. If μ is an eigenvalue of $DQ(0)$, and the eigenvector w of μ lies in the complexification of $T_0\Upsilon_0$, the tangent space of the stratum of 0, then $DP(0)|_{V^\Gamma}$ has the eigenvalue μ.*

Proof. This follows from the proof of part 1 of Theorem 37. □

Theorem 41. *Resume the setting of Theorem 37. Let μ be a real eigenvalue of S with (real) eigenvector w. Assume $w \notin T_0\Upsilon_0$.*

1. *If w is contained in the tangent cone C_0Y, then $DP(0)$ has an eigenvalue λ with $\lambda\bar{\lambda} = \mu$.*
2. *Suppose $\mathbb{R}w \cap C_0Y = \{0\}$, and assume μ_1 and $\bar{\mu}_1$ to be a pair of complex eigenvalues of S with $|\mu| = |\mu_1|$. Denote their eigenvectors by w, w_1, and $\bar{w}_1$, respectively, and assume that the space W generated by w, w_1, and $\bar{w}_1$ has a nontrivial intersection with the tangent cone. Then $DP(0)$ has eigenvalues λ and $\bar{\lambda}$ with $\lambda\bar{\lambda} = \mu$ and $\lambda^2 = \mu_1$.*

Proof. 1. w lies in the space $R \subset T_0Y$ (see Remark 9 and Proposition 32). Since R decomposes into S-invariant subspaces R_X (Remark 29), the projection w_X of w onto R_X is either the zero vector or an eigenvector of μ. w_X lies in the tangent cone C_0Y because of the product structure of C_0Y (Remark 34). Thus, we can assume $w \in R_X$, where X is not the fixed point space. There exists a point $v \in X$ with $\rho_{\mathrm{q}}(v) = w$. The space $V_1 := \mathrm{span}\,\Gamma v$ is an irreducible subrepresentation. We will show that V_1 is invariant with respect to $DP(0)$.

Let $V = V_1 \oplus \bigoplus_{i>1} V_i$ be a decomposition into irreducible subrepresentations. Choose a multihomogeneous Hilbert map σ with respect to $\bigoplus_i V_i$. Then there is one component, say σ_1, with multidegree $(2,0,\ldots)$. There is a polynomial Λ with $\sigma = \Lambda \circ \rho$, which defines a change of coordinates of the orbit space. We have $\sigma_{\mathrm{q}} = D\Lambda(0)\rho_{\mathrm{q}}$, and get $D\Lambda(0)w = \sigma_{\mathrm{q}}(v) = (\sigma_1(v),0,\ldots) =: w'$, because the quadratic components of σ except of σ_1 vanish on V_1 (Remark 29). Furthermore:

$$(\sigma_{\mathrm{l}} + \sigma_{\mathrm{q}})(x) \in \mathbb{R}w' \Longleftrightarrow x \in V_1$$

We show that V_1 is invariant with respect to $DP(0)$. The map $S' := DA(0)SDA(0)^{-1}$ has the eigenvalue μ with eigenvector w'. Remark 35 yields for $x \in V_1$

$$(\sigma_1 + \sigma_q)(DP(0)x) = S' \circ (\sigma_1 + \sigma_q)(x) = S'(cw') = \mu cw' = \mu(\sigma_1 + \sigma_q)(v) ,$$

where c is a real constant. Therefore, $DP(0)x \in V_1$. This infers that $DP(0)|_{V_1}$ has either only one eigenvalue λ or a pair of eigenvalues λ and $\bar{\lambda}$. From the proof of Theorem 6 we conclude that $\mu = \lambda\bar{\lambda}$.

2. We can assume $W \subset R_X \otimes_{\mathbb{R}} \mathbb{C}$, where $X \neq V^{\Gamma}$ is an isotypical component of V. There is a point $v \in X$ with $w := \rho_q(v) \in W$.

Let V_1 by the smallest vector space which contains v and is invariant with respect to Γ and $DP(0)$. V_1 is the sum of m equivalent irreducible subrepresentations. We show $m = 2$, and that the type of the representation is real. By Lemma 36, there is an m-dimensional space R'' generated by the iterates of w. Therefore, $m = \dim R'' \leq \dim W = 3$. If $m = 1$ or $m = 3$, then $DP(0)|_{V_1}$ has a real eigenvalue. Since $C_0Y \cap W$ does not contain any eigenvector of $S|_W$, neither $DP(0)|_{V_1}$ can have a real eigenvector nor the type of V_1 can be $\mathbb{C}$ or $\mathbb{H}$ (cf. Theorem 37 parts 2 and 3). Therefore, we have that $m = 2$, V_1 is of real type, and $DP(0)|_{V_1}$ has a pair of complex eigenvalues λ and $\bar{\lambda}$. Due to Theorem 6 either μ_1 or $\bar{\mu}_1$ must be the square of λ. □

Remark 42. Due to Theorem 37 all eigenvalues of $DP(0)$ can be discovered by Proposition 40 and Theorem 41.
If we use $DQ(0)$ instead of S, the conditions in Theorem 41 are stronger, but the assertions remain true. If there are common eigenvalues in the blocks A_1 and A_2, it may appear that not all eigenvalues of $DP(0)$ can be discovered using $DQ(0)$: Resume Example 39, considering the eigenvectors of $DQ(0)$ we cannot conclude that $DP(0)$ has the pair of eigenvalues $1 \pm \mathrm{i}$ using part 2 of Theorem 41. However, this conclusion is possible, if we consider the eigenvectors of S.
Let B be an arbitrary complement of $T_0\Upsilon_0$ in T_0Y with projection $\Pi : T_0Y \to B$. Define $S' = DQ(0)|_{T_0\Upsilon_0} + \Pi \circ DQ(0)$. The Theorems 37 and 41 remain true, if we replace S by S'.
We get the same result, if we consider generalized eigenvectors in place of genuine eigenvectors.

Example 43. Consider the matrix $\begin{pmatrix} a & -b & 0 & 0 \\ b & a & 0 & 0 \\ 0 & 0 & c & -d \\ 0 & 0 & d & c \end{pmatrix}$ with eigenvalues $\lambda_{1,2} = a \pm ib$ and $\lambda_{3,4} = c \pm id$, which is equivariant with respect to the diagonal action of SO(2) on $\mathbb{R}^4$. The orbit space is given by the relations $\rho_1\rho_2 = \rho_3^2 + \rho_4^2$ and $\rho_1 \geq 0$. Since the Hilbert basis is quadratic, we have $C_0Y = Y$. The eigenvalues of the reduced map are $\mu_1 = \lambda_1\lambda_2 = a^2 + b^2$, $\mu_2 = \lambda_3\lambda_4 = c^2 + d^2$, and $\mu_{3,4} = \lambda_{1,2}\lambda_{3,4} = ac + bd \pm \mathrm{i}(ad - bc)$ with eigenvectors $(1,0,0,0)$, $(0,1,0,0)$, and $(0,0,1,\pm\mathrm{i})$, respectively.

Even if $\mu_3 = \mu_4 \in \mathbb{R}$, none of their eigenvectors lies in the tangent cone. Applying Theorem 41 we can deduce that P has an eigenvalue λ or a pair of eigenvalues λ and $\bar{\lambda}$ with $|\lambda|^2 = a^2 + b^2$ and another eigenvalue μ or another pair of eigenvalues μ and $\bar{\mu}$ with $|\mu|^2 = c^2 + d^2$.

If we consider the symmetry of the subgroup $\mathbb{Z}_2 = \{\mathrm{id}, -\mathrm{id}\}$, then a minimal Hilbert basis consists of 10 quadratic elements: x_1^2, x_2^2, x_1x_2, x_3^2, x_4^2, x_3x_4, x_1x_3, x_2x_4, x_1x_4, and x_2x_3.

The reduced system has the eigenvalues $\mu_1 = \lambda_1^2$, $\mu_2 = \lambda_2^2$, $\mu_3 = \lambda_1\lambda_2$, $\mu_4 = \lambda_3^2$, $\mu_5 = \lambda_4^2$, $\mu_6 = \lambda_3\lambda_4$, $\mu_7 = \lambda_1\lambda_3$, $\mu_8 = \lambda_1\lambda_4$, $\mu_9 = \lambda_2\lambda_3$, and $\mu_{10} = \lambda_2\lambda_4$. Suppose we have computed the tangent cone and the eigenvectors, then we can conclude by Theorem 41 that the original system has two pairs of eigenvalues λ, $\bar{\lambda}$, μ, and $\bar{\mu}$ with $\lambda^2 = \mu_1$ and $\mu^2 = \mu_4$.

The eigenvalues $\mu_7, \ldots, \mu_{10}$ have the same modulus. If the arguments of λ_1 and λ_3 are equal, and $\lambda_1 \notin \mathbb{R}$, then $\mu_8 = \mu_9 \in \mathbb{R}$, and (μ_7, μ_{10}) is a complex pair of eigenvalues. Obviously, both triples (μ_7, μ_8, μ_{10}) and (μ_7, μ_9, μ_{10}) satisfy the first condition of part 2 of Theorem 41. However, for all triples of eigenvectors corresponding to these eigenvalues the second condition fails.

The same situation occurs, if $|\lambda_1| = |\lambda_3|$. Then we have $\mu_3 = \mu_6 \in \mathbb{R}$ and four pairs of complex conjugated eigenvalues (μ_1, μ_2), (μ_4, μ_5), (μ_7, μ_{10}), and (μ_8, μ_9) on the orbit space; all of them have the modulus μ_3. Therefore, any pair together with either μ_3 or μ_6 satisfy the first condition of part 2 of the theorem, but the second condition can only be satisfied for the triples (μ_1, μ_2, μ_3) and (μ_4, μ_5, μ_6).

Using Proposition 24 we adopt these results for an arbitrary equilibrium of Q on Y. The following theorem is an analogue to Theorem 41.

Theorem 44. *Let P and Q be as in (2). Assume Γx to be an invariant orbit of P. Then $y := \rho(x)$ is an equilibrium of Q. Let $\widetilde{P}$ and a as in Proposition 24. Let B be a complement of $T := T_y\Upsilon_y$ in T_yY with projection Π, where Υ_y is the stratum of y. Define $S = DQ(y)|_{T_y\Upsilon_y} + \Pi \circ DQ(y)$.*
Let μ be an eigenvalue of $S|_{R_y}$ (or $S|_{T_yY}$) with eigenvector w.

1. *If $w \in T \otimes_{\mathbb{R}} \mathbb{C}$, then $D\widetilde{P}(0)$ has the eigenvalue μ^a.*
2. *If $\mu \in \mathbb{R}$ and $w \in C_yY \cap B$, then $D\widetilde{P}(0)$ has an eigenvalue λ with $\lambda\bar{\lambda} = \mu^a$.*
3. *Suppose $\mu \in \mathbb{R}$ and that S has the complex eigenvalues μ_1 and $\bar{\mu}_1$ such that $|\mu| = |\mu_1|$ and the eigenvectors w, w_1 and $\bar{w}_1$, respectively, generate a space W which has a nontrivial intersection with $C_yY \cap B$. Then $D\widetilde{P}(0)$ has eigenvalues λ and $\bar{\lambda}$ with $\lambda^2 = \mu_1^a$ and $\lambda\bar{\lambda} = \mu^a$.*

Corollary 45. *The invariant orbit is normally hyperbolic, if and only if for every eigenvector w of 1 of $DQ(y)$ neither $\mathbb{R}w \cap C_yY \neq \{0\}$ nor there is a critical eigenvalue $\mu_1 \notin \mathbb{R}$ of $DQ(y)$ with eigenvector w_1 such that* $\mathrm{span}\{w, w_1, \bar{w}_1\} \cap C_yY \neq \{0\}$.

Proof. Γx is normally hyperbolic, if and only if $D\widetilde{P}(0)$ has no critical eigenvalues. It follows immediately from the above theorem that $D\widetilde{P}(0)$ has no critical eigenvalue, if and only if all eigenvectors corresponding to the eigenvalue 1 of S satisfy the condition stated above.

However, this condition holds for $DQ(y)$, if and only if it holds for S. Because if the case handled in Remark 38 arises, then A_1 has already a critical eigenvalue. □

The statements of this section are also valid for vector fields. In that context, Kœnig derived a similar criterion for hyperbolicity in [23]: Γx is normally hyperbolic, if and only if the center space of $DQ(y)$ and the tangent cone have a trivial intersection.

9 Reductive, Non-Compact Groups

Finally, we drop the assumption of compactness of Γ. To guarantee the existence of a Hilbert basis, the existence of slices, and the validity of the slice theorem and Proposition 2 we just need to assume that the group Γ is a reductive real algebraic group (see [34]), i.e., all rational representations of Γ are completely reducible.

However, then orbits are not necessarily compact any more, and it is possible that an orbit is contained in the closure of another one. Therefore, we cannot expect that a Hilbert map separates the orbits. Since each orbit contains an unique closed orbit, the space of closed orbits, called the algebraic quotient, rather than the orbit space should be embedded by a Hilbert map. Again, a set of inequalities exists which describes the embedded set ([6]). Although a Hilbert map separates the closed orbits of the complexified group $\Gamma_{\mathbb{C}}$, this is not valid for the orbits of Γ themselves. It seems reasonable to consider only those systems which respect the complexified action.

Example 46. Let $\Gamma := \mathbb{R}^*$ act on $\mathbb{R}^2$ by $(r,(x,y)) \mapsto (r^2x, r^{-2}y)$. Again $\rho(x,y) = xy$ is a Hilbert map. The orbits $\Gamma(1,1) = \{(x,y) \mid xy = 1 \wedge x > 0\}$ and $\Gamma(-1,-1) = \{(x,y) \mid xy = 1 \wedge x < 0\}$ have the same image.

The complexification of $\mathbb{R}^*$ is $\Gamma_{\mathbb{C}} := \mathbb{C}^*$. Both orbits $\Gamma(1,1)$ and $\Gamma(-1,-1)$ are subsets of $\Gamma_{\mathbb{C}}(1,1) = \{(x,y) \mid xy = 1\}$.

An important difference to the compact case is that there are representations of non-compact groups which do not have a non-constant invariant polynomial. In this case eigenvalues belonging to such a representation may not be found by the reduced system.

Example 47. Consider the action of $\mathbb{R}^*$ on $\mathbb{R}^2$ by $(r,(x,y)) \mapsto (rx, r^{-1}y)$. The algebra of invariant polynomials is generated by $\rho(x,y) = xy$. The space of closed orbits is $\mathbb{R}$.

The equivariant linear map $P(x,y) = (\lambda x, \mu y)$ is projected to $Q(z) = \lambda\mu z$. Although it is true that the eigenvalues of $DQ(0)$ are products of the eigenvalues of $DP(0)$ (Theorem 6), the stability of P is not determined by $DQ(0)$, i.e., Corollary 8 no longer holds.

Example 47 shows that the reduced map does not carry enough information about the eigenvalues of the original system.

At points with compact isotropy subgroup this problem does not arise, because the algebraic quotient has the structure of the orbit space of the compact isotropy group near such points.

References

1. Abraham, R. H., and Marsden, J. E. Foundations of Mechanics. Benjamin, Reading, Massachusetts, 1978.
2. Abud, M., and Sartori, G. The geometry of spontaneous symmetry breaking. Annals of Physics **150** (1983), 307–372.
3. Arnold, L. Random Dynamical Systems. Springer-Verlag, Berlin, 1998.
4. Bierstone, E. Lifting isotopies from orbit space. Topology **14** (1975), 245–252.
5. Bredon, G. E. Introduction in Compact Transformation Groups. Academic Press, New York, 1972.
6. Bremigan, R. J. Cohomology and real algebraic quotients. J. Algebra **159** (1993), 275–305.
7. Chossat, P. Forced reflectional symmetry breaking of an $O(2)$-symmetric homoclinic cycle. Nonlinearity **6** (1993), 723–731.
8. Chossat, P., and Dias, F. The 1:2 resonance with $O(2)$-symmetry and its application in hydrodynamics. J. Nonlinear Sci. **6** (1995), 723–732.
9. Chossat, P., Guyard, F., and Lauterbach, R. Generalized heteroclinic cycles in spherically invariant systems and their perturbations. preprint, Berlin, 1998.
10. Chossat, P., and Lauterbach, R. Le théorème de Hartman-Grobman et la réduction à l'espace des orbits. C. R. Acad. Sci. Paris **325** (1997), 595–600.
11. Derksen, H. Computation of Invariants for Reductive Groups. Universität Basel, 1997.
12. Fiedler, B., Sandstede, B., Scheel, A., and Wulff, C. Bifurcation from relative equilibria of noncompact group actions: Skew products, meanders, and drifts. Documenta Mathematica **1** (1996), 479–505.
13. Field, M. Equivariant dynamical systems. Transactions of the AMS **259** (1980), 185–205.
14. Field, M. Local Structur of Equivariant Dynamics. In Singularity Theory and its Applications II, M. Roberts and I. Stewart (Eds.), Lecture Notes in Mathematics **1463**, Springer-Verlag, Berlin, 1991, 142–166.
15. Gatermann, K. Computer Algebra Methods for Equivariant Dynamical Systems. LNM 1728, Springer-Verlag, Berlin, 2000.
16. Golubitsky, M. and Schaeffer, D. G. Singularities and Groups in Bifurcation Theory I, vol. 51 of Appl. Math. Sci. Springer-Verlag, New York, 1985.
17. Golubitsky, M., Stewart, I., and Schaeffer, D. G. Singularities and Groups in Bifurcation Theory II, vol. 69 of Appl. Math. Sci. Springer-Verlag, New York, 1988.

18. Greuel, G.-M., Pfister, G., and Schönemann, H. Singular version 1.2 User Manual. In Reports On Computer Algebra, no. 21. Centre for Computer Algebra, University of Kaiserslautern, June 1998. `http://www.mathematik.uni-kl.de/~zca/Singular`
19. Harris, J. Algebraic Geometry, vol. 133 of Graduated Texts in Mathematics. Springer-Verlag, New York, 1992.
20. Jänich, K. Differenzierbare G-Mannigfaltigkeiten. Lecture Notes in Mathematics **59**, Springer-Verlag, Berlin, Heidelberg, 1968.
21. Kato, T. Perturbation Theory for Linear Operators. Springer-Verlag, Berlin, 1966.
22. Kemper, G. Calculating invariant rings of finite groups over arbitrary fields. J. Symb. Comput. **21** (1996), 351–366.
23. Kœnig, M. Linearization of vector fields on the orbit space of the action of a compact Lie group. Math. Proc. Camb. Phil. Soc. **121** (1997), 401–424.
24. Krupa, M. Bifurcations of relative equilibria. SIAM J. Math. Anal. **21** (1990), 1453–1486.
25. Laure, P., Menck, J., and Scheurle, J. Quasiperiodic drift flow in the Couette-Taylor problem. In Bifurcation and Symmetry (Marburg 1991), Internat. Ser. Numer. Math. **104**, Birkhäuser Verlag, Basel, 1992, 191–202.
26. Lauterbach, R., and Sanders, J. Bifurcation analysis for spherically symmetric systems using invariant theory. J. Dyn. Differ. Equations **9** (1997), 535–560.
27. Mather, J. N. Differentiable Invariants. Topology **16** (1977), 145–155.
28. Melbourne, I., Lamb, J., and Wulff, C. Bifurcation from Relative Periodic Solutions. Berlin, 1999.
29. Menck, J. Analyse nichthyperbolischer Gleichgewichtspunkte in dynamischen Systemen unter Ausnutzung von Symmetrien, mit Anwendungen von Computeralgebra. Doctoral thesis, Universität Hamburg, 1992.
30. Procesi, C., and Schwarz, G. W. Inequalities defining orbit spaces. Invent. Math. **81** (1985), 539–554.
31. Rumberger, M. Lyapunov exponents on the orbit space. *Discrete Contin. Dynam. Systems* 7 (2001).
32. Rumberger, M. On the eigenvalues on the orbit space. *to appear in J. Pure and Appl. Algebra.*
33. Rumberger, M. Symmetrische dynamische Systeme: Differenzierbarkeit und linearisierte Stabilität. Doctoral thesis, Universität Hamburg, 1997.
34. Rumberger, M. Finitely differentiable invariants. Math. Zeitschrift **229** (1998), 675–694.
35. Rumberger, M., and Scheurle, J. Invariant C^r-Functions and Center Manifold Reduction. Progress in Nonlinear Differential Equations and Their Applications **19**, Birkhäuser Verlag, Basel, 1996, 145–153.
36. Scheurle, J. Some aspects of successive bifurcations in the Couette-Taylor problem. Fields Inst. Comm. **5** (1996), 335–345.
37. Schwarz, G. W. Smooth function invariant under the action of a compact Lie group. Topology **14** (1975), 63–68.
38. Schwarz, G. W. Lifting smooth homotopies of orbit spaces. Publ. Math. IHES **51** (1980), 37–135.
39. Slodowy, P. Der Scheibensatz für algebraische Transformationsgruppen. In Algebraische Transformationsgruppen und Invariantentheorie, H. Kraft, P. Slodowy, and T. A. Springer (Eds.), DMV Seminar **13**, Birkhäuser Verlag, Basel, Boston, Berlin, 1989, 89–113.

40. Sturmfels, B. Algorithms in Invariant Theory. Springer-Verlag, Wien–New York, 1993.
41. Teman, R. Infinite-Dimensional Dynamical Systems in Mechanics and Physics, vol. 68 of Appl. Math. Sci. Springer-Verlag, New York, 1988.
42. Weyl, H. The Classical Groups, 2nd ed. Princeton University Press, 1946.

Multi-Pulse Homoclinic Loops in Systems with a Smooth First Integral

Dmitry Turaev*

Weierstrass Institut für Angewandte Analysis und Stochastik, Mohrenstr. 39, 10117 Berlin, Germany
e-mail: turaev@wias-berlin.de

Abstract. We prove that the orbit-flip bifurcation in the systems with a smooth first integral (e.g. in the Hamiltonian ones) leads to appearance of infinitely many multi-pulse self-localized solutions. We give a complete description to this set in the language of symbolic dynamics and reveal the role played by special non-selflocalized solutions (e.g. periodic and heteroclinic ones) in the structure of the set of self-localized solutions. We pay a special attention to the superhomoclinic ("homoclinic to homoclinic") orbits whose presence leads to a particularly rich structure of this set.

1 Introduction

Consider a $2n$-dimensional ($n \geq 2$) dynamical system

$$\dot{x} = X(x)$$

with a smooth first integral H, i.e.,

$$H'(x)X(x) \equiv 0. \tag{1}$$

A Hamiltonian system with n degrees of freedom is a natural example but the symplectic structure is not important for our purposes.

Let X have a hyperbolic equilibrium state O at the origin (i.e. $X(0) = 0$ and the eigenvalues of the matrix $X'(0)$ do not lie on the imaginary axis). By (1)

$$H'(0)X'(0) = 0$$

so, since $X'(0)$ is non-degenerate by assumption, the linear part of H at O vanishes. Assume that the quadratic part of H at O is *a non-degeneracy quadratic form*. It is an easy exercise to check that when this non-degeneracy assumption holds, the system near O may be brought by a linear transformation of coordinates to the following form

$$\dot{u} = -Bu + \dots, \quad \dot{v} = B^{\top}v + \dots \tag{2}$$

* Project: Homoclinic Bifurcations in Hamiltonian Systems (Klaus Schneider)

where $u \in R^n$, $v \in R^n$, the dots stand for nonlinearities and B is a matrix whose eigenvalues have positive real parts. Moreover, the first integral takes the form

$$H = (v, Bu) + \dots \qquad (3)$$

where the dots stand for the third and higher order terms.

Let $\lambda_1, \dots \lambda_n$ be the eigenvalues of B, ordered in such way that $0 < \mathrm{Re}\lambda_1 \leq \dots \leq \mathrm{Re}\lambda_n$. We assume that the first two leading eigenvalues of B are real and different; precisely, we assume

$$0 < \lambda_1 < \lambda_2 < \mathrm{Re}\lambda_i \qquad (i > 2).$$

In this case the matrix B may be written in the form

$$B = \begin{pmatrix} \lambda_1 & 0 & \\ & & O \\ 0 & \lambda_2 & \\ & & \\ & O & B^0 \end{pmatrix} \qquad (4)$$

where the real parts of the eigenvalues of B^0 are strictly greater than λ_2.

The equilibrium state O is a saddle with n-dimensional stable and unstable manifolds W_O^s and W_O^u which are tangent at O to the u-space and v-space, respectively. Both the invariant manifolds lie in the $(2n-1)$-dimensional level $\{H = 0\}$ and they may intersect *transversely* in that level, producing a number of homoclinic loops, i.e. the orbits which tend to O both as $t \to +\infty$ and $t \to -\infty$ (see Fig. 1). This paper addresses the question on the possible structure of homoclinic loops in the given class of systems, in particular, on the conditions for the coexistence of infinitely many of homoclinic loops.

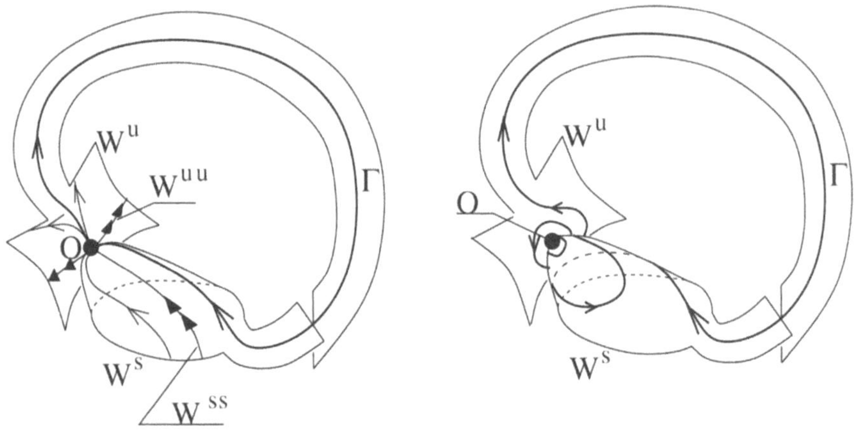

Fig. 1. A homoclinic orbit Γ of a transverse intersection of the stable and unstable manifolds of a saddle (left) or a saddle-focus (right) O.

It follows from [1–3] that (generically) there exist infinitely many homoclinic loops in an arbitrarily small neighborhood of a single homoclinic loop to a saddle-focus (this is the case where λ_1 and λ_2 are a pair of complex-conjugate numbers, we do not consider this case in this paper). On the contrary, when the equilibrium state is a saddle (i.e. λ_1 is real) no other homoclinic loops can accumulate to a homoclinic loop in general position [4]. The homoclinic loops correspond to self-localized (decaying to zero as $t \to \pm\infty$) solutions of (1). When O is a saddle, this solution tends to zero monotonically in time whereas the time dependence of any component of the self-localized solution is, typically, oscillatory when O is a saddle-focus. Thus, the cited results suggest that a self-localized solution with oscillatory tails is accompanied by infinitely many multi-pulse solutions, and self-localized solutions with monotonic tails do not form infinite series, generically. This contradicts to the fact that plenty of multi-pulse solutions with monotonic tails have been seen in different Hamiltonian systems.

To resolve this problem, a simple scenario of appearance of infinitely many homoclinic loops to a saddle was proposed in [4]: if a saddle periodic orbit L exists in the zero level of the first integral ($L \in \{H = 0\}$) and if the unstable manifold of the saddle O intersects transversely the stable manifold of L whereas the unstable manifold of L intersects transversely the stable manifold of O, then infinite sequence of homoclinic loops exists which accumulate to the union of O, L and the pair of heteroclinic connections. This statement is a simple consequence of λ-lemma: take a small cross-section S to L in $\{H = 0\}$; since $W^u_O \cap S$ intersects $W^s_L \cap S$ transversely, the infinite sequence of images of $W^u_O \cap S$ by the Poincaré map near L accumulates to $W^s_L \cap S$; each of these images must, hence, intersect $W^s_O \cap S$ transversely (as $W^s_L \cap S$ does so by assumption), producing thereby a homoclinic orbit (Fig. 2).

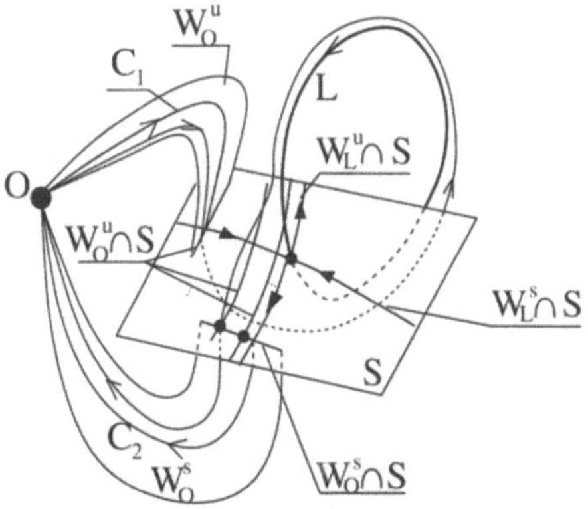

Fig. 2. Infinitely many homoclinic loops appear as a result of a transverse intersection of the invariant manifolds of O and a saddle periodic orbit L.

We start this paper with showing how such configuration appears at the so-called orbit-flip bifurcation of the homoclinic loop[1]. Namely, let the system have a transverse homoclinic loop Γ. We assume that Γ enters O as $t \to +\infty$ along *the leading direction*, i.e. it is tangent at O at $t = +\infty$ to that eigenvector of B in the u-space which corresponds to the eigenvalue λ_1. On the contrary, we require that at $t = -\infty$,

the homoclinic orbit Γ leaves O along the eigenvector of $B^\top$ in the v-space which corresponds to the eigenvalue λ_2 (the next after leading).

Note that the situation we consider here is essentially irreversible, so our orbit-flip bifurcation is different in many instances from those considered earlier in the reversible case [5,6].

The trajectories in the unstable manifold which leave O not along the leading direction form a smooth $(n-1)$-dimensional submanifold W^{uu} of W^u, transverse to the leading direction and tangent at O to the invariant subspace (in the v-space) of the matrix $B^\top$ which corresponds to the eigenvalues $\lambda_2, \ldots, \lambda_n$. The above assumption implies that $\Gamma \subset W^{uu}$. The presence of a common orbit of the n-dimensional manifold W^s and the $(n-1)$-dimensional manifold W^{uu} both lying in the $(2n-1)$-dimensional hypersurface $\{H = 0\}$ is an event of codimension one. By a small perturbation of the system (not moving it out of the class of systems with a smooth first integral) the orbit of homoclinic intersection of W^u and W^s will, generically, miss W^{uu}. To study this bifurcation we will embed our system (1) in a one-parameter family of systems with a smooth first integral, depending continuously on a parameter μ (the first integral H is assumed to depend continuously on μ as well). The original system will correspond to $\mu = 0$ and we consider the bifurcations at small μ. The system will retain its form (2), (4) (with the formula (3) still valid for H) where $\lambda_{1,2}$ and B^0 are now continuous functions of μ (as well as the terms denoted by dots in (2),(3) are).

Since the manifolds W^s and W^u depend on μ continuously and their intersection along Γ is transverse at $\mu = 0$, this intersection persists at small μ and the corresponding homoclinic orbit Γ_μ depends on μ continuously. We assume that $\Gamma_\mu \nsubseteq W^{uu}$ at $\mu \neq 0$; moreover $\Gamma_\mu \subseteq W^{u+}$ at $\mu > 0$ and $\Gamma_\mu \subseteq W^{u-}$ at $\mu < 0$ where W^{u+} and W^{u-} denote the two connected components into which W^{uu} divides W^u (Fig. 3).

Theorem 1 in the next Section shows that, generically, a saddle periodic orbit $L \in \{H = 0\}$ is born from Γ as μ passes through zero and this indeed implies the birth of infinitely many multi-pulse homoclinic loops. In the same Section we also analyze how the general structure of the set of homoclinic loops is changed due to the orbit-flip bifurcation. Namely, we establish that if

[1] Note that the orbit-flip is the only codimension-1 homoclinic bifurcation in the class of systems with a first integral which could give rise to the birth of infinite series of multi-pulse self-localized solutions with monotonic tails (the two other codimension-1 bifurcations - the tangency of stable and unstable manifolds and the transition from a saddle to a saddle-focus - are known to produce no non-oscillating multi-pulse loops).

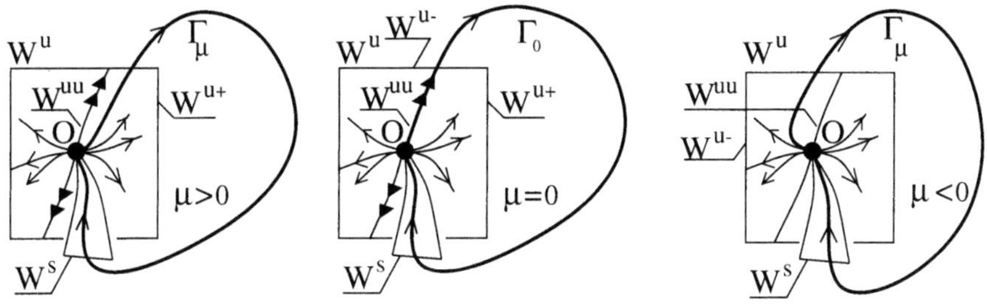

Fig. 3. The orbit-flip bifurcation: at $\mu = 0$ the homoclinic orbit Γ lies in the strong-unstable manifold of the saddle O.

a homoclinic loop $\tilde{\Gamma}$ in general position exists simultaneously with the bifurcating loop Γ, then either a double homoclinic loop close to a concatenation $\tilde{\Gamma}\Gamma$ or an infinite family of loops close to $\tilde{\Gamma}\Gamma^k$ $(k = 1, \ldots, \infty)$ is born as μ passes through zero (see theorems 2,3).

Far richer possibilities are opening when we include in the picture the so-called *superhomoclinic* (i.e. "homoclinic to homoclinic") orbits. Like the existence of a homoclinic orbit to a single periodic orbit implies the existence of infinitely many periodic orbits [7], the existence of an orbit which is homoclinic to a single homoclinic loop may imply the existence of infinitely many of loops. We show in Section 3 that at the moment of the orbit-flip bifurcation in the so-called orientable case the homoclinic loop Γ has *the unstable manifold* $W^u_\Gamma \in \{H = 0\}$ which is a smooth n-dimensional manifold with a boundary (the boundary is the manifold W^{uu}) which consists of the orbits whose limit set as $t \to -\infty$ is Γ. This manifold is the limit of the unstable manifold of the periodic orbit L_μ which tends to Γ as $\mu \to 0$ (the stable manifold of L_μ tends to the stable manifold of O). Since W^s_Γ is n-dimensional and since it lies, as a whole, in the $(2n-1)$-dimensional level $\{H = 0\}$, it may intersect transversely with W^s_O. Here, we call the orbits of such intersection the superhomoclinic orbits (see Fig. 4). We show that their presence implies immediately the existence of an infinite set of multi-pulse homoclinic loops with a nontrivial structure.

Bifurcations of superhomoclinic orbits in general (non-Hamiltonian) systems were studied in [8,9] (some cases were considered earlier in [10–12]). For systems with the smooth first integral, superhomoclinic orbits were discovered in [13] (the proofs are in [14]) in connection with the problem of the explanation of the existence of infinitely many self-localized solutions in an applied problem. Our construction here is close to that studied in [8,9] and it is quite different from that in [13,14]. However, the main idea remains the same: superhomoclinic orbits seem to play a major role in organizing the set of multi-pulse homoclinic loops in Hamiltonian systems.

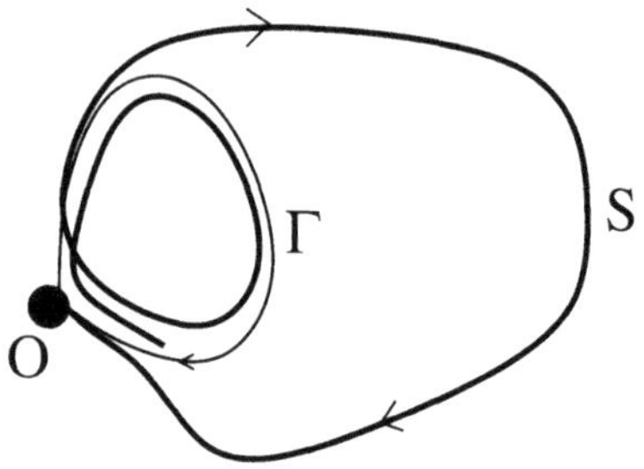

Fig. 4. A supehomoclinic orbit S is α-limit to the homoclinic loop Γ and ω-limit to the saddle O.

Finally, the author would like to acknowledge the support by the DFG-Schwerpunktprogramme DANSE and to express his gratitude to L.P.Shilnikov who proposed him this problem many years ago.

2 Orbit-Flip Bifurcation

We impose, first, some genericity assumptions on the system under consideration, which are necessary to study the orbit-flip bifurcation. The first two of them were the transversality of the intersection of W^s and W^u along Γ and the requirement that $\Gamma \notin W^{ss}$ (i.e. it enters O as $t \to +\infty$ along the leading direction).

To formulate the third genericity assumption we recall (see [15]) that an extended stable manifold W^{see} which is a smooth $(n+2)$-dimensional invariant manifold tangent at O to the direct sum of the u-space and the invariant subspace of $B^\top$ in the v-space which corresponds to the leading eigenvalues λ_1 and λ_2. Note that W^{see} contains the stable manifold W^s, so it contains the homoclinic orbit Γ (note that W^{see} is not unique but any two of such manifolds are tangent to each other at every point of W^s). We require that at $\mu = 0$, at the points of Γ the manifold W^{see} is transverse to the strong unstable manifold W^{uu} (by invariance of W^{see} and W^{uu} it is sufficient to require the transversality at an arbitrary single point on Γ).

According to [16] this kind of transversality assumption is sufficient for the result of [17] to be fulfilled; namely, it guarantees the existence of a C^1-smooth invariant repelling $(n+2)$-dimensional manifold which is transverse to W^{uu} and which contains all orbits staying in a small neighborhood of the homoclinic loop Γ for all times.

The fourth genericity assumption is

$$\lambda_2 \neq 2\lambda_1.$$

It is not a technical assumption; we will see that the cases $\lambda_2 < 2\lambda_1$ and $\lambda_2 > 2\lambda_1$ are indeed different (though the results are similar). We will also need different smoothness assumptions in these cases: the system will be assumed C^r-smooth with $r \geq 3$ at $\lambda_2 < 2\lambda_1$ and $r \geq 4$ at$\lambda_2 > 2\lambda_1$.

Most importantly, the last, fifth, genericity assumption is different in the cases $\lambda_2 < 2\lambda_1$ and $\lambda_2 > 2\lambda_1$. If $\lambda_2 < 2\lambda_1$, then in W^s there exists a special smooth (at least C^2) invariant $(n-1)$-dimensional manifold W^{s0} which is tangent at O to the eigenspace of the matrix B in the u-space which corresponds to the eigenvalues $\lambda_1, \lambda_3, \ldots, \lambda_n$ (i.e. it is transverse to the eigenvector corresponding to the eigenvalue λ_2). The existence of this manifold is proved later. We will assume that in this case

$$\Gamma \not\subset W^{s0}.$$

Basically, this means that when Γ enters O at $t = +\infty$, the coordinate u_1 (the projection onto the stable leading eigenvector) behaves asymptotically as

$$u_{11}e^{-\lambda_1 t} + u_{12}e^{-\lambda_2 t} + O(e^{-2\lambda_1 t})$$

where $u_{12} \neq 0$ (the non-vanishing of u_{11} is given by the assumption $\Gamma \not\in W^{ss}$).

When $\lambda_2 > 2\lambda_1$, the special manifold W^{u0} is not defined uniquely and, moreover, the above assumption is unnecessary. An important requirement we need in this case is that

the intersection of the extended unstable manifold W^{ue} with the stable manifold W^s along Γ is transverse in R^{2n}.

This extended unstable manifold is an $(n+1)$-dimensional smooth invariant manifold which is tangent at O to the direct sum of the v-space and the leading eigenvector in the u-space (it is the eigenvector corresponding to the leading eigenvalue λ_1 of B) (see [15]). This manifold is not unique but any two of them contain the stable manifold W^u and are tangent to each other at every point of W^u. Hence, the transversality assumption above is well posed (recall that $\Gamma \subset W^u$). Note that we speak here about the transversality in the whole phase space, not in the level $\{H = 0\}$. The intersection of W^{ue} with $\{H = 0\}$ is the union of two n-dimensional manifolds: one is W^u and the second is a smooth manifold W^{u1} which intersects W^u at the points of the manifold W^{uu} transversely in $\{H = 0\}$. Since $\Gamma_\mu \subset W^{uu}$ at $\mu = 0$, the above transversality assumption can be read as the transversality (in $\{H = 0\}$) of the intersection of W^{u1} and W^s along the homoclinic loop Γ_0. Note that this requirement is unnecessary if $\lambda_2 < 2\lambda_1$.

In both cases, the fifth non-degeneracy assumption can be expressed as a non-vanishing of some functional $A(X)$ which will be explicitly defined later. We will introduce also a functional $a(X)$ whose non-vanishing is equivalent to the transversality of W^u and W^s. The signs of A and a determine the structure of bifurcations which happened at $\mu \neq 0$.

Theorem 1. *Let U be a sufficiently small neighborhood of Γ_0 in the level $\{H = 0\}$. At $Aa\mu \geq 0$ there is no other orbit, except for Γ_μ and O, which stays in U for all times. At $Aa\mu < 0$, the set of the orbits staying in U*

for all times consists of: O, Γ_μ, a single-round periodic orbit L_μ, a pair of heteroclinic orbits $C_{1\mu}$ and $C_{2\mu}$ - the former is α-limit to O and ω-limit to L_μ whereas the latter is α-limit to L_μ and ω-limit to O, and a sequence of homoclinic loops $\Gamma_{k\mu}$ ($\Gamma_{k\mu}$ is a k-round loop, $k = 2, \ldots$, one such loop for each k) which accumulate to the union $O \cup L_\mu \cup C_{1\mu} \cup C_{2\mu}$.

Generically, in addition to Γ, the system at $\mu = 0$ may have some number of other homoclinic loops $\Gamma_+^1, \ldots, \Gamma_+^{m_+}$ and $\Gamma_-^1, \ldots, \Gamma_-^{m_-}$ which correspond to transverse intersection of W^s and W^u and which do not lie neither in W^{uu} nor in W^{ss} (i.e. they leave and enter O along the leading directions). We assume that the loops $\Gamma_+^1, \ldots, \Gamma_+^{m_+}$ lie in W^{s+} and the loops $\Gamma_-^1, \ldots, \Gamma_-^{m_-}$ lie in W^{s-} where $W^{s\pm}$ are two components into which W^{ss} divides W^s: we assume that the orbit Γ_μ belongs to W^{s+}.

Let U be a small neighborhood of the homoclinic bunch $\Gamma \cup \Gamma_+^1 \cup \ldots \cup \Gamma_+^{m_+} \cup \Gamma_-^1 \cup \ldots \cup \Gamma_-^{m_-} \cup O$ in the level $\{H = 0\}$. It is a union of a small neighborhood of O with $m_+ + m_- + 1$ handles U_0, U_{1+}, ..., U_{m_++}, U_{1-}, ..., U_{m_--} (the handle U_0 surrounds Γ). Since the fundamental group of U is nontrivial, every orbit in U gets its natural coding which describes the sequence of handles the orbit visits as time runs. Thus, the coding of O is the empty sequence, Γ is coded by 0, the loops $\Gamma_\pm^i$ are coded by $i\pm$ respectively, the periodic orbit L_μ from theorem 1 is coded by the infinite sequence of 0's, the heteroclinic orbits $C_{1\mu}$ and $C_{2\mu}$ are coded, respectively, by the infinite to the right and infinite to the left sequences of 0's; the k-round homoclinic loops from theorem 1 are coded by 0^k.

Theorem 2. *Except for the orbits given by theorem 1 and the homoclinic loops $\Gamma_\pm^i$, the set of all orbits lying entirely in U contains the following orbits (and only them): double homoclinic loops $(i-)0$ (where $i = 1, \ldots, m_-$) at $Aa\mu > 0$; nothing at $\mu = 0$; exactly one homoclinic loop $(i+)0^k$ for each $k \geq 1$ and $i = 1, \ldots, m_+$, and m_+ heteroclinic connections $(1+)0^\infty$, ..., $(m_++)0^\infty$ from O to L_μ at $Aa\mu < 0$ (as $k \to +\infty$, the limit of the sequence of loops $(i+)0^k$ is the heteroclinic connection $(i+)0^\infty$).*

Let us prove theorems 1 and 2. Choose the coordinates $(u_1, u_2, \ldots, u_n, v_1, v_2, \ldots, v_n)$ near O such that the u_1-axis will be the eigenvector of B corresponding to the leading eigenvalue λ_1, the u_2-axis will be the eigenvector of B corresponding to the next eigenvalue λ_2 and the plane $(u_1 = u_2 = 0)$ will be the eigenspace corresponding to the rest of the spectrum of B; similarly, let the v_1-axis be the eigenvector of $B^\top$ corresponding to λ_1, the v_2-axis be the eigenvector of $B^\top$ corresponding to λ_2 and the plane $(v_1 = v_2 = 0)$ be the eigenspace corresponding to the rest of the spectrum of $B^\top$. By assumption, Γ enters O at $t = +\infty$ tangent to the u_1-axis. We choose the sign of u_1 such that $u_1 > 0$ on Γ at t close to $+\infty$; i.e. the component W^{s+} of W^s corresponds to the positive direction of the u_1-axis. At $\mu = 0$ the homoclinic orbit Γ is tangent at O to the v_2-axis at $t = -\infty$. We assume that $v_2 > 0$ on

Γ at t close to $-\infty$. Moreover, we assume that Γ adjoins O at $t = -\infty$ from the side of positive v_1 at $\mu > 0$ and from the side of negative v_1 at $\mu < 0$; i.e. the component W^{u+} extends from W^{uu} towards $v_1 > 0$ and W^{u-} extends towards negative v_1.

Let us straighten the invariant manifolds W^s and W^u near O so that their equations will be, respectively, $v = 0$ and $u = 0$ locally. The system will take the following form near O:

$$\dot{u} = -Bu + f(u, v)u, \quad \dot{v} = B^\top v + g(u, v)v \tag{5}$$

where f and g are some C^{r-1}-functions vanishing at zero. The first integral is now locally written as

$$H = (v, Bu) + H_0(u, v) \tag{6}$$

where H_0 vanish identically both at $u = 0$ and $v = 0$. According to [18] (see also [19,12] and [15]), by an additional C^{r-1}-smooth transformation of coordinates system (5) is brought to the following form, where we denote $u^0 = (u_3, \ldots, u_n)$ and $v^0 = (v_3, \ldots, v_n)$:

$$\begin{aligned}
\dot{u}_1 &= -\lambda_1 u_1 + f_{11}(u_1, v)u_1 + f_{12}(u_1, u_2, v)u_2 + f_{10}(u, v)u^0, \\
\dot{u}_2 &= -\lambda_2 u_2 + f_{21}(u_1, v)u_1 + f_{22}(u_1, u_2, v)u_2 + f_{20}(u, v)u^0, \\
\dot{u}^0 &= -B^0 u^0 + f_{01}(u_1, v)u_1 + f_{02}(u_1, u_2, v)u_2 + f_{00}(u, v)u^0, \\
\dot{v}_1 &= \lambda_1 v_1 + g_{11}(u, v_1)v_1 + g_{12}(u, v_1, v_2)v_2 + g_{10}(u, v)v^0, \\
\dot{v}_2 &= \lambda_2 v_2 + g_{21}(u, v_1)v_1 + g_{22}(u, v_1, v_2)v_2 + g_{20}(u, v)v^0, \\
\dot{v}^0 &= (B^0)^\top v^0 + g_{01}(u, v_1)v_1 + g_{02}(u, v_1, v_2)v_2 + g_{00}(u, v)v^0
\end{aligned} \tag{7}$$

with the C^{r-1}-functions f_{ij}, g_{ij} vanishing at zero and satisfying the following identities

$$\begin{array}{lll}
f_{i1}(0, v) \equiv 0, & g_{i1}(u, 0) \equiv 0 & (i = 1, 2, 0), \\
f_{i2}(0, 0, v) \equiv 0, & g_{i2}(u, 0, 0) \equiv 0 & (i = 2, 0), \\
f_{11}(u_1, 0) \equiv 0, & f_{12}(u_1, u_2, 0) \equiv 0, & f_{10}(u, 0) \equiv 0, \\
g_{11}(0, v_1) \equiv 0, & g_{12}(0, v_1, v_2) \equiv 0, & g_{10}(0, v) \equiv 0
\end{array} \tag{8}$$

and, at $\lambda_2 < 2\lambda_1$, the following additional identities

$$\begin{array}{lll}
f_{12}(0, 0, v) \equiv 0, & g_{12}(u, 0, 0) \equiv 0, & \\
f_{21}(u_1, 0) \equiv 0, & f_{22}(u_1, u_2, 0) \equiv 0, & f_{20}(u, 0) \equiv 0, \\
g_{21}(0, v_1) \equiv 0, & g_{22}(0, v_1, v_2) \equiv 0, & g_{20}(0, v) \equiv 0.
\end{array} \tag{9}$$

By [12], an additional C^{r-2}-smooth coordinate transformation can be done in the case $\lambda_2 > 2\lambda_1$ which keeps the system in the form (7),(8) with f_{ij}, g_{ij}

(now C^{r-2}) satisfying the following additional identities:

$$\frac{\partial f_{1j}}{\partial v_1} \equiv 0 \quad \text{at} \quad v_1 = 0$$
$$\frac{\partial g_{1j}}{\partial u_1} \equiv 0 \quad \text{at} \quad u_1 = 0. \tag{10}$$

Hereafter we assume that the system is brought to this form. We denote the smoothness of the obtained system as q (i.e. $q = r - 1$ at $\lambda_2 < 2\lambda_1$ and $q = r - 2$ at $\lambda_2 > 2\lambda_1$, so $q \geq 2$ in both cases).

In these coordinates, the non-leading manifolds W^{ss} and W^{uu} are given by equations $\{v = 0, u_1 = 0\}$ and $\{u = 0, v_1 = 0\}$, respectively. Furthermore, identities (8) guarantee that the extended unstable manifold W^{ue} is tangent to $\{u_2 = 0, u^0 = 0\}$ at the points of the local unstable manifold $W^u_{\text{loc}} : \{u = 0\}$. Indeed, the tangents to W^{ue} at the points W^u_{loc} form a continuous field of linear spaces invariant with respect to the flow linearized along the orbits in W^u_{loc} and this field is transverse to W^{ss} at O. According to [15] such field is unique. When identities (8) are satisfied, the space $\{u_2 = 0, u^0 = 0\}$ is invariant with respect to the linearized flow and it is transverse to W^{ss} at O, hence it is the tangent to W^{ue} indeed. Thus, W^{ue} is locally given by an equation of the form

$$(u_2, u^0) = h^{ue}(u_1, v) \tag{11}$$

where h^{ue} vanishes at zero along with its first derivatives. Note that h^{ue} must vanish identically at $u_1 = 0$ because W^{ue} contains $W^u_{\text{loc}} : \{u = 0\}$ by definition. Now, it is seen that the first integral (6) on W^{ue}_{loc} is written in the form

$$H = \lambda_1 u_1 (v_1 - h^{u1}(u_1, v))$$

for some smooth h^{u1} which vanish at zero along with the first derivative. Hence, the intersection $W^{ue}_{\text{loc}} \cap \{H = 0\}$ is the union of W^u_{loc} and a C^1-manifold W^{u1}_{loc} given by (11) with the constraint

$$v_1 = h^{u1}(u_1, v). \tag{12}$$

The intersection of W^{u1} with W^u_{loc} must be an $(n-1)$-dimensional invariant submanifold of W^u_{loc}, transverse to the v_1-axis in virtue of (12). Such a submanifold is unique – it is W^{uu}_{loc}. Thus,

$$W^{u1}_{\text{loc}} \cap W^u_{\text{loc}} = W^{uu}_{\text{loc}},$$

i.e. $h^{u1}(0, v) \equiv 0$.

Analogously, the tangent to W^{see} at the points of W^s_{loc} is $v^0 = 0$.

When $\lambda_2 < 2\lambda_1$, identities (8), (9) imply that the evolution of the variables (u_1, u_2) on W^s_{loc} is independent on u^0 and is governed by the linear system

$$\dot{u}_1 = \lambda_1 u_1, \qquad \dot{u}_2 = \lambda_2 u_2.$$

Thus, for every orbit in $W^s_{loc} \backslash W^{ss}$ we have $u_2(t) = C u_1^\alpha(t)$ with $\alpha = \lambda_2/\lambda_1 < 2$. It follows that $u_2 = 0$ is a unique invariant submanifold of W^s_{loc} which is transverse to the u_2-axis and which is at least C^2-smooth. We denote this manifold as W^{s0}.

Take a small $d > 0$ and consider a pair of $(2n-2)$-dimensional cross-sections Π^{in} and Π^{out} to the homoclinic loop Γ: $\Pi^{\text{in}} = \{u_1 = d\} \cap \{H = 0\}$ and $\Pi^{\text{out}} = \{v_2 = d\} \cap \{H = 0\}$. Let $M^{\text{in}}(u^{\text{in}}, v^{\text{in}}) = \Gamma \cap \Pi^{\text{in}}$ and $M^{\text{out}}(u^{\text{out}}, v^{\text{out}}) = \Gamma \cap \Pi^{\text{out}}$. Since $M^{\text{in}} \in W^s_{loc}$ and $M^{\text{out}} \in W^u_{loc}$, it follows that $v^{\text{in}} \equiv 0$ and $u^{\text{out}} \equiv 0$. By assumption, $M^{\text{out}} \in W^{uu}_{loc}$ at $\mu = 0$, therefore $v_1^{\text{out}}|_{\mu=0} = 0$. When μ increases through zero, the value of v_1^{out} changes from negative values to positive, so we may simply assume

$$v_1^{\text{out}} = \mu. \tag{13}$$

Recall that $v_2^{\text{out}} = u_1^{\text{in}} = d$. Since $M^{\text{in}} \notin W^{s0}$ at $\lambda_2 < 2\lambda_1$, it follows that

$$u_2^{\text{in}} \neq 0 \text{ in the case } \lambda_2 < 2\lambda_1. \tag{14}$$

We take a small $\delta > 0$ and shrink Π^{in} and Π^{out} to the size δ neighborhoods of M^{in} and M^{out}, respectively. In particular, we have $\|v^0 - v^{0\text{out}}\| \leq \delta$ on Π^{out}. Since the orbit Γ is tangent to the v_2-axis at $\mu = 0$ by assumption, it follows that

$$\|v^{\text{out}}\| \ll d$$

on Π^{out}.

Orbits which lie in the level $\{H = 0\}$ in a small neighborhood of Γ must intersect $\Pi^{in,out}$, so the problem of the study of these orbits reduces to the study of the Poincaré map on these cross-sections. The flow near the global piece of the loop Γ outside the d-neighborhood of the saddle defines *the global map* T_{glo} from Π^{out} to Π^{in}. Since the corresponding flight time is bounded, this map is a diffeomorphism and it is well approximated by its Taylor expansion at the point M^{out}.

Recall that $H = 0$ on Π^{out} and $v_2 = \mathit{const} \neq 0$. Hence, by (4) and (6), u_2 is a smooth function of (u_1, v_1, u^0, v^0) for points in Π^{out}. Thus, (u_1, v_1, u^0, v^0) form a good set of coordinates on Π^{out}. Analogously, (u_2, v_2, u^0, v^0) are the coordinates on Π^{in} (here, $u_1 = \mathit{const} \neq 0$ and v_1 is found from the condition $H = 0$).

Now, we can write the map $T_{\text{glo}} : M \mapsto \bar{M}$ as

$$\begin{cases} \bar{v}_2 = a_1(v_1 - \mu) + b_1 u_1 + c_1(v^0 - v^{0\text{out}}) + d_1 u^0 + \dots \\ \bar{u}_2 - u_2^{\text{in}} = a_2(v_1 - \mu) + b_2 u_1 + c_2(v^0 - v^{0\text{out}}) + d_2 u^0 + \dots \\ \bar{v}^0 = a_3(v_1 - \mu) + b_3 u_1 + c_3(v^0 - v^{0\text{out}}) + d_3 u^0 + \dots \\ \bar{u}^0 - u^{0\text{in}} = a_4(v_1 - \mu) + b_4 u_1 + c_4(v^0 - v^{0\text{out}}) + d_4 u^0 + \dots \end{cases} \tag{15}$$

where the dots stand for non-linear (quadratic and higher order) terms.

The intersection of W^{uu}_{loc} with Π^{out} is $\{v_1 = 0, u = 0\}$, so it follows from (15) that we have

$$\bar{v}^0 = c_3(v^0 - v^{0\text{out}})$$

on the tangent to $T_{glo}(W^{uu}_{\rm loc}\cap\Pi^{\rm out})$ at $\mu=0$. The tangent to $W^{see}_{\rm loc}$ is $\bar{v}^0=0$, so the transversality of W^{uu} to W^{see} means that

$$\det c_3 \neq 0.$$

This allows for recasting (15) in the so-called cross-form: $\bar{M}=T_{glo}M$ if and only if

$$\begin{cases} \bar{v}_2 = a_1(v_1-\mu)+b_1u_1+c_1\bar{v}^0+d_1u^0+\dots \\ \bar{u}_2-u_2^{\rm in} = a_2(v_1-\mu)+b_2u_1+c_2\bar{v}^0+d_2u^0+\dots \\ v^0-v^{0\rm out} = a_3(v_1-\mu)+b_3u_1+c_3\bar{v}^0+d_3u^0+\dots \\ \bar{u}^0-u^{0\rm in} = a_4(v_1-\mu)+b_4u_1+c_4\bar{v}^0+d_4u^0+\dots \end{cases} \tag{16}$$

for some new coefficients a,b,c,d, and for some functions of $(v_1-\mu,u_1,\bar{v}^0,u^0)$ of at least second order of smallness which are denoted by dots in the right-hand sides of this formula.

When the map is written in the cross-form, it is obvious that the transversality of $T_{glo}(W^u_{\rm loc}\cap\Pi^{\rm out})$ to $W^s_{\rm loc}\cap\Pi^{\rm in}$ at the point $M^{\rm in}$ is equivalent to

$$a_1\neq 0, \tag{17}$$

and the transversality of $T_{glo}(W^{u1}_{\rm loc}\cap\Pi^{\rm out})$ to $W^s_{\rm loc}\cap\Pi^{\rm in}$ at the point $M^{\rm in}$ is equivalent to

$$b_1\neq 0. \tag{18}$$

So, our genericity assumptions are (17) and (14) in the case $\lambda_2<2\lambda_1$, and (17) and (18) in the case $\lambda_2>2\lambda_1$.

We can now introduce the quantities a and A from Theorems 1 and 2:

$$a=-a_1 \tag{19}$$

and

$$A=\begin{cases} -\dfrac{\lambda_2}{\lambda_1}a_1u_2^{\rm in} & \text{at } \lambda_2<2\lambda_1 \\ b_1d & \text{at } \lambda_2<2\lambda_1 \end{cases} \tag{20}$$

Let us now proceed to the evaluation of the *local map* from the cross-sections $\Pi^{\rm in}$ to $\Pi^{\rm out}$ which is defined by the orbits in the d-neighborhood of the saddle O. This is a much less trivial problem because an orbit starting on $\Pi^{\rm in}$ may stay near O for an unboundedly large time before reaching the cross-section $\Pi^{\rm out}$.

The regular method which allows for resolving this difficulty is based upon the study of a specific boundary value problem considered in [7]. Namely, as it follows from [7] for our particular case, if an orbit in a small neighborhood of a saddle starts at $t=0$ with some point $M_0(u_{10},u_{20},u_0^0,v_{10},v_{20},v_0^0)$ and reaches a point $M_\tau(u_{1\tau},u_{2\tau},u_\tau^0,v_{1\tau},v_{2\tau},v_\tau^0)$ at the moment $t=\tau$, then the values of (v_{10},v_{20},v_0^0) and $(u_{1\tau},u_{2\tau},u_\tau^0)$ are uniquely defined by (u_{10},u_{20},u_0^0), $(v_{1\tau},v_{2\tau},v_\tau^0)$ and τ. Moreover, such M_0 and M_τ exist for any given $\tau\geq 0$

and small (u_{10}, u_{20}, u_0^0), $(v_{1\tau}, v_{2\tau}, v_\tau^0)$; the corresponding piece of the orbit is found as the unique solution of the following system of integral equations

$$\begin{cases} v_1(t) = e^{-\lambda_1(\tau-t)}v_{1\tau} - \int_t^\tau e^{\lambda_1(t-s)}\big(g_{11}(u(s),v_1(s))v_1(s) \\ \qquad +g_{12}(u(s),v_1(s),v_2(s))v_2(s) \\ \qquad +g_{10}(u(s),v(s))v^0(s)\big)\,ds \\ v_2(t) = e^{-\lambda_2(\tau-t)}v_{2\tau} - \int_t^\tau e^{\lambda_2(t-s)}\big(g_{21}(u(s),v_1(s))v_1(s) \\ \qquad +g_{22}(u(s),v_1(s),v_2(s))v_2(s) \\ \qquad +g_{20}(u(s),v(s))v^0(s)\big)\,ds \\ v^0(t) = e^{-(B^0)^\top(\tau-t)}v_\tau^0 - \int_t^\tau e^{(B^0)^\top(t-s)}\big(g_{01}(u(s),v_1(s))v_1(s) \\ \qquad +g_{02}(u(s),v_1(s),v_2(s))v_2(s) \\ \qquad +g_{00}(u(s),v(s))v^0(s)\big)\,ds \\ u_1(t) = e^{-\lambda_1 t}u_{10} + \int_0^t e^{\lambda_1(s-t)}\big(f_{11}(u_1(s),v(s))u_1(s) \\ \qquad +f_{12}(u_1(s),u_2(s),v(s))u_2(s) \\ \qquad +f_{10}(u(s),v(s))u^0(s)\big)\,ds \\ u_2(t) = e^{-\lambda_2 t}u_{20} + \int_0^t e^{\lambda_2(s-t)}\big(f_{21}(u_1(s),v(s))u_1(s) \\ \qquad +f_{22}(u_1(s),u_2(s),v(s))u_2(s) \\ \qquad +f_{20}(u(s),v(s))u^0(s)\big)\,ds \\ u^0(t) = e^{-B^0 t}u_0^0 + \int_0^t e^{B^0(s-t)}\big(f_{01}(u_1(s),v(s))u_1(s) \\ \qquad +f_{02}(u_1(s),u_2(s),v(s))u_2(s) \\ \qquad +f_{00}(u(s),v(s))u^0(s)\big)\,ds. \end{cases} \tag{21}$$

This system is obtained by integration of (7). According to [7], the solution of (21) on the interval $t \in [0,\tau]$ is found by successive approximations. The first approximation is

$$(u(t) = 0,\ v(t) = 0).$$

Using identities (8), (9), (10) one can see (the detailed computation for a general case can be found in [18,12]) that the second and all the further approximations have the form

$$\begin{aligned} &v_1(t) = e^{-\lambda_1(\tau-t)}v_{1\tau} + O(e^{-\lambda'(\tau-t)}),\ u_1(t) = e^{-\lambda_1 t}u_{10} + O(e^{-\lambda' t}) \\ &v_2(t) = e^{-\lambda_2(\tau-t)}v_{2\tau} + O(e^{-\lambda'(\tau-t)}),\ u_2(t) = e^{-\lambda_2 t}u_{20} + O(e^{-\lambda' t}) \\ &v^0(t) = O(e^{-\lambda'(\tau-t)}), \qquad\qquad u^0(t) = O(e^{-\lambda' t}) \end{aligned} \tag{22}$$

where λ' is some constant such that

$$\lambda' > \min(2\lambda_1, \lambda_2) \tag{23}$$

(note that $\lambda' < \mathrm{Re}\lambda_3$); the $O(\cdot)$-terms in (22) are bounded uniformly, for all successive approximations. Hence, the solution of (21) has the same form. Note that up to the order $(q-1)$ the derivatives of the successive approximations with respect to the data $\{t, \tau, u_{10}, u_{20}, u_0^0, v_{1\tau}, v_{2\tau}, v_\tau^0)$ satisfy, uniformly, the estimates obtained by the formal differentiation of (22) (see [18,12]). Therefore, formulas (22) give estimates for the solution of (21) along with the derivatives up to the $(q-1)$-th order.

By (22), the following relation holds for the point M_0 and its time τ shift M_τ:

$$v_{10} = e^{-\lambda_1\tau}v_{1\tau} + O(e^{-\lambda'\tau}),\ u_{1\tau} = e^{-\lambda_1\tau}u_{10} + O(e^{-\lambda'\tau})$$

$$v_{20} = e^{-\lambda_2\tau}v_{2\tau} + O(e^{-\lambda'\tau}),\ u_{2\tau} = e^{-\lambda_2\tau}u_{20} + O(e^{-\lambda'\tau}) \tag{24}$$

$$v_0^0 = O(e^{-\lambda'\tau}), \qquad u_\tau^0 = O(e^{-\lambda'\tau}).$$

Suppose now that $M_0 \in \Pi^{\text{in}}$ and $M_\tau \in \Pi^{\text{out}}$. It means that $u_{10} = d > 0$, and $v_{2\tau} = d > 0$. Since $H = 0$ at M_0, it follows that

$$v_{10} = -\frac{\lambda_2}{\lambda_1}\frac{u_{20}}{u_{10}}v_{20} - \frac{1}{u_{10}\lambda_1}(v_0^0, B^0 u_0^0) + \ldots \tag{25}$$

where the dots stand for the terms (vanishing at $v_{20} = 0, v_0^0 = 0$) of order higher than two.

Now, it is seen that given any small u_{20}, v_τ^0, u_0^0 and sufficiently large τ the corresponding values of $v_{20}, v_{1\tau}, v_0^0$ and u_τ^0 are defined uniquely and the following estimates hold:

– $\lambda_2 < 2\lambda_1$

$$\begin{aligned}
v_{1\tau} &= -\frac{\lambda_2}{\lambda_1}\frac{v_{2\tau}}{u_{10}}u_{20}e^{(\lambda_1-\lambda_2)\tau} + O(e^{(\lambda_1-\lambda')\tau}),\\
u_{1\tau} &= e^{-\lambda_1\tau}u_{10} + O(e^{-\lambda'\tau}), \qquad u_\tau^0 = O(e^{-\lambda'\tau}),\\
v_{20} &= e^{-\lambda_2\tau}v_{2\tau} + O(e^{-\lambda'\tau}), \qquad v_0^0 = O(e^{-\lambda'\tau});
\end{aligned} \tag{26}$$

– $\lambda_2 > 2\lambda_1$

$$\begin{aligned}
v_{1\tau} &= O(e^{(\lambda_1-\lambda')\tau})\\
u_{1\tau} &= e^{-\lambda_1\tau}u_{10} + O(e^{-\lambda'\tau}), \qquad u_\tau^0 = O(e^{-\lambda'\tau}),\\
v_{20} &= O(e^{-\lambda'\tau}), \qquad v_0^0 = O(e^{-\lambda'\tau}).
\end{aligned}$$

These formulas define (implicitly) the map T_{loc} from Π^{in} to Π^{out} if we assume u_{20} close to u_2^{in}, u_0^0 close to $u^{0\text{in}}$, v_τ^0 close to $v^{0\text{out}}$ and $u_{10} = v_{2\tau} = d$.

Combining formulas (26) and (16), we arrive to the following formula for the Poincaré map $T = T_{glo} \circ T_{loc} : \Pi^{\text{in}} \to \Pi^{\text{in}}$ (we denote $\nu = \min(\lambda_1, \lambda_2 - \lambda_1)$ and $\mathcal{A} = A$ at $\lambda_2 > 2\lambda_1$ and $\mathcal{A} = A[1 + (u_2 - u_2^{\text{in}})/u_2^{\text{in}}]$ at $\lambda_2 > 2\lambda_1$):

$$\begin{cases} \bar{v}_2 = a\mu + \mathcal{A}e^{-\nu\tau} + \phi(\bar{v}^0, \mu) + o(e^{-\nu\tau}), \\ v_2 = o(e^{-\nu\tau}), \qquad v^0 = o(e^{-\nu\tau}) \\ \bar{u}_2 = u_2^{\text{in}} + \psi(\bar{v}_2, \bar{v}^0, \mu) + o(e^{-\nu\tau}), \\ \bar{u}^0 = u^{0\text{in}} + \psi^0(\bar{v}_2, \bar{v}^0, \mu) + o(e^{-\nu\tau}), \end{cases} \tag{27}$$

where ϕ, ψ, ψ^0 are some smooth functions vanishing at zero:

$$\bar{u}_2 = u_2^{\text{in}} + \psi(\bar{v}_2, \bar{v}^0, \mu), \qquad \bar{u}^0 = u^{0\text{in}} + \psi^0(\bar{v}_2, \bar{v}^0, \mu) \tag{28}$$

is the equation of the surface w^{u*} equal to $T_{glo}(W^u_{\text{loc}} \cap \Pi^{\text{out}})$ at $\lambda_2 < 2\lambda_1$ and to $T_{glo}(W^{u1}_{\text{loc}} \cap \Pi^{\text{out}})$ at $\lambda_2 > 2\lambda_1$; the subset of this surface given by the equation

$$\bar{v}_2 = a\mu + \phi(\bar{v}^0, \mu) \tag{29}$$

is $w^{uu} = T_{glo}(W^{uu}_{\text{loc}} \cap \Pi^{\text{out}})$.

Since $\mathcal{A} \neq 0$ (recall that $u_2 - u_2^{\text{in}}$ is small on Π^{in}), it follows that the first equation of (27) can be resolved with respect to τ, provided

$$A(\bar{v}_2 - a\mu - \phi(\bar{v}^0, \mu)) > 0.$$

If we make an additional change of coordinates on Π^{in}:

$$\begin{aligned} u_{2,new} &= u_2 - u_2^{\text{in}} - \psi(v_2, v^0, \mu), \\ u^0_{new} &= u^0 - u^{0\text{in}} - \psi^0(v_2, v^0, \mu), \\ v_{2,new} &= v_2 - \phi(v^0, \mu), \end{aligned} \tag{30}$$

so that equations of w^{u*} and w^{uu} become, respectively,

$$w^{u*} : \quad (u_2, u^0) = 0 \tag{31}$$

and

$$w^{uu} : \quad (u_2, u^0) = 0, \quad v_2 = a\mu, \tag{32}$$

then, after resolving (27) with respect to τ, the Poincaré map T can be written in the following form

$$(\bar{u}_2, \bar{u}^0, v_2, v^0) = \xi(u_2, u^0, \bar{v}_2, \bar{v}^0) \tag{33}$$

where ξ is a smooth function defined at

$$A(\bar{v}_2 - a\mu) > 0 \tag{34}$$

and vanishing at $\bar{v}_2 = a\mu$ along with the first derivatives, so that

$$\xi = o(\bar{v}_2 - a\mu). \tag{35}$$

If we assume $\xi = 0$ at $A(\bar{v}_2 - a\mu) < 0$, then the right-hand side of (33) will define a contracting map. Its unique fixed point $M^*(u_2^*, v_2^*, u^{0*}, v^{0*})$ will be a fixed point of the Poincaré map T if and only if v_2^* satisfies (34). By (35),

$$v_2^* = o(v_2^* - a\mu), \tag{36}$$

so it is obvious now that the map T has a fixed point if and only if $Aa\mu < 0$.

The fixed point of the Poincaré map corresponds to the periodic orbit L_μ. By (36), $v_2^* \to 0$ as $\mu \to 0$. By (33), it follows that $(u_2^*, v_2^*, u^{0*}, v^{0*}) \to 0$ as $\mu \to 0$, i.e. the periodic orbit merges into the homoclinic loop Γ_0 at $\mu = 0$.

Take some $K > 0$ and let us call as a vertical surface a surface of the kind $(u_2, u^0) = \eta(v_2, v^0)$ with $\|\eta'\| \leq K$ and let a horizontal surface be a surface of the kind $(v_2, v^0) = \nu(u_2, u^0)$ with $\|\nu'\| \leq K$. It is immediately seen from (33)-(35) that for every $K > 0$, if the range of μ and v_2 is sufficiently small, the preimage of any horizontal surface which intersects the region $A(v_2 - a\mu) > 0$ is a horizontal surface again, and the image of any vertical surface is a piece of a vertical surface (this piece is bounded by w^{uu} and lies in the region $A(v_2 - a\mu) > 0$). Moreover, when restricted to a vertical surface the map T is expanding and it is contracting on horizontal surfaces.

Thus, the map T has a hyperbolic structure and, in particular, its fixed point is a saddle (so L_μ is a saddle periodic orbit) whose stable manifold is a horizontal surface and the unstable manifold is a piece of a vertical surface. Due to the hyperbolicity, all the orbits of the map T must leave Π^{in} after a number of iterations (forward or backward), except for the fixed point. For the flow itself, this means that the only orbits which may stay in a small neighborhood U of the loop are the periodic orbit L_μ and, possibly, some orbits in $W^s(O)$ or $W^u(O)$ (such orbits correspond to finite, at least from one side, orbits of the Poincaré map T).

The orbits from $W^u(O)$ or $W^s(O)$ correspond to the orbits of the map T starting on $w^u = T_{glo}(W^u_{\text{loc}} \cap \Pi^{\text{out}})$ or, respectively, ending on $w^s = W^s_{\text{loc}} \cap \Pi^{\text{in}} = \{v_2 = 0, v^0 = 0\}$. If such an orbit is infinite to the right, it must start with a point on w^u and tend to the fixed point M^*. Thus, it must belong to the stable manifold of M^*, i.e. the starting point on w^u is defined uniquely as the intersection of $w^s(M^*) \cap w^u$ (this intersection is unique because $w^s(M^*)$ is a horizontal surface and w^u is vertical, by our assumption of the transversality of w^u and w^s). This gives us a unique heteroclinic orbit $C_{1\mu}$ which is α-limit to O and ω-limit to L_μ.

The rest are homoclinic loops and the heteroclinic orbit $C_{2\mu}$ which is α-limit to L_μ and ω-limit to O. We start with homoclinic loops. They correspond to the intersection of w^u with w^s (the original loop Γ) and with its preimages $w_k^s = T^{-k}w^s$. When exists, each of these preimages is a horizontal surface which, hence, has a unique intersection point with w^u and this intersection corresponds to the homoclinic loop $\Gamma_{k\mu}$. Thus, the problem of existence of homoclinic loops is reduced to the following question: until which k the surfaces w_k^s intersect the region $A(v_2 - a\mu) > 0$? At $Aa\mu \geq 0$, the surface w^s itself does not lie in this region so it has no preimages. Therefore, no homoclinic loops $\Gamma_{k\mu}$ exist with $k \geq 1$ (heteroclinic orbits cannot exists either because there is no periodic orbit at these μ). When $Aa\mu < 0$, the surface w^s lie in $A(v_2 - a\mu) > 0$. Hence, it has a preimage w_1^s. By (33),(35), we have $v_2 = o(\mu)$ on w_1^s, therefore $A(v_2 - a\mu) > 0$ on w_1^s, so it has a preimage as well, and so on: we obtain the infinite sequence of preimages w_k^s for all of which $v_2 = o(\mu)$ uniformly. Thus we have proved the existence of homoclinic loops $\Gamma_{k\mu}$ at $Aa\mu < 0$. Since the horizontal surfaces w_k^s stay all in a bounded region they must accumulate to the stable manifold of the saddle fixed point M^*. Therefore, they must intersect the unstable manifold of M^* which gives us the existence of the heteroclinic orbit $C_{2\mu}$ which is α-limit to L_μ and ω-limit to O (this orbit is unique because w^s can have no more than one intersection with $W^u(M^*)$ since the latter is a piece of a vertical surface). This finishes the proof of theorem 1.

To prove theorem 2, note that in a small neighborhood of O there is no orbit which starts in a small neighborhood of a point in $W^s \backslash W^{ss}$ with $\{H = 0\}$ and comes in a small neighborhood of any point in $W^u \backslash W^{uu}$. Indeed, for such an orbit we would have $v_{1\tau} \neq 0$ and $u_{10} \neq 0$ in formula (24), and this makes it clearly impossible to have $H(M_0) = 0$ or $H(M_\tau) = 0$ at sufficiently large τ (recall that the large flight time τ corresponds to the orbits starting close to the stable invariant manifold of O).

Therefore, any orbit which stays in a small neighborhood U of the homoclinic bunch $\Gamma \cup \Gamma_+^1 \cup \ldots \cup \Gamma_+^{m+} \cup \Gamma_-^1 \cup \ldots \cup \Gamma_-^{m-} \cup O$ in the level $\{H = 0\}$ and which starts close to a loop $\Gamma_\pm^i$ must enter a small neighborhood of Γ (and stay there after that) immediately after one passage near O. Thus, except for the orbits which stay all the time in a small neighborhood of Γ, the system may have in U only such orbits which start in $W_{\text{loc}}^u(O)$, make one round near one of the loops $\Gamma_\pm^i$ and then enter a small neighborhood of Γ. To stay there, these orbits must either come into $W_{\text{loc}}^s \cap \Pi^{\text{in}}$ after a number of rounds near Γ, or they must belong to the stable manifold of the periodic orbit L_μ which exists at $Aa\mu < 0$. So, to prove the theorem we must, for every loop Γ_γ^i ($\gamma = \pm$), take a small piece of $W_{\text{loc}}^u(O)$ near this loop, continue it by the orbits of the flow close to the loop back to a small neighborhood of O, then trace how it goes to the loop Γ, make one round near Γ and examine how the obtained surface intersects (on the cross-section Π^{in}) the surface $w^s = W_{\text{loc}}^s \cap \Pi^{\text{in}}$ (this intersection will correspond to a double loop $(i\gamma)0$) and, at $Aa\mu < 0$, the surfaces $w_k^s = T^{-k} w^s$ (these intersections will correspond to the loops $(i\gamma)0^k$) and the stable manifold $w^s(M^*)$ of the saddle fixed point of T (this intersection will correspond to the heteroclinic orbit $(i\gamma)0^\infty$).

Let $\Pi_{i\pm}^{\text{in}}$ be small cross-sections to the local stable manifold, intersecting the loops $\Gamma_\pm^i$, respectively. We may assume that $u_1 = d > 0$ on Π_{i+}^{in} and $u_1 = -d < 0$ on Π_{i-}^{in}. A piece of W_{loc}^u mapped by the flow near a loop Γ_γ^i on the cross-section $\Pi_{i\gamma}^{\text{in}}$ is a surface transverse to W_{loc}^s. The image of this surface by the local map on the cross-section Π^{out} to the loop Γ is found by formulas (24) where one should put $u_{10} = d > 0$ at $\gamma = +$ and put $u_{10} = -d < 0$ at $\gamma = -$ (recall that $v_{2\tau} = d > 0$ on Π^{out}). Thus, this image is a surface tangent to $W^u_{\text{loc}} \cap \Pi^{\text{out}}$ in the case $\lambda_2 < 2\lambda_1$ or to $W^{u1}_{\text{loc}} \cap \Pi^{\text{out}}$ in the case $\lambda_2 > 2\lambda_1$. In both cases this surface is bounded by $W^{uu}_{\text{loc}} \cap \Pi^{\text{out}}$.

When applying the global map (16) to this surface we will obtain a vertical surface (in the coordinates given by (30)) adjoining to w^{uu} from the side $\gamma A(v_2 - a\mu) > 0$. It is seen now immediately that this surface has an intersection (and this intersection is transverse and unique) with w^s and with any horizontal curve $o(\mu)$-close to w^s (at $Aa\mu < 0$ such are the preimages w^s_k of w^s and their limit $w^s(M^*)$; see the proof of theorem 1) if and only if $\gamma Aa\mu < 0$. This is in a complete correspondence with the statement of theorem 2. End of the proof.

Theorem 2 treats the case of a finite number of loops $\Gamma^i_\pm$, but it can be easily generalized to the case of an infinite set of loops. Namely, let a number of saddle periodic orbits $L_1, \ldots, L_m$ exists in the level $\{H = 0\}$ at $\mu = 0$ (hence, at all small μ). Suppose the unstable manifold of L_i intersects the stable manifold of L_j transversely at some number $m_{ij} \geq 0$ of heteroclinic (homoclinic at $i = j$) orbits C_{ijs} ($s = 1, \ldots, m_{ij}$ at $m_{ij} \geq 1$). Then (see [7,20]), one can take a sufficiently small neighborhood V of $L_1 \cup \ldots \cup L_m \cup_{ijs} C_{ijs}$ in the level $\{H = 0\}$ such that the set N of all orbits staying in V entirely will be a hyperbolic set topologically conjugate to a subshift of finite type, described by the following transition graph G (oriented): it has m vertices denoted as $L_1, \ldots, L_m$ and, for every $i = 1, \ldots, m$, from the vertex L_i one edge, denoted also as L_i, goes to the same vertex, plus m_{ij} edges denoted as $L_i^{\bar{k}} C_{ijs} L_j^{\bar{k}}$ ($s = 1, \ldots, m_{ij}$) go to the vertex L_j, for every $j = 1, \ldots, m$; here $\bar{k}$ is a sufficiently large integer. In other words, for every infinite oriented path in graph G, in V there exists an orbit whose natural coding is read from the consecutive edges in this path, and this correspondence between the paths in the graph and the orbits of N is one-to-one and continuous. Every orbit of N has local stable and unstable manifolds the size of which is bounded away from zero. If the codings of two forward semiorbits are close, then their stable manifolds are close as well; also, if the codings of two backward semiorbits are close, then their unstable manifolds are close.

Let $W^u(O)$ intersect transversely the stable manifolds of periodic orbits L_i at $m_{0i} \geq 0$ heteroclinic orbits C_{0is}, $s = 1, \ldots, m_{0i}$ at $m_{0i} \geq 1$, $i = 1, \ldots, m$, and let $W^s(O)$ intersect transversely the unstable manifolds of periodic orbits L_i at $m_{i0} \geq 0$ heteroclinic orbits C_{i0s}, $s = 1, \ldots, m_{i0}$ at $m_{i0} \geq 1$. Then, by λ-lemma, m_{0i} pieces of $W^u(O)$ will come sufficiently close to the local unstable manifold of L_i; hence, each of them will have one point of transverse intersection with the stable manifold of every orbit of N close to L_i. Analogously, m_{i0} pieces of $W^s(O)$ will come sufficiently close to the local stable manifold of L_i, so each of these pieces will have one point of transverse intersection with the unstable manifold of every orbit of N close to L_i. Thus, if we enlarge the neighborhood V by adding to it a small neighborhood of O and the heteroclinic orbits C_{0is} ($s \leq m_{0i}$) and C_{i0s} ($s \leq m_{i0}$, $i = 1, \ldots, m$) in the level $\{H = 0\}$, then in the new V there will exist a set $\tilde{N} \supseteq N$ of orbits for which the natural coding will give a one-to-one continuous correspondence with the set of the oriented paths (infinite, or starting at O^u, or ending in O^s) in the graph $\tilde{G}$ obtained from G by adding a pair of vertices O^s and O^u with the edges $C_{0is}L_i^{\bar{k}}$ $(s \leq m_{0i})$ aiming from O^u to L_i and $L_i^{\bar{k}}C_{0is}$ $(s \leq m_{i0})$ aiming from L_i to O^s, $i = 1, \ldots, m$. By construction, the paths starting with O^u and ending at O^s correspond to homoclinic loops, and if the graph G is nontrivial, the set of these loops will be infinite, of course.

When all the heteroclinic orbits C_{0is} and C_{i0s} are in general position, i.e. they do not lie in strong unstable or, respectively, strong stable manifolds W^{uu} and W^{ss} of O, there are no other orbits lying entirely in V except for O and those from the set $\tilde{N}$ described above. This follows from the fact we established while proving theorem 2 that in a neighborhood of O there can be no orbit which would lie in $\{H = 0\}$ and pass from a small neighborhood of a point in $W^s \backslash W^{ss}$ with $\{H = 0\}$ to a small neighborhood of any point in $W^u \backslash W^{uu}$ - hence, every positive or negative semiorbit in V which comes close to O must enter $W^s_{\text{loc}}(O)$ or, respectively, $W^u_{\text{loc}}(O)$, so it belongs to the set $\tilde{N}$ indeed.

So, we assume that C_{0is} and C_{i0s} are in general position. Moreover, we divide the orbits C_{i0s} into two groups: those lying in W^{s+} and those lying in W^{s-}. Accordingly, we change notations denoting these heteroclinics as C_{i0s+} $(s \leq m_{i0+})$ and C_{i0s-} $(s \leq m_{i0-})$ where m_{i0+} and m_{i0-} are the number of the orbits in W^{s+} and the number of the orbits in W^{s-} respectively, so that $m_{i0+} + m_{i0-} = m_{i0}$. We also change the graph $\tilde{G}$ by splitting the vertex O^s into two: O^{s+} and O^{s-}, so that the edges corresponding to the orbits C_{i0s+} end at O^{s+} and those corresponding to C_{i0s-} end at O^{s-}.

Let $\Pi^{\text{in}}_{is\pm}$ be small cross-sections to the local stable manifold, intersecting the orbits $C_{i0s\pm}$, respectively. We may assume that $u_1 = d > 0$ on Π^{in}_{is+} and $u_1 = -d < 0$ on Π^{in}_{is-}. A piece of $W^u_{\text{loc}}(L_i)$ mapped by the flow near an orbit $C_{i0s\gamma}$ $(\gamma = \pm)$ on the cross-section $\Pi^{\text{in}}_{is\gamma}$ is a surface transverse to $W^s_{\text{loc}}(O)$. Since the local unstable manifolds of the backward orbits in $\tilde{N}$ depend continuously on their coding, local unstable manifolds of all backward orbits in $\tilde{N}$ whose coding start with a sufficiently long sequence of L_i's lie close to $W^s_{\text{loc}}(L_i)$ (at least in C^1-sense). Therefore, if we took the value of $\bar{k}$ sufficiently large when constructing the set $\tilde{N}$, we will have for every path g in the graph $\tilde{G}$ which ends with the edge $C_{i0s\gamma}$ that the unstable manifold of the corresponding backward semiorbit intersects $\Pi^{\text{in}}_{is\gamma}$ at a surface w^s_g transverse to $W^s_{\text{loc}}(O)$ and the sizes of these surfaces are bounded away from zero, as well as the angles they form with $W^s_{\text{loc}}(O)$.

Let us now assume that at $\mu = 0$ there exists a homoclinic orbit Γ undergoing the orbit-flip bifurcation and the genericity assumptions of theorem 1 hold. We can now apply the arguments of theorem 2 to the surfaces w^s_g, uniformly to all of them. This will give that the images of these surfaces by the local map on the cross-section Π^{out} to the loop Γ are some surfaces, whose size is bounded away from zero, confined all in a small angle around $W^u_{\text{loc}} \cap \Pi^{\text{out}}$ in the case $\lambda_2 < 2\lambda_1$ or around $W^{u1}_{\text{loc}} \cap \Pi^{\text{out}}$ in the case $\lambda_2 > 2\lambda_1$. In both cases the surfaces are bounded by $W^{uu}_{\text{loc}} \cap \Pi^{\text{out}}$. All the surfaces coming from Π^{in}_{is+} adjoin to $W^{uu}_{\text{loc}} \cap \Pi^{\text{out}}$ from one side and the surfaces coming from Π^{in}_{is-} adjoin to $W^{uu}_{\text{loc}} \cap \Pi^{\text{out}}$ from the other side, exactly by the same rule as in theorem 2. Thus, exactly like in theorem 2, we arrive at the following statement.

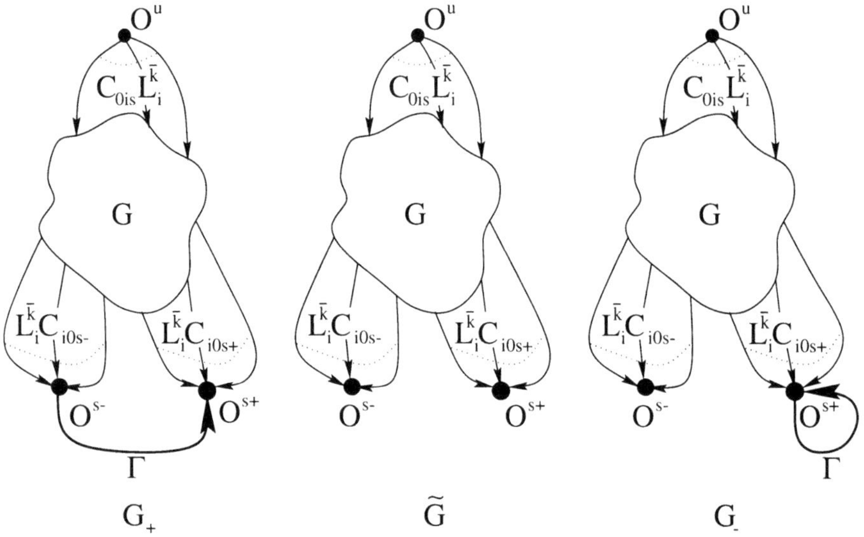

Fig. 5. The graphs G_+ and G_- are obtained from $\tilde{G}$ by adding one edge labelled Γ which ends at O^{s+} and starts at O^{s-} or O^{s+}, respectively.

Theorem 3. *Let U be the union of the neighborhood V of the set $\tilde{N}$ with a small neighborhood of Γ in $\{H = 0\}$. Then the set of all orbits lying in U entirely is (excluding O and Γ) in one-to-one continuous correspondence with the oriented paths in the graph $\tilde{G}$ at $\mu = 0$, G_+ at $Aa\mu > 0$ and G_- at $Aa\mu < 0$ where G_+ and G_- are obtained from $\tilde{G}$ by adding one more edge Γ which starts with O^{s-} or O^{s+}, respectively, and ends at O^{s+} in both cases (Fig. 5). The homoclinic loops correspond to the paths starting with O^u and ending at one of the vertices $O^{s\pm}$.*

3 Superhomoclinic Orbits

Let us now consider in more detail the behavior of orbits in a small neighborhood of the homoclinic loop Γ at the moment of the orbit-flip bifurcation (i.e. at $\mu = 0$). The problem reduces to the study of the Poincaré map T on the cross-section Π^{in}. By (33)-(35), the map T is written in the following form

$$(\bar{u}_2, \bar{u}^0, v_2, v^0) = \xi(u_2, u^0, \bar{v}_2, \bar{v}^0) = o(\bar{v}_2) \tag{37}$$

where ξ is a smooth function defined at

$$A\bar{v}_2 > 0 \tag{38}$$

and vanishing at $\bar{v}_2 = 0$ along with the first derivatives. If we define the function ξ at $A\bar{v}_2 \leq 0$ as $\xi = 0$, then the right-hand side of (37) will be a smooth function defined for all small $u_2, u^0, \bar{v}_2, \bar{v}^0$, whose first derivatives will be all small. Hence, theorem 4.4 of [15] is applied which gives the existence of a smooth attracting invariant manifold $\tilde{w}$ for the map T. Namely, this manifold $\tilde{w}$ has the form

$$(u_2, u^0) = \tilde{\eta}(v_2, v^0) \tag{39}$$

for some smooth function $\tilde{\eta}$ (the invariance of this manifold implies that $\tilde{\eta}$ vanishes at $\bar{v}_2 = 0$ along with the first derivatives), and every forward semiorbit of T which never leaves Π^{in} must tend uniformly to $\tilde{w}$. Hence, every infinite backward semiorbit of T must lie in $\tilde{w}$.

Note that it is obvious from (37) that on $\tilde{w}$ the map T^{-1} is defined and strongly contracting everywhere in the region (34). We will show that the orbits of the flow which start on $\tilde{w}$ with $v_2 \leq 0$ do not come to the cross-section Π^{out} after passing near the saddle O, so they do not return to Π^{in}. This means that the domain of the Poincaré map T on $\tilde{w}$ lies in the region $v_2 > 0$, i.e. the contracting map T^{-1} maps the region $Av_2 > 0$ inside the region $v_2 > 0$. Hence, at $A < 0$ the backward semiorbit of every point in $\tilde{w}$ leaves Π^{in} with the iterations of T^{-1}, whereas at $A > 0$ for every point in $\tilde{w}$ with positive v_2 its backward semiorbit stays in Π^{in}. Since T^{-1} is contracting, all infinite semiorbits must tend to the fixed point in the origin in Π^{in}. Thus, we have that the manifold

$$\tilde{w}^u : (u_2, u^0) = \tilde{\eta}(v_2, v^0), v_2 > 0 \tag{40}$$

is the *unstable manifold* of the origin in Π^{in} at $A > 0$. Since this point is the intersection point of Γ with Π^{in}, it follows that the orbits of the flow which pass through the points of w^u have the homoclinic loop Γ as the α-limit set. This gives us the following result:

Lemma 1. *Let $A > 0$ for the homoclinic loop Γ at the moment of the orbit-flip bifurcation. Then, the unstable set of Γ (i.e. the set of all orbits which tend to Γ as $t \to -\infty$) is non-empty and it is a smooth n-dimensional manifold $W^u(\Gamma)$ with the boundary $W^{uu}(O)$ which is tangent at the points of Γ to W^u if $\lambda_2 < 2\lambda_1$ and to W^{u1} if $\lambda_2 > 2\lambda_1$. All the orbits in $\{H = 0\}$ which do not belong to $W^u(\Gamma)$, W^s or W^u leave a small neighborhood of Γ both as $t \to +\infty$ and $t \to -\infty$.*

To prove this statement it remains to show that the orbits of the flow which start on $\tilde{w}$ with $v_2 \leq 0$ do not come to Π^{out}. Recall that we assume the transversality of the manifolds $W^{see}(O)$ and $W^{uu}(O)$ at the points of Γ, which is equivalent to the existence of an $(n+2)$-dimensional repelling smooth invariant manifold $W^{see}(\Gamma)$ which contains Γ and $W^s_{\text{loc}}(O)$ and which is transverse to W^{uu} at O [17,16] (it is tangent to $v^0 = 0$ at O, in fact).

The intersection of $W^{see}(\Gamma)$ with Π^{in} is a surface

$$w^{see} : v^0 = \varphi(u_2, u^0, v_2)$$

with some smooth function φ vanishing at $v_2 = 0$. By construction, w^{see} is invariant with respect to T. Since the derivatives of the function η in (39) are small at small v_2, it follows that w^{see} intersects the invariant manifold $\tilde{w}$ along a smooth invariant curve

$$w^* : (v^0, u_2, u^0) = \zeta(v_2)$$

where $\zeta(0) = 0$. The orbits which start on w^* lie in the invariant manifold $W^{see}(\Gamma)$; since the latter is transverse to W^{uu}, it follows that $v^0 = O(v_1, v_2)$ for every orbit starting with w^*, all the time this orbit lies in a neighborhood of O (moreover, W^{see} is tangent to $v^0 = 0$ at O, so we also have that $\|v^0\| \ll d$). Therefore, the evolution of the v_2-coordinate on this orbit is given by the equation of the form

$$\dot{v}_2 = \lambda_2 v_2 + o(v_1, v_2)$$

(see (7). By (24), the ratio u_2/u_1 remains uniformly bounded for this orbit (since $\lambda_2 > \lambda_1$ and $u_1 = d \neq 0$ initially). Hence, since the orbit lies in $\{H = 0\}$, it follows that $v_1 = O(v_2)$ and we have

$$\dot{v}_2 = \lambda_2 v_2 + o(v_2).$$

It is now obvious that the orbits which start on w^* with nonpositive v_2 can never enter the region of positive v_2 so they leave the d-neighborhood of O through the cross-section $v_2 = -d$ (the orbit Γ which pass through the point $v_2 = 0$ on w^* tends to O).

Now, take any point M on $\tilde{w}$ with $v_2 \leq 0$ and let M^* be the point of intersection of the surface $\{v_2 = const\}$ through the point M with w^*. This surface is transverse to w^{see}. Since the cone $\|u, v_1, v_2\| \leq K\|v^0\|$ is, at every K, invariant with respect to the forward flow linearized at the point O, it follows that the tangents to every surface obtained by the forward shift by the local flow near O of a surface transverse to w^{see} belong all to such cone with a sufficiently large K, provided the size d of the neighborhood of O under consideration is small enough. Thus, the forward time t shift M_t of the point M will remain in such a cone with the vertex at the time t shift M_t^* of the point M^* which makes it impossible for M_t to belong to Π^{out} (in that case both M_t and M_t^* would have $\|v^0\| \ll d$ but the difference in v_2 would be of order d which would contradict the invariant cone property). This proves the claim.

The invariant manifold $W^u(\Gamma)$ is n-dimensional and lies in the level $\{H = 0\}$. Hence, it may have orbits of the transverse (in this level) intersection with $W^s(O)$. We call such orbits *superhomoclinic.* Let S be a superhomoclinic orbit of transverse intersection of $W^u(\Gamma)$ with $W^s(O)$. Assume that S enters O at $t = +\infty$ along the leading direction, i.e. it is tangent to the u_1-axis. Moreover, we assume that S adjoins O from the side of positive u_1, i.e. $S \subseteq W^u(\Gamma) \cap W^{s+}$ (as we will see the case $S \subseteq W^u(\Gamma) \cap W^{s-}$ is trivial). Let U be a small neighborhood of $\Gamma \cup S \cup O$. It is a ball (around O) with two handles around Γ and S. We can therefore consider a natural code for the orbits in U describing the sequence of the handles visited by the orbits. Note that the codings of the orbits in $W^u(O)$ are finite to the left and the codings of the orbits in $W^s(O)$ are finite to the right, so the codings of homoclinic loops are finite to both sides; the coding of O is empty.

Let Ω be the set of sequences of symbols S and Γ constructed by the following rule: for some positive integer $\bar{k}$ take all infinite or starting with Γ and infinite to the right sequences obtained by repeated concatenation of subsequences Γ and $S\Gamma^{\bar{k}}$ in an arbitrary order; then change the infinite sequence composed of Γ's only to the one-symbol sequence $\{\Gamma\}$ and, for every other sequence which ends by the infinite string of Γ's, omit this string; the set thus obtained plus the empty sequence is the set Ω.

Theorem 4. *There exists a sufficiently large $\bar{k}$ and a small neighborhood U of $\Gamma \cup S \cup O$ such that the set of all orbits lying entirely in U is in one-to-one correspondence (provided by natural coding) with Ω.*

Proof. The intersection of $W^u(\Gamma)$ with the cross-section Π^{in} is the invariant manifold $\tilde{w}^u$ of the Poincaré map T. The manifold $\tilde{w}^u$ is given by (40) but we will change coordinates on Π^{in} such that it would have the equation

$$u = 0, \ v_2 > 0; \tag{41}$$

since the function $\tilde{\eta}$ in (40) vanishes at zero along with its derivatives, this coordinate transformation would not change the formula (37), nor it would change the formula (16) for the map $T_{glo} : \Pi^{\text{out}} \to \Pi^{\text{in}}$.

Let $P(0, v_P) \in \tilde{w}^u$ be a point of intersection of the superhomoclinic orbit S with Π^{in}. By assumption, this orbit belongs to the stable manifold of O, hence it must eventually come to $W^s_{\text{loc}}(O)$. Moreover, this orbit lies in W^{s+}. Hence, it must intersect the cross-section $\{u_1 = d\}$ at some point $Q(u_Q, 0) \in W^s_{\text{loc}}(O)$. Let $\tilde{\Pi}^{\text{in}}$ be a piece of the cross-section $\{u_1 = d\}$ around Q in the level $\{H = 0\}$. The flow near S defines a map T_S from a small neighborhood of P on Π^{in} into $\tilde{\Pi}^{\text{in}}$, so that $Q = T_S P$.

The map T_S corresponds to a finite flight time, so its derivatives are bounded and it is well approximated by its linearization at the point P, like the map T_{glo} near Γ. We can write T_S in the following form

$$\begin{cases} \tilde{v} = \tilde{a}(v - v_P) + \tilde{b}u + \dots \\ \tilde{u} - u_Q = \tilde{c}(v - v_P) + \tilde{d}u + \dots \end{cases} \tag{42}$$

where the dots stand for non-linear (quadratic and higher order) terms; $(\tilde{u}, \tilde{v})$ denote coordinates on $\tilde{\Pi}^{\text{in}}$. Note that by assumption of the transversality of $W^u(\Gamma)$ to $W^s(O)$ the image of a small piece of surface $u = 0$ around the point P by the map T_S is a surface transverse to $v = 0$ in $\tilde{\Pi}^{\text{in}}$. It means that $\tilde{a} \neq 0$ in (42).

The map from $\tilde{\Pi}^{\text{in}}$ to Π^{out} is given by formulas (24) where one should put $u_{10} = d > 0$ and $v_{2\tau} = d > 0$. Note that the flight time τ must be taken sufficiently large because $\tilde{\Pi}^{\text{in}}$ is a small neighborhood of the point Q which lies in $W^s_{\text{loc}}(O)$ and whose forward orbit stays, therefore, infinitely long time in the d-neighborhood of O. Now, combining formulas (42),(24) and (16), one can see that the map $T_{glo}T_{loc}T_S$ by the flow from a small neighborhood of P in Π^{in} close to the superhomoclinic orbit S and then close to Γ back to Π^{in} is given by the formula

$$(\bar{u}_2, \bar{u}^0, v_2 - \nu_2^s, v^0 - \nu^{0s}) = \tilde{\xi}(u_2, u^0, \bar{v}_2, \bar{v}^0) = o(\bar{v}_2) \tag{43}$$

where $\tilde{\xi}$ is a smooth function defined at sufficiently small u and sufficiently small positive[2] $\bar{v}_2$ and vanishing at $\bar{v}_2 = 0$ along with the first derivatives, and $u = \nu^s(v)$ is the preimage of $W^s_{\text{loc}} \cap \tilde{\Pi}^{\text{in}}$ on Π^{in}; by construction, $0 = \nu^s(v_P)$.

Note that we cannot control the range of $\bar{v}_2$ for which the function $\tilde{\xi}$ is defined (we only know that it is defined at sufficiently small positive $\bar{v}_2$ which corresponds to sufficiently large time τ of the flight from $\tilde{\Pi}^{\text{in}}$ to Π^{out}). In particular, the value of v_{2P} can be out of the domain of $\tilde{\xi}$. However, it is easy to see from (43) and (37) that for a sufficiently large $\bar{k}$ the map $\tilde{T} = T^{\bar{k}-1}T_{glo}T_{loc}T_S$ from a small neighborhood of P is still written in the form (43) where the function $\tilde{\xi}$ is defined for $\bar{v}_2 \in (0, \delta]$ with some $\delta > v_{2P}$ and the range of the map $(u, \bar{v}) \mapsto (\bar{u}, v)$ defined by formula (43) now lies inside its domain (the domain of $\tilde{\xi}$).

If we define the functions $\tilde{\xi}$ and ξ in formulas (43) and (37), respectively, as zero at $\bar{v}_2 = 0$, we obtain a rectangular domain in Π^{in} where a pair of maps T and $\tilde{T}$ are defined, for both of which the corresponding cross-maps $(u, \bar{v}) \mapsto (\bar{u}, v)$ take this domain into itself and they are both strongly contracting. Thus, the lemma [7] on a saddle fixed point of a sequence of saddle operators in the product of Banach spaces is applied here which gives that for every sequence $\{\sigma_i\}_{i=-\infty}^{+\infty}$ of symbols 0 and 1 there exists a unique sequence of points M_i such that $M_{i+1} = TM_i$ if $\sigma_i = 0$ and $M_{i+1} = \tilde{T}M_i$ if $\sigma_i = 1$. Moreover, the points M_i depend continuously on the corresponding sequences $\{\sigma_i\}_{i=-\infty}^{+\infty}$ and each of these points has a stable manifold which is a horizontal surface (i.e. a surface of the kind $v = \nu(u)$ where the derivative of ξ is sufficiently small). Every such surface has a unique point of the transverse intersection with the vertical surface $w^u = T_{glo}(W^u_{\text{loc}} \cap \Pi^{\text{out}})$. Thus, for every

[2] Note that if $S \in W^{s-}$, we would have $u_{10} = -d < 0$ in (24) which would give $\bar{v}_2 < 0$ in (43). Thus, the orbits starting close to P would return to that part of Π^{in} where further iterations of T or $\tilde{T}$ are not defined. Hence, in that case, no orbits other than S, Γ and O can lie in U entirely.

infinite to the right sequence $\{\sigma_i\}_{i=0}^{+\infty}$ there exists a unique sequence of points M_i such that $M_0 \in w^u$ and $M_{i+1} = TM_i$ if $\sigma_i = 0$ and $M_{i+1} = \tilde{T}M_i$ if $\sigma_i = 1$.

The obtained sequences $\{M_i\}$ correspond to the trajectories of the original maps T and $\tilde{T}$ if and only if the coordinate v_2 is not zero for every point M_i in the sequence. If $\bar{v}_2 = 0$ at some point M_{i+1}, it means that the corresponding values of $\tilde{\xi}$ or ξ are zero in, respectively, (43) or (37). Hence, M_{i+1} is the origin in Π^{in}, i.e. $M_{i+1} = \Gamma \cap \Pi^{\text{in}}$, and either $M_{i+1} = \tilde{T}M_i$ - in this case $M_i \in (T_{glo}T_{loc}T_S)^{-1}(W^s_{\text{loc}} \cap \tilde{\Pi}^{\text{in}})$, or $M_{i+1} = TM_i$ - in this case $v_2 = 0$ at the point M_i which means that $M_i = \Gamma \cap \Pi^{\text{in}}$ as well. Thus, we have that either $\{\sigma_i\}$ consists of all 0's, so all the points of the corresponding sequence $\{M_i\}$ are the same fixed point $\Gamma \cap \Pi^{\text{in}}$ of T, or all points M_i have $v_2 \neq 0$, or there is a point $M_i \in (T_{glo}T_{loc}T_S)^{-1}(W^s_{\text{loc}} \cap \tilde{\Pi}^{\text{in}})$ for which all the previous points have nonzero v_2 and $\tilde{T}M_i = M_{i+1} = \Gamma \cap \Pi^{\text{in}}$ which means that $\sigma_i = 1$ and all the further symbols are 0's. Vice versa, if the sequence $\{\sigma_i\}$ ends by an infinite sequence of 0's, some point M_i must belong to the stable manifold of the fixed point $\Gamma \cap \Pi^{\text{in}}$ (which is defined as a unique horizontal surface passing through this point and invariant with respect to T^{-1}), i.e. $M_i \in \{v = 0\}$.

Hence, the sequences $\{M_i\}$ correspond to the trajectories of the original maps T and $\tilde{T}$ if and only if the corresponding sequence $\{\sigma_i\}$ does not end with an infinite sequence of 0's. If the sequence $\{\sigma_i\}$ ends with an infinite sequence of 0's, we will cut the sequence $\{M_i\}$ at the last point to which $\sigma_i = 1$ corresponds. The new sequence $\{M_i\}$ will be a trajectory of the original maps T and $\tilde{T}$ which ends on the surface $(T_{glo}T_{loc}T_S)^{-1}(W^s_{\text{loc}} \cap \tilde{\Pi}^{\text{in}})$. All this is now in a complete correspondence with the statement of the theorem: recall that one iteration of the map T corresponds to one round of an orbit of the flow near the loop Γ and one iteration of the map T_S corresponds to one round near the superhomoclinic orbit S. End of the proof.

References

1. R.L.Devaney, Homoclinic orbits in Hamiltonian systems, J. Diff. Equat. 21, 431-438 (1976).
2. L.A.Beljakov and L.P.Shilnikov, Homoclinic curves and complex solitary waves, Methods of qualitative theory of differential equations, Gorky, 10-23 (1987).
3. L.M.Lerman, Complex dynamics and bifurcations in a Hamiltonian system having a transversal homoclinic orbit to a saddle-focus, Chaos 1(2), 174-189 (1991).
4. D.V.Turaev and L.P.Shilnikov, On Hamiltonian systems with homoclinic curves of a saddle, Soviet Math. Doclady 304, 811-81 (1989).
5. B.Sandstede, Instability of localized buckling modes in a one-dimensional strut model, Philos. Trans. R. Soc. Lond., Ser. A 355, No.1732, 2083-2097 (1997).
6. A.R.Champneys and J.F.Toland, Bifurcation of a plethora of multi-modal homoclinic orbits for autonomous Hamiltonian systems, Nonlinearity 6, 665-721 (1993).
7. L.P.Shilnikov, A problem of Poincaré-Birkhoff, Math. USSR Sbornik 74, No.3, 378-397 (1967).

8. A.J.Homburg, Some global aspects of homoclinic bifurcations of vector fields, PhD thesis, University of Groningen (1993).
9. A.J.Homburg, Global aspects of homoclinic bifurcations of vector fields, Memoirs of the AMS 121, No.578 (1996).
10. D.V.Turaev, One case of bifurcations of a contour composed by two homoclinic curves of a saddle, Methods of qualitative theory of differential equations, Gorky, 45-58 (1984).
11. J.-M.Gambaudo, Ordre, desordre, et frontiére des systeme Morse-Smale, PhD thesis, University of Nice (1987).
12. D.V.Turaev, The bifurcations of systems with a pair of homoclinic curves of a saddle, PhD thesis, Nizhny Novgorod State University (1991).
13. V.M.Eleonsky, N.E.Kulagin, D.V.Turaev and L.P.Shilnikov, On the classification of self-localized states of the electromagnetic field within a nonlinear medium, DAN SSSR 309, 848-851 (1989).
14. L.P.Shilnikov and D.V.Turaev, Superhomoclinic orbits and multi-pulse homoclinic loops in Hamiltonian systems with discrete symmetries, Regular and Chaotic dynamics 2, No. 3/4, 126-138 (1997).
15. L.Shilnikov, A.Shilnikov, D.Turaev, L.Chua, Methods of qualitative theory in nonlinear dynamics. Part I, World Scientific, Singapore (1998).
16. D.V.Turaev, On dimension of non-local bifurcational problems, Bifurcation and Chaos 6, 919-948 (1996).
17. B.Sandstede, Center manifolds for homoclinic solutions, Preprint WIAS (1995).
18. I.M.Ovsyannikov and L.P.Shilnikov, Systems with a homoclinic curve of a multidimensional saddle-focus and spiral chaos, Math. USSR Sbornik 2, 415-443 (1992).
19. V.S.Afraimovich, On smooth transformations of coordinates, Methods of qualitative theory of differential equations, Gorky, 10-21 (1984).
20. V.S.Afraimovich and L.P.Shilnikov, On attainable transitions from Morse-Smale systems to systems with many periodic motions, Math. USSR Izvestija 8, No.6, 1235-1270 (1974).

Quantum Chaos and Quantum Ergodicity

A. Bäcker and F. Steiner*

Abteilung Theoretische Physik, Universität Ulm, Albert-Einstein-Allee 11, 89069 Ulm, Germany

Abstract. We report on some of our results which have been achieved within the *DFG Schwerpunktprogramm "Ergodentheorie, Analysis und effiziente Simulation dynamischer Systeme" (1994-2000)*. One main point of our research programme has been the search for universal statistical properties of energy spectra and eigenfunctions of quantum mechanical systems whose classical dynamics is chaotic. The mode-fluctuation distribution $P(W)$ has been proposed as a universal signature of quantum chaos and a conjecture on its limit distribution has been put forward. The conjecture turns out to be mathematically equivalent to a hypothesis on the value distribution of dynamical zeta functions on the critical line and has been successfully tested for several chaotic systems. For certain systems this can be expressed in terms of the Selberg zeta function.

For a large class of ergodic systems the quantum ergodicity theorem holds, which (roughly speaking) states that almost all eigenfunctions become equidistributed in the semiclassical limit. Particular attention has been paid to the question of subsequences of exceptional, non-quantum ergodic eigenfunctions, and their counting function. Such eigenfunctions are for example bouncing-ball modes occurring in billiards with two parallel walls (like the stadium or the Sinai billiard). Also "scarred" eigenfunctions showing localization along unstable periodic orbits could give rise to a non-quantum ergodic subsequence of eigenfunctions. Furthermore, the rate by which the classical limit is approached has been studied.

We conclude by giving a short summary of the other topics studied by our group within the Schwerpunktprogramm.

1 Introduction

Today it is well–known that the generic systems in classical physics are not the standard "textbook systems", which are completely integrable, but the non–integrable ones which range from near-integrable to ergodic and even fully chaotic systems. The latter display a very strong dependence on the initial conditions, i.e., even extremely small changes in the initial conditions have exponentially growing effects on the time evolution of the system. The chaotic behaviour of such sytems arises from the non-linearity of the differential equations describing them, i.e., from the non–linearity of Hamilton's equations in the case of classical mechanics. In the following it is assumed that the classical system is described by a time-independent Hamiltonian

* Project: Ergodic Theory, Dynamical Zeta Functions and Efficient Simulation of Classical and Quantum Dynamical Systems (Frank Steiner)

$H(\mathbf{p}, \mathbf{q})$ which for systems with f degrees of freedom is defined on the phase space $(\mathbf{p}, \mathbf{q}) \in \mathbb{R}^f \times \mathbb{R}^f$. Here $\mathbf{q} \in \mathbb{R}^f$ denotes the coordinates and $\mathbf{p} \in \mathbb{R}^f$ the canonical momenta. The observables in classical mechanics are (smooth) functions $A(\mathbf{p}, \mathbf{q})$ on phase space. In the following we shall assume that the Hamiltonian flow on the energy shell is (at least) ergodic; thus the energy $E = H(\mathbf{p}, \mathbf{q})$ is the only constant of motion. Although such systems are fully deterministic, their time evolution may be practically unpredictable in the long–time limit, $t \to \infty$. Prototype examples of strongly chaotic systems are certain billiards, which are given by the free motion of a point particle with mass m inside a compact Euclidean domain Ω with elastic reflections at the boundary $\partial\Omega$, or by the geodesic motion on a compact Riemann surface with constant negative curvature, or by billiards on compact domains with constant negative curvature.

Modern physics is based on quantum mechanics which governs not only the world of atoms, molecules and elementary particles, but is also more fundamental than classical mechanics in that the latter can be obtained from it in the *semiclassical limit*, $\hbar \to 0$, where $\hbar$ denotes Planck's constant. The semiclassical limit is different from the classical limit, for which $\hbar$ is precisely equal to zero, because, in general, quantal functions are non–analytic in $\hbar$ as $\hbar \to 0$. The basic quantity in quantum mechanics is a certain complex function Ψ called probability amplitude or wave function associated with every quantum mechanical (pure) state. In the simplest case of a single particle the wave function $\Psi(\mathbf{q}, t)$ gives the probability density $|\Psi(\mathbf{q}, t)|^2$ of finding the particle at the point $\mathbf{q}$ and at time t, i.e., $\int_D |\Psi(\mathbf{q}, t)|^2 \, \mathrm{d}^f q$ is the probability to find the particle at time t in a given domain $D \subset \Omega$. The state space of quantum mechanics is the Hilbert space $L^2(\Omega, \mathrm{d}^f q)$ where $\Omega \subset \mathbb{R}^f$ denotes the configuration space. The wave function Ψ is calculated from a differential equation, which for non–relativistic systems, i.e., for particles of low velocity, is the *Schrödinger equation*

$$\mathrm{i}\hbar \frac{\partial \Psi(\mathbf{q}, t)}{\partial t} = \widehat{H} \Psi(\mathbf{q}, t) \ . \tag{1}$$

Here $\widehat{H}$ is a self-adjoint differential operator called the Hamiltonian or Schrödinger operator, which is obtained from the classical Hamiltonian $H(\mathbf{p}, \mathbf{q})$ of the associated dynamical system. Obviously, the Schrödinger equation is a deterministic equation, since knowledge of Ψ at $t = t'$ implies its knowledge at all subsequent times $t'' > t'$. However, the interpretation of $|\Psi|^2$ as the probability density of an event is an indeterministic interpretation. It is a remarkable fact that the Schrödinger equation is a *linear* differential equation even for systems where the associated classical dynamics is non–linear. In the case of billiards on a compact domain Ω, the wave function $\Psi(\mathbf{q}, t'')$ at time t'' evolving from the initial state at time t' with wave function $\Psi(\mathbf{q}, t')$ is given by

$$\Psi(\mathbf{q}, t'') = \sum_{n=1}^{\infty} c_n \psi_n(\mathbf{q}) \mathrm{e}^{-\frac{\mathrm{i}}{\hbar} E_n (t'' - t')} \ , \tag{2}$$

where $\psi_n(\mathbf{q})$ and E_n are the eigenfunctions and eigenvalues of the *stationary Schrödinger equation*

$$-\frac{\hbar^2}{2m}\Delta\psi_n(\mathbf{q}) = E_n\psi_n(\mathbf{q}) \ , \quad \mathbf{q} \in \Omega \ , \tag{3}$$

$\psi_n \in L^2(\Omega, \mathrm{d}^f q)$, with the appropriate Dirichlet, Neumann or periodic boundary conditions on $\partial\Omega$. Here Δ denotes the Laplace-Beltrami operator, which on the Euclidean plane is the usual Laplacian. Due to the compactness of Ω, the quantal energy spectrum $\{E_n\}$ is purely discrete. The complex coefficients c_n in eq. (2) are given in terms of the initial wave function $\Psi(\mathbf{q}, t')$ at time t' by

$$c_n = \int\limits_{\Omega} \psi_n^*(\mathbf{q})\Psi(\mathbf{q}, t')\, \mathrm{d}^f q \tag{4}$$

and obey the normalization condition $\sum_{n=1}^{\infty} |c_n|^2 = 1$.

For Euclidean billiards the mathematical problem defined by eq. (3) is the well–known eigenvalue problem of the Helmholtz equation, which for example in two dimensions also describes the eigenfrequencies of a vibrating membrane or of flat microwave cavities.

Whereas in classically bounded conservative systems, which are chaotic, one of the most obvious properties of chaos manifests itself in the long-time behaviour, like the exponential decay of certain correlation functions for $t \to \infty$, their quantum mechanical counterparts do not display such a behaviour since the time-dependent wave function (2) is an almost periodic function for fixed $\mathbf{q}$. Thus with respect to the long-time behaviour it is commonly believed that there is no phenomenon in quantum mechanics, that could be called temporal quantum chaos, which can be considered as a direct manifestation of the chaotic classical motion in the corresponding quantum system. (See, however, the recent paper [1].) For this reason, most work in the field of *quantum chaos* [2] has not been devoted to the long-time asymptotics $t \to \infty$, but instead to the study of energy levels and eigenfunctions, in particular of their statistical properties. Thus one of the main research lines of quantum chaos is the search for fingerprints left on the quantum system, i.e., on $\{E_n\}$ and $\{\psi_n\}$, by its classical counterpart. Since such fingerprints are expected to be seen most clearly in the semiclassical limit, one is interested in the behaviour of the energy levels $\{E_n\}$ and wave functions $\{\psi_n\}$ in this limit, which for quantum billiards corresponds to the limit $E_n \to \infty$.

2 Spectral Statistics and Quantum Chaos

One main point of our research programme was the search for universal statistical properties of energy spectra and eigenfunctions of quantum mechanical systems whose classical dynamics is chaotic. Let us start therefore by giving a short overview on the statistical behaviour of the spectrum of energy levels. Depending on the properties of the corresponding classical dynamical systems one expects a different behaviour of the spectral statistics: It has been conjectured that the energy level statistics of integrable systems can be described by a Poissonian random process [3]. In contrast, the statistics of classically strongly chaotic systems are conjectured to be given by the corresponding random matrix distributions [4], where the statistics of systems possessing only one anti-unitary symmetry (for example time-reversal invariance) should be described by the Gaussian orthogonal ensemble (GOE), and the statistics of systems without symmetry should be described by the Gaussian unitary ensemble (GUE). Random matrix theory (RMT) is by now a well established mathematical theory describing the properties of such ensembles of matrices [5]. However, although much effort has been spent on proving these conjectures, only in certain special cases rigorous results have been achieved. We will return to this point later.

Let us now define some statistical measures, which will be illustrated by numerical results for the integrable circle billiard and for the chaotic cardioid billiard (the domain Ω of these billiards is shown in figs. 12 and 13). In both cases the spectra of even symmetry of the desymmetrized systems have been used (for details see [6,7]).

The spectral staircase function $N(E)$ (mode number, integrated level density) is given by

$$N(E) := \#\{n \,|\, E_n \leq E\} \tag{5}$$

and counts the number of energy levels E_n below a given energy E. The function $N(E)$ can be separated into a mean smooth part $\overline{N}(E)$ and a fluctuating part, the *mode fluctuations* $N_{\text{fluc}}(E)$

$$N(E) = \overline{N}(E) + N_{\text{fluc}}(E) \ . \tag{6}$$

For two-dimensional billiards with domain Ω, the mean behaviour $\overline{N}(E)$ is given by the generalized Weyl formula [8] ($\hbar = 2m = 1$, $E \to \infty$)

$$\overline{N}(E) = \frac{\mathcal{A}}{4\pi}E - \frac{\mathcal{L}}{4\pi}\sqrt{E} + \mathcal{C} \ , \tag{7}$$

where $\mathcal{A}$ denotes the area of the billiard, and $\mathcal{L} := \mathcal{L}^- - \mathcal{L}^+$, where $\mathcal{L}^-$ and $\mathcal{L}^+$ are the lengths of the boundary $\partial\Omega$ with Dirichlet and Neumann boundary conditions, respectively. Fig. 1 shows $N(E)$ and $\overline{N}(E)$ for the first eigenvalues of the circle billiard and the cardioid billiard, respectively. One observes that (7) describes the mean behaviour of $N(E)$ very well even down to the ground state, despite of being an asymptotic result.

To compare the quantal spectra of different systems, one has to get rid of the system-dependent constants in $\overline{N}(E)$, which is achieved by *unfolding* the spectra by

$$x_n := \overline{N}(E_n) \ , \qquad n = 1, 2, 3, \ldots \ . \tag{8}$$

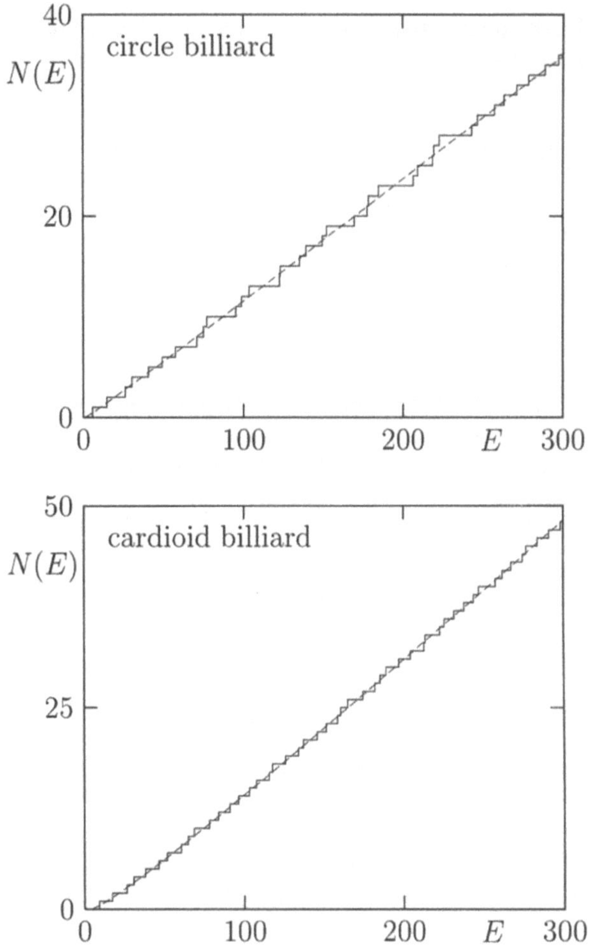

Fig. 1. Spectral staircase function $N(E)$ and its mean behaviour $\overline{N}(E)$, shown as dashed line, for the circle billiard and the cardioid billiard.

The unfolded spectrum $\{x_n\}$ has by construction a unit mean level spacing. In the sequel we will assume that the spectra are already unfolded, and in order to keep the notation simple, $N(x)$ will denote the spectral staircase of the unfolded energy spectrum, and $N_{\text{fluc}}(x)$ the corresponding fluctuating part. Thus $N(x) = x + N_{\text{fluc}}(x)$. Sometimes it is useful to include further long range oscillations in the unfolding procedure, see [7] for a detailed discussion.

To investigate the statistical distribution of quantum spectra, several statistical measures have been introduced. An important statistics is the nearest-neighbour level-spacing distribution $P(s)$ which is the probability density for the distribution of the distances $x_{n+1} - x_n$ between adjacent energy levels.

More precisely, the question is whether a limit distribution of $x_{n+1}-x_n$ exists in the following sense

$$\lim_{n\to\infty} \frac{\#\{i \leq n \mid a \leq x_{i+1} - x_i \leq b\}}{n} = \int_a^b P(s)\, \mathrm{d}s \tag{9}$$

and of what shape $P(s)$ is.

According to the above–mentioned conjectures one expects for integrable systems

$$P_{\text{Poisson}}(s) = \mathrm{e}^{-s} \quad . \tag{10}$$

Since $P(s) \to 1$ for $s \to 0$ this behaviour is called *level attraction.*

The result of RMT for the level-spacing distribution is in the case of the GOE in very good approximation described by the Wigner distribution [5]

$$P_{\text{GOE}}(s) \approx P_{\text{Wigner}}(s) = \frac{\pi}{2} s \exp\left(-\frac{\pi}{4} s^2\right) \quad . \tag{11}$$

In the case of the GUE one has in very good approximation [5]

$$P_{\text{GUE}}(s) \approx \frac{32}{\pi^2} s^2 \exp\left(-\frac{4}{\pi} s^2\right) \quad . \tag{12}$$

In both cases we have $P(s) \to 0$ for $s \to 0$, which is called *level repulsion.*

To illustrate this behaviour, we show in fig. 2 the level-spacing distribution for the circle and for the cardioid billiard, respectively. The two distributions are in good agreement with the expected behaviour of a Poissonian random process and of the GOE, respectively, which was also observed for numerous other systems, see, e.g., [2] and references therein.

For long-range correlations the situation turns out to be more complicated. A convenient measure for medium- and long-range correlations is provided by the number variance $\Sigma^2(L, \hat{x})$, which is defined by

$$\Sigma^2(L, \hat{x}) := \left\langle [N(x+L) - N(x) - L]^2 \right\rangle_{\hat{x}} \quad , \quad L > 0 \quad , \tag{13}$$

which measures the variance of the number of eigenvalues in an energy interval of length L. Here the brackets denote an energy averaging for which there are a priori different possibilities. We choose the following rectangular averaging ($c > 1$)

$$\langle f(x) \rangle_{\hat{x}} := \frac{1}{(c-1)\hat{x}} \int_{\hat{x}}^{c\hat{x}} f(x)\, \mathrm{d}x \quad . \tag{14}$$

For a Poissonian random process one has

$$\Sigma^2_{\text{Poisson}}(L) = L \quad . \tag{15}$$

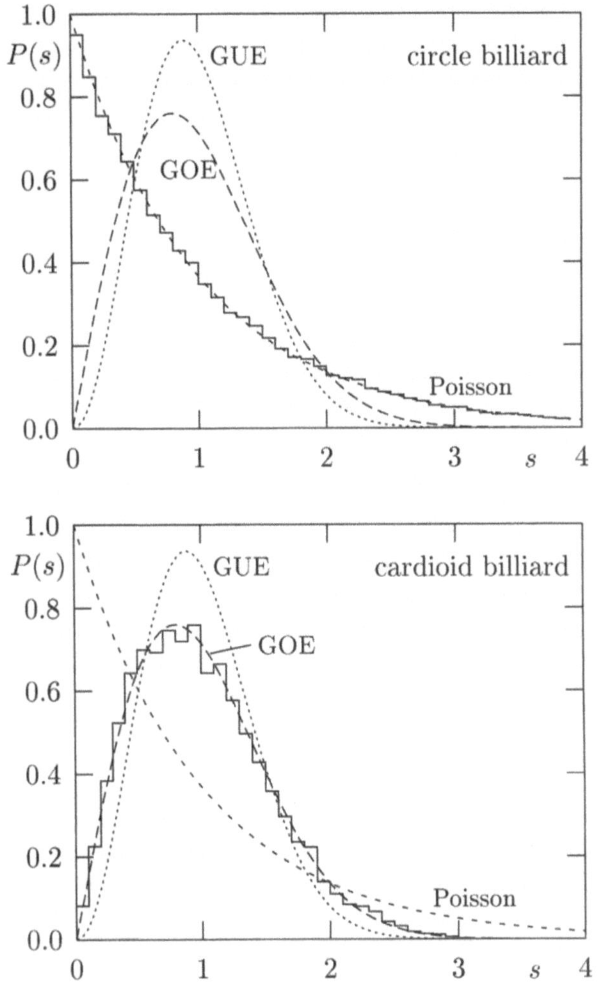

Fig. 2. Level–spacing distribution for the circle billiard (even symmetry, 100 000 eigenvalues) and the cardioid billiard (even symmetry, 11 000 eigenvalues).

In RMT one obtains in the case of the GOE [5]

$$\Sigma^2_{\mathrm{GOE}}(L) = \frac{2}{\pi^2}\left\{\ln(2\pi L) + \gamma + 1 + \frac{1}{2}\mathrm{Si}^2(\pi L) - \frac{\pi}{2}\mathrm{Si}(\pi L) \right.$$
$$\left. - \cos(2\pi L) - \mathrm{Ci}(2\pi L) + \pi^2 L\left[1 - \frac{2}{\pi}\mathrm{Si}(2\pi L)\right]\right\} , \tag{16}$$

and for the GUE [5]

$$\Sigma^2_{\mathrm{GUE}}(L) = \frac{1}{\pi^2}\{\ln(2\pi L) + \gamma + 1 - \cos(2\pi L) - \mathrm{Ci}(2\pi L) + \pi^2 L \left[1 - \frac{2}{\pi}\mathrm{Si}(2\pi L)\right]\} , \tag{17}$$

where $\gamma = 0.5772\ldots$ is Euler's constant and $\mathrm{Si}(x)$ and $\mathrm{Ci}(x)$ are the sine and cosine integral, respectively. Notice that the results (15)–(17) are "stationary", i.e., they do not depend on the energy $\hat{x}$ around which the averaging is carried out according to (14).

Fig. 3 shows the number variance for the circle billiard and the cardioid billiard, respectively. For small values of L, i.e., for short–range correlations, one observes good agreement with the conjectured Poisson (15) and GOE (16) behaviour. However, for larger values of L one observes deviations of $\Sigma^2(L, \hat{x})$ from $\Sigma^2_{\mathrm{Poisson}}(L)$ and $\Sigma^2_{\mathrm{GOE}}(L)$, respectively, and the number variance fluctuates around a saturation plateau of height $\Sigma^2_\infty(\hat{x})$. Notice that the fluctuations are much larger for the integrable system, for which also the saturation plateau is higher than for the cardioid billiard.

The RMT results for large L are $\Sigma^2_{\mathrm{GOE}}(L) \sim \frac{2}{\pi^2}\ln(2\pi L)$ and $\Sigma^2_{\mathrm{GUE}}(L) \sim \frac{1}{\pi^2}\ln(2\pi L)$ which is in contrast to the observed saturation of the number variance, see, e.g., [9–14,6]. Berry has given a semiclassical argument for the saturation behaviour of the number variance [9] and of the spectral rigidity [15]. It turns out that with $L_{\max} := a\sqrt{\hat{x}}$, where $a > 0$ is a system-dependent constant, one has the universality regime for $L \ll L_{\max}$, in which the spectral statistics are expected to be described by random matrix theory. For $L \gg L_{\max}$, however, the number variance clearly deviates from the RMT result and rather oscillates around its saturation plateau $\Sigma^2_\infty(\hat{x})$. The saturation value $\Sigma^2_\infty(\hat{x})$ increases with increasing energy $\hat{x}$ in the limit of large $\hat{x}$.

However, the above presented picture does not always hold, as there are important exceptions to the generically expected behaviour. The general behaviour of chaotic systems is violated by systems possessing arithmetical chaos [13,16–22], where, e.g., the level-spacing distribution and the two-point statistics nearly behave as for classically integrable systems. Another example of a strongly chaotic system showing non–generic spectral statistics are quantized cat maps [23].

As an example for a class of integrable systems, the eigenvalue statistics for the geodesic flow on flat tori are studied in [24]. It is proven that the pair correlation function is Poissonian for a set of full Lebesgue measure in the parameter space of tori, whereas the consecutive level-spacing distribution does not exist for a set of second Baire category (a topologically large set). Explicit examples of tori with Poissonian pair correlation function are given in [25]. A further class of integrable systems showing exceptional behaviour are harmonic oscillators with irrational ratio of frequencies, for which the nearest–neighbour level–spacing distribution and other spectral statistics do not possess a limit distribution, see, e.g., [3,26–30] and references therein.

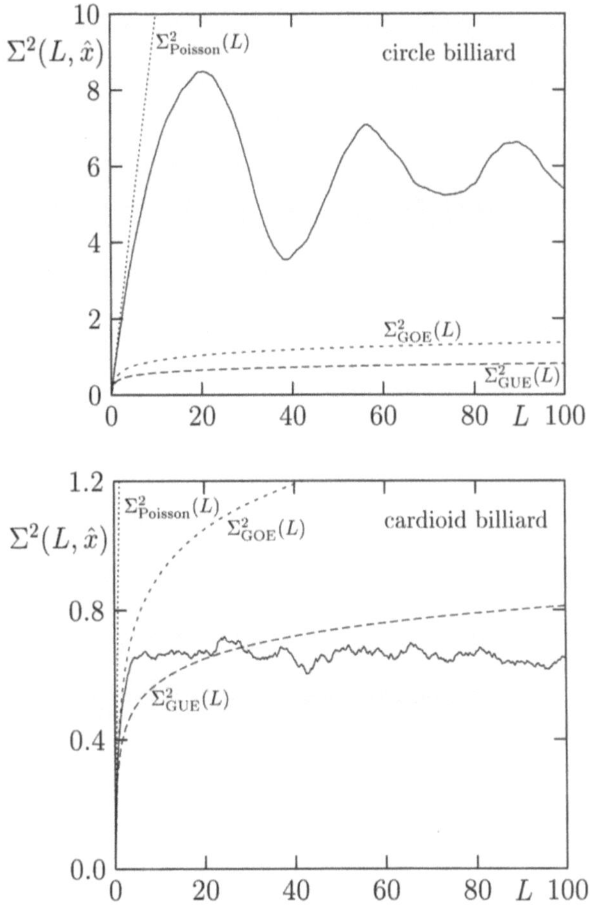

Fig. 3. Number variance for the circle and cardioid billiard for $\hat{x} = 1000$ and $c = 3$.

Recently it has been shown that for the class of quantized skew translations on the torus (also called parabolic maps) [31–33], which are ergodic but not mixing for irrational parameters, the spectral statistics do not possess a limit distribution [34].

As the standard spectral statistics do not always provide a measure to distinguish between integrable and chaotic systems, in particular even for systems as strongly chaotic as the geodesic motion on arithmetic surfaces of constant negative curvature, the distribution of the suitably normalized mode fluctuations have been proposed in [35,36] as an alternative statistical measure. Fig. 4 shows $N_{\text{fluc}}(x)$ for the first 3000 eigenvalues of the circle billiard and the cardioid billiard. Since $N_{\text{fluc}}(x)$ behaves very irregularly the idea is to consider $N_{\text{fluc}}(x)$ as a random function of x and to look for its statistical

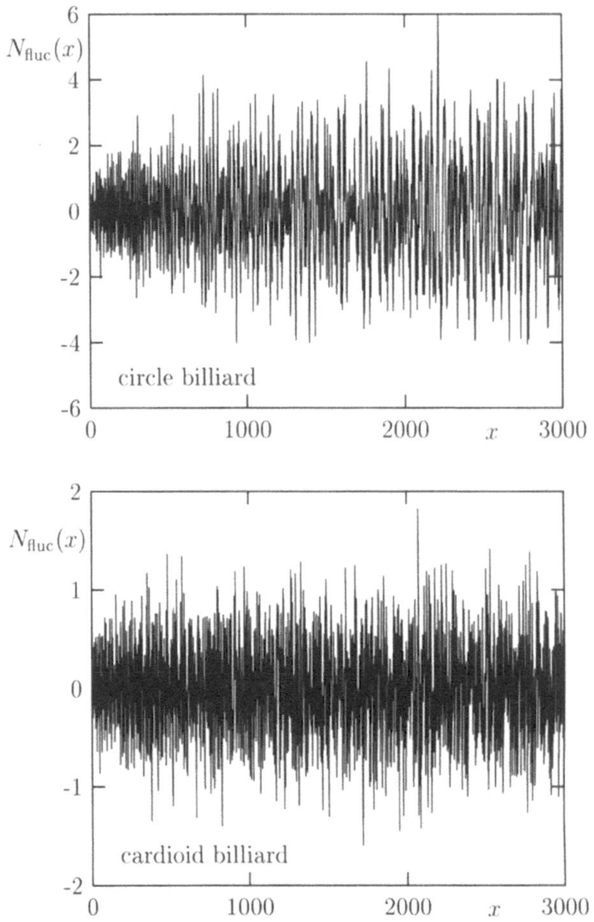

Fig. 4. Plot of N (x) for the first 3000 eigenvalues of the circle billiard and the cardioid billiard. Notice that the fluctuations are much larger for the integrable system in comparison with the chaotic system.

properties. However, in fig. 4 one observes an increase in the amplitude of the fluctuations and thus $N_{\text{fluc}}(x)$ itself cannot possess a limit distribution. Therefore, one defines the variance

$$D(x) := \frac{\Xi(c)}{(c-1)\,x} \int\limits_{x}^{cx} [N_{\text{fluc}}(y)]^2 \, \mathrm{d}y \ , \tag{18}$$

with $c > 1$, where $\Xi(c)$ is a constant ensuring normalization, see [7]. With this one can construct the following random variable

$$W(x) := \frac{N_{\text{fluc}}(x)}{\sqrt{D(x)}} \quad , \tag{19}$$

where x is randomly chosen from the interval $[T, cT]$. If a limit distribution of the random variable $W(x)$ exists for $T \to \infty$, it has by construction a second moment of one if the second moment exists. Furthermore, the first moment of this limit distribution is zero because $N_{\text{fluc}}(x)$ describes by definition the fluctuations of $N(x)$ around the mean behaviour $\overline{N}(x) = x$.

The conjecture proposed in [35,36] can now be formulated as follows:

Conjecture: *For bound conservative and scaling systems the quantity $W(x)$, eq. (19), $x \in [T, cT]$, possesses a limit distribution for $T \to \infty$. This distribution is absolutely continuous with respect to the Lebesgue measure on the real line, with a density $P(W)$*

$$\lim_{T\to\infty} \frac{1}{(c-1)\,T} \int_T^{cT} g(W(x))\,\rho(x/T)\,\mathrm{d}x = \int_{-\infty}^{\infty} g(W)\,P(W)\,\mathrm{d}W \quad , \tag{20}$$

where $g(x)$ is a bounded continuous function, and $\rho(t) \geq 0$ is a continuous density on $[1, c]$ with $\frac{1}{c-1}\int_1^c \rho(t)\,\mathrm{d}t = 1$.

Furthermore, the limit distribution has zero mean and unit variance,

$$\int_{-\infty}^{\infty} W\,P(W)\,\mathrm{d}W = 0 \quad , \quad \int_{-\infty}^{\infty} W^2\,P(W)\,\mathrm{d}W = 1 \ . \tag{21}$$

If the corresponding classical system is strongly chaotic, having only isolated and unstable periodic orbits, then $P(W)$ is universally a Gaussian, $P(W) = \frac{1}{\sqrt{2\pi}}\exp(-\frac{1}{2}W^2)$. In contrast, a classically integrable system leads to a non-Gaussian density $P(W)$.

In the following we will consider two aspects only, for a detailed discussion see [36,7,37].

For classically chaotic systems a semiclassical approximation to $N_{\text{fluc}}(x)$ is given by means of Gutzwiller's periodic-orbit theory [2], see section 3. In this theory $N_{\text{fluc}}(x)$ can be expressed in terms of the *dynamical zeta function* $Z(s)$ which involves only properties of the periodic orbits like their lengths, stabilities and Maslov indices. The non-trivial zeros of the dynamical zeta function $Z(s)$ are directly connected with the quantal levels of the corresponding system. Using the argument principle, one obtains for two–dimensional billiards

$$\widetilde{N}_{\text{fluc}}(p) = \frac{1}{\pi}\arg\, Z(\mathrm{i}p) \ . \tag{22}$$

(The tilde distinguishes $\widetilde{N}_{\text{fluc}}(p)$, which is a function of the momentum $p = \sqrt{E}$, from $N_{\text{fluc}}(x)$ which is a function of the unfolded energy x.)

An interesting number theoretical example is provided by the non-trivial zeros of the Riemann zeta function $\zeta(s)$. Their statistics behave in many respects like the eigenvalues of some hypothetical classically chaotic system without anti-unitary symmetry. Here one has

$$\widetilde{N}_{\text{fluc}}(p) = \frac{1}{\pi} \arg \zeta\left(\frac{1}{2} + \mathrm{i}p\right) , \tag{23}$$

where $\widetilde{N}_{\text{fluc}}(p)$ is the fluctuating term of the counting function $N(p)$, which counts the non-trivial Riemann zeros ρ_n, $\zeta(\rho_n) = 0$ with $0 < \operatorname{Re}\rho_n < 1$ and $0 < \operatorname{Im}\rho_n \leq p$. It can be shown that for a certain class of zeta functions, which includes the Riemann zeta function, the corresponding normalized fluctuations (given here for the Riemann zeta function)

$$\widetilde{W}(p) = \frac{\widetilde{N}_{\text{fluc}}(p)}{\sqrt{\frac{1}{2\pi^2}\ln\ln p}} \tag{24}$$

have indeed a Gaussian limit distribution [38–40].

For integrable systems a lot of results concerning the limit distribution $P(W)$ of $W(x)$ have been obtained, see [7,37,41,42] for summaries. For several completely integrable two-dimensional geodesic flows it is possible to show

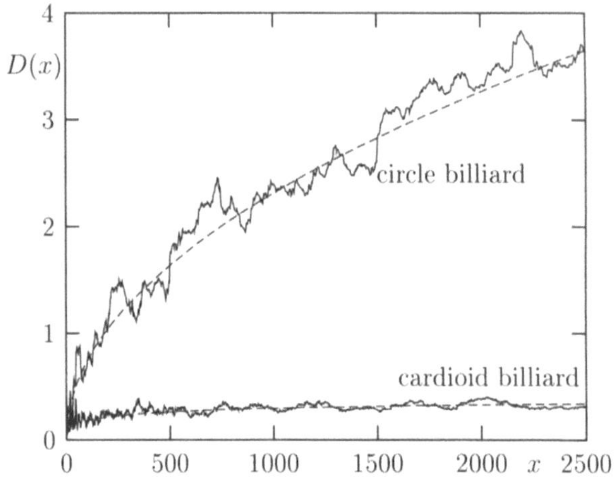

Fig. 5. Plot of the variance $D(x)$ for the circle billiard and the cardioid billiard using $c = 1.1$. Also shown (dashed curves) are the expectations based on semiclassical arguments [15], which lead [20,36] to $D(x) \sim C\sqrt{x}$ for the integrable case and $D(x) \sim \frac{1}{2\pi^2}\ln x + C'$ for the chaotic (GOE) case.

that $N(E)$ can be decomposed as follows (see eqs. (5) and (7))

$$N(E) = \frac{\mathcal{A}}{4\pi} E + E^{1/4} \, \Theta(\sqrt{E}) \ , \qquad (25)$$

where the remainder term $E^{1/4} \, \Theta(\sqrt{E})$ fluctuates around zero with an amplitude increasing as $E^{1/4}$ and is identified with $N_{\rm fluc}(E)$. The first term is the Weyl term. Since the considered systems have no boundary, there is no additional term involving $\mathcal{L}$, and also the constant $\mathcal{C}$ in (7) is zero. It could be shown that eq. (20) holds and, furthermore, that the corresponding limit distribution decays faster than a Gaussian. The basic idea for proving the existence of a limit distribution of $\Theta(\sqrt{E})$ is to express $N(E)$ in terms of a sum over periodic orbits, or (equivalently) to reduce $N(E)$ to a lattice-point problem and then to exploit the Poisson summation formula to obtain an expression of the remainder term as a trigonometric series. If the remainder term is an *almost periodic function* [43] then it has a limit distribution [44]. A simple example is given by the eigenvalues of the Laplacian on a two-dimensional torus. The eigenvalues read (in suitable units) $E_{n,m} = n^2 + m^2$ with $n, m \in \mathbb{Z}$. Thus to determine $N_{\rm fluc}(E)$ one is led to the *classical circle problem*, due to Gauß, where one considers the difference between the number of lattice points inside a circle of radius $R = \sqrt{E}$ and the corresponding area πR^2. For a long time one was concerned with finding optimal estimates of the form $O(E^\alpha)$ for the error term $N(E) - \pi \, E$. Gauß proved $\alpha = \frac{1}{2}$ which follows from the fact that the remainder term cannot grow faster than the length of the boundary. Hardy's conjecture [45] that $\alpha = \frac{1}{4} + \varepsilon$ for all positive ε is still unproven. Probably the best estimate at the present time is $|N(E) - \pi \, E| \le C_\varepsilon E^{23/73} (\log E)^{315/146}$ [46]. It was Heath-Brown [47] who had the idea to study the probabilistic properties of $\Theta(\sqrt{E})$, see eq. (25), instead of obtaining an optimal value for α. In the case of the circle problem he was then able to prove that $\Theta(R)$ possesses a limit distribution.

Before we turn to the $P(W)$ distribution, we present in figure 5 the variance $D(x)$ for the circle and cardioid billiard using the first 11 000 levels and $c = 1.1$, see eq. (18). For the integrable system one observes that the variance of $N_{\rm fluc}(x)$ is much larger than for the chaotic system, for which $D(x)$ increases only logarithmically. Fig. 6 shows the mode–fluctuation distribution $P(W)$ for the cardioid billiard and the stadium billiard (for which also the contribution of the bouncing ball orbits has been included in the unfolding procedure, see [7] for details). In both cases very good agreement with the Gaussian normal distribution (shown as dashed line) is observed. It should be emphasized that this is even true for the arithmetic systems [36,13,7], for which the usual spectral statistics, like the level-spacing distribution, show strong deviations from the expected GOE behaviour.

For comparison, fig. 7 shows the mode–fluctuation distribution $P(W)$ for a) a triangular billiard, b) the circle billiard, c) the torus and d) a rectangular billiard [7]. In accordance with the conjecture and the analytical results, for

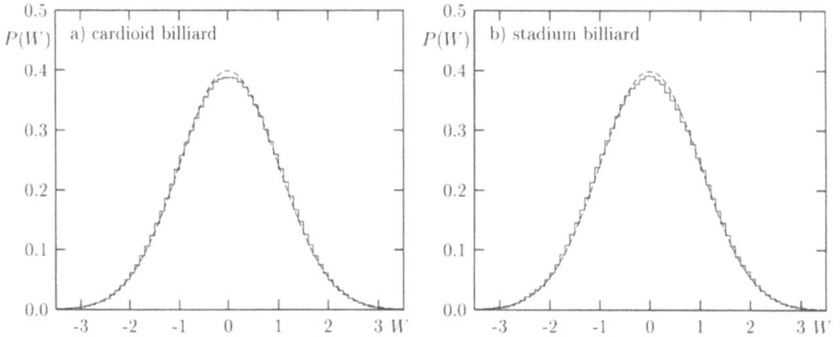

Fig. 6. Mode–fluctuation distribution $P(W)$ for the cardioid billiard and the stadium billiard in comparison with the Gaussian normal distribution (dashed curve).

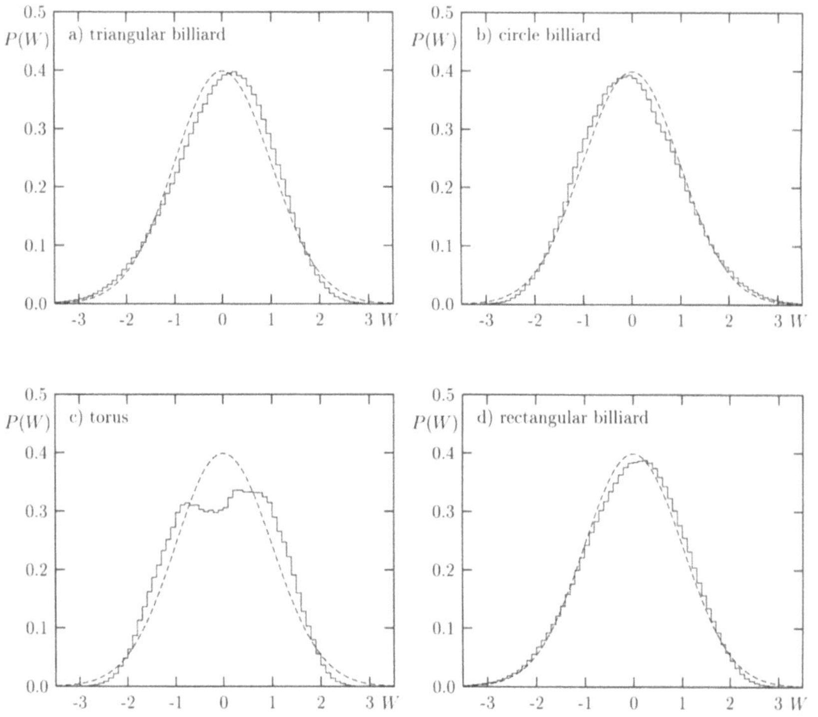

Fig. 7. Mode–fluctuation distribution $P(W)$ for four integrable billiards in comparison with the Gaussian normal distribution (dashed curve).

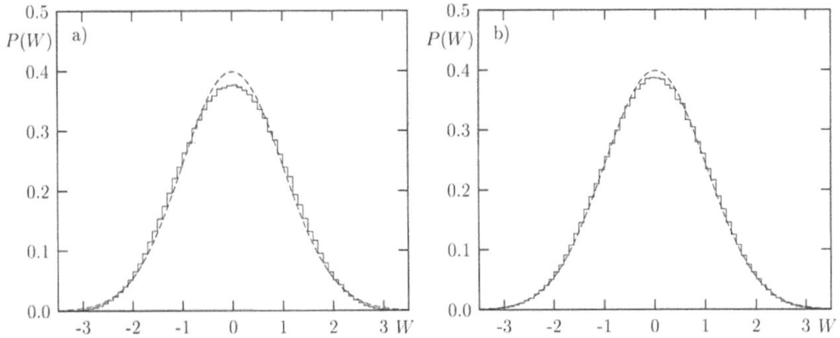

Fig. 8. Mode–fluctuation distribution for the Riemann zeta function, computed according to (24) a) for the first 50 000 zeros and b) for 50 000 zeros starting from the $10^{20} + 143\,780\,420$th zero.

these integrable systems clear deviations from the Gaussian behaviour are observed.

Finally we show in fig. 8 the mode–fluctuation distribution for the Riemann zeta function which has been computed [48] according to (24) from the non-trivial zeros [49]. Notice that the convergence to the (proven) Gaussian limit distribution is very slow.

In [50] the mode–fluctuation distribution $P(W)$ has been analyzed for experimentally measured spectra of several two– and three–dimensional superconducting microwave cavaties in the frequency range below 20 GHz. For the stadium billiard and the 3D–Sinai billiard it is found that $P(W)$ is in good agreement with our conjecture.

For the integrable geodesic flow on Riemannian manifolds all of whose geodesics are closed a trace formula has been derived in [51] and it has been shown that $P(W)$ is a box-shaped function and thus different from a Gaussian which is again in accordance with our conjecture.

3 Trace Formulae and Symbolic Dynamics

A fundamental formula which relates the energy levels of a given quantum system to properties of the corresponding classical dynamical system is provided by the Gutzwiller trace formula [52,2]. This formula gives a formal semiclassical expression for the quantum mechanical density of states as a sum over periodic orbits. With $\hbar = 2m = 1$, $d(p) := \frac{\mathrm{d}N(E=p^2)}{\mathrm{d}p} = 2pd(E=p^2) = \sum_n \delta(p - p_n)$, $p > 0$, $p_n = \sqrt{E_n}$, and the mean behaviour $\bar{d}(p) := \frac{\mathrm{d}\bar{N}(E=p^2)}{\mathrm{d}p}$, the *Gutzwiller trace formula* reads for two–dimensional, strongly chaotic billiards (i.e., all orbits are isolated and unstable)

$$d(p) \sim \bar{d}(p) + d_{\text{fluc}}(p) = \bar{d}(p) + \frac{1}{\pi} \sum_{\gamma} \sum_{k=1}^{\infty} \frac{l_\gamma \cos\left(k l_\gamma p - \frac{\pi}{2} k \nu_\gamma\right)}{\sqrt{|2 - \operatorname{Tr} M_\gamma^k|}} \tag{26}$$

for $p \to \infty$. Here γ denotes the primitive periodic orbits, and the k-summation accounts for their multiple traversals. l_γ is the geometric length of the primitive periodic orbit γ and ν_γ denotes the Maslov index, see below. $\operatorname{Tr} M_\gamma$ is the trace of the monodromy matrix M_γ of the primitive periodic orbit γ where M_γ is obtained from the linearization of the flow around the orbit.

As for chaotic systems the number of periodic orbits up to a given length is expected (and for several systems known) to grow exponentially, the sum on the right hand side in eq. (26) is not absolutely convergent, see [53,54]. For systems with only even Maslov indices ν_γ, generalized periodic orbit sum rules have been derived, which involve absolutely convergent sums and integrals only [54]. In [54] the test function $h(p)$ is assumed to fulfil the following three conditions:

a) $h(p)$ is even in p

b) $h(p)$ is analytic in the strip $|\operatorname{Im} p| \leq \sigma + \epsilon$, $\epsilon > 0$

c) $|h(p)| \leq a|p|^{-2-\delta}$ with $\delta > 0$, $a > 0$ for $|p| \to \infty$,

with $\sigma = h_{\text{top}}^{\text{flow}} - \overline{\lambda}/2$, where $h_{\text{top}}^{\text{flow}}$ is the topological entropy of the Hamiltonian flow and $\overline{\lambda}$ is a "mean Lyapunov exponent" [54]. In general one has $\sigma = P(\frac{1}{2}) > 0$, where $P(\beta)$ is the topological pressure, see [37] for details.

Including also the case of odd Maslov indices ν_γ, the generalized periodic orbit sum rules read [54,6]

$$\sum_n h(p_n) \sim \int_0^\infty h(p') \bar{d}(p') \, \mathrm{d}p' + \mathcal{C} h(0) + \sum_\gamma \sum_{k=1}^{\infty} \frac{l_\gamma}{\sqrt{|2 - \operatorname{Tr} M_\gamma^k|}} \mathcal{F}_\gamma\{h(k l_\gamma)\} \ , \tag{27}$$

where

$$\mathcal{F}_\gamma\{h(x)\} := \frac{1}{\pi} \int_0^\infty h(p') \cos\left(x p' - \frac{\pi}{2} k \nu_\gamma\right) \, \mathrm{d}p' \ , \tag{28}$$

and $\mathcal{C}$ is the constant in the generalized Weyl formula, eq. (7). Formally one can obtain this result by multiplying both sides of eq. (26) by a test function $h(p)$ and integrating over p from 0 to ∞.

In the mathematical literature similar trace formulae have been derived by Colin de Verdière [55,56], Chazarain [57] and Duistermaat and Guillemin [58] in the context of pseudodifferential and Fourier integral operators. For the connection to Schrödinger operators see [59–62].

The first example of a trace formula which is even exact and not just a semiclassical approximation is the Selberg trace formula which applies to certain manifolds with constant negative curvature. The associated dynamical systems play an important role in ergodic theory and in the study of quantum chaos. Our group has extensively studied these systems and the Selberg trace formula in a series of papers, see, e.g., [10,12,13,16,17,19,20,22,63–71].

Trace formulae play an important role in the attempts to understand the statistical properties of energy levels. Using trace formulae the spectral fluctuations can be expressed in terms of periodic orbit sums and are thus strongly linked to the correlations in the length spectra of periodic orbits.

Using appropriate test functions $h(p)$ one can get information on the classical length spectrum using the quantum mechanical energy levels. Conversely, one may use the periodic orbits to compute energy levels which thus provides a substitute for the WKB– and EBK–formulae, which can be applied to integrable systems only.

In fig. 9 we illustrate how the lengths l_γ of the shortest periodic orbits can be obtained from the quantal energy levels. This is done by a Fourier analysis of the quantum mechanical density of states. For the choice [63,64,72]

$$h(p) = \cos(pL)\,\mathrm{e}^{-p^2 t}\,; \qquad p = \sqrt{E}\ , \quad t > 0\ , \qquad L \in \mathbb{R}\ , \qquad (29)$$

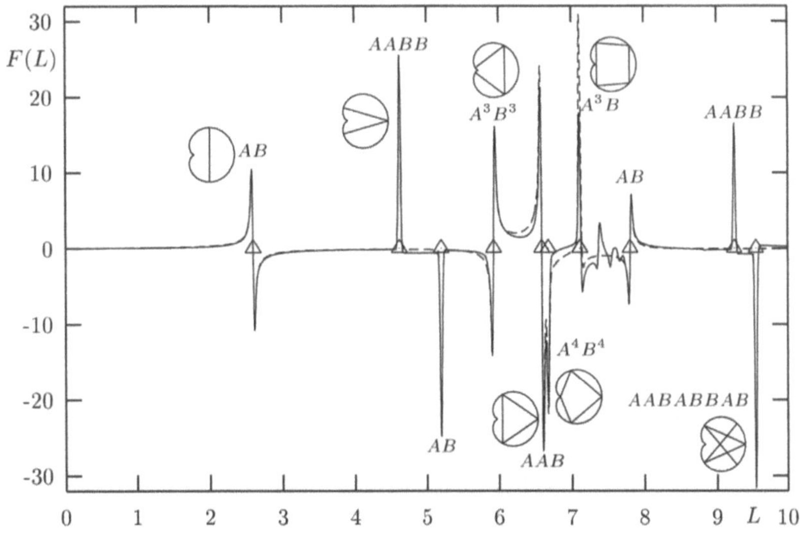

Fig. 9. The trace of the cosine–modulated heat kernel for the cardioid billiard is shown (eigenvalues of odd symmetry). The full line is $F(L)$, eq. (32), with $n = 11000$, $t = 0.0001$, and the dashed line shows the semiclassical result $F\ (L)$, eq. (33), using the periodic orbits up to geometric length 10. The triangles mark the lengths l_γ of the shortest periodic orbits in the desymmetrized system. Also shown are the corresponding orbits in the full cardioid billiard together with their symbolic code.

as test function in eq. (27) one obtains the so-called *trace of the cosine-modulated heat kernel*, $\mathrm{Tr}\{\cos\left((-\Delta)^{1/2}L\right) e^{t\Delta}\}$. The parameter t determines a smooth cut–off allowing for a suppression of the oscillations occurring for a sharp cut–off. The test function (29) yields for the right–hand side of (27) for even $k\nu_\gamma = 2m$ [63]

$$\mathcal{F}_\gamma\{h(x)\} = \frac{(-1)^m}{4} \frac{1}{\sqrt{\pi t}} \left[\exp\left(-\frac{(x-L)^2}{4t}\right) + \exp\left(-\frac{(x+L)^2}{4t}\right)\right] \ , \tag{30}$$

whereas for odd $k\nu_\gamma = 2m+1$ one obtains [6]

$$\begin{aligned}\mathcal{F}_\gamma\{h(x)\} = \frac{(-1)^m}{4} \frac{1}{\pi t} \Bigg[&(x-L)\, {}_1F_1\left(1, \frac{3}{2}, -\frac{(x-L)^2}{4t}\right) \\ &+(x+L)\, {}_1F_1\left(1, \frac{3}{2}, -\frac{(x+L)^2}{4t}\right)\Bigg] \ ,\end{aligned} \tag{31}$$

where ${}_1F_1(a,b,z)$ is Kummer's function [73].

In fig. 9 we show a plot of the trace of the cosine–modulated heat kernel where we have subtracted the mean behaviour from the left–hand side of (27). This gives [6,13]

$$F(L) := \sum_{n=1}^{n_{\max}} \cos(p_n L)\, \mathrm{e}^{-p_n^2 t} - \int_0^{p_{\max}} \cos(pL)\, \mathrm{e}^{-p^2 t} \bar{d}(p)\, \mathrm{d}p - \mathcal{C} \ , \tag{32}$$

where $n_{\max}$ is the index of the largest available eigenvalue and $p_{\max} = \sqrt{E_{n_{\max}}}$. The function $F(L)$ has to be compared with the semiclassical expression according to the trace formula (27)

$$F_{\mathrm{sc}}(L) := \sum_\gamma \sum_{k=1}^{\infty} \frac{l_\gamma}{\sqrt{|2 - \mathrm{Tr} M_\gamma^k|}} \mathcal{F}_\gamma\{h(kl_\gamma)\} \ . \tag{33}$$

One observes a very good agreement between both sides of the trace formula. The periodic orbits corresponding to the peaks are also shown, together with their symbolic code (see below).

As a second example for the application of trace formulae we consider the *cosine quantization rule*[66,74]. Using a semiclassical expression $N_{\mathrm{sc}}(E)$ for the staircase function $N(E)$ according to the semiclassical trace formula (26), an efficient quantization rule for the determination of semiclassical approximations to the eigenvalues is given by the zeros of the function $f(E)$,

$$f(E) := \cos\left(\pi N_{\mathrm{sc}}(E)\right) = 0 \ . \tag{34}$$

Fig. 10 shows $f(E)$ for the cardioid billiard (odd symmetry), where all periodic orbits up to code length 20 have been included in the sum over periodic orbits. The triangles correspond to the exact eigenvalues, and one observes very good overall agreement between the zeros of $f(E)$ and the eigenvalues; for the first 31 eigenvalues the average error in units of the mean level spacing is 5.7%, see [75] for a detailed analysis.

From the above examples it is clear that for a thorough understanding of the properties of a given quantum system a detailed knowledge of the periodic orbits of the corresponding classical system is necessary. One of the best approaches to determine the periodic orbits is to find a symbolic dynamics for the system. Roughly speaking, a symbolic dynamics gives a labelling of trajectories of a given system by means of bi-infinite symbol sequences. Periodic orbits correspond to periodic symbol sequences and can thus be searched for in a systematic way. In particular for billiards a lot of results in this direction have been obtained leading to so-called Markov partitions. For non-uniformly hyperbolic billiard systems with singularities, Markov partitions are usually countably infinite, see, e.g., [76–79]. However, if one is interested in a symbolic description of trajectories, it is already sufficient and even more appropriate to find a finite generating partition (see, e.g., [80]), without the need for the Markov property.

In [81] the existence of finite generating partitions for a large class of two-dimensional hyperbolic maps with singularities is proven. The (explicit) construction is based on a partition determined by the singularity lines of the map, which is shown to be generating. This result applies for example to the cardioid billiard, to dispersing and semidispersing billiards and to

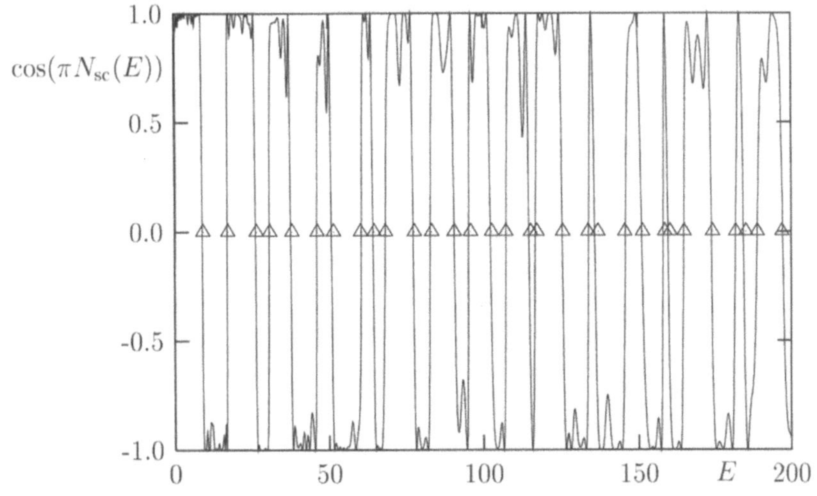

Fig. 10. Cosine quantization rule using the unsmoothed semiclassical spectral staircase function for the cardioid billiard (odd symmetry). All orbits up to code length 20 have been taken into account. The triangles mark the exact eigenvalues E_n.

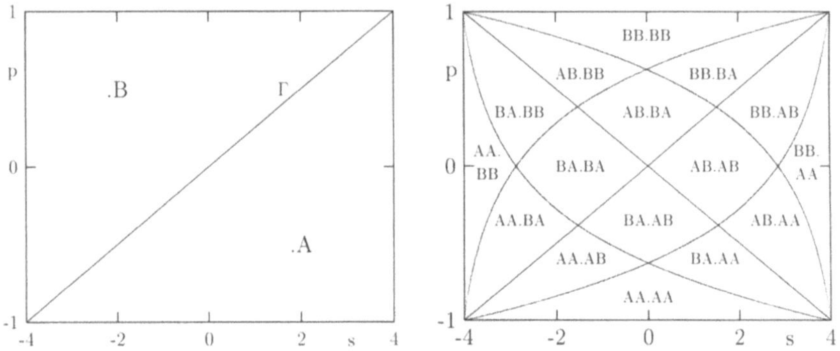

Fig. 11. Initial partition of the Poincaré section $\mathcal{P}$ for the cardioid billiard which is given by the two symbols A and B separated by the singularity line Γ. The figure to the right shows a refined partition where four symbols are fixed.

Bunimovich billiards. In particular for the stadium billiard this gives a proof of a previously conjectured symbolic dynamics [82].

Instead of giving the general statement, we would like to illustrate the result for the cardioid billiard, where the symbolic dynamics is given in terms of two symbols [83,84]. The boundary of the cardioid billiard is given in polar coordinates by $\rho(\varphi) = 1 + \cos\varphi$ with $\varphi \in [-\pi, \pi]$. The usual Poincaré section is given by $\mathcal{P} = \{(s,p) \mid s \in [-4,4],\ p \in [-1,1]\}$, where $s = 4\sin(\varphi/2)$ is the arclength coordinate along the boundary and p is the projection of the unit velocity vector (after the reflection at the boundary) on the tangent in the point of the billiard boundary corresponding to s. The normalized invariant measure of the billiard map $P : \mathcal{P} \to \mathcal{P}$ is given by $\mathrm{d}\mu = \frac{1}{16}\mathrm{d}s\,\mathrm{d}p$. In these coordinates the singularity line of P, i.e., the set of points which will hit the cusp of the cardioid in the next iteration, is simply given by $\Gamma = \{(s,p) \mid p = s/4,\ s \in [-4,4]\}$. This leads to a partition of $\mathcal{P}$ into two regions A and B as shown in fig. 11 [84]. Also shown is an example of a refined partition, where four symbols are fixed, giving altogether 16 cells. However, when fixing $2n$ symbols with $n \geq 5$ it turns out that not all symbol sequences are realized by the dynamical system, i.e., the number of cells of the refined partition is smaller than 2^{2n}. This has been called “pruning” of symbol sequences from the binary tree of possible combinations of letters A and B. A description of the set of forbidden symbol sequences has been proposed in terms of the so–called pruning front in the symbol plane [85]. This conjecture has been tested successfully for the cardioid billiard [84,75]. For certain model systems some progress in proving the pruning front conjecture has been made recently [86,87]. However, for the cardioid and other billiard systems this problem is still open.

4 Eigenfunctions and Quantum Ergodicity

The behaviour of the underlying classical system has also important consequences on the structure of the quantum eigenfunctions. A commonly used description of a quantum mechanical state is the Wigner function [88],

$$\mathrm{W}(\mathbf{p},\mathbf{q}) = \int\limits_{\mathbb{R}^2} \mathrm{e}^{\mathrm{i}\mathbf{q}'\mathbf{p}}\, \psi^\star\left(\mathbf{q}-\frac{\mathbf{q}'}{2}\right)\psi\left(\mathbf{q}+\frac{\mathbf{q}'}{2}\right)\, \mathrm{d}^2 q' \ . \tag{35}$$

The Wigner function can be seen as a phase space representation of the wave function. It is not a probability density, as it may also take negative values. According to the "semiclassical eigenfunction hypothesis" the Wigner function semiclassically concentrates on those regions in phase space, which a generic orbit explores in the long–time limit $t \to \infty$ [89–91]. Therefore, in integrable systems the Wigner function $W(\mathbf{p},\mathbf{q})$ is expected to localize on the invariant tori. In fact, WKB constructions give so–called "quasimodes" [92–95], which localize in the semiclassical limit on the invariant tori. However, it is not guaranteed that a quasimode approximates an eigenfunction of the system, as is nicely seen in the well-known example of the double well potential, see, e.g., [95, §33, pp. 236].

For ergodic systems the semiclassical eigenfunction hypothesis implies that the Wigner function should semiclassically condense on the energy surface, i.e., $W(\mathbf{p},\mathbf{q}) \sim \frac{1}{\mathrm{vol}(\Sigma_E)}\,\delta(H(\mathbf{p},\mathbf{q})-E)$, where $H(\mathbf{p},\mathbf{q})$ is the Hamilton function and $\mathrm{vol}(\Sigma_E)$ is the volume of the energy shell defined by $H(\mathbf{p},\mathbf{q}) = E$. In fact this is the correct picture for "almost all" eigenfunctions. This follows from the quantum ergodicity theorem [96–102] (see also [21,103] for general introductions), which roughly speaking states that for almost all eigenfunctions the expectation values of quantum observables tend to the mean value of the corresponding classical observable in the semiclassical limit. "Almost all" means that this holds for a subsequence $\{\psi_{n_j}\} \subset \{\psi_n\}$ of density one, i.e.,

$$\lim_{E\to\infty} \frac{\#\{n_j \mid E_{n_j} \le E\}}{\#\{n \mid E_n \le E\}} = 1 \ . \tag{36}$$

As shown in [104], the quantum ergodicity theorem is equivalent to the validity of the semiclassical eigenfunction hypothesis for a subsequence of eigenfunctions of density one if the classical system is ergodic.

As an example consider as an observable the characteristic function χ_D of a domain $D \subset \Omega$. The quantum ergodicity theorem implies that in case the classical flow is ergodic, there exists a subsequence $\{\psi_{n_j}\} \subset \{\psi_n\}$ of density one, such that

$$\lim_{j\to\infty} \int\limits_D |\psi_{n_j}(\mathbf{q})|^2 \, \mathrm{d}^2 q = \frac{\mathrm{vol}(D)}{\mathrm{vol}(\Omega)} \tag{37}$$

for every measurable subset $D \subset \Omega$. So for almost all eigenfunctions the probability of finding a particle in a certain region D of the position space Ω becomes proportional to the volume D if the energy E becomes large.

More precisely the quantum ergodicity theorem can be formulated in terms of pseudo–differential operators (see, e.g., [105–108]) as follows:

Quantum ergodicity theorem [102]:
Let $\Omega \subset \mathbb{R}^2$ be a compact two-dimensional domain with piecewise smooth boundary, and let $\{\psi_n\}$ be an orthonormal set of eigenfunctions of the Dirichlet Laplacian $-\Delta$ on Ω. If the classical billiard flow on the energy shell $\Sigma_1 = S^1 \times \Omega$ is ergodic, then there is a subsequence $\{n_j\} \subset \mathbb{N}$ of density one such that

$$\lim_{j\to\infty} \langle \psi_{n_j}, A\psi_{n_j} \rangle = \int\limits_{\Sigma_1} \sigma(A)\, \mathrm{d}\mu \ , \tag{38}$$

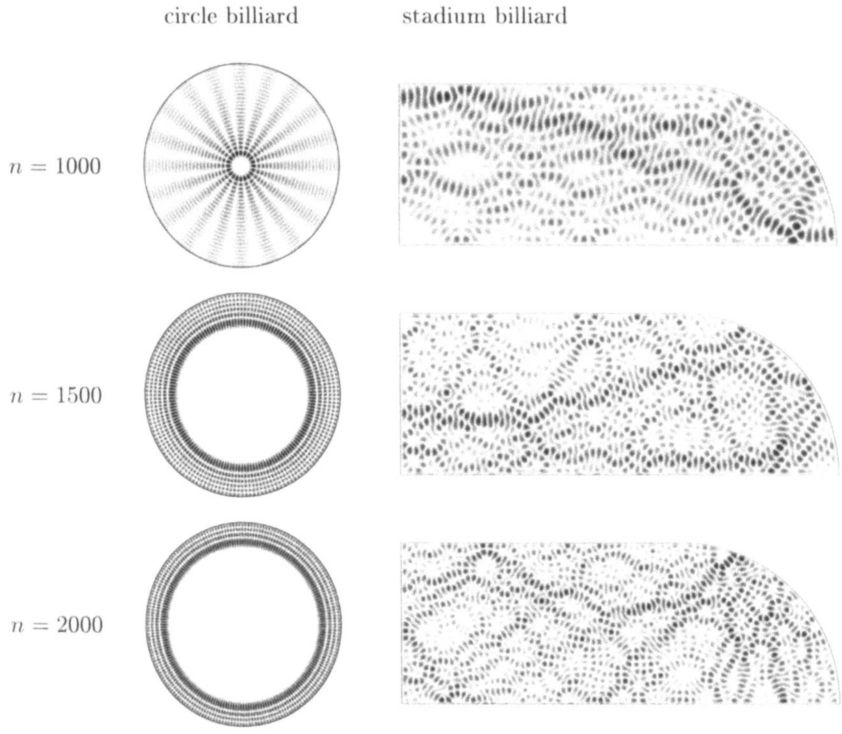

Fig. 12. Density plots of $|\psi_n(\mathbf{q})|^2$ for some eigenfunctions of the integrable circle billiard and the chaotic stadium billiard. Large values are encoded in black, and small values correspond to white.

for every classical pseudodifferential operator A of order zero, whose Schwartz kernel $K_A(\mathbf{q}, \mathbf{q}') = \langle \mathbf{q}|A|\mathbf{q}' \rangle$ has support in the interior of $\Omega \times \Omega$. Here $\sigma(A)$ is the principal symbol of A and the average of $\sigma(A)$ is taken over the energy shell Σ_1 with respect to the Liouville measure μ.

The quantum ergodicity theorem does not just hold for two–dimensional Euclidean domains, but holds also for compact Riemannian manifolds without boundary [96,99,98,97], and a similar statement is true for systems where the Hamilton operator and the observables depend explicitly on $\hbar$ [100]. Also for certain quantized ergodic area–preserving maps on the torus the equipartition of eigenfunctions is proven [109,31,110].

The sequence of expectation values $\langle \psi_{n_j}, A\psi_{n_j} \rangle$ defines (roughly speaking) a sequence of measures on the energy shell, see, e.g., [104] for details. The limit points of this sequence of measures are called *quantum limits.* The special situation that there is only one quantum limit, i.e., the Liouville measure, is called *unique quantum ergodicity.* This behaviour is conjectured to be true for the eigenfunctions of the Laplacian on a compact manifold of negative curvature [112]. This conjecture is supported by a number of previous numerical studies, see, e.g., [67,113,71,114]. Recently it has been shown that quantized skew translations, which are uniquely ergodic, but not mixing for irrational parameters, are also quantum unique ergodic [33].

Fig. 12 illustrates the situation in position space for eigenfunctions of the integrable circle billiard and of the chaotic stadium billiard, respectively. The eigenfunctions of the circle billiard resemble the caustics obtained from the projection of the tori in phase space on position space, whereas the eigenfunctions of the stadium billiard look irregular for sufficiently large n.

However, one also observes so–called scarred eigenfunctions in the stadium and the cardioid billiard, which show localization along unstable periodic orbits [115–117,111], see fig. 13 for an example. This localization is not in contradiction to the quantum ergodicity theorem, as it only makes a statement about almost all eigenfunctions and therefore does not exclude the existence of eigenfunctions behaving differently. The degree of localization of an eigenfunction is also reflected in its maximum norm $||\psi||_\infty$. This has been studied in detail in [117], where numerical results for the maximum norms for several chaotic systems were found to agree with the expectation obtained from a random wave model for the eigenfunctions. This suggests that localization phenomena in exceptional sequences of eigenfunctions are weak.

Another candidate for a non–quantum ergodic subsequence of eigenfunctions are the so-called bouncing ball (bb) modes which occur in billiards with two parallel walls (like the stadium billiard or the Sinai billiard). These eigenfunctions are localized on the rectangular part of the billiard, see fig. 14 for some examples in the stadium billiard. The quantum ergodicity theorem gives

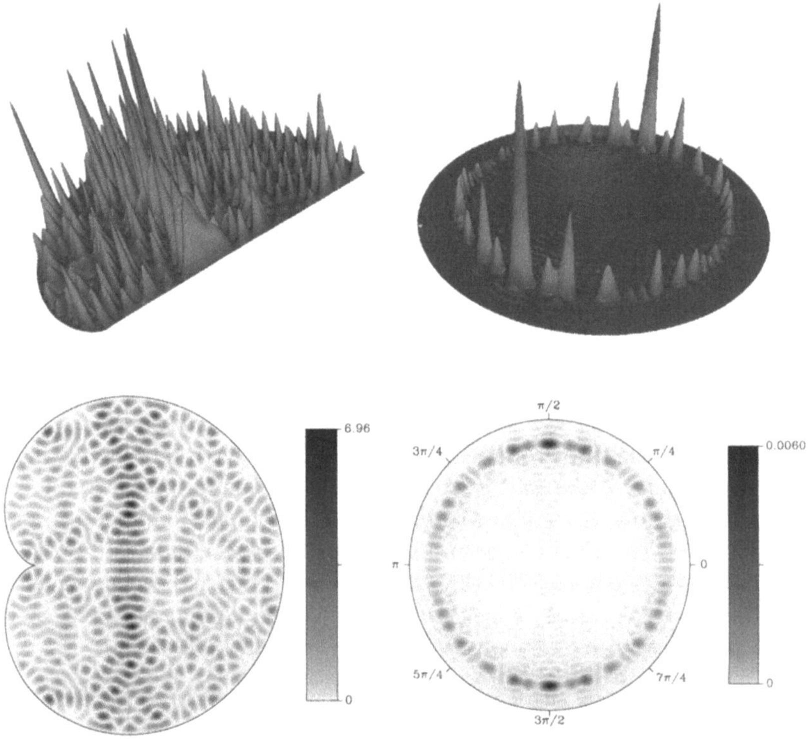

Fig. 13. Example of an eigenfunction showing localization along the shortest unstable periodic orbit AB. The left column shows a three-dimensional plot of $|\psi_{344}(\mathbf{q})|^2$ for the desymmetrized cardioid billiard (odd symmetry) and the corresponding density plot. In the right column a three dimensional plot of the momentum distribution $|\widehat{\psi}_{344}(\mathbf{p})|^2$ and the corresponding density plot is shown, see [111] for details. Here large values of $|\widehat{\psi}_{344}(\mathbf{p})|^2$ in some radial direction correspond to eigenstates with strong localization in momentum in that direction. The localization in position space is thus nicely reflected by the significant enhancements in the directions $(\varphi = \pi/2, 3\pi/2)$ of the orbit.

an upper bound for the counting function $N_{\text{bb}}(E)$ of bouncing ball modes,

$$\lim_{E\to\infty} \frac{N_{\text{bb}}(E)}{N(E)} = 0 \; . \tag{39}$$

Further information on the asymptotic behaviour of the counting function of such localized states is provided in [118], where a heuristic argument shows

$$N_{\text{bb}}(E) \sim \alpha E^{\delta} \; , \qquad E \to \infty \; . \tag{40}$$

From eq. (39) it follows that $\delta < 1$. As shown in [118] the exponent δ is determined by the local shape of the curved boundary connected with the

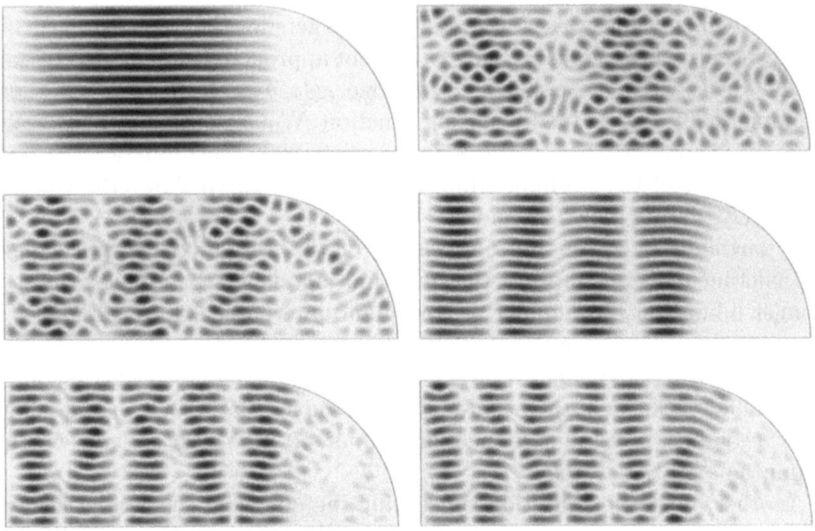

Fig. 14. Density plots of $|\psi_n(\mathbf{q})|^2$ for a consecutive series of bouncing ball modes in the stadium billiard. These eigenfunctions show localization on the rectangular part of the billiard.

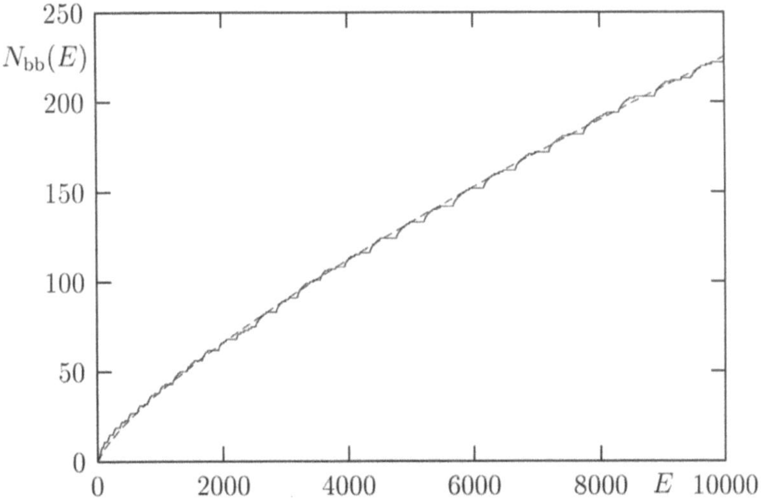

Fig. 15. The number $N\quad(E)$ of bouncing ball modes for the stadium billiard (odd–odd symmetry) together with the fit αE^{δ}, where $\alpha = 0.20$ and $\delta = 0.76$.

straight segment. For the stadium billiard one gets for the exponent $\delta = 3/4$, which was also obtained in [119] by a different approach. For the Sinai billiard on the torus with a circular scatterer one gets $\delta = 9/10$. Fig. 15 shows the numerically determined counting function $N_{\text{bb}}(E)$ obtained from visual selection of bouncing ball states from the first 2000 eigenfunctions of odd–odd symmetry of the stadium billiard. The dashed curve shows the result of a fit to αE^{δ} giving $\alpha = 0.20$ and $\delta = 0.76$ [118].

A particularly interesting application of this result concerns the question whether one can improve the quantum ergodicity theorem by giving a stronger bound on the number of non-quantum-ergodic states. In [118] it is shown explicitly that for any $\delta < 1$, arbitrarily close to 1, one can find an ergodic billiard possessing $N_{\text{bb}}(E) \sim \alpha E^{\delta}$ for the asymptotic behaviour of the number of the non-quantum-ergodic bouncing ball modes. This shows that the quantum ergodicity theorem cannot be improved without further assumptions on the system.

Another important question concerning the quantum ergodicity theorem is the rate by which quantum mechanical expectation values tend to their corresponding limit. This can be studied quantitatively by considering the quantities

$$S_m(E, A) = \frac{1}{N(E)} \sum_{E_n \le E} \left| \langle \psi_n, A\psi_n \rangle - \overline{\sigma(A)} \right|^m \quad . \tag{41}$$

Quantum ergodicity is equivalent to $S_m(E, A) \to 0$ for $E \to \infty$ and $m \ge 1$. Several arguments suggest a rate

$$S_1(E, A) \sim E^{-\frac{1}{4}} \qquad \text{as } E \to \infty \ , \tag{42}$$

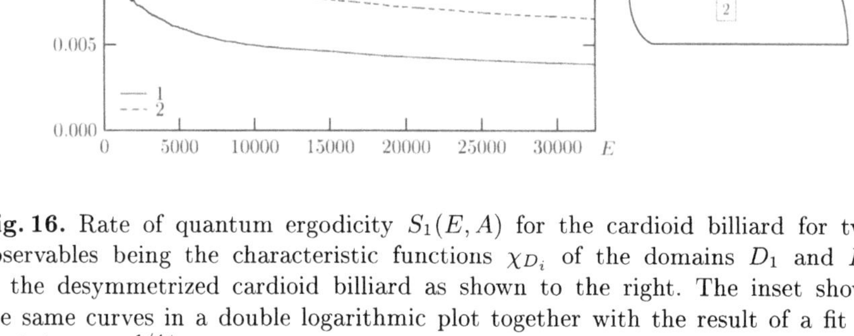

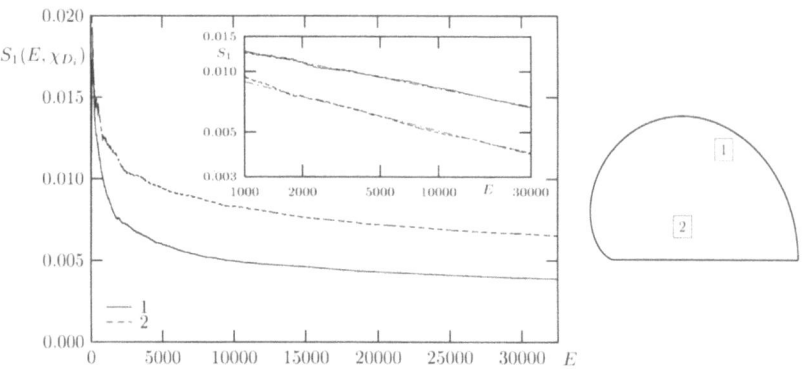

Fig. 16. Rate of quantum ergodicity $S_1(E, A)$ for the cardioid billiard for two observables being the characteristic functions χ_{D_i} of the domains D_1 and D_2 in the desymmetrized cardioid billiard as shown to the right. The inset shows the same curves in a double logarithmic plot together with the result of a fit to $S_1\ (E) = aE^{-1/4+\epsilon}$.

see [104] for a summary on the proven and conjectured behaviour of $S_m(E, A)$. Detailed studies of the rate of quantum ergodicity on surfaces of constant negative curvature and for Euclidean billiards can be found in [71,104].

In fig. 16 the numerically computed rate $S_1(E, A)$ for $A = \chi_{D_i}$ is shown. A fit to $S_1^{\mathrm{fit}}(E) = aE^{-1/4+\epsilon}$ gives $\epsilon = 0.007$ for $A = \chi_{D_1}$ and $\epsilon = 0.064$ for $A = \chi_{D_2}$. Thus in the first case there is very good agreement with the expectation (42), whereas in the second case a significant deviation is observed. This deviation is caused by a considerable number of localized eigenfunctions, like the one in fig. 13, i.e., a subsequence where the quantum limit may be different from the Liouville measure. In [104] it is shown that if the counting function for such a subsequence of localized eigenfunctions increases more strongly than $E^{3/4}$ then the rate of quantum ergodicity is slower than (42). In contrast to Euclidean billiards, for surfaces of constant negative curvature the rate of quantum ergodicity was always found to be in excellent agreement with $E^{-1/4}$ [71].

5 Other Topics

In the foregoing sections we have discussed some of our results obtained within the Schwerpunktprogramm (SPP) during the years 1994–1999. We would like to conclude by giving a very short summary of the other results achieved by our group within the same programme. (Here we do not mention our results on topics different from the research projects of the SPP.)

Several papers have been devoted to the study of quantum chaos in three-dimensional systems. In [70] an exact Selberg trace formula has been derived for compact polyhedral billiards which tesselate the three-dimensional hyperbolic space of constant negative curvature. For a tetrahedral billiard being strongly chaotic, part of the length spectrum of periodic orbits and of the energy spectrum has been computed. Recently the tetrahedral billiard has found a fascinating application in cosmology [120] where the evolution of initial metric perturbations of a small hyperbolic universe are computed from the eigenvalues and eigenfunctions of the Laplace–Beltrami operator yielding the angular power spectrum of the fluctuations of the cosmic microwave background. Asymptotic laws for the mean multiplicities of lengths of closed geodesics in arithmetic hyperbolic three-orbifolds have been derived in [69]. The Selberg trace formula, the Selberg zeta function and the cosine quantization rule have been discussed in detail [68] for Picard's billiard which is a non-compact three-dimensional arithmetical billiard defined on the fundamental domain of the Picard group $\mathrm{PSL}(2, \mathbb{Z}[\mathrm{i}])$.

Based on the theory of Hecke operators a numerical algorithm has been worked out and applied [22] to compute a large number of very precise eigenvalues of the Laplacian for several three-dimensional, non-compact billiards in the hyperbolic upper half space defined on the fundamental domains of the groups $\mathrm{PSL}(2, \mathcal{O})$, where $\mathcal{O}$ is the ring of integers of the imaginary quadratic number fields $\mathbb{Q}(\sqrt{-D})$, $D = 1$ (Picard group), 2, 3, 7, 11, 19. These systems provide further examples of arithmetic quantum chaos by possessing spectral statistics close to Poissonian [22].

The paper [121] contains a new derivation of the semiclassical trace formula for billiards in three dimensions starting from a boundary integral equation. A practical method for determining the monodromy matrix and the Maslov index in three dimensions is described.

The chaotic scattering in a two–dimensional Coulombic muffin–tin potential has been studied in [122,123]. Using the thermodynamic formalism the Hausdorff dimension, escape rate and Kolmogorov–Sinai entropy of the system has been determined. Furthermore, the statistics of the scattering phases have been investigated.

In [124] quantization rules based on Selberg's trace formula have been investigated for the geodesic motion on two compact Riemannian surfaces of genus two and on the triangular billiard $T^*(2,3,8)$.

The quantum mechanical time evolution (2) implies that the autocorrelation function

$$C(t) := \langle \Psi(0)|\Psi(t)\rangle = \sum_{n=1}^{\infty} |c_n|^2 \mathrm{e}^{-\frac{\mathrm{i}}{\hbar}E_n t} \tag{43}$$

has the mathematical form of a generalized Weyl sum. If the eigenvalues are at most quadratic in n, i.e., $E_n = an^2 + bn + c$, one obtains under certain conditions on the coefficients $|c_n|^2$ the classical Jacobi theta function. It turns out that $C(t)$ shows as a function of t quite a "chaotic" behaviour in the complex plane. Exploiting ergodic theory, it was possible [125–127] to derive several theorems on the value distribution of certain theta sums which have important applications in quantum mechanics. (See also [128].)

For sytems which are neither chaotic nor integrable but show a complicated mixture of regular and chaotic behaviour, the semiclassical periodic-orbit approximations are more complicated than in the original Gutzwiller trace formula (26). In [129–131] uniform approximations for contributions to Gutzwiller's periodic–orbit sum for the spectral density have been derived which are valid close to bifurcations of periodic orbits in systems with mixed phase space. Results are given for the tangent, the period–doubling, the period–tripling and for generic period–quadrupling bifurcations. Bifurcations occur also in parameter–dependent integrable systems. As an example, the elliptical billiard and its transition to an oval billiard has been studied in [129]. Uniform approximations for diffractive contributions to the trace formula in billiard systems with corners have been worked out in [132].

In [133,134] semiclassical methods are used to explain the mass asymmetry in nuclear fission and the properties of metal clusters and quantum dots.

Acknowledgments

We would like to thank Dr. R. Aurich, Dr. J. Bolte, Dr. S. Brandis, Prof. N. Chernov, Dr. H. Dullin, Dr. J. Marklof, Dr. C. Matthies, Dr. H. Ninnemann, R. Schubert, Dr. M. Sieber, G. Steil, Dr. P. Stifter and M. Taglieber for the fruitful collaboration in the SPP. We are grateful to Dr. R. Aurich for the kind provision of the numerical results for the mode–fluctuation distribution displayed in fig. 8. Furthermore we would like to thank Prof. M. Robnik and Dr. T. Prosen for the kind provision of the eigenvalues of the cardioid billiard. Fig. 13 is visualized using `Geomview` from *The Geometry Center* of the University of Minnesota and then rendered using `Blue Moon Rendering Tools` written by L.I. Gritz. Financial support by the Deutsche Forschungsgemeinschaft DFG is gratefully acknowledged.

References

1. R. Aurich and F. Steiner: *Temporal quantum chaos*, Int. J. Mod. Phys. B **13** (1999) 2361-2369.
2. M. C. Gutzwiller: *Chaos in Classical and Quantum Mechanics*, Springer-Verlag, New York, (1990).
3. M. V. Berry and M. Tabor: *Level clustering in the regular spectrum*, Proc. R. Soc. London Ser. A **356** (1977) 375–394.
4. O. Bohigas, M.-J. Giannoni and C. Schmit: *Characterization of chaotic quantum spectra and universality of level fluctuation laws*, Phys. Rev. Lett. **52** (1984) 1–4.
5. M. L. Mehta: *Random Matrices*, Academic Press, San Diego, revised and enlarged second edn., (1991).
6. A. Bäcker, F. Steiner and P. Stifter: *Spectral statistics in the quantized cardioid billiard*, Phys. Rev. E **52** (1995) 2463–2472.
7. R. Aurich, A. Bäcker and F. Steiner: *Mode fluctuations as fingerprints of chaotic and non-chaotic systems*, Int. J. Mod. Phys. B **11** (1997) 805–849.
8. H. P. Baltes and E. R. Hilf: *Spectra of Finite Systems*, Bibliographisches Institut, Mannheim, Wien, Zürich, (1976).
9. M. Berry: *Semiclassical formula for the number variance of the Riemann zeros*, Nonlinearity **1** (1988) 399–407.
10. R. Aurich and F. Steiner: *Energy-level statistics of the Hadamard-Gutzwiller ensemble*, Physica D **43** (1990) 155–180.
11. A. Cheng and J. L. Lebowitz: *Statistics of energy levels in integrable quantum systems*, Phys. Rev. A **44** (1991) R3399–R3401.
12. R. Aurich and F. Steiner: *Periodic–orbit theory of the number variance $\Sigma^2(L)$ of strongly chaotic systems*, Physica D **82** (1995) 266–287.
13. R. Aurich, F. Scheffler and F. Steiner: *On the subtleties of arithmetical quantum chaos*, Phys. Rev. E **51** (1995) 4173–4189.
14. P. M. Bleher and J. L. Lebowitz: *Variance of number of lattice points in random narrow elliptic strip*, Ann. Inst. Henri Poincaré: Probab. et Statist. **31** (1995) 27–58.
15. M. V. Berry: *Semiclassical theory of spectral rigidity*, Proc. R. Soc. London Ser. A **400** (1985) 229–251.

16. R. Aurich and F. Steiner: *On the periodic orbits of a strongly chaotic system*, Physica D **32** (1988) 451–460.
17. R. Aurich, E. B. Bogomolny and F. Steiner: *Periodic orbits on the regular hyperbolic octagon*, Physica D **48** (1991) 91–101.
18. E. B. Bogomolny, B. Georgeot, M.-J. Giannoni and C. Schmit: *Chaotic billiards generated by arithmetic groups*, Phys. Rev. Lett. **69** (1992) 1477–1480.
19. J. Bolte, G. Steil and F. Steiner: *Arithmetical chaos and violation of universality in energy level statistics*, Phys. Rev. Lett. **69** (1992) 2188–2191.
20. J. Bolte: *Some studies on arithmetical chaos in classical and quantum mechanics*, Int. J. Mod. Phys B **7** (1993) 4451–4553.
21. P. Sarnak: *Arithmetic quantum chaos*, Israel Math. Conf. Proc. **8** (1995) 183–236.
22. G. Steil: *Eigenvalues of the Laplacian for Bianchi groups*, in: *Emerging Applications of Number Theory* [135], pp. 617–641.
23. J. P. Keating: *The cat maps: Quantum mechanics and classical motion*, Nonlinearity **4** (1991) 309–341,
24. P. Sarnak: *Values at integers of binary quadratic forms*, in: *Harmonic analysis and number theory* (Montreal 1996), CMS Conf. Proc. **21**, American Mathematical Society, Providence, RI (1997) 181–203.
25. A. Eskin, G. A. Margulis and S. Mozes: *Quadratic forms of signature (2,2) and eigenvalue spacings on rectangular 2-tori*, preprint, `http://zaphod.uchicago.edu/~eskin/` (1998).
26. A. Pandey, O. Bohigas and M.-J. Giannoni: *Level repulsion in the spectrum of two-dimensional harmonic oscillators*, J. Phys. A **22** (1989) 4083–4088.
27. P. M. Bleher: *The energy level spacing for two harmonic oscillators with golden mean ratio of frequencies*, J. Statist. Phys. **61** (1990) 869–876.
28. P. M. Bleher: *The energy level spacing for two harmonic oscillators with generic ratio of frequencies*, J. Statist. Phys. **63** (1991) 261–283.
29. C. D. Greenman: *The generic spacing distribution of the two-dimensional harmonic oscillator*, J. Phys. A **29** (1996) 4065–4081.
30. J. Marklof: *The n-point correlations between values of a linear form*, preprint IHES/M/98/66, with an appendix *The number of solutions of simultaneous quadratic equations* by Z. Rudnick (1998). Ergodic Theory Dynam. Systems, **20** (2000) 1127–1172.
31. A. Bouzouina and S.-D. Bièvre: *Equipartition of the eigenfunctions of quantized ergodic maps on the torus*, Commun. Math. Phys. **178** (1996) 83–105.
32. S. De Bièvre, M. Degli Esposti and R. Giachetti: *Quantization of a class of piecewise affine transformations on the torus.*, Commun. Math. Phys. **176** (1996) 73–94.
33. J. Marklof and Z. Rudnick: *Quantum unique ergodicity for parabolic maps*, preprint IHES/M/99/01, math-ph/9901001 (1999). To appear in Geometric and Functional Analysis.
34. A. Bäcker and G. Haag: *Spectral statistics for quantized skew translations on the torus,*, J. Phys. A **32** (1999) L393-L398.
35. F. Steiner: *Quantum chaos*, in *Schlaglichter der Forschung. Zum 75. Jahrestag der Universität Hamburg 1994*, (ed. R. Ansorge) Festschrift published on the occasion of the 75th anniversary of the University of Hamburg, pp. 543–564, Dietrich Reimer Verlag, Berlin und Hamburg (1994).

36. R. Aurich, J. Bolte and F. Steiner: *Universal signatures of quantum chaos*, Phys. Rev. Lett. **73** (1994) 1356–1359.
37. J. Bolte: *Semiclassical trace formulae and eigenvalue statistics in quantum chaos*, Open Syst. Inf. Dynamics **6** (1999) 167–226.
38. A. Selberg: *Contributions to the theory of the Riemann zeta-function*, Arch. Math. Naturvid. **48** (1946) 89–155.
39. H. L. Montgomery: *Selberg's work on the zeta-function*, in: *Number theory, trace formulas and discrete groups (Symposium in Honor of Atle Selberg, Oslo, July 14-21, 1987)* (Eds. K. E. Aubert, E. Bombieri and D. Goldfeld), pp. 157–168, Academic Press, (1989).
40. A. Selberg: *Collected Papers, Vol. II*, Springer-Verlag, Heidelberg, (1991).
41. P. M. Bleher, F. J. Dyson and J. L. Lebowitz: *Non-Gaussian energy level statistics for some integrable systems*, Phys. Rev. Lett. **71** (1993) 3047–3050.
42. P. M. Bleher: *Trace formula for quantum integrable systems, lattice-point problem, and small divisors*, in: *Emerging Applications of Number Theory* [135], pp. 1–38.
43. A. S. Besicovitch: *Almost Periodic Functions*, Dover Pub., Cambridge, (1954).
44. P. M. Bleher: *On the distribution of the number of lattice points inside a family of convex ovals*, Duke Math. J. **67** (1992) 461–481.
45. G. H. Hardy: *The average order of the arithmetical functions $P(x)$ and $\Delta(x)$*, Proc. London Math. Soc. **15** (1915) 192–213.
46. M. N. Huxley: *Exponential sums and lattice points*, Proc. London Math. Soc. **66** (1993) 279–301.
47. D. R. Heath-Brown: *The distribution and moments of the error term in the Dirichlet divisor problem*, Acta Arithmetica **60** (1992) 389–415.
48. R. Aurich: private communication.
49. A. M. Odlyzko: private communication.
50. H. Alt, A. Bäcker, C. Dembowski, H.-D. Gräf, R. Hofferbert, H. Rehfeld and A. Richter: *Mode fluctuation distribution for spectra of superconducting microwave billiards*, Phys. Rev. E **58** (1998) 1737–1742.
51. R. Schubert: *The trace formula and the distribution of eigenvalues of Schrödinger operators on manifolds all of whose geodesics are closed*, DESY report, DESY 95–090.
52. M. C. Gutzwiller: *Periodic orbits and classical quantization conditions*, J. Math. Phys. **12** (1971) 343–358.
53. M. Sieber and F. Steiner: *Classical and quantum mechanics of a strongly chaotic billiard system*, Physica D **44** (1990) 248–266.
54. M. Sieber and F. Steiner: *Generalized periodic-orbit sum rules for strongly chaotic systems*, Phys. Lett. A **144** (1990) 159–163.
55. Y. Colin de Verdière: *Spectre du laplacien et longueurs des géodésiques périodiques I* (in French), Compositio Math. **27** (1973) 83–106.
56. Y. Colin de Verdière: *Spectre du laplacien et longueurs des géodésiques périodiques II* (in French), Compositio Math. **27** (1973) 159–184.
57. J. Chazarain: *Formule de Poisson pour les variétés Riemanniennes* (in French), Invent. Math. **24** (1974) 65–82.
58. J. J. Duistermaat and V. W. Guillemin: *The spectrum of positive elliptic operators and periodic bicharacteristics*, Invent. Math. **29** (1975) 39–79.
59. V. Guillemin and A. Uribe: *Reduction and the trace formula*, J. Differ. Geom. **32** (1990) 315–347.

60. R. Brummelhuis and A. Uribe: *A semi-classical trace formula for Schrödinger operators*, Commun. Math. Phys. **136** (1991) 567–584.
61. T. Paul and A. Uribe: *Sur la formule semi-classique des traces* (in French), C. R. Acad. Sci., Paris, Ser. I **313** (1991) 217–222.
62. E. Meinrenken: *Semiclassical principal symbols and Gutzwiller's trace formula*, Rep. Math. Phys. **31** (1992) 279–295.
63. R. Aurich, M. Sieber and F. Steiner: *Quantum chaos of the Hadamard-Gutzwiller model*, Phys. Rev. Lett. **61** (1988) 483–487.
64. R. Aurich and F. Steiner: *Periodic-orbit sum rules for the Hadamard-Gutzwiller model*, Physica D **39** (1989) 169–193.
65. R. Aurich and F. Steiner: *From classical periodic orbits to the quantization of chaos*, Proc. R. Soc. London Ser. A **437** (1992) 693–714.
66. R. Aurich and F. Steiner: *Staircase functions, spectral rigidity and a rule for quantizing chaos*, Phys. Rev. A **45** (1992) 583–592.
67. R. Aurich and F. Steiner: *Statistical properties of highly excited quantum eigenstates of a strongly chaotic system*, Physica D **64** (1993) 185–214.
68. C. Matthies: *Picards Billard. Ein Modell für Arithmetisches Quantenchaos in drei Dimensionen*, Ph.D. thesis, II. Institut für Theoretische Physik, Universität Hamburg, (1995).
69. J. Marklof: *On multiplicities in length spectra of arithmetic hyperbolic three-orbifolds*, Nonlinearity **9** (1996) 517–536.
70. R. Aurich and J. Marklof: *Trace formulae for three-dimensional hyperbolic lattices and application to a strongly chaotic tetrahedral billiard*, Physica D **92** (1996) 101–129.
71. R. Aurich and M. Taglieber: *On the rate of quantum ergodicity on hyperbolic surfaces and for billiards*, Physica D **118** (1998) 84–102.
72. M. Sieber and F. Steiner: *Quantum chaos in the hyperbola billiard*, Phys. Lett. A **148** (1990) 415–419.
73. M. Abramowitz and I. A. Stegun (eds.): *Pocketbook of Mathematical Functions*, Verlag Harri Deutsch, Thun – Frankfurt/Main, abridged edn., (1984).
74. R. Aurich, C. Matthies, M. Sieber and F. Steiner: *Novel rule for quantizing chaos*, Phys. Rev. Lett. **68** (1992) 1629–1632.
75. A. Bäcker: *Classical and Quantum Chaos in Billiards*, Ph.D. thesis, Abteilung Theoretische Physik, Universität Ulm, (1998).
76. L. A. Bunimovich and Ya. G. Sinai: *Markov partitions for dispersed billiards*, Commun. Math. Phys. **78** (1980) 247–280, erratum, ibid. **107** (1986) 357–358.
77. L. A. Bunimovich, Ya. G. Sinai and N. I. Chernov: *Markov partitions for two-dimensional hyperbolic billiards*, Russ. Math. Surveys **45** (1990) 105–152.
78. T. Krüger and S. Troubetzkoy: *Markov partitions and shadowing for non-uniformly hyperbolic systems with singularities*, Ergodic Theory Dynam. Systems **12** (1992) 487–508.
79. T. Krüger and S. Troubetzkoy: *Symbolic dynamics via shadowing for diffeomorphisms with non zero exponents*, preprint (1997).
80. V. M. Alekseev and M. V. Yakobson: *Symbolic dynamics and hyperbolic dynamic systems*, Physics Reports **75** (1981) 287–325.
81. A. Bäcker and N. Chernov: *Generating partitions for two-dimensional hyperbolic maps*, Nonlinearity **11** (1998) 79–87.
82. O. Biham and M. Kvale: *Unstable periodic orbits in the stadium billiard*, Phys. Rev. A **46** (1992) 6334–6339.

83. H. Bruus and N. D. Whelan: *Edge diffraction, trace formulae and the cardioid billiard*, Nonlinearity **9** (1996) 1023–1047.
84. A. Bäcker and H. R. Dullin: *Symbolic dynamics and periodic orbits for the cardioid billiard*, J. Phys. A **30** (1997) 1991–2020.
85. P. Cvitanović, G. Gunaratne and I. Procaccia: *Topological and metric properties of Hénon-type strange attractors*, Phys. Rev. A **38** (1988) 1503–1520.
86. A. de Carvalho: *Pruning fronts and the formation of horseshoes*, Ergodic Theory Dynam. Systems **19** (1999) 851–894.
87. Y. Ishii: *Towards the kneading theory for Lozi mappings I: A solution of the pruning front conjecture and the first tangency problem*, Nonlinearity **10** (1997) 731–747.
88. E. P. Wigner: *On the quantum correction for thermodynamic equilibrium*, Phys. Rev. **40** (1932) 749–759.
89. M. V. Berry: *Regular and irregular semiclassical wavefunctions*, J. Phys. A **10** (1977) 2083–2091.
90. A. Voros: *Semi-classical ergodicity of quantum eigenstates in the Wigner representation*, in: *Stochastic Behavior in Classical and Quantum Hamiltonian Systems*, no. 93 in Lecture Notes in Physics, 326–333, Springer-Verlag, Berlin, (1979).
91. M. V. Berry: *Semiclassical mechanics of regular and irregular motion*, in: *Comportement Chaotique des Systèmes Déterministes — Chaotic Behaviour of Deterministic Systems* (Eds. G. Iooss, R. H. G. Hellemann and R. Stora), 171–271, North-Holland, Amsterdam, (1983).
92. V. I. Arnold: *Modes and quasimodes*, Funct. Anal. Appl. **6** (1972) 94–101.
93. V. F. Lazutkin: *Asymptotics of the eigenvalues of the Laplacian and quasimodes. A series of quasimodes corresponding to a system of caustics close to the boundary*, Math. USSR Izv. **7** (1973) 439–466.
94. Y. Colin de Verdière: *Quasi-modes sur les variétés Riemanniennes* (in French), Invent. Math. **43** (1977) 15–52.
95. V. F. Lazutkin: *KAM theory and semiclassical approximations to eigenfunctions*, Springer-Verlag, Berlin, (1993).
96. A. I. Shnirelman: *Ergodic properties of eigenfunctions* (in Russian), Usp. Math. Nauk **29** (1974) 181–182.
97. A. I. Shnirelman: *On the asymptotic properties of eigenfunctions in the regions of chaotic motion*, in: V. F. Lazutkin: *KAM Theory and Semiclassical Approximations to Eigenfunctions*, Springer-Verlag Berlin (1993).
98. S. Zelditch: *Uniform distribution of eigenfunctions on compact hyperbolic surfaces*, Duke. Math. J. **55** (1987) 919–941.
99. Y. Colin de Verdière: *Ergodicité et fonctions propres du laplacien* (in French), Commun. Math. Phys. **102** (1985) 497–502.
100. B. Helffer, A. Martinez and D. Robert: *Ergodicité et limite semi-classique* (in French), Commun. Math. Phys. **109** (1987) 313–326.
101. P. Gérard and E. Leichtnam: *Ergodic properties of eigenfunctions for the Dirichlet problem*, Duke Math. J. **71** (1993) 559–607.
102. S. Zelditch and M. Zworski: *Ergodicity of eigenfunctions for ergodic billiards*, Commun. Math. Phys. **175** (1996) 673–682.
103. A. Knauf and Ya. G. Sinai (with a contribution by V. Baladi): *Classical Nonintegrability, Quantum Chaos*, DMV–Seminar 27, Birkhäuser, Basel, (1997).

104. A. Bäcker, R. Schubert and P. Stifter: *Rate of quantum ergodicity in Euclidean billiards*, Phys. Rev. E **57** (1998) 5425–5447, erratum ibid. **58** (1998) 5192.
105. L. Hörmander: *The Analysis of Linear Partial Differential Operators III*, Springer-Verlag, Berlin, Heidelberg, (1985).
106. G. B. Folland: *Harmonic Analysis in Phase Space*, vol. 122 of *Annals of Mathematics Studies*, Princeton University Press, Princeton, (1989).
107. M. E. Taylor: *Pseudodifferential Operators*, no. 34 in Princeton Mathematical Series, Princeton University Press, Princeton, New Jersey, (1981).
108. R. Schubert: *Mikrolokale Analysis und Spurformeln*, (1996), Diploma thesis, II. Institut für Theoretische Physik, Universität Hamburg.
109. M. Degli Esposti, S. Graffi and S. Isola: *Classical limit of the quantized hyperbolic toral automorphisms*, Commun. Math. Phys. **167** (1995) 471–507.
110. S. De Bièvre and M. Degli Esposti: *Egorov theorems and equidistribution of eigenfunctions for the quantized sawtooth and baker maps*, Ann. Inst. Henri Poincaré, Physique Théorique **69** (1996) 1–30.
111. A. Bäcker and R. Schubert: *Chaotic eigenfunctions in momentum space*, J. Phys. A **32** (1999) 4795–4815.
112. Z. Rudnick and P. Sarnak: *The behaviour of eigenstates of arithmetic hyperbolic manifolds*, Commun. Math. Phys. **161** (1994) 195–213.
113. R. Aurich and F. Steiner: *Quantum eigenstates of a strongly chaotic system and the scar phenomenon*, Chaos, Solitons and Fractals **5** (1995) 229–255.
114. O. M. Auslaender and S. Fishman: *Exact eigenfunctions of a chaotic system*, Physica D **128** (1999) 180–223.
115. E. J. Heller: *Bound-state eigenfunctions of classically chaotic Hamiltonian systems: Scars of periodic orbits*, Phys. Rev. Lett. **53** (1984) 1515–1518.
116. S. W. McDonald and A. N. Kaufmann: *Wave chaos in the stadium: Statistical properties of short-wave solutions of the Helmholtz equation*, Phys. Rev. A **37** (1988) 3067–3086.
117. R. Aurich, A. Bäcker, R. Schubert and M. Taglieber: *Maximum norms of chaotic quantum eigenstates and random waves*, Physica D **129** (1999) 1–14.
118. A. Bäcker, R. Schubert and P. Stifter: *On the number of bouncing-ball modes in billiards*, J. Phys. A **30** (1997) 6783–6795.
119. G. Tanner: *How chaotic is the stadium billiard? A semiclassical analysis*, J. Phys. A **30** (1997) 2863–2888.
120. R. Aurich: *The fluctuations of the cosmic microwave background for a compact hyperbolic universe*, Astrophys. J. **524** (1999) 497-503.
121. M. Sieber: *Billiard systems in three dimensions: the boundary integral equation and the trace formula*, Nonlinearity **11** (1998) 1607–1623.
122. S. Brandis: *Chaos in a Coulombic muffin-tin potential*, Phys. Rev. E **51** (1995) 3023–3031.
123. S. Brandis: *Classical and Quantum Chaotic Scattering in a Muffin–Tin Potential*, Ph.D. thesis, II. Institut für Theoretische Physik, Universität Hamburg, (1995), DESY report DESY 95–101.
124. H. Ninnemann: *Gutzwiller's octagon and the triangular billiard* $T^*(2,3,8)$ *as models for the quantization of chaotic systems by Selberg's trace formula*, Int. J. Mod. Phys B **9** (1995) 1647–1753.
125. J. Marklof: *Limit theorems for theta sums*, Duke Mathematical Journal **97** (1999) 127–153.

126. J. Marklof: *Limit Theorems for Theta Sums with Applications in Quantum Mechanics*, Shaker Verlag, Aachen, (1997), Dissertation, Universität Ulm, 1997.
127. J. Marklof: *Theta sums, Eisenstein series, and the semiclassical dynamics of a precessing spin*, in: *Emerging Applications of Number Theory* [135], pp. 405–450.
128. P. Stifter, C. Leichtle, W. P. Schleich and J. Marklof: *Das Teilchen im Kasten: Strukturen in der Wahrscheinlichkeitsdichte*, Zeitschrift für Naturforschung **52a** (1997) 377–385.
129. M. Sieber: *Semiclassical transition from an elliptical to an oval billiard*, J. Phys. A **30** (1997) 4563–4596.
130. H. Schomerus and M. Sieber: *Bifurcations of periodic orbits and uniform approximations*, J. Phys. A **30** (1997) 4537–4562.
131. M. Sieber and H. Schomerus: *Uniform approximation for period-quadrupling bifurcations*, J. Phys. A **31** (1998) 165–183.
132. M. Sieber, N. Pavloff and C. Schmit: *Uniform approximation for diffractive contributions to the trace formula in billiard systems*, Phys. Rev. E **55** (1997) 2279–2299.
133. M. Brack, S. M. Reimann and M. Sieber: *Semiclassical interpretation of the mass asymmetry in nuclear fission*, Phys. Rev. Lett. **79** (1997) 1817–1820.
134. M. Brack, P. Meier, S. M. Reimann and M. Sieber: *Manifestation of classical orbits in nuclei, metal clusters and quantum dots*, in: *Proceedings of the International Symposium on Similarities and Differences between Atomic Nuclei and Microclusters, Tsukuba, July 1-4, 1997* (Eds. Y. Abe, I. Arai, S. M. Lee and K. Yabana), pp. 17–28. American Institute of Physics, (1998).
135. D. A. Hejhal, J. Friedman, M. C. Gutzwiller and A. M. Odlyzko (eds.): *Emerging Applications of Number Theory*, The IMA Volumes in Mathematics and its Applications, Vol. **109**, Springer-Verlag, (1999).

Periodic Orbits and Attractors for Autonomous Reaction-Diffusion Systems

Matthias Büger*

Mathematisches Institut der Justus-Liebig-Universität Gießen, Arndtstraße 2, 35392 Gießen, Germany

1 Problem and Motivation

The dynamics of solutions of one autonomous reaction-diffusion equation

$$\frac{d}{dt}y = \lambda \Delta y + g(y) \tag{1}$$

with Dirichlet or other boundary conditions on an interval $\Omega = (0,1)$ has been examined by many authors [6], [8], [9], [12], [17]. In this case, all solutions converge to the set of stationary points.

In our project, we examine systems of two reaction-diffusion equations of the form

$$\frac{d}{dt}\begin{pmatrix} u \\ v \end{pmatrix} = \lambda \Delta \begin{pmatrix} u \\ v \end{pmatrix} + f\begin{pmatrix} u \\ v \end{pmatrix}, \quad t > 0, \tag{2}$$

with Dirichlet boundary conditions where the diffusion parameter λ is positive and the non-linearity $f : \mathbb{R}^2 \to \mathbb{R}^2$ is sufficiently smooth. If we have $f(0,0) = (0,0)$ and f is at least C^1, then we can write f in the form

$$f\begin{pmatrix} u \\ v \end{pmatrix} = \begin{pmatrix} u \\ v \end{pmatrix} F\begin{pmatrix} u \\ v \end{pmatrix} + \begin{pmatrix} -v \\ u \end{pmatrix} G\begin{pmatrix} u \\ v \end{pmatrix}$$

with functions $F, G : \mathbb{R}^2 \to \mathbb{R}$. We call F the radial and G the tangential part of f.

If we have $G \equiv 0$, then we can show that all (bounded) solutions of (2) tend to (the set of) stationary solutions, i.e. we get dynamics similar to the dynamics of one reaction-diffusion equation (1). Hence, a difference in the dynamics between one equation (1) and a system (2) can only be caused by a non-trivial tangential part G. Thus, the easiest system (2) in which we can expect more complicated dynamics is given by the model system

$$\frac{d}{dt}\begin{pmatrix} u \\ v \end{pmatrix} = \lambda \Delta \begin{pmatrix} u \\ v \end{pmatrix} + \begin{pmatrix} u \\ v \end{pmatrix} F\begin{pmatrix} u \\ v \end{pmatrix} + \begin{pmatrix} -v \\ u \end{pmatrix}, \quad t > 0, \tag{3}$$

where the tangential part G of f is constantly 1. The examination of the dynamics of (3) can be looked at as a first step in order to understand the dynamics of the general case (2).

* Project: Periodic Orbits and Attractors for Autonomous Reaction-Diffusion-Systems (Hans-Otto Walther)

Since the dynamics of (2) is relatively simple if $G \equiv 0$ but not at all understood in the general case, system (3) is the easiest case which has not been observed yet, and its dynamics is therefore of interest from the theoretical point of view.

We study (3) under Dirichlet boundary conditions first on an interval $(0, 1)$, later on some bounded and smooth domain $\Omega \subset \mathbb{R}^n$ with $n > 1$. It is possible to make a similar examination for Neumann or other boundary conditions, but the Dirichlet boundary conditions seem to be interesting since they are more restrictive which means that even the existence of stationary or periodic solutions besides the zero solution is a non-trivial question.

Since we are interested in the behaviour of bounded solutions only, we assume that F is chosen in a way that all solutions stay bounded. For example, this is satisfied if F itself is bounded.

Whenever we omit a proof or give only a sketch, the full proof can be found in [4].

2 New Methods: Planar Solutions and Torsion Numbers

2.1 Oscillation Numbers

Many results about the dynamics of one single reaction-diffusion equation, in particular global ones, are based on oscillation number properties which can only be applied in the case of one equation and for a one-dimensional domain Ω. First steps and basic results in this theory were made by Nickel [15] in 1962, Matano [13] in 1982 and Henry [10] in 1985. The work of Angenent [1] in 1988 brought the theory to some kind of final state. The concept of oscillation numbers works as follows: If $y \in C([0,T) \times (0,1), \mathbb{R})$ is a solution of a linear equation of the form

$$\frac{d}{dt} y = \lambda \Delta y + q(t,x) y \tag{4}$$

with Dirichlet or other boundary conditions on an interval $\Omega = (0,1)$, then the number of sign-changes $Z(t)$ of $y(t, \cdot)$ is defined by

$$Z(t) := \sup\{k \in \mathbb{N} : \exists\, 0 < x_1 < \ldots < x_k < 1 : \\ y(t, x_j) y(t, x_{j+1}) < 0 \;\forall\, 1 \le j \le k-1\}.$$

It is remarkable that $Z(t)$ is finite for all $t > 0$, no matter how many zeros the inital state $y(0, \cdot)$ has, as Angenent [1] proved. The most important result is that $Z : (0, \infty) \to \mathbb{N}_0$ is a non-increasing function. Precisely, the value of $Z(t)$ decreases whenever $y(t, \cdot)$ has a multiple zero. The following figure gives an idea how the number of zeros drops at $t = t_0$.

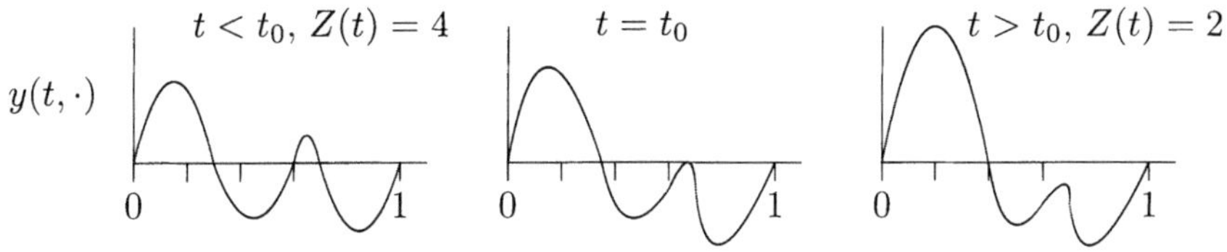

Fig. 1. Decreasing of the oscillation number

These so called oscillation number results are useful since they can be applied to the linearization of (1) at some stationary solution as well as to equation (1) itself if g in (1) can be written in the form $g(y) = y\hat{g}(y)$ (which is, for example, the case if we have $g \in C^1$ and $g(0) = 0$). The problem which we have to face is that these oscillation number results cannot be generalized to a system of two equations like our system (3). Thus, we want to use the special form of the model system (3) in order to develop a method which allows us to use the ideas of oscillation number results for our system.

2.2 Planar Solutions

It is easy to see that there is one class of solutions of system (3) which can be examined by looking only at one single equation: Solutions (u, v) which can be written in the form

$$\begin{pmatrix} u \\ v \end{pmatrix}(t) = r(t)\begin{pmatrix} \cos(t + \varphi_0) \\ \sin(t + \varphi_0) \end{pmatrix} \tag{5}$$

where r is a state function and φ_0 is some real constant. One can easily check that (5) is a solution of (3) if and only if r solves the evolution equation

$$\frac{d}{dt}r = \lambda \Delta r + rF(r\cos(t + \varphi_0), r\sin(t + \varphi_0)) \tag{6}$$

with Dirichlet boundary conditions. If (u, v) is a solution of the form (5), then for every t the states $u(t, \cdot) = r(t, \cdot)\cos(t + \varphi_0)$ and $v(t, \cdot) = r(t, \cdot)\sin(t + \varphi_0)$ are linearly dependent. This means that for all t the curve

$$\gamma_t : [0, 1] \ni x \mapsto (x, u(t, x), v(t, x)) \in [0, 1] \times \mathbb{R}^2$$

lies in a plane, which is illustrated in the next figure.

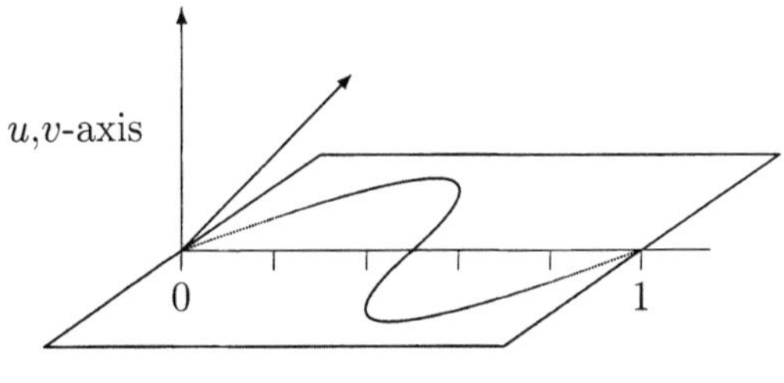

Fig. 2. Curve γ_t for a planar solution (u, v)

Since γ_t lies in a plane for all t, we call solutions of the form (5) planar solutions.

Let $X_2 := \{(u_0, v_0) \in C^1([0,1], \mathbb{R}^2) : u_0(0) = u_0(1) = v_0(0) = v_0(1) = 0\}$ be the underlying space. The index 2 indicates that X_2 is a space of functions which map into $\mathbb{R}^2$; analogously, we put $X_1 := \{u_0 \in C^1([0,1], \mathbb{R}) : u_0(0) = u_0(1) = 0\}$. If we want to prove existence and smoothness of the semiflow of solutions on some space, then Hilbert spaces such as, for example, the Sobolev space $H_0^1(\Omega, \mathbb{R}^2)$ would be best to deal with (for existence theory see [9], [11], [16]). There is, however, a chance to construct a semiflow on spaces like C^0 or C^1 (which we need here) using Sobolev imbedding theorems. This was done in [14]. The advantage of working in a C^1-space is that we can deal with C^1-functions better than with distributions.

Then, we can introduce the set $\mathcal{P}$ of all planar points formally by

$$\mathcal{P} := \{(u_0, v_0) \in X_2 : \exists\, \varphi_0 \in \mathbb{R},\ r_0 \in C^1([0,1], \mathbb{R}) \text{ such that } (u_0, v_0) = r_0(\cos\varphi_0, \sin\varphi_0)\}\,.$$

Clearly, a planar solution $(u, v) : [0, \infty) \to X_2$ has its values in the set $\mathcal{P}$. Furthermore, it is not hard to show that $(u_0, v_0) \in \mathcal{P}$ implies that the (uniquely determined) solution of (3) with initial value (u_0, v_0) is planar.

Planar solutions are interesting because they can be examined by looking at only one reaction-diffusion equation (6). We note that (6) is non-autonomous and, thus, its solutions will not tend to stationary solutions, in general; but since the non-linearity in (6) depends periodically on t, we can apply results like [2], [5] in order to determine the dynamics of (6). These results ensure that all solutions of (6) converge either to stationary solutions or to a solution which is time-periodic with period 2π. Hence, every planar solution converges either to the trivial solution or to a (planar) solution which is periodic with period 2π.

We note that the zero solution is unstable if the diffusion constant λ is sufficiently small. If we already know that every solution converges either to the trivial solution or to a (planar) periodic solution, then the fact that the trivial solution does not attract all solutions for sufficiently small λ will imply that (planar) periodic solutions exist for these λ.

2.3 Convergence to Planar Solutions

In the last section we have shown that the set $\mathcal{P}$ of planar points of the underlying space X_2 is (positively) invariant under the semiflow induced by (3), and we are able to give a good description of the dynamics in $\mathcal{P}$. The set $\mathcal{P}$, however, is a small (nowhere dense) subset of X_2, and we are interested in the dynamics of all solutions of (3).

Given an initial state $(u_0, v_0) \in X_2$ and the corresponding solution $(u, v) : [0, \infty) \to X_2$ of (3), the ω-limit set

$$\omega(u_0, v_0) := \{(\bar{u}, \bar{v}) : \exists (t_n), t_n \nearrow \infty : (u, v)(t_n) \to (\bar{u}, \bar{v})\ (n \to \infty)\}$$

describes the long-time behaviour of (u, v). The following result will be the crucial point of this paper: It gives a connection between the ω-limit set of any solution of (3) and the set of planar points.

Theorem 1. *For every* $(u_0, v_0) \in X_2$ *we get* $\omega(u_0, v_0) \subset \mathcal{P}$.

Interpretation of Theorem 1. Loosely speaking, Theorem 1 says that every solution of (3) will finally approach the planar set $\mathcal{P}$, i.e. in the long term every solution will become 'nearly' planar. This has the effect that knowledge about planar solutions can be used in order to describe the long time behaviour of all solutions.

Sketch of the proof. The proof of Theorem 1 is long and technical, but since this result is a key for all further examinations, we want to present a sketch of the proof.

First we note that if (u, v) is the solution of (3) with initial value (u_0, v_0), then (w, z) defined by

$$\begin{pmatrix} w \\ z \end{pmatrix}(t) := \begin{pmatrix} \cos t & \sin t \\ -\sin t & \cos t \end{pmatrix} \begin{pmatrix} u \\ v \end{pmatrix}(t)$$

solves the rotated system

$$\frac{d}{dt}\begin{pmatrix} w \\ z \end{pmatrix} = \lambda\Delta\begin{pmatrix} w \\ z \end{pmatrix} + \begin{pmatrix} w \\ z \end{pmatrix} F\left(\begin{pmatrix} \cos t & -\sin t \\ \sin t & \cos t \end{pmatrix}\begin{pmatrix} w \\ z \end{pmatrix}\right). \tag{7}$$

Now let us assume that $\omega(u_0, v_0)$ is not contained in $\mathcal{P}$. This means that there is a point $(\bar{u}_0, \bar{v}_0) \in \omega(u_0, v_0) \setminus \mathcal{P}$ with corresponding solution $(\bar{u}, \bar{v})$ of (3). Since $(\bar{u}_0, \bar{v}_0)$ is not a planar point, we get $(\bar{u}, \bar{v})(t) \notin \mathcal{P}$ for all $t \geq 0$. We introduce $(\bar{w}, \bar{z})$ as above. Thus, $(\bar{w}, \bar{z})$ solves system (7). In particular, this means that $\bar{w}$ and $\bar{z}$ solve equation

$$\frac{d}{dt}y = \lambda\Delta y + y h(t) \tag{8}$$

with Dirichlet boundary conditions where h is given by

$$h(t) := F\left(\begin{pmatrix} \cos t & -\sin t \\ \sin t & \cos t \end{pmatrix}\begin{pmatrix} \bar{w} \\ \bar{z} \end{pmatrix}(t)\right).$$

Of course, the function h depends on $\bar{w}$ and $\bar{z}$ and, therefore, (8) is not helpful if one wants to determine solutions of (7); but equation (8) shows that

- $\bar{w}$ as well as $\bar{z}$ solve a linear equation of the form (4) and, thus, oscillation number results can be applied;
- $\bar{w}$ and $\bar{z}$ solve the same linear equation which leads to the conjecture that they become equal (up to some multiplicative constant) in the long term.

We want to write $(\bar{w}, \bar{z})$ in polar coordinates; in particular, we want to define an angular function $\bar{\varphi}$ which satisfies

$$\frac{\bar{z}(t,x)}{\bar{w}(t,x)} = \tan\bar{\varphi}(t,x) \qquad \text{for } (\bar{w}(t,x), \bar{z}(t,x)) \neq (0,0). \tag{9}$$

Clearly, we can define $\bar{\varphi}(t,x)$ uniquely up to multiples of π as long as $(\bar{w}(t,x), \bar{z}(t,x)) \neq (0,0)$. The case $(\bar{w}(t,x), \bar{z}(t,x)) = (0,0)$ will, of course, occur, for example at the boundary $x \in \{0,1\}$. Then we can use de l' Hospital's formula and, if $\bar{w}(t,\cdot)$ and $\bar{z}(t,\cdot)$ have only simple zeros at x, define $\bar{\varphi}(t,x)$ by

$$\frac{\bar{z}_x(t,x)}{\bar{w}_x(t,x)} = \tan\bar{\varphi}(t,x)\,.$$

Since $\bar{w}$, $\bar{z}$ solve (8), oscillation number results can be applied. Hence, the number of zeros of $\bar{w}(t,\cdot)$ and $\bar{z}(t,\cdot)$ is finite and decreasing for $t > 0$; furthermore, the number of zeros decreases by at least 1 whenever a multiple zero occurs. In particular, it follows that — from some time on — neither $\bar{w}(t,\cdot)$ nor $\bar{z}(t,\cdot)$ have multiple zeros. These ideas where used in [3] in order to show that there is $t_0 > 0$ and a function $\bar{\varphi} : [t_0,\infty) \times [0,1] \to \mathbb{R}$ such that $\bar{\varphi}(t,\cdot)$ is continuous for all $t \geq t_0$ and (9) is satisfied for all $(t,x) \in [t_0,\infty) \times [0,1]$ for which the expression on the left side of (9) is defined.

Since $(\bar{w}, \bar{z})$ was said to be non-planar, $\bar{\varphi}(t,\cdot)$ will not be constant for any t. We take some $t_1 > t_0$ and assume that

$$\max_{[0,1]} \bar{\varphi}(t_1,\cdot) = 0\,.$$

We note that this maximum exists since $\bar{\varphi}(t_1,\cdot) : [0,1] \to \mathbb{R}$ is a continuous function. Let $x_1 \in [0,1]$ be chosen such that $\bar{\varphi}(t_1,x_1) = 0$. Then (9) leads to the following result:

- If $\bar{w}(t_1,x_1) \neq 0$, then $\bar{z}(t_1,\cdot)$ has a multiple zero at x_1.
- If $\bar{w}(t_1,x_1) = 0$, then we get $\bar{z}(t_1,x_1) = 0$ and

$$\frac{\bar{z}_x(t_1,x_1)}{\bar{w}_x(t_1,x_1)} = 0\,.$$

Hence, $\bar{z}(t_1,\cdot)$ must have a multiple zero at x_1.

Because $\bar{z}(t_1,\cdot)$ has a multiple zero at x_1, the number of sign changes $\bar{Z}(t)$ of $\bar{z}(t,\cdot)$ drops at $t = t_1$. This means that for any $\varepsilon > 0$ the number of sign-changes of $\bar{z}(t_1-\varepsilon,\cdot)$ is larger than the number of sign-changes of $\bar{z}(t_1+\varepsilon,\cdot)$. Since $\bar{Z}$ drops every time a multiple zero occurs and $\bar{Z}$ is finite, this can only happen finitely many times. Hence, for nearly all $\varepsilon > 0$ neither $\bar{z}(t_1-\varepsilon,\cdot)$ nor $\bar{z}(t_1+\varepsilon,\cdot)$ will have a multiple zero — and we will take ε this way. Then it is easy to show that there are neighbourhoods $U^{\pm}$ of $\bar{z}(t_1 \pm \varepsilon,\cdot)$ in X_1 such that all elements in $U^{\pm}$ have the same number of zeros (note that we are working in the C^1-topology). Since (w,z) accumulates at every element of the corresponding ω-limit set and $(\bar{w}(t_1 \pm \varepsilon, \cdot), \bar{z}(t_1 \pm \varepsilon, \cdot))$ both belong to this set, there must be sequences $(t_n^{\pm})$, $t_n^{\pm} \nearrow \infty$, such that (in the space X_2)

$$(w,z)(t_n^{\pm}) \to (\bar{w}(t_1 \pm \varepsilon, \cdot), \bar{z}(t_1 \pm \varepsilon, \cdot)) \qquad (n \to \infty)\,.$$

In particular, we can choose $(t_n^{\pm})$ in a way that

$$z(t_n^{\pm}) \in U^{\pm} \qquad \text{for all } n.$$

Since both sequences $(t_n^{\pm})$ tend to ∞, this implies that the number of sign-changes of $z(t,\cdot)$ is not monotone. But this contradicts the oscillation number results which can be applied to z as a solution of (8).

Clearly, our arguments depend on the assumption that

$$\max_{[0,1]} \bar{\varphi}(t_1, \cdot) = 0\,.$$

In general, this will not be the case. But if $\max_{[0,1]} \bar{\varphi}(t_1,\cdot) = \alpha \neq 0$, then we introduce functions $(\bar{w}_\alpha, \bar{z}_\alpha)$ which satisfy a system similar to (7) and their angular function $\bar{\varphi}_\alpha$ is given by $\bar{\varphi}_\alpha = \bar{\varphi} - \alpha$. Then we get

$$\max_{[0,1]} \bar{\varphi}_\alpha(t_1, \cdot) = 0\,,$$

and we can proceed as above. □

2.4 The Concept of Torsion

In this section we want to explain what — in the sense of dynamics — really forces the convergence of an arbitrary solution to a planar solution. Assume that $(u,v) : [0,\infty) \to X_2$ is a solution of (3) and introduce the rotated functions (w,z) as in the last section. Furthermore, suppose (for simplicity) that $w(0,\cdot)$ is positive in $(0,1)$. Then the curve

$$\gamma_0 : [0,1] \ni x \mapsto (x, w(0,x), z(0,x)) \in [0,1] \times \mathbb{R}^2$$

lies between two planes P^- and P^+.

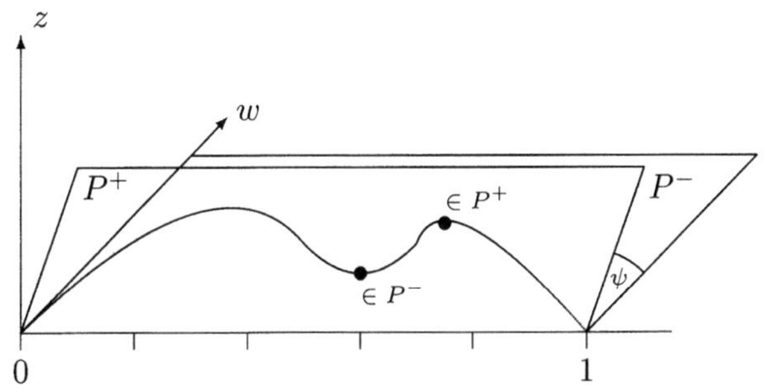

Fig. 3. Curve γ_0 between P^- and P^+

Since the curve γ_0 touches P^- and P^+, we can proceed as in the proof of Theorem 1, and we obtain that γ_t lies strictly between P^- and P^+ for every $t > 0$, i.e. γ_t is not in contact with these two planes for any $t > 0$. This means that if we denote by $P_t^\pm$ the planes which touch the curve γ_t, then the angle between P_t^+ and P_t^- is strictly decreasing. We call this angle *torsion* and denote it by $\psi(t)$. The motivation to name ψ torsion is that it measures in some sense the twist of the curve γ_t. In the notation of the last section, $\psi(t)$ is given by

$$\psi(t) = \max_{[0,1]} \varphi(t,\cdot) - \min_{[0,1]} \varphi(t,\cdot)\,.$$

We note that ψ can, in general, have values larger than 2π if the curve γ_t twists around the x-axis several times. Under some conditions (which are, for example, satisfied if $w(0,\cdot)$ is positive), there is $t_0 > 0$ such that $\psi(t)$ is strictly decreasing on $[t_0, \infty)$ (see [3]), and converges to 0, which means that the solution converges to the set of planar points.

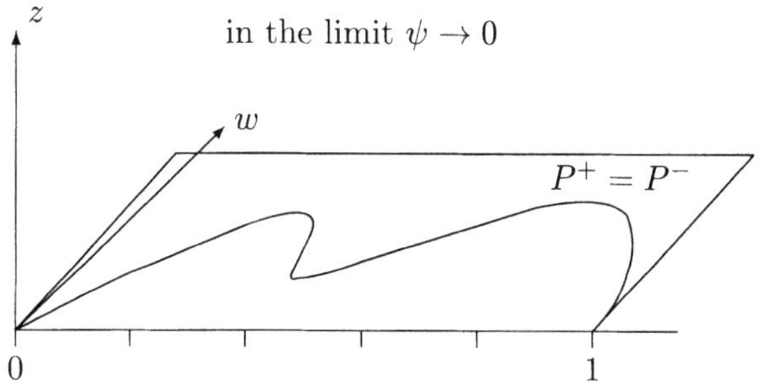

Fig. 4. Limit of γ_t for $t \to \infty$

Although the concept of torsion is closely related to polar coordinates, the proof of the underlying result Theorem 1 does not use a polar coordinates form of system (3) which would be

$$\begin{aligned} \frac{d}{dt} r &= \lambda(r_{xx} - r\varphi_x^2) + rF(r(\cos\varphi, \sin\varphi)) \\ \frac{d}{dt}\varphi &= \lambda\left(\varphi_{xx} + 2\frac{r_x\varphi_x}{r}\right) + 1\,. \end{aligned} \tag{10}$$

Even if we ignore the problems which boundary conditions we have to use for φ and how we have to deal with the singularity in the second equation of (10) if $r = 0$, it is not clear from (10) that φ approaches a function of the form $\bar\varphi(t,x) = t + \bar\varphi_0$ with $\bar\varphi_0 \in \mathbb{R}$. This convergence is in fact a consequence of Theorem 1.

3 Main Results

3.1 A Poincaré-Bendixson Result

We have seen that planar solutions play an important role in the examination of the dynamics of (3). Furthermore, there are Poincaré-Bendixson results [2], [5] (see also [6]) for the single non-autonomous equation (6) and, thus, for planar solutions. It follows from [2] that every solution of (3) converges to a stationary solution or to a solution which is periodic with period p where $2\pi/p$ is an integer. Since all solutions of (3) tend to the set of planar points, it seems easy to conclude that all solutions will converge to some periodic (and planar) solution. Clearly, each ω-limit set contains only planar points (by Theorem 1) and, thus, at least one (planar) stationary solution or (planar) periodic solution. But it is by no means obvious that the ω-limit set does not contain any other (planar) points such as connecting orbits between different planar stationary or periodic solutions. In particular, this could happen if there are heteroclinic cycles in $\mathcal{P}$ as it is shown in the following figure (for the time-2π-map, which means that periodic solutions become fixed points of this map).

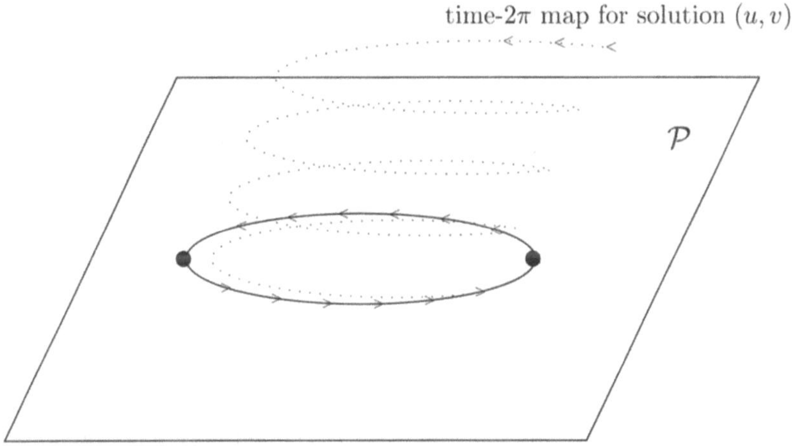

Fig. 5. Situation where a heteroclinic cycle in $\mathcal{P}$ occurs and is contained in the ω-limit set of (u, v)

Hence, the fact that all planar solutions tend to periodic or stationary solutions is not sufficient to conclude that all solution of (3) converge to a stationary or a periodic solution. We need more information about the dynamics of the planar solutions. The work of Chen and Poláčik [5] guarantees that the semiflow on their planar points $\mathcal{P}$ has a gradient like structure which means, in particular, that neither heteroclinic nor homoclinic cycles occur.

This means that a situation like it is shown in figure 5 os not possible. Using that structure on $\mathcal{P}$ and the fact that all solutions of (3) tend to the set of planar points by Theorem 1, we show the following result:

Theorem 2. *Every solution of (3) converges either to zero or to a planar and periodic solution with period 2π. Formally, this means that for every initial value (u_0, v_0) the ω-limit set $\omega(u_0, v_0)$ is contained in $\{(0,0)\} \cup \mathcal{P}_P$ where $\mathcal{P}_P$ is defined by*

$$\mathcal{P}_P := \{(\bar{u}_0, \bar{v}_0) \in \mathcal{P} : \text{the solution } (\bar{u}, \bar{v}) \text{ of (3) with initial value } (\bar{u}_0, \bar{v}_0) \\ \text{is periodic with period } 2\pi\}\,.$$

This result can be looked at as a Poincaré-Bendixson result for the whole system (3).

Proof. We omit the proof since it is long and technical. For a proof, we refer to [4].

3.2 Stability Properties of the Zero Solution and the Periodic Orbits

A stationary solution respectively a periodic orbit is called stable if solutions which start close to them stay close to these sets. In applications, stability properties play an important role since only stable states can be observed numerically.

In order to determine stability properties of stationary or periodic solutions one usually linearizes the equation at this solution and examines the dynamics of this linearization. The operator

$$L_0 : \left\{ \begin{pmatrix} \xi_0 \\ \eta_0 \end{pmatrix} \in C^2([0,1], \mathbb{R}^2) : \begin{pmatrix} \xi_0 \\ \eta_0 \end{pmatrix}(0) = \begin{pmatrix} \xi_0 \\ \eta_0 \end{pmatrix}(1) = 0 \right\} \to C([0,1], \mathbb{R}^2)$$
$$\begin{pmatrix} \xi_0 \\ \eta_0 \end{pmatrix} \mapsto \left(\lambda \Delta + F(0,0) + \begin{pmatrix} 0 & -1 \\ 1 & 0 \end{pmatrix} \right) \begin{pmatrix} \xi_0 \\ \eta_0 \end{pmatrix}$$

is called the linear variational operator at the zero solution because the linearization of system (3) at the zero solution is given by

$$\frac{d}{dt} \begin{pmatrix} \xi \\ \eta \end{pmatrix} = L_0 \begin{pmatrix} \xi \\ \eta \end{pmatrix}. \tag{11}$$

If the spectrum $\sigma(L_0)$ is contained in $\{\zeta \in \mathbb{C} : \operatorname{Re}\zeta \leq -\rho\}$ for some $\rho > 0$, then the zero solution is stable; if L_0 has at least one eigenvalue with positive real part, then the zero solution is unstable. An elementary computation gives

$$\sigma(L_0) = \{-\lambda\pi^2 n^2 + F(0,0) \pm i : n \in \mathbb{N}\}\,.$$

Thus, if we set

$$\lambda_0 := \frac{F(0,0)}{\pi^2},$$

then the zero solution is stable if $\lambda > \lambda_0$ and unstable if $\lambda < \lambda_0$. Furthermore, at $\lambda = \lambda_0$ a stable periodic solution bifurcates out of the origin.

If we assume that, in addition, F decreases along every ray starting at the origin, i.e.

$$\frac{\partial}{\partial \rho} F(\rho(\cos\varphi, \sin\varphi)) < 0 \qquad \text{for all } \rho > 0,\ \varphi \in \mathbb{R}, \tag{12}$$

then we get the following result:

Theorem 3. *Assume that F satisfies (12) and $\lambda \geq \lambda_0$. Then every solution of (3) converges to zero, and the zero solution is stable. In particular, this means that (3) has trivial dynamics.*

The examination of stability properties of a periodic solution (u_p, v_p) is more complicated since the linearization of (3) at $(u_p, v_p)(t) = r_p(t)(\cos t, \sin t)$ is given by the non-autonomous system

$$\begin{aligned} \frac{d}{dt}\begin{pmatrix} u \\ v \end{pmatrix} &= \lambda\Delta\begin{pmatrix} u \\ v \end{pmatrix} + \begin{pmatrix} u \\ v \end{pmatrix} f\left(r_p(t)\begin{pmatrix} \cos t \\ \sin t \end{pmatrix}\right) \\ &\quad + r_p(t)\begin{pmatrix} \cos t \\ \sin t \end{pmatrix}\left[Df\left(r_p(t)\begin{pmatrix} \cos t \\ \sin t \end{pmatrix}\right)\begin{pmatrix} u \\ v \end{pmatrix}\right] + \begin{pmatrix} -v \\ u \end{pmatrix}. \end{aligned} \tag{13}$$

Since (u_p, v_p) is 2π-periodic by Theorem 2, we have to examine the spectrum of the time-2π-map $T : X_2 \to X_2$ corresponding to the variational system (13). Because the time-derivative $(u_p', v_p') := \frac{d}{dt}(u_p, v_p)$ of the periodic solution (u_p, v_p) satisfies $T(u_p', v_p') = (u_p', v_p')$, 1 is contained in the spectrum $\sigma(T)$. This is, in fact, true for all periodic solutions of an autonomous system. If 1 is a simple and isolated eigenvalue of T and $\sigma(T) \setminus \{1\}$ is contained in $\{\zeta \in \mathbb{C} : |\zeta| \leq \rho\}$ for some $\rho < 1$, then the periodic solution is stable; if T has at least one eigenvalue ν with $|\nu| > 1$, then the periodic solution is unstable.

The examination of the spectrum $\sigma(T)$ is much more complicated than the examination of $\sigma(L_0)$ which can actually be calculated. Clearly, system (13) and, thus, $\sigma(T)$ depends on the underlying (planar) periodic solution (u_p, v_p). Therefore, it makes sense to ask which properties of (u_p, v_p) may effect stability or instability of the corresponding periodic orbit.

In the case of one autonomous reaction-diffusion equation (1), stability properties of non-trivial stationary solutions can be determined by looking at their number of sign-changes (i.e. at their oscillation number), provided that the non-linearity g satisfies some monotonicity condition (see [9,p.118]). In this case a stationary solution is stable if it has no sign-change and unstable if there are any. Since all periodic solutions of (3) are planar, it makes sense to examine the sign-changes of the corresponding function r_p.

Theorem 4. *Let $(u_p, v_p)(t) = r_p(t)(\cos t, \sin t)$ be a periodic solution of (3). If $r_p(0) \in X_1$ has at least one sign-change, then the corresponding periodic orbit is unstable.*

Note. Since $r_p(0)$ and $r_p(2\pi) = r_p(0)$ have the same number of sign-changes, oscillation number results ensure that the number of sign-changes of $r_p(t)$ does not depend on t.

Sketch of the proof. The main idea of the proof is not to examine the linear variational system (13) but to look at certain solutions of the original system (3) which start nearby the periodic orbit, namely solutions $(u, v)_\varepsilon$, $\varepsilon > 0$, with initial value

$$(u_0, v_0)_\varepsilon := (u_p, v_p)(0) + \varepsilon(0, \sin(\pi\cdot)) = (r_p(0), \varepsilon \sin(\pi\cdot)) .$$

We note that the curve

$$\gamma_\varepsilon : [0,1] \ni x \mapsto (x, (u_0, v_0)_\varepsilon(x)) \in [0,1] \times \mathbb{R}^2$$

is contained in an angular space

$$A(\alpha_\varepsilon, \beta_\varepsilon) := \{(\xi_0, \eta_0) \in X_2 : (\xi_0, \eta_0) := \rho_0(\cos\varphi_0, \sin\varphi_0) \text{ with } \rho_0 \in C([0,1], \mathbb{R}),\ \rho_0 \geq 0,\ \alpha_\varepsilon \leq \varphi_0 \leq \beta_\varepsilon\}$$

with $0 < \alpha_\varepsilon < \beta_\varepsilon < \pi$, i.e. with torsion $\psi_\varepsilon = \beta_\varepsilon - \alpha_\varepsilon < \pi$. In this situation, the torsion number results [3], which we have explained in the section *The concept of torsion*, ensure that $(u, v)_\varepsilon(2\pi k)$ is contained in angular spaces $A(\alpha_\varepsilon(k), \beta_\varepsilon(k))$ with the property that $(\alpha_\varepsilon(k))$ is strictly increasing while $(\beta_\varepsilon(k))$ is strictly decreasing. Thus, the torsion $\psi_\varepsilon(t)$ converges to zero and $(u, v)_\varepsilon(2\pi k)$ approaches the set

$$P^+(\varphi_0) := \{(\xi_0, \eta_0) \in X_2 : (\xi_0, \eta_0) := \rho_0(\cos\varphi_0, \sin\varphi_0) \text{ with } \rho_0 \in C([0,1], \mathbb{R}),\ \rho_0 \geq 0\} \quad \subset \mathcal{P}$$

with $\alpha_\varepsilon(k), \beta_\varepsilon(k) \to \varphi_0 \quad (k \to \infty)$. Since $r_p(t, \cdot)$ has a sign-change for all t, the periodic orbit Γ_p corresponding to (u_p, v_p) has positive distance (in X_2) to the set $P^+(\varphi_0)$ which shows that no matter how small ε is, $(u, v)_\varepsilon$ will leave some neighbourhood of Γ_p. This means, by definition, that the periodic orbit Γ_p is not stable. □

Note. Theorem 4 holds for all F and, thus, for all systems of the form (3) without further restrictions on F such as monotonicity properties. In this sense, our result is even stronger than the result which we get for one equation. The reason is, loosely speaking, that having a system of two reaction-diffusion equations instead of one equation we can perturb the periodic solution in a normal direction — we examine $(r_p(0), 0) + \varepsilon(0, \sin(\pi\cdot))$ — which cannot be done if we consider only one equation. Hence, the additional direction makes proofs of instability easier while on the other hand proofs of stability become more complicated.

In the case of one equation (1), stability of a positive stationary solution can only be shown under some monotonicity assumption on the non-linearity g. If we did not have any restriction on g, then it would be possible that three positive stationary solutions, two of them stable and one unstable, bifurcate from one positive stationary solution (which had been stable before). Since stability of a planar periodic solution of (3) implies that the corresponding radial function r_p is a stable solution of (6), it is clear that we cannot expect to prove stability of a planar periodic orbit without assuming monotonicity of F. For the rest of this section we assume that (12) is satisfied. Then we can prove

Theorem 5. *If (12) is satisfied, then every (planar) periodic solution* $(u_p, v_p)(t) = r_p(t)(\cos t, \sin t)$ *of (3) with the property that* $r_p(0) \in X_1$ *is positive in* $\Omega = (0,1)$ *is a stable solution (formally, we obtain that the periodic solution is orbitally asymptotically stable).*

Sketch of the proof. We prove stability of the periodic orbit by looking at the spectrum of the period map T of the corresponding linear system (13). We have to show that T has an isolated and simple eigenvalue 1 while all other eigenvalues are contained in some disc around the origin with radius smaller than 1.

Let us assume that $\mu \in \mathbb{C}$ is an eigenvalue of T. Then there is a, in general complex valued, function $(u_0, v_0) \in C^2([0,1], \mathbb{C}^2)$ such that

$$T(u_0, v_0) = \mu(u_0, v_0) .$$

Let $X_2^C := \{R + iI : R, I \in X_2\}$ be the complexification of X_2 and $(u, v) : [0, \infty) \to X_2^C$ the solution of (13) with initial value (u_0, v_0). Clearly, we get this solution if we look at its real and imaginary part as (real valued and, thus, uniquely determined) solutions of (13). Furthermore, we introduce the (complex valued) rotated solution

$$\begin{pmatrix} w \\ z \end{pmatrix}(t) := \begin{pmatrix} \cos t & \sin t \\ -\sin t & \cos t \end{pmatrix} \begin{pmatrix} u \\ v \end{pmatrix}(t)$$

which solves the corresponding rotated system

$$\begin{aligned} \frac{d}{dt}\begin{pmatrix} w \\ z \end{pmatrix} = \lambda\Delta\begin{pmatrix} w \\ z \end{pmatrix} + \begin{pmatrix} w \\ z \end{pmatrix} f\left(r_p(t)\begin{pmatrix} \cos t \\ \sin t \end{pmatrix}\right) \\ +r_p(t)\begin{pmatrix} 1 \\ 0 \end{pmatrix}\left[Df\left(r_p(t)\begin{pmatrix} \cos t \\ \sin t \end{pmatrix}\right)\begin{pmatrix} \cos t & -\sin t \\ \sin t & \cos t \end{pmatrix}\begin{pmatrix} w \\ z \end{pmatrix}\right] . \end{aligned} \tag{14}$$

We note that $(w, z)(0) = (u, v)(0) = (u_0, v_0)$ implies that

$$(w, z)(2\pi) = (u, v)(2\pi) = T(u_0, v_0) = \mu(u_0, v_0)$$

which means that (u_0, v_0) is also an eigenfunction for the time-2π-map associated with the rotated system (14). Although (14) is also not decoupled, it is easier to examine than the original variational system (13). In particular, it follows that v_0 is an eigenfunction of the time-2π-map corresponding to the linear equation

$$\frac{d}{dt}z = \lambda\Delta z + zf\left(r_p(t)\begin{pmatrix}\cos t\\ \sin t\end{pmatrix}\right) \tag{15}$$

which we get by looking at the second equation in (14). Using well known results for one reaction-diffusion equation in one space dimension like oscillation number properties, one can show that the eigenvalues corresponding to the time-2π-map T_z for (15) are real, and the eigenfunction ζ_1 which belongs to the largest eigenvalue μ_z of T_z is (strictly) positive. Furthermore, r_p is an eigenfunction of T_z associated to eigenvalue 1. Using, for example, torsion number results [3] to the solution of system

$$\frac{d}{dt}\begin{pmatrix}y_1\\ y_2\end{pmatrix} = \lambda\Delta\begin{pmatrix}y_1\\ y_2\end{pmatrix} + \begin{pmatrix}y_1\\ y_2\end{pmatrix} f\left(r_p(t)\begin{pmatrix}\cos t\\ \sin t\end{pmatrix}\right) \tag{16}$$

with initial state $(\zeta_1, r_p(0))$, the fact that the torsion number is non-increasing implies $\mu_z \leq 1$. Furthermore, if we have $\mu_z = 1$, then it follows that the torsion is constant and, thus, it has to be zero [3]. This implies that ζ_1 and $r_p(0)$ are linearly dependent. Hence, 1 is the largest eigenvalue and the eigenvalue 1 is simple.

We can proceed as above if we have $v_0 \neq 0$ and obtain that, in this case, the corresponding eigenvalue μ is real, $\mu \leq 1$ and, if $\mu = 1$, then $v_0 = cr_p(0)$ with some constant c. We get this result using the fact that the second equation of (14) is decoupled from the first one.

If, on the other hand, $v_0 = 0$, then u_0 solves the first equation in (14), and the fact that this equation depends on v_0 is no problem in this case. Then we can argue somewhat similarly as above and obtain that the eigenvalue μ must be strictly inside the unit disc.

If we put these results together, we obtain that if $|\mu| \geq 1$, then $v_0 \neq 0$ and $\mu = 1$; in particular, we get $v_0 = cr_p(0)$. If we keep c fixed, then u_0 is uniquely determined because if $u_0, \hat{u}_0$ were different, then (u_0, v_0) and $(\hat{u}_0, v_0)$ and, thus, also their difference $(u_0 - \hat{u}_0, 0)$ would be eigenfunctions associated with eigenvalue 1. This contradicts the result that $v_0 = 0$ implies $|\mu| < 1$.

Since the spectrum of T consists of eigenvalues which accumulate only at 0, the assertion follows. □

3.3 Closing Remarks on the Global Dynamics in One Space Variable

We have seen that every solution of (3) finally reaches a set of planar points. Furthermore, we have proved a Poincaré-Bendixson result for (3) which means that every solution tends either to zero or to a (planar) periodic solution. In the last section we have also shown that stability properties of a planar perodic solution $r_p(t)(\cos t, \sin t)$ can be determined by looking at the number of zeros of the underlying function $r_p(t) \in X_1$.

These results make us assume that only the planar points are important for the examination of the dynamics of (3). But this is only partly true: In order to describe the whole dynamics of (3), we have to examine the global attractor A, which is a maximal compact invariant set which attracts each bounded subset B of X_2. Formally, we introduce

$$\omega(B) := \bigcap_{t\geq 0} \overline{\bigcup_{s\geq t} T(s,0)B}$$

where $T(\tau_1, \tau_2)$ is the evolution operator for system (3). Then, for every bounded B, the set $\omega(B)$ is contained in the global attractor A.

In general, a global attractor need not to exist at all, but if there is a compact subset of X_2 which attracts all bounded subsets of X_2, then there is a global attractor. One can show that a global attractor for (3) exists if F is bounded. For a more detailed review on global attractors see [7].

Clearly, the ω-limit set $\omega(u_0, v_0)$ is contained in $\mathcal{P}$ for any $(u_0, v_0) \in X_2$, but if we look at $\omega(B)$ for bounded subsets B of X_2, then $\omega(B)$ will, in general, not be contained in $\mathcal{P}$. This has the consequence that the global attractor A is not contained in $\mathcal{P}$, either. We can construct a non-planar element of A as follows:

Assume that there is a periodic solution $r(t)(\cos t, \sin t)$ of (3) such that $r(0) \in X_1$ has at least one sign change in $(0, 1)$. Such a solution will exist if λ is sufficiently small. As shown in Theorem 4, solutions of (3) which start at

$$(u_0, v_0)_\varepsilon := (r(0), 0) + \varepsilon(0, \sin(\pi \cdot))$$

will not stay close to our perodic orbit. These solutions will converge to a (planar) perodic orbit without sign-changes. Hence, there is an orbit which connects these two periodic orbits. Close to $(r(0), 0)$ this connecting orbit will look like

$$(r(0), y_\varepsilon)$$

where y_ε is small but (strictly) positive. In particular, $(r(0), y_\varepsilon)$ is contained in A but not in $\mathcal{P}$ because $r(0)$ has at least one sign change in $(0, 1)$ while y_ε is positive in $(0, 1)$ and, hence, $r(0)$ and y_ε are linearly independent. Since this connecting orbit is contained in the global atractor A, we get $A \not\subset \mathcal{P}$, in general. Hence, the dynamics of system (3) are more complicated than the dynamics of one equation (6). Therefore, it is even more surprising that a major part of the dynamics, i.e. the limit set $\omega(u_0, v_0)$ for every $(u_0, v_0) \in X_2$, can be described by looking at planar solutions only.

4 Problems in Higher Dimensions

In the previous sections we examined the dynamics of solutions of system (3) in the case that the underlying domain Ω was one-dimensional. For the rest of this article we want to deal with the question what kind of methods we can use and which results we can search for if Ω is bounded and smooth, but $\dim \Omega > 1$.

The main problem which occurs if we want to proceed similarly to the case $\Omega = (0,1)$ is that the oscillation number results are only available in one space variable. Since the concept of torsion numbers and, thus, the proof of Theorem 1 were based on oscillation number results, our methods for $\dim \Omega = 1$ cannot be used if we want to deal with the case $\dim \Omega > 1$.

Up to this point it is not clear whether we only have to search for new methods in order to get results similar to case $\dim \Omega = 1$ or if the dynamics themself is different. The example which we present in the following section shows that there is in fact a fundamental difference in the dynamics between the cases $\dim \Omega = 1$ and $\dim \Omega > 1$.

4.1 A Non-Planar Periodic Solution

If $\dim \Omega = 1$, then Theorem 1 shows that all solutions converge to the set $\mathcal{P}$ of planar points. In particular, there are no non-planar periodic solutions. If we take

$$\Omega = \{(x_1, x_2) \in \mathbb{R}^2 : x_1^2 + x_2^2 < 1\}$$

to be the unit disc in $\mathbb{R}^2$ and put

$$F(u,v) := 1 - u^2 - v^2 ,$$

then we can show that there is $\lambda > 0$ such that (3) has a time-periodic solution (u, v) with the following properties:

- $u(0,(x_1,x_2)) > 0$ for $x_1 > 0$ and $u(0,(x_1,x_2)) < 0$ for $x_1 < 0$,
- $v(0,(x_1,x_2)) = u(0,(x_2,-x_1))$.

In particular, we have $(u,v)(0) \notin \mathcal{P}$ and, thus, (u,v) is not a planar solution.

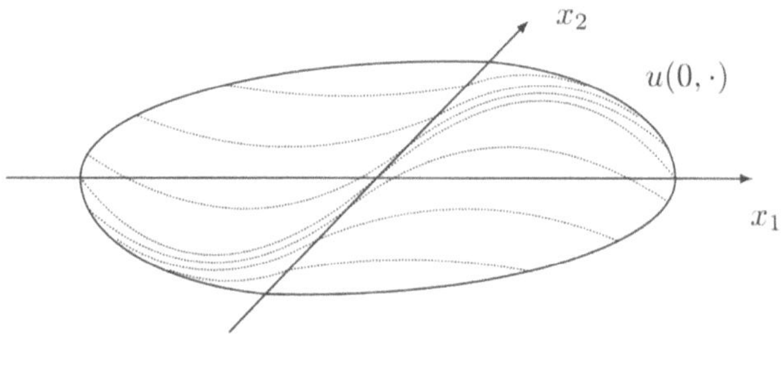

Fig. 6. $u(0,\cdot)$ for the non-planar, periodic solution (u,v)

This example is interesting because of the following reasons:

- The underlying domain Ω as well as the underlying non-linearity F are the most simple case one can think of. If we find complicated dynamics like non-planar periodic solutions in this case, then we will get problems in any case where $\dim \Omega > 1$.
- The dynamics of the non-planar periodic solution can easily be described: If t increases, $u(t, \cdot)$ can be obtained by $u(0, \cdot)$ rotated around $(0, 0)$ with angular t, and $v(t, \cdot)$ always coincides with $u(t, \cdot)$ rotated around $(0, 0)$ with angular $\pi/2$.
- Although the dynamics of (u, v) themself can easily be described, we can by no means reduce the examination of (u, v) to one single reaction-diffusion equation as it is possible in the case of a planar periodic solution.
- We can prove that (u, v) is unstable for some parameters λ, but in general stability properties are hard to examine because we have to deal with a linear variational system which cannot be (partly) decoupled like in the case of a planar periodic solution (see section 3.2).

In particular, this example shows that we cannot get a result like Theorem 1 in the case $\dim \Omega > 1$. Thus, not only the method used in the proof of Theorem 1 fails for $\dim \Omega > 1$, but the result of Theorem 1 itself is not valid in this case.

4.2 Positively Projected Solutions and Their Dynamics

Using the example constructed in the last section we know that we cannot expect results similar to those of case $\dim \Omega = 1$ here; in particular, we will not get convergence to the set of planar points, in general. However, we are able to construct a positively invariant subspace $X^+_{2,\Omega}$ of the underlying space

$$X_{2,\Omega} := \left\{(u_0, v_0) \in C^1(\overline{\Omega}, \mathbb{R}^2) : u_0|_{\partial\Omega} = v_0|_{\partial\Omega} = 0\right\}$$

and prove results similar to the case $\dim \Omega = 1$ for solutions in this subset.

If we look at one reaction-diffusion equation of the form (4) with Dirichlet boundary conditions on Ω, $\dim \Omega > 1$, then the maximum principle ensures that the subset of positive points

$$X^+_{1,\Omega} := \{u_0 \in X_{1,\Omega} : u_0 \geq 0 \text{ in } \Omega\}$$

of $X_{1,\Omega} := \left\{u_0 \in C^1(\overline{\Omega}, \mathbb{R}) : u_0|_{\partial\Omega} = 0\right\}$ is positively invariant under the semiflow defined by (4). Solutions in $X^+_{1,\Omega}$ are often called positive solutions. They play an important role since in many applications (like population models in biology) positivity is assumed.

Unfortunately, the subset of $X_{2,\Omega}$ in which u is positive, the subset in which v is positive or even the subset in which both u and v are positive are not (positively) invariant under the semiflow defined by system (3). The reason is that the 'rotation term' $(-v,u)$ forces a rotation on the values of (u,v) and, thus, the values neither of u nor of v will remain positive, in general. This effect is not really new to us since in the case $\dim \Omega = 1$ the plane which a solution finally reaches also spins around the origin, forced by the same rotation term $(-v,u)$. Hence, it is obvious to look for a positivity property relative to some plane which spins around the origin driven by the term $(-v,u)$ and, therefore, with constant angle velocity 1. Given $(u_0,v_0) \in X_{2,\Omega}$ and $\alpha \in \mathbb{R}$ we introduce

$$\begin{pmatrix} u_\alpha \\ v_\alpha \end{pmatrix} := \begin{pmatrix} \cos\alpha & \sin\alpha \\ -\sin\alpha & \cos\alpha \end{pmatrix} \begin{pmatrix} u_0 \\ v_0 \end{pmatrix}.$$

This means that we get (u_α, v_α) by rotating (u_0,v_0) with the angle α, and (u_0,v_0) can be written as

$$\begin{pmatrix} u_0 \\ v_0 \end{pmatrix} = u_\alpha \begin{pmatrix} \cos\alpha \\ \sin\alpha \end{pmatrix} + v_\alpha \begin{pmatrix} -\sin\alpha \\ \cos\alpha \end{pmatrix}.$$

Then the map $(u_0,v_0) \mapsto u_\alpha(\cos\alpha, \sin\alpha)$ can be looked at as an orthogonal projection of (u_0,v_0) onto the plane

$$P_\alpha := \{(\xi_\alpha,\eta_\alpha) \in X_{2,\Omega} : (\xi_\alpha,\eta_\alpha) \text{ has the form } (\xi_\alpha,\eta_\alpha) = r_\alpha(\cos\alpha,\sin\alpha)\}.$$

If there is some real α such that $u_\alpha \geq 0$ in Ω, then (u_0,v_0) is called *positively projected*, because it can be mapped to the element $u_\alpha(\cos\alpha,\sin\alpha)$ of the plane P_α by an orthogonal projection, where u_α satisfies $u_\alpha \geq 0$ in Ω.

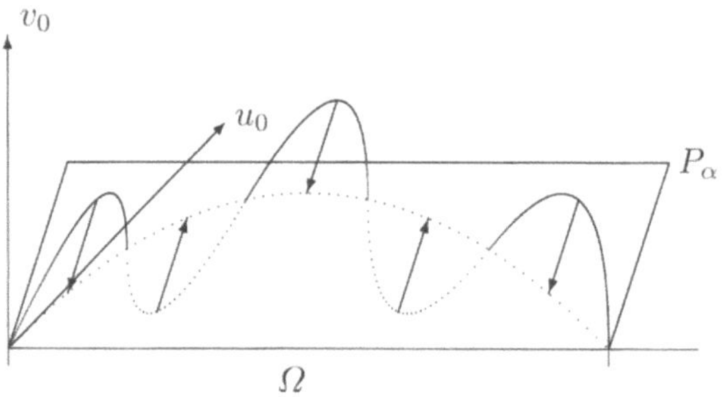

Fig. 7. A positively projected (u_0,v_0)

It is not hard to prove that the set $X^+_{2,\Omega}$ of positively projected solutions is invariant. For a proof, we look at the rotated solution

$$\begin{pmatrix} w_\alpha \\ z_\alpha \end{pmatrix}(t) := \begin{pmatrix} \cos t & \sin t \\ -\sin t & \cos t \end{pmatrix} \begin{pmatrix} u_\alpha \\ v_\alpha \end{pmatrix}(t).$$

Then w_α solves

$$\frac{d}{dt} w_\alpha = \lambda \Delta w_\alpha + w_\alpha F\left(\begin{pmatrix} \cos(t+\alpha) & -\sin(t+\alpha) \\ \sin(t+\alpha) & \cos(t+\alpha) \end{pmatrix}\begin{pmatrix} w_\alpha \\ z_\alpha \end{pmatrix}(t)\right)$$

and $w_\alpha(0,\cdot) \geq 0$. Thus, the maximum principle implies $w_\alpha(t,\cdot) \geq 0$ for all $t \geq 0$, i.e. $(u,v)(t) \in X^+_{2,\Omega}$ for all $t \geq 0$.

This invariance property is the reason why the role of positively projected solutions can be compared to the role of positive solutions in the theory of one reaction-diffusion system without a tangential part (i.e. with $G \equiv 0$).

Defining the set $\mathcal{P}_\Omega$ of planar points of $X_{2,\Omega}$ as in the case $\dim \Omega = 1$, we get the following result:

Theorem 6. *Let $(u,v) : [0,\infty) \to X^+_{2,\Omega}$ be a positively projected solution. Then (u,v) converges to the set of (positively projected and) planar points, i.e.*

$$\omega((u,v)(0)) \subset \mathcal{P}_\Omega \cap X^+_{2,\Omega}\,.$$

In the proof of Theorem 6 we can, roughly speaking, proceed similarly to the proof of Theorem 1, we only have to replace oscillation number results by maximum principle arguments.

Clearly, Theorem 6 is weaker than the result of Theorem 1 since Theorem 6 only deals with positively projected solutions. On the other hand, the example constructed in the last section gives an idea what kind of effects can happen if the underlying solution is not positively projected.

In order to understand the dynamics of all positively projected solutions, we would like to have a result analogous to the Poincaré-Bendixson result Theorem 2. Since the proof of Theorem 2 needs a result of Chen and Poláčik [5] which is based on oscillation number results, we cannot proceed as in the proof of Theorem 2 here. If we restrict ourselfs to non-linearities F which satisfy the monotonicity condition (12) — which restricts the number of periodic orbits and makes uniqueness results possible — then we can proceed as follows:

1. First of all we can show that if $r_p(t)(\cos t, \sin t)$ is a (planar and) positively projected periodic solution of (3) (which means that $r_p(t,\cdot)$ has no sign-change for any t), then the corresponding periodic orbit is stable. Here we use arguments similar to those used in the proof of Theorem 5.
2. With arguments related to the case $\dim \Omega = 1$ we obtain that there is some $\lambda_0 > 0$ such that (3) has trivial dynamics if $\lambda \geq \lambda_0$ and the zero solution is unstable if $\lambda < \lambda_0$.
3. We assume that $\lambda < \lambda_0$ and $(u,v)(t) = r(t)(\cos t, \sin t)$ is a positively projected and planar solution of (3). Using a monotonicity property we show that $r(2\pi j)$, $j = 1,2,\ldots$, will converge to some function $r_0 \not\equiv 0$ which has no sign-change. This means, in particular, that planar and positively projected solutions converge to a periodic orbit.
4. Let (u,v) be a positively projected solution of (3). By Theorem 6, $\omega((u,v)(0))$ is contained in the set $\mathcal{P}_\Omega$ of planar points. Using again monotonicity properties and our knowledge about the behaviour of solutions close to zero, we can show that (u,v) does not converge to zero.

Hence, $\omega((u,v)(0))$ contains at least one planar and positively projected element which is not zero.

5. Assume that $\lambda < \lambda_0$. We note that the ω-limit of any solution is closed and invariant. Hence, if (u,v) is a positively projected solution of (3), steps 3 and 4 together ensure that $\omega((u,v)(0))$ contains a planar and positively projected periodic orbit Γ. By step 1, this periodic orbit is stable. This means that (u,v) must stay (arbitrarily) close to Γ for all (sufficiently) large t and, hence, $\omega((u,v)(0))$ is contained in Γ. This gives $\omega((u,v)(0)) = \Gamma$, i.e. any positively projected solution converges to some planar and positively projected periodic solution.
6. Using the fact that any positively projected solution converges to some planar and positively projected periodic solution by step 5 and that each planar and positively projected periodic orbit is stable by step 1, we can show that if $\lambda < \lambda_0$ then there is exactly one such periodic orbit, which we will denote by Γ_p.

4.3 Open Problems and Further Research

The examination of the dynamics of (3) in the case $\dim \Omega > 1$ is far from being completed. Besides other interesting questions, we concentrate on the following problems:

1. Can the attractor of the positively projected solutions be described?
2. Can we get any results for non-planar (periodic) solutions?
3. What happens to our problem if we introduce different diffusion rates?
4. Can we find a larger class of non-linearities (i.e. with in general non-constant tangential part G) such that our results remain valid?

References

[1] S. Angenent, The zero set of a solution of a parabolic equation, J. reine angew. Math. **390** (1988), 79–96.

[2] P. Brunovský, P. Poláčik, B. Sandstede, Convergence in general periodic parabolic equations in one space dimension, Nonlinear Analysis **18** (1992), 209–215.

[3] M. Büger, Torsion numbers, a tool for the examination of symmetric reaction-diffusion systems related to oscillation numbers, Discr. Cont. Dyn. Sys. **4** (1998), 691–708.

[4] M. Büger, Convergence to periodic solutions in autonomous reaction-diffusion systems in one space variable, to appear in J. Differential Equations.

[5] X.-Y. Chen, P. Poláčik, Gradient-like structure and Morse decompositions for time-periodic one-dimensional parabolic equations, J. Dynam. Diff. Eq. **7** (1995), 73–107.

[6] B. Fiedler, J. Mallet-Paret, A Poincaré-Bendixson theorem for scalar reaction diffusion equations, Arch. Rat. Mech. Anal. **107:4** (1989), 325–345.

[7] J.H. Hale, Asymptotic behavior of dissipative systems, Mathematical surveys and monographs 25, Am. Math. Soc. Providence 1988.

[8] J.K. Hale, G. Raugel, Convergence in gradient-like systems with applications to PDE, Z. Angew. Math. Phys. **43** (1992), 63–124.

[9] D. Henry, Geometric theory of semilinear parabolic equations, Springer-Verlag, Berlin-Heidelberg 1981.

[10] D. Henry, Some infinite dimensional Morse Smale systems defined by parabolic differential equations, J. Differential Equations **59** (1985), 165–205.

[11] A. Lunardi, Analytic semigroups and optimal regularity in parabolic problems, Birkhäuser, Basel-Boston-Berlin 1995.

[12] H. Matano, Convergence of solutions of one-dimensional semilinear equations, J. Math. Kyoto Univ. **18** (1978), 221–227.

[13] H. Matano, Nonincrease of the lap number of a solution for a one-dimensional semi-linear parabolic equation, J. Fac. Sci. Univ. Tokyo, Sec. IA **29** (1982), 401–441.

[14] X. Mora, Semilinear parabolic problems define semiflows on C^k spaces, Trans. AMS **278** (1983), 21–55.

[15] K. Nickel, Gestaltaussagen über Lösungen parabolischer Differentialgleichungen, J. reine angew. Math. **211** (1962), 78–94.

[16] A. Pazy, Semigroups of linear operators and applications to partial differential equations, Springer-Verlag, New York-San Francisco-London 1975.

[17] T.I. Zelenyak, Stabilization of solutions of boundary value problems for a second order parabolic equation with one space variable, Differential Equations **4** (1968), 17–22.

Unconditionally Stable Explicit Schemes for the Approximation of Conservation Laws

Christiane Helzel and Gerald Warnecke*

Institut für Analysis und Numerik, Otto-von-Guericke Universität Magdeburg, 39016 Magdeburg, Germany

Abstract. We consider explicit schemes for homogeneous conservation laws which satisfy the geometric Courant-Friedrichs-Lewy condition in order to guarantee stability but allow a time step with CFL-number larger than one. A brief overview over existing unconditionally stable schemes for hyperbolic conservation laws is provided, although the focus is on LeVeque's large time step Godunov scheme. For this scheme we explore the question of entropy consistency for the approximation of one-dimensional scalar conservation laws with convex flux function and describe a possible way to extend the scheme to the two-dimensional case. Numerical calculations and analytical results show that an increase of accuracy can be obtained because the error introduced by the modified evolution step of the large time step Godunov scheme may be less important than the error due to the projection step.

1 Introduction

The long time unsteady numerical dynamics for hyperbolic conservation laws still presents quite a challenge. Most of the published work in this field deals with either steady flows, i.e. the actual limit of infinite time, or rather short time unsteady flow. The basic problem is that hyperbolic conservation laws have solutions with no dissipation in smooth regions and very little nonlinear dissipation at shocks. Numerical approximation always leads to the introduction of dissipation, which is especially strong at shocks. The numerical propagation of waves may also be distorted by a phase error, see Morton [30] for a recent discussion.

Even one-dimensional problems may provide a challenge to modern high-resolution schemes, when numerical results for long time dynamics are required. Examples of such practical applications are long pneumatic or hydraulic pipes in which piston driven pressure waves move back and forth over relatively large distances interacting with each other. Due to the accumulation of the numerical dissipation and phase error, currently available schemes produce useless results for the longer time periods that have to be considered, see Iben and Tadmor [14].

Other numerical problems arise if very different scales are modeled by the equations, for instance in low Mach number flow calculations. In [15], Klein

* Project: Large Time Dynamics of Systems of Nonlinear Hyperbolic Conservation Laws, and Their Numerical Approximation (Gerald Warnecke)

uses the unconditionally stable Godunov-type scheme of LeVeque, which is reviewed in this paper, to calculate the influence of the fast acoustic waves in such a system.

Finite volume schemes are currently the most widely used basic tool for computational fluid dynamics of compressible flows and multidimenensional schemes for hyperbolic conservation laws in general. The idea of solving Riemann problems was introduced by Godunov [6] and was developed considerably in the last two decades. The available number of variants of this numerical approach is quite large and there are many special techniques with corresponding theoretical basis for various applications.

For long time numerics all possibilities for more efficiency of the calculations have to be exploited. This includes adaptive methods, special solvers and any means of increasing the time steps used without sacrificing accuracy. That is where the large time step method of LeVeque, that will be explained in detail below, is of special interest. It provides a simple means of taking larger time steps for one-dimensional systems with Riemann solver based finite volume schemes.

In Section 2 we provide the basic theoretical and numerical concepts of conservation laws which are used in this paper. A short overview over existing unconditionally stable schemes for conservation laws is given in Section 3 followed by a detailed review of the large time step Godunov-type scheme of LeVeque in Section 4. In order to get an entropy-consistent solution for problems which contain rarefaction waves, it is necessary to use an entropy consistent Riemann solver in every time step. Such Riemann solvers are discussed in some detail. Nevertheless, the question whether the large time step Godunov scheme gives an entropy consistent approximation for 1D nonlinear systems of conservation laws is still unsolved. For the case of scalar conservation laws with convex flux function new results could be obtained, which are discussed in Section 5. An extension of the large time step method for two space dimensions is discussed in Section 6.

2 Basic Concepts for the Numerical Approximation of Conservation Laws

2.1 Hyperbolic Conservation Laws

We consider numerical schemes for solving nonlinear *hyperbolic systems of conservation laws*, i.e. equations of the general form

$$\frac{\partial}{\partial t}\mathbf{u} + \sum_{i=1}^{d} \frac{\partial}{\partial x_i}\mathbf{f}_i(\mathbf{u}) = \mathbf{0} \tag{1}$$

with given initial values

$$\mathbf{u}(\mathbf{x}, 0) = \mathbf{u}_0(\mathbf{x}) \tag{2}$$

where $\mathbf{x} \in \mathbb{R}^d$, $t \in \mathbb{R}^+ = \{t \in \mathbb{R} : t \geq 0\}$ and $\mathbf{u} : \mathbb{R}^d \times \mathbb{R}^+ \to \mathbb{R}^p$ is a vector valued function of conserved quantities, e.g. mass, momentum, and energy for the Euler equations. The vector valued functions $\mathbf{f}_i : \mathbb{R}^p \to \mathbb{R}^p$ are called flux functions.

The physical interpretation of this system of equations can be seen by rewriting (1) into the following *integral form* by using the Gauss theorem

$$\frac{\partial}{\partial t}\int_{\Omega} \mathbf{u} d\mathbf{x} + \sum_{i=1}^{d} \int_{\partial\Omega} \mathbf{f}_i(\mathbf{u}) n_i dS = \mathbf{0}, \tag{3}$$

where Ω is an arbitrary domain of $\mathbb{R}^d$ and $\mathbf{n} = (n_1, \dots, n_d)^T$ is the outward unit normal vector of the boundary $\partial\Omega$ of the considered domain Ω. Now the change of the total amount in Ω of the conserved quantities given by $\int_{\Omega} \mathbf{u} d\mathbf{x}$ is equal to the outward flux through the surface $\partial\Omega$.

Another form, the *quasilinear form*, of (1) can be obtained by carrying out the differentiations of the flux functions, then we get

$$\frac{\partial}{\partial t}\mathbf{u} + \sum_{i=1}^{d} \mathbf{A}_i(\mathbf{u}) \frac{\partial}{\partial x_i}\mathbf{u} = \mathbf{0} \tag{4}$$

where $\mathbf{A}_i(\mathbf{u}) = \mathbf{f}'(\mathbf{u})$ is the Jacobian matrix of the flux function $\mathbf{f}_i$. The d-dimensional system of conservation laws is called *hyperbolic*, if for any $\mathbf{u} \in \Omega$ and any $\mathbf{w} \in \mathbb{R}^d$, $\mathbf{w} \neq \mathbf{0}$ the matrix $\sum_{i=1}^{d} w_i \mathbf{A}_i(\mathbf{u})$ has p real eigenvalues $\lambda_1(\mathbf{u}) \leq \dots \leq \lambda_p(\mathbf{u})$ and a complete set of linearly independent right eigenvectors $\mathbf{r}_1(\mathbf{u}), \dots, \mathbf{r}_p(\mathbf{u})$.

Nice descriptions of the theory of hyperbolic conservation laws as well as the concepts of numerical schemes can be found in the text books of Godlewski, Raviart [4], [5], Kröner [16] and LeVeque [26], [28]. Here we briefly want to cover those definitions and theoretical results which will be needed for our considerations.

The solution to the initial value problem (1), (2) is in general not continuous, i.e. the solution may contain shocks. Therefore we are not only interested in classical, i.e. differentiable solutions of (1), (2) but also in *weak* solutions which allow discontinuities. Though one could develop the integral form (3) into a definition of weak solutions we prefer to use the distributional form.

Definition 1. Assume that the initial values $\mathbf{u}_0$ are in the space of locally bounded measurable functions, i.e. $\mathbf{u}_0 \in L^{\infty}_{loc}(\mathbb{R}^d)^p$. A function $\mathbf{u} \in L^{\infty}_{loc}(\mathbb{R}^d \times \mathbb{R}^+)^p$ is called a *weak solution* of the Cauchy problem (1), (2) if $\mathbf{u}$ satisfies the integral equation

$$\int_0^{\infty}\int_{\mathbb{R}^d}\left(\mathbf{u}\frac{\partial}{\partial t}\varphi + \sum_{i=1}^{d} \mathbf{f}_i(\mathbf{u})\frac{\partial}{\partial x_i}\varphi\right) d\mathbf{x}dt + \int_{\mathbb{R}^d} \mathbf{u}_0(\mathbf{x})\varphi(\mathbf{x},0)d\mathbf{x} = \mathbf{0} \tag{5}$$

for all test functions $\varphi \in C_0^{\infty}(\mathbb{R}^d, \mathbb{R}^+)^p$, i.e. vector valued test functions of compact support in $\mathbb{R}^d \times \mathbb{R}^+$ which are infinitely smooth.

Equation (5) implies that jump discontinuities in weak solutions are not completely arbitrary. At a point of discontinuity the two states $\mathbf{u}_+$, $\mathbf{u}_-$ between which the solution jumps are related to the geometry of the surface of discontinuity by the *Rankine-Hugoniot condition.* It says that the relation

$$[\mathbf{u}_+ - \mathbf{u}_-]n_t + \sum_{j=1}^{d} [\mathbf{f}_j(\mathbf{u}_+) - \mathbf{f}_j(\mathbf{u}_-)]\, n_{x_j} = \mathbf{0} \tag{6}$$

has to hold, where the vector $\mathbf{n} = (n_t, n_{x_1}, \ldots, n_{x_d})$ is the normal of the surface of discontinuity.

Even if weak solutions with a jump discontinuity satisfy the Rankine-Hugoniot condition they are not necessarily unique. We need an additional condition to characterize the physically relevant weak solutions, which is called an *entropy condition.* There is no general theory for multidimensional systems available. For scalar conservation laws and systems in one space dimension a number of mathematical criteria such as the Lax condition, various entropy inequalities or the viscosity method are in use, see [4], [5], [26], [34].

For a given numerical scheme we do not only have to know whether the sequence of approximate solutions is convergent. We also have to consider whether the limit solution is actually the physically relevant solution to the problem considered. In order to show this entropy consistency of a given numerical scheme at least for scalar conservation laws, various equivalent entropy conditions can be applied, see [26]. We will consider the question of entropy consistency for the large time step Godunov scheme of LeVeque[22] in Section 5.

2.2 Numerical Schemes for Conservation Laws

A general framework for many numerical schemes for solving systems of conservation laws with given initial values and boundary conditions consists of the following steps: A *projection* of the initial values respectively of the numerical solution at the previous time step onto piecewise constant step functions corresponding to the underlying mesh. This is followed by an *evolution step* in which the system of conservation laws with initial values corresponding to this finite dimensional subspace of step functions is solved either exactly or numerically. The evolution step can be carried out in many different ways, leading to different numerical schemes. We will restrict our considerations to Godunov-type schemes, which were introduced by Godunov [6], where the evolution is calculated by solving Riemann problems between neighboring mesh cells. Higher-resolution can be obtained by considering the evolution of piecewise linear instead of piece constant functions.

Now we first want to restrict our considerations to the case of one space dimension, i.e. we are looking for a numerical solution of the system of conservation laws

$$\mathbf{u}_t(x,t) + \mathbf{f}_x(\mathbf{u}(x,t)) = 0,$$

with given initial values $\mathbf{u}(x,0)$, where $\mathbf{u}$ and $\mathbf{f}$ are in $\mathbb{R}^p$. The numerical calculations have to be restricted to a bounded interval $[a,b] \in \mathbb{R}$ with appropriate numerical boundary conditions.

For a given discretization $a = x_{-\frac{1}{2}} < x_{1+\frac{1}{2}} < \ldots < x_{m+\frac{1}{2}} = b$ of the computational interval $[a,b]$ at time t_n, we can define a piecewise constant vector valued function $\mathbf{u}^n(x,t_n) = \mathbf{U}_i^n(x)$ for $x \in [x_{i-\frac{1}{2}}, x_{i+\frac{1}{2}})$, where

$$\mathbf{U}_i^n = \frac{1}{h}\int_{x_{i-\frac{1}{2}}}^{x_{i+\frac{1}{2}}} \mathbf{u}(x,t_n)dx$$

is the cell average of the exact or numerical solution $\mathbf{u}$ in the mesh cell $[x_{i-\frac{1}{2}}, x_{i+\frac{1}{2}})$ at time t_n. We will always use k to denote the time step and h to describe the mesh width of a one-dimensional scheme. The solution at the next time step can now be calculated by solving Riemann problems at each cell interface, i.e. solving the conservation law with initial values consisting of two constant states. We get these Riemann problems by considering the cell averages between two neighboring mesh cells. The solution of the Riemann problem consist in general of constant states which are connected by shocks, rarefaction waves or contact discontinuities, see Smoller [34].

For some conservation laws, for instance the Euler equations of gas dynamics, the Riemann problem can be solved exactly. See Chorin [2] for a scheme where the exact Riemann solution is approximated. For many applications it is not necessary or possible to get an exact solution of the Riemann problem. Therefore approximative Riemann solvers were developed. One which is often used is the Roe solver, see [31] or any text book on numerical schemes for conservation laws [5], [26], [35]. The use of the Roe Riemann solver for an unconditionally stable extension of a Godunov-type scheme is considered in Section 4.1.

As long as the waves coming from neighboring Riemann problems do not interact as shown in Figure 1 (a), we can calculate the solution $\mathbf{u}(x,t)$ for $t > t_n$ at every point x. This solution can be written in the form

$$\mathbf{u}(x,t) = \mathbf{u}^n(x,t_n) + \sum_j \left(\mathbf{u}_j^n(x,t) - \mathbf{u}_j^n(x,t_n)\right) \tag{7}$$

where

$$\mathbf{u}_j^n(x,t_n) = \begin{cases} \mathbf{U}_{j-1}^n & : x < x_{j-\frac{1}{2}} \\ \mathbf{U}_j^n & : x \geq x_{j-\frac{1}{2}} \end{cases} \tag{8}$$

is a piecewise constant function, defined over the whole x axis as the Riemann initial values corresponding to the jump across the cell interface at $x_{j-\frac{1}{2}}$. The solution $\mathbf{u}(x,t)$ calculated via (7) consists of the piecewise constant solution $\mathbf{u}^n(x,t_n)$ at the previous time step plus possibly a wave in $\mathbf{u}$ due to a Riemann problem with an influence at the point x. Here $\mathbf{u}_j^n(x,t)$ is the numerical or exact Riemann solution at a point x and time t to the initial values $\mathbf{u}_j^n(x,t_n)$. As long as there are no interactions between the waves of neighboring Riemann problems there is always at most one Riemann problem that modifies the solution $\mathbf{u}(x,t)$ at a fixed position x and time t. Therefore, we can calculate the solution $\mathbf{u}(x,t)$ at every point x by only solving Riemann problems. This classical version of a Godunov-type scheme requires a time step restriction so that the fastest wave occurring by considering all Riemann problems at a given time t_n can only move through a half mesh cell during one time step. This is guaranteed by requiring

$$CFL := \max_i \max_w |\lambda_w(\mathbf{u}_{i-1},\mathbf{u}_i)| \frac{k}{h} \leq \frac{1}{2},$$

where $\lambda_w(\mathbf{u}_{i-1},\mathbf{u}_i)$ is the w-th eigenvalue of the Jacobian matrix of the flux function at the cell interface between the mesh cells $i-1$ and i. This is a special and, as we see below, more restrictive than necessary form of the *Courant-Friedrich-Lewy (CFL) condition*, which was introduced in [3]. In general the CFL condition says, see [3], that a *necessary* condition for a numerical scheme to be stable and therefore convergent is that its numerical domain of dependence contains the true domain of dependence of the partial differential equation considered, at least in the limit as k and h go to zero.

For a linear system $\mathbf{u}_t + \mathbf{A}\mathbf{u}_x = \mathbf{0}$ with $\mathbf{A} \in \mathbb{R}^{m\times m}$ the domain of dependence of the exact solution at a given point (x,t) consists of the m points $x-\lambda_1 t, \ldots, x-\lambda_m t$. If we use a numerical scheme where the solution in one mesh cell is only influenced by the neighboring mesh cells as it is the case for the Godunov scheme, then we have to require that the exact domain of dependence has to be inside this numerical domain of dependence. For a linear system this gives us the time step restriction $|\lambda_{max}|\frac{k}{h} \leq 1$.

The wave speeds s_w arising in the Riemann problem for a linear system are equal to the eigenvalues λ_w of the matrix $\mathbf{A}$. For a nonlinear system of conservation laws the maximal eigenvalues are a upper bound of the maximal occurring wave speeds. This gives us a time step restriction $CFL \leq 1$, where now waves from neighboring Riemann problems might interact with each other. That the interaction of waves inside one mesh cell does not lead to serious problems, even if we consider nonlinear systems of equations, is due to the fact that we are only interested in the cell averages of the solution $\mathbf{U}_i^{n+1}$ for every mesh cell. This is because only the cell averages are needed as initial values for the next time step. In order to calculate the cell averages it is not necessary to calculate the solution $\mathbf{u}(x,t)$ for every point $x \in [x_{i-\frac{1}{2}}, x_{i+\frac{1}{2}})$. Instead we only need to calculate the flux through the boundaries at each mesh cell. This can be seen by considering the following integral form of the conservation law for every mesh cell, see LeVeque [26]

$$\int_{x_{i-\frac{1}{2}}}^{x_{i+\frac{1}{2}}} \mathbf{u}(x,t_{n+1})dx = \int_{x_{i-\frac{1}{2}}}^{x_{i+\frac{1}{2}}} \mathbf{u}^n(x,t_n)dx + \int_{t_n}^{t_{n+1}} \left(\mathbf{f}(\mathbf{u}^n(x_{i-\frac{1}{2}},t)) - \mathbf{f}(\mathbf{u}^n(x_{i+\frac{1}{2}},t))\right) dt.$$

This is equivalent to the *conservative form* of the Godunov scheme

$$\mathbf{U}_i^{n+1} = \mathbf{U}_i^n - \frac{k}{h}\left(\mathbf{F}_{i+\frac{1}{2}}(\mathbf{U}_i^n, \mathbf{U}_{i+1}^n) - \mathbf{F}_{i-\frac{1}{2}}(\mathbf{U}_{i-1}^n, \mathbf{U}_i^n)\right),$$

with the *numerical flux function*

$$\mathbf{F}_{i+\frac{1}{2}}(\mathbf{U}_i^n, \mathbf{U}_{i+1}^n) = \frac{1}{k}\int_{t_n}^{t_{n+1}} \mathbf{f}(\mathbf{u}^n(x_{i+\frac{1}{2}}, t))dt.$$

Therefore, we only have to determine the flux along the cell boundaries. This is relatively easy because the solution is constant at the cell boundary as long as no wave from a neighboring Riemann problem crosses the cell interface. This again shows that the time step restriction shown in Figure 1 (a), where the fastest wave can maximally move through half of the neighboring mesh cell, is more restrictive then necessary. The flux difference form of the Godunov scheme allows time steps where the fastest wave can maximally go through one mesh cell as indicated in Figure 1 (b).

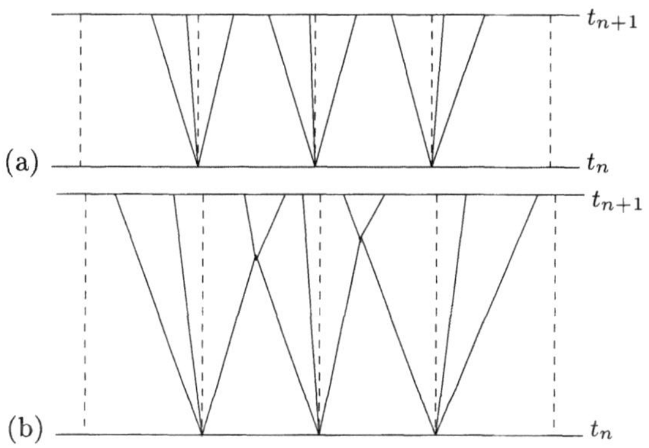

Fig. 1. Waves arising in the solution of Riemann problems for one time step of the Godunov scheme; **a:** $CFL \leq 0.5$, **b:** $CFL \leq 1$

After having defined the basic concept of the Godunov scheme we now want to review some analytical properties. For the Godunov scheme, one has a consistency result which says that if in the limit, when k and h go to zero, the scheme is convergent, then the limit function is a weak solution. This is a consequence of the Lax-Wendroff Theorem [17], which more generally holds for schemes which can be written in conservative form so that the numerical flux function is consistent with the flux function of the conservation law, i.e. $\mathbf{F}(\mathbf{u}, \mathbf{u}) = \mathbf{f}(\mathbf{u})$.

In the general form, the Lax-Wendroff Theorem does not say whether the Godunov scheme, or any conservative scheme, converges at all. Indeed, it is possible to construct conservative and consistent schemes which are unstable and therefore of no use. On the other hand, even if a scheme converges, the weak solution is not unique in general which leads to the question whether a numerical scheme is *entropy consistent*, i.e., whether the limit solution is the physical relevant solution of the problem considered. For the Godunov scheme we get a positive answer to the latter question from the Harten-Lax Theorem, see [10], [11]. There the discrete form of the entropy inequality, a special form of the entropy condition, is used to pick out the physical relevant weak solution.

The concept of the Godunov scheme is implemented in the software package CLAWPACK [18], which was used for our calculations. It may be used with any Riemann solver. There the wave propagation algorithm developed by LeVeque [27] is used to update the solution $\mathbf{u}$ at every time step. High-resolution can be obtained by considering piecewise linear reconstruction. This requires the use of limiter functions in order to avoid spurious oscillations along shock fronts, see [27]. The CLAWPACK program considers very general problems, including conservation laws with a capacity function or conservation laws with source terms in one, two, and three dimensions, see [19], [27] or [28] for details.

The simplest way to extend a one dimensional method to a multidimensional scheme is to use a *dimension splitting* approach. For a two-dimensional problem one could for instance first apply the Godunov scheme in x direction to calculate new cell averages which are then used as initial values for applying the Godunov scheme in y direction. These splitting schemes perform well for many test cases, but the solution might depend on the mesh which was used for the calculation because the mesh determines the preferred direction which is approximated by the scheme. Other schemes, for instance the finite volume scheme in CLAWPACK, were developed in order to approximate more closely the multidimensional behavior of the equations, see [27].

3 Unconditionally Stable Schemes for Conservation Laws

In contrast to the situation for numerical methods discretizing ordinary differential equations, it is possible to derive unconditionally stable explicit schemes, i.e., explicit schemes without a restriction of the time step, for the approximation of conservation laws. That this is possible, in principle, can be seen by considering the geometric CFL condition, i.e. that the domain of dependence of the exact solution must be contained in the domain of dependence for the numerical solution. This necessary condition for stability is satisfied if the space stencil of the scheme is automatically increased with a larger time step.

A generalization of the Godunov scheme which allows larger time steps was developed by LeVeque, see [20], [22]. In the following we will refer to this scheme as the LTS Godunov-type scheme. Here the interaction of waves corresponding to neighboring Riemann problems is approximated as a linear superposition. We will review this approach in the next section.

A generalization of the Glimm scheme to an unconditionally stable scheme was considered by Wang in [37]. In difference to the LTS Godunov-type scheme the piecewise constant initial values for the next time step are obtained by taking the exact or approximative solution at some randomly chosen point in the mesh cell instead of the cell average over the solution which is use by Godunov type schemes.

An unconditionally stable scheme based on front tracking was developed by Holden and Risebro [13]. The good performance of this scheme for the Euler equations is demonstrated in [12].

Billett and Toro [1] consider the weighted average flux (WAF) scheme which allows time steps with Courant numbers up to two, see [36].

Finally, we note that unconditionally stable explicit methods, based on rational Runge-Kutta methods, have been considered for parabolic equations in the past, see Hairer [7] and Satofuka [32]. They were also used in flow calculations, see Satofuka [33].

4 The Large Time Step Godunov Scheme of LeVeque

Now we want to consider the LTS Godunov scheme of LeVeque [20], [21], [22], [25] applied to one dimensional systems of conservation laws. As we have seen in Section 2.2 the classical concept of the Godunov scheme, where the solution $\mathbf{u}$ is defined as the combination of the solution of Riemann problems as long as there are no interactions of waves, can be extended to the case $CFL \leq 1$ in which no wave can cross a neighboring cell interface. In order to extend the Godunov scheme to time steps with $CFL > 1$, one also has to take the interaction of waves into consideration. In the LTS Godunov scheme this interaction is approximated by linear superposition of waves, which means that the interaction of waves will not lead to a change of the speed or strength of the waves, as indicated in Figure 2. In the nonlinear case the waves would have an influence on each other which would lead to a change of the wave speed, direction and strength as shown in Figure 1 (b).

The LTS Godunov scheme can also be defined via equation (7) with the only difference that for a given point x there may be not only one Riemann problem that leads to a change of $\mathbf{u}(x,t)$. The numerical solution $\mathbf{u}(x,t)$ at a given point x is equal to the value $\mathbf{u}(x,t_n)$ at the old time step plus the jump in $\mathbf{u}$ across all waves which have crossed the line between (x,t_n) and (x,t), see [22].

In [21], LeVeque has proved that his modification of the Godunov scheme, with an appropriate Riemann solver for each Riemann problem occurring in the discretization as it will be discussed in Section 4.1, can be written in conservative form with a numerical flux function that is consistent with the

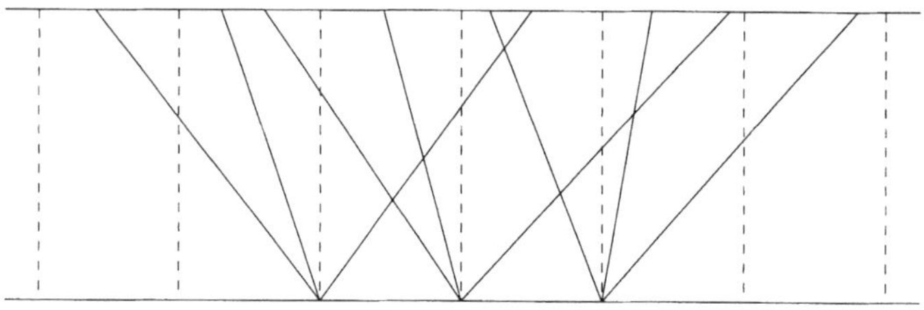

Fig. 2. Linear interaction of waves as approximated by the large time step Godunov scheme

conservation law. Hence, if the scheme converges in the limit as k and h go to zero, the limit function is a weak solution, in consequence of the Lax-Wendroff Theorem.

For scalar conservation laws he could also show that the approximate solution is total variation diminishing for arbitrary Courant numbers. This property implies via Helly's Theorem that for every subsequence of the mesh discretization, with $k_l \to 0$ and $k_l/h_l = const$, every sequence of approximate solutions contains a convergent subsequence locally in L^1. Together with the Lax-Wendroff Theorem this gives convergence of the LTS Godunov scheme to a weak solution at least in the case of scalar conservation laws.

Furthermore, he conjectured that the approximated weak solution is consistent with the entropy condition, i.e. is the physical relevant weak solution of the scalar conservation law. The proof of the entropy consistency turned out to be difficult due to the fact that the scheme is not monotone. Some preliminary results concerning the entropy consistency of the LTS Godunov scheme and the LTS Glimm scheme where shown in Wang, Warnecke [38], [39].

4.1 The Solution of the Riemann Problem

A necessary condition in order to get an entropy stable approximation is that the solution $\mathbf{u}_j^n(x,t)$ of each single Riemann problem is entropy consistent. This can always be obtained by using an exact Riemann solver. Since we are only interested in the cell averages of the numerical solution at discrete time steps, the use of entropy consistent approximative Riemann solvers is often easier and leads to the same accuracy of the numerical results.

Here we want to consider the use of Roe's Riemann solver, see [31], for the LTS scheme. Roe suggested instead of determining the Riemann solution of a given nonlinear system to calculate the solution of the Riemann problem for a corresponding *linear system*

$$\mathbf{u}_t + \mathbf{A}(\mathbf{u}_L, \mathbf{u}_R)\mathbf{u}_x = \mathbf{0}, \quad \mathbf{u}(x,0) = \begin{cases} \mathbf{u}_L : x < x_0 \\ \mathbf{u}_R : x \geq x_0 \end{cases}$$

where the matrix $\mathbf{A}(\mathbf{u}_L, \mathbf{u}_R)$ has to be a consistent approximation of the Jacobian matrix for the Riemann problem. Furthermore, the Rankine-Hugoniot condition has to be satisfied and finally the constant matrix $\mathbf{A}(\mathbf{u}_L, \mathbf{u}_R)$ has to be diagonalizable with real eigenvalues, see for instance [26]. This means that the solution of the Riemann problem for a nonlinear system of conservation laws is reduced to the solution of the Riemann problem for a linear system, which displays locally the basic properties of the nonlinear system.

The Riemann problem for the resulting linear system can be solved exactly. A disadvantage of this approach comes from the fact that the solution of a linear system only consists of discontinuities, whereas the Riemann solution of a nonlinear hyperbolic conservation law can also contain rarefaction waves. Fortunately, in most situations this does not lead to any problems because the change in the cell average of the solution caused by a rarefaction wave can usually be approximated by a single discontinuity moving with a speed that is equal to the average between the most left and the most right going part of the rarefaction wave. The only situation in which the Roe solver fails for $CFL \leq 1$ is the case of a rarefaction wave with $\lambda(\mathbf{u}_L) < 0 < \lambda(\mathbf{u}_R)$ for one eigenvalue of the Jacobian matrix, e.g. as in transonic flows. The approximation of a rarefaction wave by a discontinuity has the effect that either the cell average at the left hand side or the cell average at the right hand side will be influenced. In the case of a transonic rarefaction wave both cell averages should be changed. An entropy-fix has to be used to overcome this problem, see for instance Harten and Hyman [8].

For the LTS scheme the entropy-fix has to be extended. In order to explain this, we want to consider the approximation of a rarefaction wave for a scalar conservation law, for instance the Burgers equation. It can be extended to systems of conservation laws in an analogous way with the only difference that the solution of the Riemann problem consists of more than one wave.

If the exact solution of the Riemann problem is a *shock*, then we can calculate the shock speed using the Rankine-Hugoniot condition. The jump across the shock is equal to the jump of the Riemann initial values. This jump is added to all the cell averages of the mesh cells which were crossed by this wave during the time step. The jump is partially added to the cell averages of the mesh cells which were only touched by the wave.

If the solution of the Riemann problem with initial values u_L for $x < x_0$ and u_R for $x \geq x_0$ is a *rarefaction wave*, then in the case of a convex flux function the solution is equal to

$$u(x,t) = \begin{cases} u_L & : \quad (x-x_0)/t < \lambda(u_L) \\ u_L + \frac{(x-x_0)/t - \lambda(u_L)}{\lambda(u_R) - \lambda(u_L)}(u_R - u_L) & : \lambda(u_L) \leq (x-x_0)/t < \lambda(u_R) \\ u_R & : \quad (x-x_0)/t \geq \lambda(u_R) \end{cases}$$

Our numerical discretization of a rarefaction wave consists of a piecewise constant step function as shown in the middle of Figure 3. The most left and the most right going parts are the solid lines in Figure 3. If the rarefaction

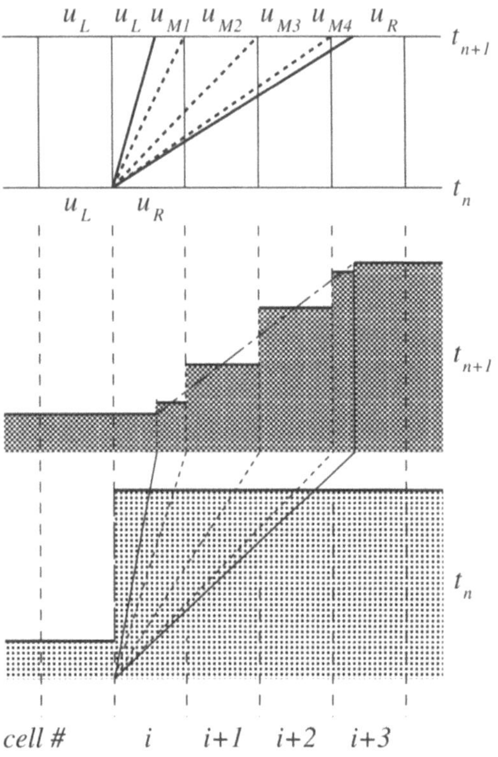

Fig. 3. Approximation of a rarefaction wave that is spread out over different mesh cells

wave at the new time step is spread out over more than one mesh cell, then we introduce new waves, shown as dashed lines in Figure 3, so that there is one jump discontinuity in every mesh cell between $x_0 + \lambda(u_L)\triangle t$ and $x_0 + \lambda(u_R)\triangle t$.

The jumps cross each of these newly introduced waves as well as across the most left and the most right going wave are chosen in such a way that the jump in the initial values is equally distributed to all of these waves. A similar approximation of rarefaction waves was used in LeVeque [23].

4.2 A Geometric Interpretation of the Entropy-Fix

In order to approximate the correct cell average for a rarefaction wave, we do not always need to use several waves. In Figure 4 different possibilities for the occurrence of rarefaction waves are shown. The most left and the most right going part of the rarefaction wave are plotted as solid lines. The dashed pointed line in each case corresponds to the speed calculated by the Rankine-Hugoniot condition, i.e. is equivalent to the wave speed used by

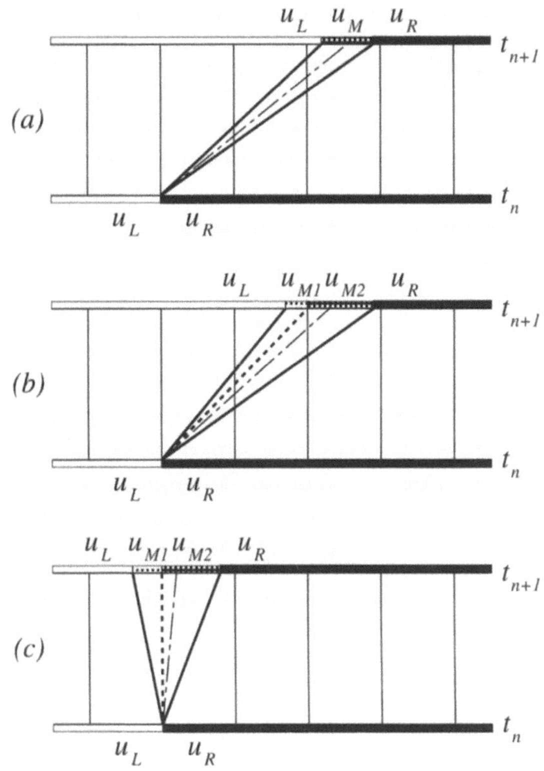

Fig. 4. Geometric interpretation of the entropy-fix for rarefaction waves

the Roe solver. The dashed lines are the waves which were introduced to approximate the rarefaction wave.

Considering Figure 4 (a) it becomes clear that if the rarefaction wave at time t_{n+1} is only extended over a single cell then we are able to calculate the change of the cell average by approximating the rarefaction wave by a single discontinuity which satisfies the Rankine-Hugoniot condition. If the rarefaction wave at time t_{n+1} is spread out over more than one mesh cell, as shown in Figure 4 (b), we need an entropy-fix, i.e. we need to introduce further waves, in order to approximate the correct cell averages.

For the Godunov scheme with Courant number smaller than one, the only possibility of a rarefaction wave that is spread out over more than one mesh cell is shown in Figure 4 (c). Hence the Godunov-type scheme with Roe Riemann solver and $CFL \leq 1$ needs an entropy-fix in the case of a transonic rarefaction wave only.

4.3 The Error of the LTS Godunov Scheme

There are two components which have an influence on the error of the LTS Godunov scheme. In each time step we make a projection onto piecewise constant functions, which are then used as initial values for the next time step. This leads to a projection error in every time step. By using larger time steps the projection error is reduced because fewer projections in comparison with the classical Godunov-type scheme are necessary in order to calculate the solution at a fixed time.

The second part of the error is generated in the evolution step. The LTS Godunov scheme handles the interaction of waves as a linear superposition which can lead to a large error, especially at discontinuities in nonlinear problems.

The reduction of the projection error can best be seen by considering linear problems, because then the linear superposition of waves is exact and the only error of the scheme is due to the projections in each time step. Here we consider the LTS approximation of a conservation law with capacity function, which is implemented in CLAWPACK, see [19, note # 6].

Example 2. We consider the conservation law with capacity function

$$\kappa(x)u_t + 2u_x = 0, \qquad \kappa(x) = \begin{cases} 2 : x < 5 \\ 1 : x \geq 5 \end{cases}$$

In CLAWPACK solutions for systems of conservation laws with capacity function are calculated using the flux difference form

$$\mathbf{U}_i^{n+1} = \mathbf{U}_i^n - \frac{\triangle t}{\kappa_i \triangle x} \cdot \left[\mathbf{F}_{i+1}^n - \mathbf{F}_i^n\right],$$

where κ_i is the cell average of $\kappa(x)$ in the mesh cell i. The capacity function causes a change of the propagation speed of the waves. This can easily be approximated by the LTS Godunov scheme. Instead of considering the superposition of waves that are straight lines as shown in Figure 2 every wave can have a different slope in each mesh cell depending on the value of $\kappa(x_i)$. The numerical results in Figure 6 show the good performance of the LTS scheme for this problem.

The good performance of the LTS scheme for this example is not astonishing. For linear systems we would always get the most accurate approximation by applying the LTS Godunov-type scheme with a Courant number large enough that we would only need one time step to calculate the solution.

Of course we are mostly interested in solving nonlinear problems. In LeVeque [25] a truncation error analysis for nonlinear systems is given. For smooth solutions the error made during one time step due to the linear superposition is of the order $\mathcal{O}(k^3)$ which gives a contribution to the global error of order $\mathcal{O}(k^2)$, see [24] for a proof.

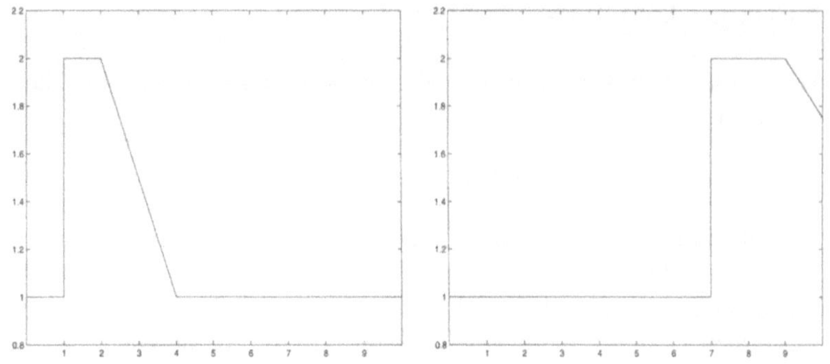

Fig. 5. Left: initial conditions, right: exact solution at time $t = 5$ for Example 2

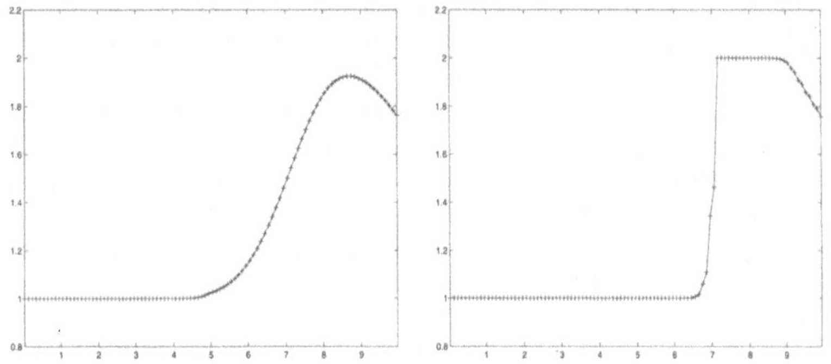

Fig. 6. Numerical solution at time $t = 5$ with $\triangle x = 0.1$ and different time steps for Example 2; Left: $CFL \leq 1$, right: $CFL \leq 10$

The projection error introduced in every time step is $\mathcal{O}(h^2)$ which gives a contribution of $\mathcal{O}(h^2/k)$ to the global error. Hence the global error of the LTS Godunov scheme is $c_1k^2 + c_2h^2/k$ with constant values c_1 and c_2. The best choice for the mesh width and time step can be obtained by setting $c_1k^2 = c_2h^2/k$ which gives a time step $k = ch^{2/3}$. Although, these analytical results are only valid for the case when h goes to zero, it indicates that we might get a more accurate solution by using time steps with Courant number larger than one at least if the solution is smooth. Numerical results confirm this. Further evidence for the advantage of using larger time steps is the error estimate of Lucier [29] for a shock-capturing scheme for scalar conservation laws of LeVeque [20], which is similar to the LTS Godunov-type scheme considered here.

5 Entropy Consistency

In this section we consider the special case of a scalar conservation law with convex flux function, i.e.

$$u_t + f(u)_x = 0, \quad f'' > 0 \tag{9}$$

with initial values $u(x,0) = u_0(x)$. For scalar conservation laws the LTS Godunov-type scheme converges to a weak solution in consequence of the Lax-Wendroff Theorem and the TVD property, see LeVeque [21]. In contrast to the classical Godunov scheme with Courant number smaller or equal one, the extension for larger time steps is not monotone. This is due to the fact that the interaction of waves is considered as a linear superposition. The linear superposition of a shock and a rarefaction wave can produce values which are smaller than the minimum of the initial values.

The total variation diminishing property of the LTS scheme, which was shown in [21], guarantees that for monotone initial values the scheme is monotone. Therefore the scheme is said to be *monotonicity preserving.*

In the scalar case the solution of (9) with monotone increasing initial values consists of rarefaction waves possibly connected by constant states. Note that it is not possible that the characteristics cross each other in this case. Nevertheless, rarefaction waves from different Riemann problems can go through one and the same mesh cell which will not lead to any numerical problems. If the initial values are monotone decreasing then the solution of the scalar conservation law at a later time consists of shocks which may cross each other.

The entropy consistency for monotone schemes was proved by Harten, Hyman and Lax [9]. Hence, it follows, that the LTS Godunov scheme leads to an entropy consistent approximation as long as the initial values are monotone. This means especially that the linear superposition of arbitrary many shocks is an entropy stable approximation if h and k go to zero.

It remains to consider the interaction of shocks and rarefaction waves. First we consider the case where a rarefaction wave is moving faster than a shock as indicated in Figure 7 (a).

As a model problem we consider Burgers equation

$$u_t + uu_x = 0$$

with initial values

$$u(x,0) = \begin{cases} 0 & : \quad x \leq 1.5 \\ (x-1.5)/5.5 & : 1.5 \leq x < 7 \\ -1 & : \quad x \geq 7. \end{cases}$$

The solution consists of a rarefaction wave followed by a shock which is initially located at $x = 7$ and which is then moving to the left hand side. After the first time step of the LTS scheme with a sufficiently coarse mesh, as shown

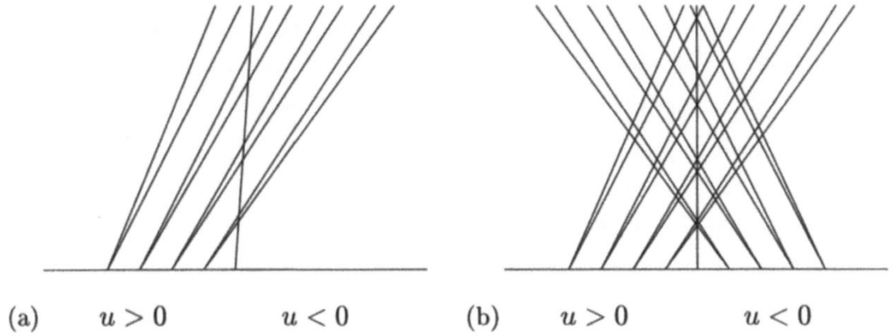

Fig. 7. Structure of the characteristics for two different initial values; left: interaction of rarefaction wave and shock, right: interaction of two rarefaction waves with a shock in between

in Figure 8 (left), we see that in the region to the right of the shock the value u is too small. Here we used a time step with $CFL \leq 10$. If we refine the mesh and take the same Courant number this error is reduced and hardly visible, see Figure 9 (left).

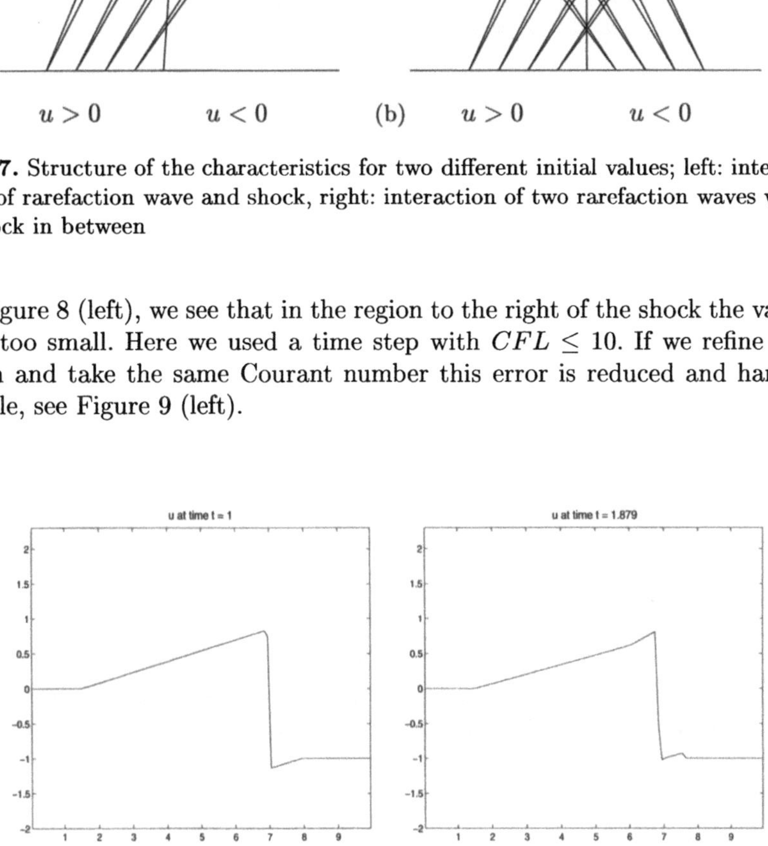

Fig. 8. Interaction of rarefaction wave and shock calculated with the LTS Godunov scheme using $\Delta x = 0.1$ and $CFL \approx 10$; Left: after first time step, right: after second time step

But also for this relatively coarse mesh the LTS Godunov scheme is correcting the error automatically by itself during the next time steps. Figure 8 (right) shows the numerical solution after the second step. Inside the region where the value of u became too small during the first time step we get a rarefaction wave which is now moving to the left hand side and which therefore corrects the error from the first time step. Also the shock is now smeared out over two mesh cells and will be approximated by one shock moving to the right and another shock moving to the left hand side. In the second time

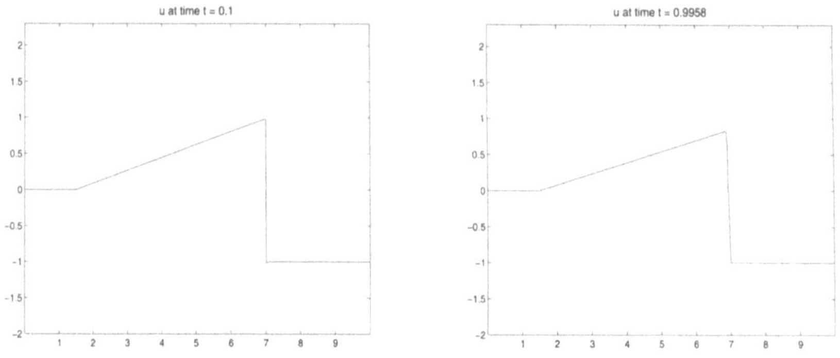

Fig. 9. Interaction of rarefaction wave and shock calculated with the LTS Godunov scheme using $\triangle x = 0.01$ and $CFL \approx 10$; Left: after first time step, right: after 10 time steps

step the shock which is moving to the right creates a small region to the right of the strong shock where the solution becomes too large. An entropy violating solution to this problem can not persist because an error inside the rarefaction wave, as can be seen in Figure 8 (right), will be smeared-out in the next time step. This coincides with the motivation for the conjecture of entropy consistency of the LTS Godunov scheme given in [21]. A similar 'self-correction' of the LTS Godunov scheme was described in LeVeque [23].

For the finer mesh we obtain a very accurate solution after ten time steps, see Figure 9 (right) which shows that the error is not accumulated for this problem. The question is whether it might be possible to construct cases where the error from one time step will increase during the next time step and where the 'self-correction' of the scheme does not work. Such a situation is considered in our next test case, where the initial values consist of two rarefaction waves which are connected by a shock.

For the numerical calculations we again consider the Burgers equation with initial values

$$u(x,0) = \begin{cases} 0 & : \; x < 2.5 \\ 0.25(x-2.5)/2.5 & : \; 2.5 \leq x < 5 \\ -0.25(7.5-x)/2.5 & : \; 5 \leq x < 7.5 \\ 0 & : \; x \geq 7.5. \end{cases} \tag{10}$$

For this example the shock is stationary at $x = 5$. This especially means that there is no numerical viscosity along the shock. For time steps with $CFL > 1$ the rarefaction waves from both sides of the shock cross the shock, compare with Figure 7 (b). On the left hand side of the shock the LTS Godunov-type scheme leads to values larger than the exact values and on the right hand side it leads to values which are to small. Now during the next

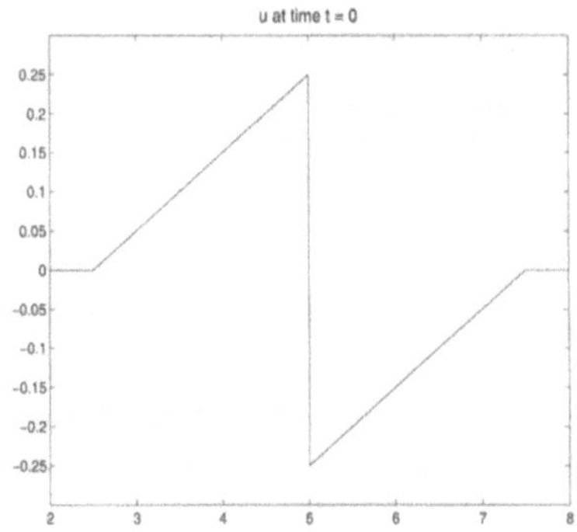

Fig. 10. Initial values as defined in Equation 10.

time steps the rarefaction waves from both sides will further contribute to this error instead of smearing it out. This happens due to the symmetry of the problem. Figure 11 shows the numerical solution for $CFL \approx 20$ and different mesh widths $\triangle x$. Note that for every small perturbation of the initial values which destroys the symmetry, this structure of the numerical solution would not persist over longer times.

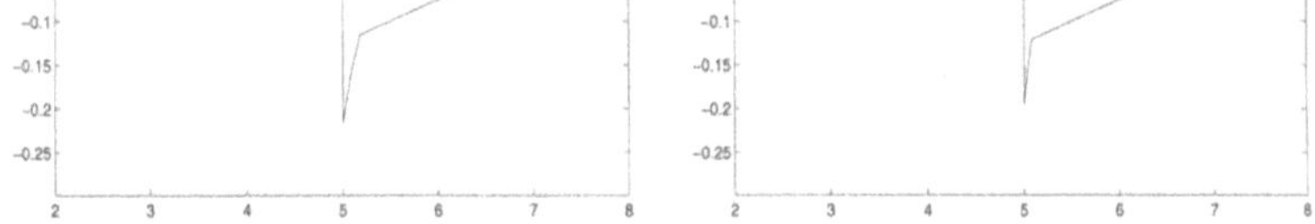

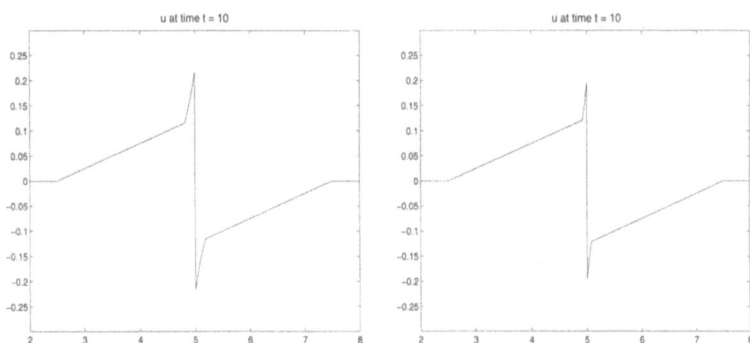

Fig. 11. Interaction of two rarefaction waves with a centred shock calculated with the LTS Godunov scheme using $CFL \leq 20$. Left: $\triangle x = 0.01$, right: $\triangle x = 0.005$

Now we want to consider the question whether the limit function would still be entropy stable. In order to do this we use Oleinik's entropy condition. It may be used for any scalar conservation law with convex flux function.

Theorem 3. *(Oleinik) Assume* $d = p = 1$ *and that the flux function* $f : \mathbb{R} \to \mathbb{R}$ *is convex. The limit function* u *of any sequence of approximations is the unique entropy solution of (9) if there is a constant* $E > 0$ *such that for all*

$a > 0$, $t > 0$ and $x \in \mathbb{R}$ the inequality

$$\frac{u(x+a,t) - u(x,t)}{a} < \frac{E}{t} \tag{11}$$

is satisfied.

That equation (11) implies uniqueness is shown in Smoller [34].

The entropy condition of Oleinik has the advantage that it can be applied directly to the piecewise constant function of approximated cell averages $u^n(x, t_n)$. In order to do this a natural choice of the parameter a would be $a = \triangle x$. A sufficient condition for the entropy consistency of the numerical scheme is that there exists a constant $E > 0$ such that

$$U_{j+1}(t) - U_j(t) < E\frac{\triangle x}{t} \tag{12}$$

for all $t > 0$ and $\triangle x \to 0$, see [26].

In Figure 12 (left) and 13 (left) we show the maximal slopes $|(U_i - U_{i-1})|/\triangle x$ occurring inside the rarefaction waves over the time t for two different discretizations with $CFL \leq 2$. For this special case the maximal discrete slope occurs between the two mesh cells to the left respectively between the two mesh cells to the right of the shock discontinuity.

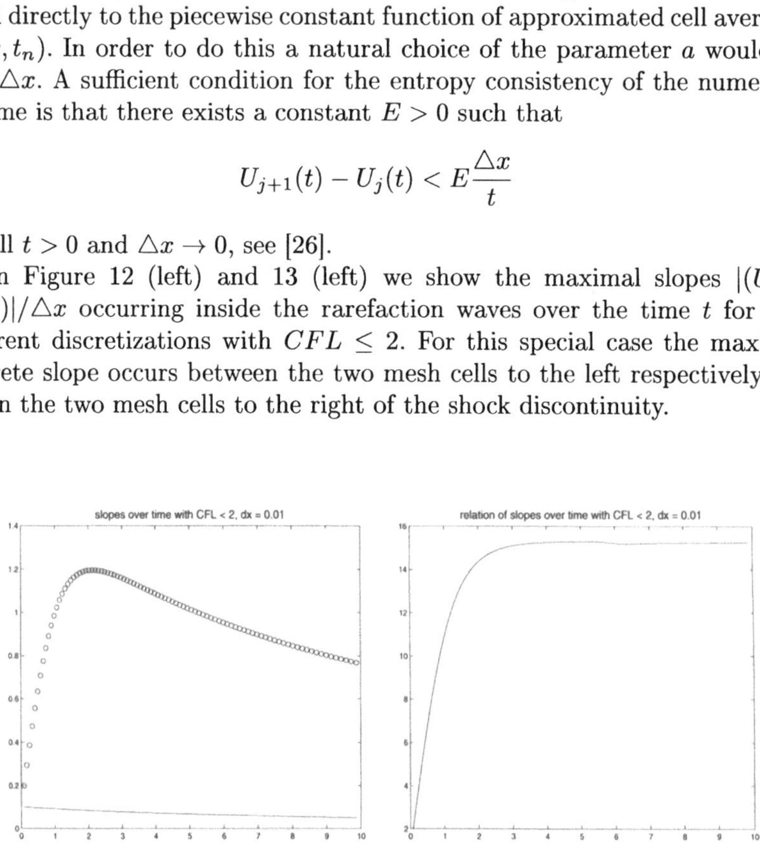

Fig. 12. Left: maximal slope (ooo) and exact slope (—) inside the rarefaction wave over time t for $\triangle x = 0.01$, $CFL \leq 2$, right: ratio between maximal slope and exact slope.

In Figure 12, 13 (left) the exact slope is also plotted, as a solid line, which is a monotone decreasing function of time. We can see that the curves of the maximal discrete slopes have a greater maximum for a finer discretization. Furthermore the function of maximal slope reaches its maximum at an earlier time if a finer mesh is used. In Figure 12 and 13 (right) we show the ratio

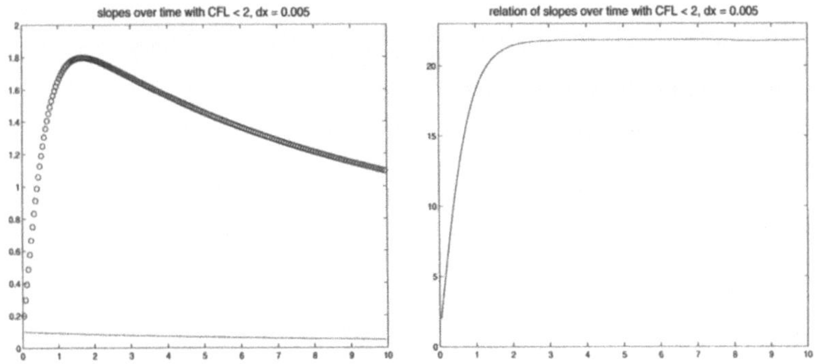

Fig. 13. Left: maximal slope inside the rarefaction wave over time t for $\Delta x = 0.005$, $CFL \leq 2$, right: ratio between maximal slope and exact slope.

between the maximal numerical slope and the exact slope. Here it can be seen that after this ratio has reached its maximum value the function remains constant. To explain this, we consider the cell averages in the two mesh cells to the left of the shock. These are the mesh cells $i+1$ and $i+2$ in Figure 14. The stationary shock for our example with initial values (10) sits at $x = 5$ which corresponds to the boundary between the mesh cells $i+2$ and $i+3$.

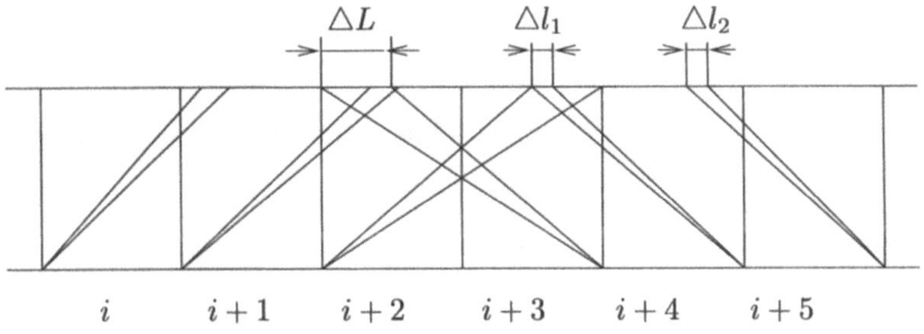

Fig. 14. Linear superposition of two rarefaction waves with an intermediate stationary shock between the mesh cells $i+2$ and $i+3$.

The cell average of the mesh cell $i+2$ at time t_{n+1} is equal to

$$U_{i+2}^{n+1} = U_{i+2}^n + (U_{i+1}^n - U_{i+2}^n) + A_{i+2}(U_i^n, U_{i+1}^n) + B_{i+2}(U_{i+3}^n, U_{i+4}^n).$$

The term $(U_{i+1}^n - U_{i+2}^n)$ is the contribution of the rarefaction wave of the Riemann problem between the mesh cells $i+1$ and $i+2$ which completely went through the cell $i+2$. The terms $A_{i+2}(U_i^n, U_{i+1}^n)$ and $B_{i+2}(U_{i+3}^n, U_{i+4}^n)$ are the contributions of the Riemann problems between the mesh cells i and $i+1$ as well as respectively $i+3$ and $i+4$ to the value of the cell average in the mesh cell $i+2$ at time t_{n+1}. For our example the term $(U_{i+1}^n - U_{i+2}^n) + A_{i+2}(U_i^n, U_{i+1}^n)$ decreases the cell average and the term $B(U_{i+3}^n, U_{i+4}^n)$ leads to an increase of the cell average.

Note that we are taking $CFL = 2$ whereby the front of the wave originating between cells $i+3$ and $i+4$ crosses exactly two mesh cells to the left. An increase of the jump between the mesh cells $i+3$ and $i+4$ increases the wave width $\triangle L$. If the slope is relatively large compared to the slope between the other mesh cells, than also the distance $\triangle L$ becomes larger, which on the other hand has the effect that the rarefaction wave arising between mesh cell $i+3$ and $i+4$ is smeared out over a larger distance. This means that the error due to the linear superposition of the waves will again be corrected by the scheme. In contrast to the example where we only had the interaction of one rarefaction wave with a shock wave we first get an increase of the jump of the cell averages between the mesh cells $i+1$ and $i+3$ during the first time steps before the scheme will balance the error. This effect can be seen very nicely from the Figures 12 and 13 (right), which show that after some time while the jump between the two mesh cells to the left of the shock is increased, we get a balanced relation compared to the exact jump which is decreasing for our example.

For smaller mesh cells we also have smaller time steps and therefore the maximal slope will be reached earlier. The decrease of the maximal slope in Figure 12, 13 (left) is proportional to the decrease of the exact slope over time which can be seen from Figure 12, 13 (right), where we have plotted the ratio between the maximal slope inside the rarefaction waves and the exact slope.

However, as we can also see from Figure 12 and 13 the maximal slope as well as the ratio between the maximal slope and the exact slope is increasing if the mesh is refined. Therefore, we do not get an upper bound for the constant E in the discrete version of the Oleinik entropy condition (12). But we can also see that the maximal slope will not be twice as large if the mesh width is reduced by the factor two. For the special case of Burgers equation, using time steps with $CFL \leq 2$, we can show that the maximal slope will increase by a factor $\sqrt{2}$ if we bisect the mesh cells. On the other hand the region where this error occurs becomes half as large. Furthermore, we get from the TVD property of the scheme that the number of discontinuities is limited. This means that we can not get infinitely many regions where such an error might arise if we refine the mesh. Therefore, in the limit as $\triangle x$ goes to zero the error in the L_1 norm goes to zero and the solution goes to the exact solution. We can also see this from Oleinik's entropy condition if we apply it in a different form to the discrete solution. Instead of setting $a = \triangle x$ we can take the value a arbitrary but fixed and then consider a sequence of approximations. If we for instance set $a = 0.005$ which is equal to $\triangle x$ in the calculations shown in Figure 13, then we can consider the maximal slope $|(U_{i-1} - U_i)/a|$ inside the rarefaction wave for different discretizations. Figure 15 shows the maximal slopes for this constant value of a and $\triangle x = 0.0025$. These maximal

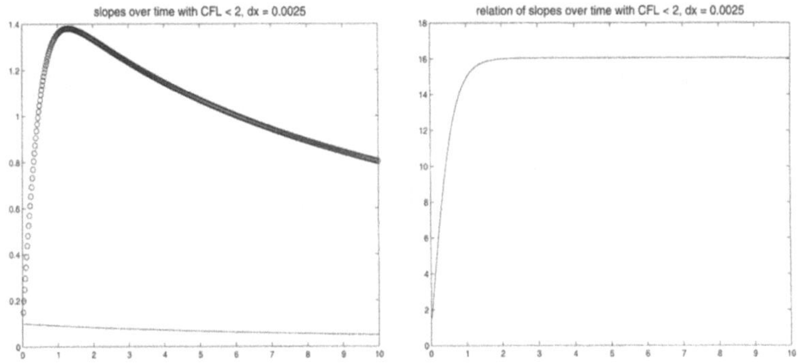

Fig. 15. Left: maximal slope for $a = 2\triangle x$ over time t with $\triangle x = 0.0025$, $CFL \leq 2$, right: ratio between this slope and the exact slope

slopes can now be directly compared with those shown in Figure 13, where the same value for the constant a was used. The numerical calculations show that the slopes corresponding to the fixed value of a are decreasing as $\triangle x$ goes to zero. Therefore we have found an upper bound of the constant E to the given value a. This upper bound is equal to the maximal slope arising in the discretization with $\triangle x = a$. The case $CFL > 2$ will not lead to further problems for the example considered.

6 The LTS Scheme in 2D

In this section we will restrict our considerations to two dimensional scalar conservation laws with convex flux function. The simplest way to extend the LTS Godunov scheme to two space dimensions can be obtained by using a dimension splitting approach, which was considered in LeVeque [20].

Here we want to consider another possibility, which was motivated by the 2D CLAWPACK algorithm, see [27], and is a generalization to $CFL > 1$. In contrast to the wave propagation algorithm in CLAWPACK we do not solve transverse Riemann problems, instead we consider for every mesh cell the transport due to the Riemann problems along all four cell boundaries, as will be described later.

For a 2D advection equation

$$u_t + au_x + bu_y = 0, \quad a, b > 0$$

the exact influence of a single mesh cell to the change of the cell average of the neighboring cells is indicated in Figure 16.

Here the value of the mesh cell (i, j) is shifted in x direction over a distance $a\triangle t$ and in y direction over a distance $b\triangle t$. In contrast to the dimension splitting approach we use the values u at the old time step, and not intermediate values, in order to calculate the evolution.

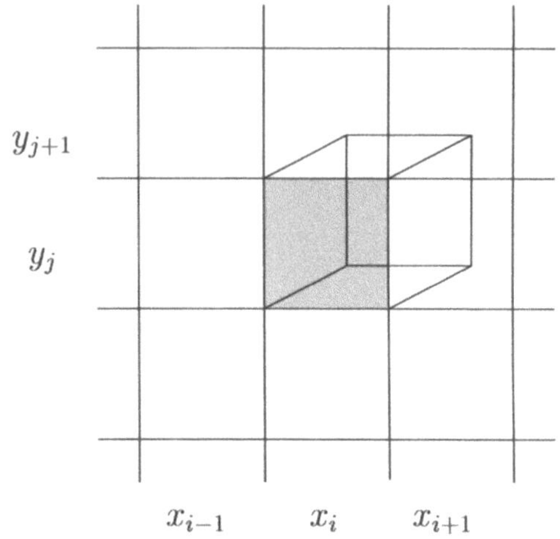

Fig. 16. Description of the flow across each cell boundary of the mesh cell (i,j) using a time step with $CFL < 1$

In order to generalize this scheme to time steps with Courant numbers larger then one, we first consider the case where the solution of the Riemann problems for both directions can be approximated by a discontinuity. This means that the solution consists either of a shock wave or of a rarefaction wave which can be approximated by a single discontinuity, i.e. no entropy-fix is necessary. The algorithm is indicated in Figure 17. If s_{ij}^x is the speed of the discontinuity for the Riemann problems between the mesh cells (i,j) and $(i+1,j)$ then the jump across the cell interface can be propagated over a distance $s_{ij}^x \triangle t$. But we also have to take into account the propagation in y direction. Let s_{ij-1}^y and s_{ij}^y denote the speeds of the Riemann problems in y direction arising on the bottom and the top of the mesh cell (i,j) then the discontinuity across the mesh cells (i,j) and $(i+1,j)$ is also propagated in y direction corresponding to these speeds. Therefore, the jump across each cell interface has an influence over a trapezoidal region as shown in Figure 17.

In order to get a conservative scheme we do not add the whole jump, for our example $u(x_{i+1},y_j) - u(x_i,y_j)$, in the area which is covered by the trapezoid. Instead we have to normalize the jump in such a way that the change due to the Riemann problem considered is equal to

$$s_{ij}^x \triangle t \triangle y [u(x_{i+1},y_j) - u(x_i,y_j)].$$

If we consider a cell interface where the corresponding Riemann problem in normal direction is a rarefaction wave then the trapezoidal area is divided into several trapezoids where we always have to add another cell average, corresponding to the 1D approximation of a rarefaction wave and the normalization in order to get a conservative scheme. If the Riemann problem for one or both of the transverse cell interfaces consists of a rarefaction wave then

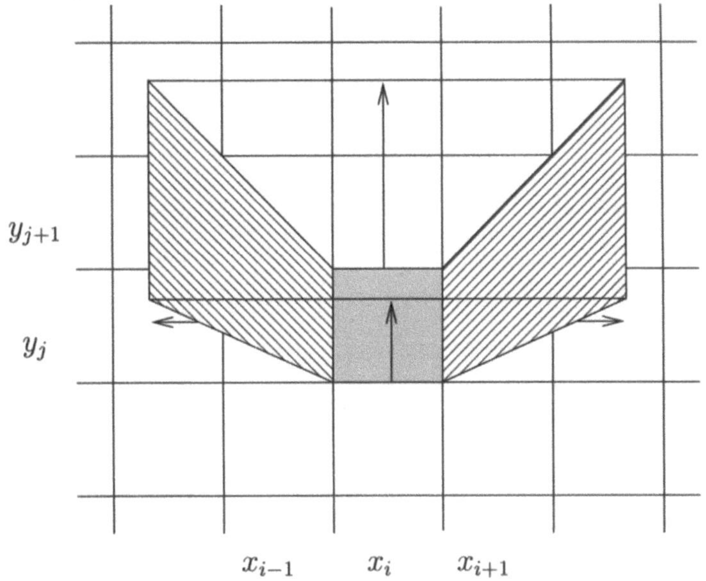

Fig. 17. Schematic description of the influence of the flow across the cell boundaries of the mesh cell (i, j) for time steps with $CFL > 1$

we use the most up-going part of the Riemann solution across the cell interface on the top of cell (i, j) and the most down-going part of the Riemann solution across the cell interface at the bottom to calculate the trapezoid area which is influenced by the jump across the mesh cells (i, j) and $(i + 1, j)$.

The area which is influenced by the flow over a cell interface is a trapezoid or consists of one or two triangles, depending on the waves of the transverse Riemann problems, see Figure 18 (a) and (b). In Figure 18(c) we show the reference area which is used to normalize the flux across the cell boundary considered, in order to get a conservative scheme.

In cases where the area consists of two triangles the scheme can lead to larger inaccuracies. As long as the area remains a trapezoid we expect a higher accuracy than by using the classical Godunov scheme, see Example 4

Example 4. We consider the 2D Burgers equation

$$u_t + uu_x + uu_y = 0$$

with piecewise constant initial values

$$u(x, y, 0) = \begin{cases} 1 : xy > 0 \\ 0 : xy < 0. \end{cases}$$

Along the sides of monotone deceasing initial values the solution will contain shocks and along the sides with monotone increasing initial values the solution consists of rarefaction waves. For the solution shown in Figure 19 we show

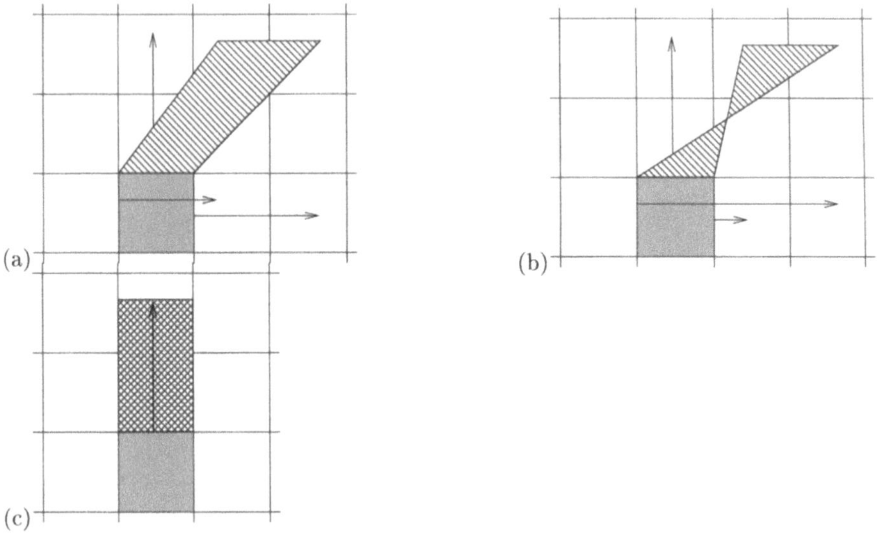

Fig. 18. Different possibilities for the area that is influenced by a Riemann problem in y-direction

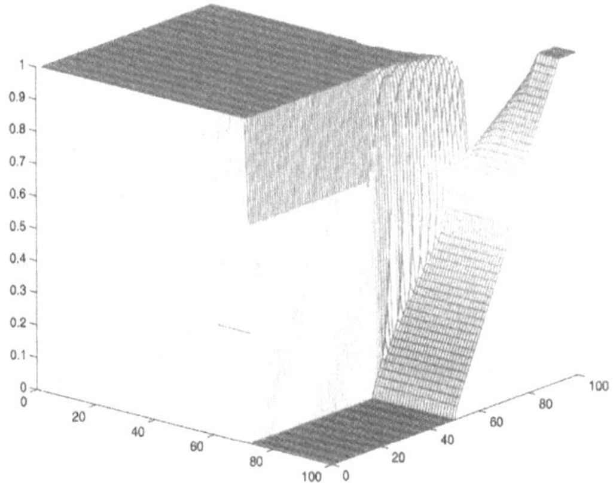

Fig. 19. Solution of Example 4 at time $t = 40$ with $\triangle x = 1$ using the LTS Godunov scheme

the numerical results which were obtained by using our first order version of the two dimensional LTS-Godunov scheme with a time step restriction $s_{i-1,j}\triangle t \leq s_{i,j}\triangle t$ and $s_{i,j-1}\triangle t \leq s_{i,j}\triangle t$, where s is the maximal wave speed for the Riemann problem between two neighboring mesh cells either in x-direction or in y-direction. With this restriction on the time step we exclude the situation shown in Figure 18 (b), in our example this time step restriction corresponds to a Courant number slightly smaller than two. We compare our solution with those obtained by using the first order version of CLAWPACK, see Figure 20. One clearly sees that the larger Courant number leads to less smearing of the solution.

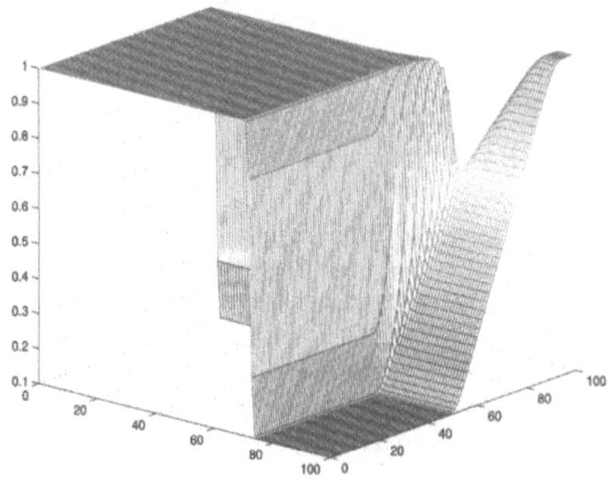

Fig. 20. Solution of Example 4 at time $t = 40$ with $\triangle x = 1$ using the first order CLAWPACK algorithm with $CFL \leq 1$

Conclusions and Outlook

The article reviewed the current state of development of the large time step scheme of LeVeque for hyperbolic conservation laws. A key open problem for over a decade was the entropy consistency, even for scalar conservation laws with convex flux function, for Courant numbers larger then two. The key difficulty in proving this is the interaction of shocks with rarefaction waves. We presented some numerical experiments that should be helpful in order to prove the entropy-consistency of the scheme.

The advantage of using Courant numbers larger than one in finite volume schemes lies not only in the fact that fewer time steps are needed to cover a given time interval. There is also a reduction of the numerical dissipation and corresponding gain in accuracy for at least moderately higher Courant numbers. Also, in general flows there are frequently large regions where the nonlinearities are very weak, so that these schemes can be accurate at least locally for quite large Courant numbers.

Though we present an improvement for scalar multidimensional conservation laws, the extension to multidimensional systems is what is really needed in the future. This would be of considerable interest to the computational fluid dynamics community, even if these type of schemes were used only locally in adaptive calculations.

Acknowledgments

This work was supported by the DFG Priority Program DANSE. We would like to thank Professor Bernold Fiedler for the coordination of this program.

References

1. S.J. Billett and E.F. Toro: On WAF-type schemes for multidimensional hyperbolic conservation laws. J. Comput. Phys. **130** (1997) 1-24.
2. A.J. Chorin: Random choice solution of hyperbolic systems. J. Comput. Phys. **22** (1976) 517-533.
3. R. Courant and K.O. Friedrichs and H. Lewy: Über die partiellen Differenzengleichungen der mathematischen Physik. Math. Ann. **100** (1928) 32-74.
4. E. Godlewski and P.-A. Raviart: Hyperbolic Systems of Conservation Laws. Ellipses, 1990.
5. E. Godlewski and P.-A. Raviart: Numerical Approximation of Hyperbolic Systems of Conservation Laws. Springer-Verlag, 1995.
6. S.K. Godunov: A difference scheme for numerical computation of discontinuous solutions of equations of fluid dynamics. Mat. Sb. **47** (1959) 271-306.
7. E. Hairer: Unconditionally stable explicit methods for parabolic equations. Numer. Math. **35** (1980) 57-68.
8. A. Harten and J.M. Hyman: Self adjusting grid methods for one-dimensional hyperbolic conservation laws. J. Comp. Phys. **50** (1983) 235-269.
9. A. Harten and J.M. Hyman and P.D. Lax: On finite difference approximations and entropy conditions for shocks. Comm. Pure Appl. Math. **29** (1976) 297-322.
10. A. Harten and P.D. Lax: A random choice finite difference scheme for hyperbolic conservation laws. SIAM J. Numer. Anal. **18** (1981) 289-315.
11. A. Harten and P.D. Lax and B. van Leer: On upstream differencing and Godunov-type schemes for hyperbolic conservation laws. SIAM Review **25** (1983) 35-61.
12. H. Holden and K.-A. Lie and N.H. Risebro: An unconditionally stable method for the Euler equations. J. Comput. Phys. **150** (1999) 76-96.
13. H. Holden and N.H. Risebro: A method of fractional steps for scalar conservation laws without the CFL condition. Math. of Comp. **60** (1993) 221-232.
14. U. Iben and E. Tadmor: High resolution schemes for conservation laws in the simulation of injection systems. Technical Report 00-06, UCLA, 2000.
15. R. Klein: Semi-implicit extension of a Godunov-type scheme based on low Mach number asymptotics I: One-dimensional flow. J. Comput. Phys. **121** (1995) 213-237.
16. D. Kröner: Numerical Schemes for Conservation Laws. Teubner-Verlag, 1997.
17. P.D. Lax and B. Wendroff: Systems of conservation laws. Comm. Pure Appl. Math. **13** (1960) 217-237.

18. R.J. LeVeque: CLAWPACK software. available from http://www.amath.washington.edu/~rjl/clawpack.html
19. R.J. LeVeque: CLAWPACK User Notes.
20. R.J. LeVeque: Large time step shock-capturing techniques for scalar conservation laws. SIAM J. Numer. Anal. **19** (1982) 1091-1109.
21. R.J. LeVeque: Convergence of a large time step generalization of Godunov's method. Comm. Pure Appl. Math. **37** (1984) 463-477.
22. R.J. LeVeque: A large time step generalization of Godunov's method for systems of conservation laws. SIAM J. Numer. Anal. **22** (1985) 1051-1073.
23. R.J. LeVeque: Some preliminary results using a large time step generalization of the Godunov method. In F. Angrand et. al, editors, Numerical methods for the Euler equations of Fluid Dynamics, SIAM, Philadelphia (1985) 32-47.
24. R.J. LeVeque: Second order accuracy of Brenier's time-discrete method for nonlinear systems of conservation laws. SIAM J. Numer. Anal. **25** (1988) 1-7.
25. R.J. LeVeque: Hyperbolic conservation laws and numerical methods. Von Karman Institute of Fluid Dynamics Lecture Series, 90-3, 1990
26. R.J. LeVeque: Numerical Methods for Conservation Laws. Birkhäuser Verlag, 1990.
27. R.J. LeVeque: Wave propagation algorithms for multidimensional hyperbolic systems. J. Comput. Phys. **131** (1997) 327-353.
28. R.J. LeVeque: Nonlinear Conservation Laws and Finite Volume Methods for Astrophysical Fluid Flow. In O. Steiner and A. Gautschy, editors, Computational Method for Astrophysical Fluid Flow, 27th Saas-Fee advanced course lecture note, Springer-Verlag, 1998.
29. B.J. Lucier: Error bounds for the methods of Glimm, Godunov and LeVeque. SIAM J. Numer. Anal. **22** (1985) 1074-1081.
30. K.W. Morton: On the analysis of finite volume methods for evolutionary problems. SIAM J. Numer. Anal. **35** (1998) 2195-2222.
31. P.L. Roe: Approximate Riemann solver, parameter vectors, and difference schemes. J. Comput. Phys. **43** (1981) 357-372.
32. N. Satofuka: A new explicit method for the numerical solution of parabolic differential equations. Numerical properties and methodologies in heat transfer, Proc. 2nd nat. Symp., College Park, 1981, 97-108, 1983.
33. N. Satofuka: Numerical solution of fluid dynamic equations using rational Runge-Kutta methods. Comp. Meth. in Appl. Sci. Eng. **VII** (1986) 249-259.
34. J. Smoller: Shock Waves and Reaction-Diffusion Equations. Springer-Verlag, 1994.
35. E.F. Toro: Riemann solvers and numerical methods for fluid dynamics. Springer-Verlag, 1997.
36. E.F. Toro and S.J. Billett: A unified Riemann-problem-based extension of the Warming-Beam and Lax-Wendroff schemes. IMA J. Numer. Anal. **17** (1997) 61-102.
37. J. Wang: Large time step generalizations of Glimm's scheme for systems of conservation laws. Chin. Ann. of Math. **913** (1988) 50-63.
38. J. Wang and G. Warnecke: On entropy consistency of large time step schemes I. The Godunov and Glimm schemes. SIAM J. Numer. Anal. **30** (1993) 1229-1251.
39. J. Wang and G. Warnecke: On entropy consistency of large time step schemes II. Approximate Riemann solvers. SIAM J. Numer. Anal. **30** (1993) 1252-1267.

Color Plates

The Algorithms Behind GAIO – Set Oriented Numerical Methods for Dynamical Systems

Michael Dellnitz, Gary Froyland, and Oliver Junge

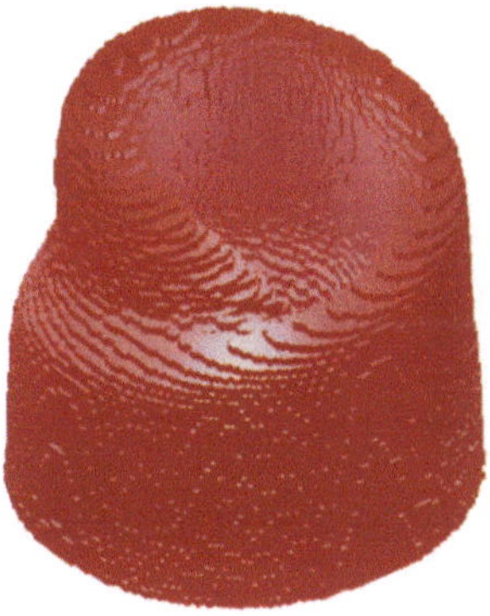

Plate 1. Covering of the relative global attractor for the knotted flow.

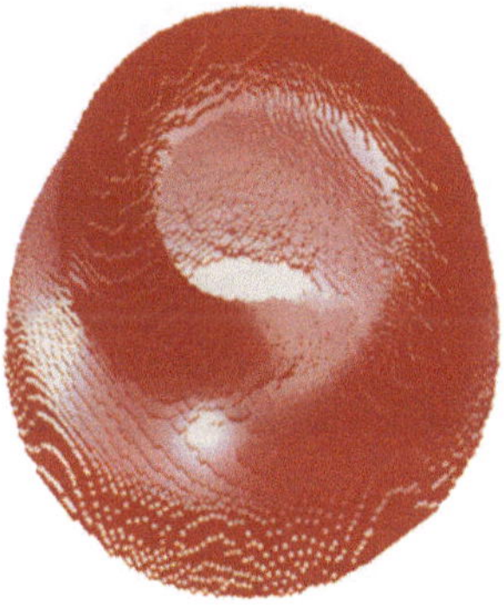

Plate 2. As for Plate 1, from another viewpoint.

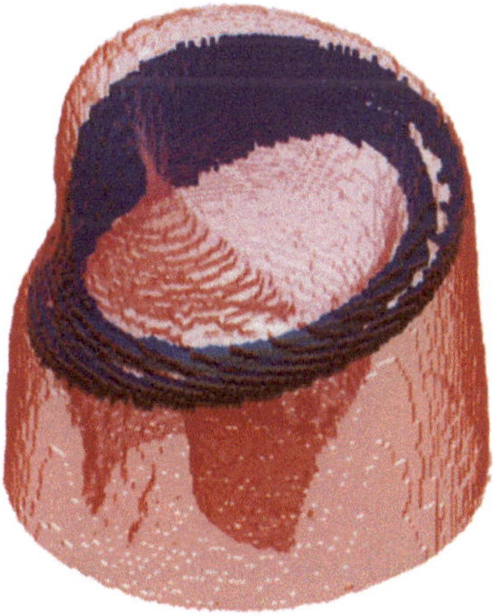

Plate 3. Coverings of the relative global attractor (red), and the chain recurrent set (blue).

Plate 4. Covering of the chain recurrent set (blue) at a deeper level of subdivision. The knot trajectory that defines the flow is shown in red.

Plate 5. Covering of the two-dimensional stable manifold of the steady state $(0, 0, 0)$.

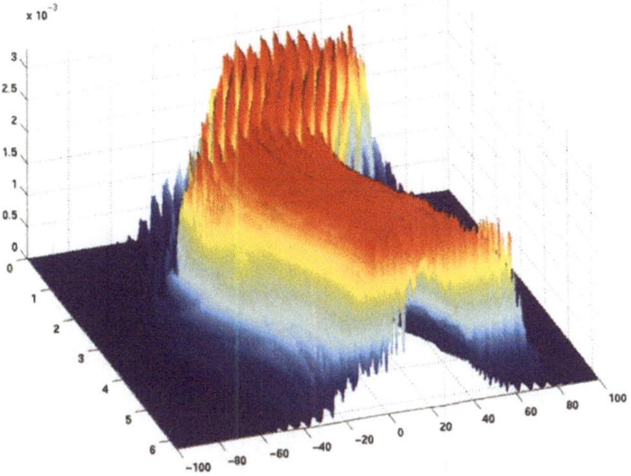

Plate 6. A three-dimensional version of Figure 5 with the density plotted in the z-coordinate, and coloured according to the density.

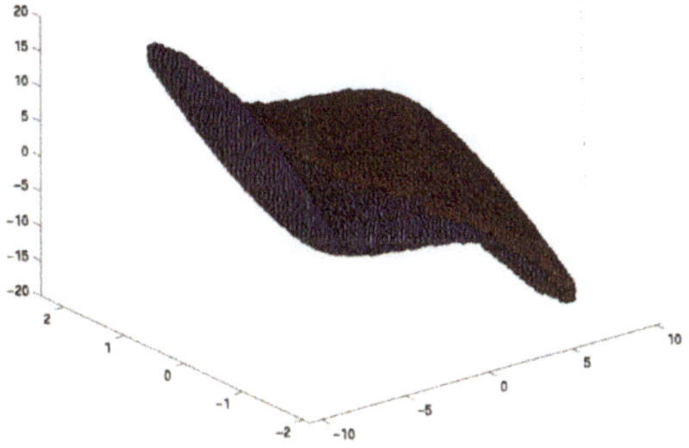

Plate 7. Separation of the invariant set into two almost-invariant pieces.

Polynomial Skew Products

Manfred Denker and Stefan-M. Heinemann

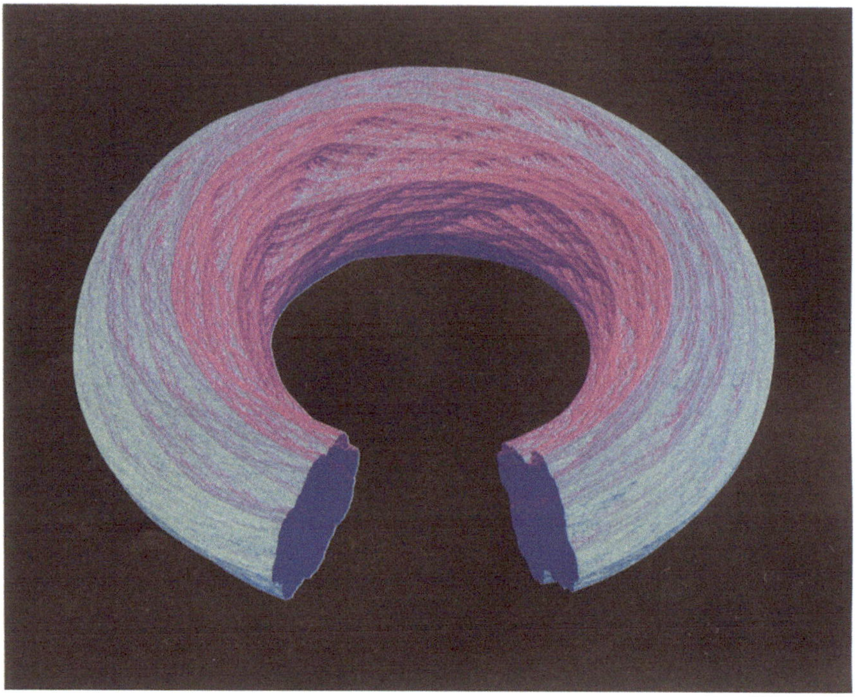

Plate 8. The Julia Set of the Doughnut Map

$$(x, y) \mapsto (x^2 + (1+i) \cdot (y^2 - y)/10 - 1/5, y^2)$$

(cut open at $\Im(y) > 9/10$).

Plate 9. The Julia Set of the Doughnut Map

$$(x, y) \mapsto (x^2 + (1 + i) \cdot (y^2 - y)/10 - 1/5, y^2)$$

with the canonical Markov partition (here, of iteration order 3) induced by the dynamics.

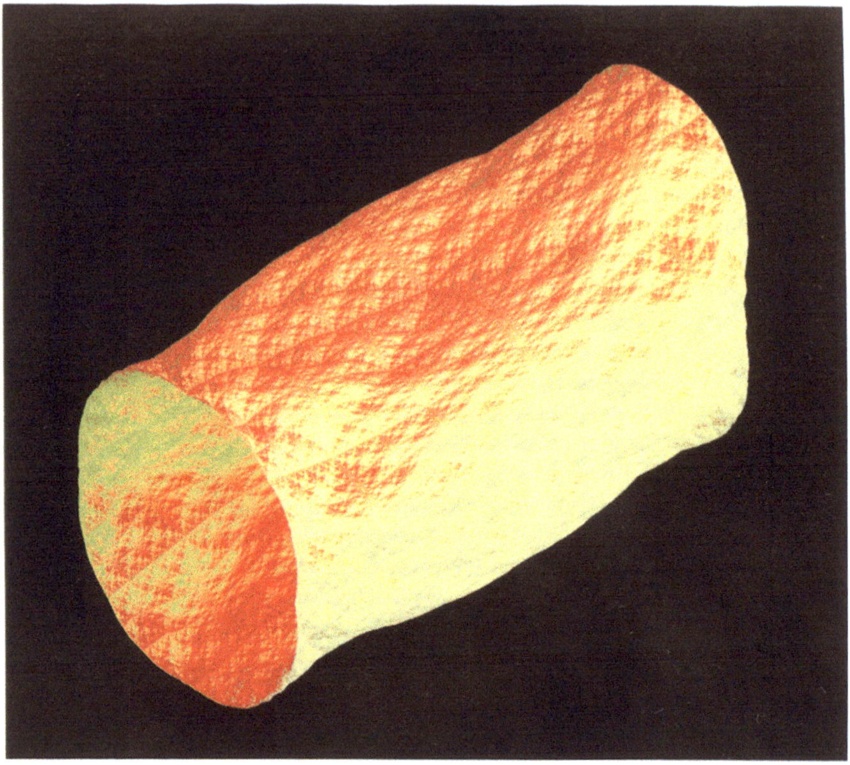

Plate 10. The Julia Set of the Skew Product of Cannellono Type

$$(x, y) \mapsto (x^2 + (y^2 - y - 1)/10, y^2 - 2) .$$

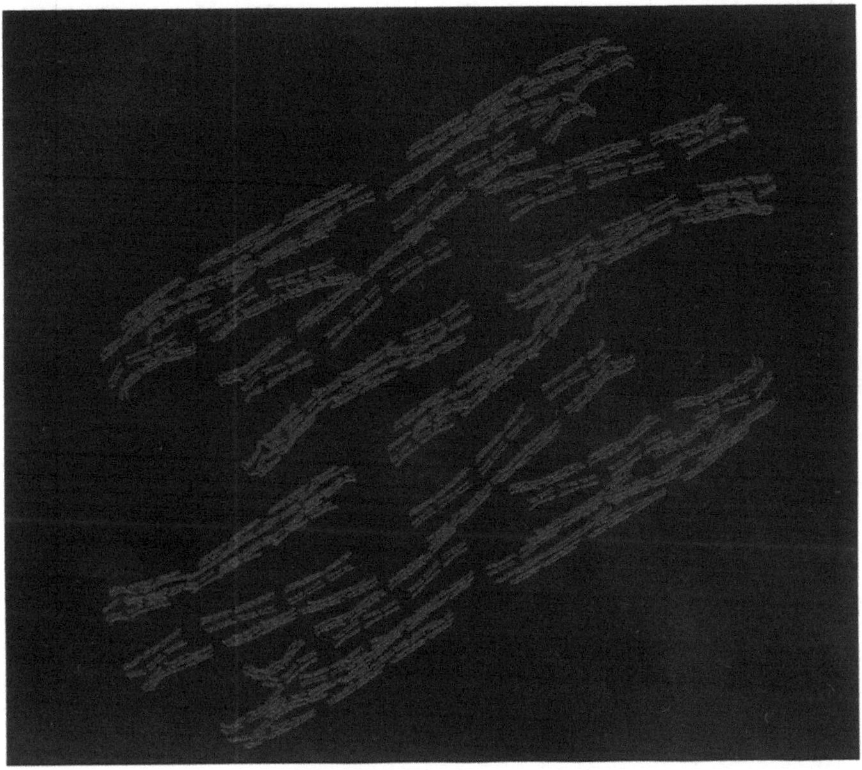

Plate 11. The Julia Set of the Skew Product

$$(x, y) \mapsto (x^2 - (y^2 - y - 1)/10, y^2 - 21/10)\ .$$

Simulation and Numerical Analysis of Dendritic Growth

Michael Fried and Andreas Veeser

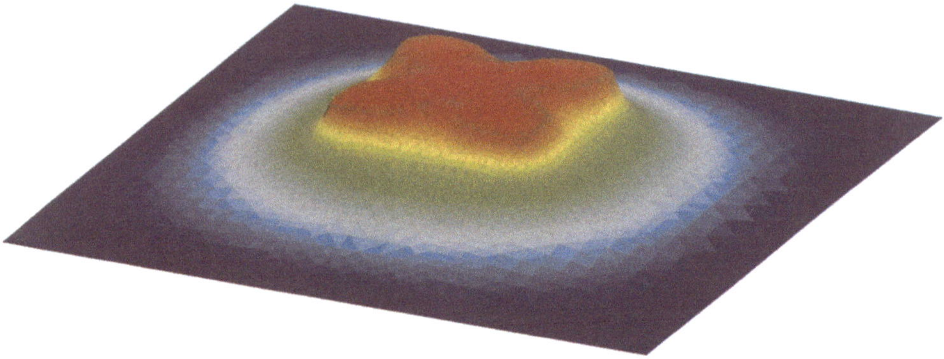

Plate 12. Fourfold Anisotropy: Graph of the Temperature Θ

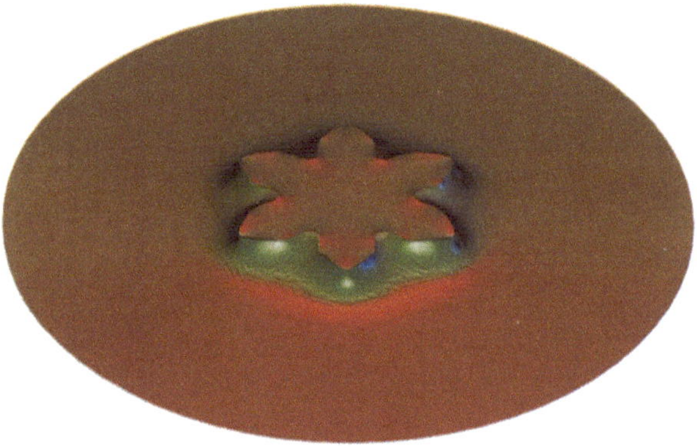

Plate 13. Sixfold Anisotropy: Graph of the Temperature with $T_0 = -1$

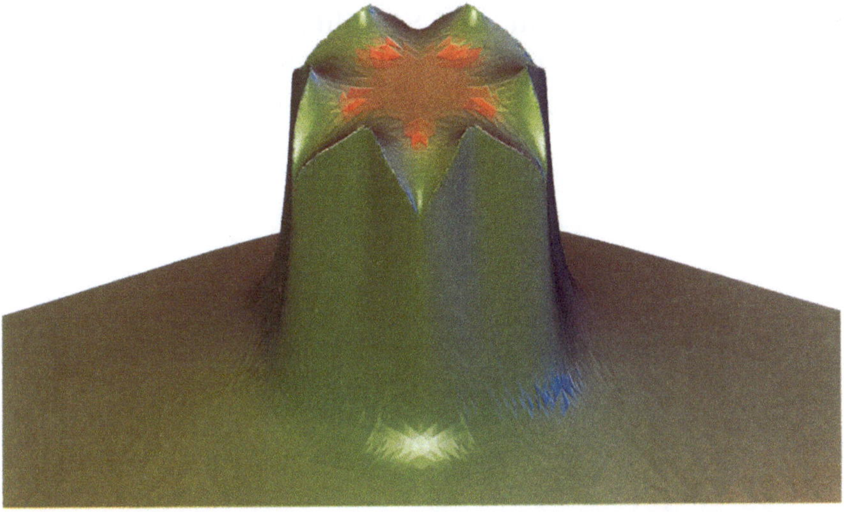

Plate 14. Sixfold Anisotropy: Graph of the Temperature with $T_0 = -2$, right: zoom to the solid phase

Existence, Bifurcation, and Stability of Profiles for Classical and Non-Classical Shock Waves

Heinrich Freistühler, Christian Fries and Christian Rohde

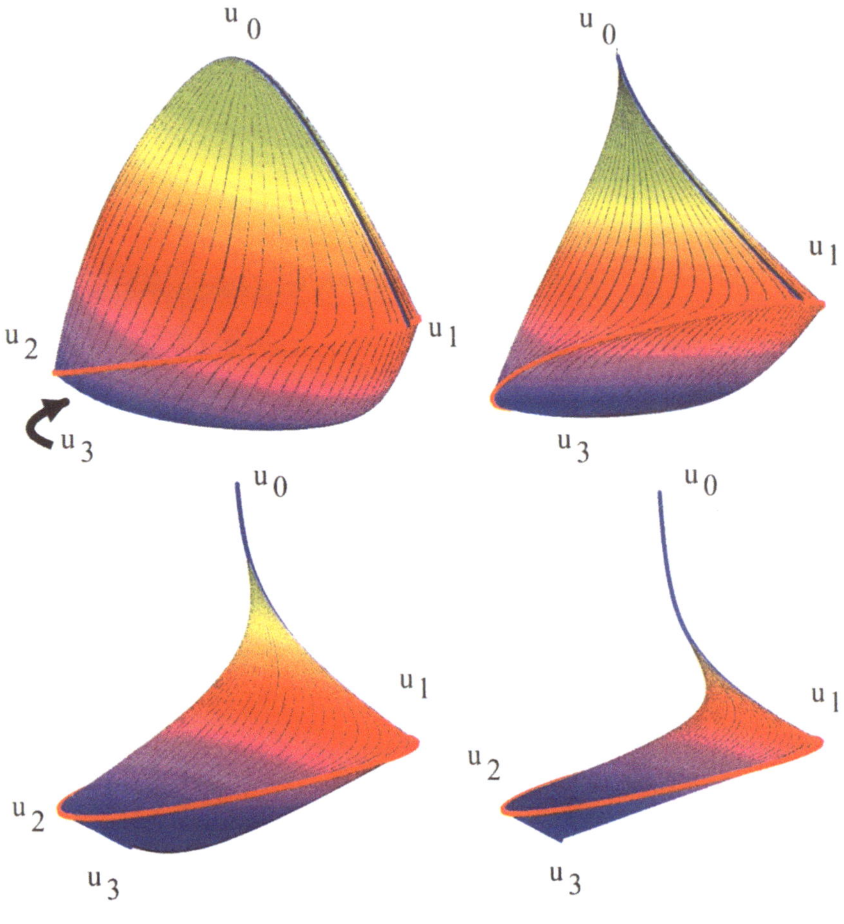

Plate 15. Projection to b_1b_2v–space for $\omega = 7.5, 1.0, 0.25, 0.06$.

Aspects on Data Analysis and Visualization for Complex Dynamical Systems

Becker, Bürkle, Happe, Preußer, Rumpf, Spielberg and Strzodka

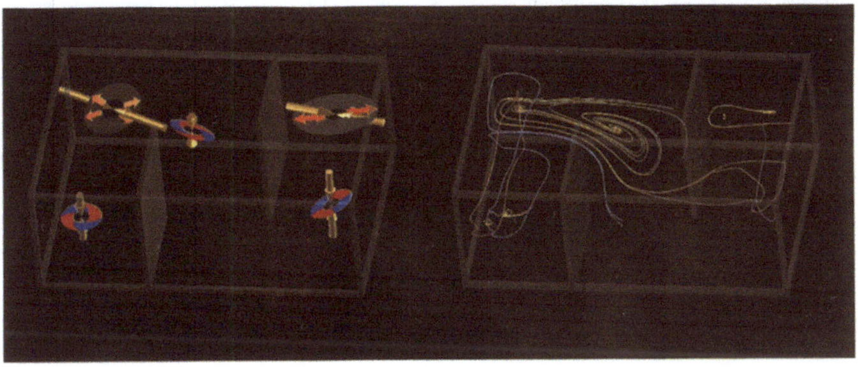

Plate 16. On the left icons placed at critical points in an incompressible flow in three dimensions, on the right the critical points are taken as starting point for particle lines.

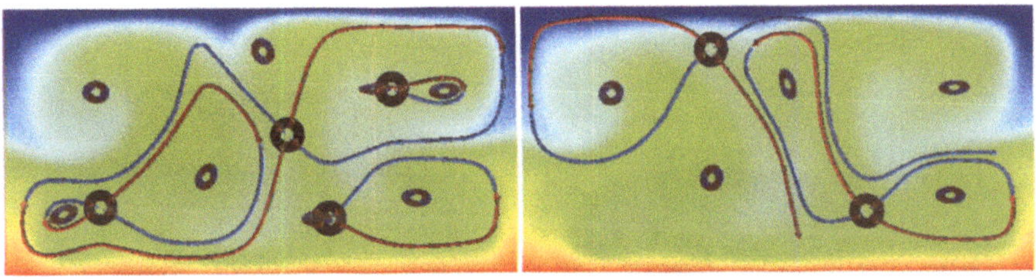

Plate 17. The topology of a convective flow in 2D depicted by icons and selected streamlines for two timesteps.

Plate 18. Several diffusion timesteps are depicted from the vector valued nonlinear anisotropic diffusion method applied to a convective flow field in a 2D box

Plate 19. The incompressible flow in a water basin with two interior walls and an inlet (*on the left*) and an outlet (*on the right*) is visualized by anisotropic nonlinear diffusion. Color is indicating the velocity

Plate 20. Texture transport in the von Kármán vortex street.

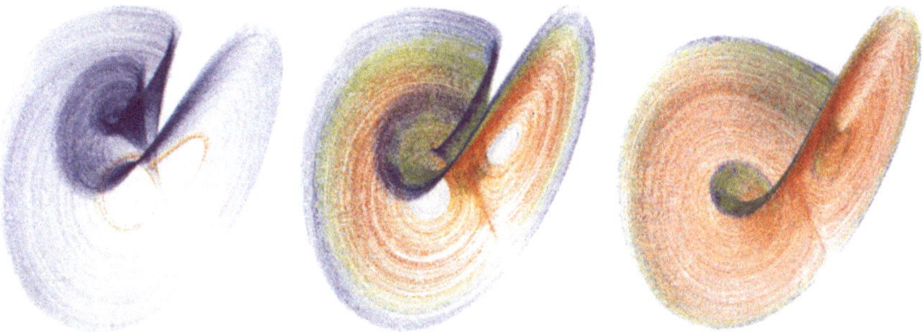

Plate 21. Invariant sets in the Lorenz family are visualized using a coverage with shaded streamlines.

Author Index

MIX
Papier aus verantwortungsvollen Quellen
Paper from responsible sources
FSC® C105338

If you have any concerns about our products,
you can contact us on
ProductSafety@springernature.com

In case Publisher is established outside the EU,
the EU authorized representative is:
Springer Nature Customer Service Center GmbH
Europaplatz 3, 69115 Heidelberg, Germany

Printed by Libri Plureos GmbH
in Hamburg, Germany